ANNUAL REVIEW
OF PHYSIOLOGY

ANNUAL REVIEW
OF PHYSIOLOGY

VOLUME 66, 2004

JOSEPH F. HOFFMAN, *Editor*
Yale University School of Medicine

DAVID L. GARBERS, *Associate Editor*
University of Texas, Southwestern Medical Center

www.annualreviews.org   science@annualreviews.org   650-493-4400

ANNUAL REVIEWS
4139 El Camino Way • P.O. Box 10139 • Palo Alto, California 94303-0139

ⱭⱤ

ANNUAL REVIEWS
Palo Alto, California, USA

International Standard Serial Number: 0066-4278
International Standard Book Number: 0-8243-0366-0
Library of Congress Catalog Card Number: 39-15404

All Annual Reviews and publication titles are registered trademarks of Annual Reviews.

⊗ The paper used in this publication meets the minimum requirements of American National Standards for Information Sciences—Permanence of Paper for Printed Library Materials. ANSI Z39.48-1992.

Annual Reviews and the Editors of its publications assume no responsibility for the statements expressed by the contributors to this *Annual Review*.

TYPESET BY TECHBOOKS, FAIRFAX, VA
PRINTED AND BOUND BY MALLOY INCORPORATED, ANN ARBOR, MI

PREFACE

I call our readers' attention to an analysis by Mallon et al. (2003) documenting changes that have occurred in basic science departments as well as in the fields in which PhDs are granted in U.S. medical schools in the past 19 years or so. I think most of us were aware of a marked increase in interdisciplinary science, but perhaps were not to the extent that the impact this increase had on departmental structure and graduate education. The bottom line is that the number of basic science departments has decreased, primarily due to mergers or new departments, but importantly with an overall increase in faculty numbers. One consequence of this change that is of primary concern is in the possible dissociation of the teaching of basic science from traditional departmental lines/disciplines. Whether this latter transition has altered the direction of graduate education is not clear, but there is no doubt that changes in the spectra of fields in which PhDs are granted has dramatically changed in the past 10 years. That is to say the number of PhDs in the fields of cell biology, neuroscience, molecular biology, and genetics has increased by approximately 100 to almost 250%, whereas that of physiology has remained *level*. I find this a surprising result that needs to be addressed by the physiology community and the relevant faculties. If this trend prevails, then the portent for future transitional/integrative physiological investigation presents serious limitations on the development of our field. It will also have a profound effect on the detailed coverage offered by the *Annual Review of Physiology*.

The format of the current volume (V66) mirrors that of previous volumes in its thematic presentation of advances in the various traditional fields of physiology. These themes change from year to year and depend on current developments and the ripeness for review. Of special note is the perspective chapter on Principles of Membrane Structure contributed by S. Jonathan Singer, as well as two Special Topic sections: one section, edited by Janos Lanyi, on Proton and Electron Transporters; the second section, edited by Stephen Smith, concerns advances in functional imaging in the service of physiology.

We are grateful to all the authors for their contributions to this volume. The Editorial Committee is always open to and encourages suggestions for future topics and authors. We can be reached at www.annualreviews.org. Access to various current and well as previous chapters can be viewed at this site as well.

Joseph F. Hoffman
Editor

Mallon WT, Biebuyck JF, Jones RF. 2003. The reorganization of basic science departments in U.S. medical schools, 1980–1999. *Acad. Med.* 78:302–6.

Annual Review of Physiology
Volume 66, 2004

CONTENTS

ERRATA

   An online log of corrections to *Annual Review of Physiology* chapters
   may be found at http://physiol.annualreviews.org/errata.shtml

OTHER REVIEWS OF INTEREST TO PHYSIOLOGISTS

From the *Annual Review of Biochemistry*, Volume 72 (2003)

The Rotary Motor of Bacterial Flagella, Howard C. Berg

The Enzymes, Regulation, and Genetics of Bile Acid Synthesis, David W. Russell

Protein-Lipid Interplay in Fusion and Fission of Biological Membranes, Leonid V. Chernomordik and Michael M. Kozlov

Signals for Sorting of Transmembrane Proteins to Endosomes and Lysosomes, Juan S. Bonifacino and Linton M. Traub

Challenges in Enzyme Mechanism and Energetics, Daniel A. Kraut, Kate S. Carroll, and Daniel Herschlag

TRK Receptors: Roles in Neuronal Signal Transduction, Eric J. Huang and Louis F. Reichardt

Dynamics of Cell Surface Molecules During T Cell Recognition, Mark M. Davis, Michelle Krogsgaard, Johannes B. Huppa, Cenk Sumen, Marco A. Purbhoo, Darrell J. Irvine, Lawren C. Wu, and Lauren Ehrlich

Proteomics, Heng Zhu, Metin Bilgin, and Michael Snyder

From the *Annual Review of Cell and Developmental Biology*, Volume 19 (2003)

Cyclic Nucleotide-Gated Ion Channels, Kimberly Matulef and William N. Zagotta

Genes, Signals, and Lineages in Pancreas Development, L. Charles Murtaugh and Douglas A. Melton

Cadherins as Modulators of Cellular Phenotype, Margaret J. Wheelock and Keith R. Johnson

Genomic Imprinting: Intricacies of Epigenetic Regulation in Clusters, Raluca I. Verona, Mellissa R.W. Mann, and Marisa S. Bartolomei

Transport Protein Trafficking in Polarized Cells, Theodore R. Muth and Michael J. Caplan

Tetraspanin Proteins Mediate Cellular Penetration, Invasion, and Fusion Events and Define a Novel Type of Membrane Microdomain, Martin E. Hemler

Plasma Membrane Disruption: Repair, Prevention, Adaptation, Paul L. McNeil and Richard A. Steinhardt

S. Jonathan Singer

Annu. Rev. Physiol. 2004. 66:1–27
doi: 10.1146/annurev.physiol.66.032902.131835
Copyright © 2004 by Annual Reviews. All rights reserved
First published online as a Review in Advance on August 18, 2003

SOME EARLY HISTORY OF MEMBRANE MOLECULAR BIOLOGY

S. Jonathan Singer

*Division of Biological Sciences, University of California, San Diego, 9500
Gilman Drive, La Jolla, California 92093-0322; email: ssinger@ucsd.edu*

Key Words thermodynamics, integral and peripheral proteins, membrane
asymmetry, transport proteins, fluid mosaic model

■ **Abstract** This article is mostly about the beginnings of the molecular biology
of membranes, covering the decade 1964–1974. It is difficult to read (or write) this
article because of a sense of déja vu. Most of the material in it is considered commonplace today, having been established experimentally since then. But at the time
this work was begun, practically nothing was known about the molecular structure
and the mechanisms of the functions of membranes. This situation existed because no
membrane proteins of the kind I called integral had as yet been isolated in a pure state,
and therefore none had had their amino acid sequence determined. The first integral
membrane protein to be so characterized was human erythrocyte glycophorin, in 1978.
It was the use of the thermodynamic reasoning that had been developed for the study
of water-soluble proteins, together with the information from several key experiments
carried out in a number of laboratories during the early decade, that led us to the
fluid mosaic model of membrane structure in 1972. Without direct evidence to confirm
the model in 1971–1972, my colleagues and I nevertheless had the confidence in it
to pursue some of the consequences of the model for a new understanding of many
membrane functions, which I present here in some detail. Finally, I discuss two recent
high-resolution X-ray crystallographic studies of integral proteins to ask how well the
structural and functional proposals that we derived from the fluid mosaic model fit
these remarkably detailed X-ray results.

INTRODUCTION

I first became interested in the molecular structure of biological membranes in
1964, coming from a research background involving the physical chemistry of
water-soluble proteins. I soon found that my ignorance about membranes was
paralleled by the generally rudimentary state of knowledge of the subject, particularly where membrane proteins were concerned. No characteristic membrane
protein had yet been isolated and its amino acid sequence determined by 1964;
this first occurred in 1978. (This was largely because membrane proteins are not
soluble in water). From 1964 to early 1971, when I published a long article on the

lipid-protein mosaic model of membrane structure (1), which was extended to include the fluid character of the mosaic in the *Science* paper of early 1972 (2), my evolving picture of membrane structure and corresponding functions was therefore largely thermodynamic and theoretical. In 2003, all of that work seems remote and old hat, largely confirmed and greatly extended by a tremendous amount of experimental work in the intervening years. It is this now largely forgotten early period, however, that is the main subject of this memoir.

MEMBRANE STRUCTURE IN 1964

Protein-Lipid Ratios

The ratio by weight of membrane-associated proteins to lipids varies from about 1.0 for the red blood cell to 3 for mitochondrial membranes. The only marked exception is myelin, which is an insulating rather than a conducting membrane: This has a ratio of only 0.23. Proteins not only constitute the major component of most membranes but also carry out most of the functions of membranes. Their structures are therefore of primary concern.

Lipid Structure

The experiments of Gorter & Grendel in 1925 (3) were carried out by measuring the surface area on a Langmuir trough of a monolayer of lipid extracted from a known number of red blood cells. They concluded that the amount of lipid was equivalent to twice the surface area of the cells and that the lipids were therefore organized as a continuous bilayer around the cell. This work is referred to today as the source of the lipid bilayer concept and is widely accepted. However, for technical reasons, these experiments turned out to be faulty, and when done correctly (4), the area occupied by the monolayer of lipids in the trough was nearly the same as, rather than double, the cell surface area. The lipid is therefore organized either as a monolayer around the entire cell (which is thermodynamically unsound), or as a bilayer covering only roughly half of the total surface area of the red blood cell. We return to this important fact below.

Electron Microscopy

Conventional transmission electron microscopy of fixed, stained, and plastic-embedded cells revealed that most types of membranes exhibited a similar railroad-track-like appearance (5). The distance between the outer edges of the track was about 90 Å, and if the lipid was regarded as a Gorter-Grendel continuous bilayer of about 70 Å average thickness, this result was interpreted (5, 6) to mean that the membrane protein was largely in the form of two continuous layers of unfolded protein of 10 Å thickness on both sides of the lipid bilayer (Figure 1A). This came to be known as the unit-membrane, or the Davson-Danielli-Robertson (DDR) model, and was the generally accepted view of membrane structure in 1964.

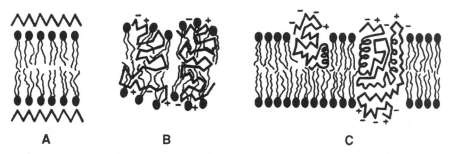

Figure 1   Schematic representations of membrane models as of 1966. (*A*) The Davson-Danielli-Robertson (DDR) model with its lipid bilayer, covered over both surfaces by unfolded protein monolayers. (*B*) The Benson model, consisting of lipoprotein subunits of intertwined fatty acyl and polypeptide chains. The ionic groups of the lipids and proteins are confined to the water surfaces. (*C*) An early version of the lipid-protein mosaic model, with amphipathic integral proteins interspersed with lipid bilayer (see text for details).

THE THERMODYNAMICS OF PROTEIN STRUCTURE

Hydrophobic Interactions

By 1964, considerable advances had been made in understanding the structures of water-soluble protein molecules. It was appreciated that a considerable (thermodynamically favorable) decrease in free energy accompanied the removal of hydrophobic amino acid residues from contact with water into the more hydrophobic milieu of the interior of the protein molecule. (This is the same principle involved in the immiscibility of oil and water, which, despite the massive efforts by owners of ocean-going petroleum tankers to prove to the contrary, still holds true). The simple model system for this hydrophobic effect, the equilibrium (1)

$$CH_4 \text{ (in water)} \rightleftharpoons CH_4 \text{ (in benzene)}, \qquad\qquad 1.$$

is accompanied by a favorable change in free energy, ΔG, of -2.6 kcal//mol (7). The overall globular structure of most water-soluble proteins provides compacted interior domains where hydrophobic residues are often sequestered away from contact with water, which greatly contributes to the stability (i.e., lowering the free energy) of the entire structure. The first water-soluble proteins to have their entire three-dimensional structures determined to 3 Å resolution by X-ray crystallography, myoglobin (8), and hemoglobin (9), demonstrated this structural principle.

Hydrogen Bonding

Many polar groups in proteins are capable of forming hydrogen bonds, the largest number of these being the peptide C=O and N-H groups. In a water milieu, the equilibrium (2) (where $\cdots$ represents a hydrogen bond)

$$\begin{matrix} \diagdown \\ \diagup \end{matrix} C{=}O\cdots H_2O + H_2O \cdots H{-}N\begin{matrix} \diagup H \\ \diagdown H \end{matrix} \rightleftharpoons \begin{matrix} \diagdown \\ \diagup \end{matrix} C{=}O\cdots H{-}N\begin{matrix} \diagup H \\ \diagdown H \end{matrix} + H_2O \cdots H_2O \qquad 2.$$

has a $\Delta G \sim 0$ (10), meaning that the sum of a $\diagup$C=O$\cdots$H–N$\diagdown$ hydrogen bond formed in the hydrophobic interior of a globular protein plus the $H_2O \cdots H_2O$ bond formed between the released H_2O molecules, is not energetically favored over the sum of the separate $\diagup$C=O$\cdots$H₂O and $\diagdown$N–H$\cdots$H₂O hydrogen bonds formed on the water-exposed surface of the protein molecule. Formation of $\diagup$C=O$\cdots$H–N bonds in the hydrophobic interior, in this sense, doesn't help to drive the formation of the globular conformation of the protein molecule in water solution. However, if the $\diagup$C=O$\cdots$H–N$\diagdown$ bonds do not form in the globular interior, burying each pair of unbonded $\diagup$C=O and H–N groups costs about 4 kcal/mol in free energy. Similar considerations apply to other hydrogen-bonding polar groups. In other words, in order to create a stable, lowest free-energy globular conformation, a substantial fraction of the groups capable of forming hydrogen bonds must do so with one another in the globular hydrophobic interior of the protein molecule in the absence of H_2O molecules with which to bond.

Some water-soluble proteins and many membrane proteins contain covalently attached saccharide residues, which have numerous hydrogen-bonding OH groups as well as ionic residues. By the same reasoning given in this and the following sections, it can be shown (1) that saccharide groups exhibit a lower, more favorable free energy in contact with water than buried in the hydrophobic interior of proteins or membranes.

Hydrophilic Interactions

Several of the amino acids have side chains that are ionized at pH 7: R, K, and H are positively charged, D and E negatively. Can these residues be buried in a thermodynamically favorable, or least unfavorable, free energy in the interior of a water-soluble globular protein away from contact with the water solvent, with its extraordinarily large dielectric constant? Several possible mechanisms to achieve such burial may be considered. Burying an isolated ion in the low dielectric constant medium ($\sim$2) of the hydrophobic interior is ruled out because it would cost a great deal of energy and is entirely unlikely. An ionic group can, however, be buried as a polar, nonionic group by discharging it first. For example, the carboxylate ion of D and E may be discharged by the reaction: $-COO^- + H_3O^+ \rightleftharpoons -COOH + H_2O$. The positive (unfavorable) ΔG for this process is given by 2.303 RT |(pH − pK)|, where R is the gas constant, T the absolute temperature, pK is log K, the dissociation constant of the group, and pH is the ambient pH. For a carboxyl group of pK = 4.5 to be protonated at pH 7.0, ΔG = +3.3 kcal/mol at 25°C and is about the same for lysine (pK 9.5); it is largest for arginine (pK $\sim$ 12), and least for histidine (pK $\sim$ 7). Thus burying an ionic residue in the protein interior by first discharging it is, with the exception of histidine, quite unfavorable energetically. [This, however, is not taking into account the favorably negative ΔG of simultaneously burying the hydrophobic remainder (mostly methylene groups) of each of these amino acid residues in the protein interior.]

In principle, another way of burying an ionic group is to include it as an ion pair $(-+)$ with an oppositely charged group. This has often been proposed, although in 1971 (1) I had already pointed out its prohibitively large free-energy cost. A model reaction for this process is the transfer of the zwitterion glycine from water to a nonpolar, or less polar, solvent

$$
\begin{array}{cc}
\text{CH}_2\ (\text{in water}) & \rightleftharpoons \quad \text{CH}_2\ (\text{in nonaqueous solvent}). \qquad 3.
\end{array}
$$

The ΔG for this process may be determined approximately from the equilibrium solubilities of glycine in water and other solvents as $RT\ln (Xw/Xs)$, where Xw and Xs are the mol fractions of glycine in the saturated solutions in water and the solvent, respectively, at the temperature T. It is a fact (11) that the solubility of glycine decreases sharply with the decreasing dielectric constant, or increasing hydrophobic nature, of the solvent: For acetone (dielectric constant ~20 and still fully miscible with water), ΔG is already $+6$ kcal/mol (1). For the globular interior of a protein molecule (dielectric constant ~2, approximately that of a liquid hydrocarbon), this positive ΔG would be much larger. The conclusion is that burying charged groups as ion pairs in a protein interior away from contact with water is energetically very costly and correspondingly quite improbable. However, this has occasionally been suggested to happen in the interior domain of membrane proteins and is almost certainly incorrect.

THE THERMODYNAMICS OF MEMBRANE STRUCTURE

The foregoing thermodynamic principles developed for water-soluble proteins apply generally to all biological systems in their particular environments. (The DNA molecule is a wonderful example.) These principles have provided the theoretical basis for our investigations of membrane structure in the period 1964–1972.

Membrane Models (1966–1972)

I have already made reference to the DDR model of membrane structure (Figure 1A) (6). In 1966, Benson (12) published a model (Figure 1B) (the lipoprotein subunit model) in which the protein chains are interspersed with and wound around the fatty acyl chains of the phospholipids in the membrane interior to produce discrete lipoprotein subunits that, packed against one another in the plane of the membrane, constitute the membrane. The lipid is not in the classical bilayer form because it is disrupted by the associated protein chains. The polar head groups of the lipids and the ionic residues of the proteins are located in contact with water on the two membrane surfaces, whereas the hydrophobic residues of the protein are located largely within the hydrophobic membrane interior, away from contact with water.

Also in 1966 (13), we published an early version of the lipid-protein mosaic model of the membrane (Figure 1*C*), based on the thermodynamics I discussed in the previous section. Similar ideas were put forward independently by Wallach & Zahler (14). The lipids and the (integral) proteins (as we later referred to them) are arranged largely independently of one another in a mosaic pattern in the plane of the membrane.[1] The hydrophobic fatty acyl chains of the lipids and a large fraction of the nonpolar amino acid residues of the integral proteins are sequestered in the membrane interior away from contact with water. The ionic and saccharide groups of the lipids are in direct contact with water. The lipids are organized as a bilayer, interrupted by the intercalated integral proteins. This is consistent with the finding (4) mentioned above that the lipid bilayer constitutes only about one half the red blood cell membrane surface area. I suggested that integral proteins occupy the remaining half. The integral protein may protrude into the water from either one or both of the membrane surfaces; in the latter case, the protein spans the membrane. These extramembranous domains contain all the charged amino acid residues of the protein in contact with water. The integral proteins are therefore proposed to be amphipathic polypeptides, having one or two hydrophilic domains exterior to the membrane, connected to a generally hydrophobic domain that is confined to the interior of the membrane.

How does each of these three models fare in meeting the thermodynamic criteria outlined in the previous section? In the DDR model (Figure 1*A*), the unfolded protein chains on the two membrane surfaces would perforce have most of their hydrophobic amino acid residues exposed to contact with water. The ionic and polar head-groups of the lipids in the bilayer, on the other hand, would be significantly shielded from contact with water by the sheets of unfolded proteins overlying them. Both of these features are thermodynamically quite unfavorable. (In addition, the structural blandness of the DDR model makes it difficult to imagine how a membrane might exhibit a wide range of enzymatic, transport, and transmembrane signaling properties. The DDR model sometimes was pictured with pores through the lipid bilayer, to account for ion permeability, without any justification, theoretical or evidential, in support.)

By stretching out the polypeptide chains and placing them in direct contact with individual hydrocarbon chains of the lipids, the Benson model (Figure 1*B*) would not allow the protein peptide groups $\rangle$C=O and $\rangle$N–H to form many hydrogen bonds with one another; likewise for other polar residues on various amino acid

---

[1]The recent evidence (cf. 16) for the existence of separate lipid-protein domains within the lipid bilayer (e.g., so-called lipid rafts and caveolae) are not at odds with the lipid-protein mosaic model. I had suggested that the model could accommodate the association of a "fraction of the membrane lipid with the protein moieties, creating a mosaic of [some] lipoproteins interspersed with the remaining lipid in a bilayer" (1, p. 198).

side chains. This would be one of the main factors contributing a large unfavorable free energy to the Benson membrane structure in water. Furthermore, the lipid is not organized as a bilayer, which does not maximize the fatty acid hydrophobic interactions.

The lipid-protein mosaic model (Figure 1*C*) was designed to be thermodynamically feasible. The amphipathic integral proteins have their hydrophobic amino acid residues largely buried in the membrane interior away from contact with water, and their polar, non-ionizable residues are distributed between both exterior and interior domains. Their ionizable residues are essentially all in contact with water on the exterior, extramembranous domains of the integral proteins. The ionic and polar saccharide moieties of the phospholipids are exposed to water on both sides of the bilayer. This structure largely maximizes both hydrophobic and hydrophilic interactions. In the membrane interior, the $\diagdown$C=O and H-N$\diagup$ peptide groups can hydrogen bond to one another (see further below), as can at least some of the side chains of the polar amino acid residues. The wide range of protein to lipid ratios in different membranes mentioned above can easily be accommodated by the lipid-protein mosaic model. (The interesting functional possibilities of the lipid-protein mosaic model were also appreciated, and are discussed below.)

Extensions of the Lipid-Protein Mosaic Model in 1971

In 1970–1971, I had the opportunity of a sabbatical leave to elaborate upon and extend the lipid-protein mosaic model (1), although there had still been no membrane proteins isolated and their structures analyzed. At about that time, however, there were other types of experiments published that were at least consistent with some features of the lipid-protein mosaic model. da Silva & Branton proved in 1970 (16), by freeze-etching experiments in electron microscopy, that protein molecules were deeply embedded in membranes and appeared to be arranged in a mosaic within the lipid bilayer. Bretscher in 1971 (17), using chemical labeling methods, demonstrated that a major protein of the red blood cell membrane was exposed to water on both sides of the membrane (i.e., spanned it). (The Bretscher result is consistent with either the Benson or the lipid-protein mosaic models, but not the DDR.) In 1968, Lenard and I (18) showed that 70% of the phosphorylamine residues of the membrane phosphoglycerides could be removed from intact red blood cell membranes by phospholipase C action, without changing the protein circular dichroism spectrum or the conformational stability of the proteins toward increasing temperature, which are characteristic of the untreated membranes, results that are most consistent with the lipid-protein mosaic model.

INTEGRAL AND PERIPHERAL PROTEINS   One extension of the lipid-protein mosaic model (1) proposed that there are two quite different types of proteins associated with membranes. One type is the membrane-embedded integral protein and the other is a peripheral protein. I anticipated that amphipathic integral proteins would generally be insoluble in water because of their predicted large transmembrane clusters of hydrophobic residues. Certain membrane proteins, however, such as cytochrome c of mitochondrial membranes and spectrin of red blood cell membranes, were exceptions to the rule. These proteins could be isolated intact and without associated lipid by mild treatments of their respective membranes, and when isolated behaved much like ordinary water-soluble proteins. Some investigators viewed these proteins as typical of all membrane proteins generally, but I suggested that these proteins were of another category: peripheral to the membrane, attached noncovalently to specific integral proteins where these protruded from the bilayer into the aqueous phase. The attachment would be via forces that operate to produce specific water-soluble protein aggregates, such as the four polypeptide chain hemoglobin molecule.

Different peripheral proteins were projected at the time to play many diverse and important roles in membrane biology (19): in signal transmission across membranes (see below); in binding extracellular matrices to the exterior face, and cytoskeletal elements to the interior face, of plasma membranes; and in limiting the diffusion of membrane components. This last function includes a role in the formation of distinctive membrane domains such as in initiating endocytic vesicles in membranes, or in synapses at neural and neuromuscular junctions, or in cell adhesion sites. These and many other essential functions have since been proposed and established for particular peripheral proteins of different membranes, but they are not discussed in depth here.

THE DOMINANCE OF THE α-HELIX IN HYDROPHOBIC DOMAINS OF INTEGRAL PROTEINS   Because of the energetic desirability, referred to above, of forming the maximum number of $\rangle$C=O $\cdots$ H-N$\langle$ peptide hydrogen bonds in the hydrophobic interior domain of the integral protein, and because of the fact that the α-helix is the most efficient structure for the formation of these bonds, I suggested that these interior domains "may all be largely in the α-helical conformation" (1, p. 201), instead of the mixed structure shown in Figure 1C. (Peptide hydrogen bonds can also be made between adjacent β-pleated sheets, but where the numbers of adjacent β chains are small (<10), the two outer chains' hydrogen-bond formation is incomplete. However, see Porins, below.) An inference drawn from this conclusion is that in order to traverse the hydrophobic interior of the bilayer ($\sim$35 Å), a single α-helix with each amino acid residue translated about 1.5 Å along the axis of the α-helix would require a stretch of $\sim$23 mainly hydrophobic residues admixed with some polar, but, few, if any, ionic residues, in a continuous sequence. Such long stretches of mainly hydrophobic sequences occur very infrequently in water soluble proteins, and this was suggested to be the critical structural distinction between water-soluble and integral membrane proteins.

DIFFERENT STRUCTURAL TYPES OF INTEGRAL PROTEINS

Monotopic Proteins

I was then also led to suggest that integral proteins might be of several structural types, each with its own functional properties. One type might have its sequence in "two distinct halves [hydrophilic and hydrophobic]... that fold up more-or-less independently of each other to give three-dimensional exterior and interior regions to the molecule," as apparently occurs with cytochrome b$_5$ of microsomes. This integral membrane protein in situ is cleaved by trypsin into two parts, "one carrying the b$_5$ activity, which is soluble in water, the other remaining membrane-bound [(20)], which confers hydrophobicity on the entire intact molecule and presumably is responsible for its attachment to the membrane" (1, pp. 199–200). This type of integral protein might have a structure like that in Figure 2A, incorporating the idea discussed above that the membrane-intercalated hydrophobic domain is likely to be an α-helix. This structure allows for only a single peptide bond cleavage to separate and solubilize the external hydrophilic domain from the membrane-intercalated hydrophobic domain. (This type of hypothetical structure is now well-known, and is referred to as monotopic.)

Polytopic Proteins

On the other hand, "it is conceivable that the single polypeptide chain could have successive segments weaving in and out of the interior and exterior regions of the protein" (1, p. 200), as depicted schematically in Figure 2B, again with the several interior hydrophobic domains in the α-helical conformation. "In [such a] case, proteolytic digestion [of the protein in situ in the membrane] would not be expected to cleave off the whole of the exterior regions of the protein" (1, p. 200). (Such hypothetical integral protein structures with varying numbers of hydrophobic transmembrane α-helices are now known to be common and are referred to as polytopic.) Some possible functional differences between monotopic and polytopic integral proteins that were deduced at the time are discussed below.

Where the exterior domains of the amphipathic integral proteins that are in contact with water are of sufficient length, it is likely that they would fold up into a conformation that is more-or-less globular, with the structural characteristics of water-soluble proteins. In the case of the polytopic integral proteins, the several hydrophilic loop regions connecting successive transmembrane helices, together with the one or both N- or C-terminal domains protruding into the aqueous phase from the same side of the membrane, would be expected to associate noncovalently to form a somewhat globular domain on each side of the membrane, each domain connected by several peptide linkages to the opposite ends of the several helices in the transmembrane domain (Figure 2B).

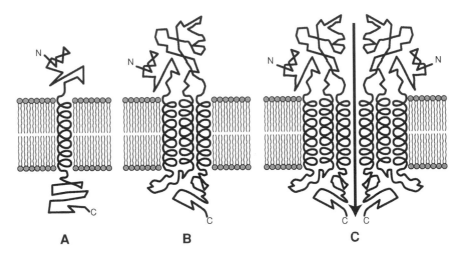

Figure 2   The lipid-protein mosaic model in 1971, with three different proposed types of integral protein structures spanning the membrane: (*A*) A monotopic protein molecule with a single stretch of mostly hydrophobic and non-ionizable amino acids in α–helical conformation within the bilayer, connected by single peptide bonds to two hydrophilic exterior domains extending into the water on both sides of the membrane. (*B*) A polytopic amphipathic integral protein molecule with, in this case, three transmembrane hydrophobic α–helical stretches connected by hydrophilic sequences extending into the two aqueous phases. The number of transmembrane helices and the membrane sidedness of the N and C termini can differ with different polytopic proteins. (*C*) A subunit aggregate of identical or homologous subunits related by a pseudo-axis of symmetry down the water-filled transmembrane channel along the central axis of the channel, serving in the specific transmembrane transport of small hydrophilic molecules and ions. The numbers of subunits in the aggregate, the numbers of transmembrane helices per subunit, and the membrane sidedness of the N and C termini can differ with different transport proteins.

Transport Proteins

Another type of integral protein structure that I proposed in 1971, depicted schematically in cross section in Figure 2*C*, was designed to provide a thermodynamically feasible mechanism whereby small ions and hydrophilic compounds might be transported across membranes. Because hydrophilic species are insoluble in hydrophobic media, such transport cannot occur by simple diffusion across the lipid of the membrane. The "rotating carrier" model of transport was widely considered in the 1960s. In this model, a transport protein is embedded in the membrane by some unspecified means, containing an active site for a specific hydrophilic species to be transported across the membrane, a site that is exposed to the aqueous phase. On receipt of some signal, the transport protein would rotate within and diffuse across the membrane, depositing the hydrophilic species on the opposite side of the membrane. I considered this transport model to be thermodynamically unsatisfactory because presumably it would require that the ionic and hydrophilic surface domain of the protein move entirely across the hydrophobic lipid bilayer to the other side. Instead, I proposed (1) that all such transport proteins consisted of a small number of specific transmembrane integral protein subunits (generally either identical or homologous to one another so as to form a stable aggregate) that were noncovalently bound to one another around an axis of rotational pseudo-symmetry, creating a continuous water-filled channel along the axis of the aggregate from one side of the membrane to the other (Figure 2C). No example of such a subunit aggregate integral membrane protein with a central water-filled channel was known at the time, but such structures were well-known among water-soluble subunit aggregate proteins such as hemoglobin (9). The thermodynamic structural constraints on the model in Figure 2C are similar to those on the integral proteins depicted in Figures 2A,B, except that the amino acid residues of the helices lining the transmembrane channel could occasionally be ionic or hydrophilic if they were in contact with the water in the channel. With an appropriately positioned active site on one of the subunits for a transportable hydrophilic ligand and some kind of gate in the channel (as represented schematically, for example, in Figure 3), a quaternary conformational change requiring only little energy could be induced in the aggregate that would then expose the active site and bound ligand in the channel to the other side of the membrane, where the ligand could be released.

In the ensuing years, many integral membrane proteins involved in transport have been studied chemically, a few have been studied structurally by electron

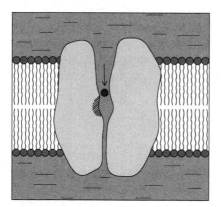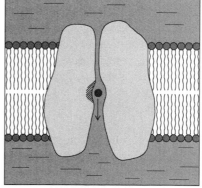

Figure 3   A schematic mechanism of transmembrane transport by a subunit aggregate integral protein with a structure like that in Figure 2C. It includes an active site (*shaded*) on a protein subunit that is specific for the small hydrophilic molecule (*dark sphere*). For active transport the aggregate can exist in two conformations, one favored either by the noncovalent binding of an allosteric modulator, a change in the voltage gradient across the membrane, or by covalent bond formation (e.g., by kinase action). This modification causes a quaternary rearrangement (*left panel* to *right panel*) of the subunits that releases the ion to the aqueous compartment on the other side of the membrane.

microscopy and X-ray crystallography, and most have been found to conform in essential features to the subunit-aggregate model of Figure 2C, with varying numbers of identical or homologous subunits in different such proteins.

Although most of the transmembrane domains of each of the subunits in such transport proteins were suggested and later found to be composed of α-helices, for the reasons previously given, there is one class of transport proteins called porins that is the exception; their general structure is discussed next.

Porins

We suggested above that the transmembrane domains of integral proteins would essentially all be α-helices because this was the most efficient way of forming the $C=O \cdots H-N$ peptide hydrogen bonds required for the formation of stable polypeptide structures in a lipid bilayer. However, in 1978 Kennedy (21) proposed that a stable membrane-intercalated barrel-like protein structure could be constructed with all of the peptide hydrogen bonds formed, which contained only β-pleated sheets and no α-helices. The requirements were that the β-sequences forming the staves of the barrel were sufficiently long to span the membrane and their numbers sufficiently large ($\sim$12–20). In such a case, a single polypeptide chain containing the $\sim$12–20 successive sequences of the long β-strand-forming stretches might produce a barrel of antiparallel β-strands in which all of the $\rangle C=O \cdots H-N \langle$ peptide hydrogen bonds could be formed between adjacent strands because the low degree of curvature of the multi-β-stranded barrel would not induce undue bending strain on the $C=O \cdots H-N$ hydrogen bonds. Every β-strand in the barrel would have successive residues facing the inside (aqueous core) or the outside (lipid bilayer) of the barrel, and the membrane-spanning portion of the β-strand would be restricted to amino acid sequences that accommodated this alternation of hydrophilic and hydrophobic interactions. The water-filled core of the β-barrel might be of significantly larger diameter that that of the water-filled channel of the subunit-aggregate proteins (Figure 2C), with each subunit containing only a small number of α-helical subunits. This would generally make the porins capable of transporting larger hydrophilic species than the subunit-aggregate proteins.

Such β-barrel integral proteins were subsequently demonstrated to be characteristic of the transport proteins called porins, by Weiss et al. (22) and Jap (23). The porins apparently are restricted to the outer membranes of gram-negative bacteria and the outer membranes of mitochondria, and are many fewer than the ubiquitous subunit-aggregate type of transport proteins represented in Figure 2C.

Diffusion Across the Plane of the Membrane

These considerations of the thermodynamics of transfer across the membrane, besides their relevance to the transport of hydrophilic small molecules and ions, also bear on the general problem of membrane asymmetry. I anticipated that once an integral protein became entirely incorporated in a membrane, with its hydrophilic domains protruding from the bilayer into the aqueous phase, it was thermodynamically highly unlikely for these domains to be transferred across the membrane during the lifetime of the protein in the membrane. If during integration of the integral protein in the bilayer, it was inserted in a directional manner, this would therefore lead to an asymmetric membrane, one face having a totally different protein domain composition and character from the other.

The asymmetrical distribution of membrane-bound oligosaccharides on the two surfaces of membranes had been demonstrated several times (cf. 24–26) before we did. We developed ferritin-conjugated plant agglutinins as specific electron stains for oligosaccharides and demonstrated their specific asymmetric binding to the exoplasmic surface of several different membranes (27, 28), including the luminal surface of the endoplasmic reticulum (29). Antigenic epitopes on integral membrane proteins were likewise each shown, by specific ferritin–conjugated antibody labeling (30, 31), to be located on hydrophilic domains of integral proteins exclusively on one surface of their respective membranes. Integral proteins do not exist in an equilibrium distribution across the membrane. (This also means that a peripheral protein that binds specifically to one hydrophilic domain of a particular integral protein will also be asymmetrically bound to its associated membrane.)

Similar considerations apply to membrane lipids that have ionic or oligosaccharide head groups. These lipids do not normally transfer across synthetic phospholipid bilayers or resting membranes at any significant rate. However, during certain biochemically active stages of the cell cycle, transfer of some phospholipids that are ionic at pH 7.0, such as phosphatidyl ethanolamine in bacterial inner membranes, are transferred from one half bilayer (where they are synthesized) to the other (32). The mechanisms for such transfers are not understood but appear to involve the mediation of special integral proteins called flippases (cf. 33). The lipids of membranes are nevertheless asymmetrically distributed in the two halves of the bilayer (see below). Whether this distribution is an equilibrium one for each of the lipids (determined, perhaps, by the nonequilibrium absolutely asymmetrical distribution of the integral proteins in the particular membrane) is not clear. If the lipid flippases are either variably present or variably inactive during the cell cycle, an equilibrium distribution of the individual lipids in the two half bilayers may exist only at certain times, but not at others.

Diffusion in the Plane of the Bilayer

The lipid-protein mosaic model, with the integral proteins embedded in the lipid bilayer, raises the question, what is the distribution of a protein in the plane of the membrane? The lipids in many membranes at 37° had already been shown by 1971 (through a variety of physical methods) to be in a fluid rather than a solid state (34–36) if their fatty acid moieties collectively contained sufficient numbers of unsaturated linkages. If a membrane was a lipid-protein mosaic consisting of individual integral proteins dispersed in a fluid lipid matrix, the membrane might in effect be a viscous two-dimensional liquid-like solution. In a seminar that I gave at the Rockefeller Institute in 1971, I raised the issue of diffusion of proteins in the plane of the membrane, and Siamon Gordon informed me of the paper that had just been published by Frye & Edidin (37). In this remarkable paper, the authors produced mouse-human cell fusion heterokaryons and followed, with time after cell fusion, the surface distributions of a mouse and a human cell surface protein antigen using the double immunofluorescent technique. Immediately after cell fusion the two antigens were confined to the two distinct halves of the heterokaryon surface, but after about 40 min at 37°, the two antigens were completely intermixed over the entire heterokaryon surface. With suitable controls, the authors concluded that global free diffusion of the two antigens in the membrane had occurred.

In our own experiments around this time, we had shown by immunoelectron microscopy that the Rho (D) antigen on human red blood cell membranes (30) and the H-2 alloantigen on mouse red blood cell membranes (31) appeared to exist in small clusters that were randomly distributed over the membrane surfaces. In the light of the Frye-Edidin experiment, these results were consistent with a limited amount of free diffusion of these proteins to form the clusters in the membrane, but we could not rule out the possibility that the clusters were stable aggregates in the membrane. As it turned out, the adult human red blood cell is unique in limiting membrane diffusion because integral proteins in its membrane are bound to an underlying continuous matrix (skeleton) of peripheral proteins, including spectrin, that severely restricts the lateral diffusion of integral proteins in the membrane (cf. 38). Our work employed red blood cells exclusively at that time, and so we missed observing the global diffusion of membrane proteins in our experiments that Frye & Edidin observed with lymphocytes. Other experiments carried out at about the same time by Taylor et al. (39) on the antibody-induced clustering of a cell surface antigen of lymphocytes (the so-called capping phenomenon), cells that do not have a continuous membrane skeleton, also attested to the global character of the movement of protein antigens in membranes. With the appearance of the Frye-Edidin paper, we incorporated the concept of global diffusion into our lipid-protein membrane model, which appeared as the fluid mosaic model in February 1972. Our three-dimensional representation (figure 3 in Reference 1) of the lipid-protein mosaic model, unchanged, was now an instantaneous snapshot of the fluid mosaic model.

The Primary Structure of Glycophorin

In 1978, Marchesi and his colleagues (40) reported the complete amino acid sequence of a major glycoprotein of the human red blood cell membrane, glycophorin A. This was the first complete sequence of an integral membrane protein to be published. The work was the culmination of a difficult experimental program of protein isolation, purification, and peptide (especially of glycopeptide) sequence analysis, before the days when protein sequences were much more readily obtained from the corresponding cDNA sequence.

Glycophorin A is 131 amino acids long, recognizably divided into three sequential domains: with an N-terminal sequence of 72 amino acids, containing 23 ionic residues and the positively charged N terminus, and all of the large number of saccharide units of the protein; a succeeding stretch of 23 residues, mostly hydrophobic, together with a few polar but not a single ionizable residue; followed by a final sequence of 36 residues containing 12 ionic residues plus the negatively charged C terminus. This structure perfectly fits the prediction of the structure of a monotopic integral protein (Figure 2A) of the plasma membrane, with an exterior hydrophilic domain containing many ionic residues and all of the oligosaccharides of the molecule (28), followed by a bilayer-spanning hydrophobic domain containing 23 amino acids (presumably in an α-helix) but no ionizable residues, followed in turn by a cytoplasmic hydrophilic domain containing many ionic residues but no oligosaccharide. The N terminus in this case faces the exoplasm, the C terminus the cytoplasm.

SOME FUNCTIONS OF MEMBRANES PREDICTED FROM THE FLUID MOSAIC MODEL

Up to this point, we have mainly discussed the structure of membranes, as the subject progressed from 1964 to 1972, leading us to formulate on a thermodynamic basis the fluid mosaic model of membrane structure. I turn now to our proposals for the mechanisms of a number of membrane functions that emerged from further consideration of this model.

Signal Transmission Across Cell Membranes

We realized that the presence of amphipathic integral proteins that spanned the plasma membrane of a cell constituted a means of transmitting chemical information (signaling) from the exterior medium into the cell, a subject that was in its infancy at the time. These signaling mechanisms could be initiated by the noncovalent specific transient binding of a water-soluble ligand molecule, say, a hormone (either small, or as large as a protein like insulin) to a site on the exterior-facing hydrophilic domain of a particular amphipathic integral protein. The sequellae to this binding would depend in part on the structure (Figure 2) of that integral protein. (Such protein hormone ligands, by the way, can be construed as one type of peripheral proteins.) The possible effects of such ligand binding to monotopic integral proteins (receptors) is considered first.

Monotopic Proteins

Monotopic proteins (Figure 2A) have a single peptide bond connecting the exterior hydrophilic domain to the α–helical hydrophobic domain, followed by another single bond to the interior hydrophilic domain. In the absence of stable aggregate formation by this monotopic protein with itself or other proteins in the membrane, I considered this structure to make it unlikely that any ligand-binding conformational change induced in the external domain would be directly transmitted across the single helical domain to the interior-facing hydrophilic domain of the same integral protein molecule. Instead, the fluid mosaic model suggested the following mechanistic variation on the Monod-Changeaux-Wyman model of protein cooperative phenomena (41). The exterior hydrophilic domain of the monotopic protein "can exist in one of two conformational states, one of which is favored by [specific] ligand-binding. In the normal unbound conformation, the integral protein is monomolecularly dispersed within the membrane, but in the conformation promoted by ligand binding, its aggregation is thermodynamically favored. The binding of a ligand molecule at one integral protein site, followed by diffusion of [a] nonliganded [molecule of the] protein to it, might then lead to [their] aggregation and spontaneous change in conformation of the [aggregate-bound nonliganded molecule]" (2, p. 729). The induced proximity of the two interior hydrophilic domains of the now dimerized receptor may then activate them. This predicted mechanism of diffusion-controlled cooperative conformational changes has since been found to explain the activation of many monotopic membrane receptors, such as the majority of monotopic receptor tyrosine kinases, where the binding of a hormone to the exterior hydrophilic domain results in a dimerization of the receptor, followed by a mutual tyrosine phosphorylation within their now dimerized and activated interior hydrophilic domains (cf. 42).

[More generally, I suggested that the newly discovered global diffusion in membranes had as its "real purpose...to permit some critical integral proteins to retain their translational mobility in the plane of the membrane, as an obligatory step in their function" (2, p. 730). This has since become clear, for example, in the formation of intracellular synapses and other adhesions, in the endocytosis of membrane proteins, and in intracellular signaling.]

Polytopic Proteins

With polytopic integral proteins (Figure 2b), the situation is quite different. The polytopic protein molecule can be a single integrated structure, unlike the monotopic molecule, which contains three structurally non-integrated domains. The exterior-facing hydrophilic domains and separately, the interior-facing hydrophilic domains are most likely to be structurally integrated at their surfaces with their neighboring hydrophobic helical sequences such that the binding of a specific ligand to the exterior facing domain can lead to a conformational change that affects the structure of the entire protein molecule in the membrane, including the interior-facing domain. The signal could thus be transmitted directly across the single polytopic molecule, and not require the prior diffusion and aggregation of the ligand-bound integral protein in the membrane. This appears to be the general mechanism for transmembrane signaling involving polytopic proteins (Figure 2B), such as with the superfamily of heterotrimeric G protein–coupled receptors (GPCR) (see 43). The noncovalent binding of a ligand (except for the case of the rhodopsins, where the retinal ligand is covalently bound) to the exterior hydrophilic domain of this seven-transmembrane helical protein receptor triggers a conformational change that apparently affects the structure of its integrated interior hydrophilic domain so as to change its binding characteristics for the components of the heterotrimeric G protein in the cytoplasm. This initiates one of several intracellular reaction cascades depending upon the particular G protein involved (44).

Transport Proteins

At a time when no thermodynamically satisfactory model of a transport protein was available and no such protein had as yet been isolated and characterized, I was the first to propose (1) a thermodynamically feasible structure and mechanism for them, namely, that transport proteins would generally consist of small aggregates of identical or homologous transmembrane amphipathic subunits forming a water-filled transmembrane channel down the central axis of the aggregate (Figure 2C). Transmembrane transport of specific ions and hydrophilic small molecules would occur by their movement through the appropriately structured water channel characteristic of each specific transport protein aggregate (Figure 3), via an appropriately stimulated conformational change in the channel. I applied these ideas to propose a mechanism for the ligand-gated ion transport protein, the acetylcholine receptor of the neuromuscular junction.

It was known by 1970 that acetylcholine binding to the acetylcholine receptor could affect the ion permeability of the membrane, but the receptor had not yet been characterized in any detail, and the activation process was not understood. I suggested in 1971 that "the binding of acetylcholine at a specific receptor site on one subunit [of the subunit aggregate constituting the receptor] could induce a quaternary rearrangement of the different subunits of the same protein molecule and thus change the ion-permeability characteristics of a pore extending down the central axis of the aggregate" (1, p. 206–7). This predicted model of the structure and function of the acetylcholine receptor has since been verified in its essential details (45–47). The receptor consists of five subunits: two identical α, and one each of β, γ, and δ subunits, each homologous in amino acid sequence to the α protein (45). The aggregate, of five-fold rotational pseudo-symmetry, has a transmembrane water-filled channel down its central axis (46) which, upon the binding of acetylcholine to the exterior domain of one of the two α subunits (45), causes the channel to undergo a conformational change (47) that increases its ion permeability. In other cases examined in recent years, such as a variety of K^+, Na^+, and Ca^{2+} voltage–gated ion transport proteins, all of which consist of four subunit aggregates, there is a more elaborate structural mechanism for opening and closing the ion channel of the subunit aggregate as determined by high-resolution X-ray crystallographic analysis (cf. 48).

Membrane Asymmetry

The asymmetry of membranes and the associated differences of protein and lipid structures in their two half layers suggested to us some novel properties and functions of membranes. I discuss briefly several of these that occurred to us during the period 1971–1974.

ON THE MEMBRANE INTERCALATION OF INTEGRAL PROTEINS   I suggested that the totally asymmetric orientation of the individual kinds of integral proteins across a membrane entailed certain consequences for the process of the initial intercalation of the integral protein molecule into the membrane. I considered that the amphipathic structure of integral proteins would render them insoluble in water solution (chaperones were unknown at the time). "A ribosome with bound messenger RNA may become attached to a cell membrane. There would have to be specificity to this attachment, so that the membrane proteins would ultimately become associated with the right membrane. . . . As the individual protein molecules are made, they would then be inserted directly into the membrane without being solubilized" (1, p. 215). This would account for a directionality to the integral protein insertion process, starting from the N terminus. These speculations preceded the discoveries of the signal peptides and signal recognition particles (cf. 49) but were generally on the right track.

THE BILAYER COUPLE HYPOTHESIS   The asymmetric distribution of the different lipids in the two halves of the lipid bilayer was first demonstrated for the red blood cell membrane (50, 51), but it is now known to be a general property of cell membranes. Of the major phospholipids of the red blood cell membrane, phosphatidyl choline and sphingomyelin are concentrated in the exterior half of the bilayer and phosphatidyl ethanolamine and phosphatidyl serine in the cytoplasmic half. The first three of these have zwitterionic head groups, the last has a net negativity charged head group. This means that the surface of the cytoplasmic half layer carries not only a chemically different structure but also a net negative charge from its lipids compared with that of the surface of the exterior half layer. We anticipated that these lipid half-layer differences (ignoring differences owing to the asymmetric distribution of different integral proteins in the membrane), could confer somewhat different properties on the two half layers and differential responses to various perturbations of the lipid bilayer of the membrane. By analogy to a bimetallic couple, we referred to this as the bilayer couple hypothesis (52).

Drug-erythrocyte interactions   As one consequence of the bilayer couple hypothesis, we proposed in 1974 (52) that amphipathic drugs, containing both a lipid-soluble hydrophobic domain and an ionically charged group, would distribute differentially into the two half layers of the membrane lipid bilayer of red blood cells, depending on the nature of their ion charge. This would result in different changes in the curvature of the membrane (Figure 4). The normal shape of the red blood cell is the well-known smooth-surfaced biconcave disc (Figure 5A). If a positively charged drug such as chlorpromazine (Figure 6), which can be discharged by the removal of its N-bound proton, is added to the cell, it first enters the exterior half layer of the bilayer by its hydrophobic domain, but it can diffuse across the bilayer as the uncharged species, and then reacquire its proton and positive charge when exposed to the water on the cytoplasmic side. There it is electrically attracted to and concentrated in the more negatively charged cytoplasmic

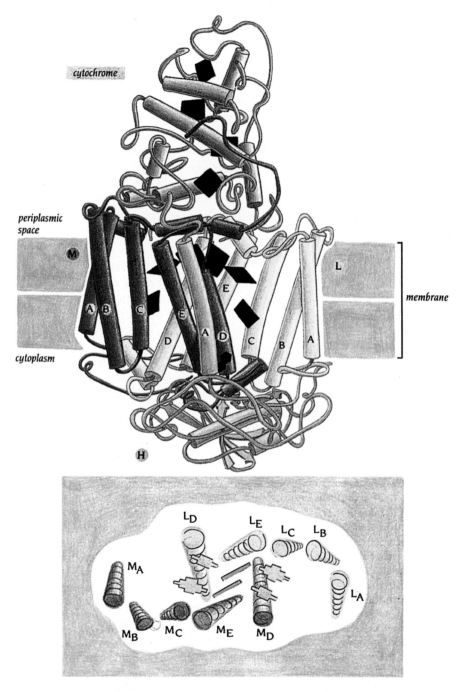

cytochrome

periplasmic space

M

L

membrane

cytoplasm

A B C E E D A D C B A

H

L_D L_E L_C L_B

M_A

L_A

M_B M_C M_E M_D

See legend on next page

Figure 7 Schematic views of the X-ray structure of PRC viewed parallel to the membrane (*top*), and only its transmembrane domain viewed perpendicularly to the membrane from one face (*bottom*). Shown are the homologous L (*yellow*) and M (*red*) integral protein molecules with their respective five transmembrane α-helices (in the order A, B, C, D, and E in the amino acid sequence); the H integral protein molecule (*green*) with its single transmembrane α-helix and large hydrophilic cytoplasmic domain, and the peripheral protein cytochrome (*purple*), along with the more or less linear arrangement of the electron transfer pigments and heme groups (*top, black*). (Taken from Reference 56, with permission).

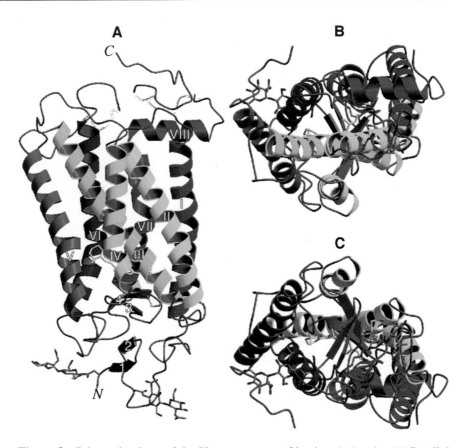

Figure 8   Schematic views of the X-ray structure of bovine rhodopsin: (*A*) Parallel to the membrane, (*B*) view into the plane of the membrane from the cytoplasmic side, and (*C*) from the exoplasmic side of the disc membrane. The seven hydrophobic transmembrane helices are labeled I through VII according to their order in the amino acid sequence. The outer surfaces of the helices in (*A*) coincide, more or less, with the outer surface of the surrounding membrane bilayer. *N* and *C* in part (*A*) represent the N and C termini of the rhodopsin molecule on its exoplasmic and cytoplasmic sides, respectively. Several of the bound pigment molecules are labeled 1–4. (From Reference 63. Copyright 2000, with permission from Routledge/Taylor & Francis Books, Inc.)

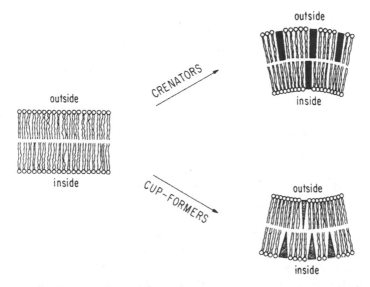

Figure 4 The mechanism of interaction of amphipathic drugs with the asymmetric halves of the lipid bilayer, according to the bilayer couple hypothesis. (See text and Figure 5) (From Reference 52. Copyright 1974, with permission from AAAS.)

half layer, where as a consequence of expanding the area of the cytoplasmic half layer compared with the exoplasmic half (Figure 4), it induces cup–shapes in the intact erythrocyte (Figure 5C). On the other hand, the very similar drug methochlorpromazine (Figure 6), because of its permanent positive charge, cannot readily diffuse across the bilayer. Therefore, it concentrates in the exterior half layer, expanding its area relative to the cytoplasmic half layer, thereby producing a crenated-shaped cell (Figure 4; Figure 5B). Negatively charged drugs such as

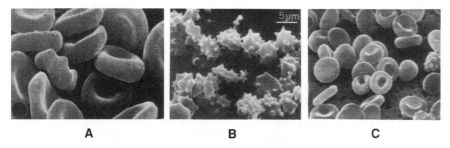

Figure 5 Intact human red blood cells as observed by scanning electron microscopy. (A) normal cells; (B) crenated cells formed after the addition of 0.2 mm methochlorpromazine (Figure 6); (C) cup-shaped cells formed after the addition of 6 μm chlorpromazine (Figure 6). The magnification of Figure 5A is slightly larger than for B and C. See text for details. (From Reference 52. Copyright 1974, with permission from AAAS.)

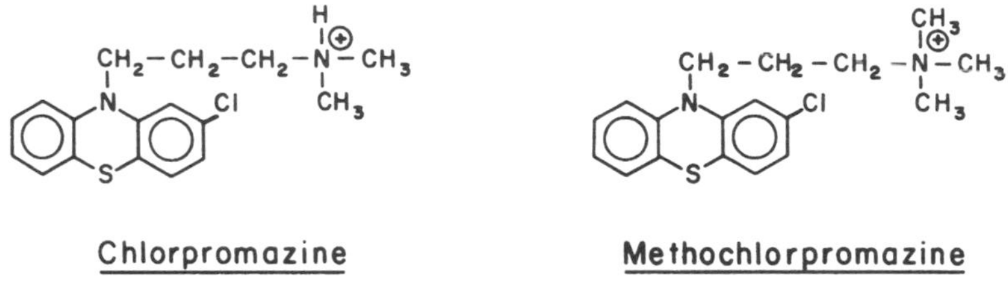

Chlorpromazine   Methochlorpromazine

Figure 6 The structures of two structurally similar amphipathic drugs used in the experiments shown in Figure 5.

2, 4-dinitrophenol do likewise (52, 53) owing to electrical repulsion from the net negatively charged lipids in the cytoplasmic half of the bilayer.

Membrane curvature The bilayer couple hypothesis is relevant to the general property of membrane curvature, a subject that has not received the attention it deserves. Many intracellular membranes of eukaryotic cells, e.g., the Golgi saccules, exhibit regions of sharp curvature at their rims that are continuous with their relatively flat domains. Similarly, regions of flat membranes that become vesiculated, as in endocytosis, or vesicles that undergo fusion with flat cell membranes, as in secretion, undergo changes in curvature. At these regions of high curvature, the half layer on the convex side occupies a larger area than its associated concave half layer, and these changes require some specific mechanism to produce. Such mechanisms are generally unknown. Factors that might be involved include the insertion or diffusion and concentration into the regions of curvature of asymmetrically shaped lipids (54); an asymmetrically distributed addition of electric charge on particular lipids (e.g., by phosphorylation); the concentration of asymmetrically shaped or asymmetrically charged membrane-spanning proteins; or the asymmetrical and localized attachment of peripheral proteins such as clathrin (55).

HIGH-RESOLUTION X-RAY CRYSTALLOGRAPHIC STRUCTURES OF THE PHOTOSYNTHETIC REACTION CENTER AND OF RHODOPSIN

This article has concentrated mostly on the early period in our study of the structure of biological membranes, particularly from 1964 to 1974, culminating in the development and exploration of the fluid mosaic model of membrane structure. At the time, the relevant experimental information, especially about the membrane proteins, was not yet available, and the model was based largely on thermodynamic theoretical considerations. Since that early period, membrane structure studies have become an industry. Many integral proteins have been isolated and characterized biochemically and physico-chemically, and the fluid mosaic model has generally been confirmed by such experiments. The most remarkably detailed information has come from X-ray crystallographic studies at 3 Å resolution, or better, of a number of single integral membrane proteins or protein complexes isolated from their native membranes and crystallized with the use of nonionic detergents. For the purposes of this article, I refer to only two of these studies; one on the photoreaction center (PRC) of photosynthetic bacterial membranes (56), published in 1985; and the other, on bovine rhodopsin (57, 58), published in 2000 and 2002, to determine whether and how well the predictions of the original fluid mosaic model, made 13 to 20 years earlier, fit these remarkably detailed structural results.

The PRC of *Rhodopseudomonas viridis*

The PRC was the first integral membrane protein to have its complete three-dimensional structure determined to better than 3 Å resolution by X-ray crystallography. PRC is a protein complex in the membranes of vesicles inside *R. viridis* cells that acts like a solar battery: It converts light energy into electron energy, and by quantum mechanical electron tunneling between the several kinds of pigments positioned at successive fixed sites within the protein complex, transfers the electrons across the membrane (59, 60). The three-dimensional structure of the PRC membrane complex is represented parallel to the bilayer in the top portion of Figure 7 (see color insert). The complex contains four noncovalently linked proteins: three, H, L, and M, that are transmembrane integral proteins and one, a cytochrome b_5 (Figure 7), that is a peripheral protein attached to the H-L-M complex.

Regarding the predictions of the original fluid mosaic model, they are largely met by the PRC structure. (*a*) The PRC exhibits both monotopic and polytopic types of transmembrane integral proteins, as well as a peripheral protein that is entirely located in the aqueous phase and is attached to the surface of the integral protein complex where the latter protrudes into the periplasmic aqueous phase. (*b*) The transmembrane domains of the integral proteins are entirely α-helical, and range from about 20 to 25 residues in length, as anticipated. (*c*) Most remarkably, of the more than 150 ionic amino acids of the PRC complex, none is located within the 11 transmembrane α–helices except for a few that are situated near either end of a helix where these residues can be in at least partial contact with the aqueous phase. This is in striking accord with the earlier discussion about hydrophobic and hydrophilic interactions.

A closer look at this cross section through the PRC, together with the view from one face of the complex (Figure 7, *bottom*) down through the transmembrane domain, reveals additional interesting features of the structure. (*a*) The homologous M and L chains are related by a pseudo twofold axis of rotation perpendicular to the membrane, but the single transmembrane helix of the H chain is not situated on this axis (Figure 7, *bottom*), as one might initially have expected. This off-axis shift of the H helix might be influenced by its interactions with the large external domain of the H chain and the peripheral cytochrome molecule, both of which are monomeric, and therefore asymmetrically associated with the L-M pseudosymmetric dimer. (*b*) Most of the L and M helices are only rather loosely packed into an overall oval configuration (Figure 7, *bottom*), with the A and D helices of both chains fairly isolated from the others. (*c*) The five helices of either the L or M chains are not arranged precisely in the same order as in the amino acid sequence, with the E and D helices inverted in their positions.

Bovine Rhodopsin

Rhodopsins are integral membrane proteins of rod and cone cells in the retina of the eye that are centrally involved in animal vision. Rhodopsin is a member of a superfamily of heterotrimeric G protein–coupled receptors (GPCR) (cf. 43), each responding to stimulation by a different specific ligand, generally binding to the receptor at its exoplasmic domain. This activates its particular G protein bound at the receptor's cytoplasmic domain, by a conformational change that is transmitted through the transmembrane domain of the polytopic receptor. All the GPCRs so far studied have seven hydrophobic transmembrane helices. Rhodopsin is unusual in having its ligand, 11-*cis* retinal, covalently bound to the receptor. This bond is within the transmembrane domain of rhodopsin: A Schiff base is formed by retinal with K296 of helix VII (61, 62). With all other GPCRs, the ligand is a water-soluble molecule that binds noncovalently to the exoplasmic or transmembrane domain of the receptor. With rhodopsin, the activation occurs by the absorption of a light photon by retinal, which converts the 11-*cis* form to the 11-*trans* isomer, thereby triggering a conformational change in the transmembrane domain of rhodopsin that is transmitted to its cytoplasmic domain.

Recently, nearly the entire three-dimensional structure of bovine rhodopsin has been determined at 2.8 Å resolution (57, 58) (Figure 8, see color insert). Some unusual features of this structure, not exhibited by the PRC complex, are of general interest. (*a*) The transmembrane domain has seven generally hydrophobic α-helices, but these helices are usually quite kinked (and therefore locally non-α-helical), often found at proline residues. (*b*) The seven helices are intertwined with one another in a complex structure in an arrangement that is largely unrelated to their order in the amino acid sequence of rhodopsin. Furthermore, the arrangement of the helices with respect to one another differs at different levels in the membrane, which is probably at least partly associated with the complex pattern of the kinking of the individual helices. (*c*) The transmembrane helices are tightly packed into a complex oval array, as seen from the cytoplasmic surface (Figure 8*B*) and the exoplasmic (intradiscal) surface (Figure 8*C*) of the membrane. This is in contrast to the looser aggregation of the PCR helices (Figure 7, *bottom*). (*d*) There is a total of 11 ionizable residues distributed among five of the seven transmembrane helices of rhodopsin. Six of these residues are near the membrane surfaces, probably in partial contact with the aqueous medium, and therefore present no thermodynamic problem. Of the remaining five, D83 is near the middle of helix II; E113 and E122 in helix III, H211 in helix V, and K296 in helix VII (which makes the Schiff base bond with retinal) are also internal. (*e*) Particularly interesting is the presence of seven individual water molecules at fixed positions within the transmembrane domain, which have to be stably bound in the structure to be detected by X-ray crystallography (58).

Because the notion that amphipathic integral transmembrane proteins even existed, and further that their transmembrane domains were composed largely of hydrophobic α-helices, were largely conjectural in 1971, I had neither the insight nor the courage to think about internally bound water molecules, helical kinking, or possible interhelix arrangements at the time, but they have now become very interesting problems. The rhodopsin structure may have features common to many polytopic integral proteins, compared with the simpler PCR structure, particularly for those proteins, which, like rhodopsin, have to undergo an extended conformational change as part of their biochemical function.

What about the few ionizable residues referred to above that appear to be located in the interior of the transmembrane domain of rhodopsin? As discussed in a previous section, such residues are more stably located if they are in the discharged state in the water-free hydrophobic interior. But some residues such as D-83 and E-113 appear to be hydrogen-bonded to one of the fixed water molecules in the transmembrane domain of rhodopsin. This may further stabilize these discharged ionizable residues in the generally hydrophobic interior. Making and breaking these relatively weak hydrogen bonds may also play a role in any conformational changes involving the membrane-interior portions of the protein.

The complex configurations of the kinked transmembrane helical domains of polytopic integral proteins such as rhodopsin may have important consequences. For example, they may produce crevices in the water-exposed surfaces of such proteins and thereby allow solvent water (and not only the stably bound water molecules as discussed above) to penetrate to some depth into portions of the interior of the transmembrane protein in one or more of its conformations. This transient interior water may make it thermodynamically favorable for certain membrane-interior ionizable residues to ionize transiently in those conformations that allow these residues to be in contact with the water. The crystal structure of rhodopsin shows E113 to be close to the Schiff base linkage, and the existence of an ion-pair between the carboxylate negative charge of the (presumed to be) ionized E-113 and the (presumed to be) positively charged Schiff base has been suggested. As we have demonstrated, however, an ion-pair has a very high free energy (i.e., is highly unfavorable) in a nonpolar environment. These ionizable groups may therefore be ionized only in those conformations that allow water into interior parts of the transmembrane domain. In other conformations in the activation cycle, the region around E-113 and the Schiff base may be closed to water permeation and be hydrophobic. In such a conformation, both ionizable groups would then likely exist in the discharged state and be hydrogen bonded to one another.

Such water penetration into transmembrane crevices could also be involved in the mysterious proteolysis of peptide bonds that seem to be located at some depth in the apparently hydrophobic interior of a polytopic protein (cf. 63). In any event, water must have access to such apparently buried proteolyzable peptide bonds, because the elements of water have to be added across the broken bond. But such crevices might also transiently admit some appropriately shaped hydrophilic domains of proteases to hydrolyze such interior peptide bonds.

Lastly, the proximity of E113 (on helix III) to the Schiff base attached to K296 (on helix VII) of rhodopsin also provides an opportunity to emphasize the obvious fact, too often ignored, that the linear, sequence-ordered topography of the hydrophobic domains of integral proteins that we usually draw from hydropathy plots (64) of polytopic integral proteins has no bearing on the real three-dimensional appositions of the helices in the transmembrane domain.

CONCLUSION

The early days of membrane molecular biology have been treated here from a personal perspective. In the absence of even one isolated and structurally characterized membrane protein, the analysis of membrane structure that I engaged in was mainly of thermodynamic nature. This, together with some critical experiments by others on the global diffusion of proteins in membranes, led me to the fluid mosaic model of membrane structure, which is still valid today as a general overview of the structures and functions of biological membranes. It has held up to the subsequent structural information obtained with many isolated membrane proteins and their complexes. A particularly exciting period in membrane molecular biology has been introduced by the successful application of X-ray crystallographic methods to crystals of individual membrane proteins and protein complexes in nonionic detergents, with a resolution of 3 Å or better. However, it is still a high art to produce usable crystals of such water-insoluble proteins. But the extraordinary insights provided by this method are well worth the effort. Appropriate chemical modification and bioengineering of specific membrane proteins may be useful in making suitable crystals of these proteins easier to produce; for example, methods may be developed such as attaching readily crystallized and already structurally analyzed water-soluble proteins to the membrane proteins in such defined ways that do not affect the structures of either. These and other technological advances can be anticipated in the future and will expedite the coming revolution in the detailed understanding of membrane molecular biology.

ACKNOWLEDGMENTS

I gratefully acknowledge the cooperation of the many outstanding postdoctoral fellows and graduate students who participated in the work carried out during the period from 1964–1974 that is, in part, presented in this review.

The *Annual Review of Physiology* is online at http://physiol.annualreviews.org

LITERATURE CITED

1. Singer SJ. 1971. The molecular organization of biological membranes. In *Structure and Function of Biological Membranes*, ed. LI Rothfield, pp. 145–222. New York: Academic
2. Singer SJ, Nicolson GL. 1972. The fluid mosaic model of the structure of cell membranes. *Science* 175:720–31
3. Gorter E, Grendel F. 1925. On bimolecular layers of lipoids on the chromocytes of the blood. *J. Exp. Med.* 41:439–43
4. Bar KS, Deamer DW, Cornwell DG. 1966. Surface area of human erythrocyte lipids: reinvestigation of experiments on plasma membrane. *Science* 153:1010–12
5. Robertson JD. 1964. Unit membranes: a review with recent new studies of experimental alterations and a new subunit structure in synaptic membranes. In *Cellular Membranes in Development*, ed. M Locke, pp. 1–81. New York/London: Academic
6. Davson H, Danielli JF. 1952. *The Permeability of Natural Membranes*. London/New York: Cambridge Univ. Press. 2nd ed.
7. Kauzmann W. 1959. Some factors in the interpretation of protein denaturation. *Advan. Protein Chem.* 14:1–63
8. Kendrew JC. 1961. The three-dimensional structure of a protein molecule. *Sci. Am.* 205:96–111
9. Perutz MF. 1964. The hemoglobin molecule. *Sci. Am.* 211:64–76
10. Klotz IM, Franzen JS. 1962. Hydrogen bonds between model peptide groups in solution. *J. Am. Chem. Soc.* 84:3461–66
11. Cohn EJ, Edsall JT. 1943. *Proteins, Amino Acids, and Peptides*. New York: Reinhold
12. Benson AA. 1966. On the orientation of lipids in chloroplast and cell membranes. *J. Am. Oil Chem. Soc.* 43:265–70
13. Lenard J, Singer SJ. 1966. Protein conformation in cell membrane preparations as studied by optical rotary dispersion and circular dichroism. *Proc. Natl. Acad. Sci. USA* 56:1828–35
14. Wallach DFH, Zahler PH. 1966. Protein conformations in cellular membranes. *Proc. Natl. Acad. Sci. USA* 56:1552–59
15. Anderson RGW, Jacobson K. 2002. A role for lipid shells in targeting proteins to caveolae, rafts, and other lipid domains. *Science* 296:1821–25
16. Pinto da Silva P, Branton D. 1970. Membrane splitting in freeze etching. *J. Cell Biol.* 45:598–605
17. Bretscher M. 1971. A major protein which spans the human erythrocyte membrane. *J. Cell Biol.* 59:351–57
18. Lenard J, Singer SJ. 1968. Structure of membranes: reaction of red blood cell membranes with phospholipase C. *Science* 159:738–39
19. Singer SJ. 1974. Molecular organization of membranes. *Annu. Rev. Biochem.* 43:805–33
20. Ito A, Sato R. 1968. Purification by means of detergents and properties of cytochrome b_5 from liver microsomes. *J. Biol. Chem.* 243:4922–23
21. Kennedy SJ. 1978. Structures of membrane proteins. *J. Membr. Biol.* 42:265–79
22. Weiss MS, Wacker T, Weckesser J, Welte W, Schulz GE. 1990. The three-dimensional structure of porin from *Rhodobacter capsulatus* at 3 Å resolution. *FEBS Lett.* 267:268–72
23. Jap BK. 1989. Molecular design of PhoE porin and its functional consequences. *J. Mol. Biol.* 205:407–19
24. Benedetti EL, Emmelot P. 1967. Studies on plasma membranes. IV. The ultrastructural localization and content of sialic acid in plasma membranes isolated from rat liver and hepatoma. *J. Cell Sci.* 2:499–512
25. Eylar EH, Madoff MA, Brody OV, Oncley JL. 1962. The contribution of sialic acid

to the surface charge of the erythrocyte. *J. Biol. Chem.* 237:1992–2000

26. Gasic GJ, Berwick L, Sorrentino M. 1968. Positive and negative colloidal iron as cell surface electron stains. *Lab Invest.* 18:63–71

27. Nicolson GL, Singer SJ. 1971. Ferritin-conjugated plant agglutinins as specific saccharide stains for electron microscopy: application to saccharides bound to cell membranes. *Proc. Natl. Acad. Sci. USA* 68:942–45

28. Nicolson GL, Singer SJ. 1974. The distribution and asymmetry of mammalian cell surface saccharides utilizing ferritin-conjugated plant agglutinins as specific saccharide stains. *J. Cell Biol.* 60:236–48

29. Hirano H, Parkhouse B, Nicolson GL, Lennox ES, Singer SJ. 1972. Distribution of saccharide residues on membrane fragments from a myeloma-cell homogenate: its implications for membrane biogenesis. *Proc. Natl. Acad. Sci. USA* 69:2945–49

30. Nicolson GL, Masouredis SP, Singer SJ. 1971. Quantitative two-dimensional ultrastructural distribution of Rho (D) antigenic sites on human erythrocyte membranes. *Proc. Natl. Acad. Sci. USA* 68:1416–20

31. Nicolson GL, Hyman R, Singer SJ. 1971. The two-dimensional topographic distribution of H-2 histocompatibility alloantigens on mouse red blood cell membranes. *J. Cell Biol.* 50:905–10

32. Rothman J, Kennedy EP. 1977. Rapid transmembrane movement of newly synthesized phospholipids during membrane assembly. *Proc. Natl. Acad. Sci. USA* 74:1821–25

33. Daleke DL, Lyles JV. 2000. Identification and purification of aminophospholipid flippases. *Biochim. Biophys. Acta* 1486:108–27

34. Hubbell WL, McConnell HM. 1968. Spin-label studies of the excitable membranes of nerve and muscle. *Proc. Natl. Acad. Sci. USA* 61:12–16

35. Wilkins MHF, Blaurock AE, Engelman DM. 1971. Bilayer structure in membranes. *Nat. New Biol.* 230:72–76

36. Melchoir DL, Morowitz HJ, Sturtevant JM, Tsong TY. 1970. Characterization of the plasma membrane of *Mycoplasma laidlawii*. VII. Phase transitions of membrane lipids. *Biochim. Biophys. Acta* 219:114–22

37. Frye CD, Edidin M. 1970. The rapid intermixing of cell surface antigens after formation of mouse-human heterokaryons. *J. Cell Sci.* 7:319–35

38. Tokuyasu KT, Schekman R, Singer SJ. 1979. Domains of receptor mobility and endocytosis in the membranes of neonatal human erythrocytes and reticulocytes are deficient in spectrin. *J. Cell Biol.* 80:481–86

39. Taylor RB, Duffus WPH, Raff MC, de Petris S. 1971. Redistribution and pinocytosis of lymphocyte surface immunoglobulin molecules induced by anti-immunoglobulin antibody. *Nat. New Biol.* 233:225–29

40. Tomita M, Furthmayr H, Marchesi VT. 1978. Primary structure of human erythrocyte glycophorin A. Isolation and characterization of peptides and complete amino acid sequence. *Biochemistry* 17:4756–70

41. Monod J, Changeux JP, Wyman J. 1963. Allosteric proteins and cellular control systems. *J. Mol. Biol.* 6:306–29

42. Schlessinger J, Ullrich A. 1992. Growth factor signaling by receptor tyrosine kinases. *Neuron* 9:383–91

43. Vassilatis DK, Hohmann JG, Zeng H, Li F, Ranchalis JE, et. al. 2003. The G protein-coupled receptor repertoires of human and mouse. *Proc. Natl. Acad. Sci. USA* 100:4903–8

44. Hur EM, Kim KT. 2002. G protein-coupled receptor signalling and cross-talk. Achieving rapidity and specificity. *Cell. Signal.* 14:397–405

45. Weill CL, McNamee MG, Karlin A. 1974. Affinity-labeling of purified acetylcholine receptor from *Torpedo californica*. *Biochim. Biophys. Res. Commun.* 61:997–1003

46. Unwin N. 1993. Nicotinic acetylcholine receptor at 9 Å resolution. *J. Mol. Biol.* 229:1101–24

47. Unwin N. 1995. Acetylcholine receptor channel imaged in the open state. *Nature* 373:37–43

48. Jiang Y, Lee A, Chen J, Cadene M, Chait BT, et al. 2002. Crystal structure and mechanism of a calcium-gated potassium channel. *Nature* 417:515–22

49. Blobel G. 2000. Protein targeting (Nobel Lecture). *Chembiochem* 1:87–102

50. Zwaal RFA, Roelofsen B, Colley CM. 1973. Localization of red cell membrane constituents. *Biochim. Biophys. Acta* 300:159–82

51. Verkleij AJ, Zwaal RFA, Roelofsen B, Comfurius P, Kastelijn D, et. al. 1973. The asymmetric distribution of phospholipids in the human red cell membrane. A combined study using phospholipases and freeze-etch electron microscopy. *Biochim. Biophys. Acta* 323:178–93

52. Sheetz MP, Singer SJ. 1974. Biological membranes as bilayer couples. A molecular mechanism of drug-erythrocyte interactions. *Proc. Natl. Acad. Sci. USA* 71:4457–61

53. Sheetz MP, Singer SJ. 1976. Equilibrium and kinetic effects of drugs on the shapes of human erythrocytes. *J. Cell Biol.* 70:247–51

54. Israelachvili J, Marcelja S, Horn R. 1980. Physical principles of membrane organization. *Q. Rev. Biophys.* 13:121–200

55. Pearse BMF. 1976. Clathrin: a unique protein associated with the intracellular transfer of membrane by coated vesicles. *Proc. Natl. Acad. Sci. USA* 73:1255–59

56. Deisenhofer J, Epp O, Miki K, Huber R, Michel H. 1985. Structure of the protein subunits in the photosynthetic reaction center of *Rhodopseudomonas viridis* at 3 Å resolution. *Nature* 318:618–24

57. Palczewski K, Kumasaka T, Hori T, Behnke CA, Motoshima H, et. al. 2000. Crystal structure of rhodopsin: a G protein-coupled receptor. *Science* 289:739–45

58. Okada T, Fujiyoshi Y, Silow M, Navarro J, Landau EM, et. al. 2002. Functional role of internal water molecules in rhodopsin revealed by X-ray crystallography. *Proc. Natl. Acad. Sci. USA* 99:5982–87

59. Brandon C, Tooze J. 1991. *Introduction to Protein Structure*, pp. 203–12. New York/London: Garland

60. Knapp EW, Fischer SF, Zinth W, Sander M, Kaiser W, et al. 1985. Analysis of optical spectra from single crystals of *Rhodopseudomonas viridis* reaction centers. *Proc. Natl. Acad. Sci. USA* 82:8463–67

61. Ovchinnikov YA. 1982. Rhodopsin and bacteriorhodopsin: structure-function relationships. *FEBS Lett.* 148:179–91

62. Dunn RJ, Hackett NR, Huang KS, Jones SS, Khorana HG. 1983. Studies on the light-transducing pigment bacterial rhodopsin. *Cold Spring Harbor Symp. Quant. Biol.* 48:853–62

63. Brown M, Ye J, Rawson RB, Goldstein JL. 2000. Regulated intramembrane proteolysis: a control mechanism conserved from bacteria to humans. *Cell* 100:391–98

64. Kyte J, Doolittle RF. 1982. A simple method for displaying the hydropathic character of a protein. *J. Mol. Biol.* 157:105–32

Annu. Rev. Physiol. 2004. 66:29–48
doi: 10.1146/annurev.physiol.66.032102.140723
Copyright © 2004 by Annual Reviews. All rights reserved
First published online as a Review in Advance on August 18, 2003

MYOCARDIAL AGING AND SENESCENCE:
Where Have the Stem Cells Gone?

Mark A. Sussman
*SDSU Heart Institute, San Diego State University, Department of Biology, LS426,
San Diego, California 98182; email: sussman@heart.sdsu.edu*

Piero Anversa
*New York Medical College, Cardiovascular Research Institute, Department of Medicine,
Valhalla, New York 10595; email: piero_anversa@nymc.edu*

Key Words   heart, cardiomyocyte, telomerase, p53, Akt, niche

■ **Abstract**   Heart failure remains a leading cause of hospital admissions and mortality in the elderly, and current interventional approaches often fail to treat the underlying cause of pathogenesis. Preservation of structure and function in the aging myocardium is most likely to be successful via ongoing cellular repair and replacement, as well as survival of existing cardiomyocytes that generate contractile force. Research has led to a paradigm shift driven by application of stem cells to generate cardiovascular cell lineages. Early controversial findings of pluripotent precursors adopting cardiac phenotypes are now widely accepted, and current debate centers upon the efficiency of progenitor cell incorporation into the myocardium. Much work remains to be done in determining the relevant progenitor cell population and optimizing conditions for efficient differentiation and integration. Significant implications exist for treatment of pathologically damaged or aging myocardium since future interventional approaches will capitalize upon the use of cardiac stem cells as therapeutic reagents.

INTRODUCTION

The primary function of the heart, to generate the force that pumps blood, depends upon the coordinated contraction of cardiac myocytes. Approximately five billion myocytes are contained within a normal human left ventricle (1), although this cell population is depleted with age by both extrinsic and intrinsic factors. A lifetime of cardiomyopathic insults combined with cellular senescence leaves the aged myocardium with lower hemodynamic performance and a diminished contractile reserve (2–10). Often, this impaired function leads to a structural remodeling downward spiral ending in failure. The aged myocardium is also more susceptible to damage following pathological insults such as reperfusion injury (11). Consequently, heart failure among the elderly remains a leading cause of morbidity and mortality in the industrialized world. Despite a pharmacopoeia based on decades of

0066-4278/04/0315-0029$14.00

research, current interventional approaches fail to treat the predominant underlying defect: cardiomyocytes that are not replaced, together with declining function of the remaining cells.

AGING, SENESCENCE, AND STEM CELL REGENERATION

Aging is a stochastic process that involves apparently random variables and manifestations; genes, environment, and probabilistic changes all contribute to define the lifespan of the organism (12, 13). Extrinsic and intrinsic factors cooperate in the determination of the rate of aging and the aging phenotype (14, 15). Genetically engineered models of aging and premature aging syndromes in humans dictated by genetic mutations indicate that senescence is at least in part genetically controlled (16–18). In spite of the presence of a systemic clock that regulates the aging of the organism as a whole, organs age at a different pace. An example is provided by the progeroid syndromes, in which the same genetic mutation hits all the cells of the organism, but has different consequences on different organs (18). The central nervous system and the immune system are unaffected by the disease for a long time, whereas other organs are severely impaired at an early age (19, 20).

Cardiomyocyte death is inextricably linked to the normal life cycle of myocardial structure and function (21). Cardiomyocyte apoptosis is no longer considered controversial, and cumulative evidence points to a contributory role of apoptosis in at least some cardiomyopathic diseases. Acceleration of cell death in the aging myocardium (22) has been proposed as a mechanism for decreased hemodynamic performance and increased risk of heart failure in the elderly (23, 24). Prevailing dogma has maintained that cardiomyocytes are not replaced and that the cumulative loss of these cells resulting from pathological insults or aging impairs contractile performance of the heart. Concurrently, aging myocytes accumulate markers of cellular senescence that typically appear as cells near the end of their useful lifespan. Cellular senescence is a complex phenotype that entails changes in both function and replicative capacity. Unlike apoptosis, which eliminates damaged cells from tissues, senescent cells remain alive despite changes in morphology, metabolism, and derangement of differentiated functions (25). In this fashion, senescence has been proposed to contribute to the deterioration of cell functions in aging.

Accumulation of senescent cardiomyocytes takes a toll on myocardial structure and function. Decreased contractile performance results from changes at the molecular level in expression of contractile protein isoforms and altered levels of calcium-handling proteins (reviewed in 26). Impaired calcium handling in senescent myocytes (27, 28) inevitably leads to reactive hypertrophy that increases cell volume without accompanying improvement of contractility (29). Aged cardiomyocytes contribute to heart failure not only with poor function, but also by promoting maladaptive rather than adaptive remodeling. Proliferation of myocyte nuclei, increased myocyte volume, and connective tissue accumulation are the major determinants of remodeling in the hypertrophic senescent myocardium (30).

Enlarged cardiomyocytes are incapable of hypertrophy and proliferation, express cell cycle inhibitors, and cannot be transcriptionally reprogrammed in response to increased workload (31, 32). Aging also elevates wall stress that is not normalized by ventricular remodeling (33), and the capacity to adaptively remodel the myocardium is lost in elderly patients (34). Optimal therapeutic interventions to inhibit heart failure and antagonize aging must not only prevent cell death, but also antagonize the development and accumulation of senescent cardiomyocytes. Ideally, this could be accomplished by a combination of delayed senescence together with promotion of regeneration of cardiomyocytes lost through time and pathological damage.

Aging is a heterogeneous process. The study of aging is simple in single-cell organisms, in which the aging of the entire body reflects the aging of its unique cell (35). When more complex animals are evaluated, the correlation among cellular aging, tissue or organ aging and, finally, organism aging is not easily established. However, eukaryotic cells age individually in vitro and in vivo, in both simple and complex organisms (36, 37). The independent lifespan of each cell suggests that a major component of aging occurs at the cellular level and that key questions about aging have to be addressed at this level. The rate of organ aging depends strictly on the pace of cell-autonomous processes resulting in the accumulation of older cells with impaired function (37). This accumulation is, in turn, linked to a decline in the capacity of the organ to replace old cells with better functioning younger cells. Therefore, aging of the heart is defined at the level of the controlling cell, i.e., the cardiac stem cell (CSC).

THE NEW PARADIGM OF MYOCARDIAL CELLULAR REPLICATION

The widely held view that cardiomyocytes are incapable of regeneration is based upon lack of evidence to the contrary: Continuous cardiomyocyte cell lines with highly differentiated features could not be established, cytokinesis of adult cardiomyocytes was a relatively rarely observed event, and prompting of adult cardiomyocytes to reenter the cell cycle by molecular alterations of cell cycle control was unsuccessful. Moreover, normal hearts challenged with pathological insults remain impaired rather than repairing the damage. Although these incontrovertible observations are true, anecdotal evidence left perplexing questions that challenged the status quo (38, 39). Space limitations preclude an in-depth discussion of this topic, which has been recently reviewed (40). Briefly, morphometric analysis indicates that the number of myocytes increases from birth to young adulthood in rodents and humans (38). The application of cytological and biochemical markers for cycling cells and DNA synthesis to the myocardium revealed a frequency of cardiomyocyte replication heretofore unsuspected and prompted further investigation into the origin of these cells, whether derived from existing myocytes or from a progenitor cardiomyocyte precursor population, previously referred to as cardiac stem cells [engendering a substantial amount of heated debate (41–43)]. This new conceptualization has forced reevaluation of regeneration and perpetuation of cardiomyocytes throughout life (Figure 1, see color insert).

A major paradigm shift has taken hold of the cardiovascular community fueled by the discovery that stem cells have the capacity to differentiate into the three major cell types of the myocardium: smooth muscle, endothelial, and cardiomyocyte. The persistence of this observation through independent validation in multiple laboratories throughout the world (44–46) bodes well for the potential of this area to have a major impact upon cardiovascular research. Despite recent disputes regarding frequency (47–49), the prevailing opinion is that such transdifferentiation events do occur. Within the past two years, coincident efforts by several groups have demonstrated integration of stem cells into the intact heart (50–53). Some studies have observed repopulation of damaged myocardium by bone marrow–derived stem cells that confer improved hemodynamic function (54–57). Thus the question no longer appears to be whether cardiomyocyte replacement occurs, but rather what are the characteristics of the progenitor population, how efficiently can such cells integrate into the functioning myocardium, and can the process be manipulated by genetic or biochemical techniques to potentiate the regenerative properties of these cells?

Much of the current debate in cardiac stem cell research revolves around issues of the frequency and efficiency of integration and differentiation of stem cells into myocardial tissue. Sorting out the specifics is clouded by multiple controversies (58, 59). First, estimates vary widely with regard to the capacity of donated cells to survive and subsequently integrate into host tissue. Second, skeptics raised concern over fusion between donated cells and the adopted home tissue creating cellular chimerism that could masquerade as integration (60–64). Third, there is an issue of whether donated stem cells promote recruitment of endogenous progenitors to assist in regeneration. Fourth, the influence of endogenous factors such as cytokines, growth factors, and the local cellular milieu upon stem cells remains poorly understood. Resolution of these key issues will ultimately determine the efficacy and feasibility of stem cell therapy as an interventional approach to repair damaged or failing myocardium. Regardless of whether one is a zealous advocate or confirmed cynic, all parties agree that improving stem cell trafficking, survival, and integration into the myocardium will eventually lead to major advances in treatment of cardiac disease and could revolutionize management of declining heart function in the aged.

STEM CELL POPULATION CHARACTERISTICS

Markers

Stem cells are defined as clonogenic cells capable of self-renewal and multilineage differentiation (65, 66). Stem cells give rise to oligolineage progenitors that lose self-renewal properties, forming progeny with limited differentiating potential. The ultimate fate of progeny is the generation of a functionally competent mature cell (65). This model of cell growth has been identified only in the hematopoietic system. However, it is commonly applied to all self-renewing organs. Recent observations made in our laboratory indicate that the heart possesses a subpopulation of cells that have the functional characteristics of stem cells (59, 67). These undifferentiated cells are clonogenic, self-renewing, and pluripotent. On the basis of in vitro and in vivo results, three classes of CSCs have been identified: c-kit, MDR1, and Sca-1-like-positive cells (59). These surface antigens, typically present in hematopoietic stem cells (HSCs), have been detected in CSCs independently or in combination. Although c-kit^{POS} cells can grow and differentiate in the various cardiac lineages (67), it is not known whether MDR1 and Sca-1-like-positive cells have the same potential. The expression of MDR1 and the ability to exclude the Hoechst dye defines a small group of HSCs, called side population, in the bone marrow (68, 69) and skeletal muscle (70, 71). MDR1^{POS} cells in the bone marrow give rise to the myeloid, lymphoid, and erythroid cell lineages (68, 69, 72, 73), whereas in skeletal muscle they regenerate muscle fibers (74). Additionally, MDR1^{POS} cells from skeletal muscle are of bone marrow origin and can repopulate a depleted bone marrow (75). Preliminary results in vivo in humans (59, 76) and in animal models (77, 78) have shown that MDR1^{POS} cells differentiate in progenitors and precursors of myocytes, endothelial cells, smooth mucle cells, and fibroblasts. A similar pattern of growth has been recognized in vitro by differentiation of clonogenic cells. Sca-1-like protein, a marker of stem cells, has been found by confocal microscopy in undifferentiated cells in mouse (78), rat (79), dog (80), and pig cardiac tissue (P. Anversa, unpublished observation), and in human heart (59, 76, 79). Sca-1-like protein is detected not only on stem cells but also on the surface of endothelial cells (76, 78), suggesting that this antigen may be particularly sensitive to growth factors active in the generation of vascular structures. MDR1^{POS} cells differentiate in skeletal muscle cells (74), and this preferential lineage commitment points to a greater potential for the formation of cardiac myocytes. The most powerful regenerating cells of the heart might be those that express both c-kit and MDR1. This cell group may divide and differentiate into myocytes and coronary-resistance arterioles and capillary profiles, essentially producing de novo myocardium. Differential expression of stem cell surface antigens may result in distinct cellular growth responses and lineage commitments, leading to a prevailing formation of myocytes, coronary vessels, or a combination of both.

Additional Phenotypic Characteristics

We have documented that the adult rat heart contains a population of undifferentiated cells that express surface antigens typically found in HSCs (65, 72, 81, 82): c-kit, MDR1, and Sca-1 (77, 78, 83, 84). Similar results have been obtained in the canine (80), pig (P. Anversa, unpublished results) and, most importantly, in the human heart (59, 76; P. Anversa, unpublished observation). A restricted pool of cells positive for a Sca-1-like epitope (59, 76) has been identified utilizing antibodies against the mouse Ly6A/E protein (76, 77). Cells with an identical phenotype have been detected in the normal human bone marrow, which also possesses cells positive for both c-kit and Sca-1-like protein (P. Anversa, unpublished observation). Lineage-negative c-kit-positive (Lin^-c-kit^{POS}) cells have been collected from the rat heart and are self-renewing, clonogenic, and pluripotent (67). Single clonogenic cells were able to differentiate into myocytes, endothelial cells, smooth muscle cells, and fibroblasts. Thus CSCs can generate all the components of the myocardium. CSCs have been isolated from the myocardium, and long-term cultures have been developed (67). Thus far only Lin^-c-kit^{POS} cells have been shown to have the three components of stemness: self-renewal, clonogenicity, and multipotentiality. The properties of MDR1 and Sca-1-like-positive cells have not been determined. In vitro, Lin^-c-kit^{POS} cells grow as a monolayer when seeded in substrate-coated dishes or form spheroids when cultured in suspension, mimicking the biology of neural stem cells (67, 86). In vivo, CSCs implanted in the infarcted ventricle or mobilized from sites of storage to the dead myocardium (77, 87) migrate, engraft, undergo multilineage commitment, and promote cardiac regeneration. Thus CSCs exist but their distribution and growth characteristics in the various regions of the heart are unknown.

NICHES   Differences in microenvironment and mechanical forces are present in the anatomical parts of the heart. These variables may play a critical role in dictating the function and fate of CSCs. Such information is crucial for understanding the effects of structural and physical factors on the behavior of CSCs in the adult heart (Figure 2, see color insert). Stem cells are usually sheltered in specialized structures called niches, a concept introduced in 1978 by Schofield, defined as "a stable microenvironment that might control hematopoietic stem cell behavior" (88). Recently, the niche has been viewed as "a subset of tissue cells and extracellular substrates that can indefinitely house one or more stem cells and control their self-renewal and progeny production in vivo" (89). Stem cell niches provide a microenvironment designed to preserve the survival and replication potential of stem cells (89). Primitive cells divide rarely, and those endowed in a niche have a much higher probability of self-renewal (89). Niches are present in all self-renewing organs (90–92). Preliminary results indicate that CSCs are stored in niches that are preferentially located in the atria and apex but are also detectable in the ventricle (77, 93, 94). The peculiar topographical distribution of CSCs in the heart suggests that a relationship may exist between the function of CSCs and level of hemodynamic stress. CSCs accumulate in niches located in the atria and apex that are anatomical areas exposed to low levels of wall stress. The preferential localization of CSCs in zones where physical forces are modest is consistent with the common sites of storage of stem cells in self-renewing organs (89–91): CSCs occupy the most protected areas of the heart.

RECEPTORS AND PARACRINE FACTORS   CSCs express two surface receptors: c-Met and IGF-1R (77). c-Met is the receptor for human growth factor (HGF) and insulin-like growth factor (IGF)-1R is the receptor for IGF-1. The HGF-cMet and

Hypothetical Cardiomyocyte Life Cycle

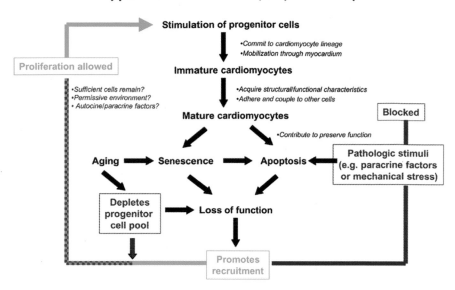

Figure 1 Flowchart of cardiomyocyte life and death. Major stages of life and influences are represented in bold type. Negative forces (*red*) and positive forces (*green*) influence progenitor cells. Features of the cell population (*italicized*) change as cells progress through life. Proliferative capacity depends upon multiple factors (*blue italicized*).

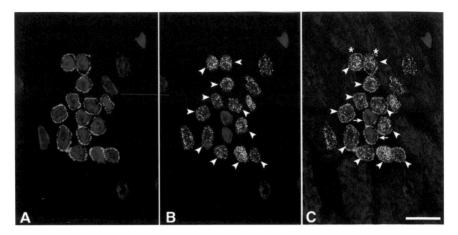

Figure 2 Clusters of primitive and early committed cells. Section of the left ventricle of an adult rat heart. Thirteen small cells are present within the myocardium. *Panel A* illustrates nuclei by the blue fluorescence of propidium iodide. The stem cell maker c-kit is shown on the surface of all cells by green fluorescence. *Panel B* demonstrates that 11 (*arrowheads*) out of 13 nuclei express the cardiac-specific transcription factor GATA-4 represented by yellow fluorescent dots. The combination of c-kit-positive GATA-4-positive cells (*arrowheads*), c-kit-positive–GATA-4-negative cells (*arrows*) and myocytes is documented in *panel C*. The myocyte cytoplasm is recognized by the red fluorescence of α-sarcomeric actin. Two of the 11 c-kit-positive GATA-4-positive cells (*asterisks*) possess a thin layer of myocyte cytoplasm. Confocal microscopy. Scale bar = 10 μm.

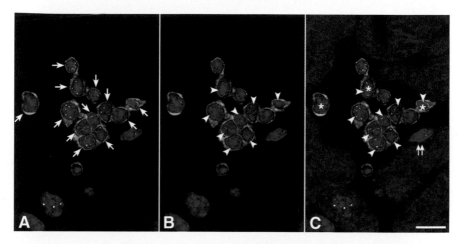

Figure 3 Clusters of cycling primitive and early committed cells. Section of left atrium of an adult mouse heart. Thirteen small cells are present within the interstitium of atrial tissue. *Panel A* illustrates nuclei by the blue fluorescence of propidium iodide. The stem cell maker c-kit is shown on the surface of 13 cells (*arrows*) by green fluorescence. Ten of the 13 c-kit positive cells are telomerase-competent. The expression of the catalytic subunit of telomerase is characterized by white fluorescent dots. *Panel B* demonstrates that 8 of the 10 c-kit-positive-telomerase-competent cells (*arrowheads*) express the cell cycle marker MCM5, represented by magenta fluorescent dots. The combination of cells positive for c-kit, telomerase and MCM5 (*arrowheads*) is documented in *panel C*. The myocyte cytoplasm is recognized by the red fluorescence of α-sarcomeric actin. Three of the 13 c-kit-positive cells (*asterisks*) possess a thin layer of myocyte cytoplasm. A small developing myocyte that lost the stem cell surface antigen is also present (*double arrows*). Confocal microscopy. Scale bar = 10 μm.

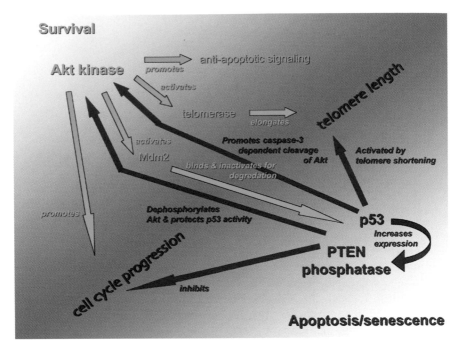

Figure 4   Flowchart of the relationship between p53 versus Akt-mediated survival signaling. Negative forces (*red*) and positive forces (*green*) influence cell survival and proliferation. Features of the cell population related to capacity for proliferation, such as telomere length and cell cycle checkpoints, are caught in the balance (*purple*).

IGF-1-IGF-1R systems are present in CSCs of mice, rats, and dogs. Binding to the c-Met receptor promotes invasion of primitive cells in vitro and in vivo (77, 87, 95, 96). IGF-1 has no locomotion property in vitro and in vivo (77) but is implicated in cell growth and differentiation (77, 87, 95–98). IGF-1 activates IGF-1R, which results in the translocation of Akt to the nucleus and, thereby, promotes cell survival (99, 100). HGF also prevents cell death through the PI3-K/Akt pathway (101, 102). c-Met binds directly to the death receptor Fas, blocking Fas-mediated apoptosis (103). Finally, HGF protects myocytes from apoptosis mediated by reactive O_2 (104). HGF and IGF-1 are synthesized in CSCs, progenitors, and precursors and can modulate CSC behavior, viability, and function.

The IGF-1-IGF-1R system plays a positive role in the migrating cells. Translocation of cells mediated by HGF is characterized by cell division when IGF-1 is present, together with labeling for nuclear proteins implicated in the cell cycle (e.g., Cdc6, MCM5, and Ki67), as well as markers of DNA replication (BrdU and mitosis) (38, 105–112). Thus IGF-1 is a powerful mitogen for primitive and early committed cells in the heart, and HGF is equally potent in stimulating the movement of the same cells to target areas within the myocardium. Myocardial aging is characterized by a decrease in the expression of IGF-1 and IGF-1R in myocytes (113), and this attenuation of the IGF-1-IGF-1R system may also involve CSCs and early committed cells. Preliminary results are consistent with the possibility that both the HGF-c-Met and the IGF-1-IGF-1R systems are downregulated with age in CSCs.

SENESCENCE IN THE AGING MYOCARDIUM

Mechanisms

There are three possibilities regarding the deficiency in myocardial regeneration with age: (*a*) CSC growth and differentiation are attenuated with age; this notion assumes that CSCs are subjected to aging effects, as are the other cells of the myocardium. (*b*) The ability of CSCs to migrate in the old heart is diminished; this implies that aged CSCs do not respond, as do young CSCs, to growth factors promoting their activation and translocation. (*c*) The number of functionally competent CSCs decreases with age by growth arrest, apoptotic and/or necrotic death of old CSCs; thus an attenuation of survival factors occurs in aged CSCs.

According to the dogma, stem cells possess unlimited self-renewal capacity or an ability to replace themselves, which exceeds the lifespan of the organ and organism (114, 115). Self-renewing, pluripotent, and clonogenic cells should escape the devastating effects of time and the deleterious changes associated with aging and senescence. Stem cells were interpreted as the fountain of youth, were considered to be immortal, and were assumed to be well equipped against any form of aging effect. However, this is not the case in the bone marrow (116–118) and is not the case in the heart. Primitive cells positive for p16[INK4a] have been detected in the heart of patients with cardiac decompensation and signs of precocious organ aging.

Apoptosis of c-kit-positive undifferentiated cells was very high in these aged hearts and was always associated with the expression of p16^{INK4a} (119). The fraction of cycling putative CSCs increased; this phenomenon is consistent with results in the bone marrow (114). Myocyte apoptosis and necrosis are evident in young adult Fischer 344 rats (38, 120, 121) and FVB mice (122) and increase with age. Myocyte hypertrophy and replication tend to preserve muscle mass and function, which deteriorate in the old animals (123, 124). With time, myocyte division constitutes the prevailing form of reactive growth (79) while myocyte hypertrophy declines (32, 113). Myocyte multiplication, however, is not sufficient to sustain cell number and ventricular performance. Cavitary dilation, decreased ventricular mass-to-chamber volume ratio, elevated wall stress, and impaired mechanical behavior of myocytes become apparent in the old heart and cardiac failure supervenes (38, 121, 124). Thus activation of old CSCs undergoing a limited number of doublings prior to cellular senescence and growth arrest may be viewed as the process underlying cardiac aging. This time-dependent reduction in CSC renewal would correspond with increased CSC lineage commitment, depleting the atrial and apical niches, thus leading to cellular senescence characterized by loss of the ability to replicate by selective alterations in specialized cell cycle regulatory functions (125, 126). Specifically, replicative senescence corresponds to an irreversible cell cycle block in G0/G1 triggered by integrated actions of the p53/p21/p19 and p16/Rb pathways (127, 128). This phase of viable cell cycle arrest is accompanied by various changes in cell physiology, morphology, and gene expression (129). The discovery that telomeres shorten with each cell multiplication in vitro suggests a mechanism by which a cell could sense its own aging.

Migration, proliferation, and differentiation of several cell types, including stem cells, are modulated by integrins (87, 130), which are transmembrane glycoprotein heterodimers composed of one α and one β chain (131, 132). These adhesion receptors interact with components of the extracellular matrix, which results in the reorganization of the cytoskeleton and in the transduction of signals from the interstitium to the intracellular compartment (133). HGF stimulates the expression, function, and cell membrane distribution of specific classes of integrins (87, 133–137). c-Met, which is present on CSCs and early committed cells (87), forms complexes with $\alpha_6\beta_4$ integrin receptor, which is activated by its ligand plasmatic fibronectin, potentiating the invasive property of the c-Met-positive cells (136, 137). Activation of integrin receptors on CSCs may initiate the migratory pathway involving the phosphorylation of FAK and paxillin and their physical association (130, 133, 138–140). Also, phospho-FAK activates the PI3-kinase-Akt survival pathway (141) further preserving cell viability. Finally, phospho-Akt increases the expression and activity of telomerase in myocytes (122), and a similar function may be operative in primitive and lineage-determined cells. Telomerase activity enhances myocyte growth and survival (142, 143). These pathways of cell migration may constitute part of the molecular mechanisms responsible for translocation of repairing cells to the damaged heart.

Telomerase

Two postulates related to the senescent phenotype prevail: (*a*) Replicative senescence is associated with telomere shortening and (*b*) p53 expression is associated with senescence. The phenomenon of replicative senescence that limits cellular mitotic activity occurs in concert with telomeric shortening. Telomerase is a cellular transcriptase that antagonizes the onset of replicative senescence by maintaining telomeres (144). For this reason, telomere length may serve as a "mitotic clock" that winds down with each cell division until the telomeres become critically short and chromosomal instability ensues (145, 146). As with mice expressing the constitutively active p53, telomerase-deficient mice also show some phenotypic changes consistent with premature aging (147). Telomere-driven senescence may be part of a system that evolved to limit tumor growth (148). The high level of telomerase activity in many human cancer cells is consistent with this postulate. In rat myocardium, telomere shortening increases with age in a subgroup of myocytes, suggesting that some cardiomyocytes were replicating chromatin throughout life with decreased telomerase activity in adults (149). Recent attention has turned to the role of telomerase in primitive cells and cardiomyocyte proliferation, survival, and myocardial repair (Figure 3, see color insert). Telomerase expression declines after birth with postnatal development (150) and aging but persists throughout life with gender-dependent differences in activity (151). Expression of telomerase was colocalized with nuclei of cycling myocytes identified by immunostaining with Ki67, a marker of dividing cells (142). Increased telomerase expression in the failing myocardium promotes preservation of telomere length and continued proliferation of myocytes. Myocyte proliferation is also apparent following overexpression of telomerase in vivo, resulting in a postnatal hyperplastic phenotype in hearts of transgenic mice (152). Hearts from these telomerase-overexpressing transgenics are also resistant to apoptotic injury; however, forced telomerase expression at abnormally high levels induces undesirable hypertrophic remodeling both in vivo and in vitro. Conversely, elimination of telomerase in knockout mice causes dilated cardiomyopathy, increased apoptosis, decreased cell proliferation, and elevation of p53 expression (143). Collectively these observations suggest that under normal circumstances telomerase activity is required for normal myocyte proliferation and maintenance of cardiac structure and function. Elevation of telomerase activity may also occur in the damaged or aged myocardium to augment repair processes, but loss of telomerase activity precipitates premature aging and a decline heart function.

p53

Changes in protein expression level, DNA-binding activity, and phosphorylation state of p53 are characteristic of the senescent state (153, 154). Mice expressing a mutation resulting in activation of p53 display an early onset of phenotypes associated with aging (155). In the context of the myocardium, p53 expression is elevated in heart failure (156), ischemia-reperfusion damage (157), and right-side pressure overload (158). Isolated cardiomyocytes exposed to hypoxic conditions accumulate p53, and adenoviral overexpression of p53 induces apoptosis (159), highlighting the point that p53-mediated effects can cross boundaries from senescence into apoptosis under certain conditions. p53 is also responsible for apoptosis of cultured cardiomyocytes in response to mechanical stretch (160). A dynamic regulation of p53 occurs via cross talk with Akt (also known as protein kinase B) as shown in Figure 4 (see color insert). In the context of p53, conflicting signals transduced by Akt (a serine/threonine kinase that promotes cell survival) and p53 are integrated via negative feedback between the two pathways. Specifically, proapoptotic stimulation leads to p53-dependent destruction of Akt, whereas Akt activation by serum stimulation leads to inhibition of p53 (161–163). Akt-dependent inhibition of p53 is mediated primarily through the action of Mdm2 (164, 165). Akt phosphorylates Mdm2, which then enters (or is retained within) the nucleus, binds to and blocks the transcriptional domain of p53, exits the nucleus as a p53/Mdm2 complex, and targets p53 for degradation. Conversely, the protein phosphatase PTEN suppresses Akt activity, thereby protecting p53 from Mdm2 (166, 167). Inactivation of PTEN leads to cardiac hypertrophy mediated through PI3-K, implicating Akt activation in the balance conferred by PTEN signaling (168). The state of this signaling network determines the fate of the cell to survive or to enter apoptosis/senescence or hypertrophy. The mechanisms for induction of senescence or apoptosis by p53 remain obscure, but evidence suggests that p53 involvement is triggered by the telomere shortening associated with declining replicative capacity of aged cells (25).

ANTAGONIZING MYOCARDIAL AGING WITH AKT

Aging is associated with increased cellular senescence, apoptosis, increased p53, decreased telomerase, and loss of the local IGF-1/IGF-1 receptor system in the myocardium that drives Akt activation (169). Aging-associated loss of Akt activity is found in senescent primary cardiac fibroblast cultures (170) and old hepatocytes (171). Aging myocardial cells exhibit depressed antiapoptotic signaling that could account for increased susceptibility to damage and death (22). Akt activation has been linked to regulation of the cell cycle, with Akt inhibition leading to senescence-like arrest (172) and activation promoting the transition from G2 to M phase in the cell cycle (173). Increased Akt activation extends the replicative lifespan of skeletal muscle satellite cells in vitro (174). Akt promotes growth and survival of stem cell populations of erythroid progenitors (175) and primordial germ cells (176). In fact, increased Akt activation in the failing heart may be a compensatory response to promote survival, although too late to rescue the heart (177). Thus Akt may promote stem cell expansion and cellular regeneration in addition to antagonizing senescence and apoptosis.

Probing myocardial Akt biology yielded surprising insights, but also revealed frustrating stumbling blocks that derailed experimental strategies designed to inhibit pathogenesis by fostering Akt activation and enhancing cell survival. Cardiac development and growth depend upon activation of Akt for early cardiomyocyte development (178) and postnatal cardiac growth (179, 180). These growth-promoting effects of Akt were evident in transgenic mice engineered for cardiac-specific expression of constitutively activated Akt, which produced enlarged hypertrophic or failing hearts (181). Hypertrophic effects were coupled with hypercontractility (182), similar to the effect observed following inactivation of PTEN phosphatase, a negative regulator of Akt activity (168). Thus the potential for Akt to benefit the myocardium by antagonizing cell death and senescence remains unfulfilled.

Drawing upon the beneficial effect of IGF-1 in the myocardium, which induces nuclear translocation of Akt kinase, we have engineered a nuclear-targeted Akt (Akt/nuc) construct. Hearts from Akt/nuc transgenics are distinct from other cardiac-targeted Akt transgenics by virtue of the following unique properties: lack of deleterious remodeling, hyperdynamic performance, resistance to pathological insult, singular genetic expression profile, and expansion of their cardiac progenitor cell population. Stimulatory effects of Akt upon telomerase activity and cell cycle progression presumably account for an increase in the number of cardiomyocytes labeled with Ki67 antibody, which identifies cycling cells (107). There are no reports of Ki67-positive cells that do not divide. Akt/nuc mice possess 1.7-fold more Ki67$^+$ cells compared with nontransgenic controls (1.54 $\pm$ 0.19 versus 0.92 $\pm$ 0.14, respectively; p $\leq$ 0.005). Akt/nuc mice possess 2.2-fold more c-kit$^+$/α-sarcomeric actin$^+$ cells compared with nontransgenic controls (59 $\pm$ 7 versus 27 $\pm$ 8 per 100 mm^2, respectively; p $=$ 0.005). Moreover, c-kit$^+$/myc-tag$^+$ cells were also found in sections from Akt/nuc mice. Such cells represent the transitional state between progenitor state to committed myocyte expressing the α-myosin heavy chain promoter that drives Akt/nuc transgene expression. Early expression of the transgene could provide an explanation for the potentiating effect of Akt/nuc upon the progenitor cell population by enhancing proliferation and survival. These diverse preliminary results provide strong support for the speculation that Akt/nuc antagonizes aging of the myocardium on a molecular, cellular, and structural level.

CONCLUSIONS

This review questions the old paradigm of cardiac growth in the adult heart viewed as the consequence of myocyte hypertrophy only. Accumulating evidence supports the notion that a stem cell pool resides in the heart and that tissue regeneration occurs by activation, mobilization, and lineage commitment of these primitive cells. Understanding the mechanisms controlling the translocation, multiplication, and differentiation of cardiac stem cells, as illustrated by manipulation of the Akt signaling cascade, could promote novel approaches for the repair of the damaged heart. This is a major challenge that several laboratories have taken in the past few years and will aggressively pursue during the next decade in order to improve the currently devastating outcome of heart failure of ischemic and nonischemic origin in the old heart.

The *Annual Review of Physiology* is online at http://physiol.annualreviews.org

LITERATURE CITED

1. Beltrami AP, Urbanek K, Kajstura J, Yan SM, Finato N, et al. 2001. Evidence that human cardiac myocytes divide after myocardial infarction. *N. Engl. J. Med.* 344:1750–57

2. Nitahara JA, Cheng W, Liu Y, Li B, Leri A, et al. 1998. Intracellular calcium, DNAse activity and myocyte apoptosis in aging Fischer 344 rats. *J. Mol. Cell. Cardiol.* 30:519–35

3. Burgess ML, McCrea JC, Hedrick HL. 2001. Age associates changes in cardiac matrix and integrins. *Mech. Ageing Dev.* 122:1739–56

4. Duque G. 2000. Apoptosis in cardiovascular aging research: future directions. *Am. J. Geriatr. Cardiol.* 9:263–64

5. Grodzicki T, Michalewicz L, Messerli FH. 1998. Aging and essential hypertension: effect of left ventricular hypertrophy on cardiac function. *Am. J. Hypertens.* 11:425–29

6. Lakatta EG, Sollott SJ, Pepe S. 2001. The old heart: operating on the edge. *Novartis Found. Symp.* 235:172–96

7. Lakatta EG. 2002. Age-associated cardiovascular changes in health: impact on cardiovascular disease in older persons. *Heart Fail. Rev.* 7:29–49

8. Patel MB, Sonnenblick EH. 1998. Age associated alterations in structure and function of the cardiovascular system. *Am. J. Geriatr. Cardiol.* 7:15–22

9. Ren J, Jefferson L, Sowers JR, Brown RA. 1999. Influence of age on contractile response to insulin-like growth factor 1 in ventricular myocytes from spontaneously hypertensive rats. *Hypertension* 34:1215–22

10. Malat Z, Fornes P, Costagliola R, Esposito B, Belmin J, et al. 2001. Age and gender effects on cardiomyocyte apoptosis in the normal human heart. *J. Gerontol. A Biol. Sci. Med. Sci.* 56:M719–23

11. Azhar G, Gao W, Liu L, Wei JY. 1999. Ischemia-reperfusion in the adult mouse heart influence of age. *Exp. Gerontol.* 34:699–714

12. Jazwinski SM, Kim S, Lai CY, Benguria A. 1998. Epigenetic stratification: the role of individual change in the biological aging process. *Exp. Gerontol.* 33:571–80

13. Takahashi Y, Kuro-o M, Ishikawa F. 2000. Aging mechanisms. *Proc. Natl. Acad. Sci. USA* 97:12407–8

14. Guarente L, Kenyon C. 2000. Genetic pathways that regulate ageing in model organisms. *Nature* 408:255–62

15. Kuro-o M. 2001. Disease model: human aging. *Trends Mol. Med.* 7:179–81

16. Kuro-o M, Matsumura Y, Aizawa H, Kawaguchi H, Suga T, et al. 1997. Mutation of the mouse *klotho* gene leads to a syndrome resembling ageing. *Nature* 390:45–52

17. Vogel H, Lim D-S, Karsenty G, Finegold M, Hasty P. 1999. Deletion of Ku86 causes early onset of senescence in mice. *Proc. Natl. Acad. Sci. USA* 96:10770–75

18. Ostler EL, Wallis CV, Sheerin AN, Faragher RGA. 2002. A model for the phenotypic presentation of Werner's syndrome. *Exp. Gerontol.* 37:285–92

19. Postiglione A, Soricelli A, Covelli EM, Iazzetta N, Ruocco A, et al. 1996. Premature aging in Werner's syndrome spares the central nervous system. *Neurobiol. Aging* 17:325–30

20. James SE, Faragher RG, Burke JF, Shall S, Mayne LV. 2000. Werner's syndrome

T lymphocytes display a normal in vitro life-span. *Mech. Ageing Dev.* 121:139–49

21. Kajstura J, Mansukhani M, Cheng W, Reiss K, Krajewski S, et al. 1995. Programmed cell death and expression of the protooncogene bcl-2 in myocytes during postnatal maturation of the heart. *Exp. Cell Res.* 219:110–21

22. Centurione L, Antonucci A, Miscia S, Grilli A, Rapino M, et al. 2002. Age-related death-survival balance in myocardium: an immunohistochemical and biochemical study. *Mech. Ageing Dev.* 123:341–50

23. Kajstura J, Chen W, Sarangarajan R, Li P, Li B, et al. 1996. Necrotic and apoptotic myocyte cell death in the aging heart of Fischer 344 rats. *Am. J. Physiol. Heart Circ. Physiol.* 271:H215–28

24. Sabbah HN, Sharov VG. 1998. Apoptosis in heart failure. *Prog. Cardiovasc. Dis.* 40:549–62

25. Itahana K, Dimiri G, Campisi J. 2001. Regulation of cellular senescence by p53. *Eur. J. Biochem.* 268:2784–91

26. Swynghedauw B, Besse S, Assayag P, Carre F, Chevalier B, et al. 1995. Molecular and cellular biology of the senescent hypertrophied and failing heart. *Am. J. Cardiol.* 76:2D–7D

27. Lim CC, Apstein CS, Colucci WS, Liao R. 2000. Impaired cell shortening and relengthening with increased pacing frequency are intrinsic to the senescent mouse cardiomyocyte. *J. Mol. Cell. Cardiol.* 32:2075–82

28. Zhou YY, Lakatta EG, Xiao RP. 1998. Age-associated alterations in calcium current and its modulation in cardiac myocytes. *Drugs Aging* 13:159–71

29. Leont'eva TA, Abuladze ZS, Semenova LA, Tsellarius IuG. 1983. Age-related changes in the cardiomyocyte population of the heart ventricles of white rats. *Ontogenez* 14:624–29

30. Olivetti G, Melissari M, Balbi T, Quaini F, Sonnenblick EH, Anversa P. 1994. Myocyte nuclear and possible cellular hyperplasia contribute to ventricular remodeling in the hypertrophic senescent heart in humans. *J. Am. Coll. Cardiol.* 24:140–49

31. Kajstura J, Pertoldi B, Leri A, Beltrami CA, Deptala A, et al. 2000. Telomere shortening is an in vivo marker of myocyte replication and aging. *Am. J. Pathol.* 156:813–19

32. Takahashi T, Schunkert H, Isoyama S, Wei JY, Nadal-Ginard B, et al. 1992. Age-related differences in the expression of proto-oncogene and contractile protein genes in response to pressure overload in the rat myocardium. *J. Clin. Invest.* 89:939–46

33. Capasso JM, Palackal T, Olivetti G, Anversa P. 1990. Severe myocardial dysfunction induced by ventricular remodeling in aging rat hearts. *Am. J. Physiol. Heart Circ. Physiol.* 259:H1086–96

34. Levine TB, Levine AB, Bolenbaugh J, Green PR. 2002. Reversal of heart failure remodeling with age. *Am. J. Geriatr. Cardiol.* 11:299–304

35. Guarente L, Kenyon C. 2000. Genetic pathways that regulate ageing in model organisms. *Nature* 408:255–62

36. Wood WB. 2000. Aging of *C. elegans*: mosaics and mechanisms. *Cell* 95:147–50

37. Campisi J. 2001. From cells to organisms: can we learn about aging from cells in culture? *Exp. Gerontol.* 36:607–18

38. Anversa P, Olivetti G. 2002. Cellular basis of physiological and pathological myocardial growth. In *Handbook of Physiology. The Cardiovascular System: the Heart*, ed. E Page, H Fozzard, JR Solaro, pp. 75–144. New York: Oxford Univ. Press

39. Rumyantsev PP. 1964. DNA synthesis and nuclear division in embyonical and postnatal histogenesis of myocardium. *Arch. Anat.* 47:59–65

40. Anversa P, Leri A, Kajstura J, Nadal-Ginard B. 2002. Myocyte growth and cardiac repair. *J. Mol. Cell. Cardiol.* 34:91–105

41. Hughes S. 2002. Cardiac stem cells. *J. Pathol.* 197:468–78
42. Anversa P, Nadal-Ginard B. 2002. Myocyte renewal and ventricular remodelling. *Nature* 415:240–43
43. Wagers AJ, Sherwood RI, Christensen JL, Weisman IL. 2002. Little evidence for developmental plasticity of adult hematopoietic stem cells. *Science* 297:2256–59
44. Boheler KR, Czyz J, Tweedie D, Yang HT, Anisimov SV, Wobus AM. 2002. Differentiation of pluripotent embryonic stem cells into cardiomyocytes. *Circ. Res.* 91:189–201
45. Mummery C, Ward D, van den Brink CE, Bird SD, Doevendans PA, et al. 2002. Cardiomyocyte differentiation of mouse and human embryonic stem cells. *J. Anat.* 200:233–42
46. Hierly AM, Seale P, Lobe CG, Rudnicki MA, Megeny LA. 2002. The postnatal heart contains a myocardial stem cell population. *FEBS Lett.* 530:239–43
47. Anversa P, Nadal-Ginard B. 2002. Cardiac chimerism: methods matter. *Circulation* 106:e129–31
48. Taylor D, Hruban R, Rodriguez ER, Goldschmidt-Clermont PJ. 2002. Cardiac chimerism as a mechanism for self-repair. Does it happen and if so to what degree? *Circulation* 106:2–4
49. Laflamme MA, Myerson S, Saffitz JE, Murry CE. 2002. Evidence for cardiomyocyte repopulation by extracardiac progenitors in transplanted human hearts. *Circ. Res.* 90:634–40
50. Malouf NN, Coleman WB, Grisham JW, Lininger RA, Madden VJ, et al. 2001. Adult-derived stem cells from the liver become myocytes in the heart in vivo. *Am. J. Pathol.* 158:1929–35
51. Jackson KA, Majka SM, Wang H, Pocius J, Hartley CJ, et al. 2001. Regeneration of ischemic cardiac muscle and vascular endothelium by adult stem cells. *J. Clin. Invest.* 107:1395–402
52. Orlic D, Kajstura J, Chimenti S, Jakoniuk I, Anderson SM, et al. 2001. Bone marrow cells regenerate infarcted myocardium. *Nature* 410:701–5
53. Behfar A, Zingman LV, Hodgson DM, Rauzier JM, Kane GC, et al. 2002. Stem cell differentiation requires a paracrine pathway in the heart. *FASEB J.* 16:1558–66
54. Min JY, Sullivan MF, Yang Y, Zhang JP, Converso KL, et al. 2002. Significant improvement of heart function by cotransplantation of human mesenchymal stem cells and fetal cardiomyocytes in postinfarcted pigs. *Ann. Thoracic Surg.* 74:1568–75
55. Orlic D, Kajstura J, Chimenti S, Bodine DM, Leri A, Anversa P. 2001. Transplanted adult bone marrow cells repair myocardial infarcts in mice. *Ann. NY Acad. Sci.* 938:221–29
56. Orlic D, Kajstura J, Chimenti S, Limana F, Jakoniuk I, et al. 2001. Mobilized bone marrow stem cells repair the infarcted heart, improving function and survival. *Proc. Natl. Acad. Sci. USA* 98:10344–49
57. Caplice NM, Gersh BJ. 2003. Stem cells to repair the heart. A clinical perspective. *Circ. Res.* 92:6–8
58. Orlic D, Hill JM, Arai AE. 2002. Stem cells for myocardial regeneration. *Circ. Res.* 91:1092–102
59. Quaini F, Urbanek K, Beltrami AP, Finato N, Beltrami CA, et al. 2002. Chimerism of the transplanted heart. *N. Engl. J. Med.* 346:5–15
60. Terada N, Hamazaki T, Oka M, Hoki M, Mastalerz DM, et al. 2002. Bone marrow cells adopt the phenotype of other cells by spontaneous cell fusion. *Nature.* 416:542–45
61. Ying Ql, Nichols J, Evans EP, Smith AG. 2002. Changing potency by spontaneous fusion. *Nature* 416:545–48
62. Vassilopoulos G, Wang PR, Russell DW. 2003. Transplanted bone marrow regenerates liver by cell fusion. *Nature* 422:901–4
63. Wang X, Willenbring H, Akkari Y, Torimaru Y, Foster M, et al. 2003. Cell fusion

is the principal source of bone marrow-derived hepatocytes. *Nature* 422:897–901

64. Spees JL, Olson SD, Ylostalo J, Lynch PJ, Smith J, et al. 2003. Differentiation, cell fusion, and nuclear fusion during ex vivo repair of epithelium by human adult stem cells from bone marrow stroma. *Proc. Natl. Acad. Sci. USA* 100:2397–402

65. Fuchs E, Segre JA. 2000. Stem cells: a new lease on life. *Cell* 100:143–55

66. Weissman IL. 2000. Stem cells: units of development, units of regeneration, and units in evolution. *Cell* 100:157–68

67. Beltrami A, Chimenti S, Limana F, Barlucchi L, Quaini F, et al. 2001. Cardiac c-kit positive cells proliferate in vitro and generate new myocardium in vivo. *Circulation* 104:II–324

68. Bunting KD, Zhou S, Lu T, Sorrentino BP. 2000. Enforced P-glycoprotein pump function in murine bone marrow cells results in expansion of side population stem cells in vitro and repopulating cells in vivo. *Blood* 96:902–909

69. Scharenberg CW, Harkey MA, Torok-Storb B. 2002. The ABCG2 transporter is an efficient Hoechst 33342 efflux pump and is preferentially expressed by immature human hematopoietic progenitors. *Blood* 99:507–12

70. Jackson KA, Mi T, Goodell MA. 1999. Hematopoietic potential of stem cells isolated from murine skeletal muscle. *Proc. Natl. Acad. Sci. USA* 96:14482–86

71. Zhou S, Schuetz JD, Bunting KD, Colapietro AM, Sampath J, et al. 2001. The ABC transporter Bcrp1/ABCG2 is expressed in a wide variety of stem cells and is a molecular determinant of the side-population phenotype. *Nat. Med.* 7:1028–34

72. Bunting KD. 2002. ABC transporters as phenotypic markers and functional regulators of stem cells. *Stem Cells* 20:11–20

73. Zhou S, Morris JJ, Barnes Y, Lan L, Schuetz JD, Sorrentino BP. 2002. Bcrp1 gene expression is required for normal numbers of side population stem cells in mice, and confers relative protection to mitoxantrone in hematopoietic cells in vivo. *Proc. Natl. Acad. Sci. USA* 99: 12339–44

74. Asakura A, Seale P, Girgis-Gabardo A, Rudnicki MA. 2002. Myogenic specification of side population cells in skeletal muscle. *J. Cell Biol.* 159:123–34

75. McKinney-Freeman SL, Jackson KA, Camargo FD, Ferrari G, Mavilio F, Goodell MA. 2002. Muscle-derived hematopoietic stem cells are hematopoietic in origin. *Proc. Natl. Acad. Sci. USA* 99:1341–46

76. Urbanek K, Quaini F, Bussani R, Silvestri F, Jakoniuk I, et al. 2002. Cardiac stem cell growth and death differs in acute and chronic ischemic heart failure in humans. *Circulation* 106:II–382

77. Chimenti S, Barlucchi L, Limana F, Jakoniuk I, Cesselli D, et al. 2003. Local mobilization of resident cardiac primitive cells by growth factors repairs the infarcted heart. *Circulation* 106:II–14

78. Cesselli D, Kajstura J, Jakoniuk I, Urbanek K, Hasahara H, et al. 2003. Cardiac stem cells are endowed in niches of the adult mouse heart and possess the ability to divide and differentiate in the various cardiac lineages. *Circulation* 106:II–286

79. Kajstura J, Leri A, Castaldo C, Quaini F, Mancarella S, et al. 2001. Cardiac stem cells mediate myocyte replication in the young and senescent rat heart. *Circulation* 104:II–220

80. Linke A, Castaldo C, Chimenti S, Leri A, Kajstura J, et al. 2002. Mobilization of cardiac stem cells by growth factors promotes repair of infarcted myocardium improving regional and global cardiac function in conscious dogs. *Circulation* 106:II–52

81. Blau HM, Brazelton TR, Weimann JM. 2001. The evolving concept of a stem cell: entity or function? *Cell.* 105:829–41

82. Sellers SE, Tisdale JF, Agricola BA, Metzger ME, Donahue RE, et al. 2001. The effect of multidrug-resistance 1 gene versus neo transduction on ex vivo and in vivo

expansion of rhesus macaque hematopoietic repopulating cells. *Blood* 97:1888–91

83. Anversa P, Nadal-Ginard B. 2002. Myocyte renewal and ventricular remodelling. *Nature* 415:240-43

84. Ma X, Ling KW, Dzierzak E. 2001. Cloning of the Ly-6A (Sca-1) gene locus and identification of a 3' distal fragment responsible for high-level gamma-interferon-induced expression in vitro. *Br. J. Haematol.* 114:724–30

85. Jackson KA, Majka SM, Wulf GG, Goodell MA. 2002. Stem cells: a minireview. *J. Cell Biochem. Suppl.* 38:1–6

86. Caldwell MA, He X, Wilkie N, Pollack S, Marshall G, et al. 2001. Growth factors regulate the survival and fate of cells derived from human neurospheres. *Nat. Biotechnol.* 19:475–79

87. Torella D, DiMeglio F, Chimenti S, Jakoniuk I, Nadal-Ginard B, et al. 2003. Molecular control of progenitor cell migration to the infarcted myocardium. *Circulation* 106:II–7

88. Schofield R. 1978. The relationship between the spleen colony-forming cell and the hemopoietic stem cells. *Blood Cells* 4:7–25

89. Spradling A, Drummond-Barbosa D, Kai T. 2001. Stem cells find their niche. *Nature* 414:98–104

90. Taylor G, Lehrer MS, Jensen PJ, Sun TT, Lavker RM. 2000. Involvement of follicular stem cells in forming not only the follicle but also the epidermis. *Cell* 102:451–61

91. Wright NA. 2000. Epithelial stem cell repertoire in the gut: clues to the origin of cell lineages, proliferative units and cancer. *Int. J. Exp. Pathol.* 81:117–43

92. Shinohara T, Orwig KE, Avarbock MR, Brinster RL. 2001. Remodeling of the postnatal mouse testis is accompanied by dramatic changes in stem cell number and niche accessibility. *Proc. Natl. Acad. Sci. USA* 98:6186–91

93. Nadal-Ginard B, Kajstura J, Anversa P, Leri A. 2003. A matter of life and death: cardiac myocyte apoptosis and regeneration. *J. Clin. Invest.* 111:1457–59

94. Anversa P, Kajstura J, Nadal-Ginard B, Leri A. 2003. Primitive cells and tissue regeneration. *Circ. Res.* 92:579–82

95. Stella MC, Comoglio PM. 1999. HGF: a multifunctional growth factor controlling cell scattering. *Int. J. Biochem. Cell Biol.* 31:1357–62

96. Powell EM, Mars WM, Levitt P. 2001. Hepatocyte growth factor/scatter factor is a motogen for interneurons migrating from the ventral to dorsal telencephalon. *Neuron* 30:79–89

97. Reiss K, Cheng W, Ferber A, Kajstura J, Li P, et al. 1996. Overexpression of insulin-like growth factor-1 in the heart is coupled with myocyte proliferation in transgenic mice. *Proc. Natl. Acad. Sci. USA* 93:8630–35

98. Butler AA, LeRoith D. 2001. Minireview: tissue-specific versus generalized gene targeting of the *igf1* and *igf1r* genes and their roles in insulin-like growth factor physiology. *Endocrinology* 142:1685–88

99. Camper-Kirby D, Welch S, Walker A, Shiraishi I, Setchell KD, et al. 2001. Myocardial Akt activation and gender: increased nuclear activity in females versus males. *Circ. Res.* 88:1020–27

100. Welch S, Plank D, Witt S, Glascock B, Schaefer E, et al. 2002. Cardiac-specific IGF-1 expression attenuates dilated cardiomyopathy in tropomodulin-overexpressing transgenic mice. *Circ. Res.* 90:641–48

101. Xiao GH, Jeffers M, Bellacosa A, Mitsuuchi Y, vande Woude GF, Testa JR. 2001. Anti-apoptotic signaling by HGF/Met via the phosphatidylinositol 3-kinase/Akt and mitogen-activated protein kinase pathways. *Proc. Natl. Acad. Sci. USA* 98:247–52

102. Mildner M, Eckart L, Lengauer B, Tschachler E. 2002. HGF/scatter factor inhibits UVB-induced apoptosis of

human keratinocytes but not of keratinocyte-derived cell lines via the phosphatidylinositol 3-kinase/Akt pathway. *J. Biol. Chem.* 277:14146–52

103. Wang X, DeFrances MC, Dai Y, Pediaditakis P, Johnson C, et al. 2002. A mechanism of cell survival: sequestration of Fas by the HGF receptor Met. *Mol. Cell.* 9:411–21

104. Kitta K, Day RM, Ikeda T, Suzuki YJ. 2001. HGF protects cardiac myocytes against oxidative stress-induced apoptosis. *Free Radical Biol. Med.* 31:902–10

105. Dolbeare F. 1996. Bromodeoxyuridine: a diagnostic tool in biology and medicine, part III. Proliferation in normal, injured and diseased tissue, growth factors, differentiation, DNA replication sites and in situ hybridization. *Histochem. J.* 28:531–75

106. Dutta A, Bell SP. 1997. Initiation of DNA replication in eukaryotic cells. *Annu. Rev. Cell Dev. Biol.* 13:293–32

107. Scholzen T, Gerdes J. 2000. The Ki-67 protein: from the known to the unknown. *J. Cell. Physiol.* 182:311–22

108. MacCallum DE, Hall PA. 2000. The location of pKi67 in the outer dense fibrillary compartment of the nucleolus points to a role in ribosome biogenesis during the cell cycle. *J. Pathol.* 190:537–44

109. MacCallum DE, Hall PA. 2000. The biochemical characterization of the DNA binding activity of pKi67. *J. Pathol.* 191:286–98

110. Stoeber K, Tisty TD, Happerfield L, Thomas GA, Romanov S, et al. 2001. DNA replication licensing and human cell proliferation. *J. Cell Sci.* 114:2027–41

111. Beltrami AP, Urbanek K, Kajstura J, Yan S-M, Finato N, et al. 2001. Evidence that human cardiac myocytes divide after myocardial infarction. *N. Engl. J. Med.* 344:1750–57

112. Leri A, Kajstura J, Anversa P. 2002. Myocyte proliferation and ventricular remodeling. *J. Cardiac Fail.* (Suppl.)8:S518–25

113. Cheng W, Reiss K, Li P, Chun MJ, Kajstura J, et al. 1996. Aging does not affect the activation of the myocyte insulin-like growth factor-1 autocrine system after infarction and ventricular failure in Fischer 344 rats. *Circ. Res.* 78:536–46

114. Morrison SJ, Wandycz AM, Akashi K, Globerson A, Weissman IL. 1996. The aging of hematopoietic stem cells. *Nat. Med.* 2:1011–16

115. Sudo K, Ema H, Morita Y, Nakauchi H. 2000. Age-associated characteristics of murine hematopoietic stem cells. *J. Exp. Med.* 192:1273–80

116. Geiger H, Van Zant G. 2002. The aging of lympho-hematopoietic stem cells. *Nat. Immunol.* 3:329–33

117. Chen J, Astle BA, Harrison DE. 1999. Development and aging of primitive hematopoietic stem cells in BALB/cBy mice. *Exp. Hematol.* 27:928–35

118. Chen J, Astle CM, Harrison DE. 2000. Genetic regulation of primitive hematopoietic stem cell senescence. *Exp. Hematol.* 28:442–50

119. Chimenti C, Kajstura J, Urbanek K, Colussi C, Maseri A, et al. 2003. Telomerase activity and myocyte regeneration are positive determinants of dilated cardiomyopathy in elderly patients. *Circulation* 106:II–383

120. Anversa P, Leri A, Beltrami CA, Guerra S, Kajstura J. 1998. Myocyte death and growth in the failing heart. *Lab. Invest.* 78:767–86

121. Anversa P, Kajstura J. 1998. Ventricular myocytes are not terminally differentiated in the adult mammalian heart. *Circ Res.* 83:1–14

122. Andreoli AM, Kajstura J, Nadal-Ginard B, Anversa P, Leri A. 2000. Myocyte proliferation in the adult heart is enhanced by IGF-1 overexpression in an age and gender dependent manner. *Circulation* 102:II–235

123. Anversa P, Palackal T, Sonnenblick EH, Olivetti G, Meggs LG, Capasso JM. 1990. Myocyte cell loss and myocyte cellular

hyperplasia in the hypertrophied aging rat heart. *Circ. Res.* 67:871–85

124. Capasso JM, Fitzpatrick D, Anversa P. 1992. Cellular mechanisms of ventricular failure: myocyte kinetics and geometry with age. *Am. J. Physiol. Heart Circ. Physiol.* 262:H1770–81

125. Campisi J. 1997. The biology of replicative senescence. *Eur. J. Cancer* 33:703–9

126. Rubin H. 2002. The disparity between human cell senescence in vitro and lifelong replication in vivo. *Nat. Biotechnol.* 20:675–81

127. Schwarze SR, Shi Y, Fu VX, Watson PA, Jarrard DF. 2001. Role of cyclin-dependent kinase inhibitors in the growth arrest at senescence in human prostate epithelial and uroepithelial cells. *Oncogene* 20:8184–92

128. Kim H, You S, Farris J, Kong BW, Christman SA, et al. 2002. Expression profiles of p53-, p16(INK4a)-, and telomere-regulating genes in replicative senescent primary human, mouse, and chicken fibroblast cells. *Exp. Cell Res.* 272:199–208

129. Faragher RGA, Kipling D 1998. How might replicative senescence contribute to human ageing? *BioEssays* 20:985–91

130. Eliceiri BP. 2001. Integrin and growth factor receptor crosstalk. *Circ. Res.* 89:1104–10

131. Schwartz MA, Schaller MD, Ginsberg MH. 1995. Integrins: emerging paradigms of signal transduction. *Annu. Rev. Cell Dev. Biol.* 11:549–99

132. Giancotti FG, Ruoslahti E. 1999. Integrin signaling. *Science* 285:1028–32

133. Petit V, Thiery J-P. 2000. Focal adhesions: structure and dynamics. *J. Biol. Cell* 92:477–94

134. Stella MC, Comoglio PM. 1999. HGF: a multifunctional growth factor controlling cell scattering. *Int. J. Biochem. Cell Biol.* 31:1357–62

135. Trusolino L, Cavassa S, Angelini P, Ando M, Bertotti A, et al. 2000. HGF/scatter factor selectively promotes cell invasion by increasing integrin avidity. *FASEB J.* 14:1629–40

136. Trusolino L, Bertotti A, Comoglio PM. 2001. A signaling adapter function by alpha6beta-4 integrin in the control of HGF-dependent invasive growth. *Cell* 107:643–54

137. Mercurio AM, Rabinovitz I, Shaw LM. 2001. The alpha 6 beta 4 integrin and epithelial cell migration. *Curr. Opin. Cell Biol.* 13:541–45

138. Sieg DJ, Hauck CR, Ilic D, Klingbeil CK, Schaefer E, et al. 2000. FAK integrates growth-factor and integrin signals to promote cell migration. *Nat. Cell Biol.* 2:249–56

139. Turner CE. 2000. Paxillin and focal adhesion signaling. *Nat. Cell. Biol.* 2:E231–36

140. Turner CE. 2000. Paxillin interactions. *J. Cell Sci.* 113:4139–40

141. Jones RJ, Brunton VG, Frame MC. 2000. Adhesion-linked kinases in cancer; emphasis on src, focal adhesion kinase and PI 3-kinase. *Eur. J. Cancer* 36:1595–606

142. Leri A, Barlucchi L, Limana F, Deptala A, Darzynkiewicz Z, et al. 2001. Telomerase expression and activity are coupled with myocyte proliferation and preservation of telomeric length in the failing heart. *Proc. Natl. Acad. Sci. USA* 98:8626–31

143. Leri A, Franco S, Zacheo A, Barlucchi L, Chimenti S, et al. 2003. Ablation of telomerase and telomere loss leads to cardiac dilatation and heart failure. *EMBO J.* 22:1–9

144. Liu J-P. 1999. Studies of the molecular mechanisms in the regulation of telomerase activity. *FASEB J.* 13:2091–104

145. Harley CB. 1991. Telomerase loss: mitotic clock or genetic time bomb? *Mutat. Res.* 256:271–82

146. Kipling D. 2001. Telomeres, replicative senescence and human ageing. *Maturitas* 38:25–37

147. Samper E, Flores JM, Blasco MA. 2001. Restoration of telomerase activity rescues chromosomal instability and premature

aging in $Terc^{-/-}$ mice with short telomeres. *EMBO Rep.* 2:800–7

148. Ahmed A, Tollefsbol T. 2001. Telomeres and telomerase: basic science implications for aging. *J. Am. Geriatr. Soc.* 49:1105–9

149. Kajstura J, Pertoldi B, Leri A, Beltrami CA, Deptala A, et al. 2000. Telomere shortening is an in vivo marker of myocyte replication and aging. *Am. J. Pathol.* 156:813–19

150. Borges A, Liew CC. 1997. Telomerase activity during cardiac development. *J. Mol. Cell Cardiol.* 29:2717–24

151. Leri A, Malhotra A, Liew CC, Kajstura J, Anversa P. 2000. Telomerase activity in rat cardiac myocytes is age and gender dependent. *J. Mol. Cell Cardiol.* 32:385–90

152. Oh H, Taffet GE, Youker KA, Entman ML, Overbeek PA, et al. 2001. Telomerase reverse transcriptase promotes cardiac muscle cell proliferation, hypertrophy, and survival. *Proc. Natl. Acad. Sci. USA* 98:10308–13

153. Oren M, Damalas A, Gottlieb T, Michael D, Taplick J, et al. 2002. Regulation of p53: intricate loops and delicate balances. *Ann. NY Acad. Sci.* 973:374–83

154. Haupt Y, Robles AI, Prives C, Rotter V. 2002. Deconstruction of p53 functions and regulation. *Oncogene* 21:8223–31

155. Tyner SD, Venkatachalam S, Choi J, Jones S, Ghebranious N, et al. 2002. p53 mutant mice that display early ageing-associated phenotypes. *Nature* 415:45–53

156. Song H, Conte JV Jr, Foster AH, McLaughlin JS, Wei C. 1999. Increased p53 protein expression in human failing myocardium. *J. Heart Lung Transplant.* 18:744–49

157. Xie Z, Koyama T, Abe K, Fuji Y, Sawa H, Nagtasima K. 2000. Upregulation of p53 protein in rat heart subjected to a transient occlusion of the coronary artery followed by reperfusion. *Jpn. J. Physiol.* 50:159–62

158. Ikeda S, Hamada M, Hiwada K. 1999. Cardiomyocyte apoptosis with enhanced expression of p53 and Bax in right ventricle after pulmonary arterial banding. *Life Sci.* 65:925–33

159. Long X, Bolyut MO, Hipolito ML, Lundberg MS, Zheng JS, et al. 1997. p53 and the hypoxia-induced apoptosis of cultured neonatal rat cardiac myocytes. *J. Clin. Invest.* 99:2635–43

160. Leri A, Liu Y, Claudio PP, Kajstura J, Wang X, et al. 1999. Insulin-like growth factor-1 induces Mdm2 and downregulates p53, attenuating the myocyte renin-angiotensin system and stretch-mediated apoptosis. *Am. J. Pathol.* 154:567–80

161. Bachelder RE, Ribick MJ, Marchetti A, Falcioni R, Soddu S, et al. 1999. p53 inhibits alpha 6 beta 4 integrin survival signaling by promoting the caspase 3-dependent cleavage of Akt/PKB. *J. Cell Biol.* 147:1063–72

162. Sabbatini P, McCormick F. 1999. Phosphoinositide 3-OH kinase (PI3K) and PKB/Akt delay the onset of p53-mediated, transcriptionally dependent apoptosis. *J. Biol. Chem.* 274:24263–69

163. Gottlieb TM, Leal JF, Seger R, Taya Y, Oren M. 2002. Cross-talk between Akt, p53 and Mdm2: possible implications for the regulation of apoptosis. *Oncogene* 21:1299–303

164. Mayo LD, Donner DB. 2001. A phosphatidylinositol 3-kinase/Akt pathway promotes translocation of Mdm2 from the cytoplasm to the nucleus. *Proc. Natl. Acad. Sci. USA* 98:11598–603

165. Ogawara Y, Kishishita S, Obata T, Isazawa Y, Suzuki T, et al. 2002. Akt enhances Mdm2-mediated ubiquitination and degradation of p53. *J. Biol. Chem.* 277:21843–50

166. Mayo LD, Dixon JE, Durden DL, Tonks NK, Donner DB. 2002. PTEN protects p53 from Mdm2 and sensitizes cancer cells to chemotherapy. *J. Biol. Chem.* 277:5484–89

167. Stambolic V, MacPherson D, Sas D, Lin Y, Snow B, et al. 2001. Regulation of PTEN transcription by p53. *Mol. Cell.* 8:317–25

168. Crackower MA, Oudit GY, Kozieradzki I, Sarao R, Sun H, et al. 2002. Regulation of myocardial contractility and cell size by distinct PI3K-PTEN signaling pathways. *Cell* 110:737–49

169. Leri A, Kajstura J, Li B, Sonnenblick EH, Beltrami CA, et al. 2000. Cardiomyocyte aging is gender-dependent: the local IGF1-IGF-1R system. *Heart Dis.* 2:108–15

170. Diez C, Nestler M, Friedrich U, Vieth M, Stolte M, et al. 2001. Down-regulation of Akt/PKB in senescent cardiac fibroblasts impairs PDGF-induced cell proliferation. *Cardiovasc. Res.* 49:731–40

171. Ikeyama S, Kokkonen G, Shack S, Wang XT, Holbrook NJ. 2002. Loss in oxidative stress tolerance with aging linked to reduced extracellular signal-regulated kinase and Akt kinase activities. *FASEB J.* 16:114–16

172. Collado M, Medema RH, Garcia-Cao I, Dubuisson ML, Barradas M, et al. 2000. Inhibition of the phosphoinositide 3-kinase pathway induces a senescence-like arrest mediated by p27Kip1. *J. Biol. Chem.* 275:21960–68

173. Kandel ES, Skeen J, Majewski N, Di Cristofano A, Pandolfi PP, et al. 2002. Activation of Akt/protein kinase B overcomes a G(2)/M cell cycle checkpoint induced by DNA damage. *Mol. Cell. Biol.* 22:7831–41

174. Chakravarthy MV, Abraha TW, Schwartz RJ, Fiorotto ML, Booth FW. 2000. Insulin-like growth factor-I extends in vitro replicative life span of skeletal muscle satellite cells by enhancing G1/S cell cycle progression via the activation of phosphatidylinositol 3′-kinase/Akt signaling pathway. *J. Biol. Chem.* 275:35942–52

175. Kapur R, Cooper R, Zhang L, Williams DA. 2001. Cross-talk between alpha(4)

beta(1)/alpha(5)beta(1) and c-Kit results in opposing effect on growth and survival of hematopoietic cells via activation of focal adhesion kinase, mitogen-activated protein kinase, and Akt signaling pathways. *Blood* 97:1975–81

176. De Miguel MP, Cheng L, Holland EC, Federspiel MJ, Donovan PJ. 2002. Dissection of the c-Kit signaling pathway in mouse primordial germ cells by retroviral-mediated gene transfer. *Proc. Natl. Acad. Sci. USA* 99:10458–63

177. Esposito G, Rapacciuolo A, Naga Prasad SV, Takaoka H, Thomas SA, et al. 2002. Genetic alterations that inhibit in vivo pressure-overload hypertrophy prevent cardiac dysfunction despite increased wall stress. *Circulation* 105:85–92

178. Klinz F, Bloch W, Addicks K, Hescheler J. 1999. Inhibition of phosphatidylinositol-3-kinase blocks development of functional embryonic cardiomyocytes. *Exp. Cell Res.* 247:79–83

179. Shioi T, Kang PM, Douglas PS, Hampe J, Yballe CM, et al. 2000. The conserved phosphoinositide 3-kinase pathway determines heart size in mice. *EMBO J.* 19:2537–48

180. Shiojima I, Yefremashvili M, Luo Z, Kureshi Y, Takahashi A, et al. 2002. Akt signaling mediates postnatal heart growth in response to insulin and nutritional status. *J. Biol. Chem.* 277:37670–77

181. Matsui T, Tao J, del Monte F, Lee KH, Li L, et al. 2001. Akt activation preserves cardiac function and prevents injury after transient cardiac ischemia in vivo. *Circulation* 104:330–35

182. Condorelli G, Drusco A, Stassi G, Bellacosa A, Roncarati R, et al. 2002. Akt induces enhanced myocardial contractility and cell size in vivo in transgenic mice. *Proc. Natl. Acad. Sci. USA* 99:12333–38

Annu. Rev. Physiol. 2004. 66:49–75
doi: 10.1146/annurev.physiol.66.032102.141555
First published online as a Review in Advance on October 15, 2003

Viral-Based Myocardial Gene Therapy Approaches to Alter Cardiac Function

Matthew L. Williams and Walter J. Koch[1]

Departments of Surgery, and Pharmacology, and Cancer Biology, Duke University Medical Center, Durham, North Carolina 27710; email: walter.koch@jefferson.edu

Key Words   adenovirus, heart failure, coronary catheterization, β-adrenergic signaling, calcium handling, arrhythmias

■ **Abstract**   In recent years there has been a rapid expansion in our understanding of the molecular biology that underpins human physiology. In the heart, elegant molecular pathways have been elucidated, and derangements in these pathways have been identified as factors in cardiac disease. However, as our understanding has grown, we have recognized that there exist only relatively crude tools to effect changes in molecular pathophysiology. The ultimate promise of gene therapy is to correct the molecular derangements that cause illness. To bring this promise to fruition in the clinical arena, many problems need to be solved, and chief among these remains reliable and robust delivery of genes to the target organ. To this end, viral vectors have been utilized with success more frequently than any other method of gene delivery. The use of these vectors in the heart has already offered promising novel benefit for human ischemic heart disease, and studies in animal models have given glimpses of hope that gene therapy may provide future therapeutic benefit in heart failure by improving cardiac function.

VIRAL VECTORS USED IN CARDIAC GENE THERAPY

Safe, efficient, and reproducible gene delivery to a target organ remains the fundamental obstacle to be overcome prior to widespread application of gene therapy in the clinical arena. The ideal vector for transduction of a gene into a diseased cell would possess several characteristics: ease of production, efficient transduction, low immunogenicity, controlled expression, organ specificity, and a low risk of oncogenesis. Unfortunately, this idealized vector remains elusive. Among those used in gene therapy research, viral vectors have demonstrated the greatest utility for applications in the cardiovascular system. Specifically for gene therapy approaches to alter cardiac physiology, three viral vector systems have been employed–the adenovirus, which has been the primary vector utilized, and

---

[1]Current address: Center for Translational Medicine, Thomas Jefferson University, 1015 Walnut Street, Room 801, Philadelphia, Pennsylvania 19107.

TABLE 1   Properties of viral vectors used for cardiac gene therapy

	Integration of viral DNA into host cell chromosome?	Length of expression	Difficulty of production	Immunogenicity	Insert size
Adenovirus	No	Up to 4 weeks	Low	Moderate	10 kB
Adeno-associated virus	Yes	Long-term	High	Low	5 kB
Lentivirus	Yes	Unknown	High	Unknown	8 kB

to a lesser extent the adeno-associated virus (AAV) and lentivirus (see Table 1). These vector systems, detailed in Table 1, all have advantages and shortcomings, and their utility for cardiac gene therapy is discussed below.

Adenoviral Vectors

The most frequently used vector in gene therapy for heart disease has been the adenovirus (1). The adenovirus lacks an envelope and has a double-stranded DNA genome of 36 kB. Adenoviruses employed in gene therapy research have been modified to prevent viral replication and limit the host immune response (1, 2). By deletion of regions of the viral genome (i.e., E1 region), these replication-deficient viruses are made safer and more efficient as vectors for gene therapy. In addition, recombinant adenoviruses have a larger cDNA (i.e., transgene) carrying capacity than most other viral systems (1, 2). Importantly, unlike other viral vectors, the adenoviral genome is not integrated into host DNA but instead persists in the cell as episomal DNA (2). Advantages of the adenovirus as a gene therapy vector are numerous. First, it is relatively easy and inexpensive to produce large quantities of adenovirus that carry a target gene. Second, it is very efficient at infecting a cell and has a broad host range (1, 2). As described in more detail below, adenoviral vectors can efficiently infect myocardium and produce robust transgene expression in cardiomyocytes. The third advantage of adenovirus is that because adenoviral DNA is not stably integrated, the worry of oncogenesis is limited, and the Adenoviridae have never been identified as a causative agent in human malignancy (2). Transient transgene expression supported by adenoviral vectors has often been viewed as a disadvantage of this vector; however, this characteristic may well prove to be an attractive quality because of the iatrogenic malignancies caused by retroviral gene therapy for children stricken with severe combined immunodeficiency (3).

Despite the positive aspects of adenovirus, it is far from the ideal vector. The transient nature of expression is indeed limiting in some cases, and adenoviruses can be a robust stimulus for host immune responses with inflammation being observed in tissues harboring the vector (1, 2). Moreover, because of immunologic memory, second or third exposures to the adenoviral vector in research animals have demonstrated reduced levels of expression (4). The receptor exploited by the adenovirus, the coxsackie-adenoviral receptor (CAR) is widely expressed in humans, limiting tissue tropism and therefore fidelity of expression (2).

Early cardiac gene delivery experiments utilized first-generation adenoviral vectors that were E1 deleted, and these were quite immunogenic and had a short duration of expression (5). Subsequently, second-generation and more advanced adenoviral vectors have become available with additional deletions (such as E3 and E4), as well as completely "gutted" vectors (6). These advanced vectors may help to advance the field of cardiac gene therapy because they appear to cause less inflammation and support significantly longer persistence of transgene expression (6). Despite the current limitations, the adenovirus has been the most commonly employed vector for cardiac gene transfer to date, serving to deliver transgenes to potentially improve ischemic heart disease, heart failure, and to alter the electrophysiological function of the heart (5).

Adeno-Associated Virus

An alternative to the adenovirus as a vector for gene therapy is the adeno-associated virus (AAV) (7). The AAV is a nonpathogenic human parvovirus. In the absence of helper virus, usually herpesvirus or adenovirus, the AAV enters a latent life cycle and integrates into chromosomal DNA (2, 7). Cellular receptors identified as targets for AAV include the heparin sulfate proteoglycan (8), fibroblast growth factor receptor 1 (9), and $\alpha V\beta 5$ integrin (10). Unlike the adenovirus, the AAV demonstrates a decreased immunologic reaction and extended periods of expression following successful gene transfection (7). AAV can be used to affect gene transfer in both dividing and nondividing cells, and AAV has been used to target gene expression chronically in the heart (11, 12).

Despite these apparent advantages, there are several drawbacks to the use of AAV vectors. The efficiency of gene transduction with the use of AAV is low, and the production of AAV in high titers is labor-intensive and costly. Early reports of cardiac gene transfer demonstrated that AAV vectors are <50% as efficient at transfecting myocardium as adenoviral vectors (11). Moreover, AAV has considerably less cloning space (5 kB) compared to other vectors (7, 13). Finally, because of the potential integration into the host chromosome, concerns regarding malignancy remain. Nonetheless, AAV vectors (along with advanced adenoviruses) may hold the key for safer human cardiac gene therapy in the near future, and a recent animal study has documented the rescue of cardiomyopathy with recombinant AAV vectors carrying a specific transgene targeting myocardial calcium (Ca^{2+}) handling (12) (more details below).

Lentiviral Vectors

A third vector employed in gene therapy research in the heart has been the lentivirus (14). Lentiviral vectors, of which HIV is the prototype, are enveloped viruses that fuse with the cell membrane following the interaction with a cellular receptor (2). These viruses are able to infect nondividing cells and deliver a gene that is integrated into the host genome, thus providing for prolonged expression but at a cost of the theoretical risk of oncogenesis. In studies to date using lentiviral vectors, the virus has been an HIV-1 type that has been genetically modified to preclude replication, and an insertion reporter gene has been used to identify cells that have been successfully infected and transduced. The lentivirus can accept constructs up to 8 kB, so it has a clear advantage over AAV. Only one report has demonstrated successful gene transduction to adult myocardium utilizing a lentiviral vector, and robust expression was measured seven days after gene transfer (14). More time and studies are needed to determine whether this vector system will become an important player in cardiac gene therapy studies.

METHODS OF CARDIAC GENE DELIVERY

To achieve robust and reliable expression in myocardium, various methods of gene delivery to the heart have been developed. In the future it might be possible to alter viral receptors to allow directed gene delivery, but at present no gene therapy studies on the heart have utilized a vector system that can be targeted to the heart via remote intravenous injection. Rather, it has been necessary to deliver the viral vector directly to the heart as an end-destination targeted organ. Various methods have been developed with success, including ex vivo delivery to cardiac isografts and allografts, invasive and noninvasive direct injection into the myocardium, and both antegrade and retrograde delivery through the coronary circulation (5). In addition, advances have been made that allow delivery through single coronary arteries without a surgical incision in a percutaneous fashion (15, 16). Delivery of the marker transgene β-galatosidase (β-Gal) to the heart by various in vivo intracoronary delivery methods is represented in Figure 1 (see color insert).

Direct Myocardial Injection

The first reports of viral gene delivery to the myocardium utilized the method of direct injection (17). Numerous studies of this type have followed, and this method continues to be utilized (5). Generally, through a thoracotomy of variable size, a small-caliber needle is used to penetrate the substance of the myocardium and inject viral particles into a localized area of the heart. Alternatively, direct myocardial delivery can be achieved via a percutaneous catheter guided into the ventricle, and from an endocardial approach the viral vector is injected directly into the myocardium (18). These direct injection methods result in reliable and robust expression of transgene in the target area. However, this method carries with it the drawbacks of localized expression (generally only in and around the needle tract) as well as the physical damage of penetrating the heart muscle and injury from the localized perfusate pressure. In fact, early studies reported inflammation in the heart owing to adenovirus, but more likely the majority of the inflammation is caused by needle injury. Importantly, for certain transgene applications, including ischemic heart disease, this methodology is more than adequate for delivering angiogenic growth factors, which are secreted proteins. As detailed in the next section, a small deposit of transgene expression can be effective for stimulating angiogenesis in ischemic myocardium (19). By contrast, the direct injection of viral vectors to improve the contractile function of the failing heart is far from ideal because global transgene expression covering a large percentage of ventricular myocardium is required to positively affect the disease (5).

Intracoronary Delivery to the Arrested or Transplanted Heart

Gene delivery to an explanted or arrested heart avoids many of the difficulties of in vivo delivery. Ex vivo gene delivery to transplanted hearts in both rat and rabbit models of heterotopic cardiac transplantation has been successfully demonstrated by perfusion of virus into the aortic root, with the vector allowed to remain in the coronary circulation while the cardiac graft was sewn into place (20–23). Our laboratory has demonstrated the feasibility of ex vivo gene delivery via the coronary arteries, and manipulation of myocardial β-adrenergic receptor (β−AR) signaling improved the function of the transplanted heart (21–23). This includes failing rabbit hearts where ex vivo intracoronary adenoviral-mediated β_2-AR gene delivery allowed for more rapid reverse remodeling of the unloaded transplanted heart (23). Adenoviral delivery to isolated hearts using a Langendorff perfusion apparatus has also been utilized as an ex vivo cardiac gene delivery strategy, and this method has proven to be a powerful way to study different intracoronary delivery parameters that can increase the efficiency of myocardial transgene expression such as perfusion pressure, Ca^{2+} concentration in the perfusate solution, and adjuvants such as serotonin, which increase vascular permeability (24). Overall, this technique of delivery could prove useful in transplantation, but would be limited to only this subset of clinical applications in human patients.

A technique applicable to a larger set of patients that are undergoing cardiac surgery is delivery of transgenes through the coronary arteries of an arrested heart that is on cardiopulmonary bypass (CPB). Davidson et al. performed the first adenoviral-mediated gene delivery to the myocardium during routine cardiopulmonary bypass in piglets (25). The two most significant findings of this study were that myocardial transgene expression was noted as early as eight hours after administration and that no evidence of extracardiac expression was measured (25). A subsequent study demonstrated the importance of certain delivery parameters and cardioplegic conditions during CPB gene delivery in the neonatal pig (26). This technique has also been tried in larger animals; Bridges et al. have delivered adenoviral transgenes to the dog heart during CPB (27). In this study, the authors used a true isolated circuit by utilizing the myocardial venous system (27). To date, in preclinical studies, CPB is the only administration method that limits delivery to the heart and avoids systemic exposure while concomitantly providing potential functional augmentation.

In Vivo Intracoronary Gene Delivery Methods
to the Beating Heart

The delivery of viral vectors via the coronary circulation in vivo was an advance in global and ventricular delivery methods. A method that perfused the entire heart through the coronary circulation was first done in rats where both the aorta and pulmonary artery were clamped while the adenovirus solution was injected into the left ventricle (LV) (28). Our laboratory was the first to demonstrate that a similar intracoronary gene delivery technique in rabbits could result in the enhancement of global cardiac function (29). In these rabbits, aortic cross-clamping and subsequent intraventricular delivery of a β_2-AR-containing adenovirus led to global and robust overexpression of β_2-ARs (29). In this study, there was a direct relationship noted between β_2-AR overexpression levels attained in the LV of rabbits and functional contractile responses to β-AR stimulation (29). This is illustrated in Figure 2. As described in more detail below in the section specifically addressing

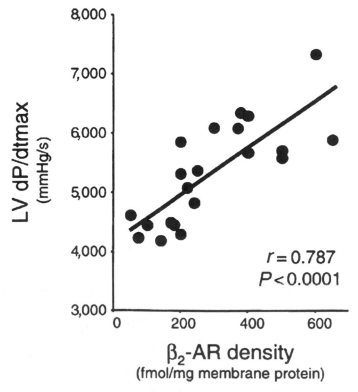

Figure 2   In vivo cardiac function in normal rabbits following intracoronary delivery of Adeno-β_2-AR. The graph depicts the correlation between the degree of left ventricular β_2-AR overexpression and the maximum LV dP/dt (measure of contractility) response to isoproterenol. A significant linear relationship was found by t-test (reprinted with permission from Reference 29).

heart failure gene therapy, this delivery method (or similar derivations) has been employed by our laboratory and others for delivery of adenoviral and AAV vectors to various animal models of cardiomyopathy (12, 30–32). Although the technique of intracoronary delivery of viral vectors through a surgical incision to directly visualize the heart has proven reliable, this method has limited clinical applicability. It would likely be limited to only those patients who were already undergoing open heart surgery for other reasons, for example, coronary artery bypass grafting or the implantation of a ventricular assist device. However, importantly, it has served as a powerful model to demonstrate critical proof of principle in animal models of heart failure.

Percutaneous Intracoronary Virus Delivery to Myocardium

The delivery method that holds the most promise for widespread clinical application is percutaneous intracoronary delivery of viral vectors to target myocardium. Barr et al. first reported this delivery system using a β-Gal adenovirus (15). Our group has improved upon this method by determining critical delivery parameters such as adenoviral purity, dose and perfusion volume, and pressure (16, 33). Moreover, by selectively catheterizing either the left circumflex coronary artery (LCx) or the right coronary artery (RCA), transgene expression can be targeted to either the LV or right ventricle, respectively (16, 34). Ventricular-specific gene delivery is demonstrated in Figure 1. Subselective coronary artery catheterization and adenoviral-mediated gene delivery of transgenes to rabbit hearts that alter β-AR signaling can enhance in vivo cardiac contractile function and performance and are effective in models of ventricular dysfunction and failure (16, 34, 35).

In addition to the above methodology, others have been able to cannulate the coronary venous system percutaneously to successfully deliver viral transgenes to the heart (36). Recently, global gene delivery to the heart has been performed by a novel percutaneous approach. In this method, balloon occlusion of the proximal aorta is followed by infusion of adenovirus into the aortic root. The infusion is supplemented with microbubbles and sonographic waves that are targeted to increase porosity of the coronary circulation; enhanced gene delivery was achieved (37). Because of the decreased morbidity of the percutaneous as opposed to surgical approach, these promising intracoronary delivery methods remain the best hope for widespread application to patients with a variety of disease states, including heart failure. Specific examples of the application of these methods of myocardial delivery to various heart diseases in both animal models and humans are described below.

APPLICATIONS OF CARDIAC GENE THERAPY

It is well known that heart disease is the leading cause of death in the United States. Over 7 million people have suffered a heart attack, 6 million people suffer with angina pectoris, 5 million people carry a diagnosis of congestive heart failure, and it is estimated that 2 million Americans have atrial arrhythmias (38). Current therapy for heart disease is improving, but the cost to society in terms of morbidity, mortality, and healthcare expenditures remains enormous. Myocardial gene transfer utilizing viral vectors holds promise as a potential novel molecular therapeutic strategy for altering cardiac physiology. Multiple molecular targets have been successfully manipulated to produce beneficial effects in animal models, and limited early trials in humans have been undertaken to treat ischemic heart disease. The molecular interventions for ischemic heart disease have primarily been growth factors designed to stimulate new blood vessel growth (i.e., angiogenesis).

In addition to ischemic heart disease, gene therapy approaches and investigation have also targeted acute and chronic heart failure. Derangements in many systems of the failing heart, such as adrenergic signaling, Ca^{2+} handling, or increased apoptosis, allow for many potential targets for intervention. Finally, targeted gene therapies for arrhythmias to alter the electrical conduction system of the heart have been undertaken.

Gene Therapy for Ischemic Heart Disease

The use of gene therapy to improve severe myocardial ischemia is the area in which clinical application of viral vectors in the heart appears to be the most close at hand (19). Clinical trials in humans employing angiogenic growth factors as transgenes are underway, and initial results appear promising. The process of angiogenesis is complex and requires multiple steps including, but not limited to, activation of endothelial cells, degradation of basement membrane and extracellular matrix, and migration and proliferation of endothelial cells. Angiogenic factors are small proteins that send a signal that is, in turn, amplified by other effector cells. Most of the research in this area has focused on the use of two secreted angiogenic factors in particular: the vascular endothelial growth factor (VEGF) and fibroblast growth factor (FGF) (39, 40). The use of gene therapy to deliver secreted gene products in the heart is attractive on many levels because many of the hurdles of cardiac gene transfer are side-stepped. For example, the problem of widespread gene delivery to a large number of cells is unnecessary provided enough of the secreted factor is produced by target cells and this paracrine signal is amplified. Because the areas affected by ischemia are often discrete, direct injection can achieve targeted myocardial therapy, and robust expression in a localized area can be achieved. In addition, the transient presence of the gene product (a characteristic of the adenoviral-based expression system) may produce a sustained therapeutic effect lasting past the loss of gene expression. It has been suggested that the ideal regimen for angiogenic factor administration would be a "sustained but transient (2–3 week) increase in local angiogenic protein concentration at a focus of myocardial ischemia" (19). The ideal technique of administration, then, would be direct myocardial gene therapy. In animal models, both adenoviral and plasmid DNA delivery have been used with success to deliver angiogenic factors to ischemic heart muscle (41–44). Below, we present some discussion on reports that have utilized viral vectors as the mode of gene delivery to the ischemic heart, including human clinical trials.

Numerous models of ischemic disease in animals have shown improvement after delivery of angiogenic factors. In a porcine model of ischemia, Mack et al. used an adenoviral vector to deliver VEGF-121 (the 121-amino-acid form of VEGF) to ischemic myocardium via direct injection with positive results (41). Adenoviral VEGF delivery by direct myocardial injection has also proved effective in a porcine-pacing model of heart failure (42). These and earlier animal models prompted several clinical trials using adenovirus-mediated VEGF delivery to the ischemic human heart (39). Early results of a phase I clinical trial that employed an adenoviral vector to deliver VEGF to patients with ischemic heart disease have been released (39). Twenty-one patients received VEGF-121 via an adenoviral vector. Fifteen patients received the virus as an adjunct to cardiac surgery, and in these patients the virus was directly injected into myocardium that could not be revascularized by a surgical conduit. Six others received the virus by direct injection through a small thoracotomy. Again, these patients were not candidates for revascularization by standard therapy (bypass surgery or percutaneous angioplasty and stent placement). Since this was a phase I trial, the data were not controlled (39). There was uniform improvement in subjective classification of angina both for those who underwent surgery and those who received Ad-VEGF-121 as sole therapy; however, there was no significant difference in coronary angiogram appearance when pre- and posttreatment films were compared, despite a trend toward improvement (39). Clearly, the lack of control groups precludes firm conclusions about these results, but because the trial was performed in humans it remains an important result and justifies further study with a phase 2/3 trial that would include satisfactory control groups. The placebo effect cannot be dismissed, and firm conclusions await further controlled studies. Importantly, there were no adverse effects related to VEGF therapy, such as hemangiomas or other vascular malformations.

In addition to VEGF, an FGF adenovirus has been used in a clinical trial for ischemic heart disease based on success in animal models (43, 44). This trial involved catheter-mediated intracoronary delivery of a replication-deficient adenovirus that contains FGF-4 (Ad-FGF-4) (40). Intracoronary delivery of adenoviral vectors has theoretical limitations in treatment of ischemia with angiogenic factors because the occluded conduits that require treatment may also block transgene delivery to target tissues. In this study, 71 patients underwent double-blind randomization to receive either Ad-FGF-4 or placebo (40). The main endpoint was an exercise stress test (ETT), although other measurements such as dobutamine stress echocardiograms and blood samples were taken to determine if an immune response to the adenovirus had occurred and to titrate for the presence of FGF-4. Safety concerns were addressed and there appeared to be no adverse events that could be attributed to Ad-FGF-4 administration. An overall comparison of placebo versus all treatment groups demonstrated no significant difference in ETT time (40). However, there were some encouraging results from subsets of patients. In particular, those patients who had low neutralizing antibody titers to adenovirus ($<1:100$)

demonstrated significant improvement when compared with those patients who received the Ad-FGF-4 who had a higher antibody titer (40). One interesting finding that clearly has been repeated in other clinical trials for angina therapy is the real and large placebo effect; 21% of patients who received placebo demonstrated improvement of greater than 30% of their ETT time at 12 weeks (40).

Gene Therapy Approaches for Heart Failure

Heart failure has become increasingly common in the United States. From 1979 to 1999, hospitalizations rose 155%, to 962,000 per year, and a recent report has described the lifetime risk of acquiring heart failure as 1 in 5 for both men and women (38). These sobering statistics highlight the need for novel therapies to treat this progressive and morbid disease. The understanding of heart failure on a molecular level has progressed, and therapies targeted to the derangements at the level of the myocyte offer optimism in the field. Two separate but connected signaling pathways that are abnormal in the failing heart have become the primary targets of heart failure gene therapy efforts. These systems, intracellular Ca^{2+} handling and the β-AR system, have become focused targets for novel therapies. Major molecular targets for these systems are illustrated in Figure 3. In addition to altering myocyte Ca^{2+} handling and β-AR signaling, other strategies have been aimed at inhibiting apoptosis, a process that many believe contributes to the progressively worsening decline in heart function and the dismal prognosis for these patients. To date, only studies in animal models of heart failure have been done using various viral vectors containing transgenes targeting these pathways.

TARGETING CA^{2+} HANDLING IN THE FAILING HEART   The Ca^{2+} cycle involved in myocyte contraction and relaxation plays a central role in normal heart function and exhibits derangements in heart failure (45) (Figure 3a). Depolarization of the myocyte allows Ca^{2+} entry into the myocyte via membrane voltage-dependent channels. This Ca^{2+} influx, in turn, activates the ryanodine receptor (RyR) to release intracellular stores of Ca^{2+} from the sarcoplasmic reticulum (SR). This Ca^{2+}-induced Ca^{2+}-release allows Ca^{2+} binding to troponin C and thus facilitates function of the contractile machinery (45). For relaxation to occur, Ca^{2+} concentration must decline, a process caused by various pumps in the SR and sarcolemmal membranes, among them the SR Ca^{2+} ATPase (SERCA2a). At the sarcolemmal level, Ca^{2+} can leave the myocytes via the $Na^+ Ca^{2+}$ exchanger (NCX) (45). Regulation of SERCA2a is important to the overall homeostasis of the cardiomyocytes. Phospholamban (PLB) is an inhibitor of SERCA2a, and phosphorylation of PLB via cAMP-dependent protein kinase A (PKA) or calmodulin kinase II relieves this inhibition. Thus the β-AR system and SR Ca^{2+} handling are linked through the actions of cAMP.

Various molecules and pathways in the myocyte Ca^{2+} cycle are altered in heart failure (Figure 3). SERCA2a activity is decreased, as are its mRNA and protein levels (46). There appears to be an alteration in the proportion of PLB

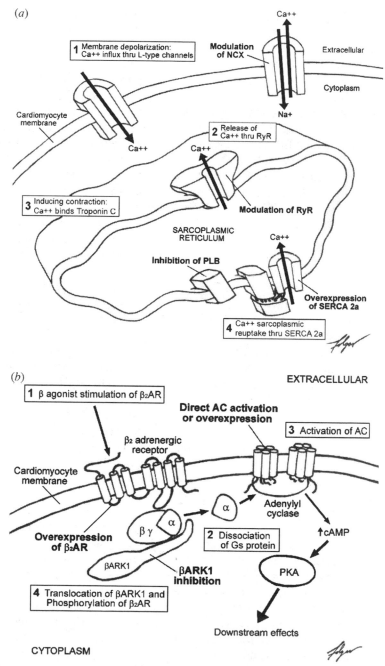

Figure 3 The (*a*) Ca^{2+} handling and (*b*) β-AR systems in the cardiomyocyte. The individual steps in the system are numbered; targets for therapeutic intervention are highlighted in bold.

that is phosphorylated in human heart failure, such that the activity of SERCA2a is depressed and SR Ca^{2+} handling is disturbed (46). As detailed below, in vitro experiments and adenoviral-mediated gene transfer in animal models of heart failure have been performed that correct these molecular defects in myocyte Ca^{2+} handling, and improvements in parameters of cardiac function have been observed.

Animal studies have demonstrated that the restoration (via overexpression) of SERCA2a improves compromised function and ameliorates heart failure. LV function was improved in an aortic banding model of heart failure when an adenovirus containing the SERCA2a transgene was administered via a catheter-based intracoronary approach (31). Survival and LV dimensions were similarly improved when the SERCA2a virus was administered via catheter in this heart failure model compared with control rats that received a β-Gal adenovirus (47). It should be noted that both the SERCA2a and NCX are responsible for removal of Ca^{2+} from the cytoplasm, either back into the SR or out of the cell. Thus one caveat to consider is the relative proportions of these mechanisms responsible for Ca^{2+} shifts that restore relaxation in the animal model studied. In the rat, the animal model in which both of the above studies were performed, over 90% of the Ca^{2+} flux out of the cytoplasm is accounted for by mechanisms of the SR of which SERCA2a is the most important (45). In the human myocyte, only about 70% of this Ca^{2+} flux is accounted for by SR mechanisms; the remainder is removed by sarcolemmal means (45, 46). Thus using rodent models of heart failure may not be ideal for studying SR manipulations. However, importantly, it has been demonstrated that contraction and relaxation of failing human myocytes isolated from patients undergoing cardiac transplantation were improved by SERCA2a gene transfer in vitro via adenoviral infection (48).

PLB has also been targeted for a viral-based gene therapy approach for heart failure. Data supporting PLB as a potential therapeutic target for heart failure first came from genetically engineered mice where the PLB gene knockout produced a model of enhanced cardiac contractility, and this mouse has rescued various murine models of heart failure (49–51). Interestingly, recent studies in other mouse models of cardiomyopathy showed improved Ca^{2+} handling but no rescue of the phenotype, demonstrating that models of heart failure in the mouse can be quite different (52). Moving beyond the mouse, PLB has been targeted using larger animal models. The use of an antisense mRNA to down-regulate the expression of PLB has been utilized to improve the contractile performance of both neonatal rat (53) and failing adult human myocytes (54). Moreover, recently, Hoshijima et al. delivered a recombinant AAV vector containing a dominant-negative mutant of PLB via an open chest intracoronary (aorta cross-clamp) delivery method to cardiomyopathic hamsters and suppressed development of heart failure (12). In this model, a mutant of PLB that contained pseudophosphorylation at Ser16, which is the site of PKA phosphorylation, was used. This mutation mimics phosphorylation and removes the baseline PLB inhibition of SERCA2a. This molecular alteration allowed improved reconstitution of SR Ca^{2+} stores and improved ventricular performance in the cardiomyopathic hamster model (12). Furthermore, the use of AAV as a vector provided sustained expression, and at 12 weeks significant amounts of the protein product of the transgene were detected (12). This represents the first report using an AAV vector in a chronic heart failure model.

PLB appears to be a useful target in animal models of heart failure; however, two recent genomic studies in selected human populations have revealed the striking finding that naturally occurring, but rare, mutations in the human PLB gene lead to cardiomyopathy (55, 56). In the first, a mutation of the protein prevents it from regulating SERCA2a but allows it to act as a sink for PKA, thus preventing adrenergic removal of PLB inhibition of SERCA (55). This has led to a heart failure phenotype and supports the current hypothesis that removal of PLB inhibition would be beneficial to the heart. However, a second human PLB mutation that introduces a premature stop codon leaves affected homozygous individuals without functional PLB protein (56). Thus, effectively, these individuals are PLB knockouts, and they actually developed progressive and severe dilated cardiomyopathy that necessitated cardiac transplantation early in life (56). These results demonstrate the importance of PLB in heart function and disease in the human, although the surprising result of a PLB null as a cardiomyopathy-inducing human mutation is difficult to reconcile with our current understanding of the role of PLB and the data in the mouse. It may be that chronic inhibition of PLB's effects in human heart failure would be deleterious, as it is in this natural experiment of a PLB human knockout. Alternatively, PLB targeting acutely via viral gene transfer could provide benefit in a chronic heart failure setting, as data from animal studies suggest.

The NCX protein appears to be an intriguing target as well as it appears to be important in the human cardiomyocyte, especially when it is increased under conditions of heart failure (57–59). In fact, NCX overexpression via adenoviral-mediated gene transfer to rabbit myocytes resulted in lowered inotropic properties (59). This potential target is now being pursued in gene therapy experiments because molecular therapies targeted to decrease NCX activity in the failing myocyte, as in using antisense, may be a novel therapeutic strategy. Importantly, the alteration of cardiac NCX activity could have profound effects on cardiac arrhythmias as well (58).

A novel molecule thought to be part of the Ca^{2+} handling and myocyte contraction-relaxation cycle is the S100A1 protein. This protein is a member of the EF-hand Ca^{2+} binding S100 family and is expressed at high levels in the heart (60). S100A1 has been found to colocalize with both the SR and contractile elements of cardiac tissue (60). The expression of S100A1 is down-regulated in end-stage heart failure (61) but increased in pressure-overload hypertrophy (62). Thus its plasticity in cardiac disease may indicate its importance. In the mouse, S100A1 has been knocked out, and these mice have impaired cardiac contractile responses to β-AR stimulation and marked decompensation in response to surgical aortic constriction (63). Interestingly, transgenic overexpression in the hearts of mice causes baseline enhancement of cardiac function that remains after β-AR stimulation (64). Consistent with these mouse studies, the transfer of S100A1 via an adenovirus to cardiac myocytes in culture and to engineered heart tissue in vitro has demonstrated enhanced Ca^{2+} transients and contractility (65). Despite this encouraging evidence, no clear mechanism for the salutary effects of this transgene has been elucidated as of yet, although S100A1 does interact with RyR in skeletal muscle and a similar interaction has been recently shown in the heart (64). Thus it appears that the function of S100A1 is at the SR level and may involve RyR Ca^{2+} release.

TARGETING THE β-AR SYSTEM IN HEART FAILURE BY GENE TRANSFER β_1- and β_2-ARs mediate the inotropic, chronotropic, and lusitropic responses in the heart to the catecholamines norepinepherine, and epinephrine (66). β_1- and β_2-ARs are the dominant forms found in the myocardium, with β_1-ARs, the most abundant subtype, accounting for 75–80% of the total (67). Agonist binding to β-ARs activates the guanine nucleotide binding protein Gs, which is located primarily within the sarcolemmal membrane of myocytes. Gs, in turn, stimulates the downstream effector enzyme adenylyl cyclase (AC), and the subsequent accumulation of cAMP activates protein kinase A (PKA), which phosphorylates several proteins that ultimately are involved in enhancing cardiac function (66, 67). Recent evidence has demonstrated that β_2-ARs can elicit qualitatively different signaling mechanisms from those of β_1-ARs within isolated cardiac myocytes, primarily through additional and alternative coupling to the heterotrimeric Gi protein (66, 68). β_1-ARs can stimulate apoptosis and elicit cellular pathology, whereas β_2-AR stimulation can be protective against apoptosis and lead to enhanced cardiac function (69). The myocardial β-AR system is illustrated in Figure 3.

β-ARs, similar to other G protein–coupled receptors (GPCRs), undergo a rapid process of regulation that leads to functional uncoupling. This homologous desensitization is initiated by phosphorylation of the activated receptor by the serine/threonine kinases known as the G protein receptor kinases (GRKs) (66, 70). Homologous desensitization of β-ARs requires the action of a second protein, β-arrestin, which binds to phosphorylated receptors and sterically interdicts further G protein-coupling (66, 70). β-arrestin binding directs the internalization of desensitized GPCRs, which leads to receptor degradation and recycling (66). The GRK that appears most important for β-ARs is the β-AR kinase (βARK1 or GRK2) (70).

Several recent data have solidified the myocardial β-AR system as being critical for normal and compromised heart function. Recent success in the treatment of heart failure with β-AR antagonists solidifies the importance of this system in cardiac disease (71, 72). Moreover, human polymorphisms in β-AR genes can influence prognosis and incidence of heart failure (73, 74). In the heart failure patient, many changes occur in the β-AR system including down-regulation of β-ARs selective for the β_1-AR subtype (75, 76). This alters the β_1-AR/β_2-AR ratio, a situation that could alter the relative importance of the two predominant β-AR subtypes, and given the qualitative signaling differences between the two subtypes (68), this is seen as particularly important. Furthermore, the AC inhibitory

G protein, Gi, is elevated (77), which not only contributes to the functional uncoupling and dysfunction of the β-AR system in heart failure but also increases the importance of the increased β_2-AR ratio.

The βARK1 enzyme is also intimately related to the derangements of the β-AR system in heart failure. βARK1 mRNA, protein, and activity are all elevated in human heart failure (76). The elevation in βARK1 activity appears to contribute to the β-AR desensitization that marks end-stage heart failure. Overall, it has been demonstrated that excessive catecholamines trigger molecular changes in the β-AR system, leading to dampening of sympathetic signaling that contributes to deterioration of ventricular function (78). This is generally thought to be an adaptive mechanism of the heart to protect compromised myocardium from catecholamine flogging. However, novel data generated from transgenic mice and adenoviral-mediated gene transfer experiments to rabbits with induced heart failure have shown that β-AR down-regulation and βARK1 up-regulation are maladaptive in the failing heart. Moreover, gene therapy strategies aimed at correcting these abnormalities may have therapeutic potential (Figure 3) (30, 66) (see below for more detail).

Differences in β-AR subtype signaling have been clearly demonstrated in transgenic mice through myocardial-targeted overexpression of β_1- and β_2-AR subtypes by use of the α-myosin heavy chain gene (αMHC) promoter (79, 80). First, the β_2-AR transgene was overexpressed in mouse hearts at nearly 200-fold greater than wild-type levels, and these mice demonstrated enhanced in vivo LV performance with surprisingly minimal pathology (79). Moreover, αMHC-β_2-AR overexpression rescued a mouse model of decompensated hypertrophy and heart failure (81). In contrast, αMHC-β_1-AR overexpression at only modest (10- to 15-fold) levels resulted in a phenotype of myocardial hypertrophy with rapid progression to failure (80).

Studies using adenoviruses containing β-AR subtypes have also demonstrated these quantitative and qualitative signaling differences. At the level of the cultured cardiac myocyte, β-AR overexpression via a replication-deficient adenovirus (Adeno-β_2-AR) enhanced adrenergic signaling in both normal and failing adult rabbit myocytes (82, 83) and has protected against β_1-AR induced pathology (i.e., apoptosis) (69, 84). In vivo gene transfer of Adeno-β_2-AR via open chest intracoronary delivery (aortic cross-clamp) to normal rabbit hearts demonstrated enhanced catecholamine contractile responsiveness (29) (Figure 2), and percutaneous LCx-mediated gene transfer of Adeno-β_2-AR to the rabbit heart resulted in enhanced in vivo LV function (16). Enhanced β-AR myocardial responsiveness has also been shown after Adeno-β_2-AR delivery to the normal rat heart (85). Furthermore, a model of cardiac transplantation in which the heart underwent Adeno-β_2-AR gene transfer prior to heterotopic transplantation demonstrated enhanced function in the isograft (21).

In addition to transfer of Adeno-β_2-AR to normal hearts and transplant grafts, β_2-AR overexpression has improved models of cardiac dysfunction. In a pressure overload model in a rat heart, β_2-AR overexpression improved catecholamine responsiveness, thus preserving myocardial function (86). More recently, we have utilized a rabbit heart failure heterotopic transplantation model to demonstrate a functional and potentially therapeutic benefit of β_2-AR overexpression (23). Heterotopic transplantation of a failing rabbit heart simulates the situation of hearts unloaded by ventricular assist devices, and these failing hearts eventually show signs of recovery after 30 days of unloading in the cervical position of donor rabbits (23). However, if the failing rabbit hearts are treated with intracoronary Adeno-β_2-AR prior to transplantation, functional recovery and an actual enhancement of function is seen within 5 days (23). Thus β_2-AR gene delivery may indeed be a novel and useful molecular adjunct to existing therapies in select heart failure patients. One particular population might be represented by heart failure patients that have polymorphic forms of the β_2-AR gene, which are negatively associated with heart failure outcomes (73). Delivery of normal β_2-ARs to these patients in the form of viral-mediated gene delivery may provide a specific benefit.

Another component of the myocardial β-AR signaling system that has shown potential importance is βARK1, which appears to play a critical role in heart function and dysfunction (66). Ample evidence of βARK1's importance in the cardiovascular system has been gleaned from transgenic mice models as well as from adenoviral-mediated gene transfer experiments in larger animals. First, βARK1 was overexpresed in mouse hearts by use of the α-MHC promoter, and these mice had markedly attenuated responses to β-AR agonists (87). In a reciprocal fashion, mice that expressed a peptide inhibitor of βARK1 in their hearts had decreased βAR desensitization, and cardiac βARK1 inhibition led to enhanced in vivo baseline LV function as well as responses to catecholamines (87, 88). This peptide inhibitor, βARKct, is made up of the final 194 amino acids of βARK1, and contains the G protein $\beta\gamma$-subunit (G$\beta\gamma$)-binding domain, which competes with endogenous βARK1 for membrane translocation and activation (89). The reciprocal nature of the physiology found in these two transgenic lines of mice indicates a critical role for βARK1 in normal cardiac regulation and function.

Further evidence supporting the idea that βARK1 inhibition may be beneficial comes from heterozygous βARK1 transgenic mice, where LV function was similar to that of βARKct mice (90). Thus by lowering cardiac βARK1 activity through either inhibition with the βARKct or the loss of one gene allele, a profound effect on in vivo heart function is seen. In addition, βARKct mice have been able to rescue several murine models of heart failure, imparting improvement in myocardial β-AR signaling, LV function and remodeling, cardiac hypertrophy and survival (91–94). Treatment of heart failure through enhancement, rather than suppression, of β-AR signaling runs counter to current conventional wisdom in the treatment of the failing heart. The finding that β-AR agonists are detrimental and that β-AR blockade is beneficial in heart failure leads many to believe that any intervention that enhances β-AR signaling will be detrimental to the already compensated heart. However, it has been demonstrated that β-AR blockade, together with βARKct expression, can act synergistically in heart failure. For example, cardiac βARK1 inhibition prolongs survival and augments β-AR antagonist therapy in a mouse model of severe heart failure (93). At a molecular level, altering β-AR density and/or inhibiting βARK1 are novel manipulations that cannot be fitted in any existing pharmacological paradigm. Thus the data generated should not be viewed with any bias to previous data from β-AR agonists or antagonist studies. βARK1 inhibition does represent a potential novel class of drugs.

Owing to this evidence from transgenic mice, it was postulated that alteration of βARK1 signaling via the βARKct peptide could improve cardiac function via gene transfer using viral vectors. Consistent with the mouse data, in vitro studies utilizing a βARKct-containing adenovirus (Adeno-βARKct) demonstrated the ability to restore βAR signaling and improve contractility of cardiac myocytes purified from failing hearts (83, 95). In vivo studies using Adeno-βARKct have also supported the hypothesis that βARK1 is a novel therapeutic target in the failing heart. First, delivery of Adeno-βARKct to rabbit hearts via an open chest global intracoronary delivery method prevented heart failure following myocardial infarction (MI) via LCx ligation (30). Adeno-βARKct delivery at the time of MI prevented β-AR down-regulation and βARK1 up-regulation, demonstrating the importance of maintaining a normal β-AR signaling system (30, 66). In this study, in addition to β-AR signaling normalization in the post-MI heart, βARKct expression led to increased in vivo LV function with improvement seen even at 21 days post-MI, when hemodynamic failure is quite evident in this model (30).

Percutaneous subselective coronary artery catheterization and adenoviral gene delivery has allowed our laboratory to deliver the βARKct transgene to post-MI rabbit hearts already in failure (34). Three weeks following MI, 5×10^{11} total viral particles (tvp) of Adeno-βARKct were delivered through the LCx, and both regional and global LV function were measured one week later, and subsequent βARKct expression enhanced LV regional systolic function (34). These results are shown in Figure 4. Moreover, desensitized β-AR signaling was reversed only in the LV, not the RV, demonstrating the ventricular-selectivity of LCx-mediated gene delivery (34).

Other applications of the βARKct transgene have been successfully undertaken. Global myocardial intracoronary delivery of Adeno-βARKct can enhance the function of hearts that have undergone cardioplegic arrest and cold ischemia with increased ex vivo function measured with βARKct expression even after 4 h of arrest (96). Moreover, donor heart contractile dysfunction following prolonged ex vivo preservation can be prevented by gene-mediated β-AR signaling modulation through intracoronary Adeno-βARKct delivery (97). Finally, using percutaneous right coronary artery–selective catheterization, Adeno-βARKct delivery to the RV has shown beneficial effects in a model of RV failure following pulmonary artery banding (35). Overall, in both transgenic mouse models and these examples of Adeno-βARKct delivery to normal and failing rabbit hearts, the data are overwhelming in demonstrating that βARK1 inhibition results in positive outcomes in heart failure. It is important to point out that the βARKct itself may affect more than β-AR signaling in the heart, and this could be part of its mechanisms of action. First, βARK1 desensitizes receptors other than β-ARs and thus

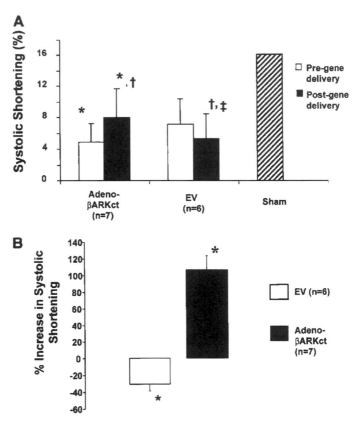

Figure 4   Ventricular-specific reversal of left ventricular cardiac dysfunction in heart failure rabbits after LCx-mediated delivery of Adeno-βARKct. After HF was induced in rabbits via myocardial infarction, animals received either Adeno-βARKct or empty virus (EV) via percutaneous subselective left circumflex coronary artery catheterization. (A) LV systolic function was assessed by sonomicrometry. (B) Also shown is the percent increase one week after βARKct gene delivery. *P < 0.05 compared with sham; †P < 0.05 compared with pregene delivery values; ‡P < 0.05 compared with Adeno-βARKct treated animals (reprinted with permission from Reference 34).

signaling through other GPCRs (that are βARK1 targets) will be increased after βARKct expression, which may account for the observed beneficial effects (66). Furthermore, other Gβγ-dependent processes may be altered in hearts expressing the βARKct.

An additional component of the β-AR system that has been exploited for beneficial effect in preclinical experiments is cardiac overexpression of adenyl cyclase (AC), most notably Type VI AC. This strategy also runs counter to dogma stating that an increase in cAMP will be detrimental to the failing heart. Evidence to the contrary was first shown in transgenic mice using the α-MHC promoter to

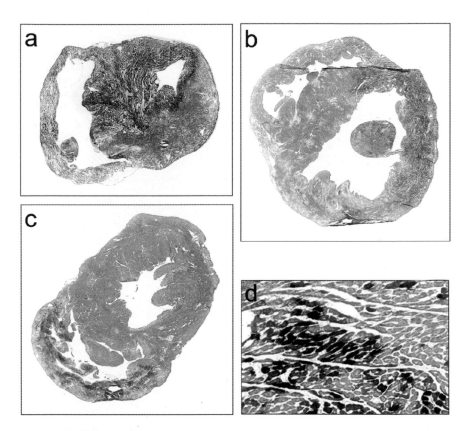

Figure 1 Representative sections of rabbit hearts that have undergone delivery of the β-galactosidase gene using an adenoviral vector. (*a*) Open-chest intracoronary delivery via aortic cross-clamp and injection into the aortic root. (*b*) Left ventricular-specific delivery via percutaneous catheterization of the left circumflex coronary artery. (*c*) Right ventricular-specific delivery via percutaneous catheterization of the right coronary artery. (*d*) High-power magnification (40×) of a ventricular section demonstrating β-galactosidase expression in myocytes.

overexpress Type VI AC (98). These mice had enhanced responsiveness to adrenergic stimulation (98). The AC transgene in these mice also improves function and survival in a murine model of cardiomyopathy (99, 100). Importantly, it has been demonstrated that percutaneous catheter-mediated intracoronary delivery of an AC Type VI adenovirus could significantly improve LV function in adult pigs, and this improvement in LV function persisted for as long as 57 days (101). In addition, cAMP generation in these hearts was also enhanced without any detected arrhythmogenic effects (101). This study is significant on several fronts because it is the first demonstration of a functional cardiac effect after intracoronary delivery to a large animal model (adult pig), and chronic effects were seen after adenoviral-mediated gene delivery in this model (101). This study was performed in normal pigs and the AC gene therapy strategy awaits application to a heart failure model.

Targeting Myocardial Apoptosis and Cardiac Arrhythmias

Another strategy to improve heart failure results from the observation that evidence of apoptosis can be found in human heart failure (102) and that inhibition of apoptosis can improve murine models of heart failure (103). It should be noted, however, that the connection between apoptosis and the progression of heart failure has yet to be definitively linked in humans. Despite the lack of conclusive evidence, promising reports within the field of gene therapy provide hope that these strategies will provide novel therapies for heart failure. These approaches seek to change the balance of apoptosis to ensure survival of post-mitotic cardiomyocytes and thus improve heart failure. Bcl-2 is a potent and well-described mediator of cell survival (104) and has been used to tip the balance against apoptosis in an effort to improve a model of post-ischemic heart failure. Interestingly, in a recent report utilizing an open chest intracoronary delivery method for an adenovirus containing Bcl-2, both function and measurements of LV wall thinning were improved in a rabbit model of post-ischemic myocardial dysfunction (32). Another antiapoptotic, prosurvival molecule, Akt, was cloned into an adenoviral vector and shown to prevent doxorubicin-induced heart failure in rats (105), as well as to improve post-ischemia LV dysfunction in a rat model (106). Moreover, intracoronary adenovirus-mediated gene transfer of a constitutive active Akt to rat hearts limited infarct size in vivo in a model of ischemia-reperfusion injury (107).

Cardiac arrhythmias present a common and often fatal medical problem affecting many individuals. Diseases of the conduction system due to defects in different ion channels of the heart account for many cases of stroke and fatal arrhythmias, and many millions are spent on pharmacotherapy, conduction pathway ablation, and implantable devices, such as the automatic defibrillators and permanent pacemakers (108). Electrical dysfunction of the heart also can contribute to heart failure (108, 109). To potentially combat this at the molecular level, viral-mediated gene transfer of different potassium (K^+) channel proteins has been utilized to target both arrhythmias and cardiac dysfunction in experimental models. Recently, Miake et al. have converted quiescent heart muscle cells into pacemaker cells via adenoviral-mediated gene transfer of an altered K^+ channel that leads to spontaneous depolarization or automaticity (110). This group used the mutant K^+ channel, Kir2.1AAA, as a dominant-negative mutant that suppresses inward K^+ currents and, subsequently, excitability (110). In guinea pigs infected with the Kir2.1AAA adenovirus, ventricular myocytes demonstrated automaticity with the development of spontaneous depolarizations, whereas myocytes from control animals did not (110). Pacemaker activity of the heart has also been altered by in vivo adenoviral-mediated gene transfer of the $G\alpha i$ protein, which inhibits AC and enhances parasympathetic control of the heart (111). In this study, the $G\alpha i$ adenovirus was delivered to the AV node of guinea pigs via percutaneous catheterization, which resulted in slower conduction and reversal of atrial fibrillation (111). These studies demonstrate a novel potential therapy for patients with diseases of the conduction system that require a pacemaker, such as the "sick sinus syndrome" or atrioventricular block.

Different studies using other K^+ channels have also led to beneficial effects concerning cardiac arrhythmias. Overexpression of the human delayed rectifier K^+ channel HERG in rabbit ventricular myocytes via adenovirus-mediated gene transfer suppressed cardiac arrhythmia early after depolarizations (112). This demonstrated the potential of using gene therapy of delayed rectifying K^+ channels to suppress arrhythmias caused by hyperexcitability and unstable repolarizations (112). Dual gene therapy has also recently been demonstrated using adenoviruses containing SERCA1 and the K^+ channel, Kir2.1, and represents a potential powerful next step in exploring molecular ways to treat the failing and dysfunctional heart (113). In guinea pigs, in vivo gene delivery of these two transgenes caused shortening of the cardiac action potential (via Kir2.1 overexpression), but the expected loss of contractile function was blocked by SERCA1 overexpression (113). These data demonstrate the potential power and therapeutic benefit of manipulating both the excitability and contractility of the failing heart via viral-mediated gene transfer.

CONCLUSIONS AND FUTURE DIRECTIONS

In experimental animals, alterations in cardiac function have been demonstrated through gene therapy, and this technology has been successfully applied to models of ischemia, heart failure, and arrythmias. As we begin to exploit our understanding of molecular pathophysiology to treat cardiac pathology, it is important to recognize the limitations in the field. Current gene delivery methods remain quite invasive. Vectors with tissue tropism that obviate the need for direct myocardial delivery would represent a significant advance. Furthermore, research has not yielded any method to control expression after gene delivery. Analogous to a physician adjusting a dose to produce a desired effect, finer exogenous control of expression would be desirable. Moreover, while viral-based methods of gene delivery to the myocardium of small mammals has been repeatedly demonstrated by a wide variet of investigators, there remain relatively few reports of global gene delivery in large animal models, a prerequisite for application of gene therapy in human heart failure.

Already, gene therapy for cardiac ischemia has reached human trials, and it is our hope that other applications of gene therapy for cardiac disease will one day be tested in the clinical arena. To achieve this objective, it will be necessary to continue to identify targets for therapy and make advances in gene delivery methods.

The *Annual Review of Physiology* is online at http://physiol.annualreviews.org

LITERATURE CITED

1. Leiden JM. 2000. Human gene therapy: the good, the bad, and the ugly. *Circ. Res.* 86:923–25
2. Mah C, Byrne BJ, Flotte TR. 2002. Viral–based delivery systems. *Clin. Pharmacokinet.* 41:901–11
3. Hacein-Bey-Abina S, von Kalle C, Schmidt M, Le Deist F, Wulffraat N, et al. 2003. A serious adverse event after successful gene therapy for X-linked severe combined immunodeficiency. *N. Engl. J. Med.* 348:255–56
4. Li JJ, Ueno H, Pan Y, Tomita H, Yamamoto H, et al. 1995. Percutaneous transluminal gene transfer into canine myocardium in vivo by replication-defective adenovirus. *Cardiovasc. Res.* 30:97–105
5. White DC, Koch WJ. 2001. Myocardial gene transfer. *Curr. Cardiol. Rep.* 3:37–42
6. Amalfitano A, Parks RJ. 2002. Separating fact from fiction: assessing the potential of modified adenovirus vectors for use in human gene therapy. *Curr. Gene Ther.* 2:111–33
7. Stilwell JL, Samulski RJ. 2003. Adeno-associated virus vectors for therapeutic gene transfer. *Biotechniques* 34:148–50
8. Summerford C, Samulski RJ. 1998. Membrane-associated heparin sulfate proteoglycan is a receptor for adeno-associated virus type 2 virions. *J. Virol.* 72:1438–45
9. Qing K, Mah C, Hansen J, Zhou S, Dwarki V, Srivastava A. 1999. Human fibroblast growth factor receptor 1 is a co-receptor for infection by adeno-associated virus 2. *Nat. Med.* 5:71–77
10. Summerford C, Bartlett JS, Samulski RJ. 1999. $\alpha V\beta 5$ integrin: a co-receptor for adeno-associated virus type 2 infection. *Nat. Med.* 5:78–82
11. Svensson EC, Marshall DJ, Woodard K, Lin H, Jiang F, et al. 1999. Efficient and stable transduction of cardiomyocytes after intramyocardial injection or intracoronary perfusion with recombinant adeno-associated virus vectors. *Circulation* 99:201–5
12. Hoshijima M, Ikeda Y, Iwanaga Y, Minamisawa S, Date MO, et al. 2002. Chronic suppression of heart-failure progression by a pseudophosphorylated mutant of phospholamban via in vivo cardiac rAAV gene delivery. *Nat Med.* 8:864–71
13. Dong JY, Fan PD, Frizzell RA. 1996. Quantitative analysis of the packaging capacity of recombinant adeno-associated virus. *Hum. Gene. Ther.* 7:2101–12
14. Zhao J, Pettigrew GJ, Thomas J, Vandenberg JI, Delriviere L, et al. 2002. Lentiviral vectors for delivery of genes into neonatal and adult ventricular cardiac myocytes in vitro and in vivo. *Basic Res. Cardiol.* 97:348–358
15. Barr E, Carroll J, Kalynych AM, Tripathy SK, Kozarsky K, et al. 1994. Efficient catheter-mediated gene transfer into the heart using replication-defective adenovirus. *Gene. Ther.* 1:51–58

16. Shah AS, Lilly RE, Kypson AP, Tai O, Hata JA, et al. 2000. Intracoronary adenovirus-mediated delivery and overexpression of the β_2-adrenergic receptor in the heart: prospects for molecular ventricular assistance. *Circulation* 101:408–14

17. Kass-Eisler A, Falck-Pedersen E, Alvira M, Rivera J, Buttrick PM, et al. 1993. Quantitative determination of adenovirus-mediated gene delivery to rat cardiac myocytes in vitro and in vivo. *Proc. Natl. Acad. Sci. USA* 90:11498–502

18. Sanborn TA, Hackett NR, Lee LY, El-Sawy T, Blanco I, et al. 2001. Percutaneous endocardial transfer and expression of genes to the myocardium utilizing fluoroscopic guidance. *Catheter Cardiovasc. Interv.* 52:260–66

19. Freedman SB, Isner JM. 2001. Therapeutic angiogenesis for ischemic cardiovascular disease. *J. Mol. Cell. Cardiol.* 33:379–93

20. Kypson AP, Peppel K, Akhter SA, Lilly RE, Glower DD, et al. 1998. Ex vivo adenovirus-mediated gene transfer to the adult rat heart. *J. Thorac. Cardiovasc. Surg.* 115:623–30

21. Kypson A, Hendrickson S, Akhter S, Wilson K, McDonald P, et al. 1999. Adenovirus-mediated gene transfer of the β_2-adrenergic receptor to donor hearts enhances cardiac function. *Gene. Ther.* 6:1298–304

22. Shah AS, White DC, Tai O, Hata JA, Wilson KH, et al. 2000. Adenoviral-mediated genetic manipulation of the myocardial β-adrenergic signaling system in transplanted hearts. *J. Thorac. Cardiovasc. Surg.* 120:581–88

23. Tevaearai HT, Eckhart AD, Walton GB, Keys JR, Wilson K, Koch WJ. 2002. Myocardial gene transfer and overexpression of β_2-adrenergic receptors potentiates the functional recovery of unloaded failing hearts. *Circulation* 106:124–29

24. Donahue JK, Kikkawa K, Thomas AD, Marban E, Lawrence JH. 1998. Accelera-tion of widespread adenoviral gene transfer to intact rabbit hearts by coronary perfusion with low calcium and serotonin. *Gene Ther.* 5:630–34

25. Davidson MJ, Jones JM, Emani SM, Wilson KH, Jaggers J, et al. 2001. Cardiac gene delivery with cardiopulmonary bypass. *Circulation* 104:131–33

26. Jones JM, Wilson KH, Koch WJ, Milano CA. 2002. Adenoviral gene transfer to the heart during cardiopulmonary bypass: effect of myocardial protection technique on transgene expression. *Euro. J. Cardiothorac. Surg.* 21:847–52

27. Bridges CR, Burkman JM, Malekan R, Konig SM, Chen H, et al. 2002. Global cardiac-specific transgene expression using cardiopulmonary bypass with cardiac ischemia. *Ann. Thorac. Surg.* 73:1939–46

28. Hajjar RJ, Schmidt U, Matsui T, Guerrero JL, Lee KH, et al. 1998. Modulation of ventricular function through gene transfer in vivo. *Proc. Natl. Acad. Sci. USA* 95:5251–56

29. Maurice JP, Hata JA, Shah AS, White DC, McDonald PH, et al. 1999. Enhancement of cardiac function after adenoviral-mediated in vivo intracoronary β_2-adrenergic receptor gene delivery. *J. Clin. Invest.* 104:21–29

30. White DC, Hata JA, Shah AS, Glower DD, Lefkowitz RJ, Koch WJ. 2000. Preservation of myocardial β_2-adrenergic receptor signaling delays the development of heart failure after myocardial infarction. *Proc. Natl. Acad. Sci. USA* 97:5428–33

31. Miyamoto MI, del Monte F, Schmidt U, DiSalvo TS, Kang ZB, et al. 2000. Adenoviral gene transfer of SERCA2a improves left-ventricular function in aortic-banded rats in transition to heart failure. *Proc. Natl. Acad. Sci. USA* 97:793–98

32. Chatterjee S, Stewart AS, Bish LT, Jayasankar V, Kim EM, et al. 2002. Viral gene transfer of the antiapoptotic factor Bcl-2 protects against chronic postischemic heart failure. *Circulation* 106:I212–17

33. Emani SM, Shah AS, Bowman MK, Emani S, Wilson K, et al. 2003. Catheter-based intracoronary myocardial adenoviral gene delivery: importance of intraluminal seal and infusion flow-rate. *Mol. Ther.* 8:306–13

34. Shah AS, White DC, Emani S, Kypson AP, Lilly RE, et al. 2001. In vivo ventricular gene delivery of a β adrenergic receptor kinase inhibitor to the failing heart reverses cardiac dysfunction. *Circulation* 103:1311–16

35. Emani SM, Shah AS, Bowman MK, White DC, Emani S, et al. 2003. Right ventricular targeted gene transfer of a β−adrenergic receptor kinase inhibitor improves ventricular performance after pulmonary artery banding. *J. Thorac. Cardiovasc. Surg.* In press

36. Boekstegers P, von Degenfeld G, Giehrl W, Heinrich D, Hullin R, et al. 2000. Myocardial gene transfer by selective pressure-regulated retroinfusion of coronary veins. *Gene Ther.* 7:232–40

37. Beeri R, Guerrero JL, Supple G, Sullivan S, Levine RA, Hajjar RJ. 2002. New efficient catheter-based system for myocardial gene delivery. *Circulation* 106:1756–59

38. Lloyd-Jones DM, Larson MG, Leip EP, Beiser A, D'Agostino RB, et al. 2002. Lifetime risk for developing congestive heart failure: the Framingham Heart Study. *Circulation* 106:3068–72

39. Rosengart TK, Lee LY, Patel SR, Sanborn TA, Parikh M, et al. 1999. Angiogenesis gene therapy: phase I assessment of direct intramyocardial administration of an adenovirus vector expressing VEGF121 cDNA to individuals with clinically significant severe coronary artery disease. *Circulation* 100:468–74

40. Grines CL, Watkins MW, Helmer G, Penny W, Brinker J, et al. 2002. Angiogenic gene therapy (AGENT) trial in patients with stable angina pectoris. *Circulation* 105:1291–97

41. Mack CA, Patel SR, Schwarz EA, Zan-zonico P, Hahn RT, et al. 1998. Biologic bypass with the use of adenovirus-mediated gene transfer of the complementary deoxyribonucleic acid for vascular endothelial growth factor 121 improves myocardial perfusion and function in the ischemic porcine heart. *J. Thorac. Cardiovasc. Surg.* 115:168–76

42. Leotta E, Patejunas G, Murphy G, Szokol J, McGregor L, et al. 2002. Gene therapy with adenovirus-mediated myocardial transfer of vascular endothelial growth factor 121 improves cardiac performance in a pacing model of congestive heart failure. *J. Thorac. Cardiovasc. Surg.* 123:1101–13

43. Safi J Jr, DiPaula AF Jr, Riccioni T, Kajstura J, Ambrosio G, et al. 1999. Adenovirus-mediated acidic fibroblast growth factor gene transfer induces angiogenesis in the nonischemic rabbit heart. *Microvasc. Res.* 58:238–49

44. Giordano FJ, Ping P, McKirnan MD, Nozaki S, DeMaria AN, et al. 1996. Intracoronary gene transfer of fibroblast growth factor-5 increases blood flow and contractile function in an ischemic region of the heart. *Nat. Med.* 2:534–39

45. Bers DM. 2002. Cardiac excitation-contraction coupling. *Nature* 415:198–205

46. Houser SR, Margulies KB. 2003. Is depressed myocyte contractility centrally involved in heart failure? *Circ. Res.* 92:350–58

47. del Monte F, Williams E, Lebeche D, Schmidt U, Rosenzweig A, et al. 2001. Improvement in survival and cardiac metabolism after gene transfer of sarcoplasmic reticulum Ca^{2+}-ATPase in a rat model of heart failure. *Circulation* 104:1424–29

48. del Monte F, Harding SE, Schmidt U, Matsui T, Kang ZB, et al. 1999. Restoration of contractile function in isolated cardiomyocytes from failing human hearts by gene transfer of SERCA2a. *Circulation* 100:2308–11

49. Minamisawa S, Hoshijima M, Chu G, Ward CA, Frank K, et al. 1999. Chronic phospholamban-sarcoplasmic reticulum calcium ATPase interaction is the critical calcium cycling defect in dilated cardiomyopathy. *Cell* 99:313–22

50. Freeman K, Lerman I, Kranias EG, Bohlmeyer T, Bristow MR, et al. 2001. Alterations in cardiac adrenergic signaling and calcium cycling differentially affect the progression of cardiomyopathy. *J. Clin. Invest.* 107: 967–74

51. Sato Y, Kiriazis H, Yatani A, Schmidt AG, Hahn H, et al. 2001. Rescue of contractile parameters and myocyte hypertrophy in calsequestrin overexpressing myocardium by phospholamban ablation. *J. Biol. Chem.* 276:9392–99

52. Song Q, Schmidt AG, Hahn HS, Carr AN, Frank B, et al. 2003. Rescue of cardiomyocyte dysfunction by phospholamban ablation does not prevent ventricular failure in genetic hypertrophy. *J. Clin. Invest.* 111:859–67

53. Eizema K, Fechner H, Bezstarosti K, Schneider-Rasp S, van der Laarse A, et al. 2000. Adenovirus-based phospholamban antisense expression as a novel approach to improve cardiac contractile dysfunction: comparison of a constitutive viral versus an endothelin-1-responsive cardiac promoter. *Circulation* 101:2193–99

54. del Monte F, Harding SE, Dec GW, Gwathmey JK, Hajjar RJ. 2002. Targeting phospholamban by gene transfer in human heart failure. *Circulation* 105:904–7

55. Schmitt JP, Kamisago M, Asahi M, Li GH, Ahmad F, et al. 2003. Dilated cardiomyopathy and heart failure caused by a mutation in phospholamban. *Science* 299:1410–13

56. Haghighi K, Kolokathis F, Pater L, Lynch RA, Asahi M, et al. 2003. Human phospholamban null results in lethal dilated cardiomyopathy revealing a critical difference between mouse and human. *J. Clin. Invest.* 111:869–76

57. Weber CR, Piacentino V, Margulies KB, Bers DM, Houser SR. 2002. Calcium influx via I(NCX) is favored in failing ventricular myocytes. *Ann. NY Acad. Sci.* 976:478–79

58. Sipido KR, Volders PG, Vos MA, Verdonck F. 2002. Altered Na/Ca exchange activity in cardiac hypertrophy and heart failure: A new target for therapy? *Cardiovasc. Res.* 53:782–805

59. Schillinger W, Janssen PM, Enami S, Henderson SA, Ross RS, et al. 2000. Impaired contractile performance of cultured rabbit ventricular myocytes after adenoviral gene transfer of Na^+-Ca^{2+} exchanger. *Circ. Res.* 87:581–87

60. Haimoto H, Kato K. 1988. S100A0 ($\alpha\alpha$) protein in cardiac muscle. Isolation from human cardiac muscle and ultrastructural localization. *Eur. J. Biochem.* 171:409–15

61. Remppis A, Greten T, Schafer BW, Hunziker P, Erne P, et al. 1996. Altered expression of the Ca^{2+}-binding protein S100A1 in human cardiomyopathy. *Biochim. Biophys. Acta* 1313:253–57

62. Ehlermann P, Remppis A, Guddat O, Weimann J, Schnabel PA, et al. 2000. Right ventricular upregulation of the Ca^{2+} binding protein S100A1 in chronic pulmonary hypertension. *Biochim. Biophys. Acta* 1500:249–55

63. Du XJ, Cole TJ, Tenis N, Gao XM, Kontgen F, et al. 2002. Impaired cardiac contractility response to hemodynamic stress in S100A1-deficient mice. *Mol. Cell. Biol.* 8:2821–29

64. Most P, Remppis A, Pleger ST, Loffler E, Ehlermann P, et al. 2003. Transgenic overexpression of the Ca^{2+}-binding protein S100A1 in the heart leads to increased in vivo myocardial contractile performance. *J. Biol. Chem.* 278:33809–17

65. Most P, Bernotat J, Ehlermann P, Pleger ST, Reppel M, et al. 2001. S100A1: a regulator of myocardial contractility. *Proc. Natl. Acad. Sci. USA* 98:13889–94

66. Rockman HA, Koch WJ, Lefkowitz RJ. 2002. Seven-transmembrane-spanning

receptors and heart function. *Nature* 415:206–12

67. Brodde OE. 1993. Beta-adrenoceptors in cardiac disease. *Pharmacol. Ther.* 60:405–30

68. Steinberg SF. 1999. The molecular basis for distinct β-adrenergic receptor subtype actions in cardiomyocytes. *Circ. Res.* 85:1101–11

69. Communal C, Singh K, Sawyer DB, Colucci WS. 1999. Opposing effects of B1 and B2 adrenergic receptors on cardiac myocyte apoptosis: role of pertussis toxin-sensitive G protein. *Circulation* 100:2210–12

70. Iaccarino G, Lefkowitz RJ, Koch WJ. 1999. Myocardial G-protein-coupled receptor kinases: implications for heart failure therapy. *Proc. Assoc. Am. Phys.* 111: 399–405

71. Clinical Trial. 1999. Effect of metoprolol CR/XL in chronic heart failure: metoprolol CR/XL randomized intervention trial in congestive heart failure (MERIT-HF). *Lancet* 353:2001–7

72. Packer M, Bristow MR, Cohn JN, Colucci WS, Fowler MB, et al. 1996. The effect of Carvedilol on morbidity and mortality in patients with chronic heart failure. U.S. Carvedilol Heart Failure Study Group. *N. Engl. J. Med.* 334:1349–55

73. Ligget SB, Wagoner LE, Craft LL, Hornung RW, Hoit BD, et al. 1998. The Ile164 β_2-adrenergic receptor polymorphism adversely affects the outcome of congestive heart failure. *J. Clin. Invest.* 102:1534–39

74. Small KM, Wagoner LE, Levin AM, Kardia SL, Ligget SB. 2002. Synergistic polymorphisms of β_1- and α_2c-adrenergic receptors and the risk of congestive heart failure. *N. Engl. J. Med.* 347:1135–42

75. Bristow MR, Ginsburg R, Umans V, Fowler M, Minobe W, et al. 1986. β_1- and β_2-adrenergic receptors subpopulations in nonfailing and failing human ventricular myocardium: coupling of both receptor subtypes to muscle contraction and selective β_2 receptor down-regulation in heart failure. *Circ. Res.* 59:297–309

76. Ungerer M, Bohm M, Elce JS, Erdmann E, Lohse MJ. 1993. Altered expression of β-adrenergic receptor kinase and β_1-adrenergic receptors in the failing human heart. *Circulation* 87:454–63

77. Feldman AM. 1993. Modulation of adrenergic receptors and G transduction proteins in failing human ventricular myocardium. *Circulation* 87:IV27–IV34

78. Cohn JN, Levine TB, Olivari MT, Garberg V, Lura D, et al. 1984. Plasma norepinephrine as a guide to prognosis in patients with chronic congestive heart failure. *N. Engl. J. Med.* 311:819–23

79. Milano CA, Allen LF, Rockman HA, Dolber PC, McMinn TR, et al. 1994. Enhanced myocardial function in transgenic mice overexpressing the β_2-adrenergic receptor. *Science* 264:582–86

80. Engelhardt S, Hein L, Wiesmann F, Lohse MJ. 1999. Progressive hypertrophy and heart failure in β_1-adrenergic receptor transgenic mice. *Proc. Natl. Acad. Sci. USA* 96:7059–64

81. Dorn GW, Tepe NM, Lorenz JN, Davis MG, Koch WJ, Ligget SB. 1999. Low and high transgenic expression of β_2-adrenergic receptors differentially affects cardiac hypertrophy and function in Gαq overexpressing mice. *Proc. Natl. Acad. Sci. USA* 96:6400–5

82. Drazner MH, Peppel KC, Dyer S, Grant AO, Koch WJ, Lefkowitz RJ. 1997. Potentiation of β-adrenergic signaling by adenoviral-mediated gene transfer in adult rabbit ventricular myocytes. *J. Clin. Invest.* 99:288–96

83. Akhter SA, Skaer CA, Kypson AP, McDonald PH, Peppel KC, et al. 1997. Restoration of β-adrenergic signaling in failing cardiac ventricular myocytes via adenoviral-mediated gene transfer. *Proc. Natl. Acad. Sci. USA* 94:12100–5

84. Zhu WZ, Zheng M, Koch WJ, Lefkowitz RJ, Koblika BK, Xiao RP. 2001. Dual modulation of cell survival and cell death

by β_2-adrenergic signaling in adult mouse cardiomyocytes. *Proc. Natl. Acad. Sci. USA* 98:1607–12

85. Kawahira Y, Sawa Y, Nishimura M, Sakakida S, Ueda H, et al. 1999. Gene transfection of β_2-adrenergic receptor into the normal rat heart enhances cardiac response to β-adrenergic agonist. *J. Thorac. Cardiovasc. Surg.* 118:446–51

86. Kawahira Y, Sawa Y, Nishimura M, Sakakida S, Ueda H, et al. 1998. In vivo transfer of a β_2-adrenergic receptor gene into the pressure-overloaded rat heart enhances cardiac response to β-adrenergic agonist. *Circulation* 98:II262–67

87. Koch WJ, Rockman HA, Samama P, Hamilton RA, Bond RA, et al. 1995. Cardiac function in mice overexpressing β-adrenergic receptor kinase or a βARK inhibitor. *Science* 268:1350–53

88. Akhter SA, Eckhart AD, Rockman HA, Shotwell KF, Lefkowitz RJ, Koch WJ. 1999. In vivo inhibition of elevated myocardial β-adrenergic receptor kinase activity in hybrid transgenic mice restores normal β-adrenergic signaling and function. *Circulation* 100:648–53

89. Koch WJ, Inglese J, Stone WC, Lefkowitz RJ. 1993. The binding site for the beta gamma subunits of heterotrimeric G proteins on the beta-adrenergic receptor kinase. *J. Biol. Chem.* 268:8256–60

90. Rockman HA, Akhter SA, Choi D-J, Jaber M, Lefkowitz RJ, et al. 1998. Regulation of myocardial contractile function by the level of β-adrenergic receptor kinase-1 in gene targeted mice. *J. Biol. Chem.* 273:18180–84

91. Rockman HA, Chien KR, Choi DJ, Iaccarino G, Hunter JJ, et al. 1998. Expression of β-adrenergic receptor kinase 1 inhibitor prevents the development of heart failure in gene targeted mice. *Proc. Natl. Acad. Sci. USA* 95:7000–5

92. Freeman K, Olsson MC, Iaccarino G, Bristow MR, Lefkowitz RJ, et al. 2001. Alterations in cardiac adrenergic signaling and calcium cycling differentially af-

fect the progression of cardiomyopathy. *J. Clin. Inv.* 107:967–74

93. Harding VB, Jones LR, Lefkowitz RJ, Koch WJ, Rockman HA. 2001. Cardiac βARK1 inhibition prolongs survival and augments β-blocker therapy in a mouse model of severe heart failure. *Proc. Natl. Acad. Sci. USA* 98:5809–14

94. Eckhart AD, Fentzke RC, Lepore J, Lang R, Lin H, et al. 2002. Inhibition of βARK1 and restoration of myocardial β-adrenergic signaling in a mouse model of dilated cardiomyopathy induced by $CREB_{A133}$ expression. *J. Mol. Cell Cardiol.* 34:669–77

95. Eckhart AD, Koch WJ. 2002. Expression of a β-adrenergic receptor kinase inhibitor reverses dysfunction in failing cardiomyocytes. *Mol. Ther.* 5:74–79

96. Tevaearai HT, Eckhart AD, Shotwell KF, Wilson K, Koch WJ. 2001. Ventricular dysfunction after cardioplegic arrest is improved after myocardial gene transfer of a β-adrenergic receptor kinase inhibitor. *Circulation* 104:2069–74

97. Tevaearai HT, Walton GB, Eckhart AD, Keys JR, Koch WJ. 2002. Donor heart contractile dysfunction following prolonged ex vivo preservation can be prevented by gene-mediated β-adrenergic signaling modulation. *Eur. J. Cardiothorac. Surg.* 22:733–37

98. Gao MH, Lai NC, Roth DM, Zhou J, Zhu J, et al. 1999. Adenylyl cyclase increases responsiveness to catecholamine stimulation in transgenic mice. *Circulation* 99:1618–22

99. Roth DM, Gao MH, Lai NC, Drumm J, Dalton N, et al. 1999. Cardiac-directed adenylyl cyclase expression improves heart function in murine cardiomyopathy. *Circulation* 99:3099–102

100. Roth DM, Bayat H, Drumm JD, Gao MH, Swaney JS, et al. 2002. Adenylyl cyclase increases survival in cardiomyopathy. *Circulation* 105:1989–94

101. Lai NC, Roth DM, Gao MH, Fine S, Head BP, et al. 2000. Intracoronary

delivery of adenovirus encoding adenylyl cyclase VI increases left ventricular function and cAMP-generating capacity. *Circulation* 102:2396–401

102. Olivetti G, Abbi R, Quaini F, Kajstura J, Cheng W, et al. 1997. Apoptosis in the failing human heart. *N. Engl. J. Med.* 336:1131–41

103. Hirota H, Chen J, Betz UA, Rajewsky K, Gu Y, et al. 1999. Loss of a gp130 cardiac muscle cell survival pathway is a critical event in the onset of heart failure during biomechanical stress. *Cell* 97:189–98

104. Adams JM, Cory S. 1998. The Bcl-2 protein family: arbiters of cell survival. *Science* 281:1322–26

105. Taniyama Y, Walsh K. 2002. Elevated myocardial Akt signaling ameliorates doxorubicin-induced congestive heart failure and promotes heart growth. *J. Mol. Cell. Cardiol.* 34:1241–47

106. Matsui T, Tao J, del Monte F, Lee KH, Li L, et al. 2001. Akt activation preserves cardiac function and prevents injury after transient cardiac ischemia in vivo. *Circulation* 104:330–35

107. Miao W, Luo Z, Kitsis RN, Walsh K. 2000. Intracoronary, adenovirus-mediated Akt gene transfer in heart failure limits infarct size following ischemia-reperfusion injury in vivo. *J. Moll. Cell. Cardiol.* 32:2397–402

108. Marban E. 2002. Cardiac channelopathies. *Nature* 415:213–18

109. Armoundas AA, Wu R, Juang G, Marban E, Tomaselli GF. 2001. Electrical and structural remodeling of the failing ventricle. *Pharmacol. Ther.* 92:213–30

110. Miake J, Marban E, Nuss HB. 2002. Biological pacemaker created by gene transfer. *Nature* 419:132–33

111. Donahue JK, Heldman AW, Fraser H, McDonald AD, Miller JM, et al. 2000. Focal modification of electrical conduction in the heart by viral gene transfer. *Nat. Med.* 6:1395–98

112. Nuss HB, Marban E, Johns DC. 1999. Overexpression of human potassium channel suppresses hyperexcitability in rabbit ventricular myocytes. *J. Clin. Invest.* 103:889–96

113. Ennis IL, Li RA, Murphy AM, Marban E, Nuss HB. 2002. Dual gene therapy with SERCA1 and Kir2.1 abbreviates excitation without suppressing contractility. *J. Clin. Invest.* 109:393–400

Annu. Rev. Physiol. 2004. 66:77–101
doi: 10.1146/annurev.physiol.66.071702.145229
First published online as a Review in Advance on September 8, 2003

DEVELOPMENTAL REGULATION OF LUNG LIQUID TRANSPORT

Richard E. Olver,[1] Dafydd V. Walters,[2] and Stuart M.Wilson[1]

[1]*Tayside Institute of Child Health, Lung Membrane Transport Group, Division of Maternal and Child Health Sciences, Ninewells Hospital and Medical School, University of Dundee, Dundee DD1 9SY, United Kingdom; email: R.E.Olver@dundee.ac.uk; S.M.Wilson@dundee.ac.uk*
[2]*Department of Child Health, St. George's Hospital Medical School, University of London SW17 0RE, United Kingdom; email: D.V.Walters@sgms.ac.uk*

Key Words lung epithelial ion transport, epithelial Na^+ channel, Cl^- channels, lung development

■ **Abstract** The developing distal lung epithelium displays an evolving liquid transport phenotype, reflecting a changing and dynamic balance between Cl^- ion secretion and Na^+ ion absorption, which in turn reflects changing functional requirements. Thus in the fetus, Cl^--driven liquid secretion predominates throughout gestation and generates a distending pressure to stretch the lung and stimulate growth. Increasing Na^+ absorptive capacity develops toward term, anticipating the switch to an absorptive phenotype at birth and beyond. There is some empirical evidence of ligand-gated regulation of Cl^- transport and of regulation via changes in the driving force for Cl^- secretion. Epinephrine, O_2, glucocorticoid, and thyroid hormones interact to stimulate Na^+ absorption by increasing Na^+ pump activity and apical Na^+ conductance (G_{Na^+}) to bring about the switch from net secretion to net absorption as lung liquid is cleared from the lung at birth. Postnatally, the lung lumen contains a small Cl^--based liquid secretion that generates a surface liquid layer, but the lung retains a large absorptive capacity to prevent alveolar flooding and clear edema fluid. This review explores the mechanisms underlying the functional development of the lung epithelium and draws upon evidence from classic integrative physiological studies combined with molecular physiology approaches.

FETAL LUNG LIQUID SECRETION

Source of Secretion

Although there are much older descriptions of liquid within the fetal lung, the narrative history of fetal lung liquid can be said to have begun in 1885 with Preyer (1), who observed that fetal lungs were filled with liquid and assumed that its origin was inhaled amniotic liquid—an error that persisted well into the 20th century. The correct interpretation by Potter & Bohlender in 1941 (2) and others (3) came from experiments of nature in which the lung distal to congenital airway atresias and stenoses was observed to be distended by liquid, indicating that the source of the liquid was distal to the obstruction. The same conclusion was reached as a result of experiments designed to give rise to hypopituitarism in fetal rabbit pups in which the trachea was incidentally occluded, resulting in distension of the lungs (4). These observations were confirmed by later experiments involving tracheal ligation in several species (e.g., 5, 6).

It is likely that lung liquid is present from the moment in embryogenesis that a lumen is formed, i.e., when the lung buds from the primitive foregut, at 5 weeks post-conception in humans. As Potter & Bohlender (2) put it, "a lumen is present, something must fill the space; it cannot be air and must be fluid." Liquid has been shown to be present within the lung lumen of human lung explants as early as 6 weeks gestation (7) and at 69 days (term = 147 days) in the lungs of the fetal sheep (8).

Although the observations described above point to a pulmonary rather than amniotic source of lung liquid, determining the cellular origin of lung liquid has been more problematic. Candidates include alveolar cells (type I and type II: ATI and ATII) and airway cells (e.g., Clara cells, ciliated surface epithelial cells, submucosal gland serous cells). In favor of an alveolar source are the observations that the alveolar epithelium constitutes 99% of the total lung epithelial surface area, lung liquid is uniform in composition throughout the respiratory tract (R.E. Olver, E.E. Schneeberger & D.V. Walters, unpublished observation) and, within the alveolar epithelium, type II cells are numerous and possess microvilli characteristic of a transporting phenotype.

Further support for an alveolar source of lung liquid comes from observations in airway (9, 10) demonstrating that the unstimulated fetal bronchial epithelium is predominantly Na^+ absorbing (as are Clara cells, see Reference 11), rather than secretory, and that proximal and distal lung ion transport are differentially regulated by β-adrenoceptor activation, with the airway epithelium exhibiting Cl^- secretion and the whole lung responding with Na^+ absorption. On the other hand, even though rodent fetal type II cells [fetal distal lung epithelial cells (FDLE)] exhibit Cl^- currents in culture in a gestation-dependent manner (12) and respond to β-adrenoceptor agonists with increased Na^+ absorption (as in the intact fetal lung) (13), they demonstrate a predominantly absorptive phenotype at rest (e.g., 12, 13–15). This may represent a species difference or loss of secretory phenotype, the result of accelerated differentiation toward a neonatal phenotype in studies that generally use hormone-supplemented culture media under hyperoxic conditions. Exposure to elevated P_{O2} provides the lung with an important, maturational stimulus (14, 16–20), and many of the early studies on FDLE were carried out in a gas mixture consisting of water-saturated air with 5% CO_2 ($P_{O2} \sim 147$ torr), which does not represent postnatal alveolar P_{O2} (90–100 torr) let alone fetal P_{O2} ($\sim$23 torr). However, elements of a Cl^- secretory mechanism can be demonstrated under appropriate conditions. Thus Ramminger et al. (15) found apical Cl^- currents in FDLE cells grown in a hormone-free minimally defined culture medium at fetal P_{O2}, which were downregulated (and Na^+ currents upregulated) by addition of a mixture of dexamethasone and triiodothyronine (T_3). However, the possibility that the properties of Cl^- secretion and Na^+ absorption reside in different alveolar cell types (i.e., ATI cells and ATII cells, respectively) cannot be discounted.

Although bronchial submucosal glands, which are important for maintaining airway hydration postnatally, may contribute to fetal lung liquid secretion, they cannot be the sole source because rodent lungs possess few such glands and yet secrete lung liquid normally.

Mechanism of Secretion

Once it had been established that the source of fetal lung liquid was the lung itself, possible mechanisms of secretion narrowed down to lung liquid being an ultrafiltrate of plasma or a secretion of the lung epithelium. Analysis of lung liquid composition (21, 22) demonstrated that lung liquid was quite unlike amniotic fluid or plasma (Table 1), indicating that it was a specialized secretion of the lung itself. This conclusion was further supported by the results of experiments in fetal sheep (23, 24) in which raising pulmonary vascular filtration pressure had no effect on the rate of lung liquid secretion.

A more definitive answer to the question of the mechanism underlying lung liquid secretion came from the work of Olver & Strang (23) that compared the ratio of the observed unidirectional fluxes of ions across the alveolar epithelium with those predicted for passive movement by the Ussing flux ratio equation (25). The most striking and, at the time, surprising finding of these studies was the transport of Cl^- ions into the lung lumen against an electrochemical gradient. Thus with the passive movement of Na^+ as a co-ion, the transport of Cl^- provides the osmotic driving force for lung liquid secretion.

At the time of Olver & Strang's observations (23) in fetal lung, there were few descriptions of active Cl^- transport in the literature, and none in mammalian lung, and the idea of what is now known as secondary active transport of Cl^- had not been conceived. Subsequent work in vivo (24, 26, 27) and in vitro (7, 28–30), variously using inhibitors of the Na^+, K^+, $2Cl^-$ cotransporter (bumetanide, furosemide)

TABLE 1 Composition of simultaneously sampled lung liquid, plasma, lung lymph, and amniotic liquid from fetal sheep of 125–147 days gestation[a]

	Na^+ (mM)	K^+ (mM)	Cl^- (mM)	HCO_3^- (mM)	Protein (g L^{-1})
Lung liquid	150	6.3	157	2.8	0.027
Plasma	150	4.8	107	24	4.09
Lymph	147	4.8	107	25	3.27
Amniotic liquid	113	7.6	87	19	0.10

[a]Reference 22.

and Na^+ pump (ouabain), have added to our understanding of the mechanism of lung liquid secretion and shown it to have the features of the consensus model for secondary active Cl^- secretion first described by Silva et al. (31). Cl^- entry into the cell at the basolateral surface via the Na^+, K^+, $2Cl^-$ cotransporter is driven by the electrochemical gradient for Na^+ generated by the Na^+ pump, raising intracellular Cl^- above its electrochemical potential and allowing it to exit passively via apically located anion-selective channels, thus generating a lumen-negative transepithelial potential difference; Na^+ follows passively and water flows down the osmotic gradient thus established. What this model does not explain is the low bicarbonate concentration in fetal lung liquid relative to plasma (2.8 versus 24 mM; see Table 1). Olver & Strang (23) observed that after the concentration of HCO_3^- in lung liquid was raised above that in plasma, it declined asymptotically to reach its original level after 4 to 5 h while failing to show saturation kinetics. The subsequent discovery of a Na^+-H^+ antiporter in fetal sheep ATII cell apical membrane vesicles (32) provides a possible explanation for this phenomenon.

Regulation of Secretion

A number of factors have been shown to be involved in regulating lung liquid absorption via upregulation of Na^+ transport (see below), but the regulation of secretion via control of Cl^- transport has received less attention. Because the fetal lung shares the property of active Cl^- transport with the gills of certain fish, Cassin & Perks (33) examined the effect of prolactin (an important ion transport regulatory hormone in fish) on lung liquid secretion in chronically catheterized fetal goats and found that, at micromolar concentrations, it stimulated secretion.

Paracrine pathways involved in the upregulation of lung liquid secretion include prostaglandins and growth factors. Evidence for a role for prostaglandins in regulating secretion based on the use of cyclo-oxygenase and prostaglandin synthase inhibitors has been conflicting, with either no effect (34) or an inhibitory effect (35) observed. However, studies in first-trimester lung explants have shown that prostaglandins PGE_2 and $PGF_{2\alpha}$ (36) and growth factors KGF and FGF-10 (37) stimulate secretion, as judged by an increase in luminal area, the effect being linked to cAMP and blocked by bumetanide. This finding is in contrast to the well-described effect of β-adrenoreceptor agonists and cAMP in inducing Na^+ absorption in the mature fetal lung and may reflect the preponderance of a columnar (i.e., surface airway-like) phenotype of the cells lining the lumen of the immature explant tubules. Chronic drainage of lung liquid leads to an increase in the lung liquid secretion rate (38), suggesting that there is a mechanism able to sense a decrease in volume and correct it. The way in which such changes in volume are sensed and mediated is unknown.

Nature of the Cl^- Efflux Pathway

The properties of the apically located Cl^- efflux pathway in fetal lung epithelium have been little studied, but such evidence as there is points to channels other than the cystic fibrosis transmembrane regulator (CFTR). Olver & Strang (23) described a halide permeability sequence of $I^- > Br^- > Cl^-$ in the intact fetal sheep lung in vivo, but it must be borne in mind that results from this type of preparation inevitably reflect hybrid properties of the paracellular shunt pathway and both cell membranes. Although rat FDLE (13) and ATII (39) cells display CFTR-like currents, KGF-stimulated lung liquid secretion is independent of CFTR in the mouse model (40, 41), an observation that may explain the absence of defective lung growth in cystic fibrosis. One of the few studies to suggest a candidate Cl^- efflux pathway (42) identified a G-protein–regulated large conductance (375 pS) maxi-Cl^- channel on the basis of patch-clamp studies in fetal guinea pig ATII cells, and a similar large conductance Cl^- channel has been described in cultured adult ATII cells (43).

Secretion and Lung Growth

There is evidence for active Cl^- transport driving lung liquid secretion in early gestation (7, 8), and the rate of secretion of lung liquid into the lung lumen has been shown in fetal sheep to rise from $1.6 \text{ ml} \cdot \text{h}^{-1} \cdot \text{kg}^{-1}$ body weight at mid-gestation (8) to peak at $3.5–5.5 \text{ ml} \cdot \text{h}^{-1} \cdot \text{kg}^{-1}$ in the third trimester (for reviews see 44, 45) before declining to under $3 \text{ ml} \cdot \text{h}^{-1} \cdot \text{kg}^{-1}$ in the days before birth (46–49). A similar pattern of diminished lung liquid secretion near term has been observed in guinea pig lungs in vitro (50). The view (48, 49) that lung volume parallels the decline in lung liquid secretion rate near term is in agreement with what would be expected from the increase in sensitivity of the fetal lung to epinephrine, and the increasing number of active Na^+ channels in the epithelium (as unmasked by the action of amiloride on the resting fetal lung; see 10). However, one group (45), using Blue Dextran as a dilution indicator, has reported a continuing rise in volume during late gestation. The discrepancy here may be methodological because the use of Blue Dextran (51), if not carefully solubilized before being added to lung liquid (52), leads to overestimation of lung liquid volume in the mature fetus (53).

Lung liquid is secreted against a resistance provided by the larynx and nasopharynx (54, 55), with the vocal cords acting as a one-way valve (47). This process generates a luminal pressure 2–3 torr above amniotic fluid pressure compared with the chest wall elastic recoil pressure of 0.2 torr (56). Thus the pressure generated by secretion is the principal expanding force acting on the lung and, with compliance of the fluid-filled lung estimated at 15–20 ml/torr, is sufficient to develop a volume of approximately 30 ml/kg body weight in the third trimester, roughly equivalent to the functional residual capacity (FRC) of the ovine and human neonatal lung (57, 58). Clinical and experimental evidence indicates that the volume of liquid within the lung, which is a function of the compliance of the lung/chest wall and distending pressure, is a crucial determinant of lung growth.

Clinical situations in which the formation of a liquid FRC is impaired, with resultant lung hypoplasia, include oligohydramnios, congenital diaphragmatic hernia, skeletal dysplasia, and neuromuscular disorders associated with diaphragmatic paralysis (59). Irrespective of the cause of lung hypoplasia, the morphological findings are surprisingly uniform, with deficits in quantitative lung growth, reduced numbers of airway generations (60), reduced pulmonary vascular bed and reduced acinar size, but apparently normal differentiation and maturation (59, 60). Animal models of oligohydramnios have provided results consistent with these findings and have demonstrated normal functional maturity as judged by surfactant phospholipid content (61).

Simple but nevertheless informative studies in fetal sheep have been undertaken to investigate the influence of lung liquid volume on lung growth and development by modulating the outflow resistance (and thus distending pressure) of the lung. Continually draining lung liquid, and so preventing the development of a liquid FRC, results in hypoplastic lungs with an excess of ATII cells and lamellar inclusion bodies (62) but normal surfactant status (63), whereas obstructing the lung outflow leads to the development of hyperplastic lungs with thin alveolar walls sparsely populated by ATII cells (62) as a result of time-dependent differentiation of ATII cells into ATI cells (64).

Work by Nardo and coworkers (65) has demonstrated that it is the change in volume, presumably acting via stretch-activated mechanisms, rather than the change in pressure, that mediates the effects of airway obstruction on lung growth, and there is a substantial body of in vitro evidence to indicate that cyclical stretch is a prerequisite for optimal growth of lung in culture (see, e.g., 66, 67), a proposal supported by in vivo experiments on the developing lung (68). It might be thought that the fetal breathing movements (FBM), first described at the beginning of the 20th century (69, 70) and later characterized by Dawes (71), would provide this physical stimulus in vivo. However, individual FBM are virtually isometric and result in shifts in lung liquid volume of no more than 1 ml (71). Nevertheless, alternating periods of FBM and apnea are associated with changes in outflow resistance (72) and significant changes in lung liquid volume, which might be expected to influence lung growth.

The observation that increasing the outflow resistance to secreted lung liquid can be used to grow the lung (5, 6, 62, 63, 73) and can correct the lung hypoplasia associated with experimental oligohydramnios (74) led to the idea of plugging the lung (i.e., obstructing outflow) in utero in order to rescue the hypoplastic lung phenotype associated with diaphragmatic hernia (75–77). However, despite the use of numerous permutations (duration and timing of plugging, administration of antenatal steroids and/or surfactant), this approach to both experimental and clinical congenital diaphragmatic hernia has been disappointing, with diminished surfactant and numbers of ATII cells, persistent pulmonary hypertension (78), and poor respiratory function as a result (79, 80). Because continuous, as opposed to cyclical, stretch or strain is cytotoxic in vitro (66), it seems likely that any successful therapeutic intervention may need to incorporate an element of periodicity in addition to other strategies designed to enhance alveolar maturation.

LIQUID TRANSPORT DURING THE PERINATAL PERIOD

Removal of Lung Liquid at Birth

Although the secretion of liquid into the lung lumen is vital for the growth of the lung, this liquid must be removed at birth to allow the newborn infant to breathe air. Indeed, the retention of liquid at birth [e.g., after Caesarean section (CS)] (81) impairs oxygenation of the blood, and severe fluid retention is an important feature of neonatal respiratory distress syndrome (RDS), the most common cause of death among premature and newborn infants in the developed world (82).

The first evidence that birth is associated with significant changes in lung fluid balance came from studies undertaken by Faure-Fremiet & Dragiou (83), who measured the water content of sheep lungs obtained from the Paris meat market at La Villette. They found that, irrespective of gestational age, fetal sheep lung tissue had a water content of 87–89%, compared with 76–78% in the lungs of newborn sheep. This observation was confirmed and extended some 40 years later by Aherne & Dawkins (84), who carried out a detailed study of lung water content of fetal and newborn rabbit lungs. Their data showed clearly that the removal of lung liquid began immediately after birth and was virtually complete within 2 h.

Early investigators generally attributed the removal of lung liquid at birth to compression of the chest wall during labor. Apart from the fact that, in humans, the widest part of the body is the head and shoulders, making it unlikely that the chest would be subject to sustained compression during normal delivery, a number of key observations effectively preclude this simple hypothesis. Liquid is cleared from the lungs of animals that have been delivered by CS after the onset of labor (as in emergency CS) just as efficiently as in those delivered vaginally (85), and in animals in which the trachea is occluded (47, 86), the volume of liquid nevertheless falls rapidly during labor. The removal of lung liquid from the lung lumen is thus linked to the process of labor, which may well explain the increased incidence of RDS among infants delivered by elective CS (87, 88).

Removal of Liquid from the Lung Parenchyma

When lung liquid is absorbed across the pulmonary epithelium, it accumulates in the interstitial space of the lung (84, 85). Pooling of liquid occurs particularly in the loose connective tissue of the perivascular areas from where it is slowly removed over 4 to 6 h. The accumulation of liquid in the interstitial space is accompanied by a rise in pressure up to an average of 6 cm water (89) at 1 to 2 h after birth before declining to the usual subatmospheric pressure observed in lung tissue by 4 to 6 h, mirroring the histological and weight changes in the lung over the same time course. The observation that interstitial pressure does not rise in those parts of the lung that are not ventilated in premature, surfactant-deficient lungs (89) is consistent with the failure of local lung liquid absorptive mechanisms. Removal of liquid from the lung interstitium is by the vascular and lymphatic systems, and estimates of the size of the lymphatic component vary between 11% (90) for animals in labor and 50% (91) for animals not in labor. Because the pulmonary lymphatic system can cope with a fivefold increase in pulmonary filtration, there is likely to be redundancy in both the lymphatic and vascular mechanisms.

Role of Epinephrine

The observation that fetal epinephrine reaches very high levels during labor (92) coincided with experiments that explored the effects of this and other hormones on lung liquid balance. Walters & Olver (93) demonstrated that epinephrine and iso-proterenol, but not norepinephrine, infused intravenously into chronically catheterized mature fetal sheep, caused rapid absorption of lung liquid that could be inhibited by prior treatment with propranolol. This and subsequent studies (47, 94) showed that the β-adrenoceptor-mediated response was developmentally regulated and could be mimicked by dibutyryl cAMP (95), a cell-permeant cAMP analogue. The physiological relevance of these findings to adaptation of the lungs at birth was confirmed by the observation that the relationship between fetal plasma epinephrine concentration and the rate of lung liquid absorption during labor was the same as that found during epinephrine infusion into sheep fetuses of comparable maturity not subject to labor (47).

The absorption of lung liquid seen during infusion of epinephrine (93) or cAMP-analogues (95), or during spontaneous labor (96), is abolished by instilling a Na^+ channel antagonist, amiloride, into the lung lumen, providing evidence that lung liquid absorption is dependent upon epinephrine-evoked Na^+ transport. The mechanisms underlying this control have since been extensively studied using fetal distal lung epithelial (FDLE) cells isolated from fetal rats and maintained in short-term culture (for review, see 97). Such experiments have confirmed that activation of β-adrenoceptors stimulates Na^+ transport in fetal alveolar epithelium and have shown that this response involves a rise in apical membrane Na^+ conductance (G_{Na^+}) (see e.g., 97–99). Whether β-adrenoceptor activation is also responsible for the increase in Na^+ pump activity seen during labor (100, 101) is unknown.

Role of Other Hormones

The surge in fetal epinephrine levels seen during labor is preceded by a slower but substantial and sustained increase in the circulating levels of thyroid and glucocorticoid hormones (see e.g., 102, 103), which are known to be important for the functional maturation of the lung's surfactant secreting mechanism (see e.g., 104). The first evidence that they may also be important to the functional maturation of the Na^+-absorbing phenotype came from studies in fetal sheep, which established that fetal thyroidectomy prevented the development of the normal, absorptive response to epinephrine (105). Although this effect was reversed by exogenous T_3, the infusion of T_3 into immature fetuses did not cause precocious maturation of this absorptive response unless given in conjunction with a glucocorticoid (105–108), thus demonstrating the hormones' synergistic mode of action.

Their effect, which is produced within ~2 h, appears to be mediated via gene expression because it is blocked by protein synthesis inhibitors (107). Interestingly, as well as blocking the development of the response to epinephrine, fetal thyroidectomy also prevents the development of an absorptive response to dibutyryl cAMP (105), suggesting that this effect of T_3 is mediated at a site downstream to β-adrenoceptor-mediated cAMP formation (105, 106). The mechanism by which epinephrine controls Na^+ transport was further explored by studying the effects of thyroid and glucocorticoids upon the ability of a β-adrenoceptor agonist, isoproterenol, to regulate the conductive properties of the FDLE cell apical plasma membrane (15). Isoproterenol had no effect upon G_{Na^+} in FDLE cells cultured in a medium devoid of all hormones and growth factors, and exposing the cells to either T_3 or dexamethasone failed to restore this response, in spite of evidence for the presence of functional Na^+ channels (13, 15). However, isoproterenol elicited clear increases in G_{Na^+} when these cells were exposed to a combination of T_3 and dexamethasone (13).

These studies, and observations on the hormonal regulation of the expression of the epithelial Na^+ channel (ENaC) (see below), establish a novel view of pulmonary physiology in which thyroid and glucocorticoid hormones synergistically prime the lung by developing a mechanism that allows epinephrine to modulate G_{Na^+} and so control lung liquid clearance during labor (15, 105–108) and postnatal life (109). Glucocorticoids are well-established regulators of Na^+ transport, but the requirement for T_3 was surprising. However, a role for T_3 in the control of pulmonary Na^+ transport provides a physiological basis for the increased incidence of RDS among hypothyroid infants (110, 111). Moreover, in guinea pigs, premature delivery by CS attenuates the normal rise in T_3 levels (103), and the observation that this hormone is important to the development of the Na^+-absorbing phenotype may, in part, explain the high incidence of RDS in babies delivered by elective CS, particularly in those babies born prematurely (87, 88). Consistent with this conclusion is the observation that infants born at or before 30 weeks gestation show evidence of suboptimal Na^+ absorptive capacity (112).

The enormous evolutionary pressure to survive the adaptation to air breathing would predict the existence of mechanisms additional to those described above to ensure the successful switch from fetal lung liquid secretion to postnatal lung liquid absorption. Other candidate regulatory hormones include arginine vasopressin (AVP), somatostatin (113), dopamine (114), and serotonin (115), which, similar to epinephrine, rise during labor. Although AVP can be shown to inhibit lung liquid secretion (116, 117), possibly acting via the V_1 receptor and protein kinase C (118), its effect has been reported to be weaker than that of epinephrine. Perks and coworkers have demonstrated in guinea pigs that somatostatin is a potent inhibitor of lung liquid secretion (113), whereas dopamine and serotonin act via activation of Na^+ transport. Additionally they have described a freely diffusible stretch factor, as yet chemically unidentified, that causes lung liquid absorption when the lung is stretched by modest volumes comparable to the first breath (119, 120). Although the physiological relevance of these findings is not clear, and their confirmation in other species has yet to be undertaken, it is reasonable to assume that some may provide backup for the β-adrenoceptor-dependent pathway and thus explain why β-adrenoceptor blockade fails to prevent lung liquid absorption during labor (121).

Nature of the Na$^+$ Entry Pathway

Studies of classical, Na$^+$-absorbing epithelia, such as the kidney tubule and amphibian bladder, have indicated that the rate of epithelial Na$^+$ transport is normally restricted by the rate at which Na$^+$ can diffuse down the inwardly directed Na$^+$ gradient across the apical membrane (122). Although it was known that the channels involved in lung liquid clearance were distinct from neuronal Na$^+$ channels, being characteristically sensitive to amiloride and certain of its analogues (123), the identity of the ion channel species that allowed this Na$^+$ entry step was unknown until the early 1990s. Canessa et al. (124), using size fractionation and oocyte expression of mRNA isolated from the colonic epithelia of salt-deficient rats, identified a novel cDNA sequence encoding an amiloride-sensitive Na$^+$ channel that they named α-ENaC. This channel was found to be expressed in kidney, colon (124), and lung (125). Two homologous cDNA sequences, β-ENaC and γ-ENaC, were subsequently identified (126). These failed to induce amiloride-sensitive currents when injected into *Xenopus* oocytes but, when coinjected (together) with α-ENaC, induced amiloride-sensitive currents that were at least 100-fold greater than those elicited by α-ENaC alone (126). It was therefore proposed that epithelial Na$^+$ channels were composed of three homologous subunits (α-, β-, and γ-ENaC), each encoded by separate genes (126). Furthermore, as α-ENaC was the only subunit that formed a conductive ion channel when expressed independently, this protein was suggested to be the critical pore-forming subunit, with β- and γ-ENaC serving as accessory subunits that modified the conductive and kinetic properties of the resultant Na$^+$ channels. When coexpressed in *Xenopus* oocytes, α-, β-, and γ-ENaC form highly selective Na$^+$ channels that are characterized by a conductance of ~4 pS and slow gating kinetics (97, 126) and are thus essentially identical to the Na$^+$ channels found in classical Na$^+$-absorbing epithelia (127, 128).

Observations on the hormonal regulation of ENaC expression in lung are consistent with the functional studies outlined above (see Roles of Other Hormones). Tchepichev and coworkers (129) demonstrated that prenatal steroids, but not thyroid hormones, could advance the timing of the increase in α-ENaC mRNA in fetal rat lung, but neither hormone had any effect on β-ENaC or γ-ENaC subunit mRNA expression. Subsequently, putative binding sites for thyroid and glucocorticoid receptors were identified in the promoter region of α-ENaC (130) and, although T_3 alone had no effect on reporter gene activity, it was shown to potentiate glucocorticoid-evoked ENaC gene expression. In the late-gestation guinea pig, α-ENaC expression rises in parallel with fetal plasma cortisol. The demonstration that metyrapone, an inhibitor of glucocorticoid synthesis, abolishes the increase in circulating cortisol and prevents the increased expression of α-ENaC after delivery by CS emphasizes the importance of this hormone in the developmental regulation of ENaC near term. Nevertheless, interpretation of the observed effects of glucocorticoids is not straightforward because, in addition to any effects on gene expression, cortisol can increase the concentration of T_3 in fetal blood, presumably through an effect on thyroid hormone-metabolizing enzymes (e.g., 131).

Compelling evidence that ENaC is essential for the clearance of liquid from the lung was provided by studies on fetal mice in which null mutations in both α-ENaC alleles precipitated a lethal phenotype in which liquid was retained within the lung, causing death from severe respiratory distress within 48 h of birth (132). Interestingly, deletion of the β- (133) or γ-ENaC (134) alleles slows lung liquid clearance but does not lead to overt pulmonary symptoms, supporting the view that these subunits act as accessory subunits rather than being directly involved in channel formation. In contrast to the effects of deletions of ENaC subunits, the results of aquaporin channel deletions in mice suggest that they do not play a significant role in perinatal lung liquid clearance (135, 136) even though several aquaporin water channels (AQP3–5) have been found in different regions of the lung epithelium and are developmentally regulated (137).

A number of mutations described in all three ENaC subunits give rise to the clinical condition of pseudohypoaldosteronism (PHA), a salt-losing state that results in severe electrolyte disturbances. At first sight, the absence of neonatal respiratory symptoms in these patients (138) would seem to provide evidence against a central role for ENaC in human lung liquid clearance. However, recent studies indicate that oocytes expressing ENaC subunit mutants found in PHA may exhibit significantly more Na^+ transport than oocytes with a null mutation (139, 140). These observations suggest that humans with PHA may have enough residual ENaC activity to avoid perinatal waterlogging of the lungs and that the requirement for ENaC function in the lungs is considerably less than in the kidney.

Although it was thought likely that highly selective Na^+ channels would be present in distal lung epithelial cells, electrophysiological studies have generally shown (but see 125) that the most abundant Na^+-permeable channel expressed in these cells is an amiloride-sensitive, nonselective cation channel that discriminates poorly between Na^+ and K^+ and has a conductance of 20–25 pS (98, 99, 141–146). There is evidence that such nonselective cation channels are formed when α-ENaC is expressed independently of other ENaC subunits (147–149). However, electrometric studies of distal lung epithelial cells grown as confluent monolayers demonstrate that cAMP-coupled agonists activate a Na^+-selective channel (13, 39). A possible resolution to this apparent contradiction has been presented by Eaton and colleagues, who showed that the conductive and kinetic properties of the channels expressed by distal lung epithelial cells are influenced by the conditions under which cells are cultured and that, if maintained on permeable supports at the air-liquid interface in medium supplemented with steroids, the predominant channel type is a small-conductance, highly selective Na^+ channel comprising α-, β-, and γ-ENaC (149) as described by Voilley et al. in lung (125).

Studies of Na^+ channels that have been biochemically purified from Na^+-absorbing epithelia and inserted into lipid bilayers reveal that their ionic selectivity and amiloride sensitivity can be altered by chemical modifications such as protein kinase A-mediated phosphorylation (150). Moreover, such channels appear to be expressed as heteromultimeric complexes containing at least six associated proteins, one of which is the pertussis toxin–sensitive GTP-binding protein subunit $G_{\alpha i\text{-}3}$. The presence of this accessory protein may account for the G protein–mediated regulation of channel activity that has been described in fetal alveolar cells (151). Moreover, pertussis toxin–mediated ADP-ribosylation can also modify the ionic selectivity and amiloride sensitivity of both phosphorylated and nonphosphorylated Na^+ channels (150), thus conferring a considerable degree of plasticity on the lung's epithelial Na^+ conductivity properties.

There is evidence for an additional, glucose-dependent Na^+ entry pathway in fetal (152) and adult (153, 154) lung. Thus in fetal sheep, glucose is rapidly removed from lung liquid by a phloridzin-inhibitable mechanism presumed to be the sodium-glucose transporter SGLT1, which has a constant K_m of 0.14 mM over the latter part of gestation and a V_{max} that increases with gestation to about 9 μmol/min at term (152). As in adult rat lung (153, 154), glucose entry into the lumen from the blood is passive and too slow (152) for the sodium-glucose symport to produce a major effect on lung liquid movement. Thus it would be unlikely to contribute in a major way to fetal lung liquid balance even if glucose were present in fetal lung liquid. The lack of effect of phloridzin on perinatal lung water clearance in fetal/newborn rats confirms this view (155). Thus Na^+-glucose cotransport is not a major contributor to Na^+ movement and its importance may be to keep the respiratory surface free of glucose and so diminish the risk of bacterial growth.

Ambient P_{O2} and the Control of Na^+ Transport

A number of early studies provided evidence that pulmonary Na^+ transport was inherently sensitive to changes in ambient P_{O2} (16, 17), raising the possibility that the rise in alveolar P_{O2} seen as the newborn infant takes its first breaths may provide a physiologically relevant stimulus for increased Na^+ transport.

Observations that O_2-evoked stimulation of Na^+ transport was associated with increased abundance of mRNA encoding α-, β-, and γ-ENaC (156), and that the α-ENaC promoter possesses a binding site for the redox-sensitive transcription factor, nuclear factor-κB (NF-κB) (130), led to the hypothesis that P_{O2} facilitates Na^+ transport by upregulating NF-κB-mediated expression of α-ENaC, which, in turn, stimulates a rise in G_{Na^+} (18, 156). Consistent with this view is the finding that this G_{Na^+} response is reduced by pharmacological inhibition of NF-κB (157). Difficulties with this hypothesis include the observations that an increase in the rate of Na^+ transport is apparent before any increase in ENaC mRNA abundance can be discerned (156) and that, under certain conditions, O_2-evoked Na^+ transport can occur with no change in ENaC mRNA abundance (158). Although physiological increases in P_{O2} can activate the α-ENaC promoter with a time course that parallels a rise in G_{Na^+} (20), these events occur only after P_{O2} has been raised for

24–48 h (159). In contrast, the O_2-evoked stimulation of Na^+ transport (measured as short circuit current, I_{SC}) is almost fully developed by $\sim$24 h. The observation by Ramminger et al. (19) that raising P_{O2} causes a detectable increase in basolateral Na^+ pump capacity in under 6 h suggests that the Na^+ pump, rather than ENaC, may be the initial O_2-responsive element in the Na^+ transport pathway, with up-regulation of α-ENaC expression occurring as an adaptation to the increased rate of Na^+ transport. Unlike the β-adrenoceptor-mediated stimulation of Na^+ transport, which is dependent upon prior exposure to thyroid/glucocorticoid hormones (15, 105–108), the stimulatory effects of increased P_{O2} occur independently of these hormones (158). Whether there is a release from nitric oxide-induced G_{Na^+} repression as P_{O2} rises at birth, as might be predicted from adult ATII cell data (159a,b), is unknown.

The temporary fall in epithelial resistance lasting several hours following elevation of P_{O2} in FDLE (156) is consistent with the increase in solute (160) and water (161) permeability noted at the onset of breathing in vivo and would be expected to facilitate liquid absorption secondary to the observed increase in Na^+ transport. Conversely, the failure of these mechanisms, particularly if coupled with failure of P_{O2}-dependent Na^+ absorptive mechanisms, as in the unventilated, surfactant-deficient lung of the preterm infant, can be expected to contribute to the waterlogging characteristic of the lung in RDS (82).

Although there is evidence to suggest that expression of the genes encoding Na^+ pump subunits can be regulated by P_{O2} (162), studies of FDLE cells suggest that physiologically relevant increases in P_{O2} do not cause any overt change in the abundance of mRNA encoding Na^+ pump subunits (S.M. Wilson & G.H. Olver, unpublished data). Moreover, the rate of Na^+ transport is normally limited by G_{Na^+} (see e.g., 122), and thus the hypothesis that changes in Na^+ pump capacity underlie the O_2-evoked stimulation of Na^+ transport presupposes the existence of mechanisms permitting rapid control over G_{Na^+} well before any rise in α-ENaC expression.

LIQUID TRANSPORT DURING POSTNATAL LIFE

Postnatal Secretion

The lung's bulk secretory capability is gradually lost in the neonatal period and the post-neonatal lung lumen contains only a small amount of liquid. Estimates of its volume, achieved by different methods (163–165), are in good agreement and give values around 0.35 ml/kg body weight, which is approximately 1% of the lung liquid volume in the fetus. Even though the volume of liquid is small, there is evidence of surface liquid resulting from Cl^- ion transport throughout the respiratory tract: in alveoli (166), airways (airway surface liquid) (9, 10), and submucosal glands, where HCO_3^- can substitute for Cl^- (167, 168) and net movement of liquid is associated with mucus secretion. Submucosal glands are the major site of CFTR expression in the lungs (169), and the secretion of liquid from these glands, in contrast to the secretion of fluid in the fetal lung, appears to be dependent upon the activity of CFTR (167, 168) and the aquaporin, AQP5 (170). Whether submucosal glands are the total source of airway and alveolar surface liquid in the postnatal lung and whether the volume of the liquid lining the lung can be sensed, and if so how, are currently unanswered questions.

Continuing Change in the Neonatal Period

The switch in lung epithelium phenotype, secretory in the fetal state and absorptive after birth, is not as sudden as it may first appear. In perfused sheep lungs in the first few weeks of life, lung liquid secretion can be restored by blocking Na^+ transport with amiloride (171). However, the effects of amiloride and epinephrine wane with postnatal age, so that by 6 months epinephrine has little absorptive effect, and amiloride, rather than causing secretion, merely slows absorption (171). Furthermore, complete β-adrenoceptor blockade does not cause complete loss of absorptive capacity in the postnatal lung (172). These findings could be interpreted as a functional change in the way ENaC interacts with amiloride or in the way ENaC is activated by epinephrine but, more likely, they may be due to the age-dependent appearance in the lung epithelium of differentially regulated ion transport systems functioning in parallel to ENaC. Cyclic nucleotide-gated (CNG) channels have been found both histologically (173) and functionally (174, 175) in adult sheep and rat lung. However, even though blocking both ENaC and CNG channels in mature sheep can induce lung liquid secretion, there is no evidence of functionally active CNG channels in the sheep fetus or newborn (176).

Na^+ Absorption in Adult Lung

The Na^+-absorbing phenotype of the lung is normally maintained throughout postnatal life (97) despite the rapid postpartum fall in circulating epinephrine levels, thus protecting lung function by preventing the accumulation of liquid in the alveolar space (177, 178). The alveolar flooding that occurs in pulmonary edema, a condition with many underlying causes, including heart failure, septic or hemorrhagic shock and ascent to high altitude (i.e., low ambient P_{O2}) (97, 177, 178), has been attributed to disturbances in pulmonary hemodynamics and epithelial damage. However, there is evidence that impaired Na^+ absorption contributes to alveolar flooding (179–181), which may be due, at least in part, to hypoxic inhibition of Na^+ transport (180, 181). Whatever its underlying cause, the eventual removal of this edema fluid from the lung is dependent upon increased Na^+ transport (177–179). Na^+ transport is thus vital to the integrated function of the respiratory tract.

CONCLUSION

The epithelium of the developing fetal lung displays a secretory phenotype up to the time of birth, with the secreted fluid functioning as a liquid framework or stent around which the developing airspaces form. During the latter part of gestation, absorptive mechanisms develop that are capable of switching the lung epithelium from secretion to absorption as the lung adapts to air breathing. The absorptive mechanisms expressed at birth persist into adult life, maintaining a dry lung interior and providing a defense against the development of pulmonary edema. Although the phenotypic switch from secretion to absorption is apparently sudden and dramatic, involving a complex interplay of physical and hormonal mechanisms, it is only one stage in a sequence of developmental changes in function that begin in the third trimester and continue into the neonatal period. Although we understand some of the mechanisms underlying several of the elements of this process, our understanding of the way in which these mechanisms integrate temporo-spatially to enable the transition to the postnatal phenotype remains incomplete.

ACKNOWLEDGMENT

The authors thank the Wellcome Trust for program and project grant support over many years.

The *Annual Review of Physiology* is online at http://physiol.annualreviews.org

LITERATURE CITED

1. Preyer W. 1885. *Specielle Physiologie des Embryo*, ed. L Fernau, p. 149. Leipzig: Th. Grieben
2. Potter EL, Bohlender GP. 1941. Intrauterine respiration in relation to development of the fetal lung. *Am. J. Obstet. Gynecol.* 42:14–22
3. Griscom NT, Harris GBC, Wohl MEB, Vawter GF, Eraklis AJ. 1969. Fluid-filled lung due to airway obstruction in the newborn. *Pediatrics* 43:383–90
4. Jost A, Policard A. 1948. Contribution expérimentale a l'étude du développement prénatal du poumon chez le lapin. *Arch. Anat. Micr.* 37:323–32
5. Carmel JA, Friedman F, Adams FH. 1965. Tracheal ligation and lung development. *Am. J. Dis. Child.* 109:452–56
6. Lanman JT, Shaffer A, Herod L, Ogawa Y, Castellanos R. 1971. Distensibility of the fetal lung with fluid in sheep. *Pediatr. Res.* 5:586–90
7. McCray PB Jr, Bettencourt JD, Bastacky J. 1992. Developing bronchopulmonary epithelium of the human fetus secretes fluid. *Am. J. Physiol. Lung Cell. Mol. Physiol.* 262:L270–79
8. Olver RE, Schneeberger EE, Walters DV. 1981. Epithelial solute permeability ion transport and tight junction morphology in the developing lung of the fetal lamb. *J. Physiol.* 315:395–412
9. Cotton CU, Lawson EE, Boucher RC, Gatzy JT. 1983. Bioelectric properties and ion transport of airways excised from adult and fetal sheep. *J. Appl. Physiol.* 55:1542–49
10. Olver RE, Robinson EJ. 1986. Sodium and chloride transport by the tracheal epithelium of fetal, new-born and adult sheep. *J. Physiol.* 375:377–90
11. van Scott MR, Davis CW, Boucher RC. 1989. Na^+ and Cl^- transport across rabbit nonciliated bronchiolar epithelial (Clara) cells. *Am. J. Physiol. Cell Physiol.* 256:C893–901
12. Rao AK, Cott GR. 1991. Ontogeny of ion transport across fetal pulmonary

epithelial cells in monolayer culture. *Am. J. Physiol. Lung Cell Mol. Physiol.* 261:L178–87

13. Collett A, Ramminger SJ, Olver RE, Wilson SM. 2002. β-adrenoceptor mediated control over apical membrane conductive properties in fetal distal lung epithelia. *Am. J. Physiol. Lung Cell. Mol. Physiol.* 282:L621–30

14. Pitkänen OM, Tanswell AK, O'Brodovich H. 1995. Fetal lung cell-derived matrix alters distal lung epithelial ion transport. *Am. J. Physiol. Lung Cell. Mol. Physiol.* 268:L762–L71

15. Ramminger SJ, Inglis SK, Olver RE, Wilson SM. 2002. Hormonal modulation of Na^+ transport in rat fetal distal lung epithelial cells. *J. Physiol.* 544:567–77

16. Barker PM, Gatzy JT. 1993. Effect of gas composition on liquid secretion by explants of distal lung of fetal rat in submersion culture. *Am. J. Physiol. Lung Cell. Mol. Physiol.* 265:L512–17

17. Acarregui MJ, Snyder JM, Mendeson CR. 1993. Oxygen modulates the differentiation of human fetal lung in vitro and its responsiveness to cAMP. *Am. J. Physiol. Lung Cell. Mol. Physiol.* 264:L465–74

18. Rafii B, Tanswell AK, Otulakowski G, Pitkänen O, Belcastro-Taylor R, et al. 1998. O_2-induced ENaC expression is associated with NF-κB activation and blocked by superoxide scavenger. *Am. J. Physiol. Lung Cell. Mol. Physiol.* 275:L764–70

19. Ramminger SJ, Baines DL, Olver RE, Wilson SM. 2000. The effects of PO_2 upon transepithelial ion transport in fetal rat distal lung epithelial cells. *J. Physiol.* 524:539–47

20. Baines DL, Ramminger SJ, Collett A, Haddad JJE, Best OG, et al. 2001. Oxygen-evoked Na^+ transport in rat fetal distal lung epithelial cells. *J. Physiol.* 532:105–13

21. Adams FH, Moss AJ, Fagan L. 1963. The tracheal fluid in the fetal lamb. *Biol. Neonate* 5:151–58

22. Adamson TM, Boyd RDH, Platt HS, Strang LB. 1969. Composition of alveolar liquid in the fetal lamb. *J. Physiol.* 204:159–68

23. Olver RE, Strang LB. 1974. Ion fluxes across the pulmonary epithelium and the secretion of lung liquid in the fetal lamb. *J. Physiol.* 241:327–57

24. Carlton DP, Cummings JJ, Chapman DL, Poulain FR, Bland RD. 1992. Ion transport regulation of lung liquid secretion in foetal lambs. *J. Dev. Physiol.* 17:99–107

25. Koefoed-Johnsen V, Ussing HH. 1953. Contribution of diffusion and flow to the passage of D_2O through living membranes. *Acta Physiol. Scand.* 27:38–48

26. Cassin S, Gause G, Perks AM. 1986. The effects of bumetanide and furosemide on lung liquid secretion in fetal sheep. *Proc. Soc. Exp. Biol. Med.* 181:427–31

27. Thom J, Perks AM. 1990. The effects of furosemide and bumetanide on lung liquid production by in vitro lungs from fetal guinea pigs. *Can. J. Physiol. Pharmacol.* 68:1131–35

28. Krochmal EM, Ballard ST, Yankaskas JR, Boucher RC, Gatzy JT. 1989. Volume and ion transport by fetal rat alveolar and tracheal epithelia in submersion culture. *Am. J. Physiol. Renal Physiol.* 256:F397–407

29. McCray PB Jr. 1990. Secretion of lung liquid by the fetal alveolar epithelium is chloride dependent and mediated by β adrenergic and cAMP-dependent pathways. *Am. Rev. Respir. Dis.* 141:A163(Abstr.)

30. McCray PB Jr, Welsh MJ. 1991. Developing fetal alveolar epithelial cells secrete fluid in primary culture. *Am. J. Physiol. Lung Cell. Mol. Physiol.* 260:L494–500

31. Silva P, Stoff J, Field M, Fine L, Forrest JN, et al. 1977. Mechanism of active chloride secretion by shark rectal gland: role of Na–K–ATPase in chloride

transport. *Am. J. Physiol. Renal Physiol.* 233:F298–306

32. Shaw AM, Steele LW, Butcher PA, Ward MR, Olver RE. 1990. Sodium-proton exchange across the apical membrane of the alveolar type II cell of the fetal sheep. *Biochim. Biophys. Acta* 1028:9–13

33. Cassin S, Perks AM. 1982. Studies of factors which stimulate lung fluid secretion in fetal goats. *J. Dev. Physiol.* 4:311–25

34. Kitterman JA. 1986. Physical factors and fetal lung growth. In *Reproduction and Perinatal Medicine III, Respiratory Control and Lung Development in the Fetus and Newborn*, ed. PW Nathaniels, pp. 63–85. Ithaca, NY: Perinatalogy Press

35. Cassin S. 1984. Effect of indomethacin on fetal lung liquid formation. *Can. J. Physiol. Pharmacol.* 62:157–59

36. McCray PB Jr, Bettencourt JD. 1993. Prostaglandins stimulate fluid secretion in human fetal lung. *J. Dev. Physiol.* 19:29–36

37. Graeff RW, Wang G, McCray PB Jr. 1999. KGF and FGF-10 stimulate liquid secretion in human fetal lung. *Pediatr. Res.* 46:523–37

38. Nardo NL, Hooper SB, Harding R. 1995. Lung hypoplasia can be reversed by short-term obstruction of the trachea in fetal sheep. *Pediatr. Res.* 38:690–96

39. Jiang X, Ingbar DH, O'Grady SM. 1998. Adrenergic stimulation of Na^+ transport across alveolar epithelial cells involves activation of apical Cl^- channels. *Am. J. Physiol. Cell Physiol.* 275:C1610–20

40. Zhou L, Graeff RW, McCray PB Jr. 1996. Keratinocyte growth factor stimulates CFTR-independent fluid secretion in the fetal lung in vitro. *Am. J. Physiol. Lung Cell. Mol. Physiol.* 271:L987–94

41. Simonet WS, DeRose ML, Bucay N, Nguyen HQ, Wert SE, et al. 1995. Pulmonary malformation in transgenic mice expressing human keratinocyte growth factor in the lung. *Proc. Natl. Acad. Sci. USA* 92:12461–65

42. Kemp PJ, McGregor GG, Olver RE. 1993. G protein-regulated large-conductance chloride channels in freshly isolated fetal type II alveolar epithelial cells. *Am. J. Physiol. Lung Cell. Mol. Physiol.* 265:L323–29

43. Schneider GT, Cook DI, Gage PW, Young JA. 1985. Voltage sensitive, high conductance chloride channels in the luminal membrane of cultured pulmonary alveolar (type II) cells. *Pflügers Arch.* 430:299–307

44. Strang LB. 1991. Fetal lung liquid: secretion and reabsorption. *Physiol. Rev.* 71:991–1016

45. Harding R, Hooper SB. 1996. Regulation of lung expansion and lung growth before birth. *J. Appl. Physiol.* 81:209–24

46. Kitterman JA, Ballard PL, Clements JA, Mescher EJ, Tooley WH. 1979. Tracheal fluid in fetal lambs: spontaneous decrease prior to birth. *J. Appl. Physiol.* 47:985–89

47. Brown MJ, Olver RE, Ramsden CA, Strang LB, Walters DV. 1983. Effects of adrenaline and of spontaneous labour on the secretion and absorption of lung liquid in the fetal lamb. *J. Physiol.* 344:137–52

48. Dickson KA, Maloney JE, Berger PJ. 1986. Decline in lung liquid volume before labor in fetal lambs. *J. Appl. Physiol.* 61:2266–72

49. Pfister RE, Ramsden CA, Neil HL, Kyriakides MA, Berger PJ. 2001. Volume and secretion rate of lung liquid in the final days of gestation and labour in fetal sheep. *J. Physiol.* 535:889–99

50. Perks AM, Dore JJ, Dyer R, Thom J, Marshall JK, et al. 1990. Fluid production by in vitro lungs from fetal guinea pigs. *Can. J. Physiol. Pharmacol.* 68:505–13

51. Lines A, Hooper SB, Harding R. 1997. Lung liquid production rates and volumes do not increase before labor in

healthy fetal sheep. *J. Appl. Physiol.* 82:927–32

52. Cassin S, Perks AM. 2003. Measuring the lung liquid volume and secretion using radioiodinated serum albumin and blue dextran. *J. Appl. Physiol.* 94:1293–94

53. Pfister RE, Ramsden CA, Neil HL, Kyriakides MA, Berger PJ. 1999. Errors in estimating lung liquid volume in fetal lambs when using radiolabeled serum albumin and blue dextran. *J. Appl. Physiol.* 87:2366–74

54. Fisk NM, Parkes MJ, Moore PJ, Hanson MA, Wigglesworth JS, et al. 1992. Mimicking low amniotic pressure by chronic pharyngeal drainage does not impair lung development in the sheep. *Am. J. Obstet. Gynecol.* 166:991–96

55. Fewell JE, Johnson P. 1983. Upper airway dynamics during breathing and during apnoea in fetal lambs. *J. Physiol.* 339:495–504

56. Vilos GA, Liggins GC. 1982. Intrathoracic pressures in fetal sheep. *J. Dev. Physiol.* 4:247–56

57. Geubelle F, Karlberg P, Koch G, Lind J, Walgreen G, et al. 1959. L'aération du poumon chez le nouveau-né. *Biol. Neonate* 1:169–210

58. Klaus M, Tooley WH, Weaver KJ, Clements JA. 1962. Lung volume in the new born infant. *Pediatrics* 30:111–16

59. Wigglesworth JS. 1988. Lung development in the second trimester. *Br. Med. Bull.* 44:894–908

60. Hislop A, Hey E, Reid L. 1979. The lungs in congenital bilateral renal agenesis and dysplasia. *Arch. Dis. Child.* 54:32–38

61. Moessinger AC, Singh M, Donnelly DF, Haddad GG, Collins H, et al. 1987. The effect of prolonged oligohydramnios on fetal lung development, maturation and ventilatory patterns in the newborn guinea pig. *J. Dev. Physiol.* 9:419–27

62. Alcorn D, Adamson TM, Lambert TF, Maloney JE, Ritchie BC. 1977. Morphological effects of chronic tracheal ligation and drainage in the fetal lamb lung. *J. Anat.* 123:649–60

63. Moessinger AC, Harding R, Adamson TM, Singh M, Kim G. 1990. Role of lung liquid volume in growth and maturation of the fetal sheep lung. *J. Clin. Invest.* 86:1270–77

64. Flecknoe S, Harding R, Maritz G, Hooper SB. 2000. Increased lung expansion alters the proportions of type I and type II alveolar epithelial cells in fetal sheep. *Am. J. Physiol. Lung Cell. Mol. Physiol.* 278:L1180–85

65. Nardo L, Hooper SB, Harding R. 1998. Stimulation of lung growth by tracheal obstruction in fetal sheep: relation to luminal pressure and lung liquid volume. *Pediatr. Res.* 43:184–90

66. Liu M, Skinner SJ, Xu J, Han RN, Tanswell AK, et al. 1992. Stimulation of fetal rat lung cell proliferation in vitro by mechanical stretch. *Am. J. Physiol. Lung Cell. Mol. Physiol.* 263:L376–83

67. Liu M, Xu J, Souza P, Tanswell B, Tanswell AK. 1995. The effect of mechanical strain on fetal rat lung cell proliferation: comparison of two- and three-dimensional culture systems. *In Vitro Cell Dev. Biol. Anim.* 31:858–66

68. Liggins GC, Vilos GA, Campos GA, Kitterman JA, Lee CH. 1981. The effects of bilateral thoracoplasty on lung development in fetal sheep. *J. Dev. Physiol.* 3:275–82

69. Ballantyne JW. 1902. *Manual of Antenatal Pathology and Hygiene.* Edinburgh: Green

70. Barcroft J. 1946. *Researches of Pre-Natal Life.* Oxford, UK: Blackwell Sci.

71. Dawes GS, Fox HE, Leduc BM, Liggins GC, Richards BT. 1972. Respiratory movements and rapid eye movement sleep in the foetal lamb. *J. Physiol.* 220:119–43

72. Harding R, Bocking AD, Sigger JN. 1986. Influence of upper respiratory tract on liquid flow to and from fetal lungs. *J. Appl. Physiol.* 61:68–74

73. Hooper SB, Han VKM, Harding R. 1993. Changes in lung expansion after pulmonary DNA synthesis and IGF-II gene expression in fetal sheep. *Am. J. Physiol. Lung Cell. Mol. Physiol.* 265:L403–9

74. Adzick NS, Harrison MR, Glick PL, Villa RL, Finkbeiner W. 1984. Experimental pulmonary hypoplasia and oligohydramnios: relative contributions of lung fluid and fetal breathing movements. *J. Pediatr. Surg.* 19:658–63

75. Hedrick MH, Estes JM, Sullivan KM, Bealer JF, Kitterman JA, et al. 1994. Plug the lung until it grows (PLUG): a new method to treat congenital diaphragmatic hernia in utero. *J. Pediatr. Surg.* 29:612–17

76. Skarsgard ED, Meuli M, VanderWall KJ, Bealer JF, Adzick NS, et al. 1996. Fetal endoscopic tracheal occlusion ('Fetendo-PLUG') for congenital diaphragmatic hernia. *J. Pediatr. Surg.* 31:1335–38

77. Papadakis K, De Paepe ME, Tackett LD, Piasecki GJ, Lus FI. 1998. Temporary tracheal occlusion causes catch-up lung maturation in a fetal model of diaphragmatic hernia. *J. Pediatr. Surg.* 33:1030–37

78. O'Toole SJ, Karamanoukian HL, Irish MS, Sharma A, Holm BA, et al. 1997. Tracheal ligation: the dark side of in utero congenital diaphragmatic hernia treatment. *J. Pediatr. Surg.* 32:407–10

79. Harrison MR, Mychaliska GB, Albanese CT, Jennings RW, Farrell JA, et al. 1998. Correction of congenital diaphragmatic hernia in utero IX: fetuses with poor prognosis (liver herniation and low lung-to-head ratio) can be saved by getoscopic temporary tracheal occlusion. *J. Pediatr. Surg.* 33:1017–22

80. Davey MG, Hedrick HL, Bouchard S, Mendoza JM, Schwarz U, et al. 2003. Temporary tracheal occlusion in fetal sheep with lung hypoplasia does not improve postnatal lung function. *J. Appl. Physiol.* 94:1054–62

81. Berger PJ, Smolich JJ, Ramsden CA, Walker AM. 1996. Effect of lung liquid volume on respiratory performance after caesarean delivery in the lamb. *J. Physiol.* 492:905–12

82. O'Brodovich H. 1996. Immature epithelial Na^+ channel expression is one of the pathological mechanisms leading to human neonatal respiratory distress syndrome. *Proc. Assoc. Am. Physicians* 108:345–55

83. Faure-Fremiet E, Dragiou J. 1923. Le développement du poumon foetal chez le mouton. *Arch. Anat. Micr.* 19:411–74

84. Aherne W, Dawkins MJR. 1964. The removal of fluid from the pulmonary airways after birth in the rabbit, and the effect on this of prematurity and prenatal hypoxia. *Biol. Neonate* 7:214

85. Bland RD, McMillan DD, Bressack MA, Dong L. 1980. Clearance of liquid from the lungs of newborn rabbits. *J. Appl. Physiol.* 49:171–77

86. Berger PJ, Kyriakides MA, Smolich JJ, Ramsden CA, Walker AM. 1998. Massive decline in lung liquid before vaginal delivery at term in the fetal sheep. *Am. J. Obstet. Gynecol.* 178:223–27

87. Fedrick J, Butler NR. 1972. Hyaline-membrane disease. *Lancet* 2(7780):768–69

88. Hales KA, Morgan MA, Thurnau GR. 1993. Influence of labor and route of delivery on the frequency of respiratory morbidity in term neonates. *Int. J. Gynaecol. Obstet.* 43:35–40

89. Miserocchi GD, Haxhiu Poskurica B, Del Fabbro M. 1994. Pulmonary interstitial pressure in anesthetized paralyzed new-born rabbits. *J. Appl. Physiol.* 74:1171–77

90. Bland RD, McMillan DD, Bressack MA, Dong L. 1982. Lung fluid balance in lambs before and after birth. *J. Appl. Physiol.* 53:992–1004

91. Humphreys PW, Normand ICS, Reynolds EOR, Strang LB. 1967. Pulmonary lymph flow and the uptake of liquid from the lungs of the lamb at the start of breathing. *J. Physiol.* 193:1–29

92. Lagercrantz H, Bistoletti P. 1977. Catecholamine release in the newborn infant at birth. *Pediatr. Res.* 11:889–93

93. Walters DV, Olver RE. 1978. The role of catecholamines in lung liquid absorption at birth. *Pediatr. Res.* 12:239–42

94. Lawson EF, Brown ER, Torday JS, Mandansky DL, Taeusch HW. 1978. The effects of epinephrine on tracheal fluid flow and surfactant efflux in fetal sheep. *Am. Rev. Respir. Dis.* 118:1023–26

95. Walters DV, Ramsden CA, Olver RE. 1990. Dibutyryl cAMP induces a gestation-dependent absorption of fetal lung liquid. *J. Appl. Physiol.* 68:2054–59

96. O'Brodovich H, Hannam V, Seear M, Mullen JB. 1990. Amiloride impairs lung water clearance in newborn guinea pigs. *J. Appl. Physiol.* 68:1758–62

97. Matalon S, O'Brodovich H. 1999. Sodium channels in alveolar epithelial cells: molecular characterization, biophysical properties and physiological significance. *Annu. Rev. Physiol.* 61:627–61

98. Ito Y, Niisato N, O'Brodovich H, Marunaka Y. 1997. The effects of brefeldin A on terbutalin-induced sodium absorption in fetal rat distal lung epithelium. *Pflügers Arch.* 434:492–94

99. Marunaka Y, Niisato N, O'Brodovich H, Eaton DC. 1999. Regulation of an amiloride-sensitive Na^+-permeable channel by a β_2 adrenergic agonist, cytosolic Ca^{2+} and Cl^- in fetal rat alveolar epithelium. *J. Physiol.* 515:669–83

100. Bland RD, Boyd CAR. 1986. Cation transport in lung epithelial cells derived from fetal, newborn and adult rabbits. *J. Appl. Physiol.* 61:507–15

101. Chapman DL, Widdicombe JH, Bland RD. 1990. Developmental differences in rabbit lung epithelial cell Na^+-K^+ ATPase. *Am. J. Physiol. Lung Cell. Mol. Physiol.* 259:L481–87

102. Polk DH. 1995. Thyroid hormone metabolism during development. *Reprod. Fertil. Dev.* 7:469–77

103. Baines DL, Folkesson HG, Norlin A, Bingle CD, Yuan HT, et al. 2000. The influence of mode of delivery, hormonal status and postnatal O_2 environment on epithelial sodium channel (ENaC) expression in guinea pig lung. *J. Physiol.* 522:147–57

104. Gross I, Dynia DW, Wilson CM, Ingleson LD, Gewolb IH, et al. 1984. Glucocorticoid-thyroid hormone interactions in fetal rat lung. *Pediatr. Res.* 18:191–96

105. Barker PM, Brown MJ, Ramsden CA, Strang LB, Walters DV. 1988. The effect of thyroidectomy in the fetal sheep on lung liquid reabsorption induced by adrenaline or cAMP. *J. Physiol.* 407:373–83

106. Barker PM, Brown MJ, Ramsden CA, Strang LB, Walters DV. 1990. Synergistic action of triiodothyronine and hydrocortisone on epinephrine-induced reabsorption of fetal lung liquid. *Pediatr. Res.* 27:588–91

107. Barker PM, Walters DV, Markiewicz M, Strang LB. 1991. Development of the lung liquid reabsorptive mechanism in fetal sheep: synergism of triiodothyronine and hydrocortisone. *J. Physiol.* 433:435–49

108. Barker PM, Strang LB, Walters DV. 1990. The role of thyroid hormones in maturation of adrenaline-sensitive lung liquid reabsorptive mechanism in fetal sheep. *J. Physiol.* 424:473–85

109. Folkesson HG, Norlin A, Wang Y, Abedinpour P, Matthay MA. 2000. Dexamethasone and thyroid hormone

pretreatment upregulate alveolar epithelial fluid clearance in adult rats. *J. Appl. Physiol.* 88:416–24

110. Redding RA, Pereira C. 1974. Thyroid function in respiratory distress syndrome (RDS) of the newborn. *Pediatrics* 54:423–28

111. Cuestas RA, Lindall A, Engel RR. 1976. Low thyroid hormones and respiratory-distress syndrome of the newborn. Studies on cord blood. *N. Engl. J. Med.* 295:297–302

112. Barker PM, Gowen CW, Lawson EE, Knowles MR. 1997. Decreased sodium ion absorption across nasal epithelium of very premature infants with respiratory distress syndrome. *J. Pediatr.* 130:373–77

113. Perks AM, Kwok YN, McIntosh CHS, Ruiz T, Kindler PM. 1992. Changes in somatostatin-like immunoreactivity in lungs from perinatal guinea pigs and the effects of somatostatin-14 on lung liquid production. *J. Dev. Physiol.* 18:151–59

114. Chua BA, Perks AM. 1998. The effect of dopamine on lung liquid production by in vitro lungs from fetal guinea pigs. *J. Physiol.* 513:283–94

115. Chua BA, Perks AM. 1999. The pulmonary neuroendocrine system and drainage of the fetal lung: effects of serotonin. *Gen. Comp. Endocrinol.* 113:374–87

116. Perks AM, Cassin S. 1989. The effects of arginine vasopressin and epinephrine on lung liquid production in fetal goats. *Can. J. Physiol. Pharmacol.* 67:491–89

117. Perks AM, Kindler PM, Marshall J, Woods B, Craddock M, et al. 1993. Lung liquid production by in vitro lungs from fetal guinea pigs: effects of arginine vasopressin and arginine vasotocin. *J. Dev. Physiol.* 19:203–12

118. Albuquerque CA, Nijland MJ, Ross MJ. 1998. Mechanism of arginine vasopressin suppression of ovine fetal lung liquid: lack of a V2-receptor effect. *J. Matern. Fetal Med.* 7:177–82

119. Nelson PG, Perks AM. 1996. Effects of lung expansion on lung liquid production in vitro by lungs from fetal guinea pigs. I. Basic studies and the effects of amiloride and propranolol. *Reprod. Fertil. Dev.* 8:335–46

120. Nelson PG, Perks AM. 1996. Effects of lung expansion on lung liquid production in vitro by lungs from fetal guinea pigs. II. Evidence for an inhibitory factor. *Reprod. Fertil. Dev.* 8:347–54

121. Chapman DL, Carlton DP, Nielson DW, Cummings JJ, Poulain FR, et al. 1994. Changes in lung liquid during spontaneous labour in fetal sheep. *J. Appl. Physiol.* 76:523–30

122. Lewis SA, Eaton DC, Clausen C, Diamond JM. 1977. Nystatin as a probe for investigating the electrical properties of a tight epithelium. *J. Gen. Physiol.* 70:427–40

123. Garty H, Benos DJ. 1988. Characteristics and regulatory mechanisms of the amiloride–blockable Na^+ channel. *Physiol. Rev.* 68:309–73

124. Canessa CM, Horisberger JD, Rossier BC. 1993. Epithelial sodium channel related to proteins involved in neurodegeneration. *Nature* 361:467–70

125. Voilley N, Lingueglia E, Champigny G, Mattei MG, Waldmann R, et al. 1994. The lung amiloride-sensitive Na^+ channel: biophysical properties, pharmacology, ontogenesis, and molecular cloning. *Proc. Natl. Acad. Sci. USA* 91:247–51

126. Canessa CM, Schild L, Buell G, Thorens B, Gautschi I, et al. 1994. Amiloride-sensitive epithelial Na^+ channel is made of three homologous subunits. *Nature* 367:463–66

127. Eaton DC, Ohara A, Ling BN. 1991. Cellular regulation of amiloride-blockable Na^+ channels. *Biomed. Res.* 12:31–35

128. Barbry P, Hofman P. 1997. Molecular biology of Na^+ absorption. *Am. J. Physiol. Gastrointest. Liver Physiol.* 273:G571–85

129. Tchepichev S, Ueda J, Canessa C, Rossier BC, O'Brodovich H. 1995. Lung epithelial channel subunits are differentially regulated during development and by steroids. *Am. J. Physiol. Cell. Physiol.* 269:C805–12

130. Otulakowski G, Raffii B, Bremner HR, O'Brodovich H. 1999. Structure and hormone responsiveness of the gene encoding the α-subunit of the rat amiloride-sensitive epithelial sodium channel. *Am. J. Respir. Cell. Mol. Biol.* 20:1028–40

131. Sensky PL, Roy CH, Barnes RJ, Heath MF. 1994. Changes in fetal thyroid hormone levels in adrenalectomized fetal sheep following continuous cortisol infusion 72 h before delivery. *J. Endocrinol.* 140:79–83

132. Hummler E, Baker P, Gatzy J, Berrmann F, Verdumo C, et al. 1996. Early death due to defective neonatal lung liquid clearance in α-ENaC-deficient mice. *Nat. Genet.* 12:325–28

133. McDonald F, Yang B, Hrstka RF, Drummond HA, Tarr DE, et al. 1999. Disruption of the β subunit of the epithelial Na+ channel in mice: hyperkalemia and neonatal death associated with a pseudohypoaldosteronism phenotype. *Proc. Natl. Acad. Sci. USA* 96:1727–31

134. Barker PM, Nguyen MS, Gatzy JT, Grubb B, Norman H, et al. 1998. Role of the γ-ENaC subunit in lung liquid clearance and electrolyte balance in newborn mice. Insights into perinatal adaptation and pseudohypoaldosteronism. *J. Clin. Invest.* 102:1634–40

135. Bai C, Fukuda N, Song Y, Ma T, Matthay MA, et al. 1999. Lung fluid transport in aquaporin-1 and aquaporin-4 knock-out mice. *J. Clin. Invest.* 103:555–61

136. Ma T, Fukuda N, Song Y, Matthay MA, Verkman AS. 2000. Lung fluid transport in aquaporin-5 knockout mice. *J. Clin. Invest.* 105:93–100

137. Borok Z, Verkman AS. 2002. Role of aquaporin water channels in fluid transport in lungs and airways. *J. Appl. Physiol.* 93:2199–206

138. Kerem E, Bistritzer T, Hanukoglu A, Hofmann T, Zhou Z, et al. 1999. Pulmonary epithelial sodium-channel dysfunction and excess airway liquid in pseudohypoaldosteronism. *N. Engl. J. Med.* 341:156–62

139. Bonny O, Chraibi A, Loffing J, Jaeger NF, Grunder S, et al. 1999. Functional expression of a pseudohypoaldosteronism type 1 mutated epithelial Na+ channel lacking the pore forming region of its alpha subunit. *J. Clin. Invest.* 104:967–74

140. Chang SS, Grunder S, Hanukoglu A, Rosler A, Mathew PM, et al. 1996. Mutations in subunits of the epithelial sodium channel cause salt wasting with hyperkalemic acidosis, pseudohypoaldosteronism type 1. *Nat. Genet.* 12:248–53

141. Orser BA, Bertlik M, Fedorko L, O'Brodovich H. 1991. Cation selective channel in fetal alveolar type II epithelium. *Biochim. Biophys. Acta* 1094:19–26

142. Marunaka Y. 1996. Amiloride-blockable Ca²⁺-activated Na⁺-permeant channels in the fetal distal lung epithelium. *Pflügers Arch.* 431:748–56

143. Yue G, Hu P, Oh Y, Jilling T, Shoemaker RL, et al. 1993. Culture-induced alterations in alveolar type II cell Na+ conductance. *Am. J. Respir. Cell. Mol. Biol.* 265:L630–40

144. Feng ZP, Clark RB, Berthiaume Y. 1993. Identification of nonselective cation channels in cultured rat alveolar type II cells. *Am. J. Respir. Cell. Mol. Biol.* 9:248–54

145. MacGregor GG, Olver RE, Kemp PJ. 1994. Amiloride-sensitive Na+ channels in fetal type II pneumocytes are regulated by G proteins. *Am. J. Physiol. Lung Cell. Mol. Physiol.* 267:L1–8

146. Tohda H, Foskett JK, O'Brodovich H, Marunaka Y. 1994. Cl⁻ regulation of a Ca²⁺-activated nonselective cation channel in β-agonist-treated fetal distal lung

epithelium. *Am. J. Physiol. Cell Physiol.* 266:C104–9

147. Kizer N, Guo XL, Hruska K. 1997. Reconstitution of stretch activated cation channels by expressing alpha-subunit of the epithelial sodium channel cloned from osteoblasts. *Proc. Natl. Acad. Sci. USA* 94:1013–18

148. Jain L, Chen X-C, Malik B, Al-Khalili O, Eaton DC. 1999. Antisense oligonucleotides against the α-subunit of ENaC decrease lung epithelial cation channel activity. *Am. J. Physiol. Lung Cell. Mol. Physiol.* 276:L1046-L51

149. Jain L, Chen X-L, Ramosevac S, Brown LA, Eaton DC. 2001. Expression of highly selective sodium channels in alveolar type II cells is determined by culture conditions. *Am. J. Physiol. Lung Cell. Mol. Physiol.* 280:L646–58

150. Benos DJ, Fuller CM, Shlyonsky VG, Berdiev BK, Ismailov II. 1997. Amiloride-sensitive Na^+ channels: insights and outlooks. *NIPS* 12:55–61

151. Kemp PJ, Olver RE. 1996. G protein regulation of alveolar ion channels: implications for lung fluid transport. *Exp. Physiol.* 81:493–504

152. Barker PM, Boyd CAR, Ramsden CA, Strang LB, Walters DV. 1989. Pulmonary glucose transport in the fetal sheep. *J. Physiol.* 409:15–27

153. Basset G, Crone C, Saumon G. 1987. Significance of active ion transport in transalveolar water absorption: a study in isolated rat lung. *J. Physiol.* 384:311–24

154. Saumon G, Martet G. 1996. Effect of changes in paracellular permeability on airspace liquid clearance: role of glucose transport. *Am. J. Physiol. Lung Cell. Mol. Physiol.* 270:L191–98

155. O'Brodovich H. 1991. Epithelial ion transport in the fetal and perinatal lung. *Am. J. Physiol. Cell Physiol.* 261:C555–64

156. Pitkänen OM, Tanswell AK, Downey G, O'Brodovich H. 1996. Increased PO_2 al-ters the bioelectric properties of the fetal distal lung epithelium. *Am. J. Physiol. Lung Cell. Mol. Physiol.* 270:L1060–66

157. Haddad JJE, Collett A, Land SC, Olver RE, Wilson SM. 2001. NF-κB blockade reduces the O_2-evoked rise in Na^+ conductance in fetal alveolar cells. *Biochem. Biophys. Res. Commun.* 281:987–92

158. Richard K, Ramminger SJ, Inglis SK, Olver RE, Land SC, et al. 2003. O_2 can raise fetal pneumocyte Na^+ conductance without affecting ENaC mRNA abundance. *Biochem. Biophys. Res. Commun.* 305:671–76

159. Baines DL, Ramminger SJ, Collett A, Best OG, Olver RE, et al. 2000. PO_2-evoked increases in Na^+ transport and epithelial Na^+ channel α subunit (α-ENaC) gene transcription in fetal distal lung epithelial (FDLE) cells. *J. Physiol.* 527:24P

159a. Jain L, Chen XJ, Brown LA, Eaton DC. 1998. Nitric oxide inhibits lung sodium transport through a cGMP-mediated inhibition of epithelial cation channels. *Am. J. Physiol. Lung Cell Mol. Physiol.* 274:L475–84

159b. Guo Y, DuVall MD, Crow JP, Matalon S. 1998. Nitric oxide inhibits Na^+ absorption across cultured type II monolayers. *Am. J. Physiol. Lung Cell Mol. Physiol.* 274:L369–77

160. Egan EA, Olver RE, Strang LB. 1975. Changes in non-electrolyte permeability of alveoli and the absorption of lung liquid at the start of breathing in the lamb. *J. Physiol.* 244:161–79

161. Carter EP, Umenishi F, Matthay MA, Verkman AS. 1997. Developmental changes in alveolar water permeability in perinatal rabbit lung. *J. Clin. Invest.* 100:1071–78

162. Wendt CH, Towle H, Sharma R, Duvick S, Kawakami K, et al. 1998. Regulation of Na-K-ATPase gene expression by hyperoxia in MDCK cells. *Am. J. Physiol. Cell Physiol.* 274:C356–64

163. Rennard SI, Basset G, Lecossier D,

O'Donell KM, Pinkston P, et al. 1986. Estimation of volume of epithelial lining fluid recovered by lavage using urea as a marker of detection. *J. Appl. Physiol.* 60:532–38

164. Stephens RH, Benjamin AR, Walters DV. 1996. Volume and protein concentration of epithelial lining in perfused in situ postnatal sheep lungs. *J. Appl. Physiol.* 80:1911–20

165. Weibel ER, Untersee P, Gil J, Zulaug M. 1973. Morphometric estimation of pulmonary diffusion capacity. VI. Effect of varying positive pressure inflation of air spaces. *Respir. Physiol.* 18:285–308

166. Neilson DW. 1988. Changes in the pulmonary alveolar subphase at birth in term and preterm lambs. *Pediatr. Res.* 23:418–22

167. Ballard S, Fountain J, Inglis S, Corboz M, Taylor A. 1995. Chloride secretion across distal airway epithelium: relationship to submucosal gland distribution. *Am. J. Physiol. Lung Cell. Mol. Physiol.* 268:L526–31

168. Inglis SK, Corboz MR, Taylor AE, Ballard ST. 1997. Effect of anion transport inhibition on mucus secretion by airway submucosal glands. *Am. J. Physiol. Lung Cell. Mol. Physiol.* 272:L372–77

169. Engelhardt JF, Yankaskas JR, Ernst ST, Yang Y, Marino CR, et al. 1992. Submucosal glands are the predominant site of CFTR expression in the human bronchus. *Nat. Genet.* 2:240–48

170. Song Y, Verkman AS. 2001. Aquaporin-5 dependent fluid secretion in airway submucosal glands. *J. Biol. Chem.* 276:41288–92

171. Ramsden CA, Markiewicz M, Walters DV, Gabella G, Parker KA, et al. 1992. Liquid flow across the epithelium of the artificially perfused lung of fetal and postnatal sheep. *J. Physiol.* 448:579–97

172. Stephens RH, Benjamin AR, Walters DV. 1998. The regulation of lung liquid absorption by endogenous cAMP in postnatal sheep lungs perfused in situ. *J. Physiol.* 511:587–97

173. Ding C, Potter ED, Qiu W, Coon SL, Levine MA, et al. 1997. Cloning and widespread distribution of the rat rod-type cyclic nucleotide-gated cation channel. *Am. J. Physiol. Cell. Physiol.* 272:C1335–44

174. Junor RW, Benjamin AR, Alexandrou D, Guggino S, Walters DV. 1999. A novel role for cyclic nucleotide gated cation channels in lung liquid homeostasis in sheep. *J. Physiol.* 520:255–60

175. Kemp PJ, Kim KJ, Borok Z, Crandall ED. 2001. Re-evaluating the Na conductance of adult rat alveolar type II pneumocytes: evidence for the involvement of cGMP-activated cation channels. *J. Physiol.* 536:693–701

176. Junor RW, Benjamin AR, Alexandrou D, Guggino S, Walters DV. 2000. Lack of a role for cyclic nucleotide gated cation channels in lung liquid absorption in fetal sheep. *J. Physiol.* 523:493–501

177. Matthay MA, Wiener-Kronish JP. 1990. Intact epithelial barrier function is critical for the resolution of pulmonary edema in man. *Am. Rev. Respir. Dis.* 142:1250–57

178. Ware LB, Matthay MA. 2001. Alveolar epithelial fluid clearance is impaired in the majority of patients with acute lung injury and acute respiratory distress syndrome. *Am. J. Respir. Crit. Care Med.* 163:1376–83

179. Rafii B, Gillie DJ, Sulowski C, Hannam V, Cheung T, et al. 2002. Pulmonary oedema fluid induces non-αENaC dependent Na^+ transport and fluid absorption in distal lung. *J. Physiol.* 544:537–48

180. Scherrer U, Sartori C, Lepori M, Allemann Y, Duplain H, et al. 1999. High-altitude pulmonary edema: from exaggerated pulmonary hypertension to

a defect in transepithelial sodium transport. *Adv. Exp. Med. Biol.* 474:93–107

181. Sartori C, Allemann Y, Trueb L, Lepori M, Maggiorini M, et al. 2000. Exaggerated pulmonary hypertension is not sufficient to trigger high-altitude pulmonary oedema in humans. *Schweiz. Med. Wochenschr. Suppl.* 130:385–89

Annu. Rev. Physiol. 2004. 66:103–29
doi: 10.1146/annurev.physiol.66.032102.150822
Copyright © 2004 by Annual Reviews. All rights reserved
First published online as a Review in Advance on October 15, 2003

MECHANISM OF RECTIFICATION IN INWARD-RECTIFIER K+ CHANNELS[1]

Zhe Lu

*Department of Physiology, University of Pennsylvania, Philadelphia,
Pennsylvania 19104; email: zhelu@mail.med.upenn.edu*

■ **Abstract**   Inward rectifiers are a class of K^+ channels that can conduct much larger inward currents at membrane voltages negative to the K^+ equilibrium potential than outward currents at voltages positive to it, even when K^+ concentrations on both sides of the membrane are made equal. This conduction property, called inward rectification, enables inward rectifiers to perform many important physiological tasks. Rectification is not an inherent property of the channel protein itself, but reflects strong voltage dependence of channel block by intracellular cations such as Mg^{2+} and polyamines. This voltage dependence results primarily from the movement of K^+ ions across the transmembrane electric field along the pore, which is energetically coupled to the blocker binding and unbinding. This mutual displacement mechanism between several K^+ ions and a blocker explains the signature feature of inward rectifier K^+ channels, namely, that at a given concentration of intracellular K^+, their macroscopic conductance depends on the difference between membrane voltage and the K^+ equilibrium potential rather than on membrane voltage itself.

INTRODUCTION

Inward-rectifier K^+ channels are a group of membrane proteins that enact important physiological tasks such as controlling the resting membrane potential, regulating cardiac and neuronal electrical activity, coupling insulin secretion to blood glucose levels, and maintaining electrolyte balance. Inward rectifiers are so named because they act as bio-diodes in the cell membrane. They selectively conduct K^+ ions and can carry much larger inward currents at membrane voltages (V_m) negative to the K^+ equilibrium potential (E_K) than outward currents at voltages positive to it, even with equal K^+ concentrations ($[K^+]$) on both sides of the membrane. The diode-like property underlies their ability to perform vital physiological functions. Excellent review articles have summarized the rich literature on inward-rectifier K^+ channels (e.g., 1–4). In the following, we focus on the mechanism of inward rectification.

---

[1]This review is dedicated to the memory of Prof. Bernhard Katz (1911–2003).

0066-4278/04/0315-0103$14.00
103

DISCOVERY OF ANOMALOUS INWARDLY RECTIFYING K^+ CONDUCTANCE

Research on inward-rectifier K^+ channels began in the mid-20th century, when Katz described (5) a novel K^+ current in skeletal muscle membrane that rectifies "anomalously" in the inward direction, compared with the already familiar outwardly rectifying K^+ current. His discovery started a long journey of exploration into the nature, function, and mechanism of the anomalously rectifying current, later referred to as inwardly rectifying K^+ current (6). Over the past half-century and more, numerous studies by many researchers using various techniques have gradually formed an increasingly clear and detailed picture of how the curious diode-like behavior of K^+ currents arises in a biological membrane. With this insight we are able to appreciate how nature has perfected the amazingly sharp inward rectification by optimizing the system's components.

Katz's seminal paper (5) reported two important observations: "One might expect under these conditions [60 mM K_2SO_4 and 50 mM sucrose bathing solution] transmembrane current to be carried exclusively by potassium ions and, with K^+ concentrations equal on both sides, membrane resistance to be independent of current polarity. In fact, however, we found definite rectifying behavior, the resistance for outwardly directed current (at the cathode) being much larger than for inwardly directed current ..."[2] (Figure 1). Checking his finding with other methods Katz reported: "In addition we measured variations in impedance at 2 kHz, ... and found that, at elevated external K^+ concentrations, impedance increased at the cathode and decreased at the anode." Based on these observations, he concluded: "Muscle membrane exhibits anomalous rectifying properties when potassium concentrations inside and outside the fiber are made approximately equal." The phenomena were subsequently confirmed and further characterized by others including Hodgkin & Horowicz (7) and Adrian & Freygang (8, 9).

EMPIRICAL QUANTITATIVE DESCRIPTION OF INWARD RECTIFICATION

Hodgkin & Horowicz further examined (7) Katz's second observation that increasing $[K^+]_{ext}$ reduces the membrane impedance to inward K^+ current. Finding that when the difference between V_m and E_K is large and positive and K^+ ions move outward, K^+ permeability (P_K) falls to about 0.05×10^{-6} cm/s but that, when the difference is negative and K^+ ions move inward, P_K rises to about 8×10^{-6} cm/s, they concluded: "... potassium permeability varies greatly with the force acting on

---

[2]Quoted passages translated by Dr. Paul De Weer.

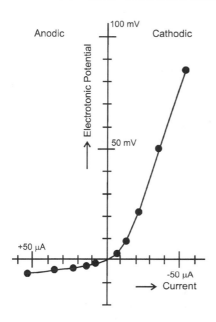

Figure 1 Rectification in frog sartorius muscle in the presence of elevated external K^+ concentration (60 mM K_2SO_4 + 50 mM sucrose). Ordinate: electrotonic potential deviation; abscissa: magnitude of polarizing current (redrawn from Reference 5).

the potassium ions." The phenomenon was elegantly illustrated by Hagiwara and colleagues (10) with a set of rectifying current-voltage (I-V) curves recorded from starfish eggs at various $[K^+]_{ext}$ (Figure 2A). The graph shows that altering $[K^+]_{ext}$ causes a seemingly parallel shift of the I-V relationship. For a sufficiently steep dependence of K^+ conductance on voltage, the resulting I-V relation will feature a negative slope at certain voltages positive to E_K, and the I-V curves at various $[K^+]_{ext}$ will cross each other (11). To analyze how G/G_{max} (normalized conductance) of inward rectifiers depends on $V_m - E_K$ (ΔV), Hagiwara & Takahashi (12) employed a Boltzmann equation similar to that used by Hodgkin & Huxley (13) to describe voltage dependence of Na^+ channel inactivation, where they replaced V_m by ΔV (Equation 1).

$$\frac{G}{G_{max}}(\Delta V) = \frac{1}{1 + e^{\left(\frac{\Delta V - \Delta V_h}{v}\right)}},$$ 1.

where ΔV_h is the ΔV at which G/G_{max} becomes $1/2$, and v characterizes the curvature of the I-V relation. Rearranging Equation 1 yields the linear form

$$\ln\left(\frac{G_{max}}{G} - 1\right) = \frac{\Delta V}{v} - \frac{\Delta V_h}{v}.$$ 2.

Hagiwara & Takahashi (12) found that the experimentally observed quantity defined by the left side of Equation 2 indeed increases linearly with ΔV over a wide range (Figure 2B). Their empirical equation therefore quantitatively describes the characteristic feature of inward rectifiers, namely, that G/G_{max} varies with ΔV

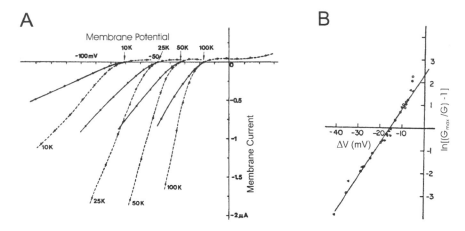

Figure 2 Dependence of inward rectification on the extracellular K^+ concentration. (A) I-V relations of starfish egg cell membrane at four different $[K^+]_{ext}$ (10, 25, 50, and 100 mM) in Na^+-free media. Continuous lines, instantaneous current; dashed lines, steady-state current. (B) Plot of $\ln[(G_{max}/G) - 1]$ against ΔV at different $[K^+]_{ext}$ (in mM). Filled circles, 10 K^+; open triangles, 25 K^+; filled triangles, 50 K^+; open circles, 100 K^+ (A modified from Reference 10; *J. Gen. Physiol.* 1976, 67:621–38 by copyright permission of the Rockefeller Univ. Press; B modified from Reference 12, with permission).

rather than with absolute V_m. Subsequent studies (14, 15) showed that the relation holds only when E_K is altered by varying $[K^+]_{ext}$ but not $[K^+]_{int}$.

INITIAL CLUES TO FUNCTION AND MECHANISM OF INWARD RECTIFICATION

Hodgkin & Horowicz wrote in 1959: "At present neither the physiological significance nor the physical nature of this rectification is understood. The matter is particularly puzzling because the rectification is in the opposite direction to that required to explain the recovery of potential and the loss of potassium during the impulse" (7). The first clue regarding function came from Noble's study of action potentials in cardiac Purkinje fibers (6, 16), where he hypothesized that inward rectification causes a decrease in potassium conductance upon membrane depolarization, producing the long plateau in the action potential, and then an increase during repolarization, accelerating the latter (see also 17).

A clue to a physical mechanism of inward rectification came from Armstrong & Binstock's study (18) of why intracellular tetraethylammonium (TEA) prolongs action potentials in squid axon, as first reported by Tasaki & Hagiwara (19). While the latter authors had concluded that TEA causes the plateau by reducing Na^+

permeability, Armstrong & Binstock countered that during the plateau, Na^+ conductance (g_{Na}) relative to both K^+ and background leak conductances (g_K and g_{leak}) must be larger than at rest because the potential is now closer to the equilibrium potential for Na^+. If g_K and g_{leak} remain unchanged between rest and plateau, then the total membrane conductance ($g_K + g_{Na} + g_{leak}$) must increase during the plateau, contrary to the conclusion of Tasaki & Hagiwara (19). They argued further that the problem could be solved by allowing g_K to be lower during the plateau than at rest, as was proposed by Noble to explain the low conductance plateau (reflecting inward rectification) in Purkinje fiber action potentials (16). Finally, they showed that the potassium I-V curve rectifies inwardly when TEA is injected into the axon (Figure 3). Inhibition of K^+ currents by TEA depends not only on V_m but also on $[K^+]_{ext}$, reminiscent of what was previously observed in inward rectifiers.

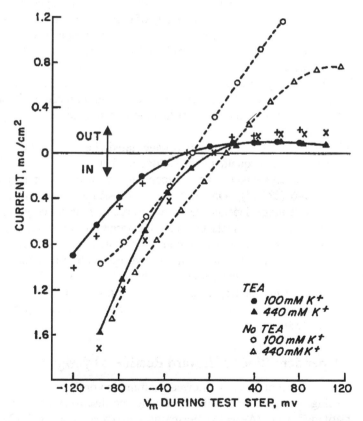

Figure 3 Current-voltage curves of squid axons in high K^+ measured 200 μs after steps from the unclamped resting potential. $\times$ and $+$ indicate membrane current of TEA-injected axon before subtraction of leakage current. TEA curves were determined on a single fiber, controls on two different fibers (from Reference 18; *J. Gen. Physiol.* 1965, 48:859–72 by copyright permission of the Rockefeller Univ. Press).

Both elevated $[K^+]_{ext}$ and membrane hyperpolarization speed dissociation of quaternary ammonium (QA) ions (20). On the basis of these and other findings, Armstrong reasoned that the transmembrane K^+ currents are mediated by pores rather than carriers as had been proposed, including for inwardly rectifying K^+ currents (21).

To explain the acceleration of QA dissociation by extracellular K^+, Armstrong proposed (20) that, in addition to the bimolecular binding-unbinding of QAs with channels, a displacement reaction between an extracellular K^+ and an intracellular QA ion also occurs, a phenomenon sometimes called the *trans* K^+ knock-off effect. Furthermore, he argued that acceleration of QA dissociation by hyperpolarization must not reflect a direct action of the transmembrane electric field on the positively charged QA ion because the effect is small when $[K^+]_{ext}$ is low. Instead, he hypothesized that it reflects the acceleration of an extracellular K^+ ion by the electric field as it enters the pore and, in turn, displaces the QA lodged in the inner pore. The hypothesis was supported by a study of French & Shoukimas (22) who found that channel block by symmetric QAs with 2- to 5-carbon alkyl chains, i.e., TEA to TPeA (tetrapentylammonium), exhibited virtually identical voltage dependence with an apparent valence of $\sim$0.15. They argued that for the valence of 0.15 to reflect QA binding at a site 15% into the electric field, i.e., for a 8 or 12 Å wide (to accommodate TEA or TPeA) seawater-filled mouth to represent 15% of the total channel resistance (taken as 10^{11} Ω), that mouth would have to be impossibly deep (380 or 850 Å).

Although TEA block of squid voltage-gated channels produces an inwardly rectifying I-V curve qualitatively similar to that of true inward rectifiers, the former has much weaker voltage dependence than the latter [valence <1 versus up to 5–6 (23, 24)]. On the basis of the steady-state conductance-voltage (G-V) curve, voltage dependence of inward rectifiers is actually as strong as that of voltage-gated K^+ channels. Owing to the absence of known endogenous blockers, intrinsic gating models were proposed to explain inward rectification (25–28). In fact, we now know that apparent inward rectification in some channel types, such as HERG, HCN, and certain K^+ channels in plants, does reflect intrinsic voltage-dependent channel gating (29–32). Additionally, the possibility that inward rectification reflects inherent asymmetry of the ion conduction pore has also been considered (33).

Voltage-Dependent Block of Inward Rectifiers by Mg^{2+}

Patch-clamp techniques (34) allowed examination of the role of intracellular ions in causing rectification in cells. Following the discovery of extracellular Mg^{2+} block of the NMDA (*N*-methyl-D-aspartic acid) receptor channels (35, 36), the first breakthrough in the search for endogenous blockers was evidence that intracellular Mg^{2+} causes rectification in classic inward rectifiers (37, 38) and in ATP-sensitive K^+ channels (39). Matsuda et al. (37) demonstrated the ohmic property of the I-V relation of single channels (Figure 4*A*), which can be blocked by intracellular

Mg^{2+}. Independently, Vandenberg (38) showed that the linear I-V relation recorded (with a very rapid voltage ramp) in the absence of intracellular Mg^{2+} becomes again inwardly rectifying when Mg^{2+} is added back (Figure $4B-G$). These studies identified Mg^{2+} as the first type of ion, present in all cells, capable of rendering the channels inwardly rectifying. The voltage dependence of channel block by Mg^{2+} was interpreted to reflect Mg^{2+} traveling across part of the transmembrane electric field to reach its site in the inner pore (1, 38), as proposed by Woodhull for proton block of voltage-gated Na^{2+} channels (40).

Woodhull (40) studied voltage dependence (apparent valence ∼0.3) of Na^+ channel block by extracellular protons (permeant blockers). Adopting an Eyring rate theory model of the kind used by Woodbury (33) to account for the I-V relation in ion channels, Woodhull explained the cause of voltage-dependent proton block as follows: "Hydrogen ions may pass through open sodium channels, but stick inside the channels for a relatively long time. While a hydrogen ion is inside the channel, sodium passage is blocked. If the site where hydrogen ions bind and block is inside the channel, a fraction of the electrical potential drop across the membrane is felt by ions in the site. A positive potential inside the axon would repel hydrogen ions from the binding site and reduce blocking." The appealing notion that voltage dependence of block reflects a direct effect of the electrical field on the blocking ion itself is intuitively straightforward and has had a profound influence on ideas regarding the cause underlying voltage dependence of channel block and rectification.

Hille & Schwarz further explored the use of Eyring models in understanding inward rectification (41). They investigated a three-site model where K^+ (coming from either side) can occupy any site, whereas the monovalent ionic blocker (coming from the internal side) can bind only at the internal two sites, and the probability of finding one or more blockers in the channel is influenced by the electric forces acting on the blocker and on the K^+ ions. Indeed, the model predicts shifts of the I-V relation with $[K^+]_{ext}$, although the computed valence of channel block (1.5) is much lower than the value (of up to 5–6) observed experimentally (8, 12, 23, 24, 42). Hille & Schwarz argued that more reasonable valence values could be obtained by increasing the number of ion-binding sites but, as they pointed out, the diagram method is not suited for those more complex cases. However, they did investigate the effect of increasing the charge on the blocker. Making it divalent increases the effective valence dramatically, but then the electrical forces on the divalent blocker and monovalent K^+ ion no longer match, and the block shifts much less than E_K when $[K^+]_{ext}$ is raised. This is a natural property of models where a di- or multivalent blocking ion is affected directly by the electric field.

Although voltage-dependent binding of intracellular Mg^{2+} renders the channels inwardly rectifying, its voltage dependence is too weak to account for the strong rectification commonly observed in many cell types. Also, even though both the single-channel and the instantaneous macroscopic I-V relation are linear in the absence of Mg^{2+}, the channel open probability and macroscopic current elicited

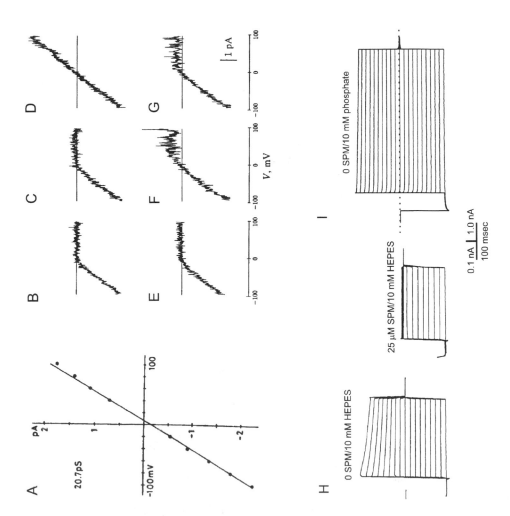

by membrane depolarization diminish with time and, consequently, the steady-state macroscopic I-V relation is still inwardly rectifying even in the nominal absence of intracellular Mg^{2+} (23, 24, 37, 43–45). The intrinsic gating hypothesis thus remained a possible reason underlying the Mg^{2+}-independent rectification.

Strong Voltage-Dependent Block of Inward Rectifiers by Polyamines

In 1993, molecular cloning of inward rectifier K^+ channels (46, 47), first ROMK1 (Kir1.1) and IRK1 (Kir2.1) and then others, rejuvenated the field and paved the way for further mechanistic studies. Shortly thereafter, in a second breakthrough in the search for endogenous blockers, intracellular polyamines [putrescine (PUT), spermidine (SPD), and spermine (SPM)] were identified as endogenous blockers of cloned inward rectifiers heterologously expressed in *Xenopus* oocytes, with such strong voltage dependence that the sharp rectification they caused resembled that in native cells (48–50) (Figure 4*H*). As expected, altering $[K^+]_{ext}$ causes a nearly parallel shift of the I-V curve in the presence of polyamines (51). Furthermore, lowering the intracellular polyamine concentration reduces inward rectification (52–54). However, even after the inside of a membrane patch is exhaustively

←———————————————————————————————————————

Figure 4   Effects of intracellular Mg^{2+} and spermine on the I-V relationship of inward rectifier K^+ channels. (*A*) Single-channel I-V curve recorded from a patch on an open (ruptured) cardiac ventricular cell, with near symmetric $[K^+]$ ($\sim$150 mM K^+ and 1 mM EGTA). (*B–G*) Single-channel I-V curves (filtered at 1 kHz) from a cell-attached and later excised inside-out patch of a ventricular cell elicited by a 115-ms voltage ramp from -96 mV to $+96$ mV. The pipette contained (in mM) 150 KCl and 5 HEPES (pH 7.3). (*B*) Cell-attached patch; bath containing (in mM) 150 KCl, 5 HEPES (pH 7.3), and 4 K_2ATP (free $[Mg^{2+}]$ was estimated at 1.2 mM). (*C*) Patch excised into the bath solution. (*D*) Bath solution changed to same solution nominally without $MgCl_2$ (actual free $[Mg^{2+}]$ estimated at 0.2 μM). (*E*) Bath solution returned to original solution (free $[Mg^{2+}]$ estimated at 1.2 mM). (*F,G*) Bath solution changed to a solution with free $[Mg^{2+}]$ estimated at 17 μM. (*H*) Macroscopic currents recorded from an inside-out membrane patch of an oocyte heterologously expressing HRK1 (Kir2.3) when V_m was stepped from the 0 mV holding potential to -80 mV and then to various V_m between -80 and 80 mV in 10 mV increments. Solutions on both sides of the membrane contain (in mM) $\sim$150 K^+ buffered with 10 mM HEPES (pH 7.3). Intracellular solution also contains 1 mM EGTA and 1 mM EDTA without (*left*) or with (*right*) 25 μM spermine (SPM). (*I*) Macroscopic currents recorded from an inside-out membrane patch of an oocyte heterologously expressing IRK1 (Kir2.1) during voltage steps from the 0 mV holding potential to -100 mV and then to voltages between -100 and 100 mV in 10 mV increments. Solutions on both sides of the membrane contain (in mM) 100 K^+, 5 EDTA, and 10 phosphate (pH 8.0). (*A* from Reference 37 with permission; *B–G* taken from 38, with permission; *H* modified from Reference 48 with permission; *I*, H.G. Shin & Z. Lu, unpublished data).

perfused with a solution nominally devoid of Mg^{2+} and polyamines, the channels still exhibit modest inward rectification (55–57). Thus the possibility remained that inward rectification results from intrinsic gating which is somehow enhanced by intracellular cations. Residual inward rectification in the absence of Mg^{2+} and polyamines appears to vary in extent between batches of ingredients used to prepare the recording solutions (58), an observation that led to identification of such contaminants as hydroxyethylpiperazine in HEPES and ethylenediamine in EDTA as the cause of residual rectification (59). Indeed, when recorded from a patch adequately perfused with blocker-free solutions, the steady-state I-V relation of IRK1 is practically linear (58, 59) (see also Figure 4*I*). Thus inward rectification reflects strong voltage dependence of channel block by intracellular cations, not intrinsic voltage-dependent channel gating.

Many complex features of channel block by intracellular polyamines have been observed. For example, a plot of the extent of channel block (at a given blocker concentration) against membrane voltage exhibits a shallow and a steep phase and thus requires an equation with two Boltzmann terms (60). At a given V_m, the extent of channel block as a function of polyamine concentration (dose-inhibition relation) also exhibits two phases, requiring an equation with two hyperbolic terms (61). These phenomena suggest formation of multiple blocked states. One proposed explanation is that the multiple blocked states reflect binding of more than one polyamine molecule in the pore, which would then help account for the remarkably strong voltage dependence. However, the binding of multiple polyamine molecules is not likely to produce much stronger voltage dependence because the charges carried by the amines do not appear to sense much of the transmembrane electric field (62). Also, the calculation by French & Shoukimas (22) shows that as little a drop as 15% in transmembrane voltage would need a great distance ($\sim$380–850 Å) in a pore 8–12 Å wide. There is no direct evidence to date either confirming or ruling out the possibility that more than one polyamine binds in the pore. Most, if not all, data can be accounted for by assuming that a single polyamine molecule bound to the pore suffices to block K^+ conduction, although it is conceivable that several polyamines bind in the wide internal vestibule of the pore and act as surface charges (63).

In some cases, multiple blocked states result from nonhomogeneous protonation of polyamines (64), an interpretation facilitated by analogy with a similar, if more robust, phenomenon in retinal cGMP-gated channels (65, 66). In the presence of a given species of long polyamines, as the membrane is depolarized the IRK1 current decreases in two phases separated by a hump and, contrary to the expectation for a nonpermeant pore blocker, a significant residual IRK1 current remains at extreme positive V_m. This multi-phasic polyamine-blocking behavior most likely reflects two blocked channel states formed by the binding of fully and partially protonated polyamines, because it is pH-titrated with a pKa value (8.1) very near that (8.3) of the tested polyamine species. In each of the blocked states, the polyamine is bound with characteristic affinity and probability of traversing the pore. The notion that polyamines traverse the pore with finite probability is supported by the observation that philanthotoxin (spermine with a bulky group attached at one end) acts as a nonpermeant ionic blocker in IRK1, a strategy first used to study polyamine block of glutamate receptor channels (67). As discussed below, certain acidic residues in the channels are critical for high-affinity binding of blocking ions. Thus just as intracellular protons titrate amine groups in blockers, they may also titrate those channel residues, and nonhomogeneous channel protonation could in principle give rise to different affinities for blocking ions. Additionally, the channel apparently becomes blocked as soon as the leading end of a polyamine chain enters the effective part of the long inner pore and begins to displace the innermost K^+, before the blocker reaches its final, deeper binding site; this may generate multiple serially related blocked states (62, 68).

MOLECULAR CHARACTERIZATION OF THE ION CONDUCTION PORE

Analysis of the amino acid sequence (deduced from the cDNA) suggests that each of the four subunits (69, 70) of inward rectifiers has two transmembrane segments (M1 and M2), approximately equivalent to the fifth and sixth transmembrane segments (S5 and S6) of voltage-gated channels (46, 47) (Figure 5). Extrapolating the insight gained from voltage-gated K^+ channels (e.g., 71–78) to inward rectifiers, the region between M1 and M2 containing the signature sequence (79, 80) would form the outer K^+-selective part of the ion conduction pore, and M2 (equivalent to S6 in voltage-gated channels) would form part of the inner pore. Consistent with this proposed topology, mutagenesis studies show that the presence of an acidic residue at a seemingly conserved site in M2 (e.g., D172 in IRK1) is crucial for the high-affinity binding of intracellular blocking ions (81–83). In the case of ROMK1, an acidic residue substituted for N171 (corresponding to IRK1's D172) in M2 confers a much higher affinity for blocking ions, hence stronger rectification (Figure 6), whereas a basic residue renders the channel essentially insensitive (82). On the basis of these observations, it was proposed that residue 171 affects the binding of blocking ions through an electrostatic mechanism (82), a proposal further supported by the findings that an acidic residue is effective at a number of different sites within M2 (62) and that a histidine substituted at position 171 confers inward rectification in a pH-dependent manner (82, 84).

Taglialatela et al. made the unexpected discovery that the cytoplasmic termini in IRK1 also underlie the affinity of IRK1 for intracellular blocking ions (85) (Figure 7A). Indeed, C-terminal residues E224 and E299 were later identified as critical for high-affinity binding of blocking ions (61, 85–87). The acidic residues in the C terminus and that in M2 affect the blocking ion binding energy in an apparently additive manner (61) (Figure 7B). These findings strongly suggest that the C terminus extends the inner pore formed by M2.

The presence of blocker binding sites in the inner pore raises the issue of the inner pore's dimension, a question tackled by several groups using cysteine

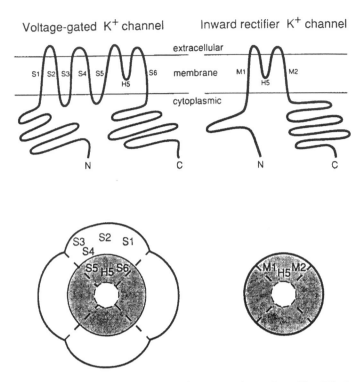

Figure 5   Proposed membrane topology of the IRK1 inward-rectifier K$^+$ channel. By analogy with the suggested membrane arrangement of voltage-gated K$^+$ channels, the M1 and M2 segments of IRK1 are shown as membrane-spanning and the intervening H5 sequence (the P-region) as dipping from the extracellular surface into the membrane. Compared with the voltage-gated K$^+$ channel, which may contain an outer shell of S1, S2, and S3 segments in contact with lipids, the inward-rectifier K$^+$ channel represents a reduced structure corresponding to only the inner core of the voltage-gated K$^+$ channel (from Reference 47 with permission).

mutagenesis-based electrophysiological approaches (88, 89). For example, Lu et al. varied the number of cysteine substitutions and correlated that number with the current inhibition caused by intracellular methanethiosulfonate-ethyltrimethylammonium (MTSET) (88). They found that, for some residues in M2 that line the inner pore, several or sometimes all four substitutions (one per subunit) were needed to completely eliminate current with MTSET and argued that the pore width at that location probably exceeded 12 Å. Further studies also suggested that cytoplasmic termini form a wide internal vestibule (90). On the other hand, a steep drop in the blocking rate constant of intracellular QAs of increasing size occurs only between tetrapropyl- and tetrabutylammonium (91), which suggests that although the inner pore may generally be wide, a ∼9 Å constriction

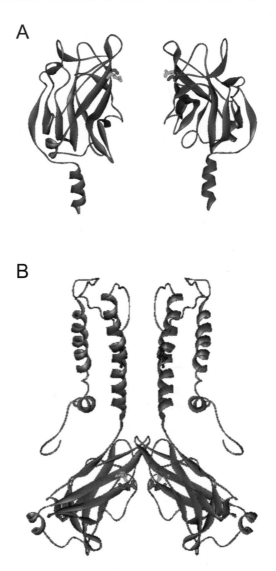

Figure 8   Ribbon representations of the structure of the cytoplasmic termini of GIRK1 [(*A*) (99) and the closed state structure of KirBac1.1 (*B*) (100)], with two subunits shown in each case. The residues whose side-chains are colored in burgundy (I138 in KirBac1.1), pink (S225 in GIRK1 and E187 in KirBac1.1), and green (E300 in GIRK1 and E258 in KirBac1.1) correspond to residues D172, E224, and E299, respectively, in IRK1.

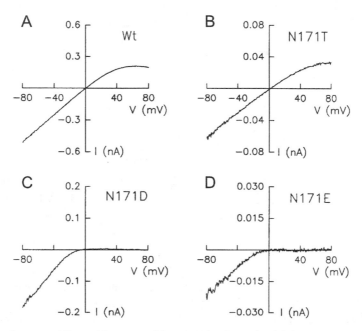

Figure 6   An acidic residue at position 171 in the M2 of ROMK1 causes strong rectification. The I-V curves were recorded from cell-attached membrane patches of oocytes heterologously expressing wild-type (N171) and mutant (N171T, N171D, and N172E) ROMK1 channels. The pipette solution contained (in mM) 100 KCl, 0.3 CaCl$_2$, 1 MgCl$_2$, and 10 HEPES (pH 7.1) (from Reference 82 with permission).

exists between the intracellular solution and the QA binding site, most probably formed by certain residues in the C terminus (62, 91). As discussed below, a constricted region for blocking ions may be critical in achieving sharp rectification.

Comparison of the membrane topology between inward-rectifier and voltage-gated K$^+$ channels also yields insight into the latter's architecture. On the basis of such a comparison, Kubo et al. proposed that a voltage-gated K$^+$ channel consists of a central ion conduction module composed of four S5-S6 (equivalent to M1-M2) units, surrounded by four voltage-sensing modules S1–S4 (47) (Figure 5). This notion of modular architecture was also articulated by others on the basis of an apparent structural conservation of the voltage-sensing module in the superfamily of voltage-gated channels (92). Additionally, the C terminus may form a module extending the inner pore (61, 85–87). The concept of modular architecture of K$^+$ channels must be correct because both inward-rectifier and voltage-gated K$^+$ channels can be built using the bacterial KcsA pore (80, 93), whose subunit has two transmembrane (M1-M2) segments as the ion conduction module. Attaching the voltage-sensing (S1-S4) module of the Shaker K$^+$ channel to the KcsA pore yields

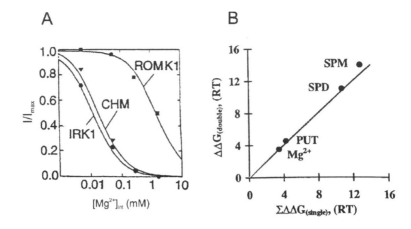

Figure 7 Cytoplasmic termini affect the binding of intracellular blocking ions to the IRK1 pore. (*A*) Normalized current (I/I_{max}) is plotted against intracellular [Mg^{2+}] for IRK1, ROMK1, and chimeric channels (CHM) in which the cytoplasmic termini of ROMK1 are replaced by those of IRK1. (*B*) Changes in blocker binding energy caused by the double mutation D172N in M2 and E244G in the C terminus of IRK1 are plotted against the sum of the changes in binding energy caused by each mutation alone (*A* from Reference 85; reprinted with permission from *Science* 264:844–47, copyright 1994; *B* reprinted with permission from Reference 61).

a voltage-gated K^+ channel (94), provided two essential sequences from Shaker are preserved to assure functional coupling between voltage sensors and channel gate (95). (For structural studies of voltage-gated K^+ channels, see 96.) Similarly, attaching the cytoplasmic termini of IRK1 to the KcsA pore yields an inward rectifier. Furthermore, substitution of an acidic residue at a proper site in M2/S6 confers strong inward rectification in both the KcsA-IRK1 and the KcsA-Shaker chimeric channels (94).

Crystallographic Studies of the Ion Conduction Pore

Determination of the atomic structure of a number of K^+ channels provided the foundation for understanding the structure-function relation underlying permeation in K^+ channels. Studies on the KcsA pore revealed how a $\sim$12 Å long K^+ selectivity filter is formed by the signature sequence, whereas the remaining ($\sim$20 Å) inner part of the pore is lined by M2, and how the structure gives rise to K^+ conduction (80, 97, 98). Structural studies on the cytoplasmic termini of GIRK1 (Kir3.1) show that they not only extend the pore intracellularly (beyond what is formed by M2) by at least another $\sim$30 Å but also form a wide internal vestibule (99) (Figure 8*A*, see color insert). Width of the pore extension varies from 7 to 15 Å (narrower at both ends), sufficiently wide to contain hydrated K^+ ions, although their presence, or that of any blocking ions, remains to be demonstrated. The structure (100) of a (closed) bacterial inward rectifier (KirBac1.1) shows that the length of the entire channel pore is $\sim$90 Å (Figure 8B). The residue corresponding to D172 in IRK1 is located somewhat internal to the K^+ selectivity filter, whereas the residues corresponding to E224 and E299 in IRK1 form a ring (80, 99, 100). The distance between the acidic residues in M2 and those in the C terminus, as inferred from the structure of KirBac1.1, would be about 35 Å (100), significantly exceeding the length ($\leq$20 Å) of a natural polyamine molecule.

THE CAUSE UNDERLYING THE VOLTAGE DEPENDENCE OF CHANNEL BLOCK

Inward rectification does not indicate intrinsic channel gating, rather it reflects voltage dependence of channel block by intracellular cations. Does the voltage dependence result from a direct influence of the transmembrane electric field on the blocking ion as hypothesized by Woodhull for proton block of Na^+ channels (40), or does it reflect an indirect influence on the blocker via an effect of the electric field on K^+ ions in the pore, as hypothesized by Armstrong for TEA block of voltage-gated K^+ channels (20)? In the latter case, the voltage dependence would literally reflect " ... the force acting on the potassium ions," as Hodgkin & Horowicz had proposed (7). Furthermore, why does the conductance of inward rectifiers vary with $V_m - E_K$ instead of with V_m itself, and what are the theoretical and physical underpinnings of the empirical equation of Hagiwara & Takahashi (12) that quantitatively describes inward rectification?

Clues came from quantitative analyses of QA block of inward rectifiers (91, 101–103). For example, while the affinity of ROMK1 for TEA decreases with increasing $[K^+]_{ext}$ (Figure 9A), the voltage dependence of channel block increases

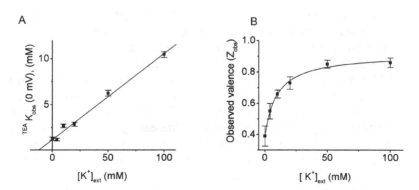

Figure 9 Extracellular K^+ dependence of ROMK1 block by intracellular TEA. $^{TEA}K_{obs}$(0 mV) (A) and observed valence (B) are plotted against $[K^+]_{ext}$ (modified from Reference 102. *J. Gen. Physiol.* 1998, 112:211–21 by copyright permission of the Rockefeller Univ. Press).

(Figure 9*B*), indicating that the voltage dependence must at least in part result from the movement of K^+ ions within the electric field along the pore (102). Similar dependence on $[K^+]_{ext}$ was also observed with other QAs of varying sizes (from TEA to TPeA) and with Mg^{2+}. Furthermore, observed valences of channel block by divalent Mg^{2+} and by monovalent TEA are nearly identical in the presence of 4-mM extracellular K^+, and both increase identically in 100-mM extracellular K^+ (102), even though some data suggest that the smaller divalent Mg^{2+} binds at a deeper site (61, 62, 81–83, 104). Nonetheless, the comparable voltage dependence suggests that neither ion traverses a significant fraction of the electric field. Any voltage dependence must therefore reflect primarily the movement of K^+ ions across the electric field, as K^+ and TEA displace one another during block-unblock of the pore (102) (For K^+ and V_m effects on QA block kinetics, see 91, 101). Such a mechanism, as shown below, predicts linear dependence of the observed dissociation constant for TEA binding ($^{TEA}K_{obs}$) on $[K^+]_{ext}$, as has been observed (Figure 9*A*).

In the kinetic scheme of Figure 10*A*, the dissociation of TEA from the internal side of the pore occurs with or without direct displacement by an extracellular K^+ ion. Intracellular $[K^+]$ is assumed constant. In such a scheme, the fraction of channels not blocked is expressed as G/G_{max}:

$$\frac{G}{G_{max}} = \frac{1}{1 + \dfrac{[TEA]_{int}}{^{TEA}K_{obs}}},$$

where

$$^{TEA}K_{obs} = {}^{TEA}K\left(1 + \frac{[K^+]_{ext}}{{}^{K}K}\right) = {}^{TEA-K}K\left({}^{K}K + [K^+]_{ext}\right). \qquad 3.$$

Therefore, the X intercept (Figure 9*A*) of a plot of $^{TEA}K_{obs}$ against the concentration of extracellular K^+ (or other alkali metal ion) estimates the apparent dissociation constant (^{K}K or $^{other}K$) for extracellular ion binding: 13 mM for K^+ (Figure 9*A*), 16 mM for Rb^+, 34 mM for Cs^+, and $\sim$1 M for Na^+ ($[K^+]_{int} = 100$ mM for all cases), consistent with the expected ion selectivity sequence (103). [A ^{K}K of 13 mM appears reasonable because 17 mM accounts for the dependence of the single-channel conductance on $[K^+]$ (105).] The channel's varying affinity for different extracellular ions results in different degrees of rectification depending on the extracellular ion species present: steepest with K^+ and shallowest with Na^+ (103). Furthermore, the apparent affinity of the pore for extracellular ions itself is significantly affected by the intracellular ion species (K^+ versus Rb^+), which suggests that significant interactions among ions in the pore are nonelectrostatic (103).

At high $[K^+]_{ext}$ the occupancy of state ChK_m^+ (Figure 10*A*) is minimal, and the reaction scheme reduces to a mechanism (Figure 10*B*) where binding-unbinding of TEA and of a K^+ ion are essentially always coupled. Equation 3 then reduces to

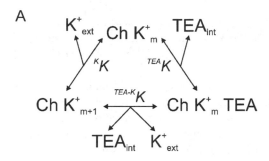

Figure 10 (*A*) Intracellular TEA binds to the pore either with or without direct displacement of a K$^+$ ion. (*B*) Reduced scheme where TEA and one extracellular K$^+$ ion obligatorily displace one another from the pore during block-unblock. (*C*) Generalized version of scheme *B*, where the binding of an amine blocker (AM) displaces n K$^+$ ions and vice versa.

$$\frac{G}{G_{max}} = \frac{1}{1 + \dfrac{[TEA]_{int}}{^{TEA}K_{obs}}}, \quad \text{where} \quad ^{TEA}K_{obs} = [K^+]_{ext}{}^{TEA-K}K. \qquad 4.$$

Assuming the electric potential energy in the reaction to be a Boltzmann function of V_m, one obtains

$$\frac{G}{G_{max}} = \frac{1}{1 + \dfrac{[TEA]_{int}}{[K^+]_{ext}{}^{TEA-K}K(0\,mV)e^{-\frac{Z_{TEA-K}FV_m}{RT}}}}, \qquad 5.$$

where Z_{TEA-K} is the valence, and F, R, and T have their usual meanings. Replacing $[K^+]_{ext}$ in Equation 5 with the Nernst equation for E_K yields

$$\frac{G}{G_{max}} = \frac{1}{1 + \dfrac{[TEA]_{int}}{\left[K^+\right]_{int}^{TEA-K} K(0\,mV)\,e^{-\frac{F(Z_{TEA-K}V_m - Z_K E_K)}{RT}}}}, \qquad 6.$$

where Z_K ($= 1$) is the valence of a K^+ ion. In the limit where the observed voltage dependence results solely from the movement of a K^+ ion across the transmembrane electric field, the numerical value of Z_{TEA-K} equals Z_K. This appears to be the case because the experimentally determined Z_{TEA-K} [called $^{TEA}(z\delta)_s$ in 102] is unity. Accordingly, Equation 6 reduces to

$$\frac{G}{G_{max}} = \frac{1}{1 + \dfrac{[TEA]_{int}}{\left[K^+\right]_{int}^{TEA-K} K(0\,mV)\,e^{-\frac{Z_K F(V_m - E_K)}{RT}}}}, \qquad 7.$$

which has a form equivalent to Equation 1, the empirical relation of Hagiwara & Takahashi (12) between G/G_{max} and $V_m - E_K$ for a given $[K^+]_{int}$.

Certain long polyamines (e.g., spermine with four amine groups) cause much sharper rectification ($Z = {\sim}5$) (58, 64, 68), raising the question whether a significant part of the strong voltage dependence derives directly from the polyamine charges themselves (e.g., 50, 60, 64, 106). The answer came from an approach pioneered by Miller with alkyl bis-quaternary ammonium ions in a study of sarcoplasmic reticulum K^+ channels (107) and first applied to inward rectifiers by Pearson & Nichols (106), who examined IRK1 block by a series of alkyl bis- and mono-amines of varying length. The most intriguing finding, in the present context, was that the apparent valence of bis-amine block increases with alkyl chain length (106, see also 64). Subsequent resolution of confounding technical problems allowed a sufficiently accurate characterization (discussed below) of kinetics and energetics of channel block by various amines to reveal the actual cause of the large apparent valence underlying sharp rectification (62, 68).

The apparent valence of channel block by alkyl bis- and mono-amines (carrying 1 or 2 positive charges; Figure 12A) increases to ${\sim}4$–5 with increasing chain length (Figure 11A), which indicates that the voltage dependence primarily reflects the movement in the electric field, not of blockers, but of K^+ ions. Effective, strongly voltage-dependent block requires a number of complementary properties of blocker and channel. The blocker apparently needs to carry at least one charge at both ends of a hydrophobic chain of a certain length. Although the trailing end always binds near the same site on the intracellular side, the leading end penetrates deeper into the pore as the chain is elongated (Figure 12B). Judging from how the energetic coupling coefficient (Ω) (108) between D172 and alkyl bis-amines varies with chain length (Figure 11B), the leading amine of bis-C9 gets closest to D172, whereas that of bis-C8 or bis-C10, respectively, falls slightly short of or overshoots D172 (additionally or alternatively, the drop-off in the Ω value may reflect buckling of the alkyl chain). Thus the longer the blocker the deeper it reaches and the more K^+ ions are displaced across the narrow selectivity filter,

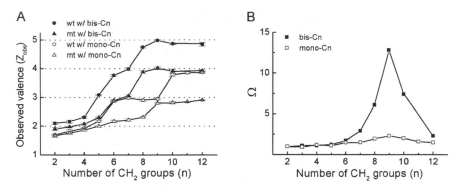

Figure 11 (*A*) Valences for block of wild-type and D172N mutant channels and (*B*) Ω [= $(^{wt}K_d{}^{C2} \times {}^{mt}K_d{}^{Cn})/({}^{mt}K_d{}^{C2} \times {}^{wt}K_d{}^{Cn})$] at 0 mV between wild-type and mutant channels, plotted for alkyl mono- and bis-amines of varying chain length (C2–C12) (from Reference 62. *J. Gen. Physiol.* 2003, 121:261–75 by copyright permission of the Rockefeller Univ. Press).

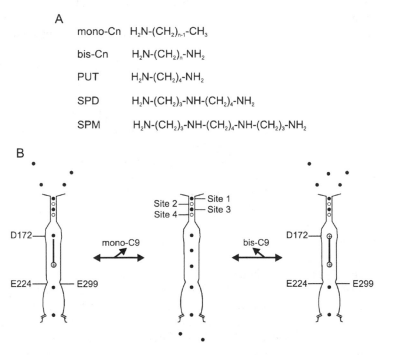

Figure 12 Blocker structures and channel models. (*A*) Chemical structures of mono-Cn, bis-Cn, PUT, SPD, and SPM. (*B*) Models for voltage dependence of channel block by mono- and bis-C9. The outer, K^+-selective part of the pore contains two K^+ ions in two possible configurations (sites 1 and 3 versus 2 and 4), whereas the inner pore contains up to five K^+ ions displaceable by an amine blocker.

thus producing stronger voltage dependence. Beyond the value of ∼5 reached at bis-C8, the apparent valence increases no further, which suggests that the leading amine rarely enters deeply into the selectivity filter itself, consistent with the very low probability of the amine blocker traversing the entire pore (64). Electrostatic interactions between D172 and the leading amine of bis-amines contribute up to ∼2 kcal/mol (in the case of bis-C9), whereas those between E224/E299 and the trailing amine contribute <1 kcal/mol. Hydrophobic interactions between the alkyl chain and the pore contribute ∼0.4 kcal/mol per methylene group. In the case of mono-amines, although their valence increases with chain length, their Ω value with respect to D172 remains near unity (Figure 11), which suggests that the sole (trailing) amine group always binds near the same internal site whereas the alkyl chain reaches deep into the pore (Figure 12*B*). Absence of a leading amine results in 20- and 600-fold reductions in the affinity and blocking rate constant, respectively.

In KirBac1.1, whose structure is known, the distance between the residues that correspond to D172 and E224/E299 in IRK1 is about 35 Å (100) (Figure 8*B*), much longer than bis-C9. If this distance also applies to (open) IRK1, the trailing amine of bis-C9 must interact with E224/E299 over some distance while the leading amine is located near the plane of D172. This appears to be the case because replacing both E224 and E299 with neutral residues affects the binding energy of bis-amines only modestly (∼0.7 kcal/mol at 0 mV) but reduces the rate constant of channel block dramatically and renders the I-V curve inwardly rectifying even in the absence of any blockers. In contrast, the electrostatic effect of D172 on the binding of cationic blockers appears more localized. First, although an acidic residue substituted at a number of sites in M2 is effective, the substitution must be made within a small region. Second, the interaction energy between D172 and the leading amine rises quite steeply over only three methylene groups (Figure 11*B*). Third, substituting a neutral residue for D172 neither reduces the apparent rate constant of channel block nor renders the macroscopic or single-channel I-V curve inwardly rectifying in the absence of blockers (59, 109).

To produce maximal voltage dependence by displacing the most K$^+$ ions, a blocker needs to be at least as long as bis-C8. Cells do not synthesize bis-C8 from PUT (bis-C4) by extending the methylene chain, but instead generate SPD or SPM by attaching one or two amino-propyl groups (Figure 12*A*). Both these long polyamines block the channels with strong voltage dependence comparable to that of bis-C8, although SPM has much higher affinity. This strong binding is apparently achieved by two amines in one half of SPM interacting primarily with D172 (∼3.5 kcal/mol) and the remaining two in the other half interacting primarily with E224 and E299 (∼2.7 kcal/mol). Effectively, each half-SPM molecule acts as a pseudodivalent cation, enhancing the electrostatic interaction, hence the binding energy. Therefore, and counterintuitively, the multiple amines in spermine serve primarily to increase the affinity of channel block, not its voltage dependence. Beyond the energetic consequences, the interactions of polyamines with E224 and E299 also help maximize the rate of channel block.

In addition to providing acidic and hydrophobic residues to interact with amine blockers' charges and methylene groups, the inner pore of the channel must harbor several K^+ ions, a characteristic enhanced significantly by the presence of the acidic residues. To achieve the requisite steep voltage dependence with an apparent valence as high as ~ 5, up to five K^+ ions must be effectively displaced by an intracellular blocker out of the inner pore across the transmembrane electric field (Figures 11A and 12B). Interestingly, valences for amine blockers increase with chain length in apparently discrete steps of ~ 1 (Figure 11A), as if one K^+ ion at a time is displaced as the leading end of amine blockers of increasing length reaches deeper into the pore. Discrete displacement will occur if K^+ ions cannot exchange position with (or bypass) the blocker, even if they can do so among themselves (i.e., mix) in the inner pore. If K^+ ions were miscible in part of the inner pore, i.e., not truly in single file, a snug fit of the blocker in a more intracellular region remains imperative for achieving strong voltage dependence: Binding of the blocker's trailing amine at a narrow, more intracellular region would prevent any K^+ from reaching the intracellular solution as the blocker's leading end "pushes" K^+ ions outwardly through the electric field across the ion selectivity filter, engendering strong voltage dependence. In this context, the value of the maximal valence of channel block need not be the same as the value of the Ussing flux ratio exponent (110) for these channels [$n' = 2.2$ (111)] because the exponent is determined by the number of K^+ ions that move along the pore in true single-file fashion, i.e., unable to exchange position with one another (110, 112). Mixing of small metal ions in the inner pore may also help explain why Mg^{2+} blocks with a much lower apparent valence (~ 1), even though it may bind at a site just internal to the narrow selectivity filter (61, 81–83, 104).

Finally, does the model of mutual displacement between K^+ ions and a multivalent cationic amine (AM) blocker account for the empirical equation of Hagiwara & Takahashi (12) for inward rectification? Scheme C in Figure 10 generalizes scheme B. Correspondingly, Equation 7 may be generalized for the case where n K^+ ions are displaced by a blocker and vice versa:

$$\frac{G}{G_{max}} = \frac{1}{1 + \dfrac{[AM]}{\left[K^+\right]_{int}^n {}^{AM-K}K(0\,mV)e^{-\frac{nZ_K F(V_m - E_K)}{RT}}}}, \qquad 8.$$

which is exactly equivalent to the equation of Hagiwara & Takahashi (12) (their Equation 3 or Equation 1 above). Comparing Equations 1 and 8 reveals the theoretical underpinnings of the empirical terms:

$$\Delta V_h = \frac{RT}{nZ_K F} \ln \frac{\left[K^+\right]_{int}^n {}^{AM-K}K(0\,mV)}{[AM]}, \quad \text{and} \quad \nu = \frac{RT}{nZ_K F} \qquad 9.$$

in Hagiwara and Takahashi's equation. The simple displacement model thus quantitatively accounts for the defining characteristic of inward rectification, i.e., that G/G_{max} depends, not on V_m itself, but on $V_m - E_K$ at constant $[K^+]_{int}$.

ACKNOWLEDGMENTS

The author thanks Dr. Paul De Weer for critical review and discussion of the manuscript. The cited experimental studies from the author's laboratory were supported by National Institutes of Health grants GM55560 and GM61929. The author was the recipient of an Independent Scientist Award from the National Institutes of Health (HL03814).

The *Annual Review of Physiology* is online at http://physiol.annualreviews.org

LITERATURE CITED

1. Matsuda H. 1991. Magnesium gating of the inwardly rectifying K$^+$ channel. *Annu. Rev. Physiol.* 53:289–98

2. Nichols CG, Lopatin AN. 1997. Inward rectifier potassium channels. *Annu. Rev. Physiol.* 59:171–91

3. Lopatin AN, Nichols CG. 2001. Inward rectifiers in the heart: an update on I$_{K1}$. *J. Mol. Cell. Cardiol.* 33:625–38

4. Stanfield PR, Nakajima S, Nakajima Y. 2002. Constitutively active and G-protein coupled inward rectifier K$^+$ channels: Kir2.0 and Kir3.0. *Rev. Physiol. Biochem. Pharmacol.* 145:50–179

5. Katz B. 1949. Les constantes électriques de la membrane du muscle. *Arch. Sci. Physiol.* 3:285–99

6. Noble D. 1965. Electrical properties of cardiac muscle attributable to inward going (anomalous) rectification. *J. Cell. Comp. Physiol.* 66:127–36

7. Hodgkin AL, Horowicz P. 1959. The influence of potassium and chloride ions on the membrane potential of single muscle fibres. *J. Physiol.* 148:127–60

8. Adrian RH, Freygang WH. 1962. The potassium and chloride conductance of frog muscle membrane. *J. Physiol.* 163:61–103

9. Adrian RH, Freygang WH. 1962. Potassium conductance of frog membrane under controlled voltage. *J. Physiol.* 163:104–14

10. Hagiwara S, Miyazaki S, Rosenthal NP. 1976. Potassium current and the effect of cesium on this current during anomalous rectification of the egg cell membrane of a starfish. *J. Gen. Physiol.* 67:621–38

11. Noble D, Tsien RW. 1968. The kinetics and rectifier properties of the slow potassium current in cardiac Purkinje fibres. *J. Physiol.* 195:185–214

12. Hagiwara S, Takahashi K. 1974. The anomalous rectification and cation selectivity of the membrane of a starfish egg cell. *J. Membr. Biol.* 18:61–80

13. Hodgkin AL, Huxley AF. 1952. The dual effect of membrane potential on sodium conductance in the giant axon of *Loligo*. *J. Physiol.* 116:497–506

14. Hagiwara S, Yoshii M. 1979. Effects of internal potassium and sodium on the anomalous rectification of the starfish egg as examined by internal perfusion. *J. Physiol.* 292:251–65

15. Leech CA, Stanfield PR. 1981. Inward rectification in frog skeletal muscle fibres and its dependence on membrane potential and external potassium. *J. Physiol.* 319:295–309

16. Noble D. 1962. A modification of Hodgkin-Huxley equations applicable to Purkinje fibre action and pace-maker potentials. *J. Physiol.* 160:317–52

17. Zaritsky JJ, Redell JB, Tempel BL, Schwarz TL. 2001. The consequence of disrupting cardiac inwardly rectifying K$^+$ current (I$_{K1}$) as revealed by the targeted deletion of the murine *Kir2.1* and *Kir2.2* genes. *J. Physiol.* 533:697–710

18. Armstrong CM, Binstock L. 1965. Anomalous rectification in the squid giant axon injected with tetraethylammonium chloride. *J. Gen. Physiol.* 48:859–72

19. Tasaki I, Hagiwara S. 1957. Demonstration of two stable potential states in the squid giant axon under tetraethylammonium chloride. *J. Gen. Physiol.* 40:858–85

20. Armstrong CM. 1971. Interaction of tetraethylammonium ion derivatives with the potassium channels of giant axons. *J. Gen. Physiol.* 58:413–37

21. Horowicz P, Gage PW, Eisenberg RS. 1968. The role of the electrochemical gradient in determining potassium fluxes in frog striated muscle. *J. Gen. Physiol.* 51:193s–203s

22. French RJ, Shoukimas JJ. 1981. Blockage of squid axon potassium conductance by internal tctra-n-alkylammonium ions of various sizes. *Biophys. J.* 34:271–91

23. Burton FL, Hutter OF. 1990. Sensitivity to flow of intrinsic gating in inwardly rectifying potassium channel from mammalian skeletal muscle. *J. Physiol.* 424:253–61

24. Silver MR, DeCoursey TE. 1990. Intrinsic gating of inward rectifier in bovine pulmonary artery endothelial cells in the presence and absence of internal Mg^{2+}. *J. Gen. Physiol.* 96:109–33

25. Ciani S, Krasne S, Miyazaki S, Hagiwara S. 1978. A model for inward rectification: electrochemical-potential-dependent gating of membrane channels. *J. Membr. Biol.* 44:103–34

26. Ohmori H. 1978. Inactivation kinetics and steady-state current noise in the anomalous rectifier of tunicate egg cell membranes. *J. Physiol.* 281:77–99

27. Gunning R. 1983. Kinetics of inward rectifier gating in the eggs of the marine polychaete, *Neanthes arenaceodentata. J. Physiol.* 342:437–51

28. Kurachi Y. 1985. Voltage-dependent activation of the inward-rectifier potassium channel in the ventricular cell membrane of guinea-pig heart. *J. Physiol.* 366:365–85

29. Smith PL, Baukrowitz T, Yellen G. 1996. The inward rectification mechanism of the HERG cardiac potassium channel. *Nature* 379:833–36

30. Santoro B, Liu DT, Yao H, Bartsch D, Kandel ER, et al. 1998. Identification of a gene encoding a hyperpolarization-activated pacemaker channel of brain. *Cell* 93:717–29

31. Anderson JA, Huprikar SS, Kochian LV, Lucas WJ, Gaber RF. 1992. Functional expression of a probable *Arabidopsis thaliana* potassium channel in *Saccharomyces cerevisiae. Proc. Natl. Acad. Sci. USA* 89:3736–40

32. Schachtman DP, Schroeder JI, Lucas WJ, Anderson JA, Gaber RF. 1992. Expression of an inward-rectifying potassium channel by the Arabidopsis KAT1 cDNA. *Science* 258:1654–58

33. Woodbury JW. 1971. Eyring rate theory model of the current-voltage relationships of ion channels in excitable membranes. In *Chemical Dynamics: Papers in Honor of Henry Eyring*, ed. JO Hirschfelder, pp. 601–17. New York: Wiley

34. Hamill OP, Marty A, Neher E, Sakmann B, Sigworth FJ. 1981. Improved patch-clamp techniques for high-resolution current recording from cells and cell-free membrane patches. *Pflügers Arch.* 391:85–100

35. Nowak L, Bregestovski P, Ascher P, Herbet A, Prochiantz A. 1984. Magnesium gates glutamate-activated channels in mouse central neurons. *Nature* 307:462–65

36. Mayer ML, Westbrook GL, Guthrie PB. 1984. Voltage-dependent block by Mg^{2+} of NMDA responses in spinal cord neurones. *Nature* 309:261–63

37. Matsuda H, Saigusa A, Irisawa H. 1987. Ohmic conductance through the inwardly rectifying K channel and blocking by internal Mg^{2+}. *Nature* 325:156–59

38. Vandenberg CA. 1987. Inward rectification of a potassium channel in cardiac ventricular cells depends on internal magnesium ions. *Proc. Natl. Acad. Sci. USA* 84:2560–64

39. Horie M, Irisawa H, Noma A. 1987. Voltage-dependent magnesium block of adenosine-triphosphate-sensitive potassium channel in guinea-pig ventricular cells. *J. Physiol.* 387:251–72

40. Woodhull AM. 1973. Ionic blockage of sodium channels in nerve. *J. Gen. Physiol.* 61:687–708

41. Hille B, Schwarz W. 1978. Potassium channels as multi-ion single-file pores. *J. Gen. Physiol.* 72:409–42

42. Hagiwara S, Miyazaki S, Krasne S, Ciani S. 1977. Anomalous permeabilities of the egg cell membrane of a starfish in K^+-Tl^+ mixtures. *J. Gen. Physiol.* 70:269–81

43. Ishihara K, Mitsuiye A, Noma A, Takano M. 1989. The Mg^{2+} block and intrinsic gating underlying inward rectification of the K^+ current in guinea-pig cardiac myocytes. *J. Physiol.* 419:297–320

44. Ishihara K, Hiraoka M. 1994. Gating mechanism of the cloned inward rectifier potassium channel from mouse heart. *J. Membr. Biol.* 142:55–64

45. Stanfield PR, Davies NW, Shelton PA, Khan IA, Brammar WJ, et al. 1994. The intrinsic gating of inward rectifier K^+ channels expressed from the murine *IRK1* gene depends on voltage, K^+ and Mg^{2+}. *J. Physiol.* 475:1–7

46. Ho K, Nichols CG, Lederer WJ, Lytton J, Vassilev PM, et al. 1993. Cloning and expression of an inwardly rectifying ATP-regulated potassium channel. *Nature* 362:31–38

47. Kubo Y, Baldwin TJ, Jan YN, Jan LY. 1993. Primary structure and functional expression of a mouse inward rectifier potassium channel. *Nature* 362:127–33

48. Lopatin AN, Makhina EN, Nichols CG. 1994. Potassium channel block by cytoplasmic polyamines as the mechanism of intrinsic rectification. *Nature* 372:366–69

49. Ficker E, Taglialatela M, Wible BA, Henley CM, Brown AM. 1994. Spermine and spermidine as gating molecules for inward rectifier K^+ channels. *Science* 266:1068–72

50. Fakler B, Brandle U, Glowatzki E, Weidemann S, Zenner HP, Ruppersberg JP. 1995. Strong voltage-dependent inward rectification of inward rectifier K^+ channels is caused by intracellular spermine. *Cell* 80:149–54

51. Lopatin AN, Nichols CG. 1996. [K^+] dependence of polyamine-induced rectification in inward rectifier potassium channels (IRK1, Kir2.1). *J. Gen. Physiol.* 108:105–13

52. Bianchi L, Roy ML, Taglialatela M, Lundgren DW, Brown AM, Ficker E. 1996. Regulation by spermine of native inward rectifier K^+ channels in RBL-1 cells. *J. Biol. Chem.* 271:6114–21

53. Shyng SL, Sha Q, Ferrigni T, Lopatin AN, Nichols CG. 1996. Depletion of intracellular polyamines relieves inward rectification of potassium channels. *Proc. Natl. Acad. Sci. USA* 93:12014–19

54. Lopatin AN, Shantz LM, Mackintosh CA, Nichols CG. 2000. Modulation of potassium channels in the hearts of transgenic and mutant mice with altered polyamine biosynthesis. *J. Mol. Cell. Cardiol.* 32:2007–24

55. Aleksandrov A, Velimirovic B, Clapham DE. 1996. Inward rectification of the IRK1 K^+ channel reconstituted in lipid bilayers. *Biophys. J.* 70:2680–87

56. Shieh RC, John SA, Lee J-K, Weiss JN. 1996. Inward rectification of the IRK1 channel expressed in *Xenopus* oocytes: effects of intracellular pH reveal an intrinsic gating mechanism. *J. Physiol.* 494:363–76

57. Lee J-K, Scott JA, Weiss JN. 1999. Novel gating mechanism of polyamine block in the strong inward rectifier K^+ channel Kir2.1. *J. Gen. Physiol.* 113:555–65

58. Guo D, Lu Z. 2000. Pore block versus intrinsic gating in the mechanism of

inward rectification in the strongly rectifying IRK1 channel. *J. Gen. Physiol.* 116:561–68

59. Guo D, Lu Z. 2002. IRK1 inward rectifier K$^+$ channels exhibit no intrinsic rectification. *J. Gen. Physiol.* 120:539–51

60. Lopatin AN, Makhina EN, Nichols CG. 1995. The mechanism of inward rectification of potassium channels: "long-pore plugging" by cytoplasmic polyamines. *J. Gen. Physiol.* 106:923–55

61. Yang J, Jan YN, Jan LY. 1995. Control of rectification and permeation by residues in two distinct domains in an inward rectifier K$^+$ channel. *Neuron* 14:1047–54

62. Guo D, Ramu Y, Klem AM, Lu Z. 2003. Mechanism of rectification in inward-rectifier K$^+$ channels. *J. Gen. Physiol.* 121:261–75

63. Xie L-H, John SA, Weiss JN. 2002. Spermine block of the strong inward rectifier potassium channel Kir2.1: dual roles of surface screening and pore block. *J. Gen. Physiol.* 120:53–66

64. Guo D, Lu Z. 2000. Mechanism of IRK1 channel block by intracellular polyamines. *J. Gen. Physiol.* 115:799–813

65. Lu Z, Ding L. 1999. Blockade of a retinal cGMP-gated channel by polyamines. *J. Gen. Physiol.* 113:35–43

66. Guo D, Lu Z. 2000. Mechanism of cGMP-gated channel block by intracellular polyamines. *J. Gen. Physiol.* 115:783–97

67. Bahring R, Bowie D, Benveniste M, Mayer ML. 1997. Permeation and block of rat GluR6 glutamate receptor channels by internal and external polyamines. *J. Physiol.* 502:575–89

68. Guo D, Lu Z. 2003. Interaction mechanisms between polyamines and IRK1 inward rectifier K$^+$ channels. *J. Gen. Physiol.* In press

69. MacKinnon R. 1991. Determination of the subunit stoichiometry of a voltage-activated potassium channel. *Nature* 350:232–35

70. Yang J, Jan YN, Jan LY. 1995. Determination of the subunit stoichiometry of an inwardly rectifying potassium channel. *Neuron* 15:1441–47

71. MacKinnon R, Miller C. 1989. Mutant potassium channels with altered binding of charybdotoxin, a pore-blocking peptide inhibitor. *Science* 245:1382–85

72. MacKinnon R, Yellen G. 1990. Mutations affecting TEA blockade and ion permeation in voltage-activated K$^+$ channels. *Science* 250:276–79

73. MacKinnon R, Heginbotham L, Abramson T. 1990. Mapping the receptor site for charybdotoxin, a pore-blocking potassium channel inhibitor. *Neuron* 5:767–71

74. Yellen G, Jurman ME, Abramson T, MacKinnon R. 1991. Mutations affecting internal TEA blockade identify the probable pore-forming region of a K$^+$ channel. *Science* 251:939–42

75. Hartmann HA, Kirsch GE, Drewe JA, Taglialatela M, Joho, RH, Brown AM. 1991. Exchange of conduction pathways between two related K$^+$ channels. *Science* 251:942–44

76. Choi KL, Mossman C, Aubé J, Yellen G. 1993. The internal quaternary ammonium receptor site of Shaker potassium channels. *Neuron* 10:533–41

77. Kirsch GE, Shieh CC, Drewe JA, Vener DF, Brown AM. 1993. Segmental exchanges define 4-aminopyridine binding and the inner mouth of K$^+$ pores. *Neuron* 11:503–12

78. Lopez GA, Jan YN, Jan LY. 1994. Evidence that the S6 segment of the Shaker voltage-gated K$^+$channel comprises part of the pore. *Nature* 367:179–82

79. Heginbotham L, Lu Z, Abramson T, MacKinnon R. 1994. Mutations in the K$^+$ channel signature sequence. *Biophys. J.* 66:1061–67

80. Doyle DA, Morais CJ, Pfuetzner RA, Kuo A, Gulbis JM, et al. 1998. The structure of the potassium channel: molecular basis of K$^+$ conduction and selectivity. *Science* 280:69–77

81. Stanfield PR, Davies NW, Shelton PA, Sutcliffe MJ, Khan IA, et al. 1994. A single aspartate residue is involved in both intrinsic gating and blockage by Mg^{2+} of the inward rectifier, IRK1. *J. Physiol.* 478: 1–6

82. Lu Z, MacKinnon R. 1994. Electrostatic tuning of Mg^{2+} affinity in an inward-rectifier K^+ channel. *Nature* 371:243–46

83. Wible BA, Taglialatela M, Ficker E, Brown AM. 1994. Gating of inwardly rectifying K^+ channels localized to a single negatively charged residue. *Nature* 371:246–49

84. Lu Z, MacKinnon R. 1995. Probing a potassium channel pore with an engineered protonatable site. *Biochemistry* 34:13133–38

85. Taglialatela M, Wible BA, Caporaso R, Brown AM. 1994. Specification of pore properties by the carboxyl terminus of inwardly rectifying K^+ channels. *Science* 264:844–47

86. Taglialatela M, Ficker E, Wible BA, Brown AM. 1995. C-terminus determinants for Mg^{2+} and polyamine block of the inward rectifier K^+ channel IRK1. *EMBO J.* 14:5532–41

87. Kubo Y, Murata Y. 2001. Control of rectification and permeation by two distinct sites after the second transmembrane region in Kir2.1 K^+ channel. *J. Physiol.* 531:645–60

88. Lu T, Nguyen B, Zhang X, Yang J. 1999. Architecture of a K^+ channel inner pore revealed by stoichiometric covalent modification. *Neuron* 22:571–80

89. Loussouarn G, Makhina EN, Rose T, Nichols CG. 2000. Structure and dynamics of the pore of inwardly rectifying K_{ATP} channels. *J. Biol. Chem.* 275:1137–44

90. Lu T, Zhu YG, Yang J. 1999. Cytoplasmic amino and carboxyl domains form a wide intracellular vestibule in an inwardly rectifying potassium channel. *Proc. Natl. Acad. Sci. USA* 96:9926–31

91. Guo D, Lu Z. 2001. Kinetics of inward-rectifier K^+ channel block by quaternary alkylammonium ions: dimension and properties of the inner pore. *J. Gen. Physiol.* 117:395–405

92. Li-Smerin Y, Swartz KJ. 1998. Gating modifier toxins reveal a conserved structural motif in voltage-gated Ca^{2+} and K^+ channels. *Proc. Natl. Acad. Sci. USA* 95:8585–89

93. Schrempf H, Schmidt O, Kümmerlen R, Hinnah S, Müller D, et al. 1995. A prokaryotic potassium ion channel with two predicted transmembrane segments from *Streptomyces lividans*. *EMBO J.* 14:5170–78

94. Lu Z, Klem AM, Ramu Y. 2001. Ion conduction pore is conserved among potassium channels. *Nature* 413:809–13

95. Lu Z, Klem AM, Ramu Y. 2002. Coupling between voltage sensors and activation gate in voltage-gated K^+ channels. *J. Gen. Physiol.* 120:663–76

96. Jiang Y, Lee A, Chen J, Ruta V, Cadene M, et al. 2003. X-ray structure of a voltage-dependent K^+ channel. *Nature* 423:33–41

97. Morais-Cabral JH, Zhou Y, MacKinnon R. 2001. Energetic optimization of ion conduction rate by the K^+ selectivity filter. *Nature* 414:37–42

98. Zhou Y, Morais-Cabral JH, MacKinnon R. 2001. Chemistry of ion coordination and hydration revealed by a K^+ channel-Fab complex at 2.0 Å resolution. *Nature* 414:43–48

99. Nishida M, MacKinnon R. 2002. Structural basis of inward rectification: cytoplasmic pore of the G protein-gated inward rectifier GIRK1 at 1.8 Å resolution. *Cell* 111:957–65

100. Kuo A, Gulbis JM, Antcliff JF, Rahman T, Lowe ED, et al. 2003. Crystal structure of the potassium channel KirBac1.1 in the closed state. *Science* 300:1922–26

101. Oliver D, Hahn H, Antz C, Ruppersberg JP, Fakler B. 1998. Interaction of permeant and blocking ions in cloned inward-rectifier K^+ channels. *Biophys. J.* 74:2318–26

102. Spassova M, Lu Z. 1998. Coupled ion movement underlies rectification in an inward-rectifier K$^+$ channel. *J. Gen. Physiol.* 112:211–21

103. Spassova M, Lu Z. 1999. Tuning the voltage dependence of tetraethylammonium block with permeant ions in an inward-rectifier K$^+$ channel. *J. Gen. Physiol.* 114:415–26

104. Fujiwara Y, Kubo Y. 2002. Ser165 in the second transmembrane region of the Kir2.1 channel determines its susceptibility to blockade by intracellular Mg^{2+}. *J. Gen. Physiol.* 120:677–93

105. Lu Z, MacKinnon R. 1994. A conductance maximum observed in an inward-rectifier potassium channel. *J. Gen. Physiol.* 104:477–86

106. Pearson WL, Nichols CG. 1998. Block of the Kir2.1 channel pore by alkylamine analogues of endogenous polyamines. *J. Gen. Physiol.* 112:351–63

107. Miller C. 1982. Bis-quaternary ammonium blockers as structural probes of the sarcoplasmic reticulum K$^+$ channel. *J. Gen. Physiol.* 79:869–91

108. Hidalgo P, MacKinnon R. 1995. Revealing the architecture of a K$^+$ channel pore through mutant cycles with a peptide inhibitor. *Science* 268:307–10

109. Oishi K, Omori K, Ohyama H, Shingu K, Matsuda H. 1998. Neutralization of aspartate residues in the murine inwardly rectifying K$^+$channel IRK1 affects the substate behaviour in Mg^{2+} block. *J. Physiol.* 510.3:675–83

110. Hodgkin AL, Keynes RD. 1955. The potassium permeability of a giant nerve fibre. *J. Physiol.* 128:61–88

111. Stampe P, Arreola J, Pérez-Cornejo P, Begenisich T. 1998. Nonindependent K$^+$ movement through the pore in IRK1 potassium channels. *J. Gen. Physiol.* 112:475–84

112. Begenisich T, De Weer P. 1980. Potassium flux ratio in voltage-clamped squid giant axons. *J. Gen. Physiol.* 76:83–98

Annu. Rev. Physiol. 2004. 66:131–59
doi: 10.1146/annurev.physiol.66.041002.142720
First published online as a Review in Advance on October 15, 2003

METABOLIC REGULATION OF POTASSIUM CHANNELS

Xiang Dong Tang,[1] Lindsey Ciali Santarelli,[1]
Stefan H. Heinemann,[2] and Toshinori Hoshi[1]

[1]Department of Physiology, University of Pennsylvania, Philadelphia, Pennsylvania 19104
[2]Research Unit Molecular and Cellular Biophysics, Medical Faculty of the Friedrich
Schiller University Jena, Drackendorfer Str. 1, D-07747, Jena, Germany;
email: tang3@mail.med.upenn.edu; ciali@mail.med.upenn.edu; ite@uni-jena.de;
hoshi@hoshi.org

Key Words metabolism, oxygen-sensitive K^+ channels, BK_{Ca} channels

■ **Abstract** Potassium (K^+) channels exist in all three domains of organisms: eubacteria, archaebacteria, and eukaryotes. In higher animals, these membrane proteins participate in a multitude of critical physiological processes, including food and fluid intake, locomotion, stress response, and cognitive functions. Metabolic regulatory factors such as O_2, CO_2/pH, redox equivalents, glucose/ATP/ADP, hormones, eicosanoids, cell volume, and electrolytes regulate a diverse group of K^+ channels to maintain homeostasis.

INTRODUCTION

Animals require matter and energy to maintain a relatively stable internal environment, as formulated by Cannon in the early twentieth century (1). Cell volume, cellular ATP/ADP ratio, and redox state are some of the well-known vital cellular homeostatic parameters that need to be carefully regulated. Cells must be able to effectively monitor these parameters and respond to the changes for survival. Homeostasis critically depends on oxygen-dependent ATP synthesis in mitochondria.

Many enzymatic pathways are involved in cellular metabolism. In theory, fluctuations in any of the metabolic substrates, intermediates, and end-products (herein referred to as metabolites) could be used to sense the metabolic state to maintain homeostasis. These metabolically relevant factors include metabolic substrates (glucose, proteins/amino acids, lipid/cholesterol), metabolic intermediates (hormones, redox equivalents, ATP/ADP, eicosanoids), metabolic end-products (urea, CO_2, water, NO, CO), and environmental factors that affect metabolism (water, O_2, pH, Na^+/K^+/Ca^{2+}/Mg^{2+}, cell volume and osmolarity, temperature, vitamins). The following issues come into play placing some constraint as to which metabolic

0066-4278/04/0315-0131$14.00

parameters could be utilized as signals. First, physiological fluctuations in the metabolite concentrations need to be large enough for the cellular machinery to detect. Second, it is more advantageous to monitor rate-limiting or high-priority metabolites. Third, the kinetics or time course of the metabolite fluctuations must be considered. Changes in the relative concentrations of some metabolites may occur very rapidly, whereas the concentrations in others may change much more slowly. A survey of the studies thus far suggests that a variety of metabolites are used to monitor the metabolic state. Once a significant change in any of the relevant metabolic parameters is detected, this information is conveyed to the effector sites to initiate appropriate adaptive responses, which may involve changes in gene expression.

Ion channels mediate electrical and chemical signaling events and are involved in both rapid nongenomic cellular signaling and often slower genomic signaling. For example, Ca^{2+} influx through Ca^{2+} permeant channels in a typical animal cell not only induces depolarization but also initiates a series of Ca^{2+}-dependent biochemical signaling cascades (1a). Intracellular Ca^{2+} signals may trigger rapid events, such as neurotransmitter or hormone release and muscle contraction, as well as changes in gene expression. Thus ion channel proteins are well suited to mediate the information flow between changes in the high-priority metabolic parameters and the effectors involved in maintaining homeostasis. This link between the channels and metabolism can be bidirectional; metabolism alters ion channels, and ion channels, in turn, can affect metabolism. Ion channel proteins themselves could contain the molecular elements required to sense changes in metabolites. Alternatively, the coupling may be indirect, and the channels may be regulated via standard second messengers such as Ca^{2+}. As illustrated below, both of these mechanisms are observed.

Potassium channels regulate numerous metabolically important physiological functions by mediating K^+ secretion and/or by regulating the action potential pattern and frequency, which in turn may affect intracellular $[Ca^{2+}]$. Fluid secretion in lacrimal tracts (2), respiratory tracts (3), digestive tracts (4), and many exocrine glands (5) and in electrolyte absorption in intestines (6) and reabsorption in kidney (7) are some of the examples in the respiratory and digestive systems. In the vascular system, K^+ channels participate in blood pressure regulation (8) and penile erection (9). Neuronal K^+ channels regulate neurotransmitter release (10) and action potential firing (11, 12) and contribute to virtually every regulatory function of the nervous system, including regulation of food, water, and salt intake (13); stress response (13a,b); and higher cognitive functions (14, 15).

POTASSIUM CHANNEL DIVERSITY

Potassium channels exist in all three domains of organisms including bacteria (16); archaea (16); and eukarya such as fungi (17, 18), protozoa (19), plants (20), and animals (21). On the basis of predicted membrane topology, K^+ channels are conventionally classified into three major groups: 6 transmembrane (6TM) K$^+$ channels, 2TM-2TM/leak K$^+$ channels, and 2TM/Kir inward rectifiers. Each group is further divided into multiple families based on sequence similarity. The 6TM group includes the EAG, KQT, SK$_{Ca}$, Slo (BK$_{Ca}$, 7TM), and voltage-gated K$^+$ channel families (Kv1–6, Kv8–9). The structural analysis of an archaebacterium voltage-gated channel, KvAP, suggests that the structure of 6TM channels deviates markedly from a complex made of 6TM helices (22). The 2TM-2TM group contains TWIK, TREK, TASK, TRAAK, and THIK. The 2TM/Kir group consists of Kir1–7. (See Supplemental Materials at http://www.annualreviews.org/supmat/ supmat.asp. K$^+$ Channels Regulated by Other Metabolic Parameters.)

POTASSIUM CHANNELS REGULATED BY METABOLIC PARAMETERS

All animals require oxygen for survival. Oxygen acts as the terminal electron acceptor in the mitochondrial electron-transport chain. The mitochondrial electron-transport system is leaky, and some electrons do escape, often forming free radicals capable of oxidizing cellular constituents. The energy derived from ATP synthesized during glycolysis and in the mitochondrial oxidative phosphorylation is then used for a variety of cellular functions, including maintenance of ion concentration gradients. Any of the metabolites involved in this cellular metabolic scheme can function as metabolic parameters and regulate K$^+$ channel function (Table 1). Here we focus on K$^+$ channel regulation by oxygen, free radical–derived reactive species/reducing equivalents, pCO$_2$/pH, and ATP/ADP.

Oxygen-Sensitive K$^+$ Channels

Given the essential role of oxygen for life, O$_2$ sensing may be considered as the most global and perhaps the most ancient form of metabolic sensing. All cells have some ability to sense O$_2$; however, higher-order animals have evolved to possess cells specialized for O$_2$ sensing: carotid body chemoreceptors, pulmonary neuroepithelial bodies, adrenal chromaffin cells, pulmonary vascular smooth muscle cells, and neurons in selected brain regions. The results from these tissues collectively indicate that acute hypoxia generally inhibits K$^+$ channels, leading to rapid membrane depolarization, greater Ca^{2+} influx, neurotransmitter/hormone release, or vasoconstriction (62–64, 64a). S-nitrosylation-mediated hypoxia sensing directly by central neurons was also reported (65).

OXYGEN-SENSITIVE K$^+$ CHANNELS IN DIFFERENT TISSUES   The carotid body regulates respiration in response to changes in oxygen tension (66). Hypoxia (<50–60 mmHg) causes carotid body glomus cells to depolarize, increase Ca^{2+} influx, and release the principal neurotransmitter dopamine, leading to sympathetic activation (64). Multiple hypoxia-sensitive K$^+$ channels are likely to work in concert to sense

TABLE 1 Homeostatic parameters, metabolic factors, and K^+ channels

Homeostatic parameters	Input	Output/clearance	Metabolic regulators	K^+ channels involved	Molecular targets
Blood O_2 (40 mmHg, tissue)	Inspiration	Oxidized to H_2O	pO_2 Hypoxia: $pO_2 < 50$ mmHg	Rat Kv (23) Rat BK_{Ca} (24, 25) hTASK3 (26).	Cys? Cys? Cys?
Redox state					
NADH	Glycolysis	NAD^+	$NADH/NAD^+$	Rat BK_{Ca} (27, 28)	Cys?
NADPH	PPP	$NADP^+$	$NADPH/NADP^+$	Rat BK_{Ca} (27, 28)	Cys?
GSH ($\sim$5–10 mM)	Synthesis	GSSG	GSH/GSSG (10/1)	Rat BK_{Ca} (27, 28)	Cys?
$O_2^{\cdot-}$	Multiple sources	H_2O_2	$O_2^{\cdot-}$	BK_{Ca} (29)	Cys?
H_2O_2	SOD	H_2O	H_2O_2	Rat BK_{Ca} (30)	Cys?
$OH^{\bullet-}$	Fenton reaction	H_2O	$OH^{\bullet-}$	Kv (31)	Cys?
$ONOO^-$	From NO, $O_2^{\bullet-}$		$ONOO^-$	K_{ATP} (29)	Cys?
HOCl	H_2O_2/Cl^-		HOCl	?	?
Chloramines	From HOCl		Chloramines	Shaker (32) hSlo1 (33)	Met Met
Blood CO_2 (46 mmHg, tissue)	Metabolically generated	Expiration Renal excretion	pCO_2/pH Alkalosis: $P_a < 38$ mmHg, Acidosis: $P_a > 44$ mmHg	Kir6.2 (34) ROMK1 (35) TASK1 (36)	Thr or Cys or His Lys or Val or Gln His or Lys
Blood pH/proton (pH 7.3–7.5)	Food intake Renal reabsorption Metabolically generated	Sweat Intestine Kidney Lung (CO_2)			
Blood glucose (75–95 mg/dl)	Food intake Glycogenesis Renal reabsorption	Catabolism (CO_2 + ATP) Glycosylation Fat store	Glucose Hypoglycemia (<50–70 mg/dl) ATP/ADP	K_{ATP} (37)	C terminus

		Consumption			
Cellular ATP	Glycolysis				
	TCA				
	Oxidative phosphorylation				
Blood amino acids/proteins	Food intake	Urea, NH_3	Heme	hSlo1 (38)	Heme-binding motif
	Renal reabsorption	Glycogenesis	CO		His
	Amino acid synthesis	Protein synthesis	Direct effect	Rat BK_{Ca} (39)	Ser/Thr
		Heme/CO	Via PKG	Rat BK_{Ca} (40)	
		NO	NO (1–10 nM)		?
		Hormones	Direct effect	Rat BK_{Ca} (41)	Ser1134
		Fat store	Via PKG	bSlo1(42)	
			Nitrosylation	?	
			Hormones		Ser922
			Via PKA	bSlo1 (42)	Ser1151/Ser1154
			Via PKC	bSlo1 (42)	Tyr124
			Via RTK	Kv2.1 (43)	
Hormones	Hypoxia	Catabolism			
	Cold/heat	Hormone reuptake			
	Other stresses				
Lipids (AA: 6–49 μM in human plasma)	Food intake	Catabolism	PIP_2	Kir (37)	Arg
	Renal reabsorption	PIP_2	PUFA/AA	Rabbit BK_{Ca} (44)	?
	Fatty acid synthesis	Eicosanoids	EETs	BK_{Ca} (45)	?
		Fat store	20-HETEs	BK_{Ca} (46)	?
		Cell membranes	PGI_2	BK_{Ca} (47)	?
			LTD_4	TASK2 (48)	?
Blood Na^+/osmolarity (280–294 mOsm/kg)	Food intake	Sweat	Hyponatremia	rSlo2 (49)	Na^+ bowl
	Renal reabsorption	Intestines	$[Na^+]_i$		
		Kidneys			
		Lungs (CO_2)			
Blood K^+	Food intake	Sweat	$[K^+]_o$	Kir2.1 (50)	?

(Continued)

TABLE 1 (*Continued*)

Homeostatic parameters	Input	Output/clearance	Metabolic regulators	K$^+$ channels involved	Molecular targets
	Renal reabsorption	Intestines Kidneys	Hyperkalemia		
Blood Ca^{2+}/Mg^{2+}	Food intake Renal, intestinal reabsorption	Bone building Sweat Intestines Kidneys	[Ca^{2+}]$_i$ Direct Via CaM [Ca^{2+}]$_o$ [Mg^{2+}]$_i$	mSlo1 (51) SK$_{Ca}$ (51a) hERG (52) mSlo1 (53) Kir2.1 (54)	Ca^{2+} bowl Asp362/Asp367 Glu518/Glu519 Glu374/Glu399/Asn 397 Asp171/Glu224
Blood volume	Water intake Na$^+$ intake	Sweat Intestines Kidneys Lungs	Stretch Direct Via AA Via Ca^{2+}	TRAAK (55) K$_{2P}$ (56, 57) BK$_{Ca}$ (58)	? ? ?
Blood pressure (100–140/60–95 mmHg)	Water intake Na$^+$ intake Cardiac pumping Vascular resistance	Sweat Intestines Kidneys Lungs Vasodilation			
Cell volume/integrity	Hypotonic shock Solute influx Stretch	RVI RVD	Cell volume Direct Via LTD$_4$ Via Ca^{2+} Via CaM	IK$_{Ca}$ (59) TASK2 (48) BK$_{Ca}$ (60) IK$_{Ca}$ (4, 61)	? ? ? ?

Abbreviations: PPP, pentose phosphate pathway; SOD, superoxide dismutase; TCA, tricarboxylic acid cycle; PTK, protein tyrosine kinases; PIP$_2$, phosphatidylinositol-4,5-biphosphate; PUFA, polyunsaturated fatty acids; AA. arachidonic acids; EET, epoxyeicosatrienoic acids; 20-HETE, hydroxyeicosatrienoic acids; LTD4, leukotriene D4; CaM, calmodulin; RVI, regulatory volume increase; RVD, regulatory volume decrease.

hypoxia and contribute to the acute hypoxia-induced depolarization. K_{Ca} (24, 25, 67, 68), Kv (69, 70), K_{2P} (71), and hERG (72) in the carotid bodies of different species are known to be acutely inhibited by hypoxia. Inhibition of these K$^+$ channels depolarizes the membrane potential, leading to an increase in the intracellular Ca^{2+}.

In pulmonary neuroepithelial bodies, both K_{Ca} and K_{2P} have been reported to be sensitive to hypoxia (73) with the latter playing a more dominant role (74). hTASK3 is expressed in neuroepithelial bodies and inhibited by hypoxia (26).

Hypoxia constricts pulmonary vasculature by inhibiting multiple K$^+$ channel types (75). Transcripts of Kv1.2, Kv1.5, Kv1.7, Kv2.1, Kv4.1, and Kv9.2, all of which were speculated to be oxygen sensitive, were detected in rat pulmonary artery (PA) cells (23). In addition, a low-threshold, depolarization-activated non-inactivating K$^+$ current in rabbit PA cells is also inhibited by hypoxia (76). It is not known which K$^+$ channel is regulated by hypoxia in human PA cells.

Chromaffin cells in the adrenal medulla increase release of catecholamines and other autocrine and paracrine molecules such as adenosine and ATP in response to hypoxia as part of the organism's stress response. I_A, I_K, K_{Ca} (77), and K_{ATP} (78) in chromaffin cells are inhibited by hypoxia. In clonal pheochromocytoma (PC12) cells derived from rat adrenal chromaffin cells, Kv1.2, Kv1.3, Kv2.1, Kv3.1, and Kv3.2 are expressed. However, only Kv1.2 is sensitive to hypoxia when expressed in *Xenopus* oocytes (79, 80).

MOLECULAR MECHANISMS OF ACUTE O$_2$ SENSITIVITY   Although the acute effects of hypoxia on several channel types are now well established, the molecular identity of the primary O$_2$ sensor to trigger the channel inhibition remains unresolved. How the O$_2$ sensing is coupled to changes in K$^+$ channel function is also unclear. The available evidence suggests that different sensing mechanisms may exist in different cell types. The channel proteins themselves or their closely associated auxiliary proteins may contain O$_2$-binding sites; binding and unbinding of oxygen could regulate the channel activity. Alternatively, O$_2$-binding sites may reside in distinct molecular entities that indirectly regulate the channels via second messengers. Yet another possibility is that the channels are not directly altered by O$_2$ but instead by those physiological changes that accompany hypoxia such as changes in redox equivalents, pH, pCO$_2$, and/or Ca^{2+} (see below).

Many studies have focused on the idea that the O$_2$ sensitivity of K$^+$ channels is mediated by free radical–derived reactive species. Some studies suggest that the membrane NADPH oxidase, a multi-subunit enzyme complex that produces the superoxide anion ($O_2^{\bullet-}$), is a critical component in hypoxia sensitivity (81, 82). The hypoxia sensitivity of the whole-cell K$^+$ current was absent in pulmonary neuroepithelial body cells of mice lacking functional gp91^{phox}, a catalytic subunit of the NADPH oxidase (73). However, in other tissues, including carotid body glomus cells and chromaffin cells, disruption of the gp91^{phox} subunit did not markedly alter their hypoxia responses (83–87). This finding is generally interpreted to argue against the NADPH oxidase involvement. Multiple gp91^{phox} homologues, NOX1, 3, 4, and 5, have been recently identified in various nonphagocytic cells (88, 89), and the absence of gp91^{phox} (NOX2) alone may not totally obliterate the hypoxia sensitivity. Other studies using mitochondrial electron-transport blockers appear to favor the idea that mitochondria may be a key O_2 sensor that is functionally linked with K^+ channels (90, 91). Hypoxia, at least transiently, increases the production of free radical–derived reactive species, probably from the mitochondrial semiubiquinone cycle (92). These reactive species produced by hypoxia may act as a second messenger and alter the K^+ channel function (this is discussed in more detail below), although global changes in the cellular reduced/oxidized glutathione (GSH/GSSG) ratio are unlikely to be involved (93).

If the plasma membrane K^+ channels are regulated by hypoxia using separate cytoplasmic O_2 sensors such as mitochondria, the hypoxia sensitivity should be diminished when the channel activity is assayed in the excised patch-clamp configurations. Heterologously expressed Slo1 BK and native BK_{Ca} channels in carotid body glomus cells are inhibited by hypoxia even in the excised patch configuration at intermediate Ca^{2+} concentrations (94, 95). Kv4.2/Kvβ1.2 channel complexes expressed in human embryonic kidney (HEK) cells also show similar hypoxia sensitivity in isolated cell-free membrane patches (96). If one assumes that excised membrane patches are largely devoid of cytoplasmic factors, this finding suggests that, at least for BK_{Ca} and Kv4.2/Kvβ1.2 channels, the O_2 sensor resides within or near the membrane in the vicinity of the channels. This assumption regarding the membrane patches, however, has been challenged (97).

The most biologically prevalent gas-sensing mechanism involves heme proteins; axial ligation of heme, iron protoporphyrin IX, is modulated by O_2, CO, and NO (98). This gas-sensing mechanism is utilized in numerous proteins including hemoglobins and soluble guanylate cyclases. If a K^+ channel contains a heme moiety, this arrangement may allow the protein to be oxygen sensitive. A recent study indicates that Slo BK channels contain a functional heme-binding domain and that the channel gating is profoundly altered by heme at a very low concentration (38). This finding keeps open the possibility that some K^+ channels are directly regulated by oxygen under certain conditions.

GENOMIC RESPONSES TO HYPOXIA   In addition to the acute depolarization mediated in part by inhibition of K^+ channels, many cells show slower genomic responses to hypoxia involving the transcription factor HIF-α (hypoxia-inducible factor) (99, 100). HIF-α is normally hydroxylated at conserved proline residues in an O_2-dependent manner and subject to rapid degradation by prolyl hydroxylases (100a). Hypoxia removes the degradation influence of the hydroxylases. Proline hydroxylation and/or the subsequent transcriptional activity of the HIFα-β complex likely regulate K^+ channels, and this issue warrants a comprehensive study. Modulation of hERG and Kv1.3 K^+ channels in response to long-term hypoxia (101, 102) may represent such phenomenon.

Redox-Regulated K+ Channels

Electrons inevitably leak out of the mitochondrial electron-transport chain, initiating a series of downstream redox reactions. Inefficient mitochondrial function leads to greater electron leak, potentially promoting oxidative damage of cellular macromolecules. In addition, xanthine oxidase, NADPH oxidase, nitric oxide synthase in the absence of either L-arginine or tetrahydrobiopterin, cytochrome P450, cyclooxygenase, and lipoxygenase also generate reactive species (102a,b). Thus the overall cellular redox status may be considered a parameter indicative of the cellular metabolic state. Once considered as mostly unwanted but unavoidable metabolic by-products (except in immune system cells), free radical–derived reactive species are now recognized to have cell signaling roles and participate in many cellular functions including O_2 sensing (see above), cell proliferation, and programmed cell death (103, 104). Furthermore, aberrant production of these reactive species is associated with accelerated aging and many diseases including hypercholesterolemia, arthrosclerosis, hypertension, diabetes mellitus, erectile dysfunction, and numerous neurodegenerative diseases (104a). Redox regulation of K+ channels may be used as a mechanism to sense and respond to changes in the cellular metabolism and its efficiency. Numerous K+ channels are in fact redox-regulated (105), at least in experimental settings. Application of oxidizing and reducing agents produces functional changes, and the effects are attributed to oxidation of the channel protein itself. Typically, the gating characteristics are altered while the permeation properties remain largely unchanged. One difficult question that is yet to be addressed adequately in many of these investigations is the nature of the physiological trigger of the oxidative changes observed. If it is a direct free-radical–mediated reaction, what oxidant is responsible? Alternatively, does an enzymatic mechanism that selectively oxidizes specific amino acid residues exist?

BK$_{Ca}$ CHANNELS AS A LINKER OF CELLULAR EXCITABILITY AND METABOLISM   BK$_{Ca}$ channels are activated by voltage and intracellular Ca^{2+} and Mg^{2+} in an allosteric manner (106, 107). Because of the intracellular Ca^{2+}-dependent activation mechanism, these K+ channels are often considered to link cellular excitability and metabolism (108). In addition, many studies have shown that BK$_{Ca}$ channels are also hypoxia-sensitive (24, 25, 67, 68) and redox-regulated; as such these channels may provide a multi-link between the cellular excitability and metabolic state.

Oxidation sensitivities of BK$_{Ca}$ channels have been examined using exogenously applied oxidizing and reducing agents. Unfortunately, the results, especially in native BK$_{Ca}$ channels, do not show any consistent effect. In some tissues, oxidizing reagents increased the channel activity (109–113), whereas in others they markedly inhibited the overall channel activity (30, 39, 114–116). The effects of oxidizing agents, whether stimulatory or inhibitory, were often but not always reversed by treatment with reducing agents such as NADH, GSH, and dithiothreitol (DTT). For example, in rabbit PA cells, reducing agents such as

NADH, GSH, and DTT inhibited BK_{Ca} channels (27, 28), but oxidants such as 5,5′-dithiobis-2-nitrobenzoic acid (DTNB), GSSG, and NAD^+ activated the channels (28). In guinea-pig taenia caeci (113), rat hippocampal neurons (116), horse trachea smooth muscle (117), and rat skeletal muscle (30) BK_{Ca} channels were inhibited by various oxidants such as DTNB, 2-(aminoethyl)methanethiosulfonate (MTSEA), NEM, and H_2O_2. GSH and DTT generally reversed the oxidant-induced inhibition, suggesting the involvement of cysteine oxidation.

Many factors can contribute to the variability in the oxidant sensitivity of BK_{Ca}. Depending on the experimental design employed, the redox manipulations may be confounded with changes in other relevant variables, such as Ca^{2+} and pH, both of which are known regulators of BK_{Ca}. Contaminating divalent and trivalent ions may also be a critical factor. Production of the powerful oxidant hydroxyl radical by the Fenton reaction involves Fe^{2+}/Fe^{3+}-dependent steps, and the degree of metal ion chelation in the recording solutions may prove to be an important confounding variable. The presence of Fe^{2+}/Fe^{3+} does facilitate oxidation of ion channels (118). Native BK_{Ca} channels in different tissues are also likely to contain differentially spliced, pore-forming α subunits and may contain auxiliary β subunits (119). Those BK_{Ca} channels containing the cysteine-rich stress axis–regulated exon (STREX) segment in the tail domain of the Slo1 subunit are more redox-sensitive (120), and the auxiliary subunit $BK\beta3$ also dramatically enhances the redox sensitivity of mSlo1 (121). Furthermore, each BK_{Ca} complex may contain multiple oxidation targets that determine the overall redox sensitivity. It is possible that different experimental reagents could attack distinct target residues, in a state-dependent manner.

Although all amino acids can be oxidized, sulfur-containing cysteine and methionine residues are reversibly and most readily oxidized by biologically relevant oxidants. Oxidized cysteine residues are reduced using reducing agents such as DTT. Those treatments that promote cysteine oxidation and modification, such as DTNB, MTSEA, pCMB, and NEM, generally inhibit the overall activity of heterologously expressed Slo1 BK_{Ca} channels even in the excised patch-clamp configuration (30, 33, 122). Cysteine oxidation decreases the overall channel activity by decreasing the number of channels available to open by driving the channels to a very long-lived closed state (30). The presence of multiple target cysteine residues is suggested by the results of Soto et al. (30), which showed two distinct inhibitory effects of H_2O_2 on hSlo1 based on the recovery time course with the reducing agent DTT. Which cysteine residue in Slo1 underlies the oxidant sensitivity is not currently known.

Additional complexity in the oxidant sensitivity of Slo1 is conferred by oxidation of methionine residues. Methionine residues are oxidized to methionine-S-sulfoxide and methionine-R-sulfoxide by biological oxidants, such as hydroxyl radicals and chloramines, and enzymatically reduced back to methionines in a stereo-specific manner by the enzyme protein methionine sulfoxide reductase (MSR) A and B (123). Reversible oxidation of methionine involving MSRA/B has been suggested to play a role both as a regulatory mechanism and also as an endogenous oxidant scavenger system (124). In hSlo1, methionine oxidation induced by the oxidizing agent chloramine-T increases the open probability by specifically altering Ca^{2+}-independent gating transitions (33). The pore-forming subunit Slo1 contains many cysteine and methionine residues, and differential oxidation of these residues in different experimental settings may, at least in part, account for the seemingly conflicting results reported (33). Considering the complexity of the BK_{Ca} channel gating and the presence of many cysteine and methionine residues, systematic investigations controlling all the relevant gating variables, such as voltage, Ca^{2+}, and Mg^{2+}, are required to completely elucidate the biophysical and molecular mechanisms.

Kv AND hERG CHANNELS AND OXIDATION   Oxidative regulation is also reported for many Kv channels both in native and heterologous systems (31, 125–127). N-type inactivation of Kv1.4 and Kv3.4 (Raw3) is reversibly redox-modulated by means of a cysteine in the distal part of the N-terminal inactivating ball domain (128). The auxiliary β subunits $Kv\beta1.1$ and $Kv\beta1.3$, which specifically coassemble with Kv1, also provide cysteine-containing N-terminal ball domains to equip Kv with redox-dependent inactivation (128–130).

Methionine oxidation also regulates Kv-type channels. In N-type inactivation of ShC/B channels, oxidation of a specific methionine residue in the N-terminal ball domain disrupts the ball-and-chain inactivation by increasing the polar nature of the distal N terminus (32, 131). Differential oxidation of this critical residue accounts for the macroscopic inactivation variability when the channels are expressed in oocytes (32, 131). P/C-type inactivation of ShB channels is also altered by methionine oxidation (132).

When expressed in *Xenopus* oocytes, Kv1.3, Kv1.4, Kv1.5, and Kv3.4 were all inhibited by photoactivation of rose bengal, which generates singlet oxygen (31). N-type inactivation of Kv1.4 and Kv3.4 channels was disrupted by tert-butyl hydroxyperoxide (31). Deactivation kinetics of Kv1.4 is also altered by oxidation (127).

The $Kv\beta$ subunit structural analysis suggests that these β subunits (also $Kv\beta2$, which does not harbor an N-terminal ball domain and often coassembles with Kv1.4) resemble oxidoreductase enzymes (133, 134). This structural similarity is tantalizingly suggestive of the notion that $Kv\beta$ subunits mediate coupling of the metabolic state and the cellular electrical activity. However, experimental evidence to support this idea is not yet available.

Thus functional properties of many Kv channels are clearly altered by oxidizing and reducing treatments. How these oxidative modifications contribute to cell function needs to be investigated. Physiological significance of oxidative modification of *h*ERG channels involved in the repolarization phase of cardiac action potentials is more readily apparent. High extracellular glucose increases production of reactive oxygen species and reduces *h*ERG-mediated currents (135). This finding may partially account for hyperglycemia-induced prolongation of the Q-T interval and ventricular arrhythmias (135).

K_{ATP} CHANNELS AND OXIDATION   Application of oxidizing agents leads to marked changes in electrophysiological properties of K_{ATP} in a variety of systems. Treatment with cysteine-oxidizing and -modifying reagents reversibly decreases the open probability of K_{ATP} channels in cardiac muscle (136), skeletal muscle (137), and pancreatic β cells (138, 139). The potential physiological relevance is suggested by the finding that topical application of xanthine/xanthine oxidase to generate superoxide anions blunted the pig pial artery dilation induced by the K_{ATP} opener cromakalim (139a). The redox regulation of K_{ATP} in pancreatic β cells may be particularly physiologically relevant considering their roles in glucose-induced insulin release (138, 139) and also in the cytoprotective preconditioning phenomenon (see below). Interestingly, the oxidant sensitivity is greater in K_{ATP} channels recorded from aged animals than from young animals (137). Whether the greater sensitivity is a result of aging or itself an aging accelerator is not known.

Potassium Channels Regulated by pCO_2/pH

Carbon dioxide (CO_2) generated by cellular metabolism is cleared through lungs by respiration controlled by the central chemoreceptors located in the medulla in the brainstem, along with the peripheral chemoreceptors in the aorta and carotid arteries. Carbonic anhydrases accelerate the conversion of cellular CO_2 to HCO_3^- and H^+. Thus intracellular pH (pH_i) may be a good indicator of the cellular metabolic state. Furthermore, cells need to initiate adaptive measures in response to intracellular acidosis following ischemia/hypoxia and trauma (140). Oxidative stress is also known to induce intracellular acidosis (141). Considering the importance of pH, it is not surprising that all major classes of K^+ channels show at least some pH_i sensitivity. Molecular and biophysical investigations of pH sensitivities of the channels are, however, often problematic because multiple properties of the channels are pH-dependent and the pK_a values of amino acids vary considerably in different environments, further complicating the investigation strategy. The histidine side chain is often involved in the pH sensitivity of K^+ channels in the physiological pH range. Other amino acids, such as cysteine, lysine, and glutamic acid, may also be involved in some cases. Although site-directed mutagenesis has been valuable, it is necessary to distinguish whether any elimination or reduction of the pH sensitivity is caused by disruption in the pH sensor mechanism or in the mechanism that couples the pH sensor and its effector.

BK_{Ca} Low pH_i inhibits BK_{Ca} channels in many native cells such as rat pancreatic β cells (142), cultured fetal hippocampal neurons (143), mouse neocortical neurons (144), rat artery smooth muscle cells (145, 146), cultured human renal proximal tubule epithelial cells (147), rabbit tracheal smooth muscle cells (148), type I cells of the neonatal rat carotid body (149), and rat skeletal muscle (150). The low pH_i-mediated inhibition of the rat skeletal BK_{Ca} channel incorporated into planar lipid bilayers involves a shift in the voltage dependence to more positive voltages in a Ca^{2+}-independent manner (150). To better elucidate the biophysical mechanism,

Avdonin et al. (151) investigated the effects of low pH_i on heterologously expressed hSlo1 in the virtual absence of divalent ions so that the hSlo1 channel essentially works as a voltage-gated channel. Low pH_i dramatically increased the open channel probability by shifting the voltage dependence to more negative voltages. The stimulatory effect was antagonized by high Ca^{2+} or Mg^{2+} and also by pretreatment of the channel with the histidine modifier DEPC, suggesting that protonation of the histidine side chain may mimic the divalent-dependent activation of the channel. The hSlo1 channel possesses an intrinsic ability to be stimulated by low pH_i, but this property appears to be somehow modified and has become inhibitory in many native systems. A contribution of the auxiliary subunit $BK\beta1$ is not likely because coexpression with mouse $BK\beta1$ does not alter the stimulatory response (151) and because rat skeletal muscle BK_{Ca} channels, which are inhibited by low pH_i (150), are not associated with any $BK\beta$ subunit (8).

SHAKER   P/C-type inactivation of ShB channels is in part dependent on the amino acid at position 449 in the external mouth of the pore (152). When a histidine residue is present at this position, C-type inactivation becomes dependent on extracellular pH (pH_o). Similar regulation is observed when a phenylalanine is present at position 425 (153). However, low pH_i does not alter C-type activation, instead it reduces the single-channel current amplitude so that 20% of the current may be blocked at physiological pH (154).

Kv1.3   As found with Shaker channels, P/C-type inactivation of Kv1.3 is also accelerated by low pH_o, and this effect is mediated by the histidine residue at the position equivalent to T449 in ShB (155–157).

Kv1.4 $K_V1.4$ undergoes rapid N-type and often slower C-type inactivation, which may be pH-regulated, from either side of the membrane (158). The N-type inactivation time course was slowed by lowering pH_i in the range 6–8, and H16 and E2 in the N terminus contribute to the pH_i dependence (159). Lowering pH_o accelerates the time courses of both N- and C-type inactivation, and the mutation H508Q in the S5-pore linker greatly diminishes the pH dependence (158, 160).

Kv1.5   Low pH_o in the physiological range ($\sim$7) decreases the Kv1.5 peak current amplitude (161–163), and this effect is markedly diminished by the mutations in the ion conduction pathway external to the selectivity filter (R487) (162, 163) or in the turret region (H463) (162). R487 in hKv1.5 is equivalent to T449 in ShB, which in part mediates P/C-type inactivation (152) and external TEA binding (164). These results are interpreted to indicate that R487-mediated P/C-type inactivation is stabilized by protonation of H463 (162). Protonation of H463 is speculated to influence pK_a of R487 (165). pH_o is also a critical determinant in the external K^+ sensitivity of hKv1.5 and mKv1.3 (165). Typically, Shaker-type channels without N-type inactivation become nonfunctional when external K^+ is removed. This phenomenon is absent in hKv1.5 and mKv1.3 at $pH_o > 7.3$ but present at $pH_o = 6.0$ (165). Using the Na^+ currents through P/C-type inactivated channels (166), Zhang et al. demonstrated that the peak current reduction is caused by dramatic acceleration of the P/C-type inactivation kinetics by low pH_o (167). Physiologically, pH-dependent properties of Kv1.5, which contribute to phase 3 repolarization, may account for the acidosis-induced prolongation of cardiac action potential (161).

Kir/K_{ATP} The weak inward rectifiers Kir1.x (ROMKx) and Kir4.1/Kir5.1 are inhibited by low pH_i with $pK_a \sim 6.7$ (35, 168–170). Kir1.1a (ROMK1) and Kir1.1b (ROMK2) expressed in *Xenopus* oocytes are sensitive to low pH_i, especially in the absence of ATP and Mg^{2+} (169). The pH_i sensitivity of Kir1.1a (ROMK1) is mediated by multiple amino acid residues located in both the N and C termini. The high pH_i sensitivity of Kir1.1b (ROMK1), compared with that of Kir4.1, is in part conferred by K66 and K80 in the N-terminal Q-region and by H342 and H354 in the C terminus (35, 171). Whether all these residues represent actual pH-sensing elements or are involved in coupling between the pH sensor and the effector is not known. In Kir1.1b (ROMK2), the pH_i sensitivity is under the control of PKA-dependent phosphorylation, increasing the versatility of the channel (172).

Kir5.1 is primarily located in tubule cells of kidney, as well as in acinar and ductal cells of pancreas and the brain (173–176). In kidney and brain, Kir5.1 coassembles with Kir4.1 to form functional heterotetramers (177) that are inhibited by protons at pH_i 6–8. However, the homomeric Kir4.1 channel is relatively unaffected in this range (173). In pancreas, Kir 5.1 coassembles with Kir4.2 (175), but the pH sensitivity of this heterotetramer is not known. The small-conductance Kir7.1 in the retinal pigment epithelium is also inhibited by pH_i (178).

Enhanced activation of K_{ATP} channels accompanies shortening of the cardiac action potential during ischemic acidosis (179) and hypercapnic acidosis (180). Although PTX-sensitive G proteins may participate in the acidosis-induced vasodilation in porcine coronary arteries (181), direct inhibition of K_{ATP} channels by low pH_I has also been reported (180–183). Pancreatic (Kir6.2/SUR1) (183) and vascular (Kir6.1/SUR2B) (180) K_{ATP} channels were also inhibited by low pH_i when expressed in HEK cells. The pH_i sensitivity of Kir6.2 within the physiological range requires T71 within the N terminus, C166 within the M2 domain, and H175 within the C terminus (34). The voltage-dependent block of K_{ATP} by intracellular polyamines is dependent on pH_i and the sensitivity is partially mediated by H216 in the channel C terminus (182).

K_{ATP} Channels Regulated by Glucose and ATP/ADP

Glucose is the major energy source in many organisms. The plasma glucose level is tightly regulated to protect critical organs such as the brain. Multiple K^+ channels respond to fluctuation in the blood glucose concentration, ensuring glucose homeostasis.

PANCREATIC β CELLS   Pancreatic β cells change the insulin secretion rate in response to fluctuations in blood glucose. The mechanism underlying the coupling between blood glucose-sensing and insulin secretion is not entirely clear, but the involvement of K$^+$ channels in this process, including K$_{ATP}$, is likely. Increased blood glucose, through multiple steps, stimulates ATP production in mitochondria. Closing of plasma membrane K$_{ATP}$ by increased ATP production is expected to promote depolarization and Ca^{2+}-dependent release of insulin. Regulation of K$_{ATP}$ by various nucleotides is exceedingly complex. The ATP/ADP ratio is generally taken as the physiological stimulus for the K$_{ATP}$ closure (184), although the channel regulation by phosphatidylinositol 4,5 bisphosphate (PIP$_2$) may also be important (185). K$_{ATP}$ channels are found abundantly in insulin secretory granules, suggesting that these channels may directly alter exocytosis (186). Pancreatic β cells also express Kv1.4, Kv1.6, and Kv2.1, and these voltage-gated K$^+$ channels modulate insulin release indirectly by altering Ca^{2+} influx through voltage-gated Ca^{2+} channels (187).

OTHER GLUCOSE-SENSING CELLS   In addition to pancreatic β cells, neurons in the ventromedial hypothalamic nucleus (VMHN) in the hypothalamus function as glucose sensors. In response to systemic hypoglycemia, some VMHN neurons release glucagons, leading to an increase in blood catecholamines released from adrenal chromaffin cells. K$_{ATP}$ appear to play a key role in this response (13, 188); in mice lacking K$_{ATP}$ (Kir6.2$^{-/-}$), the hypoglycemia sensitivity was severely blunted. These mice also exhibited reduced food intake in response to neuroglycopenia (188). Although the K$_{ATP}$ in VMHN are generally similar to those in pancreatic β cells, inhibition of rat hypothalamic K$_{ATP}$ by high glucose may be ATP-independent but involve lactate (189). VMHN is linked to control of feeding and body weight; lesions in VMHN induce hyperphagia and obesity (190). Proopiomelanocortin (POMC) neurons in the hypothalamus also work as glucose sensors. The POMC neurons possess A-type K$^+$ channels, opioid-sensitive K$^+$ channels, and K$_{ATP}$ (Kir6.2/SUR1) (191). In rat glomus cells of the carotid body, hypoxia-sensitive K$^+$ currents were also glucose sensitive (68).

Potassium Channels Regulated by CO

Carbon monoxide (CO) is a metabolic end-product of catalysis of heme by heme oxygenases. The gaseous molecule is similar to NO in its ability to relax vascular smooth muscle via activation of BK$_{Ca}$ in rat tail artery (39), pig cerebral arteriole (192), and rat PA (40). CO also activates K$^+$ channels in rat renal interlobe artery (193), rat PA (40), and rat gracilis muscle artery (194). Potassium channels in other tissues are also affected by CO: a 70-pS K$^+$ channel found in rat renal thick ascending limb (195), a cGMP-activated K$^+$ channel in rabbit corneal epithelial cells (196), and IK$_{Ca}$/SK$_{Ca}$ in pig nerve terminals and the nerve trunks in the smooth muscles of the urethra and esophagogastric junction (197). Carbon monoxide can directly modify an extracellular histidine in BK$_{Ca}$ channels in rat vascular smooth muscle (39), or work indirectly through the sGC-cGMP-PKG pathway (40, 196). In rat tail artery, pretreatment with NO blunted CO-dependent activation of BK_{Ca} (198). Carbon monoxide may also enhance the frequency of BK_{Ca} channel opening by increasing the coupling between Ca^{2+} sparks and BK_{Ca} activation (192).

Potassium Channels in Mitochondria

Cellular respiration takes place in mitochondria to synthesize ATP. Swelling of the mitochondrial matrix is an essential step in the enhanced ATP synthesis following Ca^{2+} influx to the matrix. Mitochondrial ion channels should be able to provide a rapid read-out of the metabolic function of mitochondria. A variety of ion channels, including some K^+-selective channels, are present in the outer and inner mitochondrial membranes (199, 200). Mitochondrial K_{ATP} channels (mtK_{ATP}) were first recorded from fused giant inner mitochondrial membrane vesicles (mitoplasts) prepared from rat liver mitochondria (201). Potassium-selective single-channel openings of $\sim$10 pS were inhibited when ATP was added to the matrix side in the inside-out configuration. Glybenclamide and 4-AP also inhibited the openings. These observations indicate that mtK_{ATP} channels are oriented so that the ATP-binding side faces the matrix. Healthy mitochondria maintain a large matrix-negative potential across the inner membrane caused by active extrusion of H^+ from the matrix to the cytoplasm (202). Because the matrix and cytoplasmic K^+ concentrations are similar (199), a decrease in the matrix ATP concentration would open these mtK_{ATP} channels and depolarize the inner mitochondrial membrane potential, which in turn is expected to alter the mitochondrial ATP synthesis (but see 203). Thus the information flow between mtK_{ATP} function and the mitochondrial respiration is bidirectional. The technical difficulty associated with directly recording mtK_{ATP} (204) thus far has forced most of the studies to rely on those pharmacological agents designed to specifically open or close mtK_{ATP} channels without affecting plasma membrane K_{ATP}. Numerous studies using this pharmacological approach suggest that activation of mtK_{ATP} plays a cell-protective role in ischemic preconditioning (205, 206). Activation of mtK_{ATP} channels leads to generation of nitric oxide and reactive oxygen species (207) that can, in turn, alter K_{ATP} channels (208, 209) and also initiate a series of signal transduction events. The molecular identify of mtK_{ATP} is not known. The pharmacological characteristics suggest that mtK_{ATP} may resemble Kir6.1/SUR1 (210).

Mitochondrial BK_{Ca} ($mtBK_{Ca}$) channels are also reported to exist in the inner mitochondrial membrane (211–213). Large openings expected from $mtBK_{Ca}$ channels were recorded in mitoplasts prepared from a glioma cell line (211). The channel-opening frequency was increased by Ca^{2+} and reduced by the typical BK_{Ca} channel blocker charybdotoxin (CTX) (211). Recordings from mitoplasts prepared from ventricular myocytes, which contain few plasma membrane BK_{Ca}, also revealed CTX-sensitive large openings (212). As found with mtK_{ATP}, the electrophysiological results indicate that the tail domain of the $mtBK_{Ca}$ faces the matrix side, and the normally extracellular side of the channel faces the cytoplasm. The presence of $mtBK_{Ca}$ proteins in mitochondria was further verified using an antibody against the C terminus of the channel (212). Activation of $mtBK_{Ca}$ may also have a cytoprotective role following ischemia (212). $mtBK_{Ca}$ channels are thus likely to play a key role in monitoring the mitochondrial metabolic state and also in controlling the mitochondrial function. The regulatory repertoire of BK_{Ca} is well suited for this function. Both mitochondrial respiration and $mtBK_{Ca}$ are Ca^{2+}-, redox-, and voltage-regulated. If depolarization of the inner mitochondrial membrane is important in the cytoprotective action of mtK_{ATP} activation, opening of $mtBK_{Ca}$ channels, which have a much larger amplitude, may prove to be more effective. The future investigation of $mtBK_{Ca}$ channels promises to yield exciting findings.

Potassium Channel Regulation by Other Metabolic Factors

Potassium channel function is regulated by other metabolically relevant factors, including phospholipids, fatty acids, arachidonic acid-derived metabolites, electrolytes, and cell volume as summarized in the on-line supplemental material. (See Supplemental Material at http://www.annualreviews.org/supmat/supmat.asp. K+ Channels Regulated by Other Metabolic Parameter.)

CONCLUSION

The vast diversity of K+ channels allows cells to monitor metabolic factors, such as oxygen, redox equivalents, pCO_2/pH, and ATP/ADP/glucose in a finely tuned manner. These metabolic parameters clearly regulate K+ channels, but how the metabolic changes are sensed and how the information regarding the changes is then coupled to the channel function is yet to be studied in detail. Undoubtedly, the recent advances in the channel structure elucidation and systematic genomic and proteomic approaches will facilitate our study of the intricate and reciprocal information link between K+ channel function and metabolism.

ACKNOWLEDGMENTS

Supported in part by the National Institutes of Health and the American Heart Association.

The *Annual Review of Physiology* is online at http://physiol.annualreviews.org

LITERATURE CITED

1. Cannon WB. 1932. *The Wisdom of the Body.* New York: Norton
1a. Hille B. 2001. *Ion Channels of Excitable Membranes.* Sunderland, MA: Sinauer
2. Park KP, Beck JS, Douglas IJ, Brown PD. 1994. Ca^{2+}-activated K+ channels are in-

volved in regulatory volume decrease in acinar cells isolated from the rat lacrimal gland. *J. Membr. Biol.* 141:193–201
3. Jensen BS, Strøbaek D, Christophersen P, Jørgensen TD, Hansen C, et al. 1998. Characterization of the cloned

human intermediate-conductance Ca^{2+}-activated K^+ channel. *Am. J. Physiol. Cell Physiol.* 275:C848–56

4. Rufo PA, Jiang L, Moe SJ, Brugnara C, Alper SL, Lencer WI. 1996. The antifungal antibiotic, clotriamzole, inhibits Cl^- secretion by polarized monolayers of human colonic epithelial cells. *J. Clin. Invest.* 98:2066–75

5. Sørensen JB, Nielsen MS, Gudme CN, Larsen EH, Nielsen R. 2001. Maxi K^+ channels co-localised with CFTR in the apical membrane of an exocrine gland acinus: possible involvement in secretion. *Pflügers Arch.* 442:1–11

6. Turnheim K, Plass H, Wyskovsky W. 2002. Basolateral potassium channels of rabbit colon epithelium: role in sodium absorption and chloride secretion. *Biochim. Biophys. Acta* 1560:51–66

7. Giebisch G. 2001. Renal potassium channels: function, regulation, and structure. *Kidney Int.* 60:436–45

8. Brenner R, Perez GJ, Bonev AD, Eckman DM, Kosek JC, et al. 2000. Vasoregulation by the $\beta 1$ subunit of the calcium-activated potassium channel. *Nature* 407:870–76

9. Melman A, Christ GJ. 2001. Integrative erectile biology. The effects of age and disease on gap junctions and ion channels and their potential value to the treatment of erectile dysfunction. *Urol. Clin. North Am.* 28:217–31, vii

10. Klyachko VA, Ahern GP, Jackson MB. 2001. cGMP-mediated facilitation in nerve terminals by enhancement of the spike afterhyperpolarization. *Neuron* 31:1015–25

11. Hosseini R, Benton DC, Dunn PM, Jenkinson DH, Moss GW. 2001. SK3 is an important component of K^+ channels mediating the afterhyperpolarization in cultured rat SCG neurones. *J. Physiol.* 535:323–34

12. Pedarzani P, Kulik A, Muller M, Ballanyi K, Stocker M. 2000. Molecular deter-minants of Ca^{2+}-dependent K^+ channel function in rat dorsal vagal neurones. *J. Physiol.* 527:283–90

13. Ashford ML, Boden PR, Treherne JM. 1990. Glucose-induced excitation of hypothalamic neurones is mediated by ATP-sensitive K^+ channels. *Pflügers Arch.* 415:479–83

13a. Xie J, McCobb DP. 1998. Control of alternative splicing of potassium channels by stress hormones. *Science* 280:443–46

13b. Erxleben C, Everhart AL, Romeo C, Florance H, Bauer MB, et al. 2002. Interacting effects of N-terminal variation and strex exon splicing on slo potassium channel regulation by calcium, phosphorylation, and oxidation. *J. Biol. Chem.* 277:27-45–52

14. Meiri N, Ghelardini C, Tesco G, Galeotti N, Dahl D, et al. 1997. Reversible antisense inhibition of Shaker-like Kv1.1 potassium channel expression impairs associative memory in mouse and rat. *Proc. Natl. Acad. Sci. USA* 94:4430–34

15. Giese KP, Storm JF, Reuter D, Fedorov NB, Shao LR, et al. 1998. Reduced K^+ channel inactivation, spike broadening, and after-hyperpolarization in $Kv\beta 1.1$-deficient mice with impaired learning. *Learn. Mem.* 5:257–73

16. Derst C, Karschin A. 1998. Evolutionary link between prokaryotic and eukaryotic K^+ channels. *J. Exp. Biol.* 201(Pt. 20):2791–99

17. Roberts SK. 2003. TOK homologue in *Neurospora crassa*: first cloning and functional characterization of an ion channel in a filamentous fungus. *Eukaryot. Cell* 2:181–90

18. Zhou X, Vaillant B, Loukin S, Kung C, Saimi Y. 1995. YKC1 encodes the depolarization-activated K^+ channel in the plasma membrane of yeast. *FEBS Lett.* 373:170–76

19. Jegla T, Salkoff L. 1994. Molecular evolution of K^+ channels in primitive eukaryotes. *Soc. Gen. Physiol. Ser.* 49:213–22

20. Schroeder JI, Ward JM, Gassmann W. 1994. Perspectives on the physiology and structure of inward-rectifying K$^+$ channels in higher plants: biophysical implications for K$^+$ uptake. *Annu. Rev. Biophys. Biomol. Struct.* 23:441–71

21. Jan LY, Jan YN. 1997. Cloned potassium channels from eukaryotes and prokaryotes. *Annu. Rev. Neurosci.* 20:91–123

22. Jiang Y, Lee A, Chen J, Ruta V, Cadene M, et al. 2003. X-ray structure of a voltage-dependent K$^+$ channel. *Nature* 423:33–41

23. Davies AR, Kozlowski RZ. 2001. Kv channel subunit expression in rat pulmonary arteries. *Lung* 179:147–61

24. Olschewski A, Hong Z, Linden BC, Porter VA, Weir EK, Cornfield DN. 2002. Contribution of the K$_{Ca}$ channel to membrane potential and O$_2$ sensitivity is decreased in an ovine PPHN model. *Am. J. Physiol. Lung Cell Mol. Physiol.* 283:L1103–9

25. Wyatt CN, Wright C, Bee D, Peers C. 1995. O$_2$-sensitive K$^+$ currents in carotid body chemoreceptor cells from normoxic and chronically hypoxic rats and their roles in hypoxic chemotransduction. *Proc. Natl. Acad. Sci. USA* 92:295–99

26. Hartness ME, Lewis A, Searle GJ, O'Kelly I, Peers C, Kemp PJ. 2001. Combined antisense and pharmacological approaches implicate hTASK as an airway O$_2$ sensing K$^+$ channel. *J. Biol. Chem.* 276:26499–508

27. Lee S, Park M, So I, Earm YE. 1994. NADH and NAD modulates Ca^{2+}-activated K$^+$ channels in small pulmonary arterial smooth muscle cells of the rabbit. *Pflügers Arch.* 427:378–80

28. Park MK, Lee SH, Ho WK, Earm YE. 1995. Redox agents as a link between hypoxia and the responses of ionic channels in rabbit pulmonary vascular smooth muscle. *Exp. Physiol.* 80:835–42

29. Wei EP, Kontos HA, Beckman JS. 1996. Mechanisms of cerebral vasodilation by superoxide, hydrogen peroxide, and peroxynitrite. *Am. J. Physiol. Heart Circ. Physiol.* 271:H1262–66

30. Soto MA, Gonzalez C, Lissi E, Vergara C, Latorre R. 2002. Ca^{2+}-activated K$^+$ channel inhibition by reactive oxygen species. *Am. J. Physiol. Cell Physiol.* 282:C461–71

31. Duprat F, Guillemare E, Romey G, Fink M, Lesage F, et al. 1995. Susceptibility of cloned K$^+$ channels to reactive oxygen species. *Proc. Natl. Acad. Sci. USA* 92:11796–800

32. Ciorba MA, Heinemann SH, Weissbach H, Brot N, Hoshi T. 1997. Modulation of potassium channel function by methionine oxidation and reduction. *Proc. Natl. Acad. Sci. USA* 94:9932–37

33. Tang XD, Daggett H, Hanner M, Garcia ML, McManus OB, et al. 2001. Oxidative regulation of large conductance calcium-activated potassium channels. *J. Gen. Physiol.* 117:253–74

34. Piao H, Cui N, Xu H, Mao J, Rojas A, et al. 2001. Requirement of multiple protein domains and residues for gating K$_{ATP}$ channels by intracellular pH. *J. Biol. Chem.* 276:36673–80

35. Fakler B, Schultz JH, Yang J, Schulte U, Brandle U, et al. 1996. Identification of a titratable lysine residue that determines sensitivity of kidney potassium channels (ROMK) to intracellular pH. *EMBO J.* 15:4093–99

36. Morton MJ, O'Connell AD, Sivaprasadarao A, Hunter M. 2003. Determinants of pH sensing in the two-pore domain K$^+$ channels TASK-1 and -2. *Pflügers Arch.* 445:577–83

37. MacGregor GG, Dong K, Vanoye CG, Tang L, Giebisch G, Hebert SC. 2002. Nucleotides and phospholipids compete for binding to the C terminus of K$_{ATP}$ channels. *Proc. Natl. Acad. Sci. USA* 99:2726–31

38. Tang XD, Xu R, Reynolds MF, Garcia ML, Heinemann SH, Hoshi T. 2003. Haem can bind to and inhibit mammalian

calcium-dependent Slo1 BK channels. *Nature* 425:531–35

39. Wang R, Wu L, Wang Z. 1997. The direct effect of carbon monoxide on K_{Ca} channels in vascular smooth muscle cells. *Pflügers Arch.* 434:285–91

40. Naik JS, Walker BR. 2001. Homogeneous segmental profile of carbon monoxide-mediated pulmonary vasodilation in rats. *Am. J. Physiol. Lung Cell Mol. Physiol.* 281:L1436–43

41. Thomas SL, Chmelar RS, Lu C, Halvorsen SW, Nathanson NM. 1997. Tissue-specific regulation of G-protein-coupled inwardly rectifying K^+ channel expression by muscarinic receptor activation in ovo. *J. Biol. Chem.* 272:29958–62

42. Zhou XB, Arntz C, Kamm S, Motejlek K, Sausbier U, et al. 2001. A molecular switch for specific stimulation of the BK_{Ca} channel by cGMP and cAMP kinase. *J. Biol. Chem.* 276:43239–45

43. Tiran Z, Peretz A, Attali B, Elson A. 2003. Phosphorylation-dependent regulation of Kv2.1 channel activity at tyrosine 124 by Src and by protein-tyrosine phosphatase epsilon. *J. Biol. Chem.* 278:17509–14

44. Clarke AL, Petrou S, Walsh JV Jr, Singer JJ. 2002. Modulation of BK_{Ca} channel activity by fatty acids: structural requirements and mechanism of action. *Am. J. Physiol. Cell Physiol.* 283:C1441–53

45. Li PL, Campbell WB. 1997. Epoxy-eicosatrienoic acids activate K^+ channels in coronary smooth muscle through a guanine nucleotide binding protein. *Circ. Res.* 80:877–84

46. Sun CW, Alonso-Galicia M, Taheri MR, Falck JR, Harder DR, Roman RJ. 1998. Nitric oxide-20-hydroxy-eicosatetraenoic acid interaction in the regulation of K^+ channel activity and vascular tone in renal arterioles. *Circ. Res.* 83:1069–79

47. Armstead WM. 1998. Brain injury impairs prostaglandin cerebrovasodilation. *J. Neurotrauma* 15:721–29

48. Niemeyer MI, Cid LP, Barros LF, Sepulveda FV. 2001. Modulation of the two-pore domain acid-sensitive K^+ channel TASK-2 (KCNK5) by changes in cell volume. *J. Biol. Chem.* 276:43166–74

49. Yuan A, Santi CM, Wei A, Wang ZW, Pollak K, et al. 2003. The sodium-activated potassium channel is encoded by a member of the Slo gene family. *Neuron* 37:765–73

50. Zaritsky JJ, Eckman DM, Wellman GC, Nelson MT, Schwarz TL. 2000. Targeted disruption of Kir2.1 and Kir2.2 genes reveals the essential role of the inwardly rectifying K^+ current in K^+-mediated vasodilation. *Circ. Res.* 87:160–66

51. Xia XM, Zeng X, Lingle CJ. 2002. Multiple regulatory sites in large-conductance calcium-activated potassium channels. *Nature* 418:880–84

51a. Xia XM, Fakler B, Rivard A, Wayman G, Johnson-Pais T, et al. 1998. Mechanisms of calcium gating in small-conductance calcium-activated potassium channels. *Nature* 395:5-3–7

52. Johnson JP Jr, Balser JR, Bennett PB. 2001. A novel extracellular calcium sensing mechanism in voltage-gated potassium ion channels. *J. Neurosci.* 21:4143–53

53. Shi J, Krishnamoorthy G, Yang Y, Hu L, Chaturvedi N, et al. 2002. Mechanism of magnesium activation of calcium-activated potassium channels. *Nature* 418:876–80

54. Doi T, Fakler B, Schultz JH, Ehmke H, Brandle U, et al. 1995. Subunit-specific inhibition of inward-rectifier K^+ channels by quinidine. *FEBS Lett.* 375:193–96

55. Maingret F, Fosset M, Lesage F, Lazdunski M, Honoré E. 1999. TRAAK is a mammalian neuronal mechano-gated K^+ channel. *J. Biol. Chem.* 274:1381–87

56. Kim D. 1992. A mechanosensitive K^+ channel in heart cells. Activation

by arachidonic acid. *J. Gen. Physiol.* 100:1021–40

57. Niu W, Sachs F. 2003. Dynamic properties of stretch-activated K+ channels in adult rat atrial myocytes. *Prog. Biophys. Mol. Biol.* 82:121–35

58. Hoyer J, Distler A, Haase W, Gogelein H. 1994. Ca²⁺ influx through stretch-activated cation channels activates maxi K+ channels in porcine endocardial endothelium. *Proc. Natl. Acad. Sci. USA* 91:2367–71

59. Grunnet M, MacAulay N, Jorgensen NK, Jensen S, Olesen SP, Klaerke DA. 2002. Regulation of cloned, Ca²⁺-activated K+ channels by cell volume changes. *Pflügers Arch.* 444:167–77

60. Hanaoka K, Wright JM, Cheglakov IB, Morita T, Guggino WB. 1999. A 59 amino acid insertion increases Ca²⁺ sensitivity of rbslo1, a Ca²⁺-activated K+ channel in renal epithelia. *J. Membr. Biol.* 172:193–201

61. Devor DC, Frizzell RA. 1993. Calcium-mediated agonists activate an inwardly rectified K+ channel in colonic secretory cells. *Am. J. Physiol. Cell Physiol.* 265:C1271–80

62. Lahiri S, Prabhakar NR, Forster RE. 2000. *Oxygen Sensing: Molecule to Man.* New York: Plenum

63. Prabhakar NR. 2000. Oxygen sensing by the carotid body chemoreceptors. *J. Appl. Physiol.* 88:2287–95

64. López-Barneo J, Pardal R, Ortega-Sáenz P. 2001. Cellular mechanism of oxygen sensing. *Annu. Rev. Physiol.* 63:259–87

64a. Haddad GG, Jiang C. 1997. O₂–sensing mechanisms in excitable cells: role of plasma membrane K+ channels. *Annu. Rev. Physiol.* 59:23–42

65. Lipton AJ, Johnson MA, Macdonald T, Lieberman MW, Gozal D, Gaston B. 2001. S-nitrosothiols signal the ventilatory response to hypoxia. *Nature* 413:171–74

66. López-Barneo J, López-López JR, Ureña J, González C. 1988. Chemotransduction in the carotid body: K+ current modulated by PO₂ in type I chemoreceptor cells. *Science* 241:580–82

67. Peers C. 1990. Effects of D600 on hypoxic suppression of K+ currents in isolated type I carotid body cells of the neonatal rat. *FEBS Lett.* 271:37–40

68. Pardal R, López-Barneo J. 2002. Carotid body thin slices: responses of glomus cells to hypoxia and K+-channel blockers. *Respir. Physiol. Neurobiol.* 132:69–79

69. Ganfornina MD, López-Barneo J. 1991. Single K+ channels in membrane patches of arterial chemoreceptor cells are modulated by O₂ tension. *Proc. Natl. Acad. Sci. USA* 88:2927–30

70. Sanchez D, López-López JR, Pérez-García MT, Sanz-Alfayate G, Obeso A, et al. 2002. Molecular identification of Kvα subunits that contribute to the oxygen-sensitive K+ current of chemoreceptor cells of the rabbit carotid body. *J. Physiol.* 542:369–82

71. Buckler KJ. 1999. Background leak K+-currents and oxygen sensing in carotid body type 1 cells. *Respir. Physiol.* 115:179–87

72. Overholt JL, Ficker E, Yang T, Shams H, Bright GR, Prabhakar NR. 2000. Chemosensing at the carotid body. Involvement of a HERG-like potassium current in glomus cells. *Adv. Exp. Med. Biol.* 475:241–48

73. Fu XW, Wang D, Nurse CA, Dinauer MC, Cutz E. 2000. NADPH oxidase is an O₂ sensor in airway chemoreceptors: evidence from K+ current modulation in wild-type and oxidase-deficient mice. *Proc. Natl. Acad. Sci. USA* 97:4374–79

74. O'Kelly I, Peers C, Kemp PJ. 1998. O₂-sensitive K+ channels in neuroepithelial body-derived small cell carcinoma cells of the human lung. *Am. J. Physiol. Lung Cell Mol. Physiol.* 275:L709–16

75. Michelakis ED, Archer SL, Weir EK. 1995. Acute hypoxic pulmonary

vasoconstriction: a model of oxygen sensing. *Physiol. Res.* 44:361–67

76. Osipenko ON, Evans AM, Gurney AM. 1997. Regulation of the resting potential of rabbit pulmonary artery myocytes by a low threshold, O_2-sensing potassium current. *Br. J. Pharmacol.* 120:1461–70

77. Rychkov GY, Adams MB, McMillen IC, Roberts ML. 1998. Oxygen-sensing mechanisms are present in the chromaffin cells of the sheep adrenal medulla before birth. *J. Physiol.* 509:887–93

78. Mochizuki-Oda N, Takeuchi Y, Matsumura K, Osawa Y, Watanabe Y. 1997. Hypoxia-induced catecholamine release and intracellular Ca^{2+} increase via suppression of K^+ channels in cultured rat adrenal chromaffin cells. *J. Neurochem.* 69:377–87

79. Conforti L, Millhorn DE. 1997. Selective inhibition of a slow-inactivating voltage-dependent K^+ channel in rat PC12 cells by hypoxia. *J. Physiol.* 502:293–305

80. Conforti L, Bodi I, Nisbet JW, Millhorn DE. 2000. O_2-sensitive K^+ channels: role of the Kv1.2-subunit in mediating the hypoxic response. *J. Physiol.* 524:783–93

81. Cross AR, Henderson L, Jones OT, Delpiano MA, Hentschel J, Acker H. 1990. Involvement of an NAD(P)H oxidase as a pO_2 sensor protein in the rat carotid body. *Biochem. J.* 272:743–47

82. Sanders KA, Sundar KM, He L, Dinger B, Fidone S, Hoidal JR. 2002. Role of components of the phagocytic NADPH oxidase in oxygen sensing. *J. Appl. Physiol.* 93:1357–64

83. Archer SL, Reeve HL, Michelakis E, Puttagunta L, Waite R, et al. 1999. O_2 sensing is preserved in mice lacking the gp91 phox subunit of NADPH oxidase. *Proc. Natl. Acad. Sci. USA* 96:7944–49

84. He L, Chen J, Dinger B, Sanders K, Sundar K, et al. 2002. Characteristics of carotid body chemosensitivity in NADPH oxidase-deficient mice. *Am. J. Physiol. Cell Physiol.* 282:C27–33

85. O'Kelly I, Peers C, Kemp PJ. 2001. NADPH oxidase does not account fully for O_2-sensing in model airway chemoreceptor cells. *Biochem. Biophys. Res. Commun.* 283:1131–34

86. Roy A, Rozanov C, Mokashi A, Daudu P, Al-Mehdi AB, et al. 2000. Mice lacking in gp91 phox subunit of NAD(P)H oxidase showed glomus cell $[Ca^{2+}]_i$ and respiratory responses to hypoxia. *Brain Res.* 872:188–93

87. Thompson RJ, Farragher SM, Cutz E, Nurse CA. 2002. Developmental regulation of O_2 sensing in neonatal adrenal chromaffin cells from wild-type and NADPH-oxidase-deficient mice. *Pflügers Arch.* 444:539–48

88. Cheng GJ, Cao ZH, Xu XX, van Meir EG, Lambeth JD. 2001. Homologs of gp91phox: cloning and tissue expression of Nox3, Nox4, and Nox5. *Gene* 269:131–40

89. Shiose A, Kuroda J, Tsuruya K, Hirai M, Hirakata H, et al. 2001. A novel superoxide-producing NAD(P)H oxidase in kidney. *J. Biol. Chem.* 276:1417–23

90. Archer SL, Weir EK, Reeve HL, Michelakis E. 2000. Molecular identification of O_2 sensors and O_2-sensitive potassium channels in the pulmonary circulation. *Adv. Exp. Med. Biol.* 475:219–40

91. Michelakis ED, Rebeyka I, Wu XC, Nsair A, Thebaud B, et al. 2002. O_2 sensing in the human ductus arteriosus: regulation of voltage-gated K^+ channels in smooth muscle cells by a mitochondrial redox sensor. *Circ. Res.* 91:478–86

92. Becker LB, vanden Hoek TL, Shao ZH, Li CQ, Schumacker PT. 1999. Generation of superoxide in cardiomyocytes during ischemia before reperfusion. *Am. J. Physiol. Heart Circ. Physiol.* 277:H2240–46

93. Sanz-Alfayate G, Obeso A, Agapito MT, González C. 2001. Reduced to oxidized glutathione ratios and oxygen sensing in calf and rabbit carotid body chemoreceptor cells. *J. Physiol.* 537:209–20

94. Lewis A, Peers C, Ashford ML, Kemp PJ. 2002. Hypoxia inhibits human recombinant large conductance, Ca^{2+}-activated K^+ (maxi-K) channels by a mechanism which is membrane delimited and Ca^{2+} sensitive. *J. Physiol.* 540:771–80

95. Riesco-Fagundo AM, Pérez-Garcia MT, González C, López-López JR. 2001. O_2 modulates large-conductance Ca^{2+}-dependent K^+ channels of rat chemoreceptor cells by a membrane-restricted and CO-sensitive mechanism. *Circ. Res.* 89:430–36

96. Pérez-Garcia MT, López-López JR, González C. 1999. $Kv\beta1.2$ subunit coexpression in HEK293 cells confers O_2 sensitivity to Kv4.2 but not to Shaker channels. *J. Gen. Physiol.* 113:897–907

97. Rustenbeck I, Dickel C, Herrmann C, Grimmsmann T. 1999. Mitochondria present in excised patches from pancreatic B-cells may form microcompartments with ATP-dependent potassium channels. *Biosci. Rep.* 19:89–98

98. Chan MK. 2001. Recent advances in heme-protein sensors. *Curr. Opin. Chem. Biol.* 5:216–22

99. Kaelin WG Jr. 2002. How oxygen makes its presence felt. *Genes Dev.* 16:1441–45

100. Safran M, Kaelin WG Jr. 2003. HIF hydroxylation and the mammalian oxygen-sensing pathway. *J. Clin. Invest.* 111:779–83

100a. Epstein AC, Gleadle JM, McNeill LA, Hewitson KS, O'Rourke J, et al. 2001. *C. elegans* EGL-9 and mammalian homologs define a family of dioxygenases that regulate HIF by prolyl hydroxylation. *Cell* 107:43–54

101. Fontana L, D'Amico M, Crociani O, Biagiotti T, Solazzo M, et al. 2001. Long-term modulation of HERG channel gating in hypoxia. *Biochem. Biophys. Res. Commun.* 286:857–62

102. Conforti L, Petrovic M, Mohammad D, Lee S, Ma Q, et al. 2003. Hypoxia regulates expression and activity of Kv1.3 channels in T lymphocytes: a possible role in T cell proliferation. *J. Immunol.* 170:695–702

102a. Cai H, Harrison DG. 2000. Endothelial dysfunction in cardiovascular diseases: the role of oxidant stress. *Circ. Res.* 87:840–44

102b. Kamata H, Hirata H. 1999. Redox regulation of cellular signalling. *Cell Signal.* 11:1–14

103. Schafer FQ, Buettner GR. 2001. Redox environment of the cell as viewed through the redox state of the glutathione disulfide/glutathione couple. *Free Radic. Biol. Med.* 30:1191–212

104. Finkel T, Holbrook NJ. 2000. Oxidants, oxidative stress and the biology of ageing. *Nature* 408:239–47

104a. Halliwell B, Gutteridge JMC. 1999. *Free Radicals in Biology and Medicine.* New York: Oxford Univ. Press. 936 pp.

105. Kourie JI. 1998. Interaction of reactive oxygen species with ion transport mechanisms. *Am. J. Physiol. Cell Physiol.* 275:C1–24

106. Shi J, Cui J. 2001. Intracellular Mg^{2+} enhances the function of BK-type Ca^{2+}-activated K^+ channels. *J. Gen. Physiol.* 118:589–606

107. Zhang X, Solaro CR, Lingle CJ. 2001. Allosteric regulation of BK channel gating by Ca^{2+} and Mg^{2+} through a nonselective, low affinity divalent cation site. *J. Gen. Physiol.* 118:607–35

108. Behrens MI, Vergara C, Latorre R. 1988. Calcium-activated potassium channels of large unitary conductance. *Braz. J. Med. Biol. Res.* 21:1101–17

109. Thuringer D, Findlay I. 1997. Contrasting effects of intracellular redox couples on the regulation of maxi-K channels in isolated myocytes from rabbit pulmonary artery. *J. Physiol.* 500:583–92

110. Hayabuchi Y, Nakaya Y, Matsuoka S, Kuroda Y. 1998. Hydrogen peroxide-induced vascular relaxation in porcine

coronary arteries is mediated by Ca^{2+}-activated K^+ channels. *Heart Vessels* 13:9–17

111. Barlow RS, White RE. 1998. Hydrogen peroxide relaxes porcine coronary arteries by stimulating BK_{Ca} channel activity. *Am. J. Physiol. Heart Circ. Physiol.* 275:H1283–89

112. Gong L, Gao TM, Huang H, Tong Z. 2000. Redox modulation of large conductance calcium-activated potassium channels in CA1 pyramidal neurons from adult rat hippocampus. *Neurosci. Lett.* 286:191–94

113. Lang RJ, Harvey JR, McPhee GJ, Klemm MF. 2000. Nitric oxide and thiol reagent modulation of Ca^{2+}-activated K^+ (BK_{Ca}) channels in myocytes of the guinea-pig taenia caeci. *J. Physiol.* 525(Pt. 2):363–76

114. Brzezinska AK, Gebremedhin D, Chilian WM, Kalyanaraman B, Elliott SJ. 2000. Peroxynitrite reversibly inhibits Ca^{2+}-activated K^+ channels in rat cerebral artery smooth muscle cells. *Am. J. Physiol. Heart Circ. Physiol.* 278:H1883–90

115. Vergara C, Soto M, González C, Lissi E, Latorre R. 2001. Inhibition of a Ca^{2+}-activated K^+ channel by H_2O_2. *Biophys. J.* 80:A220 (Abstr.)

116. Soh H, Jung W, Uhm DY, Chung S. 2001. Modulation of large conductance calcium-activated potassium channels from rat hippocampal neurons by glutathione. *Neurosci. Lett.* 298:115–18

117. Wang ZW, Nara M, Wang YX, Kotlikoff MI. 1997. Redox regulation of large conductance Ca^{2+}-activated K^+ channels in smooth muscle cells. *J. Gen. Physiol.* 110:35–44

118. Taglialatela M, Pannaccione A, Iossa S, Castaldo P, Annunziato L. 1999. Modulation of the K^+ channels encoded by the human ether-a-gogo-related gene-1 (hERG1) by nitric oxide. *Mol. Pharmacol.* 56:1298–308

119. Orio P, Rojas P, Ferreira G, Latorre R.

2002. New disguises for an old channel: MaxiK channel β-subunits. *News Physiol. Sci.* 17:156–61

120. Erxleben C, Everhart AL, Romeo C, Florance H, Bauer MB, et al. 2002. Interacting effects of N-terminal variation and strex exon splicing on slo potassium channel regulation by calcium, phosphorylation, and oxidation. *J. Biol. Chem.* 277:27045–52

121. Zeng XH, Xia XM, Lingle CJ. 2003. Redox-sensitive extracellular gates formed by auxiliary β subunits of calcium-activated potassium channels. *Nat. Struct. Biol.* 10:448–54

122. DiChiara TJ, Reinhart PH. 1997. Redox modulation of hslo Ca^{2+}-activated K^+ channels. *J. Neurosci.* 17:4942–55

123. Weissbach H, Etienne F, Hoshi T, Heinemann SH, Lowther WT, et al. 2002. Peptide methionine sulfoxide reductase: structure, mechanism of action, and biological function. *Arch. Biochem. Biophys.* 397:172–78

124. Hoshi T, Heinemann SH. 2001. Regulation of cell function by methionine oxidation and reduction. *J. Physiol.* 531:1–11

125. Rozanski GJ, Xu Z. 2002. Sulfhydryl modulation of K^+ channels in rat ventricular myocytes. *J. Mol. Cell Cardiol.* 34:1623–32

126. Xu Z, Patel KP, Lou MF, Rozanski GJ. 2002. Up-regulation of K^+ channels in diabetic rat ventricular myocytes by insulin and glutathione. *Cardiovasc. Res.* 53:80–88

127. Stephens GJ, Owen DG, Robertson B. 1996. Cysteine-modifying reagents alter the gating of the rat cloned potassium channel Kv1.4. *Pflügers Arch.* 431:435–42

128. Ruppersberg JP, Stocker M, Pongs O, Heinemann SH, Frank R, Koenen M. 1991. Regulation of fast inactivation of cloned mammalian IK_A channels by cysteine oxidation. *Nature* 352:711–14

129. Rettig J, Heinemann SH, Wunder F,

Lorra C, Parcej DN, et al. 1994. Inactivation properties of voltage-gated K$^+$ channels altered by presence of β-subunit. *Nature* 369:289–94

130. Heinemann SH, Rettig J, Wunder F, Pongs O. 1995. Molecular and functional characterization of a rat brain Kvβ 3 potassium channel subunit. *FEBS Lett.* 377:383–89

131. Kuschel L, Hansel A, Schönherr R, Weissbach H, Brot N, et al. 1999. Molecular cloning and functional expression of a human peptide methionine sulfoxide reductase (hMsrA). *FEBS Lett.* 456:17–21

132. Chen J, Avdonin V, Ciorba MA, Heinemann SH, Hoshi T. 2000. Acceleration of P/C-type inactivation in voltage-gated K$^+$ channels by methionine oxidation. *Biophys. J.* 78:174–87

133. Gulbis JM, Mann S, MacKinnon R. 1999. Structure of a voltage-dependent K$^+$ channel β subunit. *Cell* 97:943–52

134. Bähring R, Milligan CJ, Vardanyan V, Engeland B, Young BA, et al. 2001. Coupling of voltage-dependent potassium channel inactivation and oxidoreductase active site of Kvβ subunits. *J. Biol. Chem.* 276:22923–29

135. Zhang Y, Han H, Wang J, Wang H, Yang B, Wang Z. 2003. Impairment of human ether-à-go-go-related gene (HERG) K$^+$ channel function by hypoglycemia and hyperglycemia. Similar phenotypes but different mechanisms. *J. Biol. Chem.* 278:10417–26

136. Han J, Kim E, Ho WK, Earm YE. 1996. Sulfhydryl redox modulates ATP-sensitive K$^+$ channels in rabbit ventricular myocytes. *Biochem. Biophys. Res. Commun.* 219:900–3

137. Tricarico D, Camerino DC. 1994. ATP-sensitive K$^+$ channels of skeletal muscle fibers from young adult and aged rats: possible involvement of thiol-dependent redox mechanisms in the age-related modifications of their biophysical and pharmacological properties. *Mol. Pharmacol.* 46:754–61

138. Salgado AP, Pereira FC, Seiça RM, Fernandes AP, Flatt PR, et al. 1999. Modulation of glucose-induced insulin secretion by cytosolic redox state in clonal β-cells. *Mol. Cell Endocrinol.* 154:79–88

139. Islam MS, Berggren PO, Larsson O. 1993. Sulfhydryl oxidation induces rapid and reversible closure of the ATP-regulated K$^+$ channel in the pancreatic β-cell. *FEBS Lett.* 319:128–32

139a. Ross J, Armstead WM. 2003. Differential role of PTK and ERK MAPK in superoxide impairment of K$_{ATP}$ and K$_{Ca}$ channel cerebrovasodilation. *Am. J. Physiol. Regul. Integr. Comp. Physiol.* 285:R149–54

140. Kintner DB, Anderson MK, Fitzpatrick JH Jr, Sailor KA, Gilboe DD. 2000. ^{31}P-MRS-based determination of brain intracellular and interstitial pH: its application to in vivo H$^+$ compartmentation and cellular regulation during hypoxic/ischemic conditions. *Neurochem. Res.* 25:1385–96

141. Tsai KL, Wang SM, Chen CC, Fong TH, Wu ML. 1997. Mechanism of oxidative stress-induced intracellular acidosis in rat cerebellar astrocytes and C6 glioma cells. *J. Physiol.* 502:161–74

142. Tabcharani JA, Misler S. 1989. Ca^{2+}-activated K$^+$ channel in rat pancreatic islet B cells: permeation, gating and blockade by cations. *Biochim. Biophys. Acta* 982:62–72

143. Church J, Baxter KA, McLarnon JG. 1998. pH modulation of Ca^{2+} responses and a Ca^{2+}-dependent K$^+$ channel in cultured rat hippocampal neurones. *J. Physiol.* 511:119–32

144. Liu H, Moczydlowski E, Haddad GG. 1999. O$_2$ deprivation inhibits Ca^{2+}-activated K$^+$ channels via cytosolic factors in mice neocortical neurons. *J. Clin. Invest.* 104:577–88

145. Schubert R, Krien U, Gagov H. 2001. Protons inhibit the BK$_{Ca}$ channel of rat

small artery smooth muscle cells. *J. Vasc. Res.* 38:30–38

146. Petkova-Kirova P, Gagov H, Krien U, Duridanova D, Noack T, Schubert R. 2000. 4-aminopyridine affects rat arterial smooth muscle BK_{Ca} currents by changing intracellular pH. *Br. J. Pharmacol.* 131:1643–50

147. Hirano J, Nakamura K, Itazawa S, Sohma Y, Kubota T, Kubokawa M. 2002. Modulation of the Ca^{2+}-activated large conductance K^+ channel by intracellular pH in human renal proximal tubule cells. *Jpn. J. Physiol.* 52:267–76

148. Kume H, Takagi K, Satake T, Tokuno H, Tomita T. 1990. Effects of intracellular pH on calcium-activated potassium channels in rabbit tracheal smooth muscle. *J. Physiol.* 424:445–57

149. Peers C, Green FK. 1991. Inhibition of Ca^{2+}-activated K^+ currents by intracellular acidosis in isolated type I cells of the neonatal rat carotid body. *J. Physiol.* 437:589–602

150. Laurido C, Candia S, Wolff D, Latorre R. 1991. Proton modulation of a Ca^{2+}-activated K^+ channel from rat skeletal muscle incorporated into planar bilayers. *J. Gen. Physiol.* 98:1025–42

151. Avdonin V, Tang XD, Hoshi T. 2003. Stimulatory action of internal protons on Slo1 BK channels. *Biophys. J.* 84:2969–80

152. López-Barneo J, Hoshi T, Heinemann SH, Aldrich RW. 1993. Effects of external cations and mutations in the pore region on C-type inactivation of *Shaker* potassium channels. *Receptors Channels* 1:61–71

153. Pérez-Cornejo P. 1999. H^+ ion modulation of C-type inactivation of Shaker K^+ channels. *Pflügers Arch.* 437:865–70

154. Starkus JG, Zarga Z, Heinemann SH. 2003. Mechanisms for the inhibition of Shaker potassium channels. *Pflügers Arch.* In press

155. Busch A, Hurst R, North R, Adelman J, Kavanaugh M. 1991. Current inactiva-tion involves a histidine residue in the pore of the rat lymphocyte potassium channel RGK5. *Biochem. Biophys. Res. Commun.* 179:1384–90

156. Deutsch C, Lee SC. 1989. Modulation of K^+ currents in human lymphocytes by pH. *J. Physiol.* 413:399–413

157. Jager H, Rauer H, Nguyen AN, Aiyar J, Chandy KG, Grissmer S. 1998. Regulation of mammalian Shaker-related K^+ channels: evidence for non-conducting closed and non-conducting inactivated states. *J. Physiol.* 506:291–301

158. Li X, Bett GC, Jiang X, Bondarenko VE, Morales MJ, Rasmusson RL. 2003. Regulation of N- and C-type inactivation of Kv1.4 by pH_o and K^+: evidence for transmembrane communication. *Am. J. Physiol. Heart Circ. Physiol.* 284:H71–80

159. Padanilam BJ, Lu T, Hoshi T, Padanilam BA, Shibata EF, Lee HC. 2002. Molecular determinants of intracellular pH modulation of human Kv1.4 N-type inactivation. *Mol. Pharmacol.* 62:127–34

160. Claydon TW, Boyett MR, Sivaprasadarao A, Ishii K, Owen JM, et al. 2000. Inhibition of the K^+ channel Kv1.4 by acidosis: protonation of an extracellular histidine slows the recovery from N-type inactivation. *J. Physiol.* 526:253–64

161. Steidl JV, Yool AJ. 1999. Differential sensitivity of voltage-gated potassium channels Kv1.5 and Kv1.2 to acidic pH and molecular identification of pH sensor. *Mol. Pharmacol.* 55:812–20

162. Kehl SJ, Eduljee C, Kwan DC, Zhang S, Fedida D. 2002. Molecular determinants of the inhibition of human Kv1.5 potassium currents by external protons and Zn^{2+}. *J. Physiol.* 541:9–24

163. Trapani JG, Korn SJ. 2003. Effect of external pH on activation of the Kv1.5 potassium channel. *Biophys. J.* 84:195–204

164. Heginbotham L, MacKinnon R. 1992. The aromatic binding site for

tetraethylammonium ion on potassium channels. *Neuron* 8:483–91

165. Jäger H, Grissmer S. 2001. Regulation of a mammalian Shaker-related potassium channel, hKv1.5, by extracellular potassium and pH. *FEBS Lett.* 488:45–50

166. Starkus JG, Kuschel L, Rayner MD, Heinemann SH. 1997. Ion conduction through C-type inactivated Shaker channels. *J. Gen. Physiol.* 110:539–50

167. Zhang S, Kurata HT, Kehl SJ, Fedida D. 2003. Rapid induction of P/C-type inactivation is the mechanism for acid-induced K$^+$ current inhibition. *J. Gen. Physiol.* 121:215–25

168. Leung YM, Zeng WZ, Liou HH, Solaro CR, Huang CL. 2000. Phosphatidylinositol 4,5-bisphosphate and intracellular pH regulate the ROMK1 potassium channel via separate but interrelated mechanisms. *J. Biol. Chem.* 275:10182–89

169. McNicholas CM, MacGregor GG, Islas LD, Yang Y, Hebert SC, Giebisch G. 1998. pH-dependent modulation of the cloned renal K$^+$channel, ROMK. *Am. J. Physiol. Renal Physiol.* 275:F972–81

170. Mauerer UR, Boulpaep EL, Segal AS. 1998. Regulation of an inwardly rectifying ATP-sensitive K$^+$ channel in the basolateral membrane of renal proximal tubule. *J. Gen. Physiol.* 111:161–80

171. Xu H, Yang Z, Cui N, Giwa LR, Abdulkadir L, et al. 2000. Molecular determinants for the distinct pH sensitivity of Kir1.1 and Kir4.1 channels. *Am. J. Physiol. Cell Physiol.* 279:C1464–71

172. Leipziger J, MacGregor GG, Cooper GJ, Xu J, Hebert SC, Giebisch G. 2000. PKA site mutations of ROMK2 channels shift the pH dependence to more alkaline values. *Am. J. Physiol. Renal Physiol.* 279:F919–26

173. Tanemoto M, Kittaka N, Inanobe A, Kurachi Y. 2000. In vivo formation of a proton-sensitive K$^+$ channel by heteromeric subunit assembly of Kir5.1 with Kir4.1. *J. Physiol.* 525:587–92

174. Liu Y, McKenna E, Figueroa DJ, Blevins R, Austin CP, et al. 2000. The human inward rectifier K$^+$ channel subunit kir5.1 (KCNJ16) maps to chromosome 17q25 and is expressed in kidney and pancreas. *Cytogenet. Cell Genet.* 90:60–63

175. Pessia M, Imbrici P, D'Adamo MC, Salvatore L, Tucker SJ. 2001. Differential pH sensitivity of Kir4.1 and Kir4.2 potassium channels and their modulation by heteropolymerisation with Kir5.1. *J. Physiol.* 532:359–67

176. Cui N, Giwa LR, Xu H, Rojas A, Abdulkadir L, Jiang C. 2001. Modulation of the heteromeric Kir4.1-Kir5.1 channels by pCO$_2$ at physiological levels. *J. Cell. Physiol.* 189:229–36

177. Tucker SJ, Imbrici P, Salvatore L, D'Adamo MC, Pessia M. 2000. pH dependence of the inwardly rectifying potassium channel, Kir5.1, and localization in renal tubular epithelia. *J. Biol. Chem.* 275:16404–7

178. Shimura M, Yuan Y, Chang JT, Zhang S, Campochiaro PA, et al. 2001. Expression and permeation properties of the K$^+$ channel Kir7.1 in the retinal pigment epithelium. *J. Physiol.* 531:329-46

179. Bethell HW, Vandenberg JI, Smith GA, Grace AA. 1998. Changes in ventricular repolarization during acidosis and low-flow ischemia. *Am. J. Physiol. Heart Circ. Physiol.* 275:H551–61

180. Wang X, Wu J, Li L, Chen F, Wang R, Jiang C. 2003. Hypercapnic acidosis activates K$_{ATP}$ channels in vascular smooth muscles. *Circ. Res.* 92:1225–32

181. Ishizaka H, Gudi SR, Frangos JA, Kuo L. 1999. Coronary arteriolar dilation to acidosis: role of ATP-sensitive potassium channels and pertussis toxin-sensitive G proteins. *Circulation* 99:558–63

182. Baukrowitz T, Tucker SJ, Schulte U, Benndorf K, Ruppersberg JP, Fakler B. 1999. Inward rectification in K$_{ATP}$ channels: a pH switch in the pore. *EMBO J.* 18:847–53

183. Xu H, Cui N, Yang Z, Wu J, Giwa LR,

et al. 2001. Direct activation of cloned K_{ATP} channels by intracellular acidosis. *J. Biol. Chem.* 276:12898–902

184. Huopio H, Shyng SL, Otonkoski T, Nichols CG. 2002. K_{ATP} channels and insulin secretion disorders. *Am. J. Physiol. Endocrinol. Metab.* 283:E207–16

185. Baukrowitz T, Schulte U, Oliver D, Herlitze S, Krauter T, et al. 1998. PIP_2 and PIP as determinants for ATP inhibition of K_{ATP} channels. *Science* 282:1141–44

186. Geng X, Li L, Watkins S, Robbins PD, Drain P. 2003. The insulin secretory granule is the major site of K_{ATP} channels of the endocrine pancreas. *Diabetes* 52:767–76

187. MacDonald PE, Ha XF, Wang J, Smukler SR, Sun AM, et al. 2001. Members of the Kv1 and Kv2 voltage-dependent K^+ channel families regulate insulin secretion. *Mol. Endocrinol.* 15:1423–35

188. Miki T, Liss B, Minami K, Shiuchi T, Saraya A, et al. 2001. ATP-sensitive K^+ channels in the hypothalamus are essential for the maintenance of glucose homeostasis. *Nat. Neurosci.* 4:507–12

189. Ainscow EK, Mirshamsi S, Tang T, Ashford ML, Rutter GA. 2002. Dynamic imaging of free cytosolic ATP concentration during fuel sensing by rat hypothalamic neurones: evidence for ATP-independent control of ATP-sensitive K^+ channels. *J. Physiol.* 544:429–45

190. Sakaguchi T, Bray GA, Eddlestone G. 1988. Sympathetic activity following paraventricular or ventromedial hypothalamic lesions in rats. *Brain Res. Bull.* 20:461–65

191. Ibrahim N, Bosch MA, Smart JL, Qiu J, Rubinstein M, et al. 2003. Hypothalamic proopiomelanocortin neurons are glucose responsive and express K_{ATP} channels. *Endocrinology* 144:1331–40

192. Jaggar JH, Leffler CW, Cheranov SY, Tcheranova D, E S, Cheng X. 2002. Car-

bon monoxide dilates cerebral arterioles by enhancing the coupling of Ca^{2+} sparks to Ca^{2+}-activated K^+ channels. *Circ. Res.* 91:610–17

193. Kaide JI, Zhang F, Wei Y, Jiang H, Yu C, et al. 2001. Carbon monoxide of vascular origin attenuates the sensitivity of renal arterial vessels to vasoconstrictors. *J. Clin. Invest.* 107:1163–71

194. Zhang F, Kaide J, Wei Y, Jiang H, Yu C, et al. 2001. Carbon monoxide produced by isolated arterioles attenuates pressure-induced vasoconstriction. *Am. J. Physiol. Heart Circ. Physiol.* 281:H350–58

195. Liu H, Mount DB, Nasjletti A, Wang W. 1999. Carbon monoxide stimulates the apical 70-pS K^+ channel of the rat thick ascending limb. *J. Clin. Invest.* 103:963–70

196. Rich A, Farrugia G, Rae JL. 1994. Carbon monoxide stimulates a potassium-selective current in rabbit corneal epithelial cells. *Am. J. Physiol. Cell Physiol.* 267:C435–42

197. Werkstrom V, Ny L, Persson K, Andersson KE. 1997. Carbon monoxide-induced relaxation and distribution of haem oxygenase isoenzymes in the pig urethra and lower oesophagogastric junction. *Br. J. Pharmacol.* 120:312–18

198. Wang R, Wu L. 2003. Interaction of selective amino acid residues of K_{Ca} channels with carbon monoxide. *Exp. Biol. Med.* 228:474–80

199. Bernardi P. 1999. Mitochondrial transport of cations: channels, exchangers, and permeability transition. *Physiol. Rev.* 79:1127–55

200. Zoratti M, Szabo I. 1994. Electrophysiology of the inner mitochondrial membrane. *J. Bioenerg. Biomembr.* 26:543–53

201. Inoue I, Nagase H, Kishi K, Higuti T. 1991. ATP-sensitive K^+ channel in the mitochondrial inner membrane. *Nature* 352:244–47

202. Nicholls DG, Ward MW. 2000. Mitochondrial membrane potential and neuronal glutamate excitotoxicity: mortality and millivolts. *Trends Neurosci.* 23:166–74

203. Kowaltowski AJ, Seetharaman S, Paucek P, Garlid KD. 2001. Bioenergetic consequences of opening the ATP-sensitive K+ channel of heart mitochondria. *Am. J. Physiol. Heart Circ. Physiol.* 280:H649–57

204. Jonas EA, Buchanan J, Kaczmarek LK. 1999. Prolonged activation of mitochondrial conductances during synaptic transmission. *Science* 286:1347–50

205. Oldenburg O, Cohen MV, Yellon DM, Downey JM. 2002. Mitochondrial K$_{ATP}$ channels: role in cardioprotection. *Cardiovasc. Res.* 55:429–37

206. O'Rourke B. 2000. Pathophysiological and protective roles of mitochondrial ion channels. *J. Physiol.* 529:23–36

207. Lebuffe G, Schumacker PT, Shao ZH, Anderson T, Iwase H, Vanden Hoek TL. 2003. ROS and NO trigger early preconditioning: relationship to mitochondrial K$_{ATP}$ channel. *Am. J. Physiol. Heart Circ. Physiol.* 284:H299–308

208. Sasaki N, Sato T, Ohler A, O'Rourke B, Marban E. 2000. Activation of mitochondrial ATP-dependent potassium channels by nitric oxide. *Circulation* 101:439–45

209. Zhang DX, Chen YF, Campbell WB, Zou AP, Gross GJ, Li PL. 2001. Characteristics and superoxide-induced activation of reconstituted myocardial mitochondrial ATP-sensitive potassium channels. *Circ. Res.* 89:1177–83

210. Liu Y, Ren G, O'Rourke B, Marban E, Seharaseyon J. 2001. Pharmacological comparison of native mitochondrial K$_{ATP}$ channels with molecularly defined surface K$_{ATP}$ channels. *Mol. Pharmacol.* 59:225–30

211. Siemen D, Loupatatzis C, Borecky J, Gulbins E, Lang F. 1999. Ca^{2+}-activated K channel of the BK-type in the inner mitochondrial membrane of a human glioma cell line. *Biochem. Biophys. Res. Commun.* 257:549–54

212. Xu W, Liu Y, Wang S, McDonald T, Van Eyk JE, et al. 2002. Cytoprotective role of Ca^{2+}-activated K+ channels in the cardiac inner mitochondrial membrane. *Science* 298:1029–33

213. Yermolaieva O, Tang XD, Daggett H, Hoshi T. 2001. Calcium-activated mitochondrial K channel is involved in regulation of mitochondrial membrane potential and permeability transition. *Biophys. J.* 80:950A(Abstr.)

Annu. Rev. Physiol. 2004. 66:161–81
doi: 10.1146/annurev.physiol.66.050802.084104
First published online as a Review in Advance on September 8, 2003

STRUCTURE AND FUNCTION OF GLUTAMATE RECEPTOR ION CHANNELS[1]

Mark L. Mayer
*Laboratory of Cellular and Molecular Neurophysiology, Building 36, Room 2B28,
NICHD, NIH, DHHS, Bethesda, Maryland 20892; email: mlm@helix.nih.gov*

Neali Armstrong
*Department of Biochemistry and Molecular Biophysics, 650 West 168th Street, Columbia
University, New York, NY 10032; email: naa15@columbia.edu*

Key Words AMPA, kainate, NMDA, crystallography, channels, structure

■ **Abstract** A vast number of proteins are involved in synaptic function. Many have been cloned and their functional role defined with varying degrees of success, but their number and complexity currently defy any molecular understanding of the physiology of synapses. A beacon of success in this medieval era of synaptic biology is an emerging understanding of the mechanisms underlying the activity of the neurotransmitter receptors for glutamate. Largely as a result of structural studies performed in the past three years we now have a mechanistic explanation for the activation of channel gating by agonists and partial agonists; the process of desensitization, and its block by allosteric modulators, is also mostly explained; and the basis of receptor subtype selectivity is emerging with clarity as more and more structures are solved. In the space of months we have gone from cartoons of postulated mechanisms to hard fact. It is anticipated that this level of understanding will emerge for other synaptic proteins in the coming decade.

INTRODUCTION

The amino acid S-glutamate acts as the neurotransmitter at the majority of excitatory synapses in the brain and spinal cord of vertebrates. The pioneering work establishing the physiological role of glutamate receptor ion channels (iGluRs) in synaptic transmission relied heavily on the development of selective agonists and antagonists. By 1980 this approach had led to the recognition of three major iGluR subtypes, the AMPA, kainate, and NMDA receptors, which were named after their selective agonists (1). Each of these receptors activates a cation-selective ion

---

161

channel permeable to Na^+ and K^+, with differing degrees of permeability to, and block by, the divalent cations Ca^{2+} and Mg^{2+} (2). This functional classification remains in widespread use and indeed largely defines the biology of glutamate receptor ion channels.

A corresponding molecular classification of glutamate receptor channels arose from the application of cDNA cloning techniques. This revealed that each of the major functional subtypes of glutamate receptor ion channel was comprised of a family of genes. The AMPA receptors are assembled from the GluR1-GluR4 subunits (also called GluRA-GluRD). Two gene families, GluR5-GluR7 and KA1/KA2, encode kainate receptors; the latter form functional channels only when coassembled with members of the GluR5-GluR7 family. NMDA receptors, which typically must bind both glutamate and glycine for activation, are assembled from three gene families. All NMDA receptors contain obligate NR1 subunits that serve two roles. The NR1 subunits contain the glycine-binding site, but in addition are required for trafficking of heteromeric receptor assemblies to the plasma membrane. NR1 subunits coassemble with the NR2A-NR2D subunits that encode glutamate-binding sites; less commonly they coassemble with the NR3A/NR3B gene families that encode glycine-binding sites distinct from those on the NR1 subunit. Two orphan members of the iGluR gene family, the delta 1 and delta 2 receptors, are not well characterized because their functional activity and ligand-binding properties have defied analysis. Adding further complexity to the functional analysis of the iGluR gene family is the occurrence of alternative splicing and RNA editing. Finally, the ongoing sequencing of microbial genomes has resulted in the discovery of a growing family of bacterial ligand–gated ion channels of which GluR0 was the first to be characterized (3, 4). Several reviews summarize much of the above material and should be consulted for in-depth discussion of the physiological roles and molecular biology of glutamate receptor ion channels and their roles in synaptic function (4–9).

Domain Organization of Glutamate Receptor Ion Channels

The cloning of iGluR genes led immediately to attempts to relate their amino acid sequence to function and mechanism. Progress was at first limited by the absence of structural information, and for several years experiments were designed and interpreted on the basis of an erroneous transmembrane topology model derived from one for ligand-gated nicotinic receptor ion channels. The recognition that iGluRs had a unique topology was coupled to the discovery that their N-terminal and agonist-binding domains shared amino acid sequence homology with bacterial periplasmic-binding proteins and were thus likely to have a similar structure. The periplasmic protein homology–based model for the agonist-binding domain was, however, incompatible with the topology present in nicotinic receptor ion channels. Establishment of the correct iGluR topology was a major step forward and paved the way for a direct analysis of iGluR structure and function. Such analysis

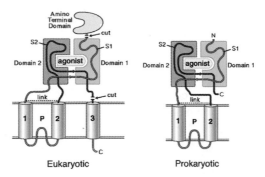

Eukaryotic   Prokaryotic

Figure 1 Domain organization of eukaryotic and prokaryotic iGluRs. The S1 peptide segment that precedes the two-TM pore loop insertion is colored cyan; the S2 peptide segment is colored pink. The S1 and S2 segments cross over into domain 2 and domain 1, respectively. Agonists and competitive antagonists bind in a fissure located between domain 1 and 2. Eukaryotic iGluRs have additional structural elements consisting of an N-terminal domain, a third TM segment, and a cytoplasmic domain. The S1S2-ligand binding core was expressed as a soluble protein by genetically excising the ion channel pore, N-terminal domain, and the C-terminal domain as required.

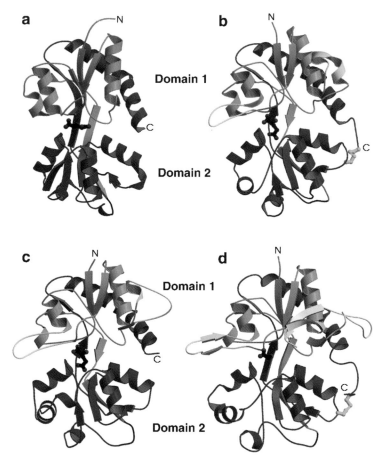

Figure 2 Crystal structures of the ligand-binding cores of (*a*) GluR0, (*b*) GluR2, (*c*) GluR6, and (*d*) NR1. The ribbon diagrams are colored using the scheme from Figure 1 with S1 colored cyan and S2 colored pink. Domain 1 is at the top and domain 2 at the bottom. Loops that are present in GluR2, GluR6, and NR1, but not in GluR0, are colored yellow. The disulfide bond that links the C terminus of S2 to domain 1 is drawn in green; in GluR6 the C terminus was disordered, and there was no electron density after the end of the last helix in domain 2. The ball and stick models show glutamate (or, for NR1, glycine) bound in the cleft between domains 1 and 2. Note the different conformation for the glutamate molecule in GluR0 compared with that in GluR2 and GluR6.

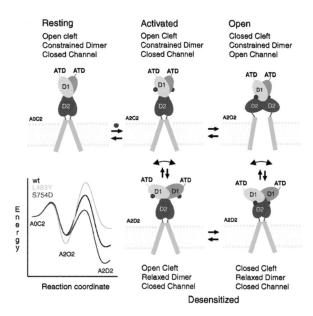

Figure 3 State diagram for activation and desensitization of glutamate receptor ion channels. Shown schematically are a pair of ligand-binding cores and their associated ion channel segments. Proceeding from the resting state, the binding of glutamate is shown by attachment of a red ball to domain 1; this triggers domain closure and ion channel activation (*top row*). The bottom row shows entry into desensitized states that results from relaxation of the dimer interface into a conformation that permits both the agonist-binding core and the ion channel to remain in closed conformations. Shown on the bottom left is a hypothetical reaction scheme for activation and desensitization with the coordinate for wild-type plotted in black. Mutations that increase the stability of the dimeric assembly (*green*) deepen the energy well for the open state of the ion channel (A2O2) and raise the barrier for entry into the desensitized state (A2D2). Mutations that destabilize the dimer interface (*red*) have the opposite effect.

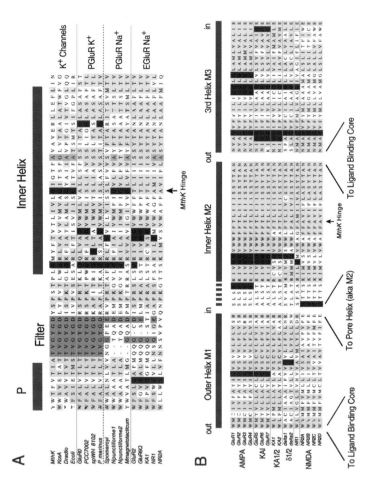

Figure 4 (A) Amino acid sequence alignments for the pore loop and inner helix of K⁺ channels and iGluRs. Conserved residues in the ion-binding site are colored orange; glycine residues are shaded in red. A dashed line separates K⁺-selective prokaryotic iGluRs from those that lack the K⁺ channel signature sequence and would be expected to be permeable to both Na⁺ and K⁺. (B) Amino acid sequence alignments of the TM segments of eukaryotic iGluRs. Conserved glycine residues are shaded in red. The location of glycine residues varies between subtypes.

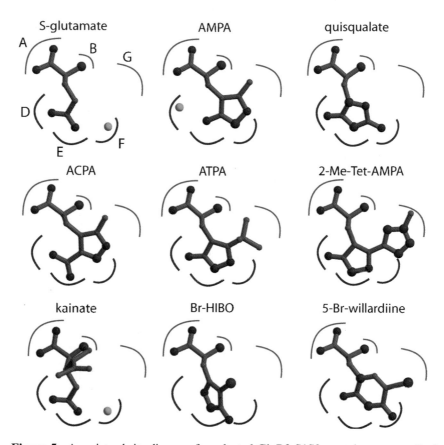

Figure 5 Agonist subsite diagram for selected GluR2 S1S2 crystal structures. Each GluR2 S1S2 agonist complex was superimposed, using only domain 1 residues, on the S1S2 glutamate structure. The α-carboxyl and α-amino groups occupy subsites A and B, respectively. Subsite C is an aliphatic region capped by Tyr450, which is not visible in the orientation shown here (53). The γ-anionic moieties occupy subsites D, E, and F. Subsite G is a large hydrophobic pocket in domain 1, which is occupied by the 5-substituents from AMPA, ACPA, ATPA, and 2-Me-Tet-AMPA. The bottom row contains the diagrams for the partial agonists, kainate, Br-HIBO, and 5-Br-willardiine. Notice that all three partial agonists have substituents located outside the subsite map. Oxygen atoms are red, nitrogen atoms are blue, bromine atoms are purple, and water molecules are green. Subsites A, B, and G are formed by residues from domain 1 and are colored blue; subsites D, E, and F are formed by residues from domain 2 and are colored pink.

revealed that iGluRs are composed of an agonist-binding domain interrupted by insertion of a two-transmembrane (TM) pore loop ion channel; in eukaryotic but not prokaryotic iGluRs there is an additional N-terminal domain, a third TM segment, and a cytoplasmic C-terminal domain (Figure 1, see color insert). In some fish, amphibian, and bird iGluRs the N-terminal domain is absent.

AGONIST-BINDING CORE   Key experiments, which defined the correct TM topology for iGluRs, combined two approaches. The first was genetic insertion of glycosylation and proteolytic enzyme recognition sites, followed by expression in eukaryotic cells and analysis of shifts in molecular weight produced by incubation with glycosidases and proteases. This approach, combined with hydropathy analysis, defined the boundaries of the extracellular and, by inference, intracellular and TM polypeptide segments (10–13). The second approach capitalized on this revised topology and utilized homology modeling based on the hypothesis that the agonist-binding domain of iGluRs was likely to be a two-domain structure similar to that in periplasmic-binding proteins, the structures of several of which had already been solved (14–16). Although the sequence homology of iGluRs to periplasmic proteins was recognized early on (17), interruption of the homologous region by insertion of the ion channel pore domain hindered recognition that the agonist-binding core recognizes ligands by a mechanism similar to that in periplasmic proteins. Once the correct topology for iGluRs was established, it was possible to convert the agonist selectivity of an acceptor protein by swapping the S1 and S2 segments between AMPA and kainate receptor subunits. This proved that in combination the S1S2 segments constituted the agonist-binding core despite the insertion of the ion channel domain (15). Further homology modeling, combined with site-directed mutagenesis, reinforced the conclusion that the S1S2 segments formed the agonist-binding site in AMPA, kainate, and NMDA receptors and that their structure was related to that in periplasmic proteins (18–20). Such studies paved the way for the establishment of the domain organization of an iGluR subunit (illustrated in Figure 1), in which the agonist-binding core is a discrete structure, joined to the N-terminal domain and the ion channel pore by polypeptide linkers of currently unknown structure. It is now widely accepted that four subunits assemble to form functional iGluRs, although direct experimental proof for this has yet to be obtained.

AMINO-TERMINAL DOMAIN   The ~400-amino acid polypeptide segment that makes up the N-terminal domain is a major, but not sole, determinant of subtype-specific assembly within iGluR gene families (21, 22). In the case of the AMPA receptor GluR4 and NR1 subunits, the N-terminal domain forms dimers in solution (22, 23), most likely related to a role in receptor assembly. Homology modeling and functional tests by site-directed mutagenesis established that this domain has a fold similar to that of leucine-isoleucine-valine-binding protein (LIVBP) and the agonist-binding domain in G–protein–coupled glutamate receptors (mGluRs). The structure of the N-terminal domain is expected to be distinct from the structure of the agonist-binding core of iGluRs because in LIVBP and mGluRs, the two globular domains are connected by three β-strands instead of the pair of β-strands found in the iGluR glutamate- and glycine-binding cores (24, 25). The two-domain fold of the N-terminal domain suggests that it might also encode a ligand-binding site, but identification of the endogenous ligands, if any, for the majority of iGluR N-terminal domains has yet to be attempted. In the case of NMDA receptors, functional experiments have established that Zn^{2+} and the noncompetitive antagonist ifenprodil bind to the N-terminal domain to modulate ion channel gating via an allosteric mechanism (26–29). The details of the coupling mechanism remain obscure, in part, because the structure of the N-terminal domain and its packing relative to the other domains in iGluRs remain to be determined.

ION CHANNEL PORE   The S1 and S2 polypeptide sequences that make up the agonist-binding core are interrupted by insertion of two, membrane-spanning, most likely α-helical segments, which in combination with a pore helix and pore loop make the narrowest part of the pore (3, 30–32). Although the ion selectivity of eukaryotic iGluRs and K^+ channels is strikingly different, they share significant amino acid sequence homology and probably have similar structures (3, 4, 30). Indeed, in the case of some of the recently discovered bacterial iGluRs, which have the required amino acid sequences to selectively bind K^+ but not Na^+ (3, 4), the structure of the selectivity filter is almost certainly like that determined for the potassium channels KcsA, MthK, and KvAP (33–35).

Because the structural basis for K^+ selectivity requires a rigid shell of main chain carbonyl oxygen atoms in the pore loop that replace water molecules of solvation during passage of K^+ ions through the lipid bilayer (36), it is easy to imagine either of two scenarios that would allow nonselective permeation of Na^+ and K^+ through a similar structure in iGluRs. Either the pore is sufficiently wide to allow Na^+ and K^+ ions to pass the narrowest segment in their hydrated state, or the pore loops in iGluRs are more flexible than those in K^+ channels, thus permitting close contact with both Na^+ and K^+ ions during permeation. Measurement of the permeability of large organic cations supports the former mechanism (37, 38). It is also conceivable that in iGluRs the narrowest part of the pore is partly lined by amino acid side chains rather than main chain peptide bonds as occurs in K^+ channels (33, 36), but determining this will require structural analysis.

In the case of bacterial iGluRs, the S2 segment of the agonist-binding domain is followed by a short extracellular C terminus (3, 39), whereas in vertebrate iGluRs, the S2 segment is linked to a third TM segment such that the C terminus is intracellular (Figure 1). The functional role of the last TM domain has not been clearly defined, but it appears, at least for NMDA receptors, to participate in channel gating because mutations in this region alter desensitization rates as well as channel open times (40). Mutants lacking this domain cannot form functional receptors (22, 41). Interestingly, expression of the deleted domain in *trans* is sufficient to rescue channel activity (41).

CARBOXYL-TERMINAL DOMAIN   The C terminus of vertebrate iGluRs varies in length from around 20 to 500 amino acids. It interacts with numerous cytoskeletal proteins and is important for receptor trafficking (42, 43). The C-terminal domain in NMDA receptors is much larger than that in AMPA and kainate receptors, suggesting that it might have as yet unknown functions. Amino acid sequence alignments do not reveal significant homology with other proteins, and direct structural studies will be required to make progress in understanding the role of this domain.

EXPRESSION AND CRYSTALLIZATION OF AGONIST-BINDING CORES

The discovery that the agonist-binding core of iGluRs is a two-domain structure, interrupted by insertion of the membrane-spanning ion channel pore, led to the realization that with protein engineering it should be possible to genetically excise the agonist-binding domain from the pore region and generate a water-soluble protein that retains ligand-binding activity. Overexpression of constructs in which the agonist-binding core encoded by the S1 and S2 segments was isolated from the N-terminal domain, pore region, and C-terminal domain permitted for the first time biochemical and subsequently structural approaches to be applied to the study of iGluR function. This approach should eventually permit a similar analysis of the N- and C-terminal domain structures, although progress to date has been slower than for the agonist-binding core.

The initial S1S2 constructs generated for the AMPA receptor GluR4 subunit established that the isolated domain retained the appropriate selectivity for AMPA receptor-preferring ligands, which bound with K_d values similar to those established for full-length receptors in membranes (44). Extensions of this approach subsequently allowed the successful expression of the agonist-binding cores for GluR2 (45), GluR6 (46), and NR1 (47, 48).

These initial attempts at over-expression of iGluR agonist-binding cores, while successful in establishing the feasibility of the approach, needed further modification before structural studies were possible. First, protein expression levels, which were on the order of 0.1 mg S1S2 protein per liter of cell culture in the initial studies on AMPA receptors, were increased 10- to 100-fold to yield sufficient protein for attempts at crystallization. This was achieved by establishing conditions for the refolding of milligram quantities of denatured GluR2 S1S2 protein expressed in *Escherichia coli* and purified from inclusion bodies (49). More recently, strains of *E. coli* engineered for expression of proteins containing disulfide bonds have allowed expression of milligram quantities of iGluR S1S2 constructs without the need for refolding (50). Second, to obtain well-ordered crystals that diffract to high resolution it was necessary to define the minimum S1S2 constructs that retained biological activity and conformational stability. Such protein engineering is typically required in crystallographic studies because the flexible polypeptide segments present at the N and C termini, and in loops extending from the core of the protein, typically adopt a range of conformations that interferes with the packing required to generate the well-ordered crystal lattice needed for high-resolution diffraction of X rays. Starting from the initial constructs that were first reported to retain AMPA-binding activity, a series of nine additional GluR2 constructs was generated and screened for binding activity and agonist-induced resistance to proteolysis. This led to the identification of a first generation construct, which when crystallized with kainate, diffracted to 1.5 Å using synchrotron radiation (51). On the basis of this approach the agonist-binding cores have been crystallized for GluR0 (39), GluR2 (52), GluR6 (M.L. Mayer, unpublished work), and NR1 (50). The similar fold of the S1S2-binding cores for GluR0 and for AMPA, kainate, and NMDA receptors, with the agonist bound in a cleft between domain 1 and domain 2, is shown in Figure 2 (see color insert) and illustrates the contributions of the S1 and S2 peptide segments to each domain.

STRUCTURAL STUDIES ON AMPA RECEPTORS

In initial experiments with the first generation GluR2 S1S2 construct, complexes with glutamate and other ligands did not give crystals with sufficient diffraction for structure determination (52). However, after a further round of protein engineering to remove disordered peptide sequences, for which electron density was missing in the first generation complex with kainate, a second generation construct was obtained that crystallized both in the ligand-free (apo) state as well as with a range of agonists and antagonists (53). At present this construct has been used to solve the crystal structures of GluR2 S1S2 complexes with 12 full and partial agonists, 2 competitive antagonists as well as some allosteric modulators, and the apo state. Complementing these static structures are studies of the protein dynamics of glutamate-bound GluR2 S1S2 (54). The use of protein engineering to obtain a well-ordered protein core, as well as expression in bacteria to prevent glycosylation, which frequently produces disorder in protein crystals, was a key determinant in the success of this strategy, and when this was not done, the crystals obtained did not diffract to high resolution (55). Studies with GluR4 do not reveal any difference in ligand binding, function, stability, or cell surface expression of protein in which S1S2 is either glycosylated or nonglycosylated (56).

   The first crystal structure of the GluR2 S1S2 kainate complex revealed a two-domain globular protein with the agonist partially buried in a cleft between the two lobes (52). The lobes are connected by a pair of antiparallel β-strands, and the overall structure is similar to the fold of GlnBP, the periplasmic-binding protein for glutamine (57). The larger of the domains is encoded by the peptide segment S1 but includes two α-helices and a β-strand that occur after the ion channel insertion and are thus encoded by S2. This was an important finding because it showed that the S1 and S2 peptide segments do not correspond to discrete structural domains within the agonist-binding core. Also of note was the observation that the extent of domain closure of the kainate complex was intermediate between that for the apo and glutamine-bound forms of GlnBP. The significance of this for determining the mechanism of partial agonism became apparent when structures of complexes with glutamate and other full agonists were solved.

MECHANISM OF ACTIVATION BY AGONISTS

Subsequent structural studies on GluR2 S1S2, using the more readily crystallized second generation construct, established that, similar to its periplasmic counterparts, GluR2 S1S2 undergoes a large conformational change upon binding agonists, switching from a relaxed, open-cleft apo conformation, to a constrained closed-cleft conformation when bound by AMPA and glutamate (53). Compared with the open-cleft apo structure, the lobes of the AMPA and glutamate complexes rotate closer together by $\sim21°$. Both X-ray crystallographic (58) and biochemical investigations (59) suggest that the mechanism for agonist binding and domain closure is a two-step process. In the docking step, the agonist binds to residues in domain 1 via its α-amino and α-carboxyl groups. With the agonist partially secured in the binding cleft, the γ-carboxyl group electrostatically attracts the base of helix F that points its N-terminal dipole into the cleft. The locking step involves rotation of domain 2 toward domain 1 and closure of the binding cleft. How this domain closure might open the ion channel was revealed by the observation that in the majority of crystal forms studied, the agonist-binding cores form dimers in which the agonist-binding clefts face outward, and the two S1S2 linkers, which replace the ion channel pore, lie on the same side of the dimer on the face opposite from the N-terminal domain. The dimer surface, which buries 1150 Å^2 of solvent-accessible surface per monomer, is formed exclusively by domain 1. As a result, when the pair of subunits in a dimeric assembly undergoes agonist-induced stabilization of the closed cleft conformation, the pair of ion channel linkers on domain 2 swing apart from each other, like a pair of opening scissors. Because these linkers are replacing the channel-forming region from the full-length receptor, it is easy to envision how agonist-induced domain closure can do work on the TM segments, pulling or twisting the ion channel to open (Figure 3, see color insert).

The dimer interface is a unique feature of iGluRs. Both GluR2 S1S2 and its bacterial homologue GluR0 S1S2 assemble as dimers in solution and crystallize as dimers, whereas the bacterial periplasmic-binding proteins do not. Amino acid sequence alignments for the agonist-binding cores of GluR0, GluR2, and three structurally related periplasmic proteins, reveal conservation of hydrophobic residues in the GluR0 and GluR2 dimer interface, whereas the corresponding surface of periplasmic proteins contains charged amino acid residues that form noncomplementary contacts when two monomers are superimposed on the GluR0 dimer (39). Thus despite folds similar to those of periplasmic proteins, the surface of the ligand-binding core of iGluRs has evolved to support the formation of dimeric assemblies that play a key role in activation, and desensitization (see below).

Recent crystallographic studies on the MthK calcium-activated potassium channel led to the proposal of a gating mechanism in which ligand-induced closing and rotation of a set of four two-domain calcium-binding cores produced a 30° bend in the inner TM helix thus opening the pore (34, 60). A conserved glycine residue in the inner helix of MthK acts as a hinge that permits this motion and is conserved in other K$^+$ channels. Because GluR0 is also a two TM ligand–gated K$^+$ channel, a gating mechanism identical to that in MthK would have been expected. However, sequence alignments show that for GluR0 and related prokaryotic iGluR K$^+$-selective channels the glycine hinge has been replaced by an alanine or serine, possibly because the hinge has moved closer to the start of M2. In this context it is interesting that in inward rectifier K$^+$ channels the glycine hinge also has moved and appears to be closer to the C terminus of the inner helix than in MthK. Curiously, for three of the prokaryotic nonselective iGluRs that lack the TVYGYD signature sequence, there is a glycine at the MthK hinge position. Eukaryotic iGluRs contain numerous conserved glycine residues in their TM segments, the location of which differs for individual iGluR subtypes (Figure 4, see color insert). It is likely that at least some of these glycine residues act as hinge regions, similar to the mechanism proposed for MthK, and that several helices must bend during gating, as suggested by the glycine-rich sequence of M3.

THE MECHANISM OF ACTION OF COMPETITIVE ANTAGONISTS

According to classical receptor theory, competitive antagonists bind to the agonist recognition site but do not activate the receptor. Many iGluR competitive antagonists are larger than their agonist counterparts, and thus a likely mechanism for antagonism would be steric interference with the activation process. Crystal structures of complexes of the GluR2 S1S2 agonist-binding core with two different antagonists, DNQX and ATPO, reveal just such a mechanism. Even though DNQX and ATPO have entirely unrelated chemical structures, the former is a quinoxalinedione and the latter is from the isoxazole family of compounds, they both appear to prevent domain closure via a foot-in-the-door mechanism (53, 61). However, as their different chemical structures suggest, the location of the foot is different. For DNQX, it is probably the interaction between the 7-nitro group and Thr686 in domain 2 that props the binding cleft open. Interestingly, ATPO, which is a derivative of AMPA, interacts with the same subsites as conventional agonists. In this case, the foot, or the 3-phosphono-methoxy group, binds to the base of helix F; agonists interact with the same location in domain 2. The extended distance between the isoxazole ring and the phosphono-methoxy group is what prevents domain closure with ATPO in the binding cleft. Of note, DNQX crystals were grown in a buffer containing ammonium sulfate, and in the receptor antagonist complex a sulfate ion acts as a bridge between the base of helix F and the antagonist molecule. Whether the binding of quinoxalines such as CNQX and

DNQX is altered by sulfate or phosphate ions in vivo has not been determined, and it is possible that a network of water molecules can support the binding of these antagonists in the absence of sulfate or phosphate. Also of interest is the finding that compared with the apo structure, both DNQX and ATPO produce between 2.5 and 6.0° of domain closure. In wild-type receptors this is insufficient to activate ion channel gating, but in iGluRs with the lurcher mutation, which destabilizes the closed state of the channel (62), quinoxalines act as partial agonists (63).

THE BINDING SITE FOR AMPA RECEPTOR-SELECTIVE AGONISTS

The GluR2 S1S2 construct used for crystallographic studies is a member of the AMPA receptor gene family for which a large series of ligands has been synthesized. At present, structures have been solved for the agonists glutamate, AMPA, and kainate (52, 53); quisqualate (64); willardiine and its 5-F, 5-Br, and 5-I derivatives (65); and a series of isoxazoles, (S)-2-Me-Tet-AMPA, (S)-ACPA, (S)-Br-HIBO (66), and ATPA (67). A key feature that emerged from these studies was the demonstration that in the active closed cleft conformation, the GluR2 agonist-binding pocket is sufficiently large to accommodate three water molecules as well as glutamate. In combination with agonists, these water molecules supply surrogate ligand atoms and contribute to a hydrogen bond network holding the agonist-binding core in its active, closed cleft conformation. Displacement of subsets of these structural water molecules allows the binding of heterocyclic glutamate analogues such as AMPA and its derivatives, quisqualate and the willardiines. Analysis of these structures also revealed that not all AMPA receptor–selective agonists act as isosteres of glutamate. A map of the occupancy of subsites within the agonist-binding pocket by individual agonists is shown in Figure 5 (see color insert). This reveals that quisqualate and the willardiines have interactions identical to those for glutamate, whereas for some of the sites occupied by AMPA and kainate, there is an interchange between water molecules and ligand atoms.

STRUCTURAL BASIS FOR PARTIAL AGONISM

Agonist efficacy, or the ability to activate once bound, is defined by the unique interactions that each agonist makes with the receptor. As discussed above, AMPA and glutamate both induce ~20° domain closure in GluR2 S1S2, and both agonists evoke similar size currents from nondesensitizing GluR2 receptors expressed in *Xenopus* oocytes (58). The AMPA and glutamate structures are essentially identical to each other, as well as to structures solved in complex with other full agonists such as quisqualate (65), ACPA, and 2-Me-Tet-AMPA (66). However, as shown in Figure 5, all these agonists occupy distinct combinations of subsites within the binding cleft and have unique contributions from water molecules. This suggests that the common conformation of the binding cleft is what defines the activity of these agonists. Indeed, the S1S2 complex with the partial agonist kainate has a conformation unlike that of the full-agonist structures: The GluR2 S1S2 kainate complex shows only 12° of domain closure relative to the apo conformation. The intermediate degree of domain closure produced by kainate, coupled with the observation that kainate also acts as a partial agonist at native AMPA receptors (68), suggests that the extent of domain closure determines the strength of ion channel activation. Although appealing, this hypothesis suffered from a number of complications, including the observation that glutamate and kainate bind to different subsites in the agonist-binding core and that only a single pair of ligands was compared. In addition, it remained unclear how differences in domain closure are related to ion channel activity. Reinforcement for the hypothesis that the extent of domain closure is related to the degree of activation by AMPA receptor agonists came from studies of a series of heterocyclic compounds related in structure to AMPA (66). One of these, (S)-Br-HIBO, produced 18° domain closure and generated 85% of the maximum response evoked by glutamate, whereas the other compounds examined, AMPA, 2-Me-tet-AMPA, and ACPA, produced the same extent of domain closure as glutamate and acted as full agonists.

In an attempt to address some of these issues, two sets of experiments were performed. The first utilized the high-resolution structure of the GluR2 S1S2 kainate complex, careful examination of which revealed that the isopropenyl group of kainate prevented full domain closure owing to collision with the side chains of Tyr450 and Leu650 (52, 53). Mutation of the hydrophobic leucine side chain to the smaller polar threonine side chain was expected to permit further domain closure and increase the efficacy of kainate. The L650T mutant behaved as predicted and increased the efficacy of kainate from 2 to 24% relative to maximum response for the reference compounds glutamate and quisqualate, which behave as full agonists in both wild-type GluR2 and the L650T mutant (58). The structure of the L650T S1S2 complex with kainate revealed a corresponding increase in domain closure from 12° for the wild-type complex to 15° for the L650T mutant complex.

An unexpected result of these experiments was the observation that AMPA, one of the reference compounds that acts as a full agonist on wild-type AMPA receptors (with efficacy similar to glutamate and quisqualate) behaved as a partial agonist for the L650T mutant with a maximum response only 38% of that produced by glutamate and quisqualate. Crystal structures of the L650T mutant S1S2 complex with quisqualate and AMPA were solved and revealed no difference from wild-type for quisqualate, whereas for AMPA, the agonist-binding core adopted both fully closed and partially closed conformations. Destabilization of the fully closed AMPA-bound conformation probably results from reduced hydrophobic contacts between the face of domain 1 with the Leu650 side chain in domain 2 and perturbation of solvent structure in the AMPA-bound complex.

The observation that the AMPA-bound complex of the L650T mutant can adopt multiple conformations suggests that the agonist-binding core has greater conformational flexibility than previously anticipated and that the degree of domain closure is a direct determinant of agonist efficacy. Nonetheless, because the mode of binding of AMPA and kainate differs from that for glutamate and quisqualate, it was desirable to perform similar experiments using a series of structurally related partial agonists that bind to the same sites as glutamate. The 5-substituted willardiines, which bind exactly like glutamate and act as partial agonists at AMPA receptors, were ideal compounds for this purpose. The willardiines act as partial agonists because close contacts between the 5-substituent of the uracil ring and the side chain of Met708 prevent complete domain closure. For glutamate and the willardiine five-position series—H, F, Br, and I—Met708 adopts one of three different rotamers as the size of the ligand increases, but because this is insufficient to fully relieve steric hindrance, there is an additional rigid body movement of domain 2, allowing the agonist-binding core to adopt a progressively more open conformation. Coupled to this variation in the extent of domain closure for the GluR2 S1S2 agonist-binding core is a corresponding reduction in the extent of activation and desensitization of the full-length receptor as the size of the 5-substituent is increased (69).

Viewed in the context of the GluR2 S1S2 dimer, domain closure leads to an increase in the distance between protomer linker regions. Because these linker regions replace the channel-forming region from the full-length receptor, linker separation not only suggests a physical mechanism for channel activation but also reveals, when plotted against agonist efficacy, how incremental increases in domain closure lead to corresponding increases in ion channel activation. Summarized in Figure 6 is the correlation between linker separation and relative agonist efficacy for all the crystal structures for which there is corresponding functional data. The relatively linear correlation for the agonist-bound points is striking and provides strong support for the initial domain closure hypothesis. The major anomaly on this plot, 5-iodo-willardiine, has a distorted linker because of crystal lattice contacts in this region; however, consistent with the trend for other AMPA receptor partial agonists, the degree of domain closure measured for 5-iodo-willardiine was intermediate between that for kainate and 5-bromo-willardiine (69). Such a trend does not exclude the existence of additional mechanisms for partial agonism as seen, for example, in the L650T complex with AMPA, which alters the stability of the closed cleft conformation such that on average partial agonist-bound receptors enter this state less frequently than full agonists (58).

Single-channel analysis reveals that the graded activation of AMPA receptors by 5-substituted willardiines occurs because partial agonists produce preferential activation of lower conductance substates, whereas full agonists preferentially activate higher conductance substates. However, the conductance of the individual substates is the same for full and partial agonists (69). This important result indicates that while the agonist-binding domain can adopt a range of conformations from open to closed, the ion channel likely populates a discrete set of conformations as defined by ion flux. The nature of the coupling mechanism that permits this will likely be understood only when structures of the agonist-binding core attached to the ion channel are solved. This remains a formidable challenge.

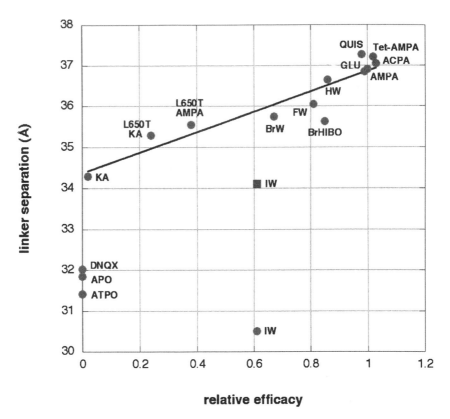

Figure 6   Correlation between agonist efficacy and GluR2 S1S2 linker separation. Agonist efficacies were measured from nondesensitizing GluR2-L483Y receptors at a saturating concentration of agonist (58, 66, 69). Efficacies are scaled relative to glutamate which is arbitrarily set to 1. The APO, DNQX, and ATPO complexes are assumed to have an efficacy of zero. Linker separation is defined as the distance between the $C\alpha$ atoms from the glycine residue in the engineered Gly-Thr S1S2 linker. The corresponding crystal structure is indicated next to each point. IW linker separation was also calculated as described by Jin et al. (69) and is indicated by a square. The linear fit to the agonist-bound points (excluding IW) yields a correlation coefficient of 0.921.

AMPA RECEPTOR DESENSITIZATION AND THE DIMER INTERFACE

A striking feature of iGluR responses, particularly for AMPA and kainate receptors, is the rapidity of onset and extent of desensitization in response to the sustained application of agonist. From the beginning of studies on iGluRs, it was recognized that desensitization resulted from a conformational change intrinsic to the receptor itself, but the nature of this process remained obscure for more than a decade. The first clues came from mapping the results from functional studies onto the GluR2 agonist-binding core crystal structure. This revealed that amino acid substitutions in the flip-flop splice variants of AMPA receptors, which regulate sensitivity to allosteric modulators of desensitization (70), lie on the dimer interface of the agonist-binding core. Likewise, the L483Y mutant, which attenuated the desensitization of GluR3 (71), was also found to lie on the dimer interface.

Establishment of the mechanism by which allosteric modulators and point mutants in the dimer interface regulate the process of desensitization was obtained from a combination of biochemical and electrophysiological approaches. Using analytical ultracentrifugation to measure the dissociation constant for dimer formation by the isolated GluR2 agonist-binding core, Sun et al. found that mutations that attenuate desensitization result in a more stable dimer complex, whereas mutations that disrupt the dimer interface accelerate the rate of onset and increase the extent of desensitization (72). A striking correlation between the extent of equilibrium desensitization in full-length receptors and the dissociation constant for dimer dissociation suggests that the processes are causally related. A simple extension of the model proposed for activation can accommodate the process of desensitization if it is assumed that the agonist binding energy is available either to open the ion channel gate or to rearrange the dimer interface, allowing the protein to relax into a low-energy state in which the agonist-binding cores remain in their active, closed cleft conformation while the ion channel closes (Figure 3). In retrospect it is clear that activation and desensitization are linked processes that use a common structural element, namely the dimer interface.

Consistent with the above picture was the discovery that cyclothiazide, an allosteric modulator of desensitization, stabilized the formation of dimers by the isolated GluR2 agonist-binding core. The crystal structure of a ternary complex of cyclothiazide, AMPA, and GluR2 S1S2 revealed that two cyclothiazide molecules were bound at the base of the dimer interface, making hydrogen bond contacts with the alternatively spliced serine side chain of the cyclothiazide-preferring flip splice variant at the base of helix J and gluing the subunits together (72).

STRUCTURAL STUDIES ON NMDA RECEPTORS

For reasons that are not yet clear, some iGluR S1S2 constructs express well and can be crystallized without difficulty, whereas others have resisted attempts at protein expression. Fortunately, the NR1 glycine-binding site of the NMDA receptor gene family has been expressed and crystallized in complex with a series of agonists and antagonists (50). As expected, glycine and other NR1-preferring ligands bind in a cleft between the two lobes of the S1S2 construct, comparable to the mechanism observed for periplasmic proteins, GluR0, GluR2, and GluR6. The fold of NR1 shows two notable differences from that of GluR2 and GluR6. First, the lengths of loop one and loop two are increased, compared with those in AMPA and kainate receptors, and in combination form a prominent ridge on the external surface of domain 1, which might be expected to form an interaction surface for contact with another subunit, or perhaps the N-terminal domain (Figure 1). Second, the orientation of one of the domain 2 helices (G) differs substantially from that in GluR2 and GluR6. Despite this, AMPA, kainate, and NMDA receptors share many common features in their ligand-binding sites.

The structure of the NR1 S1S2 agonist-binding core explains clearly why glycine is bound with high affinity, how other much larger agonists and partial agonists fit into the ligand binding pocket, and why glutamate is excluded. Despite its simple chemistry, glycine makes a total of eight hydrogen bonds and ion pair interactions with the NR1 agonist-binding site. In addition, five structural water molecules are trapped together with glycine in the closed cleft conformation. The glycine-water complex makes a network of hydrogen bond contacts that stabilizes the agonist-bound closed cleft conformation. S-glutamate is prevented from binding by a steric barrier formed by the replacement of a leucine in GluR2 by Trp731 in NR1, the large aromatic side chain of which would collide with the γ-carboxyl group of glutamate. In addition, there is loss of a hydrogen bond donor at the base of helix F, which occurs on replacement of a threonine in GluR2 by Val689 in NR1. In combination, these amino acid substitutions make the binding of glutamate to NR1 too weak to detect.

Similar to GluR2 S1S2 complexes with competitive antagonists, NR1 adopts an open cleft conformation when crystallized with the antagonist, 5,7-dichlorokynurenic acid (DCKA). DCKA acts as a wedge, or foot-in-the-door, preventing closure of the NR1 agonist-binding core. The binding of the planar bicyclic DCKA antagonist molecule is stabilized in part by stacking interactions with aromatic amino acids, reminiscent of that seen for the GluR2 complex with DNQX. Comparison of the DCKA and glycine-bound complexes reveals a 24° difference between the open and closed cleft conformations, 3° larger than that observed for GluR2 complexes. Of note, the two protomers in the asymmetric unit of the NR1 DCKA complex differ in the extent of domain closure by 6°; at present the structure of an apo complex has not been reported, but it is likely that, as observed for GluR2 DNQX and ATPO complexes, NMDA receptor glycine site antagonists can also produce moderate domain closure.

Although several lines of evidence support the conclusion that functional NMDA receptors assemble as heteromers of two NR1 and two NR2 subunits (73–75), it remains unclear how the subunits are arranged around the ion channel pore and whether they form homo- or hetero-dimers. The NR1 S1S2 construct used by Furukawa & Gouaux is monomeric in solution, and a dimerization interface similar to that observed for GluR2 and GluR0 is absent from the crystal lattice (50). The implications of this result are unclear as the literature contains conflicting reports on the association between NMDA receptor subunits. The N-terminal domain of NR1 is required for homodimerization of soluble NR1 constructs (22), and an NR1 S1S2 construct slightly longer than the one crystallized by Furukawa & Gouaux also forms dimers in solution (47). Together these results suggest that NR1 does form homodimers but that the dimerization energy was too high to compete with lattice contacts elsewhere in the crystal. Recent work using tandem-linked NMDA receptor subunits suggests a 1-1-2-2 arrangement as opposed to a positioning of subunits kitty-corner to each other (1-2-1-2) (41).

FUNCTIONAL STUDIES OF IGLURS IN THE CRYSTALLIZED STATE

An issue that frequently arises during crystallographic studies of isolated protein domains is how to relate these structures to the function of the intact receptor. There is also the valid concern that, when packed in a crystal lattice, the conformations observed may not accurately reflect those of the protein in solution. While this must always be addressed on a case-by-case basis, some elegant experiments suggest that in the case of GluR2 the behavior of the agonist-binding core in the crystal is remarkably close to that observed in solution (65). Crystal-soaking experiments reveal that GluR2 S1S2 AMPA complexes have sufficient conformational mobility to allow AMPA to unbind and be displaced by the partial agonist 5-Br-willardiine, which then stabilizes the agonist-binding core in the partially closed conformation characteristic of partial agonists. The experiments that established this used the anomalous scattering of X rays by bromine atoms to measure the binding of 5-Br-willardiine in the presence of AMPA. Crystals of the GluR2 S1S2 AMPA complex were washed in a solution containing 350 nM AMPA and then soaked in a solution containing 350 nM AMPA and various concentrations of 5-Br-willardiine. An EC_{50} of 12.9 μM was determined for development of an anomalous signal owing to binding of 5-Br-willardiine that displaced AMPA from crystallized iGluR2. In independent experiments with soluble GluR2 S1S2, an IC_{50} of 4.0 μM was measured for displacement of 350 nM [^{3}H]-AMPA by 5-Br-willardiine. Measurements of the decrease in domain closure that occurred during the crystal titration experiment gave an EC_{50} of 6.8 μM. In total, these experiments suggest that despite the constraints imposed by the crystal lattice, GluR2 S1S2 molecules have sufficient mobility to allow agonists to bind and dissociate and to undergo global conformational changes induced by the binding of full and partial agonists.

CONCLUSION

The physiological role of iGluRs is now well established, but even this maturing field remains an active area of research because of the continuing discovery of novel regulatory mechanisms. By comparison, analysis of the molecular basis of iGluR function is an emerging research area, and even though the knowledge gained from the studies reviewed here is without precedent, major questions remain to be answered. Determining the structure of a full-length vertebrate iGluR in its native oligomeric state will probably remain an elusive goal for some time. This is because techniques for the overexpression and purification of eukaryotic ion channel proteins are in their infancy. However, such structures will ultimately be required to fully understand the processes of activation, desensitization, allosteric modulation, and protein-protein interactions with other components of the synaptic membrane.

Expression of prokaryotic ion channel genes in *E. coli* has proven to be significantly more tenable than expression of eukaryotic genes. Fortunately for ion channel structural biologists, the rapidly increasing number of bacterial genome sequences has revealed that these organisms harbor homologues to virtually every family of eukaryotic ion channel. Several of these genes have been successfully overexpressed in *E. coli*, extracted from cell membrane with detergents, purified and crystallized. As proof that these homologues are likely to provide the route to a crystal structure of an intact glutamate receptor, at present every ion channel crystal structure in the Protein Data Bank is derived from a prokaryotic organism, with the exception of the nicotinic acetylcholine receptor (78). Thus far GluR homologues have been identified in at least nine bacterial genome sequences. This growing family of bacterial glutamate receptor ion channels, especially those for which sequence analysis predicts permeability to both Na^+ and K^+, are interesting and promising crystallographic targets. However, because the currently known bacterial iGluRs lack the N-terminal domain and third TM segment of their vertebrate counterparts, there are likely to be significant differences in their structure and function.

Stepping backward from the challenging goal of an intact receptor assembly, solving the structures of the N-terminal domain, and especially the N-terminal domain S1S2 complex, which makes up the whole of the extracellular portion of iGluRs, would also be a major advance. The acetylcholine-binding peptide represents an analogous case for nicotinic and related GABA, glycine, and 5-HT3 receptors. Crystallization of this protein revealed for the first time the structure of the pentameric assembly that forms the ion channel vestibule and agonist-binding site (76). In the case of iGluRs, a comparable structure would be expected to reveal the packing of the N-terminal domain and S1S2 domains, as well as the symmetry of an iGluR tetramer that currently is believed to involve assembly of the ligand-binding core as a dimer of dimers. Although this involves a symmetry violation on moving from the twofold axis of the ligand-binding core dimers to the fourfold axis of the ion channel (72), such an arrangement apparently occurs in the inward rectifier K^+ channels (77).

An immediately accessible, and ongoing research area is the determination of structures for the ligand-binding cores of additional members of the iGluR gene family; also useful would be the structures of complexes with additional agonists and antagonists because this would provide the information required for design and synthesis of novel iGluR subtype–selective ligands. The availability of such compounds would greatly advance our understanding of synaptic physiology and the complex functional roles that iGluRs play in CNS function.

The study of glutamate receptor function has advanced with unprecedented speed during the past few years, and no other neurotransmitter receptor family is as well characterized in terms of structure and mechanism, although it could be argued that nicotinic acetylcholine receptors come close (78). It is hoped that this rate of progress will continue and extend to other receptors and ligand-gated channels as well as to the complex of molecules involved in synaptic transmission.

ACKNOWLEDGMENTS

We thank Eric Gouaux for advice and discussion. Neali Armstrong is supported by the Jane Coffin Childs Memorial Fund for Medical Research. MLM is funded by the Intramural Research program of the National Institute of Child Health and Human Development, National Institutes of Health, Department of Health and Human Services.

The *Annual Review of Physiology* is online at http://physiol.annualreviews.org

LITERATURE CITED

1. Watkins JC, Evans RH. 1981. Excitatory amino acid transmitters. *Annu. Rev. Pharmacol. Toxicol.* 21:165–204
2. Mayer ML, Westbrook GL. 1987. The physiology of excitatory amino acids in the vertebrate central nervous system. *Prog. Neurobiol.* 28:197–276
3. Chen GQ, Cui C, Mayer ML, Gouaux E. 1999. Functional characterization of a potassium-selective prokaryotic glutamate receptor. *Nature* 402:817–21
4. Kuner T, Seeburg PH, Guy HR. 2003. A common architecture for K^+ channels and ionotropic glutamate receptors? *Trends Neurosci.* 26:27–32
5. McBain CJ, Mayer ML. 1994. *N*-methyl-D-aspartic acid receptor structure and function. *Phys. Rev.* 74:723–60
6. Dingledine R, Borges K, Bowie D, Traynelis SF. 1999. The glutamate receptor ion channels. *Pharmacol. Rev.* 51:7–45
7. Seeburg P. 1993. The molecular biology of mammalian glutamate receptor channels. *Trends Neurosci.* 16:359–65
8. Hollmann M, Heinemann S. 1994. Cloned glutamate receptors. *Annu. Rev. Neurosci.* 17:31–108
9. Nakanishi S. 1992. Molecular diversity of glutamate receptors and implications for brain function. *Science* 258:597–603
10. Hollmann M, Maron C, Heinemann S. 1994. N-glycosylation site tagging suggests a three transmembrane domain topology for the glutamate receptor GluR1. *Neuron* 13:1331–43
11. Bennett JA, Dingledine R. 1995. Topology profile for a glutamate receptor: Three transmembrane domains and a channel-lining re-entrant membrane loop. *Neuron* 14:373–84
12. Wo ZG, Oswald RE. 1995. A topological analysis of goldfish kainate receptors predicts three transmembrane segments. *J. Biol. Chem.* 270:2000–9
13. Wo ZG, Oswald RE. 1994. Transmembrane topology of two kainate receptor subunits revealed by N-glycosylation. *Proc. Natl. Acad. Sci. USA* 91:7154–58
14. O'Hara PJ, Sheppard PO, Thogersen H, Venezia D, Haldeman BA, et al. 1993. The ligand-binding domain in metabotropic

glutamate receptors is related to bacterial periplasmic binding proteins. *Neuron* 11:41–52

15. Stern-Bach Y, Bettler B, Hartley M, Sheppard PO, O'Hara PJ, Heinemann SF. 1994. Agonist-selectivity of glutamate receptors is specified by two domains structurally related to bacterial amino acid binding proteins. *Neuron* 13:1345–57

16. Quiocho FA, Ledvina PS. 1996. Atomic structure and specificity of bacterial periplasmic receptors for active transport and chemotaxis: variation of common themes. *Mol. Microbiol.* 20:17–25

17. Nakanishi N, Shneider NA, Axel R. 1990. A family of glutamate receptor genes: evidence for the formation of heteromultimeric receptors with distinct channel properties. *Neuron* 5:569–81

18. Swanson GT, Gereau RW, Green T, Heinemann SF. 1997. Identification of amino acid residues that control functional behavior in GluR5 and GluR6 kainate receptors. *Neuron* 19:913–26

19. Paas Y, Eisenstein M, Medevielle F, Teichberg VI, Devillers-Thiery A. 1996. Identification of the amino acid subsets accounting for the ligand binding specificity of a glutamate receptor. *Neuron* 17:979–90

20. Foucaud B, Laube B, Schemm R, Kreimeyer A, Goeldner M, Betz H. 2003. Structural model of the NMDA receptor glycine site probed by site-directed chemical coupling. *J. Biol. Chem.* 278:24011–17

21. Ayalon G, Stern-Bach Y. 2001. Functional assembly of AMPA and kainate receptors is mediated by several discrete protein-protein interactions. *Neuron* 31:103–13

22. Meddows E, Le Bourdelles B, Grimwood S, Wafford K, Sandhu S, et al. 2001. Identification of molecular determinants that are important in the assembly of N-methyl-D-aspartate receptors. *J. Biol. Chem.* 276:18795–803

23. Kuusinen A, Abele R, Madden DR, Keinanen K. 1999. Oligomerization and ligand-binding properties of the ectodomain of the alpha-amino-3-hydroxy-5-methyl-4-isoxazole propionic acid receptor subunit GluRD. *J. Biol. Chem.* 274:28937–43

24. Sack JS, Saper MA, Quiocho FA. 1989. Periplasmic binding protein structure and function. Refined X-ray structures of the leucine/isoleucine/valine-binding protein and its complex with leucine. *J. Mol. Biol.* 206:171–91

25. Kunishima N, Shimada Y, Tsuji Y, Sato T, Yamamoto M, et al. 2000. Structural basis of glutamate recognition by a dimeric metabotropic glutamate receptor. *Nature* 407:971–77

26. Paoletti P, Perin-Dureau F, Fayyazuddin A, Le Goff A, Callebaut I, Neyton J. 2000. Molecular organization of a zinc binding N-terminal modulatory domain in a NMDA receptor subunit. *Neuron* 28:911–25

27. Perin-Dureau F, Rachline J, Neyton J, Paoletti P. 2002. Mapping the binding site of the neuroprotectant ifenprodil on NMDA receptors. *J. Neurosci.* 22:5955–65

28. Masuko T, Kashiwagi K, Kuno T, Nguyen ND, Pahk AJ, et al. 1999. A regulatory domain (R1-R2) in the amino terminus of the N-methyl-D-aspartate receptor: effects of spermine, protons, and ifenprodil, and structural similarity to bacterial leucine/isoleucine/valine binding protein. *Mol. Pharmacol.* 55:957–69

29. Zheng F, Erreger K, Low CM, Banke T, Lee CJ, et al. 2001. Allosteric interaction between the amino terminal domain and the ligand binding domain of NR2A. *Nat. Neurosci.* 4:894–901

30. Panchenko VA, Glasser CR, Mayer ML. 2001. Structural similarities between glutamate receptor channels and K^+ channels examined by scanning mutagenesis. *J. Gen. Physiol.* 117:345–60

31. Kuner T, Wollmuth LP, Karlin A, Seeburg PH, Sakmann B. 1996. Structure of the NMDA receptor channel M2 segment inferred from the accessibility of substituted cysteines. *Neuron* 17:343–52

32. Kuner T, Beck C, Sakmann B, Seeburg PH. 2001. Channel-lining residues of the

AMPA receptor M2 segment: structural environment of the Q/R site and identification of the selectivity filter. *J. Neurosci.* 21:4162–72

33. Doyle DA, Cabral JM, Pfuetzner RA, Kuo A, Gulbis JM, et al. 1998. The structure of the potassium channel: molecular basis of K+ conduction and selectivity. *Science* 280:69–77

34. Jiang Y, Lee A, Chen J, Cadene M, Chait BT, MacKinnon R. 2002. Crystal structure and mechanism of a calcium-gated potassium channel. *Nature* 417:515–22

35. Jiang Y, Lee A, Chen J, Ruta V, Cadene M, et al. 2003. X-ray structure of a voltage-dependent K+ channel. *Nature* 423:33–41

36. Zhou Y, Morais-Cabral JH, Kaufman A, MacKinnon R. 2001. Chemistry of ion coordination and hydration revealed by a K+ channel-Fab complex at 2.0 Å resolution. *Nature* 414:43–48

37. Burnashev N, Villarroel A, Sakmann B. 1996. Dimensions and ion selectivity of recombinant AMPA and kainate receptor channels and their dependence on Q/R site residues. *J. Physiol.* 496:165–73

38. Villarroel A, Burnashev N, Sakmann B. 1995. Dimensions of the narrow portion of a recombinant NMDA receptor channel. *Biophys. J.* 68:866–75

39. Mayer ML, Olson R, Gouaux E. 2001. Mechanisms for ligand binding to GluR0 ion channels: crystal structures of the glutamate and serine complexes and a closed apo state. *J. Mol. Biol.* 311:815–36

40. Ren H, Honse Y, Karp BJ, Lipsky RH, Peoples RW. 2003. A site in the fourth membrane-associated domain of the *N*-methyl-D-aspartate receptor regulates desensitization and ion channel gating. *J. Biol. Chem.* 278:276–83

41. Schorge S, Colquhoun D. 2003. Studies of NMDA receptor function and stoichiometry with truncated and tandem subunits. *J. Neurosci.* 23:1151–58

42. Sheng M. 2001. Molecular organization of the postsynaptic specialization. *Proc. Natl. Acad. Sci. USA* 98:7058–61

43. Scannevin RH, Huganir RL. 2000. Postsynaptic organization and regulation of excitatory synapses. *Nat. Rev. Neurosci.* 1:133–41

44. Kuusinen A, Arvola M, Keinanen K. 1995. Molecular dissection of the agonist binding site of an AMPA receptor. *EMBO J.* 14:6327–32

45. Arvola M, Keinänen K. 1996. Characterization of the ligand-binding domains of glutamate receptor (GluR)-B and GluR-D subunits expressed in *Escherichia coli* as periplasmic proteins. *J. Biol. Chem.* 271:15527–32

46. Keinanen K, Jouppila A, Kuusinen A. 1998. Characterization of the kainate-binding domain of the glutamate receptor GluR-6 subunit. *Biochem. J.* 330:1461–67

47. Ivanovic A, Reilander H, Laube B, Kuhse J. 1998. Expression and initial characterization of a soluble glycine binding domain of the *N*-methyl-D-aspartate receptor NR1 subunit. *J. Biol. Chem.* 273:19933–37

48. Miyazaki J, Nakanishi S, Jingami H. 1999. Expression and characterization of a glycine-binding fragment of the *N*-methyl-D-aspartate receptor subunit NR1. *Biochem. J.* 340(Pt 3):687–92

49. Chen GQ, Gouaux E. 1997. Overexpression of a glutamate receptor (GluR2) ligand binding domain in *Escherichia coli*: application of a novel protein folding screen. *Proc. Natl. Acad. Sci. USA* 94:13431–36

50. Furukawa H, Gouaux E. 2003. Mechanisms of activation, inhibition and specificity: crystal structures of NR1 ligand-binding core. *EMBO J.* 22:1–13

51. Chen GQ, Sun Y, Jin R, Gouaux E. 1998. Probing the ligand binding domain of the GluR2 receptor by proteolysis and deletion mutagenesis defines domain boundaries and yields a crystallizable construct. *Protein Sci.* 7:2623–30

52. Armstrong N, Sun Y, Chen GQ, Gouaux E. 1998. Structure of a glutamate-receptor ligand-binding core in complex with kainate. *Nature* 395:913–17

53. Armstrong N, Gouaux E. 2000. Mechanisms for activation and antagonism of an AMPA-sensitive glutamate receptor: crystal structures of the GluR2 ligand binding core. *Neuron* 28:165–81

54. McFeeters RL, Oswald RE. 2002. Structural mobility of the extracellular ligand-binding core of an ionotropic glutamate receptor. Analysis of NMR relaxation dynamics. *Biochemistry* 41:10472–81

55. Fethiere J, Andersson A, Keinanen K, Madden DR. 2000. Crystallization of an AMPA receptor binding domain without agonist: importance of carbohydrate content and flash-cooling conditions. *Acta Crystallog. D* 56:1625–29

56. Pasternack A, Coleman SK, Fethiere J, Madden DR, LeCaer JP, et al. 2003. Characterization of the functional role of the N-glycans in the AMPA receptor ligand-binding domain. *J. Neurochem.* 84:1184–92

57. Sun YJ, Rose J, Wang BC, Hsiao CD. 1998. The structure of glutamine-binding protein complexed with glutamine at 1.94 Å resolution: comparisons with other amino acid binding proteins. *J. Mol. Biol.* 278:219–29

58. Armstrong N, Mayer M, Gouaux E. 2003. Tuning activation of the AMPA-sensitive GluR2 ion channel by genetic adjustment of agonist-induced conformational changes. *Proc. Natl. Acad. Sci. USA* 100: 736–41

59. Abele R, Keinanen K, Madden DR. 2000. Agonist-induced isomerization in a glutamate receptor ligand-binding domain. A kinetic and mutagenetic analysis. *J. Biol. Chem.* 275:21355–63

60. Jiang Y, Lee A, Chen J, Cadene M, Chait BT, MacKinnon R. 2002. The open pore conformation of potassium channels. *Nature* 417:523–26

61. Hogner A, Greenwood JR, Liljefors T, Lunn ML, Egebjerg J, et al. 2003. Competitive antagonism of AMPA receptors by ligands of different classes: crystal structure of ATPO bound to the GluR2 ligand-binding core, in comparison with DNQX. *J. Med. Chem.* 46:214–21

62. Zuo J, De Jager PL, Takahashi KA, Jiang W, Linden DJ, Heintz N. 1997. Neurodegeneration in Lurcher mice caused by mutation in delta2 glutamate receptor gene. *Nature* 388:769–73

63. Taverna F, Xiong ZG, Brandes L, Roder JC, Salter MW, MacDonald JF. 2000. The Lurcher mutation of an alpha-amino-3-hydroxy-5-methyl-4-isoxazolepropionic acid receptor subunit enhances potency of glutamate and converts an antagonist to an agonist. *J. Biol. Chem.* 275:8475–79

64. Jin R, Horning M, Mayer ML, Gouaux E. 2002. Mechanism of activation and selectivity in a ligand-gated ion channel: structural and functional studies of GluR2 and quisqualate. *Biochemistry* 41:15635–43

65. Jin R, Gouaux E. 2003. Probing the function, conformational plasticity, and dimer-dimer contacts of the GluR2 ligand-binding core: studies of 5-substituted Willardiines and GluR2 S1S2 in the crystal. *Biochemistry* 42:5201–13

66. Hogner A, Kastrup J, Jin R, Liljefors T, Mayer M, et al. 2002. Structural basis for AMPA receptor activation and ligand selectivity: crystal structures of five agonist complexes with the GluR2 ligand-binding core. *J. Mol. Biol.* 322:93

67. Lunn ML, Hogner A, Stensbol TB, Gouaux E, Egebjerg J, Kastrup JS. 2003. Three-dimensional structure of the ligand-binding core of GluR2 in complex with the agonist (S)-ATPA: implications for receptor subunit selectivity. *J. Med. Chem.* 46:872–75

68. Patneau DK, Vyklicky L, Mayer ML. 1993. Hippocampal neurons exhibit cyclothiazide-sensitive rapidly desensitizing responses to kainate. *J. Neurosci.* 13:3496–509

69. Jin R, Banke TG, Mayer ML, Traynelis SF, Gouaux E. 2003. Structural basis for partial agonist action at ionotropic glutamate receptors. *Nat. Neurosci.* 6:803–10

70. Partin KM, Bowie D, Mayer ML. 1995. Structural determinants of allosteric

regulation in alternatively spliced AMPA receptors. *Neuron* 14:833–43

71. Stern-Bach Y, Russo S, Neuman M, Rosenmund C. 1998. A point mutation in the glutamate binding site blocks desensitization of AMPA receptors. *Neuron* 21:907–18

72. Sun Y, Olson R, Horning M, Armstrong N, Mayer M, Gouaux E. 2002. Mechanism of glutamate receptor desensitization. *Nature* 417:245–53

73. Benveniste M, Mayer ML. 1991. Kinetic analysis of antagonist action at *N*-methyl-D-aspartic acid receptors. Two binding sites each for glutamate and glycine. *Biophys. J.* 59:560–73

74. Clements JD, Westbrook GL. 1991. Activation kinetics reveal the number of glutamate and glycine binding sites on the *N*-methyl-D-aspartate receptor. *Neuron* 7:605–13

75. Behe P, Stern P, Wyllie DJ, Nassar M, Schoepfer R, Colquhoun D. 1995. Determination of NMDA NR1 subunit copy number in recombinant NMDA receptors. *Proc. R. Soc. London Ser. B* 262:205–13

76. Brejc K, van Dijk WJ, Klaassen RV, Schuurmans M, van der Oost J, et al. 2001. Crystal structure of an ACh-binding protein reveals the ligand-binding domain of nicotinic receptors. *Nature* 411:269–76

77. Kuo A, Gulbis JM, Antcliff JF, Rahman T, Lowe ED, et al. 2003. Crystal structure of the potassium channel KirBac1.1 in the closed state. *Science* 300:1922–26

78. Miyazawa A, Fujiyoshi Y, Unwin N. 2003. Structure and gating mechanism of the acetylcholine receptor pore. *Nature* 424:949–55

Annu. Rev. Physiol. 2004. 66:183–207
doi: 10.1146/annurev.physiol.66.032102.114509
First published online as a Review in Advance on August 14, 2003

BIOCHEMICAL INDICATORS OF STRESS AND METABOLISM: Applications for Marine Ecological Studies

Elizabeth P. Dahlhoff

*Department of Biology and Environmental Studies Institute, Santa Clara University,
Santa Clara, California 95053; email: edahlhoff@scu.edu*

Key Words RNA:DNA ratio, metabolic enzymes, marine physiological ecology,
heat shock protein, climate change

■ **Abstract** Studies investigating the effects of temperature, food availability, or
other physical factors on the physiology of marine animals have led to the development of biochemical indicators of growth rate, metabolic condition, and physiological
stress. Measurements of metabolic enzyme activity and RNA/DNA have been especially valuable as indicators of condition in studies of marine invertebrates and fishes,
groups for which accurate determination of field metabolic rates is difficult. Properly
calibrated and applied, biochemical indicators have been successfully used in studies of rocky intertidal ecology, where two decades of experimentation have generated
rigorous, testable models for determining the relative influences of biotic and abiotic
factors on species distribution, abundance, and interaction. Biochemical indicators of
condition and metabolic activity (metabolic enzymes, RNA/DNA) have been used to
test nutrient-productivity models by demonstrating tight linkages between nearshore
oceanographic processes (such as upwelling) and benthic rocky intertidal ecosystems.
Indices of condition and heat stress (heat shock proteins, or Hsps) have begun to be
used to test environmental stress models by comparing condition, activity, and Hsp
expression of key rocky intertidal predator and prey species. Using biochemical indicators of condition and stress in natural systems holds great promise for understanding
mechanisms by which organisms respond to rapid environmental change.

INTRODUCTION

Understanding the mechanisms by which environmental variation impacts organisms in nature is of great interest to comparative biologists and ecologists and is
becoming increasingly important as the effects of climate change cascade through
the biosphere. As a result, there has been increasing interest in determining the
physiological condition of organisms in a natural context (1–8). Central to the success of many of these integrative studies has been the development of biochemical
indicators. Biochemical indicators of metabolic activity are key components of

0066-4278/04/0315-0183$14.00

synthetic or metabolic biochemical pathways that are directly or indirectly linked to processes important for survival or reproduction. The basic premise of their use is that adjustments in rates of physiological processes are necessary to bring metabolic demands into alignment with available energy supply. Biochemical indicators of stress are typically components of the cellular stress response, which are upregulated as a consequence of exposure to environmental conditions that perturb cellular protein structure. The concentration of biochemical components of metabolism or stress shift as a direct result of changing environmental conditions, resulting in alterations in protein synthesis or metabolism that impact performance, growth, or reproductive output. Thus measuring biochemical indicators of stress or metabolism can be used as a "snapshot" of the condition of the organism at the time it was sampled or collected.

Biochemical indicators of condition were first developed by fishery biologists to examine wild populations of economically important fish stocks (9–13). Traditional techniques for determining growth rates or nutritional condition, such as otolith size or liver-somatic index, were time-consuming and could not be used to measure changes in nutritional condition or growth rate on short (e.g., <4 week) timescales (10, 14). Furthermore, these measures were of little utility for assessing survival and growth of early larval stages upon which environmental food availability, temperature, and toxins may have especially large effects (10). These researchers needed a tool with which they could quickly and inexpensively determine the condition of wild fish. It was critical to be able to assay large numbers of individuals and work with preserved specimens or frozen tissue, since sampling typically occurred on vessels with no laboratory facilities. Subsequent studies of wild and hatchery-reared fish (3, 15–19) and marine bivalves and crustaceans in aquaculture (20–22) demonstrate the utility of a suite of biochemical indicators for determining nutritional condition and metabolic activity in situ, including metabolic enzyme activities and the ratio of ribonucleic acid to deoxyribonucleic acid (RNA:DNA ratio, or RNA/DNA). A parallel line of investigation resulted in the discovery that organisms exposed to extreme conditions up-regulate the expression of stress-inducible molecular chaperones [heat shock proteins (Hsps) and ubiquitin], which minimize tissue levels of unfolded proteins by repairing them or tagging them for destruction (23–25). As these methods became more widespread in their use, they drew the attention of marine ecologists and eco-physiologists, who have subsequently used biochemical indicators of stress and metabolism to explore mechanistic relationships between organisms and their environment in a natural context. It is these eco-physiological studies that are the focus of this review.

METABOLIC ENZYME ACTIVITIES AS INDICES OF CONDITION

Assessing the physiological condition of ecologically important marine organisms is essential for understanding how environmental change affects survival and predicting how a species or group will interact with competitors or predators.

Metabolic rate is an excellent indicator of physiological condition (26–30). Elegant techniques have been developed to measure in situ metabolic rates of mammals, birds, and large fishes such as sharks and tunas (31–36). Unfortunately, determining field metabolic rates is costly and time-consuming, and in some habitats (such as the wave-swept rocky intertidal or the deep sea) logistically challenging. In addition, many ecological studies require much larger sample sizes than are feasible by measuring metabolic rate alone.

It was primarily out of the need to assess metabolic activity in inaccessible habitats or for large numbers of individuals that the use of biochemical indicators was brought into comparative and ecological marine studies. The first indicators developed to determine physiological condition in a natural context were measurements of metabolic enzyme activities in pelagic, Antarctic, and deep-sea fishes (6, 37–39). These studies showed that metabolic rate declines rapidly as a function of minimum depth of occurrence of each species, concomitant with a steep decline in food availability with depth (6, 39). This pattern persists even when the effects of declining temperature and differences in body size are taken into account, suggesting that declining food availability with increased depth explains, in part, declines in metabolic rate. In these studies, the activities of lactate dehydrogenase (LDH), a glycolytic enzyme critical for burst swimming capacity, and citrate synthase (CS), a tricarboxylic acid (TCA) cycle enzyme critical for aerobic metabolism, were also measured and shown to directly correlate with metabolic rate, and hence food availability, for these fish species.

Many studies have now demonstrated a close link between the activity of enzymes critical to energy metabolism, metabolic rate, and food availability for fishes (19, 38, 40–42) and marine invertebrates (2, 43–47). One key to the success of these studies is selection of the appropriate enzyme-tissue combinations for each species of interest. An excellent example of this point is the use of the glycolytic enzyme LDH as an index of condition (40, 48–52). Changes in food availability strongly affect LDH activity in white muscle. However, LDH activity (and that of other metabolic enzymes) tends to remain constant in brain, independent of changes in environmental food quality or quantity (42, 53). LDH is central to burst swimming performance because its activity allows for the continuance of energy production critical for muscle contraction during functional hypoxia (54). A decrease in LDH activity because of low food availability directly impacts swimming performance, causing a decline in the ability of an individual to escape from predators or capture prey. Conversely, brain LDH activity, while low, is conserved during starvation, presumably to allow the individual to survive until conditions are more ideal for active movement and growth. Similar tissue-specific responses are observed for other metabolic enzymes (such as CS or pyruvate kinase). In addition, linkages between metabolic enzymes and condition also depend on the natural history of the organism of interest (19, 50, 52, 55, 56). Thus calibration of biochemical indicators with laboratory studies, as well as using multiple measures whenever possible (biochemical indicators, metabolic rate measurements, growth rates, etc.), is critical.

Another important point regarding the use of metabolic enzymes as biochemical indicators of metabolism is that while variation in enzyme activity may correlate with metabolic rate and environmental food availability, the underlying cause for metabolic variation may not be related to food availability per se, but to some other environmental variable. A powerful example of this point has been demonstrated by recent work conducted on deep-sea pelagic cephalopods (57, 58). For deep-sea squids and octopi, metabolic rate is highly correlated with CS activity from mantle tissue, and activities of mantle CS and octopine dehydrogenase (ODH, an invertebrate analog of LDH) decline precipitously with minimum depth of occurrence, as is observed for teleost fishes (6, 39). However, unlike many of the fishes used in those landmark studies, cephalopods are active visual predators. The fact that CS and ODH activities decline with depth implies that burst swimming ability is being lost with depth, either due to changes in food availability or to decreased availability of light for visual predation (57). Enzymatic activities of other swimming tissues (fin and arm) in these deep-sea cephalopods are actually quite high, implying no temperature-independent decline in metabolic rate with depth, but instead a shift in locomotor strategy from burst swimming (where mantle is primarily used) to sustained, aerobic swimming, where arm and fin movement is most prevalent. Thus in this case, a decline in CS and ODH activity in mantle tissue with depth is not strictly a function of changes in food availability but is also a relaxation in the need for burst locomotion (to catch prey or escape predators) in the light-limited deep sea (58). This example underscores the importance of understanding the natural history and physiology of one's study organisms when using biochemical indicators for ecological studies.

RNA/DNA AS INDICATOR OF CONDITION

Another technique used to index condition (often in combination with aforementioned enzyme activities) is the RNA:DNA ratio, or RNA/DNA. The basic principle of using RNA/DNA as a measure of condition is that total RNA content is primarily a function of ribosome number and is correlated with new protein synthesis, whereas DNA content remains constant in an individual because it is a function of chromosome number (9, 10, 59). The ratio of RNA to DNA (rather than total RNA content) is used because changes in cell size or water content alter RNA concentration independently of any change in protein synthesis. A number of studies have shown that RNA/DNA is directly related to both tissue growth and nutritional status (reviewed in 16). Altering feeding ration in the laboratory leads to significant shifts in ribosomal RNA content and protein synthesis for a number of groups, including cod, trout, sandbass, shrimp, octopi, and sea urchins (41, 60–65). Changes in protein synthesis directly correlate with RNA/DNA, although this pattern depends on tissue type and the extent of laboratory treatment. For example, studies of the barred sand bass *Paralabrax nebulifer* and the Atlantic cod *Gadus morhua* suggested that protein synthesis, measured both directly and as indicated by RNA/DNA, was more sensitive to starvation in white than in red muscle (10, 61). Published studies show that starved individuals tend to have lower growth rates, higher mortality rates, and lower RNA/DNA than fed controls. Changes in protein synthetic capacity, either measured directly or indexed by RNA/DNA, respond to changes in food availability on the order of days, illustrating the power of this technique for assaying organisms in a rapidly changing natural environment (9, 59).

Analysis of data published for three distinct groups of marine animals, teleost fishes (10, 66–68), crustaceans (7, 62), and molluscs (22, 46, 69), held in the laboratory and either food-deprived or fed ad lib on a high-quality diet demonstrate that RNA/DNA significantly correlates with nutritional condition (Figure 1A: results of nested ANCOVA given in figure caption). Determination of food availability in situ in marine systems is difficult for animals at higher trophic levels, but for suspension-feeding invertebrates and fishes, regional primary productivity is an excellent indicator of food availability (70–74). Results from published studies in

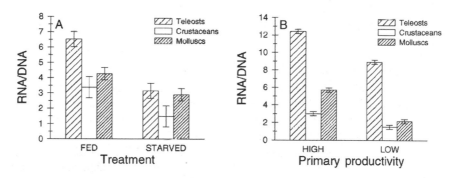

Figure 1   Correlation between food availability and RNA/DNA of muscle tissue from lab-acclimated (*A*) and field-acclimatized (*B*) fishes, crustaceans, and molluscs. Sources of data given in text. LAB: Least squares means ($\pm$SEM) were determined using nested analysis of covariance (ANCOVA), with phylogenetic group ($N \geq 2$ species, nested within group), and treatment (fed, fasted) as main effects. RNA/DNA is correlated with lab nutritional condition (ANCOVA, treatment: $F_{1,11} = 24.75$, p < 0.05) independent of phylogenetic group (food × group interaction NS). FIELD: Least squares means ($\pm$SEM) were determined using nested ANCOVA, with phylogenetic group ($N \geq 2$ species, nested within group) and food availability (high PP, low PP) as main effects. In this analysis, primary productivity variation was grouped as high or low on the basis of measurements provided by the authors in each publication. Studies were excluded from literature survey if primary productivity data were not provided or easily accessed from another publication. RNA/DNA is correlated with food availability in nature (ANCOVA, condition: $F_{1,13} = 79.83$, p < 0.0001) independent of phylogenetic group (food availability × group interaction NS), although there are significant differences in RNA/DNA between phylogenetic groups (ANCOVA, group: $F_{2,13} = 175.08$, p < 0.0001).

which primary productivity drives food-web dynamics (upwelling coastal ecosystems and the Antarctic) were compared (Figure 1B) and show that RNA/DNA is linked to food availability for planktivorous or larval fishes (43, 75), crustaceans (7, 45), and molluscs (2, 47, 76) living in a variety of natural systems. This analysis suggests that RNA/DNA is correlated with food availability in nature and is thus a good indicator of condition, whether variation in food availability is related to differences between seasons (2, 7, 45, 76) or study localities (43, 47, 75). However, there are significant differences in RNA/DNA between phylogenetic groups, emphasizing the importance of laboratory calibration using species of interest whenever possible.

There is still considerable debate about the acceptability of using metabolic enzymes or RNA/DNA as indicators of condition for animals in a natural context. First, it is difficult to distinguish between effects of food availability on protein synthesis from other factors that may influence the rates of metabolic processes. In particular, environmental temperature variation profoundly impacts metabolic rate, as most marine animals are ectotherms, leading to under or over estimation of metabolic activity when temperature is variable (53, 77–79). The use of multiple indicators, as well as close monitoring of temperature, can minimize (but not eliminate) this consideration. Second, individuals held under identical conditions in the laboratory, or exposed to similar conditions in nature, may have different metabolic enzyme activities or RNA/DNA owing to genetic variation within populations, differing availability of stored carbon, or other factors (80–83). Third, DNA content and protein expression will vary among different phylogenetic groups and between different tissue types, making comparisons between species potentially difficult. Fourth, for RNA/DNA determinations, results often depend on the technique used to determine nucleic acid content, in part because RNA is much more sensitive to enzymatic degradation than DNA. Recent advances in nucleic acid quantification techniques have minimized this final problem (84, 85), but the other points remain considerations for applying these techniques to ecological studies.

HEAT SHOCK PROTEINS AS INDICATORS
OF THERMAL STRESS

The increased importance of determining effects of environmental extremes on the physiology of animals in nature, as well as recent advances in biochemical and molecular techniques, have led to an explosion of studies investigating stress-inducible molecular chaperones (Hsps and ubiquitin) in an ecological context (47, 86–90). Temperature extremes, pH extremes, anoxia, or high heavy metal concentrations can induce molecular chaperones, and repeated exposure to sublethal temperatures may induce expression of stress proteins that enhance thermal tolerance (25, 91–93). Hsp induction benefits organisms by enhancing their ability to recover from environmental stress and to cope with subsequent stress (25, 86, 94), although there may be a fitness cost associated with over-expressing Hsps (95–97). When stress leads to accumulation of irreversibly damaged proteins, the ubiquitin response is activated. Ubiquitin is a small protein that attaches to proteins beyond repair and tags them for destruction by proteases (94). Western blots treated with antiubiquitin antibody reveal a heterogeneous population of tagged proteins or ubiquitin conjugates. Although increases in diversity of proteins conjugated to ubiquitin have been used as an index of thermal stress in situ (5, 98), differential expression of Hsps (especially the 70 kDa class) are the most commonly used biochemical indicator of heat stress in eco-physiological studies (47, 87, 99–101).

Variation in expression of Hsps in nature is correlated with differences in average environmental temperature and degree of environmental heterogeneity across large and small-scale geographic thermal gradients and through time (5, 47, 88, 89, 99, 102). The threshold at which stress-inducible forms of Hsps are upregulated in nature depends on genetic differences among populations or between species and upon recent thermal history (87, 103–106). However, as with indices of metabolic activity and condition, results must be put into informed physiological context. First, whereas some forms of Hsps are expressed only in response to conditions that unfold cellular proteins (so called stress-inducible isoforms), other forms of Hsps are present at all times as an integral part of the regular protein synthetic machinery. These constitutively expressed Hsps (cognates) are critical for folding nascent polypeptides as they are being synthesized by the ribosome (23, 24). Many techniques routinely used for Hsp quantification in natural systems (in particular, immunocytochemical analysis of Western blots) do not easily distinguish between inducible forms and cognate. Second, tissue levels of cognate Hsps may increase in response to temperature-induced elevations in metabolic rate, independently of the presence of damaged proteins. Third, genetic differences among populations may lead to differences in Hsp expression among different genotypes experiencing identical environmental conditions (89, 103, 105). Finally, although most studies are interested in one particular stressor on their study organism of interest, Hsp expression is upregulated in response to any condition that unfolds protein (23, 24). In nature, different physiological stressors often co-vary. For example, in the wave-swept rocky intertidal, where organisms living in the high zone experience both elevated body temperatures and desiccation at low tide, it is difficult to distinguish the relative importance of these two factors in upregulating Hsps. As with biochemical indicators of metabolism, these concerns can be minimized with a good understanding of natural history and ecology of the study system, careful laboratory calibration, and good experimental design.

APPLICATION OF BIOCHEMICAL INDICATORS TO FIELD STUDIES

There are a number of other biochemical indicators that are used in marine studies, including total protein content, stable isotope ratios, and nitrogen/carbon ratios (3, 107–111). However, the majority of studies using biochemical indicators to address ecological problems have used some combination of metabolic enzyme activities, RNA/DNA, and Hsp expression, and those studies are the focus of the balance of this review. One of the most compelling applications of biochemical indices in ecological studies has been in testing models that predict changes in the structure of ecological communities—species distribution, species abundance, and the strength of interactions between competitors or between predators and prey—along environmental gradients in the rocky intertidal region. Two classes of models have been developed and tested using biochemical indicators of metabolism and stress: nutrient productivity models (N/P models) and environmental stress models or ESMs (73, 112–115). These models assume that species interactions ultimately drive community structure but that communities are organized along environmental gradients, and severe habitats will have fewest number of species and lowest productivity, whereas the mildest habitats will have higher productivity, biomass, and species diversity. The two models are complementary: N/P models emphasize impacts of forces at the base of the food web on higher trophic levels (bottom-up effects), whereas ESMs assume that community structure is driven by variation in strength of interactions between basal species (algae, suspension-feeding invertebrates) and foragers along environmental stress gradients (top-down effects). Both models have been important to a qualitative understanding of rocky intertidal community dynamics, and the application of biochemical indicators to these classical ecological problems has allowed for rigorous quantification of the relationships between environmental variation and community structure.

Using Biochemical Indicators of Metabolism to Demonstrate the Importance of Bottom-Up Effects in Coastal Upwelling Ecosystems

The testing of N/P models has been in the form of comparisons between well-described rocky intertidal regions with highly distinct community structures where a clear understanding of nearshore oceanographic processes, which may be driving differences in community structure via bottom-up effects, has been developed (114–120). Studies at two sites along the Oregon coast, Strawberry Hill (SH) and Boiler Bay (BB), have been particularly illustrative (2, 46, 47, 121–123). Patterns of species distribution and abundance differ tremendously between these two sites (Figure 2, see color insert). Strawberry Hill is a site with high invertebrate biomass, high primary productivity in nearshore waters, and a potentially high rate of delivery of that productivity to nearshore suspension-feeders owing to current patterns and topography. Boiler Bay has relatively lower invertebrate biomass, lower primary productivity in nearshore waters, and oceanographic features that inhibit phytoplankton delivery to the nearshore. In the mid-low zone at SH, space is dominated by suspension-feeding invertebrates and foraging carnivores; at BB, the mid-low zone is dominated by red and brown macroalgae, as well as by foraging herbivores, principally limpets and urchins (117, 118). Suspension-feeding invertebrates, such as the mussels *Mytilus californianus* and *M. trossulus* and the barnacles *Balanus glandula* and *Chthamalus dalli*, are larger, more abundant, and have higher growth rates at SH than at BB. Foraging carnivores (the sea star *Pisaster ochraceus* and the dogwhelks *Nucella ostrina* and *N. canaliculata*) are larger and more abundant at SH than BB (121, 123, 124). Whelk foraging rates are higher at SH than BB (47). However, foraging rates of whelks and sea stars are strongly influenced by water temperature, which is not different between sites but varies tremendously during upwelling events (125–127).

If nearshore primary productivity variation is a factor driving differences in community structure between BB and SH, then this should be reflected in the condition of suspension-feeding benthic invertebrates. One example is the mussel *M. californianus*. This suspension-feeding bivalve inhabits the mid-low zone at SH and BB, is a major competitor for space, and an important prey species for foraging whelks and sea stars. It is thus central to community dynamics at these sites. Metabolic enzyme activities (CS and malate dehydrogenase, an enzyme of critical importance for anaerobic metabolism in bivalves experiencing routine environmental hypoxia) and RNA/DNA are consistently higher at SH than BB in adductor (shell) muscle of these mussels (2). This finding is generally consistent with patterns of phytoplankton availability between sites (discussed below). Reciprocal transplants of mussels between sites result in convergence of mussel condition to those native to transplant destination, suggesting a high degree of physiological plasticity in response to environmental variation in food availability between sites.

Differences in mussel condition between SH and BB persist through much of the year but are most pronounced in summer, during coastal upwelling events. Upwelling is a process that occurs along ocean shores in summer, where prevailing winds are in the same direction as prevailing currents (128, 129). With these winds (and the Coriolis effect), warm water, which has been depleted of nutrients by phytoplankton, is pushed offshore, causing cold, nutrient-rich water to upwell from the depths. During times when the winds slow (or change direction), upwelling forces dwindle, leaving the nutrient-rich water to pool at the surface. Phytoplankton in surface waters responds to upwelling relaxation by rapidly increasing growth rate until nutrients are again depleted. This phytoplankton tends to be more nitrogen-rich than phytoplankton present at other times, which means that upwelling-supported phytoplankton is of higher quality for suspension feeders than typical phytoplankton (130–132). At SH and BB, differences in surface current features are also important. During upwelling, a large eddy forms in the nearshore water off SH. This feature retains nutrient and phytoplankton-rich water in the nearshore at SH, whereas at BB, surface water is moved rapidly southward and offshore (133). Thus, more upwelling phytoplankton is delivered to suspension-feeders at SH than BB. Changes in nearshore phytoplankton concentration and quality can be rapid in intermittent upwelling ecosystems such as the SH-BB system. Mussels seem to respond rapidly to upwelling-induced influxes of phytoplankton; in some cases, mussel RNA/DNA increases within 48 h of upwelling relaxation (2, 117). Because upwelling relaxation is usually accompanied by elevated sea surface temperature, rapid increases in protein synthesis (indexed by RNA/DNA) are probably enhanced by temperature-dependent elevation of mussel metabolic rate.

Could variation in suspension feeder condition, driven by differences in near-shore primary productivity and by the delivery of high-quality phytoplankton during upwelling events, result in the radically different community structures of SH and BB? If N/P variation is driving community structure, then increased energy input into suspension-feeder growth and reproduction increases their size and abundance, effectively increasing the carrying capacity of foraging predators by increasing their food supply, assuming space is not limiting the abundance of foragers. Thus the condition of both suspension-feeders and their predators (sea stars, whelks) should be robust at high primary productivity sites. Conversely, if variation in nearshore productivity is not shaping community dynamics, then suspension-feeders may still be more robust at high productivity sites, but the condition of foraging predators, which are able to compensate for lower food quality by foraging longer or making different prey choices, would be independent of nearshore food availability.

Expanding the SH-BB California mussel comparison to other key invertebrate species at these sites, it was observed that mussels (*M. californianus, M. trossulus*), barnacles (*B. glandula*), and whelks (*N. ostrina*) are more physiologically robust at SH than BB: They have higher metabolic rates, metabolic enzyme activities, and/or tissue RNA/DNA. In contrast, metabolic activity of the sea star *Pisaster ochraceus* does not vary between sites (46, 47). If bottom-up effects were not important, there should be no differences in the metabolic activity of sea stars or whelks between sites. Conversely, if bottom-up effects were always important, then both whelks and sea stars would be more robust at SH than BB. These data suggest that bottom-up effects are important for some (whelk-barnacle, whelk-mussel), but not all (mussel-sea star) predator-prey interactions. Along with ecological studies conducted at these sites, physiological data suggest that performance of prey (and some predators) is greater in areas of high N/P input. This may ultimately negatively impact prey abundance, as a more robust predator is a more effective forager. In contrast, it may enhance prey abundance, as predator foraging may be more difficult, since robust condition of suspension feeders leads to higher growth rates (allowing prey to escape predation by getting big faster) or to more resistance to predation (the ability of a mussel to keep its shell closed during a sea star attack, for instance).

Studies of other rocky intertidal systems have supported the idea that regions with periods of high upwelling have more robust suspension-feeders and predators than regions of low upwelling. Rocky intertidal communities on the east and west coasts of the south island of New Zealand experience distinct oceanographic conditions (intermittent upwelling on the west coast, but consistently low N/P input on the east coast). West coast sites are characterized by high suspension-feeder (mussel and barnacle) recruitment and growth rates, consumer abundance, and predation rates, relative to east coast sites. As was observed in the Oregon system, upwelling input is tightly linked to suspension-feeder condition, as indexed by RNA/DNA (119, 134).

The importance of suspension-feeder condition for bringing nearshore primary productivity into the rocky intertidal food web is clearly present in upwelling-driven ecosystems. In these systems, performance and fitness of dominant suspension-feeders is linked with environmental variation via physiological condition, which varies predictably between sites or over time. But many other physical and biological factors are in play in rocky intertidal ecosystems, and the interaction between somewhat predictable environmental factors, such as temperature and food availability, can impact community dynamics in unpredictable ways. A large-scale investigation of marine ecosystems along the west coast of North America, conducted by PISCO (Partnership for Interdisciplinary Studies of Coastal Oceans) demonstrates this point (Figure 3). Fourteen rocky intertidal sites in Oregon and California were selected for study by PISCO on the basis of nearshore oceano-graphic conditions (high versus low upwelling, proximity to key features of the California current), community structure (dominated by kelps versus dominated by invertebrates in the mid-low zone), and a good understanding of community dynamics at each site. Primary productivity is highly variable between regions and sites. Generally, phytoplankton concentration in nearshore water is much higher in Oregon than in central or southern California, a pattern consistent with more intermittent upwelling events in Oregon than California, where upwelling (and offshore transport) is more persistent. Growth rates and RNA/DNA of the ecologically important suspension-feeding mussel *M. californianus* generally correlate with differences in primary phytoplankton productivity. Mussels at low phytoplankton productivity sites in central California had the lowest growth rates and RNA/DNA; mussels at high phytoplankton productivity sites in Oregon had the highest RNA/DNA and moderate growth rates.

More importantly, these data demonstrate that primary phytoplankton productivity, mussel growth, and mussel condition are not always coupled. For example, mussel growth rates at the southern California sites (Alegria and Jalama) were as high (or higher) than for mussels living at sites in Oregon rich in phytoplankton delivery (e.g., SH and Gull Haven); RNA/DNA values were high at Jalama and moderate at Alegria both years. In this case, further sampling indicated that southern California mussels were relying on an alternate food source, i.e., decaying algal matter (PISCO consortium, unpublished results). Mussels at sites in Oregon with high algal densities (e.g., BB) do not routinely have high growth rates, metabolic enzyme activities, or RNA/DNA (2, 123), so algal input alone may not drive these very high growth rates in southern California mussels. However, unlike the other sites in this study, Alegria and Jalama are south of Point Conception, and sea surface temperature is significantly higher than that for central California or Oregon. Elevated mussel body temperature will likely lead to an increase in metabolic rate. This effect, in combination with an alternate food source, means that suspension-feeders may be more robust in areas of low primary productivity input if algal cover and temperatures are both high. The fact that RNA/DNA values are higher

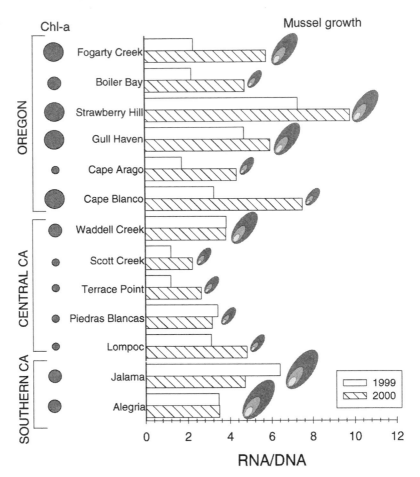

Figure 3 Nearshore primary productivity, growth rate, and RNA/DNA of adductor muscle from the dominant mussel *Mytilus californianus* at 14 rocky intertidal sites in Oregon and California. Average summer chlorophyll-a concentrations (1999, 2000) in nearshore water were determined as described in (118) and are illustrated as relative circle size graphic: low, 0.8–1.5 μg/L; moderate, 1.8–3.0 μg/L; high, 3–12 μg/L. RNA/DNA shown are for mussels collected in July of 1999 and 2000 and were determined as described in (2). Annual mussel growth rates (1999–2000) are illustrated as relative size of mussel graphic: low, 0.1–0.49 cm; moderate, 0.5–1.0 cm; high, 1.0–1.6 cm. Nearshore primary productivity predicted mussel condition, as indexed by RNA/DNA (y = 2.68,161 + 0.54788x; $F_{1,11}$ = 5.0, p < 0.05), but not as indexed by mussel growth because of very high mussel growth rates at two sites in southern California (see text). Data shown here were collected by scientists associated with the Partnership for Interdisciplinary Studies of Coastal Oceans, PISCO (B. Menge, C. Blanchette, E. Dahlhoff, P. Raimondi, S. Gaines, J. Lubchenco, D. Lohse, G. Hudson, M. Foley, J. Pamplin, C. Sorte, A. Keil, J. Freeto, J. Mello, K. Knierim & D. Ricci).

Figure 2 Contrasting patterns of community structure at rocky intertidal sites along the Pacific coast of North America have been the focus of a number of recent ecophysiological studies, including research conducted at the central Oregon sites Strawberry Hill (*upper panel*) and Boiler Bay (*lower panel*).

than expected for a low primary phytoplankton productivity site supports this assertion.

Using Biochemical Indicators of Stress and Metabolism to Test Environmental Stress Models

The rocky intertidal region is characterized by steep stress gradients created by the ebb and flow of the tide and by differences in exposure to wave splash. When low tide coincides with elevated summertime temperatures (or nighttime subzero temperatures in winter at high latitude), the amount of stress experienced by intertidal organisms can be extreme (5, 46, 47, 98, 102, 104, 135–138). Key species interactions vary along tidal or wave-exposure gradients in part because of differences in how organisms respond to physical conditions that reduce performance or compromise survival (112, 139–143). It is a classic axiom of rocky intertidal ecology that the upper tidal distribution limit of most species is determined by intolerance of physical stress, whereas the lower distribution is determined by biological interactions, such as predation or competition (112, 144–146). However, many species interactions in the rocky intertidal occur between these two extremes or between the edges of tolerance for two interacting species.

As mentioned above, ESMs are used to understand differences in community dynamics that are based on variation in strength of species interactions along environmental stress gradients (113, 115, 145). These models are typically used to describe the effects of grazers and foraging predators on the abundance of basal species (e.g., algae and suspension-feeding invertebrates), which are in competition with each other for space and other resources; thus ESMs describe top-down effects in rocky intertidal ecosystems. The predictions of ESMs depend on whether the consumer (consumer stress model) or prey (prey stress model) is most susceptible to stress. For the simple example of a two-species interaction (foraging predator, suspension-feeding prey), if the consumer is more susceptible to stress than prey (the consumer is less mobile, unable to shelter when stressful conditions arise, and/or has a less than robust physiological stress response), then in areas of high stress (e.g., higher in the intertidal, areas of low wave splash), the consumer will not control prey abundance because the majority of its energy will be used to survive the stress event. In contrast, if the prey is more susceptible to stress than the consumer (consumer is smaller than prey, able to shelter or hide from stressful conditions more easily than prey, especially if prey is sessile), in areas of high stress the prey will be weakened, leading to higher consumption rates in high stress than low stress areas. Using indices of metabolic activity (RNA/DNA, metabolic enzymes activities, metabolic rate), along with measures of the cellular stress response (Hsp expression), to study species interactions along tidal and wave-exposure gradients offers tremendous power for testing ESMs because these indicators are useful for determining physiological condition, as well as quantifying the qualitative variable stress.

Recent studies of consumers and prey at rocky intertidal sites in central Oregon (the SH-BB system discussed earlier) are a starting point for these applications. At SH, invertebrate suspension-feeders (*M. californianus*, *M. trossulus*, and *B. glandula*) are more physiologically robust (indexed by activities of metabolic enzymes and RNA/DNA) at wave-exposed (low stress) than wave-protected (high stress) areas (2, 46, 47). In addition, mussels appear to experience less heat stress (indexed by tissue levels of inducible and constitutively expressed isoforms of Hsp70) at wave-exposed than at wave-protected localities (88). Differences between wave exposures are most pronounced in summer, when mid-day low tides often coincide with relatively high air temperatures and sunny conditions, especially at SH (46, 123). One important predator-prey interaction is between the dogwhelk *Nucella ostrina* and two prey species (*M. trossulus* and *B. glandula*). Based on the premises of the prey stress model described above, the fact that dogwhelks are small, mobile, and able to avoid potentially stressful conditions would lead to the prediction that they are more resistant to stress than their prey in high-stress zones. Field experiments where both food availability and thermal stress were manipulated experimentally showed that whelks forage more actively, have higher metabolic rates, and lower levels of total Hsp70 (cognate and inducible) at wave-exposed sites, consistent with the qualitative prediction that wave-protected areas are physiologically stressful. Complete food-deprivation reduces the ability of dogwhelks to mount an effective heat shock response at wave-protected areas where heat stress is highest, suggesting that at low prey concentrations, whelks cannot tolerate elevated stress levels (47). These data suggest that although whelks are small and may be able to behaviorally avoid stressful conditions, they still experience enough stress to become physiologically compromised at the upper edge of their tidal distribution. The relative experience of stress by whelks and their prey has yet to be quantified, but preliminary studies suggest whelks may actually be more susceptible to stress than their prey [barnacles and mussels (D. McMillan & E.P. Dahlhoff, unpublished observations)], which would imply consumer stress is driving the interaction in this case.

Another important predator-prey interaction at these sites is between the foraging predator *Pisaster ochraceus* and the mussels *M. californianus* and *M. trossulus* (123, 125). As mentioned above, indices of stress and condition in mussels suggest that wave-exposed individuals are more robust than wave protected individuals. However, sea stars show no variation in metabolic activity (46) or Hsp70 expression (E.P. Dahlhoff, unpublished data) along a wave-exposure gradient. One possibility is that sea stars are so effective at behaviorally avoiding potentially stressful conditions that the time they do spend out of the water does not result in energetically significant physiological stress. In addition, recent studies of sea star (and whelk) foraging behavior suggest that sea surface temperature is a much more important determinant of ecological activity in these foragers than low-tide air temperature. During coastal upwelling, declines in sea surface temperature result in a rapid decline in sea star foraging activity. During upwelling relaxation, sea surface temperature is elevated, and foraging activity increases (125). Subsequent laboratory studies show that this decline in foraging activity is a direct consequence of temperature-induced reductions in metabolic rate (126, 127). These data suggest that foragers experiencing episodic thermal variation may have higher foraging rates at high temperature, and when temperature is low, the resulting reductions in metabolic rate lead to lower energetic demand. Thus location on the shore at low tide does not necessarily imply how much stress mobile individuals are experiencing. Again, integrating ecology with measures of cellular biochemistry and whole-organism physiology has proven highly illustrative.

FUTURE DIRECTIONS—INVESTIGATING PHYSIOLOGICAL CONSEQUENCES OF ENVIRONMENTAL CHANGE

There is an urgent need to increase our understanding of mechanisms by which animals respond to their environment, especially for organisms that occur in stressful habitats or at the edge of their range. Recent studies indicate that in both terrestrial and marine habitats, species ranges are changing substantially as a consequence of anthropogenic climate change (147–151). Whereas in many cases shifts in distribution are toward higher latitude or altitude, implying a response to overall warming of global climate, this is not uniformly the case; the impact of climate change on the distribution or abundance of given species across an environmental cline is not predictable. For example, the fact that daytime low tides tend to occur at high latitudes in summer suggests that in some ecosystems (e.g., the rocky intertidal), high-latitude sites may actually be more stressful than south temperate or subtropical localities, causing local extinction of sensitive species at high-latitude sites as a result of global warming (152). Second, although it has long been postulated that a given species reaches its highest abundance in the center of its range and that decline in abundance at the edge of a range is a result of biotic or abiotic intolerance of conditions outside that range, recent studies in the rocky intertidal region of the west coast of North America suggest that species abundance may actually be highest at the edge of a range (153, 154). Thus small changes in condition caused by shifts in climate may have profoundly negative implications for many ecologically and commercially important species that are abundant at the edge of their geographic distribution. Third, the highly chaotic nature of atmospheric and oceanic conditions during times of rapid climate flux may lead to counter-intuitive shifts in local climate. Persistent warming may elevate sea level through melting of polar sea ice and thermal expansion of water, thereby changing wave-stress patterns and increasing the frequency of extreme events (storms, droughts, spring tides), thus causing shifts in tidal distribution and abundance of important ecological species (155–158). However, in terrestrial habitats, drought conditions may lead to a decline in winter precipitation, which in turn can cause increased exposure to lethal cold stress in high-altitude or latitude regions where snow pack typically protects over-wintering animals and plants from extreme cold (159, 160).

How can biochemical indicators of stress or condition be used to gain mechanistic insight into the effects of environmental change? Recent studies of intertidal snails (genus *Tegula*) provide an excellent example. In central California, two species of *Tegula* occupy distinct vertical zones on the shore: *T. brunnea* is common in the subtidal to low intertidal, whereas *T. funebralis* is very abundant in the mid-intertidal (161). Heat stress appears to limit *T. brunnea* from occupying the mid-intertidal zone, in part due to distinct differences between species in the expression and induction of Hsps (87, 101, 104). Field-caging transplant experiments (in the presence and absence of shade) were conducted in part to mimic increased thermal stress, which would result from elevated summertime temperatures co-occurring with extreme low tides, a prediction of several climate change models (155, 156). These data show that moving the subtidal species (*T. brunnea*) into the zone occupied by the intertidal species (*T. funebralis*) rapidly increased expression of at least one isoform of Hsp70, and negatively impacted survival; the intertidal species generally did not show an increase in cellular stress, as indexed by Hsp70 expression (162). Providing shade resulted in a less extreme heat shock response for *T. brunnea* than inunshaded controls, suggesting that sun exposure (rather than desiccation or reduction in feeding time) was the most critical environmental factor causing physiological stress. Up- and down-regulation of Hsp70 expression occurred on the order of days, demonstrating the extreme physiological plasticity of the heat shock response. This study illustrates the power of using biochemical indicators of stress and condition for predicting which species will be susceptible to local extinction because of environmental change, as biochemical indicators (such as Hsp70 expression) are much more sensitive to changes in the environment than growth rate or survival rates alone. Also, it demonstrates that we can learn something about the process by which animals respond to climate change. Using these indicators concomitantly with organismal and ecological measures of responses to climate change (distribution, abundance, activity, metabolic rate, survival, fecundity, etc.) will allow for the development of more rigorous and realistic predictions of the effects of anthropogenic climate change on natural systems.

ACKNOWLEDGMENTS

I sincerely acknowledge Prof. Bruce Menge for introducing me to many of the ideas I include in this review, for enlivening and insightful conversations over the years about biochemical indicators and marine ecology, and for critically reviewing this manuscript. I also thank PISCO PIs, fellows, students, and technicians for access to unpublished results. I thank Lars Tomanek and Eric Sanford for access to unpublished studies of stress in *Tegula*, and Nathan Rank for his invaluable assistance in statistical analysis and literature review. I especially want to acknowledge my students and technicians for their diligent and tireless work that led to some of the results discussed here. Portions of this work were supported by the Andrew W. Mellon Foundation, the National Science Foundation, the Wayne and Gladys

Valley Endowment, and the David and Lucile Packard Foundation, the latter of which funds PISCO. I dedicate this review to the memory of Prof. Peter Hochachka and all the wonderful things he taught us about metabolic physiology, dogs, and living life full out.

The *Annual Review of Physiology* is online at http://physiol.annualreviews.org

LITERATURE CITED

1. Bayne BL, Widdows J. 1978. The physiological ecology of two populations of *Mytilus edulis* L. *Oecologia* 37:137–62

2. Dahlhoff EP, Menge BA. 1996. Influence of phytoplankton concentration and wave exposure on the ecophysiology of *Mytilus californianus. Mar. Ecol. Prog. Ser.* 144:97–107

3. Guderley H, Dutil JD, Pelletier D. 1996. The physiological status of Atlantic cod, *Gadus morhua*, in the wild the laboratory: estimates of growth rates under field conditions. *Can. J. Fish. Aquat. Sci.* 53:550–57

4. Hawkins AJS, Bayne BL, Bougrier S, Heral M, Iglesias JIP, et al. 1998. Some general relationships in comparing the feeding physiology of suspension-feeding bivalve molluscs. *J. Exp. Marine Biol. Ecol.* 219:87–103

5. Hofmann GE, Somero GN. 1995. Evidence for protein damage at environmental temperatures: seasonal changes in levels of ubiquitin conjugates and HSP70 in the intertidal mussel *Mytilus trossulus. J. Exp. Biol.* 198:1509–18

6. Torres JJ, Somero GN. 1988. Metabolism, enzymic activities and cold adaptation in Antarctic mesopelagic fishes. *Mar. Biol.* 98:169–80

7. Wagner M, Durbin E, Buckley L. 1998. RNA:DNA ratios as indicators of nutritional condition in the copepod *Calanus finmarchicus. Mar. Ecol. Prog. Ser.* 162:173–81

8. Wolcott TG. 1973. Physiological ecology and intertidal zonation in limpets (*Acmaea*): a critical look at 'limiting factors.' *Biol. Bull. Mar. Biol. Lab. Woods Hole* 145:389–422

9. Buckley LJ. 1984. RNA-DNA ratio: an index of larval fish growth in the sea. *Mar. Biol.* 80:291–98

10. Foster AR, Houlihan DF, Hall SJ. 1993. Effects of nutritional regime on correlates of growth rate in juvenile Atlantic cod *Gadus morula*: comparisons of morphological and biochemical measurements. *Can. J. Fish. Aquat. Sci.* 50:502–12

11. Houlihan DF, Hall SJ, Gray C, Noble BS. 1988. Growth rates and protein turnover in Atlantic cod, *Gadus morula. Can. J. Fish. Aquat. Sci.* 45:951–64

12. Mathers EM, Houlihan DF, Cunningham MJ. 1992. Nucleic acid concentrations and enzyme activities as correlates of growth rate of the Saithe *Pollachius virens*-growth rate estimates of open-sea fish. *Mar. Biol.* 112:363–69

13. Smith J, Chong C. 1982. Body weight, activities of cytochrome c oxidase and electron transport system in the liver of the American plaice *Hippoglossoides platessoides*. Can these activities serve as indicators of metabolism? *Mar. Ecol. Prog. Ser.* 9:171–79

14. Dean J, Wilson C, Haake P, Beckman D. 1983. Micro-structural features of teleost otoliths. In *Biomineralization Biological Metal Accumulation*, ed. P Westbroek, E de Jong, pp. 353–59. Amsterdam: Reidel

15. Ali M, Wootton RJ. 2003. Correlates of growth in juvenile three-spined sticklebacks: potential predictors of

growth rates in natural populations. *Ecol. Freshw. Fish* 12:87–92

16. Buckley L, Caldarone E, Ong TL. 1999. RNA-DNA ratio and other nucleic acid-based indicators for growth and condition of marine fishes. *Hydrobiologia* 401:265–277

17. Carter CG, Houlihan DF, He ZY. 2000. Changes in tissue free amino acid concentrations in Atlantic salmon, *Salmo salar* L., after consumption of a low ration. *Fish Physiol. Biochem.* 23:295–306

18. Gwak WS, Tsusaki T, Tanaka M. 2003. Nutritional condition, as evaluated by RNA/DNA ratios, of hatchery-reared Japanese flounder from hatch to release. *Aquaculture* 219:503–14

19. Pelletier D, Guderley H, Dutil JD. 1993. Effects of growth rate, temperature, season, body size on glycolytic enzyme activities in the white muscle of Atlantic cod (*Gadus morhua*). *J. Exp. Zool.* 265:477–87

20. Mayrand E, Pellerin-Massicotte J, Vincent B. 1994. Small scale variability of biochemical indices of growth in *Mya arenaria* (L.). *J. Shell. Res.* 13:199–205

21. Martinez G, Brokordt K, Aguilera C, Soto V, Guderley H. 2000. Effect of diet and temperature upon muscle metabolic capacities and biochemical composition of gonad and muscle in *Argopecten purpuratus* Lamarck 1819. *J. Exp. Mar. Biol. Ecol.* 247:29–49

22. Wright DA, Hetzel EW. 1985. Use of RNA:DNA ratios as an indicator of nutritional stress in the American oyster *Crassostrea virginica. Mar. Ecol. Prog. Ser.* 25:199–206

23. Lindquist S. 1986. The heat-shock response. *Annu. Rev. Biochem.* 55:1151–91

24. Parsell DA, Lindquist S. 1993. The function of heat-shock proteins in stress tolerance: degradation and reactivation of damaged proteins. *Annu. Rev. Genet.* 27:437–96

25. Parsell DA, Taulien J, Lindquist S. 1993. The role of heat-shock proteins in thermotolerance. *Philos. Trans. R. Soc. London Ser. B* 339:279–86

26. Childress JJ, Seibel BA. 1998. Life at stable low oxygen levels: adaptations of animals to oceanic oxygen minimum layers. *J. Exp. Biol.* 201:1223–32

27. Ferry-Graham LA, Gibb AC. 2001. Comparison of fasting and postfeeding metabolic rates in a sedentary shark, *Cephaloscyllium ventriosum. Copeia:* 1108–13

28. Bayne BL. 2000. Relations between variable rates of growth, metabolic costs and growth efficiencies in individual Sydney rock oysters (*Saccostrea commercialis*). *J. Exp. Mar. Biol. Ecol.* 251:185–203

29. Parry GD. 1984. The effect of food deprivation on seasonal changes in the metabolic rate of the limpet *Cellana tramoserica. Comp. Biochem. Physiol. A* 77:663–68

30. Torres JJ, Aarset AV, Donnelly J, Hopkins TL, Lancraft TM, Ainley DG. 1994. Metabolism of Antarctic micronektonic Crustacea as a function of depth of occurrence and season. *Mar. Ecol. Prog. Ser.* 113:207–19

31. Anava A, Kam M, Shkolnik A, Degen AA. 2000. Seasonal field metabolic rate and dietary intake in Arabian babblers (*Turdoides squamiceps*) inhabiting extreme deserts. *Func. Ecol.* 14:607–13

32. Anava A, Kam M, Shkolnik A, Degen AA. 2002. Seasonal daily, daytime and nighttime field metabolic rates in Arabian babblers (*Turdoides squamiceps*). *J. Exp. Biol.* 205:3571–75

33. Schekkerman H, Tulp I, Piersma T, Visser GH. 2003. Mechanisms promoting higher growth rate in Arctic than in temperate shorebirds. *Oecologia* 134:332–42

34. Lowe CG. 2002. Bioenergetics of free-ranging juvenile scalloped hammerhead sharks (*Sphyrna lewini*) in Kane'ohe Bay, O'ahu, HI. *J. Exp. Mar. Biol. Ecol.* 278:141–56

35. Williams TM, Haun J, Davis RW,

Fuiman LA, Kohin S. 2001. A killer appetite: metabolic consequences of carnivory in marine mammals. *Comp. Biochem. Physiol. A* 129:785–96

36. Boyd IL. 2002. Estimating food consumption of marine predators: Antarctic fur seals and macaroni penguins. *J. Appl. Ecol.* 39:103–19

37. Childress JJ, Somero GN. 1979. Depth-related enzymic activities in muscle, brain and heart of deep-living pelagic marine teleosts. *Mar. Biol.* 52:273–83

38. Sullivan KM, Somero GN. 1983. Size- and diet-related variations in enzymatic activity and tissue composition in the sablefish *Anoplopoma fimbria. Biol. Bull. Mar. Biol. Lab. Woods Hole* 164: 315–26

39. Torres JJ, Somero GN. 1988. Vertical distribution and metabolism in Antarctic mesopelagic fishes. *Comp. Biochem. Physiol. B* 90:521–28

40. Yang TH, Somero GN. 1996. Fasting reduces protein and messenger RNA concentrations for lactate dehydrogenase but not for actin in white muscle of scorpionfish (*Scorpaena guttata*, Teleostei). *Mol. Mar. Bio. Biotech.* 5:153–61

41. Dutil JD, Lambert Y, Guderley H, Blier PU, Pelletier D, Desroches M. 1998. Nucleic acids and enzymes in Atlantic cod (*Gadus morhua*) differing in condition and growth rate trajectories. *Can. J. Fish. Aquat. Sci.* 55:788–95

42. Yang TH, Somero GN. 1993. Effects of feeding and food deprivation on oxygen consumption, muscle protein concentration, and activities of energy metabolism enzymes in muscle and brain of shallow- (*Scorpaena guttata*) and deep- (*Sebastelobus alascanus*) living Scorpaenid fishes. *J. Exp. Biol.* 181:213–23

43. Geiger SP, Donnelly J, Torres JJ. 2000. Effect of the receding ice-edge on the condition of mid-water fishes in the northwestern Weddell Sea: results from biochemical assays with notes on diet. *Mar. Biol.* 137:1091–104

44. Geiger SP, Kawall HG, Torres JJ. 2001. The effect of the receding ice edge on the condition of copepods in the northwestern Weddell Sea: results from biochemical assays. *Hydrobiologia* 453:79–90

45. Cullen M, Kaufmann R, Lowery MS. 2003. Seasonal variation in biochemical indicators of physiological status in *Euphausia superba* from Port Foster, Deception Island, Antarctica. *Deep-Sea Res. II* 50:1787–98

46. Dahlhoff EP, Stillman JH, Menge BA. 2002. Physiological community ecology: variation in metabolic activity of ecologically important rocky intertidal invertebrates along environmental gradients. *Integr. Comp. Biol.* 42:862–71

47. Dahlhoff EP, Buckley BA, Menge BA. 2001. Physiology of the rocky intertidal predator *Nucella ostrina* along an environmental stress gradient. *Ecology* 82:2816–29

48. Couture P, Dutil JD, Guderley H. 1998. Biochemical correlates of growth condition in juvenile Atlantic cod (*Gadus morhua*) from Newfoundland. *Can. J. Fish. Aquat. Sci.* 55:1591–98

49. Drazen JC. 2002. A seasonal analysis of the nutritional condition of deep-sea macrourid fishes in the northeast Pacific. *J. Fish Biol.* 60:1280–95

50. Martinez M, Guderley H, Dutil JD, Winger PD, He P, Walsh SJ. 2003. Condition, prolonged swimming performance and muscle metabolic capacities of cod *Gadus morhua. J. Exp. Biol.* 206:503–11

51. Martinez M, Dutil JD, Guderley H. 2000. Longitudinal allometric variation in indicators of muscle metabolic capacities in Atlantic cod (*Gadus morhua*). *J. Exp. Zool.* 287:38–45

52. Pelletier D, Blier PU, Dutil JD, Guderley H. 1995. How should enzyme activities be used in fish growth studies? *J. Exp. Biol.* 198:1493–97

53. Kawall HG, Torres JJ, Sidell BD, Somero GN. 2002. Metabolic cold adaptation in Antarctic fishes: evidence from

enzymatic activities of brain. *Mar. Biol.* 140:279–86

54. Willmer P, Stone G, Johnston I. 2000. *Environmental Physiology of Animals.* Oxford, UK: Blackwell Sci.

55. Gibb AC, Dickson KA. 2002. Functional morphology and biochemical indices of performance: Is there a correlation between metabolic enzyme activity and swimming performance? *Integr. Comp. Biol.* 42:199–207

56. Thuesen EV, Miller CB, Childress JJ. 1998. Ecophysiological interpretation of oxygen consumption rates and enzymatic activities of deep-sea copepods. *Mar. Ecol. Prog. Ser.* 168:95–107

57. Seibel B, Thuesen E, Childress J, Gorodezky L. 1997. Decline in pelagic cephalopod metabolism with habitat depth reflects differences in locomotory efficiency. *Biol. Bull. Mar. Biol. Lab. Woods Hole* 192:262–78

58. Seibel BA, Thuesen EV, Childress JJ. 2000. Light-limitation on predator-prey interactions: consequences for metabolism and locomotion of deep-sea cephalopods. *Biol. Bull. Mar. Biol. Lab. Woods Hole* 198:284–98

59. McNamara PT, Caldarone EM, Buckley LJ. 1999. RNA/DNA ratio and expression of 18S ribosomal RNA, actin and myosin heavy chain messenger RNAs in starved and fed larval Atlantic cod (*Gadus morhua*). *Mar. Biol.* 135:123–32

60. Houlihan D, McMillan D, Agnisola C, Trara Genoino I, Foti L. 1990. Protein synthesis and growth in *Octopus vulgaris. Mar. Biol.* 106:251–59

61. Lowery MS, Somero GN. 1990. Starvation effects on protein synthesis in red and white muscle of the barred sand bass *Paralabrax nebulifer. Physiol. Zool.* 63:630–48

62. Moss SM. 1994. Use of nucleic acids as indicators of growth in juvenile white shrimp, *Penaeus vannamei. Mar. Biol.* 120:359–67

63. McMillan D, Houlihan D. 1988. The ef-fects of re-feeding on tissue protein synthesis in rainbow trout. *Physiol. Zool.* 61:429–41

64. Bishop RE, Torres JJ, Crabtree RE. 2000. Chemical composition growth indices in *Leptocephalus* larvae. *Mar. Biol.* 137:205–14

65. Frantzis A, Gremare A, Vetion G. 1992. Growth rates and RNA:DNA ratios in *Paracentrotus lividus* (Echinoidea) fed on benthic macrophytes. *J. Exp. Mar. Biol. Ecol.* 156:125–38

66. Fukuda M, Sako H, Shigeta T, Shibata R. 2001. Relationship between growth and biochemical indices in laboratory-reared juvenile Japanese flounder (*Paralichthys olivaceus*) and its application to wild fish. *Mar. Biol.* 138:47–55

67. Fieldler TJ, Clarke ME, Walsh PJ. 1998. Condition of laboratory-reared and wild-caught larval Atlantic menhaden *Brevoortia tyrannus* as indicated by metabolic enzyme activities. *Mar. Ecol. Prog. Ser.* 175:51–66

68. Gwak WS, Tanaka M. 2002. Changes in RNA, DNA and protein contents of laboratory-reared Japanese flounder *Paralichthys olivaceus* during metamorphosis and settlement. *Fish. Sci.* 68:27–33

69. Chicharo LMZ, Chicharo MA, Alves F, Amaral A, Pereira A, Regala J. 2001. Diel variation of the RNA/DNA ratios in *Crassostrea angulata* (Lamarck) *Ruditapes decussatus* (Linnaeus 1758) (Mollusca:Bivalvia). *J. Exp. Mar. Biol. Ecol.* 259:121–29

70. Worm B, Lotze HK, Sommer U. 2000. Coastal food web structure, carbon storage, and nitrogen retention regulated by consumer pressure and nutrient loading. *Limno. Oceano.* 45:339–49

71. Bustamante RH, Branch GM, Eekhout S, Robertson B, Zoutendyk P, et al. 1995. Gradients of intertidal primary productivity around the coast of South Africa and their relationships with consumer biomass. *Oecologia* 102:189–201

72. Fraser LH, Grime JP. 1997. Primary

productivity trophic dynamics investigated in a North Derbyshire, UK. *Oikos* 80:499–508

73. Oksanen L, Fretwell SD, Arruda J, Niemela P. 1981. Exploitation ecosystems in gradients of primary productivity. *Am. Nat.* 118:240–61

74. Small LF, Menzies DW. 1981. Patterns of primary productivity biomass in a coastal upwelling region. *Deep-Sea Res.* 28A:123–49

75. Garcia A, Cortes D, Ramirez T. 1998. Daily larval growth and RNA and DNA content of the NW Mediterranean anchovy *Engraulis encrasicolus* and their relations to the environment. *Mar. Ecol. Prog. Ser.* 166:237–45

76. Robbins I, Lubet P, Besnard JY. 1990. Seasonal variations in the nucleic acid content and RNA:DNA ratio of the gonad of the scallop *Pecten maximus*. *Mar. Biol.* 105:191–95

77. Kawakami Y, Mochioka N, Kimura R, Nakazono A. 1999. Seasonal changes of the RNA/DNA ratio, size, lipid contents and immigration adaptability of Japanese glass-eels, *Anguilla japonica*, collected in northern Kyushu, Japan. *J. Exp. Mar. Biol. Ecol.* 238:1–19

78. Ota AY, Landry MR. 1984. Nucleic acids as growth rate indicators for early developmental stages of *Calanus pacificus* (Brodsky). *J. Exp. Mar. Biol. Ecol.* 80:147–60

79. Saiz E, Calbet A, Fara A, Berdalet E. 1998. RNA content of copepods as a tool for determining adult growth rates in the field. *Limno. Oceano.* 43:465–70

80. Bergeron JP. 1997. Nucleic acids in ichthyoplankton ecology: a review, with emphasis on recent advances for new perspectives. *J. Fish Biol.* 51:284–302

81. Johannesson K, Johannesson B, Lundgren U. 1995. Strong natural selection causes microscale allozyme variation in a marine snail. *Proc. Natl. Acad. Sci. USA* 92:2602–6

82. Mathers EM, Houlihan DF, Burren LJ.

1994. RNA, DNA, and protein concentrations in fed and starved herring *Clupea harengus* larvae. *Mar. Ecol. Prog. Ser.* 107:223–31

83. Schmidt PS. 2001. The effects of diet and physiological stress on the evolutionary dynamics of an enzyme polymorphism. *Proc. R. Soc. London Ser. B.* 268:9–14

84. Esteves E, Chicharo MA, Pina T, Coelho ML, Andrade JP. 2000. Comparison of RNA/DNA ratios obtained with two methods for nucleic acid quantification in gobiid larvae. *J. Exp. Mar. Biol. Ecol.* 245:43–55

85. Kaplan LAE, Leamon J, Crivello JF. 2001. The development of a rapid sensitive, high-through-put protocol for RNA:DNA ratio analysis. *J. Aqua. Anim. Health* 13:276–79

86. Feder ME, Hofmann GE. 1999. Heat-shock proteins, molecular chaperones, and the heat-shock response: evolutionary and ecological physiology. *Annu. Rev. Physiol.* 61:243–82

87. Tomanek L, Somero GN. 2002. Interspecific- and acclimation-induced variation in levels of heat-shock proteins 70 (hsp70) and 90 (hsp90) and heat-shock transcription factor-1 (HSF1) in congeneric marine snails (genus *Tegula*): implications for regulation of hsp gene expression. *J. Exp. Biol.* 205:677–85

88. Roberts DA, Hofmann GE, Somero GN. 1997. Heat-shock protein expression in *Mytilus californianus*: acclimatization (seasonal tidal-height comparisons) and acclimation effects. *Biol. Bull. Mar. Biol. Lab. Woods Hole* 192:30–32

89. Dahlhoff EP, Rank NE. 2000. Functional and physiological consequences of genetic variation at phosphoglucose isomerase: heat shock protein expression is related to enzyme genotype in a montane beetle. *Proc. Natl. Acad. Sci. USA* 97:10056–61

90. Michaelak P, Minkov I, Helin A, Lerman

DN, Bettencourt BR, et al. 2001. Genetic evidence for adaptation-driven incipient speciation of *Drosophila melanogaster* along a microclimate contrast in "Evolution Canyon," Israel. *Proc. Natl. Acad. Sci. USA* 98:13195–200

91. Loeschcke V, Krebs RA, Dahlgaard J, Michalak P. 1997. High-temperature stress and the evolution of thermal resistance in *Drosophila. Exp. Suppl.* 83:175–90

92. Krebs RA, Loeschcke V. 1996. Acclimation and selection for increased resistance to thermal stress in *Drosophila buzzatii. Genetics* 142:471–79

93. Krebs RA, Feder ME. 1998. Hsp70 larval thermotolerance in *Drosophila melanogaster*: how much is enough and when is more too much? *J. Insect Physiol.* 44:1091–101

94. Hershko A, Ciechanover A. 1992. The ubiquitin system for protein degradation. *Annu. Rev. Biochem.* 61:761–807

95. Feder ME, Cartano NV, Milos L, Krebs RA, Lindquist SL. 1996. Effect of engineering Hsp70 copy number on Hsp70 expression and tolerance of ecologically relevant heat shock in larvae and pupae of *Drosophila melanogaster. J. Exp. Biol.* 199:1837–44

96. Krebs RA, Loeschcke V. 1997. Estimating heritability in a threshold trait: heat-shock tolerance in *Drosophila buzzatii. Heredity* 79:252–59

97. Krebs RA, Loeschcke V. 1999. A genetic analysis of the relationship between life-history variation and heat-shock tolerance in *Drosophila buzzatii. Heredity* 83:46–53

98. Hofmann GE, Somero GN. 1996. Protein ubiquitination and stress protein synthesis in *Mytilus trossulus* occurs during recovery from tidal emersion. *Mol. Mar. Bio. Biotech.* 5:175–84

99. Chapple JP, Smerdon GR, Berry RJ, Hawkins AJS. 1998. Seasonal changes in stress protein 70 levels reflect thermal tolerance in the marine bivalve *Mytilus*

edulis L. *J. Exp. Mar. Biol. Ecol.* 229: 53–68

100. Roberts CM. 1997. Connectivity management of Caribbean coral reefs. *Science* 278:1454–57

101. Tomanek L. 2002. The heat-shock response: its variation, regulation and ecological importance in intertidal gastropods (genus *Tegula*). *Integr. Comp. Biol.* 42:797–807

102. Helmuth BST, Hofmann GE. 2001. Microhabitats, thermal heterogeneity, and patterns of physiological stress in the rocky intertidal zone. *Biol. Bull. Mar. Biol. Lab. Woods Hole* 201:374–84

103. Neargarder GG, Dahlhoff EP, Rank NE. 2003. Variation in thermal tolerance and HSP70 expression is linked to phosphoglucose isomerase genotype in a montane leaf beetle. *Func. Ecol.* 17:213–21

104. Tomanek L, Somero GN. 1999. Evolutionary and acclimation-induced variation in the heat-shock responses of congeneric marine snails (genus *Tegula*) from different thermal habitats: implications for limits of thermotolerance and biogeography. *J. Exp. Biol.* 202:2925–36

105. Rank NE, Dahlhoff EP. 2002. Allele frequency shifts in response to climate change and physiological consequences of allozyme variation in a montane insect. *Evolution* 56:2278–89

106. Hofmann G, Buckley B, Place S, Zippay M. 2002. Molecular chaperones in ectothermic marine animals: biochemical function and gene expression. *Integr. Comp. Biol.* 42:808–14

107. Herman PMJ, Middelburg JJ, Widdows J, Lucas CH, Heip CHR. 2000. Stable isotopes as trophic tracers: combining field sampling and manipulative labeling of food resources for macrobenthos. *Mar. Ecol. Prog. Ser.* 204:79–92

108. Kreeger DA, Hawkins AJS, Bayne BL. 1996. Use of dual-labeled microcapsules to discern the physiological fates of assimilated carbohydrate, protein carbon, and protein nitrogen in

suspension-feeding organisms. *Limno. Oceanogr.* 41:208–15

109. Lorrain A, Paulet YM, Chauvaud L, Savoye N, Donval A, Saout C. 2002. Differential delta C-13 delta N-15 signatures among scallop tissues: implications for ecology and physiology. *J. Exp. Mar. Biol. Ecol.* 275:47–61

110. Pottinger TG, Carrick TR, Yeomans WE. 2002. The three-spined stickleback as an environmental sentinel: effects of stressors on whole-body physiological indices. *J. Fish Biol.* 61:207–29

111. Meyer B, Saborowski R, Atkinson A, Buchholz F, Bathmann U. 2002. Seasonal differences in citrate synthase and digestive enzyme activity in larval and post-larval Antarctic krill, *Euphausia superba. Mar. Biol.* 141:855–62

112. Menge BA, Sutherland JP. 1987. Community regulation: variation in disturbance, competition, and predation in relation to environmental stress and recruitment. *Am. Nat.* 130:730–57

113. Menge BA, Olson AM. 1990. Role of scale and environmental factors in regulation of community structure. *Trends Ecol. Evol.* 5:52–57

114. Menge BA. 2000. Bottom-up:top-down determination of rocky intertidal shorescape dynamics. In *Foodwebs at the Landscape Level,* ed. GA Polis, ME Power, G Huxel. Chicago: Univ. Chicago Press

115. Menge BA, Olson AM, Dahlhoff EP. 2002. Environmental stress, bottom-up effects, and community dynamics: integrating molecular-physiological and ecological approaches. *Integr. Comp. Biol.* 42:892–908

116. Thompson RC, Roberts MF, Norton TA, Hawkins SJ. 2000. Feast or famine for intertidal grazing molluscs: a mis-match between seasonal variations in grazing intensity and the abundance of microbial resources. *Hydrobiologia* 440:357–67

117. Menge BA, Daley BA, Wheeler PA,

Dahlhoff E, Sanford E, Strub PT. 1997. Benthic-pelagic links and rocky intertidal communities: bottom-up effects on top-down control? *Proc. Natl. Acad. Sci. USA* 94:14530–35

118. Menge BA, Daley BA, Wheeler PA, Strub PT. 1997. Rocky intertidal oceanography: an association between community structure and nearshore phytoplankton concentration. *Limno. Oceano.* 42:57–66

119. Menge BA, Daley BA, Lubchenco J, Sanford E, Dahlhoff E, et al. 1999. Top-down and bottom-up regulation of New Zealand rocky intertidal communities. *Ecol. Monogr.* 69:297–330

120. Menge BA. 2000. Top-down and bottom-up community regulation in marine rocky intertidal habitats. *J. Exp. Mar. Biol. Ecol.* 250:257–89

121. Menge BA. 1992. Community regulation: under what conditions are bottom-up factors important on rocky shores? *Ecology* 73:755–65

122. Menge BA, Farrell TM, Olson AM, vanTamelen P, Turner T. 1993. Algal recruitment and the maintenance of a plant mosaic in the low intertidal region on the Oregon coast. *J. Exp. Mar. Biol. Ecol.* 170:91–16

123. Menge BA, Berlow EL, Blanchette CA, Navarrete SA, Yamada SB. 1994. The keystone species concept: variation in interaction strength in a rocky intertidal habitat. *Ecol. Monogr.* 64:249–86

124. Sanford E, Menge BA. 2001. Spatial temporal variation in barnacle growth in a coastal upwelling system. *Mar. Ecol. Prog. Ser.* 209:143–57

125. Sanford E. 1999. Regulation of keystone predation by small changes in ocean temperature. *Science* 283:2095–97

126. Sanford E. 2002. Water temperature, predation, and the neglected role of physiological rate effects in rocky intertidal communities. *Integr. Comp. Biol.* 42:881–91

127. Sanford E. 2002. The feeding, growth,

and energetics of two rocky intertidal predators (*Pisaster ochraceus* and *Nucella canaliculata*) under water temperatures simulating episodic upwelling. *J. Exp. Mar. Biol. Ecol.* 273:199–218

128. Barth JA, Smith RL. 1998. Separation of a coastal upwelling jet at Cape Blanco, Oregon, USA. *S. Afr. J. Mar. Sci.* 19:5–14

129. Huyer AE. 1983. Coastal upwelling in the California current system. *Prog. Oceano.* 12:259–84

130. Cranford PJ, Hill PS. 1999. Seasonal variation in food utilization by the suspension-feeding bivalve molluscs *Mytilus edulis* and *Placopecten magellanicus. Mar. Ecol. Prog. Ser.* 190:223–39

131. Grant J, Bacher C. 1998. Comparative models of mussel bioenergetics and their validation at field culture sites. *J. Exp. Mar. Biol. Ecol.* 219:21–44

132. Navarro JM, Thompson RJ. 1996. Physiological energetics of the horse mussel *Modiolus modiolus* in a cold ocean environment. *Mar. Ecol. Prog. Ser.* 138:135–48

133. Kosro PM, Barth JA, Strub PT. 1997. The coastal jet: observations of surface currents along the Oregon continental shelf from HF radar. *Oceanography* 10:53–56

134. Menge BA, Sanford E, Daley BA, Freidenburg TL, Hudson G, Lubchenco J. 2002. Inter-hemispheric comparison of bottom-up effects on community structure: insights revealed using the comparative-experimental approach. *Ecol. Res.* 17:1–16

135. Hofmann GE, Somero GN. 1996. Interspecific variation in thermal denaturation of proteins in the congeneric mussel *Mytilus trossulus* and *M. galloprovincialis*: evidence from the heat-shock response and protein ubiquitination. *Mar. Biol.* 126:65–75

136. Tsuchiya M. 1983. Mass mortality in a population of the mussel *Mytilus edulis* L. caused by high temperature on rocky shores. *J. Exp. Mar. Biol. Ecol.* 66:101–11

137. Garrity SD. 1984. Some adaptations of gastropods to physical stress on a tropical rocky shore. *Ecology* 65:559–74

138. Jones KM, Boulding EG. 1999. State-dependent habitat selection by an intertidal snail: the costs of selecting a physically stressful microhabitat. *J. Exp. Mar. Biol. Ecol.* 242:149–77

139. Connell JH. 1961. The influence of interspecific competition and other factors on the distribution of the barnacle *Chthamalus stellatus. Ecology* 42:710–23

140. Connell JH. 1961. Effects of competition, predation by *Thais lapillus*, and other factors on natural populations of the barnacle *Balanus balanoides. Ecol. Monogr.* 31:61–104

141. Dayton PK. 1975. Experimental evaluation of ecological dominance in a rocky intertidal algal community. *Ecol. Monogr.* 45:137–59

142. Paine RT. 1966. Food web complexity and species diversity. *Am. Nat.* 100:65–75

143. Paine RT. 1974. Intertidal community structure: experimental studies on the relationship between a dominant competitor and its principal predator. *Oecologia* 15:93–120

144. Lewis JR. 1964. *The Ecology of Rocky Shores.* London: English Univ. Press. 323 pp.

145. Menge BA, Farrell TM. 1989. Community structure interaction webs in shallow marine hard-bottom communities: tests of an environmental stress model. *Adv. Ecol. Res.* 19:189–262

146. Menge BA, Lubchenco J, Gaines SD, Ashkenas LR. 1986. A test of the Menge-Sutherland model of community organization in a tropical rocky intertidal food web. *Oecologia* 71:75–89

147. Barry JP, Baxter CH, Sagarin RD, Gilman SE. 1995. Climate-related,

long-term faunal changes in a California rocky intertidal community. *Science* 267:672–75

148. Sagarin RD, Barry JP, Gilman SE, Baxter CH. 1999. Climate-related change in an intertidal community over short and long time scales. *Ecol. Monogr.* 69:465–90

149. Parmesan C, Root TL, Willig MR. 2000. Impacts of extreme weather and climate on terrestrial biota. *Bull. Am. Meteorol. Soc.* 81:443–50

150. Parmesan C, Yohe G. 2003. A globally coherent fingerprint of climate change impacts across natural systems. *Nature* 421:37–42

151. Walther GR, Post E, Convey P, Menzel A, Parmesan C, et al. 2002. Ecological responses to recent climate change. *Nature* 416:389–95

152. Helmuth B, Harley CDG, Halpin PM, O'Donnell M, Hofmann GE, Blanchette CA. 2002. Climate change and latitudinal patterns of intertidal thermal stress. *Science* 298:1015–17

153. Sagarin R, Gaines S. 2002. The 'abundant centre' distribution: to what extent is it a biogeographical rule? *Ecol. Lett.* 5:137–47

154. Sagarin RD, Gaines SD. 2002. Geographical abundance distributions of coastal invertebrates: using one-dimensional ranges to test biogeographic hypotheses. *J. Biogeogr.* 29:985–97

155. Easterling DR, Meehl GA, Parmesan C, Changnon SA, Karl TR, Mearns LO. 2000. Climate extremes: observations, modeling, impacts. *Science* 289:2068–74

156. Hayward TL. 1997. Pacific Ocean climate change: atmospheric forcing, ocean circulation and ecosystem response. *Trends. Ecol. Evol.* 12:150–54

157. McGowan JA, Cayan DR, Dorman LM. 1998. Climate-ocean variability and ecosystem response in the northeast Pacific. *Science* 281:210–17

158. Scavia D, Field JC, Boesch DF, Buddemeier RW, Burkett V, et al. 2002. Climate change impacts on US coastal marine ecosystems. *Estuaries* 25:149–64

159. Ayres MP, Lombardero MJ. 2000. Assessing the consequences of global change for forest disturbance from herbivores pathogens. *Sci. Total Environ.* 262:263–86

160. Inouye DW, Barr B, Armitage KB, Inouye BD. 2000. Climate change is affecting altitudinal migrants and hibernating species. *Proc. Natl. Acad. Sci. USA* 97:1630–33

161. Newell RC. 1979. *Biology of Intertidal Animals.* Kent, UK: Mar. Ecol. Surv.

162. Tomanek L, Sanford E. 2003. Heat-shock protein 70 (Hsp70) as biochemical stress indicator: an experimental field test in two congeneric intertidal gastropods (Genus: *Tegula*). *Biol. Bull. Mar. Biol. Lab. Woods Hole.* In press

Annu. Rev. Physiol. 2004. 66:209–38
doi: 10.1146/annurev.physiol.66.032102.114245
First published online as a Review in Advance on October 15, 2003

FIELD PHYSIOLOGY: Physiological Insights from Animals in Nature

Daniel P. Costa and Barry Sinervo

*Department of Ecology and Evolutionary Biology, University of California, Santa Cruz,
California 95064; email: costa@biology.ucsc.edu; sinervo@biology.ucsc.edu*

Key Words diving physiology, reproductive energetics, fish endothermy,
comparative approach, phylogeny, adaptation, evolution, field metabolic rate

■ **Abstract** Whereas comparative physiology documents the range of physiological variation across a range of organisms, field physiology provides insight into the
actual mechanisms an organism employs to maintain homeostasis in its everyday life.
This requires an understanding of an organism's natural history and is prerequisite to
developing hypotheses about physiological mechanisms. This review focuses on a few
areas of field physiology that exemplify how the underlying physiology could not have
been understood without appropriate field measurements. The examples we have chosen highlight the methods and inference afforded by an application of this physiological
analysis to organismal function in nature, often in extreme environments. The specific
areas examined are diving physiology, the thermal physiology of large endothermic
fishes, reproductive physiology of air breathing vertebrates, and endocrine physiology
of reproductive homeostasis. These areas form a bridge from physiological ecology to
evolutionary ecology.

All our examples revolve around the central issue of physiological limits as they
apply to organismal homeostasis. We view this theme as the cornerstone of physiological analysis and supply a number of paradigms on homeostasis that have been tested
in the context of field physiology.

INTRODUCTION

As a discipline, physiology examines the mechanisms by which an organism
maintains homeostasis. The broad field of comparative physiology is centered
on adaptive physiological variation among species from different environments.
Field physiology goes one step further, by measuring, with a wide spectrum of
techniques, the mechanisms employed by organisms to carry out their functions
under actual environmental conditions. Therefore, by extension, field physiology
is the examination of homeostatic mechanisms in the field. While we often find
examples of wonderful physiological mechanisms to maintain homeostasis in extreme environments, there are also excellent examples where the organism simply

0066-4278/04/0315-0209$14.00   **209** avoids the extremes by behavioral or other nonphysiological processes. For instance, small organisms take advantage of microhabitats that enable them to avoid environmental extremes. Many desert rodents avoid the daytime heat by being active at night when temperatures are moderate and remain in their burrows during the heat of day. In contrast, large mammals such as the oryx or the camel are too large and cannot escape the desert heat and thus have developed a series of well-documented physiological and anatomical adaptations to these extreme temperatures (1, 2).

A cornerstone of field physiology is an appreciation for and understanding of the natural environment and its associated history. Natural history provides a basis to develop hypotheses and questions about what physiological problems confront animals in the field and, therefore, what mechanisms an animal might require to survive (2–4). One of the innovations in recent years concerns the quantification of natural history (5) and, specifically, of the role of natural selection in shaping the evolution of physiological traits (6, 7).

Some of the earliest and best-known examples of physiology in the field started with a series of simple observations of an organism in its environment. For example, field work in the California deserts and associated observations of kangaroo rats led Schmidt-Nielsen and colleagues to ask how these animals deal with the xeric desert environment (2). Where do they get their water? Is their kidney different from other mammals? Similarly, observations of marine mammals led Scholander and Irving et al. to ask how these animals remain submerged for such prolonged periods of time (8). Do they drink sea water? Do they have an extrarenal salt gland like seabirds, or do they have a specialized kidney (9)? All these early and fundamental studies in comparative physiology started with observations of animals in their natural environment. Even though researchers made their initial observations in the field in these early studies, they almost always brought animals into the laboratory to investigate their physiology. This was because laboratory investigation was not only the obvious solution to the problem, but in many cases it was the only way possible because the available equipment was large, cumbersome, and not suitable for field use.

Nonetheless, laboratory investigation offers a precision not always possible in the field, even with the advantages of modern technology. In the laboratory, an investigator can hold all variables constant except those that are of interest. In contrast, it is difficult to control the variability in both biotic and abiotic features of the natural environment. A well-designed field study accepts the variability of the natural environment but works to insure that the variation between control and experimental groups is the same. Thus we try to create a situation where the role or impact of natural variation is reduced or at least accounted for. Finally, the issue is not whether field versus laboratory investigations are better, but how to achieve the optimum mix between both.

It would be unrealistic to review the whole range of physiological investigations that have been carried out in the field. Therefore, this review focuses on a few areas of field physiology that exemplify how the underlying physiology could not have been understood without appropriate field measurements. Notably, exploring the limits of physiological systems has been a hallmark of physiological analysis since the Krogh principle was first presented (10). The examples we have chosen highlight the methods and inference afforded by an application of this physiological analysis of extreme environments in the natural realm. Furthermore, the areas that we examine, diving physiology, the thermal physiology of large endothermic fishes, reproductive physiology of air-breathing vertebrates, and endocrine physiology of reproductive homeostasis, also bridge from the field of physiological ecology to evolutionary ecology.

In choosing these examples, we highlight the two conceptual frameworks that have served to focus our perspective on physiological processes in nature: the comparative approach applied to populations and species, and the individual approach applied to differences among individuals within a population (11). Each of these levels of biological analysis has strengths and weaknesses. For instance, the individual approach largely ignores the large-scale physiological differentiation observed at high levels of taxonomic diversity, which is the endpoint of adaptational processes (12). However, the individual approach excels at highlighting the evolutionary factors responsible for divergence in physiology at these higher levels–natural selection underlies the process of adaptation. It is also possible to carry out fine-scale analysis of causation with manipulations of physiology or those individual traits that result from physiological process. When homeostatic mechanisms fail, the outcome is invariably death. Thus differential survival or reproduction as a function of variation in organismal homeostasis among individuals is the mechanism by which physiology evolves.

In contrast, the comparative approach has classically been applied to the analysis of physiology in extreme environments and has identified those species in extreme environments that have, in some ways, extraordinary adaptations of physiology. Presumably, these species are near the functional and genetic constraints imposed on adaptational processes. Allometry is a field of study in physiological analysis that has provided a classic treatment of constraints on adaptation. Although useful in identifying such functional limits, differences in environment confound the analysis as do those biases inherent in a strictly phylogenetic analysis (12). Rarely are both approaches to the analysis of physiology in natural systems combined. We highlight a few studies that are exemplary in this regard. We do not, however, supply a comprehensive review of the evidence for these paradigms, as these have been developed over the decades in laboratory settings. Where appropriate, we supply a number of in-depth review citations to these laboratory-developed models and the extensions to a field setting.

DIVING PHYSIOLOGY

For millions of years a central component of the basic homeostatic machinery of air-breathing vertebrates has been just that, reliable access to air. The physiology of diving vertebrates is built around an enhanced capacity to function in the absence of

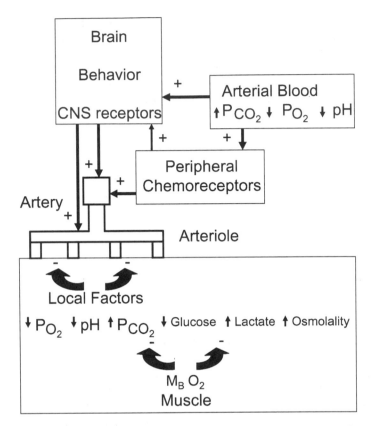

Figure 1   Proposed control of blood flow and O_2 delivery to muscle that regulates the transition from aerobic to anaerobic metabolism in a diving seal (13).

oxygen, and is marked by the tolerance by divers of considerable variation in tissue O_2, CO_2, and pH. The ability to modify or override normal control mechanisms has allowed diving reptiles, birds, and mammals to successfully reinvade and exploit the aquatic environment (Figure 1). The problems faced by diving vertebrates can be divided into adaptations that deal with pressure and adaptations to prolong their ability to remain submerged. Adaptations to pressure include mechanical effects of pressure, as well as the problems coupled with increased solubility and, in some cases, toxicity of N_2 and O_2 at high pressure. Adaptations to prolonged time spent underwater are linked to oxygen stores, how fast they are used, and whether there is any need for anaerobic metabolism. Given the difficulties of carrying out studies under high pressure, it is not surprising that the majority of research on diving has centered on metabolic processes during a dive. This review concentrates on field studies that have increased our understanding of diving physiology, a broader

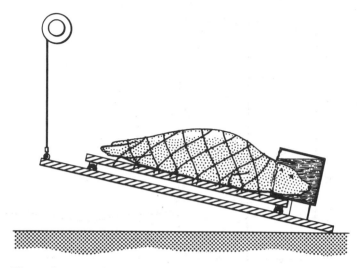

Figure 2   An example of the early experimental apparatus for experiments used a forced-dive approach (20).

appreciation of this activity can be obtained from a number of excellent reviews (13–15a).

Pioneering studies of the dive response and diving metabolism of mammals and birds were for years carried out almost exclusively in a laboratory setting (Figure 2). The early work of Scholander and Irving showed that when a captive mammal or bird was forced to dive, there was an overall reduction in metabolism coupled with an increased reliance on anaerobic metabolism as indicated by a postdive release of lactic acid (Figure 3) (16–18). This dive response was initiated by a profound bradycardia with associated reductions in cardiac output and blood flow to peripheral tissues (19). Later work using labeled microspheres confirmed that blood flow was reduced to all major organs systems except the heart, lung, and brain (20). Although this pioneering work provided insight into the maximum physiological response of a diving mammal, its relevance to the natural setting was unclear. Even Scholander (17) warned that his laboratory measurements might not adequately describe the physiology of a freely diving animal. The tools and techniques were barely available for laboratory measurements let alone for studies of freely living animals. Further evidence that the maximum dive response observed in a laboratory might not occur in nature was uncovered by Elsner et al., who studied the dive response of captive trained or unrestrained aquatic mammals. Elsner et al. found that voluntarily diving subjects exhibited a significantly reduced bradycardia compared with forced dive animals (Figure 4) (21–24).

A major breakthrough in the study of diving physiology came when Kooyman capitalized on a novel situation in the Antarctic and literally took the laboratory

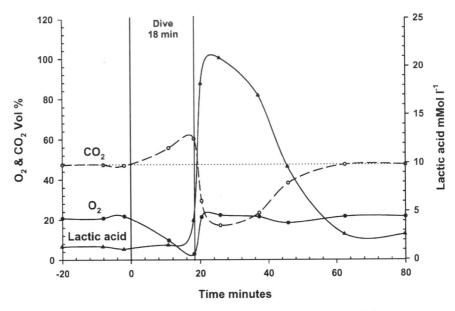

Figure 3   Effects of forced-dive submersion of a 42-kg gray seal, *Halichoerus grypus*, on arterial blood chemistry (17).

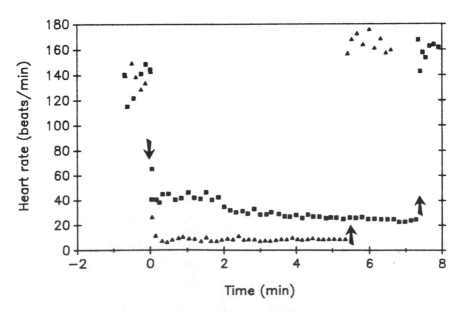

Figure 4   Heart rate in the harbor seal during trained head submersion (*solid boxes*) and forced submersion (*solid triangles*) (23).

into the field to study the diving behavior and physiology of freely diving Weddell seals, *Leptonychotes weddelli* (25). Weddell seals make a living diving in and around small openings in fast sea-ice, which is attached to the shore and in certain situations can cover the surface of the ocean for many kilometers. Because Weddell seals are adapted to living in this environment, they routinely breathe through small cracks or holes in the sea-ice. Kooyman exploited this natural ability to create a novel field laboratory. He moved a seal to an area where there was only a single sea-ice hole for many kilometers and thus created a situation where the animal had to return to the same location to breathe. Kooyman and colleagues then placed a small portable laboratory over this hole and measured the animal's pulmonary function, heart rate, and metabolic rate while the animal determined its own diving behavior (26–28). Similar to the work carried on captive unrestrained diving mammals, freely diving Weddell seals showed a moderate bradycardia. However, the degree of bradycardia increased in longer dives and was exhibited at the beginning of the dive. This indicated that the dive response varied in relation to the metabolic needs of the seal and that the seal knew whether it was going to make a long or short dive.

Although this work suggested that our understanding of the physiology of diving mammals was flawed, it was not until Kooyman's group was able to monitor blood lactic acid levels before and after a dive that they were able to show that Weddell seals dove aerobically during the majority of their dives (29, 30). This work showed that lactic acid levels remained constant for dives up to 20 min and thereafter rapidly increased with increasing dive durations (Figure 5). They concluded that animals could dive aerobically for dives lasting up to 20 min. This work defined the maximum time an animal could remain submerged without utilizing anaerobic metabolic pathways; the aerobic dive limit (ADL) was calculated as ADL (min) = total oxygen store (ml O_2)/diving metabolic rate (ml O_2 min^{-1}). An important observation was that when Weddell seals exceeded the aerobic threshold, the postdive surface interval increased disproportionately relative to dive duration (29, 30). The greater surface intervals were needed to clear the lactic acid that accumulated during the dive. One of the disadvantages of anaerobic metabolism during diving is that while a diver may increase the duration of a single dive, the total accumulated time spent underwater is reduced. This is because the animal must spend proportionately more time at the surface clearing lactic acid (13).

The tremendous potential of studying the physiology of freely diving Weddell seals in McMurdo Sound, Antarctica became apparent to other researchers as well. A variety of research teams have since used this field laboratory to examine renal function (31), lung collapse (32), blood gases and hematocrit variation (33, 34), substrate utilization (35), cardiovascular control (36), blood chemistry homeostasis during diving (37), diving metabolic rate (38), muscle and aortic temperature during diving (39), re-examination of aerobic dive limit (40), heart rate and body temperature variation, myoglobin saturation (41), hormonal control and splenic contraction (42, 43), hunting behavior (44), and locomotor mechanics (45).

While all of these studies incorporated new technologies and approaches to working with Weddell seals diving from a sea-ice hole, a few stand out for

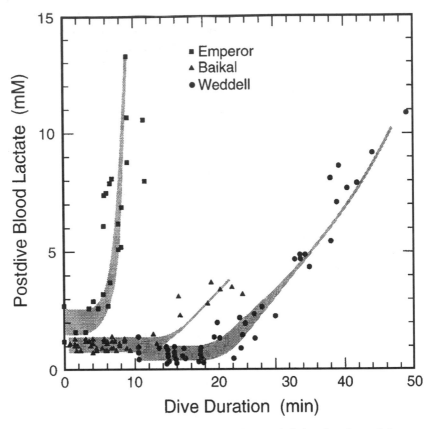

Figure 5   Postdive blood lactate concentrations and diving durations of the emperor penguin, Baikal seal, and Weddell seal (15a).

incorporating truly innovative technology. A group led by Hochachka and Zapol enlisted the engineering expertise of Hill to develop a microprocessor-controlled system to periodically sample blood during the time the seal dove (35, 36). This technology was used to document when lung collapse occurred during a dive (32); changes in blood gases and hematocrit (33); substrate utilization, cardiovascular control, heart rate, and body temperature variation, myoglobin saturation (41); hormonal control, and splenic contraction (42, 43).

Although, lung collapse had been observed in simulated dives (46) and in a freely diving trained dolphin (47), it had never been quantified in a freely diving animal. The microcomputer-controlled blood sampling system provided confirmation of the functional importance of lung collapse in a freely diving seal (Figure 6) (32). Marine mammals have an advantage in that, unlike human divers, they exclusively breath-hold dive and thus carry only a limited amount of air in their lungs during a given dive. Deep-diving marine mammals avoid problems associated with

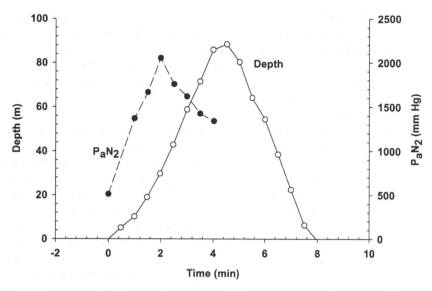

Figure 6 Depth of dive combined with serial determinations of P_aN_2 during dive. P_aN_2 values determined early during a dive when pulmonary gas exchange spaces are collapsing. Samples were collected every 30 s (32).

tissue N_2 accumulation by allowing their lungs to collapse during the initial period of the dive. As the lung collapses, air is expressed into the large bronchioles and trachea where gas uptake at elevated pressures cannot occur (13, 15a). Given that N_2 and O_2 tensions in the blood remain relatively low during the dive, nitrogen narcosis, decompression sickness, and oxygen toxicity are thought to be avoided. However, it is still unclear how penguins, small cetaceans, and sea lions avoid the bends because they can make many repetitive dives on a full lung and do not always undergo lung collapse (13, 15a, 47).

   Although Weddell seals offer a truly exceptional system to study diving physiology, they represent but a single species in a very unusual situation. The only other diving animal that has been studied in this way is the emperor penguin (48). However, comparable results have been obtained in a few studies where aerobic dive limits have been measured using freely diving trained bottlenose dolphins, *Tursiops truncatus* (49), California sea lions, *Zalophus californianus* (50), Beluga whales, *Delphinapteras leucas* (51), and captive Baikal seals, *Phoca sibirica* (52). All these studies support the ADL model as originally proposed by Kooyman et al. (29).

   One of the goals of field physiology is to understand when physiology limits behavior; therefore, it would be instructive to observe how often and under what conditions animals diving in nature stay within or exceed the ADL. Recent advances in technology have allowed simultaneous measurements of diving behavior and metabolic rate. These studies show that in certain situations diving animals may exceed their ADL, which implies the potential for anaerobic metabolism. For example, Antarctic fur seals, *Arctocephalus gazella*, rarely exceed their cADL because they make short shallow dives feeding near the surface. In contrast, the cADL is routinely exceeded in long and/or deep diving Australian *Neophoca cinerea* and New Zealand, *Phocarctos hookeri* sea lions (Figure 7) (53). Animals that feed at or near the sea bottom consume large prey that may require prolonged durations to effectively capture (53–59).

Such measurements are not just of interest to physiologists, they also have relevance to wildlife managers. For example, it is important to know whether a species in decline is operating at or close to its maximum physiological capacity, because if so, it will be less capable of compensating for normal environmental- or human-caused changes in its environment. In contrast, an animal that is operating well within its physiological capacity will be more capable of responding to environmental fluctuations. Such animals would be able to draw on a greater physiological reserve and pursue prey deeper, or dive longer or forage for greater periods (53, 56). Over the past decade many fur seals' populations (South America fur seal, *Arctocephalus australis*; New Zealand fur seal, *A. forsteri*; and California sea lion, respectively) have experienced spectacular population growth that is in marked contrast to an apparent stability or even decline in all of the sea lions that feed on or near the bottom (Steller, *Eumatopias jubatus*; Australian; southern, *Otaria flavescens*; and New Zealand sea lion), many of which are sympatric with near-surface-feeding fur seals (56).

Studies of diving behavior would not have been possible without the amazing developments in digital electronics, which have provided field biologists with a new form of biotechnology that allows the study of complex behavior and physiology in freely ranging animals. This technology has produced data loggers small enough to be attached to animals while they freely go about their activities (Figure 8, see color insert). Data from these tags are obtained when the tags are recovered (archival tags) or when transmitted via satellite (60, 61). These tags have been used extensively with marine mammals, fish, birds, and reptiles. Due to the large size and in some cases the ease of capture of marine mammals, the technology for attached instrumention was initially developed for use with these animals and has been used to record the ambient acoustic environment (62–64), heart rate (65), ventilation rate (66), swim speed (58, 67, 68) and acceleration (69, 70).

Satellite tags are used when one cannot recover the data logger from the animal. A limitation of this technology is that the satellite transmitter must be out of the water to communicate with an orbiting satellite, therefore the technology has mainly been used on air-breathing vertebrates that surface regularly such as marine birds (71, 72), sea turtles (73), marine mammals (74, 75) and most recently sharks (61, 76). An example of the kind of data that can be acquired with this technology is shown in Figure 9 (see color insert).

Future Directions

A major area of research in diving physiology that could benefit from field measurements is how deep-diving marine mammals and birds handle the effects of

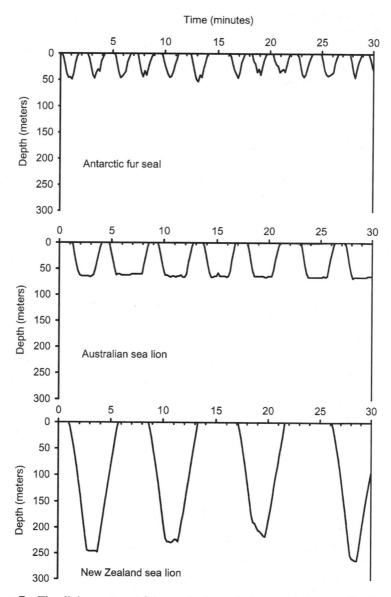

Figure 7 The diving pattern of Antarctic fur seal, *Arctocephalus gazella*, the Australian sea lion, *Neophoca cinerea*, and the New Zealand sea lion, *Phocarctos hookeri*, are compared. All axes are expressed in the same absolute units to facilitate comparison. The two sea lions feed at or near the bottom, whereas the fur seal feeds near the sea surface (53).

pressure. There are two aspects to this work: those that focus on hydrostatic effects of pressure and those associated with effects of dissolved gases such as nitrogen. Specifically, are deep divers susceptible to high-pressure nervous syndrome? How do repetitively diving marine mammals and birds that do not go through lung collapse avoid decompression sickness (15a, 77)? Further examination of the role of physiology in determining the optimal diving pattern would also be a productive area of research, as the necessary technology currently exists.

ENDOTHERMIC FISHES

A major advantage of endothermy is the ability to occupy and travel between different thermal habitats, while critical organ system, such as the nervous system and aerobic muscles can be maintained within a narrow set of biochemically optimal temperatures (78). For example, marine mammals and sea birds can go from a warm tropical island or surface waters to cold, food-rich regions of the ocean, all the while keeping their core body temperatures high and constant. However, only 27 out of 25,000 species of fish and sharks have evolved some form of regional endothermy (79). As gill breathers, fishes cannot become true endotherms because water contains only 1/40th as much oxygen, although it conducts heat 25 times faster than air (79). As a result, blood passing through the gills always equilibrates with the ambient thermal environment. Thus the vast majority of fishes remain within 1 to 2°C of the ambient environment (79).

Nevertheless, regional endothermy has developed along two different paths: one typified by tunas and sharks that conserve metabolic heat in the muscle, viscera, and brain and one recently discovered in billfishes where only the cranial cavity is warmed (79, 80). Regardless, common prerequisites for regional endothermy are large body size, a heat source, and a vascular system (counter-current heat exchanger) to conserve the heat. The physiological, biochemical, and anatomical mechanisms of tuna endothermy have been well documented (79, 81–89). Thirteen species of Scombrid tunas and 5 species of lamnid sharks exhibit regional endothermy, in some cases maintaining internal temperatures an astounding 21°C above ambient (79, 80, 90–92). Elevated temperatures in tuna are maintained by the internal location of the red oxidative muscle coupled with a vascular supply that passes blood through a counter-current heat exchanger, allowing retention of heat produced by the red muscle. In billfish, the brain and eyes are kept up to 13°C above ambient by a novel heater organ derived from the skeletal muscle around the eyes (79).

Mapping the key features of endothermy on a molecular phylogeny of the teleost fishes indicates that endothermy evolved independently at least three times in the Scombrid tunas, whereas cranial heaters evolved at least twice in the billfishes (Figure 10) (93). These independent origins of endothermy suggest that this energetically expensive strategy was under strong selection. Furthermore, the three independent origins of endothermy correspond to the independent expansion of

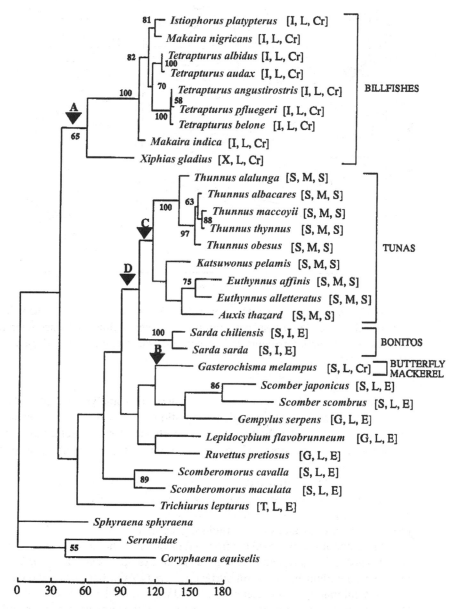

Figure 10 Phylogeny of the Scombroidei on the basis of a parsimony tree derived from a 600 base pair of cytochrome *b*. The arrows associated with specific nodes refer to (*A*) modification of the superior rectus muscle into a thermogenic organ, (*B*) modification of the lateral rectus muscle into a thermogenic organ, (*C*) systemic endothermy using vascular counter-current heat exchanges in the muscle, and (*D*) some internalization of red muscle along the horizontal septum (93).

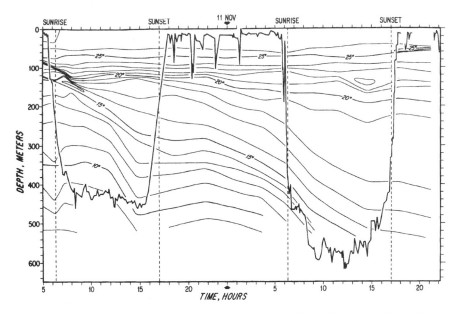

Figure 11 A depth record from a swordfish, *X. gladius*, illustrating their diurnal vertical migration. Swordfish pass through the thermocline and encounter large changes in the water temperature at dusk and at dawn. Data were derived via acoustic telemetry (94).

these three lineages into cool temperate waters. This analysis supports the role of fish endothermy as a physiological adaptation associated with niche expansion (93). However, an understanding of the importance of fish endothermy to niche expansion can be understood only by examination of the physiology of these fishes operating in nature.

Physiological measurements of freely ranging fishes were until recently carried out with acoustic tracking devices (61, 89, 94, 95). This was extremely difficult because it required real-time tracking by ship and thus tracks were limiting to just five or six days and only a single tagged animal could be followed. However, this technology provided the first real insight into the importance of regional endothermy to fishes in nature (Figure 11). For example, swordfish, *Xipias gladius*, spend the night in the warm surface waters (25°C), but during the day they vertically migrate to much deeper (400–600 m) and colder (8–9°C) waters (94). However, what is most impressive is that in spite of the large deviations in ambient temperature, the cranial temperature of the swordfish remains surprisingly constant (Figure 12).

More recently archival tags have been deployed on fishes and sharks, thereby allowing collection of data for greater time periods independent of a research vessel (61, 89). For use on fish, the archival tags are surgically implanted inside the peritoneal cavity. Pressure and internal temperature sensors are located within the body of the tag. A stalk protruding from the archival tag carries the light and water

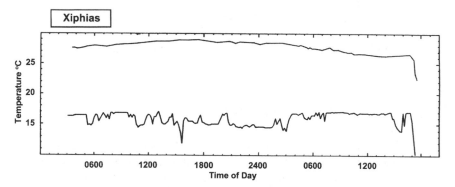

Figure 12 Temperature telemetry record from a swordfish, *X. gladius*, showing how effective the cranial heater is in keeping the brain temperature elevated and relatively constant (94).

temperature sensors externally. Data are stored in the tag until it is recovered, usually by fishermen. Information on the location, diving behavior, ambient, and internal temperature preferences can be obtained. To date blue fin tuna, *Thunnus thynnus*, have been tracked for 3.6 years (61, 89). Similarly, even though blue fin tuna repeatedly venture into cold waters, their internal body temperature remains relatively constant and elevated (Figure 13, see color insert). Finally the critical role of large body size to regional endothermy can be seen in the greater thermal variability in a small tuna compared with a larger tuna of the same species (96) (Figure 14, see color insert).

The phylogenetic analysis as used for billfish illustrates the power of the phylogenetic perspective in elucidating the origins of traits, whereas the detailed individual measurements on free-ranging fish with high-tech recording equipment provides crucial information on the actual natural history context that we noted above. More syntheses using both methods will be crucial in unraveling the origin of physiological adaptations and the environmental conditions that contributed to their evolution and maintenance. An emergent theme throughout this review is the profound role behavior has in shaping selection on physiological traits. In the case of billfish, maintenance of high cranial temperatures is thought to be the direct result of selection on the ability these fish to be effective predators across a range of thermal habitats, which requires elevated and relatively constant temperatures for efficient neurophysiological processing (the hypothesis underlying high cranial temperatures in billfish).

Future Directions

Although a significant amount of information exists on the physiological and biochemical adaptations of endothermic fishes, relatively little is known about how these processes are integrated in nature (88). Given the considerable advances in tagging technology, there is a tremendous potential to gain significant insight into the physiology and ecology of these interesting and important apex oceanic predators. Combining tags with modern satellite remote-sensing techniques will enable us to put these magnificent animals in the context of their oceanic environment (61, 89).

REPRODUCTIVE PHYSIOLOGY

The limits on reproductive physiology have been a focus of physiological ecology since early field observations noted that animals in nature can exceed three to four times basal metabolic rate for extended periods of time during reproduction (97–101). Since these early observations, researchers have focused their analyses on the short-term evolutionary consequences of such physiological effort on fitness (102).

Reproduction is thought to be costly largely because of the energetic demands imposed on parents during the acquisition of resources for progeny (77, 99, 102, 103). Two fitness consequences have been the focus: the effects of energy expenditure on progeny survival (102) and the effects of energy expenditure on adult survival to future reproductive episodes. The former issue relates to the first fundamental trade-off of life history analysis (104), referred to as the offspring size and offspring number trade-off, whereas the latter issue relates to the second trade-off, referred to as costs of reproduction in parents (105). With one possible exception (106), the relationships between energy expenditure in the wild [expressed as daily energy expenditure (DEE) or field metabolic rate (FMR)] has not been simultaneously assessed for both components of fitness, even though such information might be useful in interpreting the phase of life history (progeny versus adult fitness component) that is most sensitive to energy limitation imposed on adults. In such situations, it is possible for the unmeasured component of fitness to be related to DEE (DEE of parent affects progeny survival), while the measured component of fitness is not correlated with DEE (DEE of parent does not affect adult survival). This would result in a failure to find an association between DEE and fitness, even though such an association exists.

In brief, researchers simultaneously assess field metabolic rate with the doubly labeled water (DLW) method (107, 108) to measure DEE on reproduction of lizards, birds, and mammals (109, 110). Estimates of DEE are often made in conjunction with detailed estimates of behavioral time-budgets (77, 103, 111, 112), which when used in conjunction with laboratory estimates (e.g., O_2 consumption) of each behavior (e.g., cost-of-flight, cost-of-hovering flight, running, lactation, etc.) can be used to partition DEE to specific metabolic episodes by integrating the behavior over time. Furthermore, direct comparisons of time-budget-derived DEE and DLW-derived DEE over the entire period during which DLW is measured also can test the validity of the DLW method.

As an important experimental adjunct to methods for assessing DEE, many researchers invariably manipulate reproductive effort by litter size or brood size; such manipulations (augmentation or reduction) (99, 111, 113–116) provide a causal measure of the relationship between DEE and fitness. In addition, other researchers have used food deprivation experiments (to progeny) to increase effort expended by the parents (111, 117). This allows for the inference of direct causal effect of offspring number on parental DEE, and for the direct causal inference of the effect of offspring number (or physiological effort) on parental survival to future reproductive episodes, offspring survival, or the success of parents in producing offspring on future reproductive episodes. Other studies have compared differences in the DEE and energy investment between closely related species with different foraging behaviors (103), in different habitats (118), or between marine mammals and seabirds utilizing similar resources in the same habitat (119).

Analyses of the two life history trade-offs from a physiological perspective has had mixed results regarding the importance of physiological ceilings during reproduction as having a selective impact with cascading effects on fitness traits. Lack of a relation as a state of the real patterns must be conditioned upon the caveat noted above regarding the action of selection during adult versus progeny phases of the life history and its correlation with DEE in adults. In spite of this caveat, the discrepancies between the studies highlight the different physiological pathways in vertebrates for basic life history function [avian feeding strategies, huddling behavior on parental effort, granivory-insectivory (116, 120, 121) versus carnivory (111) or versus mammalian lactation (114, 119, 122)]. For example, in the European kestrel, *Falco tinnunculus*, manipulation of parental effort via clutch size manipulation has a clear effect on the elevation of parental effort in both parents, and the physiological effort has a clear impact on the survival of adult birds to a second clutch and future reproductive episodes (123, 124). Furthermore, food removal experiments indicate that kestrels are capable of nearly doubling energetic effort on a short-term basis (1.5 weeks) (111). Similar studies on a granivorous-insectivorous species, the great tit, *Parus major*, failed to find an associated link between DEE and fitness of cost of reproduction, even though manipulations of parental effort have demonstrated such links (125).

Similarly, in mammalian systems, detailed estimates of DEE for females during reproduction are available for the North American species the Golden-Mantled Ground Squirrel, *Spermophilus saturatus*, and red squirrels, *Tamiasciurus hudsonicus* (77, 114, 122). The most costly period of DEE for female squirrels is lactation. No correlation between litter size and DEE was found, but rather all females approached a similar physiological ceiling (114, 122). As patterns of reproductive costs and investment vary with body size, it would be useful to know how the patterns of investment vary in large mammals.

Although the energetic costs of reproduction have been studied in ungulates, primates, and pinnipeds, no study has assessed the relationship between DEE and fitness. Such fitness relations would be useful in understanding the role of physiology in shaping the evolution of life history patterns. A common feature of these highly precocial animals is that they always give birth to a single large offspring (126). Measurements of DEE during lactation in pinnipeds show two markedly different reproductive costs. The cost of reproduction in true seals (Phocidae) is quite economical (1.5–3 times BMR), whereas it is quite high (4–6 times BMR) in sea lions and fur seals (Otariidae) (126). This in part is the result of the differences in the way that females forage and provision their offspring. True seal females provision their pup primarily from stored energy reserves, often fasting during lactation, whereas sea lions and fur seals intermittently forage at sea returning to suckle the pup on shore (126).

Reproductive Endocrinology

The examples of reproductive physiology in nature illustrate how laboratory analysis of the homeostatic processes of endocrine regulation has led to a series of elegant analyses of such homeostasis in extreme environments. The advent of radioimmunoassay of reproductive hormones in free-ranging animals revolutionized our view of the endocrine system in action. In addition, the ability to insert hormone-filled implants into free-ranging animals has allowed for an analysis of causation that rivals laboratory studies, albeit in an uncontrolled field setting.

Our goal in this brief review is not to treat the endocrine axes in complete detail but rather to provide an overview of how such tools have been useful in unraveling the complexities of hormone interaction in nature. This example is useful in that endocrine homeostasis is well characterized, as are the cascading effects on target physiological traits and reproductive physiology [reviewed in (127–129)]. Endocrine homeostasis during reproduction arises from the action of protein hormones (gonadotropins) secreted by the hypothalamus and pituitary, which stimulates the gonads to produce the sex-specific steroid hormones. The up-regulation in the production of these steroid hormones in turn down-regulates the production of the gondadotropins (Figure 15). This two-part endocrine system is referred to as the hypothalamic-pituitary-gonadal (HPG) axis, and it represents a negative regulatory loop in that steroid hormones have a set point that is not exceeded owing to the negative regulation exerted by the steroids on the gonadotropins. Given that levels of steroid hormones achieve stable population-specific values and that these steroids control many physiological processes, we can view this as the elements of physiological homeostasis. The steroid hormones in turn up-regulate the transcription of specific genes related to physiology through accessory DNA-binding carrier proteins and response elements on the DNA [e.g., estrogen response elements (ERE)]. This regulation of behavioral, metabolic, and other physiological traits has been the focus of field endocrinology.

The link between individual and environmental extreme is regulated by another endocrine axis, the hypothalamic-pituitary-adrenal (HPA) axis (128, 129). Short-term and long-term regulations of sex steroids and their effects on metabolism are governed by both transient and chronic elevation of glucocorticosteroids, which have a regulatory function and a basic metabolic function. The glucocorticoids such as corticosterone can down-regulate production of the reproductive hormones (129) and also override the effects of steroid hormones on the expression of physiological and behavioral traits (128). In addition, corticosterone per se regulates gluconeogenesis and the breakdown of muscle tissue, serving as a direct regulator of metabolic physiology (130). It is for these reasons that field endocrinology

Figure 8   An Australian sea lion, *Neophoca cinerea*, mother and her pup are shown with data loggers attached. The most forward device is a satellite transmitter, the middle device is a time swim-speed and depth recorder, and the last item is a VHF radio.

Figure 9   A pseudo three-dimensional representation of the diving behavior and movement patterns of crabeater seals, *Lobodon carcinophagus*, studied in the Western Antarctic Peninsula near Adelaide Island. Data were collected with a satellite-linked data relay. Image produced by the Sea Mammal Research Unit St. Andrews Scotland (D.P. Costa, J.M. Burns & M.A. Fedak, unpublished data).

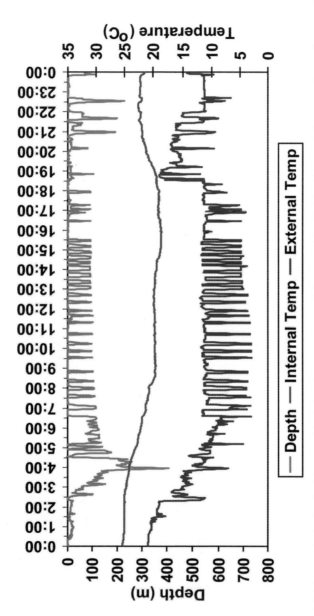

Figure 13 A 24-h record obtained with an archival data logger recovered from an Atlantic bluefin tuna. The gray line at the top of the graph shows the depth of the animal; the middle red line shows that the internal body temperature remains constant in spite of the dramatic variation in ambient water temperature. Data are unpublished and courtesy of B. Block, Stanford University.

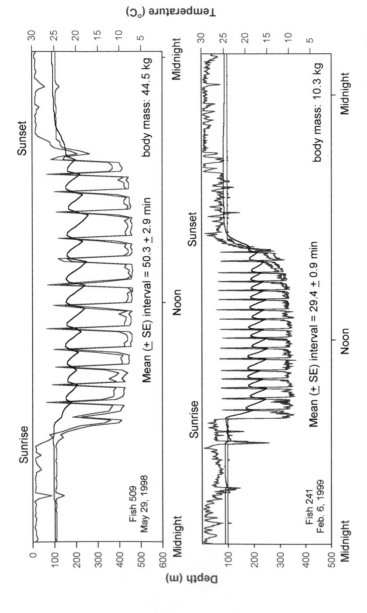

Figure 14 Two 24-h records derived from a large (44.5 kg) and small (10.3 kg) big-eye tuna, *Thunnus obesus*, off the main Hawaiian Islands, using archival data loggers. The blue line shows the animal's depth, the red line shows ambient water temperature, and the black line is the internal body temperature. Notice that due to its larger body size and greater thermal inertia, the larger fish can remain in the deeper colder water longer before it has to return to the surface to warm up (96).

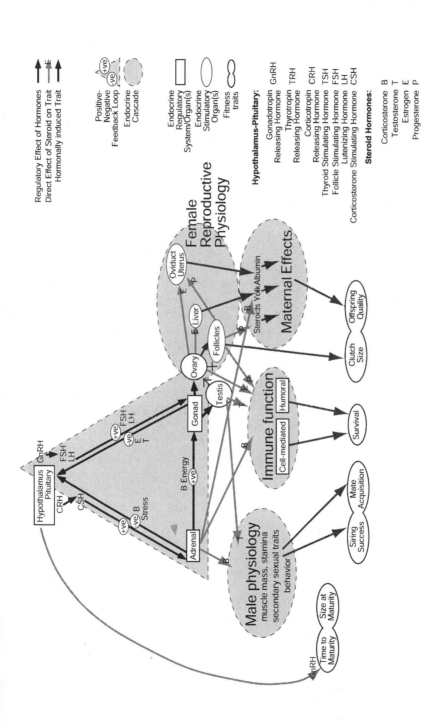

Figure 15 Endocrine homeostasis in endocrine system arises from the interaction of several endocrine axes. Endocrine homeostasis has cascading effects on a suite of physiological traits. The top triangle depicts the hypothalamic-pituitary-gonadal (HPG) axis and hypothalamic-pituitary-adrenal (HPA) axis. See text for a description of the hormone cascade. The endocrine system has control over loci that govern physiological and life history traits specific to male physiology and female reproductive physiology, which contribute to fitness (*bottom ovals*).

has focused on the action of corticosterone on a diverse array of physiological traits and how such traits are modulated when environmental conditions become either unfavorable or favorable for reproduction. In addition, corticosterone not only regulates physiology in the context of abiotic stressors (128), it also regulates physiology in the context of biotic stressors such as predators and conspecifics (129). Salient observations of nesting birds during stressful events such as snowstorms indicate that corticosterone plays a major role in modulating reproduction during stress, reducing levels of the gonadotropin lutenizing hormone (LH), which may be related to nest abandonment (131).

The actual control of reproductive physiology has turned out to be decidedly complicated (Figure 13). Initial experiments using hormone implants in nature typically involved a treatment in which a single hormone implant was tested against sham-implanted subjects (132). It was rapidly apparent that hormone interactions or hormone and environment interactions were largely responsible for the effects on physiology. Multihormone implants were developed or hormone implants were used in tandem with other manipulations ([e.g., food manipulation (133)] to further test these findings. Another basic technique was to manipulate a salient axis of the endocrine system and measure the corresponding changes in other axes [e.g., effect of gonadotropin-releasing hormone on LH and testosterone (134), effect of corticosterone of the HPA on LH of the HPG, effect of hypothalamus-pituitary control over corticosterone response (135)]. Another approach was to implant combinations of hormones to study the effect of hormone interaction on behavior and physiology (e.g., overriding effect of corticosterone over testosterone, corticosterone plus testosterone, no corticosterone, no testosterone sham-implants) (136). Manipulations of reproductive effort have also been used in tandem with measurement of plasma hormone changes (137).

The basic hypothesis concerning the role of corticosterone as a global signal of environmental stress has been invalidated by the observation that corticosterone has multiple avenues for modulating physiology. For example, the classic idea that corticosterone is immunosuppressive (129, 138) has been overturned with field implant experiments and immune challenge experiments (139). However, it is worth emphasizing that corticosterone has many effects on physiology in addition to affecting the immune system (128). Corticosterone enhances gluconeogenesis and thereby elevates blood glucose levels (128, 129), and such an effect may, at least in the short term, improve several aspects of condition including immunocompetence. Nevertheless, the role of corticosterone in modulating stressful environments has solid support from a number of studies (137, 140–142) and this information will be crucial in interpreting the action of corticosterone in stress during the expression of reproductive physiology.

Future Directions

Measurements of DEE during reproduction and its impacts on fitness are needed for other vertebrates including lizards, and more extensive assessment of fitness traits is required for mammals. While studies on lizards have focused on the measurement of DEE (107, 109), none has yet applied the techniques of experimental manipulation that are currently available. In a similar vein, mammalian studies have relied on litter manipulations to test life history theory, but only recently have litter manipulations been applied in tandem with measurements of DEE (114). Furthermore, new methods of endocrine manipulations of litter size are available (e.g., follicle stimulating hormone), which have already been used in reptilian systems. Although there are studies on the costs of reproduction in females and the impact of brood size manipulations, there are surprisingly few studies of the impact of elevated male hormones on male DEE, despite this protocol being widely used in the area of field behavioral ecology. A careful causal dissection of the impact of male reproductive hormones and energy would be most informative.

Such endocrine manipulations of life history traits are preferable over simple litter size manipulations in that hormones can trigger a physiological cascade of events and thus are more likely to capture all of the salient physiologically based costs (143). Surprisingly, the actual metabolic costs of immune challenges have not been assessed with DLW methods. It would be most informative to test animals in the wild with novel antigens and measure the cost of such immune assaults. Furthermore, such manipulations of immune function when carried out with manipulations of reproductive hormones (139) might address the synergistic impacts of immune and reproductive function on DEE and the existence of physiological ceilings during reproduction.

CONCLUDING REMARKS

We have just touched the surface of the considerable number of physiological investigations carried out in the field. However, it is apparent even from this limited review that a vast array of tools and approaches are available to address fundamental questions of physiological homeostasis and how these processes have evolved. Given that field physiology has its origins in comparative physiology, it is important to recognize the pitfalls if phylogeny is not appropriately considered (12, 110).

Recent phylogenetic methods that provide corrections for statistical biases have been developed owing to a lack of independence associated with taxonimically related species (144a). The method of phylogenetically independent contrasts is one of the most widely used methods, and it creates contrasts (e.g., the difference) between the values of extant taxa and hypothetical ancestral states that are constructed with branch lengths in the phylogeny. It is really a linear transformation of the data that constructs a data set of independent data points, removing the effect of phylogeny. Notably, these phylogenetic methods have been applied to the scaling of FMR (110). A previously reported difference in the scaling of FMR between marsupials (e.g., 0.58) and eutherian mammals (0.81) (109) is not supported when the allometric regression is corrected using the method of phylogenetically independent contrast (0.71 versus 0.82). Use of modern phylogenetic methods are preferable because the estimates of slope are unbiased by the distribution of phylogenetic data, even though some groups may be under-represented.

Field physiology has an important role in the maintenance of biodiversity and conservation. As the pace of habitat destruction, over-harvesting, and climate change increases it will be critical to understand which organism can accommodate to these human-induced alterations in their habitat. The effects of climate change are myriad (144, 145). Whereas the general trend is toward a higher mean temperature, regional climate models indicate that the changes will be more complex (145). Some habitats will become wetter, others dryer, some will get warmer, some will get more precipitation, and others may change very little. In all cases, an understanding of the physiological capability of organisms to tolerate these changes can come only from field investigations of the organism in its habitat as its response is multifaceted (144). This review has shown that animals have the capacity to respond to varying levels of food intake and/or energy expenditure during reproduction. Although a simple relationship is expected between reproductive output and cost, animals can accommodate considerable variation in energy intake and expenditure while still successfully producing offspring. Similarly, these responses are mediated by a complex endocrine system that is just beginning to be understood in the context of the environment. We have shown that some diving behaviors are fundamentally more difficult and thus likely to put specific organisms in greater jeopardy. Finally, endothermic tunas, billfish, and sharks are currently over-fished at a rate that cannot be sustained (146). All the examples used here have potential importance to the conservation of biodiversity because it is critical to know which organisms have physiological plasticity that can accommodate environmental change. Field physiology can and should have a role in solving these complex environmental issues and can at least help to identify the organisms most in need of help.

ACKNOWLEDGMENTS

This review is dedicated to the memory of Peter Hochachka. His passion and curiosity for science has been an inspiration to us all. D.P.C. thanks K. Nagy, K. Norris, and G. Kooyman for their mentorship in field physiology and natural history. S. Shaffer and M. Zavanelli provided comments on the manuscript. Special thanks to S. Cooperman and G. Somero for their patience. This review was supported by National Science Foundation grants 99-81683 and 02-13179 and the Office of Naval Research grant N000140211012.

The *Annual Review of Physiology* is online at http://physiol.annualreviews.org

LITERATURE CITED

1. Louw GN. 1993. *Physiological Animal Ecology*. New York: Wiley & Sons. 288 pp.
2. Schmidt-Nielsen K. 1998. *The Camel's Nose*. Washington DC: Island Press. 339 pp.
3. Grant PR. 2000. What does it mean to be a naturalist at the end of the

twentieth century? *Am. Nat.* 155:1–12

4. Bartholomew GA. 1986. The role of natural history in contemporary biology. *BioScience* 36:324–29

5. Greene H. 1986. Natural history and evolutionary biology. In *Predator-Prey Relationships: Perspectives and Approaches from the Study of Lower Vertebrates*, eds. ME Feder, GV Lauder, pp. 99–108. Chicago: Univ. Chicago Press

6. Kingsolver JG, Hoekstra HE, Hoekstra JM, Berrigan D, Vignieri SN, et al. 2001. The strength of phenotypic selection in natural populations. *Am. Nat.* 157:245–61

7. Hoekstra HE, Hoekstra JM, Berrigan D, Vignieri SN, Hoang A, et al. 2001. Strength and tempo of directional selection in the wild. *Proc. Nat. Acad. Sci. USA* 98:9157–60

8. Scholander PF. 1990. *Enjoying a Life in Science*. Fairbanks: Univ. Alaska Press. 226 pp.

9. Irving L, Fisher KC, McIntosh FC. 1935. The water balance of a marine mammal, the seal. *J. Cell. Comp. Physiol.* 6:387–91

10. Krogh A. 1929. Progress of physiology. *Am. J. Physiol.* 90:243–51

11. Feder ME. 1987. The analysis of physiological diversity: the prospects for pattern documentation and general questions in physiological ecology. In *New Directions in Physiological Ecology*, ed. ME Feder, AF Bennett, W Burggren, RB Huey, pp. 38–75. Cambridge, UK: Cambridge Univ. Press

12. Huey RB. 1987. Phylogeny, history, and the comparative method. In *New Directions in Physiological Ecology*, ed. ME Feder, AF Bennett, WW Burggren, RB Huey, pp. 76–101. Cambridge, UK: Cambridge Univ. Press

13. Kooyman GL. 1989. *Diverse Divers: Physiology and Behavior*. Berlin: Springer-Verlag. 200 pp.

14. Hochachka PW. 2000. Pinniped diving

response mechanism and evolution: a window on the paradigm of comparative biochemistry and physiology. *Comp. Biochem. Physiol. A* 126:435–58

15. Butler PJ, Jones DR. 1997. Physiology of diving of birds and mammals. *Physiol. Rev.* 77:837–99

15a. Kooyman GL, Ponganis PJ. 1998. The physiological basis of diving to depth: birds and mammals. *Annu. Rev. Physiol.* 60:19–32

16. Irving L. 1939. Respiration in diving mammals. *Physiol. Rev.* 19:112–34

17. Scholander PF. 1940. Experimental investigation on the respiratory function in diving mammals and birds. *Hvalrad. Skr.* 22:1–131

18. Scholander PF, Irving L, Grinnell SW. 1942. Aerobic and anaerobic changes in seal muscles during diving. *J. Biol. Chem.* 142:431–40

19. Elsner R. 1969. Cardiovascular adjustments to diving. In *The Biology of Marine Mammals*, ed. HT Andersen, pp. 117–45. New York: Academic

20. Zapol WM, Kiggins GC, Schneider RC, Qvist J, Snider MT, et al. 1979. Regional blood flow during simulated diving in the conscious Weddell seal *Leptonychotes weddelli*. *J. Appl. Physiol: Resp. Environ. Exer. Physiol.* 47:968–73

21. Elsner RW, Franklin DL, Van Citters RL. 1964. Cardiac output during diving in an unrestrained sea lion. *Nature* 202:809–10

22. Elsner R, Scholander PF. 1965. Circulatory adaptations to diving in animals and man. *Physiol. Breath Hold Diving AMA Jpn.* 1341:281–94

23. Elsner RW. 1965. Heart rate response in forced versus trained experimental dives in pinnipeds. *Hvalrad. Skr.* 48:24–29

24. Elsner R. 1966. Diving bradycardia in the unrestrained hippopotamus. *Nature* 212:408–9

25. Kooyman GL. 1968. An analysis of some behavioral and physiological characteristics related to diving in the Weddell

seal. In *Biology of the Antarctic Seas III*, ed. WL Schmitt, GA Llano, pp. 227–61. Washington DC: Am. Geophys. Union

26. Kooyman GL, Kerem DH, Campbell WB, Wright JJ. 1971. Pulmonary function in freely diving Weddell seals *Leptonychotes weddelli. Resp. Physiol.* 12:271–82

27. Kooyman GL, Campbell WB. 1972. Heart rates in freely diving Weddell seals, *Leptonychotes weddelli. Comp. Biochem. Physiol. A.* 43:31–36

28. Kooyman GL, Kerem DH, Campbell WB, Wright JJ. 1973. Pulmonary gas exchange in freely diving Weddell seals *Leptonychotes weddelli. Resp. Physiol.* 17:283–90

29. Kooyman GL, Wahrenbrock EA, Castellini MA, Davis RW, Sinnett EE. 1980. Aerobic and anaerobic metabolism during voluntary diving in Weddell seals *Leptonychotes weddelli:* evidence of preferred pathways from blood chemistry and behavior. *J. Comp. Physiol. B* 138:335–46

30. Kooyman GL, Castellini MA, Davis RW, Maue RA. 1983. Aerobic diving limits of immature Weddell seals *Leptonychotes weddelli. J. Comp. Physiol. B* 151:171–74

31. Davis RW, Castellini MA, Kooyman GL, Maue R. 1983. Renal glomerular filtration rate and hepatic blood flow during voluntary diving in Weddell seals. *Am. J. Physiol. Regul. Integr. Comp. Physiol.* 245:R743–48

32. Falke KJ, Hill RD, Qvist J, Schneider RC, Guppy M, et al. 1985. Seal lungs collapse during free diving: evidence from arterial nitrogen tensions. *Science* 229:556–58

33. Qvist J, Hill RD, Schneider RC, Falke KJ, Liggins GC, et al. 1986. Hemoglobin concentrations and blood gas tensions of free-diving Weddell seals *Leptonychotes weddelli. J. Appl. Physiol.* 61:1560–69

34. Zapol WM, Hill RD, Qvist J, Falke K, Schneider RC, et al. 1989. Arterial gas tensions and hemoglobin concentrations of the freely diving Weddell seal. *Under. Biomed. Res.* 16:363–74

35. Guppy M, Hill RD, Schneider RC, Qvist J, Liggins GC, et al. 1986. Microcomputer-assisted metabolic studies of voluntary diving of Weddell seals. *Am. J. Physiol. Regul. Integr. Comp. Physiol.* 250: R175–87

36. Hill RD, Schneider RC, Liggins GC, Schuette AH, Elliott RL, et al. 1987. Heart rate and body temperature during free diving of Weddell seals. *Am. J. Physiol. Regul. Integr. Comp. Physiol.* 253:R344–51

37. Castellini MA, Davis RW, Kooyman GL. 1988. Blood chemistry regulation during repetitive diving in Weddell seals. *Physiol. Zool.* 61:379–86

38. Castellini MA, Kooyman GL, Ponganis PJ. 1992. Metabolic rates of freely diving Weddell seals: correlations with oxygen stores, swim velocity and diving duration. *J. Exp. Biol.* 165:181–94

39. Ponganis PJ, Kooyman GL, Castellini MA, Ponganis EP, Ponganis KV. 1993. Muscle temperature and swim velocity profiles during diving in a Weddell seal, *Leptonychotes weddellii. J. Exp. Biol.* 183:341–46

40. Ponganis PJ, Kooyman GL, Castellini MA. 1993. Determinants of the aerobic dive limit of Weddell seals: analysis of diving metabolic rates, postdive end tidal PO_2s, and blood and muscle oxygen stores. *Physiol. Zool.* 66:732–49

41. Guyton GP, Stanek KS, Schneider RC, Hochachka PW, Hurford WE, et al. 1995. Myoglobin saturation in free-diving Weddell seals. *J. Appl. Physiol.* 79:1148–55

42. Hochachka PW, Liggins GC, Guyton GP, Schneider RC, Stanek KS, et al. 1995. Hormonal regulatory adjustments during voluntary diving in Weddell seals. *Comp. Biochem. Physiol. B* 112:361–75

43. Hurford WE, Hochachka PW, Schneider RC, Guyton GP, Stanek KS, et al. 1996.

Splenic contraction, catecholamine release, and blood volume redistribution during diving in the Weddell seal. *J. Appl. Physiol.* 80:298–306

44. Davis RW, Fuiman LA, Williams TM, Collier SO, Hagey WP, et al. 1999. Hunting behavior of a marine mammal beneath the Antarctic fast ice. *Science* 283:993–96

45. Williams TM, Davis RW, Fuiman LA, Francis J, Le Boeuf BJ, et al. 2000. Sink or swim: strategies for cost-efficient diving by marine mammals. *Science* 288:133–36

46. Kooyman GL, Schroeder JP, Denison DM, Hammond DD, Wright JJ, Bergman WP. 1972. Blood nitrogen tensions of seals during simulated deep dives. *Am. J. Physiol.* 223:1016–20

47. Ridgway SH, Howard R. 1979. Dolphin lung collapse and intramuscular circulation during free diving: evidence from nitrogen washout. *Science* 206:1182–83

48. Ponganis PJ, Kooyman GL, Starke LN, Kooyman CA, Kooyman TG. 1997. Postdive blood lactate concentrations in emperor penguins, *Aptenodytes forsteri. J. Exp. Biol.* 200:1623–26

49. Williams TM, Friedl WA, Haun JE. 1993. The physiology of bottlenose dolphins (*Tursiops truncatus*): heart rate, metabolic rate, plasma lactate concentration during exercise. *J. Exp. Biol.* 179:31–46

50. Ponganis PJ, Kooyman GL, Winter LM, Starke LN. 1997. Heart rate and plasma lactate responses during submerged swimming and trained diving in California sea lions, *Zalophus californianus. J. Comp. Physiol.* 167:9–16

51. Shaffer SA, Costa DP, Williams TM, Ridgway SH. 1997. Diving and swimming performance of white whales, *Delphinapterus leucas*: an assessment of plasma lactate and blood gas levels and respiratory rates. *J. Exp. Biol.* 200:3091–99

52. Ponganis PJ, Kooyman GL, Baranov EA, Thorson PH, Stewart BS. 1997. The aerobic submersion limit of Baikal seals, *Phoca sibirica. Can. J. Zool.* 75:1323–27

53. Costa DP, Gales NJ, Goebel ME. 2001. Aerobic dive limit: How often does it occur in nature? *Comp. Biochem. Physiol. A* 129A:771–83

54. Arnould JPY, Hindell MA. 2001. Dive behaviour, foraging locations, and maternal-attendance patterns of Australian fur seals (*Arctocephalus pusillus doriferus*). *Can. J. Zool.* 79:35–48

55. Costa DP, Gales NJ. 2000. Foraging energetics and diving behavior of lactating New Zealand sea lions, *Phocarctos hookeri. J. Exp. Biol.* 203:3655–65

56. Costa DP, Gales NJ. 2003. Energetics of a benthic diver: seasonal foraging ecology of the Australian sea lion, *Neophoca cinerea. Eco. Monogr.* 73:27–43

57. Croxall JP, Naito Y, Kato A, Rothery P, Briggs DR. 1991. Diving patterns and performance in the Antarctic blue-eyed shag *Phalacrocorax atriceps. J. Zool.* 225:177–200

58. Ponganis PJ, Gentry RL, Ponganis EP, Ponganis KV. 1992. Analysis of swim velocities during deep and shallow dives of two northern fur seals, *Callorhinus ursinus. Mar. Mam. Sci.* 8:69–75

59. Tremblay Y, Cherel Y. 2000. Benthic and pelagic dives: a new foraging behaviour in rockhopper penguins. *Mar. Ecol. Prog. Ser.* 204:257–67

60. Costa DP. 1993. The secret life of marine mammals: novel tools for studying their behavior and biology at sea. *Oceanography* 6:120–28

61. Block B, Costa DP, Boehlert GW, Kochevar R. 2002. Revealing pelagic habitat use: the Tagging of Pacific Pelagics Program. *Oceanolog. Acta* 25:255–66

62. Fletcher S, Le Boeuf BJ, Costa DP, Tyack PL, Blackwell SB. 1996. Onboard acoustic recording from diving northern

elephant seals. *J. Acoust. Soc. Am.* 100:2531–39

63. Burgess WC, Tyack PL, Le Boeuf BJ, Costa DP. 1998. A programmable acoustic recording tag and first results from free-ranging northern elephant seals. *Deep-Sea Res. Part II-Top. Stud. Oceanogr.* 45:1327–51

64. Costa DP, Crocker DE, Gedamke J, Webb PM, Houser DS, et al. 2003. The effect of a low-frequency sound source (acoustic thermometry of the ocean climate) on the diving behavior of juvenile northern elephant seals, *Mirounga angustirostris. J. Acoust. Soc. Am.* 113:1155–65

65. Andrews RD, Jones DR, Williams JD, Thorson PH, Oliver GW, et al. 1997. Heart rates of northern elephant seals diving at sea and resting on the beach. *J. Exp. Biol.* 200:2083–95

66. Le Boeuf BJ, Crocker DE, Grayson J, Gedamke J, Webb PM, et al. 2000. Respiration and heart rate at the surface between dives in northern elephant seals. *J. Exp. Biol.* 203:3265–74

67. Crocker DE, Gales NJ, Costa DP. 2001. Swimming speed and foraging strategies of New Zealand sea lions (*Phocarctos hookeri*). *J. Zool.* 254:267–77

68. Crocker DE, Le Boeuf BJ, Costa DP. 1997. Drift diving in female northern elephant seals: implications for food processing. *Can. J. Zool.* 75:27–39

69. Nowacek DP, Johnson MP, Tyack PL, Shorter KA, McLellan WA, Pabst DA. 2001. Buoyant balaenids: the ups and downs of buoyancy in right whales. *Proc. R. Soc. Biol. Sci. Ser. B* 268:1811–16

70. Sato K, Naito Y, Kato A, Niizuma Y, Watanuki Y, et al. 2002. Buoyancy and maximal diving depth in penguins: Do they control inhaling air volume? *J. Exp. Biol.* 205:1189–97

71. Ancel L, Kooyman GL, Ponganis PJ, Gendner J-P, Lignon J, et al. 1992. Foraging behaviour of emperor penguins as a resource detector in winter and summer. *Nature* 360:336–39

72. Bost CA, Georges JY, Guinet C, Cherel Y, Puetz K, et al. 1997. Foraging habitat and food intake of satellite-tracked king penguins during the austral summer at Crozet Archipelago. *Mar. Ecol. Prog. Ser.* 150:21–33

73. Polovina JJ, Kobayashi DR, Parker DM, Seki MP, Balazs GH. 2000. Turtles on the edge: movement of loggerhead turtles (*Caretta caretta*) along oceanic fronts, spanning longline fishing grounds in the central North Pacific, 1997–1998. *Fish. Oceanogr.* 9:71–82

74. McConnell BJ, Chambers C, Fedak MA. 1992. Foraging ecology of southern elephant seals in relation to the bathymetry and productivity of the Southern Ocean. *Ant. Sci.* 4:393–98

75. Le Boeuf BJ, Crocker DE, Costa DP, Blackwell SB, Webb PM, Houser DS. 2000. Foraging ecology of northern elephant seals. *Eco. Monogr.* 70:353–82

76. Eckert SA, Dolar LL, Kooyman GL, Perrin W, Rahman RA. 2002. Movements of whale sharks (*Rhincodon typus*) in South-east Asian waters as determined by satellite telemetry. *J. Zool.* 257:111–15

77. Kenagy GJ, Sharbaugh SM, Nagy KA. 1989. Annual cycle of energy and time expenditure in a golden-mantled ground squirrel population. *Oecologia* 78:269–82

78. Somero GN, Dahlhoff E, Lin JJ. 1996. Stenotherms and eurytherms: mechanisms establishing thermal optima and tolerance ranges. In *Society for Experimental Biology Seminar Series, 59. Animals and Temperature: Phenotypic and Evolutionary Adaptation. Meet. Soc. Exp. Biol., St. Andrews, Scotland, UK,* ed. IA Johnston, AF Bennett, pp. 53–78. Cambridge, UK/New York: Cambridge Univ. Press

79. Block BA. 1991. Endothermy in fish:

thermogenesis, ecology and evolution. In *Biochemisty and Molecular Biology of Fishes*, ed. PW Hochachka, TP Momnsen, pp. 269–311. New York: Elsevier

80. Carey FG. 1982. A brain heater in the swordfish *Xiphias gladius*. *Science* 216:1327–29

81. Block BA. 1991. Evolutionary novelties: how fish have built a heater out of muscle. *Am. Zool.* 31:726–42

82. Brill RW. 1987. On the standard metabolic rates of tropical tunas including the effect of body size and acute temperature change. *Fish. Bull.* 85:25–36

83. Graham JB, Koehrn FJ, Dickson KA. 1983. Distribution and relative proportions of red muscle in scombrid fishes: consequences of body size and relationships to locomotion and endothermy. *Can. J. Zool.* 61:2087–96

84. Hochachka PW. 1974. Enzymatic adaptations to deep sea life. In *The Biology of the Oceanic Pacific*, ed. CB Miller, pp. 107–36. Corvallis, OR: Oregon State Univ. Press

85. Stevens ED, Dizon AE. 1982. Energetics of locomotion in warm-bodied fish. *Annu. Rev. Physiol.* 44:121–32

86. Stevens ED, Lam HM, Kendall J. 1974. Vascular anatomy of the countercurrent heat exchanger of skipjack tuna. *J. Exp. Biol.* 61:145–53

87. Stevens ED, Carey FG. 1981. One why of the warmth of warm bodied fish. *Am. J. Physiol. Regul. Integr. Comp. Physiol.* 240:R151–55

88. Block BA, Stevens DE. 2001. *Tuna: Physiology, Ecology, and Evolution.* San Diego: Academic. 468 pp.

89. Gunn J, Block BA. 2001. Acoustic, archival and pop-up satellite tagging of tunas. See Ref. 88, pp. 167–224

90. Carey FG, Teal JM. 1966. Heat conservation in tuna fish muscle. *Proc. Natl. Acad. Sci. USA* 56:1464–69

91. Carey FG, Teal JM. 1969. Regulation of body temperature by the bluefin tuna. *Comp. Biochem. Physiol.* 28:205–13

92. Carey FG, Kanwisher JW, Brazier O, Gabrielson G, Casey JG, Pratt HLJ. 1982. Temperature and activities of a white shark *Carcharodon carcharias*. *Copeia* 254–60

93. Block BA, Finnerty JR, Stewart AFR, Kidd J. 1993. Evolution of endothermy in fish—mapping physiological traits on a molecular phylogeny. *Science* 260:210–14

94. Carey FG, Scharold JV. 1990. Movements of blue sharks *Prionace glauca* in depth and course. *Mar. Biol.* 106:329–42

95. Carey FG, Robison BH. 1981. Daily patterns in the activities of swordfish *Xiphias gladius* observed by acoustic telemetry. *Fish. Bull.* 79:277–92

96. Musyl MK, Brill RW, Boggs CH, Curran DS, Kazama TK, Seki MP. 2003. Vertical movements of bigeye tuna (*Thunnus obesus*) associated with islands, buoys, and seamounts near the main Hawaiian Islands from archival tagging data. *Fish. Oceanogr.* 12:152–69

97. Costa DP, Gentry RL. 1986. Free-ranging energetics of northern fur seals. In *Fur Seals: Maternal Strategies on Land and at Sea*, ed. RL Gentry, GL Kooyman, pp. 79–101. Princeton, NJ: Princeton Univ. Press

98. Davis RW, Croxall JP, O'Connell MJ. 1989. The reproductive energetics of gentoo (*Pygoscelis papua*) and macaroni (*Eudyptes chrysolophus*) penguins at South Georgia (South Atlantic Ocean). *J. Anim. Ecol.* 58:59–74

99. Hails CJ, Bryant DM. 1979. Reproductive energetics of a free living bird. *J. Anim. Ecol.* 48:471–82

100. Montevecchi WA, Birt-Friesen VL, Cairns DK. 1992. Reproductive energetics and prey harvest of Leach's storm-petrels in the Northwest Atlantic. *Ecology* 73:823–32

101. Utter JM, Lefebvre EA. 1973. Daily

energy expenditure of purple martins (*Progne subis*) during the breeding season: estimated using D_2O^{18} and time budget methods. *Ecology* 54:597–604

102. Kurta A, Bell GP, Nagy KA, Kunz TH. 1989. Energetics of pregnancy and lactation in free-ranging little brown bats *Myotis lucifugus*. *Physiol. Zool.* 62:804–18

103. Anderson RA, Karasov WH. 1988. Energetics of the lizard *Cnemidophorus tigris* and life history consequences of food-acquisition mode. *Eco. Monogr.* 58:79–110

104. Lack D. 1947. The significance of clutch size. *Ibis* 89:302–52

105. Reznick D. 1992. Measuring the costs of reproduction. *Trends Ecol. Evol.* 7:42–45

106. Reyer HU. 1983. Investment and relatedness: a cost/benefit analysis of breeding and helping in the pied kingfisher (*Ceryle rudis*). *Anim. Behav.* 32:1163–78

107. Nagy KA. 1980. CO_2 production in animals: analysis of potential errors in the doubly labeled water method. *Am. J. Physiol. Regul. Integr. Comp. Physiol.* 238: R466–73

108. Speakman JR. 1998. The history and theory of the doubly labeled water technique. *Am. J. Clin. Nutr.* 68: 932S–38S

109. Nagy KA. 1987. Field metabolic rate and food requirement scaling in mammals and birds. *Eco. Monogr.* 57:111–28

110. Nagy KA, Girard IA, Brown TK. 1999. Energetics of free-ranging mammals, reptiles, and birds. *Annu. Rev. Nutr.* 19:247–77

111. Masman D, Dijkstra C, Daan S, Bult A. 1989. Energetic limitation of avian parental effort field experiments in the Kestrel *Falco tinnunculus*. *J. Evol. Biol.* 2:435–56

112. Weathers WW, Hodum PJ, Blakesley JA. 2001. Thermal ecology and ecological energetics of California spotted owls. *Condor* 103:678–90

113. Golet GH, Irons DB, Costa DP. 2000. Energy costs of chick rearing in Black-legged Kittiwakes (*Rissa tridactyla*). *Can. J. Zool.* 78:982–91

114. Humphries Murray M, Boutin S. 2000. The determinants of optimal litter size in free-ranging red squirrels. *Ecology* 81:2867–77

115. Thomson DL, Furness RW, Monaghan P. 1998. Field metabolic rates of Kittiwakes *Rissa tridactyla* during incubation and chick rearing. *Ardea* 86:169–75

116. Tinbergen JM, Verhulst S. 2000. A fixed energetic ceiling to parental effort in the great tit? *J. Anim. Ecol.* 69:323–34

117. Arnould JPY, Boyd IL, Rawlins DR, Hindell MA. 2001. Variation in maternal provisioning by lactating Antarctic fur seals (*Arctocephalus gazella*): response to experimental manipulation in pup demand. *Behav. Ecol. Sociobiol.* 50:461–66

118. Bryant DM, Hails CJ, Tatner P. 1984. Reproductive energetics of 2 tropical bird species. *Auk* 101:25–37

119. Costa DP. 1991. Reproductive and foraging energetics of high latitude penguins, albatrosses and pinnipeds—implications for life history patterns. *Am. Zool.* 31: 111–30

120. Tinbergen JM, Dietz MW. 1994. Parental energy expenditure during brood rearing in the Great tit (*Parus major*) in relation to body mass, temperature, food availability and clutch size. *Funct. Ecol.* 8:563–72

121. Wansink D, Tinbergen JM. 1994. The influence of ambient temperature on diet in the great tit. *J. Avi. Biol.* 25:261–7

122. Kenagy GJ, Masman D, Sharbaugh SM, Nagy KA. 1990. Energy expenditure during lactation in relation to litter size in free-living golden-mantled ground squirrels. *J. Anim. Ecol.* 59:73–88

123. Daan S, Deerenberg C, Dijkstra C. 1996. Increased daily work precipitates

natural death in the kestrel. *J. Anim. Ecol.* 65:539–44

124. Deerenberg C, Pen I, Dijkstra C, Arkies BJ, Visser GH, Daan S. 1995. Parental energy expenditure in relation to manipulated brood size in the European kestrel *Falco tinnunculus*. *Zoology (Jena)* 99:39–48

125. Verhulst S, Tinbergen JM. 1997. Clutch size and parental effort in the great tit *Parus major*. *Ardea* 85:111–26

126. Costa DP. 2001. Energetics. In *Encyclopedia of Marine Mammals*, ed. WF Perrin, JGM Thewissen, B Wursig, pp. 387–94. New York: Academic

127. Sinervo B, Calsbeek RG. 2003. Physiological epistasis, ontogenetic conflict and natural selection on physiology and life history. *J. Integr. Comp. Biol.* In press

128. Wingfield JC, Manley DL, Bruenner CW, Jacobs JD, Lynn S, et al. 1998. Ecological bases of hormone-behavior interactions: the 'emergency life history stage.' *Am. Zool.* 38:191–206

129. Sapolsky R. 1992. Neuroendocrinology of the stress-response. In *Behavioral Endocrinology*, ed. JB Becker, SM Breedlove, D Crew. Cambridge, MA: MIT Press

130. Silverin B, Arvidsson B, Wingfield J. 1997. The adrenocortical responses to stress in breeding Willow Warblers *Phylloscopus trochilus* in Sweden: effects of latitude and gender. *Funct. Ecol.* 11:376–84

131. Astheimer LB, Buttemer WA, Wingfield JC. 1995. Seasonal and acute changes in adrenocortical responsiveness in an Arctic-breeding bird. *Horm. Behav.* 29:442–57

132. Marler CA, Moore MC. 1988. Evolutionary costs of aggression revealed by testosterone manipulations in free-living male lizards. *Behav. Ecol. Sociobiol.* 23:21–26

133. Marler CA, Moore MC. 1991. Supplementary feeding compensates for testosterone-induced costs of aggression in male mountain spiny lizards, *Sceloporus jarrovi*. *Anim. Behav.* 42:209–19

134. Wingfield JC, Hegner RE, Lewis DM. 1991. Circulating levels of lutenizing hormone and steroid hormones in relation to social status in the cooperatively breeding White-Browed Sparrow Weaver *Plocepasser mahali*. *J. Zool.* 225:43–58

135. Romero LM, Soma KK, Wingfield JC. 1998. Changes in pituitary and adrenal sensitivities allow the snow bunting (*Plectrophenax nivalis*), an Arctic-breeding song bird, to modulate corticosterone release seasonally. *J. Comp. Physiol. B* 168:353–58

136. Denardo DF, Sinervo B. 1994. Effects of steroid hormone interaction on activity and home-range size of male lizards. *Horm. Behav.* 28:273–87

137. Hegner RE, Wingfield JC. 1987. Effects of experimental manipulation of testosterone levels on parental investment and breeding success in male house sparrows. *Auk* 104:462–69

138. Wedekind C, Folstad I. 1994. Adaptive or non-adaptive immunosuppression by sex hormones? *Am. Nat.* 143:936–38

139. Svensson EI, Sinervo B, Comendant T. 2002. Mechanistic and experimental analysis of condition and reproduction in a polymorphic lizard. *J. Evol. Biol.* 15:1034–47

140. Ottinger MA, Adkins-Regan E, Buntin J, Cheng MF, DeVoogd T, et al. 1984. Hormonal mediation of reproductive behavior. *J. Exp. Zool.* 232:605–16

141. Silverin B, Viebke PA, Westin J. 1989. Hormonal correlates of migration and territorial behavior in juvenile willow tits during autumn. *Gen. Comp. Endocrinol.* 75:148–56

142. Wingfield JC, Silverin B. 1986. Effects of corticosterone on territorial behavior of free-living male song sparrows *Melospiza melodia*. *Horm. Behav.* 20:405–17

143. Sinervo B, Basolo Alexandra L. 1996. Testing adaptation using phenotypic manipulations. In *Adaptation*, ed. MR Rose, GV Laude, pp. 24–28. San Diego/London: Academic

144. Parmesan C, Yohe GA. 2003. Globally coherent fingerprint of climate change impacts across natural systems *Nature* 421:37–42

144a. Felsenstein J. 1985. Phylogenies and the comparative method. *Am. Nat.* 125:1–15

145. Snyder MA, Bell JL, Sloan LC, Duffy PB, Govindasamy B. 2002. Climate responses to a doubling of atmospheric carbon dioxide for a climatically vulnerable region. *Geophys. Res. Lett.* 29:1–4

146. Jackson JBC, Kirby MX, Berger WH, Bjorndal KA, Botsford LW, et al. 2001. Historical overfishing and the recent collapse of coastal ecosystems. *Science* 293:629–37

Annu. Rev. Physiol. 2004. 66:239–74
doi: 10.1146/annurev.physiol.66.032102.115105
First published online as a Review in Advance on October 15, 2003

METABOLIC RATE AND BODY TEMPERATURE REDUCTION DURING HIBERNATION AND DAILY TORPOR

Fritz Geiser

*Zoology, Center for Behavioural and Physiological Ecology, University of
New England, Armidale, NSW Australia 2351; email: fgeiser@metz.une.edu.au*

Key Words   body mass, body temperature, metabolic inhibition, temperature
effects, thermal conductance, thermoregulation

■ **Abstract**   Although it is well established that during periods of torpor heterothermic mammals and birds can reduce metabolic rates (MR) substantially, the mechanisms
causing the reduction of MR remain a controversial subject. The comparative analysis provided here suggests that MR reduction depends on patterns of torpor used, the
state of torpor, and body mass. Daily heterotherms, which are species that enter daily
torpor exclusively, appear to rely mostly on the fall of body temperature (T_b) for MR
reduction, perhaps with the exception of very small species and at high torpor T_b,
where some metabolic inhibition may be used. In contrast, hibernators (species capable of prolonged torpor bouts) rely extensively on metabolic inhibition, in addition
to T_b effects, to reduce MR to a fraction of that observed in daily heterotherms. In
small hibernators, metabolic inhibition and the large fall of T_b are employed to maximize energy conservation, whereas in large hibernators, metabolic inhibition appears
to be employed to facilitate MR and T_b reduction at torpor onset. Over the ambient
temperature (T_a) range where torpid heterotherms are thermo-conforming, the T_b-T_a
differential is more or less constant despite a decline of MR with T_a; however, in
thermo-regulating torpid individuals, the T_b-T_a differential is maintained by a proportional increase of MR as during normothermia, albeit at a lower T_b. Thermal conductance in most torpid thermo-regulating individuals is similar to that in normothermic
individuals despite the substantially lower MR in the former. However, conductance is
low when deeply torpid animals are thermo-conforming probably because of peripheral
vasoconstriction.

0066-4278/04/0315-0239$14.00

INTRODUCTION

Endothermic mammals and birds have the ability to maintain a constant high body temperature (T_b) over a wide range of ambient temperatures (T_a).[1] Because the surface area/volume ratio of animals increases with decreasing size, many small endotherms must produce substantial amounts of endogenous heat to compensate for high heat loss during cold exposure. As prolonged periods of high metabolic heat production require high food intake and food availability in the wild often fluctuates, the cost of thermoregulation may become prohibitively expensive. This is one of the reasons why not all mammals and birds are permanently homeothermic (i.e., maintain a constant high T_b), but during certain times of the day or the year enter a state of torpor (9, 92, 106). Torpor in these heterothermic endotherms is characterized by a controlled reduction of T_b, metabolic rate (MR), and other physiological functions. The T_b during torpor falls from high normothermic values of $\sim$32 to 42°C to values between -3 to < 30°C, and the minimum torpid metabolic rate (TMR) is on average reduced to 5–30% of the basal metabolic rate (BMR) (3, 50). TMR can be less than 1% of the normothermic resting metabolic rate (RMR) in some species at low T_a, emphasizing the effectiveness of torpor in reducing MR.

Although MR during torpor may be a fraction of that in normothermic individuals, regulation of T_b during torpor is not abandoned. T_b is regulated at or above a species- or population-specific minimum by a proportional increase in heat production that compensates for heat loss (30, 71). During entry into torpor, the hypothalamic set point (T_{set}) for T_b is down-regulated ahead of T_b (73). Only when T_b reaches the low T_{set} during torpor after a cooling episode is metabolic heat production used to maintain T_b at or above this minimum T_b.

However, when undisturbed and in steady-state torpor at T_a above the minimum T_b, torpid endotherms are thermo-conforming. The T_b-T_a differential in this temperature range is usually small, $\sim$1–3°C (140, 142, 159), and a reduction of T_a does not result in an increase of TMR, instead it falls with T_b until a new equilibrium is reached. Nevertheless, at any time, disturbance can result in a rise of T_{set} in torpid individuals, initiating thermoregulation. Moreover, induced torpor in the laboratory may not always result in steady-state torpor, especially when animals are not allowed to undergo their natural daily or seasonal T_b cycle, and may result in a T_{set} and TMR that are well above the minima characteristic for the species under investigation (51). Higher than characteristic minima are obviously also observed when not enough time is allowed for reaching the steady-state minima.

In addition to possessing the ability to thermoregulate during torpor, heterothermic endotherms can rewarm themselves from the low T_b during torpor by using endogenous heat production, although recent evidence from the field shows that passive rewarming also plays an important role (18, 53, 56, 97, 123, 150, 157).

---

[1]Abbreviations: T_b, body temperature; T_a, air temperature; C, apparent thermal conductance; TNZ, thermoneutral zone; T_{lc}, lower critical temperature of the TNZ; MR, metabolic rate; BMR, basal metabolic rate; TMR, torpid metabolic rate; RMR, resting metabolic rate.

In placental heterotherms, brown fat appears to be a major tissue involved in endogenous heat production during arousal via nonshivering thermogenesis, whereas in birds, monotremes, and marsupials, which appear to lack functional brown fat (128), shivering appears responsible for much of the heat produced (19, 47, 66, 111, 125).

PATTERNS OF TORPOR

Most heterothermic mammals and birds appear to use one of two common patterns of torpor: hibernation or prolonged torpor in the hibernators and daily torpor in the daily heterotherms. Hibernation is often seasonal and usually lasts from late summer/autumn to late winter/spring. However, hibernators do not remain torpid throughout the hibernation season. Bouts of torpor, during which T_b is low and bodily functions are reduced to a minimum, last for several days or weeks, but are interrupted by periodic rewarming and brief (usually less than one day) normothermic resting periods with high T_b and high energy turnover (32, 46, 88, 95, 158). Hibernators, which currently include many mammals but only a single bird species, are generally small, and most weigh between 10 and 1000 g, with a median mass of 85 g (50). However, the entire mass range of hibernators for which metabolic data are available, including black bears (*Ursus americanus*), is ~5 to 80,000 g. Nevertheless, deep torpor with a reduction of T_b by more than 10°C is restricted to species weighing less than 10 kg. Many hibernators fatten extensively before the hibernation season, refuse to hibernate when lean, and rely to a large extent on stored fat or food for an energy source in winter.

Hibernating species usually reduce T_b to below 10°C, with a minimum of -3°C in arctic ground squirrels, *Spermophilus parryii* (3), and most, with the exception of large carnivores and perhaps tropical hibernators, have T_b minima around 5°C (4, 18, 26, 50, 59). The TMR in hibernators is on average reduced to about 5% of the BMR but can be less than 1% of the RMR in normothermic individuals at low T_a. Even when the high cost of periodic arousals is considered, energy expenditure during the mammalian hibernation season is still reduced to below 15% of that the animal would have expended if it remained normothermic throughout winter (152). This enormous reduction in energy expenditure is perhaps best illustrated by the fact that many hibernating mammals can survive for 5 to 7 months entirely on body fat that has been stored prior to the hibernation season (32). Thus energy intake and expenditure in hibernators are not balanced on a daily but rather a yearly basis.

Daily torpor in the daily heterotherms is the other widely used pattern of torpor in mammals and, in contrast to hibernation, also commonly in birds (50, 114). This form of torpor is usually not as deep as hibernation, lasts only for hours rather than days or weeks, and is usually interrupted by daily foraging and feeding. On average, daily heterotherms are smaller than hibernators and most weigh between 5 and 50 g, with a median of 19 g, and a range of ~2 to 9000 g (50). In diurnal heterotherms, daily torpor is usually restricted to the night, whereas in nocturnal mammals and birds it is common in the second part of the night and the early morning. However, in the field, two bouts of torpor per day appear common in nocturnal species, and one of these is usually terminated by partially passive arousal via an increase of T_a or exposure to solar radiation (10, 58, 97). Generally, daily torpor is less seasonal than hibernation and can occur throughout the year, although its use often increases in winter. In some species from high latitudes, such as Siberian hamsters (*Phodopus sungorus*), daily torpor appears to be restricted to winter (67, 95). On the other extreme, in some warm climate species, such as subtropical nectar-eating blossom-bats (*Syconycteris australis*), daily torpor is deeper and longer in summer than in winter, and this unusual seasonal pattern appears to be explained mainly by reduced nectar availability in summer (16). Whereas daily torpor often occurs as a response to acute energy shortage, in some species it is employed regularly to balance energy budgets, even when food availability may appear favorable. For example, small arid-zone marsupials of the genus *Sminthopsis* regularly enter daily torpor in the laboratory when food is freely available (spontaneous torpor), which appears to reflect the generally low energy availability in their natural desert habitat (41, 43). In some hummingbirds, daily torpor at night is used to conserve fat stores for migration on the following day (15). Moreover, the marsupial Mulgara (*Dasycercus cristicauda*) appears to use spontaneous daily torpor during pregnancy to facilitate fat storage for the following energetically demanding period of lactation (49).

Many daily heterotherms, in contrast to most hibernators, do not exhibit extensive fattening prior to the season in which torpor is most commonly employed and typically enter torpor at times when body mass is low (39, 67, 84, 96). Large fat stores, as simulated by administration of the hormone leptin, inhibit daily torpor (55). When food is withheld from small daily heterotherms for several days they will perish (94), whereas hibernators can survive for months. The main energy supply of daily heterotherms, even in the main torpor season, remains ingested food rather than stored body fat, and they appear to balance energy expenditure and uptake on a daily basis. The T_b in daily heterotherms, such as small carnivorous marsupials (e.g., *Sminthopsis* spp.) and mice (e.g., *Peromyscus* spp.), usually falls to near 18°C, although in some hummingbirds, values below 10°C have been reported. In other, mainly large species such as tawny frogmouths (*Podargus strigoides*) or American badgers (*Taxidea taxus*), T_b is maintained just below 30°C (50, 64, 97). The TMR in daily heterotherms is on average reduced to about 30% of the BMR, although this percentage is strongly affected by body mass and other factors. When RMR at low T_a is used as a point of reference, reductions of MR during daily torpor to about 10–20% of that in normothermic individuals at the same T_a are common. Depending on the species, the duration and intensity of activity, the duration of the torpor bout, and torpor depth, overall daily energy expenditure is usually reduced 50–90% on days when daily torpor is employed, compared with days when no torpor is used (24, 83).

Thus torpor bouts in the daily heterotherms are always shorter than one day independent of food supply or prevailing ambient conditions. In contrast, although hibernators usually display bouts of torpor lasting several days or weeks, they are capable of brief torpor bouts of less than one day early and late in the hibernation season or at high T_a (32, 46, 142, 146), which superficially may appear the same as daily torpor in the daily heterotherms and, when it occurs in summer, is often referred to as aestivation. However, it appears that these short torpor bouts in hibernators are functionally nothing but brief bouts of hibernation with TMR well below those of the daily heterotherms even at the same T_b (6, 38, 51, 142, 159). Thus the term daily torpor should not be applied to describe short torpor bouts of hibernators, because it describes only the temporal pattern of T_b fluctuations without considering the apparently functional differences in the mechanism of MR reduction.

Whereas the two patterns of torpor described above appear to be the most common, it is likely that not all species conform to these but exhibit some intermediate pattern. However, in some reported cases on intermediate torpor patterns, no long-term experiments with exposure to low T_a have been conducted to clearly establish whether the species is capable of prolonged torpor bouts (e.g., 103, 161). This is an important test because, as is outlined above, hibernators commonly display test drops early in the hibernation season that superficially resemble daily torpor (146). It is interesting that even if such studies are included, heterotherms still clearly fall into two groups (50).

HYPOTHESES ON METABOLIC RATE REDUCTION

Whereas most researchers in this field agree that the reduction of MR during torpor is substantial and is pivotal for survival in many species, the mechanisms of how MR is reduced remain controversial. Several, at first glance, mutually exclusive hypotheses attempting to explain the MR reduction during torpor have been proposed. The traditional view is that as T_b and MR fall together at torpor entry and because the Q_{10} (the change in rate over a 10°C increment) between TMR and T_b often approximates 2 (which is typical of biochemical reactions), the MR reduction during torpor below BMR is explained by temperature effects (60, 62, 138). Because unexpectedly high Q_{10} (>3) have been observed in some species during torpor entry and at high T_b during torpor, it was proposed that a physiological inhibition, in addition to temperature effects, must be involved in the reduction of MR (38, 108, 145). Others have proposed that T_b may have no influence at all on TMR. They argue that MR is down-regulated at torpor entry and the fall of T_b is the consequence of and not the reason for the reduction of MR (68, 69, 127). Finally, it has been suggested that, as during normothermia, MR during torpor is a function of the T_b-T_a differential (68) or that the low TMR may be due to the low apparent thermal conductance (C) in torpid individuals (139).

The purpose of this review is to examine these hypotheses with respect to three factors that appear important in determining MR and T_b during torpor: (a) patterns of torpor, (b) state of torpor, and (c) body mass.

DATA SELECTION AND ANALYSIS

For this comparative analysis, data on MR, T_b, and body mass of heterothermic mammals and birds during normothermia and torpor were collected from the literature (Table 1). BMR was used as a reference point for the TMR of thermoconforming individuals because in both physiological states, metabolism is used only for maintenance, without extra energy expenditure for thermoregulation (12, 155). The only torpor values used in this analysis were from studies in which the following conditions pertained: The species appeared to be in steady-state torpor with a TMR below the BMR at some of the measured T_a values, and simultaneous data on T_b were available. For several species it was assumed that at steady-state torpor, T_b was $T_a + 1°C$. In some species, no BMR values were provided and for those species BMR was calculated from allometric equations. To avoid the potential problem arising from temperature corrections, TMR data were statistically analyzed in different T_b bins of 0.0–9.9°C, 10.0–14.9°C, 15.0–24.9°C, and 24.0–32.9°C, and both the TMR and the Q_{10} calculated between BMR and TMR at various T_bs were analyzed as a function of body mass. Data for torpid individuals were collected at T_a both below and above the T_{set} to examine relations between T_b and MR, MR and the apparent thermal conductance (C), and that between the T_b-T_a differential and MR (Tables 1 and 2). For most variables, data were analyzed separately for daily heterotherms and hibernators because several physiological variables differ significantly between the two groups and because all heterotherms fall into two groups when analyzed by cluster and discriminant analyses (50). Consequently, pooling of the data likely would obscure significant relationships. Data were compared using ANCOVA and t-tests as appropriate, and linear regressions were fitted using the method of least squares. Numeric values are expressed as means $\pm$ SD for n, the number of species investigated.

TORPOR ENTRY

When an animal enters torpor, the T_{set} for T_b falls faster than T_b facilitated by thermal inertia (73). As torpor entry usually occurs at low T_a, well below the thermoneutral zone (TNZ) for most species, the fall of T_{set} should theoretically result in a fall of MR from the resting metabolic rate (RMR) (energetic cost of BMR plus cost of thermoregulation) to BMR (no additional cost for thermoregulation) because heat production for normothermic thermoregulation will cease. This interpretation is supported by empirical data on sugar gliders, *Petaurus breviceps* (~120 g), which, when they become torpid in the laboratory, usually do so early in the morning, but on days when they remain normothermic, they lower T_b only slightly at the beginning of the rest phase. The transition from a nocturnal active T_{set} of ~39°C to a diurnal resting normothermic T_{set} of ~35°C results in a precipitous drop of MR that superficially appears to be a torpor entry (Figure 1). However, MR falls only from RMR to about BMR, although T_a was 10°C (~ 17°C

TABLE 1 Body mass, body temperatures (T_b), basal metabolism (BMR), torpid metabolism (TMR), and Q_{10} in heterothermic endotherms

Daily Heterotherms Group/species	Mass (g)	T_b1 (°C)	T_b2 (°C)	BMR [ml O_2/(gh)]	TMR [ml O_2/(gh)]	Q_{10}	Reference
Mammals							
Marsupialia (Didelphimorphia)							
Marmosa sp.	13	34.7	16	1.4	0.25	2.5	119
Marsupialia (Dasyuromorphia)							
Planigale gilesi	8.3	32.6	17.5	1.43	0.54	1.9	45
Planigale maculata	10	34.2	19.6	1.01	0.45	1.7	120
Ningaui yvonneae	11.6	34.4	16.6	1.35	0.26	2.5	45
Sminthopsis crassicaudata	17.3	34.1	13.7	1.2	0.25	2.2	43
	17.3	34.1	16.5	1.2	0.3	2.2	
	17.7	34.3	19	1.2	0.33	2.3	
	17.3	34.1	25.8	1.2	0.81	1.6	
Sminthopsis murina	19	35	15	1.13	0.25	2.1	57
Sminthopsis macroura	24.8	34.3	16	0.89	0.12	3.0	140
	24.8	34.3	25	0.89	0.28	3.5	
	24.8	34.3	30	0.89	0.52	3.5	
Antechinus flavipes	26	34	24.5	1.04	0.48	2.3	35, 39
Antechinus stuartii	26.1	34.1	19.9	1.06	0.66	1.4	39
	26.1	34.1	26.6	1.06[b]	0.84	1.4	
Antechinomys laniger	27.4	34.8	12	0.98	0.16	2.2	36
	27.4	34.8	16	0.98	0.18	2.5	
	27.4	34.8	25.8	0.98	0.38	2.9	
Dasycercus cristicauda	113	35.5	23	0.5	0.27	1.6	94
Dasyuroides byrnei	116	34.3	24.4	0.74	0.44	1.7	43
Marsupialia (Diprotodontia)							
Tarsipes rostratus	10	36.6	5	2.9	0.15	2.6	161
Petaurus breviceps	132	36.3	~17	0.74	0.1	2.8	27
Rodentia							
Baiomys taylori	7.3	35.5	22	1.95	~0.55	2.6	85
Reithrodontomys megalotis longicaudus	7.9	37.5	12	2.63	~0.5	1.9	148
Reithrodontomys megalotis ravus	9.5	37.3	15	2.23	~0.4	2.2	148
Gerbillus pusillus	12.6	35	16.7	1.05	0.38	1.7	14
Peromyscus eremicus	17.4	37	16	1.56	~0.3	2.2	107
Peromyscus maniculatus	18	37	21.5	1.96	0.55	2.3	115, 117, 40
Peromyscus leucopus	20	36.7	17.6	1.66	0.53	1.8	78, 22
Phodopus sungorus	25	35.6	15.5	2.01	0.7	1.7	68
	25	37	20.7	2.06	1	1.6	67
	25	35.6	27.7	2.01	1.04	2.3	68
Perognathus californicus	22	38	15	0.97	0.2	2.0	149
Perognathus hispidus	40	38	17.1	1.25	0.15	2.8	153
Steatomys pratensis	28	34.1	21	1.315	0.3	3.1	23

(Continued)

TABLE 1   (*Continued*)

Daily Heterotherms	Mass (g)	T_b1 (°C)	T_b2 (°C)	BMR [ml O_2/(gh)]	TMR [ml O_2/(gh)]	Q_{10}	Reference
Mus musculus	45.5	37.4	19	1.47	0.3	2.4	87
Saccostomus campestris	71	34	28	0.618	0.35	2.6	122
Insectivora							
Suncus etruscus	2	34.7	16	5.75	0.6	3.3	33, 34, 31
	2	34.7	22	5.75	0.8	4.7	
Notiosorex crawfordi	4	38	28	3.27	1.42	2.3	101
Crocidura russula	10	35.8	~23	2.4	0.9	2.2	124
Chiroptera							
Macroglossus minimus	16	35.3	23.1	1.29	0.7	1.7	5
Syconycteris australis	18	34.9	17.4	1.3	0.58	1.6	16, 54
	18	34.9	27	1.3	1.1	1.2	
Nyctimene albiventer	28	37	28.6	1.43	0.67	2.5	7
Carnivora							
Taxidea taxus	9000	37	28	0.3	0.13	2.5	64
Birds							
Coliiformes							
Colius striatus	51	36	18.2	0.83	0.11	3.1	113
Colius castanotus	58	38.5	18	1.2	0.1	3.4	133, 81
Trochiliformes							
Selasphorus sasin	3.1	40	23[a]	3.8	1.24	1.9	99, 130
Calypte costae	3.1	40	21[a]	3	0.39	2.9	99
Calypte anna	5.4	40	16[a]	3.85	0.17	3.7	99
	5.4	40	24[a]	3.85	0.54	3.4	
Archilochus alexandri	3.2	40	17[a]	3.5	0.2	3.5	99
	3.2	40	23[a]	3.5	0.45	3.3	
Selasphorus rufus	3.3	40	13	3.36	0.43	2.1	99, 76
	3.6	40	22.5[a]	3.35	0.48	3.0	99
Selasphorus platycercus	3.5	39	16.6	3.6	0.72	2.1	11, 12
	3.5	39	20.5	3.6	0.84	2.2	
	3.5	39	24.4	3.6	1.53	1.8	
	3.5	39	28.8	3.6	2.06	1.7	
Panterpe insignis	6	40	14	2.83[c]	0.5	1.9	162, 21
Eugenes fulgens	8	40	10	3.2	0.5	1.9	162, 21
Eulampis jugularis	8	40	18	3	1	1.6	61
Lampornis clemenciae	8	39.6	19.6	2.64	0.45	2.4	98
Strigiformes/ Caprimulgiformes							
Caprimulgus argus	75	39	29.6	0.83	0.4	2.2	20
Columbiformes							
Drepanoptila holosericea	200	37.7	25	0.72	0.27	2.2	135
Passeriformes							
Delichon urbica	18	40	26	2.4	1.2	1.6	132

(*Continued*)

TABLE 1 (Continued)

Hibernators	Mass (g)	T_b1 (°C)	T_b2 (°C)	BMR [ml O₂/(gh)]	TMR [ml O₂/(gh)]	Q_{10}	Reference
Mammals							
Monotremata							
Tachyglossus aculeatus	2800	32.2	4	0.15	0.02	2.0	1, 127
	2800	32.2	16	0.15	0.045	2.1	
Marsupialia							
(Diprotodontia)							
Acrobates pygmaeus	14	34.9	2	1.08	0.042	2.7	28, 52
	14	34.9	6	1.08	0.065	2.6	
Cercartetus lepidus	12.6	33.7	6.8	1.49	0.047	3.6	37, 38
	12.6	33.7	13	1.49	0.1	3.7	
	12.6	33.7	25.5	1.49	0.33	6.3	
Cercartetus concinnus	18.6	34.4	6.6	1.2	0.034	3.6	37, 38
	18.6	34.4	15.6	1.2	0.092	3.9	
	18.6	34.4	29	1.2	0.5	5.1	
Cercartetus nanus	36	34.3	5.9	0.66	0.019	3.5	142
	36	34.3	15	0.66	0.054	3.7	
	36	34.3	24.5	0.66	0.16	4.2	
	36	34.3	25	0.66	0.17	4.3	
	36	34.3	31.4	0.66	0.341	9.7	
Burramys parvus	50	36	2.5	0.83	0.033	2.6	29, 48
	50	36	9.3	0.83	0.043	3.0	
Rodentia							
Zapus hudsonicus	22.6	37.3	6[a]	1.5	0.043	3.1	118, 121
	25	37.3	11[a]	1.5	0.04	4.0	
Zapus princeps	33.6	37.2	5.5	1.55[b]	0.042	3.1	17
	27.7	37.2	5.5	1.66[b]	0.027	3.7	
Muscardinus avellanarius	23.5	35.8	11[a]	1.75[b]	0.04	4.6	89, 91
Eliomys quercinus	70	37	7.5[a]	1.22[b]	0.034	3.4	91
Glis glis	140	~37	~5	0.97[b]	0.017	3.5	159
	140	~37	~17	0.97[b]	0.031	5.6	
Mesocricetus auratus	90	37	5	1.19[b]	0.07	2.4	104
Cricetus cricetus	330	36.8	7.5[a]	0.88	0.032	3.1	89, 91
Tamias amoenus	60	38	1.2	1.69	0.042	2.7	44, 93
Tamias striatus	87	38.2	7	1.03	0.06	2.5	154
Spermophilus tereticaudus	125	35	11	0.72	0.048	3.1	8
	125	36	26	0.78	0.23	3.4	
Spermophilus lateralis	200	37.8	5.4	1.159	0.045	2.7	138
	200	36.2	9.3	0.598	0.068	2.2	
	200	36.5	9.5	0.82	0.064	2.6	
	200	37.2	10.9	0.629	0.068	2.3	
	200	37.4	13.9	0.867	0.084	2.7	
Spermophilus mexicanus	200	36.2	~8	0.85[b]	0.06	2.6	126
Spermophilus citellus	240	36.5	8[a]	0.79[b]	0.018	3.8	89, 90
Spermophilus saturatus	257	38	3.6	0.47[d]	0.031	2.2	46, 93
	246	38	5.3	0.47[d]	0.038	2.2	
	257	38	9.3	0.47[d]	0.048	2.2	
Spermophilus mohavensis	260	35.8	21.3	0.85	0.15	3.3	6
Spermophilus richardsonii	400	37.1	5[a]	0.535[d]	0.02	2.8	152
	400	37.1	15[a]	0.535[d]	0.04	3.2	86
Spermophilus parryii	1000	37	4.7	0.51[b]	0.012	3.2	13, 80
	1000	37	8.2	0.51[b]	0.012	3.7	
	1000	37	12.6	0.51[b]	0.014	4.4	
	1000	37	17.1	0.51[b]	0.018	5.4	
	1000	37	20.7	0.51[b]	0.047	4.3	

(Continued)

TABLE 1   (*Continued*)

Hibernators	Mass (g)	T_b1 (°C)	T_b2 (°C)	BMR [ml O_2/(gh)]	TMR [ml O_2/(gh)]	Q_{10}	Reference
Marmota flaviventris	2500	36.6	7.5	0.25	0.022	2.3	30
Marmota marmota	3100	~36	8	0.19	0.014	2.5	129
	3100	~36	21	0.19	0.024	4.0	
Marmota monax	4000	37	7[a]	0.27[b]	0.032	2.0	105
Macroscelidea							
Elephantulus rozeti	45	36.8	9	1.06	0.025	3.8	103
	45	36.8	16	1.06	0.06	4.0	
	45	36.8	27	1.06	0.27	4.0	
Elephantulus myurus	63	36.7	10	1.05	0.079	2.6	103
	63	36.7	18	1.05	0.155	2.8	
Insectivora							
Setifer setosus	270	32	16.5	0.34	0.07	2.8	91, 77
Tenrec ecaudatus	1220	33	16.5	0.27	0.025	4.2	91, 77
	360	33	16.9	0.31	0.02	5.5	
Erinaceus europaeus	700	35	5.2[a]	0.433	0.016	3.0	147
	700	35	16[a]	0.357	0.011	6.2	
Primates							
Cheirogaleus medius	250	~37	18.3	0.69[b]	0.12	2.5	18
Chiroptera							
Myotis lucifugus	6	35	5	1.53	0.06	2.9	74
	6	35	11	1.53	0.049	4.2	74, 79
	6	35	21	1.53	0.23	3.9	74, 79
	6	35	25	1.53	0.29	5.3	74
Barbastella barbastellus	7	37	4.5	2.08[b]	0.04	3.4	131
Pipistrellus pipistrellus	7.4	37	6[a]	2.05[b]	0.024	4.2	91
Nyctophilus geoffroyi	7	35.7	6.3	1.36	0.037	3.4	51
	7	35.7	15	1.36	0.09	3.7	
	7	35.7	25	1.36	0.215	5.6	
	7	35.7	29.4	1.36	0.36	8.2	
Nyctophilus gouldi	10	36	10.1	1.22	0.052	3.4	51
	10	36	16.8	1.22	0.14	3.1	
Eptesicus fuscus	10.4	36	~10	2	~0.1	3.2	75
Nyctalus noctula	23.8	37	5.3[a]	1.47[b]	0.03	3.4	91
Myotis myotis	25	37.5	4.5[a]	1.45[b]	0.04	3.0	131
Tadarida brasiliensis	16.9	36	~10	1.2	~0.1	2.6	75
Carnivora							
Ursus americanus	80,000	35	30	0.221	0.042	27.0	156, 4
Birds **Strigiformes/** **Caprimulgiformes**							
Phalaenoptilus nuttallii	35	37	10	0.788	~0.05	2.8	160
	35	37	20	0.788	0.086	3.7	

[a] - T_b calculated from T_a + 1°C.

[b] - BMR calculated from 65.

[c] - BMR calculated from 21.

[d] - BMR corrected for mass from 65.

TABLE 2   Apparent thermal conductance (C) in heterothermic endotherms during normothermia and during torpor while thermo-conforming and thermo-regulating

Group/Species	Mass (g)	C normothermia (ml O_2/g/h/°C)	C-torpor conforming [ml O_2/(gh°C)]	C-torpor-regulating [ml O_2/(gh°C)]	Reference
Marsupialia (Didelphimorphia)					
Marmosa sp.	13	0.258	0.125	0.253	119
Marsupialia (Dasyuromorphia)					
Planigale gilesi	8.3	0.33	0.145	0.38	45
Ningaui yvonneae	11.6	0.21	0.13	0.13	45
Sminthopsis crassicaudata	17.3	0.23	0.15	0.29	45
Sminthopsis macroura	24.8	0.19	0.09	0.26	140
Antechinomys laniger	27.4	0.169	0.11	0.2	36
Dasyuroides byrnei	116	0.1	0.057	0.08	45
Marsupialia (Diprotodontia)					
Cercartetus lepidus	12.6	0.21	0.052	0.17	37
Cercartetus concinnus	18.6	0.21	0.046	0.17	37
Cercartetus nanus	36	0.106	0.023	0.094	142
Petaurus breviceps	132	0.051	0.03	0.051	27, 82
Rodentia					
Phodopus sungorus	25	0.15	0.13	0.17	67, 68
Steatomys pratensis	28	0.18		0.12	23
Spermophilus lateralis	200	0.04	~0.02	0.03	144
Spermophilus parryii	1000	0.029	0.012	0.012	13, 25
Marmota marmota	3100	0.012	0.004	0.0072	129
Insectivora					
Suncus etruscus	2	0.7	0.3	0.75	33, 34
Notiosorex crawfordi	4	0.55	0.47	0.56	101
Crocidura russula	10	0.285	0.22	0.25	124
Chiroptera					
Macroglossus minimus	16	0.17	0.13	0.15	5
Syconycteris australis	18	0.16	0.13	0.14	54, 16
Birds Trochiliformes					
Selasphorus rufus	3.3	0.45	0.35	0.7	76
Panterpe insignis	6	0.32	~0.25	0.43	162, 21
Eugenes fulgens	8	0.3	~0.25	0.36	162, 21

below the T_{lc} of the TNZ). Thus the initial sharp fall of MR represents only the transient period when thermoregulatory heat production appears to be switched off to facilitate cooling, i.e., from the high normothermic T_b during the activity phase to the slightly lower normothermic T_b during the rest phase. The MR is raised again when the lowered T_{set} is approached. This transient fall in MR clearly shows that a small reduction in T_{set} can substantially reduce energy expenditure because of thermal inertia. Thermoregulatory heat production is not required during the

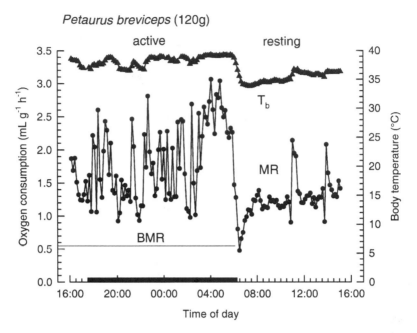

Figure 1 Metabolic rates measured as oxygen consumption of a sugar glider (*Petaurus breviceps*) during activity at night and rest during daytime. Note the precipitous drop of the metabolic rate to near BMR during the cooling phase from active to resting body temperatures (data from 82).

cooling phase and can result in a substantial MR reduction without the need for a large change of T_b. Thus the transient fall of MR is not due to the fall of T_b but the fall of T_{set}.

The initial reduction of MR at torpor onset at low T_a in most species follows a similar pattern. However, the T_{set} is reduced even further, and the substantial change in T_b that follows the reduction from RMR to BMR at torpor onset of most heterothermic species is one of the reasons why MR can fall well below BMR (Figure 2). Thus it is correct that the MR has to fall before T_b can fall, as it is often stated. However, this relationship of the initial MR decline usually explains only the reduction from RMR to BMR, not that from BMR to TMR.

Obviously, the scope of the reduction of RMR depends on size. Small mammals and birds (10 g) have a high RMR at low T_a, and a fall of T_{set} results in a large reduction of MR from RMR to BMR (Figure 3). The substantial reduction of MR, together with the large surface area of small heterotherms, results in high cooling rates (100), and the fast fall of T_b in turn affects MR.

In contrast, large heterotherms (5000 g) have a low T_{lc} of the TNZ and only a small increase of RMR over BMR at low T_a (Figure 3). Consequently, a fall of

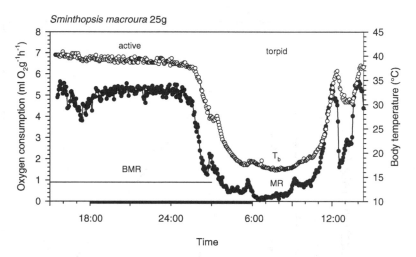

Figure 2 Metabolic rates measured as oxygen consumption of a dunnart (*Sminthopsis macroura*). Similar to that found for the glider in Figure 1, the initial fall of the metabolic rate is associated with only a small change of body temperature. Only when the metabolic rate falls below BMR does the simultaneous decline of metabolic rate and body temperature become obvious.

T_{set} and the small reduction from RMR to BMR and their low surface-to-volume ratio will result in a slow cooling rate. Even larger species, such as bears, are under thermoneutral conditions even at T_a near 0°C (137), and a fall of T_{set} should have no effect on MR. Thus physiological mechanisms employed for MR reduction during torpor entry must differ between small and large heterotherms.

TMR AND T_b IN THERMO-CONFORMING TORPID HETEROTHERMIC ENDOTHERMS

Small Heterotherms

As discussed above, small heterotherms have high RMR at low T_a and a high BMR. A reduction of T_{set} results in a precipitous drop of MR to near BMR (141) and, because not enough heat is produced for normothermic thermoregulation, T_b will follow, which in turn affects MR. Nevertheless, the reduction of MR below BMR appears to differ between small daily heterotherms and hibernators.

The main energy supply of daily heterotherms, even during the period when torpor is commonly used, remains food collected during usually daily foraging. Thus energetics differ from hibernators, which heavily rely on fat. Small daily heterotherms (body mass ~2 to 70 g) have high BMRs, and the effects of a reduction

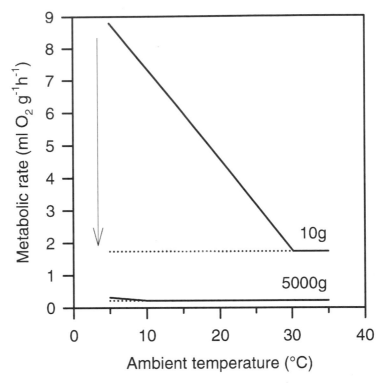

Figure 3   Resting metabolic rate within thermoneutrality (BMR) and below thermoneutrality (RMR) of a 10-g and a 5000-g endotherm. Note the wide TNZ and the small increase from BMR to RMR at low ambient temperatures in the large species and the narrow TNZ and the large increase of RMR over BMR at low ambient temperatures in the small species. This difference will strongly affect metabolic rate reduction at torpor onset.

of T_b by about 20°C, as commonly observed in daily heterotherms, results in a maximum reduction of TMR to about 25% of BMR assuming a Q_{10} of 2. This TMR may seem rather high, however because torpor in daily heterotherms is relatively shallow and brief and is usually interrupted by daily foraging, a small reduction of MR relying largely on temperature effects without major biochemical adjustments appears to be a sensible approach. Not surprisingly then, the reduction of MR in a daily heterotherm, the marsupial *Sminthopsis macroura* (Figure 4), follows closely that predicted by temperature effects ($Q_{10} = 2.5$) similar to that for many other small daily heterotherms and the minimum TMR is about 25% of BMR in many species (Table 1). Perhaps more importantly, the extrapolation of TMR as a function of T_b intercepts with BMR at the normothermic T_b of 35°C in this species. This apparent continuum between normothermia in the TNZ (BMR) and TMR as a function of T_b, which also occurs during passive rewarming (53), provides further

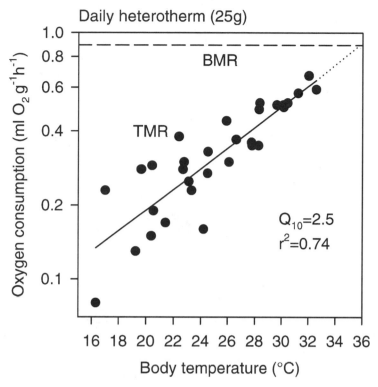

Figure 4 Metabolic rates measured as oxygen consumption of a torpid thermo-conforming small daily heterotherm, the dunnart (*Sminthopsis macroura*), as a function of body temperature. Note the intercept of the extrapolation of TMR with BMR at the normothermic body temperature of 35°C (data from 140).

evidence that the MR reduction below BMR in small daily heterotherms is largely determined by T_b.

Small hibernators undergo prolonged periods of torpor and can survive on stored fat for months. If small hibernators exhibited the same MR reduction as that of small daily heterotherms, their fuel stores would be depleted within days or weeks, well before the end of winter. Thus it is not surprising that the reduction of TMR below BMR in small hibernators (~5 to ~100 g) is much more pronounced than in daily heterotherms and that the relationship between T_b and TMR of the two groups differs. In the insectivorous bat, *Nyctophilus geoffroyi*, which is capable of prolonged torpor bouts of up to two weeks, the decline of TMR with T_b (Figure 5) is much more pronounced than in *S. macroura* (Figure 4), and the Q_{10} value for TMR of 3.0 is higher than expected for temperature effects alone. Moreover, in contrast to *S. macroura* (Figure 4), the extrapolation of TMR of *N. geoffroyi* does not intercept with BMR at the normothermic T_b of 35°C, but about 40% below

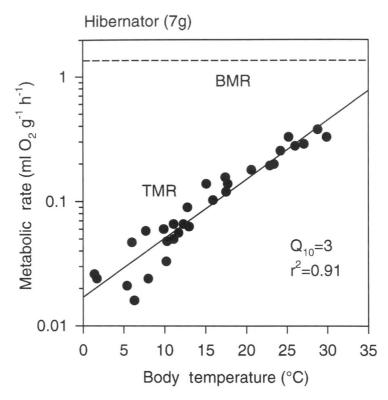

Figure 5 Metabolic rates measured as oxygen consumption of torpid thermo-conforming small hibernator, the bat (*Nyctophilus geoffroyi*), as a function of body temperature. Note the intercept with the normothermic body temperature of 35°C of the extrapolation of TMR is 40% below the BMR (data from 51).

BMR (Figure 5). Thus there is a clear break between BMR and TMR at high T_bs, suggesting that mechanisms other than temperature effects must be involved. Further evidence for largely temperature-independent metabolic inhibition comes from the marsupial hibernator, *Cercartetus nanus*, which can enter torpor within the TNZ and is able to reduce TMR to about 50% of BMR with a T_b reduction of only 2.9°C (Q_{10} = 9.7; Table 1). The minimum TMR of these small hibernators is only a fraction of that in the daily heterotherms. These observations suggest that during torpor entry, and at most T_b in steady-state torpor, small hibernators employ temperature-independent metabolic inhibition in addition to the effects of the greatly reduced T_b for MR reduction.

When the TMR as a function of T_b of thermo-conforming individuals below the BMR in a small daily heterotherm (*S. macroura*, 25 g) is compared with that of a small hibernator (*C. nanus*, 35 g), the TMR at the same T_b in the daily heterotherm

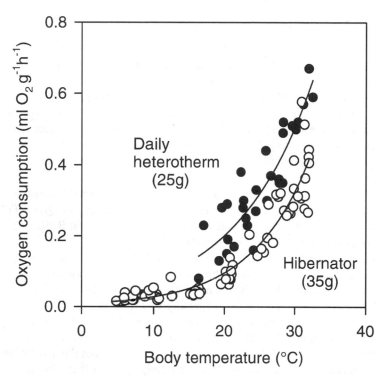

Figure 6 Metabolic rates measured as oxygen consumption of a torpid thermo-conforming small daily heterotherm, the dunnart (*Sminthopsis macroura*), compared with that of a small hibernator, the pygmy-possum (*Cercartetus nanus*), as a function of body temperature. Note that at the same body temperature the metabolic rate of the hibernator is about half of that of the daily heterotherm. The minimum metabolic rate of the daily heterotherm is about seven times that of the hibernator because of the further decline of body temperature in the latter (data from 140, 142).

is about twice that of the hibernator (Figure 6), which emphasizes the influence of metabolic inhibition in the hibernator. The minimum TMR in *S. macroura* is about sevenfold that of *C. nanus* because of the greater than 10°C lower minimum T_b in *C. nanus* (Figure 6), which emphasizes the additional effect of T_b. Log TMR is linearly related to T_b in both species, but the elevation of the regressions differs significantly (p < 0.0001, ANCOVA). The difference in TMR at the same T_b is not the result of differences in torpor duration because TMR minima in both species are reached about 3–4 hours after torpor onset. The differences in TMR at the same T_b suggest that the mechanisms of MR reduction differ between small daily heterotherms and hibernators.

Medium-Sized Heterotherms

Medium-sized heterotherms are capable of carrying more fat and have substantially lower RMR and BMR than small species. Consequently, energy constraints should be less extreme than for small species.

Information on MR reduction during torpor in medium-sized daily heterotherms is limited to three marsupials and a bird (body mass $\sim$110 to 200 g). Nevertheless, existing data suggest that TMR below the BMR in heterothermic marsupials falls as predicted by the reduction of MR via falling T_b through Q_{10} effects. In the four species examined, Q_{10} ranged between 1.6 and 2.8, suggesting that T_b plays a major role in the MR reduction below BMR (Table 1).

Similar observations have been made in medium-sized hibernators. In golden-mantled ground squirrels, *Spermophilus lateralis* (200 g), the Q_{10} for TMR between different T_b during torpor was 2.39 (62) and 2.3 between BMR and TMR (138). Similarly, in *Spermophilus saturatus* (250 g), the log TMR was a function of T_b and the Q_{10} calculated for the slope was 2.04. However, these measured Q_{10} values are restricted to low T_b during torpor ($<14°C$) and may not reflect those at high T_b. The intercept of TMR with the normothermic T_b of 38°C in *S. saturatus* was 25% below the BMR (Figure 7), suggesting that at high T_b during torpor, metabolic inhibition may generate at least some of the MR reduction in this species. Q_{10} values >3 at mainly high T_b in other medium-sized hibernators (Table 1) support this interpretation. Interestingly, at a T_b of 18.3°C, even the TMR of the fat-tailed lemur, *Cheirogaleus medius* (250 g), a tropical primate that hibernates at unusually high T_b (18), falls very close to that predicted for *S. saturatus* at the same T_b, suggesting that there are no general differences between tropical and temperate hibernators. Thus low Q_{10} values observed in medium-sized hibernators appear to be restricted to low T_b during torpor, whereas at high T_b, Q_{10} values are greater than predicted. This suggests that, unlike in small hibernators, metabolic inhibition in medium-sized hibernators predominately is used to minimize TMR at high T_b and torpor entry, whereas at low T_b these animals largely rely on T_b for TMR reduction.

Large Heterotherms

Whereas small heterotherms have to overcome the problem of having high normothermic MR and low fat stores, large heterotherms (body mass >1000 g) have to deal with a small relative surface area and only a small reduction of MR from RMR to BMR during torpor entry (Figure 3). Cooling rates consequently will be slow, and a reliance on T_b for reduction in MR will be ineffective, at least in the initial phase of torpor.

American badgers, *Taxidea taxus* ($\sim$9000 g), are the only large daily heterotherms for which metabolic data during torpor are available. Their MRs fell with T_b with a Q_{10} value of 2.15 over $\sim$10 h torpor entry (64). This suggests that despite their large size, these animals rely mainly on temperature effects for a 57% MR reduction.

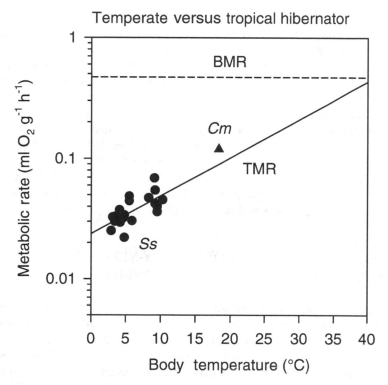

Figure 7   Metabolic rates measured as oxygen consumption of torpid thermo-conforming medium-sized hibernator, the ground squirrel (*Spermophilus saturatus, Ss*), as a function of body temperature. Note the intercept with the normothermic body temperature of 38°C of the extrapolation of TMR is 25% below the BMR. The minimum metabolism of the medium-sized tropical hibernator the lemur (*Cheirogaleus medius, Cm*) is similar to that predicted for the temperature of the ground squirrel at the T_b measured for the lemur (data from 18, 46).

Because of the small surface area and the negligible differential between TMR and BMR, large hibernators appear to employ metabolic inhibition at torpor onset, to permit cooling of the body. Echidnas, *Tachyglossus aculeatus* (body mass ~2000 g), exhibit high Q_{10} values of 6.7 between T_b and TMR during torpor entry (127), supporting this interpretation. Arctic ground squirrels, *Spermophilus parryii* (1000 g), have a constant and low TMR over a T_b range of ~5 to 13°C, suggesting that at least at the higher T_b, metabolic inhibition is involved in reducing TMR (13). Alpine marmots, *Marmota marmota* (body mass ~4000 g), appear to use metabolic inhibition during torpor entry and maintain TMR at very low levels for several hours while the T_b declines, but TMR increases somewhat later in the torpor bout (129). The likely reason why TMR returns to higher values after torpor entry is that the very low initial MR does not suffice for prolonged function.

Nevertheless, large thermal gradients may partially explain the observed patterns because the measured core T_b does not reflect peripheral temperatures during cooling. Thus, while the function of metabolic inhibition in small hibernators appears to minimize energy expenditure at all T_b to overcome the energetic constraints of small amounts of stored fat, in large hibernators this mechanism is important to allow the initial reduction of T_b that, in turn, will influence TMR to some extent.

The situation in bears is even more extreme. Because bears are under thermoneutral conditions even at very low T_a, they cannot rely on cooling through a reduction in T_{set}. Consequently, they appear to down-regulate MR to about 20% of BMR resulting in a decline of T_b of about 5°C. The $Q_{10} = 27$ for bears is very high, but this value may be inflated because the predicted BMR by Watts et al. (156) is substantially above that predicted by allometric equations for hibernators derived here, which would result in a Q_{10} of 7.2 (Figure 8B).

T_b AND THE ALLOMETRY OF STEADY-STATE BMR AND TMR OF THERMO-CONFORMING TORPID ANIMALS

As we have seen in the previous section, the fall of MR at torpor entry and consequently the fall of T_b depend on size. But even during steady-state torpor, TMR is affected by size. Small heterotherms have relatively high BMR and small fat stores, whereas large species have low BMR and relatively large energy stores. Consequently, mechanisms of MR reduction should be reflected in allometric relationships between mass and BMR and mass and TMR at different T_bs (Figure 8).

Because of the large number of birds included in the daily heterotherms, which generally have higher BMR than mammals (21), the BMR intercept differed between daily heterotherms and hibernators (p = 0.043, ANCOVA). When birds

---

→

Figure 8 Metabolic rate as a function of body mass for daily heterotherms (A) and hibernators (B). BMR (*filled circles*) and the TMR of animals with minimum T_b between 25–33°C (*unfilled squares*), 15–25°C (*filled triangles*), 10–15°C (*unfilled triangles*) and 0–10°C (*unfilled circles*) are shown.

The equations for daily heterotherms (A) were

$\log_{10} BMR = 0.678 - 0.381 \log_{10} mass, r^2 = 0.75, P < 0.001;$
$\log_{10} TMR \text{ (at } T_b \text{ 25–33°C)} = 0.320 - 0.347 \log_{10} mass, r^2 = 0.72, P < 0.001;$
$\log_{10} TMR \text{ (at } T_b \text{ 15–25°C)} = -0.037 - 0.328 \log_{10} mass, r^2 = 0.28, P < 0.001;$
$\log_{10} TMR \text{ (at } T_b \text{ 10–15°C)} = 0.070 - 0.535 \log_{10} mass, r^2 = 0.72, P = 0.034.$

The equations for hibernators (B) were

$\log_{10} BMR = 0.519 - 0.299 \log_{10} mass, r^2 = 0.77, P < 0.001;$
$\log_{10} TMR \text{ (at } T_b \text{ 25–33°C)} = -0.270 - 0.214 \log_{10} mass, r^2 = 0.78, P = 0.001;$
$\log_{10} TMR \text{ (at } T_b \text{ 15–25°C)} = -0.573 - 0.304 \log_{10} mass, r^2 = 0.49, P < 0.001;$
$\log_{10} TMR \text{ (at } T_b \text{ 10–15°C)} = -0.963 - 0.177 \log_{10} mass, r^2 = 0.24, P = 0.091;$
$\log_{10} TMR \text{ (at } T_b \text{ 0–10°C)} = -1.229 - 0.128 \log_{10} mass, r^2 = 0.22, P = 0.002.$

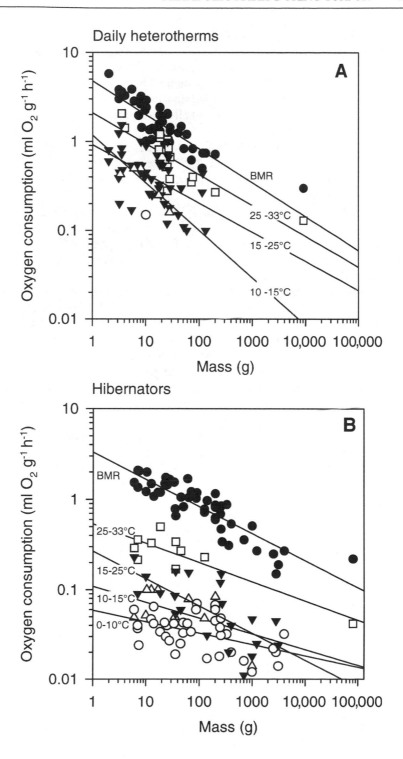

were excluded, the BMR was indistinguishable between daily heterotherms and hibernators (p = 0.49, ANCOVA). The normothermic T_b at BMR was similar (p = 0.086, t-test) between daily heterotherms ($T_b = 36.8 \pm 2.2°C$) and hibernators ($T_b = 36.2 \pm 1.5°C$), even with both birds and mammals included.

In daily heterotherms, the regression lines for TMR as a function of body mass declined in parallel with T_bs at all T_b examined (Figure 8A), and the slopes were indistinguishable among the T_b groups (p > 0.4, ANCOVA). However, the elevation differed between BMR and TMR at T_b 25–33°C, and also between TMR at T_b 25–33°C and TMR at T_b 15–25°C (p < 0.001, ANCOVA). TMRs at T_b 15–25°C and T_b 10–15°C was indistinguishable (p > 0.3, ANCOVA) likely because of the low sample size ($n = 6$) for the latter.

TMR as a function of mass of avian daily heterotherms did not differ from mammalian daily heterotherms at most T_bs examined (p > 0.1, ANCOVA). However, at T_b 10–15°C, the two groups differed in slope (p < 0.003), but again this probably reflects the low sample size ($n = 3$, in each case) rather than a biological difference.

Hibernators also reduce MR as a function of mass from BMR to TMR in parallel from normothermic T_b down to T_b 15–25°C (Figure 8B). The slope of the regression for BMR was indistinguishable from TMR at T_b 25–33°C, T_b 15–25°C, and T_b 10–15°C. However, at T_bs lower than 10°C, the slope for the regression of TMR versus mass changed significantly (p < 0.024, ANCOVA). Above T_b 15°C, the slopes for TMR versus body mass ranged from -0.214 to -0.304; below T_b 10°C, the slope was about half (-0.128) because in this T_b range the reduction of TMR relative to BMR in the small species is more pronounced than in the large species.

When TMRs based on the T_b bins of daily heterotherms (Figure 8A) and hibernators (Figure 8B) were compared, all differed significantly in elevation (p < 0.0001, ANCOVA) at the three T_b ranges that could be compared (T_b 25–33°C, 15–25°C, 10–15°C); the slopes were indistinguishable. These differences in elevation were not due to differences in T_b because mean T_bs in all T_b bins were indistinguishable (t-test; mean T_bs at T_b 25–33°C: 27.3 $\pm$ 1.6°C versus 27.3 $\pm$ 2.4°C; at T_b 15–25°C: 19.1 $\pm$ 3.0°C versus 17.8 $\pm$ 2.6°C; at T_b 10–15°C: 12.5 $\pm$ 1.5°C versus 11.1 $\pm$ 1.3°C; daily heterotherms versus hibernators, respectively).

Q_{10} BETWEEN BMR AND TMR

The relationship between TMR and body mass is reflected in the Q_{10} values. As the TMR in daily heterotherms was relatively high, the Q_{10} between BMR and TMR at various T_bs during torpor ranged between ~1.5 and 3.5, with a maximum of 4.7 in the 2-g pygmy shrew (*Suncus etruscus*) (Figure 9A). Most Q_{10} values were between 1.5 and 2.5 (average 2.3 $\pm$ 0.6, $n = 49$) and close to those typical for biochemical reactions. The Q_{10} values were not related to body mass at T_b 25–33°C and T_b 10–15°C; however, there was a weak correlation ($r^2 = 0.09$) at T_b

15–25°C. Generally, the Q_{10} values of daily heterotherms (Figure 9A) were well below those for hibernators (Figure 9B).

Hibernators at high T_b during torpor (T_b 25–33°C) have high Q_{10} values of >3 to 27 (Figure 9B) (as stated above, the value for the bear is likely to be on overestimate because of the high BMR). Whereas most of these high Q_{10}s are restricted to small species, with the exception of the bear, they are also likely to occur in other medium-sized and large hibernators, but there are no data currently available in this T_b range. At T_b 25–33°C, the Q_{10} shows a significant positive correlation with body mass ($r^2 = 0.49$) because of the bear. At intermediate T_b (T_b 15–25°C and T_b 10–15°C), Q_{10} and mass were not related despite a substantial data set. However, at all body masses, the Q_{10} values in this temperature range were generally high, with most values between 3 and 6. Only at low T_b 0–10°C was Q_{10} negatively related to body mass ($r^2 = 0.28$), reflecting a greater reduction of steady-state TMR in the small species compared with that of the large species at low T_b. The Q_{10} in most small species (<100 g) at T_b 0–10°C was >3, whereas in large hibernators (>1000 g) Q_{10} was around 2–2.5. Thus although the normal Q_{10} between BMR and TMR at low T_b in the large hibernators suggests that these animals generally rely on temperature effects for MR reduction (38), they do in fact rely heavily on metabolic inhibition. The normal Q_{10} merely reflects an average of the high Q_{10} values at high T_b during torpor and the low Q_{10} at low T_b. The overall mean Q_{10} for hibernators was 3.9 ± 3.7 ($n = 43$).

TMR AND THE T_b-T_a DIFFERENTIAL

In the T_a range where torpid heterotherms are thermo-conforming, the T_b-T_a differential is often constant or changes little, although TMR declines significantly (13, 51, 74, 79, 142, 159). These observations indicate that the T_b-T_a differential does not determine TMR above the T_{set} as has been suggested (68). The constant T_b-T_a differential, despite a declining TMR over the same T_a range, may appear surprising because at constant thermal conductance (C) the T_b-T_a differential and MR should fall together. However, it appears that C is not constant and changes from high values at high T_b to low values at low T_b, which explains the more or less constant T_b-T_a differential despite the change of TMR (142).

The lack of a functional link between the T_b-T_a differential of thermo-conforming torpid individuals and TMR also can be demonstrated by exposing torpid individuals to a decline or rise of T_a. During passive rewarming, the T_b-T_a differential often becomes very small or temporarily negative (18, 102, 136), but despite the decline in the T_b-T_a differential, the TMR increases with the rise of T_b. When the T_b-T_a differential of torpid thermo-conforming individuals is experimentally increased by a reduction of T_a, TMR does not increase, despite a rise in the T_b-T_a differential, but declines, following the decline of T_b (44). Similarly, thermal manipulations of the hypothalamus at T_b above the T_{set}, which amounts to the same as change in the T_b-T_a differential, do not elicit any increase of TMR (72).

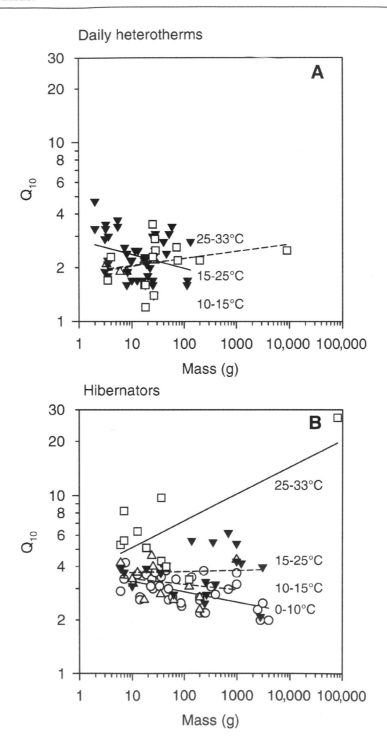

Daily heterotherms

In contrast to thermo-conforming torpid individuals, in which the T_b-T_a differential appears inconsequential for determining TMR, thermo-regulating torpid individuals below the T_{set} for T_b maintain TMR according to the T_b-T_a differential, albeit at a lower T_b. An increase of TMR in thermo-regulating torpid individuals similar to the response of the RMR has long been recognized (61). Obviously, regulation of T_b, even during torpor, will result in a proportional heat loss as occurs during normothermia, which must be compensated for by an increase in heat production. Whereas the T_b-T_a differential determines TMR in thermo-regulating torpid individuals, it has been suggested that the slope of TMR versus T_a during torpor may be shallower than that for RMR versus T_a during normothermia, perhaps because of a decrease in C at low T_b (74, 143). Although this interpretation may appear plausible, it is not supported by the empirical evidence from most species. The slope and elevation for RMR versus T_a during normothermia and that of TMR versus T_a in thermo-regulating individuals at body mass <200 g during torpor are indistinguishable for both daily heterotherms and hibernators (p = 0.144, ANCOVA), suggesting that maintenance of the T_b-T_a differential is identical between the two states and is independent of T_b. However, when species >200 g are included ($n = 3$), the slope differs significantly between normothermic and torpid thermoregulating individuals (Figure 10) suggesting that large species can reduce heat loss when thermo-regulating during torpor.

APPARENT THERMAL CONDUCTANCE AND TMR

Snyder & Nestler (139) argued that Q_{10} effects are confounded by changes in thermal conductance and consequently that Q_{10} calculations in endotherms are unsound. They found, that C is lower during torpor than during normothermia in rodents and proposed that the reduction in C is a central part of torpor, as it allows endotherms to markedly reduce levels of metabolism without abandoning regulation of T_b (139). However, this argument has several problems. First, as shown above, thermal conductance during torpor in most species is lower only

←————————————————————————————————

Figure 9 The Q_{10} between BMR and TMR as a function of body mass for daily heterotherms (A) and hibernators (B). Q_{10}s were calculated between BMR and TMR of animals with minimum T_b between 25–33°C (*unfilled squares*), 15–25°C (*filled triangles*), 10–15°C (*unfilled triangles*), and 0–10°C (*unfilled circles*).

The equation for daily heterotherms (A) was

$\log_{10} Q_{10} = 0.453 - 0.078 \log_{10} \text{mass}, r^2 = 0.09, P = 0.046$ (T_b 15–25°C).

Insignificant correlations are shown as broken lines for better identification of points. The equations for hibernators (B) were

$\log_{10} Q_{10} = 0.577 - 0.057 \log_{10} \text{mass}, r^2 = 0.28, P = 0.001$ (T_b 0–10°C);

$\log_{10} Q_{10} = 0.560 + 0.149 \log_{10} \text{mass}, r^2 = 0.49, P = 0.024$ (T_b 25–33°C).

Insignificant correlations are shown as broken lines for better identification of points.

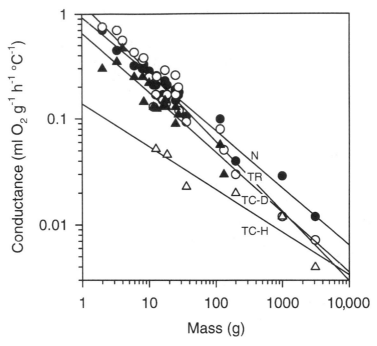

Figure 10 Conductance as a function of body mass for normothermic animals (*filled circles*, N), thermo-regulating torpid animals (*unfilled circles*, TR) and thermo-conforming torpid animals (*filled triangles*, TC-D daily heterotherms, TC-H hibernators).

The equations were

$\log_{10} C = -0.032 - 0.539 \log_{10} \text{mass}, r^2 = 0.966, P < 0.001$ (normothermia);

$\log_{10} C = 0.0992 - 0.656 \log_{10} \text{mass},$
$\quad r^2 = 0.943, P < 0.001$ (thermo-regulating torpor);

$\log_{10} C = -0.185 - 0.565 \log_{10} \text{mass},$
$\quad r^2 = 0.874, P < 0.01$ (thermo-conforming torpor, daily heterotherms);

$\log_{10} C = -0.860 - 0.402 \log_{10} \text{mass},$
$\quad r^2 = 0.915, P < 0.01$ (thermo-conforming torpor, hibernators).

when torpid animals are thermo-conforming and apparently do not regulate T_b (Table 2; Figure 9). Second, the reduction of C from RMR to TMR is small compared with the large difference in metabolism. Third, if C were important in determining MR of thermo-conforming torpid individuals, a change of C by exposure to 21% oxygen in helium, which is a more conductive atmosphere than air, should result in an increase in MR, which it does not (54). Fourth, the thermal conductance of most thermo-regulating animals is the same during torpor and normothermia (142; Figure 10), but the TMR even in thermo-regulating individuals is only a fraction of that during normothermia. Thus, as explained by Nicol et al. (127), C cannot be the reason for the low MR, but rather it is a consequence of the low TMR and peripheral vasoconstriction in thermo-conforming individuals. Nevertheless, C may play a role in reducing T_b and, consequently, TMR at high T_a or during torpor entry.

ARE Q_{10} CALCULATIONS MEANINGFUL IN ENDOTHERMS?

As is true for all other physiological measurements, common sense must be applied to the calculation of Q_{10} (155). This is especially important in heterothermic endotherms that exhibit pronounced changes in their physiological state. If meaningful calculations for Q_{10} are to be made, changes of MR that change with T_b and are based on equivalent performance at different T_bs appear most appropriate (155). Thus in the present comparison, Q_{10} was calculated between BMR and TMR in thermo-conforming individuals because both states do not include a thermo-regulatory energetic component and reflect maintenance metabolism only at different T_bs. Calculations of Q_{10} between TMR at different T_bs during torpor in thermo-conforming individuals also are appropriate, as may be comparisons between thermo-regulating individuals at the same T_a but at different T_bs because they represent equivalent performance (142).

In contrast, comparisons of thermo-regulating individuals with thermo-conforming individuals (139) are not likely to provide a meaningful Q_{10}, because a change of state rather than the effect of temperature on rates is examined. Similarly, calculations of Q_{10} during torpor entry can be meaningless if they fail to consider that the initial decline of MR is not related to a reduction of T_b, but to a reduction of T_{set} (see Figures 1 and 2). Another approach that has been applied to resolve whether T_b is a possible reason for the decline of MR has been to compare the percent change of MR with the percent change of T_b during torpor entry (69). This approach is flawed for two reasons. First, as is outlined above, the initial change of MR during torpor entry can occur without a change of T_b because MR falls from RMR to approximately BMR, and it is therefore not surprising that MR falls faster than T_b. Second, as TMR in thermo-conforming individuals is an exponential function of T_b (38, 140, 142, 159), a comparison of a percent change to assess whether T_b and TMR are related is not likely to provide meaningful information.

BIOCHEMICAL MECHANISMS

If MR reduction in many heterothermic species involves metabolic inhibition, the question arises as to the underlying biochemical mechanisms. However, considering the enormous reduction in MR that occurs especially during hibernation, in vitro data provide less clear and often contradictory results.

A potential inhibitor for MR during torpor may be a reduced pH, which may lower metabolic processes (63, 109, 110, 116). In *S. lateralis*, the respiratory quotient (RQ) falls during entry into hibernation and rises during arousal (138), suggesting that storage of CO_2 could result in a decrease of pH. This observation differs from data on the daily heterotherm, *P. sungorus*, which increases RQ during torpor entry and decreases RQ before arousal (70), suggesting that not all heterotherms store CO_2 during torpor entry.

Changes in enzyme activities at different states during torpor are other potential candidates for involvement in MR reduction. However, such measurements are often limited to a single room temperature that is representative for neither high normothermic T_b nor low torpid T_b. No major changes in enzyme activity were observed in *P. sungorus* during daily torpor with the exception of pyruvate dehydrogenase (70), but because this species appears to rely largely on T_b for MR reduction, this result is not surprising. In hibernating *Zapus hudsonicus*, several glycolytic enzymes had reduced activity by about 50%, which is similar to that observed during metabolic inhibition in some species at the whole animal level without the effect of T_b, and this change may be controlled by reversible phosphorylation (145). Similarly, mitochondrial respiration is reduced to about half during torpor in *Spermophilus* compared with that in normothermic individuals (2, 112). Furthermore, transcriptional initiation during torpor in *S. lateralis* is substantially reduced compared with that of interbout normothermia (151).

Although these are promising findings, it is clear that further in vitro work is required to fully explain what happens in vivo. Whereas Arrhenius plots of enzyme function have received considerable attention with respect to heterothermy in the past (42, 134), little recent progress has been made in this area, despite the observed increase in the Q_{10} in torpid individuals compared with that in normothermic individuals. Perhaps more emphasis on in vitro work considering cellular functions over the T_b ranges that are observed in vivo may help resolve some of the questions.

CONCLUSIONS

The above analysis suggests that mechanisms of metabolic rate reduction differ between daily heterotherms and hibernators and between small and large heterotherms. It shows that, against expectations, most of the apparently contradictive hypotheses proposed to explain metabolic rate reduction are correct. However, not all hypotheses match the published data for all species at all stages of torpor. Thus the present paper supports the view that extrapolations from one species to to another can be erroneous and underscores the strength of a comparative approach.

ACKNOWLEDGMENTS

I thank Mark Brigham and Gerhard Körtner for critical comments on the manuscript, Loren Buck for data on arctic ground squirrels, and Rebecca Drury for help with assembling the manuscript. The Australian Research Council supported the work.

The *Annual Review of Physiology* is online at http://physiol.annualreviews.org

LITERATURE CITED

1. Augee ML, Ealey EHM. 1968. Torpor in the echidna, *Tachyglossus aculeatus*. *J. Mammal.* 49:446–54
2. Barger JL, Brand MD, Barnes BM, Boyer BB. 2003. Tissue-specific depression of mitochondrial proton leak and substrate oxidation in hibernating arctic ground squirrels. *Am. J. Physiol. Regul. Integr. Comp. Physiol.* 284:R1306–13
3. Barnes BM. 1989. Freeze avoidance in a mammal: body temperatures below 0°C in an Arctic hibernator. *Science* 244:1593–95
4. Barnes BM, Toien O, Edgar DM, Grahn D, Heller C. 2000. Comparison of the hibernation phenotype in ground squirrels and bears. *Life in the Cold. Eleventh Int. Hibernation Symp.* p. 11 (Abstr.)
5. Bartels W, Law BS, Geiser F. 1998. Daily torpor and energetics in a tropical mammal, the northern blossom-bat *Macroglossus minimus* (Megachiroptera). *J. Comp. Physiol. B* 168:233–39
6. Bartholomew GA, Hudson JW. 1960. Aestivation in the Mohave ground squirrel, *Citellus mohavensis*. *Bull. Mus. Comp. Zool.* 124:193–208
7. Bartholomew GA, Dawson WR, Lasiewski RC. 1970. Thermoregulation and heterothermy in some of the smaller flying foxes (Megachiroptera) of New Guinea. *Z. Vergl. Physiol.* 70:196–209
8. Bickler PE. 1984. CO_2 balance of a heterothermic rodent: comparison of sleep, torpor, and awake states. *Am. J. Physiol. Regul. Integr. Comp. Physiol.* 246:R49–55
9. Boyer BB, Barnes BM. 1999. Molecular and metabolic aspects of mammalian hibernation. *Bioscience* 49:713–24
10. Brigham RM, Körtner G, Maddocks TA, Geiser F. 2000. Seasonal use of torpor by free-ranging Australian owlet-nightjars

(*Aegotheles cristatus*). *Physiol. Biochem. Zool.* 73:613–20
11. Bucher TL, Chappell MA. 1992. Ventilatory and metabolic dynamics during entry into and arousal from torpor in *Selasphorus* hummingbirds. *Physiol. Zool.* 65:978–93
12. Bucher TL, Chappell MA. 1997. Respiratory exchange and ventilation during nocturnal torpor in hummingbirds. *Physiol. Zool.* 70:45–52
13. Buck CL, Barnes BM. 2000. Effects of ambient temperature on metabolic rate, respiratory quotient, and torpor in an arctic hibernator. *Am. J. Physiol. Regul. Integr. Comp. Physiol.* 279:R255–62
14. Buffenstein R. 1985. The effect of starvation, food restriction, and water deprivation on thermoregulation and average daily metabolic rates in *Gerbillus pusillus*. *Physiol. Zool.* 58:320–28
15. Carpenter FL, Hixon MA. 1988. A new function of torpor: fat conservation in a wild migrant hummingbird. *Condor* 90:373–78
16. Coburn DK, Geiser F. 1998. Seasonal changes in energetics and torpor patterns in the sub-tropical blossom-bat *Syconycteris australis* (Megachiroptera). *Oecologia* 113:467–73
17. Cranford JA. 1983. Body temperature, heart rate and oxygen consumption of normothermic and heterothermic western jumping mice (*Zapus princeps*). *Comp. Biochem. Physiol.* 74A:595–99
18. Dausmann KH, Ganzhorn JU, Heldmaier G. 2000. Body temperature and metabolic rate of a hibernating primate in Madagascar: preliminary results from a field study. In *Life in the Cold. Eleventh Int. Hibernation Symp.*, ed. G. Heldmaier, M. Klingenspor, pp. 41–47. Berlin/Heidelberg/New York: Springer

19. Dawson TJ. 1983. *Monotremes and Marsupials: the Other Mammals*. London: Edward Arnold
20. Dawson WR, Fisher CD. 1969. Responses to temperature by the spotted nightjar (*Eurostopodus guttatus*). *Condor* 71:49–53
21. Dawson WR, Hudson JW. 1970. Birds. In *Comparative Physiology of Thermoregulation*, ed. GC Whittow, pp. 223–310. New York: Academic
22. Deavers DR, Hudson JW. 1981. Temperature regulation in two rodents (*Clethrionomys gapperi* and *Peromyscus leucopus*) and a shrew (*Blarina brevicaudata*) inhabiting the same environment. *Physiol. Zool.* 54:94–108
23. Ellison GTH. 1995. Thermoregulatory responses of cold-acclimated fat mice (*Steatomys pratensis*). *J. Mammal.* 76:240–47
24. Ellison GTH, Skinner JD. 1992. The influence of ambient temperature on spontaneous daily torpor in pouched mice (*Saccostomus campestris*: Rodentia Cricetidae) from southern Africa. *J. Therm. Biol.* 17:25–31
25. Erickson H. 1956. Observations on the metabolism of arctic ground squirrels (*Citellus parryii*) at different environmental temperatures. *Acta Physiol. Scand.* 36:66–74
26. Fietz J, Tataruch F, Dausmann KH, Ganzhorn JU. 2003. White adipose tissue composition in the free-ranging fat-tailed dwarf lemur (*Cheirogaleus medius*; Primates), a tropical hibernator. *J. Comp. Physiol. B* 173:1–10
27. Fleming MR. 1980. Thermoregulation and torpor in the sugar glider, *Petaurus breviceps* (Marsupialia: Petauridae). *Aust. J. Zool.* 28:521–34
28. Fleming MR. 1985a. The thermal physiology of the feathertail glider, *Acrobates pygmaeus* (Marsupialia: Burramyidae). *Aust. J. Zool.* 33:667–81
29. Fleming MR. 1985b. The thermal physiol-

ogy of the mountain pygmy-possum *Burramys parvus* (Marsupialia: Burramyidae). *Aust. Mammal.* 8:79–90
30. Florant GL, Heller HC. 1977. CNS regulation of body temperature in euthermic and hibernating marmots (*Marmota flaviventris*). *Am. J. Physiol. Regul. Integr. Comp. Physiol.* 232:R203–8
31. Fons R, Sicard R. 1976. Contribution á la conaissance du métabolisme énergétique chez deuz crocidurinae: *Suncus etruscus* (Savi, 1882) et *Crocidura russula* (Hermann, 1780) (Insectivora, Soricidae). *Mammalia* 40:299–311
32. French AR. 1985. Allometries of the duration of torpid and euthermic intervals during mammalian hibernation: a test of the theory of metabolic control of the timing of changes in body temperature. *J. Comp. Physiol. B* 156:13–19
33. Frey H. 1979. La température corporelle de *Suncus etrucus* (Sorcidae, Insectivora) au cours de l'activité, du respos normothermique et du la torpeur. *Revue Suisse Zool.* 86:653–62
34. Frey H. 1980. Le métabolisme énergétique de *Suncus etrucus* (Sorcidae, Insectivora) en torpeur. *Revue Suisse Zool.* 87:739–48
35. Geiser F. 1985. Tagesschlaflethargie bei der gelbfüssigen Breitfussbeutelspitzmaus, *Antechinus flavipes* (Marsupialia: Dasyuridae). *Z. Säugetierkunde* 50:125–27
36. Geiser F. 1986. Thermoregulation and torpor in the Kultarr, *Antechinomys laniger* (Marsupialia: Dasyuridae). *J. Comp. Physiol. B* 156:751–57
37. Geiser F. 1987. Hibernation and daily torpor in two pygmy possums (*Cercartetus* spp., Marsupialia). *Physiol. Zool.* 60:93–102
38. Geiser F. 1988a. Reduction of metabolism during hibernation and daily torpor in mammals and birds: temperature effect or physiological inhibition? *J. Comp. Physiol. B* 158:25–37
39. Geiser F. 1988b. Daily torpor and

thermoregulation in *Antechinus* (Marsupialia): influence of body mass, season, development, reproduction, and sex. *Oecologia* 77:395–99

40. Geiser F. 1991. The effect of unsaturated and saturated dietary lipids on the pattern of daily torpor and the fatty acid composition of tissues and membranes of the deer mouse *Peromyscus maniculatus. J. Comp. Physiol. B* 161:590–97

41. Geiser F. 2003. Thermal biology and energetics of carnivorous marsupials. In *Predators with Pouches: the Biology of Carnivorous Marsupials*, ed. M Jones, C Dickman, M Archer, pp. 234–49. Melbourne: CSIRO

42. Geiser F, McMurchie EJ. 1984. Differences in the thermotropic behaviour of mitochondrial membrane respiratory enzymes from homeothermic and heterothermic endotherms. *J. Comp. Physiol. B* 155:125–33

43. Geiser F, Baudinette RV. 1987. Seasonality of torpor and thermoregulation in three dasyurid marsupials. *J. Comp. Physiol. B* 157:335–44

44. Geiser F, Kenagy GJ. 1987. Polyunsaturated lipid diet lengthens torpor and reduces body temperature in a hibernator. *Am. J. Physiol. Regul. Integr. Comp. Physiol.* 252:R897–901

45. Geiser F, Baudinette RV. 1988. Daily torpor and thermoregulation in the small dasyurid marsupials *Planigale gilesi* and *Ningaui yvonneae. Aust. J. Zool.* 36:473–81

46. Geiser F, Kenagy GJ. 1988. Torpor duration in relation to temperature and metabolism in hibernating ground squirrels. *Physiol. Zool.* 61:442–49

47. Geiser F, Baudinette RV. 1990. The relationship between body mass and rate of rewarming from hibernation and daily torpor in mammals. *J. Exp. Biol.* 151:349–59

48. Geiser F, Broome LS. 1993. The effect of temperature on the pattern of torpor in a marsupial hibernator. *J. Comp. Physiol. B* 163:133–37

49. Geiser F, Masters P. 1994. Torpor in relation to reproduction in the Mulgara, *Dasycercus cristicauda* (Dasyuridae: Marsupialia). *J. Therm. Biol.* 19:33–40

50. Geiser F, Ruf T. 1995. Hibernation versus daily torpor in mammals and birds: physiological variables and classification of torpor patterns. *Physiol. Zool.* 68:935–66

51. Geiser F, Brigham RM. 2000. Torpor, thermal biology, and energetics in Australian long-eared bats (*Nyctophilus*). *J. Comp. Physiol. B* 170:153–62

52. Geiser F, Ferguson C. 2001. Intraspecific differences in behaviour and physiology: effects of captive breeding on patterns of torpor in feathertail gliders. *J. Comp. Physiol. B* 171:569–76

53. Geiser F, Drury RL. 2003. Radiant heat affects thermoregulation and energy expenditure during rewarming from torpor. *J. Comp. Physiol. B* 173:55–60

54. Geiser F, Song X, Körtner G. 1996. The effect of He-O_2 exposure on metabolic rate, thermoregulation and thermal conductance during normothermia and daily torpor. *J. Comp. Physiol. B* 166:190–96

55. Geiser F, Körtner G, Schmidt I. 1998. Leptin increases energy expenditure of a marsupial by inhibition of daily torpor. *Am. J. Physiol. Regul. Integr. Comp. Physiol.* 44:R1627–32

56. Geiser F, Goodship N, Pavey CR. 2002. Was basking important in the evolution of mammalian endothermy? *Naturwissenschaften* 89:412–14

57. Geiser F, Augee ML, McCarron HCK, Raison JK. 1984. Correlates of torpor in the insectivorous dasyurid marsupial *Sminthopsis murina. Aust. Mammal.* 7:185–91

58. Geiser F, Holloway JC, Körtner G, Maddocks TA, Turbill C, Brigham RM. 2000. Do patterns of torpor differ between free-ranging and captive mammals and birds? In *Life in the Cold. Eleventh Int. Hibernation Symp.*, ed. G Heldmaier,

M Klingenspor, pp. 95–102. Berlin/ Heidelberg/New York: Springer

59. Grigg GC, Beard LA. 2000. Hibernation by echidnas in mild climates: hints about the evolution of endothermy? In *Life in the Cold. Eleventh Int. Hibernation Symp.*, ed. G Heldmaier, M Klingenspor, pp. 5–19. Berlin/Heidelberg/New York: Springer

60. Guppy M, Withers PC. 1999. Metabolic depression in animals: physiological perspectives and biochemical generalizations. *Biol. Rev.* 74:1–40

61. Hainsworth FR, Wolf LL. 1970. Regulation of oxygen consumption and body temperature during torpor in a hummingbird, *Eulampis jugularis*. *Science* 168:368–69

62. Hammel HT, Dawson TJ, Adams RM, Anderson HT. 1968. Total calorimetric measurements on *Citellus lateralis* in hibernation. *Physiol. Zool.* 41:341–57

63. Hand SC, Somero GN. 1983. Phosphofructokinase of the hibernator *Citellus beecheyi*: temperature and pH regulation of activity via influences of the tetramer-dimer equilibrium. *Physiol. Zool.* 56:380–88

64. Harlow HJ. 1981. Torpor and other physiological adaptations of the badger (*Taxidea taxus*) to cold environment. *Physiol. Zool.* 54:267–75

65. Hayssen V, Lacy RC. 1985. Basal metabolic rates in mammals: taxonomic differences in the allometry of BMR and body mass. *Comp. Biochem. Physiol.* 81A:741–54

66. Hayward JS, Lyman CP. 1967. Nonshivering heat production during arousal and evidence for the contribution of brown fat. In *Mammalian Hibernation III*, ed. KC Fisher, AR Dawe, CP Lyman, E Schönbaum, FE South, pp. 346–55. Edinburgh: Oliver & Boyd

67. Heldmaier G, Steinlechner S. 1981. Seasonal pattern and energetics of short daily torpor in the Djungarian hamster, *Phodopus sungorus*. *Oecologia* 48:265–70

68. Heldmaier G, Ruf T. 1992. Body temperature and metabolic rate during natural hypothermia in endotherms. *J. Comp. Physiol. B* 162:696–706

69. Heldmaier G, Steiger R, Ruf T. 1993. Suppression of metabolic rate in hibernation. In *Life in the Cold*, ed. C Carey, GL Florant, BA Wunder, B Horwitz, pp. 545–48. Boulder, CO: Westview

70. Heldmaier G, Klingenspor M, Werneyer M, Lampi BJ, Brooks SPJ, Storey KB. 1999. Metabolic adjustments during daily torpor in the Djungarian hamster. *Am. J. Physiol. Endocrinol. Metab.* 276:E896–906

71. Heller HC, Hammel HT. 1972. CNS control of body temperature during hibernation. *Comp. Biochem. Physiol.* 41A:349–59

72. Heller HC, Colliver GW. 1974. CNS regulation of body temperature during hibernation. *Am. J. Physiol.* 227:583–89

73. Heller HC, Colliver GW, Beard J. 1977. Thermoregulation during entrance into hibernation. *Pflügers* 369:55–59

74. Henshaw RE. 1968. Thermoregulation during hibernation: application of Newton's law of cooling. *J. Theor. Biol.* 20:79–90

75. Herreid CF, Schmidt-Nielsen K. 1966. Oxygen consumption, temperature, and water loss in bats from different environments. *Am. J. Physiol.* 211:1108–12

76. Hiebert S. 1990. Energy costs and temporal organization of torpor in the rufous hummingbird (*Selasphorus rufus*). *Physiol. Zool.* 63:1082–97

77. Hildwein G. 1970. Capacités thermorégulatrices d'un mammifère insectivore primitif, le Tenrec; leurs variations saisonnières. *Arch. Sci. Physiol.* 24:55–71

78. Hill RW. 1975. Daily torpor in *Peromyscus leucopus* on an adequate diet. *Comp. Biochem. Physiol.* 51A:413–23

79. Hock RJ. 1951. The metabolic rates and body temperatures of hibernating bats. *Biol. Bull.* 101:289–99

80. Hock RJ. 1960. Seasonal variations in physiologic functions of Arctic ground squirrels and black bears. *Bull. Mus. Comp. Zool.* 124:155–71

81. Hoffman RA, Prinzinger R. 1984. Torpor und Nahrungsausnutzung bei 4 Mausvogelarten (Coliiformes). *J. Ornithologie* 125:225–37

82. Holloway JC. 1998. *Metabolism and thermoregulation in the sugar glider*, Petaurus breviceps (*Marsupialia*). PhD thesis. Univ. New England, Armidale, Aust. 271 pp.

83. Holloway JC, Geiser F. 1995. Influence of torpor on daily energy expenditure of the dasyurid marsupial *Sminthopsis crassicaudata*. *Comp. Biochem. Physiol.* 112A:59–66

84. Holloway JC, Geiser F. 1996. Reproductive status and torpor of the marsupial *Sminthopsis crassicaudata*: effect of photoperiod. *J. Therm. Biol.* 21:373–80

85. Hudson JW. 1965. Temperature regulation and torpidity in the pygmy mouse, *Baiomys taylori*. *Physiol. Zool.* 38:243–54

86. Hudson JW, Deavers DR. 1973. Thermoregulation at high ambient temperatures of six species of ground squirrels (*Spermophilus* spp.) from different habitats. *Physiol. Zool.* 46:95–109

87. Hudson JW, Scott JM. 1979. Daily torpor in the laboratory mouse *Mus musculus* var albino. *Physiol. Zool.* 52:205–18

88. Humphries MM, Thomas DW, Kramer DL. 2003. The role of energy availability in mammalian hibernation: a cost-benefit approach. *Physiol. Biochem. Zool.* 76:165–79

89. Kayser C. 1939. Exchanges respiratoires des hibernants réveillés. *Ann. Physiol. Physicochim. Biol.* 15:1087–219

90. Kayser C. 1961. *The Physiology of Natural Hibernation*. Oxford: Pergamon

91. Kayser C. 1964. La dépense d'énergie des mammiferes en hibernation. *Arch. Sci. Physiol.* 18:137–50

92. Kenagy GJ. 1989. Daily and seasonal uses of energy stores in torpor and hibernation. In *Living in the Cold II*, ed. A Malan, B Canguilhem, pp. 17–24. London: Libby Eurotext

93. Kenagy GJ, Vleck D. 1982. Daily temporal organisation of metabolism in small mammals: adaptation and diversity. In *Vertebrate Circadian Systems*, ed. J Aschoff, S Daan, G Groos, pp. 322–37. Berlin/Heidelberg/New York: Springer

94. Kennedy PM, MacFarlane WV. 1971. Oxygen consumption and water turnover of the fat-tailed marsupials *Dasycercus cristicauda* and *Sminthopsis crassicaudata*. *Comp. Biochem. Physiol.* 40A:723–32

95. Körtner G, Geiser F. 2000. The temporal organization of daily torpor and hibernation: circadian and circannual rhythms. *Chronobiol. Int.* 17:103–28

96. Körtner G, Geiser F. 2000. Torpor and activity patterns in free-ranging sugar gliders *Petaurus breviceps* (Marsupialia). *Oecologia* 123:350–57

97. Körtner G, Brigham RM, Geiser F. 2000. Winter torpor in a large bird. *Nature* 407:318

98. Krüger K, Prinzinger R, Schuchmann K-L. 1982. Torpor and metabolism in hummingbirds. *Comp. Biochem. Physiol.* 73A:679–89

99. Lasiewski RC. 1963. Oxygen consumption of torpid, resting, active and flying hummingbirds. *Physiol. Zool.* 36:122–40

100. Lasiewski RC, Lasiewski RJ. 1967. Physiological responses of the Blue-throated and Rivoli's hummingbirds. *Auk* 84:34–48

101. Lindstedt SL. 1980. Regulated hypothermia in the desert shrew. *J. Comp. Physiol. B* 137:173–76

102. Lovegrove BG, Lawes MJ, Roxburgh L. 1999. Confirmation of pleisiomorphic daily torpor in mammals: the round-eared elephant shrew *Macroscelides proboscideus* (Macroscelidea). *J. Comp. Physiol. B* 169:453–60

103. Lovegrove BG, Raman J, Perrin MR. 2001. Heterothermy in elephant shrews, *Elephantulus* spp. (Macroscelidea): daily torpor or hibernation? *J. Comp. Physiol. B* 171:1–10

104. Lyman CP. 1948. The oxygen consumption and temperature regulation in hibernating hamsters. *J. Exp. Zool.* 109:55–78

105. Lyman CP. 1958. Oxygen consumption, body temperature and heart rate of woodchucks entering hibernation. *Am. J. Physiol.* 194:83–91

106. Lyman CP, Willis JS, Malan A, Wang LCH. 1982. *Hibernation and Torpor in Mammals and Birds.* New York: Academic

107. MacMillen RE. 1965. Aestivation in the cactus mouse *Peromyscus eremicus. Comp. Biochem. Physiol.* 16:227–47

108. Malan A. 1986. pH as a control factor in hibernation. In *Living in the Cold*, ed. HC Heller, XJ Musacchia, LCH Wang, pp. 61–70. New York: Elsevier

109. Malan A. 1989. pH as a control factor of cell function in hibernation: the case of brown adipose tissue thermogenesis. In *Living in the Cold II*, ed. A Malan, B Canguilhem, pp. 205–15. London: Libby Eurotext

110. Malan A. 1993. Temperature regulation, enzyme kinetics, and metabolic depression in mammalian hibernation. In *Life in the Cold*, ed. C Carey, GL Florant, BA Wunder, B Horwitz, pp. 241–52. Boulder, CO: Westview

111. Marsh RL, Dawson WR. 1989. Avian adjustments to cold. In *Advances in Comparative and Environmental Physiology*, ed. LCH Wang, pp. 205–53. Berlin/Heidelberg: Springer

112. Martin SL, Maniero GD, Carey C, Hand SC. 1999. Reversible depression of oxygen consumption in isolated liver mitochondria during hibernation. *Physiol. Biochem. Zool.* 72:255–64

113. McKechnie AE, Lovegrove BG. 2001. Heterothermic responses in the speckled mousebird (*Colius striatus*). *J. Comp. Physiol. B* 171:507–18

114. McKechnie AE, Lovegrove BG. 2002. Avian facultative hypothermic responses: a review. *Condor* 104:705–24

115. McNab BK, Morrison PR. 1963. Body temperature and metabolism in subspecies of *Peromyscus* from arid and mesic environments. *Ecol. Monogr.* 33:63–82

116. Milsom WK. 1993. Metabolic depression during hibernation: the role of respiratory acidosis. In *Life in the Cold*, ed. C Carey, GL Florant, BA Wunder, B Horwitz, pp. 541–44. Boulder, CO: Westview

117. Morhardt JE. 1970. Body temperatures of white-footed mice (*Peromyscus* sp.) during daily torpor. *Comp. Biochem. Physiol.* 33:423–39

118. Morrison PR, Ryser FA. 1962. Metabolism and body temperature in a small hibernator, the meadow jumping mouse, *Zapus hudsonicus. J. Cell. Comp. Physiol.* 60:169–80

119. Morrison PR, McNab BK. 1962. Daily torpor in a Brazilian murine opossum (*Marmosa*). *Comp. Biochem. Physiol.* 6:57–68

120. Morton SR, Lee AK. 1978. Thermoregulation and metabolism in *Planigale maculata* (Marsupialia: Dasyuridae). *J. Therm. Biol.* 3:117–20

121. Muchlinski AE, Ryback EN. 1978. Energy consumption of resting and hibernating meadow jumping mice. *J. Mammal.* 59:435–37

122. Mzilikazi N, Lovegrove BG. 2002. Reproductive activity influences thermoregulation and torpor in pouched mice, *Saccostomus campestris. J. Comp. Physiol. B* 172:7–16

123. Mzilikazi N, Lovegrove BG, Ribble DO. 2002. Exogenous passive heating during torpor arousal in free-ranging rock elephant shrews, *Elephantulus myurus. Oecologia* 133:307–14

124. Nagel A. 1985. Sauerstoffverbrauch, Temperaturregulation und Herzfrequenz

bei europäischen Spitzmäusen (Soricidae). *Z. Säugetierkunde* 50:249–66

125. Nedergaard J, Cannon B. 1984. Preferential utilization of brown adipose tissue lipids during arousal from hibernation in hamsters. *Am. J. Physiol. Regul. Integr. Comp. Physiol.* 247:R506–12

126. Neumann RL, Cade TJ. 1965. Torpidity in the Mexican ground squirrel, *Citellus mexicanus parvidens* (Mears). *Can. J. Zool.* 43:133–40

127. Nicol S, Andersen NA, Mesch U. 1992. Metabolic rate and ventilatory pattern the echidna during hibernation and arousal. In *Platypus and Echidnas*, ed. ML Augee, pp. 150–9. Sydney: Royal Zool. Soc. NSW

128. Nicol S, Pavlides D, Andersen NA. 1997. Nonshivering thermogenesis in marsupials: absence of thermogenic response to β3-adrenergic agonists. *Comp. Biochem. Physiol.* 117A:399–405

129. Ortmann S, Heldmaier G. 2000. Regulation of body temperature and energy requirements of hibernating Alpine marmots (*Marmota marmota*). *Am. J. Physiol. Regul. Integr. Comp. Physiol.* 278:R698–704

130. Pearson OP. 1950. The metabolism of hummingbirds. *Condor* 52:145–52

131. Pohl H. 1961. Temperaturregulation und Tagesperiodik des Stoffwechsels bei Winterschläfern. *Z. Vergl. Physiol.* 45:109–53

132. Prinzinger R, Siedle K. 1988. Ontogeny of metabolism, thermoregulation and torpor in the house martin *Delichon u. urbica* (L.) and its ecological significance. *Oecologia* 76:307–12

133. Prinzinger R, Göppel R, Lorenz A, Kulzer E. 1981. Body temperature and metabolism in the red-backed mousebird (*Colius castanotus*) during fasting and torpor. *Comp. Biochem. Physiol.* 69A:689–92

134. Raison JK, Lyons JM. 1971. Hibernation: alteration of mitochondrial membranes as a requisite for metabolism at low temperature. *Proc. Natl. Acad. Sci. USA* 68:2092–94

135. Schleucher E. 2001. Heterothermia in pigeons and doves reduces energetic costs. *J. Therm. Biol.* 26:287–93

136. Schmid J. 1996. Oxygen consumption and torpor in mouse lemurs (*Microcebus murinus* and *M. myoxinus*): preliminary results of a study in western Madagascar. In *Adaptations to the Cold: Tenth Int. Hibernation Symp.*, ed. F Geiser, AJ Hulbert, SC Nicol, pp. 47–54. Armidale, AU: Univ. New England Press

137. Scholander PF, Hock RJ, Walters V, Johnson F, Irving L. 1950. Heat regulation in some arctic and tropical mammals and birds. *Biol. Bull.* 99:237–58

138. Snapp BD, Heller HC. 1981. Suppression of metabolism during hibernation in ground squirrels (*Citellus lateralis*). *Physiol. Zool.* 54:297–307

139. Snyder GK, Nestler JR. 1990. Relationship between body temperature, thermal conductance, Q_{10} and energy metabolism during daily torpor and hibernation in rodents. *J. Comp. Physiol. B* 159:667–75

140. Song X, Körtner G, Geiser F. 1995. Reduction of metabolic rate and thermoregulation during daily torpor. *J. Comp. Physiol. B* 165:291–97

141. Song X, Körtner G, Geiser F. 1996. Interrelations between metabolic rate and body temperature during entry into daily torpor in *Sminthopsis macroura*. In *Adaptations to the Cold: Tenth Int. Hibernation Symp.*, ed. F Geiser, AJ Hulbert, SC Nicol, pp. 63–69. Armidale, AU: Univ. New England Press

142. Song X, Körtner G, Geiser F. 1997. Thermal relations of metabolic rate reduction in a hibernating marsupial. *Am. J. Physiol. Regul. Integr. Comp. Physiol.* 273:R2097–104

143. Speakman JR, Thomas DW. 2003. Physiological ecology and energetics of bats. In *Bat Ecology*, ed. TH Kunz, M Brock Fenton, pp. 430–90. Chicago/London: Univ. Chicago Press

144. Steiger R. 1992. *Energiehaushalt im Winterschlaf vom Goldmantelziesel* (Spermophilus lateralis) *und vom Siebenschläfer* (Glis glis). Diplom thesis. Univ. Marburg. 81 pp.

145. Storey KB, Storey JM. 1990. Metabolic rate depression and biochemical adaptation in anaerobiosis, hibernation and estivation. *Q. Rev. Biol.* 65:145–74

146. Strumwasser F. 1960. Some physiological principles governing hibernation in *Citellus beecheyi. Bull. Mus. Comp. Zool. Harvard Coll.* 124:282–318

147. Thäti H. 1978. Seasonal differences in O_2 consumption and respiratory quotient in a hibernator (*Erinaceus europaeus*). *Ann. Zool. Fenn.* 15:69–75

148. Thompson SD. 1985. Subspecific differences in metabolism, thermoregulation, and torpor in the western harvest mouse *Reithrodontomys megalotis. Physiol. Zool.* 58:430–44

149. Tucker VA. 1965. Oxygen consumption, thermal conductance, and torpor in the Californian pocket mouse *Perognathus californicus. J. Cell. Comp. Physiol.* 65:393–404

150. Turbill C, Law BS, Geiser F. 2003. Summer torpor in a free-ranging bat from subtropical Australia. *J. Therm. Biol.* 28:223–36

151. van Breukelen F, Martin SL. 2002. Reversible depression of transcription during hibernation. *J. Comp. Physiol. B* 172:355–61

152. Wang LCH. 1978. Energetics and field aspects of mammalian torpor: the Richardsons's ground squirrel. In *Strategies in Cold*, ed. LCH Wang, JW Hudson, pp. 109–45. New York: Academic

153. Wang LCH, Hudson JW. 1970. Some physiological aspects of temperature regulation in normothermic and torpid hispid pocket mouse, *Perognathus hispidus. Comp. Biochem. Physiol.* 32:275–93

154. Wang LCH, Hudson JW. 1971. Temperature regulation in normothermic and hibernating eastern chipmunk, *Tamias striatus. Comp. Biochem. Physiol.* 38A: 59–90

155. Wang LCH, Lee T-F. 2000. Perspectives on metabolic suppression during mammalian hibernation and daily torpor. In *Life in the Cold: Eleventh Int. Hibernation Symp.*, ed. G Heldmaier, M Klingenspor, pp. 149–58. Berlin/Heidelberg/New York: Springer

156. Watts PD, Oritsland NA, Jonkel C, Ronald K. 1981. Mammalian hibernation and the oxygen consumption of a denning black bear (*Ursus americanus*). *Comp. Biochem. Physiol.* 69A:121–23

157. Willis CKR, Brigham RM. 2003. Defining torpor in free-ranging bats: experimental evaluation of external temperature-sensitive radiotransmitters and the concept of active temperature. *J. Comp. Physiol. B* 173:379–89

158. Willis JS. 1982. The mystery of the periodic arousal. In *Hibernation and Torpor in Mammals and Birds*, ed. CP Lyman, JS Willis, A Malan, LCH Wang, pp. 92–103. New York: Academic

159. Wilz M, Heldmaier G. 2000. Comparison of hibernation, estivation and daily torpor in the edible dormouse, *Glis glis. J. Comp. Physiol. B* 170:511–21

160. Withers PC. 1977. Respiration, metabolism, and heat exchange of euthermic and torpid poorwills and hummingbirds. *Physiol. Zool.* 50:43–52

161. Withers PC, Richardson KC, Wooller RD. 1990. Metabolic physiology of euthermic and torpid honey possums, *Tarsipes rostratus. Aust. J. Zool.* 37:685–93

162. Wolf LW, Hainsworth FR. 1972. Environmental influence on regulated body temperature in torpid hummingbirds. *Comp. Biochem. Physiol.* 41A:167–73

Annu. Rev. Physiol. 2004. 66:275–89
doi: 10.1146/annurev.physiol.66.032102.115313
Copyright © 2004 by Annual Reviews. All rights reserved
First published online as a Review in Advance on October 20, 2003

SLEEP AND CIRCADIAN RHYTHMS IN MAMMALIAN TORPOR

H. Craig Heller and Norman F. Ruby

*Department of Biological Sciences, Stanford University, Stanford,
California 94305–5020; email: hcheller@stanford.edu; ruby@stanford.edu*

Key Words sleep homeostasis, suprachiasmatic nucleus

■ **Abstract** Sleep and circadian rhythms are the primary determinants of arousal state, and torpor is the most extreme state change that occurs in mammals. The view that torpor is an evolutionary extension of sleep is supported by electrophysiological studies. However, comparisons of factors that influence the expression of sleep and torpor uncover significant differences. Deep sleep immediately following torpor suggests that torpor is functionally a period of sleep deprivation. Recent studies that employ post-torpor sleep deprivation, however, show that the post-torpor intense sleep is not homeostatically regulated, but might be a reflection of synaptic loss and replacement. The circadian system regulates sleep expression in euthermic mammals in such a way that would appear to preclude multiday bouts of torpor. Indeed, the circadian system is robust in animals that show shallow torpor, but its activity in hibernators is at least damped if not absent. There is good evidence from some species, however, that the circadian system plays important roles in the timing of bouts of torpor.

INTRODUCTION

Mammals that enter torpor undergo enormous changes in many physiological and neurophysiological variables. Deep torpor, or hibernation, involves a drop in regulated body temperature close to or even below the freezing point of water. Whole-body metabolism during deep torpor may be only 1 or 2% of basal values. Blood flow in most vascular beds falls very close to zero. Most of the nervous system becomes metabolically quiescent and electrically silent. Transcription and translation are virtually suspended. Yet, this state, which is as close to death as a living mammal can get, is spontaneously reversible. Over the winter phase of its annual cycle, a hibernator will enter repeated bouts of deep torpor that may last from an hour to weeks. Torpor bouts are punctuated by returns to normothermia that usually last a day or less. In addition to hibernators, there are many species that enter shallow torpor, also called daily torpor because it generally occurs during the inactive phase of the daily cycle. The physiological changes that occur during shallow torpor are less extreme in comparison with those in hibernation, but still dramatic in comparison with the homeostatic conditions of most mammals. Both daily torpor and hibernation raise many intriguing questions. This review addresses two areas of hibernation research: (a) the relationship between hibernation and sleep and (b) the role of the circadian system.

SLEEP AS THE EVOLUTIONARY ORIGIN OF TORPOR

Torpor has the superficial appearance of sleep. The animal retires to a nest, assumes a sleep-like posture, remains behaviorally quiescent, and exhibits elevated arousal thresholds. The assumption that torpor and sleep are homologous has been supported by numerous electroencephalographic (EEG) studies (for a review see 1). For example, hibernation in ground squirrels is entered through sleep and, as brain temperature falls, nonREM (nonrapid eye movement, NREM) sleep predominates whereas REM sleep decreases and disappears completely at a brain temperature around 25°C (2). Also, bouts of shallow torpor consist almost entirely of NREM sleep (3–5). Recently, continuous EEG recordings of black bears during the hibernation season reveal that they are predominantly asleep, but are cycling between NREM and REM sleep with very little wakefulness (B.M. Barnes, Ø. Tøien, J. Blake, D. Grahn, H.C. Heller & D.M. Edgar, unpublished data).

Characterizing the relationship between deep torpor and sleep is difficult because brain electrical activity and, therefore, the EEG changes as brain temperature falls. One attempt to overcome this methodological difficulty was based on the fact that neurons in particular parts of the brain show EEG-state selective firing rates. Long-term, single-unit recordings during continuous bouts of hibernation have shown that the spontaneous changes in firing rates of individual neurons continue to cycle through states homologous with wakefulness and NREM sleep down to brain temperatures as low as 14°C (6).

Taking into account the ecological context of mammals that undergo torpor, a logical interpretation of the EEG and single-unit studies is that the selective pressures shaping the evolution of torpor were on energy conservation and that a preadaptation was the regulated decline in mammalian body temperature associated with NREM sleep (7). The hypothesis is that selection favored lower and lower body temperatures during daily sleep periods, as well as increasing duration of those sleep periods as a means of lowering total metabolic energy needs. A constraint on the evolution of this adaptation for energy conservation is the circadian system, which in mammals serves to consolidate wakefulness during one phase of the circadian period and to allow the expression of sleep during the other phase of the circadian period (8). This constraint had to be relaxed to permit the evolution of multiday torpor characteristic of hibernation, and it may be the defining difference between daily torpor and hibernation.

Assuming that mammalian hibernation evolved as an extension of sleep and required a modification of the circadian modulation of sleep processes, the rest of this review focuses on recent studies of sleep and circadian systems of mammals that undergo daily torpor or hibernation.

SLEEP AND PERIODIC AROUSALS FROM HIBERNATION

An enduring question about hibernation has been the function and the mechanism of the periodic arousals to euthermia that occur on a fairly regular basis throughout the hibernation season. A unique hypothesis about periodic arousals was generated as a result of EEG recordings extending over the euthermic periods between bouts of torpor in ground squirrels (*Citellus* sp.). These studies showed that the animals were predominantly asleep during the arousal process and that they continued to spend most of their time asleep during the inter-bout euthermic intervals (9, 10). Of particular interest was the spectral analysis of the EEG recordings during the euthermic interval. The EEG power spectrum of NREM sleep is dominated by slow wave activity (SWA) in the 1.0 to 4.0 Hz band, called the delta band. SWA or delta power in NREM sleep is directly proportional to the duration of prior wakefulness and is therefore proposed as an indicator of sleep need and, conversely, a quantitative measure of the sleep restorative process (11). The surprising observation was that the EEG SWA was exceedingly high at the beginning of the euthermic interval and showed a monotonic decline much like the delta power profile following prolonged sleep deprivation. Even though the hibernators were apparently emerging from multiday bouts of continuous sleep, they appeared to be severely sleep deprived. It even appeared that the post-bout SWA was a function of the duration of the prior torpor bout. These observations led to the proposal that the low brain temperatures of hibernation impeded the elusive sleep restorative process and, therefore, hibernators had to periodically arouse from deep torpor for the purpose of sleep restoration (9, 10). The hypothesis was strengthened by similar EEG studies on the Djungarian hamster, which undergoes daily torpor. As was found in the studies on hibernators, torpor was followed by recovery sleep with high delta power that declined monotonically with time spent asleep. These post-bout EEG profiles were similar to those following sleep deprivation (12, 13).

The hypothesis that sleep restoration was impeded during torpor by low brain temperature was supported by studies in which golden-mantled ground squirrels (*Spermophilus lateralis*) were held at different ambient temperatures during hibernation, and EEG recordings were made during the subsequent euthermic intervals. These studies showed that for torpor bouts lasting three days or more, the magnitude of the delta power peak following a bout of hibernation was a function of the temperature of the brain during hibernation and not a function of the duration of the bout (up to seven days). In fact, when the animals were held at an ambient temperature of 21°C, it was possible to record torpor bouts lasting an average of two days, but there was no increase in delta power following arousal from the bout (14).

SWA Following Deep Torpor Does Not Reflect Sleep Deprivation

The idea that sleep following torpor was functionally homologous with sleep following prolonged wakefulness was seriously compromised by subsequent studies in which animals were sleep deprived immediately following arousal from torpor. A defining characteristic of sleep is that it is homeostatically regulated and that any NREM sleep deprivation should be followed by a SWA rebound. It was expected that sleep deprivation for several hours would shift the delta power recovery peak to a later time, and the subsequent integrated delta power expression would be greater. The opposite was found. When animals were sleep deprived for the first few hours following arousal from torpor, the expected SWA peak was completely eliminated (14–16). It appeared that whatever the recovery process was following a bout of torpor, it did not depend on sleep. However, sleep homeostasis is intact during the inter-bout intervals. Following the initial high SWA expression, sleep deprivation is followed by a proportional SWA response. Because SWA is so tightly identified with sleep homeostasis, the question of how torpor could produce a similar response that is not dependent upon sleep is intriguing.

HYPOTHESES TO EXPLAIN HIGH SWA FOLLOWING DEEP TORPOR

Three hypotheses have been put forward to explain the high SWA following emergence from torpor: the thermoregulatory hypothesis, the brain energy hypothesis, and the synaptogenesis hypothesis. The thermoregulatory hypothesis (17) proposes that the intense metabolic effort and rapid rise in core temperature associated with the arousal from torpor can create the risk of hyperthermia. Berger notes that high body temperatures prior to sleep onset are associated with higher SWA and that the higher the SWA, the greater is the fall in core temperature associated with sleep. The thermoregulatory hypothesis has been rebutted for several reasons (18). First, the SWA is expressed before the animal reaches its highest core temperature, and this core temperature characteristically does not overshoot, but approachs normal euthermic levels gradually. Second, at least over the first three days of torpor, the SWA response is proportional to duration of torpor bout even when the arousal begins at the same core temperature. Third, the brain heats up first, and the nonbrain core lags considerably behind. Therefore, the nonbrain core serves as a considerable heat sink toward the end of the arousal process, reducing the possibility of a core temperature overshoot because the regulation of body temperature in these animals is primarily dependent upon brain temperature. Rises in brain temperature induce peripheral vasodilation.

The brain energy hypothesis suggests that the high-energy demands of emergence from torpor could result in transient hypoglycemia, which is known to increase SWA. Presumably, regional hypoglycemia in the brain would increase the local release of adenosine and the depletion of astrocytic glycogen (19). Arguing against this hypothesis are measurements that show no reductions of plasma or brain glucose and brain glycogen during arousal from torpor, and no hypoglycemia is seen in the hours after arousal that could coincide with the delta power peak (20–24).

The synaptogenesis hypothesis is based on comparisons of hippocampal dendritic complexity in ground squirrel brains taken during hibernation and during euthermia. There is an apparent loss of dendritic complexity during a bout of hibernation, suggesting that a function of periodic arousal could be restoration of dendrites and synapses (25, 26). This observation of structural change is supported by a recent study of changes in synaptophysin over bouts of hibernation. Synaptophysin is associated with synaptic vesicles, so changes in the abundance of synapses and vesicles should be reflected in changes in synaptophysin and, indeed, reduced synaptophysin during bouts of hibernation was observed (27). To extend these results into an explanation of the enhanced SWA following arousal, it is necessary to assume that the torpor-related changes in dendrites and synapses that have been observed in the hippocampus also occur in thalamocortical circuits. If that assumption is correct, the following logic would apply.

Wakefulness is maintained by ascending depolarizing influences on thalamocortical circuits. These ascending influences are largely mediated by cholinergic, noradrenergic, histaminergic, and serotonergic neuromodulation. The onset of sleep is associated with a withdrawal of these activities and a resulting hyperpolarization of thalamocortical neurons, which induces the synchronized bursting activity that generates the slow waves in the cortical EEG (for a review see 19). The synaptogenesis hypothesis suggests that the loss of dendritic complexity and synapses during torpor makes it impossible for ascending influences to depolarize the thalamocortical neurons. Therefore, the cerebral cortex remains tonically hyperpolarized, which potentiates the expression of SWA. As dendrites and synapses are restored, the high SWA would gradually diminish as observed.

The synaptogenesis explanation for the high SWA following torpor gains further support from new and exciting data on reversible phosphorylation of protein tau that occurs over bouts of hibernation. Phosphorylation of protein tau is implicated in the formation of neurofibrillary tangles associated with neurodegenerative diseases such as Alzheimer's. In this type of pathology, the phosphorylation of protein tau is apparently irreversible. A study of European ground squirrels has shown formation of phosphorylated tau during bouts of torpor that is completely reversible upon arousal (28). These results were obtained from cortex as well as the hippocampus. In the hippocampus, parallel formation of phosphorylated tau with regression of synaptic contacts was observed.

SWA Following Shallow Torpor is Indistinguishable from SWA Following Sleep Deprivation

We believe that the synaptogenesis hypothesis is the most likely explanation for the high delta power following arousal from torpor in ground squirrels. The picture is less clear, however, when we extend the consideration to a species that undergoes shallow, daily torpor, the Djungarian hamster. As mentioned above, this species also shows a delta power peak following arousal from daily bouts of torpor. However, when the hamster is sleep deprived at the time that this peak normally occurs (for 1.5 h following emergence from torpor), an even larger delta power peak is expressed at the end of the sleep deprivation (29, 30). These results led to the conclusion that the delta power response following torpor in the hamster is an expression of the same phenomenon that is induced by sleep deprivation.

Several factors make it difficult to compare the results from the hamster and the ground squirrel species discussed above. First, if the ground squirrels are held at an ambient temperature that prevents their brain temperatures from falling below the levels seen in the torpid hamsters, there is no delta power response following arousal from torpor bouts lasting up to two days (14). By comparison, the longest torpor bouts in the hamster are 7 h. Also, the time course of the build up of the delta power response is much more rapid in the hamster. As a result, a 1.5-h sleep deprivation in the hamster results in a delta power peak as high as the peak seen after emergence from torpor. In contrast, a 3-h sleep deprivation in the ground squirrel produces a delta power response that is very small compared with the peak seen after a hibernation bout at a low temperature. The difference in the time constant of build up of delta power propensity makes it difficult to do comparable experiments in these two species. A sleep deprivation in the hamster that spans the time when the post torpor delta power response is expressed will itself produce a delta power response that is of a similar magnitude. We think that it is not possible currently to conclude that the delta power response seen after shallow and deep torpor have different causes, even though that may indeed be the case.

Because a major difference between shallow, daily torpor and the multiday deep torpor bouts that characterize hibernation appears to be the involvement of the circadian system, we now turn our attention to the role of the circadian system in hibernation.

HIBERNATION AND THE CIRCADIAN SYSTEM

The role of the circadian system in hibernation remains controversial. This is largely due to the lack of standardized methods for analyzing rhythms under the unique conditions of hibernation, differences among species in the extent to which the circadian system is present during hibernation, and the general paucity of studies on this topic. Several studies have examined whether the circadian system times either the entry or arousal phases of multiday torpor bouts during hibernation. Most of those studies were, however, conducted in the presence of daily cycles of illumination. This is problematic because ambient light does influence the timing of torpor entries and arousals (31). Those studies have been reviewed elsewhere and are not discussed here (32).

A few studies have demonstrated true circadian control of arousal from torpor in animals housed in constant darkness or dim light. Pocket mice are capable of multiday torpor bouts, the entries and arousals of which are timed by the circadian system. At low-ambient temperature, successive entries and arousals from torpor occur every few days. When these events occur, however, they occur at the same phase of the circadian cycle (33, 34). Similar timing of arousals from torpor have been reported for dormice (35–38). A novel approach to this issue was employed by Oklejewicz et al. (32) who used three genotypes of *tau* mutant Syrian hamsters to test for circadian rhythms in torpor entries and arousals. These animals have a point mutation that shortens their circadian period. Because the endogenous circadian period differs by $\sim$4 h between homozygotes and wild-type hamsters, circadian timing of torpor entries and arousals could be expected to differ as well. However, no evidence was found for circadian timing in either parameter in any of the animals (32). The lack of circadian timing in torpor in this species is surprising given all that is known about their circadian system in euthermic conditions.

Golden-mantled ground squirrels are probably the most intensively studied hibernators in the laboratory, and their arousal patterns have been studied under daily illumination cycles and under constant light. Arousals from torpor appear to be timed by the circadian system, but laboratory lighting may also influence timing of torpor entries and arousals. These squirrels preferentially enter torpor at night and arouse during the day when housed in a light-dark cycle (39). In squirrels in which the suprachiasmatic nucleus (SCN) had been completely ablated and, therefore, lack circadian organization, arousals occurred during the daytime just as they did for control animals (39). Under constant dim light, however, the circadian system appears to time arousals. Circadian rhythms in T_b (body temperature) persist during torpor ($T_b \sim 12°C$), and arousals are initiated at the same circadian phase of the T_b oscillation regardless of rhythm period or duration of the torpor bout (40). Thus arousals in intact squirrels can be stimulated by daytime light exposure or timed by the circadian system. Given that hibernating squirrels are not exposed to light cycles in their natural habitat, bright laboratory light may be a stronger arousal stimulus than the endogenous timing signal that would normally initiate arousal in the wild. Daytime arousals in intact squirrels may, therefore, result from positive masking effects of light.

Circadian Rhythms in Deep Torpor

The persistence of circadian rhythms during deep torpor has been well documented for golden-mantled ground squirrels, but their existence in other species remains controversial. One reason for this discrepancy is that detection and analysis of rhythms during torpor are not as straight forward as they are for euthermic animals. Rhythm period has to be analyzed on fewer cycles (typically 3–7 cycles) and on T_b rhythms of very low amplitude ($<0.1°C$) (39). Simultaneous monitoring of T_a (ambient temperature) and T_b is needed to differentiate true circadian oscillations from ones driven by T_a fluctuations because such fluctuations can interfere with endogenous T_b rhythms and lead to false positive conclusions regarding rhythm persistence (32, 39, 41, 42). A second reason for this discrepancy is that T_b rhythm amplitude decreases as T_b declines and may not be detectable below a certain T_b (39, 43). Rhythms can be detected, however, if T_a is increased so that T_b remains above that lower limit (D.A. Grahn, J.D. Miller & H.C. Heller, unpublished data). The dependence of endogenous rhythm amplitude on T_a is important because it suggests that an absence of T_b rhythms in deep torpor does not preclude rhythmicity in shallow torpor. More importantly, there has often been confusion over the detection of overt rhythms and the inference that rhythms are present based on other criteria. For example, different laboratories have demonstrated circadian control of arousal from torpor in European squirrels and hamsters but have been unable to detect circadian oscillations in T_b during deep torpor. As the authors noted, the lack of rhythmicity could be due to the presence of T_a cycles (41, 44) or the very long sampling intervals used to monitor T_b (35), both of which may have precluded detection of low-amplitude T_b rhythms. Nevertheless, circadian timing of arousals implies that the pacemaker was functional during deep torpor.

The problem of interference from T_a fluctuations in circadian studies of hibernation has been handled directly by three different studies of golden-mantled ground squirrels. Two studies documented T_b rhythms in deep torpor that were independent of T_a fluctuations (39, 40), and a third monitored hibernation in a specially constructed chamber that eliminated daily cycles in T_a entirely (45). These studies confirmed circadian rhythms in T_b and brain temperature (T_{br}) in deep torpor (40, 45). T_b and T_{br} rhythm periods were highly variable, and amplitude was extremely low ($<0.1°C$), but most periods were between 22 and 25 h. Thus rhythms not only persist during deep torpor in this species, they are also temperature compensated. A post hoc calculation for circadian rhythm period among these three studies indicates Q_{10} (the change of rate over a 10°C increment) values for circadian periods of 0.97–1.05 (31). It should be emphasized that rhythm period during torpor is highly variable and that these Q_{10}s are based on mean values; thus the Q_{10} should not be the sole means by which temperature effects on circadian function are assessed (cf 46).

Biological variables other than T_b have been monitored to assay for rhythmicity during torpor and have shown that not all rhythms present in euthermia persist during torpor. Pineal melatonin content is extremely low during torpor, but is rapidly elevated after arousal in Syrian and Turkish hamsters (47–49). The time of peak melatonin production in golden hamsters does not depend on time of arousal (47): It is stimulated by pronounced sympathetic nervous system activity that accompanies the arousal process (50). During euthermic intervals between torpor bouts, circadian sleep-wake cycles are absent and replaced by ultradian rhythms with periods of ~6 h, even though some EEG measures oscillate in a circadian pattern (45). Thus a lack of rhythmicity in output of the clock does not preclude the presence of other rhythms or a lack of rhythmicity in the SCN itself. Although T_b is of interest for hibernation studies and relatively easy to monitor, it may not be the best index of circadian organization in hibernating animals because its period and amplitude are affected by environmental light and temperature.

Seasonal Arrhythmia

Between the end of the hibernation season and beginning of the next one, hibernators maintain euthermic T_bs, and circadian rhythms in behavior and physiology are robust. In some hibernators, the transitions between seasons of hibernation and euthermia are accompanied by a damping of circadian organization. T_b and activity rhythms of European and Arctic ground squirrels are arrhythmic for many days after hibernation ends even though they are euthermic (35, 51; B.M. Barnes, personal communication). Some have suggested that the circadian system of these animals is turned off prior to the hibernation season and remains so until reemergence in the spring and that the circadian system requires several days of euthermia to return to a fully functional state.

A similar lack of rhythmicity has been observed in some golden-mantled ground squirrels for a few days after the end of hibernation, but most squirrels are completely rhythmic as soon as the hibernation season ends (39). Nevertheless, even a few days of arrhythmicity after hibernation cannot be explained easily given that these animals remain rhythmic in deep torpor. One possibility is that the circadian system functions at some minimal level during hibernation that is insufficient to sustain rhythmicity in euthermia and may require several days of euthermia to restore normal time-keeping. If the circadian system provides a daily alerting signal, as some have suggested (8), it may be toned down in winter to permit multiday torpor bouts. This phenomenon of seasonal arrhythmia is not necessarily a consequence of low T_b because circadian T_b rhythms are absent in black bears for most of hibernation even though T_b remains $>30°C$. In bears, T_b oscillates with 3–5 day cycles in hibernation (B.M. Barnes, Ø. Tøien, J. Blake, D. Grahn, H.C. Heller & D.M. Edgar, unpublished data).

THE SUPRACHIASMATIC NUCLEUS AND HIBERNATION

Because of its prominent role in circadian timing, the persistence of T_b rhythms during torpor and circadian timing of arousals implies that the SCN remains functional during hibernation. SCN activity during torpor was initially investigated in a study of brain metabolic activity in deep torpor that measured uptake of $[^{14}C]$ 2-deoxyglucose (2-DG) in over 80 brain regions at several phases of a torpor bout (52). Although overall brain metabolism was markedly reduced in torpor, the decrease in SCN metabolic activity in torpor was much less than for nearly every other brain structure examined (52–54). The relatively high metabolic activity of the SCN during torpor is also reflected in the elevated expression of c-*fos* mRNA in the SCN just prior to arousal (55).

Such high metabolic activity in the SCN during torpor implies that the SCN has a functional role in hibernation. The 2-DG studies stimulated two long-term projects that monitored hibernation patterns for several years in golden-mantled ground squirrels in which the SCN had been surgically ablated. SCN ablation did indeed have dramatic effects on hibernation patterns. In a small preliminary study, hibernation patterns were erratic after SCN ablation (56). A larger follow-up study recorded T_b telemetrically over 2.5 years in female ground squirrels and found marked disruptions in torpor patterns of SCN-lesioned squirrels (39). As expected, T_b rhythms during deep torpor and summer euthermic intervals were eliminated by SCN ablation. Most surprising, however, were other disruptions in hibernation. After SCN ablation, torpor bout duration decreased by 40%, euthermic intervals between bouts increased by 47%, and the time required to enter torpor increased by 21% (39). Moreover, the duration of the hibernation season increased by ~1 month in 50% of SCN-lesioned animals, and in the other half, there was no end to the hibernation season (39). This latter group cycled through bouts of torpor and euthermia without a summer break for the final 2 years of monitoring.

Perhaps the most important findings from that project came from animals in which the SCN was only partially ablated. Behavioral data from animals with partial SCN lesions show that these effects cannot be attributed to nonspecific sequelae of the surgery or to damage to areas adjacent to the SCN because other squirrels in this study had incomplete SCN lesions (40–90% ablation) confined within the Nissl-stained boundaries of the SCN. These animals retained circadian T_b rhythms but manifested all other changes observed in squirrels with complete SCN ablation (39); durations of their torpor bouts, euthermic intervals, and rates of torpor entry were no different from animals with complete lesions. Therefore, changes in hibernation patterns after lesion surgery were not simply the result of loss of circadian organization, but were from loss of some function served by the SCN. In addition to its known role as a circadian pacemaker, the SCN may serve a noncircadian role in hibernation related to maintenance of energy balance (31).

For the SCN to influence hibernation patterns, it must function at low tissue temperatures. Although this point seems obvious, much of the brain is electrically silent or only sporadically active in deep torpor (57). The period of circadian rhythms in neuronal firing rate from hibernating golden-mantled ground squirrels and rats is temperature compensated above 31°C ($Q_{10} = 0.95$–0.99) (46). At 25°C, however, neuronal rhythms were robust in the squirrel SCN but were eliminated in the rat SCN (46). In addition to period, rhythm amplitude is also temperature compensated in hibernators. The Q_{10} for neuronal rhythm amplitude in the rat SCN is twice that of hibernating squirrels (2.7 versus 1.3) and single units are significantly more difficult to detect in the rat than in the squirrel SCN at 25°C (46). If one extrapolates from these temperatures down to those maintained by hibernators (<10°C), it is apparent that the SCN could continue to generate action potentials during deep torpor (31). However, extracellular recordings of SCN neurons failed to detect action potentials below 16.6°C (58). Because of the background noise present in extracellular recordings, patch-clamp recordings will be necessary to resolve whether smaller, low-amplitude spikes are present below this temperature. Alternatively, humoral signals might play an important role in SCN signaling in hibernation.

There are several possible mechanisms to explain the differential impact of temperature on rat and squirrel SCN function, but the available data support the hypothesis that the loss of rhythms in rats at low temperature may be due to diminished cell synchronization. Individual SCN neurons have circadian oscillations in firing rate and must synchronize to one another to produce coherent rhythms (59, 60). When the population becomes desynchronized, the daytime peak in firing rate is lower because fewer neurons peak at the same circadian time. This appears to happen in the rat, but not in the squirrel, SCN at low temperature. The mean firing rates for individual SCN neurons measured over a 24-h period did not differ among rats and squirrels at 25°C, but the temporal variability in peak firing rate among neurons was markedly greater in rats at 25°C, but not in squirrels (46). In addition to temperature compensation of rhythm period and amplitude, cell synchrony among SCN neurons may be another property of the SCN that has been adapted to function in deep torpor.

Studies that support a role for the SCN in hibernation have not been complemented by molecular studies of core clock mechanisms. Molecular work in other brain regions and peripheral tissues in hibernation has been reviewed elsewhere (61). Studies of gene transcription have shown a marked decrease in mRNA synthesis (62–64) and an even more dramatic reduction, or even arrest, in protein synthesis in torpor (62, 65). This apparent block of protein synthesis could render the clock nonfunctional during deep torpor because oscillations in protein levels (e.g., PER, BMAL) are considered essential for circadian rhythm generation. This does not appear to be the case because the presence of circadian T_b rhythms during torpor and in arousals from torpor suggest that some proteins oscillate in the SCN during deep torpor. Molecular feedback loops may oscillate at low amplitude or in some reduced form to drive rhythms observed during hibernation.

CONCLUSIONS

Mammalian torpor reflects major evolutionary modifications of systems subserving sleep and circadian rhythms. Shallow torpor or hibernation consists primarily of a NREM sleep-like state with REM sleep disappearing as brain temperature falls. Shallow torpor expression is strongly controlled by the circadian system, whereas the circadian system is severely damped in hibernators to permit multiday bouts of torpor. Because the adaptive value of torpor is obviously energy conservation, the question of the function of periodic arousals from torpor is intriguing. In species that can forage, shallow daily torpor is adaptive, but for hibernators, the energy expenditures of periodic arousals would seem to be maladaptive. Continuous EEG studies of hibernators across bouts of torpor reveal intense sleep immediately following arousal from torpor, and this observation spawned the hypothesis that low brain temperature during torpor blocked the as yet undiscovered restorative function of sleep, and therefore hibernators had to arouse periodically to sleep. Sleep deprivation studies, however, cast doubt on this hypothesis because they revealed that intense sleep after a torpor bout was not homeostatically regulated, yet sleep later in the euthermic inter-bout interval was homeostatically regulated. A promising possible explanation of this phenomenon is that it reflects loss of dendritic and synaptic complexity during torpor and rapid restoration during the inter-bout interval. If this explanation is true, it could indicate that euthermic sleep also supports synaptic plasticity and the hibernator is a good model system for investigating that phenomenon.

Investigations of the circadian system have also revealed possible new relationships with hibernation. Clearly, there are major changes in the circadian system associated with hibernation. In the ground squirrel, which undergoes a large drop in brain temperature during hibernation, the circadian system may play a role in initiating periodic arousals. In the bear, however, which does not undergo a large drop in brain temperature during hibernation, there are no periodic arousals and no apparent circadian modulation of physiology. Thus it is likely that the circadian system is at least turned down and in some cases inactivated during the hibernation season. Lesions of the master clock in the SCN, however, have yielded evidence for critical involvement of the SCN in hibernation. Animals with SCN lesions have shorter torpor bouts and longer inter-bout intervals indicating that the circadian system in some way is defending torpor. Hibernators show the most extreme modifications of both sleep and circadian regulation and, therefore, may be a valuable model system for investigating the underlying mechanisms of sleep and the circadian control of sleep.

The *Annual Review of Physiology* is online at http://physiol.annualreviews.org

LITERATURE CITED

1. Berger RJ. 1984. Slow wave sleep, shallow torpor, and hibernation: homologous states of diminished metabolism and body temperature. *Biol. Psych.* 19:305–26
2. Walker JM, Glotzbach SF, Berger RJ, Heller HC. 1977. Sleep and hibernation in ground squirrels (*Citellus* spp): electrophysiological observations. *Am. J. Physiol. Regul. Integr. Comp. Physiol.* 233:R213–21
3. Walker JM, Garber A, Heller HC. 1979. Sleep and estivation (shallow torpor): continuous processes of energy conservation. *Science* 204:1098–100
4. Walker JM, Haskell EH, Berger RJ, Heller

HC. 1981. Hibernation at moderate temperatures: a continuation of slow wave sleep. *Experientia* 37:726–28
5. Deboer T, Tobler I. 1995. Temperature dependence of EEG frequencies during natural hypothermia. *Brain Res.* 670:153–56
6. Krilowicz BL, Glotzbach SF, Heller HC. 1988. Neuronal activity during sleep and complete bouts of hibernation. *Am. J. Physiol. Regul. Integr. Comp. Physiol.* 255:R1008–19
7. Heller HC. 1988. Sleep and hypometabolism. *Can. J. Zool.* 66:61–69
8. Edgar DM, Dement WC, Fuller CA. 1993. Effect of SCN lesions on sleep in

squirrel monkeys: evidence for opponent processes in sleep-wake regulation. *J. Neurosci.* 13:1065–79

9. Trachsel L, Edgar DM, Heller HC. 1991. Are ground squirrels sleep deprived during hibernation? *Am. J. Physiol. Regul. Integr. Comp. Physiol.* 260:R1123–29

10. Daan S, Barnes BM, Strijkstra AM. 1991. Warming up for sleep?—ground squirrels sleep during arousals from hibernation. *Neurosci. Lett.* 128:265–68

11. Borbély AA. 1982. A two process model of sleep regulation. *Human Neurobiol.* 1:195–204

12. Deboer T, Tobler I. 1996. Natural hypothermia and sleep deprivation: common effects on recovery sleep in the Djungarian hamster. *Am. J. Physiol. Regul. Integr. Comp. Physiol.* 271:R1364–71

13. Palchykova S, Deboer T, Tobler I. 2002. Selective sleep deprivation after daily torpor in the Djungarian hamster. *J. Sleep Res.* 11:313–19

14. Larkin JE, Heller HC. 1996. Temperature sensitivity of sleep homeostasis during hibernation in the golden-mantled ground squirrel. *Am. J. Physiol. Regul. Integr. Comp. Physiol.* 270:R777–84

15. Strijkstra AM, Daan S. 1998. Dissimilarity of slow-wave activity enhancement by torpor and sleep deprivation in a hibernator. *Am. J. Physiol. Regul. Integr. Comp. Physiol.* 275:R1110–17

16. Larkin JE, Heller HC. 1999. Sleep after arousal from hibernation is not homeostatically regulated. *Am. J. Physiol. Regul. Integr. Comp. Physiol.* 276:R522–29

17. García-Allegue R, Lax P, Madariaga AM, Madrid JA. 1999. Locomotor and feeding activity rhythms in a light-entrained diurnal rodent, *Octodon degus. Am. J. Physiol. Regul. Integr. Comp. Physiol.* 277:R523–31

18. Daan S, Strijkstra AM. 1998. Reply. *J. Sleep Res.* 7:71–72

19. Benington JH, Heller HC. 1995. Restoration of brain energy metabolism as the function of sleep. *Prog. Neurobiol.* 45:347–60

20. Galster W, Morrison PR. 1975. Gluconeogenesis in arctic ground squirrel between periods of hibernation. *Am. J. Physiol.* 228:325–30

21. Galster W, Morrison PR. 1970. Cyclic changes in carbohydrate concentrations during hibernation in the arctic ground squirrel. *Am. J. Physiol.* 218:1128–32

22. Lust WD, Wheaton AB, Feussner G, Passonneau J. 1989. Metabolism in the hamster brain during hibernation and arousal. *Brain Res.* 489:12–20

23. Nestler JR. 1990. Metabolic substrate change during daily torpor in deer mice. *Can. J. Zool.* 69:322–27

24. Nizielski SE, Billington CJ, Levine AS. 1989. Brown fat GDP binding and circulating metabolites during hibernation and arousal. *Am. J. Physiol. Regul. Integr. Comp. Physiol.* 257:R536–41

25. Popov VI, Bocharova LS, Bragin AG. 1992. Repeated changes of dendritic morphology in the hippocampus of ground squirrels in the course of hibernation. *Neuroscience* 48:45–51

26. Popov VI, Bocharova LS. 1992. Hibernation-induced structural changes in synaptic contacts between mossy fibres and hippocampal pyramidal neurons. *Neuroscience* 48:53–62

27. Strijkstra AM, Hut RA, de Wilde MC, Stieler J, van der Zee EA. 2003. Hippocampal synaptophysin immunoreactivity is reduced during natural hypothermia in ground squirrels. *Neurosci. Lett.* 344:29–32

28. Arendt T, Stieler J, Strijkstra AM, Hut RA, Rüdiger J, et al. 2003. Reversible paired helical filament-like phosphorylation of tau is an adaptive process associated with neuronal plasticity in hibernating animals. *J. Neurosci.* 23:6972–81

29. Deboer T, Tobler I. 2000. Slow waves in the sleep electroencephalogram after daily torpor are homeostatically regulated. *Neuroreport* 11:881–85

30. Deboer T, Tobler I. 2003. Sleep regulation in the Djungarian hamster: comparison of the dynamics leading to the slow-wave activity increase after sleep deprivation and daily torpor. *Sleep* 26:567–72

31. Ruby NF. 2003. Hibernation: when good clocks go cold. *J. Biol. Rhythms* 18:275–86

32. Oklejewicz M, Daan S, Strijkstra AM. 2001. Temporal organisation of hibernation in wild-type and *tau* mutant Syrian hamsters. *J. Comp. Physiol. B* 171:431–39

33. Lindberg RG, Gambino JJ, Hayden P. 1971. Circadian periodicity of resistance to ionizing radiation in the pocket mouse. In *Biochronometry*, ed. M. Menaker, pp. 169–85. Washington, DC: National Academy of Sciences

34. French AR. 1977. Periodicity of recurrent hypothermia during hibernation in the pocket mouse, *Perognathus longimembris*. *J. Comp. Physiol. B* 115:87–100

35. Hut RA, Barnes BM, Daan S. 2002. Body temperature patterns before, during, and after semi-natural hibernation in the European ground squirrel. *J. Comp. Physiol. B* 172:47–58

36. Daan S. 1973. Periodicity of heterothermy in the garden dormouse, *Eliomys quercinus* (L.). *Neth. J. Zool.* 23:237–65

37. Pohl H. 1967. Circadian rhythms in hibernation and the influence of light. In *Mammalian Hibernation III*, ed. KC Fisher, AR Dawe, CP Lyman, E Schönbaum Jr, FE South, pp. 140–51. New York: Elsevier

38. Wilz M, Heldmaier G. 2000. Comparison of hibernation, estivation and daily torpor in the edible dormouse, *Glis glis. J. Comp. Physiol. B* 170:511–21

39. Ruby NF, Dark J, Burns DE, Heller HC, Zucker I. 2002. The suprachiasmatic nucleus is essential for circadian body temperature rhythms in hibernating ground squirrels. *J. Neurosci.* 22:357–64

40. Grahn DA, Miller JD, Houng VS, Heller HC. 1994. Persistence of circadian rhythmicity in hibernating ground squirrels. *Am. J. Physiol. Regul. Integr. Comp. Physiol.* 266:R1251–58

41. Wollnik F, Schmidt B. 1995. Seasonal and daily rhythms of body temperature in the European hamster (*Cricetus cricetus*) under semi-natural conditions. *J. Comp Physiol. B* 165:171–82

42. Florant GL, Hill V, Ogilvie MD. 2000. Circadian rhythms of body temperature in laboratory and field marmots (*Marmota flaviventris*). In *Life in the Cold*, ed. G Heldmaier, M Klingenspor, pp. 223–31. Berlin: Springer-Verlag

43. Grahn DA, Miller JD, Heller HC. 1992. Ambient temperature (T_a) effects the amplitude, but not the tau, of circadian body temperature (T_b) rhythms in golden-mantled ground squirrels during hibernation. 3rd Meet. Soc. Res. Biol. Rhythms.101:(Abstr).

44. Waßmer T, Wollnik F. 1997. Timing of torpor bouts during hibernation in European hamsters (*Cricetus cricetus* L.). *J. Comp. Physiol. B* 167:270–79

45. Larkin JE, Franken P, Heller HC. 2002. Loss of circadian organization of sleep and wakefulness during hibernation. *Am. J. Physiol. Regul. Integr. Comp. Physiol.* 282:R1086–95

46. Ruby NF, Heller HC. 1996. Temperature sensitivity of the suprachiasmatic nucleus of ground squirrels and rats in vitro. *J. Biol. Rhythms* 11:127–37

47. Vanecek J, Jansky L, Illnerová H, Hoffmann K. 1984. Pineal melatonin in hibernating and aroused golden hamsters (*Mesocricetus auratus*). *Comp. Biochem. Physiol.* 77A:759–62

48. Darrow JM, Tamarkin L, Duncan MJ, Goldman BD. 1986. Pineal melatonin rhythms in female turkish hamsters: effects of photoperiod and hibernation. *Biol. Reprod.* 35:74–83

49. Vanecek J, Jansky L, Illnerová H, Hoffmann K. 1985. Arrest of the circadian pacemaker driving the pineal melatonin rhythm in hibernating golden hamsters,

Mesocricetus auratus. Comp. Biochem. Physiol. 80A:21–23

50. Larkin JE, Yellon SM, Zucker I. 2003. Melatonin production accompanies arousal from daily torpor in Siberian hamsters. *Physiol. Biochem. Zool.* 76:577–85

51. Hut RA, van der Zee EA, Jansen K, Gerkema MP, Daan S. 2002. Gradual reappearance of post-hibernation circadian rhythmicity correlates with numbers of vasopressin-containing neurons in the suprachiasmatic nuclei of European ground squirrels. *J. Comp. Physiol. B* 172:59–70

52. Kilduff TS, Sharp FR, Heller HC. 1982. [^{14}C] 2-deoxyglucose uptake in ground squirrel brain during hibernation. *J. Neurosci.* 2:143–57

53. Kilduff TS, Miller JD, Radeke CM, Sharp FR, Heller HC. 1990. ^{14}C2-deoxyglucose uptake in the ground squirrel brain during entrance to and arousal from hibernation. *J. Neurosci.* 10:2463–75

54. Kilduff TS, Radeke CM, Randall TL, Sharp FR, Heller HC. 1989. Suprachiasmatic nucleus: phase-dependent activation during the hibernation cycle. *Am. J. Physiol. Regul. Integr. Comp. Physiol.* 257:R605–12

55. Bitting L, Sutin EL, Watson FL, Leard LE, O'Hara BF, et al. 1994. c-fos mRNA increases in the ground squirrel suprachiasmatic nucleus during arousal from hibernation. *Neurosci. Lett.* 165:117–21

56. Dark J, Kilduff TS, Heller HC, Licht P, Zucker I. 1990. Suprachiasmatic nuclei influence hibernation rhythms of golden-mantled ground squirrels. *Brain Res.* 509:111–18

57. Gillette MU. 1991. SCN electrophysiology in vitro: rhythmic activity and endogenous clock properties. In *Suprachiasmatic Nucleus: the Mind's Clock*, ed. DC Klein,

RY Moore, SM Reppert, pp. 125–43. New York: Oxford Univ. Press

58. Miller JD, Cao VH, Heller HC. 1994. Thermal effects on neuronal activity in suprachiasmatic nuclei of hibernators and nonhibernators. *Am. J. Physiol. Regul. Integr. Comp. Physiol.* 266:R1259–66

59. Welsh DK, Logothetis DE, Meister M, Reppert SM. 1995. Individual neurons dissociated from rat suprachiasmatic nucleus express independently phased circadian firing rhythms. *Neuron* 14:697–706

60. Dudek FE, Kim YI, Bouskila Y. 1993. Electrophysiology of the suprachiasmatic nucleus: synaptic transmission, membrane properties, and neuronal synchronization. *J. Biol. Rhythms* 8:S33–7 (Suppl.)

61. Van Breukelen F, Martin SL. 2002. Invited review: molecular adaptations in mammalian hibernators: unique adaptations or generalized responses? *J. Appl. Physiol.* 92:2640–47

62. Knight JE, Narus EN, Martin SL, Jacobson A, Barnes BM, Boyer BB. 2000. mRNA stability and polysome loss in hibernating arctic ground squirrels (*Spermophilus parryii*). *Mol. Cell. Biol.* 20:6374–79

63. Bocharova LS, Gordon RY, Arkhlpov VI. 1992. Uridine uptake and RNA synthesis in the brain of torpid and awakened ground squirrels. *Comp. Biochem. Physiol.* 101B:189–92

64. Van Breukelen F, Martin SL. 2002. Reversible depression of transcription during hibernation. *J. Comp. Physiol. B* 172:355–61

65. Frerichs KU, Smith CB, Brenner M, DeGracia DJ, Krause GS, et al. 1998. Suppression of protein synthesis in brain during hibernation involves inhibition of protein initiation and elongation. *Proc. Natl. Acad. Sci. USA* 95:14511–16

Annu. Rev. Physiol. 2004. 66:291–313
doi: 10.1146/annurev.physiol.66.032802.154945
Copyright © 2004 by Annual Reviews. All rights reserved
First published online as a Review in Advance on July 23, 2003

ESTROGENS IN THE NERVOUS SYSTEM: Mechanisms and Nonreproductive Functions

Adriana Maggi, Paolo Ciana, Silvia Belcredito, and Elisabetta Vegeto

*University of Milan, Department of Pharmacological Sciences and Center of Excellence
on Neurodegenerative Diseases, Via Balzaretti 920129 Milan, Italy;
email: adriana.maggi@unimi.it, paolo.ciana@unimi.it, silvia.belcredito@unimi.it,
elisabetta.vegeto@unimi.it*

Key Words   estradiol, neuroprotection, brain, aging, transcription mechanism

■ **Abstract**   The past decade has witnessed a growing interest in estrogens and their
activity in the central nervous system, which was originally believed to be restricted to
the control of reproduction. It is now well accepted that estrogens modulate the activity
of all types of neural cells through a multiplicity of mechanisms. Estrogens, by binding
to two cognate receptors ERα and ERβ, may interact with selected promoters to initiate
the synthesis of target proteins. Alternatively, the hormone receptor complex may interfere with intracellular signaling at both cytoplasmic and nuclear levels. The generation
of cellular and animal models, combined with clinical and epidemiological studies,
has allowed us to appreciate the neurotrophic and neuroprotective effects of estrogens.
These findings are of major interest because estradiol might become an important therapeutic agent to maintain neural functions during aging and in selected neural diseases.

INTRODUCTION

The pioneering work of Gorski (1) and Arai (2) established that the central nervous
system (CNS) is a target for sex hormones. 17β-estradiol and, to some extent, several of its active metabolites were shown to be responsible for the differentiation of
sex-specific brain nuclei and for the control of functions and behaviors indispensable for reproduction. Only recently has the value of estradiol's influence on brain
functions, which are not related to reproductive activities, begun to be appreciated.
In the past decade, results of experimental and clinical studies have indicated that
estradiol may influence memory mechanisms, cognition, postural stability, fine
motor skills, mood, and affectivity. In addition, estradiol exerts a protective action
against neurodegeneration and brain injury. To achieve such a diversity of effects,
the hormone must target numerous brain nuclei and work through multifactorial
mechanisms. The aim of the present review is to discuss these newly described,
reproduction-unrelated activities of estrogens and the complexity of the underlying
mechanisms in mammalian adult brain.

0066-4278/04/0315-0291$14.00

291

HOW ESTRADIOL INFLUENCES NERVE CELL ACTIVITIES

Neuroscientists, using electrophysiological recordings, were the first to observe the rapid effects of estrogens on ion conductance, which led to the adoption of the terms nongenomic and genomic to discriminate cell responses to estrogens induced within seconds from those requiring hours to occur (3).

The long-term genomic effects have been thouroghly investigated in many estrogen target cells. It is now well established that they are mediated by the two estrogen receptors (ERs): ERα (4) and ERβ (5) [ESR1 and ESR2 according to the nomenclature proposed by the Nuclear Receptors Nomenclature Committee (5a)]. These two proteins, encoded by separate genes, are structurally and functionally distinct, and both are members of the superfamily of intracellular receptors. Upon binding the cognate ligand, ERs dimerize and bind to specific elements in the promoter of target genes [estrogen responsive elements (EREs)], where they interact with coregulators, integrators, and other proteins of the transcription machinery to regulate the synthesis of selected mRNAs and therefore the levels of estrogen-regulated proteins (6).

Less understood are the molecular mechanisms responsible for the nongenomic effects of estrogens. Recently, substantial evidence from a number of laboratories supports the concept that ERα and ERβ might also be responsible for the rapid effects of estradiol; indeed, in the cytoplasm, ER monomers can form multimeric complexes with other signaling proteins to induce rapid changes in the neuronal activity (7). On the other hand, engineered forms of ERs appear to associate with the cell membrane and cross-couple with the membrane receptor–signaling molecules to activate rapid hormonal responses (8). These receptor forms might be responsible for binding estradiol and for the ER-like immunoreactivity reported by several authors also noted in nonneural cell systems (9, 10). However, there is no clear evidence that ERα and ERβ associate with cell membrane in physiological settings. Alternatively, the existence of a specific form of membrane ER capable of rapid signaling has been suggested by a recent study (11), which needs to be confirmed by the isolation and biochemical characterization of this novel estradiol-binding protein encoded by a gene so far not predicted by genomic studies.

Finally, several splice variants of ERα and ERβ have been found in the rodent and human brain (12, 13). Typically, these variant mRNAs lack one or more of the coding exons or contain base insertions within an exon; most are translated into transcriptionally active proteins (14). The changes in expression of such variants during embryonic and postnatal development and their differential localization in adult brain nuclei (12) suggest that they contribute to the complexity of estrogen signaling in the brain.

Thus the intricacy of estradiol function in the CNS could be explained by its interaction with at least two forms of receptors and several splice variants; it is possible that all could associate with the cell membrane and reside in the cytoplasm and in the nucleus, where they could homodimerize, heterodimerize, or interact with other transcription factors or signaling molecules. These multiple combinations of interactions could generate cell-specific patterns of responses that might lead to the regulation of gene transcription in the nucleus, signal transduction pathways in the cytoplasm, or to receptors and ion channels in the cell membranes. Clearly, the effects in the cytoplasm or in the membranes could alter the genomic output of the target cells. Therefore, in the present review, to better focus on the direct target of hormone action, we provide a distinction between cytoplasmic and nuclear actions of the hormone receptor complex; furthermore, we refer to the nuclear activities of the complex as mediated by ERE or mediated by protein-protein interaction (Figure 1).

The Nucleus as a Primary Site of Action of Estrogens in Neural Cells

ERE-MEDIATED EFFECTS   The identification of genes regulated by estrogens via ERE interaction would be a major contribution toward understanding its action in neural cells. Numerous attempts have been hampered by the lack of model systems suitable for the isolation of the hormone-inducible mRNAs. The small number of

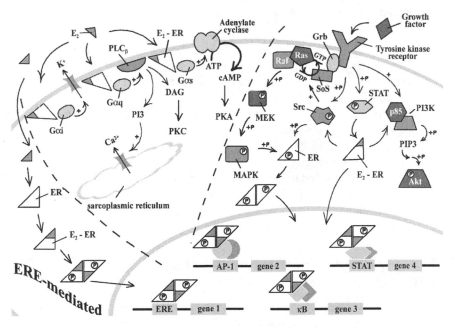

Figure 1 Estradiol, via interaction with ERs, modulates signaling in the cytoplasm and nucleus of target cells. The expression of estrogen receptor forms and other proteins interacting with the receptor, as well as the activation of other intracellular signaling pathways that cross-talk with ERs, can profoundly influence the overall response to estrogen in a given cell or tissue.

cells expressing ERs in primary nerve cells cultures and the absence of ER-positive tumors of neural origin have limited the application of technologies for the identification of differentially expressed genes. To overcome this shortcoming, several laboratories have stably transfected ERs into neuroblastoma (15) and pheocromocytoma cells (16) or into immortalized hippocampal neurons (17). These cell systems represent a valuable model for the initial identification of estrogen-regulated genes (18) and demonstrate the soundness of strategies based on the identification of estradiol-regulated genes for gaining insight into the activities of this hormone in nerve cells. For instance, the identification of BNIP-2 as a target of estradiol disclosed one of the mechanisms of estradiol anti-apoptotic activity (19–21).

Moreover, the recognition that activated ER targeted the ERE-containing promoter of a proliferative gene such as prothymosin-α revealed the role played by estrogens in neural stem cells of the adult mammals (S. Belcredito, C. Meda, G. Pollio, Y. Torrente & A. Maggi, in preparation) and led to the elucidation of the molecular bases of estrogen-induced differentiation of cells of neural origin (22).

In the forthcoming years, gene array technology combined with promoter analysis applied to the available model systems as well as to microdissected brain cells should enable the identification of the genes regulated via ERE-ER interaction in neural cells and provide a major step in the understanding of estrogen activities in the nervous system.

EFFECTS OF ESTROGEN MEDIATED BY PROTEIN-PROTEIN INTERACTION   Estrogen-coupled ERs may also interfere with the transcriptional activity of non-ERE-containing promoters by interacting with other transcription factors such as the members of the nuclear factor-kappa B (NF-κB) and activator protein 1 (AP1) families (23–25). Paradigmatic studies on these aspects have been conducted on the promoters of cytokines, such as interleukin-6 (IL-6) and tumor necrosis factor alpha (TNFα) (26–29), which show a cell type–specific interference of ERα and ERβ with NF-κB or AP1. These mechanisms, well described in transfected cells, are not clearly linked to specific functions in cells of neural origin. An exception is microglia, in which the functional interaction between estrogen-activated ERs and NF-κB could explain the strong anti-inflammatory action of the hormone recently reported (30).

The Cytoplasm as a Primary Site of Action of Estrogens in Neural Cells

In the 1970s, electrophysiological studies of neural cells demonstrated that estradiol stimulates the influx or the mobilization of intracellular Ca^{2+} within seconds, suggesting a novel action of the hormone different from the modulation of DNA transcription (31). Subsequently, several laboratories reported on estrogen effects too rapid to be mediated by activation of RNA and protein synthesis, which generated interest in a potential alternative signaling exerted through the interaction of estradiol with cytoplasmic or membrane proteins. It is now clear that these rapid responses are diversified and may involve several signaling pathways. Nonetheless, they all share the following characteristics: (*a*) The responses are too rapid to be generated by de novo protein synthesis (occur within minutes from the hormonal challenge); (*b*) they can be reproduced in the presence of inhibitors of RNA or protein synthesis; and (*c*) they can be reproduced in cells expressing a form of the estrogen receptor unable to bind the DNA or using an estrogen coupled to a membrane-impermeable molecule (e.g., estradiol-BSA).

A systematic classification of these effects is complicated by the fact that we have only a partial knowledge of the molecular interactions involved.

ESTROGENS AND G PROTEIN–SIGNALING CASCADES   It is now well accepted that estrogen modulates neural cell excitability or secretory functions by regulating the activity of ionophores or membrane receptors coupled to G proteins and their effectors (32–35). Experimental evidence demonstrates that, in neural as well as nonneural cells, estradiol may rapidly modulate the cytoplasmic content of cyclic adenosine monophosphate (cAMP; 36–40), cyclic guanosine monophosphate (cGMP; 41, 42), and inositol 1,4,5-triphosphate (PI3; 43). Furthermore, the hormone activates kinases such as protein kinase A (PKA; 38, 44, 45), protein kinase C (PKC; 46, 47), protein kinase B (PKB; 48, 49), Ca^{2+}/calmodulin-dependent kinase (CAMK; 50), and mitogen-activated protein kinase (MAPK; 51, 52) and turns on transcription factors such as cAMP-responsive element binding protein (CREB; 53, 54) and AP-1 (55). All these effects have been observed in a variety of neural cell systems: embryonic neurons in primary culture, tumors of neural origin, brain slices, and in nonneuronal cell systems. What still needs to be clarified is how the hormone interacts with so many differing molecules. A unifying hypothesis may be provided by the study of Razandi et al. (8), who showed that ERα and ERβ may physically interact with and activate various G protein α subunits, including Gαs and Gαq. This study, carried out in transfected cells, was strengthened by subsequent investigation in endothelial cells (56), here Gαi is modulated by the endogenous ERα. Thus depending on the target cell, the ERs complexed with estradiol bind and modulate the activity of a given G protein. In this way, the hormone may then interfere with functions independent of Gαs (e.g., increased cAMP), Gαq (e.g., changes phospholipase C$_\beta$ activity with alterations in Ca^{2+} and PI3 levels), or Gαi (e.g., Src signaling). This would explain the different effects reported above.

In addition, the accumulation of cyclic nucleotides could activate protein kinases, which in turn would decrease the activity of membrane receptors such as μ opioid and GABA$_B$ by phosphorylation (57) or regulate the nuclear translocation of transcription factors (53, 54).

It is important to underline that the association with G proteins cannot be the only mechanism by which the ER-bound estradiol influences cytoplasmic reactions. Indeed, the estradiol-ER complex has been shown to interact physically with other signaling molecules downstream of the G proteins such as Src or phosphatidylinositol 3-OH kinase (PI3-K), thus increasing its potential to interfere with the signaling from the membrane receptor.

ESTROGENS AND GROWTH FACTOR/CYTOKINES RECEPTOR SIGNALING   Estradiol is a well-known proliferation and differentiation factor in neural and nonneural cells. Extensive studies demonstrate the proliferative effects of estradiol in nonneural cells, but few authors have investigated the mechanisms of estrogen-dependent proliferation and differentiation of neural cells. Most likely, estrogens affect growth and differentiation of different cell systems by similar mechanisms. Indeed, genes regulated by estrogens via ERE and encoding mitotic proteins such as prothymosine α were first described in neural cell (18) and then found to be proliferation markers in breast tumor cells (58) or, conversely, cyclines originally found to be regulated by the hormone in the uterus (59) were then shown to be regulated in neural cells (60). Therefore, genomic effects are involved in the estrogen-dependent proliferation of neural cells. However, a recent investigation on neuroblastoma, where estrogens cause proliferation followed by differentiation, demonstrates that estrogen-induced proliferation is supported by the synthesis of proliferative proteins, such as prothymosin-α, whereas the ensuing estrogen-dependent differentiation depends on a cross-talk of hormone-bound ERα with the signaling molecule Stat3 (signal transducer and activator of transcription 3) (22). The interaction of ER-Stat3 potentiates Stat3 activity on the transcription of its target genes, thus determining the synthesis of the proteins indispensable for neural cell differentiation.

These data point to the relevance of nongenomic effects induced by estradiol to accomplish the full range of its action in neural cells; in the cytoplasm of target cells, estradiol bound to ER may interact with molecules involved in the signaling of growth factor and cytokine receptors. This is also true for nonneural cells, where estrogen-bound ER was shown to physically interact with and alter the action of Stat5 (61), Src (62, 63), and PI3-K (64).

Several authors reported that in neural cells estradiol may rapidly induce the phosphorylation of MAPK and Akt, the signaling molecules downstream of Src and PI3-K, respectively. These effects are physiologically relevant because they are involved in the neuroprotective effects of the hormone in primary cortical neurons (65), cerebral cortical explants (66), and neuroblastoma cell lines (67).

In addition, increased phosphorylation of Akt by PI3-K has been functionally associated with the estrogen-dependent regulation of dendritic spine growth in CA1 pyramidal neurons (68).

In cells in culture, activation of MAPK and Akt may occur in minutes (1 to 5) after estrogen treatment and can be blocked by the ER antagonist, ICI 182,780. As originally suggested by Migliaccio et al. (62), it is conceivable that these rapid changes are induced by estradiol binding to a cytoplasmic form of the ER, which by means of an adaptor protein (MNAR, modulator of nongenomic activity of estrogen receptor) interacts with the Src family of tyrosine kinases stimulating their enzymatic activity (69). This in turn attracts adaptor proteins and guanine nucleotide exchange factors, such as Grb2 and SoS, leading to the subsequent activation of p21[Ras] (70) and the downstream molecules MAPK kinase (MEK), extracellular signal-related kinase 1 and 2 (ERK1 and ERK2), and MAPK.

ERs were also shown to associate with p85, the regulatory subunit of PI3-K (64), with consequent activation of the PIP_3 (phosphatidylinositol 3,4,5-triphosphate)-dependent kinase (PDK)-mediated Akt cascade of molecular signaling. Interestingly, whereas both ERα and ERβ are capable of inducing the MAPK cascade (52), only ERα is able to interact with and activate PI3-K (64).

ERs Are the Oldest Among the Steroid Receptors: Does This Explain Their Promiscuity?

Why has estradiol acquired the capability to modulate the activity of so many signaling pathways in its target cells? A potential explanation may be found in studies on the evolution of vertebrate steroids receptors (71), in which the estrogen receptor is found to be the oldest member in the steroid receptor family. Considering that estradiol is the last to be synthesized in the metabolic pathway of steroids, it is tempting to speculate that the receptor evolved before the hormone itself, as an independent transcription factor. This hypothesis would predict that the ancestral unliganded estrogen receptor could be activated by some of the signaling molecules activated by membrane receptors. Indeed, it is now well accepted that at least ERα may be induced to activate the transcription of ERE-dependent promoters as a consequence of G protein or Ras activation (72, 73).

Therefore, the receptor, appearing before the advent of the hormone, might have served as a control integrator of cell metabolism by responding to the signaling from membrane receptors with genomic action and by interacting and regulating the activity of other cytoplasmic and nuclear signaling molecules. It is conceivable that during evolution, when estradiol was synthesized and began to control ER activity, the receptor maintained some of its original functions. This would explain why estradiol, and not the other steroids, may exert such a variety of diversified effects in so many different target tissues.

THE DEFINITION OF ESTRADIOL TARGETS IN THE NERVOUS SYSTEM: ARE ALL THE NERVE CELLS RESPONSIVE TO ESTRADIOL?

In addition to neurons, several cell types contribute to the correct functioning of the adult brain in mammals. Macroglia grant the trophic support indispensable for neuronal signaling through a network of axo-dendritic extensions, and astrocytes have a critical role in protection/survival of brain neurons. The defense against chemical and biological insults is actuated by microglia, the resident immune cells of the brain. Finally, neural stem cells have been described in the subventricular zone and hippocampus of adult rodents.

It has long been known that estradiol can accumulate in specific brain areas where it binds ERα and ERβ in neurons. Several localization studies have provided a detailed map of the neuronal expression of the two receptors in the rodent brain

(74, 75). ERβ immunoreactivity is primarily localized to cell nuclei within select regions of the brain including the olfactory bulb, cerebral cortex, septum, preoptic area, bed nucleus of the stria terminalis, amygdala, paraventricular hypothalamic nucleus, thalamus, ventral tegmental area, substantia nigra, dorsal raphe, locus coeruleus, and cerebellum. Extranuclear immunoreactivity is detected in several areas including fibers of the olfactory bulb, CA3 stratum lucidum, and CA1 stratum radiatum of the hippocampus and cerebellum. Although both receptors are generally expressed in a similar distribution through the brain, nuclear ERα immunoreactivity is the predominant subtype in the hippocampus, preoptic area, and most of the hypothalamus, whereas it is sparse or absent from the cerebral cortex and cerebellum. The expression of ERβ and ERα in the adult ovariectomized rodent brain and their significant presence in nonreproductive regions provide an anatomical framework for understanding the mechanism by which estrogen regulates specific neural systems in the mouse.

Recently, however, the presence of hormonally regulated ERs has also been found in glial cells (76, 77), in microglia (78), and in neural stem cells (S. Belcredito, C. Meda, G. Pollio, Y. Torrente & A. Maggi, in preparation; 79, 80).

Because of the intricacy of neural cell connections, the hormone can affect neuron behavior by acting on glia cells, or the effect of the hormone on microglia may have important repercussions on neurons or macroglia. The final effect of estrogens reaching the brain will be the summation of effects in the different brain regions and cell types leading to a high complexity of action difficult to understand in its completeness.

ESTRADIOL, A SEX HORMONE WITH PERVASIVE ACTIVITIES IN THE NERVOUS SYSTEM

Studies in Animals: What Have We Learned About the Nonreproductive Effects of Estrogens in the Brain?

In the 1980s, the description of a transcriptionally active ER in the mammalian hippocampus (80) challenged the traditional view of the role of estrogens in the brain as restricted to reproductive functions and suggested other domains of hormonal influence. Since then, the concept of estradiol and non-reproductive brain functions has been strengthened by (a) localization studies showing that ERs are expressed throughout the brain (74, 75); (b) behavioral studies showing sexual dimorphism (82) as well as estrogen-dependent alterations in the performance of cognitive and memory tests (83–89). For example acute rise of estradiol is associated with impaired learning on hippocampal-dependent tasks and avoidance memory tasks (90). In contrast, long-term estradiol replacement ameliorates deficits in performance of hippocampal tasks (91–93). (c) Anatomical studies showing that physiological fluctuations of estrogens or estrogen treatments are associated with dramatic changes in the plasticity of neurons in the hippocampus (94, 95). (d) The availability of ERα− and ERβ−selective knockout mice show the importance of ERβ in sustaining learning tasks and social activities (96, 97). Also ERα, which is essential for all sexually related functions, was shown to have a specific role in feeding and open field motility (98, 99). (*e*) Studies in animal and cellular models of neurodegeneration have shown a protective role of estrogens against neural cell apoptosis (20, 100–106).

On the basis of this large body of evidence, the concept that the adult brain remains highly plastic and responsive to the nonreproductive effects of estrogens is now well accepted. These neurotrophic and neuroprotective activities of the hormone provide the molecular basis for clinical and epidemiological studies in humans, which suggest the effects of the hormone in cognition and neurodegenerative disease, thus raising new perspectives for the therapeutic use of estrogens and its synthetic derivatives (Figure 2).

Clinical and Epidemiological Studies

Several studies point to a potential beneficial role of estrogens in selected brain pathologies. The role of estrogens in depression has been hypothesized for a long time on the basis that decreased plasma levels of these molecules during the menstrual phase, after parturition, or at menopause are often associated with mood disorders (107–109). Epidemiological studies have also suggested the involvement

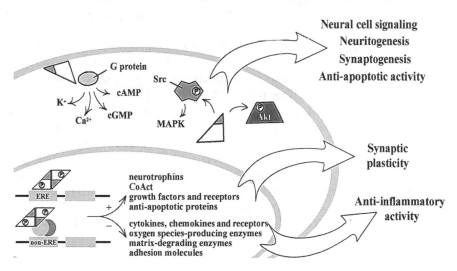

Figure 2   Cellular mediators and responses triggered by nuclear or cytoplasmic hormone-activated ERs in neuroprotection. In a variety of experimental models leading to neural cell death, estrogen functions as a protective factor. Nuclear effects on gene transcription, as well as cytoplasmic interactions with signaling molecules, result in diverse cellular responses leading to neuroprotection.

of hormonal factors in multiple sclerosis (MS): Increased levels of sex hormones during pregnancy were shown to associate with reduced severity of MS, whereas clinical symptoms are often exacerbated postpartum, when the sex hormone levels are greatly reduced (110, 111). Similarly, postmenopausal administration of estrogens may delay the onset of Alzheimer's Disease (AD) in elderly women and ameliorate Parkinson's tremors (112–114). The comparison between the different age profiles in the manifestation of psychotic episodes in the two sexes and the levels of circulating estradiol in normal and schizophrenic patients suggests a role for this hormone in schizophrenia (115). Growing evidence indicates that estradiol may influence the ability of the brain to limit the damage of ischemic insults resulting from cardiovascular disease and cerebrovascular stroke (116). In addition, clinical studies have established that estrogens influence diverse aspects of memory and cognition, movements, and fine motor skills in healthy brain as well (117).

The wealth of the above studies raises the interest in the use of estrogens as a potential powerful pharmacological tool for the prevention of diverse neurological disorders in humans, particularly in women who, with their increased life span, are now experiencing the absence of estrogens for about one third of their lives.

However, at present few prospective, adequately controlled randomized clinical trials are available that rigorously test the effects of estrogen replacement therapy on cognitive functions and on the delay of dementia. Unfortunately, very few of the studies were designed to differentiate whether estrogens can affect the onset of declining cognitive function and dementia in healthy women or can contribute to the slowing of a neurodegenerative process when it has already started. Cognitive tests performed in young women oophorectomized and treated with estrogen replacement therapy (ERT) or placebo unambiguously demonstrated that ERT improves specific components of cognition (such as immediate or delayed recall), but not spatial or visual memory (118–120).

Studies of large populations of peri- and postmenopausal women (121) found that ERT decreases the risk and severity of cognitive loss associated with AD, also after adjusting for education. Whether ERT improves memory and cognition in people suffering with AD is controversial, with laboratories reporting a modest positive effect of ERT and other studies reporting no difference between estrogen- and placebo-treated individuals. It is possible that the hormone plays a key role in maintaining normal cognition in a healthy brain, but cannot intervene when the neural circuits have already been irreparably damaged. Studies in cellular and animal models thus far seem to support this conclusion.

Estrogens and Neuroprotection: Potential Mechanisms

The neurotrophic and neuroprotective actions of estradiol might by themselves explain most of the beneficial effects on cognition, motility, and neural disorders. In several animal models, estrogens were shown to exert a trophic and neuroprotective effect and effectively prevent neuronal damage induced by neurotoxic stimuli. In ovariectomized rats, estradiol attenuates the extent of brain damage induced by permanent or transient brain ischemia (102, 103, 122), and reduces cortical lesions owing to glutamate exitotoxicity (100, 103) or secondary to status epilepticus (123), or autoimmune encephalomyelitis (124, 125).

Thus in a variety of models leading to neural death, estrogen has a protective function. Several hypotheses can be formulated to explain the mechanism of estrogen neuroprotective action: (*a*) The trophic activities of estrogens during the maturation of the central nervous system may continue to exist in the adult brain and ensure that neurons maintain the synaptic connections indispensable for neural signaling and survival; (*b*) estradiol may positively regulate the synthesis of proteins protecting neurons against apoptosis; (*c*) estradiol may influence inflammatory responses by controlling microglia reactivity and the vascular function; and (*d*) estradiol may induce proliferation of stem cells to replace neurons undergoing neurodegeneration.

INSIGHTS ON NEUROTROPHIC ACTIONS OF ESTRADIOL   The trophic actions of estrogens have been known since the pioneering work carried out by Toran-Allerand (126), who demonstrated that estradiol treatment of explant cultures of cerebral cortex and hypothalamus stimulates extensive neurite outgrowth. Since then, several studies in neuroblastoma cells or in dissociated neurons in culture showed that estradiol increases cell viability, differentiation, neurite outgrowth, and spine density (73, 127–129). Most likely these effects observed in vitro recapitulate the in vivo action of the hormone during the differentiation of the nervous system and in adult animals, where it affects the plasticity of cells such as hippocampal neurons. Thus estradiol might control the capability of neurons to extend neurites and to form synaptic connections with other cells via dendritic spines. Depending on the model system utilized, neurite extension or spine formation was observed rapidly and continued for several days after the exposure to the hormone. Altered changes in Ca^{2+} fluxes are indispensable for the increase in dendritic spines observed in hippocampal neurons. On the other hand, estrogens were shown to modulate the synthesis of growth factors such as nerve growth factor (NGF), brain-derived neurotrophic factor (BDNF), insulin-like growth factor-1 (IGF-1), transforming growth factor beta (TGFβ), and related receptors, TrkA and TrkB, in neurons and astroglia (130–135), and this de novo synthesis of growth factors is required for neurite formation. The existence of multiple mechanisms might be necessary for the generation of neurites and synaptic connections morphologically similar, but functionally distinct.

How estradiol induces neuronal cell plasticity in vivo is unclear. The observation that NGF stimulates astrocytes to function as substrates for axonal growth led to the hypothesis that an interplay between astrocytes and neurons is indispensable for estrogen effect on neuronal plasticity. This was subsequently demonstrated by a study in which estradiol, by eliciting the synthesis and release of NGF by neurons, could induce TrkA immunoreactive astrocytes to reorganize laminin into extracellular fibrillar arrays thus supporting axonal growth, structural plasticity, and synapse formation (136). On the other hand, estrogen may also cause neurite outgrowth in the absence of astrocytes, as was shown in estrogen-dependent neuritogenesis in a neuroblastoma model. In this case, neurite extension was initiated by estrogens via ER-dependent modulation of protein synthesis. Interestingly both ER isoforms contributed to neurite outgrowth even though the mechanism of action was different because ERα, but not ERβ, required a functionally active Rac 1b small G protein (137).

INSIGHTS ON ANTI-APOPTOTIC ACTIONS OF ESTRADIOL   Estradiol was shown to protect neural cells against death induced by a variety of stimuli including hypoxia, oxidative stress, excitotoxicity, glucose deprivation, exposure to amyloid β-peptide (Aβ), or other neurotoxic agents (101, 138, 139). This effect has been observed in primary neuronal cultures, tumor-derived neuronal cell lines, mixed neuron/astrocyte cell culture, and organotypic explants. The mechanism responsible for this effect is still controversial. Most of the studies in which physiological concentrations (in the namolar range) of estradiol were used support the view of an involvement of estrogen receptors and de novo protein synthesis; others using pharmacological concentrations of the hormone demonstrated its capability to directly alter the synthesis or metabolism of the toxic stimulus. Several laboratories (20, 140) demonstrated that physiological concentrations of estrogen enhance the survival of neuroblastoma cells only when transfected with ERα, thus proving the significance of this receptor in the mechanism of neuroprotection. Other studies have focused on the identification of genes modulated by estrogens in neural cells and have demonstrated that several of the known anti-apoptotic genes (Bcl-2, BclX$_L$) are transcriptionally activated by the hormone, whereas pro-apoptotic genes are down-modulated (BNIP2), thus indicating that estradiol might protect neural cells from apoptotic death by affecting the balance between apoptotic and anti-apoptotic genes (20, 141, 142).

ESTRADIOL AS AN ANTI-INFLAMMATORY AGENT   The beneficial effects of estradiol on demyelinating and neurodegenerative diseases are supported by the anti-inflammatory hypothesis, which is based upon experimental evidence on animal models and cellular systems, suggesting that estrogen suppresses the inflammatory process induced by several stimuli in the CNS. Indeed, several neural cell pathologies that benefit from hormone action, such as MS, AD, and ischemia, have a different etiology but share a strong neuroinflammatory reaction that sustains disease progression. Thus the anti-inflammatory hypothesis may represent a unifying explanation for hormone action. In addition, this mechanism is viewed as a pharmacological challenge for the prevention of these pathologies because anti-inflammatory drugs have been shown to be protective against the onset of AD (143).

The anti-inflammatory properties of female steroid hormones have been observed in vivo in animal models with CNS inflammation, i.e., experimental autoimmune encephalomyelitis (EAE), which is the animal model of MS, brain ischemia, globoid cells leukodistrophy, and experimental brain inflammation. These reports clearly demonstrated that estrogen inhibits the CNS inflammatory reaction. In fact, treatment with physiological doses of estrogen at the time of disease induction suppressed EAE by down-regulating inflammatory factors and inhibiting the migration of inflammatory cells into the CNS (107, 110, 125). Estrogen has also been shown to reduce leukocytes adhesion in the cerebral circulation of female rats subjected to transient forebrain ischemia and reperfusion (144). That this activity occurs through the specific activation of ERα (105) suggests that inflammation may also be a target for estrogen in brain vascular pathology. Similarly, it has been reported that infiltration of globoid cells in the white matter of saposin A$^{-/-}$ mice, an animal model of genetic leukodystrophy, and subsequent demyelination are strongly prevented in female animals that continually experienced pregnancy or that were treated with estrogens (145). Finally, the activation of brain macrophages induced by the bacterial endotoxin lipopolysaccharide has been blocked by administration of physiological concentrations of hormone that specifically acts through the activation of ERα (78). These animal studies provide some hints to the specific molecules that are regulated by estrogen; these include cytokines, chemokines, and their receptors (145, 146); apolipoprotein E (147); and other modulators of leukocyte migration such as matrix metalloproteinase-9 (MMP-9) and complement receptor-3 (CR3) (78).

Among the cells that are targeted by hormonal anti-inflammatory activity, microglia, the resident macrophage cells of the CNS, play a central role in brain homeostasis because they are the first cellular sensors of any local mechanical or chemical injury and immediately react by mounting the inflammatory response (148). A state of chronic microglia activation has been observed in association with the neurodegenerative process in several CNS diseases (148, 149), and it is currently believed that the production of oxygen reactive species, which inflict an oxidative stress to neighboring cells, is the major cause of neuronal cell death. Studies conducted in primary cultures of microglia cells (30) and in microglial cell lines (150) showed that nanomolar concentrations of estradiol decrease the activation of these cells toward an inflammatory stimulus. This effect has been reconciled with the blockade of the expression of pro-inflammatory factors, such as iNOS, PGE$_2$ and MMP-9, and inhibition of superoxide release and phagocytic activity. In addition, clearance of Aβ, which is impaired in AD, is enhanced by estrogen in microglia derived from the human cortex (151), a mechanism that further explains the protective role of hormone mediated by its interaction with microglia.

On the other hand, the anti-inflammatory effect of hormone has been also ascribed to its influence on vascular cells of the CNS, endothelium, and circulating leukocytes that participate in the inflammatory reaction associated with specific neuronal diseases (152, 153). The vascular activity of hormone has also been observed in the periphery and ascribed to a reduction in the expression of adhesion molecules in endothelial and leukocytic cells (154, 155).

As to the molecular mechanism of hormone action, it is clear that a reduced expression of pro-inflammatory proteins and an increased synthesis of protective factors underlie the anti-inflammatory activity of hormone. Because no EREs were found in the promoter region of these genes, it is hypothesized that the ER interfers with the signaling pathways activated by an inflammatory stimulus, such as NF-κB or AP-1, and therefore regulates their transcriptional activity.

ESTRADIOL AND NEURAL STEM CELLS   We now know that neural stem cells express both ERα and ERβ and respond to the treatment with estradiol with increased proliferation. It has been shown (80) that stem cell proliferation, at least in the hippocampus, is most active in sexually mature female mice compared with male mice, and a natural fluctuation in stem cell proliferation was also noted: Females produced more cells at proestrus (when plasma estradiol levels are highest) compared with estrus and diestrus. These cells acquired neuronal characteristics. However, further examination demonstrated that these cells have a very short survival time. Thus these neurons generated under the influence of estrogens do not seem to be of importance for the positive effects of the hormone on cognition and neuroprotection. More recently Belcredito et al. (S. Belcredito, C. Meda, G. Pollio, Y. Torrente & A. Maggi, in preparation) have demonstrated that the estrogen-induced proliferation of neural stem cells is related more to the reproductive functions by showing that the newly formed cells from the subventricular zone migrate to the accessory olfactory bulb where they exert an important function in the acquisition of odorant cues necessary for initiating sexual behavior. Therefore, the short half-life of these cells is necessary to ensure a receptive sexual behavior only after the follicle matures and ovulation occurs. These observations would rule out a role for stem cells in the described neuroprotective actions of estrogens.

CONCLUSIONS

The past decade has revealed that estradiol exerts a widespread and important regulatory function in the healthy mammalian brain through a multiplicity of mechanisms. We have also learned that estradiol may have a neuroprotective role by limiting the extent of neurodegeneration induced by injury and neural diseases. Thus estrogens might be a key element in maintaining normal brain functions particularly in aging women. We have generated a number of animal models such as ERα and ERβ knockout mice to study the contribution of each receptor subtype to estrogen functions. We have also generated ERE-luciferase reporter mice to assess the state of estrogen receptor transcriptional activity in physiopathological states. These models should be fully exploited in future studies devoted to the in-depth analysis of the mechanisms underlying the neuroprotective effects of estrogens to develop estrogen-like compounds as therapeutic agents against cognitive decline and neurodegeneration.

ACKNOWLEDGMENTS

Work in the authors' laboratory is supported by the European Community, Italian Ministry of Research, Italian National Institute for Health, and Italian Association for Cancer Research.

The *Annual Review of Physiology* is online at http://physiol.annualreviews.org

LITERATURE CITED

1. Gorski RA. 1971. Gonadal hormones and the perinatal development of neuroendocrine function. In *Frontiers in Neuroendocrinology*, ed. L Martini, WF Ganong, pp. 237–290. New York: Oxford Univ. Press

2. Arai Y. 2000. Sexual differentiation of the brain: a historical review. In *Sexual Differentiation of the Brain*, ed. A Matsumoto, pp. 1–9. Boca Raton: CRC Press

3. McEwen BS. 1981. Neural gonadal steroid actions. *Science* 211(4488):1303–11

4. Green S, Walter P, Kumar V, Krust A, Bornert JM, et al. 1986. Human oestrogen receptor cDNA: sequence, expression and homology to v-erb-A. *Nature* 320(6058):134–39

5. Kuiper GG, Enmark E, Pelto-Huikko M, Nilsson S, Gustafsson JA. 1996. Cloning of a novel receptor expressed in rat prostate and ovary. *Proc. Natl. Acad. Sci. USA* 93(12):5925–30

5a. Nuclear Receptor Nomenclature Committee. 1999. A unified nomenclature system for the nuclear receptor superfamily. *Cell* 97:161–63

6. Ciana P, Vegeto E, Beato M, Chambon P, Gustaffsson J-Å, et al. 2002. Looking at nuclear receptors from the heights of Erice. *EMBO Rep.* 3(2):125–29

7. Levin ER. 2001. Cell localization, physiology, and non-genomic actions of estrogen receptors. *J. Appl. Physiol.* 91:1860–67

8. Razandi M, Pedram A, Greene GL, Levin ER. 1999. Cell membrane and nuclear estrogen receptors (ERs) originate from a single transcript: studies of ERalpha and ERbeta expressed in Chinese hamster ovary cells. *Mol. Endocrinol.* 13(2):307–19

9. Pietras RJ, Szego CM. 1977. Specific binding sites for oestrogen at the outer surface of isolated endometrial cells. *Nature* 265(5589): 69–72

10. Pappas TC, Gametchu B, Watson CS. 1995. Membrane estrogen receptors identified by multiple antibody labeling and impeded-ligand binding. *FASEB J.* 9(5):404–10

11. Toran-Allerand CD, Guan X, MacLusky NJ, Horvath TL, Diano S, et al. 2002. ER-X: a novel, plasma membrane-associated, putative estrogen receptor that is regulated during development and after ischemic brain injury. *J. Neurosci.* 22(19):8391–401

12. Price RH Jr, Lorenzon N, Handa RJ. 2000. Differential expression of estrogen receptor beta splice variants in rat brain: identification and characterization of a novel variant missing exon 4. *Brain Res. Mol. Brain Res.* 80:260–68

13. Shupnik MA. 2002. Oestrogen receptors, receptor variants and oestrogen actions in the hypothalamic-pituitary axis. *J. Neuroendocrinol.* 14(2):85–94

14. Bolling A, Miksicek RJ. 2000. An estrogen-receptor-alpha splicing variant mediates both positive and negative effects on gene transcription. *Mol. Endocrinol.* 14:634–49

15. Ma ZQ, Spreafico E, Pollio G, Santagati S, Conti E, et al. 1993. Activated estrogen receptor mediates growth arrest and differentiation of a neuroblastoma cell line. *Proc. Natl. Acad. Sci. USA* 90(8):3740–44

16. Lustig RH, Hua P, Yu W, Ahmad FJ, Baas PW. 1994. An in vitro model for the effects of estrogen on neurons employing estrogen receptor-transfected PC12 cells. *J. Neurosci.* 14(6):3945–57

17. Fitzpatrick JL, Mize AL, Wade CB, Harris JA, Shapiro RA, Dorsa DM. 2002. Estrogen-mediated neuroprotection against beta-amyloid toxicity requires expression of estrogen receptor alpha or beta

and activation of the MAPK pathway. *J. Neurochem.* 82(3):674–82

18. Garnier M, Di Lorenzo D, Albertini A, Maggi A. 1997. Identification of estrogen-responsive genes in neuroblastoma SK-ER3 cells. *J. Neurosci.* 17(12):4591–99

19. Vegeto E, Pollio G, Pellicciari C, Maggi A. 1999. Estrogen and progesterone induction of survival of monoblastoid cells undergoing TNF-alpha-induced apoptosis. *FASEB J.* 13(8):793–803

20. Meda C, Vegeto E, Pollio G, Ciana P, Patrone C, et al. 2000. Oestrogen prevention of neural cell death correlates with decreased expression of mRNA for the pro-apoptotic protein nip-2. *J. Neuroendocrinol.* 12(11):1051–59

21. Brusadelli A, Sialino H, Piepoli T, Pollio G, Maggi A. 2000. Expression of the estrogen-regulated gene Nip2 during rat brain maturation. *Int. J. Dev. Neurosci.* 18(2–3):317–20

22. Ciana P, Ghisletti S, Mussi P, Eberini I, Vegeto E, Maggi A. 2003. ERalpha, a molecular switch converting TGFalpha-mediated proliferation into differentiation in neuroblastoma cells. *J. Biol. Chem.* In press

23. Paech K, Webb P, Kuiper GG, Nilsson S, Gustafsson J, et al. 1997. Differential ligand activation of estrogen receptors ERalpha and ERbeta at AP1 sites. *Science* 277(5331):1508–10

24. McKay LI, Cidlowski JA. 1999. Molecular control of immune/inflammatory responses: interactions between nuclear factor-kappa B and steroid receptor-signaling pathways. *Endocr. Rev.* 20(4): 435–59

25. Pfeilschifter J, Koditz R, Pfohl M, Schatz H. 2002. Changes in proinflammatory cytokine activity after menopause. *Endocr. Rev.* 2002. 23(1):90–119

26. Stein B, Yang MX. 1995. Repression of the interleukin-6 promoter by estrogen receptor is mediated by NF-kappa B and C/EBP beta. *Mol. Cell. Biol.* 15(9):4971–79

27. Galien R, Garcia T. 1997. Estrogen receptor impairs interleukin-6 expression by preventing protein binding on the NF-kappaB. *Nucleic Acids Res.* 25(12):2424–29

28. An J, Ribeiro RC, Webb P, Gustafsson JA, Kushner PJ, et al. 1999. Estradiol repression of tumor necrosis factor-alpha transcription requires estrogen receptor activation function-2 and is enhanced by coactivators. *Proc. Natl. Acad. Sci. USA* 96(26):15161–66

29. Tzagarakis-Foster C, Geleziunas R, Lomri A, An J, Leitman DC. 2002. Estradiol represses human T-cell leukemia virus type 1 Tax activation of tumor necrosis factor-alpha gene transcription. *J. Biol. Chem.* 277(47):44772–77

30. Vegeto E, Bonincontro C, Pollio G, Sala A, Viappiani S, et al. 2001. Estrogen prevents the lipopolysaccharide-induced inflammatory response in microglia. *J. Neurosci.* 21(6):1809–18

31. Kelly MJ, Moss RL, Dudley CA. 1976. Differential sensitivity of preoptic-septal neurons to microelectrophoresed estrogen during the estrous cycle. *Brain Res.* 114(1):152–57

32. Mermelstein PG, Becker JB, Surmeier DJ. 1996. Estradiol reduces calcium currents in rat neostriatal neurons via membrane receptors. *J. Neurosci.* 16:595–604

33. Wetzel CH, Hermann B, Behl C, Pestel E, Rammes G, et al. 1998. Functional antagonism of gonadal steroids at the 5-hydroxytryptamine type 3 receptor. *Mol. Endocrinol.* 12(9):1441–51

34. Valverde MA, Rojas P, Amigo J, Cosmelli D, Orio P, et al. 1999. Acute activation of Maxi-K channels (hSlo) by estradiol binding to the beta subunit. *Science* 285(5435):1929–31

35. Gu Q, Moss RL. 1996. 17beta estradiol potentiates kainate-induced currents via activation of the cAMP cascade. *J. Neurosci.* 16:3620–29

36. Minami T, Oomura Y, Nabekura J, Fukuda A. 1990. 17beta-estradiol depolarization

of hypothalamic neurons is mediated by cAMP. *Brain Res.* 519:301–7

37. Aronica SM, Kraus WL, Katzenellenbogen BS. 1994. Estrogen action via the cAMP signaling pathway: stimulation of adenylate cyclase and cAMP-regulated gene transcription. *Proc. Natl. Acad. Sci. USA* 91(18):8517–21

38. Watters JJ, Dorsa DM. 1998. Transcriptional effects of estrogen on neuronal neurotensin gene expression involve cAMP/protein kinase A-dependent signaling mechanisms. *J. Neurosci.* 18(17):6672–80

39. Kelly MJ, Levin ER. 2001. Rapid actions of plasma membrane estrogen receptors. *Trends Endocrinol. Metab.* 12(4):152–56

40. Maus M, Bertrand P, Drouva S, Rasolonjanahary R, Kordon C, et al. 1989. Differential modulation of D1 and D2 dopamine-sensitive adenylate cyclases by 17 beta-estradiol in cultured striatal neurons and anterior pituitary cells. *J. Neurochem.* 52(2):410–18

41. Russell KS, Haynes MP, Sinha D, Clerisme E, Bender JR. 2000. Human vascular endothelial cells contain membrane binding sites for estradiol, which mediate rapid intracellular signaling. *Proc. Natl. Acad. Sci. USA* 97(11):5930–35

42. Palmon SC, Williams MJ, Littleton-Kearney MT, Traystman RJ, Kosk-Kosicka D, Hurn PD. 1998. Estrogen increases cGMP in selected brain regions and in cerebral microvessels. *J. Cereb. Blood Flow Metab.* 18(11):1248–52

43. Le Mellay V, Grosse B, Lieberherr M. 1997. Phospholipase C beta and membrane action of calcitriol and estradiol. *J. Biol. Chem.* 272(18):11902–7

44. Lagrange AH, Ronnekleiv OK, Kelly MJ. 1997. Modulation of G protein-coupled receptors by an estrogen receptor that activates protein kinase A. *Mol. Pharmacol.* 51(4):605–12

45. Coleman KM, Dutertre M, El-Gharbawy A, Rowan BG, Weigel NL, Smith CL. 2003. Mechanistic differences in the activation of estrogen receptor-alpha (ERalpha)- and ERbeta-dependent gene expression by cAMP signaling pathway(s). *J. Biol. Chem.* 278(15):12834–45

46. Colin IM, Jameson JL. 1998. Estradiol sensitization of rat pituitary cells to gonadotropin-releasing hormone: involvement of protein kinase C- and calcium-dependent signaling pathways. *Endocrinology* 139(9):3796–802

47. Ansonoff MA, Etgen AM. 1998. Estradiol elevates protein kinase C catalytic activity in the preoptic area of female rats. *Endocrinology* 139(7):3050–56

48. Zhang L, Rubinow DR, Xaing G, Li BS, Chang YH, et al. 2001. Estrogen protects against beta-amyloid-induced neurotoxicity in rat hippocampal neurons by activation of Akt. *NeuroReport* 12(9):1919–23

49. Znamensky V, Akama KT, McEwen BS, Milner TA. 2003. Estrogen levels regulate the subcellular distribution of phosphorylated Akt in hippocampal CA1 dendrites. *J. Neurosci.* 23(6):2340–47

50. Hayashi T, Ishikawa T, Yamada K, Kuzuya M, Naito M, et al. 1994. Biphasic effect of estrogen on neuronal constitutive nitric oxide synthase via $Ca(2+)$-calmodulin dependent mechanism. *Biochem. Biophys. Res. Commun.* 203(2):1013–19

51. Razandi M, Pedram A, Levin ER. 2000. Estrogen signals to the preservation of endothelial cell form and function. *J. Biol. Chem.* 275(49):38540–46

52. Wade CB, Robinson S, Shapiro RA, Dorsa DM. 2001. Estrogen receptor (ER)alpha and ERbeta exhibit unique pharmacologic properties when coupled to activation of the mitogen-activated protein kinase pathway. *Endocrinology* 142(6):2336–42

53. Gu G, Rojo AA, Zee MC, Yu J, Simerly RB. 1996. Hormonal regulation of CREB phosphorylation in the anteroventral periventricular nucleus. *J. Neurosci.* 16(9):3035–44

54. Honda K, Shimohama S, Sawada H, Kihara T, Nakamizo T, et al. 2001.

Nongenomic antiapoptotic signal transduction by estrogen in cultured cortical neurons. *J. Neurosci. Res.* 64(5):466–75

55. Jezierski MK, Sturm A, Scarborough MM, Sohrabji F. 2000. NGF stimulation increases JNK phosphorylation and reduces caspase-3 activity in the olfactory bulb of estrogen-replaced animals. *Endocrinology* 142:2401–3

56. Wyckoff MH, Chambliss KL, Mineo C, Yuhanna IS, Mendelsohn ME, et al. 2001. Plasma membrane estrogen receptors are coupled to endothelial nitric-oxide synthase through Galpha(i). *J. Biol. Chem.* 276(29):27071–76

57. Kelly MJ, Qiu J, Wagner EJ, Ronnekleiv OK. 2003. Rapid effects of estrogen on G protein-coupled receptor activation of potassium channels in the central nervous system (CNS). *J. Steroid Biochem. Mol. Biol.* 83:187–93

58. Dominguez F, Magdalena C, Cancio E, Roson E, Paredes J, et al. 1993. Tissue concentrations of prothymosin alpha: a novel proliferation index of primary breast cancer. *Eur. J. Cancer* 29A(6):893–97

59. Altucci L, Addeo R, Cicatiello L, Germano D, Pacilio C, et al. 1997. Estrogen induces early and timed activation of cyclin-dependent kinases 4, 5, and 6 and increases cyclin messenger ribonucleic acid expression in rat uterus. *Endocrinology* 138(3):978–84

60. Brannvall K, Korhonen L, Lindholm D. 2002. Estrogen-receptor-dependent regulation of neural stem cell proliferation and differentiation. *Mol. Cell. Neurosci.* 21(3):512–20

61. Björnström L, Sjöberg M. 2002. Signal transducers and activators of transcription as downstream targets of nongenomic estrogen receptors actions. *Mol. Endocrinol.* 16:2202–14

62. Migliaccio A, Di Domenico M, Castoria G, de Falco A, Bontempo P, et al. 1996. Tyrosine kinase/p21ras/MAP-kinase pathway activation by estradiol-receptor complex in MCF-7 cells. *EMBO J.* 15(6):1292–300

63. Castoria G, Barone MV, Di Domenico M, Bilancio A, Ametrano D, et al. 1999. Nontranscriptional action of oestradiol and progestin triggers DNA synthesis. *EMBO J.* 18(9):2500–10

64. Simoncini T, Hafezi-Moghadam A, Brazil DP, Ley K, Chin WW, Liao JK. 2000. Interaction of oestrogen receptor with the regulatory subunit of phosphatidylinositol-3-OH kinase. *Nature* 407(6803):538–41

65. Singer CA, Figueroa-Maso XA, Batchelor RH, Dorsa DM. 1999. The mitogen-activated protein kinase pathway mediates estrogen neuroprotection after glutamate toxicity in primary cortical neurons. *J. Neurosci.* 19:2455–63

66. Singh M, Setalo G Jr, Guan X, Warren M, Toran-Allerand CD. 1999. Estrogen-induced activation of mitogen-activated protein kinase in cerebral cortical explants: convergence of estrogen and neurotrophin signaling pathways. *J. Neurosci.* 19(4):1179–88

67. Manthey D, Heck S, Engert S, Behl C. 2001. Estrogen induces a rapid secretion of amyloid beta precursor protein via the mitogen-activated protein kinase pathway. *Eur. J. Biochem.* 268(15):4285–91

68. Znamensky V, Akama KT, McEwen BS, Milner TA. 2003. Estrogen levels regulate the subcellular distribution of phosphorylated Akt in hippocampal CA1 dendrites. *J. Neurosci.* 23(6):2340–47

69. Wong CW, McNally C, Nickbarg E, Komm BS, Cheskis BJ. 2002. Estrogen receptor-interacting protein that modulates its nongenomic activity-crosstalk with Src/Erk phosphorylation cascade. *Proc. Natl. Acad. Sci. USA* 99(23):14783–88

70. Egan SE, Giddings BW, Brooks MW, Buday L, Sizeland AM, Weinberg RA. 1993. Association of Sos Ras exchange protein with Grb2 is implicated in tyrosine

kinase signal transduction and transformation. *Nature* 363(6424):45–51

71. Thornton JW. 2001. Evolution of vertebrate steroid receptors from an ancestral estrogen receptor by ligand exploitation and serial genome expansions. *Proc. Natl. Acad. Sci. USA* 98(10):5671–76

72. Kato S, Endoh H, Masuhiro Y, Kitamoto T, Uchiyama S, et al. 1995. Activation of the estrogen receptor through phosphorylation by mitogen-activated protein kinase. *Science* 270(5241):1491–94

73. Ma ZQ, Santagati S, Patrone C, Pollio G, Vegeto E, Maggi A. 1994. Insulin-like growth factors activate estrogen receptor to control the growth and differentiation of the human neuroblastoma cell line SK-ER3. *Mol. Endocrinol* 8:910–18

74. Couse JF, Lindzey J, Grandien K, Gustafsson JA, Korach KS. 1997. Tissue distribution and quantitative analysis of estrogen receptor-alpha (ERalpha) and estrogen receptor-beta (ERbeta) messenger ribonucleic acid in the wild-type and ERalpha-knockout mouse. *Endocrinology* 138(11):4613–21

75. Shughrue PJ, Merchenthaler I. 2001. Distribution of estrogen receptor beta immunoreactivity in the rat central nervous system. *J. Comp. Neurol.* 436(1):64–81

76. Jung-Testas I, Renoir M, Bugnard H, Greene GL, Baulieu EE. 1992. Demonstration of steroid hormone receptors and steroid action in primary cultures of rat glial cells. *J. Steroid Biochem. Mol. Biol.* 41(3-8):621–31

77. Santagati S, Melcangi RC, Celotti F, Martini L, Maggi A. 1994. Estrogen receptor is expressed in different types of glial cells in culture. *J. Neurochem.* 63(6):2058–64

78. Vegeto E, Belcredito S, Etteri S, Ghisletti S, Brusadelli A, et al. 2003. Estrogen receptor-alpha mediates the brain anti-inflammatory activity of estradiol. *Proc. Natl. Acad. Sci. USA.* In press

79. Brannvall K, Korhonen L, Lindholm D. 2002. Estrogen-receptor-dependent regulation of neural stem cell proliferation and differentiation. *Mol. Cell. Neurosci.* 21(3):512–20

80. Tanapat P, Hastings NB, Reeves AJ, Gould E. 1999. Estrogen stimulates a transient increase in the number of new neurons in the dentate gyrus of the adult female rat. *J. Neurosci.* 19(14):5792–801

81. Maggi A, Susanna L, Bettini E, Mantero G, Zucchi I. 1989. Hippocampus: a target for estrogen action in mammalian brain. *Mol. Endocrinol.* 3(7):1165–70

82. Fugger HN, Cunningham SG, Rissman EF, Foster TC. 1998. Sex differences in the activational effect of ERalpha on spatial learning. *Horm. Behav.* 34(2):163–70

83. Luine V, Rodriguez M. 1994. Effects of estradiol on radial arm maze performance of young and aged rats. *Behav. Neural. Biol.* 62(3):230–36

84. Singh M, Meyer EM, Millard WJ, Simpkins JW. 1994. Ovarian steroid deprivation results in a reversible learning impairment and compromised cholinergic function in female Sprague-Dawley rats. *Brain Res.* 644(2):305–12

85. Frick KM, Fernandez SM, Bulinski SC. 2002. Estrogen replacement improves spatial reference memory and increases hippocampal synaptophysin in aged female mice. *Neuroscience* 115(2):547–58

86. Lund TD, West TW, Tian LY, Bu LH, Simmons DL, et al. 2001. Visual spatial memory is enhanced in female rats (but inhibited in males) by dietary soy phytoestrogens. *BMC Neurosci.* 2(1):20

87. Sandstrom NJ, Williams CL. 2001. Memory retention is modulated by acute estradiol and progesterone replacement. *Behav. Neurosci.* 115(2):384–93

88. Markham JA, Pych JC, Juraska JM. 2002. Ovarian hormone replacement to aged ovariectomized female rats benefits acquisition of the Morris water maze. *Horm. Behav.* 42(3):284–93

89. Packard MG, Teather LA. 1997. Intrahippocampal estradiol infusion enhances memory in ovariectomized rats. *NeuroReport* 8(14):3009–13

90. Galea LA, Kavaliers M, Ossenkopp KP, Hampson E. 1995. Gonadal hormone levels and spatial learning performance in the Morris water maze in male and female meadow voles, *Microtus pennsylvanicus. Horm. Behav.* 29(1):106–25

91. Fader AJ, Hendricson AW, Dohanich GP. 1998. Estrogen improves performance of reinforced T-maze alternation and prevents the amnestic effects of scopolamine administered systemically or intrahippocampally. *Neurobiol. Learn Mem.* 69(3):225–40

92. Bimonte HA, Denenberg VH. 1999. Estradiol facilitates performance as working memory load increases. *Psychoneuroendocrinology* 24(2):161–73

93. Gibbs RB. 1999. Estrogen replacement enhances acquisition of a spatial memory task and reduces deficits associated with hippocampal muscarinic receptor inhibition. *Horm. Behav.* 36(3):222–33

94. Foy MR, Teyler TJ. 1983. 17-alpha-estradiol and 17-beta-estradiol in hippocampus. *Brain Res. Bull.* 10(6):735–39

95. Woolley CS, Gould E, Frankfurt M, McEwen BS. 1990. Naturally occurring fluctuation in dendritic spine density on adult hippocampal pyramidal neurons. *J. Neurosci.* 10(12):4035–39

96. Rissman EF, Heck AL, Leonard JE, Shupnik MA, Gustafsson JA. 2002. Disruption of estrogen receptor beta gene impairs spatial learning in female mice. *Proc. Natl. Acad. Sci. USA* 99(6):3996–4001

97. Ogawa S, Lubahn DB, Korach KS, Pfaff DW. 1997. Behavioral effects of estrogen receptor gene disruption in male mice. *Proc. Natl. Acad. Sci. USA* 94(4):1476–81

98. Geary N, Asarian L, Korach KS, Pfaff DW, Ogawa S. 2001. Deficits in E2-dependent control of feeding, weight gain, and cholecystokinin satiation in ER-alpha null mice. *Endocrinology* 142(11):4751–57

99. Nomura M, Durbak L, Chan J, Smithies O, Gustafsson JA, et al. 2002. Genotype/age interactions on aggressive behavior in gonadally intact estrogen receptor beta knockout (betaERKO) male mice. *Horm. Behav.* 41(3):288–96

100. Behl C, Skutella T, Lezoualc'h F, Post A, Widmann M, et al. 1997. Neuroprotection against oxidative stress by estrogens: structure-activity relationship. *Mol. Pharmacol.* 51(4):535–41

101. Goodman Y, Bruce AJ, Cheng B, Mattson MP. 1996. Estrogens attenuate and corticosterone exacerbates excitotoxicity, oxidative injury, and amyloid β-peptide toxicity in hippocampal neurons. *J. Neurochem.* 66(5):1836–44

102. Gridley KE, Green PS, Simpkins JW. 1997. Low concentrations of estradiol reduce beta-amyloid (25–35)-induced toxicity, lipid peroxidation and glucose utilization in human SK-N-SH neuroblastoma cells. *Brain Res.* 778(1):158–65

103. Wang Q, Santizo R, Baughman VL, Pelligrino DA, Iadecola C. 1999. Estrogen provides neuroprotection in transient forebrain ischemia through perfusion-independent mechanisms in rats. *Stroke* 30(3):630–37

104. Dubal DB, Zhu H, Yu J, Rau SW, Shughrue PJ, et al. 2001. Estrogen receptor alpha, not beta, is a critical link in estradiol-mediated protection against brain injury. *Proc. Natl. Acad. Sci. USA* 98(4):1952–57

105. Weaver CE Jr, Park-Chung M, Gibbs TT, Farb DH. 1997. 17beta-Estradiol protects against NMDA-induced excitotoxicity by direct inhibition of NMDA receptors. *Brain Res.* 761(2):338–41

106. Bebo BF Jr, Fyfe-Johnson A, Adlard K, Beam AG, Vandenbark AA, Offner H. 2001. Low-dose estrogen therapy ameliorates experimental autoimmune encephalomyelitis in two different inbred mouse strains. *J. Immunol.* 166(3):2080–89

107. Weissman MM, Bland R, Joyce RP, Newman S, Wells JE, Wittchen HU. 1993. Sex differences in rates of depression:

cross–national perspectives. *J. Affect. Disord.* 29(2–3):77–84

108. Bloch M, Schmidt PJ, Danaceau M, Murphy J, Nieman L, Rubinow RD. 2000. Effects of gonadal steroids in women with a history of postpartum depression. *Am. J. Psych.* 157(6):924–30

109. Gregoire AJ, Kumar R, Everitt B, Henderson AF, Studd JW. 1996. Transdermal oestrogen for treatment of severe postnatal depression. *Lancet* 347(9006):930–33

110. Jansson L, Holmdahl R. 1998. Estrogen-mediated immunosuppression in autoimmune diseases. *Inflamm. Res.* 47(7):290–301

111. Confavreux C, Hutchinson M, Hours MM, Cortinovis-Tourniaire P, Moreau T. 1998. Rate of pregnancy-related relapse in multiple sclerosis. Pregnancy in Multiple Sclerosis Group. *N. Engl. J. Med.* 339(5):285–91

112. Henderson VW. 1997. The epidemiology of estrogen replacement therapy and Alzheimer's disease. *Neurology* 48(5 Suppl. 7):S27–35

113. Yaffe K, Sawaya G, Lieberburg I, Grady D. 1998. Estrogen therapy in postmenopausal women: effects on cognitive function and dementia. *JAMA* 279(9):688–95

114. Tsang KL, Ho SL, Lo SK. 2000. Estrogen improves motor disability in Parkinsonian postmenopausal women with motor fluctuations. *Neurology* 54(12):2292–98

115. Huber TJ, Rollnik J, Wilhelms J, von zur Muhlen A, Emrich HM, Schneider U. 2001. Estradiol levels in psychotic disorders. *Psychoneuroendocrinology* 26(1):27–35.

116. Paganini-Hill A. 1995. Estrogen replacement therapy and stroke. *Prog. Cardiovasc. Dis.* 38(3):223–42

117. Jarvik LF. 1975. Human intelligence: sex differences. *Acta Genet. Med. Gamellol.* 24(3–4):189–211

118. Kampen DL, Sherwin BB. 1994. Estrogen use and verbal memory in healthy postmenopausal women. *Obstet. Gynecol.* 83(6):979–83

119. Sherwin BB. 1997. Estrogen effects on cognition in menopausal women. *Neurology* 48(5 Suppl. 7):S21–26

120. Robinson D, Friedman L, Marcus R, Tinklenberg J, Yesavage J. 1994. Estrogen replacement therapy and memory in older women. *J. Am. Geriatr. Soc.* 42(9):919–22

121. Kawas C, Resnick S, Morrison A, Brookmeyer R, Corrada M, et al. 1997. A prospective study of estrogen replacement therapy and the risk of developing Alzheimer's disease: the Baltimore Longitudinal Study of Aging. *Neurology* 48(6):1517–21

122. Carswell HV, Dominiczak AF, Macrae IM. 2000. Estrogen status affects sensitivity to focal cerebral ischemia in stroke-prone spontaneously hypertensive rats. *Am. J. Physiol. Heart Circ. Physiol.* 278(1):H290–94

123. Veliskova J, Velisek L, Galanopoulou AS, Sperber EF. 2000. Neuroprotective effects of estrogens on hippocampal cells in adult female rats after status epilepticus. *Epilepsia* 41(Suppl. 6):S30–35

124. Liu HY, Buenafe AC, Matejuk A, Ito A, Zamora A, et al. 2002. Estrogen inhibition of EAE involves effects on dendritic cell function. *J. Neurosci. Res.* 70(2):238–48

125. Ito A, Bebo BF Jr, Matejuk A, Zamora A, Silverman M, et al. 2001. Estrogen treatment down-regulates TNF-alpha production and reduces the severity of experimental autoimmune encephalomyelitis in cytokine knockout mice. *J. Immunol.* 167(1):542–52

126. Toran-Allerand CD, Gerlach J, McEwen B. 1980. Autoradiographic localization of [^{3}H]estradiol related to steroid responsiveness in cultures of the newborn mouse hypothalamus and preoptic area. *Brain Res.* 184:517–22

127. Blanco G, Diaz H, Carrer HF, Beauge L. 1990. Differentiation of rat hippocampal neurons induced by estrogen in vitro: effects on neuritogenesis and Na, K-ATPase activity. *J. Neurosci. Res.* 27(1):47–54

128. Brinton RD, Tran J, Proffitt P, Montoya M. 1997. 17 beta-Estradiol enhances the outgrowth and survival of neocortical neurons in culture. *Neurochem. Res.* 22(11):1339–51

129. Murphy DD, Cole NB, Greenberger V, Segal M. 1998. Estradiol increases dendritic spine density by reducing GABA neurotransmission in hippocampal neurons. *J. Neurosci.* 18(7):2550–59

130. Perez-Polo JR, Hall K, Livingston K, Westlund K. 1977. Steroid induction of nerve growth factor synthesis in cell culture. *Life Sci.* 21(10):1535–44

131. Sohrabji F, Miranda RC, Toran-Allerand CD. 1994. Estrogen differentially regulates estrogen and nerve growth factor receptor mRNAs in adult sensory neurons. *J. Neurosci.* 14(2):459–71

132. Lara HE, McDonald JK, Ojeda SR. 1990. Involvement of nerve growth factor in female sexual development. *Endocrinology* 126(1):364–75

133. Agrati P, Garnier M, Patrone C, Pollio G, Santagati S, et al. 1997. SK-ER3 neuroblastoma cells as a model for the study of estrogen influence on neural cells. *Brain Res. Bull.* 44(4):519–23

134. Sohrabji F, Miranda RC, Toran-Allerand CD. 1995. Identification of a putative estrogen response element in the gene encoding brain-derived neurotrophic factor. *Proc. Natl. Acad. Sci. USA* 92(24):11110–14

135. Cardona-Gomez GP, Chowen JA, Garcia-Segura LM. 2000. Estradiol and progesterone regulate the expression of insulin-like growth factor-I receptor and insulin-like growth factor binding protein-2 in the hypothalamus of adult female rats. *J. Neurobiol.* 43(3):269–81

136. Kawaja MD, Gage FH. 1991. Reactive astrocytes are substrates for the growth of adult CNS axons in the presence of elevated levels of nerve growth factor. *Neuron* 7(6):1019–30

137. Patrone C, Pollio G, Vegeto E, Enmark E, de Curtis I, et al. 2000. Estradiol induces differential neuronal phenotypes by activating estrogen receptor alpha or beta. *Endocrinology* 141(5):1839–45

138. Behl C, Widmann M, Trapp T, Holsboer F. 1995. 17-beta estradiol protects neurons from oxidative stress-induced cell death in vitro. *Biochem. Biophys. Res. Commun.* 216:473–82

139. Regan RF, Guo Y. 1997. Estrogens attenuate neuronal injury due to hemoglobin, chemical hypoxia and excitatory amino acids in murine cortical cultures. *Brain Res.* 764:133–40

140. Gollapudi L, Oblinger MM. 1999. Stable transfection of PC12 cells with estrogen receptor (ERalpha): protective effects of estrogen on cell survival after serum deprivation. *J. Neurosci. Res.* 56(1):99–108

141. Patrone C, Andersson S, Korhonen L, Lindholm D. 1999. Estrogen receptor-dependent regulation of sensory neuron survival in developing dorsal root ganglion. *Proc. Natl. Acad. Sci. USA.* 96(19):10905–10

142. Pike CJ. 1999. Estrogen modulates neuronal Bcl-Xl expression and β-amyloid-induced apoptosis: relevance to Alzheimer's disease. *J. Neurochem.* 72(4):1552–63

143. McGeer PL, McGeer EG. 1996. Anti-inflammatory drugs in the fight against Alzheimer's disease. *Ann. NY Acad. Sci.* 777:213–20

144. Santizo RA, Anderson S, Ye S, Koenig HM, Pelligrino DA. 2000. Effects of estrogen on leukocyte adhesion after transient forebrain ischemia. *Stroke* 31(9):2231–35

145. Matsuda J, Vanier MT, Saito Y, Suzuki K, Suzuki K. 2001. Dramatic phenotypic improvement during pregnancy in a genetic leukodystrophy: estrogen appears to be a critical factor. *Hum. Mol. Genet.* 10(23):2709–15

146. Matejuk A, Adlard K, Zamora A, Silverman M, Vandenbark AA, Offner H. 2001. 17 beta-estradiol inhibits cytokine, chemokine, and chemokine receptor mRNA expression in the central

nervous system of female mice with experimental autoimmune encephalomyelitis. *J. Neurosci. Res.* 65(6):529–42

147. Horsburgh K, Macrae IM, Carswell H. 2002. Estrogen is neuroprotective via an apolipoprotein E-dependent mechanism in a mouse model of global ischemia. *J. Cereb. Blood Flow Metab.* 22(10):1189–95

148. Gonzalez-Scarano F, Baltuch G. 1999. Microglia as mediators of inflammatory and degenerative diseases. *Annu. Rev. Neurosci.* 22:219–40

149. Kalaria RN. 1999. Microglia and Alzheimer's disease. *Curr. Opin. Hematol.* 6(1):15–24

150. Bruce-Keller AJ, Keeling JL, Keller JN, Huang FF, Camondola S, Mattson MP. 2000. Antiinflammatory effects of estrogen on microglial activation. *Endocrinology* 141(10):3646–56

151. Li R, Shen Y, Yang LB, Lue LF, Finch C, Rogers J. 2000. Estrogen enhances uptake of amyloid beta-protein by microglia derived from the human cortex. *J. Neurochem.* 75(4):1447–54

152. Gilmore W, Weiner LP, Correale J. 1997. Effect of estradiol on cytokine secretion by proteolipid protein-specific T cell clones isolated from multiple sclerosis patients and normal control subjects. *J. Immunol.* 158(1):446–51

153. Correale J, Arias M, Gilmore W. 1998. Steroid hormone regulation of cytokine secretion by proteolipid protein-specific CD4+ T cell clones isolated from multiple sclerosis patients and normal control subjects. *J. Immunol.* 161(7):3365–74

154. Nathan L, Pervin S, Singh R, Rosenfeld M, Chaudhuri G. 1999. Estradiol inhibits leukocyte adhesion and transendothelial migration in rabbits in vivo: possible mechanisms for gender differences in atherosclerosis. *Circ. Res.* 85(4):377–85

155. Caulin-Glaser T, Farrell WJ, Pfau SE, Zaret B, Bunger K, et al. 1998. Modulation of circulating cellular adhesion molecules in postmenopausal women with coronary artery disease. *J. Am. Coll. Cardiol.* 31(7):1555–60

Annu. Rev. Physiol. 2004. 66:315–60
doi: 10.1146/annurev.physiol.66.032802.155556
First published online as a Review in Advance on July 23, 2003

THE ROLE OF COREPRESSORS IN TRANSCRIPTIONAL REGULATION BY NUCLEAR HORMONE RECEPTORS

Martin L. Privalsky

Section of Microbiology, Division of Biological Sciences, University of California, Davis, California 95616; email: mlprivalsky@ucdavis.edu

Key Words SMRT, N-CoR, histone deacetylases, nuclear receptors, transcription

■ **Abstract** Nuclear receptors (also known as nuclear hormone receptors) are hormone-regulated transcription factors that control many important physiological and developmental processes in animals and humans. Defects in receptor function result in disease. The diverse biological roles of these receptors reflect their surprisingly versatile transcriptional properties, with many receptors possessing the ability to both repress and activate target gene expression. These bipolar transcriptional properties are mediated through the interactions of the receptors with two distinct classes of auxiliary proteins: corepressors and coactivators. This review focuses on how corepressors work together with nuclear receptors to repress gene transcription in the normal organism and on the aberrations in this process that lead to neoplasia and endocrine disorders. The actions of coactivators and the contributions of the same corepressors to the functions of nonreceptor transcription factors are also touched on.

OVERVIEW OF NUCLEAR RECEPTORS

Several hundred million years ago, a novel process arose among the metazoans: a transcription factor learned to bind to, and be regulated by, small hydrophobic molecules present in the cell milieu (1, 2). In this fashion, the concentration of these hydrophobic molecules could be sensed and linked to changes in gene expression. Most likely, this primordial "nuclear receptor" recognized an intracellular lipid and regulated genes involved in the synthesis and degradation of its lipid ligand. Through gene duplication and divergence, this concept was soon elaborated on, resulting in the synthesis of a series of nuclear receptors that could respond to a plethora of small hydrophobic ligands and control a corresponding assortment of target genes. A further evolutionary twist to this scheme occurred when this intracellular mechanism was adapted to allow for extracellular signaling, permitting a ligand released from one cell to diffuse into and regulate the activity of a nuclear receptor in a second cell (1, 2).

0066-4278/04/0315-0315$14.00

From these humble origins arose the foundations of modern endocrine signaling and the hundreds of nuclear receptors that today play key roles in the control of metazoan differentiation, development, reproduction, and homeostasis (3–7). The nuclear receptor family includes receptors for classic endocrine hormones, such as estrogens, androgens, glucocorticoids, T3/T4 thyroid hormones, retinoids, and vitamin D_3 (3–7). The nuclear receptor family also contains members that harken back to their primordial ancestors and respond to intermediates in lipid metabolism rather than endocrine hormones per se; examples of the latter include the peroxisome proliferator-activated receptors (PPARs), liver X receptor (LXR), and farnesoid X receptors (FXRs) (8–11). Finally, there are orphan receptors, such as the chicken ovalbumin upstream regulatory sequence transcription factors (COUP-TFs), for which no ligands have been identified (12, 13). Some orphan receptors may have a cognate ligand that is only awaiting discovery; others may have forgotten how to bind to ligand over evolutionary time and are now regulated by alterations in their expression level by covalent modifications or by interactions with other proteins.[1]

Exempting the orphan receptors, the generic nuclear receptor operates as a single-step signal transducer, transmitting an input (the binding of a small chemical ligand) into an output (a change in the transcription rate of specific target genes) (3–7). To do so, the nuclear receptor must (*a*) recognize specific DNA sequences, denoted hormone response elements (HREs), in or near the target gene, (*b*) bind to the hormone or lipid ligand, and, ultimately (*c*) mediate the molecular events that alter the rate of transcription of the target promoter.

Excellent reviews have been published describing how nuclear receptors recognize their DNA-binding sites (3, 14–16). Briefly, most nuclear receptors bind to DNA either as homodimers or as heterodimers with other members of the nuclear receptor family (Figure 1); a few can also recognize DNA as receptor monomers, or as oligomers (6, 17–27). A zinc-finger motif in each receptor monomer recognizes a six to eight nucleotide sequence on the target DNA,

---

[1]Common Acronyms: AF, activation function; APL, acute promyelocytic leukemia; AR, androgen receptor; ATRA, all-*trans* retinoic acid; CORNR box, corepressor-nuclear receptor interaction motif; COUP-TF, chicken ovalbumin upstream regulatory sequence transcription factor; Dax-1, dosage-sensitive sex reversal adrenal hypoplasia critical region on X chromosome, gene 1; DBD, DNA-binding domain; ER, estrogen receptor; FXR, farnesoid X receptor; GPS2, G protein pathway suppressor 2; GR, glucocorticoid receptor; HAT, histone acetyltransferase; HBD, hormone-binding domain; HDAC, histone deacetylase; HRE, hormone response element; LXR, liver X receptor; N-CoR, nuclear hormone receptor-corepressor; ND, nuclear receptor interaction domain; PPAR, peroxisome proliferator-activated receptor; PR, progesterone receptor; RAR, retinoic acid receptor; RD, repression domain; RXR, retinoid X receptor; SHP, short heterodimer partner; SHRM, selective hormone response modulator; SMRT, silencing mediator of retinoic acid and thyroid hormone receptors; T3R, thyroid hormone receptor; TAF, TBP-associated factor; TBL-1, transducin-like protein 1; TBLR-1, TBL-1-related protein 1; TBP, TATA-binding protein; VDR, vitamin D_3 receptor.

A. Generic nuclear receptor

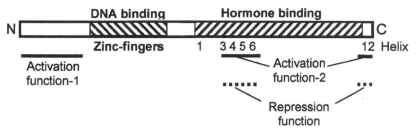

B. Nuclear receptor binding to DNA

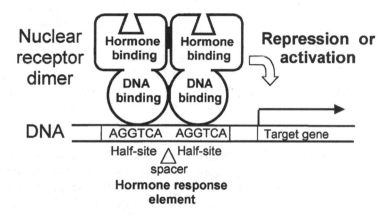

Figure 1   Nuclear receptor structure and mode of DNA binding. (*A*) Structure of a generic nuclear receptor. The primary structure of a typical nuclear receptor is presented schematically from N to C terminus. The locations of the zinc-finger/DNA-binding domain, the hormone-binding domain, and the protein domains involved in transcriptional activation (Activation Function-1 and -2) or repression (Repression Function) are indicated. (*B*) Nuclear receptor dimer bound to DNA. A representation is provided of a nuclear receptor dimer bound to a HRE upstream of a target gene. Each receptor is represented as two domains: DNA binding and hormone binding. The HRE shown here is composed of two AGGTCA half sites in a direct repeat separated by a spacer; different half-site sequences, spacings, and orientations select for the binding of different nuclear receptors.

denoted a half site (Figure 1) (6, 17–27). To recruit a receptor dimer, a functional HRE must contain two half sites arranged in a specific orientation and spacing (Figure 1*B*). For instance, thyroid hormone receptors (T3Rs) preferentially bind to two AGGTCA half sites oriented as direct repeats with a four-base spacer (DR-4s); retinoic acid receptors (RARs) bind to the same AGGTCA half sites, but oriented as a DR-5; estrogen receptors bind to AGGTCA half sites oriented as an inverted repeat with a three-base spacer (INV-3); and androgen receptors (ARs) recognize an INV-3 orientation containing AGAACA half sites (6, 17–27). This precis is necessarily a simplification: HREs in nature often contain half sites that diverge in sequence and topology from these prototypic elements. Nuclear receptors also interact with nonreceptor transcription factors, such as c-Jun and c-Fos, either to tether the receptor indirectly to the DNA or to form complexes in which both the receptor and nonreceptor contribute specific DNA contacts (28–30). These interactions can result in complex, combinatorial modes of transcriptional regulation.

Much elegant work has also been devoted to understanding how nuclear receptors recognize their hormone ligands. The operative entity in this regard is a C-terminal hormone-binding domain (HBD; Figure 1B) composed of 12 α-helical domains twisted into a triple-layered sandwich (31–36). Hormone ligand is virtually engulfed by this polypeptide sandwich, with the hormone serving as a hydrophobic core on which the receptor completes its own folding. Due to this close approximation between ligand and receptor, different hormones can invoke different conformations in the receptor (37–40). These ligand-driven receptor conformations produce distinct biological consequences. For example, ligand agonists produce receptor conformations that favor transcriptional activation, whereas ligand antagonists produce receptor conformations that favor transcriptional repression (37–40).

What molecular mechanisms allow the nuclear receptors to regulate transcription once bound to target genes? In particular, how can so many of these receptors function as both transcriptional repressors and transcriptional activators? The explanation is in the company these receptors keep. Nuclear receptors operate by recruiting an array of auxiliary polypeptides, denoted corepressors and coactivators, and it is these auxiliary proteins that mediate the molecular events that result in transcriptional repression or activation (41–53). The molecular basis of this transcriptional drama is described in greater detail below.

GENERIC INTRODUCTION TO TRANSCRIPTIONAL REGULATION BY THE NUCLEAR RECEPTORS

Nuclear receptors possess subdomains that are necessary for transcriptional regulation, yet can be distinguished from sequences required for DNA binding or hormone recognition (3, 4, 6, 7, 15). These transcriptional regulatory domains have several aliases (activation domains, activation function domains, tau domains, repression domains, silencing domains) but a common mode of operation; they represent docking surfaces on the receptor through which corepressors and coactivators are recruited (41–53). Almost all nuclear receptors possess a hormone-dependent activation domain in the receptor HBD (Figure 1A); this activation function (AF)-2 receptor domain forms a docking surface for coactivators and is assembled in three-dimensional space from portions of HBD helices 3/5/6 and 12 (37, 39, 54–56). Intriguingly, this same surface overlaps an important corepressor binding site (Figure 1A), and a yin yang mechanism operates by which hormone-induced changes in HBD helix 12 alternatively favor recruitment of one or the other class of coregulator (see below). Many, but not all, nuclear receptors possess additional activation domains within their N-terminal domains (denoted AF-1 sequences; Figure 1A) that bind coactivators, as well as less-characterized corepressor and coactivator interaction surfaces within their DNA-binding domains (41–53).

By exploiting these various docking surfaces as bait in two-hybrid or in coprecipitation experiments, researchers have compiled an increasingly thick dossier of coactivators and corepressors (41–53). The coactivators thus identified can be broadly categorized into four groups (Figure 2, see color insert): (a) histone covalent modifiers, such as the p160 family, CARM, and CBP/p300, that possess (or recruit) enzymatic activities able to modify the chromatin template, including acetylases and methylases; (b) ATP-dependent chromatin-remodeling complexes, such as the Swi/Snf family, that alter the higher-order structure and position of nucleosomes; (c) components of the mediator complex, such as TRAP/DRIP, that interact with the general transcriptional machinery to assist in assembly of the preinitiation complex; and (d) coactivators with unknown functions. These pioneer coactivators may possess cryptic histone modification or remodeling properties, may work by tethering other coactivators, or may operate through as yet unsuspected molecular mechanisms. Different coactivators can be preferentially recruited by various nuclear receptors; alternatively, a single nuclear receptor may recruit more than one coactivator, either simultaneously or sequentially (57–63).

The first corepressors identified for nuclear receptors were SMRT (silencing mediator of retinoid and thyroid hormone receptors) and N-CoR (nuclear hormone receptor-corepressor) (64–68). SMRT and N-CoR are paralogs of one another and function in similar fashions (51). A corepressor bearing some of the characteristics of SMRT, denoted SMRTER, has been identified in *Drosophila*, but may not be a true ortholog of its vertebrate namesake (69). Both N-CoR and SMRT make direct contact with the repression domain found in many nuclear receptor HBDs and nucleate the assembly of a larger array of additional corepressor polypeptides (Figure 3, see color insert) (41, 42, 45–47, 51, 53, 70–73). This array of SMRT/N-CoR and subsidiary polypeptides operates in a manner opposite of many coactivators. Whereas many coactivators possess or recruit histone acetyltransferase activities (HATs), SMRT and N-CoR recruit histone deacetylases (HDACs) (41, 42, 45–47, 51, 53, 70, 73). Similarly, whereas the mediator coactivator complex contacts and stabilizes the preinitiation complex, SMRT and N-CoR make inhibitory contacts that interfere with transcriptional initiation (44, 52, 74, 75).

It is worth noting that these same corepressors and coactivators also mediate transcriptional regulation by a variety of nonreceptor transcription factors. CBP/p300 for example, serves as a coactivator for many diverse transcription factors, including the CRE-binding protein, NF-κB and Jun/Fos (76–79). The SMRT/N-CoR corepressor complex has been implicated in transcriptional repression by NF-κB (80–84), serum response factor (84), AP-1 proteins (84), Smads (85), RBP-Jκ/CBF-1 (86–88), c-Myb (89), PLZF (90–94), BCL-6 (93–96), Pbx/Hox proteins (97–101), ETO-1 and ETO-2 (102–106), aryl hydrocarbon receptor (107), and MyoD (108) among others.

SETTING THE STAGE FOR REPRESSION: SMRT AND N-COR

The first corepressors identified for nuclear receptors were SMRT (also known as the T3R-associated cofactor, TRAC) and its close paralog, N-CoR (also known as the receptor interacting protein 13, RIP13) (64–68, 109). SMRT and N-CoR are encoded by two distinct loci but share a common molecular architecture and approximately 45% amino acid identity (65, 110, 111); additional forms of SMRT and N-CoR are generated by alternative mRNA splicing, and include SMRTτ, RIP13a, and RIP13Δ1 (Figure 4) (discussed in more detail below). Both SMRT and N-CoR can be conceptually divided into a N-terminal portion having three to four distinct transcriptional repression (or silencing) domains (RDs), and a C-terminal portion composed of two or three nuclear receptor interaction domains (NDs) (Figure 4) (64–67, 75, 110–118). The RDs are docking surfaces that recruit additional components of the corepressor complex, including histone deacetylases (HDACs), transducin-like protein 1 (TBL-1), G protein pathway suppressor 2 (GPS2), and (possibly) mSin3 and its cohorts (Figures 3 and 5) (41, 42, 45–47, 51, 53, 70, 73). SMRT and N-CoR can therefore be viewed as a protein platform that tethers to the nuclear receptors through its C-terminal domain, and on which the remainder of the corepressor complex assembles. SMRT and N-CoR are the principal point of contact between this larger corepressor complex and the nuclear receptors, and molecular events that regulate the interaction of SMRT or N-CoR with the nuclear receptor generally control the recruitment or release of the entire corepressor complex (119, 120).

SMRT contains two NDs, whereas N-CoR contains three (one of which is removed by mRNA splicing in RIP13Δ1) (Figure 4) (74, 121–123). Different nuclear receptors exhibit different affinities for the different NDs (e.g., 113, 115–117, 121, 124, 125). RARs, for example, preferentially interact with the ND2 domain in SMRT, RXRs can interact with the ND1 domains on either SMRT or N-CoR, and T3Rs interact with the ND3 and ND2 domains on N-CoR or (with lower affinity) with the ND2 and ND1 domains of SMRT (66, 115, 116, 121–123). Most nuclear receptors assemble on DNA as protein dimers, and each receptor in the dimer has been proposed to bind to a different ND on a single corepressor; as a result, each receptor dimer is believed to recruit one SMRT or N-CoR molecule (112, 116, 117, 126, 127). Receptor homodimers and heterodimers can display different N-CoR- and SMRT-binding properties. For example, T3Rs homodimers,

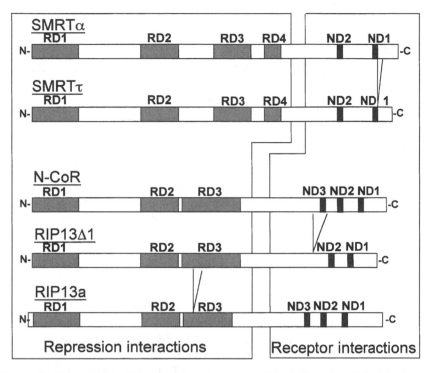

Figure 4   Comparison of the primary structures of SMRT, N-CoR, and their isoforms derived by alternative mRNA splicing. The overall architectures of the SMRT and N-CoR proteins are shown as horizontal bars, from N to C termini. The locations of the nuclear receptor interaction domains (NDs) and repression domains (RDs) are shown. A variety of labels and numbering strategies have been used in the literature for these various domains; a simplified scheme is employed here. SMRTα is the longest known version of SMRT; a shorter version, denoted SMRTτ, has a 47-amino-acid domain proximal to ND1 removed by mRNA splicing, as indicated. N-CoR has at least two additional spliced variants, denoted RIP13Δ1 and RIP13a, that create deletions in their ND3 and RD3 domains, respectively, as shown.

but not T3R/RXR heterodimers, efficiently recruit SMRT and N-CoR when bound to DNA response elements and may be important mediators of T3R repression (126, 128).

Each ND in SMRT and N-CoR contains a CORNR box (or L/I-X-X-I/V-I) motif that forms the core of the contact surface between the corepressor and nuclear receptor (40, 129–132). The CORNR motif forms an extended α-helical domain that docks into a complementary groove formed by helices 4/5/6 in the HBD of the nuclear receptor (Figures 5 and 6, see color insert) (40, 129–132). Differences in the CORNR box motifs, in adjacent amino acids, and in the sequence of the helix 3/5/6 regions of the nuclear receptors contribute to the different affinities

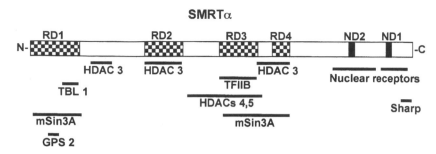

Figure 5 Protein interaction domains on SMRT. The primary structure of SMRT is depicted as a horizontal bar from N to C termini. The repression and nuclear receptor interaction domains are shown as in Figure 4. In addition, the approximate sites of interaction of other components of the corepressor complex are shown below the SMRT schematic and are further described in the text.

of various receptors for the different corepressor NDs (40, 121, 122, 129–132). Receptor HBD helix 12, located at the very C terminus, also plays an important but indirect role in the receptor/corepressor interaction. The helix 12 of most nuclear receptors can form an extended conformation in the absence of hormone that allows access of corepressor to its docking site on helices 3/5/6 (Figure 6) (40, 129–132). The binding of hormone agonist can reorient this helix 12 to a sequestered position that occludes the corepressor docking surface, resulting in release of corepressor (Figure 6) (40, 119, 129, 132–139). Interestingly, this same reorientation of helix 12 forms a new binding surface, made up of portions of helix 3/5/6 and 12, that can recruit the L-X-X-L-L motifs found in many coactivators (Figure 6) (37, 39, 54, 55). Helix 12 therefore operates as a hormone-operated molecular toggle switch that regulates the equilibrium between corepressor recruitment and coactivator recruitment (119, 120).

Which nuclear receptors can recruit SMRT or N-CoR, and which cannot? Initial studies suggested that SMRT/N-CoR corepressor binding might be limited to nuclear receptors known to repress in the unliganded state, such as T3Rs, RARs, and the COUP-TF I/II orphan receptors (64–66, 140, 141). However, further study identified a panel of additional nuclear receptors, including PPARs, VDRs, Nur 77, and hepatic nuclear factor 4α (HNF4α), that also can interact with SMRT and/or N-CoR in the absence of hormone, but had not previously been known to function as transcriptional repressors (142–150). These additional receptors are likely to function as repressors in specific contexts, perhaps on certain promoters, or in specific cell types. A third subgroup of nuclear receptors display low or no corepressor binding in the absence of hormone but gain an increased ability to bind corepressors in the presence of hormone antagonists: these include estrogen receptors (ERs), glucocorticoid receptors (GRs), progesterone receptors (PRs), and androgen receptors (ARs) (151–159). In these

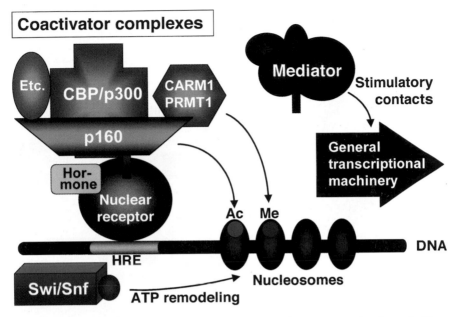

Figure 2 Coactivator complexes assembled on a nuclear receptor. A schematic illustration is shown highlighting several coactivator complexes that can be recruited by a typical nuclear receptor. The receptor is depicted as bound to a hormone response element (HRE) and to hormone agonist. A complex of the p160, CBP/p300, CARM1, and PRMT1 coactivators is shown as bound to the receptor; the ability of these coactivators to acetylate and methylate nucleosomal histones is indicated with arrows. Also shown is a mediator complex, which can interact with the same docking site on the nuclear receptor as does the p160 coactivators, and a Swi/Snf ATP-dependent chromatin remodeling complex. The composition of these different coactivator complexes and the interactions among the various components and the nuclear receptor may differ for different receptors or in different contexts.

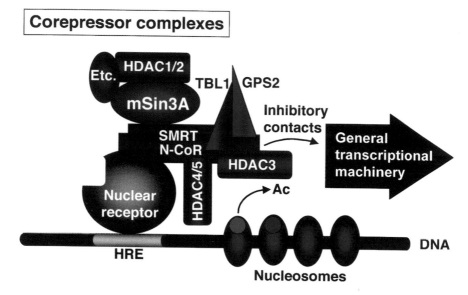

Figure 3   Corepressor complexes assembled on a nuclear receptor. A schematic is shown illustrating the SMRT/N-CoR corepressor complexes that may assemble on nuclear receptors. The receptor is depicted as bound to a hormone response element (HRE), but in the absence of hormone. The receptor is shown interacting with SMRT/N-CoR, which can recruit, in turn, additional components of the corepressor complex, such as the HDACs, mSin3, TBL-1, and GPS-2. Additional known and unknown components of the corepressor complex are indicated simply as Etc. for simplification. The ability of HDACs to remove acetyl groups from histones is indicated as an arrow, as are the inhibitory interactions between SMRT/N-CoR and the general transcriptional machinery noted in the text. Although a single complex is depicted, multiple corepressor complexes of differing compositions are likely to exist in cells.

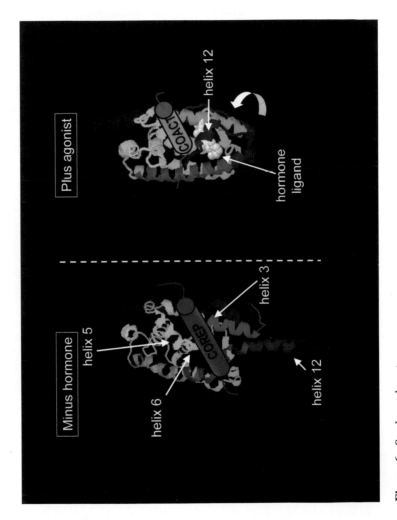

Figure 6 See legend next page

Figure 6 Corepressor docking sites, coactivator docking sites, and the helix 12 toggle switch. A ribbon diagram is shown of the hormone-binding domain of a nuclear receptor in the absence (*left panel*) versus the presence (*right panel*) of hormone agonist. Superimposed on the left ribbon diagram is a red cylinder representing the CORNR box domain of SMRT/N-CoR and emphasizing the interaction of this corepressor domain with a docking site composed of helices 3/5/6 on the nuclear receptor. Note that this docking surface is accessible to corepressor through the extended position of receptor helix 12. Superimposed on the right ribbon diagram is a green cylinder representing the L-X-X-L-L motif found in many coactivators. Note that the binding of hormone ligand (shown in *white*) has rotated receptor helix 12 to occlude the corepressor docking surface, but has simultaneously generated a new docking site for the shorter coactivator L-X-X-L-L interaction domain. To create this generic figure, the crystal structure of RXRα was employed on the left (353) and of RARγ on the right (32).

receptors, ligands such as tamoxifen or RU486 (selective hormone response modulators or SHRMs) induce unique helix 12 conformations that favor corepressor binding and are distinct from the conformations assumed in the absence of hormone or in response to hormone agonist (37–40). Indeed, the antagonist actions of these SHRMs, which are widely employed in medicine, are believed to be manifested primarily through this selective recruitment of SMRT and N-CoR corepressors; reciprocally, the different ratios of corepressors and coactivators found in different tissues may adapt the hormone response so that a given SHRM can mediate different actions in various cell types (152, 153, 155–162). This type of ligand-specific corepressor regulation is not limited to the sex steroid receptors. PPARα can recruit corepressors in the absence of hormone but displays still higher corepressor affinity in the presence of antagonists (40, 145, 163). Conversely, although RARα binds two retinoid ligands, all-*trans* and 9-*cis* retinoic acid, with equal affinities, only the former induces efficient release of SMRT corepressor (164).

Many nuclear receptors are expressed from multiple genetic loci, or by alternative mRNA splicing, to generate multiple receptor isotypes (or isoforms) that play distinct roles in development and physiology (6). These receptor isotypes can display different corepressor recruitment properties (116, 147, 165). For example, RARs are encoded by three distinct genes: α, β, and γ (166, 167). Although RARα represses target gene expression in the absence of hormone, RARβ and γ do not repress, rather they activate transcription in both the absence and presence of hormone agonist (168). These differences in transcriptional regulation reflect the corepressor binding properties of these isoforms: RARα binds corepressor strongly in vitro and in vivo, whereas RARβ and γ do not (116, 168, 169). Intriguingly, all three RAR isotypes possess comparable corepressor docking sites, but access to these docking sites is regulated differently in the various isotypes (169). Helix 12 of RARα appears to orient in the prototypic manner illustrated in Figure 6: extended in the absence of hormone (permitting corepressor binding to its helix 3/5/6 docking surface), but sequestered in the presence of hormone (displacing corepressor and recruiting coactivator). In RARβ and γ, however, hydrophobic interactions between helix 3 and helix 12 appear to gate helix 12 closed in both the absence and presence of hormone agonist; this sequestered helix 12 occludes the corepressor binding site, prevents repression, and shifts the equilibrium toward transcriptional activation (Figure 7) (169). Suitable RAR antagonists, such as AGN193109, may gate helix 12 to yet a third position to allow corepressor binding by the otherwise nonpermissive RARγ isotype (170).

The position of helix 12 of one receptor in a dimer may also be influenced by interactions with the second receptor (e.g., 56, 171). Therefore, the identity and ligand occupancy of each receptor in the dimer, as well as their spacing and orientation relative to one another, can modify the ability of the receptor dimer to recruit or to release corepressors and coactivators (117, 124, 128, 171–175). For

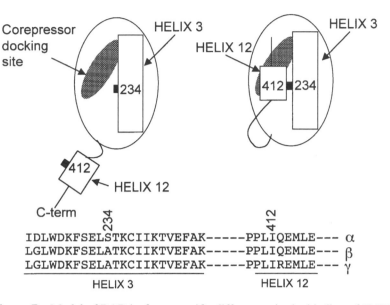

RARα minus hormone   **RARβ or γ minus hormone**

```
IDLWDKFSELSTKCIIKTVEFAK-----PPLIQEMLE--- α
LGLWDKFSELATKCIIKTVEFAK-----PPLIQEMLE--- β
LGLWDKFSELATKCIIKTVEFAK-----PPLIREMLE--- γ
```
HELIX 3   HELIX 12

Figure 7 Model of RAR isoform-specific differences in the binding of SMRT. The RARα isoform binds corepressor efficiently in the absence of hormone, whereas RARβ and RARγ do not, and this property maps to isotype-specific sequence differences in HBD helix 3. All three RAR isoforms possess strong docking sites for SMRT and N-CoR. We propose that a hydrophobic interaction between helix 12 and helix 3 in RARβ and γ, centered on Ala 234 and Ile 412, stabilizes a sequestered helix 12 conformation in these isoforms even in the absence of hormone ligand, thus preventing access of corepressor to its docking site (*right panel*). This closed conformation may mimic in some ways the agonist-bound conformation of RARα. The presence of a Ser at codon 234 in RARα is hypothesized to disrupt this hydrophobic interaction, opening the helix 12 toggle switch in the absence of hormone and permitting corepressor binding (*left panel*).

example, RXR/RARα heterodimers assembled on DR-5 HREs dissociate from corepressor in response to RARα agonists, but not in response to RXR agonists; as a result RXR is a silent partner in this context, and RXR/RARα heterodimers do not activate transcription in response to RXR agonists (174). In contrast, the RARβ isotype does not bind corepressor efficiently, and RXR/RARβ heterodimers have been reported to respond to both RAR and RXR agonists (172). Interactions with other transcription factors arrayed on a target promoter, as well as allosteric contributions of the DNA binding site itself, are also likely to influence the recruitment and/or function of corepressor complexes (100).

THE PLAYERS IN THE REPRESSION DRAMA: HISTONE DEACETYLASES AND SUPPORTING ACTORS

The RDs within SMRT and N-CoR function by recruiting more proteins that help mediate the molecular events necessary for repression (Figures 3 and 5). Best understood of these downstream recruits are the histone deacetylases (HDACs); HDACs inhibit transcription by modification of the chromatin template (detailed as the histone code, below). HDAC3, for example, interacts directly with both SMRT and N-CoR, a process that not only physically recruits HDAC3 to corepressor/receptor complexes, but also switches on the enzymatic activity of the deacetylase (Figures 3 and 5) (176–184). HDACs 4, 5, and 7 also make direct contacts with SMRT and/or N-CoR and may contribute their own blend of enzymatic specificities to the repression complex (Figures 3 and 5) (176, 185–188). Other components of the corepressor complex may serve as scaffolds, assist in substrate recognition, or regulate corepressor function. TBL-1 (transducin-like protein 1, also known as Ebi) and TBL1-R (TBL-1-related protein) complex with SMRT and N-CoR and, by making additional contacts with HDAC3, stabilize the quaternary structure of the corepressor assembly (Figures 3 and 5) (177, 180, 184, 189); TBL1 and TBLR1 also bind to histones H2B and H4 (189) and may help in chromatin substrate recognition. The GPS2 (G protein pathway suppressor 2) protein, another member of the corepressor repertoire company, interacts with both N-CoR and TBL1 (Figures 3 and 5) (184), further crosslinking and organizing the larger corepressor complex. Intriguingly, both TBL-1 and GPS2 are components of G protein–coupled signal transduction pathways and may play additional roles as regulators that couple signal transduction to transcriptional repression (see below).

N-CoR and SMRT can also interact with mSin3 (mammalian switch-independent 3 protein) (Figures 3 and 5); however, the physiological significance of this observation is controversial (75, 98, 115, 190–192). The mSin3 protein is an important corepressor for many nonreceptor transcription factors, such as Mad/Max, and forms a subcomplex with additional corepressor proteins, including HDACs 1 and 2, SAP18, SAP30, and RbAps 46/48 (98, 192–197). N-CoR and SMRT physically interact with mSin3 in vitro and in two-hybrid assays (75, 98, 115, 190–192). However, purification of SMRT or N-CoR from cells by immunoprecipitation or biochemical fractionation often fails to copurify mSin3 and its associates, even under conditions where TBL1, TBLR1, HDAC3, and GPS2 are readily detected (177, 182–184, 186, 189, 192, 198, 199). It is possible that conditions in vivo preclude access of mSin3 to SMRT and N-CoR. An alternative, more attractive explanation is that there are multiple SMRT and N-CoR complexes in the cell, some of which tether the mSin3-HDAC 1/2 subcomplex and some of which do not. Evidence for the existence of these multiple SMRT and N-CoR complexes has been reported in several independent studies, although the precise compositions and relationships of these complexes remain to be defined (98, 181, 187, 200, 201). These corepressor complexes may be preferentially recruited by different transcription factors, in different cells, or by different target promoters (e.g., 98).

These corepressor complexes might also function in a sequential fashion, or play other roles in short-term versus long-term repression.

One interesting example of how the composition of the corepressor complex may be modulated comes from studies of the T3Rβ2 isoform. T3Rβ2 is expressed primarily in the pituitary and hypothalamus and encodes a unique N-terminal domain as a result of alternative mRNA splicing (7, 202, 203). Most T3R isoforms repress transcription in the absence of hormone; T3Rβ2, however, fails to repress under these conditions, mediating instead a weak target gene activation that is further enhanced on addition of T3 hormone (204, 205). T3Rβ2 strongly binds to SMRT and N-CoR (206, 207); why does T3Rβ2 fail to repress? All T3R isoforms, including β2, interact with SMRT and N-CoR through the corepressor ND/receptor helix 3/5/6 contacts described previously (Figure 8) (207). T3Rβ2 is unique, however, in the ability of its N-terminal domain to make additional contacts with the repression domains of SMRT and N-CoR (207); these N-terminal contacts inhibit the recruitment of additional corepressor components, such as mSin3, and prevent

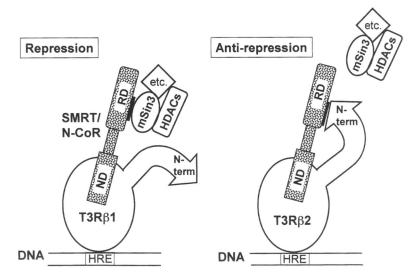

Figure 8   Model of T3R isoform-specific differences in the assembly of a functional corepressor complex. Both T3Rβ1 and T3Rβ2 bind SMRT and N-CoR corepressor strongly, but only T3Rβ1 represses transcription. All known T3R isoforms contact SMRT/N-CoR through their HBD helix 3/5/6 docking surface, as shown in both left and right panels. However, only the T3Rβ2 isoform contains an additional corepressor interaction surface on its N-terminal domain that contacts portions of the repression interaction domain within SMRT and N-CoR (*right panel*). This additional, T3Rβ2-specific interaction interferes with mSin3 recruitment by SMRT and N-CoR. We propose that this interference with corepressor assembly interferes with repression and denote this phenomenon antirepression.

repression (Figure 8) (207). Therefore, different receptors appear able to customize the composition of the corepressor complex they recruit.

In addition to the celebrity corepressor subunits described above, many additional proteins that interact or copurify with SMRT or N-CoR have been identified, including Ski (208, 209), Sno (208, 210), IR10 (189), KAP-1 (181), SHARP (211), SAP30 (98), Sun-CoR (118), and BRG1 and BAF (181, 200). Although many of these additional proteins undoubtedly contribute to corepressor function, space limitations require that these additional players be assigned nonspeaking roles and be mentioned no further in this review.

Enough of the cast of characters: what is the script through which the drama of repression is told? Much of this script is written in a histone code. Given that the work of corepressors is mediated, in part, by erasing the text written by coactivators, I must ask the indulgence of the reader in the next section, in which I discuss the histone modifications that are associated with activation, as well as those associated with repression.

HISTONE MODIFICATION: A SCRIPT USED BY COREPRESSORS AND COACTIVATORS TO MEDIATE TRANSCRIPTIONAL REGULATION

The drama of transcriptional regulation is a tale of a battle in which allied and opposing armies of histone modifiers plant their flags, recruit additional conscripts, and struggle to define domains of dominance within the chromatin template. The first glimpses of this warfare came with the observation that nucleosomal histones are often hyperacetylated in transcriptionally active chromatin, yet hypoacetylated in silent chromatin (212–215). The origins of these modifications became clearer with the discovery that many coactivators are associated with HAT activities, whereas many corepressors display HDAC activities (213–217). It was suggested initially that histone acetylation might activate gene expression by physically opening the chromatin, permitting access of the transcriptional machinery to the DNA template; deacetylation would lead to chromatin compaction and transcriptional repression. However, not all forms of histone acetylation stimulate transcription, and not all forms of deacetylation inhibit. An important insight came with the recognition that histone modifications not only alter the biophysical properties of chromatin, but also can serve as a code by which additional proteins could be recruited to, or barred from, the chromatin template (216–223). The basic tenets of this histone code have been confirmed, and substantial progress has been achieved in identifying the vocabulary of this code, its grammar, and the proteins that decipher it.

Most acetylations analyzed to date are stimulatory for transcription, such as acetylation of Lys 9 or Lys 14 on histone H3, and of Lys 5 on histone H4 (Figure 9) (212, 221–223). Acetylated lysines can serve as binding sites for proteins, such as TAF250 and Snf/Swi complexes, that contain bromo domains,

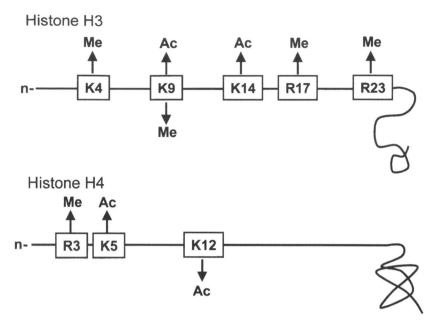

Figure 9 Simplified version of the histone code. The N-terminal domains of histone H3 and H4 are depicted as horizontal lines. Relevant lysines and arginines are shown as boxes. Acetylations (Ac) and methylations (Me) described in the text are shown with arrows. Upward arrrows indicate modifications associated with transcriptional activation; downward arrows indicate modifications associated with repression or gene silencing. Many additional modifications occur but are omitted from this diagram for simplicity.

thereby helping to recruit the general transcriptional machinery to sites of gene activation (224–227). The HDACs found in corepressors presumably reverse these stimulatory acetylations, although the interplay of the specific HAT and HDAC enzymatic activities involved in this phenomenon remains to be determined (216,217). As already noted, not all acetylation events are interpreted by the histone code as signals for transcriptional activation; acetylation of Lys 12 on histone H4, for example, is preferentially associated with inactive chromatin (Figure 9) (212, 221–223). HATs with these inhibitory substrate specificities have not yet been identified in SMRT or N-CoR corepressor complexes, and it is unclear if inhibitory histone acetylations play a role in nuclear receptor-mediated repression.

Additional entries in the chromatin lexicon include histone methylation, histone phosphorylation, histone ubiquitination, histone sumoylation, and DNA methylation (220–223, 226, 228). Histone methylation can be either stimulatory or inhibitory to transcription. Stimulatory methylations include Lys 4, Arg 17, or Arg 23 on histone H3, and Arg 3 on histone 4 (Figure 9) (220–223, 226, 228, 229).

The latter three modifications are mediated by CARM1 and PRMT1, two methyl-transferases that play a role in nuclear receptor activation by piggy-backing on the p160 coactivators (Figure 2) (230–233). Other histone methylations are strongly inhibitory for transcription. Methylation of Lys 9 on histone H3, for example, is a near-universal marker of transcriptionally repressed chromatin and plays a key role in heterochromatic silencing (Figure 9) (221, 228). Despite this potent link between specific histone methylations and transcriptional repression, the SMRT and N-CoR corepressor complexes are not known to contain methyltransferases, and the contribution of histone methylation to nuclear receptor-mediated repression remains unknown. It is important to note, however, that one histone modification can influence another. For example, acetylation of Lys 9 on histone H3 prevents methylation at this position; the HDACs recruited by SMRT/N-CoR may therefore promote histone methylation indirectly by removing this opposing acetylation (220–223, 226, 228, 234). Similarly, although nuclear receptors and their accessory proteins are not known to phosphorylate, sumoylate, or ubiquinate histones, or to methylate DNA, these additional marks of the histone code are likely to interact with the histone methylations and acetylations that are mediated by nuclear receptor coactivators and corepressors (220–222, 235–240).

BEYOND THE HISTONE CODE: TRANSCRIPTIONAL REGULATION BY PROTEIN CONTACTS, BY CHROMATIN REMODELING, AND BY MODIFYING THE MODIFIERS

Transcription is not regulated solely through chemical modifications of the chromatin template. Many nuclear receptors make direct contacts with components of the general transcriptional machinery, such as TFIIB, TBP, and the TAFs (e.g., 241–245). These contacts can either enhance or interfere with subsequent transcriptional initiation. Many nuclear receptors also recruit components of the mediator complex in the presence of hormone agonist, further contributing to the formation of a preinitiation complex on the target promoter (44, 52). Conversely, the SMRT and N-CoR corepressors not only recruit histone deacetylases to the chromatin template, but also make direct, inhibitory contacts with TFIIB and with TAF30 (74, 75). These corepressor interactions can interfere with the interaction of TFIID and TFIIB and thereby disrupt formation of the preinitiation complex and hinder target gene transcription (74).

Noncovalent changes to chromatin structure also contribute to transcriptional regulation by nuclear receptors. Many nuclear receptors physically interact with the ATP-dependent chromatin remodeling complexes, such as the Swi/Snf family; these remodeling complexes serve as coactivators by opening the nucleosome structure and permitting access of other specific and general transcription factors (246, 247). Additional chromatin remodeling complexes, such as NuRD, possess both remodeling and HDAC activities and can alter the chromatin so as to repress gene expression (248, 249).

Enzymes can have substrates beyond those implied by their nomenclature. HATs, for example, can acetylate not only histones, but also many transcription factors, including nuclear receptors (250, 251). Acetylation of the androgen receptor decreases the affinity of this receptor for corepressor (252). One coactivator can also modify and alter the function of another. The p160 coactivator, ACTR, for example, helps recruit CBP/p300 but, in turn, is acetylated by the CBP/p300 HAT and subsequently released from the target promoter (57). In this fashion, networks of interwoven modifications, mediated by nuclear receptors, corepressors, and coactivators, can invoke cyclic waves of histone and cofactor modification that play out on the order of minutes and hours (57, 59, 60, 155, 156). These cycles of modifications are paralleled by sequential changes in the recruitment of other transcription factors and in the initiation of RNA transcripts (57, 59, 60, 155, 156). The contributions of corepressors to these temporally regulated events remain unclear; however, it is very likely that corepressors and their associated HDAC activities participate in and help shape the ebb and flow of cyclic modifications already defined for coactivators.

In summary, transcriptional regulation by nuclear receptors is a dynamic process that is mediated by multiple mechanisms. Covalent and noncovalent modifications of chromatin, interactions with the transcriptional machinery, and alterations in the transcription factors themselves all contribute. Different regulatory mechanisms may be employed by different receptors, by the same receptor on different promoters, or sequentially on the same promoter. Heterodimer formation by nuclear receptors may further diversify the mechanisms available for transcriptional repression and activation.

REGULATION OF SMRT/N-COR COREPRESSOR FUNCTION

The nuclear receptors function by rendering the transcriptional machinery (described above) subservient to signals of extracellular and intracellular origin. Hormone agonists or antagonists are, of course, a key source of signal input for the nuclear receptors and operate by provoking changes in receptor helix 12, which encourage or inhibit binding of the SMRT/N-CoR complex. A quite different regulation of the receptor-SMRT/N-CoR partnership is mediated by phosphorylation of the corepressor, can occur even in the absence of hormone ligand, and can either enhance or inhibit the corepressor/nuclear receptor interaction. For example, phosphorylation of the C terminus of SMRT by casein kinase/CK2 stabilizes corepressor binding to T3Rs (253). Conversely, negative regulation of SMRT occurs in response to growth factor receptors operating through a Ras-MEKK1-MEK1 pathway (254, 255). Phosphorylation of SMRT by this kinase cascade results in a loss of affinity of the corepressor for an assortment of receptor and nonreceptor transcription factors, derepression of previously repressed target genes, and redistribution of SMRT into a cytoplasmic/perinuclear location (254, 255).

This inhibition of corepressor function by phosphorylation may contribute to the ability of epidermal growth factor receptor (EGF-R) and of HER2 (human EGF-R 2; another member of the EGF-R family) to counteract the antagonist properties of SHRMs such as tamoxifen (125, 256, 257). Conceptually similar forms of corepressor regulation can be found throughout evolution. Delta expression in *Drosophila*, for example, is derepressed in response to EGF-R-mediated phosphorylation of the SMRTER corepressor complex, resulting in release of SMRTER from the Suppressor of Hairless transcription factor and export of the corepressor into a cytoplasmic compartment (258).

A distinct mode of corepressor regulation has been identified in neural stem cells. In the absence of contrary signals, N-CoR is a strong inhibitor of astrocyte differentiation, apparently through its ability to partner with the CBF1 transcription factor and thereby repress astrocyte-specific gene expression (259). Exposure of these neural progenitor cells to ciliary neurotrophic factor induces phosphorylation of N-CoR by Akt1, resulting in redistribution of N-CoR from nucleus to cytoplasm, derepression of astrocyte-specific genes, and cell differentiation (259). Inflammatory cytokines, such as IL-1β, have also been identified as important modulators of corepressor function. For example, TAB2, a protein that operates downstream of IL-1β, has been found in a complex with N-CoR and HDAC3; intriguingly, these TAB2 complexes lack TBL1 and may be preferentially recruited by NF-κB p50 homodimers (80). IL-1β signaling results in phosphorylation of TAB2, release of the corepressor complex from the p50 homodimer, export of TAB2/N-CoR/HDAC3 out of the nucleus, and derepression of NF-κB target genes (80).

SMRT and N-CoR are not the only corepressor proteins subject to regulation by phosphorylation. Phosphorylation of class II HDACs by calmodulin-dependent protein kinases causes these deacetylases to bind to 14-3-3 proteins and be exported out of the nucleus; this process is important for relief of target gene repression by the MEF2 transcription factor during muscle cell differentiation (260–262). Phosphorylation by ERK1 and 2 acts in the converse to increase HDAC4 nuclear accumulation (263). The interactions of mSin3, HDAC1, and HDAC2 with one another, and with their transcription factor partners, are also regulated by phosphorylation (264, 265). More work will be required to understand how these modifications of the individual corepressor proteins influence the overall composition, localization, and function of the higher-order corepressor complexes that mediate nuclear receptor function.

It should be noted that phosphorylation of the nuclear receptor itself, rather than the corepressor, can also control the receptor/corepressor partnership (e.g., 125, 266, 267). Phosphorylation can similarly influence the interactions of nuclear receptors with coactivators (e.g., 266, 268–273). Furthermore, the communication between kinases and transcriptional coregulators can be bidirectional. For example, the GPS2 subunit of the TAB2/N-CoR/HDAC3 complex is a potent inhibitor of JNK1 activity, and this inhibition is enhanced by coexpression of N-CoR, but reversed by dominant-negative N-CoR constructs and by use of HDAC inhibitors (184).

In summary, corepressors are subject to an extensive panel of regulatory modifications that influence, positively or negatively, their ability to interact with their transcription partners. Acting together with ligand agonists and antagonists, these covalent modifications serve to integrate the many different signals impinging on the cell so as to produce the correct transcriptional and biological response for any given physiological context. Intriguingly, many of the known inhibitory phosphorylations of SMRT, N-CoR, and HDACs occur in response to prodifferentiation signals. Uninterrupted corepressor function, in contrast, is generally associated with maintenance of a less differentiated, more proliferative cell phenotype. It will be interesting to see if this generality holds as more is discovered about corepressor function.

SMRT VERSUS N-COR: SIMILARITIES AND DIFFERENCES

SMRT and N-CoR appear to be the products of a gene duplication event that occurred prior to the vertebrate evolutionary radiation; N-CoR and SMRT have been retained as distinct paralogs ever since. Why do vertebrates require two distinct flavors of corepressor? How do SMRT and N-CoR differ in function, and in what ways do they overlap?

Commonalties abound. The overall architecture of SMRT and N-CoR is shared, and both utilize related modes of receptor interaction and of transcriptional repression (51, 65, 110, 111). So far as has been determined, the corepressor complexes formed by N-CoR generally parallel the compositions of the analogous complexes formed by SMRT, with only modest exceptions, such as the reported associations of TAB2 and IR10 with N-CoR, but not SMRT (80, 189).

Despite these similarities, a number of functional distinctions have been detected between SMRT and N-CoR. As already noted, these two corepressors display different affinities for the different nuclear receptors (113, 115–117, 121, 122, 124, 125, 130, 132, 137). These differences tend to be quantitative, rather than absolute in nature, and can be influenced by the experimental context. For example, RARα binds to both SMRT and N-CoR in solution, yet exhibits selectivity for SMRT when the receptor is arrayed on a DNA response element (117). Differential mRNA splicing and post-translational modifications of SMRT and N-CoR also can alter the relative affinities of these corepressors for the different nuclear receptors (see below).

Additional differences in SMRT and N-CoR function have been detected in studies of the post translational modifications. Phosphorylation of SMRT in response to MAP kinase cascade signaling leads to dissociation of SMRT from its nuclear receptors partners and its nuclear export; this is not true of N-CoR (255; B. Jonas & M.L. Privalsky, unpublished observations) Conversely, IL-1β–induced phosphorylation of TAB-2 leads to nuclear export of N-CoR but not of SMRT (80). Akt signaling can phosphorylate N-CoR directly, also resulting in nuclear export, whereas the same domain of SMRT is not subject to this regulation (259). However, Akt signaling can lead to nuclear export of SMRT by phosphorylating amino acids outside of the sites characterized in this particular report (M. Mayo, personal communication). Important differences also exist between N-CoR and SMRT at the end of their lifespans. N-CoR interacts with the RING protein Siah2 and is targeted for proteosomal degradation, whereas SMRT appears to be refractory to this form of regulation (274). Different cells have different levels of Siah2, and combined with the specificity of many nuclear receptors for either SMRT or N-CoR, this differential corepressor turnover can manifest as various abilities of the same nuclear receptor to repress in different cell types (274).

Both SMRT and N-CoR are expressed as a series of additional corepressor isoforms by alternative mRNA splicing, further diversifying the functional properties of these corepressors (Figure 4). For example, the third nuclear receptor interaction domain, ND3, present in N-CoR is spliced out in the RIP13Δ1 isoform (74, 121–123). As a result, RIP13Δ1 displays a much lower affinity for T3Rs than does full-length N-CoR (74, 121–123). Conversely, RIP13Δ1 interacts more strongly with the orphan receptors COUP-TFII, Rev-erbAα, and RVR than N-CoR does (67, 113, 140). SMRT, too, undergoes alternative mRNA splicing, in this case to delete a snippet of 47 amino acids immediately following the ND1 domain (64, 66, 110, 111). We refer to this slightly smaller version of SMRT as SMRTτ (tau) and have determined that the removal of these amino acids greatly reduces the affinity of SMRTτ for certain receptors, such as T3Rβ1, but has less impact on interactions with other receptors, such as T3Rα1 (M. Goodman & M. L. Privalsky, unpublished observations).

SMRT and N-CoR are differentially expressed in different tissues and at different times in development (160, 275, 276). N-CoR, for example, is expressed at high levels in fetal liver and thymocytes, but at lower levels in thymic stroma, whereas SMRT is expressed in a reciprocal pattern (160). Although SMRT and N-CoR are both expressed in the cortical ventricular zone of the developing nervous system, N-CoR, but not SMRT, is also strongly expressed in the dorsal thalamus and in the marginal zone and cortical plate regions of the neocortex (160). Therefore, organisms appear to tailor corepressor function in a cell-specific manner by selecting from a wardrobe of distinct corepressor isoforms that differ in their affinities for different transcription factor partners and in their responses to different cell signaling pathways. Repression mediated by nuclear receptors can thereby be fine-tuned in a sophisticated fashion, adapted to yield the biologically correct response for a given target gene, cell-type, and physiological state.

COREPRESSOR DISRUPTIONS

The role of N-CoR has been explored in whole animals by gene knockout technology. A homozygous disruption of N-CoR in the mouse is embryonic lethal by day 15.5 of gestation (160); given the retention of SMRT in these N-CoR knockout animals, such experiments confirm the incomplete redundancy of the two corepressor isotypes. N-CoR −/− mice were on average 80% the size of wildtype or heterozygotic animals, with undersized livers and severe anemia. The latter appeared to be a result of defects in definitive erythropoiesis, observed also as an impairment in BFU-E colony formation (160). T cell development in N-CoR −/− mice was also disrupted, with severe reductions in total thymocytes, and many cells blocked prior to the CD25+/CD44− stage (160). Central nervous system development was aberrant in the N-CoR −/− animals, manifested as alterations in the developing thalamus, third ventricle, and the outer cortical layers (160). The prominent neurological defects observed in these knockout animals may, in part, reflect the loss of the antidifferentiation effects of N-CoR in neural stem cells, as noted above.

The N-CoR disruption engendered a variety of defects in transcriptional repression, as determined in reporter gene assays using fibroblasts derived from the N-CoR −/− animals (160). The ability of RAR to repress on a DR-5 HRE was disrupted in the N-CoR −/− fibroblasts, resulting in high levels of constitutive expression of the DR-5 reporter in the absence of cognate hormone. Defects in repression by a GAL4-T3R fusion construct and by the nonreceptor transcription factors, Mad and REST/NRSF, were also observed in the N-CoR −/− background. Tamoxifen, normally an ERα antagonist or weak agonist, functioned instead as a strong agonist in the N-CoR −/− cells, consistent with the suggestion that corepressors mediate the antagonistic actions of this and other SHRMs. Paradoxically, disruption of N-CoR also inhibited one form of transcriptional activation: the ability of RAR to induce expression of a DR-1 reporter gene in response to hormone (259). Corepressors may therefore be multifaceted in their talents and may play a role in some forms of transcriptional activation (see below).

Dominant-negative constructs have also been used to study corepressor function in transgenic mice and in Xenopus (138, 277, 278). Targeting a dominant-negative N-CoR to murine liver under hypothyroid conditions resulted in an increased expression of many genes normally subject to repression by T3Rs, including spot 14, BCL-3, glucose 6-phosphate, and 5′ deiodinase (277); for unknown reasons, malic enzyme, another T3R-regulated target gene, was not derepressed under the same conditions. Gene activation in the presence of thyroid hormone appeared normal in all cases (277). No gross anatomical differences were detected in the wild-type and transgenic mouse livers. However, a higher rate of proliferation was observed in the hepatocytes of the transgenic mice; this was paralleled by increases in the expression of genes associated with cell proliferation (277). These results suggest that N-CoR plays a key role in T3R-mediated repression in liver and also implicate this corepressor in the control of hepatocyte proliferation. Inhibition of corepressor function during Xenopus embryonic development led to defects in head development that correlated with loss of RAR-mediated repression (278); similar dominant-negative experiments during metamorphosis resulted in defects in T3R-mediated repression in tails (138). Use of trichostatin A, a potent inhibitor of HDAC activity, resulted in multiple lesions in T3-driven metamorphosis in Xenopus (279). It should be noted, however, that the dominant-negative constructs employed in both the mouse and the amphibian studies have the potential to inhibit the function of all corepressors that share the common helix 3/5/6 docking site on the nuclear receptors; HDAC inhibitors have even more pleiotropic effects.

COREPRESSORS AND COACTIVATORS: PUBLIC ENEMIES, PRIVATE FRIENDS?

As noted above, disruption of N-CoR in mouse fibroblasts causes a reporter-specific defect in RAR-mediated activation (160). Additional studies have similarly hinted at roles for corepressors in transcriptional activation by other nuclear receptors. In particular, SMRT and/or N-CoR appears to contribute to the ability of T3Rs to activate target genes that contain so-called negative HREs (i.e., DNA response elements that confer transcriptional activation in the absence, rather than in the presence, of thyroid hormone) (280–282). Overexpression of SMRT or N-CoR increases expression of these negative-response genes in the absence of hormone, whereas mutants of T3R unable to interact with corepressor are unable to mediate these effects (282). Negative regulation of this kind is important for feedback control in the thymus/pituitary/hypothalamus axis, and (possibly) in S-phase gene regulation (283). However, contradictory results have been reported, and more experiments are required to resolve these issues (284).

Corepressors may participate in transcriptional activation by physically interacting with coactivators. N-CoR has been reported to bind to p160 activators in vitro, and this interaction may be able to recruit coactivators to T3Rs in the absence of hormone ligand (285). Additional interactions between corepressors and coactivators are likely to occur, either directly or indirectly through bridging proteins. A provocative example of the latter is the SHARP protein, which was first identified as a polypeptide that can interact with the C terminus of SMRT (87, 211). However, SHARP also possesses CoRNR motifs that allow direct interactions of SHARP with nuclear receptors, as well as interaction domains for HDAC1/2, for the NuRD chromatin remodeling complex, and for an unusual RNA coactivator, denoted SRA (steroid receptor RNA coactivator) (147, 211). SHARP, therefore, has the potential to generate a wide variety of combinatorial interactions between different receptors, corepressors, chromatin remodeling complexes, and coactivators. Although the significance of these interactions in a physiological context remains to be resolved, it appears likely that the once crisp demarcation line between corepressors and coactivators is becoming increasingly blurry.

DISEASES DUE TO COREPRESSOR ABERRATIONS

Given the importance of corepressors for proper nuclear receptor function, it is not surprising that corepressor dysfunction can result in disease. One well-studied example of a corepressor disease in humans is resistance to thyroid hormone (RTH)-syndrome, an inherited endocrine disease manifested as the inability to respond correctly to elevated circulating thyroid hormone (286, 287). Most forms of RTH-syndrome map to genetic lesions in T3Rβ and are associated with the synthesis of aberrant T3Rs that fail to release corepressor properly in response to physiological concentrations of hormone (288–293). As a consequence, these mutant receptors are able to repress, but not to activate, transcription of target genes and function as dominant-negative inhibitors of transcriptional activation by wild-type T3Rs. The molecular basis for this defective corepressor release varies with the nature of the RTH-syndrome mutation: certain RTH-T3Rβ alleles produce receptors that have a reduced affinity for the T3 ligand and thus require higher than normal hormone concentrations to trigger corepressor release (288–293). Other RTH-T3Rβ mutant alleles encode receptors that bind T3 with near normal avidity but are defective in the subsequent operation of the helix 12 toggle necessary to displace the corepressor from the receptor (292, 293). Still other RTH-T3Rβ mutant receptors exhibit unusually high affinities for SMRT and N-CoR corepressors in both the absence and presence of T3 hormone (289, 292–294). Disrupting the interaction of RTH-T3Rβs with SMRT or N-CoR disrupts the dominant-negative properties of these receptors, confirming that the inability of the mutant receptors to properly release from corepressor contributes to RTH-syndrome at the molecular level (290, 292). It should be noted that many of the same RTH-mutations that cause defects in corepressor release also cause defects in coactivator binding; these coactivator acquisition defects may also contribute to the molecular failures that manifest as RTH-syndrome (295).

Mutations that result in a dominant-negative phenotype have also been found in the peroxisome-proliferator-activated receptor (PPAR)-γ (296). Wild-type PPARγ is essential for proper glucose utilization and lipid storage, whereas these dominant-negative PPARγ mutations are associated with a form of human familial type II diabetes and hypertension (286, 296). The precise role of SMRT and N-CoR corepressor in this disease remains to be established; however, as in RTH-T3Rβ, the PPARγ mutations are located in regions of the receptor anticipated to disrupt the release of SMRT/N-CoR and/or the acquisition of coactivator in response to ligand (296).

A particularly lethal example of a corepressor/nuclear receptor disease is found in human acute promyelocytic leukemia (APL). APL is associated with chromosomal translocations that replace the N terminus of the wild-type RARα with ectopic protein sequence, such as that of the promyelocytic leukemia protein (PML), promyelocytic zinc-finger protein (PLZF), nuclear mitotic apparatus protein (NuMA), STAT5, or nucleophosmin (NPM) (Figure 10) (297–300). The resulting x-RARα fusion proteins display a remarkable diversity of structure but share the ability to function as dominant-negative inhibitors of transcriptional activation by wild-type RARs. Not surprisingly, the dominant-negative properties of these x-RARα fusions are caused by aberrant corepressor interactions. PML-RARα and NPM-RARα, for example, require atypically high concentrations of all-*trans* retinoic acid (ATRA) to release SMRT corepressor and to relieve repression (90, 92, 301–307). This delayed release of corepressor correlates

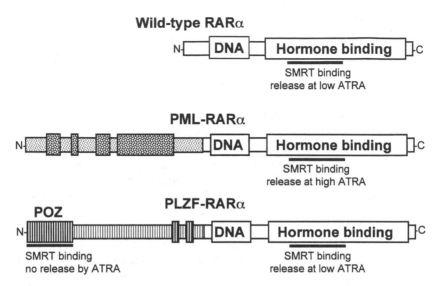

Figure 10 Schematic of wild-type RARα and of the PML-RARα and PLZF-RARα fusion proteins found in human acute promyelocytic leukemia. The primary structures of wild-type RARα, PML-RARα, and PLZF-RARα are shown from N to C terminus. The location of the DNA and hormone-binding domains of RAR are indicated. The SMRT/N-CoR docking surfaces are depicted as black bars beneath each schematic. As discussed in the text, PML-RARα requires higher ATRA concentrations to release SMRT than does wild-type RARα. PLZF-RARα does not fully release from SMRT even at high hormone concentrations owing to the presence of an additional hormone-refractory SMRT interaction surface in PLZF. Impaired corepressor release by PML-RARα and PLZF-RARα plays a key role in oncogenesis.

closely with the oncogenic properties of the x-RARα fusions: Supraphysiological concentrations of ATRA sufficient to release corepressor from PML-RARα induce differentiation of APL cells and clinical remission in mouse and human models of this disease (301, 308). Additional mutations or alternatively spliced forms of PML-RARα that impair corepressor release at even high levels of ATRA result in leukemias that are resistant to ATRA therapy (308, 309). An intriguing variation on this theme is observed with the PLZF-RARα oncoprotein. The PLZF progenitor is itself a transcriptional repressor that can recruit SMRT and N-CoR (90–92, 301, 304). As a result, the PLZF-RARα fusion contains a hormone-resistant N-CoR/SMRT interaction surface at its N terminus, in addition to the hormone-regulated corepressor interaction surface contributed by the RARα moiety at its C terminus (Figure 10) (90–92, 94, 301, 304). As might be expected from this dual form of corepressor tethering, leukemias induced by PLZF-RARα are highly resistant to differentiation therapy in response to ATRA (297–300).

A few additional points should be noted. Both PLZF and PML can recruit HDACs and mSin3 independently of SMRT or N-CoR, and these additional corepressor contacts may contribute directly or indirectly to leukemogenesis by the corresponding x-RARα fusions (93, 94, 310). The products of the reciprocal chromosomal translocations, as well as other genetic changes in the leukemic cell, may also participate in establishing the late-stage APL phenotype (311). There is an intriguing connection between the clinical treatment of APL and the protein kinase cascades mentioned above. Although most APL patients enter into remission on initial treatment with pharmacological levels of ATRA, many eventually become resistant to this therapy. These hormone-resistant leukemias nonetheless remain responsive to treatment with arsenite, which induces both apoptosis and an abortive differentiation in APL cells. Although arsenite operates in APL cells in multiple capacities, a well-known effect of arsenite is to activate MAP kinase cascades. Indeed, arsenite treatment of APL cells results in an increased phosphorylation of SMRT, its dissociation from the PML-RARα oncoprotein, and the export of the corepressor into the cytoplasm of the leukemic cell (312). Conversely, inhibitors of MAP kinase cascades block arsenite-mediated differentiation and reduce arsenite-mediated apoptosis (312). Therefore, both ATRA and arsenite may function in the treatment of APL by interfering with the interaction between PML-RARα and SMRT, the former by inducing a conformational change in helix 12 of the PML-RARα moiety, the latter by causing phosphorylation of the corepressor.

Defects in corepressor function contribute to a number of other human and animal neoplasias. A mutated version of T3Rα, denoted v-Erb A, participates in erythroid leukemogenesis by the avian erythroblastosis virus (313). v-Erb A is a virally acquired copy of the host cell T3Rα and has sustained a deletion of the helix 12 toggle switch present in the normal receptor (314, 315). As a consequence, v-Erb A binds to SMRT and N-CoR constitutively, is unable to activate reporter genes in response to T3 hormone, and functions as a dominant-negative inhibitor of the wild-type T3Rs (64–66, 182, 316, 317). v-Erb A, therefore, closely resembles the dominant-negative T3Rβ mutants found in human RTH-syndrome; why then does v-Erb A cause neoplasia, rather than endocrine malfunction? Unlike the T3Rβ mutants associated with RTH-syndrome, v-Erb A has sustained additional mutations, including amino acid substitutions within its DNA recognition domain, and is able to interfere with retinoid as well as T3 signaling (109, 318, 319). The oncogenic abilities of v-Erb A may arise from this dual ability to inhibit both T3R and RAR function (320, 321).

Although outside of the realm of nuclear receptors per se, it is interesting to note that corepressors also contribute to human acute myeloid leukemias and to human large cell lymphomas. The former are associated with chromosomal translocations that join the AML protein to the ETO protein. Native ETO is a transcriptional repressor that operates through the SMRT, N-CoR and mSin3 corepressors. The chromosomal translocation generates a fusion protein with the target gene specificity of AML, but with the repressive properties of ETO, and it is the juxtaposition of these two properties that contributes to leukemogenesis (102–106, 322). Human large cell and follicular lymphomas are associated with an overexpression of BCL-6, a zinc-finger protein closely related to PLZF. The oncogenic properties of BCL-6 parallel the ability of this transcription factor to recruit SMRT, N-CoR, and associated corepressor proteins, such as mSin3 and HDACs (94–96).

OTHER COREPRESSORS

Researchers have identified a number of nuclear receptor corepressors in addition to SMRT, N-CoR, and their immediate associates. The inclusion of these additional corepressors at the end of this review reflects only the more limited information available about these corepressors and is not meant to disparage their importance in nuclear receptor physiology.

The Alien corepressor interacts with a number of nuclear receptors, including T3Rs, VDRs and the ecdysone receptor (323, 324), but its function has best been established for the Dax-1 orphan receptor. Dax-1 (dosage-sensitive sex reversal adrenal hypoplasia critical region on the X chromosome, gene 1) can interact with a second orphan nuclear receptor, steroidogenic factor 1 (SF-1), to regulate target genes involved in adrenal and gonadal development (325, 326). The expression pattern of Alien in adrenal gland and testis parallels that of Dax-1. Mutations in Dax-1 that fail to recruit Alien fail to repress and are associated with adrenal hypoplasia congenita (327, 328). The Dax-1/SF-1 complex can also recruit N-CoR and SMRT (328), and phosphorylation of SF-1 has been reported to regulate the equilibrium between corepressor and coactivator recruitment (266).

PSF (polypyrimidine tract-binding protein-associated splicing factor) interacts with the DNA-binding domains of T3Rs and RXRs in a ligand-independent fashion (329). PSF can recruit mSin3 and associated HDACs and can contribute to transcriptional repression by these receptors (329). Hairless is another T3R-associated corepressor that can recruit HDACs independently of N-CoR or SMRT (330, 331). Hairless also associates with the RORα orphan receptor and may play specific roles in mediating T3R and RORα function in the developing brain (330–332).

The nuclear receptor orthologs identified in *Drosophila* employ a variety of corepressors. Ecdysone receptor can mediate repression through SMRTER (69). Knirps interacts with and represses target gene transcription through the C-terminal-binding protein (CtBP) corepressor (333). CtBP has been reported to mediate repression by at least two mechanisms, one involving recruitment of HDACs/mSin3, the other through interactions with the PcG/polycomb complex (334). Related proteins in vertebrates, denoted CtBP1 and 2, function as corepressors for a variety of nonreceptor transcription factors, although their contribution to nuclear receptor-mediated repression has not been established (334).

The RIP140 protein appears to serve both as a coactivator and as a corepressor of nuclear receptor function, depending on cell and promoter context and on expression level (335–339). RIP140 binds to a number of additional corepressor molecules, including HDAC 1, HDAC 3, and the vertebrate orthologs of CtBP (340). The ability of RIP140 to bind to CtBPs is regulated by acetylation of RIP140, possibly by the HAT activities of the p300/CBP coactivators (339). It is tempting to speculate that modifications of this type may switch the transcriptional properties of RIP140 from repression to activation.

Repressor of estrogen receptor activity (REA) was identified through its ability to interact with ERs bound to antiestrogens (341). Coexpression of REA augments the inhibitory actions of antiestrogens and can blunt estradiol-mediated activation (341, 342). The functions of REA may be regulated, in turn, by interaction with prothymosin α (343). Repressor of tamoxifen transcriptional activator (RTA) is a second ligand-specific negative coregulator for human ERα (344). RTA interacts directly with ERα and inhibits the partial agonist activity of SHRMs such as tamoxifen, with only minimal effects on pure agonists, such as estradiol (344). RTA also can suppress transcriptional activation by glucocorticoid and progesterone receptors in response to pure agonists, but displays no activity toward T3Rs or RARs. RTA function appears to require the integrity of an RNA-recognition motif (RRM) within RTA; mutation of this domain produces a dominant-negative RTA that allows all classes of antiestrogens to manifest partial agonist activity (344). On the male side of the equation, TGIF (5'TG3' interacting factor) has been reported to bind to androgen receptors, recruit mSin3, and repress AR-mediated gene activation in a HDAC-dependent fashion (345).

Smad4 was first characterized as the common or constitutive transcription factor partner for the regulated Smad subunits and operates as a downstream regulator of TGFβ and BMP target genes. Recently, Smad 4 was reported to selectively interact with ERα when the latter is liganded to SHRMs such as tamoxifen or raloxifene and to suppress transcriptional activity by this receptor. (346).

The last subject of this discussion is SHP (short heterodimer partner) (347). SHP is not a corepressor per se, but rather an inhibitory member of the orphan nuclear receptor family. SHP lacks an overt DNA-binding domain and may be recruited to other members of the nuclear receptor family through L-X-X-L-L motifs within the former (348). SHP interacts with and inhibits transcriptional activation by ERα, ERβ, GRs, T3Rs, RARs, and the orphan receptor, liver-homology receptor-1 (LRH-1) (347, 349–351). Perhaps the best-analyzed role of SHP is in cholesterol/bile acid homeostasis. LRH-1 is an important positive regulator of the CYC7A gene, which participates in conversion of cholesterol to bile acids. In mice, excess bile acids bind to FXR, a bile acid sensor which, in response, induces SHP expression. SHP interacts with and inhibits LRH-1 function, suppressing bile acid production in a negative feedback loop (11). SHP is, in many ways, similar to Dax-1 in its overall structure and in its ability to attenuate activation by a full-length nuclear receptor partner. However, unlike Dax-1, which confers repression by recruiting Alien, SMRT, and N-CoR (328), the precise mode of SHP-mediated inhibition has not been elucidated (352).

SUMMARY

SMRT and N-CoR were identified only eight years ago, yet an enormous amount of information has since been obtained as to how these and other corepressors help mediate nuclear receptor function. It has become clear that corepressors are essential contributors to the biological actions of nuclear receptors and that aberrations in corepressor function lead to endocrine malfunctions, neoplastic disease, or severe developmental abnormalities. The future is likely to reveal many additional, intimate links between nuclear receptors, corepressors, and key physiological and pathological processes. Furthermore, the accessibility of the corepressor/nuclear receptor partnership to experimental manipulation will permit the molecular basis of transcriptional control to be probed at ever-increasing levels of detail. The next eight years are certain to be intriguing ones.

ACKNOWLEDGMENTS

Data cited as unpublished results from the Privalsky laboratory are the results of experiments supported by Public Health Service/National Institutes of Health award DK-53528.

The *Annual Review of Physiology* is online at http://physiol.annualreviews.org

LITERATURE CITED

1. Laudet V. 1997. Evolution of the nuclear receptor superfamily: early diversification from an ancestral orphan receptor. *J. Mol. Endocrinol.* 19:207–26

2. Whitfield GK, Jurutka PW, Haussler CA, Haussler MR. 1999. Steroid hormone receptors: evolution, ligands, and molecular basis of biologic function. *J. Cell Biochem. Suppl.* 32–33:10–22

3. Apriletti JW, Ribeiro RC, Wagner RL, Feng W, Webb P, et al. 1998. Molecular and structural biology of thyroid hormone receptors. *Clin. Exp. Pharmacol. Physiol. Suppl.* 25:S2–11

4. Beato M, Klug J. 2000. Steroid hormone receptors: an update. *Hum. Reproduction Update* 6:25–36

5. Hager GL. 2001. Understanding nuclear receptor function: from DNA to chromatin to the interphase nucleus. *Prog. Nucleic Acid Res. Mol. Biol.* 66:279–305

6. Mangelsdorf DJ, Thummel C, Beato M, Herrlich P, Schütz G, et al. 1995. The nuclear receptor superfamily: the second decade. *Cell* 83:835–39

7. Zhang J, Lazar MA. 2000. The mechanism of action of thyroid hormones. *Annu. Rev. Physiol.* 62:439–66

8. Hihi AK, Michalik L, Wahli W. 2002. PPARs: transcriptional effectors of fatty acids and their derivatives. *Cell Mol. Life Sci.* 59:790–98

9. Waxman DJ. 1999. P450 gene induction by structurally diverse xenochemicals: central role of nuclear receptors CAR, PXR, and PPAR. *Arch. Biochem. Biophys.* 369:11–23

10. Kliewer SA, Lehmann JM, Milburn MV, Willson TM. 1999. The PPARs and PXRs: nuclear xenobiotic receptors that define novel hormone signaling pathways. *Recent Prog. Horm. Res.* 54:345–67; discussion 67–68

11. Repa JJ, Mangelsdorf DJ. 1999. Nuclear receptor regulation of cholesterol and bile acid metabolism. *Curr. Opin. Biotech.* 10:557–63

12. Mangelsdorf DJ, Evans RM. 1995. The RXR heterodimers and orphan receptors. *Cell* 83:841–50

13. Pereira FA, Tsai MJ, Tsai SY. 2000. COUP-TF orphan nuclear receptors in development and differentiation. *Cell Mol. Life Sci.* 57:1388–98

14. Khorasanizadeh S, Rastinejad F. 2001. Nuclear-receptor interactions on DNA-response elements. *Trends Biochem. Sci.* 26:384–90

15. Kumar R, Thompson EB. 1999. The structure of the nuclear hormone receptors. *Steroids* 64:310–19

16. Ribeiro RC, Apriletti JW, Wagner RL, West BL, Feng W, et al. 1998. Mechanisms of thyroid hormone action: insights from X-ray crystallographic and functional studies. *Recent Prog. Horm. Res.* 53:351–92; discussion 92–94

17. Forman BM, Casanova J, Raaka BM, Ghysdael J, Samuels HH. 1992. Half-site spacing and orientation determines whether thyroid hormone and retinoic acid receptors and related factors bind to DNA response elements as monomers, homodimers, or heterodimers. *Mol. Endocrinol.* 6:429–42

18. Forman BM, Evans RM. 1995. Nuclear hormone receptors activate direct, inverted, and everted repeats. *Ann. NY Acad. Sci.* 761:29–37

19. Glass CK. 1994. Differential recognition of target genes by nuclear receptor monomers, dimers, and heterodimers. *Endocr. Rev.* 15:391–407

20. Glass CK. 1996. Some new twists in the regulation of gene expression by thyroid hormone and retinoic acid receptors. *J. Endocrinol.* 150:349–57

21. Kliewer SA, Umesono K, Mangelsdorf DJ, Evans RM. 1992. Retinoid X receptor interacts with nuclear receptors in retinoic acid, thyroid hormone and vitamin D_3 signalling. *Nature* 355:446–49

22. Kurokawa R, Yu VC, Näär A, Kyakumoto S, Han Z, et al. 1993. Differential orientations of the DNA-binding domain and carboxy-terminal dimerization interface regulate binding site selection by nuclear receptor heterodimers. *Genes Dev.* 7:423–35

23. Lazar MA, Berrodin TJ, Harding HP. 1991. Differential DNA binding by monomeric, homodimeric, and potentially heteromeric forms of the thyroid hormone receptor. *Mol. Cell Biol.* 11:5005–15

24. Moore DD, Brent GA. 1991. Thyroid hormone—half-sites and insights. *New Biol.* 3:835–44

25. Näär AM, Boutin JM, Lipkin SM, Yu VC, Holloway JM, et al. 1991. The orientation and spacing of core DNA-binding motifs dictate selective transcriptional responses to three nuclear receptors. *Cell* 65:1267–79

26. Perlmann T, Rangarajan PN, Umesono K, Evans RM. 1993. Determinants for selective RAR and TR recognition of direct repeat HREs. *Genes Dev.* 7:1411–22

27. Umesono K, Murakami KK, Thompson CC, Evans RM. 1991. Direct repeats as selective response elements for the thyroid hormone, retinoic acid, and vitamin D_3 receptors. *Cell* 65:1255–66

28. Göttlicher M, Heck S, Herrlich P. 1998. Transcriptional cross-talk, the second mode of steroid hormone receptor action. *J. Mol. Med.* 76:480–89

29. Miner JN, Yamamoto KR. 1991. Regulatory crosstalk at composite response elements. *Trends Biochem. Sci.* 16:423–26

30. Saatcioglu F, Claret FX, Karin M. 1994. Negative transcriptional regulation by nuclear receptors. *Semin. Cancer Biol.* 5:347–59

31. Bourguet W, Vivat V, Wurtz JM, Chambon P, Gronemeyer H, Moras D. 2000. Crystal structure of a heterodimeric complex of RAR and RXR ligand-binding domains. *Mol. Cell* 5:289–98

32. Renaud JP, Rochel N, Ruff M, Vivat V, Chambon P, et al. 1995. Crystal structure of the RAR-gamma ligand-binding domain bound to all-*trans* retinoic acid. *Nature* 378:681–89

33. Tanenbaum DM, Wang Y, Williams SP, Sigler PB. 1998. Crystallographic comparison of the estrogen and progesterone receptor's ligand binding domains. *Proc. Natl. Acad. Sci. USA* 95:5998–6003

34. Wagner RL, Apriletti JW, McGrath ME, West BL, Baxter JD, Fletterick RJ. 1995. A structural role for hormone in the thyroid hormone receptor. *Nature* 378:690–97

35. Wagner RL, Huber BR, Shiau AK, Kelly A, Cunha Lima ST, et al. 2001. Hormone selectivity in thyroid hormone receptors. *Mol. Endocrinol.* 15:398–410

36. Williams SP, Sigler PB. 1998. Atomic structure of progesterone complexed with its receptor. *Nature* 393:392–96

37. Brzozowski AM, Pike ACW, Dauter Z, Hubbard RE, Bonn T, et al. 1997. Molecular basis of agonism and antagonism in the oestrogen receptor. *Nature* 389:753–58

38. Paige LA, Christensen DJ, Grøn H, Norris JD, Gottlin EB, et al. 1999. Estrogen receptor (ER) modulators each induce distinct conformational changes in ER alpha and ER beta. *Proc. Natl. Acad. Sci. USA* 96:3999–4004

39. Shiau AK, Barstad D, Loria PM, Cheng L, Kushner PJ, et al. 1998. The structural basis of estrogen receptor/coactivator recognition and the antagonism of this interaction by tamoxifen. *Cell* 95:927–37

40. Xu HE, Stanley TB, Montana VG, Lambert MH, Shearer BG, et al. 2002. Structural basis for antagonist-mediated recruitment of nuclear co-repressors by PPARalpha. *Nature* 415:813–17

41. Aranda A, Pascual A. 2001. Nuclear hormone receptors and gene expression. *Physiol. Rev.* 81:1269–304

42. Chen JD, Li H. 1998. Coactivation and corepression in transcriptional regulation by steroid/nuclear hormone receptors. *Crit. Rev. Eukaryot. Gene Expr.* 8:169–90

43. Horwitz KB, Jackson TA, Bain DL, Richer JK, Takimoto GS, Tung L. 1996. Nuclear receptor coactivators and corepressors. *Mol. Endocrinol.* 10:1167–77

44. Ito M, Roeder RG. 2001. The TRAP/SMCC/mediator complex and thyroid hormone receptor function. *Trends Endocrinol. Metab.* 12:127–34

45. Jenster G. 1998. Coactivators and corepressors as mediators of nuclear receptor function: an update. *Mol. Cell Endocrinol.* 143:1–7

46. Koenig RJ. 1998. Thyroid hormone receptor coactivators and corepressors. *Thyroid* 8:703–13

47. Lee JW, Lee YC, Na SY, Jung DJ, Lee SK. 2001. Transcriptional coregulators of the nuclear receptor superfamily: coactivators and corepressors. *Cell Mol. Life Sci.* 58:289–97

48. Leo C, Chen JD. 2000. The SRC family of nuclear receptor coactivators. *Gene* 245:1–11

49. McKenna NJ, O'Malley BW. 2002. Minireview: nuclear receptor coactivators—an update. *Endocrinology* 143:2461–65

50. McKenna NJ, Xu J, Nawaz Z, Tsai SY, Tsai MJ, O'Malley BW. 1999. Nuclear receptor coactivators: multiple enzymes, multiple complexes, multiple functions. *J. Steroid Biochem. Mol. Biol.* 69:3–12

51. Ordentlich P, Downes M, Evans RM. 2001. Corepressors and nuclear hormone receptor function. *Curr. Top. Microbiol. Immunol.* 254:101–16

52. Rachez C, Freedman LP. 2001. Mediator complexes and transcription. *Curr. Opin. Cell Biol.* 13:274–80

53. Xu L, Glass CK, Rosenfeld MG. 1999. Coactivator and corepressor complexes in nuclear receptor function. *Curr. Opin. Genet. Dev.* 9:140–47

54. Darimont BD, Wagner RL, Apriletti JW, Stallcup MR, Kushner PJ, et al. 1998. Structure and specificity of nuclear

receptor-coactivator interactions. *Genes Dev.* 12:3343–56

55. Mak HY, Hoare S, Henttu PM, Parker MG. 1999. Molecular determinants of the estrogen receptor-coactivator interface. *Mol. Cell Biol.* 19:3895–903

56. Nolte RT, Wisely GB, Westin S, Cobb JE, Lambert MH, et al. 1998. Ligand binding and co-activator assembly of the peroxisome proliferator-activated receptor-gamma. *Nature* 395:137–43

57. Chen H, Lin RJ, Xie W, Wilpitz D, Evans RM. 1999. Regulation of hormone-induced histone hyperacetylation and gene activation via acetylation of an acetylase. *Cell* 98:675–86

58. DiRenzo J, Shang Y, Phelan M, Sif S, Myers M, et al. 2000. BRG-1 is recruited to estrogen-responsive promoters and cooperates with factors involved in histone acetylation. *Mol. Cell Biol.* 20:7541–49

59. Shang Y, Hu X, DiRenzo J, Lazar MA, Brown M. 2000. Cofactor dynamics and sufficiency in estrogen receptor-regulated transcription. *Cell* 103:843–52

60. Sharma D, Fondell JD. 2000. Temporal formation of distinct thyroid hormone receptor coactivator complexes in HeLa cells. *Mol. Endocrinol.* 14:2001–9

61. Sharma D, Fondell JD. 2002. Ordered recruitment of histone acetyltransferases and the TRAP/Mediator complex to thyroid hormone-responsive promoters in vivo. *Proc. Natl. Acad. Sci. USA* 99:7934–39

62. Struhl K. 2001. Gene regulation. A paradigm for precision. *Science* 293:1054–55

63. Yang W, Rachez C, Freedman LP. 2000. Discrete roles for peroxisome proliferator-activated receptor gamma and retinoid X receptor in recruiting nuclear receptor coactivators. *Mol. Cell Biol.* 20:8008–17

64. Chen JD, Evans RM. 1995. A transcriptional co-repressor that interacts with nuclear hormone receptors. *Nature* 377:454–57

65. Hörlein AJ, Näär AM, Heinzel T, Torchia J, Gloss B, et al. 1995. Ligand-independent repression by the thyroid hormone receptor mediated by a nuclear receptor co-repressor. *Nature* 377:397–404

66. Sande S, Privalsky ML. 1996. Identification of TRACs (T3 receptor-associating cofactors), a family of cofactors that associate with, and modulate the activity of, nuclear hormone receptors. *Mol. Endocrinol.* 10:813–25

67. Seol W, Mahon MJ, Lee YK, Moore DD. 1996. Two receptor interacting domains in the nuclear hormone receptor corepressor RIP13/N-CoR. *Mol. Endocrinol.* 10:1646–55

68. Zamir I, Harding HP, Atkins GB, Horlein A, Glass CK, et al. 1996. A nuclear hormone receptor corepressor mediates transcriptional silencing by receptors with distinct repression domains. *Mol. Cell Biol.* 16:458–65

69. Tsai CC, Kao HY, Yao TP, McKeown M, Evans RM. 1999. SMRTER, a *Drosophila* nuclear receptor coregulator, reveals that EcR-mediated repression is critical for development. *Mol. Cell* 4:175–86

70. Jones PL, Shi YB. 2003. N-CoR-HDAC corepressor complexes: roles in transcriptional regulation by nuclear hormone receptors. *Curr. Top. Microbiol. Immunol.* 274:237–68

71. Nawaz Z, Tsai MJ, O'Malley BW. 1995. Specific mutations in the ligand binding domain selectively abolish the silencing function of human thyroid hormone receptor beta. *Proc. Natl. Acad. Sci. USA* 92:11691–95

72. Tong GX, Jeyakumar M, Tanen MR, Bagchi MK. 1996. Transcriptional silencing by unliganded thyroid hormone receptor beta requires a soluble corepressor that interacts with the ligand-binding domain of the receptor. *Mol. Cell Biol.* 16:1909–20

73. Torchia J, Glass C, Rosenfeld MG. 1998. Co-activators and co-repressors in the integration of transcriptional responses. *Curr. Opin. Cell Biol.* 10:373–83

74. Muscat GE, Burke LJ, Downes M. 1998. The corepressor N-CoR and its variants RIP13a and RIP13Delta1 directly interact with the basal transcription factors TFIIB, TAFII32 and TAFII70. *Nucleic Acids Res.* 26:2899–907

75. Wong CW, Privalsky ML. 1998. Transcriptional repression by the SMRT-mSin3 corepressor: multiple interactions, multiple mechanisms, and a potential role for TFIIB. *Mol. Cell Biol.* 18:5500–10

76. Goodman RH, Smolik S. 2000. CBP/p300 in cell growth, transformation, and development. *Genes Dev.* 14:1553–77

77. Chan HM, La Thangue NB. 2001. p300/CBP proteins: HATs for transcriptional bridges and scaffolds. *J. Cell Sci.* 114:2363–73

78. McManus KJ, Hendzel MJ. 2001. CBP, a transcriptional coactivator and acetyltransferase. *Biochem. Cell Biol.* 79:253–66

79. Blobel GA. 2002. CBP and p300: versatile coregulators with important roles in hematopoietic gene expression. *J. Leukoc. Biol.* 71:545–56

80. Baek SH, Ohgi KA, Rose DW, Koo EH, Glass CK, Rosenfeld MG. 2002. Exchange of N-CoR corepressor and Tip60 coactivator complexes links gene expression by NF-kappaB and beta-amyloid precursor protein. *Cell* 110:55–67

81. Espinosa L, Ingles-Esteve J, Robert-Moreno A, Bigas A. 2003. IkappaBalpha and p65 regulate the cytoplasmic shuttling of nuclear corepressors: cross-talk between Notch and NFkappaB pathways. *Mol. Biol. Cell* 14:491–502

82. Espinosa L, Santos S, Ingles-Esteve J, Munoz-Canoves P, Bigas A. 2002. p65-NFkappaB synergizes with Notch to activate transcription by triggering cytoplasmic translocation of the nuclear receptor corepressor N-CoR. *J. Cell Sci.* 115:295–303

83. Jang MK, Goo YH, Sohn YC, Kim YS, Lee SK, et al. 2001. Ca^{2+}/calmodulin-dependent protein kinase IV stimulates nuclear factor-kappa B transactivation via phosphorylation of the p65 subunit. *J. Biol. Chem.* 276:20005–10

84. Lee SK, Kim JH, Lee YC, Cheong J, Lee JW. 2000. Silencing mediator of retinoic acid and thyroid hormone receptors, as a novel transcriptional corepressor molecule of activating protein-1, nuclear factor-kappaB, and serum response factor. *J. Biol. Chem.* 275:12470–74

85. Luo K, Stroschein SL, Wang W, Chen D, Martens E, et al. 1999. The Ski oncoprotein interacts with the Smad proteins to repress TGFbeta signaling. *Genes Dev.* 13:2196–206

86. Kao HY, Ordentlich P, Koyano-Nakagawa N, Tang Z, Downes M, et al. 1998. A histone deacetylase corepressor complex regulates the Notch signal transduction pathway. *Genes Dev.* 12:2269–77

87. Oswald F, Kostezka U, Astrahantseff K, Bourteele S, Dillinger K, et al. 2002. SHARP is a novel component of the Notch/RBP-Jkappa signalling pathway. *EMBO J.* 21:5417–26

88. Zhou S, Hayward SD. 2001. Nuclear localization of CBF1 is regulated by interactions with the SMRT corepressor complex. *Mol. Cell Biol.* 21:6222–32

89. Li X, McDonnell DP. 2002. The transcription factor B-Myb is maintained in an inhibited state in target cells through its interaction with the nuclear corepressors N-CoR and SMRT. *Mol. Cell Biol.* 22:3663–73

90. He LZ, Guidez F, Tribioli C, Peruzzi D, Ruthardt M, et al. 1998. Distinct interactions of PML-RARalpha and PLZF-RARalpha with co-repressors determine differential responses to RA in APL. *Nat. Genet.* 18:126–35

91. Hong SH, David G, Wong CW, Dejean A, Privalsky ML. 1997. SMRT corepressor interacts with PLZF and with the PML-retinoic acid receptor alpha (RARalpha) and PLZF-RARalpha oncoproteins associated with acute promyelocytic

leukemia. *Proc. Natl. Acad. Sci. USA* 94:9028–33

92. Lin RJ, Nagy L, Inoue S, Shao W, Miller WH Jr, Evans RM. 1998. Role of the histone deacetylase complex in acute promyelocytic leukaemia. *Nature* 391: 811–14

93. Melnick A, Carlile G, Ahmad KF, Kiang CL, Corcoran C, et al. 2002. Critical residues within the BTB domain of PLZF and Bcl-6 modulate interaction with corepressors. *Mol. Cell Biol.* 22:1804–18

94. Wong CW, Privalsky ML. 1998. Components of the SMRT corepressor complex exhibit distinctive interactions with the POZ domain oncoproteins PLZF, PLZF-RARalpha, and BCL-6. *J. Biol. Chem.* 273:27695–702

95. Dhordain P, Albagli O, Lin RJ, Ansieau S, Quief S, et al. 1997. Corepressor SMRT binds the BTB/POZ repressing domain of the LAZ3/BCL6 oncoprotein. *Proc. Natl. Acad. Sci. USA* 94:10762–67

96. Dhordain P, Lin RJ, Quief S, Lantoine D, Kerckaert JP, et al. 1998. The LAZ3(BCL-6) oncoprotein recruits a SMRT/mSIN3A/histone deacetylase containing complex to mediate transcriptional repression. *Nucleic Acids Res.* 26: 4645–51

97. Asahara H, Dutta S, Kao HY, Evans RM, Montminy M. 1999. Pbx-Hox heterodimers recruit coactivator-corepressor complexes in an isoform-specific manner. *Mol. Cell Biol.* 19:8219–25

98. Laherty CD, Billin AN, Lavinsky RM, Yochum GS, Bush AC, et al. 1998. SAP30, a component of the mSin3 corepressor complex involved in N-CoR-mediated repression by specific transcription factors. *Mol. Cell* 2:33–42

99. Saleh M, Rambaldi I, Yang XJ, Featherstone MS. 2000. Cell signaling switches HOX-PBX complexes from repressors to activators of transcription mediated by histone deacetylases and histone acetyltransferases. *Mol. Cell Biol.* 20:8623–33

100. Scully KM, Jacobson EM, Jepsen K, Lunyak V, Viadiu H, et al. 2000. Allosteric effects of Pit-1 DNA sites on long-term repression in cell type specification. *Science* 290:1127–31

101. Xu L, Lavinsky RM, Dasen JS, Flynn SE, McInerney EM, et al. 1998. Signal-specific co-activator domain requirements for Pit-1 activation. *Nature* 395:301–6

102. Amann JM, Nip J, Strom DK, Lutterbach B, Harada H, et al. 2001. ETO, a target of t(8;21) in acute leukemia, makes distinct contacts with multiple histone deacetylases and binds mSin3A through its oligomerization domain. *Mol. Cell Biol.* 21:6470–83

103. Gelmetti V, Zhang J, Fanelli M, Minucci S, Pelicci PG, Lazar MA. 1998. Aberrant recruitment of the nuclear receptor corepressor-histone deacetylase complex by the acute myeloid leukemia fusion partner ETO. *Mol. Cell Biol.* 18:7185–91

104. Heibert SW, Lutterbach B, Durst K, Wang L, Linggi B, et al. 2001. Mechanisms of transcriptional repression by the t(8;21)-, t(12;21)-, and inv(16)-encoded fusion proteins. *Cancer Chemother. Pharmacol.* 48(Suppl. 1):S31–34

105. Lutterbach B, Westendorf JJ, Linggi B, Patten A, Moniwa M, et al. 1998. ETO, a target of t(8;21) in acute leukemia, interacts with the N-CoR and mSin3 corepressors. *Mol. Cell Biol.* 18:7176–84

106. Wang J, Hoshino T, Redner RL, Kajigaya S, Liu JM. 1998. ETO, fusion partner in t(8;21) acute myeloid leukemia, represses transcription by interaction with the human N-CoR/mSin3/HDAC1 complex. *Proc. Natl. Acad. Sci. USA* 95:10860–65

107. Nguyen TA, Hoivik D, Lee JE, Safe S. 1999. Interactions of nuclear receptor coactivator/corepressor proteins with the aryl hydrocarbon receptor complex. *Arch. Biochem. Biophys.* 367:250–57

108. Bailey P, Downes M, Lau P, Harris J, Chen SL, et al. 1999. The nuclear receptor corepressor N-CoR regulates differentiation: N-CoR directly interacts

with MyoD. *Mol. Endocrinol.* 13:1155–68

109. Sande S, Sharif M, Chen H, Privalsky M. 1993. v-erbA acts on retinoic acid receptors in immature avian erythroid cells. *J. Virol.* 67:1067–74

110. Ordentlich P, Downes M, Xie W, Genin A, Spinner NB, Evans RM. 1999. Unique forms of human and mouse nuclear receptor corepressor SMRT. *Proc. Natl. Acad. Sci. USA* 96:2639–44

111. Park EJ, Schroen DJ, Yang M, Li H, Li L, Chen JD. 1999. SMRTe, a silencing mediator for retinoid and thyroid hormone receptors-extended isoform that is more related to the nuclear receptor corepressor. *Proc. Natl. Acad. Sci. USA* 96:3519–24

112. Chen JD, Umesono K, Evans RM. 1996. SMRT isoforms mediate repression and anti-repression of nuclear receptor heterodimers. *Proc. Natl. Acad. Sci. USA* 93:7567–71

113. Downes M, Burke LJ, Bailey PJ, Muscat GE. 1996. Two receptor interaction domains in the corepressor, N-CoR/RIP13, are required for an efficient interaction with Rev-erbA alpha and RVR: Physical association is dependent on the E region of the orphan receptors. *Nucleic Acids Res.* 24:4379–86

114. Harding HP, Atkins GB, Jaffe AB, Seo WJ, Lazar MA. 1997. Transcriptional activation and repression by RORalpha, an orphan nuclear receptor required for cerebellar development. *Mol. Endocrinol.* 11:1737–46

115. Li H, Leo C, Schroen DJ, Chen JD. 1997. Characterization of receptor interaction and transcriptional repression by the corepressor SMRT. *Mol. Endocrinol.* 11:2025–37

116. Wong CW, Privalsky ML. 1998. Transcriptional silencing is defined by isoform- and heterodimer-specific interactions between nuclear hormone receptors and corepressors. *Mol. Cell Biol.* 18:5724–33

117. Zamir I, Zhang J, Lazar MA. 1997. Stoi-chiometric and steric principles governing repression by nuclear hormone receptors. *Genes Dev.* 11:835–46

118. Zamir I, Dawson J, Lavinsky RM, Glass CK, Rosenfeld MG, Lazar MA. 1997. Cloning and characterization of a corepressor and potential component of the nuclear hormone receptor repression complex. *Proc. Natl. Acad. Sci. USA* 94:14400–5

119. Glass CK, Rosenfeld MG. 2000. The coregulator exchange in transcriptional functions of nuclear receptors. *Genes Dev.* 14:121–41

120. Privalsky ML. 2001. Regulation of SMRT and N-CoR corepressor function. *Curr. Top. Microbiol. Immunol.* 254:117–36

121. Cohen RN, Brzostek S, Kim B, Chorev M, Wondisford FE, Hollenberg AN. 2001. The specificity of interactions between nuclear hormone receptors and corepressors is mediated by distinct amino acid sequences within the interacting domains. *Mol. Endocrinol.* 15:1049–61

122. Makowski A, Brzostek S, Cohen RN, Hollenberg AN. 2003. Determination of nuclear receptor corepressor interactions with the thyroid hormone receptor. *Mol. Endocrinol.* 17:273–86

123. Webb P, Anderson CM, Valentine C, Nguyen P, Marimuthu A, et al. 2000. The nuclear receptor corepressor (N-CoR) contains three isoleucine motifs (I/LXXII) that serve as receptor interaction domains (IDs). *Mol. Endocrinol.* 14:1976–85

124. Cohen RN, Putney A, Wondisford FE, Hollenberg AN. 2000. The nuclear corepressors recognize distinct nuclear receptor complexes. *Mol. Endocrinol.* 14:900–14

125. Lavinsky RM, Jepsen K, Heinzel T, Torchia J, Mullen TM, et al. 1998. Diverse signaling pathways modulate nuclear receptor recruitment of N-CoR and SMRT complexes. *Proc. Natl. Acad. Sci. USA* 95:2920–25

126. Cohen RN, Wondisford FE, Hollenberg

AN. 1998. Two separate NCoR (nuclear receptor corepressor) interaction domains mediate corepressor action on thyroid hormone response elements. *Mol. Endocrinol.* 12:1567–81

127. Zhang J, Zamir I, Lazar MA. 1997. Differential recognition of liganded and unliganded thyroid hormone receptor by retinoid X receptor regulates transcriptional repression. *Mol. Cell Biol.* 17: 6887–97

128. Yoh SM, Privalsky ML. 2001. Transcriptional repression by thyroid hormone receptors. A role for receptor homodimers in the recruitment of SMRT corepressor. *J. Biol. Chem.* 276:16857–67

129. Hu X, Lazar MA. 1999. The CoRNR motif controls the recruitment of corepressors by nuclear hormone receptors. *Nature* 402:93–96

130. Hu X, Li Y, Lazar MA. 2001. Determinants of CoRNR-dependent repression complex assembly on nuclear hormone receptors. *Mol. Cell Biol.* 21:1747–58

131. Marimuthu A, Feng W, Tagami T, Nguyen H, Jameson JL, et al. 2002. TR surfaces and conformations required to bind nuclear receptor corepressor. *Mol. Endocrinol.* 16:271–86

132. Perissi V, Staszewski LM, McInerney EM, Kurokawa R, Krones A, et al. 1999. Molecular determinants of nuclear receptor-corepressor interaction. *Genes Dev.* 13:3198–208

133. Baniahmad A, Leng X, Burris TP, Tsai SY, Tsai MJ, O'Malley BW. 1995. The tau 4 activation domain of the thyroid hormone receptor is required for release of a putative corepressor(s) necessary for transcriptional silencing. *Mol. Cell Biol.* 15:76–86

134. Baniahmad A, Thormeyer D, Renkawitz R. 1997. tau4/tau c/AF-2 of the thyroid hormone receptor relieves silencing of the retinoic acid receptor silencer core independent of both tau 4 activation function and full dissociation of corepressors. *Mol. Cell Biol.* 17:4259–71

135. Casanova J, Helmer E, Selmi-Ruby S, Qi JS, Au-Fliegner M, et al. 1994. Functional evidence for ligand-dependent dissociation of thyroid hormone and retinoic acid receptors from an inhibitory cellular factor. *Mol. Cell Biol.* 14:5756–65

136. Lin BC, Hong SH, Krig S, Yoh SM, Privalsky ML. 1997. A conformational switch in nuclear hormone receptors is involved in coupling hormone binding to corepressor release. *Mol. Cell Biol.* 17:6131–38

137. Nagy L, Kao HY, Love JD, Li C, Banayo E, et al. 1999. Mechanism of corepressor binding and release from nuclear hormone receptors. *Genes Dev.* 13:3209–16

138. Sachs LM, Jones PL, Havis E, Rouse N, Demeneix BA, Shi YB. 2002. Nuclear receptor corepressor recruitment by unliganded thyroid hormone receptor in gene repression during *Xenopus laevis* development. *Mol. Cell Biol.* 22:8527–38

139. Zhang J, Hu X, Lazar MA. 1999. A novel role for helix 12 of retinoid X receptor in regulating repression. *Mol. Cell Biol.* 19: 6448–57

140. Bailey PJ, Dowhan DH, Franke K, Burke LJ, Downes M, Muscat GE. 1997. Transcriptional repression by COUP-TF II is dependent on the C-terminal domain and involves the N-CoR variant, RIP13delta1. *J. Steroid Biochem. Mol. Biol.* 63:165–74

141. Shibata H, Nawaz Z, Tsai SY, O'Malley BW, Tsai MJ. 1997. Gene silencing by chicken ovalbumin upstream promoter-transcription factor I (COUP-TFI) is mediated by transcriptional corepressors, nuclear receptor-corepressor (N-CoR) and silencing mediator for retinoic acid receptor and thyroid hormone receptor (SMRT). *Mol. Endocrinol.* 11:714–24

142. Dowell P, Ishmael JE, Avram D, Peterson VJ, Nevrivy DJ, Leid M. 1999. Identification of nuclear receptor corepressor as a peroxisome proliferator-activated receptor alpha interacting protein. *J. Biol. Chem.* 274:15901–7

143. Dwivedi PP, Muscat GE, Bailey PJ, Omdahl JL, May BK. 1998. Repression of basal transcription by vitamin D receptor: evidence for interaction of unliganded vitamin D receptor with two receptor interaction domains in RIP13delta1. *J. Mol. Endocrinol.* 20:327–35

144. Herdick M, Carlberg C. 2000. Agonist-triggered modulation of the activated and silent state of the vitamin D(3) receptor by interaction with co-repressors and co-activators. *J. Mol. Biol.* 304:793–801

145. Lee G, Elwood F, McNally J, Weiszmann J, Lindstrom M, et al. 2002. T0070907, a selective ligand for peroxisome proliferator-activated receptor gamma, functions as an antagonist of biochemical and cellular activities. *J. Biol. Chem.* 277:19649–57

146. Ruse MD Jr, Privalsky ML, Sladek FM. 2002. Competitive cofactor recruitment by orphan receptor hepatocyte nuclear factor 4alpha1: modulation by the F domain. *Mol. Cell Biol.* 22:626–38

147. Shi Y, Hon M, Evans RM. 2002. The peroxisome proliferator-activated receptor delta, an integrator of transcriptional repression and nuclear receptor signaling. *Proc. Natl. Acad. Sci. USA* 99:2613–18

148. Sohn YC, Kwak E, Na Y, Lee JW, Lee SK. 2001. Silencing mediator of retinoid and thyroid hormone receptors and activating signal cointegrator-2 as transcriptional coregulators of the orphan nuclear receptor Nur77. *J. Biol. Chem* 276:43734—39

149. Tagami T, Lutz WH, Kumar R, Jameson JL. 1998. The interaction of the vitamin D receptor with nuclear receptor corepressors and coactivators. *Biochem. Biophys. Res. Commun.* 253:358–63

150. Torres-Padilla ME, Sladek FM, Weiss MC. 2002. Developmentally regulated N-terminal variants of the nuclear receptor hepatocyte nuclear factor 4alpha mediate multiple interactions through coactivator and corepressor-histone deacetylase complexes. *J. Biol. Chem.* 277:4677–87

151. Dotzlaw H, Moehren U, Mink S, Cato AC, Iniguez Lluhi JA, Baniahmad A. 2002. The amino terminus of the human AR is target for corepressor action and antihormone agonism. *Mol. Endocrinol.* 16:661–73

152. Jackson TA, Richer JK, Bain DL, Takimoto GS, Tung L, Horwitz KB. 1997. The partial agonist activity of antagonist-occupied steroid receptors is controlled by a novel hinge domain-binding coactivator L7/SPA and the corepressors N-CoR or SMRT. *Mol. Endocrinol.* 11:693–705

153. Liu Z, Auboeuf D, Wong J, Chen JD, Tsai SY, et al. 2002. Coactivator/corepressor ratios modulate PR-mediated transcription by the selective receptor modulator RU486. *Proc. Natl. Acad. Sci. USA* 99:7940–44

154. Schulz M, Eggert M, Baniahmad A, Dostert A, Heinzel T, Renkawitz R. 2002. RU486-induced glucocorticoid receptor agonism is controlled by the receptor N terminus and by corepressor binding. *J. Biol. Chem.* 277:26238–43

155. Shang Y, Brown M. 2002. Molecular determinants for the tissue specificity of SERMs. *Science* 295:2465–68

156. Shang Y, Myers M, Brown M. 2002. Formation of the androgen receptor transcription complex. *Mol. Cell* 9:601–10

157. Smith CL, Nawaz Z, O'Malley BW. 1997. Coactivator and corepressor regulation of the agonist/antagonist activity of the mixed antiestrogen, 4-hydroxytamoxifen. *Mol. Endocrinol.* 11:657–66

158. Xu J, Nawaz Z, Tsai SY, Tsai MJ, O'Malley BW. 1996. The extreme C terminus of progesterone receptor contains a transcriptional repressor domain that functions through a putative corepressor. *Proc. Natl. Acad. Sci. USA* 93:12195–99

159. Zhang X, Jeyakumar M, Petukhov S, Bagchi MK. 1998. A nuclear receptor corepressor modulates transcriptional activity of antagonist-occupied steroid hormone receptor. *Mol. Endocrinol.* 12:513–24

160. Jepsen K, Hermanson O, Onami TM, Gleiberman AS, Lunyak V, et al. 2000. Combinatorial roles of the nuclear receptor corepressor in transcription and development. *Cell* 102:753–63

161. Takimoto GS, Graham JD, Jackson TA, Tung L, Powell RL, et al. 1999. Tamoxifen resistant breast cancer: coregulators determine the direction of transcription by antagonist-occupied steroid receptors. *J. Steroid Biochem. Mol. Biol.* 69:45–50

162. Wagner BL, Norris JD, Knotts TA, Weigel NL, McDonnell DP. 1998. The nuclear corepressors NCoR and SMRT are key regulators of both ligand- and 8-bromocyclic AMP-dependent transcriptional activity of the human progesterone receptor. *Mol. Cell Biol.* 18:369–78

163. Elholm M, Dam I, Jorgensen C, Krogsdam AM, Holst D, et al. 2001. Acyl-CoA esters antagonize the effects of ligands on peroxisome proliferator-activated receptor alpha conformation, DNA binding, and interaction with co-factors. *J. Biol. Chem.* 276:21410–16

164. Hong SH, Privalsky ML. 1999. Retinoid isomers differ in the ability to induce release of SMRT corepressor from retinoic acid receptor-alpha. *J. Biol. Chem.* 274:2885–92

165. Hu X, Li S, Wu J, Xia C, Lala DS. 2003. Liver X receptors interact with corepressors to regulate gene expression. *Mol. Endocrinol.* 17:1019–26

166. Chambon P. 1996. A decade of molecular biology of retinoic acid receptors. *FASEB J.* 10:40–54

167. Sucov HM, Evans RM. 1995. Retinoic acid and retinoic acid receptors in development. *Mol. Neurobiol.* 10:169–84

168. Hauksdottir H, Farboud B, Privalsky ML. 2003. Retinoic acid receptors beta and gamma do not repress, but instead activate target gene transcription in both the absence and presence of hormone ligand. *Mol. Endocrinol.* 17:373–85

169. Farboud B, Hauksdottir H, Wu Y, Privalsky ML. 2003. Isotype-restricted corepressor recruitment: a constitutively closed helix 12 conformation in retinoic acid receptors beta and gamma interferes with corepressor recruitment and prevents transcriptional repression. *Mol. Cell Biol.* 23: 2844–58

170. Klein ES, Wang JW, Khalifa B, Gavigan SA, Chandraratna RA. 2000. Recruitment of nuclear receptor corepressor and coactivator to the retinoic acid receptor by retinoid ligands. Influence of DNA-heterodimer interactions. *J. Biol. Chem.* 275:19401–8

171. Westin S, Kurokawa R, Nolte RT, Wisely GB, McInerney EM, et al. 1998. Interactions controlling the assembly of nuclear-receptor heterodimers and co-activators. *Nature* 395:199–202

172. Germain P, Iyer J, Zechel C, Gronemeyer H. 2002. Co-regulator recruitment and the mechanism of retinoic acid receptor synergy. *Nature* 415:187–92

173. Kurokawa R, DiRenzo J, Boehm M, Sugarman J, Gloss B, et al. 1994. Regulation of retinoid signalling by receptor polarity and allosteric control of ligand binding. *Nature* 371:528–31

174. Kurokawa R, Söderström M, Hörlein A, Halachmi S, Brown M, et al. 1995. Polarity-specific activities of retinoic acid receptors determined by a co-repressor. *Nature* 377:451–54

175. Olson DP, Sun B, Koenig RJ. 1998. Thyroid hormone response element architecture affects corepressor release from thyroid hormone receptor dimers. *J. Biol. Chem.* 273:3375–80

176. Fischle W, Dequiedt F, Hendzel MJ, Guenther MG, Lazar MA, et al. 2002. Enzymatic activity associated with class II HDACs is dependent on a multiprotein complex containing HDAC3 and SMRT/N-CoR. *Mol. Cell* 9:45–57

177. Guenther MG, Lane WS, Fischle W, Verdin E, Lazar MA, Shiekhattar R. 2000. A core SMRT corepressor

complex containing HDAC3 and TBL1, a WD40-repeat protein linked to deafness. *Genes Dev.* 14:1048–57

178. Guenther MG, Barak O, Lazar MA. 2001. The SMRT and N-CoR corepressors are activating cofactors for histone deacetylase 3. *Mol. Cell Biol.* 21:6091–101

179. Guenther MG, Yu J, Kao GD, Yen TJ, Lazar MA. 2002. Assembly of the SMRT-histone deacetylase 3 repression complex requires the TCP-1 ring complex. *Genes Dev.* 16:3130–35

180. Li J, Wang J, Nawaz Z, Liu JM, Qin J, Wong J. 2000. Both corepressor proteins SMRT and N-CoR exist in large protein complexes containing HDAC3. *EMBO J.* 19:4342–50

181. Underhill C, Qutob MS, Yee SP, Torchia J. 2000. A novel nuclear receptor corepressor complex, N-CoR, contains components of the mammalian SWI/SNF complex and the corepressor KAP-1. *J. Biol. Chem.* 275:40463–70

182. Urnov FD, Yee J, Sachs L, Collingwood TN, Bauer A, et al. 2000. Targeting of N-CoR and histone deacetylase 3 by the oncoprotein v-erbA yields a chromatin infrastructure-dependent transcriptional repression pathway. *EMBO J.* 19: 4074–90

183. Wen YD, Perissi V, Staszewski LM, Yang WM, Krones A, et al. 2000. The histone deacetylase-3 complex contains nuclear receptor corepressors. *Proc. Natl. Acad. Sci. USA* 97:7202–7

184. Zhang J, Kalkum M, Chait BT, Roeder RG. 2002. The N-CoR-HDAC3 nuclear receptor corepressor complex inhibits the JNK pathway through the integral subunit GPS2. *Mol. Cell* 9:611–23

185. Fischle W, Dequiedt F, Fillion M, Hendzel MJ, Voelter W, Verdin E. 2001. Human HDAC7 histone deacetylase activity is associated with HDAC3 in vivo. *J. Biol. Chem.* 276:35826–35

186. Huang EY, Zhang J, Miska EA, Guenther MG, Kouzarides T, Lazar MA. 2000. Nuclear receptor corepressors partner with

class II histone deacetylases in a Sin3-independent repression pathway. *Genes Dev.* 14:45–54

187. Kao HY, Downes M, Ordentlich P, Evans RM. 2000. Isolation of a novel histone deacetylase reveals that class I and class II deacetylases promote SMRT-mediated repression. *Genes Dev.* 14:55–66

188. Wu X, Li H, Park EJ, Chen JD. 2001. SMRTE inhibits MEF2C transcriptional activation by targeting HDAC4 and 5 to nuclear domains. *J. Biol. Chem.* 276: 24177–85

189. Yoon HG, Chan DW, Huang ZQ, Li J, Fondell JD, et al. 2003. Purification and functional characterization of the human N-CoR complex: the roles of HDAC3, TBL1 and TBLR1. *EMBO J.* 22:1336–46

190. Alland L, Muhle R, Hou H Jr, Potes J, Chin L, et al. 1997. Role for N-CoR and histone deacetylase in Sin3-mediated transcriptional repression. *Nature* 387: 49–55

191. Heinzel T, Lavinsky RM, Mullen TM, Söderstrom M, Laherty CD, et al. 1997. A complex containing N-CoR, mSin3 and histone deacetylase mediates transcriptional repression. *Nature* 387:43–48

192. Nagy L, Kao HY, Chakravarti D, Lin RJ, Hassig CA, et al. 1997. Nuclear receptor repression mediated by a complex containing SMRT, mSin3A, and histone deacetylase. *Cell* 89:373–80

193. Alland L, David G, Shen-Li H, Potes J, Muhle R, et al. 2002. Identification of mammalian Sds3 as an integral component of the Sin3/histone deacetylase corepressor complex. *Mol. Cell Biol.* 22:2743–50

194. Ayer DE. 1999. Histone deacetylases: transcriptional repression with SINers and NuRDs. *Trends Cell Biol.* 9:193–98

195. Boehmelt G, Antonio L, Iscove NN. 1998. Cloning of the murine transcriptional corepressor component SAP18 and differential expression of its mRNA in the

hematopoietic hierarchy. *Gene* 207:267–75

196. Grandori C, Cowley SM, James LP, Eisenman RN. 2000. The Myc/Max/Mad network and the transcriptional control of cell behavior. *Annu. Rev. Cell Dev. Biol.* 16:653–99

197. Laherty CD, Yang WM, Sun JM, Davie JR, Seto E, Eisenman RN. 1997. Histone deacetylases associated with the mSin3 corepressor mediate mad transcriptional repression. *Cell* 89:349–56

198. Li J, Lin Q, Wang W, Wade P, Wong J. 2002. Specific targeting and constitutive association of histone deacetylase complexes during transcriptional repression. *Genes Dev.* 16:687–92

199. Rietveld LE, Caldenhoven E, Stunnenberg HG. 2002. In vivo repression of an erythroid-specific gene by distinct corepressor complexes. *EMBO J.* 21:1389–97

200. Downes M, Ordentlich P, Kao HY, Alvarez JG, Evans RM. 2000. Identification of a nuclear domain with deacetylase activity. *Proc. Natl. Acad. Sci. USA* 97:10330–35

201. Jones PL, Sachs LM, Rouse N, Wade PA, Shi YB. 2001. Multiple N-CoR complexes contain distinct histone deacetylases. *J. Biol. Chem.* 276:8807–11

202. Forrest D, Vennstrom B. 2000. Functions of thyroid hormone receptors in mice. *Thyroid* 10:41–52

203. Wu Y, Koenig RJ. 2000. Gene regulation by thyroid hormone. *Trends Endocrinol. Metab.* 11:207–11

204. Hollenberg AN, Monden T, Wondisford FE. 1995. Ligand-independent and -dependent functions of thyroid hormone receptor isoforms depend upon their distinct amino termini. *J. Biol. Chem.* 270:14274–80

205. Sjoberg M, Vennstrom B. 1995. Ligand-dependent and -independent transactivation by thyroid hormone receptor beta-2 is determined by the structure of the hormone response element. *Mol. Cell Biol.* 15:4718–26

206. Hollenberg AN, Monden T, Madura JP, Lee K, Wondisford FE. 1996. Function of nuclear co-repressor protein on thyroid hormone response elements is regulated by the receptor A/B domain. *J. Biol. Chem.* 271:28516–20

207. Yang Z, Hong SH, Privalsky ML. 1999. Transcriptional anti-repression. Thyroid hormone receptor beta-2 recruits SMRT corepressor but interferes with subsequent assembly of a functional corepressor complex. *J. Biol. Chem.* 274:37131–38

208. Nomura T, Khan MM, Kaul SC, Dong HD, Wadhwa R, et al. 1999. Ski is a component of the histone deacetylase complex required for transcriptional repression by Mad and thyroid hormone receptor. *Genes Dev.* 13:412–23

209. Ueki N, Hayman MJ. 2003. Signal-dependent N-CoR requirement for repression by the Ski oncoprotein. *J. Biol. Chem.* 278:24858–64

210. Shinagawa T, Dong HD, Xu M, Maekawa T, Ishii S. 2000. The *sno* gene, which encodes a component of the histone deacetylase complex, acts as a tumor suppressor in mice. *EMBO J.* 19:2280–91

211. Shi Y, Downes M, Xie W, Kao HY, Ordentlich P, et al. 2001. Sharp, an inducible cofactor that integrates nuclear receptor repression and activation. *Genes Dev.* 15:1140–51

212. Grunstein M. 1997. Histone acetylation in chromatin structure and transcription. *Nature* 389:349–52

213. Pazin MJ, Kadonaga JT. 1997. What's up and down with histone deacetylation and transcription? *Cell* 89:325–38

214. Perlmann T, Vennstrom B. 1995. The sound of silence. *Nature* 377:387–38

215. Workman JL, Kingston RE. 1998. Alteration of nucleosome structure as a mechanism of transcriptional regulation. *Annu. Rev. Biochem.* 67:545–79

216. Kuzmichev A, Reinberg D. 2001. Role of histone deacetylase complexes in the regulation of chromatin metabolism. *Curr. Top. Microbiol. Immunol.* 254:35–58

217. Urnov FD, Wolffe AP, Guschin D. 2001. Molecular mechanisms of corepressor function. *Curr. Top. Microbiol. Immunol.* 254:1–33

218. Berger SL. 2001. Molecular biology. The histone modification circus. [Comment On: Science. 2001 Apr 6;292(5514):110-3 UI: 21189485]. *Science* 292:64–65

219. Chen H, Tini M, Evans RM. 2001. HATs on and beyond chromatin. *Curr. Opin. Cell Biol.* 13:218–24

220. Freiman RN, Tjian R. 2003. Regulating the regulators: lysine modifications make their mark. *Cell* 112:11–17

221. Jenuwein T, Allis CD. 2001. Translating the histone code. *Science* 293:1074–80

222. Richards EJ, Elgin SC. 2002. Epigenetic codes for heterochromatin formation and silencing: rounding up the usual suspects. *Cell* 108:489–500

223. Strahl BD, Allis CD. 2000. The language of covalent histone modifications. *Nature* 403:41–45

224. Agalioti T, Chen G, Thanos D. 2002. Deciphering the transcriptional histone acetylation code for a human gene. *Cell* 111:381–92

225. Hassan AH, Prochasson P, Neely KE, Galasinski SC, Chandy M, et al. 2002. Function and selectivity of bromodomains in anchoring chromatin-modifying complexes to promoter nucleosomes. *Cell* 111:369–79

226. Mizzen CA, Allis CD. 2000. Transcription. New insights into an old modification. *Science* 289:2290–91

227. Zeng L, Zhou MM. 2002. Bromodomain: an acetyl-lysine binding domain. *FEBS Lett.* 513:124–28

228. Goll MG, Bestor TH. 2002. Histone modification and replacement in chromatin activation. *Genes Dev.* 16:1739–42

229. Zegerman P, Canas B, Pappin D, Kouzarides T. 2002. Histone H3 lysine 4 methylation disrupts binding of nucleosome remodeling and deacetylase (NuRD) repressor complex. *J. Biol. Chem.* 277: 11621–24

230. Daujat S, Bauer UM, Shah V, Turner B, Berger S, Kouzarides T. 2002. Crosstalk between CARM1 methylation and CBP acetylation on histone H3. *Curr. Biol.* 12:2090–97

231. Ma H, Baumann CT, Li H, Strahl BD, Rice R, et al. 2001. Hormone-dependent, CARM1-directed, arginine-specific methylation of histone H3 on a steroid-regulated promoter. *Curr. Biol.* 11: t1981–85

232. Stallcup MR. 2001. Role of protein methylation in chromatin remodeling and transcriptional regulation. *Oncogene* 20: 3014–20

233. Wang H, Huang ZQ, Xia L, Feng Q, Erdjument-Bromage H, et al. 2001. Methylation of histone H4 at arginine 3 facilitating transcriptional activation by nuclear hormone receptor. *Science* 293: 853–57

234. Rice JC, Allis CD. 2001. Histone methylation versus histone acetylation: new insights into epigenetic regulation. *Curr. Opin. Cell Biol.* 13:263–73

235. Ben-Porath I, Cedar H. 2001. Epigenetic crosstalk. *Mol. Cell.* 8:933–35

236. Bestor TH. 1998. Gene silencing. Methylation meets acetylation. *Nature* 393:311–12

237. Bird A. 2001. Molecular biology. Methylation talk between histones and DNA. *Science* 294:2113–15

238. Bird AP, Wolffe AP. 1999. Methylation-induced repression—belts, braces, and chromatin. *Cell* 99:451–54

239. Nan X, Ng HH, Johnson CA, Laherty CD, Turner BM, et al. 1998. Transcriptional repression by the methyl-CpG-binding protein MeCP2 involves a histone deacetylase complex. *Nature* 393:386–89

240. Razin A. 1998. CpG methylation, chromatin structure and gene silencing—a three-way connection. *EMBO J.* 17:4905–8

241. Baniahmad A, Ha I, Reinberg D, Tsai S, Tsai MJ, O'Malley BW. 1993. Interaction

of human thyroid hormone receptor-beta with transcription factor TFIIB may mediate target gene derepression and activation by thyroid hormone. *Proc. Natl. Acad. Sci. USA* 90:8832–36

242. Fondell JD, Brunel F, Hisatake K, Roeder RG. 1996. Unliganded thyroid hormone receptor alpha can target TATA-binding protein for transcriptional repression. *Mol. Cell Biol.* 16:281–87

243. Fondell JD, Roy AL, Roeder RG. 1993. Unliganded thyroid hormone receptor inhibits formation of a functional preinitiation complex—implications for active repression. *Genes Dev.* 7:1400–10

244. Hadzic E, Desaiyajnik V, Helmer E, Guo S, Wu SJ, et al. 1995. A 10-amino-acid sequence in the N-terminal A/B domain of thyroid hormone receptor alpha is essential for transcriptional activation and interaction with the general transcription factor TFIIB. *Mol. Cell Biol.* 15:4507–17

245. Jacq X, Brou C, Lutz Y, Davidson I, Chambon P, Tora L. 1994. Human TAF (II)30 is present in a distinct TFIID complex and is required for transcriptional activation by the estrogen receptor. *Cell* 79:107–17

246. Hager GL, Fletcher TM, Xiao N, Baumann CT, Muller WG, McNally JG. 2000. Dynamics of gene targeting and chromatin remodelling by nuclear receptors. *Biochem. Soc. Trans.* 28:405–10

247. McEwan IJ. 2000. Gene regulation through chromatin remodelling by members of the nuclear receptor superfamily. *Biochem. Soc. Trans.* 28:369–73

248. Feng Q, Zhang Y. 2003. The NuRD complex: linking histone modification to nucleosome remodeling. *Curr. Top. Microbiol. Immunol.* 274:269–90

249. Knoepfler PS, Eisenman RN. 1999. Sin meets NuRD and other tails of repression. *Cell* 99:447–50

250. Imhof A, Yang XJ, Ogryzko VV, Nakatani Y, Wolffe AP, Ge H. 1997. Acetylation of general transcription factors by histone acetyltransferases. *Curr. Biol.* 7:689–92

251. Sterner DE, Berger SL. 2000. Acetylation of histones and transcription-related factors. *Microbiol. Mol. Biol. Rev.* 64:435–59

252. Fu M, Wang C, Wang J, Zhang X, Sakamaki T, et al. 2002. Androgen receptor acetylation governs *trans* activation and MEKK1-induced apoptosis without affecting in vitro sumoylation and *trans*-repression function. *Mol. Cell Biol.* 22:3373–88

253. Zhou Y, Gross W, Hong SH, Privalsky ML. 2001. The SMRT corepressor is a target of phosphorylation by protein kinase CK2 (casein kinase II). *Mol. Cell Biochem.* 220:1–13

254. Hong SH, Wong CW, Privalsky ML. 1998. Signaling by tyrosine kinases negatively regulates the interaction between transcription factors and SMRT (silencing mediator of retinoic acid and thyroid hormone receptor) corepressor. *Mol. Endocrinol.* 12:1161–71

255. Hong SH, Privalsky ML. 2000. The SMRT corepressor is regulated by a MEK-1 kinase pathway: inhibition of corepressor function is associated with SMRT phosphorylation and nuclear export. *Mol. Cell Biol.* 20:6612–25

256. Kurokawa H, Lenferink AE, Simpson JF, Pisacane PI, Sliwkowski MX, et al. 2000. Inhibition of HER2/neu (erbB-2) and mitogen-activated protein kinases enhances tamoxifen action against HER2-overexpressing, tamoxifen-resistant breast cancer cells. *Cancer Res.* 60:5887–94

257. Kurokawa H, Arteaga CL. 2001. Inhibition of erbB receptor (HER) tyrosine kinases as a strategy to abrogate antiestrogen resistance in human breast cancer. *Clin. Cancer Res.* 7:4436s–42s; discussion 11s–12s

258. Tsuda L, Nagaraj R, Zipursky SL, Banerjee U. 2002. An EGFR/Ebi/Sno pathway promotes delta expression by inactivating Su(H)/SMRTER repression during inductive notch signaling. *Cell* 110:625–37

259. Hermanson O, Jepsen K, Rosenfeld MG.

2002. N-CoR controls differentiation of neural stem cells into astrocytes. *Nature* 419:934–39

260. Dressel U, Bailey PJ, Wang SC, Downes M, Evans RM, Muscat GE. 2001. A dynamic role for HDAC7 in MEF2-mediated muscle differentiation. *J. Biol. Chem.* 276:17007–13

261. Grozinger CM, Schreiber SL. 2000. Regulation of histone deacetylase 4 and 5 and transcriptional activity by 14-3-3-dependent cellular localization. *Proc. Natl. Acad. Sci. USA* 97:7835–40

262. McKinsey TA, Zhang CL, Olson EN. 2001. Identification of a signal-responsive nuclear export sequence in class II histone deacetylases. *Mol. Cell Biol.* 21:6312–21

263. Zhou X, Richon VM, Wang AH, Yang XJ, Rifkind RA, Marks PA. 2000. Histone deacetylase 4 associates with extracellular signal-regulated kinases 1 and 2, and its cellular localization is regulated by oncogenic Ras. *Proc. Natl. Acad. Sci. USA* 97:14329–33

264. Galasinski SC, Resing KA, Goodrich JA, Ahn NG. 2002. Phosphatase inhibition leads to histone deacetylases 1 and 2 phosphorylation and disruption of corepressor interactions. *J. Biol. Chem.* 277:19618–26

265. Tsai SC, Seto E. 2002. Regulation of histone deacetylase 2 by protein kinase CK2. *J. Biol. Chem.* 277:31826–33

266. Hammer GD, Krylova I, Zhang Y, Darimont BD, Simpson K, et al. 1999. Phosphorylation of the nuclear receptor SF-1 modulates cofactor recruitment: integration of hormone signaling in reproduction and stress. *Mol. Cell* 3:521–26

267. Juge-Aubry CE, Hammar E, Siegrist-Kaiser C, Pernin A, Takeshita A, et al. 1999. Regulation of the transcriptional activity of the peroxisome proliferator-activated receptor alpha by phosphorylation of a ligand-independent *trans*-activating domain. *J. Biol. Chem.* 274:10505–10

268. Ait-Si-Ali S, Ramirez S, Barre FX, Dkhissi F, Magnaghi-Jaulin L, et al. 1998.

Histone acetyltransferase activity of CBP is controlled by cycle-dependent kinases and oncoprotein E1A. *Nature* 396:184–86

269. Lopez GN, Turck CW, Schaufele F, Stallcup MR, Kushner PJ. 2001. Growth factors signal to steroid receptors through mitogen-activated protein kinase regulation of p160 coactivator activity. *J. Biol. Chem.* 276:22177–82

270. Puigserver P, Rhee J, Lin J, Wu Z, Yoon JC, et al. 2001. Cytokine stimulation of energy expenditure through p38 MAP kinase activation of PPARgamma coactivator-1. *Mol. Cell* 8:971–82

271. Ratajczak T. 2001. Protein coregulators that mediate estrogen receptor function. *Reprod. Fertil. Dev.* 13:221–29

272. Rowan BG, Garrison N, Weigel NL, O'Malley BW. 2000. 8-Bromo-cyclic AMP induces phosphorylation of two sites in SRC-1 that facilitate ligand-independent activation of the chicken progesterone receptor and are critical for functional cooperation between SRC-1 and CREB binding protein. *Mol. Cell Biol.* 20:8720–30

273. Wu RC, Qin J, Hashimoto Y, Wong J, Xu J, et al. 2002. Regulation of SRC-3 (pCIP/ACTR/AIB-1/RAC-3/TRAM-1) coactivator activity by IkappaB kinase. *Mol. Cell Biol.* 22:3549–61

274. Zhang J, Guenther MG, Carthew RW, Lazar MA. 1998. Proteasomal regulation of nuclear receptor corepressor-mediated repression. *Genes Dev.* 12:1775–80

275. Kurebayashi J, Otsuki T, Kunisue H, Tanaka K, Yamamoto S, Sonoo H. 2000. Expression levels of estrogen receptor-alpha, estrogen receptor-beta, coactivators, and corepressors in breast cancer. *Clin. Cancer Res.* 6:512–18

276. Martinez de Arrieta C, Koibuchi N, Chin WW. 2000. Coactivator and corepressor gene expression in rat cerebellum during postnatal development and the effect of altered thyroid status. *Endocrinology* 141:1693–98

277. Feng X, Jiang Y, Meltzer P, Yen PM. 2001.

Transgenic targeting of a dominant negative corepressor to liver blocks basal repression by thyroid hormone receptor and increases cell proliferation. *J. Biol. Chem.* 276:15066–72

278. Koide T, Downes M, Chandraratna RA, Blumberg B, Umesono K. 2001. Active repression of RAR signaling is required for head formation. *Genes Dev.* 15:2111–21

279. Sachs LM, Amano T, Shi YB. 2001. An essential role of histone deacetylases in postembryonic organ transformations in *Xenopus laevis. Int. J. Mol. Med.* 8:595–601

280. Berghagen H, Ragnhildstveit E, Krogsrud K, Thuestad G, Aprilctti J, Saatcioglu F. 2002. Corepressor SMRT functions as a coactivator for thyroid hormone receptor T3Ralpha from a negative hormone response element. *J. Biol. Chem.* 277:49517–22

281. Satoh T, Monden T, Ishizuka T, Mitsuhashi T, Yamada M, Mori M. 1999. DNA binding and interaction with the nuclear receptor corepressor of thyroid hormone receptor are required for ligand-independent stimulation of the mouse pre-prothyrotropin-releasing hormone gene. *Mol. Cell Endocrinol.* 154:137–49

282. Tagami T, Madison LD, Nagaya T, Jameson JL. 1997. Nuclear receptor corepressors activate rather than suppress basal transcription of genes that are negatively regulated by thyroid hormone. *Mol. Cell Biol.* 17:2642–48

283. Nygard M, Wahlstrom GM, Gustafsson MV, Tokumoto YM, Bondesson M. 2003. Hormone-dependent repression of the E2F-1 gene by thyroid hormone receptors. *Mol. Endocrinol* 17:79–92

284. Becker N, Seugnet I, Guissouma H, Dupre SM, Demeneix BA. 2001. Nuclear corepressor and silencing mediator of retinoic and thyroid hormone receptors corepressor expression is incompatible with T(3)-dependent TRH regulation. *Endocrinology* 142:5321–31

285. Li X, Kimbrel EA, Kenan DJ, McDonnell DP. 2002. Direct interactions between corepressors and coactivators permit the integration of nuclear receptor-mediated repression and activation. *Mol. Endocrinol.* 16:1482–91

286. Chatterjee VK. 2001. Resistance to thyroid hormone, and peroxisome-proliferator-activated receptor gamma resistance. *Biochem. Soc. Trans.* 29:227–31

287. Jameson JL. 1994. Mechanisms by which thyroid hormone receptor mutations cause clinical syndromes of resistance to thyroid hormone. *Thyroid* 4:485–92

288. Clifton-Bligh RJ, de Zegher F, Wagner RL, Collingwood TN, Francois I, et al. 1998. A novel TR beta mutation (R383H) in resistance to thyroid hormone syndrome predominantly impairs corepressor release and negative transcriptional regulation. *Mol. Endocrinol.* 12:609–21

289. Matsushita A, Misawa H, Andoh S, Natsume H, Nishiyama K, et al. 2000. Very strong correlation between dominant negative activities of mutant thyroid hormone receptors and their binding avidity for corepressor SMRT. *J. Endocrinol.* 167:493–503

290. Nagaya T, Fujieda M, Seo H. 1998. Requirement of corepressor binding of thyroid hormone receptor mutants for dominant negative inhibition. *Biochem. Biophys. Res. Commun.* 247:20–23

291. Safer JD, Cohen RN, Hollenberg AN, Wondisford FE. 1998. Defective release of corepressor by hinge mutants of the thyroid hormone receptor found in patients with resistance to thyroid hormone. *J. Biol. Chem.* 273:30175–82

292. Yoh SM, Chatterjee VK, Privalsky ML. 1997. Thyroid hormone resistance syndrome manifests as an aberrant interaction between mutant T3 receptors and transcriptional corepressors. *Mol. Endocrinol.* 11:470–80

293. Yoh SM, Privalsky ML. 2000. Resistance to thyroid hormone (RTH) syndrome

reveals novel determinants regulating interaction of T3 receptor with corepressor. *Mol. Cell Endocrinol.* 159:109–24

294. Tagami T, Gu WX, Peairs PT, West BL, Jameson JL. 1998. A novel natural mutation in the thyroid hormone receptor defines a dual functional domain that exchanges nuclear receptor corepressors and coactivators. *Mol. Endocrinol.* 12:1888–902

295. Liu Y, Takeshita A, Misiti S, Chin WW, Yen PM. 1998. Lack of coactivator interaction can be a mechanism for dominant negative activity by mutant thyroid hormone receptors. *Endocrinology* 139:4197–204

296. Barroso I, Gurnell M, Crowley VE, Agostini M, Schwabe JW, et al. 1999. Dominant negative mutations in human PPARgamma associated with severe insulin resistance, diabetes mellitus and hypertension. *Nature* 402:880–83

297. Breems-de Ridder MC, Lowenberg B, Jansen JH. 2000. Retinoic acid receptor fusion proteins: friend or foe. *Mol. Cell. Endocrinol.* 165:1–6

298. Dong S, Tweardy DJ. 2002. Interactions of STAT5b-RARalpha, a novel acute promyelocytic leukemia fusion protein, with retinoic acid receptor and STAT3 signaling pathways. *Blood* 99:2637–46

299. Minucci S, Pelicci PG. 1999. Retinoid receptors in health and disease: coregulators and the chromatin connection. *Semin. Cell Dev. Biol.* 10:215–25

300. Pandolfi PP. 1996. PML, PLZF and NPM genes in the molecular pathogenesis of acute promyelocytic leukemia. *Haematologica* 81:472–82

301. Cheng GX, Zhu XH, Men XQ, Wang L, Huang QH, et al. 1999. Distinct leukemia phenotypes in transgenic mice and different corepressor interactions generated by promyelocytic leukemia variant fusion genes PLZF-RARalpha and NPM-RARalpha. *Proc. Natl. Acad. Sci. USA* 96:6318–23

302. Dong S, Qiu J, Stenoien DL, Brinkley

WR, Mancini MA, Tweardy DJ. 2003. Essential role for the dimerization domain of NuMA-RARalpha in its oncogenic activities and localization to NuMA sites within the nucleus. *Oncogene* 22:858–68

303. Grignani F, De Matteis S, Nervi C, Tomassoni L, Gelmetti V, et al. 1998. Fusion proteins of the retinoic acid receptor-alpha recruit histone deacetylase in promyelocytic leukaemia. *Nature* 391:815–18

304. Guidez F, Ivins S, Zhu J, Soderstrom M, Waxman S, Zelent A. 1998. Reduced retinoic acid-sensitivities of nuclear receptor corepressor binding to PML- and PLZF-RARalpha underlie molecular pathogenesis and treatment of acute promyelocytic leukemia. *Blood* 91:2634–42

305. Lin RJ, Evans RM. 2000. Acquisition of oncogenic potential by RAR chimeras in acute promyelocytic leukemia through formation of homodimers. *Mol. Cell* 5:821–30

306. Minucci S, Maccarana M, Cioce M, De Luca P, Gelmetti V, et al. 2000. Oligomerization of RAR and AML1 transcription factors as a novel mechanism of oncogenic activation. *Mol. Cell* 5:811–20

307. So CW, Dong S, So CK, Cheng GX, Huang QH, et al. 2000. The impact of differential binding of wild-type RARalpha, PML-, PLZF- and NPM-RARalpha fusion proteins towards transcriptional coactivator, RIP-140, on retinoic acid responses in acute promyelocytic leukemia. *Leukemia* 14:77–83

308. Kogan SC, Hong SH, Shultz DB, Privalsky ML, Bishop JM. 2000. Leukemia initiated by PMLRARalpha: the PML domain plays a critical role while retinoic acid-mediated transactivation is dispensable. *Blood* 95:1541–50

309. Gu BW, Xiong H, Zhou Y, Chen B, Wang L, et al. 2002. Variant-type PML-RAR(alpha) fusion transcript in acute promyelocytic leukemia: use of a cryptic coding sequence from intron 2 of the RAR(alpha) gene and identification of a

new clinical subtype resistant to retinoic acid therapy. *Proc. Natl. Acad. Sci. USA* 99:7640–45

310. Khan MM, Nomura T, Kim H, Kaul SC, Wadhwa R, et al. 2001. Role of PML and PML-RARalpha in Mad-mediated transcriptional repression. *Mol. Cell* 7:1233–43

311. Zimonjic DB, Pollock JL, Westervelt P, Popescu NC, Ley TJ. 2000. Acquired, nonrandom chromosomal abnormalities associated with the development of acute promyelocytic leukemia in transgenic mice. *Proc. Natl. Acad. Sci. USA* 97:13306–11

312. Hong SH, Yang Z, Privalsky ML. 2001. Arsenic trioxide is a potent inhibitor of the interaction of SMRT corepressor with its transcription factor partners, including the PML-retinoic acid receptor alpha oncoprotein found in human acute promyelocytic leukemia. *Mol. Cell Biol.* 21:7172–82

313. Rietveld LE, Caldenhoven E, Stunnenberg HG. 2001. Avian erythroleukemia: a model for corepressor function in cancer. *Oncogene* 20:3100–9

314. Sap J, Munoz A, Damm K, Goldberg Y, Ghysdael J, et al. 1986. The c-erb-A protein is a high-affinity receptor for thyroid hormone. *Nature* 324:635–40

315. Weinberger C, Thompson CC, Ong ES, Lebo R, Gruol DJ, Evans RM. 1986. The c-erb-A gene encodes a thyroid hormone receptor. *Nature* 324:641–46

316. Ciana P, Braliou GG, Demay FG, von Lindern M, Barettino D, et al. 1998. Leukemic transformation by the v-ErbA oncoprotein entails constitutive binding to and repression of an erythroid enhancer in vivo. *EMBO J.* 17:7382–94

317. Damm K, Thompson CC, Evans RM. 1989. Protein encoded by v-erbA functions as a thyroid-hormone receptor antagonist. *Nature* 339:593–97

318. Chen HW, Privalsky ML. 1993. The erbA oncogene represses the actions of both retinoid X and retinoid A receptors but does so by distinct mechanisms. *Mol. Cell Biol.* 13:5970–80

319. Sharif M, Privalsky ML. 1991. v-erbA oncogene function in neoplasia correlates with its ability to repress retinoic acid receptor action. *Cell* 66:885–93

320. Desbois C, Pain B, Guilhot C, Benchaibi M, Ffrench M, et al. 1991. v-erbA oncogene abrogates growth inhibition of chicken embryo fibroblasts induced by retinoic acid. *Oncogene* 6:2129–35

321. Schroeder C, Gibson L, Zenke M, Beug H. 1992. Modulation of normal erythroid differentiation by the endogenous thyroid hormone and retinoic acid receptors: a possible target for v-erbA oncogene action. *Oncogene* 7:217–27

322. Chakrabarti SR, Nucifora G. 1999. The leukemia-associated gene TEL encodes a transcription repressor which associates with SMRT and mSin3A. *Biochem. Biophys. Res. Commun.* 264:871–77

323. Dressel U, Thormeyer D, Altincicek B, Paululat A, Eggert M, et al. 1999. Alien, a highly conserved protein with characteristics of a corepressor for members of the nuclear hormone receptor superfamily. *Mol. Cell Biol.* 19:3383–94

324. Polly P, Herdick M, Moehren U, Baniahmad A, Heinzel T, Carlberg C. 2000. VDR-Alien: a novel, DNA-selective vitamin D(3) receptor-corepressor partnership. *FASEB J.* 14:1455–63

325. Ikeda Y. 1996. SF-1: a key regulator of development and function in the mammalian reproductive system. *Acta Paediatr. Jpn.* 38:412–19

326. Parker KL. 1998. The roles of steroidogenic factor 1 in endocrine development and function. *Mol. Cell Endocrinol.* 140:59–63

327. Altincicek B, Tenbaum SP, Dressel U, Thormeyer D, Renkawitz R, Baniahmad A. 2000. Interaction of the corepressor Alien with DAX-1 is abrogated by mutations of DAX-1 involved in adrenal hypoplasia congenita. *J. Biol. Chem.* 275:7662–67

328. Crawford PA, Dorn C, Sadovsky Y, Milbrandt J. 1998. Nuclear receptor DAX-1 recruits nuclear receptor corepressor N-CoR to steroidogenic factor 1. *Mol. Cell Biol.* 18:2949–56

329. Mathur M, Tucker PW, Samuels HH. 2001. PSF is a novel corepressor that mediates its effect through Sin3A and the DNA binding domain of nuclear hormone receptors. *Mol. Cell Biol.* 21:2298–311

330. Potter GB, Beaudoin GM 3rd, DeRenzo CL, Zarach JM, Chen SH, Thompson CC. 2001. The hairless gene mutated in congenital hair loss disorders encodes a novel nuclear receptor corepressor. *Genes Dev.* 15:2687–701

331. Potter GB, Zarach JM, Sisk JM, Thompson CC. 2002. The thyroid hormone-regulated corepressor hairless associates with histone deacetylases in neonatal rat brain. *Mol. Endocrinol.* 16:2547–60

332. Moraitis AN, Giguere V, Thompson CC. 2002. Novel mechanism of nuclear receptor corepressor interaction dictated by activation function 2 helix determinants. *Mol. Cell Biol.* 22:6831–41

333. Nibu Y, Zhang H, Bajor E, Barolo S, Small S, Levine M. 1998. dCtBP mediates transcriptional repression by Knirps, Kruppel and Snail in the *Drosophila* embryo. *EMBO J.* 17:7009–20

334. Chinnadurai G. 2002. CtBP, an unconventional transcriptional corepressor in development and oncogenesis. *Mol. Cell* 9:213–24

335. Cavailles V, Dauvois S, L'Horset F, Lopez G, Hoare S, et al. 1995. Nuclear factor RIP140 modulates transcriptional activation by the estrogen receptor. *EMBO J.* 14:3741–51

336. Lee CH, Wei LN. 1999. Characterization of receptor-interacting protein 140 in retinoid receptor activities. *J. Biol. Chem.* 274:31320–26

337. Subramaniam N, Treuter E, Okret S. 1999. Receptor interacting protein RIP140 inhibits both positive and negative gene regulation by glucocorticoids. *J. Biol. Chem.* 274:18121–27

338. Treuter E, Albrektsen T, Johansson L, Leers J, Gustafsson JA. 1998. A regulatory role for RIP140 in nuclear receptor activation. *Mol. Endocrinol.* 12:864–81

339. Vo N, Fjeld C, Goodman RH. 2001. Acetylation of nuclear hormone receptor-interacting protein RIP140 regulates binding of the transcriptional corepressor CtBP. *Mol. Cell Biol.* 21:6181–88

340. Wei LN, Hu X, Chandra D, Seto E, Farooqui M. 2000. Receptor-interacting protein 140 directly recruits histone deacetylases for gene silencing. *J. Biol. Chem.* 275:40782–87

341. Montano MM, Ekena K, Delage-Mourroux R, Chang W, Martini P, Katzenellenbogen BS. 1999. An estrogen receptor-selective coregulator that potentiates the effectiveness of antiestrogens and represses the activity of estrogens. *Proc. Natl. Acad. Sci. USA* 96:6947–52

342. Delage-Mourroux R, Martini PG, Choi I, Kraichely DM, Hoeksema J, Katzenellenbogen BS. 2000. Analysis of estrogen receptor interaction with a repressor of estrogen receptor activity (REA) and the regulation of estrogen receptor transcriptional activity by REA. *J. Biol. Chem.* 275:35848–56

343. Martini PG, Delage-Mourroux R, Kraichely DM, Katzenellenbogen BS. 2000. Prothymosin alpha selectively enhances estrogen receptor transcriptional activity by interacting with a repressor of estrogen receptor activity. *Mol. Cell Biol.* 20:6224–32

344. Norris JD, Fan D, Sherk A, McDonnell DP. 2002. A negative coregulator for the human ER. *Mol. Endocrinol.* 16:459–68

345. Sharma M, Sun Z. 2001. 5′TG3′ interacting factor interacts with Sin3A and represses AR-mediated transcription. *Mol. Endocrinol.* 15:1918–28

346. Wu L, Wu Y, Gathings B, Wan M, Li X, et al. 2003. Smad4 as a transcription

corepressor for estrogen receptor alpha. *J. Biol. Chem.* 278:15192–200

347. Seol W, Choi HS, Moore DD. 1996. An orphan nuclear hormone receptor that lacks a DNA binding domain and heterodimerizes with other receptors. *Science* 272:1336–39

348. Johansson L, Bavner A, Thomsen JS, Farnegardh M, Gustafsson JA, Treuter E. 2000. The orphan nuclear receptor SHP utilizes conserved LXXLL-related motifs for interactions with ligand-activated estrogen receptors. *Mol. Cell Biol.* 20:1124–33

349. Borgius LJ, Steffensen KR, Gustafsson JA, Treuter E. 2002. Glucocorticoid signaling is perturbed by the atypical orphan receptor and corepressor SHP. *J. Biol. Chem.* 277:49761–66

350. Lee YK, Dell H, Dowhan DH, Hadzo-poulou-Cladaras M, Moore DD. 2000. The orphan nuclear receptor SHP inhibits hepatocyte nuclear factor 4 and retinoid X receptor transactivation: two mechanisms for repression. *Mol. Cell Biol.* 20:187–95

351. Lee YK, Moore DD. 2002. Dual mechanisms for repression of the monomeric orphan receptor liver receptor homologous protein-1 by the orphan small heterodimer partner. *J. Biol. Chem.* 277:2463–67

352. Seol W, Chung M, Moore DD. 1997. Novel receptor interaction and repression domains in the orphan receptor SHP. *Mol. Cell Biol.* 17:7126–31

353. Bourguet W, Ruff M, Chambon P, Gronemeyer H, Moras D. 1995. Crystal structure of the ligand-binding domain of the human nuclear receptor RXR-alpha. *Nature* 375:377–82

Annu. Rev. Physiol. 2004. 66:361–84
doi: 10.1146/annurev.physiol.66.032102.144149
Copyright © 2004 by Annual Reviews. All rights reserved
First published online as a Review in Advance on September 29, 2003

MOLECULAR AND INTEGRATIVE PHYSIOLOGY OF INTESTINAL PEPTIDE TRANSPORT

Hannelore Daniel

Molecular Nutrition Unit, Technical University of Munich, D-85350
Freising-Weihenstephan, Germany; email: daniel@wzw.tum.de

Key Words small intestine, protein digestion, peptide absorption, peptide transporter

■ **Abstract** Intestinal protein digestion generates a huge variety and quantity of short chain peptides that are absorbed into intestinal epithelial cells by the PEPT1 transporter in the apical membrane of enterocytes. PEPT1 operates as an electrogenic proton/peptide symporter with the ability to transport essentially every possible di- and tripeptide. Transport is enantio-selective and involves a variable proton-to-substrate stoichiometry for uptake of neutral and mono- or polyvalently charged peptides. Neither free amino acids nor peptides containing four or more amino acids are accepted as substrates. The structural similarity of a variety of drugs with the basic structure of di- or tripeptides explains the transport of aminocephalosporins and aminopenicillins, selected angiotensin-converting inhibitors, and amino acid–conjugated nucleoside-based antiviral agents by PEPT1. The high transport capacity of PEPT1 allows fast and efficient intestinal uptake of the drugs but also of amino acid nitrogen even in states of impaired mucosal functions. Transcriptional and post-transcriptional regulation of PEPT1 occurs in response to alterations in the nutritional status and in disease states, suggesting a prime role of this transporter in amino acid absorption.

GENERATION OF PEPTIDE TRANSPORTER SUBSTRATES BY THE INTESTINAL DEGRADATION OF DIETARY PROTEINS

Understanding the digestive and absorptive processes underlying human protein assimilation in the gut requires a journey almost 40 years back in time, when digestion and absorption were the focus of research by physiologists and gastroenterologists. Although today we have a much better understanding of the structure and function of the enzymes involved in hydrolysis of proteins and of the carriers responsible for intestinal uptake of free amino acids and peptides, our knowledge of what makes up the intestinal phase of protein degradation still relies on fairly old studies in experimental animals and humans. Research on protein digestion and absorption was energized in the mid-1970s by a highly controversial discussion about the existence and importance of peptide transport in mammalian intestines. The emerging new concept that not only the monomers of the proteins, the free amino acids, but also their oligomers—at least di- and tripeptides—could be taken up by intestinal epithelial cells in intact form was against the contemporary belief that only the final end products of digestion are absorbed. This historical period of research on protein digestion and absorption was comprehensively reviewed by Matthews—one of the pioneers in peptide transport—in 1991 (1). However, even after the general recognition that di- and tripeptides are taken up into intestinal epithelial cells in intact form by a transport protein for peptides designated as PEPT1, we still do not know the quantitative importance of peptide transport in overall amino acid absorption.

Before focusing on the cellular and molecular aspects of peptide transport processes, some basics on protein digestion need to be recapitulated. The average protein intake can vary considerably, but a western diet typically provides between 70 and 100 g of protein/day. In addition, endogenous protein from the secretions along the oro-gastrointestinal tract, amounts to around 50 to 60 g/day. These proteins are subjected to hydrolysis by a spectra of proteases and peptidases either secreted from stomach and pancreas or bound to the brush border membrane of enterocytes, which generates a large quantity and variety of short- and medium-sized peptides, as well as free amino acids. The jejunal contents of humans or animals, recovered after administration of model proteins such as albumin, showed that, the major fraction consisted of peptides typically with three to six amino acid residues, which, based on a mean molecular weight, corresponds to a concentration of 120 to 145 mM. The concentration of individual free amino acids in the digest varied between 0.6 and 16 mM, with a total concentration of all amino acids of 30 to 60 mM (2). Protein digestion and the absorption of the end products are generally not considered to be limited, and even quantities of 320 to 480 g of protein equivalent to 1.5 to 2.8 kg of lean meat eaten by human volunteers within 8 h did not yield evidence for a limitation in protein assimilation (1). Most proteins and oligopeptides are rapidly degraded. However, some structures are fairly resistant to hydrolysis, and the extent and the velocity by which a dietary protein is broken down to its constituents is dependent on its composition (amino acid sequence) and on post-translational modifications such as glycosylation, which render peptides more resistant to hydrolysis (3). The latter also includes thermal effects of food processing that may cause the formation of Maillard products or that lead to cross-linking reactions. In particular, glycosylated peptides and those containing multiple prolyl residues appear to resist the attack of proteases and peptidases. The presence of prolyl residues is particularly relevant, when it comes to understanding the generation and resistance of the immunodominant epitopes from alpha-gliadin, which are exceptionally rich in proline and glutamine residues (4–6). The very low activity of dipeptidyl peptidase IV and dipeptidyl carboxypeptidase I, as rate-limiting enzymes in the gastrointestinal tract, determine the digestive breakdown of these peptides (6). Other biologically active peptides containing multiple prolyl residues have also been identified in digests, for example, dietary proteins (mainly milk proteins), which led to the suggestion that those peptides released during digestion of proteins in the gut may affect body functions by their opioid, immunomodulatory or angiotensin-converting enzyme (ACE) inhibitory activity (7). Some peptides derived from dietary proteins have indeed been shown to reduce blood pressure after ingestion, which is explained by their ACE-inhibitory activity after absorption into circulation. However, studies in monolayer systems (Caco-2 cells) and intact tissue preparations have shown that the size of the peptides with identified ACE-inhibitory activity and their susceptibility to degradation are critical in determining their survival in the gut and the ability to reach the circulation in intact form (8). The most potent antihypertensive effects in spontaneously hypertensive rats were elicited by orally administered dipeptides such as Val-Tyr, Ile-Tyr, Phe-Tyr, and Ile-Trp, suggesting that the intestinal peptide transporter here allows efficient delivery into circulation (9).

The Final Stages of Extracellular and Intracellular Hydrolysis of Peptides

Although oligopeptides of medium chain length are the primary products of the luminal phase of protein digestion, they are further cleaved by a spectra of membrane anchored peptidases at the brush border of intestinal epithelial cells (Figure 1, see color insert). By measuring hydrolytic activity of peptidases in homogenates of the mucosa and in isolated brush border membranes, it became obvious that impressive differences in the activity and the subcellular location of the enzymes existed depending on the chain length of the peptide substrates (10–12). When dipeptides were used as substrates, almost 90% of total mucosal activity could be attributed to cytosolic enzymes, whereas with tripeptides, only around 50% of the hydrolytic activity originated from the soluble cytosolic fraction. In the case of tetrapeptides and those with more than four amino residues, essentially all activity was brush border membrane bound (1, 10, 11). The high activity of intracellular peptidases with the strict specificity for the hydrolysis of di- and tripeptides suggests that di- and tripeptides, but not larger ones, may be absorbed in intact form followed by the intracellular release of free amino acids. The extent to which a di- or tripeptide that is released during digestion is finally broken down at the brush border membrane or is taken up into the cell followed by intracellular hydrolysis is not known. These two pathways of course depend on the affinity and the substrate concentration of thousands of peptides, which compete simultaneously for the binding sites of either the membrane-bound enzymes or the peptide transporter. Although there is limited dipeptidase activity at the brush border membrane, luminal hydrolysis of normal di- and tripeptides occurs rapidly, in particular, when the enzymes are not overloaded with substrates that would enable peptides to bypass hydrolysis. Enzymatic cleavage is a problem whenever dipeptides are used as probes to characterize peptide transport processes, because these probes can gradually be hydrolyzed to the constitutive labeled amino acid residues and then transported by amino acid carriers. The use of glycine-proline, ß-alanine-histidine (carnosine) and glycyl-sarcosine or D-amino acids containing dipeptides—which all are more resistant to hydrolysis based on their unusual structures—reduces the hydrolysis problem and, therefore, these substrates have been employed in numerous transport studies. However, their exceptional structures not only render them more resistant to enzymatic hydrolysis but also reduce their affinity for interaction with the peptide transporter-substrate binding site. Under physiological conditions (i.e., when large scale proteolysis occurs in the lumen), the brush border peptide hydrolases are exposed to an enormous variety of substrates of variable concentrations but with a total substrate load that may be higher than their apparent V_{max} for hydrolysis. In addition, as has been shown experimentally, certain free amino acids released during hydrolysis may serve as inhibitors of the dipeptidases, and therefore a substantial quantity of di- and tripeptides may bypass enzymatic cleavage and reach the peptide transporter. This protein is then challenged again with a huge variety of substrates. On the basis of the 20 amino acids making up a protein, thousands of di- and tripeptides exist that differ in structure, molecular size, polarity, net charge, stereochemistry, and even the mesomeric structure of the peptide bond. The molecular mass for example can vary between 132 kDa for a di-glycine and 577 kDa for a tri-tryptophane, and the net charge can range (at pH 7) from neutral to trivalently anionic, as for a tri-glutamate, or trivalently cationic, as for a tri-lysine. For a transport physiologist it is hard to imagine that a single carrier protein in its thermodynamic setting would be able to transport these quite different substrates. Consequently, multiple peptide transporters for the different substrate groups were initially postulated in analogy to the different amino acid transporter classes. Indeed, some experimental findings in intact tissue preparations have suggested, on the basis of cross-inhibition studies, that there could be more than one type of peptide carrier. However, cloning of the cDNA of the intestinal di-tripeptide carrier, now designated as PEPT1, extensive analysis of mammalian genome databases, screening of intestinal tissue banks, and immunohistology have not yet provided evidence for more than one peptide carrier in the brush border membrane of intestinal epithelial cells.

THE PEPTIDE CARRIER PEPT1 IN THE APICAL MEMBRANE OF ENTEROCYTES

An important characteristic that allowed the identification of PEPT1 by expression cloning, using *Xenopus* oocytes and a cDNA library from rabbit small intestine, was its unique feature of utilizing an inwardly directed proton gradient to allow peptides to enter the cell even against a concentration gradient. The electrogenic nature of proton-peptide symport had been demonstrated unequivocally in earlier studies employing brush border membrane vesicles by Ganapathy et al. in the early 1980s (14–17). This demonstration of a H^+-peptide cotransport process shared by di- and tripeptides, but not by free amino acids or tetrapeptides, ended a long-lasting discussion on the nature of the driving forces for peptide uptake. This discussion emerged on the basis of findings of a partial Na^+ dependency of peptide uptake in intact tissue preparations. When reviewing studies with dipeptide transport in 1985, Cheeseman & Devlin found 10 papers demonstrating a Na^+ dependency and 12 papers that showed no Na^+ effects (18). Work in vesicles finally proved that peptide transport is a proton-coupled process and that the Na^+ dependency seen in vitro and in vivo is explained by the Na^+-proton antiporter activity (NHE-3) in apical membrane, which exports protons entering the cell via PEPT1 in exchange for extracellular Na^+ ions (14, 19). The acid-loading activity of PEPT1 in intestinal epithelial cells and the requirement of the apical Na^+/H^+ exchanger activity for the recovery from the acid load has been clearly demonstrated in Caco-2 cells by use of a fluorescent reporter to monitor the intracellular pH changes (19).

By screening of oocytes expressing proteins from cRNA-batches for Na^+-independent but pH-activated dipeptide uptake, the cDNA of the peptide carrier was isolated (13). Following expression cloning of the first mammalian cDNA from rabbit intestine that encoded peptide transport activity (13), the way was paved for isolation of PEPT1 clones from other species such as human, mouse, rat, chicken, cow, and *Caenorhabditis elegans* (20–25). The high-affinity isoform, PEPT2, was independently isolated by expression cloning and homology screening from a kidney cDNA library (26, 27). PEPT2 is not expressed in the intestine but in epithelial cells of the kidney tubule, lung, mammary gland, and choroid plexus, as well as in glia cells in the central nervous system and other cell types (28).

PEPT1 is a prototype transporter of the so called POT (proton-oligopeptide transporter) or PTR (peptide transporter) family with members in almost every genera (29–31). The gene encoding hPEPT1 maps to human chromosome 13q33–34 (20) and consists of 23 exons (32). The open reading frame of *hPEPT1* encodes 708 amino acids with a 50% overall sequence identity and 70% similarity to the high-affinity transporter isoform PEPT2. The amino acid sequence predicts 12 transmembrane domains in PEPT1 with both amino and carboxy termini facing the cytoplasmic side. A closely related transcript, termed *hPEPT1-RF* (for hPEPT1-regulating factor) was also cloned (34). The DNA of *hPEPT1-RF* has six exons, and the coding sequence of hPEPT1-RF shares three exons completely and two exons partially with hPEPT1. Amino acid residues 18–195 in *hPEPT1-RF* are identical to residues 8–185 in *hPEPT1*, whereas residues 1–17 and 196–208 are unique. *hPEPT1-RF* alone does not induce any peptide transport activity when expressed heterologously, but the coexpression with hPEPT1 in *Xenopus* oocytes led to a shift in the pH sensitivity profile for Gly-Sar uptake, with a more pronounced inactivation at lower pH than for PEPT1 alone (24).

A high degree of similarity in both gene-clustering and -coding sequences is found between the mammalian *PEPT1* gene orthologues. A comparison of membrane topology predictions and genomic structures indicates that human and mouse *PEPT1* genes are modular, with each transmembrane segment and loop unit encoded by a different exon, and it is thought that this modular structure may have evolved by exon shuffling and rearrangements of functional modules (32). The putative regulatory region of the *hPEPT1* gene reveals TATA boxes in unusual locations at 511 bp and 517 bp upstream from the transcription start site, whereas GC boxes are located near the start site at −29 bp and a number of others within

300 bp. This structure suggests that the GC box is a more important promoter element in the regulation of *hPEPT1* than is the TATA box with its location so far upstream from the transcription start site (32). However, there may also be more than one transcription start site.

Protein structure and membrane topology of PEPT1 have not been systematically analyzed. However, preliminary epitope insertion studies suggest that the membrane topology is identical with or similar to that predicted by hydropathy analysis (35). PEPT1 transporters with single point mutations, and functional analysis of chimeras of PEPT1 and PEPT2 with their distinctly different affinities for the same substrates, have provided evidence that the amino-terminal region with the first 4 transmembrane regions and the domains 7 to 9 play an important role in determining the substrate affinity and other characteristic features (36–39). Of central importance for PEPT1 activity is the extracellular histidine residue (H57) on the interface of the second transmembrane domain and the extracellular side that, when mutated, abolishes PEPT1 function (40–42). The histidine residue at position 121 seems to be involved in substrate binding as well (41), whereas histidine residues such as H111 and H260 did not yield evidence for an essential role in the transport process. Peptide transport was shown to be inactivated efficiently by diethlypyrocarbonate (DEPC) as a histidine-modifying agent (43, 44). However, the DEPC effects on transport can be prevented by PEPT1 substrates that contain a free terminal amino group but not by those substrate that lack the nitrogen function (45). This strongly suggests that a critical His-residue in PEPT1 lies within the substrate-binding domain and interferes with the amino function of the substrates. Adjacent to H57 is a tyrosine residue (Y56) shown to reduce the affinity for differently charged dipeptides when mutated to a phenylalanine (42). Another tyrosine residue predicted to lie in the central region of the second transmembrane domain (Y64) also appears to be involved in substrate translocation because the subtle change to a phenylalanine residue almost abolished transport (42). Moreover, Y167 in the fifth transmembrane domain also seems to be essential for PEPT1 activity (46). A mutation at residue W294 reduced substrate affinity markedly, and E595 also appears to be involved in the transport process (47). Whereas site-directed mutagenesis and the use of chimeras identified the amino-terminal half of PEPT1 as the most critical region encoding most of the functions, the role of the large extracellular domain between transmembrane domains 9 and 10 and the carboxy-terminal regions is unknown. Figure 2 (see color insert) summarizes the current understanding of PEPT1 membrane topology and depicts protein domains and individual amino acid residues that have been identified as relevant in determining the functional characteristics of the carrier.

UNIQUE FEATURES OF PEPT1

Extensive analysis of PEPT1 functions in heterologous expression systems, mainly in oocytes expressing either the rabbit or human transporters, has provided new insights into its unique features. There are two intriguing questions with regard to

PEPT1 function. The first relates to its substrate specificity or its lack of specificity because thousands of di- and tripeptides, as well as an impressive number of peptidomimetics with differing molecular size, polarity, charge and conformation, appear to be transported. The second question relates to its character as an electrogenic symporter; that regardless of the net charge of the substrate, transport with positive inward currents requires, for thermodynamic reasons, a variable substrate-to-ion (H^+ or H_3O^+) coupling ratio.

Unraveling the Substrate Specificity

Unlike other organic solute transporters, PEPT1 can be analyzed with respect to substrate specificity by the enormous number of dipeptides and tripeptides and derivatives available. Consequently, several hundred substrates have been tested in competition assays with tracer dipeptides for uptake via PEPT1 into Caco-2 cells or in yeast cells expressing the mammalian carriers. In addition, a large number of compounds has also been analyzed for electrogenicity of transport by applying the two-electrode voltage-clamp technique in *Xenopus* oocytes expressing PEPT1. The combination of competition studies and electrophysiology has proven to be valuable because it allows for easy identification of compounds that are transported and those that compete for uptake but are not transported. Although no peptide transport inhibitors were known until recently, some inhibitors where found by chance when screening for transport substrates (48), others were later synthesized and optimized by a rational approach. This has led to high-affinity type inhibitors that competitively block dipeptide transport via PEPT1, with affinities around 100-fold higher than those of normal substrates (49). Relevant inhibitors that may alter PEPT1 function in vivo in humans are the orally active sulfonylurea-antidiabetics that block transporter function with high affinity (50, 51). The conformational analysis of large sets of dipeptides, tripeptides, and amino acid derivatives, as well as peptidomimetic drugs, all shown to be transported by PEPT1, provides a solid basis for predictions of the substrate-binding template (52). Small and relatively rigid structures such as the aminocephalosporins, which can adopt only few conformations, are particularly helpful in defining the critical structural motifs in a substrate. What is important for understanding how a substrate-binding site in a transporter can accommodate such a large number of different structures is that water comes into play, with a critical role in the docking process of a substrate. The ability of water to shield electric charges of amino acid side chains in the carrier's substrate-binding domain allows both charged and polar, as well as large apolar substrates, to be accommodated at the same site. In analogy to the role of water molecules in bacterial periplasmic dipeptide (i.e., DppA) or oligopeptide binding proteins (i.e., OppA) with pleiotropic substrate-binding patterns similar to those of PEPT1 (53–55), it has to be envisaged that water acts as a versatile space-filling buffer between the substrate and the carrier-binding site, which weakens unfavorable interactions between charged substrate groups and the protein. When lysine side chains in a dipeptide are made more hydrophobic by addition of a terminal apolar ring system attached to the ε-amino group, the affinity of the compound increases substantially and renders the native substrate an efficient inhibitor of PEPT1 (49). It can be envisaged that the attached hydrophobic side chain–modifying groups expel water from the substrate-binding site and thus block the transport cycle.

The α-amino group of the peptide transporter substrates and its spatial location play a crucial role in binding affinity and translocation. This group may form a salt bridge with either an acidic residue or a histidine residue located in the substrate-binding pocket. However, there are substrates in which the α-amino group is not present, such as in some ß-lactams and ACE-inhibitors that are transported by PEPT1. Therefore, other groups in these compounds possibly interact with different residues in the binding domain, but this is not yet understood.

PEPT1 can transport simple omega-amino fatty acids (56, 57) that do not contain a peptide bond, and affinities are similar to those of normal dipeptides (Figure 3). Therefore, it is not surprising that the peptide bond can be replaced by an isosteric thioxo bond or that the peptide-bond nitrogen can be methylated with only a minor reduction in substrate affinity. The carbonyl oxygen in the peptide bond is the relevant functional component because it can form additional hydrogen bonds with the substrate-binding domain. Incorporation of this carbonyl-function into an omega-amino fatty acid increases the affinity around fivefold (Figure 3). Peptides containing proline residues possess pronounced differences in the mesomeric structure of the peptide bond (58). This led to the determination that the *trans*-conformation of the peptide bond alone allows transport of a dipeptide. In the *trans*-form, the peptide bond has a rigid, planar double-bond structure of 0.132-nm intramolecular length, and the carbonyl-oxygen—known to increase the affinity of substrates—becomes negatively charged. Peptides containing proline residues are critical substrates not only with respect to the conformation of the peptide bond but also with respect to other structural constraints. When provided in an amino-terminal position in peptides, the α-amino group is embedded into the pyrrolidine ring system, and this impairs interaction with the transporter with the consequence of a marked reduction in substrate affinity (59).

The essential and minimal structural features identified in substrates of PEPT1 that allow binding and electrogenic transport are therefore two oppositely charged head groups (i.e., amino and carboxy groups) separated by a carbon backbone with a distance of 5.5 to 6.3 Å between the centers of the head groups (56). This explains why neither free amino acids nor tetrapeptides can be bound or transported. The side chains in normal substrates formed by the 20 different amino acids most likely project into spacious and hydrated pockets in which a few contacts are made with the PEPT1 protein so that side chains differing in size and charge are accommodated with minimal adjustments of the surrounding protein structure. However, these pockets appear to be asymmetric, and therefore the protein can discriminate substrates based on the polarity and size of identical residues when those are located either in the amino-terminal or the carboxy-terminal position of a substrate. The accommodation of a peptide in the binding pocket is strongly stereoselective. D-amino acid residues in a peptide affect affinity differently when

compound	structure	apparent affinity [mM]	transport currents (I_{max} % of control)
4-aminobutyric acid	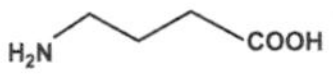	> 50	0
5-aminopentanoic acid		1.14 ± 0.06	100
5-amino-4-oxo-pentanoic acid	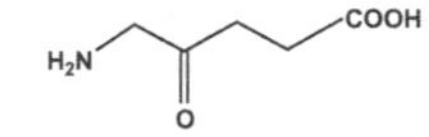	0.27 ± 0.04	100
Gly-Gly	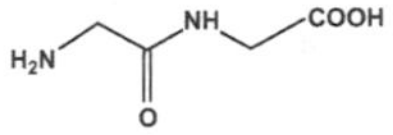	0.20 ± 0.02	100
L-Ala-L-Ala	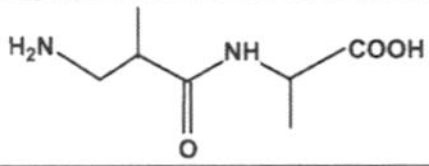	0.16 ± 0.03	100
D-Ala-L-Ala	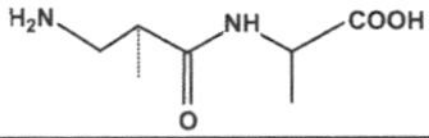	0.80 ± 0.06	70
L-Ala-D-Ala	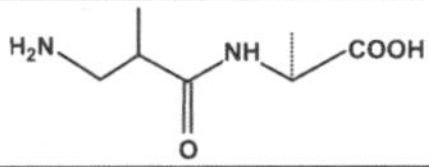	6.12 ± 0.34	30
D-Ala-D-Ala	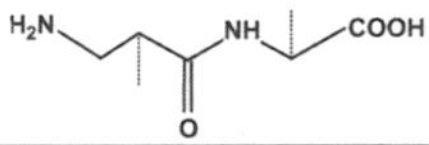	> 25	0

Figure 3   The key structural and conformational elements in PEPT1 substrates and how they affect substrate affinity and electrogenic transport. This series of model compounds has been analyzed with respect to substrate affinity and electrogenic transport under identical experimental conditions in *Pichia pastoris* cells and *Xenopus* oocytes expressing PEPT1, as described previously (56, 57). Apparent substrate affinities are derived from competition experiments with the model compounds in *P. pastoris* cells with a radioactive dipeptide serving as substrate. Inward currents generated by the compounds in *Xenopus* oocytes expressing PEPT1, determined by the two-voltage-clamp technique, are used to express the maximal transport rate. The test compounds have been applied under substrate saturation conditions and maximal transport currents are expressed as I_{max} in percent of that elicited by 10 mM Gly-(L)-Gln serving as a control in the same batch of oocytes. The comparison shows the most critical structural elements in substrates such as the intramolecular distance between the centers of the amino- and carboxy-terminal head groups and the central carbonyl function. Moreover, the stereoselective recognition of substrate side chains is demonstrated on basis of alanyl-peptides with D- and L-residues at different positions in the dipeptide.

provided in either amino- or carboxy-terminal positions. In the amino-terminal location, D-enantiomers are fairly well accepted with hydrophobic residues giving good affinity, whereas when located in the carboxy-terminal position of a dipeptide, affinity is markedly reduced. Peptides containing solely D-enantiomers of amino acids are not transported. Figure 3 provides insights into the key substrate features that determine affinity and transport by PEPT1.

Through the systematic analysis of substrate specificity, employing hundreds of compounds, PEPT1 requirements for substrate binding and transport are fairly well defined, although they can not yet be related to domains in the carrier protein. However, even without structural data for PEPT1, certain predictions can be made on binding and transport of novel compounds based on our current understanding. This has led to PEPT1 being a prime target for efficient intestinal absorption of rational designed drugs or specially designed prodrugs (60). Broad substrate specificity and also high transport rate make PEPT1 a good drug delivery system, and known drug substrates such as the aminocephalosporins or selected ACE-inhibitors show oral availability rates of up to 90% when used in doses of up to 1 g (61, 62).

The Mode of Transport

Knowledge about the transport cycle and the electrogenic nature of the transport step was derived from electrophysiological studies employing the two-electrode voltage-clamp or giant patch-clamp techniques in ooyctes expressing human or rabbit PEPT1. Both steady-state currents with various substrates as well as presteady-state currents have been analyzed (63–67). As already discussed, PEPT1 is a rheogenic carrier irrespective of the substrate's net charge. The substrate's net charge is, of course, dependent on pH and the dissociation constants of the ionizable groups, and these can be at various positions within a di- or tripeptide. Whenever PEPT1 transport activity is assessed by tracer flux studies, there is a pronounced increase in substrate uptake when the extracellular pH is decreased from neutrality to pH values of 6.5 to 6.0. This influx is completely independent of the luminal presence of Na^+, K^+, or Cl^-, only the proton-motive force that accounts for more then -50 mV at a luminal pH of 6.0 enables uphill uptake of peptides. A further lowering of the extracellular pH (<6.0) decreases transport activity, and this is generally explained as a pH-dependent inactivation of the carrier. However, the bell-shaped pH-dependence is observed only in the case of zwitterionic substrates (zwitterionic at neutral pH) and at low substrate concentrations. When anionic or even dianionic substrates are used, the transport rate increases only at low external pH (<6.0), and even at a pH of 4.5, the uptake optimum was not yet reached (68). Therefore, PEPT1 possesses a high acid resistance. But what causes the decline of transport at low pH when neutral substrates are employed? This observation can be taken as the first indicator that the substrate net charge affects the transport mode of PEPT1. More generally, it has been observed that anionic substrates are transported with higher transport rate at more acidic pH and cationic substrates at

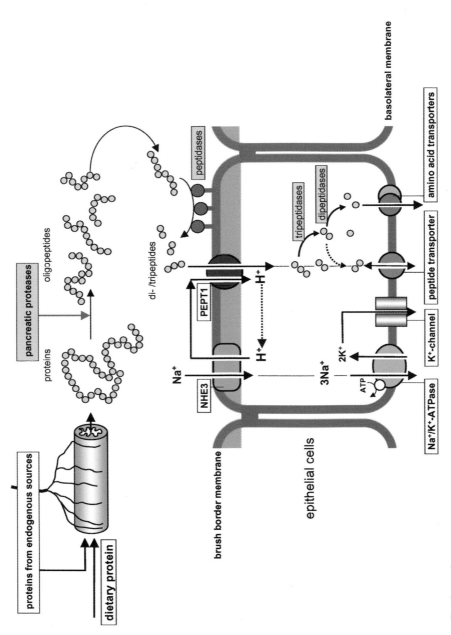

See legend on next page

Figure 1 Integrated model depicting the generation of di- and tripeptides from the hydrolysis of luminal proteins and the pathways involved in cellular uptake of peptides mediated by PEPT1 in the apical membrane of intestinal epithelial cells. Following apical influx, di- and tripeptides are sequentially hydrolyzed by multiple cytosolic hydrolases followed by basolateral efflux of the amino acids via different amino acid–transporting systems. Peptides not undergoing hydrolysis can exit the cell by a basolateral peptide-transporting system not yet identified on a molecular basis.

hPEPT1

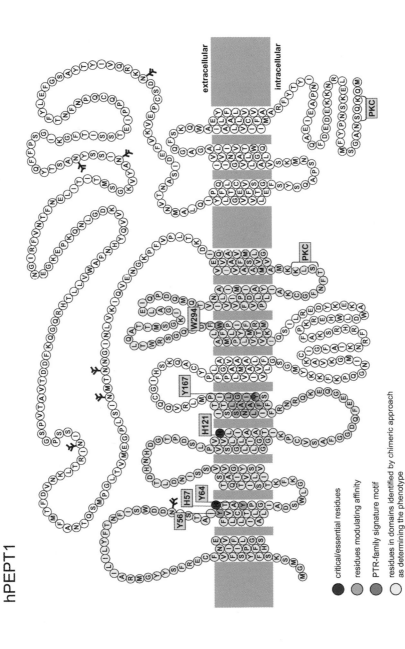

extracellular

intracellular

● critical/essential residues
● residues modulating affinity
● PTR-family signature motif
○ residues in domains identified by chimeric approach
   as determining the phenotype
⋏ putative glycosylation sites

See legend on next page

Figure 2   Model of the human PEPT1 secondary structure highlighting amino acid residues and protein domains with importance for the transport process. Individual amino acid residues identified by site-directed mutagenesis in PEPT1 proteins of various species and found to be either essential for transport or to alter substrate interaction with the protein are indicated by their distinct position and color code. Protein domains that are important for the phenotypical characteristics of PEPT1, as identified by the analysis of PEPT1/PEPT2 chimeras, are highlighted in yellow. The signature motif of the PTR-family of proton-dependent peptide transporters is projected into hPEPT1 and marked in blue. Putative glycosylation sites and PKC recognition sides, although not yet experimentally verified as used for post-translational modification, are also marked.

more neutral or slightly alkaline extracellular pH. In the latter case, however, the decline in the proton-motive force in alkaline conditions counteracts the effects on the substrate charge, and transport rates do not increase proportionally. From the detailed electrophysiological analysis of PEPT1 transporting charged and neutral substrates, the following conclusions have been derived: There is clearly a preference in any given peptide for transport of the substrate species that carries no net charge. Compounds with no net charge are translocated with a 1:1 stoichiometry in proton to substrate flux (65). In the case of dipeptides with glutamate or aspartate residues, two protons enter the cell with each substrate, but the second proton appears to be carried by the side chain carboxyl group, which is protonated prior to transport (65, 67). Cationic peptides containing lysine or arginine groups are also transported preferentially in their neutral form but in addition also in the charged form, in both cases with a 1:1 stoichiometry. However, when the positive charge is located in the carboxy-terminal side chain (i.e., Gly-Lys), the charged species can not bind to the transporter, and only the neutral form is transported in a 1:1 stoichiometry (67). Transport of the non-ionized species of cationic peptides occurs only at neutral or slightly alkaline luminal pH. Presteady-state current analysis has provided the basis for a model of the transport cycle in which protons bind first to the transporter in its outward facing conformation, followed by a change in substrate-binding affinity, and after substrate accommodation, the final conformational change follows with the translocation step (63). This model can explain most but not all of the experimental findings on the transport process. Electrophysiology, using the giant patch configuration and applied to rabbit PEPT1, has demonstrated that PEPT1 can be forced to transport bidirectionally and symmetrically (69). At saturating substrate concentrations, transport occurs in both directions with the same velocity, and only the membrane potential determines the direction and rate of transport. However, substrate binding to PEPT1 occurs in its outward facing state with a 5- to 100-fold higher affinity than on the inside (under identical experimental conditions). The differences depend on the substrate and are much more pronounced with charged dipeptides that have almost no affinity when provided from the cytosolic side. This asymmetry in the conformation of the substrate-binding domain favors the intracellular release of the substrate, which is independent of the fact that under normal physiological conditions an outward-directed transport mode with a cell inside negative membrane potential is unlikely.

TAKING PEPT1 FUNCTION BACK INTO PHYSIOLOGY

For over 40 years, starting with the first report in 1959 suggesting that dipeptides may be absorbed intact into the intestinal epithelium (70), numerous studies in experimental animals and also in humans have been carried out to characterize peptide transport processes. With the use of brush border membrane vesicles in the mid-1980s, the fundamentals of the transport process were elucidated, and the post-cloning research efforts have delivered substantial knowledge on the kinetics of PEPT1, its substrate specificity, flux coupling, and the regulation at the mRNA or protein expression level. Therefore, it is a challenge to combine knowledge gathered with the cloned PEPT1 with the physiological processes that characterize the intestinal absorption of peptides released from breakdown of dietary proteins. Although PEPT1 displays most of the expected functional characteristics, some of its features are not necessarily mirrored by findings obtained in tissue preparations or even in perfusion studies in humans. This holds true for pH dependence and substrate specificity. However, the pH dependence is difficult to relate to the intact epithelium where there is a fairly stable but slightly acidic surface pH compartment (71–73), and any changes in luminal pH are not transmitted in a linear fashion to the pH in the vicinity of the membrane. In Caco-2 cells (a model for the human small intestinal epithelium), peptide uptake across the apical membrane closely resembles the phenotypical characteristics of the cloned PEPT1 studied in heterologous expression systems. Even the dependence on the luminal pH—most likely due to a lack in a significant microclimate in Caco-2 cells—is similar to that found for PEPT1 in other expression systems. In Caco-2 cells, the interdependence of peptide uptake and the Na^+/H^+ exchanger activity has been demonstrated. The recovery of the epithelial cells from the peptide transport–induced acid load is mediated by the apical NHE-3 exchanger isoform (19, 74, 75) and the Na^+ dependence of this pH recovery process is taken as the explanation for the findings on the Na^+ dependence of peptide transport in intact tissue preparations.

The differences in substrate specificities and apparent affinities for identical substrates as obtained in intact tissues or with the cloned PEPT1 are difficult to interpret. In the intact epithelium, unstirred layer effects and hydrolysis of the peptides can lead to an underestimation of the true affinity of a substrate. For example, the apparent affinity for glycyl-glycine disappearance from the in situ perfused human intestine was determined as 43 mM (76), whereas the affinity of heterologously expressed hPEPT1 for the same substrate is at least 100-fold higher. However, what can be defined as a common theme is that the intestinal peptide transporter has a sufficiently broad substrate specificity that allows essentially every possible di- and tripeptide released during protein digestion to be transported in intact form. Whether and to which extent hydrolysis takes place at the membrane surface or in the intracellular compartment depends on the structure of the substrate and the concentration presented to the enzymes and transporters that compete for the same substrates.

Studies in human volunteers with perfusion of a segment of the small intestine with a series of glycyl-peptides with different carboxy-terminal residues have demonstrated that dipeptides are usually absorbed faster than a corresponding mixture of the same amino acids provided in free form (77). This may be attributed to the very high capacity of PEPT1 resulting from its high turn-over number (64) and high expression level. These findings on the efficient uptake of peptide-bound amino acids in human gut and the fact that peptide solutions have a lower osmolarity provide the rationale for using protein hydrolysates with short-chain peptides rather than free amino acids in enteral nutrition solutions (78, 79).

Although a rapid luminal disappearance of di- and tripeptides has been shown in human intestine, it is not known to what extent the peptides are undergoing enzymatic cleavage and what finally leaves the intestinal epithelial cell. Transepithelial movement of di- and tripeptides can be observed in tissue preparations and in monolayer cultures (i.e., Caco-2 cells) with hydrolysis-resistant dipeptides, and functional data from basolateral membrane vesicles (80) and Caco-2 cells (81, 82) have demonstrated an exit pathway for di- and tripeptides that is shared by peptidomimetics (83, 84), with characteristics similar but not identical to PEPT1. However, the molecular nature of the basolateral transporter is not known. Recently, photolabeling studies identified a protein for which peptide mapping of a fragment yielded the first sequence information that could not be related to any of the known peptide transporter genes (85). Although significant quantities of intact di- and tripeptides may leave the epithelial cell, peptide transfer from the gut lumen into the peripheral blood could be detected only in humans when large quantities of fairly stable glycyl-peptides were administered orally (86, 87). The importance of PEPT1 in the transfer of peptides from the gut lumen into circulation can most convincingly be demonstrated by the pharmacokinetics of the aminocephalosporins or ACE-inhibitors that utilize PEPT1 for intestinal absorption. These compounds are resistant to hydrolysis, and therefore their rapid appearance in circulation and almost complete absorption demonstrate the efficiency of PEPT1 for oral delivery of its substrates.

One of the most interesting question regarding peptide transport in the gut is the importance of PEPT1 in overall amino acid absorption from the diet. Demonstration of its general importance comes from studies in patients suffering from cystinuria or from Hartnup disease (88–92). In both disease types, intestinal absorption of certain amino acid is lacking or is markedly reduced by the malfunction of the underlying transporters. Whereas in case of cystinuria, a variety of missense and frameshift mutations have been identified that cause the malfunction of the two proteins, rBAT and its light chain, $b^{0,+}AT$, which operate as heterodimers to mediate transport of dibasic and neutral amino acids in kidney and gut (92); the molecular basis of Hartnup disease is not yet defined. Patients suffering from cystinuria lack sufficient intestinal transport of cystine and amino acids such as arginine and lysine (93). The latter is as an essential amino acid but, surprisingly, the patients do not develop a lysine deficiency. That they receive the essential lysine from the diet by the peptide transporter was substantiated in situ by perfusion of the jejunum of the patients providing lysine and leucine in free form or as leucyl-lysine. Whereas the absorption of lysine but not of leucine was drastically reduced in all cystinuria patients, absorption rates of lysine, when provided as a lysyl-dipeptide, form did not different from those in healthy controls (89). The same findings were obtained for lack of absorption of free arginine but efficient uptake from arginyl-leucine (90, 91). Similarly, in Hartnup patients with impaired absorption of amino acids such as histidine, tryptophan, and phenylalanine, absorption rates in vivo were far greater when the amino acids were provided in dipeptides than when perfused in free form (88). This clearly establishes that PEPT1 is important or even essential to life in humans suffering from inherited gene defects causing malfunctions of individual amino acid transporters. Another link between transport of peptides and that of free amino acids was demonstrated in Caco-2 cells (94). A dose-dependent stimulation of uptake of free arginine, but not of alanine, was observed with dipeptides containing neutral (alanine) or cationic amino acids (lysine, arginine). This phenomenon is best explained by a rapid intracellular hydrolysis of the absorbed peptides with the free amino acids binding to the $b^{0,+}$ system that operates as an obligatory exchanger and in turn enhances the apical influx of system $b^{0,+}$ substrates such as arginine. The opposite, an inhibition of dipeptide uptake by free amino acids—particularly by leucine— was also shown in different types of tissue preparations (95, 96), but this could not be demonstrated with the cloned PEPT1. Therefore it is not known how the free amino acids reduced dipeptide uptake into the intact epithelium.

REGULATION AND ADAPTATION OF PEPT1 TRANSPORT ACTIVITY

Chronic as well as acute regulation of PEPT1 activity has been demonstrated in numerous studies employing Caco-2 cells, animal models, and in human small and large intestine. Acute changes in transport activity can be induced by affecting intracellular signaling pathways with agonists or antagonists of protein kinase A (PKA) and C (PKC). An activation of PKC causes a rapid decline in apical peptide uptake based on a decreased V_{max} without changes in substrate affinity. Increasing intracellular cAMP levels by cholera toxin or forskolin treatment also inhibits peptide transport (97), whereas a decrease in the cAMP-level, for example, by the adrenoreceptor agonist clonidine stimulates transport as shown in Caco2–3B cells and in the rat jejunum (98, 99); a sigma-receptor agonist causes upregulation of PEPT1 (100). A reciprocal control of activity of PEPT1 is observed by alterations in intracellular free calcium concentration. An increase in free calcium rapidly decreases activity, whereas a decrease in free calcium causes an increased V_{max} of PEPT1 (101). These regulatory effects of second messengers on transporter activity can not yet be related to the PEPT1–protein, which in the human variant contains two canonical PKC recognition sites but no PKA site, whereas rat and the mouse PEPT1 possess one PKA and PKC consensus site. None of these motifs has been analyzed yet with respect to functionally transmitting the observed effects of second messengers at the protein level.

A regulation of PEPT1 abundance and activity in epithelial cells can also be observed by hormones such as insulin, leptin, growth hormone, and thyroid hormone (102–104). Dipeptide uptake into Caco-2 cells was stimulated by treatment with 5-nM insulin with an observed increase in the V_{max} and no alteration of K_m (102). Similarly, leptin, which is secreted into the stomach and gut lumen and binds to an apical membrane receptor, was shown to enhance dipeptide uptake into Caco-2 cells and in the rat intestine (103). Both hormones appear to increase the PEPT1 protein density in the apical membrane by recruitment of preformed transporters, because the colchicine-mediated disruption of the microtubular structures, but not inhibition of protein synthesis, prevented the incorporation of transporters into the membrane compartment in the presence of the hormones. Whereas insulin and leptin increase peptide transport activity without changes at the transcriptional level, treatment of Caco-2 cells with 3,5,3'-triiodothyronine (T3) reduced the V_{max} for dipeptide influx significantly with a concomitant reduction in the PEPT1 mRNA to 30% of control level (104). Recombinant human growth hormone has recently been shown to increase peptide transport activity in Caco-2 cells (105) and in combination with epidermal growth factor also increased PEPT1 expression in the rabbit small intestine after midgut resection (106).

Several studies have demonstrated that the expression level of PEPT1 and its function is altered by dietary treatments and in disease states. PEPT1 also shows a pronounced circadian alteration in protein expression followed by less pronounced but still significant changes in transport activity in rats (107). Protein concentration in the intestinal tissue was shown to be more then three times higher at 8:00 PM than at 8:00 AM in animals kept under a fixed light/dark cycle.

In rats, a brief fast (108), as well as sustained starvation and malnourishment (109), has been shown to be associated with a high expression level of PEPT1. The latter is important with respect to protein assimilation in states of impaired jejunal structure and function. Although other transporters such as SGLT1 displayed reduced expression levels, PEPT1 expression was sustained at a high level. Similarly, intestinal tissue damage by the chemotherapeutic agent 5-fluorouracil in rats revealed that PEPT1 protein levels in the brush border membrane remained unaffected, whereas those of sucrase and SGLT1 decreased (110). Animals treated with 5-fluorouracil showed higher PEPT1 mRNA levels, suggesting that the high resistance of the intestinal peptide transport activity to tissue damage is based on an increased protein synthesis rate. Diabetes, induced by treatment of rats with streptozotocin, also caused an increased protein expression level of PEPT1 in the small intestine and increased almost twofold the maximal peptide transport activity (111).

Dietary protein levels as well as certain free amino acids and peptides are able to change PEPT1 expression in the small intestine and its maximal transport activity (112). In rats fed increasing quantities of dietary protein for three days, the abundance of PEPT1 in the brush border membrane increased almost proportionally to the protein intake followed by a concomitant increase in peptide transport activity. Moreover, free phenylalanine and a dipeptide (Gly-Phe) administered in the diet could also induce the same changes, and a rat *pept1*-promotor-driven luciferase reporter assay applied in transfected cells with different deletion constructs identified certain elements in the promotor that also responded to phenylalanine and Gly-Phe with increased activity. Although several other free amino acids and selected dipeptides had some minor activity, the effects of Phe and Gly-Phe were most pronounced. It was suggested for rat *PEPT1* that the promotor region responsive to dietary protein involved the AP-1 binding site (TGACTCAG, nt −295), the AARE-like element-binding site (CATGGTG, nt −277) identified next to an AP-2 binding site, and an octamer-binding protein site for Oct1/Oct2. AP-1 is a transcription factor associated with the regulation of gene expression under amino acid deprivation conditions (113), and a similar AARE-like element was shown to control asparagine synthetase gene expression under essential amino acid deprivation (114). PEPT1 protein expression levels and transport activity significantly increased in Caco-2 cells that were cultured in a medium containing dipeptides (115, 116), suggesting that the substrates per se can alter the function of PEPT1 in the Caco-2 cells at the transcriptional and translation level.

PEPT1 is expressed throughout the small intestine but is virtually absent or expressed at very low levels in the human colonic mucosa under normal conditions (117). However, in patients suffering from short-bowel syndrome (SBS), interesting changes in intestinal PEPT1 expression are observed. Northern blot analysis, in situ hybridization, and immunohistochemistry indicated a high level of PEPT1 mRNA and protein confined to the epithelial cells in colonic samples obtained from patients with SBS (118). In these patients, the compensatory PEPT1 expression in colon may allow effective absorption of dietary amino acids from the diet or for treatment of SBS in enteral nutrition solutions as protein hydrolysates.

The unusual high colonic expression of PEPT1 has also been demonstrated in colon mucosa samples of patients with chronic ulcerative colitis and Crohn's disease (119). The aberrant expression of PEPT1 in chronic disease states has raised interest in a potential role of PEPT1 in inflammatory bowel diseases. PEPT1 was shown to transport the *n*-formylated peptide fMLP (formyl-Met-Leu-Phe), a major neutrophil chemotactic factor produced by *Escherichia coli* (119, 120). The transport of fMLP by hPEPT1 in Caco-2 monolayers showed a close link to neutrophil transepithelial migration and epithelial expression of immune molecules such as MHC-1 (119, 120) suggesting that the transporter may be involved in inflammatory processes (120, 121).

Bacterial infections have also been demonstrated to alter PEPT1 expression level and function in the gut. Endotoxin treatment in rats was shown to reduce PEPT1 expression at the protein level, which could be almost fully reversed by dexamethasone administration. It was suggested that the downregulation of PEPT1 by lipopolysaccharide involves proinflammatory cytokines such as tumor necrosis factor alpha and various interleukins (122). More recently, studies in suckling rats infected with *Cryptosporidium parvum*, which causes diarrhea in children and may lead to malnutrition, was studied with respect to the effects on PEPT1 expression. The data obtained suggested a transcriptional upregulation of the peptide transporter during acute cryptosporidiosis in response to malnutrition and the parasite implantation (123). Taken together, regulation of intestinal peptide transport activity occurs by transcriptional changes and altered protein translocation to the brush border membrane induced by a variety of hormones, second messengers, and in various disease states but PEPT1 shows an impressive resistance and sustained high expression level in the gut with impaired morphology and function.

CONCLUSIONS AND PERSPECTIVES

The post-cloning efforts in characterization of PEPT1 functions have substantially increased our understanding of some of its unique features. Recently, numerous studies have demonstrated that PEPT1 is subject to profound changes in gene and protein expression involving numerous hormones and different intracellular signaling pathways. There is also a growing scientific interest in intestinal peptide transport in pharmacology, with PEPT1 as a prime target for efficient oral delivery of drugs. Moreover, on the basis of the proposed role of the transporter in inflammatory processes and its increased colonic expression in disease states, a possible application to clinical medicine has appeared. Although demonstrated as essential to life in patients suffering from inherited diseases of amino acid transporters, the role of PEPT1 in overall amino acid absorption is not fully understoon, and animal models lacking the gene encoding PEPT1 are not yet available. We also do not know how polymorphisms in the *pept1*-gene in different populations affect the function of the transporter. Programs under way in pharmacogenomics with a focus on PEPT1 will eventually provide information on the importance of this transporter for interindividual variations in oral drug availability, with an expected spin-off for its significance in protein nutrition. Meanwhile, we wait for the successful identification of the molecular nature of the basolateral peptide transporter.

The *Annual Review of Physiology* is online at http://physiol.annualreviews.org

LITERATURE CITED

1. Matthews DM. 1991. Protein absorption. Development and present state of the subject. New York: Willey-Liss
2. Adibi SA, Mercer DW. 1973. Protein digestion in human intestine as reflected in luminal, mucosal, and plasma amino acid concentrations after meals. *J. Clin. Invest.* 52:1586–94
3. Kuwata H, Yamauchi K, Teraguchi S, Ushida Y, Shimokawa Y, et al. 2001. Functional fragments of ingested lactoferrin are resistant to proteolytic degradation in the gastrointestinal tract of adult rats. *J. Nutr.* 131:2121–27
4. McLachlan A, Cullis PG, Cornell HJ.

2002. The use of extended amino acid motifs for focussing on toxic peptides in coeliac disease. *J. Biochem. Mol. Biol. Biophys.* 6:319–24
5. Shan L, Molberg O, Parrot I, Hausch F, Filiz F, et al. 2002. Structural basis for gluten intolerance in celiac sprue. *Science* 297:2275–79
6. Hausch F, Shan L, Santiago NA, Gray GM, Khosla C. 2002. Intestinal digestive resistance of immunodominant gliadin peptides. *Am. J. Physiol. Gastrointest. Liver Physiol.* 283:G996–1003
7. Meisel H, Bockelmann W. 1999. Bioactive peptides encrypted in milk proteins:

proteolytic activation and thropho-functional properties. *Antonie Van Leeuwenhoek.* 76:207–15

8. Vermeirssen V, Deplancke B, Tappenden KA, Van Camp J, Gaskins HR, Verstraete W. 2002. Intestinal transport of the lactokinin Ala-Leu-Pro-Met-His-Ile-Arg through a Caco-2 Bbe monolayer. *J. Pept. Sci.* 8:95–100

9. Sato M, Hosokawa T, Yamaguchi T, Nakano T, Muramoto K, et al. 2002. Angiotensin I-converting enzyme inhibitory peptides derived from wakame (*Undaria pinnatifida*) and their antihypertensive effect in spontaneously hypertensive rats. *J. Agric. Food Chem.* 9:6245–52

10. Sterchi EE, Woodley JF. 1980. Peptide hydrolases of the human small intestinal mucosa: distribution of activities between brush border membranes and cytosol. *Clin. Chim. Acta* 14:49–56

11. Kania RK, Santiago NA, Gray GM. 1977. Intestinal surface amino-oligopeptidases. II. Substrate kinetics and topography of the active site. *J. Biol. Chem.* 252:4929–34

12. Piggott CO, O'Cuinn G, Fottrell PF. 1976. Similarities between a dipeptide hydrolase from brush-border and cytosol fractions of guinea-pig intestinal mucosa. *Biochem. J.* 1:715–17

13. Fei YJ, Kanai Y, Nussberger S, Ganapathy V, Leibach FH, et al. 1994. Expression cloning of a mammalian proton-coupled oligopeptide transporter. *Nature.* 7:563–66

14. Ganapathy V, Leibach FH. 1985. Is intestinal peptide transport energized by a proton gradient? *Am. J. Physiol. Gastrointest. Liver Physiol.* 249:G153–60

15. Ganapathy V, Burckhardt G, Leibach FH. 1985. Peptide transport in rabbit intestinal brush-border membrane vesicles studied with a potential-sensitive dye. *Biochim. Biophys. Acta* 27:234–40

16. Ganapathy V, Burckhardt G, Leibach FH. 1984. Characteristics of glycylsarcosine transport in rabbit intestinal brush-border membrane vesicles. *J. Biol. Chem.* 25:8954–59

17. Ganapathy V, Leibach FH. 1983. Role of pH gradient and membrane potential in dipeptide transport in intestinal and renal brush-border membrane vesicles from the rabbit. Studies with L-carnosine and glycyl-L-proline. *J. Biol. Chem.* 10:14189–92

18. Cheeseman CI, Devlin D. 1985. The effect of amino acids and dipeptides on sodium-ion transport in rat enterocytes. *Biochim. Biophys. Acta* 14:767–73

19. Kennedy DJ, Leibach FH, Ganapathy V, Thwaites DT. 2002. Optimal absorptive transport of the dipeptide glycylsarcosine is dependent on functional Na^+/H^+ exchange activity. *Pflügers Arch.* 445:139–46

20. Liang R, Fei YJ, Prasad PD, Ramamoorthy S, Han H, et al. 1995. Human intestinal H^+/peptide cotransporter. Cloning, functional expression, and chromosomal localization. *J. Biol. Chem.* 270:6456–63

21. Chen H, Pan Y, Wong EA, Bloomquist JR, Webb KE JR. 2002. Molecular cloning and functional expression of a chicken intestinal peptide transporter (cPepT1) in *Xenopus* oocytes and Chinese hamster ovary cells. *J. Nutr.* 132:387–93

22. Pan Y, Wong EA, Bloomquist JR, Webb KE JR. 2001. Expression of a cloned ovine gastrointestinal peptide transporter (oPepT1) in *Xenopus* oocytes induces uptake of oligopeptides in vitro. *J. Nutr.* 131:1264–70

23. Deleted in proof

24. Saito H, Okuda M, Terada T, Sasaki S, Inui K. 1995. Cloning and characterization of a rat H^+/peptide cotransporter mediating absorption of beta-lactam antibiotics in the intestine and kidney. *J. Pharmacol. Exp. Ther.* 275:1631–37

25. Fei YJ, Fujita T, Lapp DF, Ganapathy V, Leibach FH. 1998. Two oligopeptide transporters from *Caenorhabditis elegans*: molecular cloning and functional expression. *Biochem. J.* 1:565–72

26. Liu W, Liang R, Ramamoorthy S, Fei YJ, Ganapathy ME, et al. 1995. Molecular cloning of PEPT 2, a new member of the H$^+$/peptide cotransporter family, from human kidney. *Biochim. Biophys. Acta* 4:461–66

27. Boll M, Herget M, Wagener M, Weber WM, Markovich D, et al. 1996. Expression cloning and functional characterization of the kidney cortex high-affinity proton-coupled peptide transporter. *Proc. Natl. Acad. Sci. USA* 9:284–89

28. Rubio-Aliaga I, Daniel H. 2002. Mammalian peptide transporters as targets for drug delivery. *Trends Pharmacol. Sci.* 23:434–40

29. Steiner HY, Naider F, Becker JM. 1995. The PTR family: a new group of peptide transporters. *Mol. Microbiol.* 16:825–34

30. Paulsen IT, Skurray RA. 1994. The POT family of transport proteins. *Trends Biochem. Sci.* 19:404

31. Stacey G, Koh S, Granger C, Becker JM. 2002. Peptide transport in plants. *Trends Plant Sci.* 7:257–63

32. Urtti A, Johns SJ, Sadee W. 2001. Genomic structure of proton-coupled oligopeptide transporter hPEPT1 and pH-sensing regulatory splice variant. *AAPS Pharm. Sci.* 3:E6

33. Fei YJ, Sugawara M, Liu JC, Li HW, Ganapathy V, et al. 2000. cDNA structure, genomic organization, and promoter analysis of the mouse intestinal peptide transporter PEPT1. *Biochim. Biophys. Acta* 21:145–54

34. Saito H, Motohashi H, Mukai M, Inui K. 1997. Cloning and characterization of a pH-sensing regulatory factor that modulates transport activity of the human H$^+$/peptide cotransporter, PEPT1. *Biochem. Biophys. Res. Commun.* 28:577–82

35. Covitz KM, Amidon GL, Sadee W. 1998. Membrane topology of the human dipeptide transporter, hPEPT1, determined by epitope insertions. *Biochemistry* 27:15214–21

36. Doring F, Martini C, Walter J, Daniel H. 2002. Importance of a small N-terminal region in mammalian peptide transporters for substrate affinity and function. *J. Membr. Biol.* 15:55–62

37. Terada T, Saito H, Sawada K, Hashimoto Y, Inui K. 2000. N-terminal halves of rat H$^+$/peptide transporters are responsible for their substrate recognition. *Pharm. Res.* 17:15–20

38. Fei YJ, Liu JC, Fujita T, Liang R, Ganapathy V, Leibach FH. 1998. Identification of a potential substrate binding domain in the mammalian peptide transporters PEPT1 and PEPT2 using PEPT1-PEPT2 and PEPT2-PEPT1 chimeras. *Biochem. Biophys. Res. Commun.* 8:39–44

39. Doring F, Dorn D, Bachfischer U, Amasheh S, Herget M, Daniel H. 1996. Functional analysis of a chimeric mammalian peptide transporter derived from the intestinal and renal isoforms. *J. Physiol.* 15:773–79

40. Fei YJ, Liu W, Prasad PD, Kekuda R, Oblak TG, et al. 1997. Identification of the histidyl residue obligatory for the catalytic activity of the human H$^+$/peptide cotransporters PEPT1 and PEPT2. *Biochemistry* 14:452–60

41. Terada T, Saito H, Mukai M, Inui KI. 1996. Identification of the histidine residues involved in substrate recognition by a rat H$^+$/peptide cotransporter, PEPT1. *FEBS Lett.* 30:196–200

42. Chen XZ, Steel A, Hediger MA. 2000. Functional roles of histidine and tyrosine residues in the H$^+$-peptide transporter PepT1. *Biochem. Biophys. Res. Commun.* 16:726–30

43. Kato M, Maegawa H, Okano T, Inui K, Hori R. 1989. Effect of various chemical modifiers on H$^+$ coupled transport of cephradine via dipeptide carriers in rabbit intestinal brush-border membranes: role of histidine residues. *J. Pharmacol. Exp. Ther.* 251:745–49

44. Brandsch M, Brandsch C, Ganapathy ME, Chew CS, Ganapathy V, Leibach FH.

1997. Influence of proton and essential histidyl residues on the transport kinetics of the H⁺/peptide cotransport systems in intestine (PEPT 1) and kidney (PEPT 2). *Biochim. Biophys. Acta* 13:251–62

45. Terada T, Saito H, Inui K. 1998. Interaction of beta-lactam antibiotics with histidine residue of rat H⁺/peptide cotransporters, PEPT1 and PEPT2. *J. Biol. Chem.* 6:5582–85

46. Yeung AK, Basu SK, Wu SK, Chu C, Okamoto CT, et al. 1998. Molecular identification of a role for tyrosine 167 in the function of the human intestinal proton-coupled dipeptide transporter (hPepT1). *Biochem. Biophys. Res. Commun.* 8:103–7

47. Bolger MB, Haworth IS, Yeung AK, Ann D, von Grafenstein H, et al. 1998. Structure, function, and molecular modeling approaches to the study of the intestinal dipeptide transporter PEPT1. *J. Pharm. Sci.* 87:1286–91

48. Meredith D, Boyd CA, Bronk JR, Bailey PD, Morgan KM, et al. 1998. 4-aminomethylbenzoic acid is a nontranslocated competitive inhibitor of the epithelial peptide transporter PepT1. *J. Physiol.* 1:629–34

49. Knutter I, Theis S, Hartrodt B, Born I, Brandsch M, et al. 2001. A novel inhibitor of the mammalian peptide transporter PEPT1. *Biochemistry* 10:4454–58

50. Sawada K, Terada T, Saito H, Hashimoto Y, Inui K. 1999. Effects of glibenclamide on glycylsarcosine transport by the rat peptide transporters PEPT1 and PEPT2. *Br. J. Pharmacol.* 128:1159–64

51. Terada T, Sawada K, Saito H, Hashimoto Y, Inui K. 2000. Inhibitory effect of novel oral hypoglycemic agent nateglinide (AY4166) on peptide transporters PEPT1 and PEPT2. *Eur. J. Pharmacol.* 24:11–17

52. Bailey PD, Boyd CA, Bronk JR, Collier ID, Meredith D, et al. 2000. How to make drugs orally active: a substrate template for peptide transporter PepT1.

Angew. Chem. Int. Ed. Engl. 39:505–8

53. Tame JR, Dodson EJ, Murshudov G, Higgins CF, Wilkinson AJ. 1995. The crystal structures of the oligopeptide-binding protein OppA complexed with tripeptide and tetrapeptide ligands. *Structure* 15:1395–406

54. Nickitenko AV, Trakhanov S, Quiocho FA. 1995. 2 Å resolution structure of DppA, a periplasmic dipeptide transport/chemosensory receptor. *Biochemistry* 26:16585–95

55. Wang T, Wade RC. 2002. Comparative binding energy (COMBINE) analysis of OppA-peptide complexes to relate structure to binding thermodynamics. *J. Med. Chem.* 24:4828–37

56. Doring F, Will J, Amasheh S, Clauss W, Ahlbrecht H, Daniel H. 1998. Minimal molecular determinants of substrates for recognition by the intestinal peptide transporter. *J. Biol. Chem.* 4:23211–18

57. Doring F, Walter J, Will J, Focking M, Boll M, et al. 1998. Delta-aminolevulinic acid transport by intestinal and renal peptide transporters and its physiological and clinical implications. *J. Clin. Invest.* 15:2761–67

58. Brandsch M, Thunecke F, Kullertz G, Schutkowski M, Fischer G, Neubert K. 1998. Evidence for the absolute conformational specificity of the intestinal H⁺/peptide symporter, PEPT1. *J. Biol. Chem.* 13:3861–64

59. Brandsch M, Knutter I, Thunecke F, Hartrodt B, Born I, et al. 1999. Decisive structural determinants for the interaction of proline derivatives with the intestinal H⁺/peptide symporter. *Eur. J. Biochem.* 266:502–8

60. Han HK, Amidon GL. 2000. Targeted prodrug design to optimize drug delivery. *AAPS Pharm. Sci.* 2:E6

61. Mazzei T, Dentico P. 2000. The pharmacokinetics of oral cephalosporins. *Clin. Microbiol. Infect.* 6:53–54

62. Zhang H, Zhang J, Streisand JB. 2002.

Oral mucosal drug delivery: clinical pharmacokinetics and therapeutic applications. *Clin. Pharmacokinet.* 41:661–80

63. Mackenzie B, Loo DD, Fei Y, Liu WJ, Ganapathy V, et al. 1996. Mechanisms of the human intestinal H^+-coupled oligopeptide transporter hPEPT1. *J. Biol. Chem.* 8:5430–37

64. Nussberger S, Steel A, Trotti D, Romero MF, Boron WF, Hediger MA. 1997. Symmetry of H^+ binding to the intra- and extracellular side of the H^+-coupled oligopeptide cotransporter PepT1. *J. Biol. Chem.* 21:7777–85

65. Steel A, Nussberger S, Romero MF, Boron WF, Boyd CA, Hediger MA. 1997. Stoichiometry and pH dependence of the rabbit proton-dependent oligopeptide transporter PepT1. *J. Physiol.* 1:563–69

66. Amasheh S, Wenzel U, Boll M, Dorn D, Weber W, et al. 1997. Transport of charged dipeptides by the intestinal H^+/peptide symporter PepT1 expressed in *Xenopus laevis* oocytes. *J. Membr. Biol.* 1:247–56

67. Kottra G, Stamfort A, Daniel H. 2002. PEPT1 as a paradigm for membrane carriers that mediate electrogenic bidirectional transport of anionic, cationic, and neutral substrates. *J. Biol. Chem.* 6:32683–91

68. Wenzel U, Gebert I, Weintraut H, Weber WM, Clauss W, Daniel H. 1996. Transport characteristics of differently charged cephalosporin antibiotics in oocytes expressing the cloned intestinal peptide transporter PepT1 and in human intestinal Caco-2 cells. *J. Pharmacol. Exp. Ther.* 277:831–39

69. Kottra G, Daniel H. 2001. Bidirectional electrogenic transport of peptides by the proton-coupled carrier PEPT1 in *Xenopus laevis* oocytes: its asymmetry and symmetry. *J. Physiol.* 15:495–503

70. Newey H, Smyth DH. 1960. Intracellular hydrolysis of dipeptides during intestinal absorption. *J. Physiol.* 152:367–80

71. Lucas M. 1983. Determination of acid surface pH in vivo in rat proximal jejunum. *Gut* 24:734–39

72. Daniel H, Neugebauer B, Kratz A, Rehner G. 1985. Localization of acid microclimate along intestinal villi of rat jejunum. *Am. J. Physiol. Gastrointest. Liver Physiol.* 248:G293–98

73. McEwan GT, Daniel H, Fett C, Burgess MN, Lucas ML. 1988. The effect of *Escherichia coli* STa enterotoxin and other secretagogues on mucosal surface pH of rat small intestine in vivo. *Proc. R. Soc. London Ser. B* 22:219–37

74. Thwaites DT, Kennedy DJ, Raldua D, Anderson CM, Mendoza ME, et al. 2002. H/dipeptide absorption across the human intestinal epithelium is controlled indirectly via a functional Na/H exchanger. *Gastroenterology* 122:1322–33

75. Thwaites DT, Ford D, Glanville M, Simmons NL. 1999. H^+/solute-induced intracellular acidification leads to selective activation of apical Na^+/H^+ exchange in human intestinal epithelial cells. *J. Clin. Invest.* 104:629–35

76. Adibi SA, Soleimanpour MR. 1974. Functional characterization of dipeptide transport system in human jejunum. *J. Clin. Invest.* 53:1368–74

77. Adibi SA. 1986. Kinetics and characteristics of absorption from an equimolar mixture of 12 glycyl-dipeptides in human jejunum. *Gastroenterology* 90:577–82

78. Steinhardt HJ, Wolf A, Jakober B, Schmuelling RM, Langer K, et al. 1989. Nitrogen absorption in pancreatectomized patients: protein versus protein ydrolysate as substrate. *J. Lab. Clin. Med.* 113:162–67

79. Boza JJ, Moennoz D, Vuichoud J, Jarret AR, Gaudard-de-Weck D, Ballevre O. 2000. Protein hydrolysate vs free amino acid-based diets on the nutritional recovery of the starved rat. *Eur. J. Nutr.* 39:237–43

80. Dyer J, Beechey RB, Gorvel JP, Smith RT, Wootton R, Shirazi-Beechey SP. 1990. Glycyl-L-proline transport in rabbit enterocyte basolateral-membrane vesicles. *Biochem. J.* 1:565–71

81. Saito H, Hashimoto Y, Inui K. 1999. Functional characteristics of basolateral peptide transporter in the human intestinal cell line Caco-2. *Am. J. Physiol. Gastrointest. Liver Physiol.* 276:G1435–41

82. Saito H, Inui K. 1993. Dipeptide transporters in apical and basolateral membranes of the human intestinal cell line Caco-2. *Am. J. Physiol. Gastrointest. Liver Physiol.* 265:G289–94

83. Matsumoto S, Saito H, Inui K. 1994. Transcellular transport of oral cephalosporins in human intestinal epithelial cells, Caco-2: interaction with dipeptide transport systems in apical and basolateral membranes. *J. Pharmacol. Exp. Ther.* 270:498–504

84. Irie M, Terada T, Sawada K, Saito H, Inui K. 2001. Recognition and transport characteristics of nonpeptidic compounds by basolateral peptide transporter in Caco-2 cells. *J. Pharmacol. Exp. Ther.* 298:711–17

85. Shepherd EJ, Lister N, Affleck JA, Bronk JR, Kellett GL, et al. 2002. Identification of a candidate membrane protein for the basolateral peptide transporter of rat small intestine. *Biochem. Biophys. Res. Commun.* 30:918–22

86. Adibi SA. 1971. Intestinal transport of dipeptides in man: relative importance of hydrolysis and intact absorption. *J. Clin. Invest.* 50:2266–75

87. Adibi SA, Morse EL, Masilamani SS, Amin PM. 1975. Evidence for two different modes of tripeptide disappearance in human intestine. Uptake by peptide carrier systems and hydrolysis by peptide hydrolases. *J. Clin. Invest.* 56:1355–63

88. Asatoor AM, Cheng B, Edwards KD, Lant AF, Matthews DM, et al. 1970. Intestinal absorption of dipeptides and corresponding free amino acids in Hartnup disease. *Clin. Sci.* 39:1P

89. Hellier MD, Perrett D, Holdsworth CD, Thirumalai C. 1971. Absorption of dipeptides in normal and cystinuric subjects. *Gut* 12:496–97

90. Asatoor AM, Harrison BD, Milne MD, Prosser DI. 1972. Intestinal absorption of an arginine-containing peptide in cystinuria. *Gut* 13:95–98

91. Silk DB, Perrett D, Clark ML. 1975. Jejunal and ileal absorption of dibasic amino acids and an arginine-containing dipeptide in cystinuria. *Gastroenterology* 68:1426–32

92. Chillaron J, Roca R, Valencia A, Zorzano A, Palacin M. 2001. Heteromeric amino acid transporters: biochemistry, genetics, and physiology. *Am. J. Physiol. Renal Physiol.* 281:F995–1018

93. De Sanctis L, Bonetti G, Bruno M, De Luca F, Bisceglia L, et al. 2001. Cystinuria phenotyping by oral lysine and arginine loading. *Clin. Nephrol.* 56:467–74

94. Wenzel U, Meissner B, Doring F, Daniel H. 2001. PEPT1-mediated uptake of dipeptides enhances the intestinal absorption of amino acids via transport system $b^{0,+}$. *J. Cell. Physiol.* 186:251–59

95. Radhakrishnan AN. 1977. Intestinal dipeptidases and the dipeptide transport in the monkey and in man. *Ciba Found. Symp.* 50:37–59

96. Himukai M, Kano-Kameyama A, Hoshi T. 1982. Mechanisms of inhibition of glycylglycine transport by glycyl-L-leucine and L-leucine in guinea-pig small intestine. *Biochim. Biophys. Acta* 7:170–78

97. Muller U, Brandsch M, Prasad PD, Fei Y-J, Ganapathy V, Leibach FH. 1996. Inhibition of the H^+/peptide cotransporter in the human intestinal cell line Caco-2 by cyclic AMP. *Biochem. Biophys. Res. Commun.* 218:461–65

98. Berlioz F, Maoret JJ, Paris II, Laburthe M, Farinotti R, Roze C. 2000. Alpha$_2$-adrenergic receptors stimulate oligopeptide transport in a human intestinal cell line. *J. Pharmacol. Exp. Ther.* 294:466–72

99. Berlioz F, Julien S, Tsocas A, Chariot J, Carbon C, et al. 1999. Neural modulation of cephalexin intestinal absorption through the di- and tripeptide brush

border transporter of rat jejunum in vivo. *J. Pharmacol. Exp. Ther.* 288:1037–44

100. Fujita T, Majikawa Y, Umehisa S, Okada N, Yamamoto A, et al. 1999. sigma Receptor ligand-induced up-regulation of the H⁺/peptide transporter PEPT1 in the human intestinal cell line Caco-2. *Biochem. Biophys. Res. Commun.* 2:242–46

101. Wenzel U, Kuntz S, Diestel S, Daniel H. 2002. PepT1-mediated cefixime uptake into human intestinal epithelial cells is increased by Ca^{2+} channels blockers. *Antimicrob. Agents Chemother.* 46:1375–80

102. Thamotharan M, Bawani SZ, Zhou X, Adibi SA. 1999. Hormonal regulation of oligopeptide transporter Pept-1 in a human intestinal cell line. *Am. J. Physiol. Cell Physiol.* 276:C821–26

103. Buyse M, Berlioz F, Guilmeau S, Tsocas A, Voisin T, et al. 2001. PepT1-mediated epithelial transport of dipeptides and cephalexin is enhanced by luminal leptin in the small intestine. *J. Clin. Invest.* 108:1483–94

104. Ashida K, Katsura T, Motohashi H, Saito H, Inui K. 2002. Thyroid hormone regulates the activity and expression of the peptide transporter PEPT1 in Caco-2 cells. *Am. J. Physiol. Gastrointest. Liver Physiol.* 282:G617–23

105. Sun BW, Zhao XC, Wang GJ, Li N, Li JS. 2003. Hormonal regulation of dipeptide transporter (PepT1) in Caco-2 cells with normal and anoxia/reoxygenation management. *World J. Gastroenterol.* 9:808–12

106. Avissar NE, Ziegler TR, Wang HT, Gu LH, Miller JH, et al. 2001. Growth factors regulation of rabbit sodium-dependent neutral amino acid transporter ATB0 and oligopeptide transporter 1 mRNAs expression after enteretomy. *J. Parenter. Enteral. Nutr.* 25:65–72

107. Pan X, Terada T, Irie M, Saito H, Inui K. 2002. Diurnal rhythm of H⁺-peptide cotransporter in rat small intestine. *Am. J. Physiol. Gastrointest. Liver Physiol.* 283:G57–64

108. Thamotharan M, Bawani SZ, Zhou X, Adibi SA. 1999. Functional and molecular expression of intestinal oligopeptide transporter (Pept-1) after a brief fast. *Metabolism* 48:681–84

109. Ihara T, Tsujikawa T, Fujiyama Y, Bamba T. 2000. Regulation of PepT1 peptide transporter expression in the rat small intestine under malnourished conditions. *Digestion* 61:59–67

110. Tanaka H, Miyamoto KI, Morita K, Haga H, Segawa H, et al. 1998. Regulation of the PepT1 peptide transporter in the rat small intestine in response to 5-fluorouracil-induced injury. *Gastroenterology* 114:714–23

111. Gangopadhyay A, Thamotharan M, Adibi SA. 2002. Regulation of oligopeptide transporter (Pept-1) in experimental diabetes. *Am. J. Physiol. Gastrointest. Liver Physiol.* 283:G133–8

112. Shiraga T, Miyamoto K, Tanaka H, Yamamoto H, Taketani Y, et al. 1999. Cellular and molecular mechanisms of dietary regulation on rat intestinal H⁺/peptide transporter PepT1. *Gastroenterology* 116:354–62

113. Pohjanpelto P, Holtta E. 1990. Deprivation of a single amino acid induces protein synthesis-dependent increase in c-jun, c-myc and ornithine decarboxylase mRNAs in Chinese hamster ovary cells. *Mol. Cell Biol.* 10:5814–21

114. Guerrini L, Gong SS, Mangasarian K, Basilico C. 1993. *Cis-* and *trans-*acting elements involved in amino acid regulation of asparagine synthetase gene expression. *Mol. Cell Biol.* 13:3202–12

115. Walker D, Thwaites DT, Simmons NL, Gilbert HJ, Hirst BH. 1998. Substrate up-regulation of the human small intestinal peptide transporter, hPepT1. *J. Physiol.* 15:697–706

116. Thamotharan M, Bawani SZ, Zhou X, Adibi SA. 1998. Mechanism of dipeptide stimulation of its own transport in a human intestinal cell line. *Proc. Assoc. Am. Physicians* 110:361–68

117. Ford D, Howard A, Hirst BH. 2003. Expression of the peptide transporter hPepT1 in human colon: a potential route for colonic protein nitrogen and drug absorption. *Histochem. Cell Biol.* 119:37–43

118. Ziegler TR, Fernandez-Estivariz C, Gu LH, Bazargan N, Umeakunne K, et al. 2002. Distribution of the H$^+$/peptide transporter PepT1 in human intestine: upregulated expression in the colonic mucosa of patients with short-bowel syndrome. *Am. J. Clin. Nutr.* 75:922–30

119. Merlin D, Si-Tahar M, Sitaraman SV, Eastburn K, Williams I, et al. 2001. Colonic epithelial hPepT1 expression occurs in inflammatory bowel disease: transport of bacterial peptides influences expression of MHC class 1 molecules. *Gastroenterology* 120:1666–79

120. Merlin D, Steel A, Gewirtz AT, Si-Tahar M, Hediger MA, Madara JL. 1998. hPepT1-mediated epithelial transport of bacteria-derived chemotactic peptides enhances neutrophil-epithelial interactions. *J. Clin. Invest.* 1:2011–18

121. Buyse M, Tsocas A, Walker F, Merlin D, Bado A. 2002. PepT1-mediated fMLP transport induces intestinal inflammation in vivo. *Am. J. Physiol. Cell Physiol.* 283:C1795–800

122. Shu HJ, Takeda H, Shinzawa H, Takahashi T, Kawata S. 2002. Effect of lipopolysaccharide on peptide transporter 1 expression in rat small intestine and its attenuation by dexamethasone. *Digestion* 65:21–29

123. Barbot L, Windsor E, Rome S, Tricottet V, Reynes M, et al. 2003. Intestinal peptide transporter PepT1 is over-expressed during acute cryptosporidiosis in suckling rats as a result of both malnutrition and experimental parasite infection. *Parasitol. Res.* 89:364–70

Annu. Rev. Physiol. 2004. 66:385–417
doi: 10.1146/annurev.physiol.66.032902.134726
Copyright © 2004 by Annual Reviews. All rights reserved
First published online as a Review in Advance on October 15, 2003

ORAL REHYDRATION THERAPY:
New Explanations for an Old Remedy

Mrinalini C. Rao
Department of Physiology and Biophysics, University of Illinois at Chicago, Chicago, Illinois 60612; email: meenarao@uic.edu

Key Words diarrhea, glucose, short-chain fatty acid, sodium absorption, chloride secretion

■ **Abstract** Diarrheal diseases are among the most devastating illnesses globally, but the introduction of oral rehydration therapy has reduced mortality due to diarrhea from >5 million children, under the age of 5, in 1978 to 1.3 million in 2002. Variations of this simple therapy of salts and sugars are prevalent in traditional remedies in cultures world-wide, but only in the past four decades have the scientific bases for these remedies begun to be elucidated. This review aims to provide a broad understanding of the cellular basis of oral rehydration therapy. The features integral to the success of oral rehydration therapy are active glucose transport in the small intestine, commensal bacteria, and short-chain fatty acid transport in the colon. The review examines these processes and their regulation and considers new approaches that might supplement oral rehydration therapy in controlling diarrheal diseases.

INTRODUCTION

Infectious diarrheas continue to be a major health problem across the world, having widespread and devastating effects in developing countries and contributing to the health burden in developed countries. The morbidity and mortality associated with infectious diarrheas result from severe dehydration. The consequences of diarrhea are more severe in the young than in the adult because a fasting child not suffering from diarrhea can lose 1 to 2% of body weight daily (1, 2). Because the young have low nutritional reserves and children in developing countries are exposed to multiple episodes of diarrhea, malnutrition in these regions becomes commonplace. The consequences of diarrhea are similarly severe in the very elderly or immunocompromised patient (2).

Although numerous pathogens can cause diarrheas with varying etiologies, the most notorious is *Vibrio cholerae*. The seminal work of the London surgeon/ epidemiologist John Snow drew attention to the transmission of the disease, the importance of sanitation, and even the underlying cause as dehydration (3)[1]. In the century following these observations, improved sanitary conditions greatly attenuated the disease. However, in the last quarter of the twentieth century, the formulation and dissemination of oral rehydration therapy brought a tremendous improvement to the control of diarrheal diseases and in 1978 was acclaimed to be the "most important medical advance this century" (4). While there is much debate on what is an ideal oral rehydration solution (discussed below), the basic premise is that by providing the patient with an oral supplement comprised of sugar or starches, salt, and water, fluid replacement should be achieved (2, 5, 6). This seemingly simple remedy is deeply rooted in traditional medicine worldwide. Remarkably, only in the past four decades have the scientific bases of these remedies begun to be elucidated, and their success underscores why these remedies have stood the test of time.

WHAT IS ORAL REHYDRATION THERAPY?

Oral Rehydration Therapy in Traditional Medicine

The use of oral fluid replacements in the treatment of diarrhea can be documented in perhaps every culture. Diarrhea was known to Hippocrates, and elaborate treatment for diarrhea is provided in the Vedic Suśruta Saṁhitā (7). In rural Mississippi, for example, a mixture of flour and water is recommended, and another folk remedy suggests apple extracts as a binding agent for mild diarrhea prior to a visit to the pediatrician. In tropical countries, the formulations range from watery rice gruel to extracts of maize, yams, taro root, green bananas, the water of tender green coconuts (a sterile medium), a mixture of molasses and rock salt, and extracts of mung beans (2, 8, 9). In the developing world during the nineteenth and early twentieth centuries, there were opposing views on the value of traditional remedies vis-à-vis Western medicine. For example, in India, European colonizers often regarded tropical diseases "as diseases peculiar to India" and believed that the response of the European and the native to therapy would be different. Thus in colonial Bombay in the mid-1800s, even as Snow was making his prescient observations on the Broad Street epidemic (3), British medical and sanitation officers were of the firm belief that cholera was a poison developed in the soil (miasmatic theory) (10). Furthermore, medical officers in the Bombay Presidency considered Western medicine to be more efficacious and rejected the traditional use of "large draughts of salt and water" to treat cholera as it would cause inflammation and would work, if at all, only with Indians (10). The Indian response was equally unhelpful as local practitioners of Ayurveda claimed that cholera arose from cow slaughter

---

[1]As a lasting tribute to Dr. Snow, Dr. R.R. Frerichs, of the University of California at Los Angeles, has created an informative website, http://www.ph.ucla.edu/epi/snow.html, which includes access to Dr. Snow's 1855 book.

by the Europeans! However, as deciphered by Snow, improved sanitation was critical in stemming cholera (3), and Western approaches to sanitation resulted in a reduction in the number of cholera-related deaths in heavily populated Indian urban centers from 19,996 during 1857–1865 to 10,509 during 1866–1886 (10). Recently, the common practice by Bangladeshi villagers of using old cotton saris to filter home-made drinks was put to a field test, involving 65 villages and 45,000 participants. The cotton sari folded eightfold (mesh size ~20 μm) sieved out the zooplankton that harbor *Vibrio*, and filtration led to a 48% reduction in cholera (11). These improvements show that although beneficial, traditional medicines cannot be implemented in a vacuum, they are even more effective when coupled with other factors such as sanitation, education, and implementation.

Oral Rehydration Therapy in Modern Medicine

The terms oral rehydration salts or solutions (both abbreviated ORS) and oral rehydration therapy (ORT) are generally used interchangeably. The World Health Organization's (WHO) original definition of ORS was confined to the use of oral rehydration salt packets dissolved in water. However, ORT encompasses both ORS as well as recommended home fluids (RHF), comprised of Na^+ and a source of carbohydrate, ranging from rice water to cereal-based solutions and traditional soups.

Original references to the modern history of oral rehydration can be found in a number of reports (1, 2). In the late 1940s and early 1950s, a balanced salt solution was found to help correct the acidosis in children with acute diarrhea (11a). Much like the traditional remedies, these formulations included glucose more as a source of nutrition than as the major driving force for fluid absorption. Parallel to these clinical studies, seminal work by a number of physiologists, including Ussing, Curran, Crane, Fisher, and Schultz, culminated in the demonstration that the transport of a number of solutes, in particular glucose and amino acids, was coupled to the movement of Na^+ across the apical membrane (reviewed in 12–14). In 1968, the first successful clinical trials of ORS were conducted by the Cholera Research Laboratory, Dhaka,[2] in collaboration with Johns Hopkins Institute for Medical Research and Training, Baltimore, and the All India Institute for Tropical Medicine, Calcutta (15, 16). The 1971 war on the Indian subcontinent not only resulted in the creation of the nation of Bangladesh but was pivotal in making ORS available to the general public. Hitherto, administration of ORS was believed to be the purview of medical personnel, but the rampant spread of cholera and other diseases in the aftermath of the war led the director of a medical center camp to distribute ORS to the general public for administration (17). The reported death rates in the camps using conventional intravenous fluid was 6- to 16-fold greater than in those using ORS (20–50% versus 3%) (17, 18). This reaction to

---

[2]This laboratory was established in 1960 in East Pakistan (Bangladesh in 1971). In 1978, the Government of Bangladesh established the International Center for Diarrheal Disease Research B.

adversity was fortuitous and resulted in the general acceptance of nonmedical personnel administering ORS, a major step in providing care in regions where access to medical help is limited. Although the validity of this simple solution was questioned, these clinical efforts led to the launching of ORS as a therapy by the WHO in 1978.

A number of centers around the world contributed to the development of improved oral rehydration formulations, whereas the WHO and UNICEF were primarily and continue to be instrumental in the effective dissemination of the formulations. The number of OR packets distributed in developing countries has risen from 51 million in 1979–80 to 800 million in 1991–1992 (19). The effect of these interventions is best seen in the mortality rates; in 1978, approximately 5 million children under the age of 5 years died from acute diarrhea; these figures steadily dropped to ~4.6 million in 1980, 3.3 million in 1990, 1.5 million in 2000, and 1.3 million in 2002 (17, 19). Boosted by this success, WHO's goals are targeted to reducing deaths attributable to diarrhea by a further 50% by 2010 (17). The success of ORT is due to the interaction of multiple variables, including caretaker compliance, adequate supplies, effective distribution, and improved sanitation. Great strides were made in the improvement of home management from 1990–1995 (19). Assessing the appropriate management of ORT[3] is a gargantuan task as it involves surveying different regions of the world, with information from nonclinical and clinical settings, and includes multiple indicators such as accessibility to ORT, continued feeding during diarrheal episodes, and the ensuing fluid intake (19). Despite these challenges, ORT has had far-reaching repercussions globally on the management of one of the most devastating diseases. It is fitting that the first Pollin Prize for Paediatric Research was awarded to Drs. N. Hirschhorn, N.F. Pierce, D. Mahalanabis, and D.R. Nalin for their key contributions to the discovery and implementation of ORT.

Why Is ORT Lifesaving? The Composition of Oral Rehydration Fluids

The key to the success of ORT is that it replaces the fluid being lost, circumventing the need for intravenous replacement in 80 to 90% of the cases of mild-moderate diarrhea and is lifesaving in acute diarrheal diseases. It also minimizes the malnourishing effects of acute diarrhea. The composition of oral rehydration fluids has undergone some changes over the past 25 years.

DEFINED FORMULATIONS   The composition of WHO packets distributed in 1978 is provided in Table 1 (17). Although highly beneficial, it soon became clear that the therapy had some limitations (20). These can be broadly categorized into compliance, questions related to osmolarity, and determination of what should

---

[3]Survey examples include the Control of Diarrheal Diseases survey of WHO, the Multiple Indicator Cluster Surveys and Demographic and Health Surveys of UNICEF.

TABLE 1   Sugar and electrolyte composition of commonly utilized oral rehydration solutions

(CH₂O)ₙ	Standard ORS[a]	Reduced pre-osmolarity ORS[a]	Rice-based ORS[b] 50 g cooked rice powder	Amylase-resistant starch[c] 50 g high-amylose starch	Pedialyte[d]
Glucose (mM)	111	75	—	111	139
Sodium (mM)	90	75	90	90	45
Chloride (mM)	80	65	80	80	35
Potassium (mM)	20	20	20	20	20
Citrate (mM)	10*	10	10	10	—
Other Anion (mM)	—	—	—	—	30
Total Osmolarity (mOsM)	311	245	280	327	269

*Citrate varies 10–30 mM, and earlier formulations used HCO_3^-.
[a] (17).
[b] (24).
[c] (27).
[d] (11a).

accompany and follow ORT. While dehydration was averted by the standard ORS (Table 1), the near-isotonic solution neither reduced the duration of the illness nor the rate of stool loss; in some cases, an increase in stool loss was even reported. In addition, there was danger of hypernatremia (6, 20). These shortcomings made the therapy less attractive to the patient and the caretaker, leading to poor compliance. As recently as 2000, there was concern about the under-utilization of ORT in developing and developed countries (5, 21, 22). In developing countries, the problem is one of misplaced expectations; i.e., that the therapy should be a "magical" cure involving a "prestigious" drug and therefore patients preferred "drugs" to the WHO version of traditional remedies. In Europe and the United States, despite the potential saving from ORT of $1 billion, in terms of hospital care and follow-up visits, the acceptance rate of ORT was disappointingly low. The causes were varied but ironically related to the poorly informed support staff, who preferred intravenous fluids to the frequent feedings (every few minutes) required in ORT, and the uncertainty over whether ORT would be reimbursed by insurance companies (5).

The criteria for improving ORS are to decrease stool output, duration of diarrhea and emesis, and to reduce the need for intravenous therapy. There has been extensive debate over the ideal osmolarity for ORS, and there may be a difference in what is needed in developed countries compared with developing countries and the nature of the diarrheas (5, 6). Fecal Na^+ concentration in cholera is ~90 mM and replacing this was the basis of the original ORS. However, in rotaviral diarrheas, the major type of diarrhea in children in developed countries, fecal Na^+ is closer to 40 mM and therefore there is danger of hypernatremia. Pediatric societies, such as ESPGHAN[4], have recommended a multistep protocol for the treatment of acute diarrheas in developed countries, including reduced osmolarity ORS (RO-ORS), fast rehydration, and rapid return to normal feeding (22). In a multicenter trial in Brazil, India, Mexico, and Peru, administration of RO-ORS (224 mOsM) decreased stool output, continued need for ORS intake, and duration of diarrhea compared with the use of WHO-ORS (23). However, a risk in reducing osmolarity in rehydrating cholera patients is hyponatremia. When RO-ORS (75 mM each Na^+ and glucose, 65 mM Cl^-, 245 mOsM) was tested in adults and children with cholera, it was found to be as effective as standard ORS, with a chance of symptomatic hyponatremia in adults [reduced osmolarity ORS formulation in (17)]. In a meta-analysis of noncholera diarrheas in children, the effects were more dramatic, i.e., reduced stool volume and decreased need for follow-up intravenous therapy. On the basis of such meta-analyses, the WHO-UNICEF released a new reduced osmolarity ORS formula in May 2002 (17; Table 1). However, grave reservations were expressed by pioneers in the field that this move was ill-advised as the scale of the study was not large enough and the risks of hyponatremia were not fully explored (23a).

Numerous studies on improving ORT also examine what should accompany or follow ORT. The importance of continuous feeding, especially breast-feeding, during ORT and follow-up of the treatment with a regimen to replace lost fluid (2, 5, 22) is now well accepted. Feeding is critical to promote intestinal growth and to increase net absorption. In general, breast-fed children fared better, with a 35% greater energy intake and a 250% greater protein intake than formula-fed children (2).

RICE-BASED ORS   Along with the strategies described above, the quest for alternate, inexpensive improved home fluid therapy continued. Answers were sought again in traditional remedies, such as use of starchy water left after over-cooking rice to treat diarrheas in India, Bangladesh, Egypt, and Mexico. In a meta-analysis of 13 clinical trials, adults and children with cholera, when fed with rice-based (50–80 gm/L) ORS (280 mOsM, Table 1), showed ~32–36% decrease in stool output (24). In other studies, ORS was shown to decrease the duration of diarrhea and rate of purging. Not surprisingly, rice-based hypoosmolar ORS was better than standard ORS or RO-ORS in cholera patients (25). Rice has the advantage of providing glucose, amino acids, and more calories than an equivalent amount of glucose and is hypoosmolar to standard WHO-ORS. It is also cheaper and readily available in cholera-endemic regions. However, for noncholera diarrhea with feeding, there was no difference between rice-based ORS and standard ORS (26).

---

[4]European Society of Pediatric Gastroenterology, Hepatology and Nutrition.

AMYLASE-RESISTANT STARCHES   Whereas rice is readily digested by amylase in the upper small intestine, depending on the content of pectins, dextrins, glycans, and cellulose, a significant portion of starches can be amylase resistant. Recognition of the importance of short-chain fatty acids (SCFAs) in colonic absorption has led to a heightened interest in the use of amylase-resistant starches in ORT.

In a randomized, well-designed study of 48 adolescent and adult cholera patients, the efficacy of an amylase-resistant, high-amylose maize starch (Table 1) was compared with that of rice flour and standard WHO-ORS. The starch and rice (50 g each) were resuspended in WHO-ORS, rendering them somewhat hyperosmolar. When compared with WHO-ORS and rice-ORT, the resistant starch significantly decreased fecal weight and reduced the duration of diarrhea; it also caused an increased fecal excretion of starch (27). Another randomized, double-blind trial involving 62 infants suffering from persistent diarrhea elegantly demonstrated the efficacy of amylase-resistant starches contained in green bananas (28). Three groups were given equivalent caloric amounts (54 kcal/dL) for 7 days of a diet comprised of rice ORS alone, or pectin or cooked banana, made up in a rice-ORS. The last two groups showed a decrease in diarrhea as early as day 2 of treatment, and by day 4, over 80% of them had no diarrhea in contrast to 20% in the rice-ORS group. Pectin- or green banana-based ORS reduced stool output and emesis and required less rehydration solutions and fewer return medical visits. Similarly, partially hydrolyzed guar gum (Benefiber®) administered in WHO-ORS reduced duration of diarrhea and stool output in children suffering from acute, noncholera diarrhea (29).

To summarize, ORT is here to stay, and exploration of amylase-resistant starches and rice as effective sources of carbohydrates to improve ORT is of tremendous therapeutic and economic benefit.

WHAT IS THE CELLULAR BASIS FOR ORAL REHYDRATION THERAPY?

The development of ORT over the past 40 years has paralleled the huge leaps in our understanding of the molecular basis of intestinal ion and fluid transport in health and disease. A brief overview is provided (for reviews, see 14, 30–34).

Intestinal Fluid and Ion Transport

DESIGN AND REGULATION   Together, the small and large intestines process >9 liters of fluid a day; approximately 7 liters come from other exocrine glands and food intake, and the intestines themselves contribute 1.5–2 liters. In a healthy individual, over 80% of this fluid is reabsorbed in the small intestine and >18% in the colon, with less than 2% being lost in the stool. The architecture of the intestines and their lining epithelia are uniquely geared to accomplish the opposing functions of absorption and secretion. There are, however, species, age, and segmental differences along the cephalocaudal axis of the intestine in terms of morphology, expression of hydrolases, types of ion transporters, their regulation, and the luminal milieu (30, 35–37). The accepted model is that the villar/surface cells are the chief sites of absorption and the crypts are the main sites of secretion, with a few exceptions (see below) (14, 32, 38, 39). A disruption in the balance of absorption and secretion, either by decreased absorption and/or increased secretion, results in diarrhea.

Absorptive and secretory functions are tightly coordinated by a plethora of neurohumoral factors. The heterogeneous epithelial cells themselves contribute modulators, and the rest of the regulatory pathways arise from either luminal stimuli or from neural, immune, or systemic stimuli from the intestinal wall (Table 2). The multiple smooth muscle layers govern transit time. The complexities of intestinal ion transport regulation have been underscored by the discovery of multiple signaling pathways and abundant cross-talk between these cascades. Multiple isoforms for all components of the canonical signal cascades, including receptors, cyclases, phosphodiesterases, protein kinases (PK), phosphatases, and target transporters, add to this complexity (30, 31, 34, 40, 41, 46, 47). Another level of regulation is achieved by compartmentalization of transporters and their regulatory molecules into subcellular domains, such as on lipid rafts (42), on scaffolding proteins (43), or in vesicular compartments (44). The pleiotropic ability of transporters

TABLE 2   Intestinal fluid transport regulators (categorized by intracellular mediator)

	Promoters of secretion		Promoters of absorption
$\uparrow Ca^{2+}$	$\uparrow$ **Cyclic AMP**	**Tyrosine kinase and MEK cascades**	$\downarrow$ **Cyclic AMP**
Acetylcholine	Adenosine	Epidermal growth factor	α_2-adrenergic agonists
Adenosine	Bradykinin	*B. fragilis*	Enkephalins
ATP	Histamine	Enteropathogenic	Lipoxygenase
Bile acids	Prostaglandins	*E. coli*	product 12-HETE
Bradykinin	Vasoactive intestinal	*Salmonella* species	Neuropeptide Y
Galanin	peptide	*Shigella* species	Somatostatin
Gastrin-releasing	*E. coli* LT		
peptide	*V. cholerae*		**Others**
Histamine			Aldosterone
Neurotensin	$\uparrow$ **Cyclic GMP**		Dopamine
Prostaglandins	Guanylin		Glucocorticoid
Serotonin	Nitric Oxide		
Substance P	Uroguanylin		
Rotavirus	*E. coli* EAST		
C. difficile	*E. coli* STa		
V. parahaemolyticus	*Y. enterocolitica*		

EAST, enteroaggregative heat-stable toxin; LT, heat-labile enterotoxin; STa, heat-stable enterotoxin.

such as the cystic fibrosis transmembrane conductance regulator (CFTR) to influence the expression of other transporters adds yet another layer of complexity (reviewed in 45). Generally, stimuli that increase intracellular cAMP, cGMP, or Ca^{2+} (Table 2) cause net luminal fluid accumulation by either stimulating secretion and/or inhibiting absorption (46, 47). Agents that decrease cAMP promote absorption. Cross-talk exists between these cascades and the tyrosine kinase/phosphatase and the mitogen-activated kinase cascades (30, 34, 40, 41).

PARACELLULAR PATHWAYS   Transepithelial transport occurs via paracellular or transcellular pathways. The tight junction at the apical end of the cells largely defines paracellular movement in terms of charge selectivity, ion and solute movement, and tissue resistance. Transepithelial resistance increases down the cephalo-caudal axis (duodenum is $<25 \ \Omega \ cm^2$, ileum $\sim 90 \ \Omega \ cm^2$, and colon $\sim 200 \ \Omega \ cm^2$) (30). Intercellular contacts farther down the lateral space also contribute to the complex geometry and therefore modify flow. Whereas paracellular transport is passive, the tight junction is a dynamic lattice structure of 14 or more proteins, ranging from the transmembrane occludins and claudins to the plaque-associated zonula occludens (ZO) proteins and cytoskeletal actin (48). These are subject to regulation by cellular signals and by other membrane proteins further along the lateral space. Thus the interaction of E-cadherin with Ca^{2+} triggers a chain of events involving recruitment of catenins, vinculin, and the activation of phospholipase C and PKC and ZO proteins, resulting in tight junction formation (49). The apical perijunctional actomyosin ring is in close contact with the tight junction and can modify paracellular permeability (50). Of relevance to this review is the observation that agents that cause secretion [e.g., cAMP, *V. cholera* ZO toxin (ZOT), *Clostridium perfringens* toxin, and immune mediators], as well as those that promote absorption (e.g., glucose), increase paracellular permeability (40).

TRANSCELLULAR TRANSPORT IN THE SMALL INTESTINE   The study of two devastating and seemingly disparate diseases, e.g., cholera and cystic fibrosis, in the 1960s–1980s revolutionized our thinking about anion secretory and solute-independent Na^+ absorptive processes in the intestine (14, 51). This understanding followed on the heels of the discovery of the cellular basis for solute transport (see below) (13). The P-type Na^+/K^+ ATPase, located on the basolateral membrane (BLM), is the sine qua non of all transepithelial active transport processes (Figure 1, see color insert). Regulation of the pump is one way to modulate fluid loss. For example, diarrhea associated with T cell activation was recently shown to be because of increased epithelial permeability and decreased Na^+/K^+ ATPase activity, resulting in decreased solute-dependent and electroneutral Na^+ absorption in wild-type and $CFTR^{-/-}$ animals (52).

Fluid secretion   The concerted action of at least three transporters, in addition to the Na^+/K^+ pump, is needed to drive secretion; regulation of any one of these will alter net anion secretion. Na^+ moves passively via the paracellular pathway (Figure 1). The pump energizes the uphill entry of Cl^- via the BLM Na^+-K^+-$2Cl^-$ cotransporter (NKCC-1) (53). NKCC-1 contains consensus sequences for a number of kinases, and, in some cell types, an increase in NKCC-1 phosphorylation is associated with a stimulation of transport (54), whereas in others, NKCC-1 activity is governed by $[Cl^-]_i$ and/or retrieval from intracellular vesicles (53). In $NKCC^{-/-}$ mice, although cAMP-stimulated Cl^- secretion is impaired in the jejunum, the cGMP-stimulated fluid secretion is unimpaired, implying that other anion entry mechanisms are present (55). CFTR decreases NKCC-1 expression and activity in pancreatic cells (56) but not in the colon (57). The other essential BLM transporters are K^+ channels, which repolarize the cell and maintain the driving force for Cl^- exit. There are at least two types of K^+ channels in the intestine, one regulated by cAMP (KVLQTI) and the other by Ca^{2+} (14, 32).

Chloride exits the apical membrane (BBM) of the enterocyte via CFTR (reviewed in 45). CFTR expression is greater in the crypts, but there is evidence for its presence in the villus (58). Support for the critical role of CFTR in intestinal Cl^- secretion comes from the $\Delta F508$ and $CFTR^{-/-}$ mice, where the primary defect is in intestinal secretion, and the intestines are refractory to Ca^{2+}-dependent secretagogues (45). Cyclic AMP and cGMP, via PKA and PKGII, respectively, increase the open probability of CFTR by directly phosphorylating CFTR on the regulatory R domain. In addition, cAMP increases the recruitment of CFTR to the apical membrane from the subapical membrane compartment (45). There are at least two other classes of Cl^- channels in the intestine, the ClC family, and the Ca^{2+}-activated Cl^- channels (CLCA) (32, 59, 60). The intestine mainly contains ClC-2, and its localization to the BLM of villus enterocytes suggests a role in absorption (61). The members of the CLCA family are DIDS-sensitive Cl^- channels, and some members are chiefly found in the goblet cells and crypts of the small intestine (59). Germane to this review is whether CLCAs can play a role in rotaviral infections (see below).

Bicarbonate secretion is necessary for luminal alkalization, especially in the duodenum, and there are at least three exit routes for HCO_3^-: two anion exchangers, the down-regulated in adenoma protein (DRA) (62, 63), and the putative anion transporter PAT1 (64) in the villus cells and CFTR in the crypts (65). PAT1 shows higher expression in the duodenum than in the colon, whereas DRA shows the reverse pattern. Recent evidence shows that DRA and the Na^+/H^+ exchangers (NHE), NHE-2 and NHE-3, are colocalized to specific lipid microdomains and are functionally coupled (66). Entry of HCO_3^- into the enterocyte across the BLM occurs via the Na^+-HCO_3^- (NBC) family of cotransporters. The BLM also expresses the AE-2 anion exchanger proteins.

Fluid absorption   Net fluid absorption largely occurs in the jejunum and ileum and is inexorably linked to the movement of Na^+, either in conjunction with solutes (see below), or via NHEs. The bulk of nonsolute-coupled Na^+ transport across the BBM is achieved via the NHE-2 and NHE-3 isoforms, whereas the NHE-1 isoform is localized to the BLM and is involved in pH and cell volume regulation (reviewed in 30, 67, 68). There are clear distinctions in the regulation of NHE-2 and NHE-3, and the latter appears to be responsible for the bulk of intestinal Na^+ absorption because the NHE-3$^{-/-}$, but not the NHE-2$^{-/-}$, mouse has severe diarrhea (69). The NHEs, especially NHE-3, are inhibited by cAMP, cGMP, or Ca^{2+} and therefore contribute to net fluid loss. Glucocorticoids increase NHE-3 but not NHE-2 activity (70), but proximal bowel resection increases both NHE-2 and NHE-3 expression and activity (71). The molecular mechanisms underlying the regulation of NHEs involve sequestration in lipid rafts, vesicle recycling, binding to the regulatory proteins, and perhaps direct phosphorylation of the exchanger (72). The identification of the PDZ domain-containing NHE regulatory factors (NHERF) as proteins critical for the cAMP-mediated inhibition of NHE-3 was a major breakthrough in transport physiology (73). These proteins interact with a number of transporters critical to transepithelial salt and water movement, including CFTR and DRA, and are also linked to β-adrenergic receptors and to the PKA-anchoring protein ezrin. The concerted interaction of these proteins enables cAMP to simultaneously activate CFTR while inhibiting NHEs and perhaps DRA (74).

The excessive Cl^- secretion and reduced NHE activity are the bane of ORT, and interference with any of their interlinked signaling pathways would be of benefit.

TRANSCELLULAR TRANSPORT IN THE MAMMALIAN COLON   The healthy colon reabsorbs 1.5 liters of fluid/day but can triple its capacity. However, if the colon's absorptive capacity is compromised, as seen by the neurally mediated action of cholera toxin (see below), then the diarrhea is exacerbated. Some features of colonic ion transport that are distinct from the small intestine are highlighted below (Figure 2, see color insert) (reviewed in 32, 33, 75). The adult colon does not express the Na^+-dependent glucose transporter, SGLT1, and therefore Na^+ glucose is not a major mechanism for Na^+ uptake; rather, the distal colon expresses conductive Na^+ transport. The colon is the only part of the intestine where K^+ can be actively secreted and absorbed via apical K^+ channels and a K^+/H^+-ATPase, respectively. Finally, the major anions in the colonic lumen are SCFA (detailed below in the section on Role of SCFA and Commensal Bacterial Microflora).

The proximal colon largely expresses NHE-2 and NHE-3, and depending on the hormonal status, the distal colon expresses NHE-2, NHE-3, and the epithelial Na^+ channel (ENaC) (32). CFTR down-regulates the expression and activity of epithelial Na^+ channels (60). In the NHE-3$^{-/-}$ mice there is an upregulation of ENaC and DRA, which together probably compensate for the loss in Na^+ absorption (76). Recent evidence suggests that Na^+ absorption can also take place in the crypts (38, 39), perhaps even by a novel Cl^--dependent NHE (38). Equally intriguing are the observations that, by generating a hypertonic absorbate, the crypts play a critical role in fecal dehydration (77). ClC-2 may be involved in Cl^- absorption in the rat, but in the human it is present in subapical compartments (61). The bulk of Cl^- secretion can be attributed to CFTR (33). Colonic crypts and goblets express hCLCA1, and evidence in T-84 cells, but not in CFTR$^{-/-}$ mice, suggests a role for CLCAs in colonic Cl$^-$ secretion (34, 59, 60). Critical to this review is the fact that colonic absorptive processes can play a major role in rehydration therapy.

Major Perpetrators of Diarrhea

Space does not permit doing justice to the fascinating ways in which ingenious microorganisms adapt to hostile environments and coopt their host and their own colonizers (phages) to survive and proliferate. When microbial survival occurs at the expense of the host, the organisms are pathogenic and elicit their diarrheagenic effects by interfering with various host cell signaling cascades, leading to an inhibition of Na$^+$ absorption and/or stimulation of Cl$^-$ secretion. For example, *Clostridium difficile* affects the cytoskeleton whereas *Shigella dysenteriae* affects protein synthesis. Enterohemorrhagic *Escherichia coli* and *Salmonella* utilize NFκB pathways that induce the expression of galanin-1 receptors on the host cell. Endogenous galanin then activates a Ca^{2+}-mediated Cl$^-$ secretion (78). Three examples of diseases effectively treated with ORT are given below (40, 41).

CHOLERA AND RELATED DIARRHEAS   The world is currently witnessing the eighth cholera pandemic, which emerged in India and Bangladesh in 1992. The different pandemics have been associated with specific strains of *V. cholerae* (serogroups O1 biotype "classical" for the fifth and sixth, O1 El Tor for the seventh, and O139 for the eighth). Substantial progress has been made in understanding the molecular basis of *V. cholerae* virulence and survival (reviewed in 79, 80), culminating in the sequencing of the genome. This gram-negative curved rod bacterium can survive in an aquatic ecosystem where it receives some protection from the hostile milieu by associating with zooplankton and forming specific biofilms. *V. cholerae* has two chromosomes. Chromosome II appears to contain most of the genes needed for survival in an aquatic system. Chromosome I contains genes needed for metabolic function and for virulence. There are two major clusters of virulence genes, the Vibrio pathogenicity island (VPI) and the filamentous phage CTXϕ, that are viral in origin and integrated into the host chromosome. The VPI bears the genes for toxin-coregulated pili (TCPs) required for colonization, whereas the CTXϕ bears the genes for cholera toxin A and B subunits, ZOT, and accessory cholera toxin (ACE). Although many aquatic strains of *V. cholerae* are nonpathogenic, pathogenicity can be rapidly acquired by horizontal gene transfer, first of VPI, which then provides the receptor for transfer of CTXϕ. The fact that horizontal gene transfer can occur in the intestine, that *V. cholerae* utilizes quorum sensing, and that the host milieu influences the types of genes expressed in the bacteria reinforces the challenges this bacterium provides (79, 80).

Effective colonization of the bacterium in the intestine and expression of the necessary virulence genes involve unknown luminal cues and a complex regulatory dialog between VPI- and CTXϕ-coded proteins and those coded for by the "ancestral" *Vibrio* chromosomes (79). However, once the bacterium has colonized the surface cells of the small intestine, it elaborates a heteromeric toxin containing five B subunits that form a ring around the A subunit. The B subunit binds to GM1 ganglioside on the BBM and is sorted via lipid rafts into a retrograde trafficking path via the Golgi cisternae into the lumen of the endoplasmic reticulum (ER). The A subunit is then unfolded and nicked in the lumen of the ER, and the enzymatically active A1 is released via the sec61 complex. By mechanisms that are still unclear, the A1 subunit accesses the BLM G-protein, $G\alpha S$, and ADP-ribosylates it. ADP ribosylation of $G\alpha S$ inhibits its nascent GTPase activity and therefore allows $G\alpha S$ to permanently activate adenylate cyclase and generate cAMP for the life of the enterocyte (reviewed in 81). Although *V. cholerae* chiefly colonize the small intestine, by activating enteric neural pathways involving release of serotonin, PGs and VIP, it can also stimulate colonic secretion (82, 83).

Other factors produced by *V. cholerae* may contribute to the diarrhea. Most intriguing of these was the observation that ZOT, via a specific receptor, ZOT-R, increased paracellular permeability in the small intestine, but not the colon, and that ACE stimulated Cl^- transport. Although the plasmids *zot* and *ace* are now proven to be involved in the morphogenesis of CTXϕ, recent evidence suggests that C-terminal ZOT can bind to ZOT-R (reviewed in 40). The mammalian ligand for ZOT-R, zonulin, is also a potent regulator of paracellular permeability (84). Finally, strains of enterotoxigenic *E. coli* produce a heat-labile toxin (LT), which also elaborates an AB toxin, with mechanisms of action similar to CT (40, 41).

TRAVELER'S DIARRHEA AND HEAT-STABLE ENTEROTOXINS   Enterotoxigenic *E. coli* and *Yersinia enterocolitica* (85) elaborate a heat-stable enterotoxin (STa), which acts via the cGMP signaling cascade (86, 87). The STas have 18 amino acid peptides with 6 cysteine residues and are synthesized as precursor proteins and exported via the Sec pathway. STa binds to a specific BBM receptor-guanylate cyclase (GCC) in enterocytes and colonocytes. The action of STa is reversible and involves activation of a specific cGMP-dependent PKGII (47). Whereas STa acts via PKGII in the human colon, there are also suggestions that it could inhibit a phosphodiesterase and thereby stimulate cAMP production in the colon (87, 88). Specific scaffolding proteins associated with GCC and STa signaling have been identified (89). STa-induced secretion also appears to involve the enteric nervous system (82). STa stimulates Cl^- secretion and inhibits Na^+ absorption (31, 90), although recently Lucas (91) suggested that in vivo STa may chiefly be affecting Na^+ absorption rather than Cl^- secretion. The guanylin family members are the mammalian homologues of STa (Table 2). Enteroaggregative *E. coli* produces a guanylin-like toxin, EAST-1, which interacts with GCC but has a different pathogenicity than STa (92).

ROTAVIRUS-INDUCED DIARRHEAS   Rotavirus is the major cause of diarrhea in children in developed countries, but is also prevalent in developing countries, contributing (20%–25%) to overall mortality. The virus mainly affects the small intestine, is self-limiting, and shows a strong age-dependence in a variety of species, causing diarrhea in the very young and often being asymptomatic in the adult. The double-stranded RNA virus has at least six viral proteins and five nonstructural proteins (NSP), of which one, NSP4, is the first known viral enterotoxin (reviewed in 93). The mechanisms underlying rotaviral infections appear to be a potpourri of the processes utilized by various bacterial pathogens. Thus in the early stages, the virus, NSP4, and its C-terminal peptide (114–135) increase $[Ca^{2+}]_i$ in the enterocytes of all ages, but elicit only Cl^- secretion in the young animal. A pertussis toxin–sensitive activation of PLC increases $[Ca^{2+}]_i$ initially from extracellular stores and later from the ER. NSP4, but not cAMP or carbachol, stimulates Cl^- secretion only in young $CFTR^{-/-}$ mice (94). This finding led to the suggestion that NSP4 is activating CLCA in the young. The increase in $[Ca^{2+}]_i$ unleashes a host of other effects ranging from a decrease in junctional resistance to actin reorganization and a disruption in cell morphology, thereby reducing the absorptive surface area. A second stage of infection involves the host enteric system and is associated with villus damage, although not necessarily with inflammation. Although rotaviral infections curtail Na^+-glucose absorption, ORT is effective, perhaps by promoting residual glucose absorption.

Why Does ORT Work?

The cellular underpinnings of ORT rest in the digestion and absorption of nutrients; the faster the absorption of sugar, the faster the absorption of water and Na^+.

TRANSPORT OF GLUCOSE AND OTHER NUTRIENTS   Carbohydrates, which constitute 45–60% of the Western diet, are comprised of polymers of glucose, amylose (α-1,4), amylopectin (α-1,6), complex polysaccharides, and disaccharides. Salivary α–amylases start the process of digestion, but the bulk of the common polysaccharides is hydrolyzed by the pancreatic α–amylase, at the α-1,4 linkages, to release oligosaccharides. The oligosaccharides are broken down further by surface hydrolases such as maltase-glucoamylase and isomaltase, which cleave the α-1,4, and α-1,6, linkages, respectively. Similarly, surface disaccharidases break down sucrose, lactose, and maltose. Pancreatic proteases and enteral brush border peptidases cause the breakdown of proteins to release oligopeptides, dipeptides, and amino acids. There are adequate hydrolases to ensure complete digestion of amyloses, with the surface hydrolases constituting $> 10\%$ of the total BBM protein. Although pancreatic α–amylase may be low in the infant, salivary α–amylase and BBM glucoamylase appear sufficient for digestion (reviewed in 36, 95, 96).

The elucidation of the mechanism of glucose absorption in the small intestine has revolutionized the field of transport physiology from many perspectives. The key features are that hexoses can be transported both actively and passively and that active glucose transport is coupled to Na^+ movement at the BBM and to an ATP-independent step at the BLM (reviewed in 13). These observations led to the tenet of secondary active transport and vectorial movement, not only of sugars, but also of amino acids, vitamins, and ions. The major route for BBM amino acid entry now appears to be the dipeptide carrier (see Daniel, this volume; 97).

Marking another milestone in the field were the yeoman efforts of Wright and colleagues (98), leading to the molecular identification of the Na^+-glucose transporter, SGLT1, on the BBM, both by protein purification techniques and expression cloning. SGLT1 transports Na: D-glucose or D-galactose with a stoichiometry of 2:1. Site-directed mutagenesis and structure-function studies of SGLT1 have mapped the regions of Na^+ and glucose binding to the N- and C-terminal transmembrane domains, respectively (99). Mutations in SGLT1 result in the rare and potentially fatal congenital diarrhea of glucose-galactose malabsorption, which emphasizes the importance of SGLT1 as the major mechanism for active glucose uptake (100). Cloning of SGLT1 led to the identification of other distinct sugar transporters in the enterocyte; GLUT5 on the BBM, which facilitates fructose entry, and GLUT2 on the BLM, which facilitates glucose, galactose, and fructose exit from the cell (Figure 1). SGLT1 activity may influence other routes of glucose entry. Transport through SGLT1 activates PKCßII and recruits GLUT2 to the BBM; this apical GLUT2 then functions as a high-capacity, low-affinity route for sugar entry (101, 102). Activation of SGLT1 was also shown to alter perijunctional actomyosin contraction and increase tight junctional permeability and thereby paracellular glucose transport (50). The relative roles of these pathways to overall glucose transport remain to be elucidated. A role for SGLT1 as a conduit for water is discussed below under water transport.

Regulation of hexose transporters have been extensively studied, especially during development and in response to dietary changes (103). The most critical finding was the observation that elevation in cAMP (104, 105) or cGMP (106) does not alter Na^+-dependent glucose or amino acid transport. These important observations form the linchpin of the success of ORT. Other features of carbohydrate digestion also contribute to the efficacy of this treatment. First, neither SGLT1 nor GLUT2 is rate limiting; the hydrolysis of polysaccharides is the rate-limiting step. This explains the advantage of rice-based ORTs. Rice provides more molecules of glucose per osmole, but does not add substantially to the osmotic load because as the oligosaccharides are hydrolyzed, the glucose does not accumulate in the lumen, but rather is rapidly transported via SGLT1 and GLUT2. Short-chain glucose polymers reduce cAMP-induced water secretion much better than D-glucose in vitro (107). Second, carbohydrates stimulate SGLT1 expression and therefore ORT treatment will help increase glucose absorption. Third, SGLT1 and GLUT2 are expressed in late gestation, whereas GLUT5 appears at weaning (103). Thus, young infants, the most adversely affected age group in diarrhea, have the necessary cellular machinery to respond to ORT.

WATER TRANSPORT   The key element to ORT is rehydration and therefore water movement. The human small intestine can absorb up to 8 liters of water a day in the absence of any overt hydrostatic or osmotic gradients. The molecular mechanisms and physical forces underlying water transport across epithelia are hotly debated and have spurred an international symposium (108). Study in gastrointestinal epithelia is confounded by heterogeneous cell types, complex geometry of lateral spaces, and increasing tissue resistance from the leaky small intestine to the tighter colonic epithclium (6, 109). The relative contributions of the paracellular and the transcellular routes to net water movement have not been resolved (109). Paracellular transport of solutes and water across the tight junction is thought to be governed by passive driving forces, although the nature of these forces is subject to debate. A complicating feature is the dynamic nature of the tight junction and the lateral spaces discussed earlier. Transcellular water transport could occur either via water diffusing across the lipid bilayer or via membrane water transport molecules. A favored explanation for transepithelial water transport is the modified standing gradient hypothesis (based on the earlier compartment model of Curran & Macintosh and standing gradient model of Diamond & Bossert, whereby the creation of relatively small osmotic gradients is sufficient to drive water movement (reviewed in 108, 109).

Another subject of debate is whether transcellular movement of water always involves transporters (108). Two discoveries of the past decade have revolutionized our thinking of water transport in the intestine. First was the seminal discovery in 1991 of the ubiquitous aquaporin (AQP) family of integral membrane proteins with a high water selectivity (110). Permeation through aquaporins occurs in response to osmotic and hydraulic gradients, and at least 10 aquaporins have been identified in mammals (110). Considerable information is available regarding AQPs localization in a variety of tissues, regulation, ability to transport CO_2 and glycerol, association with disease states, and the atomic structure of AQP1. The identification of AQPs in the intestine have been slower, and although AQP2, AQP3, AQP4, AQP7, AQP8, and AQP10 have now been localized to various regions of the intestine (111–113), their role in diarrheal diseases remains to be elucidated. AQP10 is abundantly found in the duodenum and jejunum, and AQP4 is present on the BLM of colonic surface but not crypt epithelia (112, 113). The $AQP4^{-/-}$ mouse shows a slight increase in fecal stool water content but no impairment in the ophylline-mediated Cl^- secretion, suggesting a role in transcolonic water permeability but not in secretion (113). AQP3 has been localized to the apical membrane of colonocytes, and VIP/cAMP upregulate AQP3 mRNA expression in HT29 cells (111). If borne out in in vivo studies, this finding may have an implication for the role of cAMP in increasing water transport in diarrhea.

The second breakthrough in water transporters was the demonstration that SGLT1 expressed in *Xenopus* oocytes can transport water with a rate of 249 molecules of water:2 Na:1 glucose (reviewed in 114). This could account for the transport of 4.5–5 liters of water a day in the intestine, which therefore may not need AQPs. Based on sensitive cell-volume and voltage measurements, Loo et al. contend that this "solute-water cotransport" is specific and distinct from transport due to osmotic gradients and that both "secondary active" transport and osmotic gradients contribute to water transport (114). However, this explanation has been questioned by others who arrive at similar stoichiometries of water influx into oocytes but attribute it to osmotic flow (115). If SGLT-mediated water transport is demonstrated in epithelia, it might explain rehydration via ORT in the small intestine, but not in the colon. Perhaps in the colon, where water needs to be absorbed against larger osmotic gradients, AQPs may play a more prominent role, as suggested by their prevalence in surface colonocytes. Thus the conduit for water along the cephalocaudal axis of the intestine needs to be identified, but osmotic gradients clearly play a major role in driving water flow.

THE GENERATION OF SCFA AND ROLE OF COMMENSAL/SYMBIOTIC BACTERIAL MICROFLORA   Although the devastation caused by enteropathogens is great, symbiotic and commensal[5] microflora, largely anaerobes, play a critical role in water conservation in the gut and are essential for the efficacy of complex polysaccharides in ORT. Bacterial fermentation of undigested carbohydrates and proteins releases aliphatic SCFA. Four decades ago, SCFA were considered to contribute to carbohydrate-induced diarrhea, but their importance as a fuel source in herbivores and ruminants led to a detailed examination of their role in human and in laboratory models (116, 117). These studies reveal that SCFA, the major luminal anions, could well be the "wonder molecules" of the colon affecting a variety of functions from fluid absorption to growth, differentiation, and defense.

Source and generation   Most SCFA are generated in the cecum and proximal colon. Considering that the colon has 10^{11}–10^{12} bacteria/gm tissue, with ~40 species making up the microbiota, the nature of the fermentation depends on the microflora as well as on the composition of the undigested carbohydrates. The carbohydrates range from plant polysaccharides, such as cellulose, glycans, pectin, psyllium, and lignins, to amylase-resistant starches, and sugars, such as stacchyose. In the human colon, over 90% of pectin but only 20% of cellulose are digested. Undigested dietary fiber contributes to fecal bulk. The colonic bacteria also salvage nutrients from the sloughed off contributions of the intestinal lumen including mucins, cell debris, and secretions and synthesize vitamins (37, 116–118).

In humans on a Western diet, resistant carbohydrates make up 10 to 12% of the daily diet, yielding 0.5–0.6 moles of SCFA, which contribute 10% of the total daily energy requirement (compared with 70% in ruminants and horses) (31, 119). Acetate, propionate, and butyrate, in a ratio of 69:21:10 (70–100 mM) contribute up to 90–95% of total SCFA, with the rest being made of isobutyrate, valerate, isovalerate, and caproate (116). The colon derives 60 to 70% of its metabolic energy from SCFA, with butyrate providing >50% (120). Propionate is chiefly utilized in the liver and acetate in the muscle and peripheral tissues.

How are SCFA formed? Elaborate functional studies culminating in the recent cloning of the complete genome (121) of the abundant commensal anaerobe *Bacteroides thetaiotaomicron* illustrate how SCFA may be generated via host-commensal microbial interaction, which is briefly described here (reviewed in 37).

---

[5]Symbiosis implies benefit to both host and microbe, and commensal implies benefit to one without detriment to the other, but the distinction can be blurry.

The bacterium has distinct metabolic pathways for different nutrients and conserves energy by expressing the requisite genes only when it senses the presence of the nutrient. For example, to digest complex polysaccharides, the bacterium expresses a set of eight genes, termed the starch utilization system (sus) (*sus*A–G and *sus*R). SusR is a transcriptional regulator, which senses di- or oligosaccharides and turns on the *sus*A–G genes. The gene products, SusC–SusF, reside in the outer bacterial membrane and bind to and harness complex polysaccharides. SusG, an outer membrane $\alpha(1,4,$ or $1,6)$–amylase, starts the process of hydrolysis to oligosaccharides, which are then rapidly transported through the outer membrane porins into the periplasmic space. Here, the oligosaccharides encounter SusA, another α–amylase. Hydrolysis is completed by the SusB protein, an α-glucosidase in the cytoplasm. The sugars are utilized by the bacterium for its own energy needs and through anaerobic fermentation release pyruvate and ATP. The pyruvate can be further fermented to SCFA with the release of an additional ATP. The *Bacteroides* similarly has evolved different factories to effectively utilize host-derived glycans and mucins, the benefits of which are reaped by itself and the host (37, 122).

Mechanisms of transport   Mammals have evolved mechanisms to absorb SCFA rapidly; this process is linked to Na^+ absorption and therefore fluid conservation in the colon. Butyrate is the favored substrate, and SCFA can traverse the epithelial layer either paracellularly or transcellularly via diffusion or carrier-mediated routes. SCFA are weak electrolytes and can exist in the protonated (SCFA-H^+) or anionic forms (SCFA$^-$). The relative contributions of the transcellular routes are hotly debated because small changes in the transmembrane pH can shift the protonated:anion equilibrium (117, 123, 124). There is, however, clearly a lumen-to-serosa pH gradient and SCFA gradient that helps in net SCFA absorption. Elegant studies have tried to dissect the contributors to and the role of the pH microclimate around the apical membrane (AM) (123, 125). Protonated SCFA are lipophilic and can diffuse into the colonocyte by a process that is not saturable, not HCO_3^- dependent, is pH dependent, and increases with SCFA chain length (117, 123, 125).

To understand carrier-mediated SCFA transport, a brief summary of anion absorptive processes in the colon is appropriate (32, 75, 117, 123, 124) (Figure 2). Despite species and segmental variability, at least four carrier-mediated transporters for anions have been identified functionally on the AM of surface colonocytes and at least two of these are related to SCFA transport. The apical transporters are Cl^-/HCO_3^-, Cl^-/OH^-, $Cl^-/SCFA^-$, and $HCO_3^-/SCFA^-$ exchangers, distinguishable on the basis of their kinetics, ion specificities, and sensitivity to inhibitors such as stilbenes (DIDS, SITS), niflumic acid, or α-cyano-4-hydroxycinnamate. There is evidence for $Cl^-/SCFA$ but not $HCO_3^-/SCFA$ transporters in the crypt (124). Thus SCFA is transported into the cell, perhaps in exchange for HCO_3^-, resulting in intracellular acidification. Once inside the cell, a major portion of SCFA is metabolized. However, some could be recycled to the AM and some is transported across the BLM into the portal circulation. There is evidence for $HCO_3^-/SCFA^-$ transport

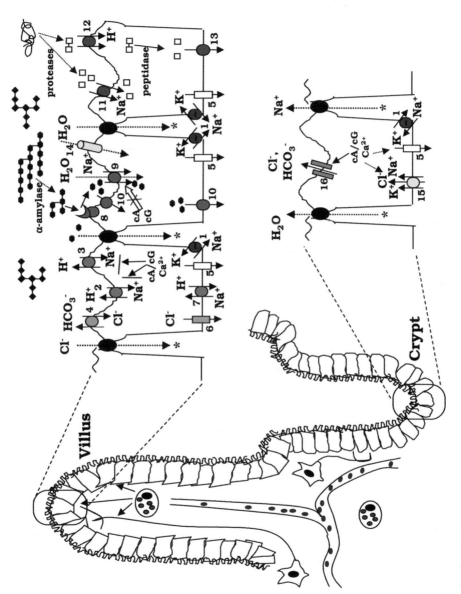

See legend on next page

Figure 1 Transport mechanisms in the small intestine. Transepithelial transport is driven by the BLM Na$^+$/K$^+$-ATPase pump [1]. In the villus, apical NHE2 and NHE3 [2 and 3] and Cl$^-$/HCO$_3^-$ exchangers [4; DRA, PAT1] drive NaCl entry and HCO$_3^-$ exit. Basolateral K$^+$ channels [5], Cl$^-$ channels [6, CLC-2], and the basolateral NHE1 [7] regulate volume and cell pH. Digestion of nutrients occurs by the action of luminal α-amylases and proteases and surface hydrolases such as isomaltase [8] and peptidase (not shown). Apical entry of glucose occurs mainly via SGLT1 [9] and recent reports also suggest GLUT-2 [10], whereas amino acid entry occurs via Na$^+$-coupled cotransporters [11] and H$^+$/dipeptide cotransporters [12]. Glucose exits the BLM via GLUT-2 [10] and amino acid via various amino acid transporters [13]. Cl$^-$ and glucose transport can occur via paracellular routes (*). Absorption of water can occur either via paracellular routes (*), via SGLT1 [9] or perhaps via aquaporin10 [14]. In the crypt, secretion of fluid is governed by the BLM Na$^+$/K$^+$-ATPase [1] and maintained by basolateral K$^+$ channels [5], the basolateral NKCC-1 [15], and CFTR [16]. Movement of Na$^+$ and water occurs via paracellular routes (*). Extracellular modulation of epithelial function results from the actions of systemic factors, immune cells, nerve cells, and the endocrine system; intracellular regulation of transport may be mediated by cAMP (**cA**), cGMP (**cG**), and calcium (Ca^{2+}).

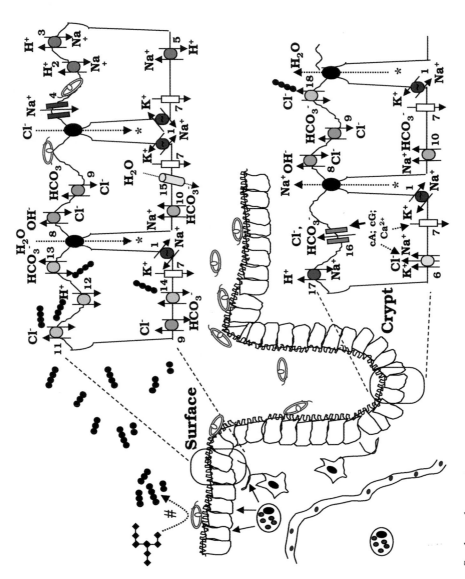

See legend on next page

Figure 2 Transport mechanisms in the colon. Transepithelial transport is driven by the BLM Na^+/K^+-ATPase pump [**1**]. On the surface cells, NHE2 and NHE3 [**2** and **3**] and ENaC channels [**4**] drive Na^+ entry, whereas apical Cl^-/OH^- exchangers [**8**] and Cl^-/HCO_3^- exchangers [**9**; DRA] drive Cl^- entry. Basolateral NHE1 [**5**], K^+ channels [**7**], and Na^+/HCO_3^- cotransporters [**10**] help maintain absorption. Water can move paracellularly (*) and via BLM aquaporins [**15**]. Commensal bacteria (*colored ovals*) break down undigested polysaccharides and proteins (*chains of solid circles*) and generate SCFA such as butyrate, propionate, and acetate. The SCFA are absorbed by surface colonocytes via apical Cl^-/butyrate exchangers [**11**], HCO_3^-/butyrate exchangers [**13**], and monocarboxylate cotransporter MCT-1 [**12**] and BLM HCO_3^-/butyrate exchangers [**14**]. In the crypt, secretion of fluid is governed by the BLM Na^+/K^+-ATPase [**1**] and maintained by basolateral K^+ channels [**7**], the basolateral NKCC-1 [**6**], and apical CFTR [**16**]. Na^+ and water follow Cl^- exit via paracellular routes (*). The apical Cl^--dependent NHE [**17**] and Cl^-/OH^- exchangers [**8**] may play a role in crypt absorption. The crypt cells extrude SCFA via apical Cl^-/butyrate exchangers [**18**]. Apical Cl^-/HCO_3^- exchangers [**9**] and basolateral Na^+/HCO_3^- cotransporters [**10**] may contribute to HCO_3^- extrusion.

across the BLM (124, 126), which may account for the transepithelial absorption of SCFA as well as SCFA-dependent HCO_3^- secretion and luminal alkalinization. The BLM also has a Cl^-/HCO_3^- exchanger and a Na^+-HCO_3^- symporter (32, 124).

There are promising leads as to the molecular identity of these transporters. First, the Cl^-/HCO_3^- anion exchanger AE-1 has been localized to the AM of surface colonocytes, and AE-2 has been localized to the BLM of surface and crypt cells. Second, the DRA protein, defective in congenital chloridorrhea, is highly expressed in the colon and can function as a Cl^-/HCO_3^-, Cl^-/Cl^-, and/or SO_4^-/Cl^- exchanger (63). Its role in fluid absorption is underscored by the fact that it is upregulated in NHE-3$^{-/-}$ mice (76). Finally, the H^+-monocarboxylate transporter MCT1 in the colon, which transports butyrate and lactate and is reduced in carcinoma states, has recently been identified (127). Butyrate transport is Na^+-independent, DIDS insensitive, and blocked by antisense mRNA inhibition of MCT1 (128). Thus MCT1 could be a major SCFA transporter in colonocytes.

Also of note is the profound effect SCFA absorption has on other mechanisms, which results in net fluid absorption. First, acidification of the cell by SCFA entry activates apical NHE, thereby increasing Na^+ absorption. SCFA can also promote Cl^- absorption either directly by recycling to the apical membrane via $Cl^-/SCFA^-$ exchangers or indirectly by being metabolized to HCO_3^-, which then exits the cell. The net result is an increase in NaCl and therefore fluid absorption (117, 129). Second, the effects of SCFA on Na^+ absorption are not blocked by cAMP and therefore, much like Na^+-dependent glucose transport in the ileum, SCFA transport is beneficial when cAMP levels are pathologically elevated, as in cholera. Third, SCFA inhibits basal and cAMP-, but not Ca^{2+}-, stimulated secretion (130, 131). Different steps in the cAMP to Cl^- secretion signaling cascade from an inhibition of adenylate cyclase (131) to that of NKCC-1 (132) or of apical membrane Cl^- transport (123) have been suggested to be the sites of SCFA action. Fourth, SCFA play a role in volume regulation and activate volume-sensitive Cl^- channels on the BLM (123, 133). Fifth, SCFA acts via PYY to alter colonic motility and increase transit time (134). Taken together all these processes promote fluid absorption.

Equally relevant to ORT are the long-term effects of SCFA. A well-designed study (135) demonstrated that long-term exposure to SCFA in vitro (in Caco2/bbe cell lines) or in vivo (colon of pectin-fed rats) results in the specific activation of NHE-3 protein and activity, but not of NHE-2 or other marker proteins. Pectin feeding did not alter ileal NHE-3, suggesting SCFA specifically promote colonic Na^+ absorption. The small intestine, however, exhibits SCFA/HCO_3^- transport (136), and SCFA stimulates ileal proglucagon mRNA as well as Glut-2 expression in a total parenteral nutrition rat model (137). Long-term treatment with SCFA increases expression of its own transporter, MCT-1 (138), which in turn promotes further Na^+ absorption.

Other benefits of commensal bacteria and SCFA   Commensal bacteria and SCFA have potent effects on numerous other colonic functions, and although probably not all the microbial effects are due to SCFA, the actions attributable to SCFA are more easily delineated. There is a strong correlation between lower levels of butyrate and disease states of the colon, including cancer, and therefore with growth and differentiation (116, 139). Both SCFA and commensal bacteria alter blood flow, stimulate the enteric nervous system, and alter gene expression in the colonocyte (37). SCFA combat oxidant-mediated injury by stimulating the production of heat shock proteins (140). Equally intriguing is the increasing evidence that microorganisms can influence and change protein expression in the host cell [e.g., SGLT1 (122)] and that the host can direct bacterial growth to the distal intestine by restricting expression of the host glycoproteins needed for colonization to that region (37). Such a symbiotic relationship has an evolutionary advantage: The anaerobe is able to survive in a hostile and competitive milieu, and the host is able to salvage nutrients from ingested material that it could not digest on its own. These relationships are being exploited for therapeutic use (see section below on Prebiotics and Probiotics).

Further Considerations

Integral to the success of ORT are active glucose transport in the small intestine, commensal bacteria, and SCFA transport in the colon, and the fact that cAMP and cGMP do not affect glucose or SCFA transport. Because a major target for ORT is children under the age of 5, understanding developmental physiology is crucial. Extensive information is available on the developmental regulation of nutrient digestion and absorption (96, 141), but less is known about ion transport processes (35, 36). The neonate appears more susceptible to diarrhea, as seen in the rotaviral studies (93), but are there any inherent defense mechanisms? The newborn rabbit colon is refractory to cGMP (142) and Ca^{2+}-dependent secretagogues in terms of Cl^- transport (143), suggesting that there are inherent pathways to prevent excess fluid loss in the neonate. The gut flora change from facultative anaerobes in the weanling to obligate anaerobes in the adult; how this influences host cell responses awaits further examination.

RECENT DEVELOPMENTS AND FUTURE DIRECTIONS

The effectiveness of ORT rests in the ability to distribute it to wide segments of the affected population, even when health care is not readily accessible. Any simple additives to this formulation that would help control diarrhea would clearly be of further benefit.

Naturally Occuring Compounds as Clues for Therapy

RICE FACTOR   Mathews et al. (144) demonstrated that a specific low M_r (< 1.5 kDa) factor in rice, termed RF, has potent antisecretory effects. RF is neither a peptide nor a glycoprotein and is hydrophobic with a net negative charge. RF inhibits intestinal cAMP-mediated Cl^- secretion by specifically decreasing the open probability of CFTR and has no effect on volume-sensitive Cl^- channels. Thus rice-based ORT could present a triple advantage: increased glucose availability, increased production of SCFA, and a factor that inhibits Cl^- secretion.

HERBAL AND OTHER NATURAL MEDICATIONS   In traditional medicine such as Ayurvedic (7) and Yunani of the Indian subcontinent and the Kampo formulations of China and Japan, the use of herbal extracts to treat diarrhea is recommended in addition to oral fluid replacement. An understanding of the biochemical basis of such formulations may help develop potent antidiarrheal additives to ORT. In Eastern Europe and in France, smectite, an unabsorbable clay, has been shown to reduce the duration of diarrhea (reviewed in 5). The efficacy of poppy seeds as an antidiarrheal agent was recognized by the ancient Egyptians long before their analgesic properties were appreciated. Opium, camphor, and asafoetida were recommended in the traditional treatment of cholera in India (10) (see next section for description of their use in modern therapy). Extracts of different parts of tropical plants ranging from *Tamarindus indica* to *Curcuma longa* have been used as antidiarrheal agents in traditional Thai medicine (8). The latex of the ornamental tree *Croton lechleri* has been used in the indigenous communities of South America to treat watery diarrheas. This latex contains proanthocyanidin oligomers, of which one, SP-303, was shown to be effective in inhibiting cholera toxin and cAMP-mediated Cl^- secretion in a mouse model and in cell lines (145).

Recently, Oi et al. (146) identified the active ingredients of Daio-kanzo-to, a Kampo medication, as coming from *Rhei rhizoma* (Daio). The most effective compound was identified as rhubarb galloyltannin and was demonstrated to inhibit three important actions of cholera toxin: ADP ribosylation, elongation of Chinese hamster ovary cells, and fluid accumulation in a rabbit ileal loop assay. Equally important, chemically synthesized analogs of the galloyltannin had similar effects. The natural and chemically synthesized analogs of the galloyltannin attenuated cholera toxin–induced fluid secretion when given before, with, or after the toxin. Given their effect on ADP ribosylation, the galloyltannins may be of benefit only in the treatment of LT- and CT-associated diarrheas.

Prebiotics and Probiotics

The past decade has seen an explosion in the exploitation of beneficial bacteria, in the form of prebiotics and probiotics, to treat diseases ranging from inflammatory bowel disease (147), antibiotic-associated and acute diarrheas (148), to colon cancer (149). The origins of this therapy also lie in traditional medicine, but the scientific bases are being elucidated (149). Prebiotics are defined as "nondigestible food ingredients that beneficially affect the host by selectively stimulating the growth and/or activity of one or a limited number of bacteria in the colon that can improve host health" (150). Prebiotics such as lactulose and fructo-oligosaccharides, which cannot be digested or absorbed in the small intestine, make their way to the colon where they promote the growth of commensal species, as well as suppress the growth of pathogens. The therapeutic use of prebiotics in acute infectious diarrheas has not been extensively investigated. Probiotics are defined as "live microbial feed supplements which beneficially affect the host animal by improving its intestinal microbial balance" (150). Probiotics have been used to study acute diarrheas. The most popular of the probiotic bacteria are *Lactobacillus* species, *Bifidobacterium bifidum*, *Streptococcus thermophilus*, and the yeast *Saccharomyces boulardii*.

The efficacy of probiotics in treating acute diarrheas is clearly dependent on the species used and on the etiology of the diarrhea. For example, *Lactobacillus* GG (L-GG) had a positive effect on adults with traveler's diarrhea, whereas in a smaller study, Lactinex (*L. acidophilus* + *L. bulgaricus*) had no beneficial effect (148). With respect to ORT, one of the most detailed studies performed on children with acute diarrhea was a multicenter, randomized, placebo-controlled, double-blinded study sponsored by ESPGHAN (151). *Lactobacillus* GG given ad libitum significantly decreased duration of acute diarrhea. Interestingly, the effect was most profound on children with rotavirus and not with bacterial infections, and L-GG may have prophylactic effects (151, 152). Antibiotic-associated diarrhea is often ($\sim$33%) caused by *C. difficile*, and both *S. boulardii* and L-GG have independently proven to be effective in reversing this diarrhea (152). Much attention is being paid to the potency and reliability of commercial formulations. VSL#3^{TM} (VSL Pharmaceuticals, Inc.), for example, contains four strains of *Lactobacilli*, three of *Bifidobacteria*, and one of *S. thermophilus*. In double-blind, placebo-controlled trials, the efficacy of VSL#3 as a prophylactic and therapeutic in pouchitis was demonstrated (153). The VSL#3 formulation restored mucosal epithelial barrier function in IL-10$^{-/-}$ mice, and a factor from these bacteria provided resistance to *Salmonella* infection (154).

With the development of probiotics of known composition, their mechanism of action is being elucidated. Probiotics could produce antibacterial agents; they could compete with the endogenous microflora for adhesion, for receptors on the epithelial cells, and for essential nutrients; they could stimulate and/or suppress the gut-associated lymphoid tissue; and they could alter the types of mucin secreted by goblet cells to interfere with host-pathogen interactions. Because microbial ecology varies with geographical location, of relevance to ORT is the question whether a globally applicable, inexpensive probiotic can be developed.

Pharmacological Approaches

Attempts to identify high-potency antagonists of intestinal fluid secretion have met with varying success.

ANTIBIOTICS AND VACCINES   Although prescribed for bacterial pathogens, antibiotics are ineffective against viruses. Attempts to develop vaccines have met with some success, but the challenges are compounded by the ingenuity of the microorganisms and the complexities of horizontal gene transfers (80).

SIGNAL TRANSDUCTION BLOCKERS   An alternate strategy is to develop drugs targeted to various steps of the fluid transport process, ranging from a direct inhibition of Cl^- channels to blocking proximal steps of the signal transduction cascade, including neural pathways. Several promising agents were effective in animal model and in vitro studies, but performed poorly in clinical trials. For example, recombinant growth hormone was a potent proabsorptive agent in vivo in the rat but failed to have the same effect in humans (155). Although serotonin receptor antagonists show some promise because serotonin release appears to be pathogen- and toxin-specific, these drugs are not applicable to all forms of diarrhea. Blockers of VIP-mediated secretion such as igmesine have greater possibilities of success because they can block CT- and LT-mediated secretion (156).

AGENTS THAT PROMOTE ABSORPTION   Much attention has focused on developing agents that promote absorption. Somatostatin, enkephalins, and α-adrenergic agents promote absorption largely by acting via the $G_{\alpha i}$ cascade to decrease cAMP production (30). Alpha2-adrenergic agonists such as clonidine have limited clinical value because of their adverse side effects. Greater success has been obtained with octreotide, the long-acting somatostatin analog. In the treatment of secretion related to VIPomas, octreotide appears to act by inhibiting release of VIP and fluid secretion (156). Recently, octreotide was demonstrated to increase net epithelial absorption rate in the jejunum and ileum of human volunteers, largely by inhibiting secretion (157). Opiate agonists such as loperamide are widely used in the treatment of mild diarrheas, largely for their effects on decreasing motility and increasing transit time. However, in severe diarrheas, loperamide can cause enteropooling. Therefore, targeting another arm of this cascade has proven to be more effective in the treatment of severe diarrhea. Met- and leu-enkephalin-containing neurons can be localized in the lamina propria in the vicinity of the enterocytes and act via δ-receptors to promote Cl^- and fluid absorption. However, enkephalins are rapidly degraded by endogenous membrane-associated metalloproteinases. A potent enkephalinase inhibitor, racecadotril, has proven to be effective in the treatment of acute diarrheas in adults and children without adversely affecting motility (156). Equally important, it has been demonstrated to be a good adjunct to ORT in a clinical trial in Peru (158).

Cl^- CHANNEL INHIBITORS   A direct inhibitor of CFTR would be ideal as the bulk of intestinal Cl^- transport is driven by this channel. Pharmacological inhibitors of Cl^- channels (45) such as 5-nitro-2(3-phenylpropyl-amino)benzoate (NPPB) and glibenclamide are not useful because they can be nonspecific. The most exciting advances in this regard are the recent reports from Verkman et al. on the development of compounds to activate CFTR as a therapeutic approach to CF (159), and on compounds that will inhibit CFTR, to combat diarrheal diseases. These authors have developed a highly sensitive method to measure CFTR activity in intact epithelial cells, using a halide-sensitive modified yellow fluorescent protein (YFP-H148Q) (160). Through combinatorial chemistry they developed 50,000 compounds, which were analyzed by high-throughput screening (161). Screening was carried out in the presence of a cocktail of forskolin, isobutylmethyl xanthine, and apigenin to ensure that screening was restricted to compounds affecting CFTR directly rather than the signaling cascade. This screening yielded six thiazolidinone compounds that inhibited cAMP-activated CFTR function. The most promising was CFTR$_{inh}$-172, which inhibited CFTR with a K_i of 300 nM, but had no effect on K^+ channels, Ca^{2+}-activated Cl^- channels, or the multidrug-resistant protein, MDR-1 (161). Intraperitoneal infusion of this nontoxic compound blocked cholera toxin–induced Cl^- secretion in a mouse model. Although direct blockers of CFTR may not be effective against rotaviral infections, such screening processes for specific drugs are nevertheless a very promising approach.

SUMMARY

Analysis of the cellular and molecular basis of time-honored remedies for the effective, accessible, and inexpensive treatment of devastating diseases such as diarrhea can be very useful. The serendipitous phenomena of glucose and SCFA transport being unaffected by secretagogues that elevate cAMP and cGMP is the crux of the success of ORT. The simplicity of ORT also contributes to its success, but can the one-size-fits-all formula continue? Issues of hypo- versus hypernatremia, hypo- versus isoosmolar, and the fact that rotavirus and cholera toxins affect different types of Cl^- channels must be considered in designing additives to ORT. Clearly, improving the formulations with either inhibitors of secretion or probiotics, as long as they are inexpensive and readily available, would be highly beneficial. Its successes will be heightened with accessible education and improved sanitation. The tremendous global impact of ORT is due to the ingenuity of numerous investigators in unraveling the secrets of the microorganisms and determining the molecular basis of secretion and absorption, as well as performing complex clinical studies under challenging conditions. With the continued multiprong collaboration of pharmacognosists, ecologists, epidemiologists, microbiologists, cell physiologists, and physicians, strategies such as oral rehydration therapies will continue to be a marvel of modern medicine.

ACKNOWLEDGMENTS

The work cited from the author's laboratory is supported by NIH DK 58135. The author is deeply indebted to Ms. Roli Prasad for her artistic and editorial help. The author thanks Ms. Regina Coleman for her help in literature search and members of her laboratory for their editorial help. The author is grateful to Drs. S. Guandalini and R.V. Benya for their critical review of the manuscript. Because of space limitations, the references cited are restricted to a broad overview, with emphasis on comprehensive reviews. Finally, the author thanks her wise rice-gruel-making grandmothers.

The *Annual Review of Physiology* is online at http://physiol.annualreviews.org

LITERATURE CITED

1. Farthing MJ. 2002. Oral rehydration: an evolving solution. *J. Pediatr. Gastroenterol. Nutr.* 34 (Suppl. 1):S64–67
2. Greenough WB 3rd, Khin-Maung-U. 1991. Oral rehydration therapy. In *Diarrheal Diseases; Current Topics in Gastroenterology*, ed. M Field, pp. 485–500. New York: Elsevier
3. Snow J. 1855. On the mode of communication of cholera. London: Churchill. 139 pp.
4. Editorial series. 1978. Water with sugar and salt. *Lancet* 2:300–1
5. Guarino A, Albano F, Guandalini S, et al. 2001. Oral rehydration: toward a real solution. *J. Pediatr. Gastroenterol. Nutr.* 33(Suppl. 2):S2–12
6. Thillainayagam AV, Hunt JB, Farthing MJ. 1998. Enhancing clinical efficacy of oral rehydration therapy: Is low osmolality the key? *Gastroenterology* 114:197–210
7. Dwivedi L, ed. 1999. Uttaratantrum. In *Suśruta Samhita: Kalpasthana-Uttartantra*, III:370–409. Varanasi, India: Chowkhamba (transl. KK Bhishagratna)
8. Farnsworth NR, Bunyapraphatsara, N. 1992. *Thai Medicinal Plants: Recommended for Primary Health Care System*. Bangkok, Thail.: Med. Plant Inf. Cent., Mahidol Univ. 402 pp.
9. Rolston DK. 1990. Oral rehydration therapy in developing countries. In *Textbook of Secretory Diarrhea*, ed. E Lebenthal, ME Duffey, pp. 371–82. New York: Raven
10. Ramanna M. 2002. *Western Medicine and Public Health in Colonial Bombay 1845–1895. Chapter 4, Coping With Diseases*, pp. 123–61. New Dehli: Orient Longman. 270 pp.
11. Colwell RR, Huq A, Islam MS, Aziz KM, Yunus M, et al. 2003. Reduction of cholera in Bangladeshi villages by simple filtration. *Proc. Natl. Acad. Sci. USA* 100:1051–55
11a. Ghishan FK. 1990. Oral rehydration solutions. In *Textbook of Secretory Diarrhea*, ed. E Lebenthal, ME Duffey, pp. 365–70. New York: Raven
12. Field M. 1974. Intestinal secretion. *Gastroenterology* 66:1063–84
13. Reuss L. 2000. One-hundred years of inquiry: the mechanism of glucose absorption in the intestine. *Annu. Rev. Physiol.* 62:939–46
14. Field M. 2003. Intestinal ion transport and the pathophysiology of diarrhea. *J. Clin. Invest.* 111:931–43
15. Pierce NF, Banwell JG, Rupak DM, Mitra RC, Caranasos GJ, et al. 1968. Effect of intragastric glucose-electrolyte infusion upon water and electrolyte balance in Asiatic cholera. *Gastroenterology* 55:333–43
16. Hirschhorn N, Kinzie JL, Sachar DB, Northrup RS, Taylor JO, et al. 1968. Decrease in net stool output in cholera during intestinal perfusion with glucose-containing solutions. *N. Engl. J. Med.* 279:176–81
17. http://www.who.int/child-adolescent-health. 2002
18. http://www.icddrb.org. 2003

19. Victora CG, Bryce J, Fontaine O, Monasch R. 2000. Reducing deaths from diarrhoea through oral rehydration therapy. *Bull. World Health Organ.* 78:1246–55

20. Greenough WB 3rd. 1980. Oral replacement therapy in diarrhea: consensus and controversy. In *Secretory Diarrhea*, ed. M Field, JS Fordtran, SG Schultz, pp. 179–85. Baltimore: Am. Physiol. Soc.

21. Guandalini S. 2002. The treatment of acute diarrhea in the third millennium: a pediatrician's perspective. *Acta Gastroenterol. Belg.* 65:33–36

22. Davidson G, Barnes G, Bass D, Cohen M, Fasano A, et al. 2002. Infectious diarrhea in children: Working Group Report of the First World Congress of Pediatric Gastroenterology, Hepatology, and Nutrition. *J. Pediatr. Gastroenterol. Nutr.* 35(Suppl. 2):S143–50

23. International Study Group Reduced-Osmolarity ORS Solutions. 1995. Multicentre evaluation of reduced-osmolarity oral rehydration salts solution. *Lancet* 345:282–85

23a. Hirschhorn N, Nalin DR, Cash RA, Greenough WB 3rd. 2002. Formulation of oral rehydration solution. *Lancet* 360:340–41

24. Gore SM, Fontaine O, Pierce NF. 1992. Impact of rice based oral rehydration solution on stool output and duration of diarrhoea: meta-analysis of 13 clinical trials. *Brit. Med. J.* 304:287–91

25. Dutta D, Bhattacharya MK, Deb AK, Sarkar D, Chatterjee A, et al. 2000. Evaluation of oral hypo-osmolar glucose-based and rice-based oral rehydration solutions in the treatment of cholera in children. *Acta Paediatr.* 89:787–90

26. Fayad IM, Hashem M, Duggan C, Refat M, Bakir M, et al. 1993. Comparative efficacy of rice-based and glucose-based oral rehydration salts plus early reintroduction of food. *Lancet* 342:772–75

27. Ramakrishna BS, Venkataraman S, Srinivasan P, Dash P, Young GP, Binder HJ. 2000. Amylase-resistant starch plus oral rehydration solution for cholera. *N. Engl. J. Med.* 342:308–13

28. Rabbani GH, Teka T, Zaman B, Majid N, Khatun M, Fuchs GJ. 2001. Clinical studies in persistent diarrhea: dietary management with green banana or pectin in Bangladeshi children. *Gastroenterology* 121:554–60

29. Alam NH, Meier R, Schneider H, Sarker SA, Bardhan PK, et al. 2000. Partially hydrolyzed guar gum-supplemented oral rehydration solution in the treatment of acute diarrhea in children. *J. Pediatr. Gastroenterol. Nutr.* 31:503–7

30. Rao MC. 2000. Absorption and secretion of water and electrolytes. In *Small Bowel Disorders*, ed. RN Ratnaike, pp. 116–33. New York: Arnold/Oxford Univ. Press

31. Chang EB, Rao MC. 1994. Intestinal water and electrolyte transport. In *Physiology of the Gastrointestinal Tract*, ed. LR Johnson, DH Alpers, J Christensen, ED Jacobson, JH Walsh, 2:2028–81. New York: Raven

32. Kunzelmann K, Mall M. 2002. Electrolyte transport in the mammalian colon: mechanisms and implications for disease. *Physiol. Rev.* 82:245–89

33. Greger R. 2000. Role of CFTR in the colon. *Annu. Rev. Physiol.* 62:467–91

34. Barrett KE, Keely SJ. 2000. Chloride secretion by the intestinal epithelium: molecular basis and regulatory aspects. *Annu. Rev. Physiol.* 62:535–72

35. Venkatasubramanian J, Sahi J, Rao MC. 2000. Ion transport during growth and differentiation. *Ann. NY Acad. Sci.* 915:357–72

36. Pacha J. 2000. Development of intestinal transport function in mammals. *Physiol. Rev.* 80:1633–67

37. Hooper LV, Midtvedt T, Gordon JI. 2002. How host-microbial interactions shape the nutrient environment of the mammalian intestine. *Annu. Rev. Nutr.* 22:283–307

38. Sangan P, Rajendran VM, Geibel JP,

Binder HJ. 2002. Cloning and expression of a chloride-dependent Na⁺-H⁺ exchanger. *J. Biol. Chem.* 277:9668–75

39. Chu J, Chu S, Montrose MH. 2002. Apical Na⁺/H⁺ exchange near the base of mouse colonic crypts. *Am. J. Physiol. Cell Physiol.* 283:C358–72

40. Fasano A. 2002. Toxins and the gut: role in human disease. *Gut* 50(Suppl. 3):9–14

41. Berkes J, Viswanathan VK, Savkovic SD, Hecht G. 2003. Intestinal epithelial responses to enteric pathogens: effects on the tight junction barrier, ion transport, and inflammation. *Gut* 52:439–51

42. Edidin M. 2003. The state of lipid rafts: from model membranes to cells. *Annu. Rev. Biophys. Biomol. Struct.* 16:16

43. Dorn GW II, Mochly-Rosen D. 2002. Intracellular transport mechanisms of signal transducers. *Annu. Rev. Physiol.* 64:407–29

44. Jilling T, Kirk KL. 1997. The biogenesis, traffic, and function of the cystic fibrosis transmembrane conductance regulator. *Int. Rev. Cytol.* 172:193–241

45. Frizzell RA, Quinton PM, Sheppard DN, Welsh MJ, Dawson DC, et al. 1990. Ten years with CFTR. 1999. *Physiol. Rev.* 79(Suppl. 1):S1–S246

46. Rao MC, deJonge HR. 1990. Ca and phospholipid-dependent protein kinases. In *Textbook of Secretory Diarrhea*, ed. E Lebenthal, M Duffey, pp. 209–32. New York: Raven

47. deJonge HR, Rao MC. 1990. Cyclic nucleotide-dependent kinases. See Ref. 46, pp. 191–207

48. Mitic LL, Van Itallie CM, Anderson JM. 2000. Molecular physiology and pathophysiology of tight junctions I. Tight junction structure and function: lessons from mutant animals and proteins. *Am. J. Physiol. Gastrointest. Liver Physiol.* 279:G250–54

49. Cereijido M, Shoshani L, Contreras RG. 2000. Molecular physiology and pathophysiology of tight junctions. I. Biogenesis of tight junctions and epithelial polarity. *Am. J. Physiol. Gastrointest Liver Physiol.* 279:G477–82

50. Nusrat A, Turner JR, Madara JL. 2000. Molecular physiology and pathophysiology of tight junctions. IV. Regulation of tight junctions by extracellular stimuli: nutrients, cytokines, and immune cells. *Am. J. Physiol. Gastrointest Liver Physiol.* 279:G851–57

51. Field M, Semrad CE. 1993. Toxigenic diarrheas, congenital diarrheas, and cystic fibrosis: disorders of intestinal ion transport. *Annu. Rev. Physiol.* 55:631–55

52. Musch MW, Clarke LL, Mamah D, Gawenis LR, Zhang Z, et al. 2002. T cell activation causes diarrhea by increasing intestinal permeability and inhibiting epithelial Na⁺/K⁺-ATPase. *J. Clin. Invest.* 110:1739–47

53. Haas M, Forbush B 3rd. 2000. The Na-K-Cl cotransporter of secretory epithelia. *Annu. Rev. Physiol.* 62:515–34

54. Selvaraj NG, Omi E, Gibori G, Rao MC. 2000. Janus kinase 2 (JAK2) regulates prolactin-mediated chloride transport in mouse mammary epithelial cells through tyrosine phosphorylation of Na⁺-K⁺-Cl⁻ cotransporter. *Mol. Endocrinol.* 14:2054–65

55. Flagella M, Clarke LL, Miller ML, Erway LC, Giannella RA, et al. 1999. Mice lacking the basolateral Na-K-2Cl cotransporter have impaired epithelial chloride secretion and are profoundly deaf. *J. Biol. Chem.* 274: 26946–55

56. Soleimani M, Ulrich CD 2nd. 2000. How cystic fibrosis affects pancreatic ductal bicarbonate secretion. *Med. Clin. N. Am.* 84:641–55

57. Bachmann O, Wuchner K, Rossmann H, Leipziger J, Osikowska B, et al. 2003. Expression and regulation of the Na⁺-K⁺-2Cl⁻ cotransporter NKCC1 in the normal and CFTR-deficient murine colon. *J. Physiol.* 549:525–36

58. Ameen NA, Martensson B, Bourguinon L, Marino C, Isenberg J, McLaughlin GE. 1999. CFTR channel insertion to the

apical surface in rat duodenal villus epithelial cells is upregulated by VIP in vivo. *J. Cell Sci.* 112:887–94

59. Fuller CM, Benos DJ. 2000. Ca^{2+}-activated Cl^- channels: a newly emerging anion transport family. *News Physiol. Sci.* 15:165–71

60. Jentsch TJ, Stein V, Weinreich F, Zdebik AA. 2002. Molecular structure and physiological function of chloride channels. *Physiol. Rev.* 82:503–68

61. Lipecka J, Bali M, Thomas A, Fanen P, Edelman A, Fritsch J. 2002. Distribution of ClC-2 chloride channel in rat and human epithelial tissues. *Am. J. Physiol. Cell Physiol.* 282:C805–16

62. Alrefai WA, Tyagi S, Mansour F, Saksena S, Syed I, et al. 2001. Sulfate and chloride transport in Caco-2 cells: differential regulation by thyroxine and the possible role of DRA gene. *Am. J. Physiol. Gastrointest. Liver Physiol.* 280:G603–13

63. Jacob P, Rossmann H, Lamprecht G, Kretz A, Neff C, et al. 2002. Down-regulated in adenoma mediates apical Cl^-/HCO_3^- exchange in rabbit, rat, and human duodenum. *Gastroenterology* 122:709–24

64. Wang Z, Petrovic S, Mann E, Soleimani M. 2002. Identification of an apical Cl^-/HCO_3^- exchanger in the small intestine. *Am. J. Physiol. Gastrointest. Liver Physiol.* 282:G573–79

65. Seidler U, Blumenstein I, Kretz A, Viellard-Baron D, Rossmann H, et al. 1997. A functional CFTR protein is required for mouse intestinal cAMP-, cGMP- and Ca^{2+}-dependent HCO_3^- secretion. *J. Physiol.* 505:411–23

66. Musch M, Arvans D, Field M, Wu G, Chang E. 2003. DRA is an apical intestinal anion exchanger functionally coupled to NHE2 and NHE3 to mediate electroneutral NaCl absorption. *Gastroenterology* 124:299(A-40)

67. Counillon L, Pouyssegur J. 2000. The expanding family of eucaryotic Na^+/H^+ exchangers. *J. Biol. Chem.* 275:1–4

68. Wakabayashi S, Shigekawa M, Pouysse-gur J. 1997. Molecular physiology of vertebrate Na^+/H^+ exchangers. *Physiol. Rev.* 77:51–74

69. Gawenis LR, Stien X, Shull GE, Schultheis PJ, Woo AL, et al. 2002. Intestinal NaCl transport in NHE2 and NHE3 knockout mice. *Am. J. Physiol. Gastrointest. Liver Physiol.* 282:G776–84

70. Cho JH, Musch MW, DePaoli AM, Bookstein CM, Xie Y, et al. 1994. Glucocorticoids regulate Na^+/H^+ exchange expression and activity in region- and tissue-specific manner. *Am. J. Physiol. Cell Physiol.* 267:C796–803

71. Musch MW, Bookstein C, Rocha F, Lucioni A, Ren H, et al. 2002. Region-specific adaptation of apical Na/H exchangers after extensive proximal small bowel resection. *Am. J. Physiol. Gastrointest. Liver Physiol.* 283:G975–85

72. Li X, Galli T, Leu S, Wade JB, Weinman EJ, et al. 2001. Na^+-H^+ exchanger 3 (NHE3) is present in lipid rafts in the rabbit ileal brush border: a role for rafts in trafficking and rapid stimulation of NHE3. *J. Physiol.* 537:537–52

73. Lamprecht G, Weinman EJ, Yun CH. 1998. The role of NHERF and E3KARP in the cAMP-mediated inhibition of NHE3. *J. Biol. Chem.* 273:29972–78

74. Lamprecht G, Heil A, Baisch S, Lin-Wu E, Yun CC, et al. 2002. The down regulated in adenoma (dra) gene product binds to the second PDZ domain of the NHE3 kinase A regulatory protein (E3KARP), potentially linking intestinal Cl^-/HCO_3^- exchange to Na^+/H^+ exchange. *Biochemistry* 41:12336–42

75. Binder HJ, Sandle GI. 1994. Electrolyte transport in the mammalian colon. See Ref. 31, pp. 2133–71

76. Melvin JE, Park K, Richardson L, Schultheis PJ, Shull GE. 1999. Mouse down-regulated in adenoma (DRA) is an intestinal Cl^-/HCO_3^- exchanger and is upregulated in colon of mice lacking the NHE3 Na^+/H^+ exchanger. *J. Biol. Chem.* 274:22855–61

77. Naftalin RJ, Pedley KC. 1999. Regional crypt function in rat large intestine in relation to fluid absorption and growth of the pericryptal sheath. *J. Physiol.* 514:211–27

78. Matkowskyj KA, Danilkovich A, Marrero J, Savkovic SD, Hecht G, Benya RV. 2000. Galanin-1 receptor up-regulation mediates the excess colonic fluid production caused by infection with enteric pathogens. *Nat. Med.* 6:1048–51

79. Reidl J, Klose KE. 2002. *Vibrio cholerae* and cholera: out of the water and into the host. *FEMS Microbiol. Rev.* 26:125–39

80. Sougioultzis S, Pothoulakis C. 2003. Bacterial infections: small intestine and colon. *Curr. Opin. Gastroenterol.* 19:23–30

81. Lencer WI. 2001. Microbes and microbial toxins: paradigms for microbial-mucosal toxins. *V. cholera*: invasion of the intestinal epithelial barrier by a stably folded protein toxin. *Am. J. Physiol. Gastrointest. Liver Physiol.* 280:G781–86

82. Lundgren O. 2002. Enteric nerves and diarrhoea. *Pharmacol. Toxicol.* 90:109–20

83. Nocerino A, Iafusco M, Guandalini S. 1995. Cholera toxin-induced small intestinal secretion has a secretory effect on the colon of the rat. *Gastroenterology* 108:34–39

84. Fasano A. 2001. Intestinal zonulin: open sesame! *Gut* 49:159–62

85. Rao MC, Guandalini S, Laird WJ, Field M. 1979. Effects of heat-stable enterotoxin of *Yersinia enterocolitica* on ion transport and cyclic guanosine $3',5'$-monophosphate metabolism in rabbit ileum. *Infect. Immun.* 26:875–78

86. Rao MC. 1985. Toxins which activate guanylate cyclase: heat-stable enterotoxins. *Ciba Found. Symp.* 112:74–93

87. Vaandrager AB. 2002. Structure and function of the heat-stable enterotoxin receptor/guanylyl cyclase C. *Mol. Cell. Biochem.* 230:73–83

88. Selvaraj NG, Prasad R, Goldstein JL, Rao MC. 2000. Evidence for the presence of cGMP-dependent protein kinase-II in human distal colon and in T84, the colonic cell line. *Biochim. Biophys. Acta* 1498:32–43

89. Scott RO, Thelin WR, Milgram SL. 2002. A novel PDZ protein regulates the activity of guanylyl cyclase C, the heat-stable enterotoxin receptor. *J. Biol. Chem.* 277:22934–41

90. Field M, Rao MC, Chang EB. 1989. Intestinal electrolyte transport and diarrheal disease. *N. Engl. J. Med.* 321:800–6, 79–83

91. Lucas ML. 2001. A reconsideration of the evidence for *Escherichia coli* STa (heat stable) enterotoxin-driven fluid secretion: a new view of STa action and a new paradigm for fluid absorption. *J. Appl. Microbiol.* 90:7–26

92. Menard LP, Dubreuil JD. 2002. Enteroaggregative *Escherichia coli* heat-stable enterotoxin 1 (EAST1): a new toxin with an old twist. *Crit. Rev. Microbiol.* 28:43–60

93. Morris AP, Estes MK. 2001. Microbes and microbial toxins: paradigms for microbial-mucosal interactions. VIII. Pathological consequences of rotavirus infection and its enterotoxin. *Am. J. Physiol. Gastrointest. Liver Physiol.* 281:G303–10

94. Morris AP, Scott JK, Ball JM, Zeng CQ, O'Neal WK, Estes MK. 1999. NSP4 elicits age-dependent diarrhea and Ca^{2+}-mediated I^- influx into intestinal crypts of CF mice. *Am. J. Physiol. Gastrointest. Liver Physiol.* 277:G431–44

95. Alpers DH. 1994. Digestion and absorption of carbohydrates and proteins. See Ref. 31, pp. 1723–49

96. Henning S, Rubin D, Shulman R. 1994. Ontogeny of the intestinal mucosa. See Ref. 31, pp. 571–610

97. Daniel H. 2004. Molecular and integrative absorption. *Annu. Rev. Physiol.* 66: In press

98. Wright EM, Hirayama BA, Loo DDF, Turk E, Hager K. 1994. Intestinal sugar transport. See Ref. 31, pp. 1751–72

99. Wright EM, Loo DD, Panayotova-Heiermann M, Hirayama BA, Turk E, et al. 1998. Structure and function of the Na⁺/glucose cotransporter. *Acta Physiol. Scand. Suppl.* 643:257–64

100. Wright EM, Turk E, Martin MG. 2002. Molecular basis for glucose-galactose malabsorption. *Cell Biochem. Biophys.* 36:115–21

101. Cheeseman CI. 2002. Intestinal hexose absorption: transcellular or paracellular fluxes. *J. Physiol.* 544:336

102. Kellett GL. 2001. The facilitated component of intestinal glucose absorption. *J. Physiol.* 531:585–95

103. Ferraris RP. 2001. Dietary and developmental regulation of intestinal sugar transport. *Biochem. J.* 360:265–76

104. Field M. 1971. Ion transport in rabbit ileal mucosa. II. Effects of cyclic 3′,5′-AMP. *Am. J. Physiol.* 221:992–97

105. Field M, Fromm D, al-Awqati Q, Greenough WB 3rd. 1972. Effect of cholera enterotoxin on ion transport across isolated ileal mucosa. *J. Clin. Invest.* 51:796–804

106. Guandalini S, Migliavacca M, de Campora E, Rubino A. 1982. Cyclic guanosine monophosphate effects on nutrient and electrolyte transport in rabbit ileum. *Gastroenterology* 83:15–21

107. Rabbani GH, Lu RB, Horvath K, Lebenthal E. 1991. Short-chain glucose polymer and anthracene-9-carboxylic acid inhibit water and electrolyte secretion induced by dibutyryl cyclic AMP in the small intestine. *Gastroenterology* 101:1046–53

108. Reuss L, Hirst BH. 2002. Water transport controversies—an overview. *J. Physiol.* 542.1:1–87

109. Spring KR. 1998. Routes and mechanism of fluid transport by epithelia. *Annu. Rev. Physiol.* 60:105–19

110. Agre P, King LS, Yasui M, Guggino WB, Ottersen OP, et al. 2002. Aquaporin water channels—from atomic structure to clinical medicine. *J. Physiol.* 542:3–16

111. Itoh A, Tsujikawa T, Fujiyama Y, Bamba T. 2003. Enhancement of aquaporin-3 by vasoactive intestinal polypeptide in a human colonic epithelial cell line. *J. Gastroenterol. Hepatol.* 18:203–10

112. Ma T, Verkman AS. 1999. Aquaporin water channels in gastrointestinal physiology. *J. Physiol.* 517:317–26

113. Wang KS, Ma T, Filiz F, Verkman AS, Bastidas JA. 2000. Colon water transport in transgenic mice lacking aquaporin-4 water channels. *Am. J. Physiol. Gastrointest. Liver Physiol.* 279:G463–70

114. Loo DD, Wright EM, Zeuthen T. 2002. Water pumps. *J. Physiol.* 542:53–60

115. Lapointe JY, Gagnon M, Poirier S, Bissonnette P. 2002. The presence of local osmotic gradients can account for the water flux driven by the Na⁺-glucose cotransporter. *J. Physiol.* 542:61–62

116. Cook SI, Sellin JH. 1998. Review article: short-chain fatty acids in health and disease. *Aliment. Pharmacol. Ther.* 12:499–507

117. Sellin JH. 1999. SCFAs: the enigma of weak electrolyte transport in the colon. *News Physiol. Sci.* 14:58–64

118. Said HM. 2004. Water-soluble vitamin absorption. *Annu. Rev. Physiol.* 66: In press

119. Russell JB, Rychlik JL. 2001. Factors that alter rumen microbial ecology. *Science* 292:1119–22

120. Roediger WE. 1980. Role of anaerobic bacteria in the metabolic welfare of the colonic mucosa in man. *Gut* 21:793–98

121. Xu J, Bjursell MK, Himrod J, Deng S, Carmichael LK, et al. 2003. A genomic view of the human-Bacteroides thetaiotaomicron symbiosis. *Science* 299:2074–76

122. Hooper LV, Wong MH, Thelin A, Hansson L, Falk PG, Gordon JI. 2001. Molecular analysis of commensal host-microbial relationships in the intestine. *Science* 291:881–84

123. Charney AN, Dagher PC. 1996. Acid-base effects on colonic electrolyte transport revisited. *Gastroenterology* 111:1358–68

124. Rajendran VM, Binder HJ. 2000. Characterization and molecular localization of anion transporters in colonic epithelial cells. *Ann. NY Acad. Sci.* 915:15–29

125. Maouyo D, Chu S, Montrose MH. 2000. pH heterogeneity at intracellular and extracellular plasma membrane sites in HT29-C1 cell monolayers. *Am. J. Physiol. Cell Physiol.* 278:C973–81

126. Tyagi S, Venugopalakrishnan J, Ramaswamy K, Dudeja PK. 2002. Mechanism of *n*-butyrate uptake in the human proximal colonic basolateral membranes. *Am. J. Physiol. Gastrointest. Liver Physiol.* 282:G676–82

127. Ritzhaupt A, Wood IS, Ellis A, Hosie KB, Shirazi-Beechey SP. 1998. Identification and characterization of a monocarboxylate transporter (MCT1) in pig and human colon: its potential to transport L-lactate as well as butyrate. *J. Physiol.* 513:719–32

128. Hadjiagapiou C, Schmidt L, Dudeja PK, Layden TJ, Ramaswamy K. 2000. Mechanism(s) of butyrate transport in Caco-2 cells: role of monocarboxylate transporter 1. *Am. J. Physiol. Gastrointest. Liver Physiol.* 279:G775–80

129. Binder HJ, Mehta P. 1989. Short-chain fatty acids stimulate active sodium and chloride absorption in vitro in the rat distal colon. *Gastroenterology* 96:989–96

130. Dagher PH, Egnor RW, Taglietta-Kohlbrecher A, Charney AN. 1996. Short-chain fatty acids inhibit cAMP-mediated chloride secretion in rat colon. *Am. J. Physiol. Cell Physiol.* 271:C1853–60

131. Resta-Lenert S, Truong F, Barrett KE, Eckmann L. 2001. Inhibition of epithelial chloride secretion by butyrate: role of reduced adenylyl cyclase expression and activity. *Am. J. Physiol. Cell Physiol.* 281:C1837–49

132. Matthews JB, Hassan I, Meng S, Archer SY, Hrnjez BJ, Hodin RA. 1998. Na-K-2Cl cotransporter gene expression and function during enterocyte differentiation.

Modulation of Cl⁻ secretory capacity by butyrate. *J. Clin. Invest.* 101:2072–79

133. Sellin J, Shelat H. 1996. Short-chain fatty acid (SCFA) volume regulation in proximal and distal rabbit colon is different. *J. Membr. Biol.* 150:83–88

134. Cherbut C, Ferrier L, Roze C, Anini Y, Blottiere H, et al. 1998. Short-chain fatty acids modify colonic motility through nerves and polypeptide YY release in the rat. *Am. J. Physiol. Gastrointest. Liver Physiol.* 275:G1415–22

135. Musch MW, Bookstein C, Xie Y, Sellin JH, Chang EB. 2001. SCFA increase intestinal Na absorption by induction of NHE3 in rat colon and human intestinal C2/bbe cells. *Am. J. Physiol. Gastrointest. Liver Physiol.* 280:G687–93

136. Harig JM, Soergel KH, Barry JA, Ramaswamy K. 1991. Transport of propionate by human ileal brush-border membrane vesicles. *Am. J. Physiol. Gastrointest. Liver Physiol.* 260:G776–82

137. Tappenden KA, Drozdowski LA, Thomson AB, McBurney MI. 1998. Short-chain fatty acid-supplemented total parenteral nutrition alters intestinal structure, glucose transporter 2 (GLUT2) mRNA and protein, and proglucagon mRNA abundance in normal rats. *Am. J. Clin. Nutr.* 68:118–25

138. Cuff MA, Lambert DW, Shirazi-Beechey SP. 2002. Substrate-induced regulation of the human colonic monocarboxylate transporter, MCT1. *J. Physiol.* 539:361–71

139. Bai L, Merchant JL. 2000. Transcription factor ZBP-89 cooperates with histone acetyltransferase p300 during butyrate activation of p21waf1 transcription in human cells. *J. Biol. Chem.* 275:30725–33

140. Ren H, Musch MW, Kojima K, Boone D, Ma A, Chang EB. 2001. Short-chain fatty acids induce intestinal epithelial heat shock protein 25 expression in rats and IEC 18 cells. *Gastroenterology* 121:631–39

141. Ferraris RP, Diamond J. 1997. Regulation of intestinal sugar transport. *Physiol. Rev.* 77:257–302

142. Selvaraj NG, Prasad R, Stephan B, Vidyasagar D, Rao MC. 1999. Segmental differences in cGMP signaling in rabbit colon is due to the cGMP protein kinase-II (PKG-II). *Gastroenterology* 116:G4040

143. Venkatasubramanian J, Selvaraj N, Carlos M, Skaluba S, Rasenick MM, Rao MC. 2001. Differences in Ca^{2+} signaling underlie age-specific effects of secretagogues on colonic Cl^- transport. *Am. J. Physiol. Cell Physiol.* 280:C646–58

144. Mathews CJ, MacLeod RJ, Zheng SX, Hanrahan JW, Bennett HP, Hamilton JR. 1999. Characterization of the inhibitory effect of boiled rice on intestinal chloride secretion in guinea pig crypt cells. *Gastroenterology* 116:1342–47

145. Gabriel SE, Davenport SE, Steagall RJ, Vimal V, Carlson T, Rozhon EJ. 1999. A novel plant-derived inhibitor of cAMP-mediated fluid and chloride secretion. *Am. J. Physiol. Gastrointest. Liver Physiol.* 276:G58–63

146. Oi H, Matsuura D, Miyake M, Ueno M, Takai I, et al. 2002. Identification in traditional herbal medications and confirmation by synthesis of factors that inhibit cholera toxin-induced fluid accumulation. *Proc. Natl. Acad. Sci. USA* 99:3042–46

147. Shanahan F. 2001. Inflammatory bowel disease: immunodiagnostics, immunotherapeutics, and ecotherapeutics. *Gastroenterology* 120:622–35

148. Vanderhoof JA, Young RJ. 1998. Use of probiotics in childhood gastrointestinal disorders. *J. Pediatr. Gastroenterol. Nutr.* 27:323–32

149. Fedorak RN, Madsen KL. 2000. Application of probiotics for GI diseases. *Monograph*

150. Fuller R, Gibson GR. 1997. Modification of the intestinal microflora using probiotics and prebiotics. *Scand J. Gastroenterol. Suppl.* 222:28–31

151. Guandalini S, Pensabene L, Zikri MA, Dias JA, Casali LG, et al. 2000. Lactobacillus GG administered in oral rehydration solution to children with acute diarrhea: a multicenter European trial. *J. Pediatr. Gastroenterol. Nutr.* 30:54–60

152. Guandalini S, Gupta P. 2002. The role of probiotics in gastrointestinal disorders of infancy and childhood. In *Infant Formula: Closer to the Reference*, ed. NCR Raiha, F Rubaltelli, pp. 29–45. Philadelphia: Vevey/Lippincott, Williams & Wilkins

153. Gionchetti P, Rizzello F, Helwig U, Venturi A, Lammers KM, et al. 2003. Prophylaxis of pouchitis onset with probiotic therapy: a double-blind, placebo-controlled trial. *Gastroenterology* 124:1202–9

154. Madsen K, Cornish A, Soper P, McKaigney C, Jijon H, et al. 2001. Probiotic bacteria enhance murine and human intestinal epithelial barrier function. *Gastroenterology* 121:580–91

155. Hogenauer C, Santa Ana CA, Porter JL, Fordtran JS. 2000. Discrepancies between effects of recombinant human growth hormone on absorption and secretion of water and electrolytes on the human jejunum compared to results reported on rat jejunum. *Dig. Dis. Sci.* 45:457–61

156. Farthing MJ. 2002. Novel targets for the control of secretory diarrhoea. *Gut* 50(Suppl. 3):15–18

157. Hogenauer C, Aichbichler B, Santa Ana C, Porter J, Fordtran J. 2002. Effect of octreotide on fluid absorption and secretion by the normal human jejunum and ileum in vivo. *Aliment. Pharmacol. Ther.* 16:769–77

158. Salazar-Lindo E, Santisteban-Ponce J, Chea-Woo E, Gutierrez M. 2000. Racecadotril in the treatment of acute watery diarrhea in children. *N. Engl. J. Med.* 343:463–67

159. Galietta LJ, Springsteel MF, Eda M, Niedzinski EJ, By K, et al. 2001. Novel CFTR chloride channel activators

identified by screening of combinatorial libraries based on flavone and benzoquinolizinium lead compounds. *J. Biol. Chem.* 276:19723–28

160. Galietta LV, Jayaraman S, Verkman AS. 2001. Cell-based assay for high-throughput quantitative screening of CFTR chloride transport agonists. *Am.* *J. Physiol. Cell Physiol.* 281:C1734–42

161. Ma T, Thiagarajah JR, Yang H, Sonawane ND, Folli C, et al. 2002. Thiazolidinone CFTR inhibitor identified by high-throughput screening blocks cholera toxin-induced intestinal fluid secretion. *J. Clin. Invest.* 110:1651–58

Annu. Rev. Physiol. 2004. 66:419–46
doi: 10.1146/annurev.physiol.66.032102.144611
First published online as a Review in Advance on July 30, 2003

Recent Advances in Carrier-Mediated Intestinal Absorption of Water-Soluble Vitamins

Hamid M. Said

*University of California School of Medicine, Irvine and VA Medical Center, Long Beach,
Long Beach, California 90822; email: hmsaid@uci.edu*

Key Words   transport mechanism, transport regulation

■ **Abstract**   Significant progress has been made in recent years toward understanding the mechanisms and regulation of intestinal absorption of water-soluble vitamins from the diet, especially those that are transported by a specialized carrier-mediated mechanism (i.e., ascorbic acid, biotin, folate, riboflavin, thiamin, and pyridoxine). The driving force involved in the uptake events and the molecular identity of the systems involved have been identified for a number of these vitamins. In addition, information about regulation of the uptake process of these micronutrients by intracellular and extracellular factors has been forthcoming. Furthermore, the 5′ regulatory region of the genes that encode a number of these transporters has been characterized, thus providing information about transcriptional regulation of the transport events. Also of interest is the identification of existence of carrier-mediated mechanisms in human colonocytes that are capable of absorbing some of the vitamins that are synthesized by normal microflora of the large intestine. Although the contribution of the latter source of vitamins toward overall host nutrition is not clear and requires further investigations, it is highly likely that it does contribute toward the cellular homeostasis of these vitamins in the localized colonocytes.

INTRODUCTION

Water-soluble vitamins represent a group of structurally and functionally unrelated compounds that share a common feature of being essential for normal health and well-being (Figure 1). These micronutrients play critical roles in maintaining normal metabolic, energy, differentiation, and growth status of mammalian cells. Because humans and other mammals cannot synthesize these compounds (except for some synthesis of niacin), they must obtain them from exogenous sources via intestinal absorption. Thus the intestine plays an important role in maintaining and regulating normal body homeostasis of these micronutrients. Impairment in intestinal absorption of these compounds can lead to states of vitamin deficiency. Such impairment occurs in a variety of conditions including intestinal diseases,

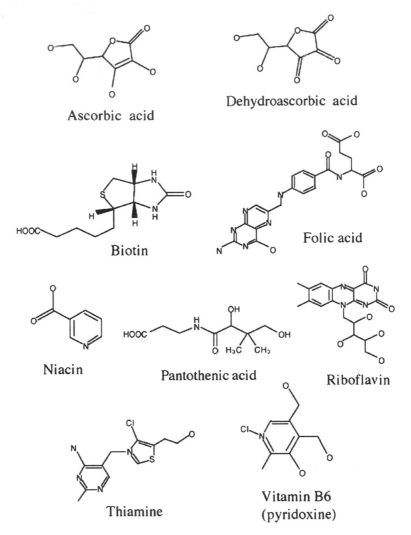

Figure 1 Chemical structure of water-soluble vitamins described in this review.

congenital disorders in the involved transport systems, drug interactions, chronic alcohol consumption, and intestinal resection. The aim of this review is to provide an update of our understanding of the cellular and molecular mechanisms and regulation of the intestinal absorption process of these essential micronutrients, with special emphasis on those vitamins that are transported by carrier-mediated mechanisms [the discussion does not include vitamin B_{12}, cobalamin, because this vitamin is transported by a receptor-mediated mechanism (reviewed in 1)]. It is hoped that such information will lead to a stimulation of interest in further studies into the mechanism and regulation of intestinal vitamin absorption in health and disease.

ASCORBIC ACID

Vitamin C exists in the diet in two forms, the reduced form, ascorbic acid, and the oxidized form, dehydroascorbic acid (DHAA) (Figure 1). Ascorbic acid acts as a cofactor for a number of important enzymes that are involved in maintaining transition metal ions in their reduced forms, in the biosynthesis of extracellular matrix proteins and neurotransmitters, and in the regulation of iron uptake. DHAA is structurally different from ascorbic acid and is similar to glucose. Its toxicity is similar to alloxan, which has long been used by endocrine investigators to induce diabetes in laboratory animals. Deficiency of vitamin C leads to a variety of clinical abnormalities including scurvy, poor wound healing, and bone and connective tissue disorders.

Primates and guinea pigs have lost the ability to synthesize ascorbic acid from glucose owing to a deficiency in the enzyme L-gluono-γ-lactone oxidase. These species must, therefore, obtain the vitamin from the diet via intestinal absorption. The mechanism of uptake of ascorbic acid has been investigated using a variety of intestinal tissue preparations. These investigations have concluded that intestinal ascorbic acid uptake occurs via a concentrative, carrier-mediated Na$^+$-dependent mechanism (1a–3). These findings were confirmed in studies with purified intestinal brush border membrane vesicle (BBMV) preparations (4, 5). As to the exit of ascorbic acid from the enterocyte, i.e., transport across the basolateral membrane of the polarized enterocyte, this process was studied using purified intestinal basolateral membrane vesicles (BLMV) preparations and found to also occur by a carrier-mediated system (4); this system, however, was found to be Na$^+$-independent. Little metabolic alterations occur in the transported ascorbic acid in the enterocytes.

Our understanding of the molecular mechanisms involved in ascorbic acid transport in mammalian cells, including those of the small intestine, has increased significantly in recent years following the cloning of two distinct isoforms of Na$^+$-dependent ascorbic acid transporters from rat and human tissues: the Na$^+$-dependent vitamin C transporter 1 and 2 (SVCT1 and SVCT2, respectively; in the case of humans these isoforms are referred to as hSVCT1 and hSVCT2) (6, 7–11). These two isoforms were found to share a high degree of homology with one another, but they did not display homology with other mammalian membrane transporters. Functional characterization of both isoforms has been performed by expression in *Xenopus* oocytes, COS-7 cells, and other cellular systems. The results showed that both SVCT1 and SVCT2 transport ascorbic acid (but not dehydroascorbic acid) and that both have higher selectivity for L-ascorbic acid compared with D-isoascorbic acid. SVCT2, however, was reported to have a higher affinity for ascorbic acid compared with that of SVCT1 (8). Both transporters were also found to transport L-ascorbic acid via an electrogenic, Na^+-dependent process with a stoichiometric ratio of Na^+ to ascorbic acid of 2:1. In addition, both SVCT1 and SVCT2 were found to have multiple potential N-glycosylation and protein kinase C (PKC) phosphorylation sites. A unique feature of both ascorbate transporters was their ability to also act as Na^+ uniporters in the absence of ascorbate, allowing Na^+ leakage to take place across the cell membrane. In regard to the distribution of the two isoforms, striking differences were observed. In both human and rat tissues, SVCT1 was found to be expressed mainly in epithelial tissues, whereas SVCT2 was found to be expressed in most tissues except the lung and skeletal muscles. In the small intestine, both SVCT1 and SVCT2 were found to be expressed with the former being the predominant form as shown by Northern blot analysis and by hybrid depletion (with antisense oligonucleotides) of intestinal poly $(A)^+$ RNA following microinjection into *Xenopus* oocytes. Recent studies have shown that flavonoids (such as quercetin) are reversible and noncompetitive inhibitors of ascorbic acid transport by the SVCT1 system, although they themselves are not transported by this system (12).

As to the absorption of DHAA, the enterocyte takes up this form of vitamin C and metabolizes it to the reduced form by the action of DHAA-reductase (4, 13–15). It is through this mechanism that the intracellular level of DHAA is believed to be maintained at low nontoxic levels. Studies on the cellular uptake of DHAA have shown that the enterocyte takes up this compound across the brush border membrane by a Na^+-independent process (4). Substantial uptake of DHAA was also reported to occur from the serosal surface of the intestinal tissue, i.e., across the basolateral membrane (16). The serosal uptake of DHAA might occur in exchange with the reduced form of vitamin C as it leaves the cell. Recently it has been shown that the systems involved in DHAA uptake at the basolateral membrane are the facilitated glucose transporters GLUT 1, 3, and 4 (17–19).

The intestinal ascorbic acid uptake process was found to be regulated by extracellular substrate levels and by an intracellular PKC-mediated pathway (20–22). Ascorbic acid supplementation to guinea pigs was found to lead to a downregulation in intestinal ascorbate absorption (20, 21). More recently, evidence has been presented to show that both SVCT1 and SVCT2 are regulated by an intracellular PKC-mediated pathway. This regulation, however, appears to involve different mechanisms. While PKC appears to act on SVCT1 by affecting its distribution at the cell membrane, its effect on SVCT2 appears to be mediated via decreasing the catalytic transport efficiency of the carrier itself at the cell membrane (22).

BIOTIN

Biotin (vitamin H; Figure 1) acts as a coenzyme for five mammalian carboxylases that catalyze essential steps in four pathways involving fatty acid biosynthesis, gluconeogenesis, and catabolism of certain amino acids and fatty acids. Deficiency of biotin leads to a variety of clinical abnormalities including neurological disorders, growth retardation, and dermal abnormalities. Biotin deficiency and suboptimal levels have been reported with increased frequency in recent years and have been shown to occur in a variety of conditions including inborn errors of biotin metabolism/transport, following chronic use of anticonvulsant drugs and following long-term use of parenteral nutrition.

Humans and other mammals cannot synthesize biotin and thus must obtain the vitamin from exogenous sources. The intestine is exposed to two sources of biotin: a dietary source and a bacterial source where the vitamin is produced by the normal microflora of the large intestine. Dietary biotin exists in both the free and protein-bound forms (23). Protein bound biotin is digested by gastrointestinal proteases and peptidases to biocytin (N-biotinyl-L-lysine) and biotin-containing short peptides (24). These forms of biotin are then converted to free biotin by the action of biotinidase (24). This hydrolysis step appears to be essential for efficient absorption and optimal bioavailability of dietary biotin (25). The source of intestinal biotinidase is believed to be the pancreatic juice (24), and the enzyme has been cloned and shown to be the product of a single-copy gene (26).

The absorption mechanism of free dietary biotin in the small intestine has been the subject of intense investigation using a variety of intestinal preparations (see 27, 28 and references therein). It is now well recognized that a specialized, Na^+-dependent, carrier-mediated mechanism, more active in the proximal than the distal part of the small intestine, is involved in absorption of dietary biotin. Studies with purified intestinal BBMV and BLMV preparations have shown that the Na^+-dependent carrier-mediated process is localized at the apical membrane domain of the polarized enterocyte (29, 30). In addition, it has been shown that the inwardly directed Na^+ gradient, and not the presence of Na^+ per se, is responsible for driving the movement of biotin against a concentration gradient in BBMV. As to the transport of biotin out of the enterocyte, i.e., transport across the basolateral membrane, this event occurs via a Na^+-independent, carrier-mediated mechanism (30).

Recent studies have shown that the intestinal biotin transport system is also utilized by two other unrelated micronutrients, namely, the water-soluble vitamin pantothenic acid and the metabolically important substrate lipoate (28). A similar interaction between biotin and the other two substrates has also been reported in cell types, such as those of the colon (31), the blood brain barrier (32), the heart (33), and the placenta (34). For these reasons, the biotin transport system is now referred to as the Na^+-dependent multivitamin transporter (SMVT). The physiological and nutritional significance of interaction between these three micronutrients at the level of membrane transport is unclear, as is how such a transport system is regulated, and both require further investigation.

The normal microflora of the large intestine synthesize a considerable amount of free biotin (35, 36), and the large intestine is capable of absorbing luminally introduced biotin (37–39). The mechanism involved in colonic biotin uptake has been recently delineated in studies using the human-derived, nontransformed colonic epithelial cell line NCM460 as a model system (31). The colonic biotin uptake process was found to occur via a specialized, Na^+-dependent, carrier-mediated mechanism.

This mechanism is shared with pantothenic acid and lipoate. These findings are similar to those observed for biotin uptake in the small intestine and suggest that a common uptake system is involved in the two areas of the intestinal tract.

Recent studies have determined the molecular identity of the intestinal biotin uptake system, the SMVT (40, 41). These studies have also reported the existence of a significant heterogeneity in the 5' untranslated region of the rat SMVT and have identified four distinct variants (I, II, III, IV) (40). Variant II was found to be the predominant form expressed in the rat small and large intestine. Functional identification of the cloned intestinal cDNAs was established by expression in different cellular systems and demonstrated a marked and specific increase in biotin and pantothenic acid uptake. The induced vitamin uptake was Na^+-dependent and saturable with an apparent K_m similar to that found in native intestinal epithelial cells. Message distribution along the vertical and longitudinal axes of the intestinal tract was also determined. Villus cells expressed significantly higher SMVT message level than crypt cells, a finding that correlates with the higher carrier-mediated biotin uptake in villus cells compared with crypt cells (40). As to the longitudinal distribution of SMVT, a similar level of expression was found along the length of the small intestine and colon (40, 41). A recent study (42) has reported the existence of another biotin transporter in addition to SMVT in peripheral blood monocytic cells. No evidence for such a system, however, was found in the intestine as shown by recent studies using the approach of small interfering RNA (siRNA) to silence the human *SMVT* (hSMVT) gene (43). In the latter study, treatment with the gene-specific siRNA led to an almost complete inhibition in carrier-mediated uptake of a physiological concentration of biotin by the human-derived intestinal epithelial Caco-2 cells. In addition, uptake by these cells as a function of concentration within the nano-molar range was found to be linear (43).

Regulation of the intestinal biotin uptake process has also been the subject of investigation in recent years. The intestinal biotin uptake process was found to be regulated by ontogeny, extracellular substrate level, and by specific intracellular protein kinase–mediated pathways (28, 44–46). Previous studies have shown that the intestinal *trans*-epithelial biotin transport process undergoes clear ontogenic changes during early stages of life (44). These changes were recently found to involve the entry step of the vitamin at the level of the brush border membrane and were mediated via changes in the V_{max} and apparent K_m of the biotin uptake process (46). Parallel changes were also seen in protein, mRNA, and the transcription rate of SMVT in the intestine during development as indicated by results of Western blotting, RT-PCR, and nuclear run-on assays, respectively (46). These findings suggest that the ontogenic regulation of the intestinal biotin uptake process involves transcriptional mechanisms. Further studies are required to identify the nature of these mechanisms. It is interesting that the renal biotin uptake process, unlike the intestinal biotin uptake process, does not appear to be affected by ontogeny (46). These findings suggest that the two biotin uptake processes are regulated differently during early stages of life.

Other studies have shown that the intestinal biotin uptake process is regulated by extracellular substrate levels (45). Biotin deficiency in rats was shown to lead to a specific and significant up-regulation in intestinal biotin uptake, whereas biotin over-supplementation results in a specific and significant down-regulation in intestinal biotin uptake. These adaptive changes were found to be mediated mainly via changes in the V_{max} of the intestinal biotin uptake process, suggesting that the regulation is mediated by changes in the number and/or activity of the biotin transporters. Further studies are required to determine the molecular mechanisms involved in this regulation.

Evidence also suggests that the intestinal biotin uptake process is under the regulation of intracellular PKC- and Ca^{2+}/calmodulin(CaM)-mediated pathways (28). Activation of PKC was found to lead to a significant inhibition in biotin uptake, whereas PKC inhibition led to a stimulation in the vitamin uptake by Caco-2 cells. This regulatory effect was found to be mediated via a decrease in the V_{max} (with no changes in the apparent K_m) of the biotin uptake process, suggesting a decrease in the activity and/or the number of the biotin uptake carriers. It is noteworthy that SMVT contains two putative PKC phosphorylation sites (40, 41), but it is not known if these sites are involved in the observed PKC regulation of biotin uptake. A role for an intracellular Ca^{2+}/CaM-mediated pathway in the regulation of intestinal biotin uptake has also been reported. Specific inhibitors of this pathway were found to cause a significant inhibition in biotin uptake (28). This inhibition was mediated via an inhibition in the V_{max} (and not the apparent K_m) of the biotin uptake process, suggesting changes in the activity and/or number of the functional biotin carriers. Thus the PKC and Ca^{2+}/CaM-mediated pathways appear to exert their effects on the biotin uptake process through different mechanisms. This conclusion is based on the observation that simultaneous activation of PKC and inhibition of Ca^{2+}/CaM pathways lead to an additive inhibition in biotin uptake (28).

Insight into transcriptional regulation of the *SMVT* gene has emerged following the cloning and characterization of the 5'-regulatory region of the rat and human genes (47, 48). As mentioned above, studies on *SMVT* expression in rat intestine have identified four distinct transcript variants owing to heterogeneity at the 5'-untranslated region (40). This finding has raised the possibility that multiple promoters may be involved in the regulation of transcription of the *SMVT* gene. Indeed, this is what we found in studies involving genome-walking (47). These studies have identified three distinct putative promoters (P1, P2, and P3) that are separated by exons of the four identified variants (Figures 2). P1 was found to contain multiple putative *cis*-elements such as GATA-1, AP-1, AP-2, and C/EBP, which included several repeats of purine-rich regions and two TATA-like elements. P2 and P3 were determined to be GC-rich and also contained many putative *cis*-regulatory elements. The functional identity of each promoter and the minimal region required for its basal activity were established by fusing the promoter constructs to the firefly luciferase reporter gene, transfecting the constructs into the rat-derived cultured intestinal epithelial IEC-6 cells, and assaying for luciferase

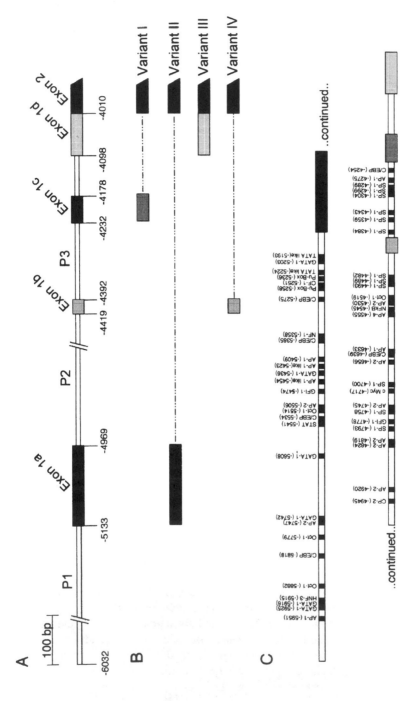

Figure 2 Diagrammatic representation of the 5′-regulatory region of the rat SMVT gene. The numbers represent nucleotides relative to the translation start site (A of ATG is considered as 1). (*A*) The relative positions of the first exons and promoters. Exons are shaded differently. (*B*) The exon composition of the four identified variants of the rat SMVT. (*C*) The positions of several identified putative regulatory elements (adapted from Reference 47).

activity. P1 was found to be the most active promoter in these cells compared with P2, whose activity was higher than that of P3.

The 5'-regulatory region of the human *SMVT* gene has also been cloned and characterized (48). Two distinct promoters (P1 and P2) were identified and both were found to be TATA-less, CAAT-less, contained highly GC-rich sites, and had multiple, putative *cis*-regulatory elements (e.g., AP1, AP2, C/EBP, SP1, NF1, and GATA). Activity of both promoters and the minimal region required for their basal activity were established using the firefly luciferase reporter gene assay following transfection into the human-derived intestinal epithelial Caco-2 cells. The minimal region required for basal activity of P1 was encoded by a sequence between -5846 and -5313, whereas that for P2 was encoded in a sequence between -4417 and -4244 relative to the translation initiation codon. No study is currently available that examines the potential role of the putative *cis*-elements in transcription regulation of the *SMVT* gene and whether the identified promoters and their minimal region are also active in vivo in transgenic animals.

FOLATE

Folate (Figure 1), an essential micronutrient, acts as a coenzyme in the synthesis of DNA and RNA and in the metabolism of several amino acids including homocystein. Folate deficiency leads to a variety of clinical abnormalities (megaloblastic anemia, growth retardation, etc.), whereas optimization of folate body homeostasis leads to promotion of good health and prevention of certain diseases (e.g., neural tube defects). Folate deficiency is a highly prevalent vitamin deficiency throughout the world and occurs for a variety of reasons including impairment in intestinal absorption of the vitamin.

Humans and other mammals have lost their ability to synthesize folate and thus must obtain the vitamin from exogenous sources via intestinal absorption. The intestine is exposed to two sources of folate: a dietary source and a bacterial source where the vitamin is synthesized by the normal microflora of the large intestine. Dietary folate exists mainly in the polyglutamate form that must be converted to the monoglutamate form prior to absorption. This conversion step involves the enzyme folate hydrolase (also referred to as folylpoly-γ-glutamate carboxypeptidase). This enzyme exists in two forms: a brush border form and an intracellular (lysosomal) form (49–52). The brush border form of the folate hydrolase is expressed mainly in the proximal part of the small intestine, whereas the intracellular form is uniformly expressed along the entire length of the small gut (53, 54). The intestinal brush border form of the folate hydrolase has been cloned (54, 55) and has been recently shown to be up-regulated in dietary folate deficiency (56).

The mechanism of absorption of folate monoglutamates in the small intestine has been the subject of intense investigations over the past three decades. Using a variety of intact intestinal tissue and cellular preparations, important features of the intestinal folate absorption process have been delineated (reviewed in 57). It is now well established that the proximal part of the small intestine is the preferred site of absorption of dietary folate monoglutamates and that the process involves a specific, carrier-mediated system. This system is highly pH dependent, with higher uptake at acidic pH compared with neutral or alkaline pH. Studies with purified intestinal BBMV have also shown the folate uptake system is concentrative, electroneutral, inhibited by the anion transport inhibitors 4,4'-diisothiocyano-2-2-disulfonic acid stilbene (DIDS) and acetamidoisothiocyanostilbene-2,2' disulfonic acid (SITS), and has a similar affinity for reduced (e.g., 5-methyltetrahydrofolate, 5-MTHF) and oxidized (e.g., folic acid) folate derivatives (58–61). These observations have led to the conclusion that a folate$^-$:OH$^-$ exchanger or a folate$^-$:H$^+$ cotransport is involved in intestinal folate uptake process (58, 59). The fact that the intestinal folate uptake process has similar affinity for reduced and oxidized folate derivatives is unique to the gut and is different from the widely investigated folate uptake system of the mouse leukemia cells. In the latter cell types, the folate uptake system has a higher preference for reduced over oxidized folate derivatives, and hence it is referred to as the reduced folate carrier (RFC) (62–64). As to the mechanism of exit of folate out of the enterocyte, i.e., transport across the basolateral membrane, studies with purified intestinal BLMV have shown the involvement of a specific carrier-mediated system that appears to be shared by oxidized and reduced folate derivatives and is DIDS sensitive (65). Functional, molecular, and immunological studies have shown that intestinal epithelial cells do not express or utilize the alternative folate uptake mechanism via the folate receptor (66, 67).

As mentioned above, the intestine is exposed to a second source of folate, i.e., the folate that is synthesized by the normal microflora of the large intestine. Significant amounts of this folate source have been shown to exist in the absorbable monoglutamate form. Studies have also shown that the large intestine is capable of absorbing some of this folate (68). Using purified colonic apical membrane vesicles isolated from the colon of human organ donors and the human-derived, nontransformed colonic epithelial NCM460 cells, recent studies have shown that folate uptake in the colon is efficient and occurs via a specific, carrier-mediated, pH-dependent, DIDS-sensitive and electroneutral mechanism similar to that found in the small intestine (69, 70). The studies with the colonic epithelial NCM460 cells have also shown that the colonic folate uptake is similar to that in the small intestine in that it is regulated by an intracellular protein tyrosine kinase (PTK)- and cAMP-mediated pathways (see below). These similarities in the mechanism and regulation of folate uptake by colonic and small intestinal epithelial cells suggest that a common mechanism is involved in folate uptake in the two regions of the intestinal tract. The identification of an efficient carrier-mediated mechanism for folate uptake in the human colon further suggests that this source of folate may contribute toward body homeostasis of folate, or at least toward cellular homeostasis of the local colonocytes. Such findings may also lead to a better understanding of the causes of localized folate deficiency described in colonic epithelia and believed to be associated with premalignant changes in colonic mucosa (71, 72).

Molecular cloning of the intestinal folate uptake system has identified the system as the RFC (in case of human it is referred to as hRFC) (73, 74). Significant homology was found in the RFC of different species, and the protein appears to carry a net positive charge, which may be important for its interaction with the negatively charged folate. Expression of RFC in *Xenopus* oocytes led to a significant and specific increase in 5-MTHF uptake (73, 74). The induced folate uptake was found to be saturable as a function of increasing the substrate concentration in the incubation medium and showed an apparent K_m value similar to that reported for folate uptake in the native intestine. Differences, however, were found in pH-dependence profile and in the degree of inhibition by oxidized and reduced folate derivatives compared with folate transport in the native intestine. Although folate uptake in the native intestine is considerably higher at acidic buffered pH compared with neutral or alkaline pH (58, 60), no such pH dependence was seen with cRNA-induced folate uptake in *Xenopus* oocytes (73, 74). In addition, the cRNA-induced folate uptake in *Xenopus* oocytes was preferentially inhibited by reduced rather than oxidized folate derivatives, whereas the folate transport system of the native small intestine handles both oxidized and reduced folate derivatives with similar affinity (see above) (58, 60). Subsequent studies on the causes of these differences have shown that the characteristics displayed by RFC depend on the cell context (75), which could be attributed to differences in membrane composition, cell-specific post-translational modifications, and/or involvement of cell-specific auxiliary protein that modulates RFC activity in the different cell types (75). The distribution of the RFC message along the vertical axis of the intestine was shown by Northern blot analysis and in situ hybridization to be significantly higher in mature epithelial cells of the villus tip compared with the immature epithelial cells of the crypt (73, 74). Also, both the small and large intestine were found to express RFC, but at different levels (73, 74).

Mutations in *hRFC* have been recently reported in two siblings with the so-called folate malabsorption syndrome (76). Those patients were found to respond favorably to oral pharmacological (but not physiological) doses of folinic acid (77). Mutational screening analysis has identified a total of four missense mutations that are located in well-conserved positions in human, mouse, rat, and hamster RFC. When experimentally introduced into the open reading frame of *hRFC* followed by expression in HeLa cells, these mutations were found to lead to a severe inhibition in folate uptake compared with cells expressing the wild-type hRFC. This inhibition in folate uptake was found to be the result of malfunctioning of the folate uptake system itself and not because of impairment in the expression of hRFC at the cell membrane (76).

Regulation of the intestinal folate absorption process by extracellular substrate levels, by intracellular protein kinase-mediated pathways, and by ontogeny has recently been reported (56, 66, 78). Dietary folate deficiency appears to lead to a significant and specific up-regulation in intestinal carrier-mediated folate uptake (56). This increase is mediated via a significant increase in the V_{max} (but not the apparent K_m) of the folate uptake process and is associated with a parallel increase in intestinal RFC protein and RNA levels. The latter findings suggest the possible involvement of a transcriptional regulatory mechanism in the up-regulation event. Indeed, more recent studies have shown that expression of the *hRFC* promoter B (the promoter that drives the expression of the intestinal variant I of hRFC), which was fused to a firefly luciferase reporter gene in intestinal epithelial Caco-2 cells grown in a folate-deficient growth medium, led to a higher promoter activity compared with expression in Caco-2 cells grown in a folate-sufficient (control) growth medium (79). The intestinal folate uptake process was also found to be under the regulation of intracellular PTK and cAMP-mediated pathways, but the mechanisms through which these pathways exert their regulatory effects on intestinal folate uptake are not clear (66). Other studies have reported that the intestinal folate absorption process is ontogenically regulated (78). This has been recently shown to involve the initial step in intestinal folate uptake, i.e., transport across the brush border membrane, and it is mediated via a decrease in the V_{max} of the folate uptake process (78a). A parallel decline in intestinal RFC protein and mRNA levels and in the rate of *RFC* transcription was also observed with maturation, indicating the involvement of a transcriptional mechanism in the ontogenic regulation of the intestinal folate uptake process (78a). Information regarding transcriptional regulation of the *RFC* gene has been forthcoming with the identification of multiple promoters that drive transcription of numerous variants of this gene in different tissues (64). In humans, promoter B appears to be responsible for driving the expression of variant I, the predominant hRFC variant expressed in the human intestine (64, 80).

Recent studies have also identified the molecular determinants that dictate plasma membrane expression of hRFC protein, as well as the cellular mechanisms that deliver the carrier protein to the cell surface (81, 82). Using a series of truncated fusion proteins of hRFC with enhanced green fluorescent protein (GFP) to image the targeting and trafficking dynamics of hRFC in living intestinal epithelial cells, the molecular determinants that dictate hRFC plasma membrane expression were shown to reside within the hydrophobic backbone of the polypeptide and not within the cytoplasmic NH_2- or COOH-terminal domains of the protein (Figure 3, see color insert). These studies have also shown that the integrity of the hRFC backbone is critical for export of the polypeptide from the endoplasmic reticulum to the cell surface. Numerous intracellular vesicular-like structures that appear to contain the hRFC protein and to be involved in intracellular trafficking were identified. This trafficking event was found to be critically dependent on intact microtubules as disruption of the microtubule network led to inhibition in motility of hRFC-containing vesicles as well as final expression of the hRFC protein at the plasma membrane.

Niacin

Niacin (nicotinic acid; also known as vitamin B_3; see Figure 1 for structure) acts as a precursor for the synthesis of NAD and NADP, two metabolically important coenzymes involved in maintaining the redox state of cells. These coenzymes are involved in catabolic reactions such as the pentose phosphate shunt and glycolysis.

Niacin also appears to have lipid-lowering effects and is in clinical use for such a purpose. Severe deficiency of niacin leads to pellagra, a disease characterized by skin lesions and inflammation of mucous membranes, diarrhea, and weight loss. Niacin deficiency occurs in alcoholics and in patients with Hartnup's disease. Patients with the latter disorder have mutations in their tryptophan transport gene, and tryptophan is a precursor for endogenous production of nicotinic acid.

The human body obtains niacin from two sources: an endogenous source in which the vitamin is produced by metabolic conversion of the amino acid tryptophan to niacin, and an exogenous source from the diet via absorption in the intestine. The mechanism of intestinal niacin transport is not well defined. Studies with rat intestinal brush border membrane vesicles have reported the involvement of a pH-dependent, carrier-mediated system for niacin uptake (83). However, the reported apparent K_m of the system is very high (millimolar range), raising questions about the relevance of such a system in intestinal absorption of physiological concentrations of niacin that exist in the lumen in the micromolar range (84). Niacin uptake by rat renal brush border membrane vesicles was reported to be Na^+-dependent, concentrative, and carrier-mediated in nature (85). With regard to the molecular identity of the intestinal niacin uptake system and its regulation, there is nothing currently known on these subjects. Also nothing is know about the mechanism of exit of niacin out of the intestinal absorptive cells, i.e., transport across the basolateral membrane. In the enterocytes, some of the transported niacin undergoes metabolism to intermediates of NAD biosynthesis.

Pantothenic Acid

Pantothenic acid (Figure 1), a member of the B-complex family of vitamins, is involved in the metabolic synthesis of coenzyme A (CoA) and acyl carrier proteins and, therefore, is important in the metabolism of carbohydrate, fat, and to a lesser extent protein. The intestine is exposed to two sources of pantothenic acid: a dietary source and a bacterial source (36). In the diet, pantothenic acid exists mainly in the form of CoA, which is hydrolyzed to free pantothenic acid in the intestinal lumen prior to absorption (86). This is then followed by transport of free pantothenic acid into the absorptive cells via the SMVT (40, 41, 87; also see the biotin section). Transport of the bacterially synthesized pantothenic acid in the large intestine also occurs by the SMVT (31). Interaction between pantothenic acid, biotin, and lipoate at the level of membrane transport is not unique to the intestine; it also occurs in other cellular systems (32–34). It is unclear how cells regulate transport of the individual vitamin via this common transport system and how the substrate level of the individual substrate affects SMVT function in these cells. Further studies are required to address these issues.

Riboflavin

Riboflavin (vitamin B_2; Figure 1), in its riboflavin-5-phosphate (FMN) and flavin adenosine dinucleotide (FAD) forms, is involved in key metabolic reactions including carbohydrate, amino acid, and lipid metabolism and in the conversion of folic acid and vitamin B_6 into their coenzyme forms. Deficiency of riboflavin occurs in humans in a variety of conditions and leads to a several clinical abnormalities including degenerative changes in the nervous system, endocrine dysfunction, skin disorders, and anemia.

Humans and other mammals can not synthesize riboflavin and thus must obtain the vitamin from exogenous sources via the intestine. Two sources of riboflavin are available to the host: a dietary source and a bacterial source where the vitamin is produced by the normal microflora in the large intestine (36). Dietary riboflavin exists mainly in the form of FMN and FAD. These forms are enzymatically hydrolyzed to free riboflavin in the intestinal lumen prior to absorption (88). Intestinal absorption of dietary riboflavin has been extensively studied using a variety of intestinal preparations that include intact tissue preparations, isolated purified intestinal membrane vesicles, and more recently cultured intestinal epithelial cell lines (89–95). Results of these studies have shown that riboflavin uptake occurs mainly in the proximal part of the small intestine and involves a specialized, Na^+-independent carrier-mediated system. This system was found to be competitively inhibited by structural analogues of riboflavin and by the Na^+/H^+ exchange inhibitor amiloride (90). In addition, the system was found to be regulated by extracellular substrate levels and by specific intracellular protein kinase-mediated pathways (90, 93, 96). Growing Caco-2 cells in a riboflavin-deficient growth medium led to an up-regulation in riboflavin uptake by these cells, whereas growing them in a medium supplemented with high pharmacological concentrations of riboflavin led to down-regulation in the vitamin uptake (90). These adaptive changes were found to be mediated via changes in the V_{max} (and not the apparent K_m) of the riboflavin uptake process, suggesting that the changes are mediated through an increase in the number (and/or activity) of the carriers involved (90). Similar adaptive changes were observed when lumiflavin (a riboflavin structural analogue that utilizes the riboflavin uptake system but cannot be phosphorylated or used as a riboflavin-like vitamin by cells) was used in place of riboflavin in such experiments (90). The latter findings suggest that it is the availability of a transportable form of the substrate at the apical surface, recognizable by the riboflavin-carrier system, that triggers these adaptive changes in the riboflavin-uptake system. Further studies are needed to delineate the cellular/molecular mechanisms involved in this adaptive regulation in riboflavin uptake. Similarly, studies in rats have shown that riboflavin deficiency leads to a significant and specific up-regulation in intestinal riboflavin uptake, whereas riboflavin over supplementation of rats leads to a significant and specific down-regulation in intestinal riboflavin uptake (96). Other studies have shown that an increase in intracellular cAMP levels leads to a down-regulation in riboflavin uptake and that this effect is mediated via a decrease in the activity of the riboflavin uptake carriers (93). It has been suggested that a PKA-mediated pathway is involved in the regulation of intestinal riboflavin uptake process (93). Some degree of riboflavin metabolism (phosphorylation) has also been reported in the transported substrate inside the enterocytes (97).

As to the bacterially synthesized riboflavin, the amount of the vitamin produced by the normal microflora in the large intestine depends on the type of diet; it is higher following intake of vegetable-based compared with meat-based diets (98). Studies have shown that considerable amounts of this riboflavin exist in the lumen of the large intestine in the free absorbable form (98, 99) and that the large intestine is capable of absorbing luminally introduced free riboflavin (39, 100). However, the mechanism involved in riboflavin uptake in the large intestine was not known until recently. Using human-derived colonic epithelial NCM460 cells as an in vitro experimental model system for colonocytes, evidence shows the involvement of a specialized, carrier-mediated mechanism in riboflavin uptake in the colon (101). This system was found to be similar to that described for the small intestine in being temperature- and energy-dependent, Na^+-independent, and inhibited by riboflavin structural analogs and by the membrane transport inhibitor amiloride (101). The colonic riboflavin uptake process was also found to be adaptively regulated by extracellular riboflavin levels and modulated by an intracellular Ca^{2+}/CaM-mediated pathway. Subsequent studies in rats have confirmed the existence of a specialized carrier-mediated system for riboflavin uptake in the colon (102). There is nothing known at present about the molecular identity of the intestinal riboflavin uptake system or its gene in any species.

Thiamine

Thiamine (vitamin B_1; Figure 1) plays a critical role in normal carbohydrate metabolism where it participates in the decarboxylation of pyruvic and α-ketoglutamic acids and in the utilization of pentose in the hexose monophosphate shunt. Thiamine deficiency in humans leads to a variety of clinical abnormalities, including neurological and cardiovascular disorders, and occurs in a high percentage of alcoholics, diabetic patients, and in patients with thiamine-responsive megaloblastic anemia (TRMA; also known as Rogers syndrome). TRMA is an autosomal-recessive disorder characterized by manifestations that include megaloblastic anemia, sensorineural deafness, and diabetes mellitus (103, 104). A genetic defect in the recently cloned thiamine transporter *SLC19A2* is believed to be the cause of TRMA (49, 105–107).

Humans and other mammals cannot synthesize thiamine; rather they obtain the vitamin from exogenous sources via intestinal absorption. The host is exposed to two sources of thiamine: dietary source and bacterial source where the vitamin is synthesized by the normal microflora of the large intestine (36). Dietary thiamine exists predominantly in the form of thiamine pyrophosphate, which is hydrolyzed by intestinal phosphatases to free thiamine in the intestinal lumen (108). Absorption of free thiamine then takes place mainly in the proximal part of the small intestine and occurs via a specialized carrier-mediated system (109, 110). Studies with purified jejunal BBMV isolated from organ donors have shown the involvement of a Na^+-independent, pH-dependent, electroneutral, amiloride-sensitive, carrier-mediated mechanism for thiamine uptake in the human intestine

(111). This process is capable of transporting thiamine against an intravesicular concentration gradient in the presence of an outwardly directed H^+ gradient. Similar findings have been reported for thiamine transport in rat intestinal BBMV (112). These findings on the thiamine uptake mechanism confirm and extend earlier findings with intact human intestinal tissue preparations (113–115). Inside the intestinal epithelial cells, some of the absorbed thiamine undergoes metabolism mainly to thiamine pyrophosphate; however, only free thiamine is transported across the basolateral membrane into the blood (serosal) side (109, 110). The exit of thiamine out of the enterocyte has also been examined using purified jejunal BLMV isolated from organ donors and appears to involve a specialized, pH-dependent, electroneutral carrier-mediated mechanism (116). Similar findings were reported for thiamine transport by rat small intestinal BLMV (117), with the exception that in the rat study, thiamine uptake was reported to be Na^+-K^+-ATPase-dependent, whereas no such dependence was found in the human study.

Recent studies of bacterially synthesized thiamine in the large intestine, using the human-derived colonic epithelial NCM460 cells, have demonstrated the existence of an efficient and specific carrier-mediated mechanism for thiamine uptake in these cells (118). This system was found to transport the vitamin via a thiamine$^+$:H^+ exchange mechanism and appears to be under the regulation of an intracellular Ca^{2+}/CaM-mediated pathway. These findings are similar to those observed with small intestinal epithelial cells, raising the possibility that a common system is involved in thiamine uptake in the two regions of the intestinal tract.

Insight into the molecular nature of the intestinal thiamine transport system has also recently emerged following the cloning of two human thiamine transporters, *SLC19A2* and *SLC19A3* (49, 105–107, 119). These thiamine transporters share 48% identity and 64% similarity with one another. They also share 40 and 39% identity at the amino acid level, respectively, with the human reduced folate carrier (hRFC) (107, 120). However, neither transporter transports folate, and hRFC does not transport thiamine (119). Messages of both thiamine transporters were found in the human intestine (120, 121; H.M. Said, unpublished observations). The *SLC19A2* message was found to be expressed in all gastrointestinal tissues in the following order: liver → stomach → duodenum → jejunum → colon → cecum → rectum → ileum (121). *SLC19A2* was also found to be expressed at the protein level in the native human intestine (121). Little, however, is currently known about the distribution of the *SLC19A3* message along the gastrointestinal tract. Using the approach of siRNA to silence the individual thiamin transporter genes (*SLC19A2* or *SLC19A3*), recent studies with Caco-2 cells have shown that both of these transporters have a role in normal intestinal thiamine uptake (H.M. Said, unpublished observations). In line with the latter observations is the recent finding that knockout mice with targeted disruption of *Slc19a2* (the mouse orthologue of the human *SLC19A2*) display a normal phenotype and have normal plasma thiamine levels (122). This finding suggests normal intestinal thiamine absorption in these mice and the involvement of a compensatory

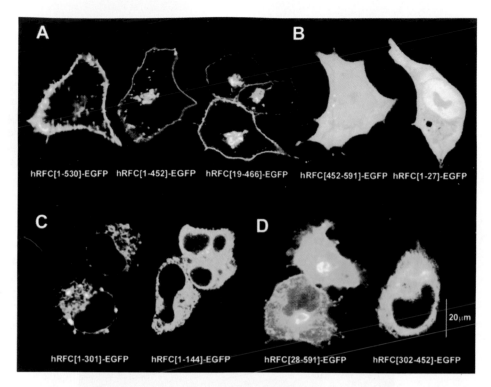

Figure 3 Cellular distribution of constructs of the hRFC fused to the enhanced GFP in the human-derived intestinal epithelial HuTu-80 cells. Images shown are of a series of representative lateral (x–y) confocal sections from cells transfected with the indicated constructs (A–D). Panel (E) demonstrates colocalization of EGFP and the red-emitting lipophilic dye FM4–64, which selectively labels the plasma membrane as a method of quantifying the extent of expression of different hRFC constructs at the plasma membrane. The *right-hand panels* show overlays of the red and green images, with regions of colocalization appearing as yellow. (F) The bar graphs show the values of colocalization of FM4–64 and EGFP fluorescence for each of the indicated constructs. Schematic diagrams on the right illustrate the structures of each construct (NH$_2$ terminus, *white dotted line*; transmembrane domain, *yellow bar*; COOH terminus, *white line*; GFP, *green arrow*). This figure is adapted from Reference 82.

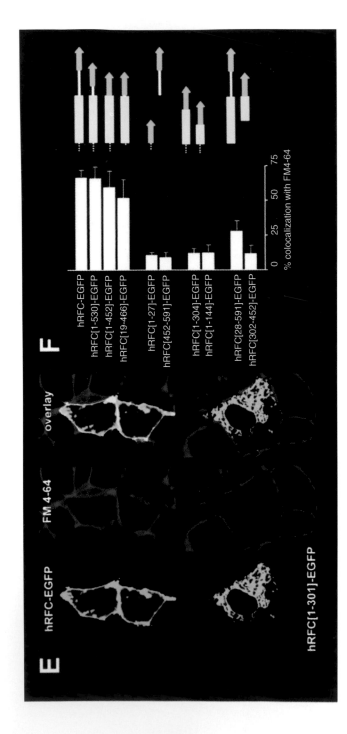

Figure 3 *(Continued)*

thiamine uptake mechanism in the intestine, most probably *Slc19a3* (the orthologue of the human SLC19A2; 120). Further studies are required to confirm this suggestion.

Regulation of the small intestinal thiamine uptake process has also been investigated using Caco-2 cells as an in vitro model system for the absorptive enterocytes. These cells possess a thiamine uptake mechanism that is similar to that of native human intestine, i.e., specific, pH-dependent, amiloride-sensitive, and carrier-mediated in nature. Using these cells, evidence suggests that the thiamine uptake process is under the regulation of an intracellular Ca^{2+}/CaM-mediated pathway (123). In another study involving intestinal mucosal biopsies, up-regulation in thiamine uptake was found in a thiamine-deficient human compared with control subjects, suggesting possible adaptive regulation of the intestinal thiamine uptake process by substrate level (113). Information regarding transcriptional regulation of the *SLC19A2* gene in intestinal epithelial cells has also recently begun to emerge. Studies have reported the cloning of the 5′-regulatory region of the *SLC19A2* gene and have functionally confirmed and characterized its promoter in Caco-2 cells (121). In addition, the minimal promoter region required for basal activity of the *SLC19A2* promoter in these cells has been determined and shown to be mainly encoded in a sequence between −356 and −36 and to include multiple *cis*-regulatory elements (Figure 4) (121). Mutational analysis, oligonucleotide competition assays, electromobility shift, and supershift assays have documented an important role for GKLF, NF-1, and SP-1 in the regulation of the *SLC19A2* promoter (124). The functionality of the full-length and minimal *SLC19A2* promoter-luciferase

```
          AatII
          -356                        NF1
CCGCGCTTAGCCCTGGGACAGAGGCCTGGCGTGCTGCCACCGCCCCAGTTC


      AP1                       GKLF
ACGTCGTGTGACCCACGACCAATGGAAGAGCAGGCAAGTATTCCCGGCGTC
       -317   F6        -305                         -275


                                          SP1
CGCTGTGATTGGTTCCCGGAGTGGAGGCGGTGGCAGAGGGTGGGCCTTAGG
F3


ACGGGTCTCCCTTAAACTGGGCGATCAGGCAGCGACCCTAGAGGCGTCTGT
-220                            -192        -183


AGGGTAAAGCTGGGGGTTCTGTAGCCGGAGGCGGCGGCGAGTCCAGAAC
                                -139      F4
```

Figure 4 Diagrammatic representation of the minimal 5′-region of the human *SLC19A2* gene required for basal promoter activity of the promoter. Bold type indicates the positions of several identified putative regulatory elements. The nucleotides of the three transcription initiation sites are underlined. Numbers represent nucleotides relative to the translational start site A of ATG as number 1. This figure is adapted from Reference 121.

constructs has also been confirmed in vivo in transgenic mice (124). Tissue distribution of the luciferase activity in the transgenic mice was found to mimic the expression pattern reported for *SLC19A2* in different human tissues. These studies suggest that human tissues differentially regulate expression levels of *SLC19A2* through the promoter and that *cis*-elements for GKLF, NF-1, and/or SP-1 play a role in regulating expression of this gene. Other studies have determined the transcription initiation sites for the *SLC19A2* gene in intestinal epithelial Caco-2 cells (121). Three such sites have been identified by 5′-rapid amplification of cDNA ends, with one at position −183, another at −192, and the third at −220 (A of the initiator ATG sequence was considered as position 1). A recent study has reported that the human thiamine transporter *SLC19A2* is a target for activation by the p53 tumor suppressor transcription factor (122). This study, however, was performed in murine erythroleukemia cells, and it is not known if a similar activation of the *SLC19A2* gene also occurs in intestinal epithelial cells.

Insight into the structure/function relationship of the *SLC19A2* protein has also been coming from both clinical and experimental findings. Clinically, 14 distinct mutations have been identified in patients with TRMA, with 3 of these mutations being missense and the rest nonsense mutations that lead to an early truncation of the protein (126). Direct demonstration of a defect in transport function of the 3 missense mutations has been experimentally established (127). Using site-directed mutagenesis (127), studies have shown that introducing the missense mutations individually into *SLC19A2* expressed in HeLa cells led to severe impairment in transport function of the carrier protein (Figure 5). A critical role for amino acid 138 (the only conserved anionic residue in any of the transmembrane domains of the SLC19A2 protein) in the function of the carrier protein in uptake of the cationic thiamine has also been demonstrated (127). Furthermore, it has been shown that both of the predicted N-glycosylation sites of the SLC19A2 protein (N63 and N314) are indeed glycosylated (127). This glycosylation, however, does not appear to have any functional significance on the ability of the carrier protein to transport thiamine and appears to have no effect on its expression at the cell membrane (127).

Studies have also delineated the mechanisms that control the intracellular trafficking and membrane targeting of the human thiamine transporter (hTHTR1) in human-derived intestinal epithelial Hutu-80 cells (128). This was performed by generating a series of hTHTR1 truncations fused with the enhanced GFP and imaging the targeting and trafficking dynamics of each construct in living cells. These studies have shown that although the full-length fusion protein is functionally expressed at the plasma membrane of Hutu-80 cells, analysis of the truncated mutants demonstrates an essential role for both the NH_2-terminal sequence and the integrity of the backbone polypeptide for cell surface expression. Most notably, truncation of hTHTR1 within a region where several TMRA truncations are clustered resulted in intracellular retention of the mutant protein. Numerous hTHTR-1 vesicle structures, which presumably are involved in

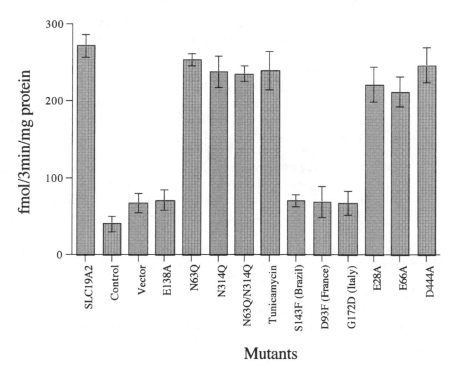

Figure 5 Functional expression (determined by means of thiamin uptake) of wild-type SLC19A2 open reading frame and missense mutants identified in patients with thiamine-responsive megaloblastic anemia (S143F, D93F, G172D), together with con-tructs that have their putative N-glycosylation sites mutated (N63Q and N314Q) and the construct with a mutation in the only conserved negatively charged amino acid in any *trans*-membrane of the protein (E138A) in HeLa cells. This figure is adapted from Reference 127.

hTHTR-1 intracellular trafficking, were also observed within the cytoplasm. Finally, confocal imaging of the dynamics of intracellular hTHTR1 vesicles revealed a critical role for microtubules—but not microfilaments—in hTHTR1 trafficking.

Vitamin B_6

Vitamin B_6 (Figure 1) represents a group of structurally related compounds, namely pyridoxine, pyridoxal, and pyridoxamine. These compounds exist in the diet in phosphorylated and unphosphorylated forms. The most biologically active form of the vitamin is pyridoxal phosphate. Vitamin B_6 acts as a cofactor in a number of metabolic reactions that include amino acid metabolism. Deficiency of vitamin B_6 leads to a variety of clinical abnormalities and occurs in numerous, conditions, including alcoholism and diabetics, and in patients with long-term use of isoniazid and other hydrazines. Low levels of vitamin B_6 have also been reported in patients with vitamin B_6-dependent seizure, an autosomal-recessive disorder suggested to result from an abnormality in pyridoxine transport into cells (129).

Humans and other mammals can not synthesize vitamin B_6 and thus most obtain the vitamin via intestinal absorption. The intestine is exposed to two sources of vitamin B_6: a dietary source and a bacterial source where the vitamin is produced by the normal microflora of the large intestine (36). Absorption of dietary vitamin B_6 occurs following hydrolysis of the phosphorylated forms of the vitamin in the intestinal lumen (130–132). Previous studies have reported no net *trans*-epithelial transport or intracellular accumulation of vitamin B_6 in the intestine (133). In addition, studies with guinea pig jejunal BBMV and isolated vascularly perfused rat intestine in vivo have failed to show saturation in the vitamin B_6 uptake process (130, 134). Thus uptake of vitamin B_6 was believed to occur via simple diffusion. This concept, however, has been recently challenged by studies in our laboratory with Caco-2 cells, which show clear evidence for the existence of a specialized, carrier-mediated system for pyridoxine uptake (134a). This system is Na^+-independent but highly dependent on acidic buffered pH and is amiloride-sensitive as evidenced in Caco-2 cells. The findings on vitamin B_6 uptake with intestinal epithelial Caco-2 cells, although similar to those reported with renal epithelial cells (135) with regard to the involvement of a carrier-mediated system, are different in the effect of pH on the uptake process. The reason for this difference is not clear and requires further investigation. The molecular identity of the intestinal vitamin B_6 uptake system and its gene is yet to be elucidated.

Concluding Remarks

Our knowledge related to the mechanisms and regulation of intestinal transport processes of water-soluble vitamins has been significantly expanded in recent years. It is now well established that absorption of these essential micronutrients occurs via specialized and regulated transport systems. In the case of a number of such vitamins, the molecular identity of the systems involved, characteristics of their 5′-regulatory region, and aspects of their regulation and cell biology (intracellular trafficking and membrane targeting) have been described. For others, such information is still lacking and in need of investigations. With the mapping of the human genome and the current availability of powerful cell and molecular biology approaches, it is expected that the recent advances made in these areas will continue and expand in the coming years. This should ultimately further our understanding of the mechanisms of absorption of these essential nutrients under normal physiological conditions and of how aberration in their absorption occurs under certain pathophysiological conditions. Such knowledge could ultimately help in the design of effective strategies to optimize the normal body homeostasis of these essential micronutrients especially in cases of deficiency and suboptimal levels.

The *Annual Review of Physiology* is online at http://physiol.annualreviews.org

LITERATURE CITED

1. Seetharam B. 1994. Gastric intrinsic factor and cobalamin absorption. In *Physiology of the Gastrointestinal Tract*, ed. L Johnson, pp. 1997–2026. New York: Raven

1a. Mellors AJ, Nahrwold DL, Rose RC. 1977. Ascorbic acid flux across the mucosal border of guinea pig and human ileum. *Am. J. Physiol. Endocrinol. Metab.* 233:E374–79

2. Patterson LT, Nahrwold DL, Rose RC. 1982. Ascorbic acid uptake in guinea pig intestinal mucosa. *Life Sci.* 31:2783–91

3. Stevenson N, Bush M. 1969. Existence and characteristics of Na^+-dependent active transport of ascorbic acid in guinea pig. *Am. J. Clin. Nutr.* 22:318–26

4. Bianchi J, Wilson FA, Rose RC. 1986. Dehydroascorbic acid and ascorbic acid transport in the guinea pig ileum. *Am. J. Physiol. Gastrointest. Liver Physiosl.* 250:G461–68

5. Siliprandi L, Vanni P, Kessler M, Semenza G. 1979. Na^+-dependent electroneutral L-ascorbate transport across brush-border membrane vesicles from guinea pig small intestine. *Biochim. Biophys. Acta* 552:129–42

6. Tsukaguchi H, Tokui T, Mackenzie B, Berger UV, Chen X, et al. 1999. A family of mammalian Na^+-dependent L-ascorbic acid transporters. *Nature* 399:70–75

7. Daruwala R, Song J, Koh WS, Rumsey SC, Levine M. 1999. Cloning and functional characterization of the human sodium-dependent vitamin Ca transporters hSVCT1 and hSVCT2. *FEBS Lett.* 460:480–84

8. Liang WJ, Johnson D, Jarvis SM. 2001. Vitamin C transport systems of mammalian cells. *Mol. Membr. Biol.* 18:87–95

9. Rajan DP, Huang W, Dutta B, Devoe LD, Leibach FH, et al. 1999. Human placental sodium-dependent vitamin C transporter (SVCT2): molecular cloning and transport function. *Biochem. Biophys. Res. Commun.* 262:762–58

10. Wang H, Dutta B, Huang W, Devoe LD, Leibach FH, et al. 1999. Human Na^+-dependent vitamin C transporter 1 (hSVCT1): primary structure, functional characteristics and evidence for a nonfunctional splice variant. *Biochim. Biophys. Acta* 1461:1–9

11. Wang H, Mackenzie B, Tsukaguchi H, Weremowicz S, Morton CC, Hediger MA. 2000. Human vitamin C (L-ascorbic acid) transporter SVCT1. *Biochim Biophys. Res. Commun.* 267:488–94

12. Song J, Kwon O, Chen S, Daruwala R, Eck PM. et al. 2001. Flavonoid inhibition of sodium-dependent vitamin C transporter 1 (SVCT1) and glucose transporter isoform 2 (GLUT2), intestinal transporters for vitamin C and glucose. *J. Biol. Chem.* 277:15252–60

13. Choi JL, Rose RC. 1989. Regeneration of ascorbic acid by rat colon. *Proc. Soc. Exp. Bio. Med.* 190:369–74

14. Rose RC, Choi JL, Koch MJ. 1988. Intestinal transport and metabolism of oxidized ascorbic acid (dehydroascorbic acid). *Am. J. Physiol.Gastrointest. Liver Physiol.* 254:G824–28

15. Schell DA, Bode AM. 1993. Measurement of ascorbic acid and dehydroascorbic acid in mammalian tissue utilizing HPLC and electrochemical detection. *Biomed. Chromatogr.* 7:267–72

16. Rose RC, Choi JL. 1990. Intestinal absorption and metabolism of ascorbic acid in rainbow trout. *Am. J. Physiol. Regul. Integr. Comp. Physiol.* 258:R1238–41

17. Vera JC, Rivas CI, Fishbarg J, Golde

DW. 1993. Mammalian facilitative hexose transporters mediates the transport of dehydroascorbic acid. *Nature* 364:79–82

18. Rumsey SC, Daruwala R, Al-Hasani H, Zarnowski MJ, Simpson IA, Levine M. 2000. Dehydroascorbic acid transport by GLUT4 in *Xenopus* oocytes and isolated rat adipocytes. *J. Biol. Chem.* 275:28246–53

19. Rumsey SC, Kwon O, Xu GW, Burant CF, Simpson I, Levine M. 1997. Glucose transport isoforms GLUT1 and GLUT3 transport dehydroascorbic acid. *J. Biol. Chem.* 272:18982–89

20. Karasov WH, Darken BW, Bottum MC. 1991. Dietary regulation of intestinal ascorbate uptake in guinea pigs. *Am. J. Physiol. Gastrointest. Liver Physiol.* 260:G108–18

21. Rose RC, Nahrwold DL. 1978. Intestinal ascorbic acid transport following diets of high or low ascorbic acid content. *Int. J. Vitam. Nutr. Res.* 48:382–86

22. Liang WJ, Johnson D, Ma LS, Jarvis SM. 2002. Regulation of the human vitamin C transporters expressed in COS-1 cells by protein kinase C. *Am. J. Physiol. Cell Physiol.* 283:C1696–704

23. Lampen J, Hahler G, Peterson W. 1942. The occurrence of free and bound biotin. *J. Nutr.* 23:11–21

24. Wolf B, Heard GS, Secor-McVoy JR, Raetz HM. 1984. Biotinidase deficiency: the possible role of biotinidase in the processing of dietary protein-bound biotin. *J. Inherit. Metab. Dis.* 7, 121–22

25. Said HM, Thuy LP, Sweetman L, Schatzman B. 1993. Transport of the biotin dietary derivative biocytin (N-biotinyl-L-lysine) in rat small intestine. *Gastroenterology* 104:75–79

26. Cole H, Reynolds TR, Lockyer JM, Buck GA, Denson T, et al. 1994. Human serum biotinidase: cDNA cloning, sequence, and characterization. *J. Biol. Chem.* 269:6566–70

27. Dyer DL, Said HM. 1997. Biotin uptake in cultured cell lines. *Methods Enzymol.* 279:393–405

28. Said HM. 1999. Cellular uptake of biotin: mechanisms and regulation. *J. Nutr.* 129:490S–493S

29. Said HM, Redha R, Nylander W. 1987. A carrier-mediated Na$^+$-gradient dependent transport for biotin in human intestinal brush-border membrane vesicles. *Am. J. Physiol. Gastrointest. Liver Physiol.* 253:G631–36

30. Said HM, Redha R, Nylander W. 1988. Biotin transport in basolateral membrane vesicles of human intestine. *Gastroenterology* 94:1157–63

31. Said HM, Ortiz A, McCloud E, Dyer D, Moyer MP, Rubin SA. 1998. Biotin uptake by the human colonic epithelial cells NCM460: A carrier-mediated process shared with pantothenic acid. *Am. J. Physiol. Cell Physiol.* 275: C1365–71

32. Spector R, Mock D. 1987. Biotin transport through the blood brain barrier. *J. Neurochem.* 48:400–4

33. Beinlich CJ, Naumovitz RD, Song WO, Neely JR. 1990. Myocardial metabolism of pantothenic acid in chronically diabetic rats. *J. Mol. Cell. Cardiol.* 22:323–32

34. Grassl SM. 1992. Human placental brush-border membrane Na$^+$-pantothenate cotransport. *J. Biol. Chem.* 267: 22902–6

35. Burkholder PR, McVeigh I. 1942. Synthesis of vitamins by intestinal bacteria. *Proc. Natl. Acad. Sci. USA* 28:285–89

36. Wrong OM, Edmonds CJ, Chadwich VS, eds. 1981. *The Large Intestine; Its Role In Mammalian Nutrition and Homeostasis.* New York: Wiley & Sons

37. Barth CA, Frigg M, Hogemeister H. 1986. Biotin absorption from the hindgut of the pig. *J. Anim. Physiol. Anim. Nutr.* 55:128–34

38. Brown BB, Rosenberg JH. 1987. Biotin absorption by distal rat intestine. *J. Nutr.* 117:2121–26

39. Sorrell MF, Frank O, Thomson AD, Aquino A, Baker H. 1971. Absorption of vitamins from the large intestine. *Nutr. Res. Int.* 3:143–48

40. Chatterjee NS, Kumar CK, Ortiz A, Rubin SA, Said HM. 1999. Molecular mechanism of the intestinal biotin transport process. *Am. J. Physiol. Cell Physiol.* 277:C605–13

41. Prasad PD, Wang H, Huang W, Fei Y, Leibach FH, et al. 1999. Cloning and functional characterization of the intestinal Na$^+$-dependent multivitamin transporter. *Arch. Biochem. Biophys.* 366: 95–106

42. Mardach R, Zempleni J, Wolf B, Cannon MJ, Jennings ML, et al. 2002. Biotin dependency due to a defect in biotin transport. *J. Clin. Invest.* 109:1617–23

43. Balamurugan K, Ortiz A, Said HM. 2003. Biotin uptake by human intestinal and liver epithelial cells: Role of the sodium-dependent multivitamin transport system, SMVT. *Am. J. Physiol. Gastrointest. Liver Physiol.* 285:673–77

44. Said HM, Redha R. 1988. Ontogenesis of the intestinal transport of biotin in the rat. *Gastroenterology* 94:68–72

45. Said HM, Mock DM, Collins J. 1989. Regulation of intestinal biotin transport in the rat: effect of biotin deficiency and supplementation. *Am. J. Physiol. Gastrointest. Liver Physiol.* 256:G306–11

46. Nabokina SM, Subramanian VS, Said HM. 2003. Comparative analysis of ontogenic changes in renal and intestinal biotin transport in the rat. *Am. J. Physiol. Renal Physiol.* 284:F737–42

47. Chatterjee NS, Rubin SA, Said HM. 2001. Molecular characterization of the 5′ regulatory region of rat sodium-dependent multivitamin transporter gene. *Am. J. Physiol. Cell Physiol.* 280:C548–55

48. Dey S, Subramanian VS, Chatterjee NS, Rubin SA, Said HM. 2002. Characterization of the 5′ regulatory region of the human sodium-dependent multivitamin transporter, hSMVT. *Biochim. Biophys. Acta* 1574:187–92

49. Chandler CJ, Wang T, Halsted CH. 1986. Pteroylpolyglutamate hydrolase from human jejunal brush borders: purification and characterization. *J. Biol. Chem.* 261:928–33

50. Reisenauer AM, Krumdieck CL, Halsted CH. 1971. Folate conjugase: two separate activities in human jejunum. *Science* 198:196–97

51. Wang T, Reisenauer AM, Halsted CH. 1986. Comparison of folate conjugase activities in human, pig, rat and monkey intestine. *J. Nut.* 115:814–19

52. Wang T, Reisenauer AM, Halsted CH. 1986. Intracellular pteroylpolyglutamate hydrolase from human jejunal mucosa: isolation and characterization. *J. Biol. Chem.* 261:13551–55

53. Chandler CJ, Harrison DA, Buffington CA, Santiago NA, Halsted CH. 1991. Functional specificity of jejunal brush-border pteroylpolyglutamate hydrolase in pig. *Am. J. Physiol. Gastrointest. Liver Physiol.* 260:G865–72

54. Halsted CH, Ling E, Carter RL, Villanueva JA, Gardner JM, Coyle JT. 1998. Folylpoly-γ-glutamate carboxypeptidase from pig jejunum molecular characterization and relation to glutamate carboxypeptidase II. *J. Biol. Chem.* 273:20417–24

55. Devlin AM, Ling EH, Peerson JM, Fernando S, Clarke R, et al. 2000. Glutamate carboxypeptidase II: a polymorphism associated with lower levels of serum folate and hyperhomocysteinemia. *Hum. Mol. Genet.* 9:2837–44

56. Said HM, Chatterjee H, Haq RU, Subramanian VS, Ortiz A, et al. 2000. Adaptive regulation of intestinal folate uptake: effect of dietary folate deficiency. *Am. J. Physiol. Cell Physiol.* 279:C1889–95

57. Rose RC. 1987. Intestinal absorption of water-soluble vitamins. In *Physiology of*

the *Gastrointestinal Tract*, ed. LR Johnson, pp. 1581–96. New York: Raven

58. Said HM, Ghishan FK, Redha R. 1987. Folate transport by human intestinal brush-border membrane vesicles. *Am. J. Physiol. Gastrointest. Liver Physiol.* 252:G229–36

59. Schron CM, Washington C, Blitzer B. 1985. The *trans*-membrane pH gradient drives uphill folate transport in rabbit jejunum. *J. Clin. Invest.* 76:2030–33

60. Selhub J, Rosenberg JH. 1981. Folate absorption in isolated brush border membrane vesicles from rat intestine. *J. Biol. Chem.* 256:4489–93

61. Schron CM, Washington C, Blitzer B. 1988. Anion specificity of the jejunal folate carrier: effects of reduced folate analogues on folate uptake and effect. *J. Membr. Biol.* 102:175–83

62. Goldman ID, Lichenstein NS, Oliverio VT. 1968. Carrier-mediated transport of the folic acid analogue, methotrexate, in the L1210 leukemia cell. *J. Biol. Chem.* 243:5007–17

63. Henderson GB, Zevely EM. 1984. Transport routes utilized by L1210 cells for the influx and efflux of methotrexate. *J. Biol. Chem.* 259:1526–31

64. Sirotnak FM, Tolner B. 1999. Carrier-mediated membrane transport of folates in mammalian cells. *Annu. Rev. Nutr.* 19:91–122

65. Said HM, Redha R. 1987. A carrier-mediated transport for folate in basolateral membrane vesicles of rat small intestine. *Biochem. J.* 247:141–46

66. Said HM, Ma TY, Ortiz A, Tapia A, Valerio CK. 1997. Intracellular regulation of intestinal folate uptake: studies with cultured IEC-6 epithelial cells. *Am. J. Physiol. Cell Physiol.* 272:C729–36

67. Witman SD, Lark RH, Coney LR, Fort DW, Frasca V, et al. 1992. Distribution of the folate receptor in normal and malignant cell lines and tissues. *Cancer Res.* 52:3396–401

68. Rong NI, Selhub J, Goldin BR, Rosenberg I. 1991. Bacterially synthesized folate in rat large intestine is incorporated into host tissue folylpolyglutamates. *J. Nutr.* 121:1955–59

69. Dudeja PK, Torania SA, Said HM. 1997. Evidence for the existence of an electroneutral, pH-dependent, DIDS-sensitive carrier-mediated folate uptake mechanism in the human colonic luminal membrane vesicles. *Am. J. Physiol. Gastrointest. Liver Physiol.* 272:G1408–15

70. Kumar CK, Moyer MP, Dudeja PK, Said HM. 1997. A protein-tyrosine kinase regulated, pH-dependent carrier-mediated uptake system for folate by human normal colonic epithelial cell line NCM 460. *J. Biol. Chem.* 272:6226–31

71. Cravo ML, Mason JB, Selhub J, Rosenberg IH. 1991. Use of deoxyuridine suppression test to evaluate localized folate deficiency in rat colonic epithelium. *Am. J. Clin. Nutr.* 53:1450–54

72. Lashner BA, Heidenreich PA, Su GL, Kane SV, Hanauer SB. 1989. The effect of folate supplementation on the incidence of displasia and cancer in chronic ulcerative colitis: a case-control study. *Gastroenterology* 97:255–59

73. Nguyen TT, Dyer DL, Dunning DD, Rubin SA, Said HM. 1997. Human intestinal folate transport: cloning, expression and distribution of complementary RNA. *Gastroenterology* 1112:783–91

74. Said HM, Nguyen TT, Dyer DL, Cowan KH, Rubin SA. 1996. Intestinal transport of folate: identification of a mouse intestinal cDNA and localization of its mRNA. *Biochim. Biophys. Acta* 1281:164–72

75. Kumar CK, Nguyen TT, Gonzales FB, Said HM. 1998. Comparison of intestinal folate carrier clone expressed in IEC-6 cells and in *Xenopus* oocytes. *Am. J. Physiol. Cell Physiol.* 274:C289–94

76. Balamurugan K, Sandoval C, Said HM. 2003. Mutations in the human reduced

folate carrier in patients with folate malabsorption syndrome. *Gastroenterology* 124:A434 (Abstr.)

77. Geller J, Kronn D, Jayabose S, Sandoval C. 2002. Hereditary folate malabsorption: family report and review of the literature. *Medicine* 81:51–68

78. Said HM, Ghishan FK, Murrell JE. 1985. Ontogenesis of intestinal transport of 5-methyltetrahydrofolate in the rat. *Am. J. Physiol. Gastrointest. Liver Physiol.* 249:G567–71

78a. Balamurugan K, Said HM. 2003. Autogenic regulation of folate transport across rat jejunal brush border membrane. *Am. J. Physiol. Gastrointest. Liver Physiol.* In press

79. Subramanian VS, Chatterjee N, Said HM. 2003. Folate uptake in the human intestine: promoter activity and effect of folate deficiency. *J. Cell Physiol.* 196:403–8

80. Gong M, Cowan KH, Gudas J, Moscow JA. 1999. Isolation and characterization of genomic sequences involved in the regulation of the human reduced folate carrier gene (RFC1). *Gene* 133:21–31

81. Subramanian VS, Marchant JS, Parker I, Said HM. 2001. Intracellular trafficking/membrane targeting of human reduced folate carrier expressed in Xenopus oocytes. *Am. J. Physiol. Gastrointest. Liver Physiol.* 281:G1477–86

82. Marchant JS, Subramanian VS, Parker I, Said HM. 2002. Intracellular trafficking and membrane targeting mechanisms of the human reduced folate carrier in mammalian epithelial cells. *J. Biol. Chem.* 277:33325–33

83. Simanjuntak MT, Tamai I, Terasaki T, Tsugi A. 1990. Carrier-mediated uptake of nicotinic acid by rat intestinal brush border membrane vesicles and relation to monocarboxylic acid transport. *J. Pharmacobio-Dyn* 13:301–9

84. Guilarte TR, Pravlik K. 1983. Radiometric-microbiologic assay of niacin using *Kloeckera brevis*: analysis

of human blood and food. *J. Nutr.* 113:2587–94

85. Schuette S, Rose RC. 1986. Renal transport and metabolism of nicotinic acid. *Am. J. Physiol. Cell Physiol.* 250:C694–703

86. Shibata K, Gross CJ, Henderson LM. 1983. Hydrolysis and absorption of pantothenate and its coenzymes in the rat small intestine. *J. Nutr.* 113:2107–15

87. Prasad PD, Wang H, Kekuda R, Fujita T, Feis YJ, et al. 1998. Cloning and functional expression of a cDNA encoding a mammalian-sodium dependent vitamin transporter mediating the uptake of pantothenate, biotin and lipoate. *J. Biol. Chem.* 273:7501–6

88. Daniel H, Binninger E, Rehner G. 1983. Hydrolysis of FMN and FAD by alkaline phosphatase of the intestinal brush border membrane. *Int. J. Vitam. Nutr. Res.* 53:109–14

89. Daniel H, Wille U, Rehner G. 1983. In vitro kinetics of the intestinal transport of riboflavin in rats. *J. Nutr.* 113:636–43

90. Said HM, Ma TY. 1994. Mechanism of riboflavin uptake by Caco-2 human intestinal epithelial cells. *Am. J. Physiol. Gastrointest. Liver Physiol.* 266:G15–21

91. Said HM, Hollander D, Khani R. 1993. Uptake of riboflavin by intestinal basolateral membrane vesicles: a specialized carrier-mediated process. *Biochim. Biophys. Acta* 1148:263–68

92. Said HM, Khani R, McCloud E. 1993. Mechanism of transport of riboflavin in rabbit intestinal brush border membrane vesicles. *Proc. Soc. Exp. Biol. Med.* 202:428–34

93. Said HM, Ma TY, Grant K. 1994. Regulation of riboflavin intestinal uptake by protein kinase A: studies with Caco-2 cells. *Am. J. Physiol. Gastro intest. Liver Physiol.* 267:G955–59

94. Hegazy E, Schwenk M. 1983. Riboflavin uptake by isolated enterocytes of guinea pigs. *J. Nutr.* 113:1702–7

95. Middleton HM. 1985. Uptake of riboflavin by rat intestinal mucosa in vitro. *J. Nutr.* 120:588–93

96. Said HM, Khani R. 1993. Uptake of riboflavin across the brush border membrane of rat intestine: regulation by dietary vitamin levels. *Gastroenterology* 105:1294–98

97. Gastaldi G, Rerrari G, Verri A, Casirola D, Orsenigo MN, Laforenza N. 2000. Riboflavin phosphorylation is the crucial event in riboflavin transport by isolated rat enterocytes. *J. Nutr.* 130:2556–61

98. Iinuma S. 1955. Synthesis of riboflavin by intestinal bacteria. *J. Vitam.* 2:6–13

99. Ocese O, Pearson PB, Schwiegert BS. 1948. The synthesis of certain B vitamins by the rabbit. *J. Nutr.* 35:577–90

100. Kasper H. 1970. Vitamin absorption in the colon. *Am. J. Protocol.* 21:341–45

101. Said HM, Ortiz A, Moyer MP, Yanagawa N. 2000. Riboflavin uptake by human-derived colonic epithelial NCM460 cells. *Am. J. Physiol. Cell Physiol.* 278:C270–76

102. Yuasa H, Hirobe M, Tomei SA, Watanabe J. 2000. Carrier-mediated transport of riboflavin in the rat colon. *Biopharm. Drug Dispos.* 21:77–82

103. Mandel H, Beratn M, Hazani A, Naveh Y. 1984. Thiamine-dependent beriberi in the thiamine-responsive anemia syndrome. *N. Engl. J. Med.* 311:836–38

104. Rogers LE, Porter FS, Sidbury JB Jr. 1969. Thiamine-responsive megaloblastic anemia. *J. Pediatr.* 74:494–504

105. Diaz GA, Banikazemai M, Oishi K, Desnick RJ, Gelb BD. 1999. Mutations in a new gene encoding a thiamine transporter cause thiamine-responsive megaloblastic anaemia syndrome. *Nat. Genet.* 22:309–12

106. Labay V, Raz T, Baron D, Mandel H, Williams H, et al. 1999. Mutations in *SLC19A2* cause thiamine-responsive

megaloblastic anaemia associated with diabetes mellitus and deafness. *Nat. Genet.* 22:300–4

107. Dutta B, Huang W, Molero M, Kekuda R. Liebach FH, et al. 1999. Cloning of the human thiamine transporter, a member of the folate transporter family. *J. Biol. Chem.* 45:31925–29

108. Sklan D, Trostler N. 1977. Site and extend of thiamin absorption in the rat. *J. Nutr.* 107:353–56

109. Rindi G. 1984. Thiamin absorption by small intestine. *Acta Vitaminol. Enzymol.* 6:47–55

110. Rindi G, Laforenza U. 2000. Thiamine intestinal transport and related issues: recent aspects. *Proc. Soc. Exp. Biol. Med.* 224:246–55

111. Dudeja PK, Tyagi S, Kavilaveettil RJ, Gill R, Said HM. 2001. Mechanism of thiamine uptake by human jejunal brush-border membrane vesicles. *Am. J. Physiol. Cell Physiol.* 281:C786–92

112. Laforenza U, Orsenigo MN, Rindi G. 1998. A thiamin: H$^+$ antiport mechanism for thiamin entry into brush border membrane vesicles from rat small intestine. *J. Membr. Biol.* 161:151–61

113. Laforenza U, Patrini C, Alvisi C, Faelli A, Licandro A, Rindi G. 1997. Thiamin uptake in human intestinal biopsy specimens, including observations from a patient with acute thiamin deficiency. *Am. J. Clin. Nutr.* 66:320–26

114. Rindi G, Ferrari G. 1977. Thiamin transport by human intestine in vitro. *Experentia* 33d:211–13

115. Thomson AD, Leevy CM. 1972. Observations on the mechanism of thiamin hydrochloride absorption man. *Clin. Sci.* 43:153–63

116. Dudeja PK, Tyagi S, Gill R, Said HM. 2003. Evidence for carrier-mediated mechanism for thiamine transport to human jejunal basolateral membrane vesicles. *Dig. Dis. Sci.* 48:109–15,

117. Laforenza U, Gastaldi G, Rindi G. 1993. Thiamin outflow from the enterocyte: A

study using basolateral membrane vesicles from rat small intestine. *J. Physiol.* 468:401–12

118. Said HM, Ortiz A, Subramanian VS, Neufeld EJ, Moyer MP, Dudeja PK. 2001. Mechanism of thiamine uptake by human colonocytes: studies with cultured colonic epithelial cell line NCM460. *Am. J. Physiol. Gastrointest. Liver Physiol.* 281:G144–50

119. Rajgopal A, Edmondson A, Goldman D, Zhao R. 2001. SLC19A3 encodes a second thiamine transporter ThTr2. *Biochim. Biophys. Acta* 1537:175–78

120. Eudy JD, Spiegelstein O, Baber RC, Wlodarczk BJ, Talbot J, et al. 2000. Identification and characterization of the human and mouse SLC19A3 gene: a novel member of the reduced folate family of micronutrient transporter genes. *Mol. Gen. Metab.* 71:581–90

121. Reidling JC, Subramanian VS, Dudeja PK, Said HM. 2002. Expression and promoter analysis of SLC19A2 in the human intestine. *Biochim. Biophys. Acta* 1561:180–87

122. Oishi K, Hofmann S, Diaz GA, Brown T, Manwani D, et al. BD. 2002. Targeted disruption of *Slc19a2*, the gene encoding the high-affinity thiamin transporter Thtr-1, causes diabetes mellitus, sensorineural deafness and megaloblastosis in mice. *Hum. Mol. Genet.* 11:2951–60

123. Said HM, Ortiz A, Kumar CK, Chatterjee N, Dudeja PK, Rubin SA. 1999. Transport of thiamine in the human intestine: mechanism and regulation in intestinal epithelial cell model Caco-2. *Am. J. Physiol: Cell Physiol.* 277:C645–51

124. Reidling JC, Said HM. 2003. In vitro and in vivo characterization of the minimal promoter region of the human thiamin transporter *SLC19A2. Am. J. Physiol. Cell Physiol.* In press

125. Lo PK, Chen JY, Tang PP, Lin J, Lin CH,

et al. 2001. Identification of a mouse thiamine transporter gene as a direct transcriptional target for p53. *J. Biol. Chem.* 276:37186–93

126. Neufeld EJ, Fleming JC, Tyartaglini E, Steinkamp MP. 2001. Thiamine-responsive megaloblastic anemia syndrome: a disorder of high-affinity thiamine transport. *Blood Cells Mol. Dis.* 27:135–38

127. Balamurugan K, Said HM. 2002. Functional role of specific amino acid residues in human thiamine transporter SLC19A2: mutational analysis. *Am. J. Physiol. Gastrointest. Liver Physiol.* 283:G37–43

128. Subramanian VS, Marchant JS, Parker I, Said HM. 2003. Cell biology of the human thiamine transporter-1 (hTHTR1): intracellular trafficking and membrane targeting mechanisms. *J. Biol. Chem.* 278:3976–84

129. Gospe SM. 2002. Pyridoxine-dependent seizures: finding from recent studies pose new questions. *Pediatr. Neurol.* 26:181–85

130. Hamm MW, Hehansho H, Henderson LM. 1979. Transport and metabolism of pyridoxamine and pyridoxamine phosphate in the small intestine. *J. Nutr.* 109:1552–59

131. Middleton HM. 1979. Intestinal absorption of pyridoxal-5′ phosphate disappearance from perfused segments of rat jejunum in vivo. *J. Nutr.* 109:975–81

132. Middleton HM. 1985. Uptake of pyridoxine by in vivo perfused segments of rat small intestine: a possible role for intracellular vitamin metabolism. *J. Nutr.* 115:1079–88

133. Serebro HA, Solomon HM, Johnson JH, Hendrix TR. 1966. The intestinal absorption of vitamin B_6 compounds by the rat and hamster. *Johns Hopkins Hosp. Bull.* 119:166–71

134. Yoshida S, Hayashi K, Kawasaki T. 1981. Pyridoxine transport in brush

border membrane vesicles of guinea pig jejunum. *J. Nutr. Sci. Vitaminol.* 27:311–17

134a. Said HM, Ortiz A. 2003. A carrier-mediated mechanism for pyrodoxine uptake by human intestinal epithelial Caco-2 cells: regulation by a PKA-mediated pathway. *Am. J. Physiol. Cell Physiol.* In press

135. Said HM, Ortiz A, Vaziri ND. 2002. Mechanism and regulation of vitamin B_6 uptake by renal tubular epithelia: studies with cultured OK cells. *Am. J. Physiol. Renal Physiol.* 282:F465–71

Annu. Rev. Physiol. 2004. 66:447–75
doi: 10.1146/annurev.physiol.66.032102.112534
Copyright © 2004 by Annual Reviews. All rights reserved
First published online as a Review in Advance on September 22, 2003

LEARNING MECHANISMS IN ADDICTION:
Synaptic Plasticity in the Ventral Tegmental Area as a Result of Exposure to Drugs of Abuse

Julie A. Kauer
*Department of Pharmacology, Physiology and Biotechnology, Brown University,
Providence, Rhode Island 02912; email: Julie_Kauer@brown.edu*

Key Words   VTA, LTP, LTD, sensitization, dopamine, psychostimulant

■ **Abstract**   One of the central questions in neurobiology is how experience modifies neural function, and how changes in the nervous system permit an animal to adapt its behavior to a changing environment. Learning and adaptation to a host of different environmental stimuli exemplify processes we know must alter the nervous system because the behavioral output changes after experience. Alterations in behavior after exposure to addictive drugs are a striking example of chemical alterations of nervous system function producing long-lasting changes in behavior. The alterations produced in the central nervous system (CNS) by addictive drugs are of interest because of their relationship to human substance abuse but also because these CNS alterations produce dramatic, easily observed alterations in behavior in response to discrete stimuli. Considerable study has been given to behavioral and biochemical correlates of addiction over the past 50 or more years; however, our understanding of the cellular physiological responses of affected CNS neurons is in its infancy. This review focuses on alterations in cellular and synaptic physiology in the ventral tegmental area (VTA) in response to addictive drugs.

INTRODUCTION

Drug addiction is a complex behavioral phenomenon characterized by compulsion to continue seeking and administering drugs despite severe consequences to work, social relationships, and health. Although many individuals try addictive drugs, most do not become addicted to a given substance. However, with repeated exposure, and in individuals with certain genetic risk factors, a shift occurs from volitional drug-taking to compulsive drug-seeking and drug-taking. One intriguing feature of addiction is that it accompanies exposure to drugs that elicit very different behavioral responses and that bind to distinct target molecules. For example, cocaine and amphetamine block dopamine and norepinephrine transporters throughout the brain and produce strong euphoric and psychomotor stimulant effects. In sharp contrast, nicotine activates nicotinic acetylcholine receptors in the brain and produces a mild sense of well-being, alertness, and relaxation, and a relatively mild rush. Morphine activates opiate receptors located on neurons throughout the brain, produces a dreamy euphoric state, and blocks the negative sensation of pain. All three drug classes can be highly addictive, although some individuals appear to be more susceptible than others. In considering the physiological changes that underlie the development of addiction, clues are provided when common neuroadaptations result from all addictive drugs.

One of the most exciting hypotheses proposed in recent years is that neuronal alterations thought to form the cellular basis for learning (for example in the hippocampus) may also occur in the mesolimbic reward system after drug exposure, producing persistent alterations of neural function that ultimately lead to addiction. This field is in its infancy, and as yet we have only preliminary evidence in support of the hypothesis. This review attempts to integrate the current knowledge of ventral tegmental area (VTA) synaptic plasticity with the more established literature on addiction and behavioral sensitization. Other brain regions also exhibit long-lasting synaptic plasticity that contributes importantly to the development of addiction, and several excellent recent reviews cover this subject (1–3).

THE MESOLIMBIC DOPAMINE CIRCUIT

Nearly all addictive drugs share the ability to cause dopamine release at nerve terminals in the nucleus accumbens (4–6). This finding has focused attention on the VTA dopamine neurons and the nucleus accumbens, key components mediating reinforcement and reward (7–9). In primates, VTA dopamine neurons are activated by primary rewards, by reward-predicting appetitive stimuli, and by novel, behavior-orienting stimuli (10). Thus when an unanticipated primary reward is presented, dopamine neurons increase their firing rate. However, with repeated pairings of a reward-predicting stimulus with the reward, firing rate increases to the reward predictor with no alteration in firing rate to the primary reward. If, however, the primary reward is unexpectedly not delivered, dopamine cell firing frequency drops. Thus the dopamine neuron acts as an error signal for the quality of environmental events relative to prediction (7).

The VTA consists of dopaminergic and GABAergic neurons (Figure 1) (11–13). Both neuron types project to the nucleus accumbens and to the prefrontal cortex. The GABAergic neurons thus are projection neurons and also have local connections onto dopamine cells (14). In addition to ordinary synaptic release of dopamine from nerve terminals, dopamine is also released from somatic and dendritic sites within the VTA (15, 16). Dopamine can diffuse many microns from the site of release, binding not only to the D2 subtype of dopamine receptor expressed by neighboring dopaminergic neurons but also to D1 and D2 receptors on GABAergic neurons and local nerve terminals (17–20). The VTA, prefrontal cortex, and nucleus accumbens are synaptically interconnected.

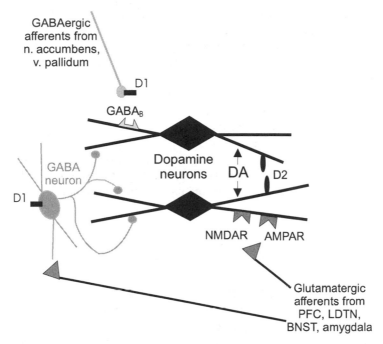

Figure 1   Diagram of the VTA microcircuit. The VTA consists of dopaminergic and GABAergic neurons. PFC, prefrontal cortex; LDTN, laterodorsal tegmental nucleus; BNST, bed nucleus of the stria terminalis; n. accumbens, nucleus accumbens; v. pallidum, ventral pallidum.

The primary excitatory afferents to the VTA include a direct glutamatergic input from prefrontal cortex (21–23). This input forms synapses on both dopamine and nondopamine cells in the VTA (24). The VTA also receives important excitatory input from the amygdala, laterodorsal tegmental nucleus, and bed nucleus of the stria terminalis (25–27). Excitatory inputs to the VTA have recently been defined in an elegant series of experiments demonstrating that prefrontal cortex–projecting dopamine neurons receive a reciprocal input from the prefrontal cortex, whereas nucleus accumbens-projecting dopamine neurons do not (28). In contrast, excitatory afferents from the laterodorsal tegmental nucleus preferentially innervate nucleus accumbens-projecting VTA dopamine neurons (S. Sesack, personal communication).

Dopamine neurons in the VTA express metabotropic and ionotropic glutamate receptors, including both NMDA and AMPA subtypes (NMDARs and AMPARs) (13, 29–31). As in other tissues, glutamate acting at these receptors excites dopamine neurons in the VTA (30, 32–37).

Acute effects of drugs of different classes on VTA neuron firing in anesthetized animals are not homogeneous and thus do not correlate well with their addictive potential. For example, activation of opiate receptors in the VTA increases firing of dopamine neurons, whereas exposure to cocaine depresses dopamine neuron firing (38, 39). Amphetamine increases firing rate of nondopamine (presumed GABAergic) neurons in the VTA. These changes in firing rate are thought to occur via a combination of direct and circuit-mediated drug effects. Firing rate in vivo, at least at a very early time point, therefore does not represent a common effect of addictive drugs of different classes. In the past several years, more investigators have turned their attention to in vitro slice preparations of the VTA to more easily characterize cellular responses to drugs of abuse. These studies, although taken out of the context in vivo in which drugs normally act, provide important building blocks in understanding the components that make up these linked circuits. Ultimately, cellular physiology and its alteration by addictive drugs will provide evidence to explain the molecular actions of these drugs on the circuit.

BEHAVIORAL SENSITIZATION

In the search for a cellular change common to multiple addictive drugs, the model of behavioral sensitization has become extremely useful. Daily drug administration results in a progressive enhancement (sensitization) to many of the behavioral effects of the drug (38, 40). In rats the enhancement can be observed as a progressive augmentation of spontaneous locomotor activity, so that rats exposed to psychostimulants will react far more strongly to a subsequent dose of psychostimulant than a drug-naive animal. The rewarding and motor-activating effects of addictive drugs may be mediated by common cellular mechanisms (9). Robinson & Berridge (41–43) have proposed the incentive salience theory of sensitization, which posits that drug addiction involves an increase in "drug-wanting" leading to drug-craving and compulsive drug-seeking behavior, despite a decreasing level of enjoyment ("drug-liking"). Through associative learning, contextual cues surrounding drug-taking become especially strong incentives in themselves. Two features of sensitization have linked it to the gradual enhancement of drug-craving, a central symptom of addiction and a major cause of drug relapse. First, nearly all drugs of abuse elicit behavioral sensitization, indicating an important common target in the brain. Second, sensitization is extremely long-lasting, for example, lasting for up to a year in rats (half of the rat's life span) (41–45). The persistence of sensitization may be comparable to the persistence of drug-craving in humans, a devastating feature of addiction that makes lasting recovery very difficult (1, 39, 41, 46).

The circumstances surrounding drug administration have potent effects on the behavioral and subjective effects of addictive drugs (43, 47–49). As a likely consequence, there appear to be behavioral differences between context-dependent sensitization (in which drug treatment is experienced in a novel environment in which sensitization is also tested) and noncontext-dependent sensitization. A striking example of this is that rats administered amphetamine intravenously without any contextual cues to predict drug administration did not exhibit psychomotor activation, nor did they exhibit sensitization five days after a nine-day daily drug treatment (50). In this review I focus on data from context-dependent sensitization because it is a robust experimental phenomenon that may well have cellular underpinnings similar to learning. As substance abusers describe strong context-dependent increases in craving levels, context-dependent behavioral effects of addictive drugs may be relevant to human addiction as well.

THE VTA IS REQUIRED FOR THE INDUCTION OF BEHAVIORAL SENSITIZATION

Psychostimulants such as amphetamine and cocaine increase dopamine concentrations by increasing dopamine release and/or by blocking reuptake of dopamine (51, 52). Administration of an acute dose of cocaine or amphetamine increases dopamine release from VTA terminals in the nucleus accumbens. Increased dopamine release in the nucleus accumbens promotes an increase in locomotor activity, whereas destruction of the VTA-accumbens pathway (mesoaccumbens pathway) causes hypoactivity and abolishes the locomotor response to amphetamine (53–55). For this reason, in attempting to identify the cellular targets of psychostimulants in the initiation of behavioral sensitization, efforts have focused on the mesoaccumbens dopamine pathway.

Behavioral sensitization is not initiated by direct effects in the nucleus accumbens. One might have expected drug delivery into the nucleus accumbens to produce sensitization because locomotion is elicited by dopamine released here (44). Instead, however, repeated amphetamine injections into the nucleus accumbens produce locomotion, but not sensitization (56, 57). Sensitization is produced by daily microinjection of amphetamine into the VTA. In fact, intra-VTA amphetamine sensitizes animals to peripherally delivered amphetamine and cocaine (57–59); interestingly, animals also become sensitized to systemic morphine after this regimen (60). Conversely, microinjection of certain compounds (D1 receptor antagonists, NMDA receptor antagonists) directly into the VTA blocks the development of behavioral sensitization to repeated peripheral administration of amphetamine (40). These data strongly suggest that the VTA is the site of initiation of sensitization.

One remarkable point about the VTA is that not only psychostimulants but also the chemically distinct opioid compounds trigger sensitization by direct effects within the VTA. Repeated microinjection of μ-opioids into the VTA, but not into nucleus accumbens, results in behavioral sensitization to peripherally administered morphine (62), and intra-VTA injections of naltrexone block sensitization to peripherally administered opioids (63). Nicotine, phencyclidine, MDMA, and ethanol have all been reported to induce sensitization as well (43). Thus sensitization represents an example of behavior in which abused drugs with quite distinct mechanisms of action share a common target in the VTA. Taken together, these results demonstrate that processes occurring in the VTA are both necessary and sufficient for the initiation of behavioral sensitization and suggest that the long-lasting plastic change that triggers sensitization is localized to the VTA.

RELATIONSHIP BETWEEN SENSITIZATION AND ADDICTION

In addition to its clear involvement in sensitization, the VTA is also linked with drug addiction, in particular drug-craving and reinstatement of drug-seeking (relapse). Treatment with repeated intermittent morphine, cocaine, or amphetamine, like treatments that produce sensitization, all produce an apparent increase in drug-induced rewarding effects measured by conditioned place preference tests (64). Animals also show conditioned place preference for sites where morphine was directly injected into VTA (65). Application of morphine to VTA neurons reinstates self-administration of either heroin or cocaine, indicating that opiate receptors in the VTA can activate circuits controlling the tendency to relapse (65). Furthermore, glutamatergic input to the VTA has recently been implicated in reinstatement of cocaine use (66). Rats were given trains of stimuli to the ventral subiculum after extinction of cocaine self-administration. These stimuli elicited cocaine-seeking behavior that was entirely blocked when glutamate receptors in the VTA were pharmacologically blocked, indicating that glutamatergic stimulation of the VTA can trigger relapse.

After long-term extinction, reinstatement of heroin- or cocaine-seeking behavior is correlated with behavioral sensitization (67). Cross-sensitization between different drugs also correlates with reinstatement of drug-seeking. For example, heroin-seeking behavior was reinstated by injection of low doses of heroin, amphetamine, or cocaine, and any one of these drugs also elicited a sensitized locomotor response. Conversely, cocaine-seeking behavior was reinstated by cocaine or amphetamine, but not heroin, and only the psychomotor stimulants (but not heroin) elicited locomotor sensitization. This study shows that self-administration of drugs, as well as injection by the experimenter, elicits sensitization and that the tendency to relapse after low drug doses correlates with sensitization.

Although locomotor sensitization may not be equivalent to craving, these data indicate that the same circuits may be employed during both processes and that the same drugs trigger neuroadaptations underlying both processes. Therefore, a more detailed understanding of cellular processes that underlie behavioral sensitization may reasonably provide useful information in understanding addictive effects of drugs of abuse.

All evidence points to the VTA being the gateway for initiation of sensitization and perhaps control of relapse, although it remains unknown whether the VTA is the site of neural changes that underlie the persistence of craving and sensitization over months and years. Considerable evidence supports the idea that repeated exposure of the VTA to addictive drugs allows long-lasting alterations in target brain regions. The nucleus accumbens, neostriatum, amygdala, and prefrontal cortex are altered after drug exposure. These areas are involved in associative appetitive learning, fear conditioning, and working memory, and each of these regions is known to undergo activity-dependent synaptic plasticity. These brain regions may also be the repositories of longer-term drug-induced alterations occurring during the development of addiction (3, 68, 69).

GLUTAMATE RECEPTORS IN THE VTA PLAY A PIVOTAL ROLE IN THE INITIATION OF SENSITIZATION

A growing body of evidence suggests that the initiation of sensitization depends on glutamatergic transmission within the VTA (70). First, selective antagonists of the NMDA subclass of glutamate receptors block sensitization to cocaine, amphetamine, and morphine (71–76). Importantly, the acute behavioral response to amphetamine or cocaine is not altered by pretreatment with NMDA receptor antagonists, so that the animal experiences motor excitation, but sensitization does not develop with repeated treatment. Local injection of an NMDA receptor antagonist directly into the VTA also blocks behavioral sensitization to peripherally administered cocaine, demonstrating that NMDA receptors within the VTA are essential (77). Second, lesions of the prefrontal cortex, which provides glutamatergic afferent input to the VTA and to the laterodorsal tegmental nucleus, block behavioral sensitization to amphetamine (78, 79). Third, repeated electrical stimulation of glutamatergic prefrontal afferents by itself sensitizes animals to subsequent cocaine administration. In contrast, identical electrical stimulation delivered to the hippocampus does not initiate sensitization (80).

NMDA RECEPTOR-DEPENDENT SYNAPTIC PLASTICITY: HYPOTHESIS FOR INITIATION OF BEHAVIORAL SENSITIZATION

These data suggested the intriguing idea that exposure to an addictive drug may elicit pathological NMDA receptor-dependent long-term potentiation (LTP) at excitatory synapses in the mesolimbic dopamine system (1, 38, 68, 70, 80–82). NMDA receptor-mediated synaptic plasticity has been well described at excitatory synapses in several brain regions (83–87). The best-understood example of NMDA receptor-dependent plasticity is hippocampal LTP (88). In hippocampus, as in the VTA and most brain regions, glutamate is released onto postsynaptic spines that contain both NMDA and non-NMDA receptors. Ordinarily, synaptic currents are carried primarily by non-NMDA receptors, since NMDA receptors are blocked by Mg^{2+} when the postsynaptic neuron is near its resting potential. When the postsynaptic cell is depolarized, however, Mg^{2+} is expelled from the NMDA receptor channel and current can flow into the cell (89, 90). When NMDA receptors pass current, Ca^{2+} enters the postsynaptic cell and the synapse becomes potentiated.

NMDA receptor-dependent LTP has been demonstrated at excitatory synapses on midbrain dopamine neurons (36, 91, 92). Importantly, while excitatory synapses on dopamine cells undergo LTP, it appears that excitatory synapses on VTA GABAergic neurons do not (Figure 2) (36). The presence of LTP at excitatory synapses in VTA strengthens the hypothesis that this process could play a crucial role in the onset of sensitization.

AMPA RECEPTOR-MEDIATED SYNAPTIC TRANSMISSION IS ENHANCED IN HIPPOCAMPUS DURING LTP AND IN THE VTA AFTER ADDICTIVE DRUGS

Although NMDA receptors are required during the initiation of hippocampal LTP, established LTP is unaffected by NMDAR antagonists. Similarly, NMDAR antagonists that block behavioral sensitization to psychostimulants are ineffective once sensitization is established (78, 81, 93). During hippocampal LTP, potentiation is apparent as an increase in non-NMDA receptor-mediated glutamate transmission (94–97). Recent work demonstrates that LTP depends critically upon expression of the AMPA receptor subunit, GluR1 (98), and, in fact, results from insertion of AMPA receptors at synaptic sites (99, 100).

The early development of sensitization (over a period of 1 to 2 weeks) may be a consequence of enhanced glutamatergic synaptic transmission in the VTA. For example, viral-mediated transfection of VTA neurons in vivo with GluR1 produced behavioral sensitization to locomotor effects of opiates (101). Furthermore, in amphetamine-sensitized rats, dopamine neurons recorded in vivo are more sensitive to ionophoretically applied glutamate or AMPA than are cells in control animals (102, 103). These data support the idea that AMPA receptor-mediated synaptic transmission onto VTA dopamine neurons is enhanced in an LTP-like fashion during early stages of sensitization. Whether mRNA and protein levels of various AMPAR subunits are increased after exposure to morphine or cocaine is controversial (104–106). However, changes in these measures are clearly not necessary for rapid increases in sensitivity to glutamate assayed electrophysiologically during LTP. Thus, to trigger LTP, it is not essential that total levels of AMPARs increase, simply that AMPAR levels at synapses increase.

IN VIVO EXPOSURE TO DIFFERENT DRUGS OF ABUSE TRIGGERS LTP AT VTA SYNAPSES

Recent work provides the best evidence yet in support of the hypothesis that exposure to drugs of abuse elicits LTP at VTA synapses (107). Excitatory synapses in VTA brain slices cut 24 h after a single exposure to cocaine were examined

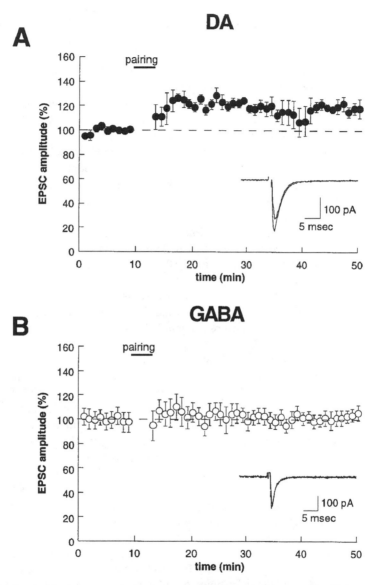

Figure 2   Long-term potentiation at excitatory synapses on VTA dopamine neurons but not GABAergic neurons. LTP was induced by pairing 1-Hz afferent stimulation with depolarization of the postsynaptic neuronal membrane to +10 mV for 200 stimuli. (*A*) LTP in the average of nine dopamine cells. (*B*) No LTP after pairing in the average of six GABAergic cells. Experiments carried out in the presence of 100 μM picrotoxin. Reprinted with permission from Bonci & Malenka (36).

for signs that LTP had taken place. LTP could not be induced at synapses on dopamine cells, as if synapses were already maximally potentiated. Moreover, the ratio between the AMPAR-mediated EPSC and the NMDAR-mediated EPSC (the AMPA/NMDA ratio) was increased in slices from the cocaine-treated animals. The AMPA/NMDA ratio is also increased in hippocampal CA3-CA1 synapses after LTP induction and is thought to result from the insertion of new AMPA receptors (but not new NMDA receptors) at potentiated synapses (100). Excitatory synapses on dopamine neurons from cocaine-treated animals had twofold higher AMPAR/NMDAR ratios when compared with those from saline-treated animals. Moreover, AMPAR/NMDAR ratios in slices from mice treated with an NMDAR antagonist before cocaine exposure were unchanged, showing that NMDAR activation is necessary for the plasticity to occur. In contrast, AMPAR/NMDAR ratios at excitatory synapses on neighboring VTA GABAergic neurons were unchanged. This is particularly important because it indicates that drug treatment increases excitation in the region relative to inhibition. Thus excitatory inputs activate dopamine neurons two times more strongly, without increasing their activation of GABA neurons, which provide important local inhibition to these cells, particularly via $GABA_B$ receptors (108). Synapses in hippocampus showed no differences between saline- and cocaine-treated mice. Animals exhibited behavioral sensitization after the same drug treatment, which correlated sensitization with LTP in the VTA for the first time. Together these findings demonstrated that a single cocaine exposure triggers LTP at excitatory synapses on dopamine neurons.

If LTP at excitatory VTA synapses contributes to the development of addiction and sensitization, LTP should follow not only cocaine treatment but also treatment with several different addictive drugs. Representatives of four distinct classes of addictive drugs all increase AMPAR/NMDAR ratios in VTA slices 24 h after drug treatment in vivo (109). Morphine, nicotine, ethanol, and either cocaine or amphetamine—drugs with little in common beyond their abuse potential— all have the same effect. In contrast, fluoxetine or carbamazepine, two centrally acting drugs used to treat psychiatric illnesses, do not increase the AMPAR/ NMDAR at VTA synapses. This work suggests that excitatory synapses on all VTA dopamine cells are potentiated because the experimenter recording from a brain slice cannot determine whether a given dopamine cell projects to nucleus accumbens or to prefrontal cortex. The global nature of this potentiation is somewhat surprising because different VTA dopamine cells clearly have distinct anatomical projections (28). It is also not clear whether excitatory synapses made by all glutamatergic inputs to VTA dopamine neurons are potentiated by drug exposure. It is a limitation of brain slice electrophysiology in the VTA that a single discrete afferent input cannot be stimulated with certainty; thus all recordings are likely to include EPSC components from a variety of synaptic afferents. One possibility is that all excitatory synapses on dopamine neurons are potentiated. Alternatively, only a subset of synapses may be potentiated, but these synapses would have to be significantly potentiated to provide the twofold potentiation observed even if the EPSC is diluted by other unchanged synapses.

It will be of interest to determine whether the effects of all these drugs are blocked, similar to effects of cocaine, when coadministered with an NMDAR antagonist. If they are, this would suggest that the major addictive drug classes produce NMDAR-dependent LTP at VTA synapses (although the requirement for local NMDARs within the VTA would still not be proven). It is possible that a decrease in NMDARs could result after drug exposure. It will also be of interest to determine whether absolute levels of synaptic NMDARs increase after drugs; if so, this might provide a mechanism to make the system even more susceptible to further LTP with repeated drug exposure. In the case of cocaine, independent measures were made indicating that the overall response of dopamine neurons to bath-applied AMPA was increased, whereas that to NMDA was not. This supports the idea that the increase in AMPAR/NMDAR ratio results from a relative increase in AMPARs. Whether this is true for the other drugs remains to be tested.

These studies measured AMPAR/NMDAR ratios 24 h after drug treatment in vivo. How rapidly are the changes initiated at synapses? Intraperitoneal injection of amphetamine leads to brain absorption of the drug within 30 min; are AMPAR/NMDAR ratios increased as soon as amphetamine is present, or does the increase require longer-term exposure to the drug (which probably remains in the brain for over 8 h) or neuroadaptations on a longer time scale? Recent work indicates that an increase in the AMPAR/NMDAR ratio can be detected when brain slices are prepared 2 h after an intraperitoneal injection of amphetamine (Figure 3) (110). The increase in the ratio is close to twofold, approximately the level seen at 24 h. This finding suggests that alterations in excitatory synapses on dopamine cells occur very rapidly after treatment with at least one addictive drug. Furthermore, the data suggest that these rapid alterations are maintained at approximately the same levels for at least 24 h. The rapid effects of amphetamine most likely require an intact circuit because preliminary data suggest that exposure of brain slices to amphetamine for 2 h does not increase the AMPAR/NMDAR ratios (L. Faleiro & J. Kauer, unpublished observations).

STRESS AND THE VTA AND DRUGS

It is commonly accepted that relapse to drug or alcohol abuse is more likely in individuals under high levels of life stress, although the correlation is not well understood (111, 112). In rats, stress or glucocorticoid receptor activation trigger many of the changes observed after treatment with addictive drugs (113). For example, in repeatedly stressed rats, the increase in extracellular dopamine levels in the nucleus accumbens is indistinguishable from that observed in rats treated repeatedly with cocaine (114). Stress increases dopamine release in striatum (115). Foot shock or food deprivation stressors are as effective as addictive drugs at

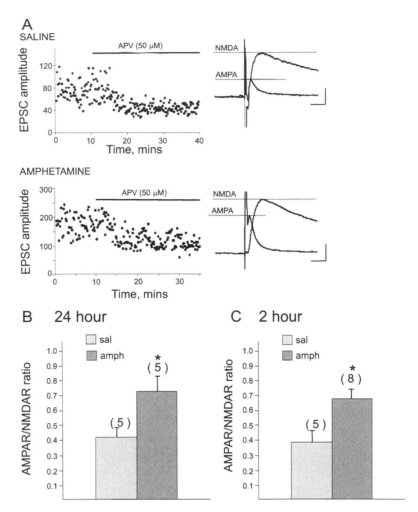

Figure 3 AMPAR/NMDAR ratios increase after in vivo treatment with amphetamine.
(*A*) Animals were injected with either saline or 2.5 mg/kg amphetamine 24 h before
brain slices were prepared. Dual-component EPSCs were evoked at +40 mV. The
NMDAR antagonist d-APV (50 μM) was then added to isolate the AMPAR component. Subtraction of the AMPAR component from the dual-component EPSC provided the NMDAR component. (*Right panels*) NMDAR and AMPAR components. *Top*
EPSC calibration: 25 pA, 10 ms; *Bottom* EPSC calibration: 50 pA, 10 ms. *Top panel*,
data from a saline-treated rat. *Bottom panel*, data from an amphetamine-treated rat.
(*B*) Averaged AMPAR/NMDAR ratios from slices prepared 24 h (*left*) and 2 h (*right*)
after amphetamine injection. Ratios were calculated from the peak currents. Open bar,
saline-treated rats; filled bar, amphetamine-treated rats. Experiments carried out in the
presence of 100 μM picrotoxin. Adapted from Faleiro & Kauer (110).

eliciting reinstatement (116), and these effects appear to be context-dependent. Adrenalectomy blocks amphetamine sensitization and this effect is reversed, and sensitization is even potentiated when dexamethasone is given after adrenalectomy (117). Other intriguing bits of the puzzle are emerging. For example, it has long been known that brief stress can "cross-sensitize" with drugs of abuse (40, 118). Thus, for example, exposure to stress leads to an enhanced locomotor response to amphetamine.

Glucocorticoid receptors are present on midbrain dopamine neurons, and repeated stress elevates GluR1 levels in VTA over several days (105). A single stress exposure also increases the AMPAR/NMDAR ratio in animals 24 h later, and this effect is blocked by a glucocorticoid receptor antagonist (109). The mechanism by which stress elicits LTP at VTA synapses is not understood, nor is it known how rapidly stress triggers LTP. Glucocorticoids act at conventional steroid receptors to activate hormone-specific gene transcription. However, more rapid effects of steroids have also been described and may result from activation of membrane receptors (119); glucocorticoids and other steroids inhibit neuropeptide release, enhance reuptake of glutamate or glycine, inhibit voltage-gated Ca^{2+} channels, and decrease the release of Ca^{2+} from intracellular stores (119). None of these cellular actions easily explains an enhancement of AMPAR/NMDAR ratios; however, NMDAR currents have also been reported to increase rapidly in the presence of glucocorticoid receptor agonists (120), and this effect could promote VTA LTP. The increased AMPAR/NMDAR ratio caused by stress was blocked when stress followed administration of an NMDAR antagonist, suggesting that NMDAR activation is downstream from glucocorticoid action (109).

The potentiation of AMPAR/NMDAR ratios was as large after stress as that recorded after any of five addictive drugs. This response may provide a key to understanding a physiological role of potentiation of VTA synapses and indicates that excitatory synapses in the VTA may be the common site modified by stress and addictive drugs, relevant to both sensitization and relapse.

HOW DO DRUGS OF ABUSE PROMOTE LTP AT EXCITATORY SYNAPSES IN THE VTA?

Psychostimulants Increase VTA Glutamate by Blocking Glutamate Reuptake

Sensitization-inducing drugs promote dopamine release into somatodendritic fields of midbrain dopamine neurons (121–124). What is the link between administration of these compounds and the activation of NMDA receptors, which are required to initiate sensitization? Using microdialysis techniques in vivo, others have demonstrated glutamate efflux in the VTA following exposure to psychostimulants (20, 125, 126). Glutamate efflux following psychostimulant treatment in vivo is not of synaptic origin because it is not Ca^{2+}-dependent (79); this finding is in agreement with direct observations in VTA brain slices that amphetamine does not enhance excitatory synaptic transmission (127). It is not clear what mechanism mediates the increase in extracellular glutamate seen after psychostimulant, but glial glutamate transporters are a likely target. After a sensitizing regimen of amphetamine, basic fibroblast growth factor (bFGF) is up-regulated in astrocytes in the VTA (but not in nucleus accumbens), suggesting that astrocytes are influenced by psychostimulant administration (128). Furthermore, antibodies to bFGF delivered into the VTA blocked amphetamine-induced behavioral sensitization, indicating an obligatory involvement of this growth factor in sensitization, at least to amphetamine (129). bFGF is also up-regulated in response to neuronal injury, suggesting that repeated exposure to amphetamine elicits cellular responses appropriate to neuronal injury but perhaps maladaptive in the uninjured nervous system (130, 131). Because astrocytes are largely responsible for glutamate reuptake, astrocytic reuptake of glutamate may be impaired after stimulant administration, perhaps as a result of bFGF upregulation, and this impairment could then increase extracellular glutamate. In theory, the enhanced extracellular glutamate could promote LTP at synapses on VTA dopamine neurons.

Psychostimulants Block LTD at VTA Synapses

Another possible link between local dopamine release and increased NMDA receptor-dependent plasticity is that long-term depression (LTD) is blocked by drugs of abuse, making it easier to induce LTP. In hippocampus and other brain regions expressing LTP, synaptic weight can be decreased by an opposing process, LTD (132–134). LTD is considered to be a normal brake mechanism preventing saturating LTP. Excitatory synapses on dopamine neurons of the VTA exhibit LTD as well as LTP (Figure 4A) (135–137).

If LTD is related to the development of sensitization or to the increased AMPAR/NMDAR ratios observed in the VTA, one might expect drugs of abuse to block LTD. In fact, the one drug examined to date, amphetamine, effectively prevents LTD at VTA synapses (Figure 4B). The block of LTD is mediated by dopamine acting on D2 receptors on dopamine neurons. LTD can also be blocked by inhibitors of protein kinase A or by chelation of intracellular Ca^{2+} (137–139). This form of LTD is not NMDAR-dependent but appears to result from Ca^{2+} entry via voltage-gated Ca^{2+} channels (135, 139). The lack of LTD to act as a brake on potentiation coupled with the documented increase in extracellular glutamate are hypothesized to trigger unusually effective LTP of excitatory synapses on VTA dopamine neurons after amphetamine or cocaine treatment.

Nicotine Promotes LTP at Excitatory Synapses on VTA Dopamine Neurons

Nicotine is an interesting drug in that it elicits very mild behavioral effects in comparison with cocaine or heroin but can be intensely addictive. Nicotine binds to and

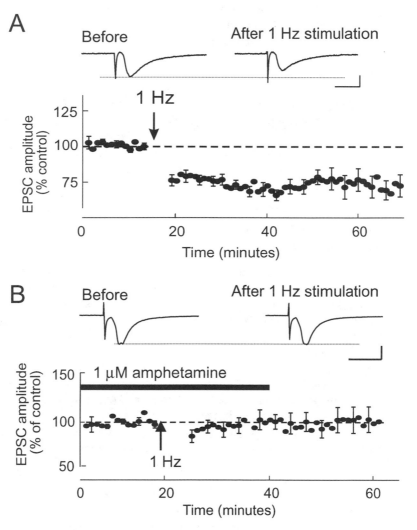

Figure 4 LTD is induced at excitatory synapses on VTA neurons, and amphetamine blocks LTD. (*A*) LTD is induced by pairing 1 Hz stimulation for 6 min with depolarization of the postsynaptic neuron to –40 mV (at the arrow; $n = 30$). Calibration: 50 pA, 10 ms. (*B*). Amphetamine (1 μM) blocks LTD following the same stimulus protocol ($n = 9$). Calibration: 100 pA, 10 ms. (*Insets*) average of five EPSCs before and 5 min after the pairing protocol. Experiments carried out in the presence of 100 μM picrotoxin. Adapted from Jones et al. (135).

activates nicotinic acetylcholine receptors (nAChRs). Like other addictive drugs, nicotine causes the release of dopamine in the nucleus accumbens (4, 140) and causes prolonged excitation of VTA dopamine neurons in vitro (141–143). One curious feature of these findings is that nAChRs desensitize within seconds after nicotine exposure, yet dopamine neurons appear to be activated for much longer periods after nicotine exposure. It was recently shown that nAChRs are present on presynaptic glutamatergic nerve terminals in the VTA. When these receptors are activated by nicotine, glutamate release is enhanced and LTP is induced at excitatory synapses on dopamine neurons (92). Interestingly, nicotine desensitization of nAChRs on GABAergic nerve terminals decreases GABAergic inhibition of dopamine neurons (144). Thus nicotine may induce LTP and increase dopamine neuron activity by simultaneously increasing glutamate release and diminishing GABAergic inhibition onto dopamine cells.

Alterations in GABAergic Synaptic Transmission May Also Promote LTP

Opiate receptors are present in the VTA only on GABAergic neurons. Thus the potent sensitizing and reinstatement-inducing effects of local opiates in the VTA (65) must be mediated by alterations in activity of GABAergic neurons. Opiates acutely hyperpolarize GABAergic neurons, preventing their firing, and effectively disinhibiting VTA dopamine neurons (145). These neurons also project to the nucleus accumbens. The precise cellular mechanisms by which morphine administration triggers an increase in the AMPAR/NMDAR ratio at VTA synapses within 24 h remain to be defined. One long-term modification of the GABA system by addictive drugs has been worked out, however. D1 receptors normally increase inhibitory postsynaptic potentials mediated by $GABA_B$ receptors, increasing local inhibition of dopamine neurons (146). After daily treatment with morphine for one week, however, D1 receptor activation decreases $GABA_B$-mediated inhibition owing to an alteration in adenosine levels (147). Furthermore, daily treatment with cocaine for 14 days produced the identical result–$GABA_B$-mediated inhibition is decreased by D1 receptor activation. The effect is most robust 1 week after the last drug treatment. Thus in animals treated with either of two distinct addictive drugs, GABAergic inhibition is decreased with local dopamine release. Similar to the block of LTD by amphetamine, decreased GABAergic inhibition also may play a key role in promoting LTP at excitatory synapses by removing a brake on dopamine cell excitation. This finding predicts that if VTA dopamine release were enhanced a week after withdrawal from either morphine or cocaine (e.g., by another drug exposure), the loss of $GABA_B$ inhibition would drive further excitation of dopamine cells. This mechanism could therefore contribute to an excitatory feedback loop responsible for greater VTA vulnerability to drugs after initial drug exposures.

Ethanol is thought to exert most of its actions via potentiation of $GABA_A$ channels, and the mesolimbic dopamine system is implicated in ethanol self-administration (148, 149). The mechanisms are not entirely clear, but during ethanol withdrawal there is a marked reduction in activity of VTA dopamine neurons (150). In vitro, ethanol excites dopamine neurons, increasing their spontaneous firing rate, but this effect does not appear to require synaptic transmission (151, 152). Little is known about ethanol's actions on excitatory synapses in the VTA, but 24 h after ethanol injection, AMPAR/NMDAR ratios are increased (109). A single in vivo ethanol exposure also alters GABAergic function in the VTA (153). Twenty-four hours after ethanol treatment, paired-pulse facilitation, ordinarily seen at inhibitory synapses onto VTA dopamine neurons, becomes paired-pulse depression. Furthermore, ethanol increases spontaneous GABA release. Together, these findings indicate that GABA synapses exhibit an increased probability of release after ethanol exposure. At first glance, it would seem that this would increase inhibition to the area generally, but in fact, the increased probability of release will reduce GABA release during trains of stimuli. Thus ethanol produces a striking alteration in the VTA that may contribute to LTP induction in vivo. The increase in paired-pulse depression lasts for more than 1 week, indicating a fairly long-lasting action. It will be important to determine whether these effects of ethanol on inhibition lead to the observed potentiation of glutamate synapses on the same neurons.

LTP IS A COMMON MECHANISM PROMOTED BY DRUGS OF ABUSE

It has been hypothesized that the increased synaptic and/or surface expression of GluR1 is sufficient to produce sensitization (154). The hypothesis includes the concept that elevated levels of GluR1 would lead to AMPARs containing only GluR1 on the surface of VTA dopamine neurons. Such homomeric GluR1 AMPARs are significantly Ca^{2+}-permeable and thus could play an important role in initiating signaling cascades that might lead to LTP induction. However, electrophysiological data do not support the idea that any significant fraction of synaptic AMPARs are GluR1 homomers on the basis of their current-voltage relationships (L. Faleiro & J. Kauer, unpublished observations). A more parsimonious hypothesis might be that any alteration of the local circuit in the VTA that promotes dopamine neuron excitability will lead to the increase in AMPAR/NMDAR ratios, to increased dopamine cell firing, and to the initiation of sensitization. Thus experimentally driven overexpression of GluR1 will promote dopamine cell excitability (101), but so would the mechanistically distinct inhibition of GABAergic VTA neurons by opioids because opioids disinhibit dopamine neurons (14). I propose that different drugs of abuse and stress promote sensitization via different mechanisms, all of which may converge to trigger LTP at excitatory synapses on VTA dopamine neurons (Figure 5). Thus overexpression of GluR1, high-frequency electrical stimulation of prefrontal cortex, and decreased GABA receptor function, etc., will also trigger LTP at dopamine neuron synapses thereby triggering sensitization.

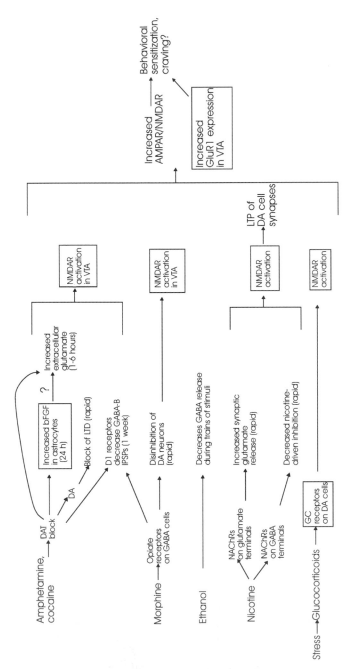

Figure 5 Diagram of possible steps linking different drugs and stress to LTP at VTA synapses. Boxed items indicate that this step has been shown to be necessary to link the drug/stress to an increased AMPAR/NMDAR ratio or sensitization.

WHAT IS THE FUNCTIONAL SIGNIFICANCE OF LTP AT VTA SYNAPSES?

Brain Regions Downstream of the VTA

LTP and LTD at excitatory synapses onto VTA cells look very similar to synaptic plasticity observed at other excitatory synapses throughout the brain and may use similar cellular mechanisms (36, 91, 92, 135–137). However, the VTA may behave functionally quite differently from cortical regions and is unlikely to be the site of cue-specific conditioning. Dopamine neurons appear to fire more or less in unison, as might be expected if dopamine release is to act as an error prediction signal (7, 156). In this model, the synaptic plasticity in the VTA may not be used for fine-tuning of distinct inputs onto each dopamine neuron but instead may be used to up- or down-regulate the excitability of the entire ensemble of dopaminergic neurons (135).

An increased global response to glutamate following stress or drug treatment will increase excitability of all VTA dopamine cells, releasing more dopamine for periods of hours. The firing of VTA neurons appears to be an error signal, increasing when environmental stimuli are better than expected, and decreasing when environmental stimuli are worse than expected. It is reasonable to think of dopamine release as a "teaching" signal, to strengthen associations between those environmental stimuli and approach behaviors that take advantage of the positive environment. Structures downstream of dopamine neurons (such as the nucleus accumbens, neostriatum, and prefrontal cortex) may use this teaching signal to modify synaptic weights and the overall output of circuits involved in motivation. When drugs of abuse or stress increase dopamine release in target areas, this is presumably interpreted by the system as environmental stimuli better than expected and may pathologically reinforce drug-taking and drug-seeking behavior. The abnormal potentiation of all excitatory synapses onto dopamine neurons (but not those on local GABA cells) by drugs will allow cortical and other glutamate inputs to enhance dopamine release potently. Beyond NMDAR involvement in generating LTP, NMDAR activation in the VTA also promotes a bursting firing pattern in dopamine neurons, resulting in significantly greater dopamine release compared with regular firing (157, 158). It is thus easy to see how LTP at synapses on dopamine neurons may be a first necessary step in a chain of responses that could contribute to addiction-related modification of downstream brain regions receiving dopamine release.

VTA

It is too early to assign a narrow significance to LTP in the VTA. The VTA is essential for the initiation of sensitization, and all addictive drugs so far tested cause synaptic potentiation similar to LTP within 24 h. It is not clear, however, if these early alterations in VTA function act solely as a transient gate to permit downstream changes or play a more significant long-term role in the persistent changes that underlie drug-craving. Several alterations in the VTA as a result of drug exposure last only a week or two, after which the system appears to reset (39, 40, 102, 103). Enhanced glutamatergic transmission at VTA synapses may also produce long-term changes in the VTA itself that contribute to relapse to drug use. This effect may be particularly strong after repeated drug exposures. One reason for the difficulty in speculating on a role of VTA plasticity is that so much more remains to be tested. Very few studies to date have examined the cellular physiology of VTA neurons in vitro at times longer than 24 h after drug treatment or after repeated drug treatment. The work of Bonci & Williams (showing that D1 receptor modulation of $GABA_B$ conductances is reversed after either chronic morphine or cocaine exposure) suggests that significant alterations in synaptic function occur a week or more after the last drug treatment (147). Similarly, a single cocaine injection increases AMPAR/NMDAR ratios for over a week (109), and a single exposure to ethanol increases the probability of GABA release for over a week (153). These late time points are important, since at these times withdrawal and tolerance mechanisms will be diminished, allowing drug-craving-related neuroadaptations to stand out more clearly. Unfortunately, technical limitations of the in vitro system have thus far limited nearly all studies to very young rats or mice; visualized patch-clamp recordings become quite difficult after 4 weeks of age. This is particularly relevant as sensitization is less robust and less long-lasting at these ages (160), indicating that examining the cellular physiology of older animals is likely to add information on the persistence of synaptic changes.

QUESTIONS

Many questions arise from recent work. Does the block of LTD by dopamine-releasing drugs in fact promote LTP? Does the local administration of each addictive drug, which is sufficient to trigger locomotor sensitization, produce LTP at dopamine cell synapses? The peripheral administration of MK-801 blocks the increased AMPAR/NMDAR ratios in response to cocaine. Are the NMDARs blocked in this experiment the same NMDARs at VTA synapses required for LTP, or are there other NMDARs in connected brain regions also necessary for LTP induction at VTA synapses? Are NMDARs required for increased AMPAR/NMDAR ratios after morphine, ethanol, and nicotine administration? Can any of these changes be induced in vitro, i.e., can increased AMPAR/NMDAR ratios be observed after exposure of a VTA slice preparation to any of the drugs? Local administration of either amphetamine or morphine causes sensitization. What is the time course over many hours of dopamine neuron firing to these and other addictive drugs? Is an increase in dopamine neuron firing and/or increased bursting activity required to induce LTP or sensitization? The similarities between stress and drugs seem an important line of inquiry at the cellular level. It is possible that synaptic alterations seen after drug use evolved as adaptive mechanisms to cope with stressful situations. The duration of stress-related effects is important to determine. Furthermore, if the same neural changes follow treatment with either drugs of abuse or stress, it seems crucial to determine why stress does not routinely elicit compulsive behavior as severe as that seen with drugs. Perhaps interfering with the pathways mediating the stress response might provide a useful approach to countering the effects of drugs of abuse. Although recent cellular physiology studies raise as many questions as they answer, we now have several avenues from which to approach rapid cellular effects of addictive drugs in the VTA. Cellular approaches both in vivo and in vitro in the near future are sure to contribute important information about how the brain and behavior are altered by drugs.

The *Annual Review of Physiology* is online at http://physiol.annualreviews.org

LITERATURE CITED

1. Hyman SE, Malenka RC. 2001. Addiction and the brain: the neurobiology of compulsion and its persistence. *Nat. Rev. Neurosci.* 2:695–703

2. Nestler EJ. 2001. Molecular basis of long-term plasticity underlying addiction. *Nat. Rev. Neurosci.* 2:119–28

3. Gerdeman GL, Partridge JG, Lupica CR, Lovinger DM. 2003. It could be habit forming: drugs of abuse and striatal synaptic plasticity. *Trends Neurosci.* 26:184–92

4. Di Chiara G, Imperato A. 1988. Drugs abused by humans preferentially increase synaptic dopamine concentrations in the mesolimbic system of freely moving rats. *Proc. Natl. Acad. Sci. USA* 85:5274–78

5. Koob GF. 1992. Drugs of abuse: anatomy, pharmacology, and function of reward pathways. *Trends Pharmacol. Sci.* 13:177–84

6. Nestler EJ. 1992. Molecular mechanisms of drug addiction. *J. Neurosci.* 12:2439–50

7. Schultz W. 1998. Predictive reward signal of dopamine neurons. *J. Neurophysiol.* 80:1–27

8. Wise RA. 1996. Neurobiology of addiction. *Curr. Opin. Neurobiol.* 6:243–51

9. Wise RA, Bozarth MA. 1987. A psychomotor stimulant theory of addiction. *Psychol. Rev.* 94:469–92

10. Schultz W. 1997. Dopamine neurons and their role in reward mechanisms. *Curr. Opin. Neurobiol.* 7:191–97

11. Grace AA, Onn S-P. 1989. Morphology and electrophysiological properties of immunocytochemically identified rat dopamine neurons recorded in vitro. *J. Neurosci.* 9:3463–81

12. Lacey MG, Mercuri NB, North RA. 1989. Two cell types in rat substantia nigra zona compacta distinguished by membrane properties and the actions of dopamine and opioids. *J. Neurosci.* 9:1233–41

13. Johnson SW, North RA. 1992. Two types of neurone in the rat ventral tegmental area and their synaptic inputs. *J. Physiol.* 450:455–68

14. Johnson SW, North RA. 1992. Opioids excite dopamine neurons by hyperpolarization of local interneurons. *J. Neurosci.* 12:483–88

15. Beart PM, McDonald D, Gundlach AL. 1979. Mesolimbic dopaminergic neurones and somatodendritic mechanisms. *Neurosci. Lett.* 15:165–70

16. Kalivas PW, Duffy P, Barrow J. 1989.

Regulation of the mesocorticolimbic dopamine system by glutamic acid receptor subtypes. *J. Pharmacol. Exp. Ther.* 251:378–87

17. Aghajanian GK, Bunney BS. 1977. Dopamine "autoreceptors:" pharmacological characterization by microiontophoretic single unit recording studies. *Naunyn Schmiedebergs Arch. Pharmacol.* 297:1–7

18. Garris PA, Ciolkowski EL, Pastore P, Wightman RM. 1994. Efflux of dopamine from the synaptic cleft in the nucleus accumbens of the rat brain. *J. Neurosci.* 14(10):6084–93

19. Gonon F. 1997. Prolonged and extrasynaptic excitatory action of dopamine mediated by D1 receptors in the rat striatum in vivo. *J. Neurosci.* 17:5972–78

20. Kalivas PW, Duffy P. 1995. D1 receptors modulate glutamate transmission in the ventral tegmental area. *J. Neurosci.* 15:5379–88

21. Beckstead RM. 1979. An autoradiographic examination of corticocortical and subcortical projections of the mediodorsal-projection (prefrontal) cortex in the rat. *J. Comp. Neurol.* 184:43–62

22. Christie MJ, Bridge S, James LB, Beart PM. 1985. Excitotoxic lesions suggest an aspartatergic projection from rat medial prefrontal cortex to ventral tegmental area. *Brain Res.* 333:169–72

23. Thierry AM, Chevalier G, Ferron A, Glowinski J. 1983. Diencephalic and mesencephalic efferents of the medial prefrontal cortex in the rat: electrophysiological evidence for the existence of branched axons. *Exp. Brain Res.* 50:273–82

24. Sesack SR, Pickel VM. 1992. Prefrontal cortical efferents in the rat synapse on unlabeled neuronal targets of catecholamine terminals in the nucleus accumbens septi and on dopamine neurons in the ventral tegmental area. *J. Comp. Neurol.* 320:145–60

25. Semba K, Fibiger HC. 1992. Afferent connections of the laterodorsal and the pedunculopontine tegmental nuclei in the rat: a retro- and antero-grade transport and immunohistochemical study. *J. Comp. Neurol.* 323:387–410

26. Garzon M, Vaughan RA, Uhl GR, Kuhar MJ, Pickel VM. 1999. Cholinergic axon terminals in the ventral tegmental area target a subpopulation of neurons expressing low levels of the dopamine transporter. *J. Comp. Neurol.* 410:197–210

27. Georges F, Aston-Jones G. 2001. Potent regulation of midbrain dopamine neurons by the bed nucleus of the stria terminalis. *J. Neurosci.* 21:RC160

28. Carr DB, Sesack SR. 2000. Projections from the rat prefrontal cortex to the ventral tegmental area: target specificity in the synaptic associations with mesoaccumbens and mesocortical neurons. *J. Neurosci.* 20:3864–73

29. Seutin V, Verbanck P, Massotte L, Dresse A. 1990. Evidence for the presence of NMDA receptors in the ventral tegmental area of the rat: an electrophysiological in vitro study. *Brain Res.* 514:147–50

30. Mercuri NB, Stratta F, Calabresi P, Bernardi G. 1992. A voltage-clamp analysis of NMDA-induced responses on dopaminergic neurons of the rat substantia nigra zona compacta and ventral tegmental area. *Brain Res.* 593:51–56

31. Mercuri NB, Stratta F, Calabresi P, Bonci A, Bernardi G. 1993. Activation of metabotropic glutamate receptors induces an inward current in rat dopamine mesencephalic neurons. *Neuroscience* 56:399–407

32. Wang T, French ED. 1993. L-glutamate excitation of A10 dopamine neurons is preferentially mediated by activation of NMDA receptors: extra- and intracellular electrophysiological studies in brain slices. *Brain Res.* 627:299–306

33. Wang T, French ED. 1993. Electrophysiological evidence for the existence of NMDA and non-NMDA receptors on

rat ventral tegmental dopamine neurons. *Synapse* 13:270–77

34. Overton PG, Clark D. 1997. Burst firing in midbrain dopaminergic neurons. *Brain Res. Rev.* 25:312–34

35. Mercuri NB, Grillner P, Bernardi G. 1996. *N*-methyl-D-aspartate receptors mediate a slow excitatory postsynaptic potential in the rat midbrain dopaminergic neurons. *Neuroscience* 14:785–91

36. Bonci A, Malenka RC. 1999. Properties and plasticity of excitatory synapses on dopaminergic and GABAergic cells in the ventral tegmental area. *J. Neurosci.* 19:3723–30

37. Jones S, Kauer JA. 1999. Amphetamine depresses excitatory synaptic transmission in the ventral tegmental area via serotonin receptors. *J. Neurosci.* 19:9780–87

38. White FJ. 1996. Synaptic regulation of mesocorticolimbic dopamine neurons. *Annu. Rev. Neurosci.* 16:405–36

39. White FJ, Kalivas PW. 1998. Neuroadaptations involved in amphetamine and cocaine addiction. *Drug Alcohol Depend.* 51:141–53

40. Kalivas PW, Stewart J. 1991. Dopamine transmission in the initiation and expression of drug- and stress-induced sensitization of motor activity. *Brain Res. Rev.* 16:223–44

41. Robinson TE, Berridge KC. 1993. The neural basis of drug craving: an incentive-sensitization theory of addiction. *Brain Res. Rev.* 18:247–91

42. Robinson TE, Berridge KC. 2000. The psychology and neurobiology of addiction: an incentive-sensitization view. *Addiction* 95(Suppl.)2:S91–117

43. Robinson TE, Berridge KC. 2001. Incentive-sensitization and addiction *Addiction* 96:103–14

44. Paulson PE, Robinson TE. 1991. Sensitization to systemic amphetamine produces an enhanced locomotor response to a subsequent intra-accumbens amphetamine challenge in rats. *Psychopharmacology* 104:140–41

45. Valadez A, Schenk S. 1994. Persistence of the ability of amphetamine preexposure to facilitate acquisition of cocaine self-administration. *Pharmacol. Biochem. Behav.* 47:203–5

46. Self DW, Nestler EJ. 1998. Relapse to drug-seeking: neural and molecular mechanisms. *Drug Alcohol Depend.* 51: 49–60

47. Badiani A, Anagnostaras SG, Robinson TE. 1995. The development of sensitization to the psychomotor stimulant effects of amphetamine is enhanced in a novel environment. *Psychopharmacology* 117:443–52

48. Badiani A, Oates MM, Day HEW, Watson SJ, Akil H, Robinson TE. 1998. Amphetamine-induced behavior, dopamine release, and c-fos mRNA expression: modulation by environmental novelty. *J. Neurosci.* 18:10579–93

49. Robinson TE, Browman KE, Crombag HS, Badiani A. 1998. Modulation of the induction or expression of psychostimulant sensitization by the circumstances surrounding drug administration. *Neurosci. Biobehav. Rev.* 22:347–54

50. Crombag HS, Badiani A, Robinson TE. 1996. Signalled versus unsignalled intravenous amphetamine: large differences in the acute psychomotor response and sensitization. *Brain Res.* 722:227–31

51. Kuzcenski R. 1983. Biochemical actions of amphetamine and other stimulants. In *Stimulants, Neurochemical, Behavioral, and Clinical Perspectives*, ed. I Creese, pp. 31–61. New York: Raven

52. Reith MEA, Meisler BE, Sershen H, Lajtha A. 1986. Structural requirements for cocaine congeners to interact with dopamine and serotonin uptake sites in mouse brain and to induce stereotyped behavior. *Biochem. Pharmacol.* 35:1123–29

53. Koob GF, Stinus L, LeMoal M. 1981. Hyperactivity and hypoactivity produced by lesions to the mesolimbic dopamine system. *Behav. Brain Res.* 3:341–59

54. Clarke PBS, Jakubovic A, Fibiger HC. 1988. Anatomical analysis of the involvement of mesolimbocortical dopamine in the locomotor stimulant actions of D-amphetamine and apomorphine. *Psychopharmacology* 96:511–20

55. Kelley AE, Throne LC. 1992. NMDA receptors mediate the behavioral effects of amphetamine infused into the nucleus accumbens. *Brain Res. Bull.* 29:247–54

56. Dougherty GG, Ellinwood EH. 1981. Chronic D-amphetamine in nucleus accumbens: lack of tolerance or reverse tolerance of locomotor activity. *Life Sci.* 28:2295–98

57. Kalivas PW, Weber B. 1988. Amphetamine injection into the ventral mesencephalon sensitizes rats to peripheral amphetamine and cocaine. *J. Pharmacol. Exp. Ther.* 245:1095–101

58. Vezina P. 1993. Amphetamine injected into the ventral tegmental area sensitizes the nucleus accumbens dopaminergic response to systemic amphetamine: an in vivo microdialysis study in the rat. *Brain Res.* 605:332–37

59. Perugini M, Vezina P. 1994. Amphetamine administered to the ventral tegmental area sensitizes rats to the locomotor effects of nucleus accumbens amphetamine. *J. Pharmacol. Exp. Ther.* 270: 690–96

60. Vezina P, Stewart J. 1990. Amphetamine administration to the ventral tegmental area but not to the nucleus accumbens sensitizes rats to systemic morphine: lack of conditioned effects. *Brain Res.* 516:99–106

61. Deleted in proof

62. Vezina P, Kalivas PW, Stewart J. 1987. Sensitization occurs to the locomotor effects of morphine and the specific mu opioid receptor agonist, DAGO, administered repeatedly to the VTA but not to the nucleus accumbens. *Brain Res.* 417:51–58

63. Kalivas PW, Duffy P. 1987. Sensitization to repeated morphine injection in the rat: possible involvement of A10 dopamine neurons. *J. Pharmacol. Exp. Ther.* 241:204–12

64. Lett BT. 1989. Repeated exposures intensify rather than diminish the rewarding effects of amphetamine, morphine, and cocaine. *Psychopharmacology* 98:357–62

65. Stewart J. 1983. Conditioned and unconditioned drug effects in relapse to opiate and stimulant drug self-administration. *Prog. Neuropsychopharmacol. Biol. Psychiatry* 7:591–97

66. Vorel SR, Liu X, Hayes RJ, Spector JA, Gardner EL. 2001. Relapse to cocaine-seeking after hippocampal theta burst stimulation. *Science* 292:1175–78

67. De Vries TJ, Schoffelmeer AN, Binnekade R, Mulder AH, Vanderschuren LJ. 1998. Drug-induced reinstatement of heroin- and cocaine-seeking behaviour following long-term extinction is associated with expression of behavioural sensitization. *Eur. J. Neurosci.* 10:3565–71

68. Nestler EJ. 2001. Molecular basis of long-term plasticity underlying addiction. *Nat. Rev. Neurosci.* 2:119–28

69. Nestler EJ. 2002. Common molecular and cellular substrates of addiction and memory. *Neurobiol. Learn. Mem.* 78:637–47

70. Wolf ME. 1998. The role of excitatory amino acids in behavioral sensitization to psychomotor stimulants. *Prog. Neurobiol.* 54:1–42

71. Karler R, Calder LD, Chaudhry IA, Turkanis SA. 1989. Blockade of "reverse tolerance" to cocaine and amphetamine by MK-801. *Life Sci.* 45:599–606

72. Stewart J, Druhan JP. 1993. Development of both conditioning and sensitization to the behavioral activating effects of amphetamine is blocked by the non-competitive NMDA receptor antagonist, MK-801. *Psychopharmacology.* 110:125–32

73. Wolf ME, Jeziorski M. 1993. Coadministration of MK-801 with amphetamine, cocaine or morphine prevents rather than transiently masks the development

of behavioral sensitization. *Brain Res.* 613:291–94

74. Jeziorski M, White FJ, Wolf ME. 1994. MK-801 prevents the development of behavioral sensitization during repeated morphine administration. *Synapse* 16:137–47

75. Ohmori T, Abekawa T, Muraki A, Koyama T. 1994. Competitive and noncompetitive NMDA antagonists block sensitization to methamphetamine. *Pharmacol. Biochem. Behav.* 48:587–91

76. Li Y, Wolf ME. 1999. Can the "state-dependency" hypothesis explain prevention of amphetamine sensitization in rats by NMDA receptor antagonists? *Psychopharmacology* 141:351–61

77. Kalivas PW, Alesdatter JE. 1993. Involvement of NMDA receptor stimulation in the ventral tegmental area and amygdala in behavioral sensitization to cocaine. *J. Pharmacol. Exp. Ther.* 267:486–95

78. Wolf ME, Dahlin SL, Hu X-T, Xue C-J, White K. 1995. Effects of lesions of prefrontal cortex, amygdala, or fornix on behavioral sensitization to amphetamine: comparison with NMDA antagonists. *Neuroscience* 69:417–39

79. Wolf ME, Xue CJ. 2000. Amphetamine increases glutamate efflux in the rat ventral tegmental area by a mechanism involving glutamate transporters and reactive oxygen species. *J. Neurochem.* 75:1634–44

80. Schenk S, Snow S. 1994. Sensitization to cocaine's motor activating properties produced by electrical kindling of the medial prefrontal cortex but not of the hippocampus. *Brain Res.* 659:17–22

81. Karler R, Calder LD, Turkanis SA. 1991. DNQX blockade of amphetamine behavioral sensitization. *Brain Res.* 552:295–300

82. Tong Z-Y, Overton PG, Clark D. 1995. Chronic administration of (+)-amphetamine alters the reactivity of midbrain dopaminergic neurons to prefrontal cortex stimulation in the rat. *Brain Res.* 674:63–74

83. Collingridge GL, Kehl SJ, McLennan H. 1983. Excitatory amino acids in synaptic transmission in the Schaffer collateral-commissural pathway of the rat hippocampus. *J. Physiol.* 334:33–46

84. Artola A, Singer W. 1987. Long-term potentiation and NMDA receptors in rat visual cortex. *Nature* 330:649–52

85. Dudek SM, Bear MF. 1992. Homosynaptic long-term depression in area CA1 of hippocampus and effects of N-methyl-D-aspartate receptor blockade. *Proc. Natl. Acad. Sci. USA* 89:4363–67

86. Mulkey RM, Malenka RC. 1992. Mechanisms underlying induction of homosynaptic long-term depression in area CA1 of the hippocampus. *Neuron* 9:967–75

87. Crair MC, Malenka RC. 1995. A critical period for long-term potentiation at thalamocortical synapses. *Nature* 375:325–28

88. Bliss TVP, Lomo T. 1973. Long-lasting potentiation of synaptic transmission in the dentate area of the anaesthetized rabbit following stimulation of the perforant path. *J. Physiol.* 232:331–56

89. Mayer ML, Westbrook GL, Guthrie PB. 1984. Voltage-dependent block by Mg^{2+} of NMDA responses in spinal cord neurones. *Nature* 309:261–63

90. Nowak L, Bregestovski P, Ascher P, Herbert A, Prochiantz A. 1984. Magnesium gates glutamate-activated channels in mouse central neurones. *Nature* 307:462–65

91. Overton PG, Richards CD, Berry MS, Berry DC. 1999. Long-term potentiation at excitatory amino acid synapses on midbrain dopamine neurons. *NeuroReport* 10:221–26

92. Mansvelder HD, McGehee DS. 2000. Long-term potentiation of excitatory inputs to brain reward areas by nicotine. *Neuron* 27:349–57

93. Karler R, Chaudhry IA, Calder LD, Turkanis SA. 1990. Amphetamine

behavioral sensitization and the excitatory amino acids. *Brain Res.* 537:76–82

94. Kauer JA, Malenka RC, Nicoll RA. 1988. A persistent postsynaptic modification mediates long-term potentiation in the hippocampus. *Neuron* 1:911–17

95. Isaac JTR, Nicoll RA, Malenka RC. 1995. Evidence for silent synapses: implications for the expression of LTP. *Neuron* 15:427–34

96. Liao D, Hessler NA, Malinow R. 1995. Activation of postsynaptically silent synapses during pairing-induced LTP in CA1 region of hippocampal slice. *Nature* 375:400–4

97. Oliet SHR, Malenka RC, Nicoll RA. 1996. Bidirectional control of quantal size by synaptic activity in the hippocampus. *Science* 271:1294–97

98. Zamanillo D, Sprengel R, Hvalby O, Jensen V, Burnashev N, et al. 1999. Importance of AMPA receptors for hippocampal synaptic plasticity but not for spatial learning. *Science* 284:1805–11

99. Shi SH, Hayashi Y, Petralia RS, Zaman SH, Wenthold RJ, et al. 1999. Rapid spine delivery and redistribution of AMPA receptors after synaptic NMDA receptor activation. *Science* 284:1811–16

100. Malinow R, Malenka RC. 2002. AMPA receptor trafficking and synaptic plasticity. *Annu. Rev. Neurosci.* 25:103–26

101. Carlezon WA Jr, Boundy VA, Haile CN, Lane SB, Kalb RG, et al. 1997. Sensitization to morphine induced by viral-mediated gene transfer. *Science* 277:812–14

102. White FJ, Hu X-T, Zhang X-F, Wolf ME. 1995. Repeated administration of cocaine or amphetamine alters neuronal responses to glutamate in the mesoaccumbens dopamine system. *J. Pharmacol. Exp. Ther.* 273:445–54

103. Zhang X-F, Hu X-T, White FJ, Wolf ME. 1997. Increased responsiveness of ventral tegmental area dopamine neurons to glutamate after repeated administration of cocaine or amphetamine is transient and

selectively involves AMPA receptors. *J. Pharmacol. Exp. Ther.* 281:699–706

104. Churchill L, Swanson CJ, Urbina M, Kalivas PW. 1999. Repeated cocaine alters glutamate receptor subunit levels in the nucleus accumbens and ventral tegmental area of rats that develop behavioral sensitization. *J. Neurochem.* 72:2397–403

105. Fitzgerald LW, Ortiz J, Hamedani AG, Nestler EJ. 1996. Drugs of abuse and stress increase the expression of GluR1 and NMDAR1 glutamate receptor subunits in the rat ventral tegmental area: common adaptations among cross-sensitizing agents. *J. Neurosci.* 16:274–82

106. Wells DG, Dong X, Quinlan EM, Huang YS, Bear MF, et al. 2001. A role for the cytoplasmic polyadenylation element in NMDA receptor-regulated mRNA translation in neurons. *J. Neurosci.* 21:9541–48

107. Ungless MA, Whistler JL, Malenka RC, Bonci A. 2001. Single cocaine exposure in vivo induces long-term potentiation in dopamine neurons. *Nature* 411:583–87

108. Cameron DL, Williams JT. 1993. Dopamine D1 receptors facilitate transmitter release. *Nature* 366:344–47

109. Saal D, Dong Y, Bonci A, Malenka RC. 2003. Drugs of abuse trigger a common synaptic adaptation in dopamine neurons. *Neuron* 37:577–82

110. Faleiro L, Kauer JA. 2003. An increase in AMPAR/NMDAR ratios in the ventral tegmental area (VTA) is detectable within two hours of amphetamine injection. *Ann. NY Acad. Sci.* In press

111. Brown SA, Vik PW, Patterson TL, Grant I, Schuckit MA. 1995. Stress, vulnerability and adult alcohol relapse. *J. Stud. Alcohol* 56:538–45

112. McFall ME, Mackay PW, Donovan DM. 1992. Combat-related posttraumatic stress disorder and severity of substance abuse in Vietnam veterans. *J. Stud. Alcohol* 53:357–63

113. Marinelli M, Piazza PV. 2002. Interaction between glucocorticoid hormones, stress

and psychostimulant drugs. *Eur. J. Neurosci.* 16:387–94

114. Kalivas PW, Duffy P. 1990. Effect of acute and daily cocaine treatment on extracellular dopamine in the nucleus accumbens. *Synapse* 5:48–58

115. Curzon G, Hutson PH, Knott PJ. 1979. Voltammetry in vivo: effect of stressful manipulation and drugs on the caudate nucleus of the rat. *Br. J. Pharmacol.* 66:127P–28

116. Shaham Y, Erb S, Stewart J. 2000. Stress-induced relapse to heroin and cocaine seeking in rats: a review. *Brain Res. Rev.* 33:13–33

117. Rivet JM, Stinus L, LeMoal M, Mormede P. 1989. Behavioral sensitization to amphetamine is dependent on corticosteroid receptor activation. *Brain Res.* 498:149–53

118. Piazza PV, Le Moal M. 1998. The role of stress in drug self-administration. *Trends Pharmacol. Sci.* 19:67–74

119. Chen YZ, Qiu J. 1999. Pleiotropic signaling pathways in rapid, nongenomic action of glucocorticoid. *Mol. Cell Biol. Res. Commun.* 2:145–49

120. Takahashi T, Kimoto T, Tanabe N, Hattori TA, Yasumatsu N, Kawato S. 2002. Corticosterone acutely prolonged *N*-methyl-D-aspartate receptor-mediated Ca^{2+} elevation in cultured rat hippocampal neurons. *J. Neurochem.* 83:1441–51

121. Mercuri NB, Calabresi P, Bernardi G. 1989. The mechanism of amphetamine-induced inhibition of rat substantia nigra compacta neurones investigated with intracellular recording in vitro. *Br. J. Pharmacol.* 98:127–34

122. Brodie MS, Dunwiddie TV. 1990. Cocaine effects in the ventral tegmental area: evidence for an indirect dopaminergic mechanism of action. *Naunyn Schmiedebergs Arch. Pharmacol.* 342:660–65

123. Lacey MG, Mercuri NB, North RA. 1990. Actions of cocaine on rat dopaminergic neurones in vitro. *Br. J. Pharmacol.* 99:731–35

124. Seutin V, Verbanck P, Massotte L, Dresse A. 1991. Acute amphetamine-induced subsensitivity of A10 dopamine autoreceptors in vitro. *Brain Res.* 558:141–44

125. Xue C-J, Ng JP, Li Y, Wolf ME. 1996. Acute and repeated systemic amphetamine administration: effects on extracellular glutamate, aspartate, and serine levels in rat ventral tegmental area and nucleus accumbens. *J. Neurochem.* 67:352–63

126. Wolf ME, Xue C-J. 1998. Amphetamine and D1 dopamine receptor agonists produce biphasic effects on glutamate efflux in rat ventral tegmental area: modification by repeated amphetamine administration. *J. Neurochem.* 70:198–209

127. Deleted in proof

128. Flores C, Rodaros D, Stewart J. 1998. Long-lasting induction of astrocytic basic fibroblast growth factor by repeated injections of amphetamine: blockade by concurrent treatment with a glutamate antagonist. *J. Neurosci.* 18:9547–55

129. Flores C, Samaha AN, Stewart J. 2000. Requirement of endogenous basic fibroblast growth factor for sensitization to amphetamine. *J. Neurosci.* 20:RC55

130. Chadi G, Cao Y, Pettersson RF, Fuxe K. 1994. Temporal and spatial increase of astroglial basic fibroblast growth factor synthesis after 6-hydroxydopamine-induced degeneration of the nigrostriatal dopamine neurons. *Neuroscience* 61:891–910

131. Kawamata T, Dietrich W, Schallert T, Gotts J, Cocke R, et al. 1997. Intracisternal basic fibroblast growth factor enhances functional recovery and upregulates the expression of a molecular marker of neuronal sprouting following focal cerebral infarction. *Proc. Natl. Acad. Sci. USA* 94:8179–84

132. Malenka RC, Nicoll RA. 1993. NMDA-receptor-dependent synaptic plasticity: multiple forms and mechanisms. *Trends Neurosci.* 16:521–27

133. Heynen AJ, Abraham WC, Bear MF.

1996. Bidirectional modification of CA1 synapses in the adult hippocampus in vivo. *Nature* 381:163–66

134. Nicoll R, Oliet S, Malenka R. 1998. NMDA receptor-dependent and metabotropic glutamate receptor-dependent forms of long-term depression coexist in CA1 hippocampal pyramidal cells. *Neurobiol. Learn. Mem.* 70:62–67

135. Jones S, Kornblum JL, Kauer JA. 2000. Amphetamine blocks long-term synaptic depression in the ventral tegmental area. *J. Neurosci.* 20:5575–80

136. Thomas MT, Malenka RC, Bonci A. 2000. Modulation of long-term depression by dopamine in the mesolimbic system. *J. Neurosci.* 20:5581–86

137. Gutlerner JL, Chapin-Penick E, Snyder EM, Kauer JA. 2002. Novel protein kinase A-dependent long-term depression of excitatory synapses. *Neuron* 36:921–31

138. Deleted in proof

139. Thomas MT, Malenka RC, Bonci A. 2000. Modulation of long-term depression by dopamine in the mesolimbic system. *J. Neurosci.* 20:5581–86

140. Koob GF. 1992. Drugs of abuse: anatomy, pharmacology, and function of reward pathways. *Trends Pharmacol. Sci.* 13:177–84

141. Calabresi P, Lacey MG, North RA. 1989. Nicotinic excitation of rat ventral tegmental neurones in vitro studied by intracellular recording. *Br. J. Pharmacol.* 98:135–40

142. Pidoplichko VI, DeBiasi M, Williams JT, Dani JA. 1997. Nicotine activates and desensitizes midbrain dopamine neurons. *Nature* 390:401–4

143. Picciotto MR, Zoli M, Rimondini R, Lena C, Marubio LM, et al. 1998. Acetylcholine receptors containing the β2 subunit are involved in the reinforcing properties of nicotine. *Nature* 391:173–77

144. Mansvelder HD, McGehee DS. 2002. Cellular and synaptic mechanisms of nicotine addiction. *J. Neurobiol.* 53:606–17

145. Johnson SW, North RA. 1992. Opioids excite dopamine neurons by hyperpolarization of local interneurons. *J. Neurosci.* 12:483–88

146. Deleted in proof

147. Bonci A, Williams JT. 1996. A common mechanism mediates long-term changes in synaptic transmission after chronic cocaine and morphine. *Neuron* 16:631–39

148. Koob GF, Roberts AJ, Schulteis G, Parsons LH, Heyser CJ, et al. 1998. Neurocircuitry targets in ethanol reward and dependence. *Alcohol Clin. Exp. Res.* 22:3–9

149. McBride WJ, Murphy JM, Ikemoto S. 1999. Localization of brain reinforcement mechanism: intracranial self-administration and intracranial place-conditioning studies. *Behav. Brain Res.* 101:129–52

150. Bailey CP, Manley SJ, Watson WP, Wonnacott S, Molleman A, Little HJ. 1998. Chronic ethanol administration alters activity in ventral tegmental area neurons after cessation of withdrawal hyperexcitability. *Brain Res.* 803:144–52

151. Brodie MA, Shefner SA, Dunwiddie TV. 1990. Ethanol increases the firing rate of dopamine neurons of the rat ventral tegmental area in vitro. *Brain Res.* 508:65–69

152. Brodie MS, Pesold C, Appel SB. 1999. Ethanol directly excites dopaminergic ventral tegmental area reward neurons. *Alcohol Clin. Exp. Res.* 23:1848–52

153. Melis M, Camarini R, Ungless MA, Bonci A. 2002. Long-lasting potentiation of GABAergic synapses in dopamine neurons after a single in vivo ethanol exposure. *J. Neurosci.* 22:2074–82

154. Carlezon WA Jr, Nestler EJ. 2002. Elevated levels of GluR1 in the midbrain: a trigger for sensitization to drugs of abuse? *Trends Neurosci.* 25:610–15

155. Deleted in proof

156. Schultz W, Apicella P, Ljungberg T. 1993. Responses of monkey dopamine neurons to reward and conditioned stimuli during

successive steps of learning a delayed response task. *J. Neurosci.* 13:900–13

157. Gonon FG. 1988. Nonlinear relationship between impulse flow and dopamine released by rat midbrain dopaminergic neurons as studied by in vivo electrochemistry. *Neuroscience* 24:19–28

158. Cooper DC. 2002. The significance of action potential bursting in the brain reward circuit. *Neurochem. Int.* 41:333–40

159. Deleted in proof

160. Kolta MG, Scalzo FM, Ali SF, Holson RR. 1990. Ontogeny of the enhanced behavioral response to amphetamine in amphetamine-pretreated rats. *Psychopharmacology* 100:377–82

Annu. Rev. Physiol. 2004. 66:477–519
doi: 10.1146/annurev.physiol.66.032102.113328
First published online as a Review in Advance on October 20, 2003

LOCALIZATION OF VOLTAGE-GATED ION CHANNELS IN MAMMALIAN BRAIN

James S. Trimmer
Department of Pharmacology, School of Medicine, University of California, Davis, California 95616-8635; email: jtrimmer@ucdavis.edu

Kenneth J. Rhodes
Neurological Disorders, Johnson & Johnson Pharmaceutical Research and Development L.L.C., Raritan, New Jersey 08869-1425; email: KRhodes2@prdus.jnj.com

Key Words   excitability, neuron, hippocampus, cortex, cerebellum

■ **Abstract**   The intrinsic electrical properties of neurons are shaped in large part by the action of voltage-gated ion channels. Molecular cloning studies have revealed a large family of ion channel genes, many of which are expressed in mammalian brain. Much recent effort has focused on determining the contribution of the protein products of these genes to neuronal function. This requires knowledge of the abundance and distribution of the constituent subunits of the channels in specific mammalian central neurons. Here we review progress made in recent studies aimed at localizing specific ion channel subunits using in situ hybridization and immunohistochemistry. We then discuss the implications of these results in terms of neuronal physiology and neuronal mechanisms underlying the observed patterns of expression.

INTRODUCTION

Voltage-gated ion channels underlie the intrinsic electrical properties of neurons. These integral membrane proteins form highly selective pores for the passage of specific ions across the plasma membrane (1). The pore itself can exist in either an open or closed state. Transitions between these states, referred to as channel gating, are based on conformational changes in the channel protein governed by changes in the membrane electrical potential (2). Because these proteins permit selective ion flux in response to voltage-dependent gating, they are referred to as voltage-dependent ion channels. In the open state, the channel pore allows the selective flux of ions down their electrochemical gradients at very high rates (up to 10^6 ions per second) (1). Voltage-dependent channels selective for sodium, calcium, and potassium ions are the major determinants of the active electrical properties of neurons (3). These channels are composed of polytopic, transmembrane, pore-forming, and voltage-sensing α or α_1 subunits (4), which are encoded by members of a large and diverse, yet homologous, gene family. In many but not all cases transmembrane and/or cytoplasmic auxiliary subunits, which in themselves do not form functional channels, can exert profound effects on channel function upon association with their corresponding α or α_1 subunits (5).

Over the past 50 years, ion channel biophysicists have used electrophysiological methods as the primary means to study the distribution of voltage- and ligand-gated ion channels in mammalian brain. There have been two primary goals of these studies. One goal has been to explore the contribution of specific ion channel classes, e.g., those specific for sodium, calcium, and potassium ions, to the complex, integrated firing properties of central neurons. In recent years, this effort has used voltage- and/or patch-clamp recordings to study the contribution of specific channels to membrane excitability at the level of a single neuron, dendrite, and even axon terminal. Although the resulting discoveries have provided a wealth of information about the biophysical signature of native ion channels and have led to major advances in our understanding of the complex interplay of channel subtypes underlying neuronal excitability, these approaches provide only limited information about the molecular composition of the channels underlying the activity of the recorded, native currents.

A more recent goal of ion channel biophysics has been to understand the molecular composition of channels underlying each distinct native current. This effort began with the biochemical purification of acceptors for neurotoxins and other pharmacological agents known to affect neuronal excitability via direct modulation of ion channel function. These biochemical approaches ultimately led to the cloning and expression of cDNAs encoding the principal, pore-forming α and α_1 subunits of voltage-gated ion channels and unearthed a large family of ion channel genes that together control the intrinsic electrical excitability of neurons. Heterologous expression of these cloned channel subunits led to insights into how the primary structure of the encoded polypeptides relates to function and also allowed for direct comparison of the properties of expressed channel subunits with their native counterparts. Taken together with data from in situ localization of transcripts encoding individual subunits, these studies provided the first opportunity to correlate the expression of molecularly defined subunits and measurable native currents. More recently, native channel complexes have been purified and their major component subunits and interacting proteins identified using biochemical approaches based upon the use of subunit-specific immunological reagents. In parallel, studies employing patch-clamp recordings, coupled to single cell analyses of mRNA expression, have been used to correlate expression of individual channel subunits with specific currents (6). However, the multitude of genes encoding ion channel subunits that operate within similar ranges of membrane voltage and have similar biophysical properties. The relative paucity of selective pharmacological tools available to dissect the function of specific channel subtypes, and the acceptance that all voltage-gated ion channels are heteromeric, multi-subunit macromolecular complexes, underscore the difficulties inherent in attempts to link molecular biology, protein biochemistry, and biophysical analyses to define native channels.

It is clear from studies of mammalian neurons and from extensive computer analyses of model cells that the intrinsic electrical properties of neurons depend not only on the overall expression level of specific channels but also on their discrete placement in the cell membrane. Given the tremendous compartmentalization of the neuronal membrane and the discrete localization of specific membrane events critical in information reception, i.e., signal transduction and propagation, the nonuniform placement of ion channels dramatically impacts neuronal function. Recognizing this, several laboratories have used biochemical, immunological, and neuroanatomical approaches to map the patterns of expression, association, and subcellular localization of specific ion channel subunits and their associated proteins in neurons. This review brings together information about the cellular and subcellular localization of individual ion channel subunits in an attempt to summarize the current state of knowledge about the distribution and molecular composition of native channel complexes in adult mammalian brain.

Molecular cloning of ion channel genes has resulted in the design of isoform-specific immunological reagents that allow for the high-resolution immunohistochemical analyses of the distribution of specific channel subunits. Used in combination with electrophysiological and modern neuroanatomical approaches, these studies provide a wealth of information about the distribution, subcellular localization, and subunit composition of ion channel complexes.

Here we review recent studies employing these approaches to define at high resolution where ion channels are expressed in mammalian brain. Where possible and appropriate, we attempt to correlate the results of anatomical and biochemical analyses with published work on channel pharmacology and biophysics, recognizing that this effort is handicapped by the technical limitations inherent in the techniques used, as described below. A detailed summary of ion channel subunit association and localization across all brain regions is well beyond the scope of even this detailed review. We therefore focus on three brain regions: the hippocampal formation, neocortex, and cerebellar cortex. We chose these regions for several reasons. First, for many years these brain regions have been the focus of intense pharmacological and electrophysiological study, owing, in large part, to the accessibility of the principal cell types to microelectrode recording, the development of in vitro slice preparations that permit visualization of individual cells and their processes at high magnification, and the critical technical advantages conferred by the exquisite cytoarchitectural organization of these structures, which greatly facilitates activation of afferent and efferent pathways while simultaneously recording pre- and/or postsynaptic responses. We explore first the diversity and classification of channel subunits and then summarize the distribution and localization of each class of voltage-gated ion channel subtype revealed using immunological and neuroanatomical approaches. We next attempt to relate the observed expression patterns to electrophysiological studies of the underlying currents. Finally, we describe mechanisms that contribute to the generation and maintenance of the observed patterns of channel localization in mammalian brain, and how these may be modulated to contribute to the plasticity of neuronal signaling.

MOLECULAR DIVERSITY AND CLASSIFICATION OF ION CHANNELS

Nomenclature of Ion Channels and Ion Channel Genes

The complexity of the ion channel gene family, combined with a dizzying array of names given to the genes and/or cDNAs as they were isolated, has resulted in a degree of confusion in the molecular nomenclature of voltage-dependent ion channels. As a response to the growing number of homegrown appellations for cloned voltage-gated potassium channels, Chandy (7) proposed a systematic nomenclature system that was widely accepted. This system was later adopted for calcium (8) and sodium (9) channels by consortia of involved investigators. Together, this has resulted in a clear, systematic, and unambiguous nomenclature for nearly all voltage-gated ion channels. The nomenclature system is based on the chemical symbol for the principal physiologically permeant ion (K for potassium, Ca for calcium, and Na for sodium), followed by the abbreviation of the ligand, which, in the case of this review, is always voltage (v). Thus we focus this review on voltage-dependent potassium (Kv), calcium (Cav), and sodium (Nav) channels. The remainder of the nomenclature relates to the gene families within these ion channel groups. For example, the prototypical Kv channels have been divided into four families (Kv1–Kv4) based on sequence homology and on their sequence similarity to single gene orthologues in *Drosophila* (name given in italics): Kv1 (*Shaker*), Kv2 (*Shab*), Kv3 (*Shaw*), and Kv4 (*Shal*). Cav channels have been divided into three groups based on both sequence homology and function: Cav1 (L-type high-voltage-activated channels), Cav2 (P/Q-, N-, and R-type low-voltage-activated channels), and Cav3 (T-type low-voltage-activated channels). Nav channels are primarily in the Nav1 family, which contains all of the classical voltage-gated sodium channels. The outlying second family, both in terms of sequence homology and function (termed Nax), contains channels whose sequences share some similarity to Nav1 sequences but have never been functionally expressed to verify their physiological function as Nav channels.

A parallel nomenclature for ion channel genes has been developed by mammalian geneticists, who assigned official UCL/HGNC/HUGO Human Gene Nomenclature symbols in conjunction with the human genome project. The Kv α subunit genes are named KCN*, with the four gene families assigned the letters A–D (i.e., Kv1–Kv4 = KCNA–KCND). The specific gene number is derived from the Kv nomenclature, such that Kv1.1 = KCNA1, Kv1.4 = KCNA4, Kv2.1 = KCNB1, Kv4.2 = KCND2, etc. Similarly, Nav channel genes are for the most part simply assigned the name SCN, and the numerical designation used in the Nav1 nomenclature followed by an A for α subunit (e.g., Nav1.1 = SCN1A, Nav1.2 = SCN2A, Nav1.3 = SCN3A, etc.). Unfortunately, some inconsistencies have crept into the Nav gene naming system, for example, Nav1.6 = SCN8A. The nomenclature for Cav genes is even more complex because the names are based on the classification system that was in place before the more systematic nomenclature was adopted, such that the gene for Cav1.1 (nee α_{1S}) is CACNA1S, Cav3.1(nee α_{1G}), is CACNA1G, etc. Thus, some knowledge of the physiological and pharmacological characteristics of the channel, as well as its history, is needed to accurately sort through the HUGO nomenclature for Cav genes. Throughout this review, we predominantly use the nomenclature system developed by ion channel researchers (Kv, Cav, and Nav), and refer to the HUGO system only when referring to the gene itself.

Voltage-Gated Ion Channel Genes

The transmembrane, pore-forming, and voltage-sensing α (for Nav and Kv) or α_1 (for Cav) subunits of mammalian voltage-gated ion channels are encoded by a large and diverse family of homologous genes. These polypeptides exist in two general leitmotifs: individual small-format subunits with six transmembrane segments (termed S1–S6), which assemble post-translationally to form tetrameric complexes, and large-format polypeptides containing four internally repeated pseudosubunit (S1–S6) domains that make up a single 24-transmembrane-segment subunit. In each case, the transmembrane S1–S6 segments present in each of the four subunits or pseudosubunits assemble to form the major portions of both the voltage-sensing apparatus and ion-selective pore. Although high-resolution structural details of these domains of mammalian voltage-gated ion channel α or α_1 subunits have not yet been obtained (10), recent elegant structural analyses of bacterial ion channels have provided tremendous insights into molecular mechanisms that may underlie ion selectivity and voltage-dependent gating (11, 12).

Genes encoding the large-format α/α_1 subunits arose from gene duplication and fusion of the coding regions of genes encoding small-format α subunits (13, 14). Sequence homology analyses reveal that these gene duplication and fusion events occurred before the divergence of eukaryotes and prokaryotes. The pore-forming and voltage-sensing functionalities of Kv channels are formed as tetrameric assemblies of four independent small-format α subunits, whereas Nav and Cav channels are formed by a single large format α/α_1 subunit. Both small- and large-format α/α_1 subunits are typically associated with smaller auxiliary subunits, which can be either cytoplasmic or single- or dual-pass transmembrane polypeptides (5). The auxiliary subunit composition of the native channel complexes is diverse and can dramatically influence channel function, intracellular trafficking, post-translational modification, stability, and localization (15).

Voltage-gated ion channel α/α_1-subunit genes are dispersed throughout the genome, although certain channel genes are found clustered in multigene complexes. Examples of ion channel gene clustering in the mammalian genome include Nav1.1, Nav1.2, Nav1.3, and Nav1.7 (the SCN1A, SCN2A, SCN3A, and SCN9A loci, respectively) on 2q24 (16, 17), and Kv1.1, Kv1.5, and Kv1.6 (KCNA1, KCNA5, and KCNA6) on 12p13 (18). These clusters presumably represent the products of ancient gene duplication events that led to the diversity of Nav and

Kv genes observed in higher mammals (14). The genes themselves can range from the quite simple to the amazingly complex. On the simple end of the spectrum, the entire coding regions of six of the seven *Shaker*-related Kv1 α-subunit genes are contained in a single exon (19). At the opposite end of the spectrum is the CACNA1E gene (encoding the Cav2.3 channel α_1 subunit) that encompasses >315 kB of genomic sequence on human chromosome 1. Even genes encoding small, cytoplasmic auxiliary subunits can be quite complex. The KCNA1B gene is over 250 kB in length, yet encodes small Kv channel auxiliary subunits on the order of only 400 amino acids ($\sim$40 kDa) in size (20).

Voltage-Gated Sodium (Nav) Channels

As a direct result of the Herculean efforts of Numa and his colleagues in 1984 (21), molecular characterization of Nav channels, and in fact of all voltage-dependent ion channels, began with the cloning of a cDNA encoding the electric eel electroplax Nav channel α subunit. The α subunit of the electroplax Nav channel was the first subunit of a voltage-dependent ion channel to be cloned, and early analyses of the deduced amino acid sequence led to a number of important insights into how ion channel primary structure might relate to function. Moreover, this team used their eel Nav channel cDNA to isolate, in rapid succession, cDNAs encoding three different rat brain Nav channel α subunits, now known as Nav1.1 (SCN1A), Nav1.2 (SCN2A) (22), and Nav1.3 (SCN3A) (23). The availability of these cDNAs led to the first studies of the regional and developmental patterns of Nav gene expression in mammalian brain (24–27). Moreover, using deduced amino acid sequences, isoform-specific antibodies were generated that allowed for analyses of the cellular and subcellular distribution of the corresponding rat brain Nav channel proteins (28, 29). These initial studies showed that the Nav1.1 mRNA was highly expressed in cerebellum and at lower levels in the striatum, hippocampus, and thalamus, and was undetectable in cortex (26, 27). The Nav1.1 polypeptide was found predominantly in the cell bodies and dendrites of expressing cells (28). Nav1.1 channels are presumably involved in propagating synaptic signals from dendrites to soma and in integration of electrical signals within the soma prior to the initiation of axonal action potentials.

Although Nav1.2 mRNA has a pattern of cellular expression somewhat similar to that observed for Nav1.1 (26), the Nav1.2 polypeptide was found specifically localized in axons and terminals (28). These early studies revealed that the highly related rat brain Nav1.1 and Nav1.2 polypeptides ($\sim$88% amino acid identity) could be targeted to the distinct subcellular domains (somatodendritic versus axonal) of rat central neurons. This localization also suggested that in central neurons, Nav1.2 is a major component of the axonal action potential conductance mechanism, as well as a regulator of neurotransmitter release in presynaptic terminals. In contrast to Nav1.1 and Nav1.2 transcripts, in rodents Nav1.3 mRNA was found at highest levels in embryonic and early postnatal brain. To date, detailed analyses of the localization of Nav1.3 polypeptide in rodent brain have not been reported (although, see 30). In human brain, however, Nav1.3 has a somatodendritic localization, similar to that observed for Nav1.1 (31).

More recently, cDNAs encoding the Nav1.6 (SCN8A) channel have been isolated from mammalian brain libraries (32). The Nav1.6 mRNA is highly expressed in cerebellar granule cells and in pyramidal and granule cells of the hippocampus (33). In many central neurons, immunoreactivity for Nav1.6 is found in the soma and proximal dendrites of cells expressing the corresponding mRNA (34, 35). Interestingly, Nav1.6 is also found as the principal Nav channel at axon initial segments (36). However, in adult neurons with myelinated axons, Nav1.6 is expressed at high densities at nodes of Ranvier (35, 37). When nodes of Ranvier initially form in response to developmental myelination, Nav1.2 channels are found in high densities at most if not all nodes, and Nav1.6 is not found (37). However, with further maturation of the node, Nav1.6 becomes the predominant Nav isoform such that Nav1.2 is seen in only a small subpopulation of nodes, where it is always found in conjunction with Nav1.6. In neurons such as retinal ganglion cells, where the myelinated portion of the axon within the optic nerve lie distal to extensive non-myelinated portions within the retina, it is clear that Nav1.2 is highly expressed in the nonmyelinated portions in the absence of Nav1.6 (37). However, in the axon initial segment, Nav1.6 is again found at high levels in conjunction with Nav1.2 (36). These findings point to the fact that a number of distinct membrane domains can exist along the length of mammalian axons. Some of these domains are defined by highly specific cell-cell interactions, such as that which occurs between neurons and glia at the node of Ranvier of myelinated axons. However, intrinsic neuronal machinery must also exist to restrict the expression of Nav isoforms to different locations, such as axon initial segments, in the nonmyelinated portions of axons.

Interestingly, Nav1.5, which is the predominant cardiac sodium channel, has recently been shown to be expressed at both the mRNA (38) and protein (39) level in mammalian brain. In mouse brain, Nav1.5 is localized to axons (39). Other Nav channels (Nav1.4, Nav1.7–Nav1.12) do not appear to be appreciably expressed in mammalian central neurons.

Voltage-Gated Calcium (Cav) Channels

The original cloning of the rabbit skeletal muscle dihydropyridine receptor/Cav channel α_1 subunit (40) led to the isolation of the first mammalian brain Cav channel cDNAs (41, 42). Subsequent analyses led to the isolation of a large family of brain α_1 subunit cDNAs. To date, nine α_1 subunit genes have been identified in the human genome (CACNA1A–H, S). These have been divided into the Cav1 (L-type), Cav2 (P/Q-, N- and R-type), and Cav3 (T-type) families (8). Native Cav channels contain auxiliary $\alpha_2\delta$, β, and γ subunits, such that the subunit composition of many native Cav channels is $\alpha_1\alpha_2\delta\beta\gamma$ (43). The diversity among these auxiliary subunits leads to isoform-specific effects that are crucial in determining the expression level and functional characteristics of the associated α_1 subunits.

L-type Cav channels play a critical role in calcium influx in the somata and dendrites of many mammalian central neurons and are involved in dendritic calcium signaling resulting from back-propagating action potentials, synaptic plasticity, active amplification of synaptic signals, and excitation-transcription coupling (44). In mammalian neurons, both Cav1.2 (α_{1C}) and Cav1.3 (α_{1D}) underlie prominent somatodendritic L-type high-voltage-activated Cav currents, as shown by pharmacological studies using dihydropyridine inhibitors (45). Cav1.1 (α_{1S}) and Cav1.4 (α_{1F}) do not appear to be expressed at appreciable levels in mammalian neurons.

P/Q- and R-type channels play critical roles in calcium flux, regulating transmitter release from presynaptic nerve terminals. Products of the Cav2.1 (α_{1A}) gene underlie P/Q-type channels, which are inhibited by the funnel web spider toxin ω-agatoxin-IVA. The Cav2.2 (α_{1B}) gene encodes N-type channels, which are inhibited by the cone snail toxin ω-GVIA-conotoxin. Pharmacological studies with these toxins suggest that Cav2.1 and Cav2.2 channels mediate much of the calcium influx that triggers neurotransmitter release from presynaptic terminals. Recombinant Cav2.3 (α_{1E}) channels are inhibited by nickel and by SNX-482, a tarantula toxin that selectively blocks R-type channels, suggesting that these Cav2.3 subunits mediate R-type currents in some tissues.

T-type channels underlie small, transient currents that activate at subthreshold membrane potentials. These channels are crucial to regulating plasma membrane calcium permeability near resting membrane potentials and during action potentials. Cav3 family members have diverse expression patterns in the mammalian brain. Cav3.1 (α_{1G}) is highly expressed in the cerebellum, where it is present on Purkinje cell dendrites in human, rat (46), and mouse (47) brain. Cav3.2 (α_{1H}) channels have not been studied at the protein level, but in rat hippocampus Cav3.2 mRNA is found at high levels in dentate granule cells and pyramidal neurons (46). Cav3.3 (α_{1I}) appears to be a major T-type Cav channel in interneurons and in parallel fibers in cerebellar cortex (47).

Voltage-Gated Potassium (Kv) Channels

Molecular characterization of Kv channels initially lagged behind that of Nav and Cav channels. Due to the diversity of Kv channel α subunit genes, the potential for oligomerization, and the complex repertoire of Kv channels in any given excitable cell, the biochemical approach that was so successful for Nav and Cav channels did not yield amounts of Kv channel subunits amenable for protein sequencing. The breakthrough for Kv channel molecular characterization came from concerted genetic and molecular analyses of potassium channel mutants in the fruit fly *Drosophila melanogaster* (48). These efforts resulted in the isolation of cDNAs encoding the Kv channel α subunit encoded at the *Shaker* gene locus (49, 50). It was immediately clear from the deduced *Shaker* amino acid sequence that this Kv channel α subunit strongly resembled one of the four internally repeated homologous pseudosubunit domains of a Nav or Cav channel. This led to the proposal, later substantiated by direct experimentation (51), that Kv channels have functional tetramers of individual α subunits. In *Drosophila*, Salkoff and colleagues showed that in addition to *Shaker*, three other Kv channel genes were present: *Shab*, *Shal*, and *Shaw* (52).

cDNAs encoding multiple members of each of the corresponding mammalian gene families (*Shab* = Kv2 or KCNB; *Shaw* = Kv3 or KCNC; and *Shal* = Kv4 or KCND) have now been isolated and expressed. Using the Shaker cD-NAs as probes, Tempel and coworkers isolated the first mammalian Kv channel cDNA, Kv1.1 (53). In rapid succession, cDNAs encoding other Kv1 (i.e., *Shaker*-related) family members (Kv1.2–Kv1.7, the products of the KCNA1–7 genes) were isolated (reviewed in 54). These different mammalian Kv1 family members had distinct functional properties when expressed alone (i.e., as homotetramers) in heterologous cells (55). Different Kv1 family members could also coassemble into channels with mixed subunit composition, and such heterotetrameric channels exhibited functional properties intermediate between those of channels formed from homotetramers of the constituent subunits (56–58). Certain Kv1 channels contain strong trafficking determinants, whereas others do not (59). Heteromeric assembly of different subunits yields channels with intermediate trafficking characteristics (60). Thus the subunit composition of Kv1 channels not only determines their gating and kinetic properties, but also dramatically affects their expression and localization.

Coincident with the identification of mammalian Kv1 α subunits, a cDNA encoding the rat brain Kv2.1 α subunit was isolated by expression cloning in *Xenopus* oocytes (61). The manner of cloning Kv2.1 is noteworthy as it reflects the high-level expression of Kv2.1 in mammalian brain. Low-stringency hybridization screening of mammalian brain cDNA libraries led to the isolation of cDNAs encoding Kv3 (62–65) and Kv4 (66, 67) subunits.

The human genome contains a total of 16 KCNA–D or Kv α subunit genes. Some of these genes generate messages that are subject to alternative splicing. In mammalian brain, the expression of many of these Kv channel α subunits is restricted to neurons, although glial cells may express a subset of the neuronal repertoire. In general, Kv channels exhibit subfamily-specific patterns of subcellular localization. Kv1 channels are found predominantly on axons and nerve terminals; Kv2 channels on the soma and dendrites; Kv3 channels can be found in dendritic or axonal domains, depending on the subunit and cell type; and Kv4 channel are concentrated in somatodendritic membranes. However, the specific locations and degree of formation of channels with heteromeric subunit combinations are highly variable and are discussed in detail below.

Kv channels also contain auxiliary subunits. The best characterized of these are the cytoplasmic Kvβ subunits associated with Kv1 family members (68). The bulk of Kv1 channel complexes in mammalian brain have associated Kvβ subunits (69). Four Kvβ subunit genes exist in the human genome, and alternative splicing can generate a number of functionally distinct isoforms (68). Inclusion of the Kvβ1.1 subunit in Kv channel complexes containing Kv1.1 or Kv1.2 dramatically alters the channel gating properties, converting the channels from sustained, or delayed-rectifier type, to rapidly inactivating, or A-type. Moreover, the specific α and β subunit composition of native complexes can dramatically impact both the expression level and function of Kv1 channels in mammalian neurons (15). Accessory subunits for Kv4 channels have also been identified recently and are encoded by two distinct sets of proteins. One set is a family of calcium-binding proteins, called KChIPs, that are members of the neuronal calcium sensor gene family (70). At least four KChIP genes have been reported to exist in mammals (70, 71), and multiple alternatively spliced isoforms of each KChIP gene product have been reported. With the exception of the KChIP4a splice variant, in heterologous expression systems all KChIP isoforms increase the surface density (72), slow the inactivation gating, and speed the kinetics of recovery from inactivation (70) of Kv4 channels. Recently, Nadal and colleagues (73) reported the identification of a dipeptidyl-peptidase-like protein (DPPX) as an accessory subunit for Kv4 channels. Co-expression of DPPX, KChIPs, and Kv4 subunits gives rise to A-type currents whose biophysical properties closely match the properties of native somatodendritic A-type currents.

In addition to these stereotypical auxiliary subunits, there exist in the genome a number of "electrically silent" α subunit–like polypeptides (74). In heterologous expression systems, these can coassemble with and functionally modify bona fide Kv α subunits (75). However, while there is some information available as to the expression pattern of these genes in mammalian tissues (74, 76), very little is known of their expression at the protein level or their contribution to native mammalian brain Kv channels.

MAPPING OF ION CHANNEL EXPRESSION IN MAMMALIAN BRAIN

Technical Considerations

The most detailed information on the distribution and subcellular localization of ion channel subunits in brain has come from the combination of in situ hybridization histochemistry and immunohistochemical or immunofluorescence techniques. As in all studies of this type, the quality of the resulting data depends on not only the technical competence of the investigators but also on the quality of the reagents and tissues used for the analyses. In general, the greater the care that is taken in generating, characterizing, and evaluating hybridization probes or antibody reagents, the higher the quality of the resulting images and greater the comfort level in interpreting the expression and staining patterns. This issue is particularly critical for immunohistochemical analyses, where antibody quality and specificity are absolutely critical. In contrast to immunoblotting, where electrophoretic mobility can be used to distinguish the protein of interest from among other labeled proteins, no independent markers of identity and specificity are available in immunohistochemistry and immunofluorescence. Reaction product reveals all sites at which antibody reagents have bound, leaving considerable room for interpretation and confusion as to which staining is specific (i.e., represents the presence of the target antigen) and which is nonspecific. Antibodies are, after all, affinity reagents, and in spite of their tremendous affinity for their target antigen, low-affinity cross-reactivity with closely related sequences is common and contributes to the signal. It is clear that there is wide variability in the extent of characterization of the antibody reagents used in published reports of channel localization. Particularly problematic, in our experience, are purified rabbit polyclonal antibodies raised de novo by commercial vendors and sold as general purpose reagents suitable for all applications, including immunohistochemistry. Frequently these reagents are purchased off-the-shelf and used without further characterization. However, in our experience, these antibodies are typically not tested extensively enough to prove either their specificity or broad utility. As a result, these antibodies typically cross-react with a range of other proteins and therefore give high nonspecific or background staining. Evidence for this comes from an appreciation that it can be difficult to reconcile the reported immunohistochemical staining patterns observed with these antibodies with what is expected based on the areal and laminar expression of the corresponding mRNA, revealed by in situ hybridization histochemistry.

As described previously (77), generating antibodies that work well for immunohistochemistry in brain is an extremely challenging endeavor. The best quality antibodies for this application are ones that have been extensively tested, using a multi-pronged testing strategy. Our approach, which we apply to every antibody reagent, is to evaluate the antibodies by immunofluorescence versus cells transiently transfected with cDNA encoding the cognate antigen, by immunoprecipitation in native or transfected cells, and by immunoblotting versus brain membranes or lysates prepared from transfected cells. Moreover, we also evaluate the immunohistochemical staining in parallel with an analysis of expression of the corresponding mRNA and ensure that the staining and expression patterns can be reconciled on the basis of neuroanatomical connections and criteria. Whenever possible, we also make antibodies to two distinct, nonoverlapping epitopes on the same channel subunit protein and show that the staining patterns observed with the two independent antibodies overlap extensively. This latter approach is rarely seen in the published literature, but we now employ this approach routinely, both as a check of the specificity of immunoreactivity and in appreciation that for some applications it is desirable to have antibodies directed at more than one epitope on a single protein. A second approach that we have exploited extensively is to generate panels of mouse monoclonal antibodies generated using synthetic peptides or GST-fusion proteins encoding unique portions of the channel or subunit sequence as the immunogens. This approach offers considerable flexibility in that subunit-specific and sometimes also multi-subunit-reactive antibodies can be obtained from a single immunization (69, 78), Moreover, this approach provides the opportunity to select antibodies based on their isotype (i.e., IgG_1, IgG_{2a}, etc.) as well as specificity, enabling us to perform multiple-label analyses in native tissues (79). The inherent advantages of monoclonal antibodies (e.g., homogeneous preparation, unlimited supply, etc.) contribute further to the attractiveness of this approach.

Nav Channel α Subunits

At least 11 of the 12 Nav channel α subunit genes present in the mammalian genome are expressed somewhere in the nervous system. When expressed in heterologous systems, these α subunits generate Nav currents whose functional properties are quite conserved, compared with the heterogeneity of function observed among the different Cav and Kv family members. Certain subtypes, such as Nav1.7, Nav1.8, Nav1.9, and Navx, are expressed predominantly in the peripheral nervous system and are not discussed here. Nav1.4 is the predominant Nav channel in adult skeletal muscle, and Nav1.5 is present in adult cardiac and embryonic skeletal muscle. The remaining four α subunits (Nav1.1, Nav1.2, Nav1.3, and Nav1.6) are expressed at high levels in brain, although in rodents Nav1.3 is predominantly found in embryonic and neonatal, but not adult, brain. A number of groups have used antipeptide rabbit polyclonal antibodies raised against sequences in the intracellular interdomain I-II linker region or the cytoplasmic C terminus to define the localization of Nav1.1, Nav1.2, Nav1.3, and Nav1.6 in rat and human brain. Recently, we also used isoform-specific monoclonal antibodies for this purpose.

Together, these studies showed that in the adult rat and human hippocampus, strong Nav1.1 staining is present on the soma of dentate granule cells and of interneurons dispersed throughout the dentate hilus (28, 29). Intense staining is also seen in stratum (s.) pyramidale of all hippocampal subfields (CA1, CA2, and CA3) with staining restricted to the pyramidal cell somata and proximal dendrites. Additional staining is present in a subpopulation of interneurons dispersed throughout s. radiatum and s. oriens. The identity of this subpopulation of interneurons has not been experimentally determined. Robust Nav1.1 staining is also observed on the soma and apical dendrites of pyramidal cells in layer V of the cerebral cortex. In the cerebellum, granule cells and Purkinje cells have strong Nav1.1 staining on the soma, and clear staining is seen on Purkinje cell dendrites. It should be noted that strikingly similar patterns of Nav1.1 staining were observed in human brain (31), suggesting that this distinct pattern of localization is evolutionarily conserved and that somatodendritic targeting of Nav1.1 is essential for normal response properties across species.

High levels of Nav1.1 and Nav1.2 mRNA are expressed in many of the same cells. However, the subcellular staining pattern of Nav1.2 is in many instances a reciprocal image of that for Nav1.1. The contrasting subcellular localizations of Nav1.1 and Nav1.2 are especially clear in the hippocampus, where the exquisitely laminar cytoarchitecture lends itself to such comparisons. In many cells expressing robust somatodendritic Nav1.1 staining, very little if any Nav1.2 staining is observed on the somatic or dendritic membrane. Thus the principal cell layers of the hippocampus are virtually lacking Nav1.2 staining, whereas areas rich in axons and nerve terminals exhibit robust staining. In the dentate gyrus, Nav1.2 is present at high levels in the middle third of the molecular layer, site of termination efferents from the entorhinal cortex to the hippocampus (the so-called medial perforant path), and absent from dentate granule cell bodies. Additional strong staining is seen in s. lucidum of CA3, the site of densely packed mossy fiber axons of dentate granule cells, whereas CA3 pyramidal cell bodies are not stained. Nav1.2 is also concentrated in s. radiatum and s. oriens of CA1 and appears to be associated with axons and terminals of the Schaffer collaterals (s. radiatum), and the commissural/associational pathway (s. oriens). As in other pyramidal cell layers, Nav1.2 is not present on pyramidal cell bodies in CA1. In these regards, the staining pattern of Nav1.2 appears virtually identical to that of the Kv1.4 α subunit (79, 80), for which experimental lesions (81) and immuno-electron microscopy (82) studies have strongly supported an axonal localization. However neither lesion studies nor immuno-electron microscopy has been utilized to verify that the Nav1.2 staining is in fact localized to axons and terminals. It should be noted that in the large pyramidal cells of hippocampal CA1, and of layer V of the neocortex, some apical dendritic staining is apparent (29). However, this staining appears to be intracellular and presumably represents a pool of newly synthesized Nav1.2 α subunits transiting through the rough endoplasmic reticulum and/or Golgi apparatus.

Similar contrasting localization of Nav1.1 and Nav1.2 is observed in the cerebellar cortex, where strong Nav1.1 staining is seen in the granule cell layer, and strong Nav1.2 staining is seen in the molecular layer, consistent with disparate polarized trafficking of granule cell-derived Nav channels. Neither subunit is strongly expressed in Purkinje cells.

The subcellular localization of Nav1.3 has only recently been reported. Early mRNA analyses showed that in rats this Nav channel was expressed predominantly in embryonic and neonatal brain, with much lower levels in the adult (24). A detailed immunohistochemical study of Nav1.3 localization has not been performed in the rat (although, see 30). However, recent studies of Nav1.3 mRNA (83) and protein (31) in human brain have revealed that the distribution of Nav1.3 in adult human brain is extensive. Certain human brain regions express moderate to high levels of Nav1.3 staining, which in all cases (and as in the rat) (30) is associated with neuronal somata and dendrites. Especially notable is staining in neocortex, where robust staining of layer III pyramidal cells is apparent (31). This strong staining extends some distance along the large apical dendrites next to these layer III pyramidal cells, whereas in other neocortical laminae, staining is less intense and restricted to somata. In rat, both layer III and layer V pyramidal cells exhibit strong Nav1.3 staining extending along the length of the apical dendrites (30). In the human hippocampus, robust staining is observed in a subpopulation of interneurons scattered throughout s. oriens of CA1 and CA2, s. lucidum of CA3, and the polymorphic layer of the dentate gyrus. Less intense somatodendritic staining is also observed in pyramidal cells, but little or no staining is seen in dentate granule cells. In cerebellar cortex, moderate staining is observed in granule cells and in the molecular layer. Thus in human brain, Nav1.3 may act together with Nav1.1 in determining the active properties of neuronal somata and dendrites.

Immunolocalization studies of Nav1.6 in mammalian brain led to the discovery that this is the major Nav channel at nodes of Ranvier of myelinated axons (34, 35, 37) and at axon initial segments (36). Nav1.6 is also robustly expressed in Purkinje cells, where it is present in the somata and throughout the dendritic arbor in rat (34) and human (31). This robust Purkinje cell staining is consistent with the finding that mutant mice lacking Nav1.6 expression are missing a Purkinje cell resurgent Nav current (84). Outside of Purkinje cells, other somatodendritic Nav1.6 staining is less impressive, with neurons in many brain regions, such as cortex and hippocampus, exhibiting somatodendritic staining that is somewhat intracellular in appearance (34, 35). This may simply represent newly synthesized Nav1.6 prior to its targeting to initial segments and/or nodes of Ranvier of myelinated axons, as opposed to a functional somatodendritic pool of Nav1.6.

Cav Channel α_1 Subunits

Molecular isoforms of Cav channel α_1 subunit polypeptides have been extensively localized in the mammalian central nervous system. In general these studies have employed immunoperoxidase staining and light level microscopy, although in certain cases autoradiographic analyses of radiolabeled toxin binding, or immuno-electron microscopy analyses have been performed.

Immunolocalization studies of Cav channels began in 1990 using monoclonal antibodies raised by the Catterall laboratory against the auxiliary $\alpha_2\delta$ subunits of purified brain L-type Cav1 channels (85). These studies found staining associated with dendrites of principal cells in cortex and hippocampus. These same investigators (86) showed that staining with this antibody was located in the cell bodies and proximal dendrites of hippocampal pyramidal cells and was clustered in high density at the base of major dendrites. This local concentration of L-type $\alpha_2\delta$ subunits led to the proposal that these channels may be involved in proximal dendritic calcium entry in response to synaptic activity (87). Subsequent studies using isoform-specific antibodies raised against peptides derived from the Cav1.2 (α_{1C}) and Cav1.3 (α_{1D}) α_1 subunits revealed that these two components of neuronal L-type channels had distinct subcellular localizations within the soma and dendrites of principal cells in the hippocampus (88, 89). Cav1.2 immunoreactivity was found to make up $\sim$75% of the total dihydropyridine binding activity in brain membranes, whereas Cav1.3 contributed $\sim$20% (89). Cav1.2 staining was found associated with the soma and proximal dendrites and in many respects resembled the staining pattern obtained with pan-L-type reagents, consistent with the role of Cav1.2 as the major component of the brain L-type CAV pool. In contrast, Cav1.3 staining was found more distally and was present in clusters (89). These clusters were later found to be at or near axosomatic and axodendritic synapses and often associated with a β2-adrenergic receptor signaling complex (90). The determinants that direct the distinct subcellular localizations of L-type Cav1.2 and Cav1.3 channels have not been investigated.

In contrast to the constrained somatodendritic localization of Cav1 channels, studies on the localization of the Cav2-family α_1 subunits have defined both axonal and somatodendritic localizations for these P/Q- and R-type channels. Westenbroek and coworkers (91) initially found prominent Cav2.1 staining in areas rich in synaptic terminals, specifically the molecular layer of the cerebellum and the mossy fiber zone of hippocampal CA3. However, certain cells (e.g., cerebellar Purkinje cells) also exhibited prominent somatodendritic Cav2.1 staining (91, 92). Other workers have found prominent somal staining for Cav2.1, although on close inspection the bulk of this staining appears intracellular and may reflect a biosynthetic, as opposed to a functional, somal pool (93). The Cav2.1 nerve terminal staining is consistent with the role of P/Q-type channels in regulating transmitter release through interaction with release machinery (94, 95).

In globus pallidus neurons, immuno-electron microscopy revealed that in addition to prominent staining in nerve terminals, immunoperoxidase reaction product was also observed in dendrites (96). Moreover, in the dendrites of cerebellar Purkinje cells, Cav2.1 is colocalized and associated through a direct physical interaction with metabotropic glutamate subtype 1 receptors (mGluR1) (97). Numerous examples of P/Q channels on dendrites, where they are susceptible to neurotransmitter regulation, have been obtained (98–102). Thus P/Q channels encoded by the Cav2.1 gene have both axonal and somatodendritic localizations. Cav2.1 transcripts are extensively processed by alternative splicing (103), raising the possibility that it may be different Cav2.1 splice variants that are differentially targeted to axons versus dendrites.

The initial studies of the localization of Cav2.2/α_{1B} were accomplished using autoradiography with the Cav2.2-specific toxin ^{125}I-ω-GVIA-conotoxin (104) to define the localization of these N-type Cav channels in mammalian brain (105–107). These studies revealed a high density of sites in striatum, hippocampus, cortex, and cerebellum in locations that corresponded to high densities of synapses. Later studies in human hippocampus and cerebellum yielded qualitatively similar results (108). More recent ^{125}I-ω-GVIA-conotoxin binding analyses have revealed a high density of binding sites for this N-type/Cav2.2-specific toxin in cortex, striatum, and hippocampus in gerbil brain (109). Higher-resolution analyses in the hippocampus revealed intense signals in regions rich in axons and presynaptic terminals, with the highest signals in s. lucidum of CA3, which contains the mossy fibers axons and terminals. The infragranular polymorphic region of the dentate gyrus, and s. radiatum and lacunosum moleculare of CA1, which contains Schaffer collaterals, also exhibited strong signals for ^{125}I-ω-GVIA-conotoxin binding (109). Together these studies suggest a correlation between high densities of N-type Cav2.2 channels and high densities of synapses, as opposed to areas rich in cell bodies.

Studies of Cav2.2 (α_{1B}) immunoreactivity yielded similar results. Westenbroek and her colleagues (88) used an antipeptide antibody to define prominent Cav2.2 immunoreactivity in punctate/synaptic structures in rat brain. They highlighted prominent staining on the dendrites of pyramidal cells in layers II/III and V of dorsal cerebral cortex, and of layers CA1–CA3 of hippocampus. Especially dramatic was the robust staining of mossy fiber expansions (presynaptic terminals) in s. lucidum of CA3, which exhibit much more pronounced staining and stand out remarkably relative to the staining in the preceding axon shafts (88). More uniform Cav2.2 staining on dendrites is also apparent, although at lower levels than in terminals. Similar results were obtained in organotypic cultures of rat hippocampal CA3, although stronger Cav2.2 staining was found associated with CA3 pyramidal cell dendrites than with mossy fiber terminals, which in this case exhibited dense staining for Cav2.1 (99). Similar analyses in human hippocampus with a Cav2.2-specific antibody (raised against a GST-Cav2.2 fusion protein) revealed much less dramatic mossy fiber terminal staining than was obtained in the studies in rat, although the Cav2.2 staining corresponded well with that for calbindin, which in CA3 is a useful marker for mossy fiber terminals (110). It is not clear if these differences reflect real variations between rodent and human brain or are simply from the different antibodies used in the various studies or from other technical aspects of the experiments (e.g., postmortem autolysis of the human tissue prior to fixation).

Cav2.2 mRNA is subjected to alternative splicing at the C terminus, generating long (Cav2.2a/α_{1b-1}) and short (Cav2.2b/α_{1b-2}) channel isoforms (111). A recent study of the subcellular localization of exogenously expressed Cav2.2 subunits in cultured hippocampal neurons suggests that the long Cav2.2a variant is specifically targeted to presynaptic terminals, whereas the shorter form is localized in the somatodendritic domain (112). This difference in targeting of exogenously expressed Cav2.2 variants may underlie the apparent nonpolarized expression of the total neuronal pool of Cav2.2 channels in brain (88). However, the finding that different splice variants of the same channel subunit have different subcellular localizations complicates efforts to map channel expression using antibodies directed against sites conserved among the products of the alternative splicing events.

Immunohistochemical staining for Cav2.3/α_{1E} (R-type) subunits is distinguished from that of Cav2.1 and Cav2.2 in that it is more prominent in the midbrain and hindbrain than in forebrain structures (113). Strong Cav2.3 somatic staining was observed in all cells examined, although upon close inspection the staining in certain cells (principal cells in neocortex and hippocampus) appeared to be perinuclear/intracellular, perhaps representing biosynthetic pools of inefficiently trafficked channels. In other neurons, such as those in globus pallidus, thalamus, anterior amygdala, subthalamic nuclei, and hypothalamus, the Cav2.3 somal staining appeared to be more plasma membrane associated and extended well into the dendrites where it appeared associated with the plasma membrane. In the cerebellar cortex, prominent staining was observed in the soma and dendrites of Purkinje cells (113). Interestingly, immuno-electron microscopy revealed that in rat globus pallidus neurons, Cav2.3 reaction product was clearly present both in dendritic structures and in a subpopulation of presynaptic boutons (96). Taken together, these results suggest that Cav2 family members exhibit distinct patterns of cellular and subcellular localization in mammalian brain. However, in some cases their expression may overlap, suggesting complex functional requirements for multiple family members at a specific site (e.g., certain presynaptic terminals).

Cav3/α_{1G-I} family members underlie T-type Cav channels. Cav3.1 (α_{1G}) is prominently expressed in hindbrain (cerebellum and medulla) with lower levels in forebrain and midbrain regions in both rats (114) and mice (47). In the cerebellar cortex, Cav3.1-specific antibodies display strong staining to thick processes in the molecular layer, with (114) or without (47) additional staining to Purkinje cell bodies. All this staining is presumably associated with Purkinje cell somata and dendrites, given the high levels of Cav3.1 mRNA in Purkinje cell somata in rats (46, 114) and relatively low levels in granule cells and basket cells. In the medulla, prominent somatodendritic staining was observed in the hypoglossal nucleus, the inferior olive, and the lateral reticular nucleus. In some neurons, the Cav3.1 staining appeared to be intracellular, whereas in others it was present uniformly on the somal and proximal dendritic membrane (114).

Only one report exists on the localization of Cav3.3 (α_{1I}) protein in mammalian brain. This study revealed that Cav3.3 staining was distinct from that for Cav3.1 (47). High levels of Cav3.3 protein were found in mouse olfactory bulb and midbrain. In the olfactory bulb, intense staining was observed in the olfactory nerve layer and glomerular layer. In hippocampus, staining was observed in a subset of interneurons dispersed throughout s. lucidum of the dentate gyrus, CA3, and CA2, with the highest density in CA2. In the dentate gyrus, strong labeling of the inner and outer thirds of the molecular layer was observed, suggesting the presence of Cav3.3 on terminals of the associational and commissural pathways of the dentate gyrus and of the lateral perforant pathway of entorhinal cortex, respectively. Alternatively, this staining could represent input-specific localization to specific subdomains of dentate granule cell dendrites.

In the cerebellar cortex, Cav3.3 staining in the molecular layer appeared diffuse and associated with parallel fibers and/or their terminals. Consistent with this model is the high level of Cav3.3 mRNA in the granule cell layer, but not in the Purkinje cell or molecular layer, in adult rats (46). The localization of Cav3.2 (α_{1H}) protein in mammalian brain has not been performed, although analysis of mRNA distribution in adult rat brain reveals high levels of Cav3.2 mRNA in hippocampal dentate granule and CA pyramidal cells, in olfactory bulb neurons, in pituitary and pineal glands, and in sensory neurons (46). Whether Cav3.2 is playing a role in mediating calcium influx into axons and terminals, or dendrites and somata, in these cells is not known.

Kv Channel α Subunits

When expressed in heterologous systems, the neuronal Kv1 α subunits can generate either transient (Kv1.4) or sustained (Kv1.1–Kv1.3, Kv1.5, Kv1.6) Kv currents. Moreover, heteromeric assembly with one another and coassembly with auxiliary Kvβ subunits can generate a diversity of function from the resultant $\alpha_4\beta_4$ channel complexes. Kv1 family members also exhibit diverse expression patterns in mammalian brain. The predominant Kv1 cellular staining pattern throughout the brain is neuronal, and subcellularly, axonal. The three most abundant Kv1 subunits expressed in mammalian brain, Kv1.1, Kv1.2, and Kv1.4, are found predominantly localized to axons and nerve terminals. In many cases these subunits are components of heteromeric channel complexes, as Kv1.1, Kv1.2, and Kv1.4 exhibit precise patterns of colocalization, and extensive association as shown by copurification. However, the subunit composition of channels containing these subunits varies across brain regions. For example, the Kv1.1 α subunit in mammalian brain appears to be segregated into two major subpopulations, one associated with Kv1.2 and one associated with Kv1.4. Kv1.1 and Kv1.4 are found robustly expressed in the relative absence of Kv1.2 within the striatal efferents in globus pallidus and pars reticulata of substantia nigra (79, 80). Kv1.1 and Kv1.2 are found in the absence of Kv1.4 in cerebellar basket cell terminals (78, 79, 115–119), and the juxtaparanodal membrane adjacent to nodes of Ranvier (78, 79, 117, 118, 120–123). Within the excitatory circuitry of the hippocampus, a number of patterns for expression for these three Kv1 α subunits emerge, providing a striking example of the complex heterogeneity of subunit association. Kv1.1, Kv1.2, and Kv1.4 are highly expressed in the middle third of the molecular layer of the dentate gyrus where they are associated with axons and terminals of the medial perforant path (78–81, 117, 118, 124, 125). Kv1.1, Kv1.2, and Kv1.4 are also found in Schaffer collateral axons, whereas Kv1.1 and Kv1.4 colocalize, in the absence of Kv1.2, in mossy fiber axons (80, 82, 117, 118, 124). However, in spite of their colocalization, it is not clear whether heteromeric channel complexes containing coassociated Kv1.1, Kv1.2, and Kv1.4 are present on perforant path and Schaffer collateral axons. Lesions in entorhinal cortex have distinct effects on the distribution of Kv1.2 and Kv1.4 in the middle third of the dentate molecular layer, suggesting that while these subunits may colocalize at the light microscope level, they may be expressed on different components of the perforant path. Similar results have been obtained with lesions placed in other subfields. For example, in CA3 predominant Kv1 channels appear to be composed of Kv1.1 with Kv1.4, whereas in the Schaffer collateral pathway, Kv1.1 is likely to be associated with Kv1.2 and Kv1.4 (81).

Electron microscopic immunohistochemical studies have demonstrated that Kv1.1, Kv1.2, and Kv1.4 are concentrated along axons and in the axonal membrane immediately preceding or within axon terminals (82, 117, 118). The immunoreactivity for Kv1.1 and Kv1.2 has been localized to the preterminal axonal membrane in s. radiatum; immunoreactivity for Kv1.4 has been localized to the preterminal extensions of mossy fiber axons (82). In these positions, activation of Kv1 channels can play a critical role in regulating the extent of nerve terminal depolarization and thereby regulate neurotransmitter release.

The other Kv1 α subunits appear to be expressed at lower levels in mammalian brain. Kv1.6 is found predominantly in interneurons, although some dendritic staining is seen throughout the brain on principal cell dendrites (79). Kv1.3 is highly expressed in the cerebellar cortex. The bulk of expression is in the parallel fiber axons of cerebellar granule cells, as a strong in situ hybridization signal is present in the granule cell layer (126), whereas strong immunostaining (125) and ^{125}I-margatoxin (specific for Kv1.2 and Kv1.3) binding (127) are found in the molecular layer. The molecular layer also contains high levels of staining for Kv1.1 (79, 125), suggesting that Kv1.1 and Kv1.3 can form heteromeric channels on parallel fibers. The expression of Kv1.5 in the brain is overall quite low (128). What Kv1.5 expression there is may be restricted to glial cells because numerous reports place Kv1.5 and Kv1.3 as important components of glial-delayed rectifier currents (129, 130). In mammals, Kv1.7 is expressed in skeletal muscle, heart, and pancreatic islets, but not brain (131).

Recent in situ hybridization, immunoprecipitation, and immunohistochemical analyses have also localized sites of expression of the Kvβ1 and Kvβ2 in mammalian brain (69, 78, 79, 132). Kvβ2 appears to be a component of many, if not all, Kv1-containing channel complexes in mammalian brain, and immunoreactivity for Kvβ2 is present in each and every location where immunoreactivity for a Kv1-family α subunit is observed (69, 79). The Kvβ1 subunit, which exerts dramatic effects on the inactivation kinetics of Kv1 channels, appears to be included more selectively. Interestingly, the pattern of immunoreactivity for Kvβ1 closely matches the expression pattern for Kv1.1 and Kv1.4, in that Kvβ1 is found to colocalize with Kv1.4 in the medial perforant path, mossy fiber pathway, and in striatal efferents to the globus pallidus (69, 79). One exception to this pattern of colocalization is in the cerebellar basket cell terminals, where there is a high density of immunoreactivity for Kv1.1 but no detectable staining for Kvβ1 (79).

Kv2 family members form delayed rectifier Kv channels that are prominently expressed in mammalian brain, where they are localized in the somatodendritic domain of neurons. Kv2.1 was the first member of this family isolated by molecular cloning approaches and is unique in that it was identified and isolated by expression cloning (61). Thus it was not surprising when immunostaining revealed that Kv2.1 was highly expressed and has an extensive distribution throughout the mammalian brain (133–135). In fact, the cellular distribution in neurons is so broad that in many regions the cellular staining pattern of Kv2.1 resembles that of a Nissl stain. However, in spite of this broad neuronal expression, within individual neurons the staining for Kv2.1 is highly restricted and is present on only the somatic and proximal dendritic membrane and absent from axons and nerve terminals. Immuno-electron microscopy (135) and excitotoxic lesion studies (136) have unambiguously confirmed the somatodendritic localization of Kv2.1. The striking subcellular distribution is accentuated by the fact that within these domains Kv2.1 is present in large clusters (133–135, 137, 138), which are present on the cell surface membrane immediately facing astrocytic processes, and over subsurface cisterns underlying the plasma membrane facing astrocytes (135). The physiological basis for the highly clustered, discrete localization of Kv2.1 to these specialized membrane domains is not known.

In spite of its widespread cellular distribution, certain cells stand out for having especially prominent Kv2.1 expression. In the cortex, pyramidal cells in layers

II/III and layer V are especially striking for their high levels of Kv2.1 expression. Kv2.1 is also present in high levels throughout the hippocampus, although the levels in dentate granule cells and CA1 pyramidal cells exceed those found in CA3 and CA2 pyramidal cells in both rat and mouse. However, it should be stressed that Kv2.1 is found on both principal cells and interneurons throughout the hippocampus (135). Among interneurons, Kv2.1 is found in the majority of cortical and hippocampal parvalbumin, calbindin, and somatostatin-containing inhibitory interneurons (135).

Kv2.2 is expressed in many of the same cells that express Kv2.1. However, unlike other Kv channels (Kv1, Kv3, and Kv4 family members), the two members of the mammalian Kv2 family apparently do not readily form heteromultimers in native neurons (although, see 139), as the subcellular localizations of Kv2.1 and Kv2.2 expressed in the same cells are distinct (140, 141). Kv2.2 is uniformly present on dendrites and along the entire length of the dendrite. The clustered, proximal dendritic localization of Kv2.1 is not observed for Kv2.2. Kv2.2 is present at high levels in olfactory bulb neurons and in cortical pyramidal neurons.

Kv3 family members have unique functional characteristics, including fast activation at voltages positive to -10 mV and very fast deactivation rates. These properties are thought to facilitate sustained high-frequency firing, and Kv3 subunits are highly expressed in fast-spiking neurons, such as neocortical and hippocampal interneurons, as well as in midbrain auditory neurons (142). Kv3 currents can have either sustained (Kv3.1, Kv3.2) or transient (Kv3.3, Kv3.4) characteristics and can form hetero-oligomeric channels with intermediate gating characteristics (142). Kv3 mRNAs are somewhat unusual among Kv α subunits in that they are subjected to extensive alternative splicing to generate subunits that differ only at their cytoplasmic carboxyl termini (63). This complicates studies of localization of these subunits because one needs to distinguish among reports on the basis of the reagents used (i.e., their specificity for sequences conserved among the alternative forms or unique to individual variants).

Initial in situ hybridization analyses revealed that unlike many other Kv subunits, Kv3.1 and Kv3.2 transcripts are expressed in only a small subset of cells in the cerebral cortex and hippocampus (143, 144). Interestingly, the in situ hybridization patterns of Kv3.1 and Kv3.2 were distinct, suggesting a strict cellular specificity to expression of these highly related Kv channel α subunits (144). Initial immunolocalization studies were performed using antibodies raised against the major splice variant of Kv3.1, termed Kv3.1b, which has a longer C terminus than the less abundant Kv3.1a variant. The studies revealed that Kv3.1b was highly expressed in interneurons and that expression was very low or undetectable in principal cells, such as neocortical and hippocampal pyramidal cells and dentate granule cells (145, 146). Kv3.1b is also robustly expressed in fast-spiking cells in the cochlear nucleus (143). Double labeling experiments (146) revealed that the subset of cortical cells labeled with anti-Kv3.1b antibodies corresponded to GABAergic interneurons (distinguished by their fast-spiking properties), which express the calcium-binding protein parvalbumin. Interestingly, Kv3.2 α subunits were found in nonfast-spiking, somatostatin-, and calbindin-containing interneurons (147). Thus, the expression patterns of Kv3.1 and Kv3.2 α subunits can distinguish different populations of interneurons, thus raising the possibility that interneuron firing patterns rely to some extent on the subtype of Kv3 channels expressed (142, 148, 149).

Studies on the exogenous expression of three different Kv3.2 splice variants (Kv3.2a, b, and c) in polarized epithelial cells revealed that alternative splicing led to differences in subcellular localization. The Kv3.2a variant was localized to the basolateral membrane, whereas the Kv3.2b and Kv3.2c isoforms were found apically (150). The epithelial cell:neuron analogous membrane hypothesis (151) predicts that as such, in neurons, Kv3.2a would be localized to the somatodendritic domain, and Kv3.2b and Kv3.2c would be localized to the axon. A recent study has now revealed that, in the case of mammalian brain Kv3.1, alternative splicing does in fact lead to a difference in the polarized expression of Kv3.1 variants in mammalian brain (152). The initial studies of Kv3.1b localization in mammalian brain revealed staining that was present in the soma, proximal dendrites, unmyelinated axons, and axon terminals of the parvalbumin-positive interneurons (145–147) and neurons in the cochlear nucleus (153). In contrast, Kv3.1a proteins were prominently expressed in the axons of some of the same neuronal populations, but there was little or no Kv3.1a protein expression in somatodendritic membrane. Thus, as for the case of Cav2.3 discussed above, alternative splicing of ion channel transcripts can generate functionally similar variants of the same channel with altered subcellular distributions (152).

Kv3.3 α subunits are also widely expressed at the mRNA level in brain (142). Both Purkinje cells in cerebellar cortex and deep cerebellar nuclei contain high levels of Kv3.3b message (154). Most brainstem auditory neurons also express Kv3.3 mRNA, where it may coassemble with Kv3.1 in a subset of cells (155). Immunofluorescence staining reveals prominent staining for Kv3.3 in Purkinje cell somata and dendrites (156) where a Kv3 channel complex of Kv3.3 and Kv3.4 may play a role in shaping large depolarizing events. Unlike Kv3.1 and Kv3.2, in neocortex and hippocampus Kv3.4 is present in principal cells (144, 157). Moreover, Kv3.4 appears to be localized to axons and nerve terminals of these cells, such that in a number of brain regions Kv3.4 is found colocalized with Kv1 family members. Combined in situ hybridization and immunohistochemistry showed that Kv3.4 is found in terminals of the perforant path; high levels of Kv3.4 mRNA, but not protein, are found in entorhinal cortex, and high levels of Kv3.4 immunostaining are present in the middle third of the molecular layer of the dentate gyrus (125, 157). Intense Kv3.4 staining is also observed in s. lucidum of CA3 and appears to be associated with mossy fiber axons and/or terminals (125, 157). In these regions, Kv3.4 may be present in the same axons and terminals as Kv1.1 and Kv1.4. Kv3.4 is also found in cerebellar basket cell terminals, which contain high levels of Kv1.1 and Kv1.2. Immuno-electron microscopy (119) revealed that although the localization of staining of Kv1.1, Kv1.2, and Kv3.4 overlaps at the light microscope level, these subunits have distinct ultrastructural localizations.

Kv1.1 and Kv1.2 are present in septate-like junctions formed between basket cell terminals and Purkinje cell axons, whereas Kv3.4 is found in nonjunctional regions of the terminals (119). These findings highlight the extent to which different highly related ion channel subunits can be precisely localized in neuronal membrane domains. It should be noted that recent data suggest certain fast-spiking cells Kv3.4 may also associate with Kv3.1 and/or Kv3.2 α subunits to generate a fast-delayed rectifier current (158). However, examples of neurons in which Kv3.4 was found colocalized with Kv3.1 and Kv3.2 were not provided. As noted above, Purkinje cell dendrites may contain a Kv3.3/Kv3.4 heteromeric channel that is important in shaping responses to certain, relatively strong depolarizing events (156).

The Kv4 α subunits Kv4.1, Kv4.2, and Kv4.3 form transient or A-type Kv channels. Experimental knockdown of Kv4 α subunit expression in mammalian neurons results in suppression of A-type Kv channels (159, 160). Kv4.1 is expressed at very low levels in mammalian brain (161), and expression that can be detected in neurons does not correlate with A-type current density (162). In contrast, Kv4.2 and Kv4.3 are expressed at relatively high levels (161), and the expression of these subunits correlates well with neuronal A-type current density in a number of neuronal types (162–166). In situ hybridization analyses show that the expression of Kv4.2 and Kv4.3 is widespread throughout the brain and that whereas in many brain regions the cellular expression of these two Kv4 genes is reciprocal or complementary, there are cells in which Kv4.2 and Kv4.3 are coexpressed (161).

Immunoreactivity for Kv4 subunits is concentrated primarily in the dendrites of central neurons. Kv4.2 is expressed at high levels in many principal cells, whereas Kv4.3 is found in a subset of principal cells and in many interneurons. In the hippocampus, dentate granule cells express high levels of Kv4.2 and Kv4.3 mRNA, and robust staining for both of these subunits is present in the outer third of the molecular layer, presumably in the distal granule cell dendrites that receive input from the lateral perforant path. A similar colocalization of Kv4.2 and Kv4.3 is seen in the distal basal and apical dendrites (in s. oriens and s. radiatum, respectively) of CA3 pyramidal neurons. Interestingly, CA2 and CA1 pyramidal cells express Kv4.2 in the absence of Kv4.3, suggesting a unique role for Kv4.2 homotetramers in the distal dendrites of these cells relative to CA3. Conversely, throughout the hippocampus, Kv4.3 is found in the absence of Kv4.2 in many interneurons. Thus different hippocampal cell types appear to have different requirements for homotetrameric and heterotetrameric Kv4.2 and Kv4.3 channels. The specific targeting of Kv4 channels to dendrites of principal cells in the hippocampus has been confirmed by excitotoxic lesion studies (136). In all cases the staining of Kv4 channels on dendrites tends to be quite uniform, with little evidence of local concentrations of clustered immunoreactivity associated with dendritic spines, postsynaptic densities, or other subcellular domains.

Principal cells and interneurons in neocortex also exhibit specific patterns of Kv4.2 and Kv4.3 staining. Pyramidal cells in layer II exhibit high levels of Kv4.3 staining, and Kv4.2 staining predominates in those in layer V. Kv4.3 is also found in interneurons scattered throughout layers II–VI. The identity of these interneurons has not been confirmed by double labeling experiments for markers of different interneuron populations (parvalbumin, somatostatin, etc.).

Although a detailed report of the expression and localization of Kv4 accessory subunits, KChIPs and DPPX, has not yet been published, images of KChIP localization have been published in abstract form and within the scope of broader papers (70, 73, 136). These studies have revealed that immunoreactivity for KChIPs 2, 3, and 4 is concentrated in somatodendritic membranes where their distribution corresponds closely with that described for Kv4.2. In contrast, the distribution of KChIP1 closely matches the distribution of Kv4.3, particularly in neocortical, hippocampal, and striatal interneurons and in the dendrites of cerebellar granule cells.

CHANNELS UNDERLYING FUNCTION IN SPECIFIC SUBCELLULAR DOMAINS

For the reasons elucidated above, it is often extremely difficult to tie the biophysical properties of recorded, native currents to the expression of individual molecularly defined channel subunits. The difficulties are magnified in the case of Kv channels, where, for example, individual subunits of the Kv1 subfamily coassemble to generate functional channels whose biophysical properties are a mixture of those displayed by each subunit expressed individually. Add to this the selective incorporation of modulatory accessory subunits, such as Kvβ1, into the channel complex, as well as the fundamental limitations on spatial and biochemical resolution inherent in the techniques currently available to study channel localization and subunit association, and the problem is further magnified. Acknowledging these technical limitations, however, it is informative (and irresistibly tempting) to try and match patterns of channel subunit immunoreactivity with native currents revealed using pharmacological or electrophysiological techniques.

Channels Targeted to Axons and Axon Terminals

Much of the data describing the properties of native channels localized to axon and axon terminals is indirect because the comparatively small size of these subcellular domains generally makes them inaccessible to patch-clamp electrodes. There are a few notable exceptions, however, and in these locations high-fidelity recordings of the native channels have been made directly. One such exception is the large (3–5 μm) diameter hippocampal mossy fiber terminal, a specialized excitatory synapse between dentate granule cells and CA3 pyramidal neurons. Jonas and colleagues have performed patch-clamp recordings from identified mossy fiber terminals and explored the contributions of voltage-gated ion channels to action potential propagation and neurotransmitter (glutamate) release from individual boutons (167, 168). Using outside-out patches, Geiger & Jonas (167) have shown that within mossy fiber terminals, Kv channels underlie frequency-dependent spike broadening, a membrane biophysical property that allows the nerve terminal to tune the extent and duration of neurotransmitter release as a function of action potential frequency. Using tetraethylammonium ion (TEA) and α dendrotoxin (DTX) as pharmacological tools, these investigators were able to establish that a rapidly inactivating, DTX- and TEA-sensitive Kv channel conductance underlies frequency-dependent spike broadening. Given this profile of pharmacological sensitivity as well as published immunohistochemical analyses demonstrating a high density of immunoreactivity for Kv1.1, Kv1.4, and Kvβ1 along mossy fiber axons (79–82, 117), the authors concluded that Kv1 channels underlie this conductance. Moreover, they concluded that the likely subunits involved include Kv1.1, coassembled with either Kv1.4 and/or the Kvβ1 Kvβ subunit (167). Although immunoreactivity for Kv3.4, a *Shal*-related α subunit that activates and inactivates rapidly, is also concentrated along mossy fiber axons (169), it is not clear whether or to what extent this channel plays a role in frequency-dependent spike broadening.

Using a similar tissue preparation, Jonas and colleagues also explored the dynamics of calcium entry at mossy fiber terminals (168). Although it is clear from these elegant studies that presynaptic Cav channel activity is critical for synchronous release of neurotransmitter from mossy fiber boutons, they did not perform pharmacological analyses to dissect the contribution of individual Cav channel types. Nonetheless, P/Q- and R-type channels are known to play an important role in neurotransmitter release within the hippocampus, and neurotransmission at the mossy fiber synapse specifically can be attenuated by $NiCl_2$ and SNX-482 (170), suggesting that Cav2.3 channels are important in regulating calcium dynamics in mossy fiber terminals (and in other fast, excitatory hippocampal synapses).

A second location where channels underlying neurotransmission have been recorded is in the cerebellar cortex. Using cerebellar slice preparations, Southan & Robertson (171, 172) have successfully recorded from large basket cell pinceau terminals. Using classical pharmacological tools and channel-specific peptide toxins, in combination with patch-clamp recordings, Southan & Robertson have demonstrated that DTX-sensitive Kv1 channel subunits underlie the low-threshold Kv conductance that modulates GABA release from the basket cell terminals. In addition, on the basis of biophysical analyses, a second, high-voltage-activated conductance has been recorded at these terminals (172) corresponding to a channel containing Kv3 α subunits. Overall, the electrophysiological and pharmacological assessment of basket cell terminals corresponds well with published immunohistochemical data, which have shown a high density of immunoreactivity for the DTX-sensitive subunits Kv1.1 and Kv1.2 (together with Kvβ2), as well as Kv3.4, at cerebellar basket cell terminals (69, 78, 79, 116, 117, 119, 124, 125). Although the beautiful work by Southan & Robertson has not been extended to explore the roles played by specific Nav or Cav channels, the basket cell terminal complex is unique in that it is one of the rare inhibitory synaptic terminals accessible to both detailed anatomical and electrophysiological analyses.

Channels Targeted to Somatodendritic Membranes

Channels that are concentrated in the somatodendritic membranes of hippocampal CA1 and neocortical layer V pyramidal cells, as well as those expressed on neocortical and hippocampal interneurons, are readily accessible for electrophysiological, pharmacological, and neuroanatomical analysis. A wealth of literature describes the role of individual Kv, Nav, and Cav conductances to excitability in somatodendritic membranes of these cell types (see 173–175). On the basis of these extensive studies, it is clear that Cav and Nav channels mediate postsynaptic responses to excitatory input, effectively regulating the active membrane properties of these cells. In large cortical pyramidal cells, Cav and Nav channels play a particularly critical role in distal dendrites, where these channels are responsible for initiating and propagating action potentials and amplify signals received on distal dendritic branches (e.g., 176–180). In neocortical and hippocampal pyramidal cells, action potentials initiated at the soma or axon initial segment frequently travel retrogradely, or back-propagate into the dendrites. This phenomenon is critical for input-specific synaptic plasticity and also for amplification of weak, distal excitatory synaptic inputs that occur in close temporal proximity with somatic action potentials discharged by the same neuron (181, 182). The contributions of dendritic Cav and Nav channels to somatodendritic excitation are extensively and effectively modulated by a diverse array of dendritic Kv channels. The density, subcellular distribution, and biophysical properties of Kv channels along the apical and basal dendrites of hippocampal and cortical pyramidal cells are fine-tuned to control subthreshold excitatory responses and control the amplitude and duration of back-propagating action potentials (182, see also 174). Somatic Kv channels, such as Kv2.1 are also exquisitely positioned as a final brake on somatic excitation, and their activation or inhibition could play a critical role in determining whether the summed level of excitation reaching the soma is sufficient to initiate an action potential (183).

As described above, immunohistochemical analyses have shown that the majority of postsynaptic, dendritic Nav channels are made up of Nav1.1 or Nav1.3 subunits. In hippocampal and neocortical neurons, Nav channel immunoreactivity is concentrated within the proximal apical dendrite, but staining is also observed along the apical dendrite and apical and basal dendritic branches (30, 31). Unfortunately, selective pharmacological tools are not available to help determine the individual contributions of these two Nav channel subunits to currents recorded in native cells. Nonetheless, although the density of recorded Nav currents along the dendritic arbors of hippocampal and neocortical pyramidal cells appears to remain constant (184, 185) or perhaps increases slightly close to the soma (180), it is clear that the biophysical properties of Nav currents change with increasing distance from the soma. In hippocampal pyramidal cells, Nav currents elicited distally in the apical dendrite (200–350 μm away) possess slow inactivation (184, 185), a property that has a major impact on the contribution of Nav channels to excitability along the dendrite because it can lead to long-lasting depolarization.

On the basis of immunohistochemical localization of Nav1.1 and Nav1.3 subunits in hippocampal and neocortical pyramidal cells, it seems plausible that Nav1.1 channels mediate the high density, more rapidly inactivating somatic and proximal dendritic sodium conductance, whereas Nav1.3 channels, perhaps, mediate the more slowly inactivating Nav conductance located more distally on the dendritic tree.

Almost all types of Cav channels are present along the dendrites of mammalian central neurons. Although the density of Cav currents appears to remain constant along the dendritic membrane, the local concentration of individual Cav channel subtypes varies at different points along the dendritic tree (87, 174). As described above, over 70% of the dihydropyridine-sensitive Cav channels in brain are encoded by the Cav1.2 subunit. Because immunoreactivity for this subunit is present along the soma and dendrites of pyramidal cells, Cav1.2 is likely to underlie a major component of the dendritic Cav current passed through L-type channels. Interestingly, a minor component of the dendritic L-type channel conductance is likely to be carried by Cav1.3 subunits. Immunoreactivity for Cav1.3 is concentrated more distally in the dendritic membrane and is found in association with punctuate structures, most likely dendritic spines (90). Convincing immunohistochemical staining in somatodendritic membranes of hippocampal or neocortical pyramidal cells has not been reported for other Cav subunits.

Somatodendritic Kv currents have been most extensively studied by patch-clamp recording in hippocampal pyramidal cells and in the large layer V pyramidal neurons in motor and somatosensory cortex (e.g., 175, 182, 186–191). In CA1 pyramidal cells, A-type currents play a critical role in synaptic integration and plasticity by controlling subthreshold excitation and the amplitude of back-propagating action potentials (182, 188). It has been reported that a DTX-sensitive channel, perhaps encoded by Kv1.1 or Kv1.2 in combination with accessory Kvβ subunits, may underlie a component of the somatic Kv current (189). However, in mammalian central neurons, the somatodendritic A-type potassium conductance is predominantly mediated by Kv4-family α subunits, which coassemble and coassociate with KChIPs (70) and perhaps also DPPX (73) as accessory subunits. In neocortical and hippocampal pyramidal cells, the density of A-type currents increases with distance from the pyramidal cell soma (182), and thus the impact of these currents in dampening postsynaptic excitation increases as a function of distance from the cell soma. Published reports of immunostaining for Kv4 subunits in the hippocampus have not described a gradient in Kv4 immunoreactivity that would match the gradient in A-type current observed in electrophysiological experiments. Nonetheless, it is clear from these studies that there is a dramatic decrease in the density of Kv4 immunoreactivity at the junction of s. radiatum (containing the proximal apical dendrites of CA1 pyramidal cells) compared with that in s. moleculare (containing the distal apical dendritic tufts of CA1 pyramidal cells) (80). The decrease in Kv4 immunoreactivity in s. moleculare indicates that there is a very low density of Kv4-mediated A-type current in the very distal dendrites of CA pyramidal cells and suggests that these distal dendrites will respond strongly to excitatory stimuli. Thus one might expect that excitatory inputs that terminate on the distal dendrites of CA1 pyramidal cells in s. moleculare would evoke large postsynaptic EPSPs because there is little A-type current to attenuate subthreshold excitation. This would contrast with the expectation of smaller EP-SPs in the more proximal apical and basal dendrites, where the very high density of Kv4 channels (and A-type current) would effectively dampen excitation and the back-propagation of action potentials (182, 187).

Electrophysiological studies have demonstrated that in neocortical layer V pyramidal cells a gradient of A-current along the apical dendrite does not appear and that there is a far lower density of A-type current in layer V pyramidal cells compared with that in hippocampal CA1 pyramidal cells (175, 182, 190, 191). These findings suggest that the dendritic A-type current does not exert as dominant an influence on excitatory input in cortical pyramidal cells as it does in CA1 (175). In published studies and in our own, unpublished work it appears that immunoreactivity for Kv4.2 (as well as some KChIP isoforms) can be followed along the entire apical dendrite of cortical layer V pyramidal cells, and that there appears to be a fairly uniform density of immunoreactivity along the dendritic shaft. However, it is important to note that Kv4.2 (and KChIP) immunoreactivity can be followed into the most distal dendritic branches and the apical dendritic tufts of neocortical pyramidal cells. Thus although there is a comparatively lower density of A-type current in the apical dendrites of neocortical pyramids, the presence of Kv4 channels in the most distal dendritic regions indicates that these channels modulate excitatory input along the entire dendrite. Thus neocortical pyramidal cell A-type currents are not compartmentalized within discrete dendritic domains as they are in the hippocampus. The functional significance of this differential distribution in hippocampus versus neocortex remains to be explored.

Some large, inhibitory, GABAergic interneurons in cortex and hippocampus have the ability to fire action potentials at extremely high frequencies without adaptation. This characteristic firing pattern is critically dependent on the expression of specific ion channels whose biophysical properties permit rapid integration of excitatory inputs and allow for sustained high-frequency propagation of action potentials. The most intensively studied fast-spiking interneurons are the large GABAergic and calbindin-immunoreactive interneurons scattered throughout s. oriens and s. radiatum of the hippocampal CA subfields and throughout all layers of the neocortex. Part of the ability of these cells to fire high-frequency action potentials comes from their expression of an extremely high density of voltage-gated Nav channels whose inactivation kinetics are slower than those observed hippocampal pyramidal cells (192, 193). Based on published immunohistochemical analyses, the best candidate for the somatodendritic Nav current in these fast-spiking interneurons is Nav1.1 (28, 29). Although there is no extensive data describing the patterns of immunoreactivity for Cav channel subtypes in interneurons, as defined using selective toxins and pharmacological tools, it is clear that a mixture of Cav currents contribute to dendritic excitability in these cells because there is a high density of N- and L-type currents, and a lower density of P/Q currents in interneurons (194). Interestingly, GABA release from hippocampal interneurons is modulated by distinct sets of Cav channels, in a cell-type and synapse-specific manner (195).

Kv channels play a key role in determining the firing properties of hippocampal and cortical interneurons. Major contributions to our understanding of Kv currents and channel subunits controlling the firing properties and excitability of these cells have come from the laboratories of McBain, Rudy, and Jonas, and their colleagues. Using electrophysiological, pharmacological, immunohistochemical, and genetic approaches, investigators in these laboratories have characterized in detail the contribution of individual Kv channel subtypes and α subunits to the complex response properties of interneurons (135, 142, 145–149, 152, 193, 196–200). From this work, it is clear that Kv3 channel subunits play a major role in determining the fast-spiking properties of these neurons, although somatodendritic Kv4 channels, as well as clustered Kv2.1 channels, also shape the action potential waveform in these cells. The electrophysiological characterization of Kv currents in interneurons is consistent with published immunohistochemical analyses, which have demonstrated a high density of immunoreactivity for Kv3 α subunits, Kv4.3, KChIP1, and Kv2.1 in hippocampal and neocortical interneurons (see also 69, 70, 79).

MECHANISMS UNDERLYING SUBCELLULAR LOCALIZATION

Biosynthetic Mechanisms

Mammalian neurons express a diverse repertoire of voltage-gated ion channels that act in concert to shape their reception, processing, integration, and transmission of information. The functional expression of channels in the plasma membrane of these neurons is regulated at a number of different levels. Expression of ion channel genes is highly regulated, with specific promoter elements acting in concert with transcriptional machinery to achieve precise temporal and spatial cellular patterns of expression (201). Although evidence for dynamic regulation of ion channel translation has not been provided, it is clear that multiple post-translational processes dramatically impact the abundance and distribution of ion channel proteins in the neuronal membrane.

Many ion channel subunits carry specific intracellular trafficking signals that regulate their exit from the rough endoplasmic reticulum, where they are synthesized, and their stepwise transit through the biosynthetic endomembrane system. Polarized sorting of channel proteins to axon- or dendrite-directed cargo vesicles in the *trans*-Golgi network, targeted insertion at discrete sites within axonal or dendritic plasma membrane subdomains, and active retention at these sites to maintain discrete patterns of channel localization in the face of membrane fluidity are also involved in determining which channels are present at precise sites in neurons. The resulting compartmentalized, nonuniform distribution of ion channels can dramatically impact neuronal signaling through local regulation of the amplitude and spread of membrane depolarization.

Ion Channel Targeting Signals

A number of recent studies have attempted to define the determinants of the differing subcellular localization of highly related ion channels in mammalian neurons. These studies have employed both polarized epithelial cell lines, such as MDCK cells, and primary cultures of neurons as dissociated cells or in organotypic culture. The most progress to date has been in the study of Kv channels, perhaps due to the fact that they are composed of small format α subunits and in many cases lack an absolute requirement for auxiliary subunit coexpression. Some of the earliest work was aimed at defining the determinants of the polarized and clustered localization of Kv2.1. The initial studies, performed in MDCK cells, showed that the cytoplasmic C terminus of Kv2.1 was necessary for both polarized expression and clustering (138). These findings were later extended to studies in cultured hippocampal neurons. These revealed that a small segment within the 410–amino acid Kv2.1 cytoplasmic tail was found to be necessary and sufficient for polarized and clustered localization (141). Interestingly, three of the four critical amino acids in this segment were serines, suggesting that changes in phosphorylation state may dynamically regulate Kv2.1 localization.

A recent study (202) has used a similar approach to define the determinants of the dendritic localization of Kv4.2. Analyses of chimeras between axonal Kv1.4 and dendritic Kv4.2 revealed a dendritic targeting determinant in the cytoplasmic C terminus of Kv4.2. Further mutational studies showed the critical sequence is a 16–amino acid motif present in all Kv4 family members. This targeting signal apparently underlies the dendritic localization of Kv4.2 and Kv4.3 observed in principal cells and in the somatodendritic localization of Kv4.3 in interneurons, observed throughout the mammalian central nervous system.

The C-terminal cytoplasmic domain also appears to control the polarized localization of Nav channels in neurons. A small sequence (CLDILFAFT) was identified that mediates both axonal targeting and somatodendritic endocytosis of Nav1.2 (203). Interestingly, this sequence is 100% conserved in all neuronal Nav channels (Nav1.1, Nav1.2, Nav1.3, and Nav1.6) that exhibit very different subcellular localizations in mammalian brain (Nav1.1 and Nav1.3 are somatodendritic, and Nav1.2 and Nav1.6 are axonal). As such, the contribution of this signal in the establishment and maintenance of the contrasting patterns of subcellular localization of different Nav isoforms in mammalian central neurons is unclear. This same group recently identified a second Nav1 targeting determinant that directs localization of these channels to the axon initial segment (204). Interestingly, this initial segment-targeting domain, as well as a recently described motif for ankyrin binding contained within this domain (205), is found in all Nav1 channels. As such, determinants of the subtype-specificity of subcellular localization of Nav family members in neurons remain to be found.

Only one study has investigated neuronal subcellular targeting signals in Cav channels. As described above, Cav2.2 exists in both long- (Cav2.2a/α_{1b-1}) and short- (Cav2.2b/α_{1b-2}) channel isoforms through alternative splicing of Cav2.2 mRNA (111). A recent study of the subcellular localization of exogenously expressed Cav2.2 subunits in cultured hippocampal neurons suggests that the long Cav2.2a variant is specifically targeted to presynaptic terminals, whereas the shorter form is localized in the somatodendritic domain (112). The synaptic localization of the larger Cav2.2a variant appears to be regulated by interaction with PDZ- and SH3-containing proteins, i.e., the long (but not short) variant contains canonical binding motifs for these modular protein-protein interaction domains (206), and these motifs are necessary for the synaptic localization of Cav2.2a (112). The specific binding partners for these canonical interaction domains of Cav2.2 and the mechanism whereby they regulate synaptic localization have not yet been determined.

CHALLENGES AND PROSPECTS

The molecular diversity of voltage-gated ion channels combined with the cellular and morphological complexity of the central nervous system poses a considerable challenge in attempts to map the contributions of specific ion channels to processes controlling neuronal excitability. The availability of cDNA clones encoding each of the members of this large gene family has allowed for the recent generation of molecularly defined antibodies, and these reagents are being used to map the locations of individual channel subunits within neuronal populations and specific subcellular domains. Most of these studies have employed immunoperoxidase staining and analyses by light microscopy. Although such approaches are powerful as a first pass, more definitive studies employing parallel excitotoxic lesions or immuno-electron microscopy are needed before more definitive conclusions as to specific subcellular localizations can be drawn. These studies will allow for a more precise correlation between channel localization and discrete functional domains (e.g., active zones of terminals, postsynaptic membranes, dendritic branch points, etc.), which may provide additional inferences for function.

A major deficit in the study of many mammalian brain ion channels is the lack of definitive pharmacology with which one can acutely modulate the function of specific molecularly defined channels. For example, specific blockers of individual neuronal sodium channel subtypes (i.e., that would discriminate among Nav1.1–Nav1.3 and Nav1.6) are not available. Structural studies of these proteins may provide information that would enable the design of pharmacological tools that could be used to discriminate between these highly related proteins. Another possibility is that subtype-specific antibodies could be used to modulate the activity of specific channels. While this has worked in some cases (for example, see 183), the general applicability of such an approach is not proven. One problem is that the external mouth of the channel pore, the most attractive site for channel blockers, is highly conserved among members of the same channel family. For this reason most antibodies directed at channel ectodomains have been generated against the S1-S2 extracellular domain. Our experience has been that although these antibodies bind to native channels, they do not affect channel function. Perhaps future attempts will yield subtype-specific blockers that will allow for the pharmacological dissection of the role of individual subunits in neuronal function.

More recently, genetic approaches have been applied to help match expression of individual channel subunits to physiologically identified currents. This effort has taken primarily two forms. First, targeted deletion of ion channel genes has been used to ablate expression in individual subunits, with the effects on the phenotype of native currents then recorded. For brain Kv channels, this approach has been used to link expression of Kv3.2 to a component of the somatodendritic Kv conductance in cortical interneurons (148). Similarly, this approach has been used to determine the effects of Kv1.1 (207–209), Kvβ1 (210), and Kvβ2 (211) removal on Kv1-family Kv currents. One limitation of the gene knockout approach is that deletion of many ion channel subunit genes results in a profound behavioral phenotype or to perinatal death. As a means to circumvent this problem, a technically elegant strategy using transient expression of dominant-negative pore mutants has been used by Nerbonne and colleagues (159, 160, 212, 213) to link individual channel subunits to specific native currents. By tagging these dominant-negative mutants with green fluorescent protein, one should be able not only to "poison" the activity of a specific channel subtype (the tagged channel must coassemble with native channel subunits in order to have its effect) but also to localize subunit expression to specific subcellular domains. One limitation of this approach is that it cannot be used to study the contributions of individual Kv α subunits within a family, given their propensity to coassemble, and of many accessory subunits because it has proven difficult to develop dominant-negative versions of these proteins.

The combinatorial diversity of subunit composition of native channels remains a particular challenge for studies of Kv and Cav channels. While some progress has been made on defining the subunit composition of native channels (214, 215), linking these biochemical populations with specific functions has not been achieved. A major goal is to define the subunit components channels and to discern their contribution to neuronal physiology. One challenge is whether pharmacological tools—small molecules, toxins, or antibodies—that modulate only specific subunit combinations of channels can be identified. These would offer advantages as selective tools for the dissection of the role of a specific channel in neuronal physiology and could also yield agents with high cellular specificity due to the large number of subunit combinations that occur in native cells. These and other challenges will drive future investigations into which ion channels are performing which specific tasks in mammalian central, and may provide a promising area for the development of therapeutics for human neurological or psychiatric disorders.

The *Annual Review of Physiology* is online at http://physiol.annualreviews.org

LITERATURE CITED

1. Hille B. 2001. *Ionic Channels of Excitable Membranes*. Sunderland, MA: Sinauer
2. Bezanilla F, Perozo E. 2003. The voltage sensor and the gate in ion channels. *Adv. Protein Chem.* 63:211–41
3. Llinas RR. 1988. The intrinsic electrophysiological properties of mammalian neurons: insights into central nervous system function. *Science* 242:1654–64
4. Catterall WA. 1995. Structure and function of voltage-gated ion channels. *Annu. Rev. Biochem.* 64:493–531
5. Hanlon MR, Wallace BA. 2002. Structure and function of voltage-dependent ion channel regulatory beta subunits. *Biochemistry* 41:2886–94
6. Sucher NJ, Deitcher DL, Baro DJ, Warrick RM, Guenther E. 2000. Genes and channels: patch/voltage-clamp analysis and single-cell RT-PCR. *Cell Tissue Res.* 302:295–307
7. Chandy KG. 1991. Simplified gene nomenclature. *Nature* 352:26
8. Ertel EA, Campbell KP, Harpold MM, Hofmann F, Mori Y, et al. 2000. Nomenclature of voltage-gated calcium channels. *Neuron* 25:533–35
9. Goldin AL, Barchi RL, Caldwell JH, Hofmann F, Howe JR, et al. 2001. Nomenclature of voltage-gated sodium channels. *Neuron* 28:365–68
10. Yellen G. 1999. The bacterial K^+ channel structure and its implications for neuronal channels. *Curr. Opin. Neurobiol.* 9:267–73
11. Doyle DA, Morais Cabral J, Pfuetzner RA, Kuo A, Gulbis JM, et al. 1998. The structure of the potassium channel: molecular basis of K^+ conduction and selectivity. *Science* 280:69–77
12. Jiang Y, Ruta V, Chen J, Lee A, MacKinnon R. 2003. The principle of gating charge movement in a voltage-dependent K^+ channel. *Nature* 423:42–48
13. Strong M, Chandy KG, Gutman GA. 1993. Molecular evolution of voltage-sensitive ion channel genes: on the origins of electrical excitability. *Mol. Biol. Evol.* 10:221–42
14. Goldin AL. 2002. Evolution of voltage-gated Na^+ channels. *J. Exp. Biol.* 205:575–84
15. Trimmer JS. 1998. Regulation of ion channel expression by cytoplasmic subunits. *Curr. Opin. Neurobiol.* 8:370–74
16. Malo D, Schurr E, Dorfman J, Canfield V, Levenson R, Gros P. 1991. Three brain sodium channel alpha-subunit genes are clustered on the proximal segment of mouse chromosome 2. *Genomics* 10:666–72
17. Beckers MC, Ernst E, Belcher S, Howe J, Levenson R, Gros P. 1996. A new sodium channel alpha-subunit gene (*Scn9a*) from Schwann cells maps to the Scn1a, Scn2a, Scn3a cluster of mouse chromosome 2. *Genomics* 36:202–5
18. Albrecht B, Weber K, Pongs O. 1995. Characterization of a voltage-activated K-channel gene cluster on human chromosome 12p13. *Receptors Channels* 3:213–20
19. Chandy KG, Williams CB, Spencer RH, Aguilar BA, Ghanshani S, et al. 1990. A family of three mouse potassium channel genes with intronless coding regions. *Science* 247:973–75
20. Leicher T, Roeper J, Weber K, Wang X, Pongs O. 1996. Structural and functional characterization of human potassium channel subunit beta 1 (KCNA1B). *Neuropharmacology* 35:787–95
21. Noda M, Shimizu S, Tanabe T, Takai T, Kayano T, et al. 1984. Primary structure of *Electrophorus electricus* sodium channel deduced from cDNA sequence. *Nature* 312:121–27
22. Noda M, Ikeda T, Kayano T, Suzuki H,

Takeshima H, et al. 1986. Existence of distinct sodium channel messenger RNAs in rat brain. *Nature* 320:188–92

23. Kayano T, Noda M, Flockerzi V, Takahashi H, Numa S. 1988. Primary structure of rat brain sodium channel III deduced from the cDNA sequence. *FEBS Lett.* 228:187–94

24. Beckh S, Noda M, Lubbert H, Numa S. 1989. Differential regulation of three sodium channel messenger RNAs in the rat central nervous system during development. *EMBO J.* 8:3611–16

25. Beckh S. 1990. Differential expression of sodium channel mRNAs in rat peripheral nervous system innervated tissues. *FEBS Lett.* 262:317–22

26. Brysch W, Creutzfeldt OD, Luno K, Schlingensiepen R, Schlingensiepen KH. 1991. Regional temporal expression of sodium channel messenger RNAs in the rat brain during development. *Exp. Brain Res.* 86:562–67

27. Black JA, Yokoyama S, Higashida H, Ransom BR, Waxman SG. 1994. Sodium channel mRNAs I, II and III in the CNS: cell-specific expression. *Brain Res. Mol. Brain Res.* 22:275–89

28. Westenbroek RE, Merrick DK, Catterall WA. 1989. Differential subcellular localization of the RI and RII Na⁺ channel subtypes in central neurons. *Neuron* 3:695–704

29. Gong B, Rhodes KJ, Bekele-Arcuri Z, Trimmer JS. 1999. Type I and type II Na⁺ channel alpha-subunit polypeptides exhibit distinct spatial and temporal patterning, and association with auxiliary subunits in rat brain. *J. Comp. Neurol.* 412:342–52

30. Westenbroek RE, Noebels JL, Catterall WA. 1992. Elevated expression of type II Na⁺ channels in hypomyelinated axons of shiverer mouse brain. *J. Neurosci.* 12:2259–67

31. Whitaker WR, Faull RL, Waldvogel HJ, Plumpton CJ, Emson PC, Clare JJ. 2001. Comparative distribution of voltage-gated sodium channel proteins in human brain. *Brain Res. Mol. Brain Res.* 88:37–53

32. Schaller KL, Krzemien DM, Yarowsky PJ, Krueger BK, Caldwell JH. 1995. A novel, abundant sodium channel expressed in neurons glia. *J. Neurosci.* 15:3231–42

33. Schaller KL, Caldwell JH. 2000. Developmental regional expression of sodium channel isoform NaCh6 in the rat central nervous system. *J. Comp. Neurol.* 420:84–97

34. Krzemien DM, Schaller KL, Levinson SR, Caldwell JH. 2000. Immunolocalization of sodium channel isoform NaCh6 in the nervous system. *J. Comp. Neurol.* 420:70–83

35. Caldwell JH, Schaller KL, Lasher RS, Peles E, Levinson SR. 2000. Sodium channel Na$_v$1.6 is localized at nodes of Ranvier, dendrites, and synapses. *Proc. Natl. Acad. Sci. USA* 97:5616–20

36. Boiko T, Van Wart A, Caldwell JH, Levinson SR, Trimmer JS, Matthews G. 2003. Functional specialization of the axon initial segment by isoform-specific sodium channel targeting. *J. Neurosci.* 23:2306–13

37. Boiko T, Rasband MN, Levinson SR, Caldwell JH, Mandel G, et al. 2001. Compact myelin dictates the differential targeting of two sodium channel isoforms in the same axon. *Neuron* 30:91–104

38. Hartmann HA, Colom LV, Sutherland ML, Noebels JL. 1999. Selective localization of cardiac SCN5A sodium channels in limbic regions of rat brain. *Nat. Neurosci.* 2:593–95

39. Wu L, Nishiyama K, Hollyfield JG, Wang Q. 2002. Localization of Nav1.5 sodium channel protein in the mouse brain. *NeuroReport* 13:2547–51

40. Tanabe T, Takeshima H, Mikami A, Flockerzi V, Takahashi H, et al. 1987. Primary structure of the receptor for calcium channel blockers from skeletal muscle. *Nature* 328:313–18

41. Mori Y, Friedrich T, Kim MS, Mikami A,

Nakai J, et al. 1991. Primary structure and functional expression from complementary DNA of a brain calcium channel. *Nature* 350:398–402

42. Starr TV, Prystay W, Snutch TP. 1991. Primary structure of a calcium channel that is highly expressed in the rat cerebellum. *Proc. Natl. Acad. Sci. USA* 88:5621–25

43. Catterall WA. 2000. Structure and regulation of voltage-gated Ca^{2+} channels. *Annu. Rev. Cell Dev. Biol.* 16:521–55

44. Johnston D, Christie BR, Frick A, Gray R, Hoffman DA, et al. 2003. Active dendrites, potassium channels and synaptic plasticity. *Philos. Trans. R. Soc. London Ser. B* 358:667–74

45. Snutch TP, Sutton KG, Zamponi GW. 2001. Voltage-dependent calcium channels—beyond dihydropyridine antagonists. *Curr. Opin. Pharmacol.* 1:11–16

46. Talley EM, Cribbs LL, Lee JH, Daud A, Perez-Reyes E, Bayliss DA. 1999. Differential distribution of three members of a gene family encoding low voltage-activated (T-type) calcium channels. *J. Neurosci.* 19:1895–911

47. Yunker AM, Sharp AH, Sundarraj S, Ranganathan V, Copeland TD, McEnery MW. 2003. Immunological characterization of T-type voltage-dependent calcium channel Ca$_V$3.1 (alpha1G) and Ca$_V$3.3 (alpha1I) isoforms reveal differences in their localization, expression, and neural development. *Neuroscience* 117:321–35

48. Jan LY, Jan YN. 1997. Cloned potassium channels from eukaryotes and prokaryotes. *Annu. Rev. Neurosci.* 20:91–123

49. Kamb A, Iverson LE, Tanouye MA. 1987. Molecular characterization of *Shaker*, a *Drosophila* gene that encodes a potassium channel. *Cell* 50:405–13

50. Papazian DM, Schwarz TL, Tempel BL, Jan YN, Jan LY. 1987. Cloning of genomic complementary DNA from *Shaker*, a putative potassium channel gene from *Drosophila*. *Science* 237:749–53

51. MacKinnon R. 1991. Determination of the subunit stoichiometry of a voltage-activated potassium channel. *Nature* 350:232–35

52. Salkoff L, Baker K, Butler A, Covarrubias M, Pak MD, Wei A. 1992. An essential 'set' of K$^+$ channels conserved in flies, mice and humans. *Trends Neurosci.* 15:161–66

53. Tempel BL, Jan YN, Jan LY. 1988. Cloning of a probable potassium channel gene from mouse brain. *Nature* 332:837–39

54. Chandy KG, Gutman GA. 1995. Voltage-gated potassium channel genes. In *Ligand-Voltage-Gated Ion Channels*, ed. RA North, pp. 1–71. Boca Raton, FL: CRC Press

55. Stuhmer W, Ruppersberg JP, Schroter KH, Sakmann B, Stocker M, et al. 1989. Molecular basis of functional diversity of voltage-gated potassium channels in mammalian brain. *EMBO J.* 8:3235–44

56. Isacoff EY, Jan YN, Jan LY. 1990. Evidence for the formation of heteromultimeric potassium channels in *Xenopus* oocytes. *Nature* 345:530–34

57. Ruppersberg JP, Schroter KH, Sakmann B, Stocker M, Sewing S, Pongs O. 1990. Heteromultimeric channels formed by rat brain potassium-channel proteins. *Nature* 345:535–37

58. Hopkins WF, Allen ML, Houamed KM, Tempel BL. 1994. Properties of voltage-gated K$^+$ currents expressed in *Xenopus* oocytes by mKv1.1, mKv1.2 and their heteromultimers as revealed by mutagenesis of the dendrotoxin-binding site in mKv1.1. *Pflügers Arch.* 428:382–90

59. Manganas LN, Wang Q, Scannevin RH, Antonucci DE, Rhodes KJ, Trimmer JS. 2001. Identification of a trafficking determinant localized to the Kv1 potassium channel pore. *Proc. Natl. Acad. Sci. USA* 98:14055–59

60. Manganas LN, Trimmer JS. 2000. Subunit composition determines Kv1 potassium channel surface expression. *J. Biol. Chem.* 275:29685–93

61. Frech G, VanDongen AMJ, Schuster G,

Brown AM, Joho RH. 1989. A novel potassium channel with delayed rectifier properties isolated from rat brain by expression cloning. *Nature* 340:642–45

62. Yokoyama S, Imoto K, Kawamura T, Higashida H, Iwabe N, et al. 1989. Potassium channels from NG108-15 neuroblastoma-glioma hybrid cells. Primary structure and functional expression from cDNAs. *FEBS Lett.* 259:37–42

63. Luneau CJ, Williams JB, Marshall J, Levitan ES, Oliva C, et al. 1991. Alternative splicing contributes to K$^+$ channel diversity in the mammalian central nervous system. *Proc. Natl. Acad. Sci. USA* 88:3932–36

64. McCormack T, Vega-Saenz de Miera EC, Rudy B. 1991. Molecular cloning of a member of a third class of *Shaker*-family K$^+$ channel genes in mammals. *Proc. Natl. Acad. Sci. USA* 88:4060

65. Rudy B, Sen K, Vega-Saenz de Miera E, Lau D, Ried T, Ward DC. 1991. Cloning of a human cDNA expressing a high voltage-activating, TEA-sensitive, type-A K$^+$ channel which maps to chromosome 1 band p21. *J. Neurosci. Res.* 29:401–12

66. Baldwin TJ, Tsaur ML, Lopez GA, Jan YN, Jan LY. 1991. Characterization of a mammalian cDNA for an inactivating voltage-sensitive K$^+$ channel. *Neuron* 7:471–83

67. Roberds SL, Tamkun MM. 1991. Cloning tissue-specific expression of five voltage-gated potassium channel cDNAs expressed in rat heart. *Proc. Natl. Acad. Sci. USA* 88:1798–802

68. Pongs O, Leicher T, Berger M, Roeper J, Bahring R, et al. 1999. Functional molecular aspects of voltage-gated K$^+$ channel beta subunits. *Ann. NY Acad. Sci.* 868:344–55

69. Rhodes KJ, Monaghan MM, Barrezueta NX, Nawoschik S, Bekele-Arcuri Z, et al. 1996. Voltage-gated K$^+$ channel beta subunits: expression and distribution of Kv beta 1 and Kv beta 2 in adult rat brain. *J. Neurosci.* 16:4846–60

70. An WF, Bowlby MR, Betty M, Cao J, Ling HP, et al. 2000. Modulation of A-type potassium channels by a family of calcium sensors. *Nature* 403:553–56

71. Holmqvist MH, Cao J, Hernandez-Pineda R, Jacobson MD, Carroll KI, et al. 2002. Elimination of fast inactivation in Kv4 A-type potassium channels by an auxiliary subunit domain. *Proc. Natl. Acad. Sci. USA* 99:1035–40

72. Shibata R, Misonou H, Campomanes CR, Anderson AE, Schrader LA, et al. 2003. A fundamental role for KChIPs in determining the molecular properties and trafficking of Kv4.2 potassium channels. *J. Biol. Chem.* 278:36445–54

73. Nadal MS, Ozaita A, Amarillo Y, de Miera EV, Ma Y, et al. 2003. The CD26-related dipeptidyl aminopeptidase-like protein DPPX is a critical component of neuronal A-type K$^+$ channels. *Neuron* 37:449–61

74. Drewe JA, Verma S, Frech G, Joho RH. 1992. Distinct spatial and temporal expression patterns of K$^+$ channel mRNAs from different subfamilies. *J. Neurosci.* 12:538–48

75. Patel AJ, Lazdunski M, Honore E. 1997. Kv2.1/Kv9.3, a novel ATP-dependent delayed-rectifier K$^+$ channel in oxygen-sensitive pulmonary artery myocytes. *EMBO J.* 16:6615–25

76. Salinas M, Duprat F, Heurteaux C, Hugnot JP, Lazdunski M. 1997. New modulatory alpha subunits for mammalian Shab K$^+$ channels. *J. Biol. Chem.* 272:24371–79

77. Bekele-Arcuri Z, Matos MF, Manganas L, Strassle BW, Monaghan MM, et al. 1996. Generation characterization of subtype-specific monoclonal antibodies to K$^+$ channel alpha- and beta-subunit polypeptides. *Neuropharmacology* 35:851–65

78. Rhodes KJ, Keilbaugh SA, Barrezueta NX, Lopez KL, Trimmer JS. 1995. Association and colocalization of K$^+$ channel alpha- and beta-subunit polypeptides in rat brain. *J. Neurosci.* 15:5360–71

79. Rhodes KJ, Strassle BW, Monaghan MM, Bekele-Arcuri Z, Matos MF, Trimmer JS. 1997. Association and colocalization of the Kvbeta1 and Kvbeta2 beta-subunits with Kv1 alpha-subunits in mammalian brain K$^+$ channel complexes. *J. Neurosci.* 17:8246–58

80. Sheng M, Tsaur ML, Jan YN, Jan LY. 1992. Subcellular segregation of two A-type K$^+$ channel proteins in rat central neurons. *Neuron* 9:271–84

81. Monaghan MM, Trimmer JS, Rhodes KJ. 2001. Experimental localization of Kv1 family voltage-gated K$^+$ channel alpha and beta subunits in rat hippocampal formation. *J. Neurosci.* 21:5973–83

82. Cooper EC, Milroy A, Jan YN, Jan LY, Lowenstein DH. 1998. Presynaptic localization of Kv1.4-containing A-type potassium channels near excitatory synapses in the hippocampus. *J. Neurosci.* 18:965–74

83. Whitaker WR, Clare JJ, Powell AJ, Chen YH, Faull RL, Emson PC. 2000. Distribution of voltage-gated sodium channel alpha-subunit and beta-subunit mRNAs in human hippocampal formation, cortex, and cerebellum. *J. Comp. Neurol.* 422:123–39

84. Raman IM, Sprunger LK, Meisler MH, Bean BP. 1997. Altered subthreshold sodium currents and disrupted firing patterns in Purkinje neurons of Scn8a mutant mice. *Neuron* 19:881–91

85. Ahlijanian MK, Westenbroek RE, Catterall WA. 1990. Subunit structure localization of dihydropyridine-sensitive calcium channels in mammalian brain, spinal cord, and retina. *Neuron* 4:819–32

86. Westenbroek RE, Ahlijanian MK, Catterall WA. 1990. Clustering of L-type Ca^{2+} channels at the base of major dendrites in hippocampal pyramidal neurons. *Nature* 347:281–84

87. Magee JC, Johnston D. 1995. Synaptic activation of voltage-gated channels in the dendrites of hippocampal pyramidal neurons. *Science* 268:301–4

88. Westenbroek RE, Hell JW, Warner C, Dubel SJ, Snutch TP, Catterall WA. 1992. Biochemical properties and subcellular distribution of an N-type calcium channel alpha 1 subunit. *Neuron* 9:1099–115

89. Hell JW, Westenbroek RE, Warner C, Ahlijanian MK, Prystay W, et al. 1993. Identification and differential subcellular localization of the neuronal class C and class D L-type calcium channel alpha 1 subunits. *J. Cell Biol.* 123:949–62

90. Davare MA, Avdonin V, Hall DD, Peden EM, Burette A, et al. 2001. A beta2 adrenergic receptor signaling complex assembled with the Ca^{2+} channel Cav1.2. *Science* 293:98–101

91. Westenbroek RE, Sakurai T, Elliott EM, Hell JW, Starr TV, et al. 1995. Immunochemical identification and subcellular distribution of the alpha 1A subunits of brain calcium channels. *J. Neurosci.* 15:6403–18

92. Restituito S, Thompson RM, Eliet J, Raike RS, Riedl M, et al. 2000. The polyglutamine expansion in spinocerebellar ataxia type 6 causes a beta subunit-specific enhanced activation of P/Q-type calcium channels in *Xenopus* oocytes. *J. Neurosci.* 20:6394–403

93. Craig PJ, McAinsh AD, McCormack AL, Smith W, Beattie RE, et al. 1998. Distribution of the voltage-dependent calcium channel alpha(1A) subunit throughout the mature rat brain and its relationship to neurotransmitter pathways. *J. Comp. Neurol.* 397:251–67

94. Rettig J, Sheng ZH, Kim DK, Hodson CD, Snutch TP, Catterall WA. 1996. Isoform-specific interaction of the alpha1A subunits of brain Ca^{2+} channels with the presynaptic proteins syntaxin and SNAP-25. *Proc. Natl. Acad. Sci. USA* 93:7363–68

95. Catterall WA. 1999. Interactions of presynaptic Ca^{2+} channels and snare proteins in neurotransmitter release. *Ann. NY Acad. Sci.* 868:144–59

96. Hanson JE, Smith Y. 2002. Subcellular distribution of high-voltage-activated

calcium channel subtypes in rat globus pallidus neurons. *J. Comp. Neurol.* 442: 89–98

97. Kitano J, Nishida M, Itsukaichi Y, Minami I, Ogawa M, et al. 2003. Direct interaction and functional coupling between metabotropic glutamate receptor subtype 1 and voltage-sensitive Cav2.1 Ca^{2+} channel. *J. Biol. Chem.* 278:25101–8

98. Llinas R, Sugimori M, Lin JW, Cherksey B. 1989. Blocking isolation of a calcium channel from neurons in mammals and cephalopods utilizing a toxin fraction (FTX) from funnel-web spider poison. *Proc. Natl. Acad. Sci. USA* 86:1689–93

99. Elliott EM, Malouf AT, Catterall WA. 1995. Role of calcium channel subtypes in calcium transients in hippocampal CA3 neurons. *J. Neurosci.* 15:6433–44

100. Kavalali ET, Zhuo M, Bito H, Tsien RW. 1997. Dendritic Ca^{2+} channels characterized by recordings from isolated hippocampal dendritic segments. *Neuron* 18:651–63

101. Mouginot D, Bossu JL, Gahwiler BH. 1997. Low-threshold Ca^{2+} currents in dendritic recordings from Purkinje cells in rat cerebellar slice cultures. *J. Neurosci.* 17:160–70

102. Magee JC, Carruth M. 1999. Dendritic voltage-gated ion channels regulate the action potential firing mode of hippocampal CA1 pyramidal neurons. *J. Neurophysiol.* 82:1895–901

103. Soong TW, DeMaria CD, Alvania RS, Zweifel LS, Liang MC, et al. 2002. Systematic identification of splice variants in human P/Q-type channel alpha1(2.1) subunits: implications for current density Ca^{2+}-dependent inactivation. *J. Neurosci.* 22:10142–52

104. Kerr LM, Yoshikami D. 1984. A venom peptide with a novel presynaptic blocking action. *Nature* 308:282–84

105. Kerr LM, Filloux F, Olivera BM, Jackson H, Wamsley JK. 1988. Autoradiographic localization of calcium chan-nels with ^{125}Iomega-conotoxin in rat brain. *Eur. J. Pharmacol.* 146:181–83

106. Takemura M, Kiyama H, Fukui H, Tohyama M, Wada H. 1988. Autoradiographic visualization in rat brain of receptors for omega-conotoxin GVIA, a newly discovered calcium antagonist. *Brain Res.* 451:386–89

107. Takemura M, Kiyama H, Fukui H, Tohyama M, Wada H. 1989. Distribution of the omega-conotoxin receptor in rat brain. An autoradiographic mapping. *Neuroscience* 32:405–16

108. Albensi BC, Ryujin KT, McIntosh JM, Naisbitt SR, Olivera BM, Fillous F. 1993. Localization of ^{125}Iomega-conotoxin GVIA binding in human hippocampus cerebellum. *NeuroReport* 4:1331–34

109. Azimi-Zonooz A, Kawa CB, Dowell CD, Olivera BM. 2001. Autoradiographic localization of N-type VGCCs in gerbil hippocampus and failure of omega-conotoxin MVIIA to attenuate neuronal injury after transient cerebral ischemia. *Brain Res.* 907:61–70

110. Day NC, Shaw PJ, McCormack AL, Craig PJ, Smith W, et al. 1996. Distribution of alpha 1A, alpha 1B and alpha 1E voltage-dependent calcium channel subunits in the human hippocampus and parahippocampal gyrus. *Neuroscience* 71:1013–24

111. Williams ME, Brust PF, Feldman DH, Patthi S, Simerson S, et al. 1992. Structure and functional expression of an omega-conotoxin-sensitive human N-type calcium channel. *Science* 257:389–95

112. Maximov A, Bezprozvanny I. 2002. Synaptic targeting of N-type calcium channels in hippocampal neurons. *J. Neurosci.* 22:6939–52

113. Yokoyama CT, Westenbroek RE, Hell JW, Soong TW, Snutch TP, Catterall WA. 1995. Biochemical properties and subcellular distribution of the neuronal class E calcium channel alpha 1 subunit. *J. Neurosci.* 15:6419–32

114. Craig PJ, Beattie RE, Folly EA, Banerjee MD, Reeves MB, et al. 1999.

Distribution of the voltage-dependent calcium channel alpha1G subunit mRNA and protein throughout the mature rat brain. *Eur. J. Neurosci.* 11:2949–64

115. McNamara NM, Muniz ZM, Wilkin GP, Dolly JO. 1993. Prominent location of a K$^+$ channel containing the alpha subunit Kv 1.2 in the basket cell nerve terminals of rat cerebellum. *Neuroscience* 57:1039–45

116. McNamara NM, Averill S, Wilkin GP, Dolly JO, Priestley JV. 1996. Ultrastructural localization of a voltage-gated K$^+$ channel alpha subunit (KV 1.2) in the rat cerebellum. *Eur. J. Neurosci.* 8:688–99

117. Wang H, Kunkel DD, Schwartzkroin PA, Tempel BL. 1994. Localization of Kv1.1 and Kv1.2, two K channel proteins, to synaptic terminals, somata, and dendrites in the mouse brain. *J. Neurosci.* 14:4588–99

118. Wang H, Kunkel DD, Martin TM, Schwartzkroin PA, Tempel BL. 1993. Heteromultimeric K$^+$ channels in terminal juxtaparanodal regions of neurons. *Nature* 365:75–79

119. Laube G, Roper J, Pitt JC, Sewing S, Kistner U, et al. 1996. Ultrastructural localization of Shaker-related potassium channel subunits and synapse-associated protein 90 to septate-like junctions in rat cerebellar Pinceaux. *Brain Res. Mol. Brain Res.* 42:51–61

120. Rasband MN, Trimmer JS, Peles E, Levinson SR, Shrager P. 1999. K$^+$ channel distribution clustering in developing and hypomyelinated axons of the optic nerve. *J. Neurocytol.* 28:319–31

121. Rasband MN, Trimmer JS, Schwarz TL, Levinson SR, Ellisman MH, et al. 1998. Potassium channel distribution, clustering, and function in remyelinating rat axons. *J. Neurosci.* 18:36–47

122. Vabnick I, Trimmer JS, Schwarz TL, Levinson SR, Risal D, Shrager P. 1999. Dynamic potassium channel distributions during axonal development prevent aberrant firing patterns. *J. Neurosci.* 19:747–58

123. Rasband MN, Park EW, Zhen D, Arbuckle MI, Poliak S, et al. 2002. Clustering of neuronal potassium channels is independent of their interaction with PSD-95. *J. Cell Biol.* 159:663–72

124. Sheng M, Tsaur ML, Jan YN, Jan LY. 1994. Contrasting subcellular localization of the Kv1.2 K$^+$ channel subunit in different neurons of rat brain. *J. Neurosci.* 14:2408–17

125. Veh RW, Lichtinghagen R, Sewing S, Wunder F, Grumbach IM, Pongs O. 1995. Immunohistochemical localization of five members of the Kv1 channel subunits: contrasting subcellular locations and neuron-specific co-localizations in rat brain. *Eur. J. Neurosci.* 7:2189–205

126. Kues WA, Wunder F. 1992. Heterogeneous expression patterns of mammalian potassium channel genes in developing adult rat brain. *Eur. J. Neurosci.* 4:1296–308

127. Koch RO, Wanner SG, Koschak A, Hanner M, Schwarzer C, et al. 1997. Complex subunit assembly of neuronal voltage-gated K$^+$ channels. Basis for high-affinity toxin interactions and pharmacology. *J. Biol. Chem.* 272:27577–81

128. Felix JP, Bugianesi RM, Schmalhofer WA, Borris R, Goetz MA, et al. 1999. Identification and biochemical characterization of a novel nortriterpene inhibitor of the human lymphocyte voltage-gated potassium channel, Kv1.3. *Biochemistry* 38:4922–30

129. Khanna R, Roy L, Zhu X, Schlichter LC. 2001. K$^+$ channels and the microglial respiratory burst. *Am. J. Physiol. Cell Physiol.* 280:C796–806

130. Chittajallu R, Chen Y, Wang H, Yuan X, Ghiani CA, et al. 2002. Regulation of Kv1 subunit expression in oligodendrocyte progenitor cells and their role in G1/S phase progression of the cell cycle. *Proc. Natl. Acad. Sci. USA* 99:2350–55

131. Kalman K, Nguyen A, Tseng-Crank J, Dukes ID, Chandy G, et al. 1998.

Genomic organization, chromosomal localization, tissue distribution, and biophysical characterization of a novel mammalian *Shaker*-related voltage-gated potassium channel, Kv1.7. *J. Biol. Chem.* 273:5851–57

132. Rettig J, Heinemann SH, Wunder F, Lorra C, Parcej DN, et al. 1994. Inactivation properties of voltage-gated K+ channels altered by presence of beta-subunit. *Nature* 369:289–94

133. Trimmer JS. 1991. Immunological identification and characterization of a delayed rectifier K+ channel polypeptide in rat brain. *Proc. Natl. Acad. Sci. USA* 88:10764–68

134. Hwang PM, Fotuhi M, Bredt DS, Cunningham AM, Snyder SH. 1993. Contrasting immunohistochemical localizations in rat brain of two novel K+ channels of the *Shab* subfamily. *J. Neurosci.* 13:1569–76

135. Du J, Tao-Chang J-H, Zerfas P, McBain CJ. 1998. The K+ channel, Kv2.1, is apposed to astrocytic processes and is associated with inhibitory postsynaptic membranes in hippocampal and cortical principal neurons and inhibitory interneurons. *Neuroscience* 84:37–48

136. Rhodes KJ, Carroll KI, Sung MA, Doliveira LC, Monaghan MM, et al. Co-localization and co-association of KChIPs and Kv4 α subunits as integral components of A-type potassium channels in mammalian brain. Submitted

137. Maletic-Savatic M, Lenn NJ, Trimmer JS. 1995. Differential spatiotemporal expression of K+ channel polypeptides in rat hippocampal neurons developing in situ and in vitro. *J. Neurosci.* 15:3840–51

138. Scannevin RH, Murakoshi H, Rhodes KJ, Trimmer JS. 1996. Identification of a cytoplasmic domain important in the polarized expression and clustering of the Kv2.1 K+ channel. *J. Cell Biol.* 135:1619–32

139. Blaine JT, Ribera AB. 1998. Heteromultimeric potassium channels formed by members of the Kv2 subfamily. *J. Neurosci.* 18:9585–93

140. Hwang PM, Cunningham AM, Peng YW, Snyder SH. 1993. CDRK DRK1 K+ channels have contrasting localizations in sensory systems. *Neuroscience* 55:613–20

141. Lim ST, Antonucci DE, Scannevin RH, Trimmer JS. 2000. A novel targeting signal for proximal clustering of the Kv2.1 K+ channel in hippocampal neurons. *Neuron* 25:385–97

142. Rudy B, Chow A, Lau D, Amarillo Y, Ozaita A, et al. 1999. Contributions of Kv3 channels to neuronal excitability. *Ann. NY Acad. Sci.* 868:304–43

143. Perney TM, Marshall J, Martin KA, Hockfield S, Kaczmarek LK. 1992. Expression of the mRNAs for the Kv3.1 potassium channel gene in the adult and developing rat brain. *J. Neurophysiol.* 68:756–66

144. Weiser M, Vega-Saenz de Miera E, Kentros C, Moreno H, Franzen L, et al. 1994. Differential expression of *Shaw*-related K+ channels in the rat central nervous system. *J. Neurosci.* 14:949–72

145. Weiser M, Bueno E, Sekirnjak C, Martone ME, Baker H, et al. 1995. The potassium channel subunit KV3.1b is localized to somatic axonal membranes of specific populations of CNS neurons. *J. Neurosci.* 15:4298–314

146. Sekirnjak C, Martone ME, Weiser M, Deerinck T, Bueno E, et al. 1997. Subcellular localization of the K+ channel subunit Kv3.1b in selected rat CNS neurons. *Brain Res.* 766:173–87

147. Chow A, Erisir A, Farb C, Nadal MS, Ozaita A, et al. 1999. K+ channel expression distinguishes subpopulations of parvalbumin- and somatostatin-containing neocortical interneurons. *J. Neurosci.* 19:9332–45

148. Lau D, Vega-Saenz de Miera EC, Contreras D, Ozaita A, Harvey M, et al. 2000. Impaired fast-spiking, suppressed cortical inhibition, and increased susceptibility to seizures in mice lacking Kv3.2 K+ channel proteins. *J. Neurosci.* 20:9071–85

149. Lien CC, Jonas P. 2003. Kv3 potassium

conductance is necessary and kinetically optimized for high-frequency action potential generation in hippocampal interneurons. *J. Neurosci.* 23:2058–68

150. Ponce A, Vega-Saenz de Miera E, Kentros C, Moreno H, Thornhill B, Rudy B. 1997. K$^+$ channel subunit isoforms with divergent carboxy-terminal sequences carry distinct membrane targeting signals. *J. Membr. Biol.* 159:149–59

151. Dotti CG, Simons K. 1990. Polarized sorting of viral glycoproteins to the axon dendrites of hippocampal neurons in culture. *Cell* 62:63–72

152. Ozaita A, Martone ME, Ellisman MH, Rudy B. 2002. Differential subcellular localization of the two alternatively spliced isoforms of the Kv3.1 potassium channel subunit in brain. *J. Neurophysiol.* 88:394–408

153. Perney TM, Kaczmarek LK. 1997. Localization of a high threshold potassium channel in the rat cochlear nucleus. *J. Comp. Neurol.* 386:178–202

154. Goldman-Wohl DS, Chan E, Baird D, Heintz N. 1994. Kv3.3b: a novel *Shaw* type potassium channel expressed in terminally differentiated cerebellar Purkinje cells and deep cerebellar nuclei. *J. Neurosci.* 14:511–22

155. Li W, Kaczmarek LK, Perney TM. 2001. Localization of two high-threshold potassium channel subunits in the rat central auditory system. *J. Comp. Neurol.* 437:196–218

156. Martina M, Yao GL, Bean BP. 2003. Properties and functional role of voltage-dependent potassium channels in dendrites of rat cerebellar Purkinje neurons. *J. Neurosci.* 23:5698–707

157. Rettig J, Wunder F, Stocker M, Lichtinghagen R, Mastiaux F, et al. 1992. Characterization of a *Shaw*-related potassium channel family in rat brain. *EMBO J.* 11:2473–86

158. Baranauskas G, Tkatch T, Nagata K, Yeh JZ, Surmeier DJ. 2003. Kv3.4 subunits enhance the repolarizing efficiency of Kv3.1

channels in fast-spiking neurons. *Nat. Neurosci.* 6:258–66

159. Malin SA, Nerbonne JM. 2000. Elimination of the fast transient in superior cervical ganglion neurons with expression of KV4.2W362F: molecular dissection of IA. *J. Neurosci.* 20:5191–99

160. Malin SA, Nerbonne JM. 2001. Molecular heterogeneity of the voltage-gated fast transient outward K$^+$ current, I(Af), in mammalian neurons. *J. Neurosci.* 21:8004–14

161. Serodio P, Rudy B. 1998. Differential expression of Kv4 K$^+$ channel subunits mediating subthreshold transient K$^+$ (A-type) currents in rat brain. *J. Neurophysiol.* 79:1081–91

162. Hattori S, Murakami F, Song WJ. 2003. Quantitative relationship between Kv4.2 mRNA and A-type K$^+$ current in rat striatal cholinergic interneurons during development. *J. Neurophysiol.* 90:175–83

163. Song WJ, Tkatch T, Baranauskas G, Ichinohe N, Kitai ST, Surmeier DJ. 1998. Somatodendritic depolarization-activated potassium currents in rat neostriatal cholinergic interneurons are predominantly of the A-type and attributable to co-expression of Kv4.2 and Kv4.1 subunits. *J. Neurosci.* 18:3124–37

164. Shibata R, Wakazono Y, Nakahira K, Trimmer JS, Ikenaka K. 1999. Expression of Kv3.1 and Kv4.2 genes in developing cerebellar granule cells. *Dev. Neurosci.* 21:87–93

165. Tkatch T, Baranauskas G, Surmeier DJ. 2000. Kv4.2 mRNA abundance and A-type K$^+$ current amplitude are linearly related in basal ganglia and basal forebrain neurons. *J. Neurosci.* 20:579–88

166. Liss B, Franz O, Sewing S, Bruns R, Neuhoff H, Roeper J. 2001. Tuning pacemaker frequency of individual dopaminergic neurons by Kv4.3L and KChip3.1 transcription. *EMBO J.* 20:5715–24

167. Geiger JRP, Jonas P. 2000. Dynamic control of presynaptic Ca^{2+} inflow by fast-inactivating K$^+$ channels in

hippocampal mossy fiber boutons. *Neuron* 28:927–39

168. Bischofberger J, Geiger JR, Jonas P. 2002. Timing efficacy of Ca^{2+} channel activation in hippocampal mossy fiber boutons. *J. Neurosci.* 22:10593–602

169. Riazanski V, Becker A, Chen J, Sochivko D, Lie A, et al. 2001. Functional and molecular analysis of transient voltage-dependent K^+ currents in rat hippocampal granule cells. *J. Physiol.* 537:391–406

170. Gasparini S, Kasyanov AM, Pietrobon D, Voronin LL, Cherubini E. 2001. Presynaptic R-type calcium channels contribute to fast excitatory synaptic transmission in the rat hippocampus. *J. Neurosci.* 21:8715–21

171. Southan AP, Robertson B. 1998. Patch-clamp recordings from cerebellar basket cell bodies and their presynaptic terminals reveal an asymmetric distribution of voltage-gated potassium channels. *J. Neurosci.* 18:948–55

172. Southan AP, Robertson B. 2000. Electrophysiological characterization of voltage-gated K^+ currents in cerebellar basket and Purkinje cells: Kv1 and Kv3 channel subfamilies are present in basket cell nerve terminals. *J. Neurosci.* 20:114–22

173. Magee JC, Johnston D. 1995. Characterization of single voltage-gated Na^+ and Ca^{2+} channels in apical dendrites of rat CA1 pyramidal neurons. *J. Physiol.* 487:67–90

174. Magee JC, Hoffman DA, Colbert C, Johnston D. 1998. Electrical calcium signalling in dendrites of hippocampal pyramidal neurons. *Annu. Rev. Physiol.* 60:327–46

175. Storm JF. 2000. K^+ channels and their distribution in large cortical pyramidal neurones. *J. Physiol.* 525:565–66

176. Larkum ME, Zhu JJ, Sakmann B. 1999. A new cellular mechanism for coupling inputs arriving at different cortical layers. *Nature* 398:338–41

177. Schiller J, Schiller Y, Stuart G, Sakmann B. 1997. Calcium action potentials restricted to distal apical dendrites of rat neocortical pyramidal neurons. *J. Physiol.* 505:605–16

178. Schwindt PC, Crill WE. 1997. Modification of current transmitted from apical dendrite to soma by blockade of voltage- and Ca^{2+}-dependent conductances in rat neocortical pyramidal neurons. *J. Neurophysiol.* 78:187–98

179. Golding NL, Spruston N. 1998. Dendritic sodium spikes are variable triggers of axonal action potentials in hippocampal CA1 pyramidal neurons. *Neuron* 21:1189–200

180. Larkum ME, Zhu JJ, Sakmann B. 2001. Dendritic mechanisms underlying the coupling of the dendritic with the axonal action potential initiation zone of adult rat layer 5 pyramidal neurons. *J. Physiol.* 533:447–66

181. Stuart GJ, Sakmann B. 1994. Active propagation of somatic action potentials into pyramidal cell dendrites. *Nature* 367:69–72

182. Hoffman DA, Magee JC, Colbert CM, Johnston D. 1997. K^+ channel regulation of signal propagation in dendrites of hippocampal pyramidal cells. *Nature* 387:869–75

183. Murakoshi H, Trimmer JS. 1999. Identification of the Kv2.1 K^+ channel as a major component of the delayed rectifier K^+ current in rat hippocampal neurons. *J. Neurosci.* 19:1728–35

184. Magee JC, Christofi G, Miyakawa H, Christie B, Lasser-Ross N, Johnston D. 1995. Subthreshold synaptic activation of voltage-gated Ca^{2+} channels mediates a localized Ca^{2+} influx into the dendrites of hippocampal pyramidal neurons. *J. Neurophysiol.* 74:1335–42

185. Mickus T, Jung H, Spruston N. 1999. Properties of slow, cumulative sodium channel inactivation in rat hippocampal CA1 pyramidal neurons. *Biophys. J.* 76:846–60

186. Wu RL, Barish ME. 1992. Two pharmacologically and kinetically distinct transient

potassium currents in cultured embryonic mouse hippocampal neurons. *J. Neurosci.* 12:2235–46

187. Hoffman DA, Johnston D. 1998. Down-regulation of transient K$^+$ channels in dendrites of hippocampal CA1 pyramidal neurons by activation of PKA and PKC. *J. Neurosci.* 18:3521–28

188. Johnston D, Hoffman DA, Magee JC, Poolos NP, Watanabe S, et al. 2000. Dendritic potassium channels in hippocampal pyramidal neurons. *J. Physiol.* 525:75–81

189. Golding NL, Mickus T, Spruston N. 1999. Dendritic calcium spike initiation and repolarization are controlled by distinct potassium channel subtypes in CA1 pyramidal neurons. *J. Neurosci.* 19:8789–98

190. Bekkers JM. 2000. Distribution and activation of voltage-gated potassium channels in cell-attached and outside-out patches from large layer 5 cortical pyramidal neurons of the rat. *J. Physiol.* 525:611–20

191. Korngreen A, Sakmann B. 2000. Voltage-gated K$^+$ channels in layer 5 neocortical pyramidal neurones from young rats: subtypes and gradients. *J. Physiol.* 525:621–39

192. Martina M, Jonas P. 1997. Functional differences in Na$^+$ channel gating between fast-spiking interneurones and principal neurones of rat hippocampus. *J. Physiol.* 505:593–603

193. Martina M, Vida I, Jonas P. 2000. Distal initiation and active propagation of action potentials in interneuron dendrites. *Science* 287:295–300

194. Lambert NA, Wilson WA. 1996. High-threshold Ca^{2+} currents in rat hippocampal interneurones and their selective inhibition by activation of GABA(B) receptors. *J. Physiol.* 492:115–27

195. Poncer JC, McKinney RA, Gahwiler BH, Thompson SM. 1997. Either N- or P-type calcium channels mediate GABA release at distinct hippocampal inhibitory synapses. *Neuron* 18:463–72

196. Du J, Haak LL, Phillips-Tansey E, Russell JT, McBain CJ. 2000. Frequency-dependent regulation of rat hippocampal somato-dendritic excitability by the K$^+$ channel subunit Kv2.1. *J. Physiol.* 522: 19–31

197. Du J, Zhang L, Weiser M, Rudy B, McBain CJ. 1996. Developmental expression and functional characterization of the potassium-channel subunit Kv3.1b in parvalbumin-containing interneurons of the rat hippocampus. *J. Neurosci.* 16:506–18

198. Hernandez-Pineda R, Chow A, Amarillo Y, Moreno H, Saganich M, et al. 1999. Kv3.1–Kv3.2 channels underlie a high-voltage-activating component of the delayed rectifier K$^+$ current in projecting neurons from the globus pallidus. *J. Neurophysiol.* 82:1512–28

199. Erisir A, Lau D, Rudy B, Leonard CS. 1999. Function of specific K$^+$ channels in sustained high-frequency firing of fast-spiking neocortical interneurons. *J. Neurophysiol.* 82:2476–89

200. Martina M, Schultz JH, Ehmke H, Monyer H, Jonas P. 1998. Functional and molecular differences between voltage-gated K$^+$ channels of fast-spiking interneurons and pyramidal neurons of rat hippocampus. *J. Neurosci.* 18:8111–25

201. Mandel G, McKinnon D. 1993. Molecular basis of neural-specific gene expression. *Annu. Rev. Neurosci.* 16:323–45

202. Rivera JF, Ahmad S, Quick MW, Liman ER, Arnold DB. 2003. An evolutionarily conserved dileucine motif in *Shal* K$^+$ channels mediates dendritic targeting. *Nat. Neurosci.* 6:243–50

203. Garrido JJ, Fernandes F, Giraud P, Mouret I, Pasqualini E, et al. 2001. Identification of an axonal determinant in the C terminus of the sodium channel Na(v)1.2. *EMBO J.* 20:5950–61

204. Garrido JJ, Giraud P, Carlier E, Fernandes F, Moussif A, et al. 2003. A targeting motif involved in sodium channel clustering at the axonal initial segment. *Science* 300:2091–94

205. Lemaillet G, Walker B, Lambert S. 2003. Identification of a conserved ankyrin-binding motif in the family of sodium channel α subunits. *J. Biol. Chem.* 278: 27333–39

206. Maximov A, Südhof TC, Bezprozvanny I. 1999. Association of neuronal calcium channels with modular adaptor proteins. *J. Biol. Chem.* 274:24453–56

207. Smart SL, Lopantsev V, Zhang CL, Robbins CA, Wang H, et al. 1998. Deletion of the K(V)1.1 potassium channel causes epilepsy in mice. *Neuron* 20:809–19

208. Zhou L, Zhang CL, Messing A, Chiu SY. 1998. Temperature-sensitive neuromuscular transmission in Kv1.1 null mice: role of potassium channels under the myelin sheath in young nerves. *J. Neurosci.* 18:7200–15

209. Zhou L, Messing A, Chiu SY. 1999. Determinants of excitability at transition zones in Kv1.1-deficient myelinated nerves. *J. Neurosci.* 19:5768–81

210. Giese KP, Storm JF, Reuter D, Fedorov NB, Shao LR, et al. 1998. Reduced K⁺ channel inactivation, spike broadening, and after-hyperpolarization in Kvbeta1.1-deficient mice with impaired learning. *Learn Mem.* 5:257–73

211. McCormack K, Connor JX, Zhou L, Ho LL, Ganetzky B, et al. 2002. Genetic analysis of the mammalian K⁺ channel beta subunit Kvbeta 2 (Kcnab2). *J. Biol. Chem.* 277:13219–28

212. Barry DM, Xu H, Schuessler RB, Nerbonne JM. 1998. Functional knockout of the transient outward current, long-QT syndrome, and cardiac remodeling in mice expressing a dominant-negative Kv4 alpha subunit. *Circ. Res.* 83:560–67

213. Malin SA, Nerbonne JM. 2002. Delayed rectifier K⁺ currents, IK, are encoded by Kv2 alpha-subunits and regulate tonic firing in mammalian sympathetic neurons. *J. Neurosci.* 22:10094–105

214. Wang FC, Parcej DN, Dolly JO. 1999. alpha subunit compositions of Kv1.1-containing K⁺ channel subtypes fractionated from rat brain using dendrotoxins. *Eur. J. Biochem.* 263:230–37

215. Coleman SK, Newcombe J, Pryke J, Dolly JO. 1999. Subunit composition of Kv1 channels in human CNS. *J. Neurochem.* 73:849–58

Annu. Rev. Physiol. 2004. 66:521–45
doi: 10.1146/annurev.physiol.66.032102.112842
Copyright © 2004 by Annual Reviews. All rights reserved
First published online as a Review in Advance on September 29, 2003

MYOSIN-1C, THE HAIR CELL'S ADAPTATION MOTOR

Peter G. Gillespie[1] and Janet L. Cyr[2]

[1]Oregon Hearing Research Center and Vollum Institute, Oregon Health and Science University, Portland, Oregon 97239; email: gillespp@ohsu.edu
[2]Sensory Neuroscience Research Center, and Departments of Otolaryngology and Biochemistry and Molecular Pharmacology, West Virginia University School of Medicine, Morgantown, West Virginia 26506; email: jcyr@hsc.wvu.edu

Key Words myosin, adaptation, microvilli, stereocilia, transduction

■ **Abstract** Given their prominent actin-rich subcellular specializations, it is no surprise that mechanosensitive hair cells of the inner ear exploit myosin molecules—the only known actin-dependent molecular motors—to carry out exotic but essential tasks. Recent experiments have confirmed that an unconventional myosin isozyme, myosin-1c, is a component of the hair cell's adaptation-motor complex. This complex carries out slow adaptation, provides tension to sensitize transduction channels, and may participate in assembly of the transduction apparatus. This review focuses on the detailed operation of the adaptation motor and the functional consequences of the incorporation of this specific myosin isozyme into the motor complex.

MECHANOTRANSDUCTION BY HAIR CELLS

Sensory cells of the inner ear, hair cells, detect sound and movements of the head and transmit signals representing those stimuli to the central nervous system. Hair cells transduce mechanical energy into electrical impulses using a profoundly nonlinear mechanism, which enables the auditory and vestibular systems to report stimuli differing more than a million-fold in intensity. Hair-cell mechanotransduction is still poorly understood and, as a result, has attracted considerable attention.

Projecting from the top of a hair cell, ~100 actin-filled stereocilia and an axonemal kinocilium constitute the hair bundle, the hair cell's mechanically sensitive organelle (Figure 1A, see color insert). Stereocilia are arranged in ranks of increasing heights, an anatomical arrangement that reflects the bundle's physiological responsiveness. Accordingly, deflection of the bundle toward the tallest stereocilia (an excitatory or positive displacement) opens cation-selective transduction channels, which are located at stereociliary tips. Deflection toward the shortest stereocilia (an inhibitory or negative displacement) closes channels; by contrast, perpendicular deflections are without effect (1).

The prevailing model for hair-cell transduction (2), the subject of several comprehensive reviews (3–5), has proven resilient and predictive. The key feature of the model, depicted in Figure 1B, is that transduction is direct: Movement of the hair bundle directly opens or closes transduction channels by modulating tension felt by the channels (2). The structure of the bundle dictates its movement in response to a stimulus. Interconnected by several varieties of crosslinks, stereocilia move together when the bundle is deflected. Because they are stiff along their lengths but flexible at their bases, deflected stereocilia pivot at their bases, where they insert into the top of the hair cell and slide along each other. Such stereociliary shearing stretches or slackens gating springs, hypothetical mechanical elements that apply tension to transduction channels. Tension in gating springs directly controls both the extent and speed of transduction-channel gating, as depicted in Figure 1B (2). Channels open when tension is high and close when tension is low; large positive deflections open channels faster and to a greater final open probability (p_{open}). Because channel opening is rapid, dependence on second-messenger cascades for transduction is precluded (2).

Tip links, extracellular filaments that traverse from the top of one stereocilium to the side of its taller neighbor, and occur only along the axis of the bundle's mechanical sensitivity (Figure 1A, *inset*), are essential for transduction-channel gating (6). Because the gating springs disappear when tip links are severed (6), tip links are either the gating springs themselves or are connected with them in series. For the purposes of this review, we use the terms tip links and gating springs interchangeably, although the tip link itself may indeed be inelastic (7).

During transduction, deflections applied to the top of the hair bundle stretch tip links by a distance smaller than the size of the mechanical stimulus, reflecting a mechanical gain of less than one. This gain factor, called γ, is approximately equal to the stereociliary height divided by the separation between adjacent stereocilia. In frog saccular hair cells, $\gamma = 0.14$ (8); a 100-nm bundle deflection therefore elongates the gating spring by only 14 nm. Due to differences in bundle morphology, $\gamma = 0.05$ in mouse vestibular hair cells (9).

ADAPTATION OF HAIR-CELL TRANSDUCTION

When a hair bundle receives a sustained excitatory deflection, the transduction current is initially large, then declines as transduction channels close, i.e., the current adapts. Responding to the new location of the hair bundle, adaptation serves to reset the p_{open} of transduction channels near the rest value of ~0.1. After returning close to this value, channels are poised near their position of optimal sensitivity along the displacement-response relationship (Figure 2). During and following adaptation, transduction channels can be fully reopened if a sufficiently large stimulus is applied to the bundle (10), a feature that distinguishes adaptation from desensitization.

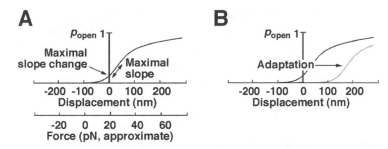

Figure 2   Resting tension and adaptation. (*A*) The relationship between bundle displacement and channel open probability (p_{open}) is described by a Boltzmann relationship. At rest (0-nm displacement), the hair cell is near the points of maximal slope (most sensitive point) and maximal slope change (greatest rectification). A substantial amount of force, $\sim$20 pN, is required to position the channel at this point. (*B*) During adaptation, the displacement-open probability relationship shifts in the direction of the stimulus (*gray line*). This example depicts the response to a 200-nm adapting stimulus. Although a 150-nm displacement is nearly saturating prior to adaptation, afterward it is not.

Two Phases of Adaptation

Depending on the size and speed of hair-bundle displacement, the decline in transduction current during adaptation can be fast (time constant, τ, of a few milliseconds or less) or slow ($\tau > 10$ ms). All hair cells studied in depth simultaneously exhibit both fast and slow adaptation, with fast adaptation most extensively characterized in turtle cochlear hair cells (11) and slow adaptation thoroughly examined in bullfrog saccular hair cells (12). The two forms of adaptation, although both dependent on Ca^{2+} entering through open transduction channels, have distinct mechanisms that are most easily distinguished by their mechanical consequences for hair bundles.

During fast adaptation, Ca^{2+} ions that enter a stereocilium through an open transduction channel bind rapidly to a site on or near the channel and induce channel closure. When channels close, tension increases in the tip link, pulling the bundle in the negative direction, opposite the direction of the stimulus. The term fast adaptation may be misleading, however, as the process seems to be more important for tuning and sensitivity of a hair cell than for adaptation per se. Fast channel reclosure is a more accurate, albeit wordy, description of this process.

If the kinetics of the stimulus and the fast channel reclosure are matched, mechanical amplification of the signal can ensue (13). As a bundle is deflected in the excitatory direction, the coordinated opening of transduction channels and the subsequent relaxation of gating springs augment bundle movement. When channels close, they reset the tension of the gating spring; if this resetting occurs on the returning phase of bundle displacement, movement in that direction will also be enhanced. This form of amplification works much like a child on a swing, who enhances the swing's oscillations by pumping his legs at the appropriate moments. Evidence for hair bundle-based mechanical amplification in lower vertebrates is consistent with use of this mechanism (14).

Slow adaptation uses a different mechanism. The prevailing model suggests that slow adaptation occurs when the upper insertion point of the tip link slides down the stereocilium in response to elevated tension imparted on it during a positive bundle displacement (6, 15). The resultant decreased tension in the tip link permits the bundle to move yet farther in the positive direction, in direct contrast to the movement induced by fast adaptation. As tension decreases, channels close, producing the decline in transduction current that is the outward manifestation of the adaptive process.

If the stimulus is sufficiently fast, fast adaptation can dominate the response of the hair cell for stimuli well below saturation (16, 17). On the other hand, because fast adaptation is often much faster than the stimulus used to excite a hair cell, fast channel reclosure can occur simultaneously with the stimulus and hence may not be readily apparent (18). With a relatively slow mechanical stimulus, hair cells appear dominated by slow adaptation and also require much larger mechanical stimuli to open channels to the same p_{open}. The type of stimulus presented to the hair bundle can influence the speed of adaptation. Deflection of the bundle with a fluid jet, which delivers a force step, results in larger time constants for adaptation than stimuli of the same rise time administered by a stiff probe, which delivers a displacement step (16). This difference in adaptation rate between the two types of stimuli can readily be accounted for by modeling of bundle mechanics (16).

Although hair-cell adaptation has been studied intensively in both auditory and vestibular organs, slow adaptation is most prominent in vestibular hair cells. This form of adaptation may be more physiologically relevant to these hair cells, which respond to stimuli of much lower frequency than do auditory hair cells. Over the past decade, the underlying molecular basis of slow adaptation in vestibular hair cells has been unveiled. We restrict the focus of this review to slow adaptation in vestibular hair cells.

Slow Adaptation's Mechanism

The motor model for slow adaptation was initially proposed by Howard & Hudspeth (15) and elaborated by Assad & Corey (12). In this model, the tension in the gating spring is governed by an adaptation motor, which controls the position of the transduction complex along the stereocilium in a Ca^{2+}-dependent manner. During this process, Ca^{2+} serves the role of a feedback signal in a negative feedback loop, which strives to maintain p_{open} at an optimal point. When tension is high, Ca^{2+} influx through an open transduction channel causes the motor to slip down the actin cytoskeleton, reducing tip-link tension and permitting channel closure. By contrast, when tension is low and Ca^{2+} entry is blocked, the motor ascends the cytoskeleton until it exerts sufficient tension to restore the transduction channel's p_{open} to ~0.1 and allow modest Ca^{2+} entry. For the remainder of this review, we often refer to this mechanism as adaptation, dropping the adjective slow. Furthermore, we consider adaptation to be the underlying movement of the transduction apparatus that leads to the decline in transduction current (10). This movement is reflected in the shift of the displacement-open probability relationship in the direction of the stimulus (Figure 2B). By this definition, adaptation can take place even if the transduction channel's p_{open} does not change. For example, during a large inhibitory displacement that slackens the tip link, the adaptation motor moves but the channel does not detect the movement until tip-link slack is taken up. This movement can be assayed, however, by applying a large excitatory displacement at different times during the adaptive process and using the transduction current to monitor the position of the transduction apparatus.

To maintain appropriate tension on the system, the motor model for slow adaptation invokes an adaptation motor, coupled mechanically to the transduction channel (or channels) at the upper end of a single tip link (15). The predominance of actin filaments in stereocilia suggests that an actin-based molecular motor, i.e., a myosin, must power adaptation. Considerable data indicate that 100–200 clustered myosin molecules make up the adaptation motor (detailed below). A structure at the upper end of the tip link, the insertional plaque (19), is appropriately located to house the adaptation motor. Consistent with that hypothesis, the insertional plaque moves up or down a stereocilium in response to extracellular Ca^{2+}-chelator application (which breaks tip links) or large excitatory stimuli (20).

Because stereocilia actin filaments are arranged with their barbed ends pointing toward stereociliary tips, the favored direction of movement of nearly all myosins, the myosin molecules within the motor continuously strive to ascend the stereocilium and therefore exert a constant tension on the gating spring. At rest, the force exerted by the adaptation motor is equal and opposite to the force in the gating spring. Converging independent estimates of the stiffness and extension of the gating spring indicate that its resting tension is relatively high, in the range of 5–20 pN (3). Maintenance of resting tension is usually attributed to the adaptation motor, although the fast adaptation mechanism may also contribute.

When the bundle is at rest in the presence of a physiologically relevant extracellular Ca^{2+} concentration (100 μM), bundle mechanical noise is very low (21). Because bundle noise arises, in part, from the stochastic progression of individual myosin molecules through their ATPase cycles (see below), the myosin molecules exerting resting tension must be catalytically inhibited. Noise increases dramatically when Ca^{2+} entry is blocked, suggesting that the low-Ca^{2+} state increases the rate at which motors traverse their ATPase cycles and carry out power-strokes, thereby resulting in mechanical noise (and bundle movement).

Because of the mechanical nature of transduction, both the forces applied to the system and the force production of the adaptation motor are relevant. When a bundle is deflected in the excitatory direction, tension in the gating spring immediately exceeds that exerted by the adaptation motor. Slipping adaptation then proceeds: Increased tension pulls the adaptation motor down the stereocilium, toward the pointed end of actin filaments and opposite the direction of active myosin travel. While slipping, the adaptation motor may still traverse its ATPase cycle, but the larger force exerted by the stimulus causes the motor to lose a tug-of-war battle. Under these circumstances, the behavior of the motor is analogous to the behavior of a contracting muscle that is pulled hard opposite its contraction; even though its constituent myosin molecules labor to shorten the muscle, the larger external force overcomes the effort of the myosins and instead lengthens it (22). In the hair cell, rather than ascending the actin filament, the motor is pulled in the opposite direction. In another view, the elevated tension may reverse the mechanical steps in the ATPase cycle, moving the adaptation-motor myosins to the conformation seen when ADP and P_i (inorganic phosphate) are bound to the protein. In this conformation, myosins might detach rapidly (M.A. Geeves, R. Milligan, J. Molloy & L. Coluccio, submitted). Consistent with both models, the rate of slipping adaptation is linearly dependent on the size of the stimulus for a broad range of stimuli (12).

Climbing adaptation occurs when the external force (from bundle displacement or opposing extension of the gating spring) is less than the force exerted by the motor and the adaptation motor moves toward the tip of a stereocilium. In contrast to slipping adaptation's rate, which depends on the stimulus size, the rate of climbing adaptation plateaus once a bundle is deflected sufficiently far in the negative direction (12). Under these conditions, the gating spring slackens, and the adaptation motor climbs with a rate that reflects the unloaded rate of movement of its constituent myosin molecules. When the slack in the gating spring is taken up, the ensuing tension begins to oppose motor movement and the motor movement slows.

Adaptation's Significance

Adaptation plays an important role in hair-cell function in both the vestibular and auditory systems. In vestibular cells, which typically have both immediate phasic responses to sustained stimuli and prolonged tonic responses, adaptation allows a cell to reduce its response to a sustained stimulus, yet still respond robustly to a new stimulus. This role is not, however, important for auditory hair cells.

By contrast, a second role for the adaptation motor appears to be universal in all hair cells. In order to be maximally sensitive, a hair cell must sit at or near the steepest part of the displacement-response relationship (Figure 2). Because transduction channels respond to tension in the gating spring, operating at this point of maximal sensitivity requires that an internal source of tension be applied to transduction channels. The adaptation motor is perfectly situated to control this tension (3); a cluster of myosin molecules could clearly supply sufficient force to maintain the necessary resting tension on the transduction channel.

A final role for the adaptation motor is more hypothetical: The motor may mediate assembly of the transduction apparatus (24). Little is known about the assembly of the transduction apparatus or its molecular makeup. It is plausible that the adaptation motor also serves to transport and assemble transduction components into a functional mechanosensitive complex.

MYOSIN IN HAIR CELLS

Of the ~40 myosin isozymes present in mammalian genomes (25), at least 6 are essential for auditory and vestibular function, including myosin-VI (Myo6), myosin-VIIa (Myo7a), myosin-1a (Myo1a), myosin-XV (Myo15), nonmuscle myosin-IIa (MYH9), and myosin-IIIa (Myo3a). Mutations in these genes result in hearing and vestibular deficits of various degrees in both mice and humans (26–34).

Many more myosin isozymes than just these 6 are expressed in auditory and vestibular tissues. For example, 13 myosin gene products were identified in a degenerate RT-PCR experiment within frog sacculus (35). RT-PCR and immunocytochemistry experiments in mouse utricles and cochlea directed toward the 8 vertebrate myosin-I isozymes identified (Myo1b, Myo1c, and Myo1e) expressed at high levels, with lower expression of the remaining 5 (36).

Although several myosin isozymes arc present in hair bundles, only myosin-1c (Myo1c; formerly known as myosin Iβ) has been shown to be located at the tips of stereocilia and exhibit biochemical properties consistent with a role as the adaptation motor. Substantial evidence indicates that Myo1c mediates slow adaptation and sets the resting tension applied to transduction channels, which ensures their optimal sensitivity.

MYOSIN-1C IS THE ADAPTATION MOTOR

Identification of the adaptation-motor myosin has been a goal of the field for over 15 years. Although Myo1c was established as the leading candidate for the hair cell's adaptation motor in the early 1990s, it took almost a decade for definitive evidence to appear demonstrating its participation in adaptation in vestibular hair cells.

Myo1c Location in Vestibular Hair Bundles

To play a role in adaptation, the adaptation motor must be found within the hair bundle and located at or near the site of transduction. Early biochemical experiments demonstrated the presence of Myo1c in purified vestibular hair bundles, making it a candidate for the adaptation motor (37). In addition, photoaffinity labeling experiments (37–39) and direct ATPase assays (40) indicated that bundle Myo1c is catalytically active.

Immunocytochemical localization of Myo1c within the hair bundle placed it at stereociliary tips, the site of transduction (37, 41). Moreover, immunogold electron microscopy more precisely localized Myo1c at or near tip-link anchors (42, 43), although this result was not universally found (44). Because the adaptation motor is thought to control tension in tip links, localization near this structure would seem to be essential.

Biochemically, stereocilia have been shown to contain 100–200 molecules of Myo1c (37, 38). Immunogold electron microscopy suggests a substantially higher number of Myo1c molecules, perhaps 500 (42). The apparent discrepancy between these values may not be as incongruous as it initially appears; the number of Myo1c molecules may have been underestimated in biochemical experiments owing to inadequate standard proteins used for quantification, or overestimated in the immunogold electron microscopy experiments if all gold particles counted did not faithfully represent individual Myo1c molecules. Nevertheless, structural and biochemical data collectively suggest that each stereocilium has several hundred Myo1c molecules, a value consistent with modeling estimates (see below).

Myo1c Participates in Adaptation

Because localization in stereocilia only circumstantially implicates Myo1c in adaptation, an experimental approach that more directly connects the Myo1c protein to adaptation was developed (45). In these studies, an inhibitor-sensitized Myo1c mutant was designed in which tyrosine-61 was changed to glycine (Y61G) to accommodate an ADP analog, N^6(2-methylbutyl) ADP (NMB-ADP) (46). An ADP analog was chosen because it was expected to maintain mutant Myo1c in a tight interaction with actin (47) and immobilize adaptation motors. Because NMB-ADP binds wild-type Myo1c poorly and has minimal effects on wild-type ATPase and motility activities, this nucleotide analog served as a specific inhibitor of the mutant Myo1c (46).

When Y61G-Myo1c was expressed in mouse hair cells using a transgenic approach, adaptation became sensitive to NMB-ADP. In cells expressing Y61G-Myo1c, the rate of adaptation to positive deflections fell by fivefold when the nucleotide analog was present, and adaptation to negative deflections disappeared completely in nearly every cell. Although wild-type Myo1c is also expressed in these cells and presumably can power adaptation, even in the presence of NMB-ADP, a few inhibited mutant molecules were nevertheless sufficient to seize the motor complex in its tracks when the analog was present. Because these results were seen only when NMB-ADP delivery was paired with Y61G-Myo1c expression, the mutant Myo1c must participate in adaptation when the nucleotide analog is absent (45). Wild-type Myo1c is almost identical in structure; it too must power adaptation in vestibular hair cells.

Although the mutant-inhibitor approach demonstrates that Myo1c is essential for adaptation, it does not prove that Myo1c is the only myosin involved in this process. Two additional experiments could address this issue. First, the inhibitor-sensitized myosin strategy could be extended to other myosin isozymes expressed in hair cells. Myo7a is a particularly attractive candidate; in Myo7a's absence, auditory hair cells require very large stimuli to begin to open their transduction channels, a result that is consistent with Myo7a applying resting tension in these cells (48). This result suggests the possibility that in auditory hair cells, Myo7a plays a role equivalent to Myo1c's role in vestibular hair cells. Because most myosin isozymes (including Myo7a) have a Tyr residue in the position equivalent to Y61G in Myo1c, the approach used by Holt et al. (45) may apply generally. The second, more direct experiment would be to test whether Y61G-Myo1c could mediate adaptation when all other hair-cell myosins are inhibited. In this experiment, native myosin motility could be blocked by ADP or ADPβS (49); under these conditions, a nucleoside triphosphate analog such as N^6(2-methylbutyl) ATP should power adaptation in hair cells containing the Y61G allele.

PROPERTIES OF MYOSIN-1C

Myosin-1c is related structurally to other members of the myosin superfamily. Although kinetic and structural measurements have been made with several myosin isozymes, we focus here on biochemical and mechanical properties of Myo1c itself (Table 1).

All myosin isozymes work by essentially the same mechanism (Figure 3), although kinetic details vary significantly among the isozymes. When free of ATP, myosin binds strongly to actin (rigor). ATP binds rapidly and dramatically reduces the affinity of myosin for actin; while in the ATP-bound state, myosin alters its conformation, essentially cocking its lever arm—the domain responsible for movement—to a position where it can exert force. While still in a weakly bound state, myosin hydrolyzes ATP to ADP and P_i, which are released sequentially. P_i leaves first, tunneling out the "back door" of the enzyme (50), which triggers a

TABLE 1   Myo1c biochemical and mechanical constants (rat Myo1c)

Constant	Value (low Ca^{2+})	Source
ATPase rate	1 s^{-1}	(46)
K_m (ATP hydrolysis)	10 μM	(46)
K_m (motility)	200 μM	(46)
V_{Myo1c} (sliding rate)	100 nm · s^{-1}	(M. Geeves, R. Milligan, J. Molloy & L. Coluccio, submitted)
κ_{Myo1c} (crossbridge stiffness)	0.5 mN · m^{-1}	(M. Geeves, R. Milligan, J. Molloy & L. Coluccio, submitted)
δ_{Myo1c} (2-step power stroke)	4 nm	(M. Geeves, R. Milligan, J. Molloy & L. Coluccio, submitted)
p (duty ratio or attachment probability)	0.05	$= K_M$ (ATPase)$/K_M$ (motility)
t_{on} (attachment time)	50 ms	$= p/$ATPase rate
f_{max} (maximal force)	2 pN	$= \kappa_{Myo1c} \delta_{Myo1c}$

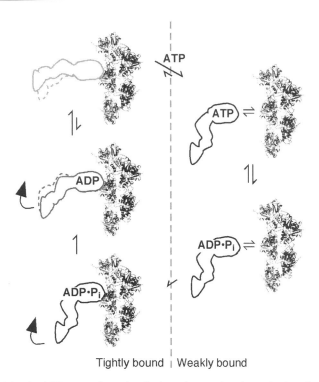

Tightly bound ┊ Weakly bound

Figure 3 Myo1c ATPase and mechanical cycle. Depicted are the head and neck region of Myo1c, drawn to scale from the data of Arthur et al. (M. Geeves, R. Milligan, J. Molloy & L. Coluccio, submitted), and a ribbon diagram of actin. Starting in the nucleotide-free state at upper left, Myo1c binds ATP very rapidly, which causes detachment of the motor (*upper right*). ATP is hydrolyzed in a readily reversible step (*lower right*). Product release occurs simultaneously with tight binding to actin; release of P_i is associated with the first step of the working stroke (*lower left*), and ADP release is associated with the second (*middle left*). Differences in sizes of the working strokes measured with mechanical and structural experiments suggest that ADP-bound and nucleotide-free Myo1c may display considerable flexibility in its lever arm (*dashed lines*).

tight interaction of myosin with actin. Simultaneously with P_i release, myosin carries out a power stroke, moving its lever arm and exerting force, thereby straining an internal elastic element. Movement occurs as the internal strain is relieved. As long as ADP is bound, myosin interacts tightly with actin, even if other associated myosin molecules strain the bound myosin in the opposite direction.

An important parameter that describes a motor protein's behavior is the duty ratio, the fraction of time during ATP hydrolysis that the motor spends tightly attached to an actin filament (51). As long as the number of myosin molecules working together (as in the adaptation motor) is substantially greater than the inverse of

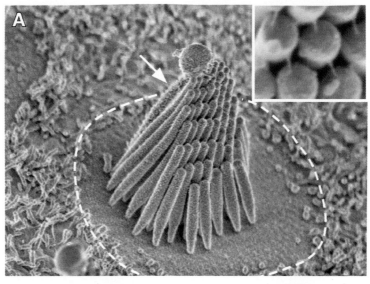

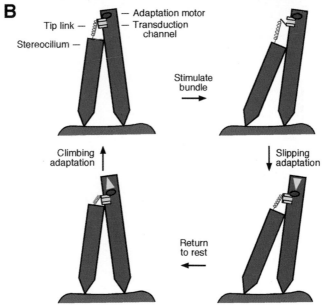

See legend on next page

Figure 1   Hair cells and hair-cell transduction. (*A*) Scanning electron micrograph of bullfrog sacculus hair cell. The apical surface of one hair cell is outlined by the dotted line. Stereocilia range in height in this bundle from ~4 to ~8 μm. The kinocilium is indicated by an arrow. *Inset*: Higher magnification of another bundle shows tip links, which interconnect stereocilia. (*B*) Model for adaptation. Starting at upper left, at rest, the adaptation motor maintains a resting tension on the transduction channels. An excitatory stimulus stretches tip links and opens transduction channels (*upper right*). Increased tension in the tip links and Ca^{2+} entry triggers slipping adaptation, which reduces tension and allows channels to close (*lower right*). When the bundle is returned to rest (*bottom left*), tip-link tension is low; the adaptation motor climbs to restore tension to the resting level (*upper left*).

A

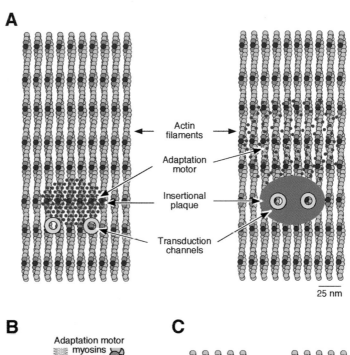

Actin
filaments

Adaptation
motor

Insertional
plaque

Transduction
channels

25 nm

B

Adaptation motor
myosins

Transduction
channel

Coupling
protein

Tip
link

Actin
filament

C

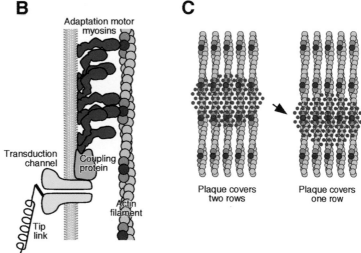

Plaque covers
two rows

Plaque covers
one row

See legend on next page

Figure 5   Models for adaptation motor structure. (*A*) Plaque model (*left*) versus dispersed model (*right*). Actin and myosin are approximately to scale. The view is through the membrane (not indicated) at the parallel actin filaments of the stereocilia. Sites available for myosin binding to actin filaments are indicated in red. The tip link (not shown) terminates at the insertional plaque; transduction channels (*yellow rings*) presumably are located nearby. The adaptation-motor myosin molecules may make up the plaque (*left*) or be dispersed nearby (*right*). (*B*) Transduction apparatus. A plausible model is indicated; adaptation-motor myosins are coupled together (mechanism not specified) and are coupled to the transduction channel (via hypothetical green coupling protein). (*C*) The plaque model should show variable interaction with actin-binding sites. The plaque is drawn to cover two rows of filaments longitudinally (74 nm), an area larger than the plaque size measured in electron micrographs. Only a fraction of actin monomers are accessible to myosins in a given orientation; fimbrin-crosslinked actin filaments will be in register as indicated. As the plaque moves, it should interact with two (*left*) or one (*right*) row of actin-binding sites, which should lead to variations in force production. Scale bar in *A* also applies to *C*.

TABLE 2 Adaptation constants (mouse hair cells)

Constant	High Ca^{2+}	Source	Low Ca^{2+}	Source
κ_g (gating-spring stiffness)	*1 mN · m^{-1} (frog)	note 1	*1.5 mN · m^{-1}	note 1
γ (geometrical gain)	0.05	(9)	0.05	(9)
S (slipping rate constant)	44 s^{-1}	(9)	N.D.	
ζ (frictional coefficient)	34 μN s · m^{-1}	$= \kappa_g S^{-1}$	170 μN · s · m^{-1}	note 2
C (climbing rate re bundle tip)	6 μm · s^{-1}	(45)	N.D.	
c (climbing rate re adaptation motor)	300 nm · s^{-1}	$= \gamma C$	100 nm · s^{-1}	note 3
δ_{motor} (working stroke, motor)	N.D.	N.D.	4 nm	(21)
x_r (gating-spring extension)	5 nm	(64)	*7 nm	
Resting tension	5 pN	$= \kappa_g x_r$	*10 pN	

*Estimated. N.D., not determined. Values are for mouse hair cells unless otherwise indicated.

Note 1   The value measured consistently in frog hair cells is used here. We also assume that stiffness increases in low Ca^{2+} to 150% of the high-Ca^{2+} value, as in frog hair cells (84).

Note 2   Assuming a fivefold reduction in slipping rate by lowering Ca^{2+}.

Note 3   We assume that by lowering Ca^{2+}, C is reduced threefold, as in frog hair cells.

the duty ratio ($\sim$20 molecules for Myo1c; Table 2), the velocity of an ensemble of myosins (and hence the climbing velocity in hair-cell adaptation) will be constant, no matter the number of myosin molecules constituting the ensemble (51).

Biochemical Properties of Myo1c

Myo1c is a $\sim$118-kDa member of the myosin-I class of motor proteins. This protein contains three domains (Figure 4): a globular head domain, which hydrolyzes ATP and binds actin; a neck domain, which contains four calmodulin-binding IQ domains; and a tail domain, which binds phospholipids with negatively charged head groups.

Head (1-700)	Neck (701-791)	Tail (792-1028)
ATP, actin binding	CaM binding	Anionic phospholipid binding

Figure 4 Myo1c structure. Myo1c consists of three domains, delineated by amino acid number in parentheses. The motor head generates force while hydrolyzing ATP. The neck provides an extended lever arm to amplify movements within the head; it binds calmodulin (CaM) and may also function to couple Myo1c to other proteins. The tail binds anionic phospholipids such as phosphatidylserine and phosphatidylinositol-4,5-bisphosphate.

MYO1C PURIFICATION   Myo1c has been purified from a variety of tissues, including bovine adrenal gland (52), rat liver (53), and porcine smooth muscle (54); moreover, the bovine (55), rat (46), and frog (S. Jean & P.G. Gillespie, unpublished data) orthologs have been heterologously expressed in Sf9 cells using baculovirus expression. Values for enzymatic properties of Myo1c experimentally determined using these different myosin sources largely agree, at least within an order of magnitude. We focus here on properties of rodent Myo1c, as those are directly comparable to the electrophysiological properties of rodent hair cells.

ATP TURNOVER   Myo1c hydrolyzes ATP at a rate of $0.1-1$ s^{-1}; the likelihood of enzyme inactivation during purification suggests that the higher value may be closer to that seen in cells. As with other vertebrate myosin-I isozymes, the actin-independent ATPase activity is relatively high; actin activates Myo1c by 2- to 10-fold. The K_m for ATP is relatively low, ~ 10 μM; moreover, ADP is not a potent inhibitor of Myo1c as it is for other myosin isozymes (56). Under normal cellular conditions of ATP (>1 mM) and ADP (<50 μM), Myo1c activity should therefore be near maximal.

MOTILITY   Measured rates of in vitro motility, where myosin adsorbed to a surface translocates fluorescently labeled actin filaments, vary depending on the source of Myo1c. At one extreme, baculovirus-expressed bovine Myo1c moved actin filaments as fast as 500 nm $\cdot$ s^{-1} (57). By contrast, baculovirus-expressed rat Myo1c managed only ~ 30 nm $\cdot$ s^{-1} at room temperature (46), whereas rat Myo1c purified from liver translocated actin at ~ 100 nm $\cdot$ s^{-1} at 37°C (M. Geeves, R. Milligan, J. Molloy & L. Coluccio, submitted). It is unclear whether these differences in motility rates reflect experimental variability between laboratories or whether they reflect true differences in the state of Myo1c purified from different sources.

Ca^{2+} REGULATION   Details of Myo1c regulation by Ca^{2+} vary between laboratories. In some hands, Ca^{2+} stimulates actin-activated ATPase activity (52, 57), whereas others find no effect of the ion (54). Moreover, few significant effects of Ca^{2+} were seen in the one study of Myo1c that used transient kinetics to measure individual steps of the cycle (M. Geeves, R. Milligan, J. Molloy & L. Coluccio, submitted). Ca^{2+} could instead affect Myo1c mechanical activity, which would not be observed in kinetics experiments where no mechanical load is applied to Myo1c. The effects of Ca^{2+} on in vitro motility are equally confounding. Some laboratories find that motility is blocked when Ca^{2+} is elevated (55), whereas others find a stimulation of motility (M. Geeves, R. Milligan, J. Molloy & L. Coluccio, submitted). It is unclear why such disparate results have been obtained in these in vitro assays. Such discrepancies are important, as the rate of adaptation is clearly affected strongly by Ca^{2+}; we expect that this effect of the ion derives from its modulation of Myo1c.

LIPID BINDING   As with other myosin-I isozymes, Myo1c binds anionic phospholipids (58). Although the basic tail domain confers lipid binding, the resulting interaction is transient (59). When Myo1c is treated with Ca^{2+}, which dissociates at least one calmodulin molecule from the IQ domains, the interaction of the enzyme with anionic phospholipids strengthens dramatically (59). One interpretation of these results is that one of the IQ sites can, when not occupied by calmodulin, form a second lipid-binding domain. This result suggests that tight interaction of Myo1c with membranes may occur only where Ca^{2+} entry occurs.

Mechanical Properties of Myo1c

Myo1c behaves in a mechanically similar way to other vertebrate myosin-I isozymes, which have been characterized more thoroughly (60). An attached Myo1c-actin crossbridge has a stiffness of $\sim$0.5 mN $\cdot$ m^{-1} and carries out a power-stroke of $\sim$4 nm (M. Geeves, R. Milligan, J. Molloy & L. Coluccio, submitted). Interestingly, as with Myo1b, the power-stroke occurs in two phases; the first phase of $\sim$3 nm corresponds to P_i release, whereas the second phase of $\sim$1 nm corresponds to ADP release (M. Geeves, R. Milligan, J. Molloy & L. Coluccio, submitted). This mechanical cycle is illustrated in Figure 3. Consistent with these results, actin filaments decorated with Myo1c show, by electron microscopy, a large difference in the position of the light-chain binding domain between the ADP-bound and nucleotide-free states of the motor protein (M. Geeves, R. Milligan, J. Molloy & L. Coluccio, submitted).

A two-phase power-stroke suggests that Myo1c is able to bear significant tension without substantial ATP hydrolysis. If movement through both phases of the power-stroke is necessary for ADP release, tension borne by attached Myo1c molecules (opposite the direction that the lever arm needs to swing) may prevent those Myo1c molecules from completing their ATPase cycles and detaching from actin. As in Myo1a, Myo1b, and smooth-muscle myosin II, this feature could allow for substantial tension development without continuous cycling on and off of actin (61, 62). Consistent with that result, hair-bundle mechanical noise, an indicator of myosin cycling, is very low at rest (21). Such low noise could improve the bundle's sensitivity at rest.

STRUCTURE AND FUNCTION OF ADAPTATION-MOTOR MYOSINS

Interpretation of hair-cell physiology data (Table 2) in the context of the slow-adaptation model (Figure 1*B*) permits a deeper understanding of adaptation-motor behavior. Here we compare the in vitro biochemical characteristics of purified Myo1c with the physiological measurements of adaptation, seeking to establish properties of Myo1c that account for force production at rest, rates of climbing and slipping adaptation, and regulation of adaptation. We also use this model to estimate the number of myosin molecules that participate in adaptation, as well as to picture their location within the hair bundle.

Because the most thorough set of Myo1c in vitro parameter values are those obtained in low Ca^{2+}, we largely restrict this comparison of Myo1c molecular properties to physiology experiments performed under low Ca^{2+} conditions. Moreover, because more in vitro properties of Myo1c have been measured with the rat isozyme, presumably similar to that from mouse, we focus on mouse hair-cell physiology when appropriate parameter values are available. In some cases, only frog hair-cell data are available; in these instances, we extrapolate the frog data to the mouse. Our goal is to explain the physiological properties of the adaptation motor from the biochemical and mechanical properties of Myo1c.

Simple Model for the Adaptation Motor

A single adaptation motor controls tension in one tip link—the number of channels controlled is unclear—and consists of m myosin molecules. Each myosin traverses through its ATPase cycle, spending time attached (t_{on}) in a strong-binding state out of a total cycle time of t_{cycle}. The probability of attachment of each myosin molecule is p, so that $t_{on} = pt_{cycle}$. While bound strongly, an individual motor exerts a force, f, that propels the motor toward the barbed end of actin filaments, e.g., the tip of a stereocilium. The product of p and m is the total number of myosin molecules attached at any given time, and the isometric force is pmf. We hope to determine the value of m with our model; in some cases, however, pm is more easily measured.

Resting Tension

We can use measurements of resting tension to estimate how many Myo1c molecules are active in the adaptation motor. Assuming a gating-spring stiffness (κ_g) of 1 mN · m^{-1} (63), the resting tension in frog hair cells in high Ca^{2+} is about 20 pN. Reduction of Ca^{2+} sensed by the motor increases this value at least twofold. Resting tension in mouse hair cells appears to be less than that in frog. We assume that the mouse κ_g is also 1 mN · m^{-1} (Table 2). The tension in the gating spring is equal to the spring's stiffness multiplied by its resting extension. Using a high-Ca^{2+} gating-spring extension (x_r) of 5 nm (64) and assuming that lowered Ca^{2+} produces a doubling of resting tension as in frog, $\kappa_g x_r = 10$ pN in mouse hair cells when Ca^{2+} is low. Because the adaptation motor generates this force, pmf is also 10 pN. With $f = 2$ pN (M. Geeves, R. Milligan, J. Molloy & L. Coluccio, submitted), this resting tension could be produced by ~5 attached myosin molecules.

Motor Velocity and Limits

The adaptation motor responds very differently to excitatory and inhibitory displacements; accordingly, dissecting climbing and slipping adaptation provides further insights into Myo1c's behavior during adaptation.

CLIMBING RATE   In mouse hair cells, the climbing rate (v_{motor}) when Ca^{2+} is high (at a cell membrane potential of –80 mV) is 300 nm s^{-1} (45). Although v_{motor} has not been measured under low Ca^{2+} conditions, in frog hair cells the rate is one third as great as the high Ca^{2+} value; we presume that the mouse adaptation motor will move similarly slower under low Ca^{2+} conditions (i.e., 100 nm $\cdot$ s^{-1}).

The attachment time can be extracted from the velocity and motor step size (δ) from $v_{motor} = \delta t_{on}^{-1}$. With a 4-nm step size, $t_{on} = 40$ ms. These values match well the in vitro properties of Myo1c; for example, a duty ratio of 0.05 and an ATPase rate of 1 s^{-1} predicts $t_{on} = 50$ ms. Similarly, a 4-nm step and v_{Myo1c} of 100 nm $\cdot$ s^{-1} (M. Geeves, R. Milligan, J. Molloy & L. Coluccio, submitted) predicts $t_{on} = 40$ ms, albeit at 37°C. This attachment time estimate is useful in determining the behavior of myosin molecules during slipping adaptation (see below).

SLIPPING FRICTION   We can use the properties of slipping adaptation to estimate *pm*, the number of attached myosin molecules. The concept of molecular friction was initially developed to explain the load exerted by weakly attached myosin crossbridges during rapid filament sliding in muscle (65). We apply the idea simplistically here, suggesting that myosin crossbridges undergoing tension during and after their power-strokes, while still tightly attached to actin, will slow the movement of the whole adaptation-motor ensemble (66). The frictional coefficient (ζ) is related to the number of attached motors (*pm*) by $\zeta = pm\kappa_m\tau_{on}$, where κ_m is the myosin-actin stiffness (per molecule). Assad & Corey (12) measured the initial rate of slipping adaptation in frog vestibular hair cells and fit their data to $V = SX$, where V is the velocity of the motor (measured at the tip of the bundle), X is the applied displacement, and S is the rate constant for adaptation (81 s^{-1} at -80 mV and 15 s^{-1} at $+80$ mV). Alternatively, the velocity can be set in terms of the force applied (F) and the macroscopic frictional coefficient (Z), $V = FZ^{-1}$. Because $F = K_gX$ (K_g is the macroscopic gating-spring stiffness, as measured at the tip of the bundle), $Z = K_gS^{-1}$. The relationship between Z and the frictional coefficient at the level of an individual adaptation motor is $\zeta = ZN^{-1}\gamma^{-2}$, where N is the number of transduction units; because $\kappa_g = K_gN^{-1}\gamma^{-2}$, we obtain the useful relationship $\zeta = \kappa_gS^{-1}$. With $\kappa_g = 1$ mN $\cdot$ m^{-1}, the frictional ratios for frog adaptation are 12 μN $\cdot$ s $\cdot$ m^{-1} at -80 mV and 65 μN $\cdot$ s $\cdot$ m^{-1} at $+80$ mV.

In mouse hair cells, the slipping rate constant of 44 s^{-1} at -80 mV (9) and $\kappa_g = 1.5$ mN $\cdot$ m^{-1} (Table 2) suggests that $\zeta = 34$ μN $\cdot$ s $\cdot$ m^{-1}; if reducing Ca^{2+} entry reduces the rate constant by fivefold as in frog hair cells (12), in low Ca^{2+}, $\zeta \approx 170$ μN $\cdot$ s $\cdot$ m^{-1}. This value is most directly comparable to in vitro properties of rat Myo1c; applying the parameter values of Table 1 gives the number of attached myosin molecules as $\sim$7 under low Ca^{2+} conditions. This number is very approximate, however, as it depends on the model chosen for slipping adaptation. Moreover, our calculation assumes that there is no strain-dependent increase in myosin dissociation, which is unlikely.

SLIPPING VIA ATP-INDEPENDENT RELEASE   An ATP-independent route of Myo1c release from actin filaments provides an alternative mechanism for slipping adaptation (J. Molloy, personal communication). Although the mechanistic details of

ATP-independent release are unknown, a plausible mechanism is that the myosin power-stroke can be reversed through stochastic fluctuation of the lever-arm position. With this mechanism, external forces delivered by a stimulus would promote power-stroke reversal. Motor detachment and the rate of slipping adaptation should depend on the size of the stimulus, a consequence that matches the experimental data (12).

FORCES FELT BY THE ADAPTATION MOTOR   Under low Ca^{2+} conditions, adaptation is linearly related to displacement size of 750 nm or less in frog hair cells (12); such a stimulus applies ~100 pN to each adaptation motor. In laser-tweezer experiments with conventional myosin, attached individual myosin molecules can resist up to 9 pN (67). If the adaptation-motor myosins have similar properties, 11 attached myosins can account for the hair-cell resistance to force.

LIMITS TO ADAPTATION   A common phenomenon observed is that slow adaptation is incomplete; in frog hair cells, the shift of the displacement-response relation is only ~80% of the stimulus amplitude (68). Incomplete adaptation has been modeled by incorporating an intracellular spring mechanical element, called the extent spring (of ~0.14 mN · m^{-1}), located parallel to the adaptation motor. In frog hair cells, the extent spring operates over an extremely broad range; linearity over bundle displacements of 1000–2000 nm indicates that it can stretch 100–200 nm. In mice, the extent spring may exhibit nonlinear behavior such that its stiffness is inversely related to the displacement step size (16). The extent spring apparently can also serve as a force to restore the adaptation motor to its rest position (39).

Climbing adaptation is also limited; the adaptation motor cannot respond to negative displacements larger than 100–500 nm (68), which correspond to a motor movement of 14–70 nm. Consistent with that result, in frog hair cells, the insertional plaque—the likely location of the adaptation motor—moves only 50–100 nm after tip links are cut with the Ca^{2+}-chelator BAPTA (20). Moreover, the extent spring, because it will either assist (68) or oppose (39) motor climbing when gating springs slacken, significantly complicates interpretation of the climbing-rate data (68).

A plausible model that explains both limits to adaptation posits that the adaptation motor contains actin-binding molecules that bind tightly to actin, yet cannot be released by ATP. These crossbridges could make up the extent spring if they included a linear elastic element of the appropriate stiffness that could sustain substantial tension without detachment, and that could elongate (perhaps by unfolding) 100 nm or more. Myosin molecules, either alone or bound to a hypothetical myosin-binding protein, could serve this role if their ATP-dependent release from actin was blocked. Elongation would have to derive from the motor's tail, as any elongation of the head would presumably destroy its actin-binding ability. Consistent with this model, hair-bundle Myo1c molecules are recalcitrant to extraction with ATP, Ca^{2+} chelators, and detergent, which would be expected to disrupt interactions between the adaptation motor and the rest of the transduction apparatus (37).

Modulation by Ca^{2+} and Calmodulin

A successful model for the adaptation motor must explain the striking effects of Ca^{2+} on adaptation. The effect on slipping rate is most clear; depolarization of a hair cell to $+80$ mV, which nearly eliminates Ca^{2+} entry, reduces the initial rate of slipping by about fivefold relative to the rate at -80 mV (12). Although it is difficult to accurately determine the Ca^{2+} concentration at the adaptation motor under these two conditions with channels fully open, a reasonable estimate would be ~ 25 μM at -80 mV and 0.1 μM when Ca^{2+} entry is blocked (69). With the bundle at rest at -80 mV, channels are open $\sim 10\%$ of the time, so the Ca^{2+} concentration is intermediate between these extremes. Although similar experiments suggest that the unloaded climbing rate slows by threefold when a hair cell is depolarized (12), the Ca^{2+} concentrations at -80 mV and $+80$ mV are not likely to be the same as those attained during measurement of slipping rate. At both membrane potentials, the transduction channels should be fully closed; hence, the difference in rate may reflect contamination of the response by fast adaptation.

The effect of Ca^{2+} is most likely due to a decrease in t_{on} or a reduction in κ_m (66). As indicated above, transient-kinetic experiments indicate little effect of Ca^{2+} on the reaction mechanism of Myo1c (M. Geeves, R. Milligan, J. Molloy & L. Coluccio, submitted), suggesting a reduction in crossbridge stiffness as the most likely locus of Ca^{2+} action. An estimate of the magnitude of this effect can come from comparing slipping rates under depolarized and resting conditions. Because the frictional coefficient decreases by about fivefold following depolarization, if Ca^{2+} simply affected κ_m, it would be only 20% the stiffness seen under low Ca^{2+} conditions. A full characterization of Myo1c's mechanical properties in the presence of high and low Ca^{2+} should be very illuminating.

In frog hair cells, the divalent-cation dependence of adaptation's slipping rate (70) concords with the Ca^{2+} sensor being calmodulin, which is present at a free concentration of ~ 35 μM in hair bundles (71). Ba^{2+} and Mg^{2+}, which bind poorly to calmodulin, do not substitute for Ca^{2+}. Sr^{2+}, which binds to but poorly activates calmodulin, acts as a competitive inhibitor of Ca^{2+} (72). The influence of Ca^{2+} on adaptation is likely due to the modulation of calmodulin binding to each of the four calmodulin-binding IQ domains located in the Myo1c neck region. Indeed, dialysis of calmodulin antagonists into frog hair cells abolishes adaptation (38). This effect on adaptation, however, may be secondary owing to the effects of such antagonists on other calmodulin-binding proteins within the hair bundle, including the plasma membrane Ca^{2+}-ATPase.

Regulation by Phosphorylation

Adaptation is also apparently regulated by phosphorylation. Introduction of cAMP through a patch-recording pipette produces a rightward shift in the displacement-response relation in turtle hair cells (73). Because the same shift can be induced with an inhibitor of phosphodiesterases (PDE), the enzyme that hydrolyzes cAMP, both PDE and the cAMP-synthetic enzyme adenylyl cyclase must both be tonically active under these experimental conditions. In mouse hair cells, agents that activate cAMP-dependent protein kinase (PKA) cause a modest rightward shift of the displacement-response curve, resulting in a decreased p_{open} at rest. By contrast, inhibitors of PKA cause a larger leftward shift (G. Géléoc & D.P. Corey, personal communication), suggesting that there is tonic PKA phosphorylation of a site that affects resting tension.

Among other possible causes, the shift in resting channel open probability could be caused by reduced Myo1c force production. Myo1c has one PKA site (Ser-701), located between the motor head and the neck domain by the first calmodulin-binding site. Ser-701 is phosphorylated by PKA in vitro; the biochemical consequences of phosphorylation are unclear, although calmodulin binding is unaffected (E. Miller & P.G. Gillespie, unpublished data). A plausible role for phosphorylation of Myo1c by PKA is to influence the motor's mechanical properties, which provides a mechanism for the observed effects of modulation of cAMP levels on transduction. Myo1c is also phosphorylated by protein kinase C (PKC) in vitro, reducing the binding of calmodulin (74). Activation of PKC, although not yet demonstrated in hair bundles, could also control the rate of adaptation by affecting Myo1c's mechanical properties.

Location and Size of the Adaptation Motor

The adaptation-motor model described here is most useful for determining the location of the motor and the number of myosin molecules contained within it. Although the mutant-inhibitor experiment described earlier clearly demonstrates that Myo1c participates in adaptation in vestibular hair cells, this experiment does not indicate where the myosin molecules are located and how many are present.

LOCATION   As diagrammed in Figure 5A (left; see color insert), the adaptation-motor myosins might be housed completely within the insertional plaque (19, 20). In favorable cross sections of the insertional plaque, individual crossbridges, perhaps corresponding to single Myo1c molecules, can be seen (19). The insertional plaque, 70 nm wide × 50 nm tall × 20 nm thick (19), is sufficient to accommodate 200 close-packed myosin-1 molecules of dimensions 5 nm tall × 3.5 nm wide × 22 nm long (75). Because close-packed molecules would not have the freedom to operate independently, it is unlikely that a plaque of this size would contain so many myosin molecules. On the other hand, uncompensated shrinkage during processing for electron microscopy could have made the plaque appear substantially smaller than its actual size.

Physiological measurements suggest that, out of the many myosins molecules in the adaptation motor, only 5–10 are bound to actin at any given time. Because of the helical pitch on actin filaments, myosin-binding sites appear only every 37 nm (76). Even if myosin molecules were well packed in the plaque, just a fraction could interact with binding sites on the actin filaments (Figures 5A,C).

Actin filaments crosslinked with fimbrin, the actin-bundling protein present in high abundance in stereocilia, are spaced laterally by 12 nm (77), so 7–8 filaments would be covered by a plaque of dimensions $\sim$25% greater than those measured by electron microscopy. More binding sites must be available for adaptation-motor myosins, however, as steric constraints would prevent continuous binding to a single row of binding sites on actin filaments. The insertional plaque does not seem to be tall enough, however, to ensure that two or more rows of binding sites are always occupied; for this coverage, twice the spacing is required (74 nm). Either the actin filaments are not in register, allowing a mixture of filaments with one binding site covered in some and others with two, or there are periodic interactions with one, then two binding sites. This latter possibility is problematic, as force exerted by the adaptation motor would fluctuate as the motor moved up or down the stereocilium. One possibility is that the plaque is large enough in vivo to always cover two sets of myosin-binding sites on exposed actin filaments; a plaque this size (74 nm in diameter) would permit continuous access of myosin molecules to 10–20 actin-binding sites.

Alternatively, the insertional plaque may not be the location of the adaptation-motor myosins. A discrepancy between the notion that the plaque contains the adaptation motor and experimental data comes from measuring the distance Ca^{2+} diffuses to affect resting p_{open} and the rate of slow adaptation (17, 78). In those experiments, the Ca^{2+}-binding site that controls those two parameters was estimated to be 100–200 nm away from the transduction channels, rather than $\sim$25 nm predicted if both channels and motors are present at the insertional plaque. These results suggest that the adaptation-motor myosins could be considerably more distant from transduction channels.

An additional inconsistency with the plaque model for adaptation-motor location is that antibody labeling shows only a modest elevation of Myo1c density at the insertional plaque (42, 43). Moreover, Myo1c density appears to be substantially higher >100 nmabove the insertional plaque (42). [A slightly different picture was portrayed by Steyger et al. (43) in frog utricle, where Myo1c density was high within 75 nm of the plaque.] The discrepancy between antibody localization and the expected Myo1c location in the insertional plaque might arise because antibodies could have limited access to tightly packed Myo1c molecules within the plaque. Alternatively, the plaque may not contain the myosin molecules (although perhaps the channels are located there). Instead, the adaptation motor myosins may be more broadly distributed above the plaque (Figure 5A, *right*), presumably coupled mechanically to the channels through an elongated elastic linkage.

MYOSIN NUMBER   As noted above, a variety of calculations suggest that *pm*, the number of attached myosin molecules, is 5–10. *p* can be calculated from the ratio of the K_m for ATP hydrolysis to the K_m for in vitro motility (51), giving $\sim$0.05 for rat Myo1c (46; P.G. Gillespie, unpublished data). Thus *m*, the total number of Myo1c molecules in each active adaptation motor, is 100–200, in good accordance with biochemical and structural estimates. As noted above, stereocilia likely contain more Myo1c molecules than this; not all will be connected to transduction channels.

Direct experimental verification of this estimated number, as well as the specific location of adaptation-motor myosins, is essential. Such detection of Myo1c in live hair cells should be feasible with modern fluorescence methods (79, 80). Because myosins participating in adaptation should move substantial distances, adaptation-dependent rearrangements should be detectable using sensitive optical methods. The distances moved (e.g., 120 nm following a 1 μm stimulus in a bullfrog hair cell, given 80% adaptation) are below the limits of optical resolution but not below the limit of optical measurement.

Coupling of Myo1c Molecules to the Transduction Apparatus

An unresolved question is how the myosin molecules are coupled to the rest of the transduction apparatus (Figure 5B). Somehow, 100–200 molecules of Myo1c must be coupled to a single gating spring. Models can be divided conceptually into two extremes. In the first model, all myosins of an adaptation motor could be coupled together by Myo1c-Myo1c self-association; the Myo1c cluster would then be attached to transduction channels by one or a few coupling molecules. In the second model, Myo1c molecules might not interact with each other, but instead would be rafted together by Myo1c-binding molecules, which in turn would interact directly or indirectly with channels.

A potential clue to the coupling mechanism comes from experiments using Myo1c or its fragments to localize its binding sites in hair cells (81). Myo1c-binding sites are found where Myo1c is located, including at stereociliary tips and in the periculticular necklace, a vesicle-rich cytoplasmic region at the bundle's base. Myo1c binding to these sites requires a calmodulin-free IQ domain (probably IQ2; K. Phillips & J.L. Cyr, unpublished data) and a calmodulin-bound IQ domain. As expected, owing to the involvement of the Myo1c IQ domains in binding, the interaction is modulated by calmodulin, which may provide a means for the Ca^{2+}-dependent regulation of the interaction (81).

Although the identity of the Myo1c-interacting proteins is not known, the conditions of the binding assay suggest that IQ domains do not mediate Myo1c-Myo1c self-association (81). Alternatively, we favor the interpretation that the Myo1c neck region interacts with other stereociliary constituents, which are probably associated with the cytoskeleton. Although interaction of Myo1c with another stereociliary protein seems most likely, it is also possible that Myo1c interacts only with lipids employing its two lipid-binding domains. Coupling of the adaptation motor to the transduction channel exclusively through lipids would require an unusual local lipid environment in the hair cell; such a model would differ significantly from those usually proposed (e.g., Figure 5B).

It is unclear exactly how interactions of the Myo1c neck region, which also serves as the lever arm, with other stereociliary constituents will influence the mechanochemical properties of the motor molecule. The identification of the region of Myo1c that is involved in the interaction, however, provides the necessary information to initiate biochemical and molecular-biological approaches for identifying the interaction partners; such studies should elucidate other components of the transduction apparatus and may lead to the identification of the transduction channel itself.

FUTURE DIRECTIONS

Slow Adaptation in Auditory Hair Cells

As noted above, the most compelling evidence for Myo1c participation in slow adaptation comes from mouse vestibular organs. By contrast, much less is known about Myo1c in auditory transduction. Myo1c is present in cochlear stereocilia, particularly in those from inner hair cells, although it appears to be present throughout stereocilia, in contrast to its punctate localization in vestibular organs near stereociliary tips (36). In addition, slow adaptation in hair cells of the mammalian cochlea, the auditory organ, is less apparent than it is in vestibular hair cells (18, 82, 83). Nevertheless, Myo1c might play a crucial role in cochlear transduction by maintaining appropriate resting tension on the channel thereby maximizing channel sensitivity. Application of the Y61G-Myo1c mutant-inhibitor approach to auditory hair cells could determine the functional role of Myo1c in these cells.

Reconciling Myo1c Properties with Hair-Cell Physiology

Although comparison of Myo1c biochemical and mechanical properties with biophysical features of adaptation has been profitable, our picture of the adaptation motor remains incomplete. For example, we still do not understand how Ca^{2+} regulates adaptation; fortunately, measurements of Myo1c mechanics in the presence and absence of Ca^{2+} may clear this issue up. In addition, the precise location of the Myo1c molecules that participate in adaptation remains unclear. We look forward to the development of sophisticated approaches to observe active Myo1c molecules in hair bundles, which should clarify some of these issues.

Mechanical measurements of mouse vestibular hair bundles are also incomplete. Better estimates of κ_g, the gating-spring stiffness, will be particularly important to obtain. Uncertainty in calibration of fluid-jet stimulators indicates that an alternative approach to measuring κ_g, using deflection of bundles with flexible glass fibers, will be more informative. Moreover, measurement of the Ca^{2+} dependence of the mechanical properties and rate of adaptation in mouse hair bundles will allow better comparison with biochemical measurements.

ACKNOWLEDGMENTS

We thank Lynne Coluccio, Mike Geeves, Justin Molloy, and Ron Milligan for communicating unpublished data. In addition, we thank Ruth Anne Eatock and David Corey for communication of manuscripts prior to publication, and Rachel

Dumont, Kevin Nusser, Diane Ronan, John Scarborough, Meredith Strassmaier, and Richard Walker for critical reading of the manuscript. Research in the authors' laboratories is supported by NIDCD grants to P.G.G. and to J.L.C., by the American Hearing Research Foundation (J.L.C.), and by a WVU Research Development Grant (J.L.C.).

The *Annual Review of Physiology* is online at http://physiol.annualreviews.org

LITERATURE CITED

1. Shotwell SL, Jacobs R, Hudspeth AJ. 1981. Directional sensitivity of individual vertebrate hair cells to controlled deflection of their hair bundles. *Ann. NY Acad. Sci.* 374:1–10

2. Corey DP, Hudspeth AJ. 1983. Kinetics of the receptor current in bullfrog saccular hair cells. *J. Neurosci.* 3:962–76

3. Hudspeth AJ. 1992. Hair-bundle mechanics and a model for mechanoelectrical transduction by hair cells. *Soc. Gen. Physiol. Ser.* 47:357–70

4. Hudspeth AJ, Choe Y, Mehta AD, Martin P. 2000. Putting ion channels to work: mechanoelectrical transduction, adaptation, and amplification by hair cells. *Proc. Natl. Acad. Sci. USA* 97:11765–72

5. Markin VS, Hudspeth AJ. 1995. Gating-spring models of mechanoelectrical transduction by hair cells of the internal ear. *Annu. Rev. Biophys. Biomol. Struct.* 24:59–83

6. Assad JA, Shepherd GMG, Corey DP. 1991. Tip-link integrity and mechanical transduction in vertebrate hair cells. *Neuron* 7:985–94

7. Kachar B, Parakkal M, Kurc M, Zhao Y, Gillespie PG. 2000. High-resolution structure of hair-cell tip links. *Proc. Natl. Acad. Sci. USA* 97:13336–41

8. Jacobs RA, Hudspeth AJ. 1990. Ultrastructural correlates of mechanoelectrical transduction in hair cells of the bullfrog's internal ear. *Cold Spring Harbor. Symp. Quant. Biol.* 55:547–61

9. Holt JR, Corey DP, Eatock RA. 1997. Mechanoelectrical transduction adaptation in hair cells of the mouse utricle, a low-frequency vestibular organ. *J. Neurosci.* 17:8739–48

10. Eatock RA, Corey DP, Hudspeth AJ. 1987. Adaptation of mechanoelectrical transduction in hair cells of the bullfrog's sacculus. *J. Neurosci.* 7:2821–36

11. Ricci AJ, Crawford AC, Fettiplace R. 2000. Active hair bundle motion linked to fast transducer adaptation in auditory hair cells. *J. Neurosci.* 20:7131–42

12. Assad JA, Corey DP. 1992. An active motor model for adaptation by vertebrate hair cells. *J. Neurosci.* 12:3291–309

13. Hudspeth AJ. 1997. Mechanical amplification of stimuli by hair cells. *Curr. Opin. Neurobiol.* 7:480–86

14. Manley GA. 2001. Evidence for an active process and a cochlear amplifier in non-mammals. *J. Neurophysiol.* 86:541–49

15. Howard J, Hudspeth AJ. 1987. Mechanical relaxation of the hair bundle mediates adaptation in mechanoelectrical transduction by the bullfrog's saccular hair cell. *Proc. Natl. Acad. Sci. USA* 84:3064–68

16. Vollrath MA, Eatock RA. 2003. Time course and extent of mechanotransducer adaptation in mouse utricular hair cells: comparison with frog saccular hair cells. *J. Neurophysiol.* e-pub 10.1152/jn.00893.2002

17. Wu YC, Ricci AJ, Fettiplace R. 1999. Two components of transducer adaptation in auditory hair cells. *J. Neurophysiol.* 82:2171–81

18. Kennedy HJ, Evans MG, Crawford MC, Fettiplace R. 2003. Fast adaptation of

mechanoelectrical transducer channels in mammalian cochlear hair cells. *Nat. Neurosci.* 6:832–36

19. Hudspeth AJ, Gillespie PG. 1994. Pulling springs to tune transduction: adaptation by hair cells. *Neuron* 12:1–9

20. Shepherd GMG, Assad JA, Parakkel M, Kachar B, Corey DP. 1992. Movement of the tip-link attachment is correlated with adaptation in bullfrog saccular hair cells. *J. Gen. Physiol.* 98:25a (Abstr.)

21. Frank JE, Markin V, Jaramillo F. 2002. Characterization of adaptation motors in saccular hair cells by fluctuation analysis. *Biophys. J.* 83:3188–201

22. Katz B. 1939. The relation between force and speed in muscular contraction. *J. Physiol.* 96:45–64

23. Deleted in proof

24. Zhao Y, Yamoah EN, Gillespie PG. 1996. Regeneration of broken tip links and restoration of mechanical transduction in hair cells. *Proc. Natl. Acad. Sci. USA* 93: 15469–74

25. Berg JS, Powell BC, Cheney RE. 2001. A millennial myosin census. *Mol. Biol. Cell* 12:780–94

26. Avraham KB, Hasson T, Steel KP, Kingsley DM, Russell LB. 1995. The mouse Snell's waltzer deafness gene encodes an unconventional myosin. *Nat. Genet.* 11:369–75

27. Donaudy F, Ferrara A, Esposito L, Hertzano R, Ben-David O, et al. 2003. Multiple mutations of MYO1A, a cochlear-expressed gene, in sensorineural hearing loss. *Am. J. Hum. Genet.* 72:1571–77

28. Gibson F, Walsh J, Mburu P, Varela A, Brown KA, et al. 1995. A type VII myosin encoded by the mouse deafness gene *shaker-1. Nature* 374:62–64

29. Lalwani AK, Goldstein JA, Kelley MJ, Luxford W, Castelein CM, Mhatre AN. 2000. Human nonsyndromic hereditary deafness DFNA17 is due to a mutation in nonmuscle myosin MYH9. *Am. J. Hum. Genet.* 67:1121–28

30. Melchionda S, Ahituv N, Bisceglia L, Sobe T, Glaser F, et al. 2001. MYO6, the human homologue of the gene responsible for deafness in Snell's waltzer mice, is mutated in autosomal dominant nonsyndromic hearing loss. *Am. J. Hum. Genet.* 69:635–40

31. Probst FJ, Fridell RA, Raphael Y, Saunders TL, Wang A, et al. 1998. Correction of deafness in shaker-2 mice by an unconventional myosin in a BAC transgene. *Science* 280:1444–47

32. Walsh T, Walsh V, Vreugde S, Hertzano R, Shahin H, et al. 2002. From flies' eyes to our ears: mutations in a human class III myosin cause progressive nonsyndromic hearing loss DFNB30. *Proc. Natl. Acad. Sci. USA* 99:7518–23

33. Wang A, Liang Y, Fridell RA, Probst FJ, Wilcox ER, et al. 1998. Association of unconventional myosin MYO15 mutations with human nonsyndromic deafness DFNB3. *Science* 280:1447–51

34. Weil D, Blanchard S, Kaplan J, Guilford P, Gibson F, et al. 1995. Defective myosin VIIA gene responsible for Usher syndrome type 1B. *Nature* 374:60–61

35. Solc CK, Derfler BH, Duyk GM, Corey DP. 1994. Molecular cloning of myosins from bullfrog saccular macula: a candidate for the hair cell adaptation motor. *Auditory Neurosci.* 1:63–75

36. Dumont RA, Zhao YD, Holt JR, Bahler M, Gillespie PG. 2002. Myosin-I isozymes in neonatal rodent auditory and vestibular epithelia. *J. Assoc. Res. Otolaryngol.* 3: 375–89

37. Gillespie PG, Wagner MC, Hudspeth AJ. 1993. Identification of a 120 kD hair-bundle myosin located near stereociliary tips. *Neuron* 11:581–94

38. Walker RG, Hudspeth AJ. 1996. Calmodulin controls adaptation of mechanoelectrical transduction by hair cells of the bullfrog's sacculus. *Proc. Natl. Acad. Sci. USA* 93:2203–7

39. Yamoah EN, Gillespie PG. 1996. Phosphate analogs block adaptation in hair cells by inhibiting adaptation-motor force production. *Neuron* 17:523–33

40. Burlacu S, Tap WD, Lumpkin EA, Hudspeth AJ. 1997. ATPase activity of myosin in hair bundles of the bullfrog's sacculus. *Biophys. J.* 72:263–71

41. Hasson T, Gillespie PG, Garcia JA, MacDonald RB, Zhao Y, et al. 1997. Unconventional myosins in inner-ear sensory epithelia. *J. Cell Biol.* 137:1287–307

42. Garcia JA, Yee AG, Gillespie PG, Corey DP. 1998. Localization of myosin-Ib near both ends of tip links in frog saccular hair cells. *J. Neurosci.* 18:8637–47

43. Steyger PS, Gillespie PG, Baird RA. 1998. Myosin Ib is located at tip link anchors in vestibular hair bundles. *J. Neurosci.* 18:4603–15

44. Metcalf AB. 1998. Immunolocalization of myosin Ibeta in the hair cell's hair bundle. *Cell Motil. Cytoskelet.* 39:159–65

45. Holt JR, Gillespie SK, Provance DW, Shah K, Shokat KM, et al. 2002. A chemical-genetic strategy implicates myosin-1c in adaptation by hair cells. *Cell* 108:371–81

46. Gillespie PG, Gillespie SK, Mercer JA, Shah K, Shokat KM. 1999. Engineering of the myosin-Iβ nucleotide-binding pocket to create selective sensitivity to N^6-modified ADP analogs. *J. Biol. Chem.* 274:31373–81

47. Cooke R, Pate E. 1985. The effects of ADP and phosphate on the contraction of muscle fibers. *Biophys. J.* 48:789–98

48. Kros CJ, Marcotti W, van Netten SM, Self TJ, Libby RT, et al. 2002. Reduced climbing increased slipping adaptation in cochlear hair cells of mice with Myo7a mutations. *Nat. Neurosci.* 5:41–47

49. Gillespie PG, Hudspeth AJ. 1993. Adenine nucleoside diphosphates block adaptation of mechanoelectrical transduction in hair cells. *Proc. Natl. Acad. Sci. USA* 90:2710–14

50. Yount RG, Lawson D, Rayment I. 1995. Is myosin a "back door" enzyme? *Biophys. J.* 68:44S–7S

51. Howard J. 2001. *Mechanics of Motor Proteins and the Cytoskeleton.* Sunderland, MA: Sinauer

52. Barylko B, Wagner MC, Reizes O, Albanesi JP. 1992. Purification and characterization of a mammalian myosin I. *Proc. Natl. Acad. Sci. USA* 89:490–94

53. Coluccio LM, Conaty C. 1993. Myosin-I in mammalian liver. *Cell Motil. Cytoskelet.* 24:189–99

54. Chacko S, Jacob SS, Horiuchi KY. 1994. Myosin I from mammalian smooth muscle is regulated by caldesmon-calmodulin. *J. Biol. Chem.* 269:15803–7

55. Zhu T, Sata M, Ikebe M. 1996. Functional expression of mammalian myosin I beta: analysis of its motor activity. *Biochemistry* 35:513–22

56. De La Cruz EM, Sweeney HL, Ostap EM. 2000. ADP inhibition of myosin V ATPase activity. *Biophys. J.* 79:1524–29

57. Zhu T, Beckingham K, Ikebe M. 1998. High affinity Ca^{2+} binding sites of calmodulin are critical for the regulation of myosin Ibeta motor function. *J. Biol. Chem.* 273:20481–86

58. Reizes O, Barylko B, Li C, Südhof TC, Albanesi JP. 1994. Domain structure of a mammalian myosin I. *Proc. Natl. Acad. Sci. USA* 91:6349–53

59. Tang N, Lin T, Ostap EM. 2002. Dynamics of myo1c (myosin-Ibeta) lipid binding and dissociation. *J. Biol. Chem.* 277:42763–68

60. Veigel C, Coluccio LM, Jontes JD, Sparrow JC, Milligan RA, Molloy JE. 1999. The motor protein myosin-I produces its working stroke in two steps. *Nature* 398:530–33

61. Coluccio LM, Geeves MA. 1999. Transient kinetic analysis of the 130-kDa myosin I (MYR-1 gene product) from rat liver. A myosin I designed for maintenance of tension? *J. Biol. Chem.* 274:21575–80

62. Geeves MA, Perreault-Micale C, Coluccio LM. 2000. Kinetic analyses of a truncated mammalian myosin I suggest a novel isomerization event preceding nucleotide binding. *J. Biol. Chem.* 275:21624–30

63. Martin P, Bozovic D, Choe Y, Hudspeth AJ. 2003. Spontaneous oscillation by hair bundles of the bullfrog's sacculus. *J. Neurosci.* 23:4533–48

64. van Netten SM, Kros CJ. 2000. Gating energies and forces of the mammalian hair cell transducer channel and related hair bundle mechanics. *Proc. R. Soc. London Ser. B* 267:1915–23

65. Leibler S, Huse DA. 1993. Porters versus rowers: a unified stochastic model of motor proteins. *J. Cell Biol.* 121:1357–68

66. Gillespie PG, Corey DP. 1997. Myosin and adaptation by hair cells. *Neuron* 19:955–58

67. Nishizaka T, Miyata H, Yoshikawa H, Ishiwata S, Kinosita KJ. 1995. Unbinding force of a single motor molecule of muscle measured using optical tweezers. *Nature* 377:251–54

68. Shepherd GMG, Corey DP. 1994. The extent of adaptation in bullfrog saccular hair cells. *J. Neurosci.* 14:6217–29

69. Lumpkin EA, Hudspeth AJ. 1998. Regulation of free Ca^{2+} concentration in hair-cell stereocilia. *J. Neurosci.* 18:6300–18

70. Hacohen N, Assad JA, Smith WJ, Corey DP. 1989. Regulation of tension on hair-cell transduction channels: displacement and calcium dependence. *J. Neurosci.* 9:3988–97

71. Walker RG, Hudspeth AJ, Gillespie PG. 1993. Calmodulin and calmodulin-binding proteins in hair bundles. *Proc. Natl. Acad. Sci. USA* 90:2807–11

72. Chao SH, Suzuki Y, Zysk JR, Cheung WY. 1984. Activation of calmodulin by various metal cations as a function of ionic radius. *Mol. Pharmacol.* 26:75–82

73. Ricci AJ, Fettiplace R. 1997. The effects of calcium buffering and cyclic AMP on mechano-electrical transduction in turtle auditory hair cells. *J. Physiol.* 501:111–24

74. Williams R, Coluccio LM. 1995. Phosphorylation of myosin-I from rat liver by protein kinase C reduces calmodulin binding. *Biochem. Biophys. Res. Commun.* 216:90–102

75. Jontes JD, Milligan RA. 1997. Brush bor-der myosin-I structure and ADP-dependent conformational changes revealed by cryo-electron microscopy and image analysis. *J. Cell Biol.* 139:683–93

76. Steffen W, Smith D, Simmons R, Sleep J. 2001. Mapping the actin filament with myosin. *Proc. Natl. Acad. Sci. USA* 98:14949–54

77. Volkmann N, DeRosier D, Matsudaira P, Hanein D. 2001. An atomic model of actin filaments cross-linked by fimbrin and its implications for bundle assembly and function. *J. Cell Biol.* 153:947–56

78. Ricci AJ, Wu YC, Fettiplace R. 1998. The endogenous calcium buffer and the time course of transducer adaptation in auditory hair cells. *J. Neurosci.* 18:8261–77

79. Adams SR, Campbell RE, Gross LA, Martin BR, Walkup GK, et al. 2002. New biarsenical ligands and tetracysteine motifs for protein labeling in vitro and in vivo: synthesis and biological applications. *J. Am. Chem. Soc.* 124:6063–76

80. Tsien RY. 1998. The green fluorescent protein. *Annu. Rev. Biochem.* 67:509–44

81. Cyr JL, Dumont RA, Gillespie PG. 2002. Myosin-1c interacts with hair-cell receptors through its calmodulin-binding IQ domains. *J. Neurosci.* 22:2487–95

82. Géléoc GS, Lennan GW, Richardson GP, Kros CJ. 1997. A quantitative comparison of mechanoelectrical transduction in vestibular and auditory hair cells of neonatal mice. *Proc. R. Soc. London Ser. B* 264:611–21

83. Kros CJ, Rüsch A, Richardson GP. 1992. Mechano-electrical transducer currents in hair cells of the cultured neonatal mouse cochlea. *Proc. R. Soc. London Ser. B* 249:185–93

84. Marquis RE, Hudspeth AJ. 1997. Effects of extracellular Ca^{2+} concentration on hair-bundle stiffness and gating-spring integrity in hair cells. *Proc. Natl. Acad. Sci. USA* 94:11923–28

Annu. Rev. Physiol. 2004. 66:547–69
doi: 10.1146/annurev.physiol.66.032102.112025
First published online as a Review in Advance on July 30, 2003

REGULATION OF RENAL K TRANSPORT BY DIETARY K INTAKE

WenHui Wang

*Department of Pharmacology, New York Medical College, Valhalla, New York 10595;
email: wenhui_wang@nymc.edu*

Key Words ROMK, protein tyrosine kinase, protein tyrosine phosphatase, renal K secretion, collecting duct

■ **Abstract** Extracellular K must be kept within a narrow concentration range for the normal function of neurons, skeletal muscle, and cardiac myocytes. Maintenance of normal plasma K is achieved by a dual mechanism that includes extrarenal factors such as insulin and β-adrenergic agonists, which stimulate the movement of K from extracellular to intracellular fluid and modulate renal K excretion. Dietary K intake is an important factor for the regulation of K secretion: An increase in K intake stimulates secretion, whereas a decrease inhibits K secretion and enhances absorption. This effect of changes in dietary K intake on tubule K transport is mediated by aldosterone-dependent and -independent mechanisms. Recently, it has been demonstrated that the protein tyrosine kinase (PTK)-dependent signal transduction pathway is an important aldosterone-independent regulatory mechanism that mediates the effect of altered K intake on K secretion. A low-K intake stimulates PTK activity, which leads to increase in phosphorylation of cloned inwardly rectifying renal K (ROMK) channels, whereas a high-K intake has the opposite effect. Stimulation of tyrosine phosphorylation also suppresses K secretion in principal cell by facilitating the internalization of apical K channels in the collecting duct.

INTRODUCTION

K homeostasis is vital for the function of many body cells: A high intracellular K concentration is critical for cell growth, whereas maintenance of extracellular K in a normal range is essential for the function of neurons, cardiac myocytes, and skeletal muscle. Either high K (hyperkalemia) or low K (hypokalemia) induces cardiac arrhythmias and interferes with normal nerve function and muscle contraction. Because the total K content in the extracellular fluid is approximately 70 meq, the normal daily K intake could easily double this value if K were exclusively distributed in the extracellular fluid. However, a high-K intake does not cause hyperkalemia in the healthy subject. Safeguarding plasma K is achieved by synchronized extrarenal and renal K handling (1–4). An increase in dietary K intake is accompanied by stimulation of insulin secretion in response to an elevated

0066-4278/04/0315-0547$14.00

glucose uptake (3). Insulin activates Na-K-ATPase in skeletal muscle and transfers K from extracellular to intracellular fluid. This shift minimizes the rise of plasma K after each meal. K stored in skeletal muscle and liver is eventually released into the extracellular fluid, filtered, and also secreted into the urine by the kidney. In addition, high-K intake stimulates K secretion in the distal colon, especially when the renal K secretion function is compromised (5, 6). When K intake is restricted, the kidney is also able to actively reabsorb K in the distal nephron segment (1). Moreover, skeletal muscle plays an important role in keeping plasma K within a relative normal range during K restriction by decreasing the surface number of Na-K-ATPase. Thus K is shifted from the intracellular to the extracellular fluid (7, 8). K depletion has also been shown to attenuate the response of skeletal muscle to insulin. As a consequence, skeletal muscle becomes insulin resistant, i.e., shifting K from extracellular to intracellular fluid is significantly diminished during hypokalemia (9). K depletion has also been demonstrated to stimulate the expression of H-K-ATPase in the colon and to increase intestinal K absorption (10). The present review is focused on the effects of changes in K diet on renal K secretion. The role of skeletal muscle in maintaining K homeostasis during varying dietary K intake has been discussed in a recent review (2).

CELLULAR MECHANISM OF K TRANSPORT ALONG THE NEPHRON

Proximal Tubules

Figure 1 provides information regarding the cellular transport functions along the nephron segments. The proximal tubules reabsorb approximately 60–70% of filtered K in similar proportion to Na and water absorption (1, 11). K reabsorption in the proximal tubule is largely passive and proceeds through paracellular pathways (12–14). However, a small fraction of K may also be reabsorbed via transcellular pathways because addition of Ba^{2+}, an inhibitor of K channels, to the lumen has been shown to diminish transepithelial K absorption (13). The action of this process has not been defined. The main driving force for passive paracellular K absorption is solvent drag and the lumen-positive transepithelial potential created by transepithelial Cl reabsorption in the late portion of the proximal tubule (15, 16). Paracellular K reabsorption is also supported by a favorable K concentration gradient from the lumen to the peritubular fluid as a consequence of the absorption of water. Thus K diffuses along a favorable electrochemical gradient through the proximal tubule. This notion is further supported by the observation that K absorption in the proximal tubule is highly sensitive to alterations in transepithelial voltage and K concentrations in the lumen (17). In addition to diffusion, paracellular K reabsorption may also be driven by solvent drag because a significant fraction of filtered K has been shown to be reabsorbed by solute-dependent water movement via a high-K permeable paracellular pathway (17–19).

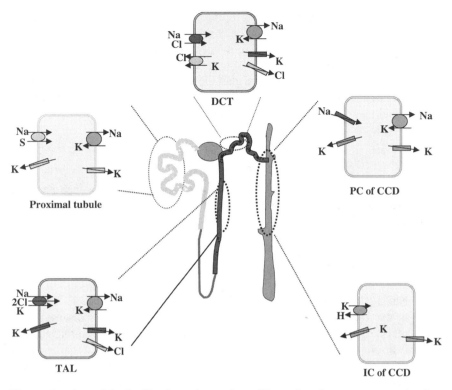

Figure 1 A model of cells along the nephron illustrating the transport mechanism of proximal tubule, thick ascending limb (TAL), distal convoluted tubule (DCT), and cortical collecting duct (CCD). PC and IC are abbreviation for principal and intercalated cell.

The mechanism by which K is absorbed via transcellular pathways is incompletely understood. It is possible that K is reabsorbed from lumen into the cell by H-K-ATPase in the apical membrane (20). However, this possibility is not supported by the finding that no significant K reabsorption occurs after inhibition of passive K transport in proximal tubules (12). Several studies have demonstrated that both a Ca^{2+}-dependent large-conductance K channel and a voltage-gated K channel, such as KCNA10, are expressed in the apical membrane (21, 22). It has also been observed that the electrochemical gradient for K across the apical membrane is above K equilibrium potential (23, 24) and favors K secretion. However, it is unlikely that a significant K secretion occurs in the proximal tubule because neither Ca^{2+}-dependent large-conductance K channels nor voltage-gated K channels have a high channel open probability under physiological conditions (21, 22). Ca^{2+}-dependent large-conductance K channels may be involved in the regulation of cell volume (22). Because the voltage-gated K channel in the apical membrane is voltage sensitive, it may contribute to stabilization of the apical membrane potential following depolarization during action of electrogenic Na-coupled glucose/amino acid transporters. K channels in the basolateral membrane play an important role in generating cell negative membrane potential, and they mediate K recycling across the basolateral membrane (25). Because several Na-coupled transport processes are electrogenic, an alteration in cell membrane potential may affect the Na-coupled transport across the apical membrane and the fluid absorption that could indirectly influence K absorption.

Thick Ascending Limb

K is also reabsorbed in the thick ascending limb (TAL), and such K reabsorption increases in K-depleted animals and decreases in high-K-adapted animals (26). Figure 1 depicts a model of TAL, indicating its transport mechanisms. K enters the cell across the apical membrane via the Na/K/2Cl cotransporter and across the basolateral membrane by Na-K-ATPase (27–30). K can move across the basolateral membrane by KCl and $KHCO_3^-$ cotransporters (31, 32). Apical K channels play a key role in K recycling, which is essential for maintaining the function of the Na/K/2Cl cotransporter (29, 33). Inhibition of K recycling blocks Na reabsorption in the TAL (1, 33, 34) and leads to salt wasting (Bartter's disease) (35, 36). K recycling not only provides an adequate supply of K for the cotransporter but also is responsible for the lumen-positive transepithelial potential that is an important driving force for passive paracellular K, Na, and divalent cation transport (1, 33). Three types of K channels have been identified in the apical membrane of the TAL of the rat kidney (37–40): a Ca^{2+}-dependent large-conductance channel and 30 and 70 pS channels. Because the Ca^{2+}-dependent large-conductance K channel has a low open probability, it is unlikely that it contributes significantly to K recycling. There is a general agreement that the 30 and 70 pS K channels are responsible for K recycling in the rat kidney. It is most likely that ROMK, a cloned inwardly rectifying renal K channel, is a key component of both 30 and 70 pS K channels because both are absent in the TAL from ROMK knockout mice (35).

Distal Convoluted Tubule

The distal convoluted tubule (DCT) secretes K, although to a lesser extent than that of the cortical collecting duct (CCD) (41). The mechanism by which K is secreted in the DCT is shown in Figure 1 and involves parallel action of KCl cotransporter (42) and the thiazide-sensitive NaCl cotransporter. Because they operate in opposite direction in the apical membrane (43, 44), this results in a net K secretion and Na reabsorption, whereas Cl is recycled across the apical membrane. The observation that infusion of $NaSO_4$ stimulates K secretion in the DCT, which is suppressed by thiazide diuretics, further supports the notion that K secretion occurs by a Na/K exchange mechanism in the DCT when luminal Cl concentration is low (42, 44).

Cortical Collecting Duct

The CCD is responsible for the final regulation of K secretion and urinary K excretion under normal conditions (1). However, the CCD can also reabsorb K when K intake is restricted (1). Two morphological distinct cells, principal and intercalated cells, are present in the CCD (1, 21, 45). Principal cells are responsible for K secretion, whereas intercalated cells reabsorb K (1, 45, 46). K secretion takes place by a two-step process: It enters the cell via the basolateral Na-K-ATPase (Figure 1) and is secreted into the lumen through apical K channels along a favorable electrochemical gradient (11). This K electrochemical gradient increases when Na transport is enhanced, for instance, by aldosterone or following augmented Na delivery (47). K can also enter the cell across the basolateral membrane via K channels in deoxycorticosterone acetate (DOCA)-treated animals because mineralocorticoids stimulate the activity of Na-K-ATPase and hyperpolarize the basolateral membrane exceeding the K equilibrium potential (48).

Two types of K channels are present in the apical membrane of the CCD: a ROMK-like small-conductance K (SK) and a Ca^{2+}-dependent large-conductance K channel (49–51). The SK channel has been shown to play a key role in K secretion under normal conditions (50–56). It is likely that the Ca^{2+}-dependent large-conductance K channel is also involved in K secretion when the tubule flow rate increases (57).

The medullary collecting duct has the ability to reabsorb K: K enters the cell across the apical membrane by an ouabain-insensitive H-K-ATPase and leaves the cell across the basolateral membrane along a favorable K electrochemical gradient (58–61). This K reabsorption pathway is regulated by dietary K intake because K depletion stimulates H-K-ATPase activity (20, 62, 63).

EFFECT OF LOW-K INTAKE ON TRANSPORT IN THE PROXIMAL TUBULE CELL

K restriction significantly decreases glomerular filtration rate and single-nephron filtration rate in rats (64). Also, fractional reabsorption in the proximal tubule is enhanced in animals on K-deficient diet. As a consequence, fluid delivery to the loop of Henle is markedly diminished. Micropuncture studies performed in chronically K-depleted animals have also demonstrated that hypokalemia inhibits NaCl transport in the rabbit proximal tubule (65, 66).

EFFECT OF K INTAKE ON K TRANSPORT IN THE TAL

Both urinary concentrating ability and transepithelial Na transport are impaired in K-depleted animals (26, 67–69). Although the mechanism by which K depletion compromises the transepithelial Na transport is not completely understood, it is possible that decreased Na transport results partially from functional and structural changes in the TAL (26, 69, 70). It has been reported that protein levels of Na/K/2Cl cotransporter in the cell membrane are significantly diminished in the TAL from K-depleted animals (26, 70–72). Also, K depletion decreases the ROMK level in the membrane of the outer medulla in which the TAL is predominant (72). Because both ROMK and Na/K/2Cl cotransporter are essential for transepithelial Na absorption in the TAL, a decrease in cotransporter and ROMK channels in the apical membrane is expected to compromise the urinary concentrating ability and transepithelial Na transport. In contrast to the Na/K/2Cl cotransporter, K depletion increases the activity of basolateral Na-K-ATPase (69). This increase may be important for maintaining intracellular K concentration at normal range.

K depletion not only causes changes in the protein levels of ion transporters that are involved in transepithelial Na reabsorption but also changes in the regulatory mechanisms of ion transport in the TAL. Stimulation of the Ca^{2+}-sensing receptor by raising external Ca^{2+} inhibits the apical 70 pS K channel (73), which is the result of increasing the activity of phospholipase A2 (PLA2) and enhancing the generation of arachidonic acid (AA) (74). Although AA can be converted to 20-hydroxyeicosatetraenoic acid (20-HETE) by cytochrome P450 monooxygenase (P450) or to PGE_2 by cyclooxygenase (75), the effect of stimulating the Ca^{2+}-sensing receptor is mediated by 20-HETE because inhibition of P450-dependent ω oxidation abolishes the effect of external Ca^{2+} (73). Interestingly, the inhibitory effect on K channel of stimulating the Ca^{2+}-sensing receptor is significantly enhanced in K-depleted rats and attenuated in animals on a high-K diet (76). The different response of the apical 70 pS K channel to external Ca^{2+} is the result of alteration of nitric oxide synthase (iNOS) expression: A high-K intake increases the expression of iNOS in the kidney, whereas a low-K intake has the opposite effect (76). Because nitric oxide (NO) interacts with P450 and inhibits P450-dependent ω oxidation (77), a decrease in NOS activity is expected to enhance P450 activity and increase the production of 20-HETE, which inhibits apical 70 pS K channel in the TAL (78). Indeed, this notion is confirmed by the findings that 20-HETE concentrations in the mTAL are doubled in rats on a low-K diet compared with rats on a high-K diet (76). Figure 2A is a model of a TAL cell showing the mechanism by which interaction between NO and 20-HETE regulates apical 70 pS K channels during K depletion.

In contrast, a high-K intake increases not only the expression of iNOS but also the expression of heme oxygenase 2 (HO-2) in the renal cortex and outer medulla (79). Figure 2B is a cell model demonstrating the role of iNOS and HO in the regulation of apical 70 pS K channels. HO is responsible for converting heme molecules to carbon monoxide (CO), bilirubin, and iron (80). It has been previously demonstrated that CO can stimulate the activity of the 70 pS K channel (81). Thus increase in CO production is expected to stimulate K recycling across the apical membrane and the Na/K/2Cl cotransporter. This view is supported by experiments in which inhibition of HO not only decreases the activity of the 70 pS K channel but also diminishes Na reabsorption along the loop of Henle (79). Moreover, inhibition of HO caused a larger decrease in Na absorption in rats on a high-K diet than that observed in animals on a normal K diet. Because the

A.   B.

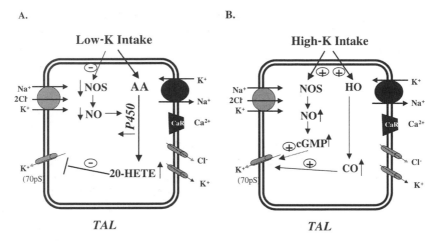

Figure 2   A model of TAL cell showing the regulatory mechanism of the apical 70 pS K channel in rats on a low-K (*A*) or a high-K diet (*B*). Abbreviations: nitric oxide synthase (NOS), heme oxygenase (HO), nitric oxide (NO), carbon monoxide (CO), 20-hydroxyeicosatetraenoic acid (20-HETE), arachidonic acid (AA) and CaR (calcium sensing receptor).

expression of HO-2 is suppressed in K-depleted animals (79), it is likely that a decrease in CO concentration may also be responsible for suppressing the activity of the apical K channels and transepithelial Na transport during K depletion.

EFFECT OF LOW-K INTAKE ON K HANDLING IN THE COLLECTING DUCT

A decrease in dietary K intake is promptly followed by reduced urinary K excretion (82–85). The decrease in K excretion is the result of suppression of K secretion in principal cells and stimulation of K absorption in intercalated cells (11). K conservation is impaired by exogenous application of mineralocorticoids. It has been shown that DOCA can still enhance K secretion in K-depleted animals (82). The effect of dietary K intake on K secretion is achieved by regulation of apical ROMK-like SK channels in principal cells and of the activity of K-H-ATPase in intercalated cells (1, 86).

REGULATION OF H-K-ATPase AND K-ATPase

Two types of H-K-ATPase are expressed in the kidney: colonic H-K-ATPase, sensitive to ouabain; and gastric H-K-ATPase, inhibited by Sch-28080 (58, 87–91). Amino acid sequence analysis shows that both isoforms share 60–70% sequence homology. Two lines of evidence indicate that gastric H-K-ATPase plays a key role in reabsorption of K in the collecting duct (62, 92). First, an ouabain-insensitive and Sch-28080-sensitive K uptake mechanism is present in the CCD and outer medullary collecting duct (OMCD). Second, the activity of gastric H-K-ATPase is markedly stimulated in OMCD from K-depleted animals (20, 62, 63). It has been reported that Rb influx, an index of K transport, increased in the OMCD from rabbits on a K-deficient diet and that this effect on Rb transport is abolished by inhibition of gastric H-K-ATPase (93, 94). Although H-K-ATPase is mainly expressed in intercalated cells, it has also been demonstrated in the apical membrane of principal cells during hypokalemia (59). An important feature of the H-K exchange process during K depletion is that the activity of H-K-ATPase is not sensitive to the inhibition of apical K channels. This is in sharp contrast to that observed in the OMCD from K-repleted animals in which inhibition of apical K channels suppresses H-K-ATPase (61). Figure 3 is a model of an intercalated cell illustrating the mechanism of K reabsorption: K enters the cell via H-K-ATPase across the apical membrane and exits across the basolateral membrane via K channels. Because apical K channel activity is suppressed in the OMCD during K restriction, the function of H-K-ATPase does not require K recycling across the apical membrane.

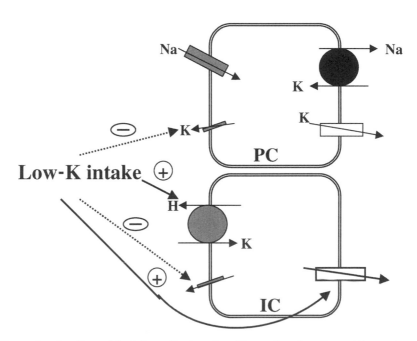

Figure 3   A cell model of the collecting duct illustrating the effect of low-K intake on apical K channels in a principal cell (PC) and the H-K exchange process in an intercalated cell (IC). The dotted line and solid line mean inhibition and stimulation, respectively.

In addition to H-K-ATPase, the kidney expresses three other types of K-ATPases: the Sch-28080-sensitive and ouabain-insensitive (Type I), the Sch-28080- and ouabain-sensitive (Type II), and the Sch-28080-insensitive and ouabain-sensitive (Type III) (91). Type II K-ATPase is expressed in the proximal tubule and the TAL, whereas Types I and III are expressed in the CCD (91). Interestingly, low-K intake significantly increases the expression of Type III K-ATPase and decreases the number of Type I K-ATPase (91). Therefore, it is conceivable that Type III K-ATPase may also be involved in K reabsorption in the CCD and OMCD during K restriction.

REGULATION OF Na-K-ATPase

K depletion has been reported to increase the activity of Na-K-ATPase and mRNA levels of $\alpha 1$ and $\beta 1$ subunits in the OMCD but not in the CCD (95). Stimulation of Na-K-ATPase facilitates NH_4 uptake because NH_4 accumulates in the interstitial space when the extracellular K concentrations fall. Thus NH_4 can substitute for K by occupying the K-binding site of Na-K-ATPase (96). This leads to an increase in proton supply and promotion of K-H exchange via H-K-ATPase in the OMCD (96).

REGULATION OF ROMK-LIKE SK CHANNELS

The ROMK-like SK channel plays a major role in K secretion across the apical membrane of principal cells (25). A low-K intake has been shown to decrease the apical K conductance (11, 72). Although patch-clamp experiments did not observe a significant decrease in channel activity in the apical membrane of the CCD in K-depleted rats compared with that of control animals (97), it is possible that channel activity in the CCD from normal animals is underestimated. We have demonstrated that the apical SK channel activity is closely coupled to the turnover rate of Na-K-ATPase (cross talk): Inhibition of Na-K-ATPase decreases the apical SK channel activity, whereas stimulation increases it (98). Because patch-clamp experiments are conducted in split-open tubules, such cross-talk mechanism may not be effective. Also, low plasma K (hypokalemia) is expected to cause inhibition of Na-K-ATPase, which leads to attenuation of the apical K channel activity. However, the patch-clamp experiments have been performed with the same bath solution in both groups of animals. Thus, it is possible that K channel activity in the CCD from K-depleted rats is overestimated.

A large body of evidence indicates that the apical SK channel is a ROMK (25, 99) because they have similar biophysical properties and regulatory mechanisms (25).The notion that the SK channel is a ROMK is further supported by experiments in which patch-clamp studies performed in tubules from ROMK-knockout mice did not observe SK channels in the TAL and the CCD (35). Although it is possible that other subunits, such as the sulfonylurea-agent receptor or the cystic fibrosis transmembrane conductance regulator (CFTR), are required to form the SK channels in the native tissue (99, 100–103), it is virtually certain that the ROMK is the pore-containing component of the native SK channels.

ROMK has three alternative splicing isoforms: ROMK1, 2, and 3 (104). The differences among the three ROMK isoforms are located in their N termini: the N terminus of ROMK2 is 19 amino acid shorter than that of ROMK1, whereas ROMK3 has an unique 21-amino-acid N terminus, which is different from that of ROMK1. The different N termini of ROMK isoforms have a functional consequence. It has been demonstrated that arachidonic acid inhibits ROMK1 but has no effect on ROMK2 or ROMK3 (105). Moreover, the distribution of ROMK isoforms differ: ROMK1 is exclusively expressed in the CCD, whereas ROMK3 is present only in the TAL. On the other hand, ROMK2 has been demonstrated in both TAL and CCD (104).

PTK MEDIATES THE INHIBITORY EFFECT OF LOW-K INTAKE ON ROMK CHANNELS IN THE CCD

ROMK channels have a putative PTK phosphorylation site in the C terminus that is at tyrosine residue 337 for ROMK1. When purified, ROMK1 is incubated with [^{32}P]ATP and active c-Src, radio-labeled ATP is incorporated into ROMK1 protein (106). Also, the fact that no significant radio-labeled ATP is incorporated into ROMK1 mutants (R1Y337A), in which tyrosine residue was mutated to alanine, suggests that tyrosine residue 337 is the PTK phosphorylation site. Moreover, the tyrosine phosphorylation of ROMK is regulated by dietary K intake: a low-K intake increases the level of the tyrosine-phosphorylated ROMK, whereas a high-K intake decreases it. This suggests that PTK is involved in mediating the inhibitory effect of low dietary K intake on the ROMK-like SK channels. This hypothesis is also supported by two important observations. First, both ROMK and c-Src, a Src family PTK, are expressed in the TAL and CCD (107). Second, the level of Src family PTKs, such as c-Src and c-Yes, increased in the renal cortex and outer medulla obtained from rats on a K-deficient diet and significantly decreased with high-K intake (108).

The role of PTK in mediating the effect of low-K intake on K secretion is further supported by experiments in which blocking PTK with herbimycin A significantly increased the activity of the ROMK-like SK channels in CCDs harvested from rats on a K-deficient diet (108). Moreover, the stimulatory effect of herbimycin A is enhanced by prolonging K restriction and is almost completely absent in high-K-adapted animals (108). However, the inhibitory effect of PTK on ROMK-like SK channels is not a direct consequence of tyrosine phosphorylation because addition of exogenous c-Src does not inhibit the channel activity in excised patches (109). Also, the observation that inhibition of microtubule formation by colchicine abolished the effect of herbimycin A on the SK channels in CCDs suggests that the effect of PTK inhibition results from the stimulation of exocytosis (110). The notion that stimulation of tyrosine dephosphorylation increases the insertion of ROMK channels into the apical membrane is also supported by two lines of evidence:

(*a*) Inhibition of PTK, a maneuver that enhanced tyrosine dephosphorylation, increases the surface density of ROMK1 detected with biochemical and confocal microscope in HEK cells transfected with ROMK1 and c-Src(111); (*b*) The effect of inhibition of PTK on ROMK1 trafficking is abolished by tetanus toxin, an agent that blocks exocytosis (112).

Several lines of evidence indicate that stimulation of tyrosine phosphorylation of ROMK1 facilitates internalization of ROMK-like SK channels. First, inhibition of PTP decreases ROMK-like SK channel activity. This effect is absent in the presence of a sucrose-containing bath or in CCDs treated with concanavalin A, a maneuver that inhibits endocytosis (113), which suggests that inhibition of PTP increases the internalization of ROMK-like SK channels. Second, inhibition of PTP significantly increases tyrosine phosphorylation of ROMK1 and reduces the number of ROMK1 detected by confocal microscopic image and surface biotin labeling in HEK293 cells transfected with ROMK1 and c-Src (114). Third, inhibition of PTP has no effect on K channel activity in cells transfected with the ROMK1 mutant, R1Y337A, indicating that phosphorylation of tyrosine residue 337 is essential for initiating the internalization of ROMK1 (114). Figure 4 is a cell model illustrating the mechanism by which a low-K intake decreases the number of apical ROMK1

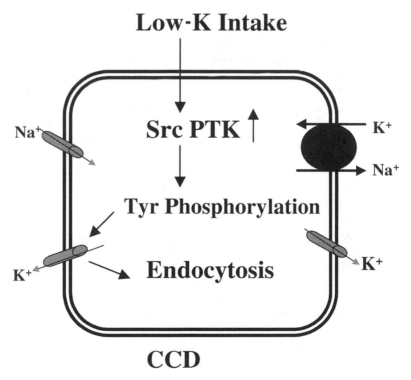

Figure 4   A cell model of the CCD showing the mechanism by which low-K intake decreases the apical K channel activity.

channels in the CCD: low-K intake stimulates the activity of PTK, which enhances the tyrosine phosphorylation of ROMK1 and the internalization of K channels.

Although the tyrosine residue is present in all three ROMK isoforms, available evidence indicates that tyrosine phosphorylation of ROMK2 and ROMK3 does not lead to endocytosis. Inhibition of PTK does not increase the activity of the 30 pS K channel in the TAL isolated from K-restricted animals (115). Although inhibition of PTK increases the activity of the 70 pS K channel in the TAL, the effect is also observed in excised patches. This suggests that the regulation of ROMK2 and ROMK3 by PTK is different from that of ROMK1. This hypothesis is also supported by experiments in which inhibition of PTK did not increase the surface ROMK2 channel number in HEK cells transfected with ROMK2 and c-Src (W.H. Wang, unpublished observation). This indicates that the PTK-dependent signaling pathway is specifically responsible for the regulation of the number of apical ROMK1 channels in the CCD.

EFFECT OF HIGH-K INTAKE ON K TRANSPORT IN THE COLLECTING DUCT

An increase in the K content of the diet leads to stimulation of K secretion and augmentation of the apical K conductance in the CCD (53, 97, 116). Figure 5 is a cell model illustrating the mechanism by which high-K intake regulates transport functions in principal and intercalated cells. The stimulatory effect of high-K intake on K secretion has two phases: an early (hours) and late phase (days). Augmentation in K secretion can be observed within several hours after a K-rich meal. This effect is mainly achieved by enhancing the function of existing Na-K-ATPase and apical K channels (117). The observation that a high-K intake for 24 h increases only Rb uptake but not ouabain-binding sites (117) indicates that the early stimulatory effect of high K is mainly mediated by stimulation of existing Na-K-ATPase. In contrast, in the late phase, the increase in K secretion is accompanied by enlargement of the basolateral membrane area and by an increase in the amount of Na-K-ATPase in the CCD (118, 119).

A high-K diet inhibits the ouabain-insensitive H-K-ATPase and K reabsorption in the OMCD (120). Although the H-K-ATPase activity is still present in the apical membrane, it may serve largely as a mechanism for proton secretion. In addition, the finding that the proton secretion is sensitive to inhibition of apical K channels suggests that K recycling is important for maintaining H-K-ATPase in the OMCD for an animal on a high-K diet (121). This hypothesis is supported further by experiments in which a K-permeable cation channel has been identified in the apical membrane of the rabbit OMCD (122). Figure 5 shows a cell model illustrating the mechanism by which K recycling across the apical membrane is coupled to H-K-ATPase in K-adapted animal. Thus K entering the cell via H-K-ATPase is secreted into the lumen with no net K absorption occurring under such conditions.

Several factors have been identified to contribute to the activation of K secretion by high-K intake (1, 123, 124). The early stimulatory effect of high-K intake on K

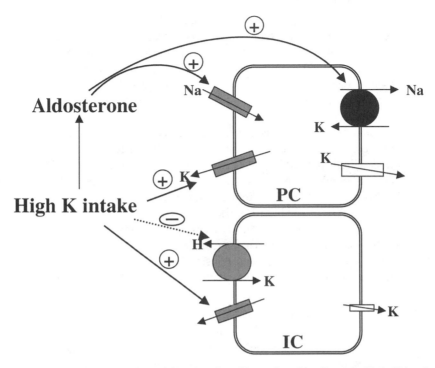

Figure 5   A cell model of the collecting duct illustrating the effect of a high-K intake on Na absorption and K secretion in the principal cell (PC) and the H-K exchange process in an intercalated cell (IC). The dotted line and solid line mean inhibition and stimulation, respectively.

secretion may result from an increase in plasma K, which stimulates Na-K-ATPase. Also, high-plasma K has been shown to inhibit the Na and water transport in the proximal tubule and increase the Na delivery to the distal tubule and tubule flow rate, which may activate Ca^{2+}-dependent large-conductance K channels and enhance K secretion (57). In addition, high K stimulates aldosterone secretion and increases Na reabsorption in the distal nephron. This leads to augmentation of the electrochemical gradient for K movement across the apical membrane.

REGULATION OF ROMK-LIKE SK CHANNEL

High-K intake significantly increases the number of apical ROMK-like SK channels in the CCD, but the mechanisms of stimulation are not completely understood. An increase in dietary K intake does not stimulate transcription of ROMK (125). Western blot analysis has also demonstrated that the total ROMK proteins are not affected by a high-K diet (72). These results support the view that the stimulatory effect of high-K intake on the apical ROMK-like SK channels is the result of their functional modulation. An increase in plasma K following high-K intake may stimulate Na-K-ATPase. Because apical K conductance is closely coupled to the basolateral Na-K-ATPase, stimulation of Na-K-ATPase is expected to increase the apical K channel activity as demonstrated in the proximal tubules (126–128) and the CCD (97, 98, 129, 108). Although high-K intake increases plasma aldosterone levels, aldosterone appears to have no direct effect on the ROMK-like SK channels because neither infusion of aldosterone nor low-Na diet (a maneuver that stimulates aldosterone secretion) increases the number of the SK channels in the CCD (108, 130). In contrast, infusion of aldosterone or low-Na diet increases the number of epithelial Na channels in the CCD of rat kidney. This suggests that an aldosterone-independent mechanism is responsible for mediating the stimulatory effect of high-K intake on the ROMK-like SK channels in the CCD. This notion is also supported by previous studies performed on the isolated CCD (129, 131). It has been reported that K secretion along the perfused initial collecting tubule rises sharply in response to high-K intake in adrenalectomized rats with clamped, circulating aldosterone levels (129, 131). This indicates that an increase in plasma K per se stimulates K secretion. The mechanism by which high-K intake stimulates the ROMK-like SK channels by an aldosterone-independent signal transduction pathway may involve several factors such as changes in PTP, PKC, and PKA activity.

ROLE OF PTK AND PTP

Although PTK plays a key role in suppressing activity of ROMK-like SK channels during K restriction, a decrease in PTK activity is only partially responsible for the stimulatory effect of high-K intake on K channel activity in the CCD. It has been demonstrated that activity of the apical ROMK-like SK channel doubles in the CCD from rats on a high-K diet for only 4 h. In contrast, a significant decrease in Src PTK expression cannot be detected in the kidney from rats on a high-K diet until 24 h (108). Because tyrosine phosphorylation and dephosphorylation of ROMK1 depend on the interaction between PTK and PTP, it is possible that high-K intake may stimulate the activity of PTP and potentiate the dephosphorylation of ROMK1. This possibility is supported by the observation that high-K intake significantly decreased tyrosine phosphorylated ROMK (106). Also, inhibition of PTP decreased the activity of ROMK-like SK channels in the CCD in rats on a high-K diet (113). However, inhibition of PTK did not increase the channel activity in control animals to the extent observed in rats fed with a high-K diet for 4 h, suggesting that factors other than stimulation of PTP or inhibition of PTK are also involved in mediating the effect of high K on the activity of ROMK-like SK channels in the CCD (108).

ROLE OF PKC

ROMK1 has three putative PKC phosphorylation sites (serine residues 4, 183, and 201), whereas ROMK2 and ROMK3 have only two putative PKC phosphorylation sites in the C terminus (132). We previously demonstrated that ROMK1 is a substrate of PKC, and serine residues 4 and 201 are two main PKC phosphorylation sites (133). Also, it has been shown that PKC is involved in the ROMK1 export from the endoplasmic reticulum (ER) to the cell membrane because deletion of serine residues 4 and 201 resulted in retention of ROMK1 in the perinuclear region and a decrease in K current owing to a low density of ROMK1 (133). Thus if high K selectively stimulates the activity of PKC, it could increase the insertion of ROMK-like SK channels from the ER to membranes in the CCD.

ROLE OF PKA/SGK

ROMK channels have three putative PKA phosphorylation sites (132). It has been shown that stimulation of PKA increased the activity of ROMK channels (134, 135). Moreover, deletion of one putative PKA phosphorylation site increased the channel closed time, and deletion of any two PKA sites abolished K current (136). In addition, PKA-induced phosphorylation may enhance the sensitivity of ROMK channels to PIP_2 (137, 138). The interaction between PKA and PIP_2 has also been observed in the mouse CCD, in which PKA potentiates the stimulatory effect of PIP_2 on the SK channel (139). Therefore, if high-K intake stimulates PKA activity, this may increase the activity of ROMK-like SK channels in the CCD. Recently, several studies have suggested that serum and glucocorticoid-dependent kinase (SGK) can increase the activity of ROMK1 channels in the oocytes expressing ROMK1 and SGK (140, 141). Because high K increases plasma aldosterone, which in turn stimulates SGK activity, SGK may be involved in mediating the effect of high-K diet on ROMK-like SK channels. Further experiments are needed to explore the role of PKA/SGK in mediating the effect of high K on ROMK1 channel activity.

FUTURE DIRECTION

Although some progress has been made in understanding the mechanisms by which dietary K intake regulates K secretion in the kidney, several important issues remain unresolved. It is still not clear how low-K intake can activate the PTK-dependent signal transduction pathway. Moreover, because Src family PTK has been shown to interact with a variety of signaling molecules (142), it would be interesting to determine the downstream signaling molecules of PTK that are involved in the regulation of ROMK1 trafficking. Also, more experiments are required to delineate the mechanisms by which high-K intake increases the apical ROMK-like K channels in the apical membrane of the CCD. However, it is most likely that an aldosterone-independent mechanism plays an important role in mediating the effect of dietary K intake on renal K secretion.

ACKNOWLEDGMENTS

The author thank Drs. Gerhard Giebisch and Charles Stier for their valuable suggestions in the preparation of the manuscript. The work is supported by National Institutes of Health grant DK47,402, DK54,983 and HL34,300.

The *Annual Review of Physiology* is online at http://physiol.annualreviews.org

LITERATURE CITED

1. Giebisch G. 1998. Renal potassium transport: mechanisms and regulation. *Am. J. Physiol. Renal Physiol.* 274:F817–33

2. McDonough AA, Thompson CB, Youn JH. 2002. Skeletal muscle regulates extracellular potassium. *Am. J. Physiol. Renal Physiol.* 282:F9667–74

3. Bia MJ, DeFronzo RA. 1981. Extrarenal potassium homeostasis. *Am. J. Physiol. Renal Physiol.* 240(9):F257–68

4. Young DB. 1988. Quantitative analysis of aldosterone's role in potassium regulation. *Am. J. Physiol. Renal Physiol.* 255:F811–22

5. Fisher KA, Binder HJ, Hayslett JP. 1976. Potassium secretion by colonic mucosal cells after potassium adaptation. *Am. J. Physiol.* 231:987–94

6. Bastl C, Hayslett JP, Binder HJ. 1977. Increased large intestinal secretion of potassium in renal insufficiency. *Kidney Int.* 12:9–16

7. Thompson CB, McDonough AA. 1996. Skeletal muscle Na,K-ATPase alpha and beta subunit protein levels respond to hypokalemic challenge with isoform and muscle type specificity. *J. Biol. Chem.* 271:32653–58

8. Azuma KK, Hensley CB, Putnam DS, McDonough AA. 1991. Hypokalemia decreases Na+-K+-ATPase alpha2- but not alpha1-isoform abundance in heart, muscle, and brain. *Am. J. Physiol. Cell Physiol.* 260:C958–64

9. Choi CSC, Thompson CB, Leong PKK, McDonough AA, Youn JH. 2001. Short-term K+ deprivation provokes insulin resistance of cellular K+ uptake revealed with the K+ clamp. *Am. J. Physiol. Renal Physiol.* 280:F95–102

10. Codina J, Pressley TA, DuBose TD Jr. 1997. Effect of chronic hypokalemia on H-K-ATPase in rat colon. *Am. J. Physiol. Renal Physiol.* 272:F22–30

11. Stanton BA, Giebisch GH. 1992. Renal potassium transport. In *Handbook of Physiology-Renal Physiology*, ed. E Windhager, pp. 813–74. New York: Oxford Univ. Press

12. Kaufman JS, Hamburger RJ. 1985. Passive potassium transport in the proximal convoluted tubule. *Am. J. Physiol. Renal Physiol.* 248:F228–32

13. Kibble JD, Wareing M, Wilson RW, Green R. 1995. Effect of barium on potassium diffusion across the proximal convoluted tubule of the anesthetized rat. *Am. J. Physiol. Renal Physiol.* 268:F778–83

14. Weinstein AM, Windhager EE. 2001. The paracellular shunt of proximal tubule. *J. Membr. Biol.* 184:241–45

15. Fromter E. 1974. Free-flow potential profile along rat kidney proximal tubule. *Pflügers Arch.* 351:69–83

16. Shirley DG, Walter SJ, Folkerd EJ, Unwin RJ, Bailey MA. 1998. Transepithelial electrochemical gradients in the proximal convoluted tubule during potassium depletion in the rat. *J. Physiol.* 513:551–57

17. Bomsztyk K, Wright FS. 1986. Dependence of ion fluxes on fluid transport by rat proximal tubule. *Am. J. Physiol. Renal Physiol.* 250:F680–89

18. Weinstein AM. 1986. A mathematical model of the rat proximal tubule. *Am. J. Physiol. Renal Physiol.* 250:F860–73

19. Wareing M, Wilson RW, Kibble JD, Green R. 1995. Estimated potassium reflection coefficient in perfused proximal convoluted tubules of the anaesthetized rat in vivo. *J. Physiol.* 488:153–61

20. Younes-Ibrahim M, Barlet-Bas C, Buffin-Meyer B, Cheval L, Rajerison R, Doucet A. 1995. Ouabain-sensitive and -insensitive K-ATPases in rat nephron: effect of K depletion. *Am. J. Physiol. Renal Physiol.* 268:F1141–47

21. Yao X, Tian S, Chan HY, Biemesderfer D,

Desir GV. 2002. Expression of KCNA10, a voltage-gated K channel, in glomerular endothelium and at the apical membrane of the renal proximal tubule. *J. Am. Soc. Nephrol.* 13:2839

22. Zweifach A, Desir GV, Aronson PS, Giebisch G. 1992. Inhibition of Ca-activated K channels from renal microvillus membrane vesicles by amiloride analogs. *J. Membr. Biol.* 128:115–22

23. Biagi B, Kubota T, Sohtell M, Giebisch G. 1981. Intracellular potentials in rabbit proximal tubules perfused in vitro. *Am. J. Physiol. Renal Physiol.* 240:F200–10

24. Biagi B, Sohtell M, Giebisch G. 1981. Intracellular potassium activity in the rabbit proximal straight tubule. *Am. J. Physiol. Renal Physiol.* 241:F677–86

25. Wang WH, Hebert SC, Giebisch G. 1997. Renal K channels: structure and function. *Annu. Rev. Physiol.* 59:413–36

26. Unwin R, Capasso G, Giebisch G. 1994. Potassium and sodium transport along the loop of Henle: effects of altered dietary potassium intake. *Kidney Int.* 46:1092–99

27. Greger R, Schlatter E, Lang F. 1983. Evidence for electroneutral sodium chloride cotransport in the cortical thick ascending limb of Henle's loop of rabbit kidney. *Pflügers Arch.* 396:308–14

28. Greger R, Schlatter E. 1983. Properties of the basolateral membrane of the cortical thick ascending limb of Henle's loop of rabbit kidney. *Pflügers Arch.* 396:325–34

29. Hebert SC, Andreoli TE. 1984. Control of NaCl transport in the thick ascending limb. *Am. J. Physiol. Renal Physiol.* 246:F745–56

30. Gamba G, Miyanoshita A, Lombardi M, Lytton J, Lee WS, et al. 1994. Molecular cloning, primary structure, and characterization of two members of the mammalian electroneutral sodium-(potassium)-chloride cotransporter family expressed in kidney. *J. Biol. Chem.* 269:17713–22

31. Leviel F, Borensztein P, Houillier P, Pail-lard M, Bichara M. 1992. Electroneutral K^+/HCO_3^- cotransport in cells of medullary thick ascending limb of rat kidney. *J. Clin. Invest.* 90:869–78

32. Borensztein P, Leviel F, Froissart M, Houillier P, Poggioli J, et al. 1991. Mechanisms of H^+/HCO_3^- transport in the medullary thick ascending limb of rat kidney. *Kidney Int. Suppl.* 33:S43–46

33. Greger R. 1985. Ion transport mechanisms in thick ascending limb of Henle's loop of mammalian nephron. *Physiol. Rev.* 65:760–97

34. Huang DY, Osswald H, Vallon V. 2000. Sodium reabsorption in thick ascending limb of Henle's loop: effect of potassium channel blockade in vivo. *Br. J. Pharmacol.* 130:1255

35. Lu M, Wang T, Yan Q, Yang X, Dong K, et al. 2002. Absence of small conductance K channel (SK) activity in apical membrane of thick ascending limb and cortical collecting duct in ROMK (Bartter's) knockout mice. *J. Biol. Chem.* 277:37881–87

36. Simon DB, Karet FE, Rodriguez J, Hamdan JH, DiPietro A, et al. 1996. Genetic heterogeneity of Bartter's syndrome revealed by mutations in the K channel, ROMK. *Nat. Genet.* 14:152–56

37. Bleich M, Schlatter E, Greger R. 1990. The luminal K^+ channel of the thick ascending limb of Henle's loop. *Pflügers Arch.* 415:449–60

38. Taniguchi J, Guggino WB. 1989. Membrane stretch: a physiological stimulator of Ca^{2+}-activated K^+ channels in thick ascending limb. *Am. J. Physiol. Renal Physiol.* 257(26):F347–52

39. Wang WH, White S, Geibel J, Giebisch G. 1990. A potassium channel in the apical membrane of the rabbit thick ascending limb of Henle's loop. *Am. J. Physiol. Renal Physiol.* 258:F244–53

40. Wang WH. 1994. Two types of K^+ channel in thick ascending limb of rat kidney. *Am. J. Physiol. Renal Physiol.* 267:F599–605

41. Schnermann J, Steipe B, Briggs JP. 1987. In situ studies of distal convoluted tubule in rat. II. K secretion. *Am. J. Physiol.* 252:970–76

42. Ellison DH, Velazquez H, Wright FS. 2003. Mechanisms of sodium, potassium and chloride transport by the renal distal tubule. *Miner. Elect. Metab.* 13:422–32

43. Ellison DH, Velazquez H, Wright FS. 1987. Thiazide-sensitive sodium chloride cotransport in early distal tube. *Am. J. Physiol. Renal Physiol.* 253:F546–54

44. Velazquez H, Wright FS. 1986. Effects of diuretic drugs on Na, Cl, and K transport by rat renal distal tubule. *Am. J. Physiol. Renal Physiol.* 250:F1013–23

45. O'Neil RG, Hayhurst AR. 1985. Functional differentiation of cell types of cortical collecting duct. *Am. J. Physiol.* 248:449–53

46. O'Neil RG. 1981. Potassium secretion by the cortical collecting tubule. *Fed. Proc.* 40:2403–7

47. Schafer JA, Troutman SL, Schlatter E. 1990. Vasopressin and mineralocorticoid increase apical membrane driving force for K^+ secretion in rat CCD. *Am. J. Physiol. Renal Physiol.* 258:F199–10

48. Sansom SC, O'Neil RG. 1986. Effects of mineralocorticoids on transport properties of the cortical collecting duct basolateral membrane. *Am. J. Physiol.* 251:743–57

49. Frindt G, Palmer LG. 1987. Ca-activated K channels in apical membrane of mammalian CCT, and their role in K secretion. *Am. J. Physiol. Renal Physiol.* 252:F458–67

50. Frindt G, Palmer LG. 1989. Low-conductance K channels in apical membrane of rat cortical collecting tubule. *Am. J. Physiol. Renal Physiol.* 256:F143–51

51. Hunter M, Lopes AG, Boulpaep EL, Giebisch G. 1984. Single channel recordings of calcium-activated potassium channels in the apical membrane of rabbit cortical collecting tubules. *Proc. Natl. Acad. Sci. USA* 81:4237–39

52. Satlin LM, Palmer LG. 1997. Apical K^+ conductance in maturing rabbit principal cell. *Am. J. Physiol. Renal Physiol.* 272:F397–404

53. Wang W, Schwab A, Giebisch G. 1990. Regulation of small-conductance K channel in apical membrane of rat cortical collecting tubule. *Am. J. Physiol. Renal Physiol.* 259:F494–502

54. Satlin LM. 1994. Postnatal maturation of potassium transport in rabbit cortical collecting duct. *Am. J. Physiol. Renal Physiol.* 266:F57–65

55. Ling BN, Hinton CF, Eaton DC. 1991. Potassium permeable channels in primary cultures of rabbit cortical collecting tubule. *Kidney Int.* 40:441–52

56. Schlatter E, Lohrmann E, Greger R. 1992. Properties of the potassium conductances of principal cells of rat cortical collecting ducts. *Pflügers Arch.* 420:39–45

57. Woda CB, Bragin A, Kleyman T, Satlin LM. 2001. Flow-dependent K secretion in the cortical collecting duct(CCD) is mediated by a TEA-sensitive channel. *Am. J. Physiol. Renal Physiol.* 280:F786–793

58. DuBose TD Jr, Gitomer J, Codina J. 1999. H^+,K^+-ATPase. *Curr. Opin. Nephrol. Hypertens.* 8:597–602

59. Guntupalli J, Onuigbo M, Wall S, Alpern RJ, DuBose TD Jr. 1997. Adaptation to low-K^+media increases H^+-K^+-ATPase but not H^+-ATPase-mediated pHi recovery in OMCD1 cells. *Am. J. Physiol. Cell Physiol.* 273:C558–71

60. Doucet A, Marsy S. 1987. Characterization of K-ATPase activity in distal nephron: stimulation by potassium depletion. *Am. J. Physiol. Renal Physiol.* 253:F418–23

61. Zhou X, Lynch IJ, Xia SL, Wingo CS. 2000. Activation of H-K-ATPase by CO_2 requires a basolateral Ba^{2+}-sensitive pathway during K restriction. *Am. J. Physiol. Renal Physiol.* 279:F153–60

62. Garg LC. 1991. Respective role of H-ATPase and H-K-ATPase in ion transport in the kidney. *J. Am. Soc. Nephrol.* 2(5):949–60

63. Buffin-Meyer B, Younes-Ibrahim M, Barlet-Bas C, Cheval L, Marsy S, Doucet A. 1997. K depletion modifies the properties of Sch-28080-sensitive K-ATPase in rat collecting duct. *Am. J. Physiol. Renal Physiol.* 272:F124–31

64. Walter SJ, Shore AC, Shirley DG. 1988. Effect of potassium depletion on renal tubular function in the rat. *Clin. Sci.* 75: 621–28

65. Burg MB, Gree N. 1976. Role of monovalent ions in the reabsorption of fluid by isolated perfused proximal renal tubules of the rabbit. *Kidney Int.* 10:221–28

66. Cardinal J, Duchesneau D. 1978. Effect of potassium on proximal tubular function. *Am. J. Physiol. Renal Physiol.* 234:F381–85

67. Rutecki GW, Cox JW, Robertson GW, Francisco LL, Ferris TF. 1982. Urinary concentrating ability and antidiuretic hormone responsiveness in the potassium-depleted dog. *J. Lab. Clin. Med.* 100:53–60

68. Senba S, Konishi K, Saruta T, Ozawa Y, Kato E, et al. 1984. Hypokalemia and prostaglandin overproduction in Bartter's syndrome. *Nephron* 37:257–63

69. Buffin-Meyer B, Marsy S, Barlet-Bas C, Cheval L, Younes-Ibrahim M, et al. 1996. Regulation of renal Na^+,K^+-ATPase in rat thick ascending limb during K^+ depletion: evidence for modulation of Na^+ affinity. *J. Physiol.* 490:623–32

70. Luke RG, Booker BB, Galla JH. 1985. Effect of potassium depletion on chloride transport in the loop of Henle in the rat. *Am. J. Physiol. Renal Physiol.* 248:F682–87

71. Kunau RTJ. 1970. Function of ascending loop of Henle in the potassium-deficient dog. *Am. J. Physiol.* 219:1071–79

72. Mennitt PA, Frindt G, Silver RB, Palmer LG. 1999. Potassium restriction downregulates ROMK expression in the rat kidney. *Am. J. Physiol. Renal Physiol.* 278:F916–24

73. Wang WH, Lu M, Hebert SC. 1996.

74. Wang WH, Lu M, Balazy M, Hebert SC. 1997. Phospholipase A2 is involved in mediating the effect of extracellular Ca on apical K channels in rat TAL. *Am. J. Physiol. Renal Physiol.* 273:F421–29

Cytochrome P-450 metabolites mediate extracellular Ca^{2+}-induced inhibition of apical K channels in the TAL. *Am. J. Physiol. Cell Physiol.* 270:C103–11

75. McGiff JC. 1991. Cytochrome P-450 metabolism of arachidonic acid. *Annu. Rev. Pharmacol. Toxicol.* 31:339–69

76. Gu RM, Wei Y, Jiang HL, Lin DH, Sterling H, et al. 2002. K depletion enhances the extracellular Ca-induced inhibition of the apical K channels in the mTAL of rat kidney. *J. Gen. Physiol.* 119:33–44

77. Oyekan AO, Youseff T, Fulton D, Quilley J, McGiff JC. 1999. Renal cytochrome P450 omega-hydroxylase and epoxygenase activity are differentially modified by nitric oxide and sodium chloride. *J. Clin. Invest.* 104:1131–37

78. Gu RM, Wei Y, Jiang H, Balazy M, Wang WH. 2001. The role of 20-HETE in mediating the effect of dietary K intake on the apical K channels in the mTAL. *Am. J. Physiol. Renal Physiol.* 280:F223–30

79. Wang T, Sterling H, Shao W, Yan QS, Bailey MA, et al. 2003. Inhibition of heme oxygenase decreases sodium and fluid absorption in the loop of Henle. *Am. J. Physiol. Renal Physiol.* 285:484–90

80. Maines MD. 1997. The heme oxygenase system: a regulator of second messenger gases. *Annu. Rev. Pharmacol. Toxicol.* 37:517–54

81. Liu HJ, Mount DB, Nasjletti A, Wang WH. 1999. Carbon monoxide stimulates the apical 70 pS K channel of the rat thick ascending limb. *J. Clin. Invest.* 103:963–70

82. Linas SL, Peterson LN, Anderson RJ, Aisenbrey GA, Simon FR, Berl T. 1979. Mechanism of renal potassium conservation in the rat. *Kidney Int.* 15:601–11

83. Malnic G, Klose RM, Giebisch G. 1964.

Micropuncture study of renal potassium excretion in the rat. *Am. J. Physiol.* 206(4):674–86

84. Malnic G, Klose RM, Giebisch G. 1966. Micropuncture study of distal tubular potassium and sodium transport in rat nephron. *Am. J. Physiol.* 211(3):529–47

85. Rabinowich L, Sarason RL, Yamauchi H, Yamanaka KK, Tzendzalian PA. 1984. Time course of adaptation to altered K intake in rats and sheep. *Am. J. Phyiol. Renal Physiol.* 247:F607–17

86. Wingo CS. 1987. Potassium transport by medullary collecting tubule of rabbit: effects of variation in K intake. *Am. J. Physiol. Renal Physiol.* 253:F1136–41

87. Kone BC. 1996. Renal H,K-ATPase: structure, function and regulation. *Miner. Elect. Metab.* 22:349–65

88. Wingo CS, Cain BD. 1993. The renal H-K-ATPase: physiological significance and role in potassium homeostasis. *Annu. Rev. Physiol.* 55:323–47

89. Armitage FE, Wingo CS. 1994. Luminal acidification in K-replete OMCDi: contributions of H-K-ATPase and bafilomycin-A1-sensitive H-ATPase. *Am. J. Physiol. Renal Physiol.* 267:F450–58

90. Codina J, Wall SM, DuBose TD Jr. 1999. Contrasting functional and regulatory profiles of the renal H+,K+-ATPases. *Semin. Nephrol.* 19:399–404

91. Silver RB, Soleimani M. 1999. H-K-ATPase regulation and role in pathophysiological states. *Am. J. Physiol. Renal Physiol.* 276:F799–811

92. Cheval L, Barlet-Bas C, Khadouri C, Féraille E, Marsy S, Doucet A. 1991. K+-ATPase-mediated Rb+ transport in rat collecting tubule: modulation during K+ deprivation. *Am. J. Physiol. Renal Physiol.* 260:F800–805

93. Zhou X, Wingo CS. 1992. Mechanisms of rubidium permeability by rabbit cortical collecting duct during potassium restriction. *Am. J. Physiol. Renal Physiol.* 263:F1134–41

94. Zhou X, Wingo CS. 1992. H-K-ATPase enhancement of Rb efflux by cortical collecting duct. *Am. J. Physiol. Renal Physiol.* 263:F43–48

95. McDonough AA, Magyar CE, Komatsu Y. 1994. Expression of Na+-K+-ATPase alpha- and beta-subunits along rat nephron: isoform specificity and response to hypokalemia. *Am. J. Physiol. Cell. Physiol.* 267:C901–8

96. Frank AE, Wingo CS, Weiner ID. 2000. Effects of ammonia on bicarbonate transport in the cortical collecting duct. *Am. J. Physiol. Renal Physiol.* 278:F219–26

97. Palmer LG, Antonian L, Frindt G. 1994. Regulation of apical K and Na channels and Na/K pumps in rat cortical collecting tubule by dietary K. *J. Gen. Physiol.* 105:693–710

98. Wang WH, Geibel J, Giebisch G. 1993. Mechanism of apical K+ channel modulation in principal renal tubule cells. *J. Gen. Physiol.* 101:673–94

99. Palmer LG, Choe H, Frindt G. 1997. Is the secretory K channel in the rat CCT ROMK? *Am. J. Physiol. Renal Physiol.* 273:F404–10

100. McNicholas CM, Nason MW, Guggino WB, Schwiebert EM, Hebert S, et al. 1997. The functional CFTR-NBF1 is required for ROMK2-CFTR interaction. *Am. J. Physiol. Renal Physiol.* 273:F843–48

101. McNicholas CM, Guggino WB, Schwiebert EM, Hebert SC, Giebisch G, Egan ME. 1996. Sensitivity of a renal K+ channel (ROMK2) to the inhibitory sulfonylurea compound glibenclamide is enhanced by coexpression with the ATP-binding cassette transporter cystic fibrosis transmembrane regulator. *Proc. Natl. Acad. Sci. USA* 93:8083–88

102. Dong K, Xu J, Vanoye CG, Welch R, MacGregor GG, et al. 2001. An amino acid triplet in the NH$_2$ terminus of rat ROMK1 determines interation with SUR2B. *J. Biol. Chem.* 276:44347–53

103. Ruknudin A, Schulze DH, Sullivan SK,

Lederer WJ, Welling P. 1998. Novel subunit composition of a renal epithelial K_{ATP} channel. *J. Biol. Chem.* 273:14165–71

104. Boim MA, Ho K, Schuck ME, Bienkowski MJ, Block JH, et al. 1995. The ROMK inwardly rectifying ATP-sensitive K channel. II. Cloning and intrarenal distribution of alternatively spliced forms. *Am. J. Physiol. Renal Physiol.* 268:F1132–40

105. Macica CM, Yang YH, Hebert SC, Wang WH. 1996. Arachidonic acid inhibits activity of cloned renal K channel, ROMK1. *Am. J. Physiol. Renal Physiol.* 271:F588–94

106. Lin DH, Sterling H, Lerea KM, Welling P, Jin L, et al. 2002. K depletion increases the protein tyrosine-mediated phosphorylation of ROMK. *Am. J. Physiol. Renal Physiol.* 283:F671–77

107. Lin DH, Sterling H, Wang WH. 2003. Protein tyrosine kinase and tyrosine phosphatase regulate the membrane location of ROMK channles in the cortical collecting duct. *FASEB J.* (17):A1214 (Abstr.)

108. Wei Y, Bloom P, Lin DH, Gu RM, Wang WH. 2001. Effect of dietary K intake on the apical small-conductance K channel in the CCD: role of protein tyrosine kinase. *Am. J. Physiol. Renal Physiol.* 281:F206–12

109. Wang WH, Lerea KM, Chan M, Giebisch G. 2000. Protein tyrosine kinase regulates the number of renal secretory K channel. *Am. J. Physiol. Renal Physiol.* 278:F165–71

110. Wei Y, Wang WH. 2001. The role of cytoskeleton in mediating the effect of vasopressin and herbimycin A on the secretory K channels in the CCD. *Am. J. Physiol. Renal Physiol.* 282:F680–86

111. Moral Z, Deng K, Wei Y, Sterling H, Deng H, et al. 2001. Regulation of ROMK1 channels by protein tyrosine kinase and tyrosine phosphatase. *J. Biol. Chem.* 276:7156–63

112. Sterling H, Lin DH, Wei Y, Wang WH.
2003. Tetanus toxin abolishes exocytosis of ROMK1 induced by inhibition of protein tyrosine kinase. *Am. J. Physiol. Renal Physiol.* 284:F510–17

113. Wei Y, Bloom P, Gu RM, Wang WH. 2000. Protein-tyrosine phosphatase reduces the number of apical small conductance K channels in the rat cortical collecting duct. *J. Biol. Chem.* 275:20502–7

114. Sterling H, Lin DH, Gu RM, Dong K, Hebert SC, Wang WH. 2002. Inhibition of protein-tyrosine phosphatase stimulates the dynamin-dependent endocytosis of ROMK1. *J. Biol. Chem.* 277:4317–23

115. Gu RM, Wei Y, Falck JR, Krishna UM, Wang WH. 2001. Effects of protein tyrosine kinase and protein tyrosine phosphatase on the apical K channels in the thick ascending limb. *Am. J. Physiol. Cell Physiol.* 281:C1185–95

116. Palmer LG. 1999. Potassium secretion and the regulation of distal nephron K channels. *Am. J. Physiol. Renal Physiol.* 277:F821–25

117. Fujii Y, Mujais SK, Katz AI. 1989. Renal ptassium adaptation: role of the Na-K pump in rat cortical collecting tubules. *Am. J. Physiol. Renal Physiol.* 256:F279–84

118. Stanton BA. 1989. Renal potassium transport: morphological and functional adaptation. *Am. J. Physiol. Regul. Integr. Physiol.* 257:R989–97

119. Hayhurst RA, O'Neil RG. 1988. Time-dependent actions of aldosterone and amiloride on Na^+-K^+-ATPase of cortical collecting duct. *Am. J. Physiol. Renal Physiol.* 254:F689–96

120. Garg LC, Narang N. 1989. Suppression of ouabin-insensitive K-ATPase activity in rabbit nephron segments during chronic hyperkalemia. *Renal Physiol. Biochem* 12:295–301

121. Zhou X, Wingo CS. 1994. Stimulation of total CO_2 flux by 10% CO_2 in rabbit CCD: role of an apical Sch-28080- and Ba-sensitive mechanism. *Am.*

J. Physiol. Renal Physiol. 267(36):F114–20

122. Xia SL, Gelband CH, Wingo CS. 1999. Cation channels of the rabbit outer medullary collecting duct. *Semin. Nephrol.* 19:472–76

123. Field MJ, Stanton BA, Giebisch GH. 1984. Differential acute effects of aldosterone, dexamethasone, and hyperkalemia on distal tubular potassium secretion in the rat kidney. *J. Clin. Invest.* 74:1792–802

124. Field MJ, Giebisch GH. 1984. Hormonal control of renal potassium excretion. *Kidney Int.* 27:379–87

125. Frindt G, Zhou H, Sackin H, Palmer LG. 1998. Dissociation of K channel density and ROMK mRNA in rat cortical collecting tubule during K adaptation. *Am. J. Physiol. Renal Physiol.* 274:F525–31

126. Beck JS, Laprade R, Lapointe JY. 1994. Coupling between transepithelial Na transport and basolateral K conductance in renal proximal tubule. *Am. J. Physiol. Renal Physiol.* 266(35):F517–27

127. Beck JS, Mairbaurl H, Laprade R, Giebisch G. 1991. Relationship between sodium transport and intracellular ATP in isolated perfused rabbit proximal convoluted tubules. *Am. J. Physiol. Renal Physiol.* 261:F634–39

128. Tsuchiya K, Wang W, Giebisch G, Welling PA. 1992. ATP is a coupling modulator of parallel Na,K-ATPase-K-channel activity in the renal proximal tubule. *Proc. Natl. Acad. Sci. USA* 89:6418–22

129. Muto S, Sansom S, Giebisch G. 1988. Effects of a high potassium diet on electrical properties of cortical collecting duct from adrenalectomized rabbits. *J. Clin. Invest.* 81:376–80

130. Palmer LG, Frindt G. 2000. Aldosterone and potassium secretion by the cortical collecting duct. *Kidney Int.* 57:1324–28

131. Muto S, Giebisch G, Sansom S. 1988. An acute increase of peritubular K stimulates K transport through cell pathways of CCT.

Am. J. Physiol. Renal Physiol. 255:F108–14

132. Ho K, Nichols CG, Lederer WJ, Lytton J, Vassilev PM, et al. 1993. Cloning and expression of an inwardly rectifying ATP-regulated potassium channel. *Nature* 362:31–38

133. Lin DH, Sterling H, Lerea KM, Giebisch G, Wang WH. 2002. Protein kinase C (PKC)-induced phosphorylation of ROMK1 is essential for the surface expression of ROMK1 channels. *J. Biol. Chem* 277:44332–38

134. MacGregor GG, Xu JZ, McNicholas CM, Giebisch G, Hebert SC. 1998. Partially active channels produced by PKA site mutation of the cloned renal K channel. *Am. J. Physiol. Renal Physiol.* 275:F415–22

135. McNicholas CM, Wang W, Ho K, Hebert SC, Giebisch G. 1994. Regulation of ROMK1 K^+ channel activity involves phosphorylation processes. *Proc. Natl. Acad. Sci. USA* 91:8077–81

136. Xu ZC, Yang Y, Hebert SC. 1996. Phosphorylation of the ATP-sensitive, inwardly rectifying K channel, ROMK, by cyclic AMP-dependent protein kinase. *J. Biol. Chem.* 271:9313–19

137. Huang CL, Feng S, Hilgemann DW. 1998. Direct activation of inward rectifier potassium channels by PIP_2 and its stabilization by $G\beta\gamma$. *Nature* 391:803–6

138. Liou HH, Zhou SS, Huang CL. 1999. Regulation of ROMK1 channel by protein kinase A via a phosphatidylinositol 4,5-bisphosphate-dependent mechanism. *Proc. Natl. Acad. Sci. USA* 96:5820–25

139. Lu M, Hebert SC, Giebisch G. 2002. Hydrolyzable ATP and PIP_2 modulate the small-conductance K channel in apical membrane of rat cortical collecting duct. *J. Gen. Physiol.* 120:603–15

140. Yoo D, Kim Y, Campo C, Nance L, King A, et al. 2003. SGK/PKA phosphorylate of Ktr1.1 regulates surface expression. *FASEB J.* 17:A470

141. Yun CC, Palmada M, Embark HM, Fedorenko O, Feng Y, et al. 2002. The serum and glucorticoid-inducible kinase SGK1 and Na/H exchange regulating factor NHERF2 synergize to stimulate the renal outer medullary K channel ROMK1. *J. Am. Soc. Nephrol.* 13:2823–30

142. Pawson T. 1995. Protein modules and signalling networks. *Nature* 373:573–79

Annu. Rev. Physiol. 2004. 66:571–99
doi: 10.1146/annurev.physiol.66.032102.111604

THE EXTRACELLULAR CYCLIC AMP-ADENOSINE PATHWAY IN RENAL PHYSIOLOGY

Edwin K. Jackson[1] and Dubey K. Raghvendra[1,2]

[1]Center for Clinical Pharmacology, Departments of Pharmacology and Medicine, University of Pittsburgh School of Medicine, Pittsburgh, Pennsylvania 15261; email: edj+@pitt.edu
[2]Department of Obstetrics and Gynecology, Clinic for Endocrinology, University Hospital Zurich, Zurich, Switzerland

Key Words ecto-phosphodiesterase, ecto-5′-nucleotidase, adenosine receptors, renal function, pancreatohepatorenal

■ **Abstract** Many cell types in the kidney express adenosine receptors, and adenosine has multiple effects on renal function. Although adenosine is produced within the kidney by several biochemical reactions, recent studies support a novel mechanism for renal adenosine production, the extracellular cAMP-adenosine pathway. This extracellular cAMP-adenosine pathway is initiated by efflux of cAMP from cells following activation of adenylyl cyclase. Extracellular cAMP is then converted to adenosine by the serial actions of ecto-phosphodiesterase and ecto-5′-nucleotidase. When extracellular cAMP is converted to adenosine near the biophase of cAMP production and efflux, this local extracellular cAMP-adenosine pathway permits tight coupling of the site of adenosine production to the site of adenosine receptors. cAMP in renal compartments may also be formed by tissues/organs remote from the kidney. For example, stimulation of hepatic adenylyl cyclase by the pancreatic hormone glucagon increases circulating cAMP, which is filtered at the glomerulus and concentrated in the tubular lumen as water is extracted from the ultrafiltrate. Conversion of hepatic-derived cAMP to adenosine in the kidney completes a pancreatohepatorenal cAMP-adenosine pathway that may serve as an endocrine link between the pancreas, liver, and kidney.

ADENOSINE RECEPTORS AND SIGNAL TRANSDUCTION

There are several excellent and recent reviews of adenosine receptors (1–4), and the reader is referred to these reviews for references to the extensive primary literature regarding the pharmacology of adenosine receptors. Adenosine receptors, also called P1 receptors, are classic heptahelical G protein–coupled receptors with four known subtypes, namely A_1, A_{2A}, A_{2B}, and A_3. A_1 receptors have a high affinity for adenosine (K_d in the low nanomolar range), whereas the affinity of A_{2A} receptors for adenosine is approximately threefold less compared with that of A_1 receptors. A_{2B}

and A_3 receptors are low-affinity adenosine receptors requiring high nanomolar concentrations for activation.

Human A_1 receptors consist of 326 amino acids, have a nominal molecular mass of 36.5 kDa, and signal via activation of G_i and G_o. The primary mode of signaling for A_1 receptors is inhibition of adenylyl cyclase; however, A_1 receptors also may stimulate phospholipase C, increase influx of potassium, and inhibit influx of calcium. Human A_{2A} receptors consist of 412 amino acids, have a nominal molecular mass of 44.7 kDa, and signal by engaging G_s, G_{olf}, and p21^{ras}. Increases in adenylyl cyclase activity mediate many of the effects of A_{2A} receptors. Human A_{2B} receptors have 332 amino acids and a nominal molecular mass of 36.3 kDa. Similar to A_{2A} receptors, A_{2B} receptors are largely coupled to stimulation of adenylyl cyclase; however, A_{2B} receptors are also known to stimulate phospholipase C via G_q. Human A_3 receptors consist of 318 amino acids, have a nominal molecular mass of 36.2 kDa, and similar to A_1 receptors, inhibit adenylyl cyclase via G_i. A_3 receptors may also stimulate phospholipase via G_q.

REGULATION OF RENAL FUNCTION BY ADENOSINE

As summarized in Figure 1, A_1 receptors importantly regulate renal function. Infusion of A_1 receptor agonists into the renal interstitium diminishes blood flow to both superficial and deep nephrons (5). In the outer cortex, A_1 receptor-induced vasoconstriction is most likely caused by contraction of preglomerular, rather than postglomerular, microvessels (6–9); however, in juxtaglomerular nephrons, A_1 receptors mediate vasoconstriction by contracting preglomerular microvessels (10), efferent arterioles (10), and outer medullary descending vasa recta (11). Angiotensin II strongly enhances A_1 receptor-induced preglomerular vasoconstriction (12–14). Contraction of preglomerular microvascular smooth muscle cells by A_1 receptors underlies the mediator role of adenosine in tubuloglomerular feedback (for review see 15), explains the ability of adenosine to decrease glomerular filtration rate (for review see 16), and potentiates postjunctional vasoconstrictor responses to renal sympathetic neurotransmission in the kidney (17, 18).

Inasmuch as juxtaglomerular cells are modified preglomerular vascular smooth muscle cells, it is not surprising that A_1 receptors also influence renin release. In this regard, A_1 receptors may restrain renin release responses, a theory known as the adenosine-brake hypothesis (19). Importantly, A_1 receptors are coupled to G_i and therefore inhibit adenylyl cyclase (20, 21). Because stimulation of renin release from juxtaglomerular cells by many stimuli involves activation of adenylyl cyclase (22), activation of juxtaglomerular A_1 receptors attenuates renin release, and antagonism of renal A_1 receptors increases renin release (15).

In proximal tubular epithelial cells, A_1 receptors mediate enhancement of transport. For example, stimulation of A_1 receptors in cultured epithelial cells that express a proximal tubular phenotype enhances both Na^+-glucose symport and Na^+-phosphate symport (23). Also, stimulation of A_1 receptors in microperfused proximal convoluted tubules increases basolateral Na^+-$3HCO_3^-$ symport (24), and

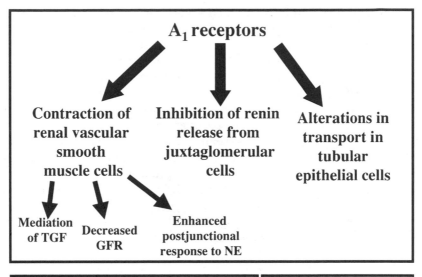

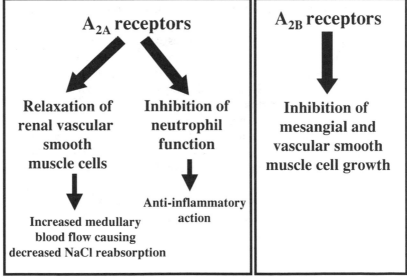

Figure 1   Regulation of renal function by adenosine. TGF, tubuloglomerular feedback; GFR, glomerular filtration rate; NE, norepinephrine.

selective antagonism of A_1 receptors decreases Na^+-dependent phosphate transport in renal proximal tubular cells via a cAMP-mediated mechanism (25, 26). Although it is clear that A_1 receptors affect epithelial transport in proximal tubules, in fact there are many more A_1 receptors per milligram of protein in the collecting duct than in the proximal tubule (27, 28). This relatively rich expression of A_1 receptors in the collecting duct strongly suggests that A_1 receptors importantly affect epithelial transport in this nephron segment. In support of this hypothesis, A_1 receptor antagonists do not increase K^+ excretion despite the fact that A_1 receptor antagonists markedly increase Na^+ excretion. Because all classes of diuretics that act proximal to the collecting duct increase K^+ excretion (29), the finding that A_1 receptor antagonists do not increase K^+ excretion, combined with the high level of expression of A_1 receptors in the collecting duct, suggests that A_1 receptors alter Na^+ and K^+ transport in epithelial cells in the collecting duct. In vivo, selective blockade of A_1 receptors always elicits an impressive natriuretic response (30, 31) without triggering a tubuloglomerular feedback-induced reduction in glomerular filtration rate (32) and without increasing K^+ excretion. Consequently, selective A_1 receptor antagonists are being developed as eukaluretic natriuretics in Na^+-retaining states such as heart failure (33). A_1 receptors may also be involved in pathophysiological mechanisms of drug-induced nephrotoxicity. For example, selective antagonism of A_1 receptors attenuates nephropathy caused by such nephrotoxins as cisplatin (34–36), gentamicin (37), cephaloridine (38), glycerol (39–41), and radiocontrast media (42, 43).

Selective A_{2A} receptor agonists augment renal blood flow (44, 45), a response caused by vasodilation of the medullary renal microcirculation (5). In this regard, A_{2A} receptor-induced increases in medullary blood flow are caused by vasodilation of afferent and efferent arterioles of juxtaglomerular nephrons (10) and vasodilation of the outer medullary descending vasa recta (11). In contrast to A_1 receptors, which directly stimulate Na^+ reabsorption by increasing epithelial transport mechanisms, A_{2A} receptors indirectly reduce Na^+ reabsorption by enhancing renal medullary blood flow. Importantly, activation of A_{2A} receptors in microvessels of juxtaglomerular nephrons (10) and in the medullary microcirculation (5) enhances medullary blood flow (46), thus altering peritubular forces that modulate Na^+ reabsorption. The net result is an increase in Na^+ excretion (46).

A_{2A} receptors also exert important anti-inflammatory actions that may protect the kidneys from injury. Activation of A_{2A} receptors strongly inhibits neutrophil-endothelial cell interactions in vitro (47), and in vivo, A_{2A} receptor activation reduces neutrophil-endothelial interactions following ischemia/reperfusion injury (48). Selective activation of A_{2A} receptors markedly decreases the renal infiltration of neutrophils and attenuates renal dysfunction in kidneys subjected to ischemia/reperfusion injury (49).

Freshly isolated preglomerular microvessels express A_{2B} receptors (50), and studies in cultured mesangial cells (51) and vascular smooth muscle cells (52–56) indicate that activation of A_{2B} receptors attenuates proliferation of, and extracellular matrix production by, mesangial and vascular smooth muscle cells. It is likely, therefore, that A_{2B} receptors importantly influence renal structure and function.

CLASSICAL PATHWAYS OF ADENOSINE PRODUCTION

The classical paradigm is that three major pathways produce adenosine. The first mechanism is the intracellular ATP pathway, which is characterized by sequential dephosphorylation of intracellular ATP to adenosine. In cells that rapidly consume

ATP and/or have a low capacity to generate ATP, enhanced utilization of ATP increases the intracellular production rate of adenosine via the intracellular breakdown of ATP to adenosine (57). Because nucleoside transporters permit movement of adenosine across cell membranes (58), stimulation of the intracellular ATP pathway results in increased exposure of nearby cells to adenosine owing to translocation of intracellular adenosine into the extracellular space.

In the kidney there are two prime examples of the intracellular ATP pathway. Increased delivery of Na^+ to the thick ascending limb of Henle stimulates Na^+-K^+-ATPase activity in the basolateral membrane of epithelial cells through the increased flux of Na^+ across the luminal membrane. Oxygen availability in the renal medulla is marginally adequate (59); consequently, increased Na^+-K^+-ATPase activity may deplete ATP levels and lead to dephosphorylation of adenine nucleotides to form adenosine. Evidence supporting this mechanism is that a high Na^+ concentration increases adenosine production by thick ascending limbs of Henle in vitro (60), and administration of hypertonic radiocontrast agents (61) and a high Na^+ intake (62, 63) augments intrarenal levels of adenosine in vivo. The intracellular ATP pathway of adenosine formation mediates, in part, the coupling of Na^+ concentrations in the loop of Henle to increases in preglomerular vascular resistance and decreases in the vascular resistance of the vasa recta, processes that are mediated by A_1 receptors and A_{2A} receptors, respectively. The increase in preglomerular vascular resistance decreases single nephron glomerular filtration and, therefore, decreases Na^+ delivery to the distal nephron—a process known as tubuloglomerular feedback (for review see 15). Reductions in the vascular resistance of the vasa recta decrease Na^+ reabsorption and thereby enhance the urinary excretion of Na^+ (46).

Reactive ischemia is a second example of the intracellular ATP pathway in the kidney. A short period of ischemia in the kidney triggers a brief increase in renal vascular resistance, a phenomenon known as reactive ischemia. Renal ischemia activates the intracellular ATP pathway of adenosine production (64–66), and adenosine causes reactive ischemia via activation of A_1 receptors in the preglomerular microvessels (67, 68).

The extracellular ATP pathway is yet another mechanism of adenosine production in the kidney (69, 70). Release of adenine nucleotides into the extracellular compartment from renal sympathetic nerve terminals, intrarenal platelets, renal endothelial cells, renal vascular smooth muscle cells and/or renal epithelial cells exposes extracellular ATP to ecto-ATPases, ecto-ADPases, and ecto-5′-nucleotidases, and these enzymes metabolize adenine nucleotides to adenosine. Endothelial cells and vascular smooth muscle cells are particularly capable of releasing adenine nucleotides, for example, in response to thrombin (71) and neutrophil-derived proteases (72), and converting adenine nucleotides to adenosine (73–75). It is unclear whether the extracellular ATP pathway importantly contributes to the renal production of adenosine; however, the release of adenine nucleotides significantly contributes to activation of renal P2 receptors (69, 70).

The transmethylation pathway of adenosine production is the hydrolysis of S-adenosyl-L-homocysteine to L-homocysteine and adenosine (76). The rate of transmethylation reactions determines the rate of adenosine biosynthesis by the transmethylation pathway because transfer of a methyl group from S-adenosyl-L-methionine to methyl acceptors forms S-adenosyl-L-homocysteine. Approximately one third of the adenosine release to the extracellular space by cardiomyocytes is through the transmethylation pathway (77), but the importance of this pathway in the kidney is unclear.

EXTRACELLULAR cAMP-ADENOSINE PATHWAY

The transmethylation pathway of adenosine production occurs at a relatively fixed rate because the rate of cellular methylation reactions is fairly constant. In contrast, the intracellular and extracellular ATP pathways for adenosine biosynthesis are activated by crisis events (for example, ischemia, platelet activation, inflammation). Therefore, the classical pathways of adenosine biosynthesis do not appear suitable for physiological regulation (that is, moment to moment fine tuning) of extracellular adenosine levels.

However, the putative cAMP-adenosine pathway could provide a mechanism for hormonally mediated fine tuning of extracellular adenosine levels in the biophase of adenosine receptors. As illustrated in Figure 2, adenosine biosynthesis by this putative pathway occurs with activation of hormone receptors coupled to adenylyl cyclase via stimulatory G proteins. This hypothesis posits that the cAMP-adenosine pathway has both intracellular and extracellular branches. The intracellular cAMP pathway occurs via conversion of intracellular cAMP to AMP by cytosolic phosphodiesterase, followed by conversion of intracellular AMP to adenosine by cytosolic 5'-nucleotidase. Adenosine transporters in the cell membrane would allow for movement of adenosine formed by the intracellular cAMP-adenosine pathway into the extracellular compartment. Under most circumstances, however, the intracellular cyclic AMP-adenosine pathway may be quantitatively insignificant because of the efficient conversion of AMP to ADP by AMP kinase. Indeed, at low concentrations, most intracellular AMP is rephosphorylated rather than dephosphorylated (78). However, in the extracellular compartment, dephosphorylation, rather than rephosphorylation, is the fate of extracellular AMP. In this regard, ecto-5'-nucleotidase is a ubiquitous enzyme fastened to the extracellular aspect of the cell membrane by a lipid-sugar linkage, and ecto-5'-nucleotidase rapidly metabolizes AMP to adenosine (79–80, for review see 81).

Hormonal activation of adenylyl cyclase causes robust efflux of intracellular cAMP into the extracellular space. Consequently, if sufficient activity of ecto-phosphodiesterase is present in the extracellular compartment, hormonal activation of adenylyl cyclase would be expected to result in extracellular production of adenosine by conversion of extracellular cAMP to AMP by ecto-phosphodiesterase, followed by conversion of extracellular AMP to adenosine by ecto-5'-nucleotidase, a sequence of reactions now known as the extracellular cAMP-adenosine pathway.

A crucial aspect of the extracellular cAMP-adenosine pathway is the efflux of intracellular cAMP. cAMP egress was described approximately four decades ago

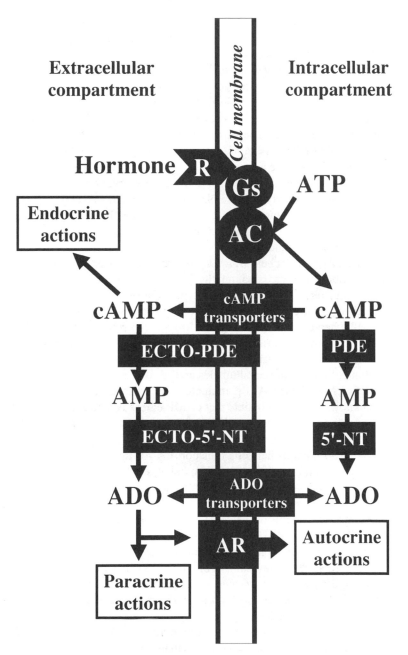

Figure 2 Extracellular cAMP-adenosine pathway. ATP, adenosine triphosphate; AMP, adenosine monophosphate; cAMP, cyclic adenosine monophosphate; ADO, adenosine; AC, adenylyl cyclase; Gs, stimulatory G-protein coupled to AC; R, receptor coupled to Gs; PDE, phosphodiesterase; 5′-NT, 5′-nucleotidase; AR, adenosine receptor.

in pigeon erythrocytes (82), yet surprisingly little is known regarding how this process occurs. cAMP egress evolved several hundred million years ago as part of a survival mechanism for the slime mold *Dicytostelium discoideum*. When faced with inadequate nutrients, *D. discoideum* secretes cAMP into the environment, and the extracellular cAMP attaches to cell surface cAMP receptors on nearby cells to trigger chemotaxis, which leads to aggregation of single-celled amoebae into a migrating slug (83). It is now known that efflux of cAMP occurs in most cells and tissues including the rat superior cervical ganglia (84), rat glioma cells in culture (85), fibroblasts (86), perfused rat livers (87), perfused rat hearts (88), rat adipose tissue (89), and swine adipocytes (90). The rate of cAMP efflux is proportional to the intracellular levels of cAMP (91, 92), occurs rapidly (within minutes) following stimulation of adenylyl cyclase (93), occurs through an energy-dependent process (94) that is temperature sensitive (95), is inhibited by the organic anion transport inhibitor probenecid (82), and is blocked partially by certain prostaglandins, in particular prostaglandin A (91).

Recently, van Aubel et al. (96) discovered that cAMP transport across apical membranes of proximal tubules is mediated by an integral membrane protein called multidrug resistance protein type 4 (MRP4; also called ABCC4). MRP4 is highly expressed in apical membranes of proximal tubules and actively transports cAMP when expressed in Sf9 cells. Moreover, probenecid inhibits both proximal tubular transport of cAMP and MRP4-mediated transport of cAMP. The integral membrane protein organic acid transporter type 1 (OAT1) is present in the basolateral membrane of proximal tubules and, similar to MRP4, transports cAMP (97). Whether MRP4 and OAT1 participate in the transport of cAMP in other cell types is unknown, and it is possible that cAMP egress is mediated by a family of transporters, the members of which variably contribute to efflux of cAMP depending on the tissue.

Why do mammalian cells transport cAMP into the extracellular space? One possibility is that such a mechanism helps regulate intracellular levels of cAMP. However, both adenylyl cyclase (98) and phosphodiesterase (99) contain large families of enzymes that are modulated in complex signaling networks to regulate intracellular cAMP. From a teleological point of view, it makes little sense for cells to reduce intracellular levels of cAMP by pumping cAMP into the extracellular compartment, thus depleting the intracellular purine pool, when these other effective systems are operational. An alternative explanation is that cAMP efflux was conserved by evolution because of its usefulness as a mechanism for regulating extracellular adenosine levels. If so, then the extracellular cAMP pathway must have biological functions.

The biological function of the extracellular cAMP pathway is that it could provide for adenosine production to attenuate, augment and/or enrich the response to hormonal stimulation of adenylyl cyclase. Inasmuch as adenosine could be synthesized in the unstirred water layer by spatially linked enzymatic reactions taking place on the cell surface (cAMP transport onto the cell surface, metabolism of cAMP to AMP by ecto-phosphodiesterase, and conversion of AMP to adenosine by membrane-bound ecto-5'-nucleotidase), quantitatively small increases in cAMP production could give rise to significant concentrations of adenosine in the biophase of cell surface adenosine receptors. This would optimize the autocrine and paracrine effects of adenosine. In addition, as discussed below, cAMP secreted into the systemic circulation could provide for an endocrine role for the extracellular cAMP-adenosine pathway. In this regard, although the half-life of adenosine in human blood is less than one second (100), cAMP is stable in blood and could function as a prohormone.

Most of the components of the putative extracellular cAMP pathway have now been well characterized. For example, adenylyl cyclase (101), ecto-5′-nucleotidase (79), and at least some cAMP transporters (96, 97) have been described at the molecular level. During the past several years, investigators have discovered and cloned numerous ecto-enyzmes that metabolize purine nucleotides including ecto-nucleoside 5′-triphosphate diphosphohydrolases (E-NTPDases 1, 2, 3, 4, 5, and 6), ecto-nucleotide pyrophosphatases (E-NPP 1, 2, and 3), alkaline phosphatases and NAD-glycohydrolases (102), and an increasing amount of information is accumulating to support the existence of ecto-phosphodiesterases that, combined with adenylyl cyclases, cAMP transporters and ecto-5′-nucleotidase, produce an operational cAMP-adenosine pathway.

Early experiments by Gorin & Brenner (103), Smoake et al. (104), Rosenberg & Dichter (105), and Kather (106) offered preliminary evidence that an extracellular cAMP-adenosine pathway is operational. In those early days, however, the hypothesis was not well articulated and evidence was limited. In the mid-1990s, the hypothesis became better formulated, and several studies now support the existence of the cAMP-adenosine pathway. For example, in cultured rat aortic vascular smooth muscle cells (107) and cardiac fibroblasts (108), exogenous cAMP is converted to AMP, adenosine, and inosine in a concentration-dependent and time-dependent manner. Steady-state levels of adenosine are reached in the culture medium within five minutes after adding exogenous cAMP, and significant increases in extracellular adenosine levels occur with concentrations of cAMP in the medium as low as 1 μmol/liter.

Several pharmacological inhibitors can be used to establish the existence of a cAMP-adenosine pathway. 3-Isobutyl-1-methylxanthine (IBMX) is a broad spectrum inhibitor of phosphodiesterases that penetrates cell membranes (109) and thus inhibits both intracellular and extracellular phosphodiesterases. 1,3-Dipropyl-8-p-sulfophenylxanthine (DPSPX), similar to IBMX, is a xanthine, but unlike IBMX, DPSPX is restricted to the extracellular compartment by a negative charge at physiological pH (110). At low concentrations, DPSPX is a nonselective adenosine receptor antagonist (111), and at high concentrations, DPSPX blocks ecto-phosphodiesterase (112, 113). α,β-Methyleneadenosine-5′-diphosphate (AMPCP) is a selective inhibitor of ecto-5′-nucleotidase, but does not inhibit cytosolic 5′-nucleotidase (81).

In vascular smooth muscle and cardiac fibroblasts, the conversion of cAMP to AMP, adenosine, and inosine (a metabolite of adenosine) is inhibited by IBMX and DPSPX (107, 108). Also, in these cells, the metabolism of cAMP to adenosine and inosine, but not to AMP, is blocked by AMPCP. Importantly, activation of adenylyl cyclase in cardiac fibroblasts with norepinephrine, isoproterenol, or forskolin increases extracellular levels of endogenous cAMP (114). This response is accompanied by an increase in extracellular levels of adenosine (114), and $2',5'$-dideoxyadenosine, an inhibitor of adenylyl cyclase, blocks the increase in both extracellular endogenous cAMP and adenosine.

Occupancy of A_2 receptors markedly increases the first-order rate constant for the efflux of cAMP in vascular smooth muscle cells (93). Thus in some cell types, the extracellular cAMP-adenosine pathway may be involved in a positive-feedback loop in which increases in extracellular adenosine by the cAMP-adenosine pathway may cause further activation of the pathway by stimulating adenylyl cyclase and by increasing the efficiency of cAMP efflux.

Studies with brain cells provide additional evidence for the cAMP-adenosine pathway. Isoproterenol, a beta-adrenoceptor agonist coupled to adenylyl cyclase via stimulatory G proteins, increases extracellular levels of cAMP in rat cerebral cortex in dissociated cell culture (105). Moreover, in this cell culture model system, exogenous cAMP is converted to AMP and adenosine (105). In astrocyte-rich cerebral cortex in culture, isoproterenol augments extracellular levels of cAMP and adenosine (115); however, this effect is not observed in neuron-enriched cultures (115). In cerebral cortical cultures, norepinephrine (116), epinephrine (116), and forskolin (117) also increase extracellular cAMP and adenosine. Importantly, in rat cerebral cortex in dissociated cell culture, the increase in adenosine biosynthesis following activation of adenylyl cyclase is inhibited by blockade of cAMP transport with probenecid (115), phosphodiesterase with IBMX plus RO 20-1724 (115) or RO 20-1724 alone (117), and ecto-$5'$-nucleotidase with GMP (117). Studies in the hippocampus also support the existence of the extracellular cAMP pathway (118) in the brain. In this regard, forskolin and exogenous cAMP increase extracellular adenosine levels and induce adenosine-mediated electrophysiological effects in superfused hippocampal slices (118). Therefore, it appears that the extracellular cAMP-adenosine pathway exists in the brain primarily as a result of egress of cAMP from astrocytes.

There is also evidence for the extracellular cAMP pathway in the cerebral vasculature. In rats equipped with an implanted closed cranial window, suffusion of pial vessels with cAMP enhances the biosynthesis of extracellular adenosine, and this phenomenon is blocked by IBMX, DPSPX, and AMPCP (119).

Investigators have also observed evidence for the cAMP-adenosine pathway in fat cells and hepatocytes. In human adipocytes, catecholamines increase both cAMP production and extracellular levels of the adenosine metabolites, inosine and hypoxanthine (120). Also, cell membranes isolated from swine adipocytes convert cAMP to AMP and adenosine (113). In swine adipocytes, the cell membrane-bound ecto-phosphodiesterase and the microsomal membrane-bound phosphodiesterase are pharmacologically distinguishable. For example, the plasma membrane ecto-phosphodiesterase is inhibited by DPSPX, but is not inhibited by selective blockers of types 1, 2, 3B, 4, or 5 phosphodiesterase. Conversely, the microsomal enzyme is blocked by inhibitors of type 3B phosphodiesterase (113). Freshly isolated hepatocytes (103, 104) and hepatocytes in culture (104) convert exogenous cAMP to AMP and nucleosides, and this activity is blocked by aminophylline and trypsin (104), suggesting the involvement of an ecto-phosphodiesterase.

In summary, current data obtained from aortic vascular smooth muscle cells, cardiac fibroblasts, cerebral cortical cells, hippocampal tissue, cerebral microvessels, adipocytes and hepatocytes confirm the existence of the extracellular cAMP-adenosine pathway.

RENAL EXTRACELLULAR cAMP-ADENOSINE PATHWAY

Although studies in nonrenal cells/tissues have provided much of the evidence supporting the existence of the extracellular cAMP-adenosine pathway, the concept was first explicitly articulated as a "transmembrane negative-feedback loop" mechanism in which adenosine formed from cAMP egressing from juxtaglomerular cells was proposed to restrain the renin release response to factors that stimulated renin release by activating adenylyl cyclase (19). A number of studies now provide direct evidence for the extracellular cAMP-adenosine pathway in the kidney.

In the isolated, perfused rat renal vascular bed, cAMP added to the perfusate markedly increases the renal secretion rates of AMP, adenosine, and inosine (112), and the metabolism of exogenous cAMP to AMP, adenosine, and inosine is blocked by IBMX and DPSPX. Also, AMPCP inhibits the metabolism of cAMP to adenosine and inosine, but does not inhibit the metabolism of cAMP to AMP in the perfused kidney (112). These data not only support the existence of the cAMP-adenosine pathway in the kidney but also suggest that the metabolism of exogenous cAMP to adenosine in the kidney occurs mostly in the extracellular compartment. In this regard, cAMP is hydrophilic, and exogenous cAMP would not be expected to penetrate cell membranes. Also, AMPCP inhibits ecto-5′-nucleotidase, but not cytosolic 5′-nucleotidase (81). Consequently, the blockade of cAMP metabolism to adenosine and inosine by AMPCP indicates an extracellular site of conversion. Finally, DPSPX is restricted to the extracellular space (110) and yet blocks the metabolism of cAMP to AMP, adenosine, and inosine, a finding consistent with an extracellular site of cAMP metabolism to AMP and adenosine. These data do not rule out the intracellular cAMP-adenosine pathway but do indicate the existence of a renal extracellular cAMP-adenosine pathway.

Additional studies in the isolated perfused rat kidney (121) demonstrate that isoproterenol increases adenosine and inosine secretion and that this effect is blocked by propranolol, a beta-adrenoceptor antagonist. These experiments suggest that stimulation of renal adenylyl cyclase by beta-adrenoceptors increases renal adenosine production. Blockade of phosphodiesterase with IBMX and ecto-5′-nucleotidase with AMPCP also inhibits isoproterenol-induced adenosine production in the isolated, perfused rat kidney, a finding consistent with endogenous cAMP being converted to adenosine. These studies provide strong support for the renal extracellular cAMP-adenosine pathway in the kidney.

In isolated perfused kidneys, inhibition of cAMP transport by the renal tubules with probenecid increases the recovery of adenosine in the venous effluent following addition of cAMP to the perfusate (122). This finding, while not ruling out an extracellular cAMP-adenosine pathway in the tubules, indicates that exogenous cAMP added to the perfusate is metabolized to adenosine in part in the vascular compartment and that this vascular adenosine is recovered in the venous drainage. The conclusion that blood vessels in the kidney express an extracellular cAMP-adenosine pathway is also supported by studies in freshly isolated, preglomerular microvessels (123). Incubation of preglomerular microvessels with cAMP augments extracellular adenosine (up to 60-fold), and this effect is inhibited by blockade of phosphodiesterase with IBMX or ecto-phosphodiesterase with DPSPX. Incubation of preglomerular microvessels with isoproterenol plus IBMX increases extracellular cAMP (up to 30-fold), a finding that indicates robust cAMP egress following activation of adenylyl cyclase in preglomerular microvessels. This conclusion is confirmed by the linear relationship between intracellular and extracellular cAMP concentrations in preglomerular microvessels. As in perfused rat kidneys, in preglomerular microvessels isoproterenol stimulates extracellular adenosine levels, and this effect of isoproterenol is blocked by propranolol, IBMX, and DPSPX. These results are fully consistent with the concept that preglomerular microvessels transport endogenous cAMP to the extracellular compartment and convert extracellular cAMP to adenosine. Therefore, preglomerular microvessels are one site in the kidney that expresses an extracellular cAMP-adenosine pathway.

In cultured rat preglomerular vascular smooth muscle cells, addition of cAMP stimulates extracellular adenosine concentrations (up to 40-fold), and this effect is blocked by IBMX (122). Evidence suggests that mesangial cells also express the extracellular cAMP-adenosine. In mesangial cells, exogenous cAMP augments extracellular concentrations of adenosine (up to 25-fold), and this effect is blocked by inhibition of phosphodiesterase with IBMX, ecto-phosphodiesterase with DPSPX, and ecto-5'-nucleotidase with AMPCP (124). Recent studies from our laboratory (E.K. Jackson, Z. Mi, C. Zhu & R.K. Dubey, unpublished data) indicate that freshly isolated proximal convoluted tubules and collecting ducts, but not thick ascending limbs of Henle, convert exogenous cAMP to AMP, adenosine, and inosine in a time-dependent and concentration-dependent manner. Moreover, the conversion of cAMP to adenosine is blocked by inhibition of phosphodiesterase with IBMX, ecto-phosphodiesterase with DPSPX, and ecto-5'-nucleotidase with AMPCP. Current evidence thus suggests the existence of the extracellular cAMP-adenosine pathway in preglomerular vascular smooth muscle cells, glomerular mesangial cells, and epithelial cells in the proximal convoluted tubules and collecting ducts.

There is also evidence that the extracellular cAMP-adenosine pathway exists in the kidney in vivo. Although IBMX is an inhibitor of the extracellular cAMP-adenosine pathway, when administered systemically at appropriate doses, IBMX causes cardiovascular effects (125) that preclude a straightforward interpretation of the data. However, using a microdialysis probe, IBMX can be placed locally into the renal cortical interstitial space, thus providing effective high local concentrations of IBMX without inducing confounding hemodynamic effects. These experiments reveal that IBMX decreases the recovery of both adenosine and inosine from the interstitial space of the renal cortex by approximately 50% (126) and support the conclusion that cAMP is metabolized to adenosine and contributes importantly to the concentrations of adenosine in the renal cortical interstitium in vivo. Finally, in rats (E.K. Jackson, W.A. Herzer, Z. Mi & R.K. Dubey, unpublished data) intrarenal and intravenous infusions of cAMP and intrarenal infusions of isoproterenol increase urinary excretion rates of both cAMP and adenosine.

PANCREATOHEPATORENAL EXTRACELLULAR cAMP-ADENOSINE PATHWAY

In addition to an autocrine/paracrine role in the kidney, the extracellular cAMP-adenosine pathway may also function as an endocrine system by which the pancreas and liver regulate renal function. The pancreas releases glucagon directly into the portal circulation in response to appropriate stimuli (127, 128), thus avoiding dilution of glucagon and maximizing the concentration of glucagon delivered to the hepatocytes. Glucagon is a powerful stimulant of hepatic adenylyl cyclase (129, 130), and activation of hepatic adenylyl cyclase causes release of large quantities of cAMP into the systemic circulation (130, 131). Inasmuch as the liver is the largest organ in the body, about 1.5 kg in a normal adult, the liver has the capacity to release significant amounts of cAMP into the hepatic vein. The kidney freely filters cAMP into the tubular lumen (132), and the concentration of cAMP in the proximal tubules would be increased by reabsorption of water by the tubules, a process that concentrates cAMP in the tubular lumen.

Freshly isolated, proximal convoluted tubules readily convert cAMP into adenosine in a concentration-dependent and time-dependent manner, and the metabolism of cAMP to adenosine is blocked by IBMX, DPSPX, and AMPCP (E.K. Jackson, Z. Mi, C. Zhu & R.K. Dubey, unpublished data). Proximal tubules also express adenosine receptors, and adenosine enhances electrolyte transport in the proximal tubule (133, 134). Taken together, the aforementioned facts suggest the endocrine linkage summarized in Figure 3. Glucagon released from the pancreas into the portal circulation stimulates hepatic adenylyl cyclase, which results in secretion of cAMP by hepatocytes into the venous circulation. Liver-derived cAMP circulates to the kidney where it is filtered into the proximal tubule. As water is extracted from the proximal tubules, cAMP becomes concentrated. Because proximal tubules can metabolize cAMP to adenosine, this process would proceed to generate adenosine. Adenosine would then engage A_1 adenosine receptors in epithelial cells to enhance electrolyte transport. It is important to note that in contrast to cAMP, which is stable in blood, adenosine has a half-life in human blood of less than one second (100). Therefore, if the pancreas and liver communicate with the kidneys via the adenosine system, the liver could not do so by releasing adenosine, but

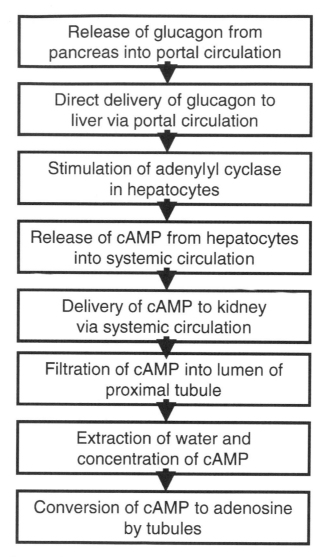

Figure 3   Pancreatohepatorenal extracellular cAMP-adenosine pathway. cAMP, cyclic adenosine monophosphate.

would have to do so by secreting a stable prohormone that could be manufactured into adenosine at the level of the target tissue. In this perspective, then, cAMP is an extracellular prohormone.

To test this putative pancreatohepatorenal extracellular cAMP-adenosine pathway, we recently examined the effects of intraportal administration of glucagon on plasma and urine levels of cAMP and adenosine in rats (135). Intraportal infusions of glucagon caused a ninefold increase in plasma levels of cAMP, and caused a ninefold and threefold increase in the urinary excretion rate of cAMP and adenosine, respectively (135). The increase in adenosine was renal in origin because glucagon did not alter plasma levels of adenosine. In additional studies, we found that intraportal, but not intravenous, infusion of glucagon caused a marked increase in renal cortical interstitial levels of cAMP and adenosine (E.K. Jackson, Z. Mi, W.A. Herzer & R.K. Dubey, unpublished data). Further studies are needed to determine whether the pancreatohepatorenal extracellular cAMP-adenosine pathway exists.

PHYSIOLOGICAL AND PATHOPHYSIOLOGICAL ROLES OF THE EXTRACELLULAR cAMP-ADENOSINE PATHWAY

Physiological Roles at Extrarenal Sites

In vascular smooth muscle cells, the cAMP-adenosine pathway may influence nitric oxide (NO) production (136). Administration of adenosine, 2-chloroadenosine and agents that elevate endogenous adenosine [such as erythro-9-(2-hydroxy-3-nonyl)adenine (EHNA), an inhibitor of adenosine deaminase, and iodotubercidin, an inhibitor of adenosine kinase] to cultured aortic vascular smooth muscle cells increases nitrite/nitrate (stable metabolites of NO) levels in the medium and enhances the metabolism of arginine to citrulline by cytosolic extracts from smooth muscle cells pretreated with these test agents. On the other hand, CPA and CGS21680, selective A_1 and A_{2A} receptor agonists, respectively, do not increase NO production. The ability of 2-chloroadenosine and EHNA plus iodotubercidin to stimulate NO production is inhibited by KF17837 (blocks A_{2A} and A_{2B} receptors) and DPSPX (nonselective adenosine receptor antagonist), but not by 1,3-dipropyl-8-cyclopentylxanthine (DPCPX; a selective A_1 receptor blocker). Incubation of vascular smooth muscle cells with exogenous cAMP at concentrations that enhance adenosine concentrations in the medium also augment nitrite/nitrate levels and citrulline formation, and these effects of exogenous cAMP are blocked by DPSPX and KF17837, but not by DPCPX. These findings indicate that in vascular smooth muscle, extracellular cAMP is metabolized to adenosine, which activates A_{2B} adenosine receptors and induces NO synthesis.

The extracellular cAMP-adenosine pathway may also contribute to estradiol-induced inhibition of vascular smooth muscle cell growth (137). Inhibition of vascular smooth muscle cell proliferation and protein and collagen synthesis by estradiol is reduced by DPSPX, KF17837, and 2',5'-dideoxyadenosine (an adenylyl cyclase inhibitor), but not by DPCPX. Moreover, the inhibitory effects of estradiol are augmented by stimulation of adenylyl cyclase with forskolin and by blockade of adenosine metabolism with EHNA plus iodotubercidin. Estradiol raises extracellular concentrations of cAMP and adenosine, and these actions of estradiol are inhibited by 2',5'-dideoxyadenosine. These findings are consistent with the hypothesis that the extracellular cAMP-adenosine pathway importantly contributes to the effects of estradiol in vascular biology.

Evidence suggests that the extracellular cAMP-adenosine pathway also participates in regulation of vascular tone, at least in some vascular beds. For example, suffusion of rat pial arteries with cAMP increases the concentrations of extracellular adenosine on the cortical surface, and this effect is inhibited by IBMX, DPSPX, and AMPCP (119). Moreover, suffusion of rat pial arteries with cAMP and adenosine produces concentration-dependent vasodilation, whereas cAMP analogues that are not converted to adenosine have little effect on pial artery vascular resistance. Also, the reductions in pial vascular resistance promoted by cAMP, adenosine and hypotension are significantly reduced by 3,7-dimethyl-1-propargylxanthine (an A_2 receptor antagonist), IBMX, DPSPX, and AMPCP, but not by 8-cyclopentyltheophylline (an A_1 antagonist). These findings strongly indicate that the cAMP-adenosine pathway is an important contributor to hypotension-induced cerebral autoregulation via activation of A_2 receptors.

The extracellular cAMP-adenosine pathway may also influence cardiac fibroblast biology, including proliferation and collagen production. In cultured cardiac fibroblasts, adenosine (both exogenous and endogenous) and adenosine analogues with A_{2B} receptor agonist activity attenuate cardiac fibroblast proliferation and collagen and protein synthesis (138–140). Studies with a broad range of adenosine receptor antagonists and agonists and with antisense oligodeoxynucleotides against the A_{2B} receptor indicate that the antigrowth effects of adenosine on cardiac fibroblasts are mediated by A_{2B} receptors (138–140). Application of exogenous cAMP to cardiac fibroblasts augments extracellular concentrations of adenosine and inosine, and these effects are blocked by IBMX, DPSPX, and AMPCP (108). Exogenous cAMP attenuates DNA synthesis, cell proliferation, and protein synthesis, and antagonism of A_2 receptors with KF17837, but not A_1 receptors with DPCPX, blocks the antigrowth effects of exogenous cAMP, but not the antigrowth effects of 8-bromo-cAMP (108). Also, the antigrowth effects of exogenous cAMP are augmented by EHNA, an inhibitor of adenosine deaminase, plus iodotubercidin, an inhibitor of adenosine kinase (108). Activation of adenylyl cyclase with either isoproterenol or forskolin augments cAMP production and concomitantly increases extracellular concentrations of adenosine in cardiac fibroblasts, and these actions are blocked by inhibition of adenylyl cyclase with 2′,5′-dideoxyadenosine (114). Forskolin and isoproterenol reduce DNA synthesis, and these actions are enhanced by EHNA plus iodotubercidin (114). Also, blockade of adenylyl cyclase with 2′,5′-dideoxyadenosine or A_2 receptors with KF17837 prevents the effects of forskolin and isoproterenol on DNA synthesis. Forskolin attenuates protein synthesis and cell proliferation, and these actions of forskolin are inhibited by antagonism of A_2 receptors with KF17837 or blockade of adenylyl cyclase with 2′,5′-dideoxyadenosine (114). Thus the evidence suggests that the extracellular cAMP-adenosine pathway is expressed in cardiac fibroblasts and functions to attenuate cardiac fibroblast collagen synthesis and proliferation via A_{2B} receptors.

Physiological Roles at Renal Sites

Renin release from juxtaglomerular cells is stimulated in response to increases in intracellular cAMP (141), and increases in intracellular cAMP cause efflux of cAMP, which activates the extracellular cAMP-adenosine pathway. Because adenosine via A_1 receptors attenuates renin release (15), the extracellular cAMP-adenosine pathway acts as a negative-feedback control on renin release.

Several lines of evidence support the concept that the extracellular cAMP-adenosine pathway functions to modulate renin release. First, as noted above, the extracellular cAMP-adenosine pathway exists in preglomerular microvessels (123) and in vascular smooth muscle cells cultured from preglomerular microvessels (122). Because juxtaglomerular cells are derived from preglomerular vascular smooth muscle cells (142), it is likely that the extracellular cAMP-adenosine pathway also exists in juxtaglomerular cells. Second, administration of cAMP (143), AMP (143), and adenosine (144, 145) directly into renal artery inhibits renin release. Third, inhibition of A_1 adenosine receptors with nonselective (theophylline, caffeine, and DPSPX) or A_1 selective (DPCPX and FK453) adenosine receptor antagonists stimulates renin release (110, 146–160).

The extracellular cAMP-adenosine pathway may also regulate proliferation and extracellular matrix production by vascular smooth muscle cells, both renal preglomerular as well as extrarenal vascular smooth muscle cells. As mentioned above, evidence indicates that the extracellular cAMP-adenosine pathway exists in both renal and nonrenal vascular smooth muscle cells. Exogenous and endogenous adenosine and adenosine analogues with A_{2B} receptor agonist activity attenuate vascular smooth muscle cell proliferation and collagen and protein synthesis (52–56). Studies with a number of adenosine receptor agonists and antagonists and with antisense oligodeoxynucleotides against the A_{2B} receptor indicate that the growth inhibitory effects of adenosine on vascular smooth muscle cells are mediated by A_{2B} receptors (52–56). Exogenous cAMP inhibits DNA synthesis in cultured aortic vascular smooth muscle cells, and EHNA and dipyridamole (an inhibitor of adenosine transport) enhance this effect (107). Antagonism of A_2 receptors with KF17837 and DPSPX decrease the inhibitory effects of cAMP on DNA synthesis, yet these adenosine receptor antagonists do not attenuate the inhibitory effects of 8-bromo-cAMP (a cAMP analogue that is not metabolized to adenosine) on DNA synthesis (107). These results suggest that cAMP-derived adenosine attenuates vascular smooth muscle cell growth and that the extracellular cAMP-adenosine pathway may importantly contribute to the regulation of vascular biology, both within and outside the kidney.

In addition, the extracellular cAMP-adenosine pathway may regulate glomerular mesangial cell (GMC) extracellular matrix production and proliferation. As reviewed above, GMCs express the extracellular cAMP-adenosine pathway (124). Exogenous cAMP reduces proliferation of GMCs, and this effect is attenuated by A_2 receptor antagonism with either DPSPX or KF17837 (124). Nonselective A_{2B} receptor agonists such as adenosine, 2-chloroadenosine, 5′-N-methylcarboxamidoadenosine and 5′-N-ethylcarboxamidoadenosine, but not selective

A_1(N^6-cyclopentyladenosine) or A_{2A} (CGS21680) receptor agonists, attenuate GMC collagen synthesis and proliferation (51). Inhibition of adenosine deaminase with EHNA or adenosine kinase with iodotubercidin augments endogenous adenosine levels in GMCs and attenuates cell proliferation and DNA and collagen synthesis (51). The inhibitory actions of 2-chloroadenosine, EHNA, and iodotubercidin on GMC growth are attenuated by DPSPX, but not DPCPX. Together, these data support the conclusion that the extracellular cAMP-adenosine pathway exists in GMCs and functions via A_{2B} receptors to inhibit GMC proliferation and collagen.

In the kidney, adenosine potentiates renal vascular responses to sympathetic nerve stimulation by activating postjunctional A_1 receptors (17, 18), and stimulation of renal sympathetic nerves releases adenosine and adenosine metabolites via a mechanism requiring intact beta-adrenoceptors (161). It is possible that noradrenergic neurotransmission in the renal vasculature is enhanced by the cAMP-adenosine pathway. Specifically, exocytosis of norepinephrine from renal sympathetic nerve terminals may activate postjunctional beta-adrenoceptors, stimulate adenylyl cyclase, and engage the cAMP-adenosine pathway to increase renal vascular levels of adenosine. Adenosine could then enhance renal vascular responses to sympathetic nerve stimulation by sensitizing renal preglomerular vascular smooth muscle cells to norepinephrine.

The extracellular cAMP-adenosine pathway may be importantly involved in the modulation of transport by renal epithelial cells. Studies by Friedlander et al. (162) demonstrate that exogenous cAMP is rapidly converted to extracellular adenosine in cultured opossum kidney cells (a cell model system with a proximal epithelial phenotype) by a mechanism that is blocked by inhibition of either ecto-5′-nucleotidase or phosphodiesterase. As mentioned above, our studies demonstrate that exogenous cAMP is converted to AMP and adenosine in freshly isolated proximal convoluted tubules and collecting ducts, and this response is blocked by inhibition of ecto-5′-nucleotidase or ecto-phosphodiesterase. Thus epithelial cells in the proximal tubule and collecting duct apparently express ecto-phosphodiesterase and rapidly convert extracellular cAMP to adenosine.

In proximal tubules, activation of A_1 receptors enhances Na^+ transport. For example, activation of A_1 receptors in cultured epithelial cells that express a proximal tubular phenotype augments both Na^+-glucose symport and Na^+-phosphate symport (23), and activation of A_1 receptors enhances basolateral Na^+-$3HCO_3^-$ symport in microperfused proximal convoluted tubules (24). These effects of adenosine most likely are mediated by inhibition of adenylyl cyclase because the natriuretic/diuretic effects of A_1 receptor antagonists are abolished by blockade of inhibitory G proteins with pertussis toxin (30). Thus it is conceivable that a negative-feedback mechanism exists in which stimulation of adenylyl cyclase attenuates epithelial transport, and this effect is moderated by adenosine formation via a local cAMP-adenosine pathway.

However, our most recent studies strongly suggest that an autocrine/paracrine extracellular cAMP-adenosine pathway in fact does not participate in the regulation of Na^+ transport by proximal epithelial cells. Although isolated proximal tubules metabolize exogenous cAMP to adenosine, even marked stimulation of adenylyl cyclase does not cause efflux of cAMP by these isolated tubules (E.K. Jackson, Z. Mi, C. Zhu & R.K. Dubey, unpublished data). It appears, therefore, that the pancreatohepatorenal extracellular cAMP pathway is more important in modulating proximal tubular function. As explained above, cAMP released from the liver in response to glucagon is concentrated in the renal cortex, and this cAMP is readily converted to adenosine (135). Whereas intrarenal infusions of glucagon do not reduce Na^+ excretion (163), intraportal infusions of glucagon in sheep cause a marked antidiuresis (164).

Although speculative, it is conceivable that the pancreatohepatorenal cAMP-adenosine pathway participates in physiological adjustments of renal transport, as well as in pathophysiological processes. In normal mammals, both hypoglycemia and exercise are powerful stimulants to glucagon release (165, 166). Activation of the pancreatohepatorenal cAMP-adenosine pathway by glucagon in response to hypoglycemia might increase Na^+-glucose symport in proximal tubules and thus increase the efficiency of glucose transport, an adaptive mechanism to combat hypoglycemia. Activation of the pancreatohepatorenal cAMP-adenosine pathway by glucagon during exercise might enhance Na^+ transport in the proximal tubules and thus increase the efficiency of Na^+ reabsorption, an adaptive mechanism to avoid volume depletion during sustained physical exertion.

The pancreatohepatorenal cAMP-adenosine pathway might be overly activated in Syndrome X (i.e., the metabolic syndrome characterized by obesity, insulin resistance, hyperlipidemia, and hypertension). Although oral glucose normally strongly inhibits glucagon secretion by the pancreas, in animals (167) and people (168) with Syndrome X, an oral glucose challenge markedly stimulates pancreatic glucagon secretion by approximately 200%. If the pancreatohepatorenal cAMP-adenosine pathway exists, each time a patient with Syndrome X ingests a high-carbohydrate meal the renal tubules would be exposed to a wave of excess adenosine production. Because adenosine causes increased reabsorption of Na^+ and vasoconstriction of the preglomerular microcirculation (15), this could contribute to the pathophysiology of hypertension in Syndrome X. Importantly, adenosine receptors also inhibit lipolysis in fat cells (169) and may reduce insulin sensitivity in skeletal muscle (170). Therefore, if adenosine is synthesized by adipocytes and skeletal muscle from liver-derived cAMP, adenosine could be a common denominator linking obesity, insulin resistance, and hypertension. However, at this time both the physiological and pathophysiological roles of the putative pancreatohepatorenal cAMP-adenosine pathway are speculative.

In the proximal tubule, the pancreatohepatorenal extracellular cAMP pathway appears to be more important than the autocrine/paracrine extracellular cAMP-adenosine pathway in determining the exposure of proximal tubules to adenosine. In the collecting duct, the opposite may be true. In this regard, our most recent studies indicate not only that the collecting duct is highly efficient at converting cAMP to adenosine (E.K. Jackson, Z. Mi, C. Zhu & R.K. Dubey, unpublished data), but also that stimulation of adenylyl cyclase activity in collecting duct epithelial cells in culture markedly increases extracellular levels of cAMP and adenosine. However, in contrast to the renal cortex, intraportal glucagon does not increase adenosine levels in the renal medullary interstitium. Studies by Ma & Ling (135a) indicate that in A6 cells expressing a distal nephron phenotype, activation of luminal A_1 receptors increases the product of the number of Na^+ channels times the open probability of Na^+ channels. Thus it is likely that autocrine/paracrine extracellular cAMP-adenosine modulates Na^+ reabsorption by the collecting duct.

SUMMARY

Substantial evidence now exists for a physiological role of the extracellular cAMP-adenosine in juxtaglomerular cells, GMCs, renal microvessels and vascular smooth muscle cells, cerebral microvessels, and cardiac fibroblasts. More limited evidence supports the existence of the extracellular cAMP-adenosine pathway in the cerebral cortex, in the hippocampus, and in adipocytes. Recent data suggest that cAMP may be a prohormone released by the liver in response to pancreatic glucagon and may be concentrated and converted to adenosine in the renal cortex. This putative pancreatohepatorenal cAMP-adenosine might be involved in such important disease states as Syndrome X. Future studies will no doubt clarify the role of the extracellular cAMP-adenosine pathway in health and disease, inside and outside of the kidney.

The *Annual Review of Physiology* is online at http://physiol.annualreviews.org

LITERATURE CITED

1. Ralevic V, Burnstock G. 1998. Receptors for purines and pyrimidines. *Pharmacol. Rev.* 50:413–92
2. Lorenzen A, Schwabe U. 2001. P1 Receptors. In *Handbook of Experimental Pharmacology; Purinergic and Pyrimidinergic Signalling I: Molecular, Nervous and Urogenitary System Function*, ed. MP Abbracchio, M Williams, pp. 19–45. Berlin: Springer-Verlag. Vol. 151/I
3. Jacobson KA, Van Rhee AM. 1997. Development of selective purinoceptor agonists and antagonists. In *Purinergic Approaches in Experimental Therapeutics*, ed. KA Jacobson, MF Jarvis, pp. 101–28. New York: Wiley-Liss
4. Feoktistov I, Polosa R, Holgate ST, Biaggioni I. 1998. Adenosine A_{2B} receptors: a novel therapeutic target in asthma? *Trends Pharmacol. Sci.* 19:148–53
5. Agmon Y, Dinour D, Brezis M. 1993. Disparate effects of adenosine A_1- and A_2-receptor agonists on intrarenal blood flow. *Am. J. Physiol. Renal Physiol.* 265: F802–6
6. Holz FG, Steinhausen M. 1987. Renovascular effects of adenosine receptor agonists. *Renal Physiol.* 10:272–82
7. Joyner WL, Mohama RE, Myers TO, Gilmore JP. 1988. The selective response to adenosine of renal microvessels from hamster explants. *Microvasc. Res.* 35: 122–31
8. Murray RD, Churchill PC. 1984. Effects of adenosine receptor agonists in the isolated, perfused rat kidney. *Am. J. Physiol. Heart Circ. Physiol.* 247:H343–48

9. Murray RD, Churchill PC. 1985. Concentration dependency of the renal vascular and renin secretory responses to adenosine receptor agonists. *J. Pharmacol. Exp. Ther.* 232:189–93

10. Nishiyama A, Inscho EW, Navar LG. 2001. Interactions of adenosine A_1 and A_{2a} receptors on renal microvascular reactivity. *Am. J. Physiol. Renal Physiol.* 280:F406–14

11. Sildorff EP, Kreisberg MS, Pallone TL. 1996. Adenosine modulates vasomotor tone in outer medullary descending vasa recta of the rat. *J. Clin. Invest.* 98:18–23

12. Munger KA, Jackson EK. 1994. Effects of selective A_1 receptor blockade on glomerular hemodynamics: involvement of renin-angiotensin system. *Am. J. Physiol. Renal Physiol.* 267:F783–90

13. Traynor T, Yang T, Huang YG, Arend L, Oliverio MI, et al. 1998. Inhibition of adenosine-1 receptor-mediated preglomerular vasoconstriction in AT_{1A} receptor-deficient mice. *Am. J. Physiol. Renal Physiol.* 275:F922–27

14. Weihprecht H, Lorenz JN, Briggs JP, Schnermann J. 1994. Synergistic effects of angiotensin and adenosine in the renal microvasculature. *Am. J. Physiol. Renal Physiol.* 266:F227–39

15. Jackson EK. 2001. P1 and P2 receptors in the renal system. In *Handbook of Experimental Pharmacology; Purinergic and Pyrimidinergic Signalling II: Cardiovascular, Respiratory, Immune, Metabolic and Gastrointestinal Tract Function*, ed. MP Abbracchio, M Williams, pp. 33–71. Berlin: Springer-Verlag. Vol. 151/II

16. Jackson EK. 1997. Renal actions of purines. In *Purinergic Approaches in Experimental Therapeutics*, ed. KA Jacobson, MF Jarvis, pp. 217–50. New York: Wiley-Liss

17. Hedqvist P, Fredholm BB. 1976. Effects of adenosine on adrenergic neurotransmission; prejunctional inhibition and postjunctional enhancement. *Naunyn Schmiedebergs Arch. Pharmacol.* 293:217–23

18. Hedqvist P, Fredholm BB, Olundh S. 1978. Antagonistic effects of theophylline and adenosine on adrenergic neuroeffector transmission in the rabbit kidney. *Circ. Res.* 43:592–98

19. Jackson EK. 1991. Adenosine: a physiological brake on renin release. *Annu. Rev. Pharmacol. Toxicol.* 31:1–35

20. Leaney JL, Tinker A. 2000. The role of members of the pertussis toxin-sensitive family of G proteins in coupling receptors to the activation of the G protein-gated inwardly rectifying potassium channel. *Proc. Natl. Acad. Sci. USA.* 97:5651–56

21. Wise A, Sheehan M, Rees S, Lee M, Milligan G. 1999. Comparative analysis of the efficacy of A_1 adenosine receptor activation of Gi/o α G proteins following coexpression of receptor and G protein and expression of A_1 adenosine receptor-Gi/o α fusion proteins. *Biochemistry* 38:2272–78

22. Kurtz A, Wagner C. 1999. Cellular control of renin secretion. *J. Exp. Biol.* 202:219–25

23. Coulson R, Johnson RA, Olsson RA, Cooper DR, Scheinman SJ. 1991. Adenosine stimulates phosphate and glucose transport in opossum kidney epithelial cells. *Am. J. Physiol. Renal Physiol.* 260:F921–28

24. Takeda M, Yoshitomi K, Imai M. 1993. Regulation of Na^+-$3HCO_3^-$ cotransport in rabbit proximal convoluted tubule via adenosine A_1 receptor. *Am. J. Physiol. Renal Physiol.* 265:F511–19

25. Cai H, Batuman V, Puschett DB, Puschett JB. 1994. Effect of KW-3902, a novel adenosine A_1 receptor antagonist, on sodium-dependent phosphate and glucose transport by the rat renal proximal tubular cell. *Life Sci.* 55:839–45

26. Cai H, Puschett DB, Guan S, Batuman V, Puschett JB. 1995. Phosphate transport

inhibition by KW-3902, an adenosine A_1 receptor antagonist, is mediated by cyclic adenosine monophosphate. *Am. J. Kidney Dis.* 26:825–30

27. Yamaguchi S, Umemura S, Tamura K, Iwamoto T, Nyui N, et al. 1995. Adenosine A_1 receptor mRNA in microdissected rat nephron segments. *Hypertension* 26:1181–85

28. Smith JA, Sivaprasadarao A, Munsey TS, Bowmer CJ, Yates MS. 2001. Immunolocalisation of adenosine A_1 receptors in the rat kidney. *Biochem. Pharmacol.* 61:237–44

29. Jackson EK. 2001. Diuretics. In *Goodman and Gilman's The Pharmacological Basis of Therapeutics*, ed. JG Hardman, LE Limbird, Chpt. 29, pp. 757–87. New York: McGraw-Hill. 10th ed.

30. Kost CK Jr, Herzer WA, Rominski BR, Mi Z, Jackson EK. 2000. Diuretic response to adenosine A_1 receptor blockade in normotensive and spontaneously hypertensive rats: role of pertussis toxin-sensitive G-proteins. *J. Pharmacol. Exp. Ther.* 292:752–60

31. Kuan CJ, Herzer WA, Jackson EK. 1993. Cardiovascular and renal effects of blocking A_1 adenosine receptors. *J. Cardiovasc. Pharmacol.* 21:822–28

32. Wilcox CS, Welch WJ, Schreiner GF, Belardinelli L. 1999. Natriuretic and diuretic actions of a highly selective adenosine A_1 receptor antagonist. *J. Am. Soc. Nephrol.* 10:714–20

33. Gottlieb SS, Skettino SL, Wolff A, Beckman E, Fisher ML, et al. 2000. Effects of BG9719 (CVT-124), an A_1-adenosine receptor antagonist, and furosemide on glomerular filtration rate and natriuresis in patients with congestive heart failure. *J. Am. Coll. Cardiol.* 35:56–59

34. Knight RJ, Collis MG, Yates MS, Bowmer CJ. 1991. Amelioration of cisplatin-induced acute renal failure with 8-cyclopentyl-1,3-dipropylxanthine. *Br. J. Pharmacol.* 104:1062–68

35. Nagashima K, Kusaka H, Karasawa A.

1995. Protective effects of KW-3902, an adenosine A_1-receptor antagonist, against cisplatin-induced acute renal failure in rats. *Jpn. J. Pharmacol.* 67:349–57

36. Suzuki F, Shimada J, Mizumoto H, Karasawa A, Kubo K, et al. 1992. Adenosine A_1 antagonists. 2. Structure-activity relationships of diuretic activities and protective effects against acute renal failure. *J. Med. Chem.* 35:3066–75

37. Yao K, Kusaka H, Sato K, Karasawa A. 1994. Protective effects of KW-3902, a novel adenosine A_1-receptor antagonist, against gentamicin-induced acute renal failure in rats. *Jpn. J. Pharmacol.* 65:167–70

38. Nagashima K, Kusaka H, Sato K, Karasawa A. 1994. Effects of KW-3902, a novel adenosine A_1-receptor antagonist, on cephaloridine-induced acute renal failure in rats. *Jpn. J. Pharmacol.* 64:9–17

39. Kellett R, Bowmer CJ, Collis MG, Yates MS. 1989. Amelioration of glycerol-induced acute renal failure in the rat with 8-cyclopentyl-1,3-dipropylxanthine. *Br. J. Pharmacol.* 98:1066–74

40. Panjehshahin MR, Munsey TS, Collis MG, Bowmer CJ, Yates MS. 1992. Further characterization of the protective effect of 8-cyclopentyl-1,3-dipropylxanthine on glycerol-induced acute renal failure in the rat. *J. Pharm. Pharmacol.* 44:109–13

41. Shimada J, Suzuki F, Nonaka H, Karasawa A, Mizumoto H, et al. 1991. 8-(Dicyclopropylmethyl)-1,3-dipropylxanthine: a potent and selective adenosine A_1 antagonist with renal protective and diuretic activities. *J. Med. Chem.* 34:466–69

42. Arakawa K, Suzuki H, Naitoh M, Matsumoto A, Hayashi K, et al. 1996. Role of adenosine in the renal responses to contrast medium. *Kidney Int.* 49:1199–206

43. Erley CM, Heyne N, Burgert K, Langanke J, Risler T, Osswald H. 1997.

Prevention of radiocontrast-induced nephropathy by adenosine antagonists in rats with chronic nitric oxide deficiency. *J. Am. Soc. Nephrol.* 8:1125–32

44. Levens N, Beil M, Jarvis M. 1991. Renal actions of a new adenosine agonist, CGS 21680A selective for the A_2 receptor. *J. Pharmacol. Exp. Ther.* 257:1005–12

45. Levens N, Beil M, Schulz R. 1991. Intrarenal actions of the new adenosine agonist CGS 21680A, selective for the A_2 receptor. *J. Pharmacol. Exp. Ther.* 257:1013–19

46. Zou AP, Nithipatikom K, Li PL, Cowley AW Jr. 1999. Role of renal medullary adenosine in the control of blood flow and sodium excretion. *Am. J. Physiol. Regul. Integr. Comp. Physiol.* 276:R790–98

47. Cronstein BN. 1994. Adenosine, an endogenous anti-inflammatory agent. *J. Appl. Physiol.* 76:5–13

48. Nolte D, Lorenzen A, Lehr HA, Zimmer FJ, Klotz KN, Messmer K. 1992. Reduction of postischemic leukocyte-endothelium interaction by adenosine via A_2 receptor. *Naunyn Schmiedebergs Arch. Pharmacol.* 346:234–37

49. Okusa MD, Linden J, Macdonald T, Huang L. 1999. Selective A_{2A} adenosine receptor activation reduces ischemia-reperfusion injury in rat kidney. *Am. J. Physiol. Renal Physiol.* 277:F404–12

50. Jackson EK, Zhu C, Tofovic SP. 2002. Expression of adenosine receptors in the preglomerular microcirculation. *Am. J. Physiol. Renal Physiol.* 283:F41–51

51. Dubey RK, Gillespie DG, Mi Z, Jackson EK. 1997. Exogenous and endogenous adenosine inhibits fetal calf serum-induced growth of rat glomerular mesangial cells via A_{2b} receptors. *Hypertension* 30:509

52. Dubey RK, Gillespie DG, Mi Z, Jackson EK. 1998. Adenosine inhibits growth of human aortic smooth muscle cells via A_{2B} receptors. *Hypertension* 31:516–21

53. Dubey RK, Gillespie DG, Jackson EK.

1999. Adenosine inhibits collagen and total protein synthesis in vascular smooth muscle cells. *Hypertension* 33:190–94

54. Dubey RK, Gillespie D, Mi Z, Suzuki F, Jackson EK. 1996. Smooth muscle cell-derived adenosine inhibits cell growth. *Hypertension* 27:766–63

55. Dubey RK, Gillespie D, Osaka K, Suzuki F, Jackson EK. 1996. Adenosine inhibits growth of rat aortic smooth muscle cells: possible role of A_{2b} receptor. *Hypertension* 27:786–93

56. Dubey RK, Gillespie DG, Shue H, Jackson EK. 2000. A_{2B} receptors mediate antimitogenesis in vascular smooth muscle cells. *Hypertension* 35:267–72

57. Schrader J. 1991. Formation and metabolism of adenosine and adenine nucleotides in cardiac tissue. In *Adenosine and Adenine Nucleotides as Regulators of Cellular Function*, ed. JW Phillis, pp. 55–69. Boca Raton: CRC Press

58. Cass CE, Young JD, Baldwin SA. 1998. Recent advances in the molecular biology of nucleoside transporters of mammalian cells. *Biochem. Cell Biol.* 76:761–70

59. Pallone TL, Silldorff EP, Turner MR. 1998. Intrarenal blood flow: microvascular anatomy and the regulation of medullary perfusion. *Clin. Exp. Pharmacol. Physiol.* 25:383–92

60. Baudouin-Legros M, Badou A, Paulais M, Hammet M, Teulon J. 1995. Hypertonic NaCl enhances adenosine release and hormonal cAMP production in mouse thick ascending limb. *Am. J. Physiol. Renal Physiol.* 269:F103–9

61. Katholi RE, Taylor GJ, McCann WP, Woods WT Jr, Womach KA, et al. 1995. Nephrotoxicity from contrast media: attenuation with theophylline. *Radiology* 195:17–22

62. Siragy HM, Linden J. 1996. Sodium intake markedly alters renal interstitial fluid adenosine. *Hypertension* 27:404–7

63. Zou A-P, Wu F, Li P-L, Cowley AW Jr.

1999. Effect of chronic salt loading on adenosine metabolism and receptor expression in renal cortex and medulla in rats. *Hypertension* 33:511–16

64. Miller WL, Thomas RA, Berne RM, Rubio R. 1978. Adenosine production in the ischemic kidney. *Circ. Res.* 43:390–97

65. Osswald H, Schmitz HJ, Kemper R. 1977. Tissue content of adenosine, inosine and hypoxanthine in the rat kidney after ischemia and postischemic recirculation. *Pflügers Arch.* 371:45–49

66. Ramos-Salazar A, Baines AD. 1986. Role of 5'-nucleotidase in adenosine-mediated renal vasoconstriction during hypoxia. *J. Pharmacol. Exp. Ther.* 236: 494–99

67. Pflueger AC, Schenk F, Osswald H. 1995. Increased sensitivity of the renal vasculature to adenosine in streptozotocin-induced diabetes mellitus rats. *Am. J. Physiol. Renal Physiol.* 269: F529–35

68. Sakai K, Akima M, Nabata H. 1979. A possible purinergic mechanism for reactive ischemia in isolated, cross-circulated rat kidney. *Jpn. J. Pharmacol.* 29:235–42

69. Inscho EW. 2001. P2 receptors in the regulation of renal microvascular function. *Am. J. Physiol. Renal Physiol.* 280:F927–44

70. Schwiebert EM, Kishore BK. 2001. Extracellular nucleotide signaling along the renal epithelium. *Am. J. Physiol. Renal Physiol.* 280:F945–63

71. Pearson JD, Gordon JL. 1979. Vascular endothelial and smooth muscle cells in culture selectively release adenine nucleotides. *Nature* 281:384–86

72. LeRoy EC, Ager A, Gordon JL. 1984. Effects of neutrophil elastase and other proteases on porcine aortic endothelial prostaglandin I₂ production, adenine nucleotide release, and responses to vasoactive agents. *J. Clin. Invest.* 74:1003–10

73. Gordon EL, Pearson JD, Dickinson ES,

Moreau D, Slakey LL. 1989. The hydrolysis of extracellular adenine nucleotides by arterial smooth muscle cells. Regulation of adenosine production at the cell surface. *J. Biol. Chem.* 264:18986–95

74. Gordon EL, Pearson JD, Slakey LL. 1986. The hydrolysis of extracellular adenine nucleotides by cultured endothelial cells from pig aorta. Feed-forward inhibition of adenosine production at the cell surface. *J. Biol. Chem.* 261:15496–507

75. Pearson JD, Carleton JS, Gordon JL. 1980. Metabolism of adenine nucleotides by ectoenzymes of vascular endothelial and smooth-muscle cells in culture. *Biochem. J.* 190:421–29

76. Lloyd HG, Deussen A, Wuppermann H, Schrader J. 1988. The transmethylation pathway as a source for adenosine in the isolated guinea-pig heart. *Biochem. J.* 252:489–94

77. Deussen A, Lloyd HG, Schrader J. 1989. Contribution of *S*-adenosylhomocysteine to cardiac adenosine formation. *J. Mol. Cell Cardiol.* 21:773–82

78. Kroll K, Decking UK, Dreikorn K, Schrader J. 1993. Rapid turnover of the AMP-adenosine metabolic cycle in the guinea pig heart. *Circ. Res.* 73:846–56

79. Misumi Y, Ogata S, Hirose S, Ikehara Y. 1990. Primary structure of rat liver 5'-nucleotidase deduced from the cDNA presence of the COOH-terminal hydrophobic domain for possible post-translational modification by glycophospholipids. *J. Biol. Chem.* 265:2178–83

80. Pearson JD, Coade SB, Cusack NJ. 1985. Characterization of ectonucleotidases on vascular smooth muscle cells. *Biochem. J.* 230:503–7

81. Zimmermann H. 1992. 5'-Nucleotidase: molecular structure and functional aspects. *Biochem. J.* 285:345–65

82. Davoren PR, Sutherland EW, Maxwell AM. 1963. The effect of *l*-epinephrine and other agents on the synthesis and release of adenosine 3',5'-phosphate

by whole pigeon erythrocytes. *J. Biol. Chem.* 238:3009–15

83. Alberts B, Bray D, Lewis J, Raff M, Roberts K, Watson JD. 1989. In *Molecular Biology of the Cell*, pp. 825–26. New York/London: Garland. 2nd ed.

84. Cramer H, Lindl T. 1974. Release of cyclic AMP from rat superior cervical ganglia after stimulation of synthesis in vitro. *Nature* 249:380–82

85. Doore BJ, Bashor MM, Spitzer N, Mawe RC, Saier MH Jr. 1975. Regulation of adenosine 3':5'-monophosphate efflux from rat glioma cells in culture. *J. Biol. Chem.* 250:4371–72

86. Kelly LA, Butcher RW. 1974. The effects of epinephrine and prostaglandin E-1 on cyclic adenosine 3':5'-monophosphate levels in WI-38 fibroblasts. *J. Biol. Chem.* 249:3098–102

87. Kuster J, Zapf J, Jakob A. 1973. Effects of hormones on cyclic AMP release in perfused rat livers. *FEBS Lett.* 32:73–77

88. O'Brien JA, Strange RC. 1975. The release of adenosine 3':5'-cyclic monophosphate from the isolated perfused rat heart. *Biochem. J.* 152:429–32

89. Zumstein P, Zapf J, Froesch ER. 1974. Effects of hormones on cyclic AMP release from rat adipose tissue in vitro. *FEBS Lett.* 49:65–69

90. Finnegan RB, Carey GB. 1998. Characterization of cyclic AMP efflux from swine adipocytes in vitro. *Obesity Res.* 6:292–98

91. Barber R, Butcher RW. 1981. The quantitative relationship between intracellular concentration and egress of cyclic AMP from cultured cells. *Mol. Pharmacol.* 19:38–43

92. King CD, Mayer SE. 1974. Inhibition of egress of adenosine 3',5'-monophosphate from pigeon erythrocytes. *Mol. Pharmacol.* 10:941–53

93. Fehr TF, Dickinson ES, Goldman SJ, Slakey LL. 1990. Cyclic AMP efflux is regulated by occupancy of the adeno-

sine receptor in pig aortic smooth muscle cells. *J. Biol. Chem.* 265:10974–80

94. Rindler MJ, Bashor MM, Spitzer N, Saier MH. 1978. Regulation of adenosine 3',5'-monophosphate efflux from animal cells. *J. Biol. Chem.* 253:5431–36

95. Brunton LL, Mayer SE. 1979. Extrusion of cAMP from pigeon erythrocytes. *J. Biol. Chem.* 254:9714–20

96. van Aubel RAMH, Smeets PHE, Peters JGP, Bindels RJM, Russel FGM. 2002. The *MRP4/ABCC4* gene encodes a novel apical organic anion transporter in human kidney proximal tubules: putative efflux pump for urinary cAMP and cGMP. *J. Am. Soc. Nephrol.* 13:595–603

97. Sekine T, Watanabe N, Hosoyamada M, Kanai Y, Endou H. 1997. Expression cloning and characterization of a novel multispecific organic anion transporter. *J. Biol. Chem.* 272:18526–29

98. Hanoune J, Defer N. 2001. Regulation and role of adenylyl cyclase isoforms. 2001. *Annu. Rev. Pharmacol. Toxicol.* 41:145–74

99. Soderling SH, Beavo JA. 2000. Regulation of cAMP and cGMP signaling: new phosphodiesterases and new functions. *Curr. Opin. Cell Biol.* 12:174–79

100. Moser GH, Schrader J, Deussen A. 1989. Turnover of adenosine in plasma of human and dog blood. *Am. J. Physiol. Cell Physiol.* 256:C799–806

101. Krupinski J, Coussen F, Bakalyar HA, Tang W-J, Feinstein PG, et al. 1989. Adenylyl cyclase amino acid sequence: possible channel- or transporter-like structure. *Science* 244:1558–64

102. Zimmermann H. 2001. Ecto-nucleotidases. In *Handbook of Experimental Pharmacology; Purinergic and Pyrimidinergic Signalling I: Molecular, Nervous and Urogenitary System Function*, ed. MP Abbracchio, M Williams, pp. 209–50. Berlin: Springer-Verlag Vol. 151/I

103. Gorin E, Brenner T. 1976. Extracellular metabolism of cyclic AMP. *Biochim. Biophys. Acta* 451:20–28

104. Smoake JA, McMahon KL, Wright RK, Solomon SS. 1981. Hormonally sensitive cyclic AMP phosphodiesterase in liver cells. An ecto-enzyme. *J. Biol. Chem.* 256:8531–35

105. Rosenberg PA, Dichter MA. 1989. Extracellular cyclic AMP accumulation and degradation in rat cerebral cortex in dissociated cell culture. *J. Neurosci.* 9: 2654–63

106. Kather H. 1990. Beta-adrenergic stimulation of adenine nucleotide catabolism and purine release in human adipocytes. *J. Clin. Invest.* 85:106–14

107. Dubey RK, Mi Z, Gillespie DG, Jackson EK. 1996. cyclic AMP-adenosine pathway inhibits vascular smooth muscle cell growth. *Hypertension* 28:765–71

108. Dubey RK, Gillespie DG, Mi Z, Jackson EK. 2000. Cardiac fibroblasts express the cAMP-adenosine pathway. *Hypertension* 36:337–42

109. Beavo JA, Reifsnyder DH. 1990. Primary sequence of cyclic nucleotide phosphodiesterase isozymes and the design of selective inhibitors. *Trends Pharmacol. Sci.* 11:150–55

110. Tofovic SP, Branch KR, Oliver RD, Magee WD, Jackson EK. 1991. Caffeine potentiates vasodilator-induced renin release. *J. Pharmacol. Exp. Ther.* 256:850–60

111. Daly JW, Jacobson KA. Adenosine receptors: selective agonists and antagonists. In *Adenosine and Adenine Nucleotides: From Molecular Biology to Integrative Physiology*, ed. L Belardinelli, A Pelleg, pp. 157–66. Boston: Kluwer

112. Mi Z, Jackson EK. 1995. Metabolism of exogenous cyclic AMP to adenosine in the rat kidney. *J. Pharmacol. Exp. Ther.* 273:728–33

113. Zacher LA, Carey GB. 1999. cAMP metabolism by swine adipocyte microsomal and plasma membranes. *Comp. Biochem. Physiol. B* 124:61–71

114. Dubey RK, Gillespie DG, Mi Z, Jackson EK. 2001. Endogenous cyclic AMP-adenosine pathway regulates cardiac fibroblast growth. *Hypertension* 37: 1095–100

115. Rosenberg PA, Knowles R, Knowles KP, Li Y. 1994. β-Adrenergic receptor-mediated regulation of extracellular adenosine in cerebral cortex in culture. *J. Neurosci.* 14:2953–65

116. Rosenberg PA, Li Y. 1995. Adenylyl cyclase activation underlies intracellular cyclic AMP accumulation, cyclic AMP transport, and extracellular adenosine accumulation evoked by β-adrenergic receptor stimulation in mixed cultures of neurons and astrocytes derived from rat cerebral cortex. *Brain Res.* 692:227–32

117. Rosenberg PA, Li Y. 1996. Forskolin evokes extracellular adenosine accumulation in rat cortical cultures. *Neurosci. Lett.* 211:49–52

118. Brundege JM, Diao L, Proctor WR, Dunwiddie TV. 1997. The role of cAMP as a precursor of extracellular adenosine in the rat hippocampus. *Neuropharmacology* 36:1201–10

119. Hong KW, Shin HK, Kim HH, Choi JM, Rhim BY, et al. 1999. Metabolism of cAMP to adenosine: role in vasodilation of rat pial artery in response to hypotension. *Am. J. Physiol. Heart Circ. Physiol.* 276:H379–82

120. Kather H. 1990. Beta-adrenergic stimulation of adenine nucleotide catabolism and purine release in human adipocytes. *J. Clin. Invest.* 85:106–14

121. Mi Z, Jackson EK. 1998. Evidence for an endogenous cAMP-adenosine pathway in the rat kidney. *J. Pharmacol. Exp. Ther.* 287:926–30

122. Jackson EK, Mi Z, Gillespie DG, Dubey RK. 1997. Metabolism of cAMP to adenosine in the renal vasculature. *J. Pharmacol. Exp. Ther.* 283:177–82

123. Jackson EK, Mi Z. 2000. Preglomerular microcirculation expresses the cyclic AMP-adenosine pathway. *J. Pharmacol. Exp. Ther.* 295:23–28

124. Dubey RK, Gillespie DG, Mi Z, Jackson EK. 1997. cyclic AMP-adenosine pathway inhibits glomerular mesangial cell growth. *Hypertension* 30:506

125. Sakamoto T, Chen C, Lokhandwala MF. 1994. Attenuation of adenylate cyclase-induced increases in renal sodium excretion by the dopamine D-2 receptor agonist SK&F 89124. *J. Auton. Pharmacol.* 14:295–306

126. Mi Z, Herzer WA, Zhang Y, Jackson EK. 1994. 3-isobutyl-1-methylxanthine decreases renal cortical interstitial levels of adenosine and inosine. *Life Sci.* 54: PL277–82

127. Radziuk J, Barron P, Najm H, Davies J. 1993. The effect of systemic venous drainage of the pancreas on insulin sensitivity in dogs. *J. Clin. Invest.* 92:1713–21

128. Taborsky GJ Jr, Ahren B, Havel PJ. 1998. Autonomic mediation of glucagon secretion during hypoglycemia: implications for impaired alpha-cell responses in type 1 diabetes. *Diabetes* 47:995–1005

129. Houslay MD. 1986. Insulin, glucagon and the receptor-mediated control of cyclic AMP concentrations in liver. Twenty-second Colworth medal lecture. *Biochem. Soc. Trans.* 14:183–93

130. Broadus AE, Kaminsky NI, Northcutt RC, Hardman JG, Sutherland EW, et al. 1970. Effects of glucagon on adenosine 3′,5′-monophosphate and guanosine 3′,5′-monophosphate in human plasma and urine. *J. Clin. Invest.* 49:2237–45

131. Kuster J, Zapf J, Jakob A. 1973. Effects of hormones on cyclic AMP release in perfused rat livers. *FEBS Lett.* 32:73–77

132. Broadus AE, Kaminsky NI, Hardman JG, Sutherland EW, Liddle GW. 1970. Kinetic parameters and renal clearances of plasma adenosine 3′,5′-mono-phosphate and guanosine 3′,5′-monophosphate in man. *J. Clin. Invest.* 49: 2222–36

133. Jackson EK. 2002. A_1 receptor antagonists as diuretic/natriuretic agents. *Drugs Future* 27:1057–69

134. Lee HT, Emala CW. 2002. Characterization of adenosine receptors in human kidney proximal tubule (HK-2) cells. *Exp. Nephrol.* 10:383–92

135. Jackson EK, Mi Z, Zhu C, Herzer WA, Dubey RK. 2003. The pancreatohepatorenal cAMP-adenosine pathway. A potential endocrine system linking the pancreas, liver, and kidney. *Hypertension* (Abstr.) In press

135a. Ma H, Ling BN. 1996. Luminal adenosine receptors regulate amiloride-sensitive Na^+ channels in A6 distal nephron cells. *Am. J. Physiol. Renal Physiol.* 270: F798–805

136. Dubey RK, Gillespie DG, Jackson EK. 1998. cyclic AMP-adenosine pathway induces nitric oxide synthesis in aortic smooth muscle cells. *Hypertension* 31: 296–302

137. Dubey RK, Gillespie DG, Mi Z, Rosselli M, Keller PJ, Jackson EK. 2000. Estradiol inhibits smooth muscle cell growth in part by activating the cAMP-adenosine pathway. *Hypertension* 35:262–66

138. Dubey RK, Gillespie DG, Jackson EK. 1998. Adenosine inhibits collagen and protein synthesis in cardiac fibroblasts: potential role of A_{2B} receptors. *Hypertension* 31:943–48

139. Dubey RK, Gillespie DG, Mi Z, Jackson EK. 1997. Exogenous and endogenous adenosine inhibits fetal calf serum-induced growth of rat cardiac fibroblasts: role of A_{2B} receptors. *Circulation* 96: 2656–66

140. Dubey RK, Gillespie DG, Zacharia LC, Mi Z, Jackson EK. 2001. A_{2B} receptors mediate the antimitogenic effects of adenosine in cardiac fibroblasts. *Hypertension* 37:716–21

141. Friis UG, Jensen BL, Hansen PB, Andreasen D, Skott O. 2000. Exocytosis and endocytosis in juxtaglomerular cells. *Acta Physiol. Scand.* 168:95–99

142. Hackenthal E, Paul M, Ganten D, Taugner R. 1990. Morphology, physiology, and molecular biology of renin secretion. *Physiol. Rev.* 70:1067–116

143. Tagawa H, Vander AJ. 1970. Effects of adenosine compounds on renal function and renin secretion in dogs. *Circ. Res.* 26:327–38

144. Deray G, Branch RA, Herzer WA, Ohnishi A, Jackson EK. 1987. Adenosine inhibits β-adrenoceptor but not DBcAMP-induced renin release. *Am. J. Physiol. Renal Physiol.* 252:F46–52

145. Deray G, Branch RA, Ohnishi A, Jackson EK. 1989. Adenosine inhibits renin release induced by suprarenal aortic constriction and prostacyclin. *Naunyn-Schmiedebergs Arch. Pharmacol.* 339:590–95

146. Langård O, Holdaas H, Eide I, Kiil F. 1983. Conditions for augmentation of renin release by theophylline. *Scand. J. Clin. Lab Invest.* 43:9–14

147. Reid IA, Stockigt JR, Goldfien A, Ganong WF. 1972. Stimulation of renin secretion in dogs by theophylline. *Eur. J. Pharmacol.* 17:325–32

148. Cannon ME, Twu BM, Yang CS, Hsu CH. 1989. The effect of theophylline and cyclic adenosine 3′,5′-monophosphate on renin release by afferent arterioles. *J. Hypertens.* 7:569–76

149. Viskoper RJ, Maxwell MH, Lupu AN, Rosenfeld S. 1977. Renin stimulation by isoproterenol and theophylline in the isolated perfused kidney. *Am. J. Physiol. Renal Physiol.* 232:F248–53

150. Peart WS, Quesada T, Tenyl I. 1975. The effects of cyclic adenosine 3′,5′-monophosphate and guanosine 3′,5′-monophosphate and theophylline on renin secretion in the isolated perfused kidney of the rat. *Br. J. Pharmacol.* 54:55–60

151. Deray G, Branch RA, Jackson EK. 1989. Methylxanthines augment the renin response to suprarenal-aortic constriction. *Naunyn-Schmiedebergs Arch. Pharmacol.* 339:690–96

152. Paul S, Jackson EK, Robertson D, Branch RA, Biaggioni I. 1989. Caffeine potentiates the renin response to furosemide in rats. Evidence for a regulatory role of endogenous adenosine. *J. Pharmacol. Exp. Ther.* 251:183–87

153. Tseng CJ, Kuan CJ, Chu H, Tung CS. 1993. Effect of caffeine treatment on plasma renin activity and angiotensin I concentrations in rats on a low sodium diet. *Life. Sci.* 52:883–90

154. Brown NJ, Porter J, Ryder D, Branch RA. 1991. Caffeine potentiates the renin response to diazoxide in man. Evidence for a regulatory role of endogenous adenosine. *J. Pharmacol. Exp. Ther.* 256:56–61

155. Kuan CJ, Wells JN, Jackson EK. 1989. Endogenous adenosine restrains renin release during sodium restriction. *J. Pharmacol. Exp. Ther.* 249:110–16

156. Kuan CJ, Wells JN, Jackson EK. 1990. Endogenous adenosine restrains renin release in conscious rats. *Circ. Res.* 66:637–46

157. Pfeifer CA, Suzuki F, Jackson EK. 1995. Selective A_1 adenosine receptor antagonism augments β-adrenergic-induced renin release *in vivo*. *Am. J. Physiol. Renal Physiol.* 269:F469–79

158. Balakrishnan VS, Coles GA, Williams JD. 1993. A potential role for endogenous adenosine in control of human glomerular and tubular function. *Am. J. Physiol. Renal Physiol.* 265:F504–10

159. van Buren M, Bijlsma JA, Boer P, van Rijn HJM, Koomans HA. 1993. Natriuretic and hypotensive effect of adenosine-1 blockade in essential hypertension. *Hypertension* 22:728–34

160. Albinus M, Finkbeiner E, Sosath B, Osswald H. 1998. Isolated superfused juxtaglomerular cells from rat kidney: a

model for study of renin secretion. *Am. J. Physiol. Renal Physiol.* 275:F991–97

161. Mi Z, Jackson EK. 1999. Effects of α- and β-adrenoceptor blockade on purine secretion induced by sympathetic nerve stimulation in the rat kidney. *J. Pharmacol. Exp. Ther.* 288:295–301

162. Friedlander G, Couette S, Coureau C, Amiel C. 1992. Mechanisms whereby extracellular adenosine 3′,5′-monophosphate inhibits phosphate transport in cultured opossum kidney cells and in rat kidney. Physiological implication. *J. Clin. Invest.* 90:848–58

163. Briffeuil P, Thu TH, Kolanowski J. 1996. A lack of direct action of glucagon on kidney metabolism, hemodynamics, and renal sodium handling in the dog. *Metab. Clin. Exp.* 45:383–88

164. Faix S, Leng L. 1997. The renal response of sheep to intraportal infusion of glucagon. *Exp. Physiol.* 82:1007–13

165. Gerich JE, Charles MA, Grodsky GM. 1976. Regulation of pancreatic insulin and glucagon secretion. *Annu. Rev. Physiol.* 38:353–88

166. Sperling MA. 1979. Glucagon: secretion and actions. *Adv. Exp. Med. Biol.* 124:29–61

167. Velliquette RA, Koletsky RJ, Ernsberger P. 2002. Plasma glucagon and free fatty acid responses to a glucose load in the obese spontaneous hypertensive rat (SHROB) model of metabolic syndrome X. *Exp. Biol. Med.* 227:164–70

168. Iannello S, Campione R, Belfiore F. 1998. Response of insulin, glucagon, lactate, and nonesterified fatty acids to glucose in visceral obesity with and without NIDDM: relationship to hypertension. *Mol. Genet. Metab.* 63:214–23

169. Foley JE, Anderson RC, Bell PA, Burkey BF, Deems RO, et al. 1997. Pharmacological strategies for reduction of lipid availability. *Ann. NY Acad. Sci.* 827:231–45

170. Thong FSL, Graham TE. 2002. The putative roles of adenosine in insulin- and exercise-mediated regulation of glucose transport and glycogen metabolism in skeletal muscle. *Can. J. Appl. Physiol.* 27:152–78

Annu. Rev. Physiol. 2004. 66:601–23
doi: 10.1146/annurev.physiol.66.032102.134711
Copyright © 2004 by Annual Reviews. All rights reserved
First published online as a Review in Advance on October 20, 2003

ALTERATIONS IN SP-B AND SP-C EXPRESSION IN NEONATAL LUNG DISEASE

Lawrence M. Nogee

Division of Neonatology, Department of Pediatrics, Johns Hopkins University School of Medicine, Baltimore, Maryland 21287-3200; email: lnogee@jhmi.edu

Key Words   respiratory distress syndrome, lung development, mutation, genetic basis of disease, interstitial lung disease

■ **Abstract**   The hydrophobic surfactant proteins, SP-B and SP-C, have important roles in surfactant function. The importance of these proteins in normal lung function is highlighted by the lung diseases associated with abnormalities in their expression. Mutations in the gene encoding SP-B result in severe, fatal neonatal lung disease, and mutations in the gene encoding SP-C are associated with chronic interstitial lung diseases in newborns, older children, and adults. This work reviews the current state of knowledge concerning the lung diseases associated with mutations in the SP-B and SP-C genes, and the potential roles of abnormal SP-B and SP-C expression and genetic variation in these genes in other lung diseases.

INTRODUCTION

Respiratory disease remains a common cause of neonatal morbidity and mortality, including the neonatal respiratory distress syndrome (RDS),[1] which occurs principally in premature infants (1, 2). The principal cause of RDS is a deficiency of pulmonary surfactant, the mixture of lipids and proteins needed to reduce alveolar surface tension and prevent end-expiratory atelectasis owing to pulmonary immaturity. Whereas surfactant phospholipids are critical in their ability to lower surface tension, specific proteins have been identified that also have important roles in surfactant function and homeostasis (3, 4). These include two glycoproteins, surfactant proteins A and D, which are part of the collectin family, and two smaller, very hydrophobic proteins, surfactant proteins B (SP-B) and SP-C (5, 6). Either SP-B or SP-C, when mixed with surfactant phospholipids, produces a surfactant

---

[1]Abbreviations: RDS, respiratory distress syndrome; SP-B, surfactant protein B; SP-C, surfactant protein C; proSP-B, surfactant protein B precursor protein; proSP-C, surfactant protein C precursor protein; SNP, single-nucleotide polymorphism; BAL, bronchoalveolar lavage; CFTR: cystic fibrosis transmembrane regulator; PAP, pulmonary alveolar proteinosis; GM-CSF, granulocyte macrophage colony stimulating factor.

preparation that readily lowers surface tension and is effective in treating RDS in experimental animals, and both SP-B and SP-C are found (in varying amounts) in commercially prepared, mammalian-derived exogenous surfactant preparations, such as Survanta™, Infasurf™, or Curosurf™, used to treat premature infants with RDS. The importance of the roles of the hydrophobic proteins in surfactant function and homeostasis is illustrated by the lung diseases resulting from abnormal expression of SP-B and SP-C due to mutations in the genes encoding these proteins. Herein we review current knowledge concerning the normal gene structures for SP-B and SP-C, allelic variants, and known mutations associated with human lung disease.

GENETICS AND PRODUCTION OF SP-B AND SP-C

SP-B is encoded by a single gene (the locus is referred to as *SFTPB*) spanning approximately 10 kilobases on the short arm of human chromosome 2 (7, 8). The gene contains 11 exons and is transcribed into an approximately 2000 base pair (bp) mRNA, with the approximately 800 bp 11th exon being untranslated (9, 10). The mRNA is translated into a 381-amino acid preprotein with the first 23 amino acids comprising a signal peptide that is cotranslationally removed. The remaining proprotein (proSP-B) is proteolytically processed at both the amino- and carboxy-terminal ends to yield the 79 amino acid mature SP-B peptide found in the airspaces, which corresponds to amino acids 201 (phenylalanine) to 279 (methionine) of the preprotein, encoded in exons 6 and 7 of the gene (10) (Figure 1, see color insert). ProSP-B is organized into three tandem domains having structural homology to saposins, proteins that interact with phospholipids and activate lysosomal hydrolases, with mature SP-B corresponding to the middle domain (11, 12). Most of the amino-terminal domain is removed first in a nonspecific step, with the subsequent removal of the carboxy terminus in a distal cellular compartment in a step thought to be specific for alveolar type II cells (13–17). Final proteolysis at the amino terminus, likely involving the cysteine protease, cathepsin H, yields the 79 amino acid (~8700 kDa) peptide found in the airspaces (18, 19). ProSP-B is post-translationally modified by N-linked glycosylation in the carboxy-terminal domain. A second glycosylation site is also present in the amino-terminal domain of proSP-B derived from SP-B alleles, which encode threonine at codon 131. Alternative splicing resulting in a 12 bp deletion at the beginning of exon 8 occurs in a small percentage of transcripts and could result in a 4 amino acid deletion from the carboxy-terminal domain of the proprotein (20–22). Mature SP-B forms oligomers in part through intermolecular disulfide bonds involving the cysteine at position 48 (10, 12, 23–25).

The roles of the amino- and carboxy-terminal proSP-B domains are not completely known. A role for the amino-terminal domain in the intracellular routing of proSP-B is supported by experiments involving deletion of this domain, which interfered with the routing of proSP-B to secretory granules (26, 27). The role of the carboxy-terminal domain is less well understood. This domain may not be essential, as genetically engineered mice that had both normal SP-B alleles inactivated and expressed a truncation mutant lacking the carboxy-terminal proSP-B domain, were viable and expressed mature SP-B (28). Increased amounts of disaturated phosphatidylcholine in the lung tissue and large lamellar bodies in the alveolar type II cells of these mice suggest a possible role in lipid metabolism for this domain.

A number of relatively frequent variations in the SP-B gene have been identified. A complex variable nucleotide tandem repeat region is located in intron 4, and a number of alleles have been characterized (29, 30). Single-nucleotide polymorphisms (SNPs) have also been identified, including an A/C transversion in the 5′ untranslated region at the −18 position from the transcription initiation site, an A/C transversion 8 bp 5′ to the exon 3 splice acceptor site, and a nonsynonymous C/T transition in codon 131 that results in encoding either threonine or isoleucine (9, 20, 31–34). This last polymorphism would alter a potential site for N-linked glycosylation and thus could have functional significance. Multiple other SNPs have also been identified in the 3′ untranslated region and introns. The functional significance of any of these SNPs is unknown.

SP-C is encoded by a single gene (referred to as *SFTPC*) on the short arm of human chromosome 8 that spans approximately 3500 bp (35, 36). The gene contains 6 exons and is transcribed into an ~900 bp mRNA, with the sixth exon being untranslated (9, 10). The mRNA is translated into a 191 or 197 amino acid proprotein (proSP-C), with mature SP-C encoded in exon 2 and corresponding to codons 24 to 58 of the mRNA (Figure 2, see color insert). ProSP-C is an integral membrane protein in a type II orientation, with the amino terminus directed toward the cytoplasm and with the region that corresponds to mature SP-C serving as the signal peptide and anchoring it in the membrane (37, 38). A domain in the amino-terminal portion of proSP-C appears to be critically important for proper routing and sorting of SP-C in the secretory pathway (39, 40). ProSP-C is processed first from the carboxy terminus and subsequently from the amino terminus in a distal cellular compartment; cathepsin H was recently implicated as being involved in proSP-C processing (38, 41–44). ProSP-C is extensively post-translationally modified, including palmitoylation of cysteine residues at positions 5 and 6 of the mature peptide, such that SP-C is a proteolipid (45–48). SP-C contains a stretch of valine residues that form a tight alpha helix, which is thought to imbed it in the membrane (49, 50). This valine-rich domain can also result in the polymerization of mature SP-C into amyloid-like fibrils (51–53).

Human SP-C mRNA is alternatively spliced at the beginning of exon 5 and yields two transcripts that differ by 18 nucleotides, with the larger transcript yielding a 6-amino acid insertion into proSP-C (54, 55). It is not known if both alternatively spliced transcripts are translated and whether there are any functional differences imparted by the 6-amino acid insertion. Such alternative splicing leading to this 6-amino acid insertion has not been observed in other species. Alternative splicing also occurs at the end of exon 5 distal to the termination codon, which results in either the insertion or deletion of 9 bp in the 3′ untranslated region (55).

Two SNPs altering the SP-C coding sequence have been identified, one in codon 138 (C or A) that encodes either threonine or asparagine, and another in codon 186 (G or A) that encodes either serine or asparagine at this location (9, 54, 55). The relative frequencies of these allelic variants, or any effects they may have on proSP-C processing or function, are unknown. Other variations observed among the reported SP-C cDNA and gene sequences may be due to allelic variation (35, 54–56). A list of known allelic variants for *SFTPB* and *SFTPC* may be found in on-line databases, such as that maintained by the National Center for Biotechnology Information (http://www.ncbi.nlm.nih.gov).

HEREDITARY SP-B DEFICIENCY

The inability to produce SP-B because of loss-of-function mutations on both alleles results in severe neonatal lung disease. The disease was first recognized in a family in which three full-term infants developed respiratory disease with symptoms and radiographic findings of surfactant deficiency, which proved fatal despite all interventions. Immunologic studies demonstrated a complete lack of SP-B protein in lung tissue and fluid of the affected infants, as well as undetectable levels of mRNA when assayed (57). Affected children in this kindred were subsequently shown to be homozygous for a frameshift mutation that accounted for the lack of SP-B, confirming SP-B deficiency as the basis for their disease (20). The same mutation was also found on five of six SP-B alleles of three other unrelated infants who had died from similar lung disease, indicating that this was a relatively common mutation responsible for the phenotype. The mutation involved the substitution of three bases for one in exon 4 of the gene, corresponding to codon 121 of the SP-B mRNA, and has been termed 121ins2 for the net two bp insertion (Figure 1).

Hereditary SP-B deficiency has since been reported as the basis for lung disease in over 75 children from unrelated families (31, 34, 58–69). The 121ins2 mutation has accounted for approximately two thirds of the mutant alleles identified to date, but numerous other mutations have been identified (Table 1). Many of the mutations identified have been nonsense or frameshift mutations that would completely preclude any SP-B production, but missense mutations or small in-frame deletions or insertions have also been identified that would allow for production of proSP-B. In these situations the proSP-B containing the altered amino acid sequences is not properly processed to mature SP-B (34).

The usual phenotype of SP-B deficiency is a full-term infant who develops diffuse parenchymal lung disease that clinically and radiographically resembles RDS in a premature infant (57, 70–73). The lung disease may be quite severe, with affected infants usually requiring mechanical ventilation, and often support with cardio-pulmonary by-pass when available. Not all affected infants have such severe lung disease initially, however, with many appearing to have milder respiratory symptoms, and X rays and a clinical picture consistent with more benign diagnoses such as transient tachypnea of the newborn (74). Factors that modify

TABLE 1 SP-B gene (SFTPB) mutations

Exon	Location in gene	Nucleotide substitution	Effect on protein	RE site created	RE site deleted	Reference
1	40	G > A	Nonsense	Alu I		
1	52	T > C	Missense: Leucine > Proline	Aci I		(34)
2	441	G > A	G > A		Ban I	(34)
2	457	Delete C	Frameshift		Cac 8I, Mwo I	(65)
2	469	T > C	Missense: Cysteine > Arginine	HhaI		(34)
2	504	G > A	Nonsense	Dde I		(34)
2	523	A > G	Splice–skip exon 2	Ban I		(34)
2	524	G > C	Splice–skip exon 2		Rsa I	
4	1454	Delete A	Splice–skip exon 4	Msp I		(34)
4	1486	T > G	Missense: Cysteine > Glycine	Mnl I		(34)
4	1549	C > GAA (121ins2)	Frameshift	Sfu I, Taq I		(20)
4	1552	Delete C	Frameshift			(67)
5	1553	Delete T	Frameshift	Msp I	Bst NI	(31, 34)
5	2415	Insert AA	Frameshift			(66)
5	2417	G > A	Missense: Glycine > Serine	Alw NI		(64)
5	2479	G > T	Create new splice site in exon			(68)
5	2595	Delete A	Frameshift	Dra III		
6	2913	G > A	Splice		Hph I	(34)
7	4334	C > A	Create new splice site	Dde I		
7	4378	T > C	Missense: Cysteine > Arginine	Hha I		(34)

(Continued)

TABLE 1   (*Continued*)

Exon	Location in gene	Nucleotide substitution	Effect on protein	RE site created	RE site deleted	Reference
7	4381	C > T	Missense: Arginine > Cysteine		Bst UI	(60)
7	4419	C > A	Nonsense		Bst NI	(34)
7	4429	C > T	Missense: Arginine > Cysteine		Hae II	(34)
7	4440	Duplicate 5 bp	Frameshift			
8	4723	G >A	Nonsense			
8	4729	Duplicate 18 bp	Insert 6 amino acids			
8	4730	C > T	Nonsense		Bst UI	
8	4751	T > G	Missense: Cysteine > Glycine			
9	6114	Insert 3 bp	Insert Proline			(34)
9	6119	Delete 12bp	Delete 4 amino acids		Pvu II	(34)
9	6122	T > C	Missense: Leucine > Proline		Pvu II	

Gene sequence numbering as per Reference (7). RE: restriction endonuclease.

the course are not clear, as even infants homozygous for null mutations have had relatively mild initial disease. However the disease resulting from SP-B deficiency is progressive, with escalating need for respiratory support and progressive alveolar and interstitial infiltrates seen on chest radiographs. Transient and inconsistent responses have been observed to therapeutic interventions such as exogenous surfactant administration and high-dose corticosteroids, and the disease is usually rapidly fatal, with the only effective treatment being lung transplantation (74–76).

Milder disease allowing for more prolonged survival has been observed in a few children with mutations that allow for some SP-B production, and has been termed partial deficiency (60, 68). Mature SP-B levels in the lung tissue of such infants have generally been approximately 5 to 10% of control values. Presumably the residual amount of mature SP-B mitigates the severity of lung disease, but the possibility that some retained function of proSP-B may also be important cannot be excluded. Children with mutations on one allele with decreased levels of SP-B but who eventually recovered from their lung disease as SP-B levels increased have also been recognized (64). Such transient deficiency may thus result from haploinsufficiency in combination with some other perinatal event that further reduces SP-B expression. The observations of partial and transient deficiency thus indicate that there is a critical level of SP-B needed for proper lung function and that decreased expression below this level results in lung disease. This hypothesis is supported by experiments with genetically engineered mice. Adult mice that could produce SP-B only under control of the tetracycline regulatable promoter developed lung disease when doxycycline was withdrawn and SP-B levels fell below approximately 25% of that observed in control animals; in addition, SP-B haploinsufficient mice were more susceptible to pulmonary oxygen toxicity than their wild-type littermates (77–79). Thus DNA sequence variants that alter SP-B expression could be important in the pathogenesis of other lung diseases such as RDS, and SP-B is thus a reasonable candidate gene for lung diseases beyond the neonatal period that involve surfactant dysfunction.

Hereditary SP-B deficiency has resulted from mutations involving the coding exons of the SP-B gene or mutations likely or proven to result in aberrant splicing. Mutations in the untranslated regions affecting gene transcription, mRNA stability, or splicing could also result in disease, but have not yet been identified. As SP-B oligomerizes in the airway, mutations in the region corresponding to mature SP-B that interfere with oligomerization or mature SP-B structure or function could also result in SP-B deficiency from a dominant-negative mechanism. Again, such a mechanism has not yet been identified, although is plausible given observations in genetically engineered animals. Expression of a construct that eliminated an intramolecular disulfide bridge between cysteines at positions 235 and 246 in transgenic mice resulted in neonatal lethality in some mice also expressing wild-type SP-B, with survival related to expression of the abnormal transgene, which indicated the feasibility of a dominant-negative mechanism resulting in SP-B deficiency (23). However, impaired processing of proSP-B was also implicated, as abnormal proSP-B containing the mutations accumulated in endoplasmic reticulum. Similarly, a synthetic peptide containing a mutation in the mature SP-B domain (R236C) was able to augment surface tension lowering normally in an in vitro system, suggesting that the disease in infants with this mutation also resulted from impaired processing of the abnormal proSP-B (80). Finally, SP-B deficiency could also result from mutations in other genes, such as those needed for the processing of proSP-B to mature SP-B, or for its secretion, and result in a similar phenotype. Such locus heterogeneity has not been demonstrated, but reduced SP-B expression has been observed in at least one large kindred in which no SP-B gene mutations were identified (21).

SP-B deficiency is an extremely rare disorder. The frequency of the 121ins2 mutation in the population has been estimated at approximately 1 in 1000 individuals in the United States (81, 82). The 121ins2 mutation has accounted for approximately 60% of mutant alleles identified and thus the carrier frequency for any SP-B mutation would be about 1 in 600. As SP-B deficiency is inherited in an autosomal-recessive fashion, the predicted disease incidence would thus be in the range of 1 in 1.5 million births. The frequency of the 121ins2 mutation, or any other

SP-B gene mutation, in or outside the United States has not been determined. The 121ins2 mutation has been identified principally in individuals of northern European descent. Mutations in unrelated individuals have also been identified in other ethnic groups. These include the 122delT (Middle Eastern), c.1043ins3 (Asian), c.479G > T (French-Canadian), and R295X (Hispanic) mutations (31, 34, 68). The frequency of these mutations in these subgroups is unknown, and estimates of disease frequency in other countries or subpopulations are not available.

Aside from the lack of SP-B, a number of changes in the metabolism of other surfactant components have been identified. The most striking of these is an apparent block in the processing of proSP-C to mature SP-C. The lung tissue and BAL fluid from SP-B-deficient infants contain large amounts of proteins that are recognized by antisera directed against either the amino-terminal epitopes of proSP-C or full-length proSP-C, and migrate at a higher molecular mass (between 6,000 and 12,000 kDa) than mature SP-C (83) (Figure 2). The presence of these abnormally migrating proSP-C-related proteins correlates with intense immunohistochemical staining of extracellular proteinaceous material in the airspaces observed in lung tissue from SP-B-deficient infants using the same antisera directed against proSP-C (34) (Figure 3, see color insert). These findings appear to result from incomplete processing of proSP-C, with secretion of the partially processed protein. To date, incompletely processed proSP-C has been observed only in association with SP-B deficiency and appears to be specific for the disorder.

The lung tissue and fluid from SP-B-deficient infants also contain a paucity of phosphatidylglycerol compared with samples from controls, although phosphatidylcholine content or metabolism has not necessarily been altered (76, 84). On ultrastructural exam, the type II cells of SP-B-deficient infants contain few if any normal-appearing lamellar bodies, the storage organelle for surfactant, and instead contain inclusions with multiple small vesicles and poorly packed lamellae (85). As the final processing steps of SP-B and SP-C take place in lamellar bodies, the inability to form this organelle could account for the incomplete processing of SP-C (10, 17).

Targeted disruption of the SP-B gene in mice results in perinatal lethality in the homozygous SP-B-deficient mice owing to their inability to inflate their lungs and establish respiration (86, 87). The lungs of such animals contain no SP-B, and the cellular phenotype is similar to that observed in human infants with aberrant processing of SP-C and a paucity of lamellar bodies. Adult mice heterozygous for the null allele have subtle abnormalities of pulmonary function and are more susceptible to pulmonary oxygen toxicity, supporting the hypothesis that haploinsufficiency for SP-B may predispose to lung disease (79). However, no abnormalities of pulmonary function could be demonstrated in one small study of adults heterozygous for the 121ins2 mutation, and such individuals generally do not have a history of lung disease (88).

Animal models have also proven useful for the evaluation of functional epitopes of proSP-B or mature SP-B by evaluating whether transgenic animals expressing SP-B gene variants can rescue the lethal SP-B-deficient phenotype. The viability of animals that expressed a SP-B gene construct encoding the amino-terminal domain and mature SP-B domains but lacking the carboxy-terminal domain in the SP-B null background suggests that nonsense mutations in the region encoding the carboxy-terminal domain might not result in lung disease (28). However, several such mutations have been identified in association with fatal lung disease, and it is likely that these nonsense mutation are also associated with reduced SP-B mRNA levels (89, 90), in contrast to the high level of expression resulting from the construct used in transgenic animals. Similarly, mice expressing a modified SP-B gene construct that substituted serine for cysteine at position 48 of the mature peptide, which is important for formation of an intermolecular disulfide bridge in dimeric SP-B, were also able to rescue the lethal SP-B phenotype, indicating that the formation of this disulfide bridge is not critical for proSP-B to be processed to mature SP-B or for mature SP-B function. However, the hysteresis of the lungs of mice expressing the mutant form of SP-B that could not form this disulfide bridge was reduced, as was the surface tension lowering ability of surfactant extracted from the lungs of such animals (24).

SP-C DEFICIENCY

When SP-C is mixed with surfactant phospholipids, it also forms an effective surfactant that rapidly lowers surface tension and can be used to treat animals with experimental RDS. A surfactant preparation combining phospholipids with a synthetic modified form of SP-C has been extensively studied (91–95). Given its importance in surfactant function, it would seem reasonable that a complete deficiency of SP-C would also result in severe neonatal lung disease resembling RDS. However, genetically engineered SP-C-deficient mice do not develop neonatal lung disease such that SP-C does not appear to be critical for normal neonatal adaptation, although the surfactant isolated from such animals is unstable at low volumes (96).

Abnormalities in SP-C expression have been linked to lung disease in older children and adults. Absent SP-C in BAL fluid and markedly reduced proSP-C expression in lung tissue were observed in a mother and two daughters with interstitial lung disease inherited in an apparent autosomal-dominant pattern (97). Analysis of the SP-C gene in this kindred did not reveal any deviations from the known SP-C sequence that would account for the lack of SP-C, and the underlying mechanism(s) responsible for the lack of SP-C and lung disease in these individuals remains unknown.

Subsequently, mutations in the SP-C gene have been clearly associated with lung disease. A substitution in the donor splice site of exon 4 of the SP-C gene that resulted in the in-frame skipping of an exon and a shortened SP-C mRNA and proprotein (that was also reduced in amount) was identified in a child with familial interstitial lung disease inherited in an autosomal-dominant fashion (98). The mutation segregated with disease, but was present on only one allele in affected individuals, with no sequence deviations observed on the other allele. Mature SP-C was also undetectable in the lung tissue from the affected child, suggesting a dominant-negative effect of the mutant allele on the normal allele. Such a mechanism is supported by the observations that proSP-C self-associates in the secretory pathway of the alveolar type II cell (39, 99) (Figure 2). Further support for this mechanism has come from in vitro experiments with constructs directing expression of wild-type SP-C or SP-C containing the exon 4 deletion transfected into A549 cells, a lung epithelial cell line. Transfected wild-type SP-C was observed in cytoplasmic vesicles in a diffuse pattern in the cells, whereas SP-C containing the exon 4 deletion was concentrated in perinuclear aggregates. Cotransfection with both wild-type and abnormal proSP-C resulted in a pattern similar to the mutant SP-C constructs, i.e., localized in perinuclear aggregates, supporting a dominant-negative mechanism (100).

A large kindred with familial pulmonary fibrosis in which a missense mutation (L188Q) was identified in the SP-C gene and segregated with disease has also been reported, further supporting the association of SP-C abnormalities and chronic lung disease (101). The onset of lung disease in the affected individuals in this kindred was highly variable, ranging from early infancy to the fifth decade, and some individuals with the mutation were symptom-free at the time of publication, consistent with incomplete penetrance. The histopathology of lung tissue from affected members also varied, including elements of both nonspecific interstitial pneumonitis and usual interstitial pneumonitis and pulmonary fibrosis. Several of the younger infants in this pedigree developed symptoms after a viral infection, indicating that environmental factors may be important in modifying the course of the disease. In support of genetic modifiers, while the initial strain of SP-C-deficient mice had relative subtle pulmonary abnormalities (96), when SP-C-deficient mice were generated in a different genetic background, severe lung disease characterized by progressive air space enlargement, macrophage infiltration, and pulmonary fibrosis developed as the mice aged (102). Identification of the modifier genes in these animals as well as environmental factors will be important in understanding the pathophysiology of the disease and in developing effective therapies.

The pathophysiology of lung disease associated with SP-C mutations is incompletely understood. The lack of mature SP-C due to a dominant-negative effect on SP-C metabolism may be important in terms of leading to alveolar instability with recurrent atelectasis, inflammation, and eventual fibrosis. However, all SP-C mutations associated with lung disease identified to date are ones that would result in production of an abnormal proSP-C molecule in contrast to precluding SP-C production (98, 101, 103). Given the hydrophobic nature of proSP-C and mature SP-C, it is also likely that the abnormal proSP-C resulting from the mutated allele is toxic to the alveolar epithelial cells. The perinuclear accumulation observed with transfection of constructs expressing mutant SP-C in vitro is consistent with aggresome formation, and an inability to clear such aggregates, contributing to type II cell injury (100, 104, 105). The potential toxicity of abnormal proSP-C was also demonstrated in experiments in which constructs containing the L188Q mutation, transfected into a murine lung epithelial cell line, caused LDH release, consistent

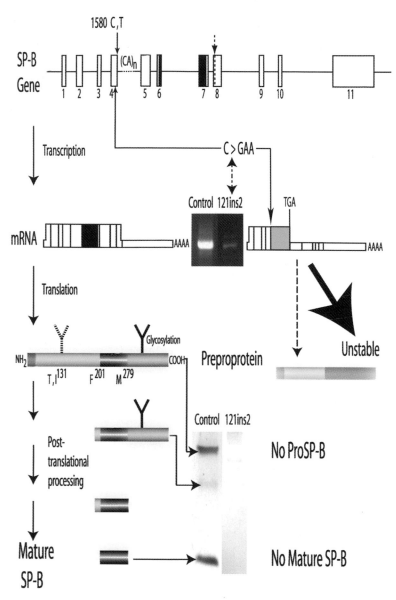

See legend on next page

Figure 1 SP-B gene, mRNA, proprotein and post-translational processing. The SP-B gene is depicted on top, with the exons represented by rectangles, introns represented by lines, and untranslated regions in the mRNA regions indicated by narrow rectangles. Mature SP-B is encoded in exons 6 and 7 (*black*), corresponding to codons 201 (phenylalanine, F) to 279 (methionine, M) of the mRNA. The locations of a single nucleotide polymorphism (C or T at genomic position 1580) in exon 4 that results in either isoleucine (I) or threonine (T) in codon 131 and a variable nucleotide tandem repeat region [$(CA)_n$] in intron 4 are indicated. A small arrow and dashed line indicate the site of alternative splicing at the beginning of exon 8. The normal transcription, translation, and post-translational processing of the SP-B gene translation product are outlined on the left and the effects of the net two base pair insertion resulting from the 121ins2 mutation are shown on the right. The net insertion alters the reading frame (*pink*) and introduces a new codon for the termination of translation after codon 214. The resulting portion of proprotein encoded after the frameshift would be unrelated to normal proSP-B (*pink*) and would not encode mature SP-B; the frameshifted transcript is also unstable leading to markedly reduced SP-B mRNA. The insets show the relative amounts of SP-B transcripts as analyzed by RT-PCR, and proSP-B and mature SP-B by Western blotting.

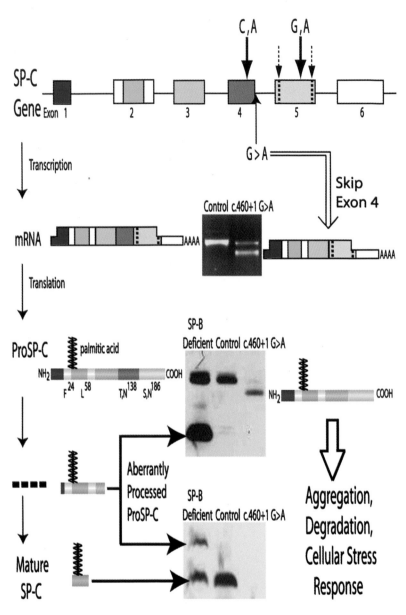

See legend on next page

Figure 2 SP-C gene, mRNA, proprotein, and post-translational processing. The SP-C gene is depicted on top, with the exons represented by rectangles, introns are represented by lines, and untranslated regions in the mRNA regions are indicated by narrow rectangles. Mature SP-C is encoded in exon 2 (*light blue*) corresponding to codons 24 (phenylalanine, F) to 58 (leucine, L) of the mRNA. Large arrows indicate the sites of single-nucleotide polymorphisms in exons 4 and 5; small arrows and dashed lines indicate the sites of alternative splicing in exon 5. The normal transcription, translation, and post-translational processing of the SP-C gene translation product are outlined on the left. The consequences of a G to A transition in the first base of intron 4 (c.460 + 1G > A) of one allele are shown on the right, with the skipping of exon 4 (*red*) in the mRNA. The insets show the relative sizes of SP-C transcripts as analyzed by RT-PCR, and proSP-C (*top*) and mature SP-C (*bottom*) as analyzed by Western blotting. With the c.460 + 1G > A mutation, a smaller transcript corresponding to the deletion of exon 4 is observed (along with the normal-sized transcript). ProSP-C from the mutant allele migrates at a lower molecular weight, normal proSP-C is reduced in amount compared with the control, and no mature SP-C is detected despite one normal allele. The dashed line at the left indicates the presumed site of the block in proSP-C processing associated with SP-B deficiency, with arrows pointing to the aberrantly processed proSP-C peptides in the insets.

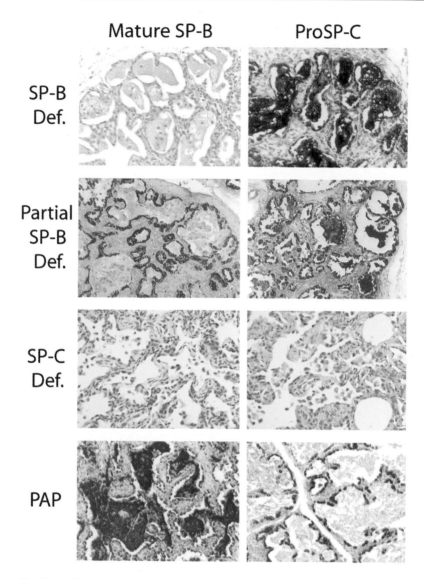

See legend on next page

Figure 3 Patterns of immunohistochemical staining of lung tissue for the surfactant proteins. Biopsy or autopsy lung sections were immuno-stained for mature SP-B (*left*) and proSPC (*right*). The top row is from a newborn compound heterozygote for two frameshift mutations (121ins2/134ins2). No SP-B staining is observed, but intense staining of the alveolar epithelium and extracellular material is observed with antisera directed against proSP-C. The second row is from an infant with partial SP-B deficiency. Positive staining for SP-B is observed, as well as for proSP-C in both the alveolar epithelium and the extracellular material. The child was homozygous for a splicing mutation in exon 5 of the SP-B gene. The third row is from an infant with a SP-C mutation on one allele that resulted in the skipping of exon 4. Reduced staining for both mature SP-B and proSP-C is observed. The bottom panels are from a neonate with respiratory failure and neonatal alveolar proteinosis due to an unknown mechanism. The proteinaceous material stains intensely for mature SP-B, but proSP-C staining is confined to the alveolar epithelium. Immuno-stained sections courtesy of S. Wert, Children's Hospital Medical Center, Cincinnati, Ohio.

with cytotoxicity (101). Collectively, these observations support a model whereby SP-C mutations cause lung disease owing to chronic cellular stress and injury resulting from an accumulation of misfolded proSP-C. Therefore, the disease related to SP-C mutations may be thought of as a conformational disease (106). In this model, the interstitial pneumonitis may be analogous to the hepatitis observed in some infants with α_1-antitrypsin deficiency, where there is accumulation of abnormal α_1-antitrypsin in the endoplasmic reticulum of the hepatocytes (107–109). The pulmonary fibrosis that develops in patients with SP-C mutations would be analogous to the hepatic cirrhosis observed in association with α_1-antitrypsin deficiency. Important unanswered questions concerning this model include how specific this process is for SP-C, or whether other proteins can misfold in the secretory pathway of the type II cell and result in similar pathology. Another question concerns what other factors, genetic or environmental, could either precipitate misfolding or facilitate proper folding of SP-C. In addition to producing and secreting surfactant, alveolar type II cells also serve as the progenitor cells for type I pneumocytes during development and following lung injury. Such hypertrophic type II cells markedly increase their expression of proSP-C. Misfolding of proSP-C during this process could result in ongoing lung injury or delayed healing from lung injury, indicating a possible role for SP-C in recovery from lung disease.

This model for the pathogenesis of lung disease due to SP-C mutations has therapeutic implications. Agents that act as chemical chaperones or promote proper folding or routing through the secretory pathway may have therapeutic value for patients with SP-C mutations or in disease resulting from misfolded SP-C or other proteins in the secretory pathway of the type II cell (110). One such agent is sodium phenylbutyrate, which has been shown to facilitate the processing of abnormal CFTR and α_1-antitrypsin (111–113). Phenylbutyrate treatment of lung epithelial cells transfected with constructs expressing mutant SP-C with the exon 4 deletion resulted in a more normal distribution of the SP-C staining pattern, suggesting that phenylbutyrate did improve the trafficking of abnormal proSP-C (100).

The potentially toxic effects of abnormal SP-C expression during development could also result in neonatal lung disease. In order to bypass the block in SP-C processing in SP-B deficiency, Conkright et al. generated transgenic mice using a construct that contained the mature SP-C sequence without the sequences corresponding to the flanking domains of the proprotein, and would thus not require proteolytic processing for mature SP-C expression. Their goal was to breed these mice with animals heterozygous for a null SP-B allele to determine whether mature SP-C could compensate for the lack of SP-B by rescuing the lethal SP-B phenotype. Unexpectedly, no viable mice were obtained, and analysis of animals expressing the transgene revealed marked abnormalities of lung development in association with high levels of expression of the transgene (114). Hence, expression of high levels of mature SP-C with its hydrophobic epitopes early in development had a toxic effect on lung development. Mutations that result in misfolded proSP-C could have similar effects, and hinder normal lung development resulting in early–onset and severe lung disease.

ABNORMAL SP-B AND SP-C EXPRESSION
IN OTHER LUNG DISEASES

Respiratory Distress Syndrome

Given its essential role in surfactant function and its developmental regulation, SP-B is a reasonable candidate gene to be involved in the pathophysiology of more common lung diseases such as RDS. The expression of SP-B is developmentally regulated, and the appearance of SP-B in human amniotic fluid increases with advancing gestation (115). SP-C mRNA and proprotein expression are also developmentally regulated in human lung (116, 117), although mature SP-C expression has been more difficult to study owing to the difficulty in developing antibodies to mature SP-C because of its extreme hydrophobicity. Low levels of SP-B and SP-C have been observed in tracheal effluent fluid of infants with RDS, although the methods used did not distinguish SP-B from SP-C (118). Haploinsufficiency for thyroid transcription factor 1 (TTF-1, also termed Nkx2.1), which is important in SP-B and SP-C gene expression, has been recognized in neonates with diffuse respiratory disease, thus indicating a potential role for decreased SP-B and/or SP-C gene transcription in RDS (119–121).

The association of several SP-B gene variants with the risk of RDS in premature infants has been examined. Several studies have examined a variable nucleotide tandem repeat region in intron 4 and found higher rates of variant alleles in populations of infants with RDS compared with control infants, suggesting a role in RDS (30, 122, 123). However, such associations have generally been with all variant alleles grouped and compared with the most common allele, rather than a risk associated with a specific allele, and have not always controlled for ethnic background. Moreover, there is no known association of these intronic variants with variation in SP-B gene expression, proper mRNA splicing, proprotein processing, or mature SP-B function. Thus the mechanism by which such alieles would influence the risk of RDS remains uncertain.

A single-nucleotide polymorphism (C or T at position 1580 in the SP-B gene) results in either isoleucine or threonine being encoded in codon 131 and alters a potential glycosylation site in proSP-B (7, 31, 34, 124, 125). Although evidence indicates that this polymorphism does affect the glycosylation of proSP-B (126), there is no evidence that such glycosylation is important in proper proSP-B folding, routing through the cell, processing to mature SP-B, or function. The corresponding codon encodes isoleucine in most other species in which the proSP-B sequence has been determined, which would not result in a glycosylation site and indicates that glycosylation at this site is not critical (5, 127). Several studies have examined the potential role of this polymorphism in RDS without clear results. The codon 131 threonine allele, when combined with certain SP-A alleles, was weakly associated with an increased risk for RDS (122, 128–133). The codon 131 threonine allele was also been found to be associated with an increased risk for ARDS in adult subjects in one study (33).

Little data are currently available concerning the role of SP-C in neonatal RDS. The SP-C gene was sequenced from a small ($n = 12$) cohort of premature infants with RDS in one study, but no deviations in the known SP-C-coding sequence were identified, nor were SP-C mRNA levels reduced, as determined by in situ hybridization (56). It is unclear whether both alleles from each patient were represented, as the sequencing analyses involved subcloning into plasmid vectors, and the coding sequence from all patients was identical, although two relatively frequent coding polymorphisms have been identified in the SP-C gene. A possible relationship between SP-C expression and neonatal lung disease was suggested by the finding of absent or markedly reduced SP-C expression in a strain of Belgian calves that died from neonatal lung disease resembling RDS (134). The mechanism for the decreased SP-C expression in these animals is unknown. The lack of SP-C in these animals might be secondary to damage to the alveolar epithelium, rather than a direct cause of disease. In a preliminary report, absent staining for proSP-C was observed in a cohort of full-term infants who died from hyaline membrane disease, although no mutations in the SP-C gene were identified that would account for the lack of proSP-C (135). The lack of SP-C expression was striking in that both SP-C mRNA and proSP-C staining in type II cells have been observed as early as 15-weeks gestation in human lung, with staining in alveolar type II cells increasing from 25-weeks gestation to term (117). It is unknown whether the lack of proSP-C in these full-term infants reflects a similar deficiency of proSP-C in premature infants with severe RDS, or was a specific finding related to either a genetic disorder or the nature of the lung injury in this cohort.

Pulmonary Alveolar Proteinosis

A striking aspect of the pathology of hereditary SP-B deficiency can be the accumulation of granular eosinophilic material filling distal airspaces, with entrapped alveolar macrophages (57, 58). This material represents secreted surfactant material, and stains intensely for SP-A and proSP-C (34) (Figure 3). The histologic appearance is similar to that of pulmonary alveolar proteinosis (PAP) in adults, which is a disease of more insidious onset and gradual progression. However, these findings of alveolar proteinosis may be variable or even absent in SP-B-deficient infants (70, 76). Additionally, severe neonatal lung disease resembling RDS with histopathologic features of alveolar proteinosis has been observed in neonates who clearly had abundant SP-B in BAL fluid and in their lung tissue, as demonstrated by immunohistochemical staining, such that the mechanism for their alveolar proteinosis was not SP-B deficiency (21, 58). Generally, these infants also had very high levels of proSP-B and proSP-C in their lung tissue, although proSP-C staining has been confined to alveolar cells and not found in the extracellular material (Figure 3). The term congenital alveolar proteinosis should therefore not be used synonymously with hereditary SP-B deficiency. Although levels of the surfactant proteins are elevated in older patients with alveolar proteinosis, this is likely related to impaired clearance of the surfactant proteins as opposed to increased production similar to what has been observed in animal models (136–139); however, this has not been shown directly in human neonates with alveolar proteinosis due to mechanisms other than SP-B deficiency.

The mechanisms responsible for disease in these neonates with features of alveolar proteinosis remain unknown. Accumulation of macrophages and variable amounts of proteinacous material in the airspaces is also part of the pathology observed in older infants with SP-C mutations. Deficiency of granulocyte-macrophage colony-stimulating factor (GM-CSF) or its receptor can result in alveolar proteinosis in experimental animals, and adult patients with PAP frequently have anti-GM-CSF antibodies (140–146). Deficiency of the common beta chain (βc) of the receptors for GM-CSF, IL-3, and IL-5 has been identified as a potential mechanism for PAP in young infants. Absent βc expression on monocytes and a lack of response of monocytes to GM-CSF were observed in four young infants with alveolar proteinosis (147). However, no definitive mutations have been identified in the GM-CSF βc receptor gene to account for these observations, and it is unclear whether the βc receptor deficiency was the primary cause of disease or a secondary phenomenon in these infants.

SUMMARY

Clinical features associated with SP-B and SP-C gene mutations are summarized in Table 2. The severe lung disease resulting from the inability to produce SP-B demonstrates the essential role for SP-B in normal lung function. The secondary disturbances in surfactant metabolism in such infants indicates an important

TABLE 2   Surfactant protein gene abnormalities and lung disease

	Hereditary SP-B deficiency	**SP-C deficiency**
Age of onset	Neonatal	Variable
Typical aymptoms	Surfactant deficiency	Tachypnea
	Respiratory failure	Cyanosis
	Airleak and pulmonary hypertension common	Failure to thrive
Inheritance	Recessive	Dominant (sporadic)
Cause	Absent SP-B	Abnormal ProSP-C
		Reduced mature SP-C
Pathology	Nonspecific injury	Alveolar accumulation of
	Alveolar proteinosis	macrophages with variable
	Interstitial fibrosis	amounts of proteinaceous material
		Interstitial thickening

intracellular role for SP-B or proSP-B beyond its role in enhancing the surface tension–lowering properties of surfactant phospholipids and demonstrates a mechanism for genetic contribution to neonatal lung disease. Unanswered questions concerning SP-B deficiency include a more complete understanding of the pathophysiology of the disorder, such as the impaired processing of proSP-C, and the contributions of partial deficiency to chronic lung diseases and SP-B gene variants to more common lung diseases. An effective treatment is also needed for the children afflicted with this fatal disorder.

Whereas loss-of-function mutations result in SP-B deficiency and an understandable phenotype of surfactant deficiency, the lung disease associated with heterozygous SP-C mutations is less well understood. The relative roles of a toxic gain-of-function from expression of abnormal proSP-C and deficiency of mature SP-C in the pathophysiology of lung disease associated with such mutations are not known. Currently there is no information concerning the epidemiology of lung disease associated with SP-C mutations, the role of abnormal SP-C expression in other lung diseases, or the association of SP-C gene variants with lung diseases. Animal models should aid in the understanding of the pathogenesis of lung disease associated with SP-C mutations and evaluating novel therapies.

ACKNOWLEDGMENTS

This work was supported by grants from the National Institutes of Health and the Eudowood Foundation. The author gratefully acknowledges the collaboration of Drs. Susan E. Wert, Timothy E. Weaver, and Jeffrey A. Whitsett of Children's Hospital Medical Center and the University of Cincinnati, Cincinnati, Ohio, and Drs. Aaron Hamvas and F. Sessions Cole, of St. Louis Children's Hospital and Washington University, St. Louis, Missouri.

The *Annual Review of Physiology* is online at http://physiol.annualreviews.org

LITERATURE CITED

1. Farrell PM, Avery ME. 1975. Hyaline membrane disease. *Am. Rev. Respir. Dis.* 111:657–88
2. Usher RH, Allen AC, McLean FH. 1971. Risk of respiratory distress syndrome related to gestational age, route of delivery, and maternal diabetes. *Am. J. Obstet. Gynecol.* 111:826–32
3. Whitsett JA, Hull WM, Ohning B, Ross G, Weaver TE. 1986. Immunologic identification of a pulmonary surfactant-associated protein of molecular weight = 6000 daltons. *Pediatr. Res.* 20:744–49
4. Whitsett JA, Ohning BL, Ross G, Meuth J, Weaver T, et al. 1986. Hydrophobic surfactant-associated protein in whole lung surfactant and its importance for biophysical activity in lung surfactant extracts used for replacement therapy. *Pediatr. Res.* 20:460–67
5. Weaver TE, Conkright JJ. 2001. Function of surfactant proteins B and C. *Annu. Rev. Physiol.* 63:555–78
6. Crouch E, Wright J. 2001. Surfactant proteins a and d and pulmonary host defense. *Annu. Rev. Physiol.* 63:521–54

7. Pilot-Matias TJ, Kister SE, Fox JL, Kropp K, Glasser SW, Whitsett JA. 1989. Structure and organization of the gene encoding human pulmonary surfactant proteolipid SP-B. *DNA* 8:75–86

8. Vamvakopoulos NC, Modi WS, Floros J. 1995. Mapping the human pulmonary surfactant-associated protein B gene (SFTP3) to chromosome 2p12 → p11.2. *Cytogenet. Cell Genet.* 68:8–10

9. Nogee LM. 1998. Genetics of the hydrophobic surfactant proteins. *Biochim. Biophys. Acta* 1408:323–33

10. Weaver TE. 1998. Synthesis, processing and secretion of surfactant proteins B and C. *Biochim. Biophys. Acta* 1408:173–79

11. Patthy L. 1991. Homology of the precursor of pulmonary surfactant-associated protein SP-B with prosaposin and sulfated glycoprotein 1. *J. Biol. Chem.* 266:6035–37

12. Hawgood S, Derrick M, Poulain F. 1998. Structure and properties of surfactant protein B. *Biochim. Biophys. Acta* 1408:150–60

13. O'Reilly MA, Weaver TE, Pilot-Matias TJ, Sarin VK, Gazdar AF, Whitsett JA. 1989. In vitro translation, post-translational processing and secretion of pulmonary surfactant protein B precursors. *Biochim. Biophys. Acta* 1011:140–48

14. Voorhout WF, Veenendaal T, Haagsman HP, Weaver TE, Whitsett JA, et al. 1992. Intracellular processing of pulmonary surfactant protein B in an endosomal/lysosomal compartment. *Am. J. Physiol. Lung Cell Mol. Physiol.* 263:L479–86

15. Weaver TE, Lin S, Bogucki B, Dey C. 1992. Processing of surfactant protein B proprotein by a cathepsin D-like protease. *Am. J. Physiol. Lung Cell Mol. Physiol.* 263:L95–103

16. Hawgood S, Latham D, Borchelt J, Damm D, White T, et al. 1993. Cell-specific post-translational processing of the surfactant-associated protein SP-B. *Am. J. Physiol. Lung Cell Mol. Physiol.* 264:L290–99

17. Voorhout WF, Weaver TE, Haagsman HP, Geuze HJ, Van Golde LM. 1993. Biosynthetic routing of pulmonary surfactant proteins in alveolar type II cells. *Microsc. Res. Tech.* 26:366–73

18. Guttentag SH, Beers MF, Bieler BM, Ballard PL. 1998. Surfactant protein B processing in human fetal lung. *Am. J. Physiol. Lung Cell Mol. Physiol.* 275:L559–66

19. Guttentag S, Robinson L, Zhang P, Brasch F, Buhling F, Beers M. 2003. Cysteine protease activity is required for surfactant protein B processing and lamellar body genesis. *Am. J. Respir. Cell Mol. Biol.* 28:69–79

20. Nogee LM, Garnier G, Dietz HC, Singer L, Murphy AM, et al. 1994. A mutation in the surfactant protein B gene responsible for fatal neonatal respiratory disease in multiple kindreds. *J. Clin. Invest.* 93:1860–63

21. Lin Z, deMello DE, Batanian JR, Khammash HM, DiAngelo S, et al. 2000. Aberrant SP-B mRNA in lung tissue of patients with congenital alveolar proteinosis (CAP). *Clin. Genet* 57:359–69

22. Lin Z, Wang G, deMello DE, Floros J. 1999. An alternatively spliced surfactant protein B mRNA in normal human lung: disease implication. *Biochem. J.* 343 Pt:145–49

23. Beck DC, Na CL, Whitsett JA, Weaver TE. 2000. Ablation of a critical surfactant protein B intramolecular disulfide bond in transgenic mice. *J. Biol. Chem.* 275:3371–76

24. Beck DC, Ikegami M, Na CL, Zaltash S, Johansson J, et al. 2000. The role of homodimers in surfactant protein B function in vivo. *J. Biol. Chem.* 275:3365–70

25. Ikegami M, Takabatake N, Weaver TE. 2002. Intersubunit disulfide bridge is not required for the protective role of SP-B against lung inflammation. *J. Appl. Physiol.* 93:505–11

26. Lin S, Akinbi HT, Breslin JS, Weaver TE. 1996. Structural requirements for targeting of surfactant protein B (SP-B) to

secretory granules in vitro and in vivo. *J. Biol. Chem.* 271:19689–95

27. Lin S, Phillips KS, Wilder MR, Weaver TE. 1996. Structural requirements for intracellular transport of pulmonary surfactant protein B (SP-B). *Biochim. Biophys. Acta* 1312:177–85

28. Akinbi HT, Breslin JS, Ikegami M, Iwamoto HS, Clark JC, et al. 1997. Rescue of SP-B knockout mice with a truncated SP-B proprotein. Function of the C-terminal propeptide. *J. Biol. Chem.* 272:9640–47

29. Todd S, Naylor SL. 1991. Dinucleotide repeat polymorphism in the human surfactant-associated protein 3 gene (SFTP3). *Nucleic Acids Res.* 19:3756

30. Floros J, Veletza SV, Kotikalapudi P, Krizkova L, Karinch AM, et al. 1995. Dinucleotide repeats in the human surfactant protein-B gene and respiratory-distress syndrome. *Biochem. J.* 305:583–90

31. Lin Z, deMello DE, Wallot M, Floros J. 1998. An SP-B gene mutation responsible for SP-B deficiency in fatal congenital alveolar proteinosis: evidence for a mutation hotspot in exon 4. *Mol. Genet. Metab.* 64:25–35

32. Guo X, Lin HM, Lin Z, Montano M, Sansores R, et al. 2000. Polymorphisms of surfactant protein gene A, B, D, and of SP-B-linked microsatellite markers in COPD of a Mexican population. *Chest* 117:249S–50S

33. Lin Z, Pearson C, Chinchilli V, Pietschmann SM, Luo J, et al. 2000. Polymorphisms of human SP-A, SP-B, and SP-D genes: association of SP-B Thr131Ile with ARDS. *Clin. Genet.* 58:181–91

34. Nogee LM, Wert SE, Proffit SA, Hull WM, Whitsett JA. 2000. Allelic heterogeneity in hereditary surfactant protein B (SP-B) deficiency. *Am. J. Respir. Crit. Care Med.* 161:973–81

35. Glasser SW, Korfhagen TR, Perme CM, Pilot-Matias TJ, Kister SE, Whitsett JA. 1988. Two SP-C genes encoding human pulmonary surfactant proteolipid. *J. Biol. Chem.* 263:10326–31

36. Fisher JH, Emrie PA, Drabkin HA, Kushnik T, Gerber M, et al. 1988. The gene encoding the hydrophobic surfactant protein SP-C is located on 8p and identifies an EcoRI RFLP. *Am. J. Hum. Genet.* 43:436–41

37. Keller A, Eistetter HR, Voss T, Schafer KP. 1991. The pulmonary surfactant protein C (SP-C) precursor is a type II transmembrane protein. *Biochem. J.* 277:493–99

38. Russo SJ, Wang W, Lomax CA, Beers MF. 1999. Structural requirements for intracellular targeting of SP-C proprotein. *Am. J. Physiol. Lung Cell Mol. Physiol.* 277:L1034–44

39. Conkright JJ, Bridges JP, Na CL, Voorhout WF, Trapnell B, et al. 2001. Secretion of surfactant protein C, an integral membrane protein, requires the N-terminal propeptide. *J. Biol. Chem.* 276:14658–64

40. Johnson AL, Braidotti P, Pietra GG, Russo SJ, Kabore A, et al. 2001. Post-translational processing of surfactant protein-C proprotein: targeting motifs in the NH(2)-terminal flanking domain are cleaved in late compartments. *Am. J. Respir. Cell Mol. Biol.* 24:253–63

41. Beers MF, Lomax C. 1995. Synthesis and processing of hydrophobic surfactant protein C by isolated rat type II cells. *Am. J. Physiol. Lung Cell Mol. Physiol.* 269:L744–53

42. Beers MF, Kim CY, Dodia C, Fisher AB. 1994. Localization, synthesis, and processing of surfactant protein SP-C in rat lung analyzed by epitope-specific antipeptide antibodies. *J. Biol. Chem.* 269:20318–28

43. Vorbroker DK, Voorhout WF, Weaver TE, Whitsett JA. 1995. Posttranslational processing of surfactant protein C in rat type II cells. *Am. J. Physiol. Lung Cell Mol. Physiol.* 269:L727–33

44. Brasch F, Ten Brinke A, Johnen G, Ochs

M, Kapp N, et al. 2002. Involvement of cathepsin H in the processing of the hydrophobic surfactant-associated protein C in type II pneumocytes. *Am. J. Respir. Cell Mol. Biol.* 26:659–70

45. Curstedt T, Johansson J, Persson P, Eklund A, Robertson B, et al. 1990. Hydrophobic surfactant-associated polypeptides: SP-C is a lipopeptide with two palmitoylated cysteine residues, whereas SP-B lacks covalently linked fatty acyl groups. *Proc. Natl. Acad. Sci. USA* 87:2985–89

46. Vorbroker DK, Dey C, Weaver TE, Whitsett JA. 1992. Surfactant protein C precursor is palmitoylated and associates with subcellular membranes. *Biochim. Biophys. Acta* 1105:161–69

47. Creuwels LA, Demel RA, van Golde LM, Benson BJ, Haagsman HP. 1993. Effect of acylation on structure and function of surfactant protein C at the air-liquid interface. *J. Biol. Chem.* 268:26752–58

48. Qanbar R, Possmayer F. 1995. On the surface activity of surfactant-associated protein C (SP-C): effects of palmitoylation and pH. *Biochim. Biophys. Acta* 1255:251–59

49. Vandenbussche G, Clercx A, Curstedt T, Johansson J, Jornvall H, Ruysschaert JM. 1992. Structure and orientation of the surfactant-associated protein C in a lipid bilayer. *Eur. J. Biochem.* 203:201–9

50. Johansson J. 1998. Structure and properties of surfactant protein C. *Biochim. Biophys. Acta* 1408:161–72

51. Gustafsson M, Thyberg J, Naslund J, Eliasson E, Johansson J. 1999. Amyloid fibril formation by pulmonary surfactant protein C. *FEBS Lett.* 464:138–42

52. Gustafsson M, Griffiths WJ, Furusjo E, Johansson J. 2001. The palmitoyl groups of lung surfactant protein C reduce unfolding into a fibrillogenic intermediate. *J. Mol. Biol.* 310:937–50

53. Johansson J. 2001. Membrane properties and amyloid fibril formation of lung surfactant protein C. *Biochem. Soc. Trans.* 29:601–6

54. Glasser SW, Korfhagen TR, Weaver TE, Clark JC, Pilot-Matias T, et al. 1988. cDNA, deduced polypeptide structure and chromosomal assignment of human pulmonary surfactant proteolipid, SPL(pVal). *J. Biol. Chem.* 263:9–12

55. Warr RG, Hawgood S, Buckley DI, Crisp TM, Schilling J, et al. 1987. Low molecular weight human pulmonary surfactant protein (SP5): isolation, characterization, and cDNA and amino acid sequences. *Proc. Natl. Acad. Sci. USA* 84:7915–19

56. Hatzis D, Deiter G, deMello DE, Floros J. 1994. Human surfactant protein-C: genetic homogeneity and expression in RDS; comparison with other species. *Exp. Lung Res.* 20:57–72

57. Nogee LM, deMello DE, Dehner LP, Colten HR. 1993. Brief report: deficiency of pulmonary surfactant protein B in congenital alveolar proteinosis. *N. Engl. J. Med.* 328:406–10

58. deMello DE, Nogee LM, Heyman S, Krous HF, Hussain M, et al. 1994. Molecular and phenotypic variability in the congenital alveolar proteinosis syndrome associated with inherited surfactant protein B deficiency. *J. Pediatr.* 125:43–50

59. Ball R, Chetcuti PA, Beverley D. 1995. Fatal familial surfactant protein B deficiency. *Arch. Dis. Child Fetal Neonatal Ed.* 73:F53

60. Ballard PL, Nogee LM, Beers MF, Ballard RA, Planer BC, et al. 1995. Partial deficiency of surfactant protein B in an infant with chronic lung disease. *Pediatrics* 96:1046–52

61. Chetcuti PA, Ball RJ. 1995. Surfactant apoprotein B deficiency. *Arch. Dis. Child Fetal Neonatal Ed.* 73:F125–27

62. de la Fuente AA, Voorhout WF, deMello DE. 1997. Congenital alveolar proteinosis in the Netherlands: a report of five cases with immunohistochemical and genetic studies on surfactant apoproteins. *Pediatr. Pathol. Lab. Med.* 17:221–31

63. Sleight E, Coombs RC, Gibson AT, Primhak RA. 1997. Neonatal respiratory

distress in near-term infants–consider surfactant protein B deficiency. *Acta Paediatr.* 86:428–30

64. Klein JM, Thompson MW, Snyder JM, George TN, Whitsett JA, et al. 1998. Transient surfactant protein B deficiency in a term infant with severe respiratory failure. *J. Pediatr* 132:244–48

65. Tredano M, van Elburg RM, Kaspers AG, Zimmermann LJ, Houdayer C, et al. 1999. Compound SFTPB 1549C→GAA (121ins2) and 457delC heterozygosity in severe congenital lung disease and surfactant protein B (SP-B) deficiency. *Hum. Mutat.* 14:502–9

66. Williams GD, Christodoulou J, Stack J, Symons P, Wert SE, et al. 1999. Surfactant protein B deficiency: clinical, histological and molecular evaluation. *J. Paediatr. Child Health* 35:214–20

67. Somaschini M, Wert S, Mangili G, Colombo A, Nogee L. 2000. Hereditary surfactant protein B deficiency resulting from a novel mutation. *Intensive Care Med.* 26:97–100

68. Dunbar AE 3rd, Wert SE, Ikegami M, Whitsett JA, Hamvas A, et al. 2000. Prolonged survival in hereditary surfactant protein B (SP-B) deficiency associated with a novel splicing mutation. *Pediatr. Res.* 48:275–82

69. Andersen C, Ramsay JA, Nogee LM, Shah J, Wert SE, et al. 2000. Recurrent familial neonatal deaths: hereditary surfactant protein B deficiency. *Am. J. Perinatol.* 17:219–24

70. Hamvas A, Nogee LM, deMello DE, Cole FS. 1995. Pathophysiology and treatment of surfactant protein-B deficiency. *Biol. Neonate* 67:18–31

71. Whitsett JA, Nogee LM, Weaver TE, Horowitz AD. 1995. Human surfactant protein B: structure, function, regulation, and genetic disease. *Physiol. Rev.* 75:749–57

72. Hamvas A. 1997. Inherited surfactant protein-B deficiency. *Adv. Pediatr.* 44:369–88

73. Nogee LM. 1997. Surfactant protein-B deficiency. *Chest* 111:129S–35S

74. Nogee LM, Wert SE, Hamvas A, Hull W, Whitsett JA. 2000. Phenotypic variability in hereditary surfactant protein B (SP-B) deficiency. *Am. J. Respir. Crit. Care Med.* 161:A523 (Abstr.)

75. Hamvas A, Cole FS, deMello DE, Moxley M, Whitsett JA, et al. 1994. Surfactant protein B deficiency: antenatal diagnosis and prospective treatment with surfactant replacement. *J. Pediatr.* 125:356–61

76. Hamvas A, Nogee LM, Mallory GB Jr, Spray TL, Huddleston CB, et al. 1997. Lung transplantation for treatment of infants with surfactant protein B deficiency. *J. Pediatr.* 130:231–29

77. Melton KR, Nesslein LL, Ikegami M, Tichelaar JW, Clark JC, et al. 2003. SP-B deficiency causes respiratory failure in adult mice. *Am. J. Physiol. Lung Cell Mol. Physiol.* 285:L573–79

78. Tokieda K, Ikegami M, Wert SE, Baatz JE, Zou Y, Whitsett JA. 1999. Surfactant protein B corrects oxygen-induced pulmonary dysfunction in heterozygous surfactant protein B-deficient mice. *Pediatr. Res.* 46:708–14

79. Tokieda K, Iwamoto HS, Bachurski C, Wert SE, Hull WM, et al. 1999. Surfactant protein-B-deficient mice are susceptible to hyperoxic lung injury. *Am. J. Respir. Cell Mol. Biol.* 21:463–72

80. Mbagwu N, Bruni R, Hernandez-Juviel JM, Waring AJ, Walther FJ. 1999. Sensitivity of synthetic surfactants to albumin inhibition in preterm rabbits. *Mol. Genet. Metab.* 66:40–48

81. Cole FS, Hamvas A, Rubinstein P, King E, Trusgnich M, et al. 2000. Population-based estimates of surfactant protein B deficiency. *Pediatrics* 105:538–41

82. Hamvas A, Trusgnich M, Brice H, Baumgartner J, Hong Y, et al. 2001. Population-based screening for rare mutations: high-throughput DNA extraction and molecular amplification from Guthrie cards. *Pediatr. Res.* 50:666–68

83. Vorbroker DK, Profitt SA, Nogee LM, Whitsett JA. 1995. Aberrant processing of surfactant protein C in hereditary SP-B deficiency. *Am. J. Physiol. Lung Cell Mol. Physiol.* 268:L647–56

84. Beers MF, Hamvas A, Moxley MA, Gonzales LW, Guttentag SH, et al. 2000. Pulmonary surfactant metabolism in infants lacking surfactant protein B. *Am. J. Respir. Cell Mol. Biol.* 22:380–91

85. deMello DE, Heyman S, Phelps DS, Hamvas A, Nogee L, et al. 1994. Ultrastructure of lung in surfactant protein B deficiency. *Am. J. Respir. Cell Mol. Biol.* 11:230–39

86. Clark JC, Weaver TE, Iwamoto HS, Ikegami M, Jobe AH, et al. 1997. Decreased lung compliance and air trapping in heterozygous SP-B- deficient mice. *Am. J. Respir. Cell Mol. Biol.* 16:46–52

87. Tokieda K, Whitsett JA, Clark JC, Weaver TE, Ikeda K, et al. 1997. Pulmonary dysfunction in neonatal SP-B-deficient mice. *Am. J. Physiol. Lung Cell Mol. Physiol.* 273:L875–82

88. Yusen RD, Cohen AH, Hamvas A. 1999. Normal lung function in subjects heterozygous for surfactant protein-B deficiency. *Am. J. Respir. Crit. Care Med.* 159:411–14

89. Carothers AM, Steigerwalt RW, Urlaub G, Chasin LA, Grunberger D. 1989. DNA base changes and RNA levels in *N*-acetoxy-2-acetylaminofluorene-induced dihydrofolate reductase mutants of Chinese hamster ovary cells. *J. Mol. Biol.* 208:417–28

90. Urlaub G, Mitchell PJ, Ciudad CJ, Chasin LA. 1989. Nonsense mutations in the dihydrofolate reductase gene affect RNA processing. *Mol. Cell Biol.* 9:2868–80

91. Davis AJ, Jobe AH, Hafner D, Ikegami M. 1998. Lung function in premature lambs and rabbits treated with a recombinant SP-C surfactant. *Am. J. Respir. Crit. Care Med.* 157:553–59

92. Hafner D, Germann PG, Hauschke D. 1998. Effects of rSP-C surfactant on oxygenation and histology in a rat-lung-lavage model of acute lung injury. *Am. J. Respir. Crit. Care Med.* 158:270–78

93. Hafner D, Germann PG, Hauschke D. 1998. Comparison of rSP-C surfactant with natural and synthetic surfactants after late treatment in a rat model of the acute respiratory distress syndrome. *Br. J. Pharmacol.* 124:1083–90

94. Hawgood S, Ogawa A, Yukitake K, Schlueter M, Brown C, et al. 1996. Lung function in premature rabbits treated with recombinant human surfactant protein-C. *Am. J. Respir. Crit. Care Med.* 154:484–90

95. Lewis J, McCaig L, Hafner D, Spragg R, Veldhuizen R, Kerr C. 1999. Dosing and delivery of a recombinant surfactant in lung-injured adult sheep. *Am. J. Respir. Crit. Care Med.* 159:741–47

96. Glasser SW, Burhans MS, Korfhagen TR, Na CL, Sly PD, et al. 2001. Altered stability of pulmonary surfactant in SP-C-deficient mice. *Proc. Natl. Acad. Sci. USA* 98:6366–71

97. Amin RS, Wert SE, Baughman RP, Tomashefski JF Jr, Nogee LM, et al. 2001. Surfactant protein deficiency in familial interstitial lung disease. *J. Pediatr.* 139:85–92

98. Nogee LM, Dunbar AE, Wert SE, Askin F, Hamvas A, Whitsett JA. 2001. A mutation in the surfactant protein C gene associated with familial interstitial lung disease. *N. Engl. J. Med.* 344:573–79

99. Wang WJ, Russo SJ, Mulugeta S, Beers MF. 2002. Biosynthesis of surfactant protein C (SP-C). Sorting of SP-C proprotein involves homomeric association via a signal anchor domain. *J. Biol. Chem.* 277:19929–37

100. Wang WJ, Mulugeta S, Russo SJ, Beers MF. 2003. Deletion of exon 4 from human surfactant protein C results in aggresome formation and generation of a dominant negative. *J. Cell Sci.* 116:683–92

101. Thomas AQ, Lane K, Phillips J 3rd, Prince M, Markin C, et al. 2002. Heterozygosity for a surfactant protein C gene mutation

associated with usual interstitial pneumonitis and cellular nonspecific interstitial pneumonitis in one kindred. *Am. J. Respir. Crit. Care Med.* 165:1322–28

102. Glasser SW, Detmer EA, Ikegami M, Na CL, Stahlman MT, Whitsett JA. 2003. Pneumonitis and emphysema in sp-C gene targeted mice. *J. Biol. Chem.* 278:14291–98

103. Nogee LM, Dunbar AE 3rd, Wert S, Askin F, Hamvas A, Whitsett JA. 2002. Mutations in the surfactant protein C gene associated with interstitial lung disease. *Chest* 121:20S–21S

104. Johnston JA, Ward CL, Kopito RR. 1998. Aggresomes: a cellular response to misfolded proteins. *J. Cell Biol.* 143:1883–98

105. Kabore AF, Wang WJ, Russo SJ, Beers MF. 2001. Biosynthesis of surfactant protein C: characterization of aggresome formation by EGFP chimeras containing propeptide mutants lacking conserved cysteine residues. *J. Cell Sci.* 114:293–302

106. Nogee LM. 2002. Abnormal expression of surfactant protein C and lung disease. *Am. J. Respir. Cell Mol. Biol.* 26:641–44

107. Perlmutter DH. 1998. Alpha-1-antitrypsin deficiency. *Semin Liver Dis.* 18:217–25

108. Perlmutter DH. 1999. Misfolded proteins in the endoplasmic reticulum. *Lab. Invest* 79:623–38

109. Lomas DA, Mahadeva R. 2002. Alpha1-antitrypsin polymerization and the serpinopathies: pathobiology and prospects for therapy. *J. Clin. Invest* 110:1585–90

110. Burrows JA, Willis LK, Perlmutter DH. 2000. Chemical chaperones mediate increased secretion of mutant alpha 1-antitrypsin (alpha 1-AT) Z: a potential pharmacological strategy for prevention of liver injury and emphysema in alpha 1-AT deficiency. *Proc. Natl. Acad. Sci. USA* 97:1796–801

111. Rubenstein RC, Egan ME, Zeitlin PL. 1997. In vitro pharmacologic restoration of CFTR-mediated chloride transport with sodium 4-phenylbutyrate in cystic

fibrosis epithelial cells containing delta F508-CFTR. *J. Clin. Invest.* 100:2457–65

112. Rubenstein RC, Zeitlin PL. 1998. A pilot clinical trial of oral sodium 4-phenylbutyrate (Buphenyl) in deltaF508-homozygous cystic fibrosis patients: partial restoration of nasal epithelial CFTR function. *Am. J. Respir. Crit. Care Med.* 157:484–90

113. Zeitlin PL. 2000. Pharmacologic restoration of delta F508 CFTR-mediated chloride current. *Kidney Int.* 57:832–37

114. Conkright JJ, Na CL, Weaver TE. 2002. Overexpression of surfactant protein-C mature peptide causes neonatal lethality in transgenic mice. *Am. J. Respir. Cell Mol. Biol.* 26:85–90

115. Pryhuber GS, Hull WM, Fink I, McMahan MJ, Whitsett JA. 1991. Ontogeny of surfactant proteins A and B in human amniotic fluid as indices of fetal lung maturity. *Pediatr. Res.* 30:597–605

116. Whitsett JA, Weaver TE, Clark JC, Sawtell N, Glasser SW, et al. 1987. Glucocorticoid enhances surfactant proteolipid Phe and pVal synthesis and RNA in fetal lung. *J. Biol. Chem.* 262:15618–23

117. Khoor A, Stahlman MT, Gray ME, Whitsett JA. 1994. Temporal-spatial distribution of SP-B and SP-C proteins and mRNAs in developing respiratory epithelium of human lung. *J. Histochem. Cytochem.* 42:1187–99

118. Chida S, Fujiwara T, Konishi M, Shimada S, Takahashi A. 1997. Surfactant proteins and stable microbubbles in tracheal aspirates of infants with respiratory distress syndrome: relation to the degree of respiratory failure and response to exogenous surfactant. *Eur. J. Pediatr.* 156:131–38

119. Iwatani N, Mabe H, Devriendt K, Kodama M, Miike T. 2000. Deletion of NKX2.1 gene encoding thyroid transcription factor-1 in two siblings with hypothyroidism and respiratory failure. *J. Pediatr.* 137:272–76

120. Krude H, Schutz B, Biebermann H, von Moers A, Schnabel D, et al.

2002. Choreoathetosis, hypothyroidism, and pulmonary alterations due to human NKX2-1 haploinsufficiency. *J. Clin. Invest.* 109:475–80

121. Pohlenz J, Dumitrescu A, Zundel D, Martine U, Schonberger W, et al. 2002. Partial deficiency of thyroid transcription factor 1 produces predominantly neurological defects in humans and mice. *J. Clin. Invest* 109:469–73

122. Kala P, Ten Have T, Nielsen H, Dunn M, Floros J. 1998. Association of pulmonary surfactant protein A (SP-A) gene and respiratory distress syndrome: interaction with SP-B. *Pediatr. Res.* 43:169–77

123. Veletza SV, Rogan PK, TenHave T, Olowe SA, Floros J. 1996. Racial differences in allelic distribution at the human pulmonary surfactant protein B gene locus (SP-B). *Exp. Lung Res.* 22:489–94

124. Jacobs KA, Phelps DS, Steinbrink R, Fisch J, Kriz R, et al. 1987. Isolation of a cDNA clone encoding a high molecular weight precursor to a 6-kDa pulmonary surfactant-associated protein. *J. Biol. Chem.* 262:9808–11. Erratum. 1998. *J. Biol. Chem.* 263(2):1093

125. Glasser SW, Korfhagen TR, Weaver T, Pilot-Matias T, Fox JL, Whitsett JA. 1987. cDNA and deduced amino acid sequence of human pulmonary surfactant-associated proteolipid SPL(Phe). *Proc. Natl. Acad. Sci. USA* 84:4007–11

126. Wang G, Christensen ND, Wigdahl B, Guttentag SH, Floros J. 2003. Differences in N-linked glycosylation between human surfactant protein-B variants of the C or T allele at the single-nucleotide polymorphism at position 1580: implications for disease. *Biochem. J.* 369:179–84

127. Emrie PA, Shannon JM, Mason RJ, Fisher JH. 1989. cDNA and deduced amino acid sequence for the rat hydrophobic pulmonary surfactant-associated protein, SP-B. *Biochim. Biophys. Acta* 994:215–21

128. Haataja R, Ramet M, Marttila R, Hallman M. 2000. Surfactant proteins A and B as interactive genetic determinants of neona-

tal respiratory distress syndrome. *Hum. Mol. Genet.* 9:2751–60

129. Floros J, Fan R. 2001. Surfactant protein A and B genetic variants and respiratory distress syndrome: allele interactions. *Biol. Neonate* 80(Suppl.)1:22–25

130. Hallman M, Haataja R, Marttila R. 2002. Surfactant proteins and genetic predisposition to respiratory distress syndrome. *Semin. Perinatol.* 26:450–60

131. Hallman M, Haataja R. 2003. Genetic influences and neonatal lung disease. *Semin. Neonatol.* 8:19–27

132. Marttila R, Haataja R, Ramet M, Lofgren J, Hallman M. 2003. Surfactant protein B polymorphism and respiratory distress syndrome in premature twins. *Hum. Genet* 112:18–23

133. Makri V, Hospes B, Stoll-Becker S, Borkhardt A, Gortner L. 2002. Polymorphisms of surfactant protein B encoding gene: modifiers of the course of neonatal respiratory distress syndrome? *Eur. J. Pediatr.* 161:604–8

134. Danlois F, Zaltash S, Johansson J, Robertson B, Haagsman HP, et al. 2000. Very low surfactant protein C contents in newborn Belgian White and Blue calves with respiratory distress syndrome. *Biochem. J.* 351:779–87

135. Wert SE, Proffit SA, Whitsett JA, Nogee LM. 1998. Reduced surfactant protein C expression in full-term infants with respiratory distress syndrome. *Pediatr. Res.* 43:303A (Abstr.)

136. Ikegami M, Jobe AH, Huffman Reed JA, Whitsett JA. 1997. Surfactant metabolic consequences of overexpression of GM-CSF in the epithelium of GM-CSF-deficient mice. *Am. J. Physiol. Lung Cell Mol. Physiol.* 273:L709–14

137. Reed JA, Whitsett JA. 1998. Granulocyte-macrophage colony-stimulating factor and pulmonary surfactant homeostasis. *Proc. Assoc. Am. Physicians* 110:321–32

138. Reed JA, Ikegami M, Robb L, Begley CG, Ross G, Whitsett JA. 2000. Distinct

changes in pulmonary surfactant home-ostasis in common beta-chain- and GM-CSF-deficient mice. *Am. J. Physiol. Lung Cell Mol. Physiol.* 278:L1164–71

139. Yoshida M, Ikegami M, Reed JA, Chroneos ZC, Whitsett JA. 2001. GM-CSF regulates protein and lipid catabolism by alveolar macrophages. *Am. J. Physiol. Lung Cell Mol. Physiol.* 280:L379–86

140. Dranoff G, Crawford AD, Sadelain M, Ream B, Rashid A, et al. 1994. Involvement of granulocyte-macrophage colony-stimulating factor in pulmonary homeostasis. *Science* 264:713–16

141. Stanley E, Lieschke GJ, Grail D, Metcalf D, Hodgson G, et al. 1994. Granulocyte/macrophage colony-stimulating factor-deficient mice show no major perturbation of hematopoiesis but develop a characteristic pulmonary pathology. *Proc. Natl. Acad. Sci. USA* 91:5592–96

142. Nishinakamura R, Nakayama N, Hirabayashi Y, Inoue T, Aud D, et al. 1995. Mice deficient for the IL-3/GM-CSF/IL-5 beta c receptor exhibit lung pathology and impaired immune response, while beta IL3 receptor-deficient mice are normal. *Immunity* 2:211–22

143. Robb L, Drinkwater CC, Metcalf D, Li R, Kontgen F, et al. 1995. Hematopoietic and lung abnormalities in mice with a null mutation of the common beta subunit of the receptors for granulocyte-macrophage colony-stimulating factor and interleukins 3 and 5. *Proc. Natl. Acad. Sci. USA* 92: 9565–69

144. Kitamura T, Tanaka N, Watanabe J, Uchida, Kanegasaki S, et al. 1999. Idiopathic pulmonary alveolar proteinosis as an autoimmune disease with neutralizing antibody against granulocyte/macrophage colony-stimulating factor. *J. Exp. Med.* 190:875–80

145. Bonfield TL, Kavuru MS, Thomassen MJ. 2002. Anti-GM-CSF titer predicts response to GM-CSF therapy in pulmonary alveolar proteinosis. *Clin. Immunol.* 105:342–50

146. Seymour JF, Doyle IR, Nakata K, Presneill JJ, Schoch OD, et al. 2003. Relationship of anti-GM-CSF antibody concentration, surfactant protein A and B levels, and serum LDH to pulmonary parameters and response to GM-CSF therapy in patients with idiopathic alveolar proteinosis. *Thorax* 58:252–57

147. Dirksen U, Nishinakamura R, Groneck P, Hattenhorst U, Nogee L, et al. 1997. Human pulmonary alveolar proteinosis associated with a defect in GM-CSF/IL-3/IL-5 receptor common beta chain expression. *J. Clin. Invest.* 100:2211–17

Annu. Rev. Physiol. 2004. 66:625–45
doi: 10.1146/annurev.physiol.66.032102.135749
First published online as a Review in Advance on September, 29 2003

EPITHELIAL-MESENCHYMAL INTERACTIONS IN THE DEVELOPING LUNG

John M. Shannon and Brian A. Hyatt

*Division of Pulmonary Biology, Department of Pediatrics, Cincinnati Children's
Hospital Medical Center, Cincinnati, Ohio 45229-3039; email: john.shannon@chmcc.org*

Key Words   lung development, tissue interactions, gene expression, growth factors

■ **Abstract**   Classical experiments in embryology have shown that normal growth,
morphogenetic patterning, and cellular differentiation in the developing lung depend
on interactive signaling between the endodermal epithelium and mesenchyme derived
from splanchnic mesoderm. These interactions are mediated by a myriad of diffusible
factors that are precisely regulated in their temporal and spatial expression. In this
review we first describe factors regulating formation of the embryonic foregut. We
then discuss the experiments demonstrating the importance of tissue interactions in
lung patterning and differentiation. Finally, we detail the roles that a few key signaling
systems—fibroblast growth factors and their receptors, *sonic hedgehog* and Gli genes,
Wnt genes and *β-catenin*, and *BMP4*—play as mediators of epithelial-mesenchymal
interactions in the developing lung.

INTRODUCTION

A benchmark in the adaptation of vertebrates to living on land was the development of a gas exchange system that provided oxygen to meet the metabolic load
of cellular respiration. As organisms increased in size, the surface area required
for adequate gas exchange became quite large, estimated at 70 m^2 in the adult
human (1). The problem of generating sufficient surface area for gas exchange
in a manageable space/volume has been solved in the basic structure of the lung,
where a series of ramifying epithelial tubules conduct air to millions of alveoli
(an average of 300 million in human adults) in the lung periphery. Alveolar type I
and type II cells constitute the epithelial component of the alveoli. Type I cells,
which cover approximately 95% of the surface area of the peripheral lung, are
highly attenuated cells that interface with pulmonary capillaries and serve as the
site of gas exchange (2). The type II cell is a multifunctional unit that synthesizes
and secretes pulmonary surfactant, serves as a progenitor cell for type I cells, di-
rectionally transports sodium from apical to basolateral cell surfaces to minimize
alveolar fluid, and produces molecules involved in innate host defense (3). The
epithelium proximal to the alveoli also contains a number of highly specialized

cell types that function in a variety of capacities, such as protection against inhaled particulates and pathogens.

In humans, the embryonic lung originates as a bifurcation at the posterior end of the laryngotracheal groove during the third to fourth week of gestation. Lung initiation is slightly different in rodents, where the two buds that will form the right and left lungs evaginate from the floor of the foregut endoderm; initiation of tracheal development begins later (4). In both humans and rodents, the endoderm then undergoes a series of dichotomous and lateral branchings during the pseudoglandular stage of development to give rise to the conducting airways. Expansion and maturation of the gas exchange region occurs later in gestation during the canalicular and saccular stages. Alveolarization begins at the very end of gestation and continues postnatally, when millions of alveoli are added. The extensive branching of the lung epithelium is accompanied by the parallel development of the pulmonary vasculature. The interfacing of the distal lung epithelium and vasculature generate what will postnatally become the functional gas exchange system.

Lung development is a highly regulated process. An extensive literature has documented that both morphogenesis and differentiation of many organs are controlled by short-range secondary inductions, commonly referred to as epithelial-mesenchymal interactions. We focus here on the role of epithelial-mesenchymal interactions in the developing lung. First we describe the observations made using classical techniques in embryology that documented the importance of tissue interactions to lung development. Then we discuss the recent advances in molecular developmental biology that have provided a wealth of information on the molecular basis of epithelial-mesenchymal interactions. Whereas much information has been gained about these regulatory mechanisms, it is clear that much more remains to be learned.

Endoderm Formation

During mammalian gastrulation, totipotent cells from the epiblast migrate through the primitive streak to form the definitive mesodermal and endodermal layers. The factors that regulate whether the cells adopt a mesodermal or endodermal fate are not yet known, and the precise point in development at which this cell fate decision is made has not been elucidated. The presumptive endoderm cells migrate into the visceral endoderm layer and over time displace these cells (5), forming a two-dimensional sheet on the surface of the embryo when gastrulation has been completed. Folding over of the rostral and caudal regions of the definitive endoderm then gives rise to the foregut and hindgut diverticula, respectively. Embryo turning closes off the midgut region, effectively completing the primitive gut tube. Shortly thereafter, the endodermal anlage that will form the thyroid, liver, dorsal and ventral pancreas, and lung emerge at specific points along the primitive gut tube. Therefore, the genes that regulate formation of the endodermal gut tube and, specifically, the foregut necessarily affect lung development.

Nodal, a member of the transforming growth factor β (TGFβ) superfamily, is one gene expressed in early development that affects formation of the endoderm.

Mice homozygous for a retroviral insertion mutation in *Nodal* lack a primitive streak, do not form a morphologically identifiable mesoderm, and show developmental arrest (6, 7). This phenotype precludes assessment of the role of *Nodal* in foregut formation. Mice with a hypomorphic *Nodal* allele, however, progress further in development. These mice show impaired development along the entire gut tube, including the foregut, and the extent of these defects is correlated with the level of Nodal expression (8). Elucidation of the genes activated by Nodal signaling in endoderm formation is an area of intense current interest.

Members of the GATA family of zinc finger transcription factors also play important roles in early endoderm formation; indeed, transfection of non-endodermal cells with GATA genes induces transcription of endodermal markers (9). Mice null for *GATA4* show developmental arrest between embryonic (E) days E7 and E9.5 because of multiple developmental abnormalities, most notably the lack of a heart tube and foregut (10, 11). *GATA6* appears to act even earlier in development. Mice with a targeted deletion of *GATA6* die between days E6.5 and E7.5, with defects in differentiation of endodermal cells, and do not show expression of *GATA4* (12).

The Sox family of genes encodes transcription factors that contain a high-mobility group (HMG) DNA-binding domain. Sox factors have been shown to regulate the specification of a variety of cell types and tissue differentiation (13). Sox genes were first implicated in endoderm formation from studies in *Xenopus* (14), where disrupting *Sox17* activity in those cells destined to form endoderm causes them to change fate to ectodermal or mesodermal lineages (15). Mice null for *Sox17* are deficient in definitive gut endoderm. The foregut develops normally until day E8.0, but thereafter elevated apoptosis depletes the foregut of definitive endoderm cells (16). In chimeric mice, *Sox17* null embryonic stem cells are able to populate ectodermal and mesodermal lineages without restriction, but they colonize the foregut sparsely and the mid- and hindguts not at all.

Forkhead genes have been shown to be involved in endoderm formation in all species studied thus far. *Foxa1* (*HNF3α*) and *Foxa3* (*HNF3γ*) are expressed throughout the developing endoderm, but targeted deletion of these genes does not affect early gut tube formation (17). Inactivation of *Foxa2* (*HNF3β*), which is normally expressed in endodermal precursors from the onset of gastrulation, results in embryonic lethality on day E7.5 to E9.5 from multiple abnormalities. Definitive endoderm cells are present in mutant embryos but fail to form a foregut and midgut (18, 19). *Foxa2*, GATA genes, and *Sox17* all exert their effects by the transcriptional activation of other genes. The identification of these gene targets and determination of how they affect endoderm formation and patterning remain great challenges in developmental biology.

Early Lung Development Requires Epithelial-Mesenchymal Interactions

Once the gut tube has formed, outpocketing of the endodermal epithelium at specific points produces the diverticula that will eventually form the thyroid, liver, pancreas, and lung. Many studies have documented that the formation of these diverticula requires a specific interaction between the endodermal epithelium and its subtending mesenchyme. The requirement for mesenchyme in the induction of lung branching morphogenesis was first demonstrated 70 years ago by Rudnick (20). In these experiments, embryonic chicken lung rudiments were grafted onto chorioallantoic membranes and cultured in ovo. If primary and secondary branches had been established in the rudiments, then normal branching morphogenesis ensued. If the mesenchyme was removed from the rudiments prior to branching, however, all further development was blocked. The development of suitable culture techniques confirmed this basic observation in both birds (21) and mammals (4, 22, 23). Lung branching morphogenesis specifically requires lung mesenchyme, and the extent of induction is dependent on the amount of mesenchyme present (24). Wessells showed that non-lung mesenchyme was capable of inducing formation of a bud in gut endoderm, but these buds did not branch further (25). Whether the endodermal buds induced by non-lung mesenchyme represented true lung induction is not known because it was not determined whether they expressed lung-specific differentiation markers. One exception to the specific requirement for lung mesenchyme may be salivary gland mesenchyme, which could support weak morphogenesis of lung epithelium (26, 27).

Lung mesenchyme is a strong inducer of branching morphogenesis in non-lung epithelium such as salivary gland (28). An impressive example of the morphogenetic capabilities of lung mesenchyme was seen when mouse lung mesenchyme was grafted onto a portion of the embryonic trachea that had been denuded of its own mesenchyme (25, 29). In these recombinants the lung mesenchyme induced a supernumerary bud that continued to branch in a typical lung-like pattern. Because pattern formation and cytodifferentiation may occur independently of each other (30), these experiments did not determine if the tracheal epithelium that had been induced to undergo lung morphogenesis had also been reprogrammed to a lung epithelial phenotype. A subsequent study (31), however, showed that the induced E13 rat tracheal epithelium exhibited specific markers of type II cell differentiation, including expression of surfactant protein C (*SP-C*), as well as the presence of osmiophilic lamellar bodies, which are the storage organelles of pulmonary surfactant. The competence of tracheal epithelium to respond to lung-inductive cues was restricted, however, because epithelium from E16 fetuses was nonresponsive to induction by E13 lung mesenchyme. Furthermore, not all endoderm was able to respond to lung mesenchyme induction: E13 esophageal and intestinal endoderm did not branch when grafted with lung mesenchyme.

The ability of lung mesenchyme to reprogram embryonic tracheal epithelium may not be particularly surprising given that the trachea arises from the endoderm immediately rostral to that which gives rise to the lung. However, a recent study by Lin et al. (32) showed that lung mesenchyme was also able to induce lung-like patterning in an unrelated epithelium, the ureteric bud. When recombined with E11.5 mouse ureteric bud epithelium, same-stage lung mesenchyme dictated a pattern of lung-like branching morphogenesis. Furthermore, these recombinants expressed *SP-C* in their most distal tips, indicating that patterning of the proximal-distal axis normally seen in the developing lung had also been induced.

The epithelium of the early respiratory tract has significant plasticity in its response to varying inductive cues. This was shown in experiments using tissue recombinants made from pseudoglandular stage lung and trachea. As noted above, tracheal epithelium grafted with lung mesenchyme was reprogrammed to undergo lung morphogenesis and epithelial differentiation. The same studies also showed that pseudoglandular stage tracheal mesenchyme completely inhibited branching when grafted onto distal lung tips that had their mesenchyme removed (25), demonstrating that branching of lung epithelium required continuous input from the associated mesenchyme. A more recent study (33) has shown that mesenchyme dictates not only the patterning of the epithelium but also its differentiated phenotype. When recombined with tracheal mesenchyme, distal lung epithelium formed a cyst made up of cells expressing markers of the proximal lung epithelium, such as cilia and mucous production. These observations, along with those demonstrating the ability of lung mesenchyme to reprogram tracheal epithelium, indicate that the entire embryonic respiratory tract epithelium, from the trachea to the distal tips, can express either a proximal or distal differentiated phenotype depending on the instructive cues it receives from its adjacent mesenchyme.

Diffusible Factors Mediate Lung Epithelial Growth and Differentiation

Pioneering studies done 50 years ago by Grobstein demonstrated that branching morphogenesis in the salivary gland was the result of short-range interactions that were mediated by diffusible signals. In the lung, Taderera (23) demonstrated that E12.5 mouse lung epithelium disintegrated when cultured by itself on a 0.45-μm filter, but branched actively when mouse lung mesenchyme was apposed to it on the other side of the filter. Chick lung mesenchyme was also effective in supporting lung epithelial branching, indicating that the requisite diffusible factors are conserved across species lines. Importantly, he also observed that the differentiation of smooth muscle and vascular elements within the lung mesenchyme was dependent on the presence of epithelium, demonstrating that inductions in lung organogenesis occur in both directions between the epithelium and mesenchyme.

These experiments were performed using purified lung epithelium, which is already specified toward lung differentiation. Transfilter recombinations prepared from embryonic rat lung and trachea have shown that specification of epithelial fate is determined by diffusible signals from the mesenchyme. When lung epithelium was enrobed in Matrigel and cultured transfilter to lung mesenchyme, it branched in a lung-like pattern (Figure 1*B*, see color insert), demonstrating that the signaling center controlling dichotomous branching of the epithelium was established across the filter. When tracheal mesenchyme was substituted for lung mesenchyme, the epithelium grew (Figure 1*C*) but did not branch, confirming that lung epithelium has no inherent capacity to branch (25). Furthermore, lung epithelium cultured transfilter to tracheal mesenchyme did not express *SP-C*; because this epithelium was *SP-C* positive at the time of isolation, these results indicate that distal lung epithelial differentiation requires continuous input from lung mesenchyme.

Tracheal epithelium cultured transfilter to tracheal mesenchyme grew and formed a cyst (Figure 1*F*) identical to that seen when lung epithelium was used as the responding epithelium. When cultured transfilter to lung mesenchyme, however, tracheal epithelium branched in a pattern identical to that seen with lung epithelium (Figure 1*E*). Furthermore, the tracheal epithelium was induced to express *SP-C* mRNA (Figure 1*G,H*), showing that induction of the distal epithelial cell phenotype did not require direct epithelial-mesenchymal contact. This observation was important because other studies have indicated that cellular processes from epithelial cells traverse the basal lamina to contact subtending fibroblasts, and the number of contacts was correlated with the extent of distal epithelial differentiation (34–36).

Two observations from these transfilter recombinant experiments indicated that close proximity of epithelium and mesenchyme was required for the induction of the lung phenotype. First, the epithelium, be it lung or trachea, never grew beyond the borders of the lung mesenchyme on the other side of the filter. Second, positioning the epithelium slightly off to the side of the lung mesenchyme resulted in no induction (Figure 1*I*). Taken together, these observations indicated that a gradient of inductive molecules produced by lung mesenchyme has a very restricted range. Such a restriction is likely necessary for limiting the number of active signaling centers in the developing lung, thus ensuring proper patterning.

Soluble Mediators of Epithelial-Mesenchymal Interactions

The complex architecture of the lung, along with its variety of differentiated cell types, estimated at over 40 (1), predicts multiple pathways of molecular regulation. Indeed, numerous soluble factors have been shown to affect various aspects of lung development. In this review we do not attempt to detail the effects of all these factors, many of which have been covered in recent work (37–39); rather we focus on a limited number of factors that have been shown to profoundly affect lung development.

Fibroblast Growth Factors and Their Receptors

The first description of a fibroblast growth factor (FGF) was made over 25 years ago by Gospodarowicz et al. (40) who discovered that *basic FGF* was a potent mitogen for 3T3 cells. Subsequent studies revealed that this FGF was but one member of a family of at least 24 related ligands. Evidence from many systems has shown that FGFs play a critical role in the development of organs, affecting cells from all three germ layers. FGF family members cause changes in cell proliferation, morphology, differentiation, and migration during organogenesis. All these responses elicited by FGFs have been demonstrated in the developing lung. In addition, all the known members of the FGF family share significant protein sequence identity within a core of approximately 120 amino acids that is responsible for one characteristic of this family—the ability to bind heparin (41). *Acidic FGF* and *basic FGF* (hereafter *FGF1* and *FGF2*, respectively) were the first FGF family members to be identified and differ from most other FGFs in that they do not contain a signal sequence for classical protein secretion; *FGF9* also does not contain a signal sequence. The receptors for *FGF1*, *FGF2*, and *FGF9* are expressed on cell surfaces, however, suggesting that these ligands act extracellularly. Evidence has shown that *FGF1* and *FGF2* can be secreted by a nonclassical, energy-dependent pathway that does not entail cell death (42).

FGF1, *FGF2*, *FGF7*, *FGF9*, *FGF10*, and *FGF18* are all expressed in the developing lung. Of these, only *FGF10* has been shown to be absolutely necessary for the initiation of lung development. Expression of *FGF10* mRNA has been localized to mesenchymal cells around distal lung epithelial tips (43). *FGF10* binds the *FGFR2IIIb*, which is found throughout the embryonic respiratory epithelium (44), with high affinity. Because *FGF10* is produced by the lung mesenchyme and its receptor is on epithelial cells, it was quickly identified as a potential mediator of epithelial mesenchymal interactions. *FGF10* appears to be the evolutionarily conserved homologue of the *Drosophila* gene *branchless*, which guides tracheal epithelial buds by chemoattraction to specific sites within the larva, thereby playing a primary role in patterning the *Drosophila* respiratory system (45). Similarly, *FGF10*-soaked beads induced dramatic chemoattraction of embryonic lung epithelium (46, 47), supporting its role in determining the spatial coordinates of the early lung. The importance of *FGF10* to lung development was most strikingly demonstrated in *FGF10* null mice, which had no lung development below the trachea (48, 49). Considered with the in vitro bead data, this result is consistent with a model in which a high local concentration of *FGF10* induces outgrowth of the two primary lung buds by chemoattraction. Since none of the other FGFs found in the lung have been shown to elicit lung epithelial chemoattraction, primary lung bud initiation cannot occur in the absence of *FGF10*. Thus *FGF10* plays a unique role in the induction of lung development that cannot be compensated by other FGF ligands.

Although a primary role for *FGF10* in lung patterning is apparent, its role in specification of epithelial phenotype has been unclear. In vitro, *FGF10* caused budding of embryonic lung epithelium in mesenchyme-free culture, but the ability of *FGF10* to sustain expression of *SP-C* was not determined (43). Similarly, a study using mesenchyme-free culture of mouse tracheal epithelium showed that a high (5 μg/ml) concentration of *FGF10* produced buds, but this study did not examine the expression of lung-specific markers (50). Another study using rat tracheal epithelium in mesenchyme-free culture was unable to demonstrate induction of *SP-C* with 1 μg/ml *FGF10* (51). It appears, however, that the local concentration of *FGF10* provided by a ligand-soaked bead, which is able to induce chemoattraction of mouse tracheal epithelium, is also able to induce *SP-C* expression (Figure 2, see color insert). How the local concentration of *FGF10* in the bead cultures relates to the level of bioactive *FGF10* that is produced by lung mesenchyme in vivo is not known. However, tightly regulated local expression of *FGF10* would be an efficient way in which to meet both the patterning and cytodifferentiation dictates of the embryonic lung.

Given the pivotal role that *FGF10* plays in lung morphogenesis, its regulation has become a topic of great interest. One proposed regulatory molecule is sonic hedgehog (*Shh*), which plays a role in the patterning of many organs. *Shh* is present throughout the epithelium in developing lung, with expression being most intense in the distal tips (52, 53). Transgenic mice in which the human *SP-C* promoter was used to overexpress Shh in the lung epithelium had hypoplastic lungs and showed reduced expression of *FGF10* mRNA (43). Mice null for *Shh* also had abnormally patterned, hypoplastic lungs (54), and the domain of *FGF10* expression was expanded such that most of the mesenchyme was positive (55). In vitro treatment of E16.5 mouse lung mesenchymal cells with *Shh* reduced *FGF10* expression by 50% (56), approximately the same extent seen in mice overexpressing *Shh* in the lung. These observations have led to a model (38) in which *Shh* produced by the most distal epithelium acts on nearby mesenchymal cells to downregulate *FGF10* levels, thereby suppressing further chemoattraction, and hence bud formation, at that location. Initiation of *FGF10* expression at other sites produces new signaling centers that give rise to the next generation of buds.

Recent evidence suggests that another factor controlling *FGF10* expression in the early lung mesenchyme is the transcription factor *Tbx4*. In the early chick embryo, *Tbx4* is expressed in mesenchymal cells adjacent to the area of the foregut from which the primary lung buds will emerge. When *Tbx4* was electroporated in ovo into foregut mesenchymal cells outside of its normal expression domain, *FGF10* expression was induced and an ectopic bud emerged (57). Furthermore, electroporation of a dominant-negative form of *Tbx4* in the region of presumptive lung bud formation suppressed both bud outgrowth and *FGF10* expression. Similarly, antisense suppression of *Tbx4* and *Tbx5* in mouse lung explants both inhibited branching and decreased *FGF10* expression (58).

FGF9 is the only other member of the FGF family that has been shown to play a critical role in lung development. In the embryonic mouse lung, *FGF9* is initially expressed in both the endodermal epithelium and mesothelium, then exclusively in the mesothelium from the pseudoglandular stage onward (59). Targeted deletion of *FGF9* resulted in decreased branching morphogenesis and pulmonary hypoplasia, which was attributed to reduced mesenchymal proliferation resulting from a decrease in *FGF10* expression (60). Recent in vitro data suggest that *FGF9* inhibits *Shh*-induced differentiation of peripheral lung mesenchymal cells, thereby maintaining a defined population of multipotent progenitor cells (61).

Similar to *FGF10*, *FGF7* (*KGF*) is secreted by mesenchymal cells in the lung. *FGF7* binds and activates only *FGFR2IIIb* on epithelial cells, thus making it another potential mediator of epithelial-mesenchymal interactions in the lung. Indeed, *FGF7* has potent effects on both the prenatal and postnatal lung epithelium. Targeted overexpression of *FGF7* in the developing lung epithelium using the *SP-C* promoter resulted in abnormal morphogenesis that resembled cystadenomatoid malformation (62). Similarly, increasing *FGF7* levels in the postnatal lung by instillation (63) or by conditional, targeted overexpression (64) caused adenomatous hyperplasia. In vitro lung explants treated with *FGF7* showed epithelial hyperplasia and cystic dilation (65, 66). In mesenchyme-free culture of embryonic mouse lung epithelium, *FGF7* by itself induced cystic dilation and patchy expression of *SP-C* (67). When given with other competence factors, however, *FGF7* stimulated widespread proliferation and expression of *SP-C* (68). Competence factors are critical for the *FGF7* response. When *FGF7* alone was given to embryonic tracheal epithelium, there was no response. In the presence of serum and elevated cyclic AMP, however, *FGF7* induced the reprogramming of tracheal epithelium to an alveolar type II cell phenotype (51). Because the competence factors by themselves were not able to elicit this response, it was concluded that *FGF7* was necessary but not sufficient to induce distal lung epithelial differentiation. *FGF7* has also been shown to stimulate type II cell differentiation in late gestation (69) and postnatally (70). However, mesenchyme-derived factors other than *FGF7* must drive type II cell differentiation because inhibiting *FGF7* activity in cocultures of adult type II cells and lung fibroblasts only partially inhibited differentiation (71). Targeted deletion of the *FGF7* gene showed no lung abnormalities (72), suggesting functional compensation by another FGF, most likely *FGF10*.

Compelling data demonstrating a primary role of FGFs in lung development have also come from studies in which FGF signaling pathways have been disrupted by the alteration or elimination of FGF receptors. At present, four FGFR have been identified, each encoded by a separate gene, and are all found in the lung. Using the *SP-C* promoter to target overexpression of a dominant-negative *FGFR2IIIb* to the surface of lung epithelial cells, Peters et al. (73) demonstrated that lung development did not progress beyond the generation of the trachea and two unbranched bronchi. Celli et al. (74) used the metallothionein promoter to express a soluble dominant-negative *FGFR2IIIb* throughout the entire developing embryo and also observed lung bud initiation but no branching morphogenesis. Targeted deletion of the *FGFR2* gene resulted in embryonic lethality shortly after implantation due to trophectoderm defects (75). When these defects were circumvented by using tetraploid fusion chimeras, *FGFR2* null mice showed no development of the respiratory tree below the trachea (76). Using the Cre-loxP recombination system, De Moerlooze et al. showed that pulmonary agenesis occurred when the DNA coding for the IIIb exon was deleted (77). This phenotype was identical to that seen in *FGF10* null mice, indicating that FGF10 signaling transduced through *FGFR2IIIb* is absolutely required for lung development.

The complete lack of lungs in mice in which signaling through *FGFR2IIIb* was abrogated, however, precluded examination of the role of FGFR-mediated signaling in later aspects of development, such as alveolarization and epithelial maturation. A recent study (78) has used bitransgenic mice to conditionally express a soluble dominant-negative FGF receptor (FGFR-Fc) in the lung. In accordance with earlier results (74), severe fetal lung hypoplasia resulted when FGFR-Fc was expressed from E6.5 to E18.5. When the transgene was activated postnatally, however, no effects on lung size, morphology, and alveolarization were seen. Similarly, Perl et al. (79) used lung-targeted conditional expression of the FGF receptor antagonist *Sprouty-4* to demonstrate that abrogating FGF signaling early in development caused severe hypoplasia and patterning defects, whereas expressing *Sprouty-4* postnatally had only a mild effect on alveolarization. Taken together, these data suggest that although FGF signaling is crucial for lung patterning during the embryonic and pseudoglandular periods, its role in postnatal development is much less critical.

FGFR2 is not the only FGF receptor signaling system used in the developing lung. Although mice null for *FGFR3* or *FGFR4* have no abnormal lung phenotype, mice null for both *FGFR3* and *FGFR4* show defective alveolar septation and abnormally elevated elastin (80). The role of *FGFR1*-mediated signaling in lung development is at present unknown because animals null for *FGFR1* die early in development (81).

The relevant interaction of FGFs with heparin occurs through their low-affinity binding of the related molecule, heparan sulfate. Heparan sulfate is found on cell surfaces and in the ECM in the form of heparan sulfate proteoglycan (HSPG). HSPGs bind FGFs and induce or stabilize formation of FGF dimers, or a ternary complex composed of ligand plus high- and low-affinity receptors (82). Beyond acting as coreceptors for FGFs on the cell surface, the HSPG that exists in the extracellular matrix can bind FGFs and thus may act as a reservoir for FGFs in the extracellular space (83). Treating lung explants with heparitinase caused decreased branching morphogenesis (84), as did adding excess heparin (85). A recent study (86) has refined the role of HSPGs in lung development, suggesting that varying levels of O-sulfation in heparan sulfate glycosaminoglycans mediate the effects of FGF10 in lung patterning.

HSPGs may not be alone in their ability to facilitate FGF binding. A recent report (87) has shown that oversulfated forms of chondroitin sulfate (CS) proteoglycan are also able to bind heparin-binding growth factors, including *FGF2*, *FGF10*, and *FGF18*, all of which are found in the developing lung. These observations gain importance in view of earlier studies showing that inhibiting proteoglycan synthesis in lung explants with β-xyloside disrupted lung branching and differentiation (88, 89) and that it was CSPG (and not HSPG) synthesis that was affected (90). Furthermore, a recent study has shown that treating lung explants and tissue recombinants with chondroitinase abrogated branching morphogenesis but did not affect the induction of *SP-C* (90a).

Sonic Hedgehog, Hedgehog-Interacting Protein, and Gli Transcription Factors

The Hedgehog signaling pathway plays a important role in the development of multiple organs (91, 92). Because *Shh* is expressed in the developing lung epithelium (52, 53) and its primary receptor, *Patched-1* (*Ptc*), is found in mesenchymal cells (52, 61, 93), *Shh* signaling is a model of epithelium-to-mesenchyme

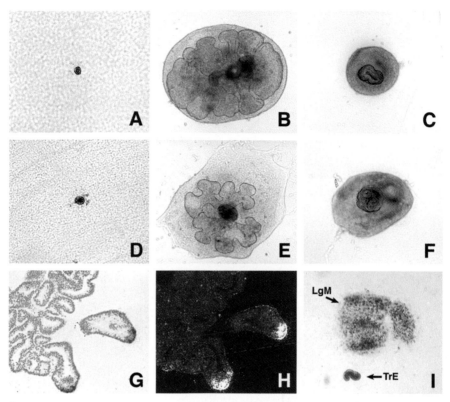

Figure 1 Diffusible factors mediate epithelial-mesenchymal interactions. No growth occurs when either embryonic rat lung (*A*) or tracheal (*D*) epithelium are enrobed in Matrigel and cultured transfilter in the absence of mesenchyme. Lung epithelium cultured transfilter to lung mesenchyme (*B*) shows substantial growth and branching at its distal tips. Lung epithelium cultured transfilter to tracheal mesenchyme (*C*) shows significant growth and no branching, a pattern identical to that seen when tracheal epithelium is cultured transfilter to tracheal mesenchyme (*F*). Tracheal epithelium cultured transfilter to lung mesenchyme (*E*) grows and branches in a lung-like pattern. This reprogramming of morphogenesis is accompanied by respecification of fate because in situ hybridization shows that cells in the distal tips of these rudiments express mRNA for the lung-specific marker *SP-C* (*G, bright field*; *H*, corresponding *dark field*). The inductive cues produced by lung mesenchyme are effective only over a short distance, because positioning the tracheal epithelium (TrE) off to the side of the transfilter lung mesenchyme (LgM) is sufficient to prevent induction (*I*). Note also that the margins of the epithelial rudiments do not extend beyond the border of the mesenchyme. Panels A–F and I are at the same magnification, as are panels G and H.

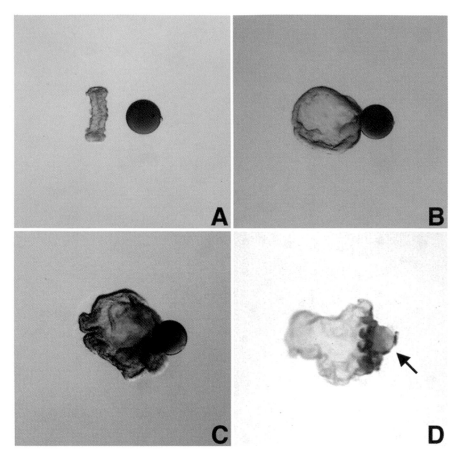

Figure 2 *FGF10* both chemoattracts and reprograms tracheal epithelium. When purified embryonic mouse tracheal epithelium is enrobed in Matrigel and placed near a
heparin bead that has been soaked in *FGF10* (*A*), the rudiment moves toward the bead,
contacting it after 48 h (*B*). By 96 h (*C*) the rudiment has expanded in size and has
begun to engulf the bead. Whole mount in situ hybridization (*D*) shows that those cells
closest to the bead (*arrow*) express *SP-C* (*dark purple*), indicating that a locally high
concentration of *FGF10* is capable of inducing a lung-specific marker.

induction. In the conventional model (see 91, 92 for reviews) of *Shh* signaling, *Ptc* directly interacts with another protein, *Smoothened (Smo)*, within the plasma membrane, and represses its signaling activity; the actual nature of this repression is unknown. Binding of *Shh* to *Ptc* releases this repression, thereby allowing *Smo*, which is closely related to the Frizzled family of Wnt gene receptors, to activate its intracellular target(s).

The expression of *Shh* in the developing lung is dynamic: It is initially expressed throughout the epithelium but becomes restricted to subsets of cells from day E16.5 onward (94). The observation that *Shh* null mice have severely hypoplastic lungs (54, 55) indicates that *Shh* is not required for lung bud formation but rather is involved in regulating branching morphogenesis. Although *Shh* deletion profoundly affects endoderm patterning, specification of epithelial cell types appears to be unaffected (55). Whereas a lack of *Shh* disrupts lung development, excessive amounts of *Shh* are equally deleterious. When Bellusci and colleagues (52) used the human *SP-C* promoter to overexpress *Shh* in the lung epithelium, they also observed pulmonary hypoplasia, albeit not as severe as that seen in *Shh* null mice. Levels of *Shh* expression, therefore, must be tightly regulated in order for normal lung development to ensue.

How are *Shh* levels modulated? One way is through binding to *Ptc* (95). *Shh* upregulates *Ptc* expression, and any *Ptc* in excess of that involved in signaling binds *Shh* and sequesters it near its source of production. This creates a negative feedback loop that restrains the spread of *Shh*. Another molecule regulating *Shh* levels is *Hedgehog-interacting protein 1 (Hip1)*, a membrane-bound protein that binds all mammalian hedgehog proteins and, similar to *Ptc*, is upregulated in response to *Shh* (96). The expression domains of *Hip1* and *Ptc* in the lung overlap (97), suggesting a functional redundancy. Targeted deletion of *Hip1* resulted in lung hypoplasia with a failure in secondary bud formation (97), which was attributed to a loss of *FGF10* expression at the prospective sites of bud formation. These data, along with those described in the previous section (55, 56), underscore the importance of *Shh* levels in determining the correct spatial localization of *FGF10*. A paradox that awaits explanation is how *FGF10* levels remain highest in the distal tip mesenchyme, which is adjacent to distal tip epithelium, the site of highest expression of *Shh*, which antagonizes *FGF10* expression.

An in vivo study (52) showed that *Shh* increased proliferation of lung mesenchymal cells. In vitro treatment of E11.5 purified mouse lung mesenchyme rudiments with *Shh* also increased proliferation and, additionally, upregulated expression of *Ptc*, *BMP-4*, the BMP antagonist *Noggin*, and smooth muscle actin and myosin (61). The same study demonstrated that *FGF9* antagonizes the effects of *Shh* on mesenchymal cell differentiation but does not affect proliferation. From these observations the investigators proposed a model in which mesenchymal cells are sandwiched between epithelium and the mesothelium that covers the surface of the lung. *Shh* produced by the epithelium stimulates mesenchymal cell proliferation and differentiation, whereas *FGF9* from the mesothelium antagonizes differentiation, thereby maintaining a population of multipotent cells in the lung periphery.

Lung mesenchyme cells exposed to exogenous *Shh* upregulate expression of *Foxf1*, and lungs of *Shh* null embryos do not express *Foxf1* in the mesenchyme (98). Furthermore, the morphology of lungs from embryos with a haploinsufficiency of *Foxf1* resembled those in *Shh* null mice and showed decreased expression of *Ptc* and *Gli3* (99). Taken together, these data suggest *Foxf1* is a key target gene for *Shh* signaling. The genes that are controlled by *Foxf1* are at present unknown.

The three Gli genes (1, 2, and 3) code for zinc finger transcription factors that are the vertebrate counterparts of the gene *cubitus interruptus*, the principal effector of *hedgehog* signaling in *Drosophila*. All three Gli genes are expressed in distinct but overlapping domains in lung mesenchyme, with expression being highest in the distal tips (100, 101). Mice in which the DNA-binding domain has been removed from *Gli1* were viable and appeared normal (102). Mice null for *Gli3* were also viable, but their lungs showed reductions in the size and shape of the different lobes (101). $Gli2^{-/-}$ mice died at birth with hypoplastic lungs that exhibited severe patterning defects (93). The right lung, which normally has four lobes, formed only one. BrdU labeling index was decreased by 40% in the lung mesenchyme of $Gli2^{-/-}$ embryos, consistent with the demonstrated ability of *Shh* to stimulate proliferation.

The complexity of the involvement of Gli genes in lung development has been demonstrated by the analysis of compound mutant mice. Embryos expressing different combinations of Gli genes showed a range of lung defects, the most striking of which was the absence of lungs, trachea, and esophagus in $Gli2^{-/-}$, $Gli3^{-/-}$ compound mutants (93). The presence of a single *Gli3* allelle ($Gli2^{-/-}$, $Gli3^{+/-}$) was enough to effect formation of a hypoplastic lungs, but the left and right lungs did not separate, and the embryos developed tracheoesophageal fistula. The phenotype seen in *Gli2/Gli3* double null embryos was more severe than that seen in *Shh* null animals, suggesting that the Gli genes may be involved in signaling pathways other than *Shh*. The complete lack of endodermal structures between the thymus and stomach in *Gli2/Gli3* double null animals suggests a role for Gli genes in formation of the early foregut. Downstream target genes regulated by the Gli genes in the developing lung have yet to be identified; it has been suggested that Glis may regulate expression of *Foxf1* (98), but direct evidence of this is lacking.

Wnt Genes and β-Catenin

The vertebrate Wnt growth factor family contains at least 21 secreted cysteine-rich glycoproteins that regulate cell-cell interactions in many embryonic tissues (103). Wnts signal through multiple pathways, the most well-characterized being the canonical *β-catenin/LEF-TCF* pathway (104). In this pathway, secreted Wnt proteins bind to Frizzled (Fzd) receptors on cell membranes, activating *Disheveled* protein, which in turn inactivates GSK-3β, thereby inhibiting phosphorylation of *β-catenin*. Hypophosphorylation stabilizes *β-catenin*, which accumulates in the cytoplasm, and is then transported to the nucleus where it heterodimerizes with members of the Lef/TCF family of transcription factors to activate downstream target genes. At least nine Wnt genes and six Fzd genes are expressed in the lung.

Wnt7b, which is expressed throughout the lung and tracheal epithelium, signals through the canonical pathway. Mice null for *Wnt7b* had hypoplastic lungs and died at birth of respiratory failure (105). Whereas differentiation of Clara cells and alveolar type II cells were normal in these animals, alveolar type I cell differentiation was impaired. Vascular development was also disrupted, particularly in the large pulmonary vessels, which were surrounded by areas of hemorrhage. *Wnt7b* was shown to be a mitogen for lung mesenchyme cells in early (E12.5) development, and the lack of *Wnt7b* increased vascular smooth muscle cell apoptosis.

The overall importance of the canonical Wnt signaling pathway to lung development has been recently demonstrated by Mucenski et al. (105a) who used targeted, conditional expression of Cre recombinase to excise loxP flanked *β-catenin* from the developing mouse lung epithelium. These animals had lungs in which deficits in lung patterning were apparent by day E13.5 and which were hypoplastic on day E18.5. Bronchiolar tubules were enlarged and elongated, and their constitutive cells expressed predominantly proximal epithelial markers such as *CC10* and *Foxj1*. Epithelial cells expressing the distal marker *SP-C* were infrequently observed. These observations suggest that *β-catenin* may be involved in the specification of the distal epithelial cell fate.

Wnt5a, which signals through a noncanonical pathway (106), was shown to be expressed in both the lung epithelium and mesenchyme early in development, with expression appearing highest in the distal tips. By the end of gestation, *Wnt5a* expression was restricted almost exclusively to the epithelium. Mice null for *Wnt5a* died shortly after birth from respiratory failure (107). The lungs of *Wnt5a* $^{-/-}$ mice had the correct number of lobes, but were slightly larger than controls owing to increased proliferation in both the epithelial and mesenchymal compartments. In addition, distal branching was increased and the interstitium was thickened. Epithelial cell differentiation occurred in these mice but appeared somewhat delayed. Tracheal development was compromised, with the tracheae being stunted and having a reduced number of cartilage rings. Because *Wnt5a* null embryos showed increased expression of *Shh*, it was proposed that *Wnt5a* signaling is required for the normal downregulation of Shh, and in the absence of *Wnt5a*, Shh-induced mesenchymal proliferation continues in late gestation. Expression of *FGF10* was also increased in *Wnt5a* null animals, which is somewhat paradoxical in light of the increased *Shh* levels. Taken as a whole, however, these data suggest that *Wnt5a* may interact with other signaling pathways critical for lung development.

BMP4, Noggin, and Gremlin

Bone morphogenetic protein 4 (BMP4), a member of the TGF$β$ superfamily of proteins, has been shown to play a role in diverse developmental processes (108). Using a transgenic mouse strain in which lacZ was inserted into the *BMP4* locus, Weaver et al. (109) showed that early in mouse development (19 somites) *BMP4* is expressed in the splanchnic mesenchyme surrounding the gut tube where the future lung buds will form. After lung buds have formed, *BMP4* continues to be expressed in the more proximal mesenchyme, but is also intensely expressed in the epithelium of the distal tips. Its localization in morphogenetically active signaling centers suggested a role for *BMP4* as a mediator of epithelial-mesenchymal interactions.

The precise level of *BMP4* expression, similar to that of *Shh* expression, appears to be critical for normal lung development. Overexpression of *BMP4* in the distal lung epithelium resulted in lung hypoplasia and distension of terminal airspaces, which was accompanied by a decrease in epithelial cell proliferation and the number of *SP-C*-positive cells (110). When the *SP-C* promoter was used to drive misexpression of either *Noggin*, a direct antagonist of *BMP4*, or a dominant-negative form of the *BMP4* receptor, *Alk6*, the lungs were also hypoplastic. Additionally, these lungs showed a reduced number of *SP-C*-positive cells and an expansion of epithelial cells expressing the proximal markers *CC10* and *Foxj1* (109). Similar results were obtained in vivo (111) and in vitro (112) using another *BMP4* antagonist, *Gremlin*. These data suggest that *BMP4* is involved in proximal-distal patterning of the lung epithelium.

Several studies have shown that expression of *BMP4* in respiratory epithelium is induced by FGF family members (47, 56, 113). The effect of exogenous *BMP4* on either lung (47) or tracheal (113) epithelium in mesenchyme-free culture was to inhibit proliferation. These observations appear to conflict with reports that showed increased branching of E11.5 mouse lung explants when *BMP4* was added to the medium (112, 114). In one study, however (114), direct injection of *BMP4* into the lung lumen had no effect on branching, suggesting that the effects of *BMP4* added to the medium were mediated by the mesenchyme. In this case, *BMP4* might induce the mesenchyme to produce factors that either antagonize the inhibitory effects of *BMP4* on epithelial proliferation or directly stimulate epithelial proliferation. Mesenchymal expression of *BMP4* appears to be controlled by *Shh* (54, 61). This regulation may be restricted to the early period of lung development because mice overexpressing *Shh* under control of the *SP-C* promoter showed no difference in *BMP4* expression (52). Taken together, these data support a model in which *BMP4* inputs from both the epithelial and mesenchymal tissue compartments, which are differentially regulated, effect the precisely controlled levels of ligand necessary for normal morphogenesis.

Although there is a decreased number of *SP-C*-positive cells in lungs of mice in which *BMP4* is antagonized by *Noggin* overexpression, *BMP4* itself does not appear to specify the distal lung epithelial phenotype. Hyatt et al. (113) showed that addition of *BMP4* to mesenchyme-free cultures of E11.5 mouse tracheal epithelium not only inhibited FGF-induced proliferation but also decreased its induction of *SP-C*. Furthermore, antagonizing *BMP4* with *Noggin* in these cultures had no significant effect on *SP-C* expression but greatly increased expression of the proximal markers *CC10* and *Foxj1*. Thus the role for *BMP4* in the distal tips may not be to induce distal differentiation, but rather to prevent epithelial cells from adopting a proximal phenotype, which would allow the cells to remain competent to respond to other inductive cues (e.g., FGFs) that specify the distal phenotype.

CONCLUDING REMARKS

The renaissance in the study of lung development of the last decade has shown that the epithelial-mesenchymal interactions controlling morphogenesis and differentiation are vastly complex. In this review we have covered only some of the soluble mediators of these interactions. Others, such as TGFβs, retinoic acid, PDGFs, EGF, VEGFs, and IGFs, have also been shown to be important for normal lung development, and details on their activities can be found in recent reviews (38, 39). Additional molecular regulators are now being identified through the application of cDNA microarray technology (115, 116). Identification of these key molecular regulators, however, is just the first step in understanding how they affect lung development. Further questions need to be addressed before the importance of any given factor can be fully ascertained. What is its level of expression, and are gradients formed? When does expression begin, and how long does it last? Where is it expressed, and does this change over the course of development? Does it interact with other factors and, if so, which ones? Our understanding of epithelial-mesenchymal interactions in the lung will increase as these questions are answered for an expanding list of molecular mediators. With this understanding should come new strategies for the prevention and treatment of pathologies resulting from abnormal lung development.

ACKNOWLEDGMENTS

The authors thank Jeff Whitsett, Mike Mucenski, Anne Perl, Ann Akeson, Jim Greenberg, and Tim LeCras for helpful discussions. We also thank lab members, past and present, who have contributed to the work presented here. These include Dennis Schellhase, Robin Deterding, Larry Nielsen, Moshe Kalina, Sarah Gebb, Tianli Pan, Sui Lin, Mike Burhans, Tyler Martin, Kathy Shannon, Kalpana Srivastava, and Xiaofei Shangguan. Brian Hyatt is a Parker B. Francis Fellow. This work was supported by a Franklin Delano Roosevelt Fellowship from the March of Dimes, and by grants P50HL56387 and R01HL071898 from the NIH/NHLBI.

The *Annual Review of Physiology* is online at http://physiol.annualreviews.org

LITERATURE CITED

1. Crapo JD, Barry BE, Gehr P, Bachofen M, Weibel ER. 1982. Cell number and cell characteristics of the normal human lung. *Am. Rev. Respir. Dis.* 125:332–37

2. Williams MC. 2003. Alveolar type I cells: phenotype and development. *Annu. Rev. Physiol.* 65:669–95

3. Fehrenbach H. 2001. Alveolar epithelial

type II cell: defender of the alveolus revisited. *Respir. Res.* 2:33–46

4. Spooner BS, Wessells NK. 1970. Mammalian lung development: interactions in primordium formation and bronchial morphogenesis. *J. Exp. Zool.* 175:445–54

5. Wells JM, Melton DA. 1999. Vertebrate endoderm development. *Annu. Rev. Cell Dev. Biol.* 15:393–410

6. Iannaccone PM, Zhou X, Khokha M, Boucher D, Kuehn MR. 1992. Insertional mutation of a gene involved in growth regulation of the early mouse embryo. *Dev. Dyn.* 194:198–208

7. Conlon FL, Lyons KM, Takaesu N, Barth KS, Kispert A, et al. 1994. A primary requirement for nodal in the formation and maintenance of the primitive streak in the mouse. *Development* 120:1919–28

8. Lowe LA, Yamada S, Kuehn MR. 2001. Genetic dissection of nodal function in patterning the mouse embryo. *Development* 128:1831–43

9. Gao X, Sedgwick T, Shi YB, Evans T. 1998. Distinct functions are implicated for the GATA-4, -5, and -6 transcription factors in the regulation of intestine epithelial cell differentiation. *Mol. Cell Biol.* 18:2901–11

10. Kuo CT, Morrisey EE, Anandappa R, Sigrist K, Lu MM, et al. 1997. GATA4 transcription factor is required for ventral morphogenesis and heart tube formation. *Genes Dev.* 11:1048–60

11. Molkentin JD, Lin Q, Duncan SA, Olson EN. 1997. Requirement of the transcription factor GATA4 for heart tube formation and ventral morphogenesis. *Genes Dev.* 11:1061–72

12. Morrisey EE, Tang Z, Sigrist K, Lu MM, Jiang F, et al. 1998. GATA6 regulates HNF4 and is required for differentiation of visceral endoderm in the mouse embryo. *Genes Dev.* 12:3579–90

13. Wegner M. 1999. From head to toes: the multiple facets of Sox proteins. *Nucleic Acids Res.* 27:1409–20

14. Hudson C, Clements D, Friday RV, Stott D, Woodland HR. 1997. Xsox17-α and -β mediate endoderm formation in *Xenopus*. *Cell* 91:397–405

15. Clements D, Woodland HR. 2000. Changes in embryonic cell fate produced by expression of an endodermal transcription factor, Xsox17. *Mech. Dev.* 99:65–70

16. Kanai-Azuma M, Kanai Y, Gad JM, Tajima Y, Taya C, et al. 2002. Depletion of definitive gut endoderm in Sox17-null mutant mice. *Development* 129:2367–79

17. Kaestner KH, Katz J, Liu Y, Drucker DJ, Schutz G. 1999. Inactivation of the winged helix transcription factor HNF3alpha affects glucose homeostasis and islet glucagon gene expression in vivo. *Genes Dev.* 13:495–504

18. Weinstein DC, Ruiz i Altaba A, Chen WS, Hoodless P, Prezioso VR, et al. 1994. The winged-helix transcription factor HNF-3 beta is required for notochord development in the mouse embryo. *Cell* 78:575–88

19. Ang SL, Wierda A, Wong D, Stevens KA, Cascio S, et al. 1993. The formation and maintenance of the definitive endoderm lineage in the mouse: involvement of HNF3/forkhead proteins. *Development* 119:1301–15

20. Rudnick D. 1933. Developmental capacities of the chick lung in chorioallantoic grafts. *J. Exp. Zool.* 66:125–54

21. Dameron F. 1961. Etude de la morphogénèse de la bronche de l'embryon de Poulet associeé à differents mesenchymes en culture in vitro. *C. R. Acad. Sci.* 262:1642–45

22. Sampaolo G, Sampaolo L. 1959. Observations histologiques sur le poumon de foetus de Cobaye, cultivé in vitro. *C. R. Assoc. Anat.* 45:707–14

23. Taderera JT. 1967. Control of lung differentiation in vitro. *Dev. Biol.* 16:489–512

24. Masters JRW. 1976. Epithelial-mesenchymal interaction during lung

development: the effect of mesenchymal mass. *Dev. Biol* 51:98–108

25. Wessells N. 1970. Mammalian lung development: interactions in formation and morphogenesis of tracheal buds. *J. Exp. Zool.* 175:455–66

26. Lawson KA. 1972. The role of mesenchyme in the morphogenesis and functional differentiation of rat salivary epithelium. *J. Embryol. Exp. Morphol.* 27:497–513

27. Ball WD. 1974. Development of the rat salivary glands. III. Mesenchymal specificity in the morphogenesis of the embryonic submaxillary and sublingual glands of the rat. *J. Exp. Zool.* 188:277–88

28. Lawson KA. 1974. Mesenchyme specificity in rodent salivary gland development: the response of salivary epithelium to lung mesenchyme in vitro. *J. Embryol. Exp. Morphol.* 32:469–93

29. Alescio T, Cassini A. 1962. Induction in vitro of tracheal buds by pulmonary mesenchyme grafted on tracheal epithelium. *J. Exp. Zool.* 150:83–94

30. Sakakura T, Nishizuka Y, Dawe CJ. 1976. Mesenchyme-dependent morphogenesis and epithelium-specific cytodifferentiation in mouse mammary gland. *Science* 194:1439–41

31. Shannon JM. 1994. Induction of alveolar type II cell differentiation in fetal tracheal epithelium by grafted distal lung mesenchyme. *Dev. Biol.* 166:600–14

32. Lin Y, Zhang S, Rehn M, Itaranta P, Tuukkanen J, et al. 2001. Induced repatterning of type XVIII collagen expression in ureter bud from kidney to lung type: association with sonic hedgehog and ectopic surfactant protein C. *Development* 128:1573–85

33. Shannon JM, Nielsen LD, Gebb SA, Randell SH. 1998. Mesenchyme specifies epithelial differentiation in reciprocal recombinants of embryonic lung and trachea. *Dev. Dyn.* 212:482–94

34. Grant M, Cutts N, Brody J. 1983. Alterations in lung basement membrane dur-

ing fetal growth and type 2 cell development. *Dev. Biol.* 97:173–83

35. Adamson I, King G. 1984. Sex-related differences in cellular composition and surfactant synthesis of developing fetal rat lungs. *Am. Rev. Respir. Dis.* 129:130–34

36. Bluemink JG, Van Maurik P, Lawson KA. 1976. Intimate cell contacts at the epithelial/mesenchymal interface in embryonic mouse lung. *J. Ultrastruct. Res.* 55:257–70

37. Shannon JM, Deterding RR. 1997. Epithelial-mesenchymal interactions in lung development. In *Lung Growth and Development*, ed. JA McDonald, pp. 81–118. New York: Dekker

38. Cardoso WV. 2001. Molecular regulation of lung development. *Annu. Rev. Physiol.* 63:471–94

39. Warburton D, Schwarz M, Tefft D, Flores-Delgado G, Anderson KD, Cardoso WV. 2000. The molecular basis of lung morphogenesis. *Mech. Dev.* 92:55–81

40. Gospodarowicz D, Ferrara N, Schweigerer L, Neufeld G. 1987. Structural characterization and biological functions of fibroblast growth factor. *Endocr. Rev.* 8:95–114

41. McKeehan WL, Wang F, Kan M. 1998. The heparan sulfate-fibroblast growth factor family: diversity of structure and function. *Prog. Nucleic Acid Res. Mol. Biol.* 59:135–76

42. Guillonneau X, Regnier-Ricard F, Dupuis C, Courtois Y, Mascarelli F. 1997. FGF2-stimulated release of endogenous FGF1 is associated with reduced apoptosis in retinal pigmented epithelial cells. *Exp. Cell Res.* 233:198–206

43. Bellusci S, Grindley J, Emoto H, Itoh N, Hogan BL. 1997. Fibroblast growth factor 10 (FGF10) and branching morphogenesis in the embryonic mouse lung. *Development* 124:4867–78

44. Urtreger AO. 1993. Developmental localization of the splicing alternatives

of fibroblast growth factor receptor-2 (FGFR2). *Dev. Biol.* 158:475–86

45. Sutherland D, Samakovlis C, Krasnow MA. 1996. *branchless* encodes a *Drosophila* FGF homolog that controls tracheal cell migration and the pattern of branching. *Cell* 87:1091–101

46. Park WY, Miranda B, Lebeche D, Hashimoto G, Cardoso WV. 1998. FGF-10 is a chemotactic factor for distal epithelial buds during lung development. *Dev. Biol.* 201:125–34

47. Weaver M, Dunn NR, Hogan BL. 2000. Bmp4 and Fgf10 play opposing roles during lung bud morphogenesis. *Development* 127:2695–704

48. Min H, Danilenko DM, Scully SA, Bolon B, Ring BD, et al. 1998. Fgf-10 is required for both limb and lung development and exhibits striking functional similarity to *Drosophila branchless*. *Genes Dev.* 12:3156–61

49. Sekine K, Ohuchi H, Fujiwara M, Yamasaki M, Yoshizawa T, et al. 1999. Fgf10 is essential for limb and lung formation. *Nat. Genet.* 21:138–41

50. Ohtsuka N, Urase K, Momoi T, Nogawa H. 2001. Induction of bud formation of embryonic mouse tracheal epithelium by fibroblast growth factor plus transferrin in mesenchyme-free culture. *Dev. Dyn.* 222:263–72

51. Shannon JM, Gebb SA, Nielsen LD. 1999. Induction of alveolar type II cell differentiation in embryonic tracheal epithelium in mesenchyme-free culture. *Development* 126:1675–88

52. Bellusci S, Furuta Y, Rush MG, Henderson R, Winnier G, Hogan BLM. 1997. Involvement of Sonic hedgehog (*Shh*) in mouse enbryonic lung growth and morphogenesis. *Development* 124:53–63

53. Urase K, Mukasa T, Igarashi H, Ishii Y, Yasugi S, et al. 1996. Spatial expression of *Sonic hedgehog* in the lung epithelium during branching morphogenesis. *Biochem. Biophys. Res. Commun.* 225:161–66

54. Litingtung Y, Lei L, Westphal H, Chiang C. 1998. Sonic hedgehog is essential to foregut development. *Nat. Genet.* 20:58–61

55. Pepicelli CV, Lewis PM, McMahon AP. 1998. Sonic hedgehog regulates branching morphogenesis in the mammalian lung. *Curr. Biol.* 8:1083–86

56. Lebeche D, Malpel S, Cardoso WV. 1999. Fibroblast growth factor interactions in the developing lung. *Mech. Dev.* 86:125–36

57. Sakiyama J, Yamagishi A, Kuroiwa A. 2003. Tbx4-Fgf10 system controls lung bud formation during chicken embryonic development. *Development* 130:1225–34

58. Cebra-Thomas JA, Bromer J, Gardner R, Lam GK, Sheipe H, Gilbert SF. 2003. T-box gene products are required for mesenchymal induction of epithelial branching in the embryonic mouse lung. *Dev. Dyn.* 226:82–90

59. Colvin JS, Feldman B, Nadeau JH, Goldfarb M, Ornitz DM. 1999. Genomic organization and embryonic expression of the mouse fibroblast growth factor 9 gene. *Dev. Dyn.* 216:72–88

60. Colvin JS, White AC, Pratt SJ, Ornitz DM. 2001. Lung hypoplasia and neonatal death in Fgf9-null mice identify this gene as an essential regulator of lung mesenchyme. *Development* 128:2095–106

61. Weaver M, Batts L, Hogan BL. 2003. Tissue interactions pattern the mesenchyme of the embryonic mouse lung. *Dev. Biol.* 258:169–84

62. Simonet WS, DeRose ML, Bucay N, Nguyen HQ, Wert SE, et al. 1995. Pulmonary malformation in transgenic mice expressing human keratinocyte growth factor in the lung. *Proc. Natl. Acad. Sci. USA* 92:12461–65

63. Ulich TR, Yi ES, Longmuir K, Yin S, Bilta R, et al. 1994. Keratinocyte growth factor is a growth factor for type II pneumocytes in vivo. *J. Clin. Invest.* 93:1298–306

64. Tichelaar JW, Lu W, Whitsett JA. 2000. Conditional expression of fibroblast growth factor-7 in the developing and mature lung. *J. Biol. Chem.* 275:11858–64

65. Shiratori M, Oshika E, Ung LP, Singh G, Shinozuka H, et al. 1996. Keratinocyte growth factor and embryonic rat lung morphogenesis. *Am. J. Respir. Cell Mol. Biol.* 15:328–38

66. Zhou L, Graeff RW, McCray PB Jr, Simonet WS, Whitsett JA. 1996. Keratinocyte growth factor stimulates CFTR-independent fluid secretion in the fetal lung in vitro. *Am. J. Physiol. Lung Cell Mol. Physiol.* 271:L987–94

67. Cardoso WV, Itoh A, Nogawa H, Mason I, Brody JS. 1997. FGF-1 and FGF-7 induce distinct patterns of growth and differentiation in embryonic lung epithelium. *Dev. Dyn.* 208:398–405

68. Deterding RR, Jacoby CR, Shannon JM. 1996. Acidic fibroblast growth factor and keratinocyte growth factor stimulate fetal rat pulmonary epithelial growth. *Am. J. Physiol. Lung Cell Mol. Physiol.* 271:L495–505

69. Chelly N, Mouhieddine-Gueddiche OB, Barlier-Mur AM, Chailley-Heu B, Bourbon JR. 1999. Keratinocyte growth factor enhances maturation of fetal rat lung type II cells. *Am. J. Respir. Cell Mol. Biol.* 20:423–32

70. Sugahara K, Rubin JS, Mason RJ, Aronsen EL, Shannon JM. 1995. Keratinocyte growth factor increases mRNAs for SP-A and SP-B in rat alveolar type II cells in culture. *Am. J. Physiol. Lung Cell Mol. Physiol.* 269:L344–50

71. Shannon JM, Pan T, Nielsen LD, Edeen KE, Mason RJ. 2001. Lung fibroblasts improve differentiation of rat type II cells in primary culture. *Am. J. Respir. Cell Mol. Biol.* 24:235–44

72. Guo L, Degenstein L, Fuchs E. 1996. Keratinocyte growth factor is required for hair development but not for wound healing. *Genes Dev.* 10:165–75

73. Peters K, Werner S, Liao X, Wert S, Whitsett J, Williams L. 1994. Targeted expression of a dominant-negative FGF receptor blocks branching morphogenesis and epithelial differentiation of the mouse lung. *EMBO J.* 13:3296–301

74. Celli G, LaRochelle WJ, Mackem S, Sharp R, Merlino G. 1998. Soluble dominant-negative receptor uncovers essential roles for fibroblast growth factors in multi-organ induction and patterning. *EMBO J.* 17:1642–55

75. Arman E, Haffner-Krausz R, Chen Y, Heath JK, Lonai P. 1998. Targeted disruption of fibroblast growth factor (FGF) receptor 2 suggests a role for FGF signaling in pregastrulation mammalian development. *Proc. Natl. Acad. Sci. USA* 95:5082–87

76. Arman E, Haffner-Krausz R, Gorivodsky M, Lonai P. 1999. Fgfr2 is required for limb outgrowth and lung-branching morphogenesis. *Proc. Natl. Acad. Sci. USA* 96:11895–99

77. De Moerlooze L, Spencer-Dene B, Revest J, Hajihosseini M, Rosewell I, Dickson C. 2000. An important role for the IIIb isoform of fibroblast growth factor receptor 2 (FGFR2) in mesenchymal-epithelial signalling during mouse organogenesis. *Development* 127:483–92

78. Hokuto I, Perl AK, Whitsett JA. 2003. Prenatal, but not postnatal, inhibition of fibroblast growth factor receptor signaling causes emphysema. *J. Biol. Chem.* 278:415–21

79. Perl AK, Hokuto I, Impagnatiello MA, Christofori G, Whitsett JA. 2003. Temporal effects of Sprouty on lung morphogenesis. *Dev. Biol.* 258:154–68

80. Weinstein M, Xu X, Ohyama K, Deng CX. 1998. FGFR-3 and FGFR-4 function cooperatively to direct alveogenesis in the murine lung. *Development* 125:3615–23

81. Ciruna BG, Schwartz L, Harpal K, Yamaguchi TP, Rossant J. 1997. Chimeric

analysis of fibroblast growth factor receptor-1 (Fgfr1) function: a role for FGFR1 in morphogenetic movement through the primitive streak. *Development* 124:2829–41

82. Ornitz DM. 2000. FGFs, heparan sulfate and FGFRs: complex interactions essential for development. *BioEssays* 22:108–12

83. Mason IJ. 1994. The ins and outs of fibroblast growth factors. *Cell* 78:547–52

84. Toriyama K, Muramatsu H, Hoshino T, Torii S, Muramatsu T. 1997. Evaluation of heparin-binding growth factors in rescuing morphogenesis of heparitinase-treated mouse embryonic lung explants. *Differentiation* 61:161–67

85. Roman J, Schuyler W, McDonald JA, Roser S. 1998. Heparin inhibits lung branching morphogenesis: potential role of smooth muscle cells in cleft formation. *Am. J. Med. Sci.* 316:368–78

86. Izvolsky KI, Shoykhet D, Yang Y, Yu Q, Nugent MA, Cardoso WV. 2003. Heparan sulfate-FGF10 interactions during lung morphogenesis. *Dev. Biol.* 258:185–200

87. Deepa SS, Umehara Y, Higashiyama S, Itoh N, Sugahara K. 2002. Specific molecular interactions of oversulfated chondroitin sulfate E with various heparin-binding growth factors. Implications as a physiological binding partner in the brain and other tissues. *J. Biol. Chem.* 277:43707–16

88. Smith C, Hilfer S, Searls R, Nathanson M, Allodoli M. 1990. Effects of ß-D-xyloside on differentiation of the respiratory epithelium in the fetal mouse lung. *Dev. Biol.* 138:42–52

89. Spooner BS, Bassett KE, Spooner BS Jr. 1993. Embryonic lung morphogenesis in organ culture: experimental evidence for a proteoglycan function in the extracellular matrix. *Trans. Kans. Acad. Sci.* 96:46–55

90. Smith C, Webster E, Nathanson M,

Searls R, Hilfer S. 1990. Altered patterns of proteoglycan deposition during maturation of the fetal mouse lung. *Cell Diff. Dev.* 32:83–96

90a. Shannon JM, McCormick-Shannon K, Burhans MS, Shangguan X, Srivastava K, Hyatt BA. 2003. Chondroitin sulfate proteoglycans are required for lung growth and morphogenesis in vitro. *Am. J. Physiol. Lung Cell Mol. Physiol.* In press

91. Ingham PW, McMahon AP. 2001. Hedgehog signaling in animal development: paradigms and principles. *Genes Dev.* 15:3059–87

92. McMahon AP, Ingham PW, Tabin CJ. 2003. Developmental roles and clinical significance of hedgehog signaling. *Curr. Top. Dev. Biol.* 53:1–114

93. Motoyama J, Liu J, Mo R, Ding Q, Post M, Hui CC. 1998. Essential function of Gli2 and Gli3 in the formation of lung, trachea and oesophagus. *Nat. Genet.* 20:54–57

94. Miller LA, Wert SE, Whitsett JA. 2001. Immunolocalization of sonic hedgehog (Shh) in developing mouse lung. *J. Histochem. Cytochem.* 49:1593–604

95. Chen Y, Struhl G. 1996. Dual roles for patched in sequestering and transducing Hedgehog. *Cell* 87:553–63

96. Chuang PT, McMahon AP. 1999. Vertebrate Hedgehog signalling modulated by induction of a Hedgehog-binding protein. *Nature* 397:617–21

97. Chuang PT, Kawcak T, McMahon AP. 2003. Feedback control of mammalian Hedgehog signaling by the Hedgehog-binding protein, Hip1, modulates Fgf signaling during branching morphogenesis of the lung. *Genes Dev.* 17:342–47

98. Mahlapuu M, Enerback S, Carlsson P. 2001. Haploinsufficiency of the forkhead gene *Foxf1*, a target for sonic hedgehog signaling, causes lung and foregut malformations. *Development* 128:2397–406

99. Lim L, Kalinichenko VV, Whitsett JA, Costa RH. 2002. Fusion of lung lobes

and vessels in mouse embryos heterozygous for the forkhead box f1 targeted allele. *Am. J. Physiol. Lung Cell Mol. Physiol.* 282:L1012–22

100. Platt KA, Michaud J, Joyner AL. 1997. Expression of the mouse *Gli* and *Ptc* genes is adjacent to embryonic sources of hedgehog signals suggesting a conservation of pathways between flies and mice. *Mech. Dev.* 62:121–35

101. Grindley JC, Bellusci S, Perkins D, Hogan BLM. 1997. Evidence for the involvement of the *Gli* gene family in embryonic mouse lung development. *Dev. Biol.* 188:337–48

102. Park HL, Bai C, Platt KA, Matise MP, Beeghly A, et al. 2000. Mouse *Gli1* mutants are viable but have defects in SHH signaling in combination with a *Gli2* mutation. *Development* 127:1593–605

103. Wodarz A, Nusse R. 1998. Mechanisms of Wnt signaling in development. *Annu. Rev. Cell Dev. Biol.* 14:59–88

104. Huelsken J, Birchmeier W. 2001. New aspects of Wnt signaling pathways in higher vertebrates. *Curr. Opin. Genet. Dev.* 11:547–53'

105. Shu W, Jiang YQ, Lu MM, Morrisey EE. 2002. Wnt7b regulates mesenchymal proliferation vascular development in the lung. *Development* 129:4831–42

105a. Mucenski ML, Wert SE, Natian JM, Loudy DE, Huelsken J, et al. 2003. Beta-catenin is required for the specification of proximal/distal cell fate during lung morphogenesis. *J. Biol. Chem.* In press

106. Slusarski DC, Corces VG, Moon RT. 1997. Interaction of Wnt a Frizzled homologue triggers G-protein-linked phosphatidylinositol signalling. *Nature* 390:410–13

107. Li C, Xiao J, Hormi K, Borok Z, Minoo P. 2002. Wnt5a participates in distal lung morphogenesis. *Dev. Biol.* 248:68–81

108. Hogan BL. 1996. Bone morphogenetic proteins: multifunctional regulators of vertebrate development. *Genes Dev.* 10:1580–94

109. Weaver M, Yingling JM, Dunn NR, Bellusci S, Hogan BL. 1999. Bmp signaling regulates proximal-distal differentiation of endoderm in mouse lung development. *Development* 126:4005–15

110. Bellusci S, Henderson R, Winnier G, Oikawa T, Hogan BLM. 1996. Evidence from normal expression and targeted misexpression that *Bone Morphogenetic Protein-4* (*Bmp-4*) plays a role in mouse embryonic lung morphogenesis. *Development* 122:1693–702

111. Lu MM, Yang H, Zhang L, Shu W, Blair DG, Morrisey EE. 2001. The bone morphogenic protein antagonist gremlin regulates proximal-distal patterning of the lung. *Dev. Dyn.* 222:667–80

112. Shi W, Zhao J, Anderson KD, Warburton D. 2001. Gremlin negatively modulates BMP-4 induction of embryonic mouse lung branching morphogenesis. *Am. J. Physiol. Lung Cell Mol. Physiol.* 280:L1030–39

113. Hyatt BA, Shangguan X, Shannon JM. 2002. BMP4 modulates fibroblast growth factor-mediated induction of proximal and distal lung differentiation in mouse embryonic tracheal epithelium in mesenchyme-free culture. *Dev. Dyn.* 225:153–65

114. Bragg AD, Moses HL, Serra R. 2001. Signaling to the epithelium is not sufficient to mediate all of the effects of transforming growth factor beta and bone morphogenetic protein 4 on murine embryonic lung development. *Mech. Dev.* 109:13–26

115. Liu Y, Hogan BL. 2002. Differential gene expression in the distal tip endoderm of the embryonic mouse lung. *Gene Expr. Patterns* 2:229–33

116. Lin S, Shannon JM. 2002. Microarray analysis of gene expression in the embryonic lung. *Chest* 121:80S–81

Annu. Rev. Physiol. 2004. 66:647–63
doi: 10.1146/annurev.physiol.66.032102.134301
First published online as a Review in Advance on September 15, 2003

GENETICALLY ENGINEERED MOUSE MODELS FOR LUNG CANCER

I. Kwak, S.Y. Tsai, and F.J. DeMayo

Department of Molecular and Cellular Biology, Baylor College of Medicine, Houston, Texas 77030; email: fdemayo@bcm.tmc.edu

Key Words   oncogene, transgenic, knockout, gene switch, conditional gene ablation

■ **Abstract**   The lung is a complex organ consisting of numerous cell types that function to ensure sufficient gas exchange to oxygenate the blood. In order to accomplish this function, the lung must be exposed to the external environment and at the same time maintain a homeostatic balance between its function in gas exchange and the maintenance of inflammatory balance. During the past two decades, as molecular methodologies have evolved with the sequencing of entire genomes, the use of in vivo models to elucidate the molecular mechanisms involved in pulmonary physiology and disease have increased. The mouse has emerged as a potent model to investigate pulmonary physiology due to the explosion in molecular methods that now allow for the developmental and tissue-specific regulation of gene transcription. Initial efforts to manipulate gene expression in the mouse genome resulted in the generation of transgenic mice characterized by the constitutive expression of a specific gene and knockout mice characterized by the ablation of a specific gene. The utility of these original mouse models was limited, in many cases, by phenotypes resulting in embryonic or neonatal lethality that prevented analysis of the impact of the genetic manipulation on pulmonary biology. Second-generation transgenic mouse models employ multiple strategies that can either activate or silence gene expression thereby providing extensive temporal and spatial control of the experimental parameters of gene expression. These highly regulated mouse models are intended to serve as a foundation for further investigation of the molecular basis of human disease such as tumorigenesis. This review describes the principles, progress, and application of systems that are currently employed in the conditional regulation of gene expression in the investigation of lung cancer.

INTRODUCTION

The lung is a complex organ composed of over 40 different cell types, whose primary function is the exchange of gases between body fluids and the external environment (1). This intimate relationship of pulmonary epithelium and external environment results in a high-risk situation, whereby the epithelium is especially vulnerable to malignant transformation secondary to repetitive exposure to injury and oncogenic pollutants. Unfortunately, early detection of resultant lung tumors is difficult. The inability to detect lung cancer at early stages makes lung cancer a deadly disease, which is borne by the fact that lung cancer is the leading cause of cancer-related death in both sexes in the United States (2). Mouse models afford the opportunity for investigators to experimentally manipulate the mouse genome in order to investigate the molecular interactions involved in oncogenic transformation and disease progression, as well as test potential therapeutic interventions (3, 4). The success of establishing such an animal model is dependent upon targeting a specific genetic modification to the appropriate pulmonary epithelial cell type at the appropriate time in lung development.[1]

Lung cancer can be divided into four major, histologically identifiable subtypes (5): adenocarcinoma, squamous cell carcinoma, small cell carcinoma, and large cell carcinoma. The diversity in subtypes of lung cancer is due to the cellular diversity of the pulmonary epithelium. Depending upon the specific cell type that is transformed, a distinct type of lung cancer will develop. The type of lung cancer that develops in an individual depends upon the individual's genetic susceptibility, gender, and exposure to environmental carcinogens (6). Currently, squamous cell carcinoma and adenocarcinoma are the most common type of lung cancer in the United States, with adenocarcinoma being the most common form of lung cancer in women (6). In fact, adenocarcinoma has increased to 45% of bronchogenic carcinomas with declines in squamous and large cell carcinomas, whereas small cell has remained at 20% (7). Therefore, the success of establishing mouse models for lung cancer will depend upon the ability to target the specific oncogenic genetic modification in a cell-specific fashion.

The complexity of developing animal models for lung cancer is not limited to the cellular diversity of lung cancer. Another complication in establishing animal models for lung cancer is controlling the timing of the genetic modification. The mouse lung begins as a protrusion from the oral endoderm on embryonic day 9.5. The lung then proceeds through four phases of development: the pseudoglandular phase, the cannalicular phase, the saccular phase, and the alveolar phase. During the pseudoglandular phase, embryonic day 9.5 to 16.5, the outpocketing of oral endoderm forms and the trachea and two lung buds continue to grow and branch. Also during this phase, the primordial lung structure is established. In the cannalicular phase, day 16.5 to 17.5, the smaller bronchiolar passages are formed. On day 17.5 to neonatal day 5, the lung enters the saccular phase in which the distal branches expand into alveolar sacs. Finally, during

---

[1]Abbreviations: AAV, adeno-associated viral vector; CCSP, Clara cell-specific protein; Dox, doxycycline; ER, estrogen receptor; ES, embryonic stem; FRT, Flp recombination target; GR, glucocorticoid receptor; LBD, ligand-binding domain; NSCLC, non-small cell lung cancer; POI, protein of interest; PR, progesterone receptor; rTS, tetracycline-controlled silencer; rtTA, reverse *tet* transactivator; SCLC, small cell lung cancer; SP-C, surfactant protein C; Tag, Simian virus large T antigen; *tetO*: *tet* operon; *tetR*, *tet* repressor; *tTA*, tetracycline-transactivator.

neonatal day 5 to 30, the lung enters the alveolar phase in which the alveolar sacs develop into mature alveoli (8). These four stages of lung development culminate first in the establishment of the pulmonary architecture and then in the differentiation of the pulmonary epithelium to establish the specific cell types needed to execute pulmonary function. Disruption of this developmental process can impair pulmonary function and result in lethality. Genetic modification, which drastically alters pulmonary cell growth and differentiation, results in neonatal lethality (9, 10). Therefore, genetic manipulation must not only target the appropriate cell type in the lung, but also must not interfere with pulmonary development in order to allow the impact of such manipulation to be investigated in the adult mouse lung.

The advent of transgenic technology has significantly improved the ability to define the role of specific genes in the process of transformation and disease progression. The first generation of transgenic mouse models for lung cancer constitutively expressed regulatory genes in the pulmonary epithelium. The subsequent generations of transgenic mouse models further enhanced the ability to clarify the specific molecular mechanisms by allowing for the cell-specific-regulated ablation or expression of genes in the lung (reviewed in 11, 12). The present review provides a brief overview of conventional and conditional mouse transgenic technologies. In particular, we focus on conditional transgenic and gene-targeting techniques for the study of lung cancer.

TRANSGENIC MOUSE MODELS

Conventional Transgenic Mouse Models

Conventional transgenic animals have revolutionized our ability to study gene function in vivo (13). The transgene DNA construct is generated by the fusion of a cell-specific promoter to direct transcription of the gene of interest. This transgene DNA is subsequently microinjected into fertilized oocytes, and the transgene undergoes random integration into the mouse genome. The viable oocytes are then transferred into pseudopregnant mothers, and the DNA obtained from progeny is assessed for integration of the transgene into the mouse genome (14, 15). Using this technology, gene expression has been directed in a cell-specific fashion to the lung. Two promoters have been extensively used to target gene expression to the mouse lung (1). The DNA fragments used are the promoters for the surfactant protein C (SP-C) and the Clara cell secretory protein (CCSP) genes. The former targets gene expression primarily to the alveolar type II cells of the distal airways (16). The latter targets transgene expression to the nonciliated secretory cells of the airways (17, 18). These promoters have been used to target many regulatory proteins specifically to the lungs to investigate the role of these proteins in lung development, physiology, and oncogenic transformation (3). The first oncogene targeted specifically to the lung was the Simian virus large T antigen (Tag). Tag was targeted with both the *SP-C* (19, 20) and *CCSP* (21, 22) promoters. Both models resulted in adenocarcinoma of the lung, a rapid transformation of the pulmonary epithelium with tumors appearing in the first few months of life. The *SPC-Tag* model resulted in the transformation of the alveolar type II cells and distal Clara cells (19). The *CCSP-Tag* model resulted in the transformation of the Clara cells of the airways (22). These models have been useful in the investigation of the regulatory pathways disrupted during tumor progression, as well as the generation of cell-specific cell lines for the investigation of the molecular regulation of lung gene expression (22, 23). These promoters have also been used to target other tumor-promoting genes to the airways. The development of pulmonary oncogenic transformation has been generated in transgenic mice by using the *SP-C* promoter to express c-*myc* (24), epidermal growth factor (24), the receptor tyrosine kinase (RON) (25, 26), and *Raf-1* (27). The *CCSP* promoter has been used to a lesser extent to target oncogenes to the lung. Pulmonary oncogenic transformation has been generated in mice expressing a dominant-negative p53 transgene (28), c-*myc* (29), and the human achaete schuete gene (30).

Although the *SP-C* and *CCSP* promoters are powerful tools in the establishment of animal models for lung cancer, the cell types to which these promoters direct transgene expression do not encompass all the cell types in the pulmonary epithelium. These models have been useful only for the generation of animal models for adenocarcinoma of the lung. Promoters are still needed to target genes specifically to the upper respiratory tract to generate animal models for pulmonary squamous cell carcinoma, as well as to direct transgene expression to the neuroendocrine cells of the lung to generate animal models for small cell carcinoma. Attempts have been made to target oncogene expression to the neuroendocrine cells. Using the calcitonin promoter, the *v-Ha-ras* gene was expressed in the lungs (31). The *v-H-ras* expression induced hyperplasia of both neuroendocrine cells and non-neuroendocrine cells. The tumors that progressed in these mice were mostly of non-neuroendocrine origin. The lack of neuroendocrine cell specificity of the development of these tumors may be due in part to the promiscuity of transgene expression early in lung development (31).

The limitations of using the above conventional transgenic approach to generate transgenic models for lung cancer is the timing of the initiation of expression of these genes under the control of these promoters. Transcription under the control of the *SP-C* promoter is initiated at embryonic day 10 (32). *CCSP*-regulated transgene expression occurs later in development in the mouse, at about embryonic day 17 (33). The fact that these promoters activate gene expression early in development limits the use of these models in the establishment of animal models of lung cancer. Expression of genes that disrupt the developmental processes during branching morphogenesis and alveologenesis result in an embryonic lethal phenotype that is of limited value for the investigation of the progression of lung cancer (9, 34). The ability to initiate transgene expression at specific time points would allow the model to be more effective in identifying the sequence of events governing tumor progression by more effectively exploiting advances in high-density gene expression analysis. The combination of inducible transgenic models with the advancing field of high-density DNA microarray expression analysis should allow the identification of early events in cancer progression by detecting changes in gene expression after acute induction of the transgene. However, the ability to conduct this analysis in the conventional model is not practical. The second limitation of the conventional transgenic approach is that once transcription of the transgene is initiated in these mice, it is irreversible. The ability to have tight control over the window of transgene expression would allow a more precise determination of the molecular events regulating cancer progression. The ability to terminate transgene expression would allow identification of the stages of tumor progression independent of the growth-promoting properties of the transgene. Conditional transgenic models have overcome these limitations.

Conditional Transgenic Mouse Models

Currently, the most effective regulatory systems for conditional transgenic mice are the ligand-inducible binary transgenic systems (35, 36). These systems are illustrated in Figure 1 and consist of using at least two transgene constructions, a regulator transgene and a target transgene, to confer regulated expression of the desired gene. The regulator transgene encodes a transcription factor whose

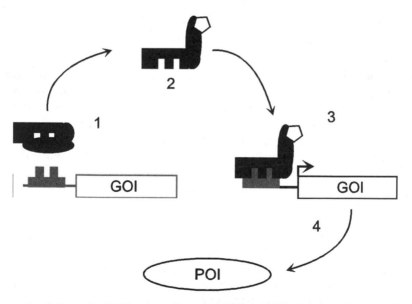

Figure 1   Schematic of a bitransgenic gene switch. 1, In the absence of ligand, the regulator (*black*) is inactive, and the gene of interest (GOI) is not expressed from the target transgene. 2, The presence of ligand (*pentagon*) converts the regulator into the active configuration. 3, The activated regulator binds to the response element in the target transgene and activates transcription of the GOI. 4, Activation of the GOI allows expression of the protein of interest (POI).

transcriptional activity is controlled by the administration of an exogenous compound. The regulator is placed under the control of a tissue-specific promoter in order to express the transcription factor in the tissue of interest. The regulator does not activate transcription of the target transgene until the animal is administered the exogenous compound. Upon administration of the exogenous compound, the regulator activates only the target transgene. The second transgenic construct, the target construct, contains the coding sequence of a protein of interest (POI) under the transcriptional control of cis-acting DNA elements that are responsive to the DNA-binding domain of the regulator transgene. The target transgene is silent until the regulator transgene is activated by the administration of the exogenous compound. In summary, the sequence of events in the ligand-inducible binary transgenic system is (a) the exogenous compound induces the transcriptional activity of the regulator transgene, (b) the regulator binds to the cis-acting DNA elements in the promoter of the target transgene, and (c) the POI encoded by the target transgene is expressed. Currently, three ligand-inducible binary transgenic systems have been established: the tetracycline transactivator-inducible system, the mifepristone gene switch, and the ecdysone regulatory system. Onlg tTA and mifepristone have been successfully established in the lung and are discussed below.

Tetracycline-Transactivator, *tTA*-Inducible System

Currently, the most common ligand-inducible binary transgenic system is the tetracycline transactivator-inducible (*tTA*) regulatory system, which has been extremely useful for generating ligand-dependent and reversible transgenic mouse models (37). The *tTA* systems are based on the Tn10-specific tetracycline resistance operon in *Escherichia coli* (38). This system utilizes the *tet* repressor (*tetR*) gene to activate target genes that are under the control of promoter elements containing the *tet* response element, the *tet* operon (*tetO*). In bacteria, *tet*-responsive genes are repressed by the *tetR*. In the presence of the antibiotic tetracycline, the *tetR* does not bind to the *tetO*, and tetracycline-resistant genes are expressed. Current engineering of the *tet* gene for the ligand-inducible binary transgenic systems has generated two forms of this regulator, the *tTA* and *rtTA* (known as *tet* off and *tet* on, respectively). The *tTA* acts as a repressor in the presence of tetracycline. In the absence of tetracycline, the *tTA* causes the transcription of the *tet*-responsive target transgene. The administration of tetracycline or commercially available doxycycline (Dox) induces a conformational change in the *tTA* that prevents DNA binding and leads to the termination of target gene expression (39). The major drawback to this system is that the transcription of the target transgene is constitutive and has to be terminated by the administration of Dox. This makes investigation of acute changes due to expression of the target transgene difficult because of the problems of the kinetics of gene repression versus gene activation expression.

The second and more commonly used generation of tetracycline-dependent transactivators is the reverse *tet* transactivator (*rtTA*). The *rtTA* does not bind DNA and activate target gene transcription in the absence of ligand. However, upon the administration of Dox, the *rtTA* will bind to the *tetO* and activate transgene expression (40). The *rtTA* system has been applied in the lung with either the *CCSP* promoter or the *SP-C* promoter used to drive expression of the *rtTA* (41, 42).

One final modification on the *tTA* system was the incorporation of a tetracycline-controlled silencer (*tTS*) to avoid any leaky expression of transgene in the absence of Dox administration (43). The modified *tTA* consists of the *tet* repressor fused to the KRAB-AB repression domain of the Kid-1 protein (44). In the absence of Dox, the *tTS* binds the *tet*-responsive element and represses target transgene transcription. In this triple transgenic model, target mice were generated along with mice bearing the *rtTA* regulator transgene and the *tTS* silencer transgene under the control of the *CCSP* promoter (45). This triple transgenic system, consisting of the target, regulator, and silencer transgene, has tighter control of transgene expression. Although complicated, this system has given tighter control of *IL-13* gene expression. The tighter control afforded by this system may be important for regulating the expression of potent oncogenes in the lungs where basal expression may impart a phenotype.

The systems have been used to drive the expression of cytokines IL-11 (41) and IL-13 (45); growth factors FGF-7 (42), FGF-10 (46), FGF-18 (47); an activated K-*ras* oncongene (48); and Cre recombinase (49). These systems have been effectively used to investigate the regulation of pulmonary cell proliferation, differentiation, and transformation. Regulated expression of FGF-7 (42) and FGF-10 (46) caused increased cell proliferation upon administration of Dox with the induction of FGF-10, which resulted in the development of adenomas. Both the development of epithelial cell proliferation and adenomas were reversible upon withdrawal of Dox. However, the most informative animal model for lung cancer was one that used the *rtTA* to induce the expression of an activated K-*ras* oncogene (48). Activation of K-*ras* using a *CCSP*-driven *rtTA* regulator resulted in the development of pulmonary adenocarcinoma. Upon withdrawal of Dox, K-*ras* expression was terminated, and the tumors regressed as a result of apoptosis. This K-*ras* model was crossed with mice deficient for either the *p53* or *Ink4A/Arf* tumor suppressor genes in order to investigate the interaction of these tumor suppressor genes with the development and regression of the K-*ras*-dependent tumors. Induction of K-*ras* expression in the presence of these null alleles caused the tumors to arise faster. However, the absence of these tumor suppressor genes did not prevent the tumors from regressing upon termination of K-*ras* expression. This ability to activate and terminate oncogene expression in the lungs allows for a dissection of the molecular mechanisms regulating pulmonary epithelial cell transformation.

Mifepristone Gene Switch

The second binary transgenic system used to regulate transgene expression in mice is based on a mutated human progesterone receptor (PR). A deletion of the carboxy terminus ligand-binding domain (LBD) of the PR alters the specificity of ligand binding of this receptor. This deletion prevents the receptor from binding the endogenous ligand progesterone, but retains the ability to bind antiprogestins such as mifepristone (50). This mutant LBD was then used to create a chimeric receptor consisting of the LBD mutant, the DNA-binding domain of the yeast Gal4 transcription factor, and the activation domain of either the herpes simplex virus, VP-16 (VP16-AD), or the activation domain of the nuclear factor kappa B p65 protein (p65-AD) (51, 52). The mifepristone regulators were called GLVP16 or GLp65 depending on whether they contained the VP16-AD or p65-AD, respectively. These regulators strongly activate a promoter containing the upstream activating sequences of the yeast Gal4 gene (UAS$_{Gal4}$) upon the addition of mifepristone (53).

The GLVP and GLp65 regulators were targeted to the lung in transgenic mice using the *SP-C* promoter (12, 54). Mice expressing the *SPC-GLVP* regulator were then bred to mice bearing the human growth hormone target transgene (*UAS$_{Gal4}$-hGH*). Binary transgenic mice containing the *SPC-GLVP* regulator and the *UAS$_{Gal4}$-hGH* transgene were administered mifepristone. Mifepristone was effective in regulating the expression of the *hGH* transgene in the alveolar type II cells from an undetectable level to a significant signal only in the binary transgenic mice. The *SPC-GLp65* regulator was used to control the expression of the FGF-3 (also called int-2) oncogene in mice. Transgene activation in response to mifepristone resulted in two phenotypes, depending on the levels of induction of FGF-3 transgene expression. Low-induction levels of FGF-3 expression induced massive free alveolar macrophage infiltration, whereas high-induction levels of FGF-3 expression resulted in diffuse alveolar type II cell hyperplasia. More importantly, both phenotypes were reversible after the withdrawal of mifepristone (54). Thus these GLVP/GLp65 systems can be utilized to activate or repress expression of a target transgene by using mifepristone, a synthetic ligand.

GENE ABLATION

Conventional Knockout Mouse Models

Endogenous genes can be mutated or ablated in vivo by using homologous recombination in embryonic stem (ES) cells (55, 56). Specific genes can be targeted in vitro for genetic modification. The ES cells with the appropriate target event can be injected into the blastoceol of the day 3.5 mouse embryo, and mice born from this manipulation contain cells from both the host embryo and targeted ES cells. ES cells can incorporate themselves into the murine germline (57), and the mutation can be transmitted to future generations, with mice homozygous for the specific genetic modification. This technology is powerful and has been used to investigate the role of many genes in pulmonary development and physiology. However, the disadvantage of this technology is that if the genetic modification affects animal viability, the role of this gene in the adult mouse model and in the development of lung cancer is not possible. One specific example was the use of homologous recombination in ES cells to delete a region of chromosome 9F1.

This region is syntactic with human chromosome 3p21.2 (58) and contains several tumor suppressor genes that are lost in human lung cancer (59). The phenotype of this mutation was embryo lethality (58). One approach to overcome this limitation was to engineer a targeting event in which somatic recombination would allow activation of a mutant oncogene. This approach was used to express a mutant K-*ras* gene (60). A "hit and run strategy" was used to generate mice in which one K-*ras* allele contained an oncogenic mutation and the other allele was wild-type. The only way for the mutant K-*ras* gene to be expressed was if somatic recombination occurred between the targeted allele and the wild-type allele. This approach overcame any potential embryonic lethality that may have occurred if the mutation occurred in every cell during development. Mice with the targeted K-*ras* mutation developed pulmonary adenocarcinomas because of the sporadic activation of the mutant allele. This approach had the advantage over transgenic approaches in that the transgene was expressed under the control of the endogenous K-*ras* promoter and the levels of expression reflected more the physiological situation. However, the cancers were not limited to the lung and were observed in other tissues, which is another limitation of ES-cell-targeted mutagenesis. The genetic modification occurs in all cells of the mouse. Therefore, phenotypes observed in extrapulmonary tissue may confound the pulmonary phenotype. This has been particularly true with the ablation of tumor suppressor genes and oncogene. Ablation of tumor suppressor genes may result in cancer in other tissues before lung cancer develops. This limitation has been overcome by the use of recombinases to generate tissue-specific or conditional knockout mouse models.

Conditional Knockout Mouse Models

Conditional knockout models use tissue-specific expression of recombinases to ablate genes in a tissue- and cell-specific fashion (36, 61). Site-specific recombinases splice DNA sequences, which are flanked by specific sequences recognized by the recombinases. Homologous recombination in ES cells is used to insert recombinase recognition sequences around an entire gene or around exons coding for regulatory domains. Transgenic technology is then used to express site-specific recombinase in a tissue-specific manner. This combination of ES cell technology with transgenic technology allows the gene of interest, flanked by the recombinase sites, to be deleted in a tissue-specific fashion. This strategy is outlined in Figure 2*A*.

In addition to being used to ablate gene expression in a tissue-specific fashion, recombinases can also be used to activate gene expression. In this case, a reporter gene or stop codons are flanked by recombinase recognition sites and inserted between a promoter and the coding region of the gene to be activated. The gene to be activated can be endogenous or a transgene expressed under the control of a ubiquitous promoter. Expression of the recombinase excises the stop codons of the reporter gene and allows transcription of the gene of interest (62). This strategy is shown in Figure 2*B*. These approaches have greatly improved the in vivo investigation of specific regulatory proteins in pulmonary development, physiology, and

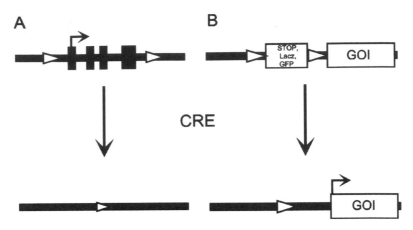

Figure 2 Ablation and activation of genes using Cre recombinase. (*A*) The gene of interest (GOI) has loxP sites (*triangles*) flanking coding regions of the gene. Cre recombinase activity deletes the coding regions, leaving one loxP site in its place. (*B*) LoxP sites flank either a reporter gene or stop codons, which are inserted between the promoter region and the coding region of a GOI preventing the transcription or translation of the GOI. Cre recombinase activity removes the sequences between the loxP site allowing expression of the GOI.

cancer. Currently, two recombinases are used to generate conditional ablation of genes.

Cre and Flp Recombinases

The most common recombinase system used to induce gene ablation in mice is the Cre–loxP system. The Cre (causes recombination) protein of bacteriophage P1 directs recombination between loxP (locus of crossover P1) sites and ablates gene sequences that are flanked by two 34-bp loxP sites (floxed genes) (63). The second recombinase is the *Saccharomyces cerevisiae* recombinase, Flp recombinase. Flp recombinase causes recombination between FRT (Flp recombination target) recognition sites (64). LoxP and FRT recognition sequences are 34-bp DNA elements consisting of an 8-bp core region flanked by palindromic sequences of 13 bp (65, 66). To date, these recombinases have not been used in a direct transgenic approach to ablate genes specifically in the lung. The limitation of this approach using the current lung promoters, *SP-C* and *CCSP*, is that cell-specific ablation would still occur during development and that the ablation of the desired gene would disrupt lung development, thus resulting in an embryonic lethal phenotype. However, this limitation can be overcome by regulating the expression of activity of the recombinase. Currently, there are three approaches to regulate recombinase activity in a tissue-specific fashion: generating a ligand-dependent recombinase, using a transgene regulatory system to regulate recombinase expression, and using gene therapy to deliver recombinases to the lung.

Ligand-Dependent Cre Recombinase

In an ideal conditional mouse model system, the control over the knockout phenotype should activate or ablate a transgene when and where investigators choose. This is possible by using Cre recombinases whose activity is directly regulated by ligand, which binds to the Cre recombinase and causes changes in conformation that allow the recombinase to edit floxed genes (reviewed in 15). The LBD of steroid hormone receptors was fused to Cre recombinase to activate Cre expression in a ligand-dependent manner. Several systems have been developed for the direct ligand-dependent regulation of Cre recombinase using mutations in the LBD of the progesterone receptor (PR) (67), estrogen receptor (ER) (68, 69), and glucocorticoid receptor (GR) (70), which bind to the antagonists mifepristone (67) and tamoxifen (71), or the agonist dexamethasone, respectively (70). These systems are called the Cre-PR, the Cre-ER, and the Cre-GRdex regulatory systems, respectively. The Cre-PR system has been successfully used to conditionally ablate genes in several tissues of mice such as heart (72), brain (67, 73), and skin (74, 75). The Cre-ER system has been used to conditionally ablate genes in the skin (76, 77), muscle (78), B lymphocytes (79), hepatocytes (80), adipocytes (81), and the embryo (82) in mice. Although potentially valuable, this technology has not been successfully reported in the lung.

rtTA Regulated-Cre Recombinase

Another approach of the regulated gene ablation is placing Cre recombinase under the control of an *O-tet*-responsive promoter. This system requires crossing mice with floxed genes with mice containing a *tet*-responsive Cre recombinase target transgene and with mice expressing the *rtTA* regulator in the lungs. The *rtTA*-regulated Cre expression has been accomplished using SPC-*rtTA* and CCSP-*rtTA* mice (49). This approach has been used to edit loxP sites and activate the expression of alkaline phosphatase and GFP reporter genes in the lungs of mice. The Cre activity was regulated for a defined window. Activation of these reporter genes was permanent in the cells expressing the Cre recombinase, as well as in all descendents of these cells. This approach was used to map pulmonary cell lineage during lung development (49). The next step is to use this approach to activate or ablate genes proposed to be in the oncogenic pathway and investigate their impact on lung development.

Gene Therapeutic Delivery of Cre Recombinase

Gene therapy has been used to deliver genes to pulmonary epithelial cells (reviewed in 83). One such approach is to deliver Cre recombinase to the lung in the form of an aerosol. Several studies use adenovirus to deliver functional Cre recombinase to the lung in vivo (84–86). Replication-defective human adenovirus containing the gene for the Cre recombinase under control of the human cytomegalovirus promoter (Ad-Cre) has been constructed (87, 88) and used for a gene switch system in mammalian cells. Injections of adeno-associated viral (AAV) vectors expressing a Cre recombinase (AAV-Cre), which have all viral coding sequences removed to avoid toxicity, can mediate recombination in mice (89, 90). Ad-Cre- or AAV-Cre-mediated delivery of Cre recombinase offers a useful tool for selective ablation of genes in adult mice (91). The Ad-Cre system has been used to initiate tumorigenesis in the lungs of animals by activating the expression of a mutated K-*ras* oncogene. In one case, gene targeting was used to insert a transcription stop sequence flanked by loxP sites, into the endogenous mutated K-*ras* gene (92). In a second instance, a mutated K-*ras* gene was placed under the transcriptional control of the β-actin promoter in which a GFP reporter gene flanked by loxP sites was inserted between the K-*ras* coding region and the promoter (93). Adenoviral delivery of Cre recombinase in both cases edited out the sequences blocking K-*ras* transcription and allowed K-*ras* expression. The results were adenocarcinomas in the airways of the mice. This approach has the major advantage of allowing ablation of genes at specific times without having the additional breeding to generate Cre transgenic and loxP homozygous mice. The disadvantage of this approach is that there is loss of the control over the cell specificity of the gene ablation.

CONCLUSIONS

The ability to genetically engineer mouse models for lung cancer has evolved from targeting transgenes constitutively to the lungs to regulate the expression and ablation of genes in the lung. Currently, this technology is limited by the number of tools available to direct these genetic events to many of the cell types clinically important for lung cancer. However, the sequencing of the mouse and human genomes and high-density DNA expression analysis will allow the identification of tools for targeting these genetic events to more lung cell types. This will allow the generation of animal models that can be used for the development of diagnostic and therapeutic approaches for the treatment of lung cancer.

ACKNOWLEDGMENTS

This manuscript was prepared with the assistance of Janet DeMayo, Patricia L. Ramsay, Kevin Lee, and John Ellsworth. This work was supported by a National Institutes of Health grant HL#61406.

The *Annual Review of Physiology* is online at http://physiol.annualreviews.org

LITERATURE CITED

1. Bloom WF, Fawcett DW. 1975. Respiratory systems. In *A Textbook of Histology*, pp. 743–65. Philadelphia: Saunders
2. Jemal A, Murray T, Samuels A, Ghafoor A, Ward E, Thun MJ. 2003. Cancer statistics, 2003. *CA Cancer J. Clin.* 53:5–26
3. Whitsett JA, Glasser SW, Tichelaar JW,

Perl AK, Clark JC, Wert SE. 2001. Transgenic models for study of lung morphogenesis and repair: Parker B. Francis lecture. *Chest* 120:27S–30

4. Herzig M, Christofori G. 2002. Recent advances in cancer research: mouse models of tumorigenesis. *Biochim. Biophys. Acta* 1602:97–113

5. el-Torky M, el-Zeky F, Hall JC. 1990. Significant changes in the distribution of histologic types of lung cancer. A review of 4928 cases. *Cancer* 65:2361–67

6. Alberg AJ, Samet JM. 2003. Epidemiology of lung cancer. *Chest* 123:21S–49

7. Thun MJ, Lally CA, Flannery JT, Calle EE, Flanders WD, Heath CW Jr. 1997. Cigarette smoking and changes in the histopathology of lung cancer. *J. Natl. Cancer Inst.* 89:1580–86

8. Ten Have-Opbroek AA. 1991. Lung development in the mouse embryo. *Exp. Lung. Res.* 17:111–30

9. Peters K, Werner S, Liao X, Wert S, Whitsett J, Williams L. 1994. Targeted expression of a dominant negative FGF receptor blocks branching morphogenesis and epithelial differentiation of the mouse lung. *EMBO J.* 13:3296–301

10. Kimura S, Hara Y, Pineau T, Fernandez-Salguero P, Fox CH, et al. 1996. The T/ebp null mouse: thyroid-specific enhancer-binding protein is essential for the organogenesis of the thyroid, lung, ventral forebrain, and pituitary. *Genes Dev.* 10:60–69

11. Jackson EL, Jacks T. 2003. Lung cancer models. http://emice.nci.nih.gov/mouse_models/organ_models/lung_models

12. Zhao B, Magdaleno S, Chua S, Wang YL, Burcin M, et al. 2000. Transgenic mouse models for lung cancer. *Exp. Lung. Res.* 26:567–79

13. Gordon JW, Scangos GA, Plotkin DJ, Barbosa JA, Ruddle FH. 1980. Genetic transformation of mouse embryos by microinjection of purified DNA. *Proc. Natl. Acad. Sci. USA* 77:7380–84

14. Palmiter RD, Brinster RL. 1986. Germ-line transformation of mice. *Annu. Rev. Genet.* 20:465–99

15. Bockamp E, Maringer M, Spangenberg C, Fees S, Fraser S, et al. 2002. Of mice and models: improved animal models for biomedical research. *Physiol. Genom.* 11:115–32

16. Glasser SW, Korfhagen TR, Wert SE, Bruno MD, McWilliams KM, et al. 1991. Genetic element from human surfactant protein SP-C gene confers bronchiolar-alveolar cell specificity in transgenic mice. *Am. J. Physiol. Lung Cell Mol. Physiol.* 261:L349–56

17. Stripp BR, Sawaya PL, Luse DS, Wikenheiser KA, Wert SE, et al. 1992. *cis*-Acting elements that confer lung epithelial cell expression of the CC10 gene. *J. Biol. Chem.* 267:14703–12

18. Ray MK, Magdaleno SW, Finegold MJ, DeMayo FJ. 1995. *cis*-Acting elements involved in the regulation of mouse Clara cell-specific 10-kDa protein gene. In vitro and in vivo analysis. *J. Biol. Chem.* 270:2689–94

19. Wikenheiser KA, Clark JC, Linnoila RI, Stahlman MT, Whitsett JA. 1992. Simian virus 40 large T antigen directed by transcriptional elements of the human surfactant protein C gene produces pulmonary adenocarcinomas in transgenic mice. *Cancer Res.* 52:5342–52

20. Wikenheiser KA, Whitsett JA. 1997. Tumor progression and cellular differentiation of pulmonary adenocarcinomas in SV40 large T antigen transgenic mice. *Am. J. Respir. Cell Mol. Biol.* 16:713–23

21. DeMayo FJ, Finegold MJ, Hansen TN, Stanley LA, Smith B, Bullock DW. 1991. Expression of SV40 T antigen under control of rabbit uteroglobin promoter in transgenic mice. *Am. J. Physiol. Lung Cell Mol. Physiol.* 261:L70–76

22. Magdaleno SM, Wang G, Mireles VL, Ray MK, Finegold MJ, DeMayo FJ. 1997. Cyclin-dependent kinase inhibitor expression in pulmonary Clara cells transformed

with SV40 large T antigen in transgenic mice. *Cell Growth Differ.* 8:145–55

23. Ikeda K, Clark JC, Bachurski CJ, Wikenheiser KA, Cuppoletti J, et al. 1994. Immortalization of subpopulations of respiratory epithelial cells from transgenic mice bearing SV40 large T antigen. *Am. J. Physiol. Lung Cell Mol. Physiol.* 267:L309–17

24. Ehrhardt A, Bartels T, Klocke R, Paul D, Halter R. 2003. Increased susceptibility to the tobacco carcinogen 4-(methylnitrosamino)-1-(3-pyridyl)-1-butanone in transgenic mice overexpressing c-myc and epidermal growth factor in alveolar type II cells. *J. Cancer Res. Clin. Oncol.* 129:71–75

25. Chen YQ, Zhou YQ, Fu LH, Wang D, Wang MH. 2002. Multiple pulmonary adenomas in the lung of transgenic mice overexpressing the RON receptor tyrosine kinase. Recepteur d'origine nantais. *Carcinogenesis* 23:1811–19

26. Chen YQ, Zhou YQ, Fisher JH, Wang MH. 2002. Targeted expression of the receptor tyrosine kinase RON in distal lung epithelial cells results in multiple tumor formation: oncogenic potential of RON in vivo. *Oncogene* 21:6382–86

27. Kerkhoff E, Fedorov LM, Siefken R, Walter AO, Papadopoulos T, Rapp UR. 2000. Lung-targeted expression of the c-Raf-1 kinase in transgenic mice exposes a novel oncogenic character of the wild-type protein. *Cell Growth Differ.* 11:185–90

28. Tchou-Wong KM, Jiang Y, Yee H, LaRosa J, Lee TC, et al. 2002. Lung-specific expression of dominant-negative mutant p53 in transgenic mice increases spontaneous and benzo(a)pyrene-induced lung cancer. *Am. J. Respir. Cell Mol. Biol.* 27:186–93

29. Geick A, Redecker P, Ehrhardt A, Klocke R, Paul D, Halter R. 2001. Uteroglobin promoter-targeted c-MYC expression in transgenic mice cause hyperplasia of Clara cells and malignant transformation of T-lymphoblasts and tubular epithelial cells. *Transgenic Res.* 10:501–11

30. Linnoila RI, Zhao B, DeMayo JL, Nelkin BD, Baylin SB, et al. 2000. Constitutive achaete-scute homologue-1 promotes airway dysplasia and lung neuroendocrine tumors in transgenic mice. *Cancer Res.* 60:4005–9

31. Sunday ME, Haley KJ, Sikorski K, Graham SA, Emanuel RL, et al. 1999. Calcitonin driven v-Ha-ras induces multilineage pulmonary epithelial hyperplasias and neoplasms. *Oncogene* 18:4336–47

32. Wert SE, Glasser SW, Korfhagen TR, Whitsett JA. 1993. Transcriptional elements from the human SP-C gene direct expression in the primordial respiratory epithelium of transgenic mice. *Dev. Biol.* 156:426–43

33. Ray MK, Wang G, Barrish J, Finegold MJ, DeMayo FJ. 1996. Immunohistochemical localization of mouse Clara cell 10-kD protein using antibodies raised against the recombinant protein. *J. Histochem. Cytochem.* 44:919–27

34. Simonet WS, DeRose ML, Bucay N, Nguyen HQ, Wert SE, et al. 1995. Pulmonary malformation in transgenic mice expressing human keratinocyte growth factor in the lung. *Proc. Natl. Acad. Sci. USA* 92:12461–65

35. DeMayo FJ, Tsai SY. 2001. Targeted gene regulation and gene ablation. *Trends Endocrinol. Metab.* 12:348–53

36. Lewandoski M. 2001. Conditional control of gene expression in the mouse. *Nat. Rev. Genet.* 2:743–55

37. Shockett PE, Schatz DG. 1996. Diverse strategies for tetracycline-regulated inducible gene expression. *Proc. Natl. Acad. Sci. USA* 93:5173–76

38. Hillen W, Berens C. 1994. Mechanisms underlying expression of Tn10 encoded tetracycline resistance. *Annu. Rev. Microbiol.* 48:345–69

39. Furth PA, St Onge L, Boger H, Gruss P, Gossen M, et al. 1994. Temporal control of gene expression in transgenic mice by a tetracycline-responsive promoter. *Proc. Natl. Acad. Sci. USA* 91:9302–6

40. Gossen M, Freundlieb S, Bender G, Muller

G, Hillen W, Bujard H. 1995. Transcriptional activation by tetracyclines in mammalian cells. *Science* 268:1766–69

41. Ray P, Tang W, Wang P, Homer R, Kuhn C 3rd, et al. 1997. Regulated overexpression of interleukin 11 in the lung. Use to dissociate development-dependent and -independent phenotypes. *J. Clin. Invest.* 100:2501–11

42. Tichelaar JW, Lu W, Whitsett JA. 2000. Conditional expression of fibroblast growth factor-7 in the developing and mature lung. *J. Biol. Chem.* 275:11858–64

43. Freundlieb S, Schirra-Muller C, Bujard H. 1999. A tetracycline controlled activation/repression system with increased potential for gene transfer into mammalian cells. *J. Gene. Med.* 1:4–12

44. Witzgall R, O'Leary E, Leaf A, Onaldi D, Bonventre JV. 1994. The Kruppel-associated box-A (KRAB-A) domain of zinc finger proteins mediates transcriptional repression. *Proc. Natl. Acad. Sci. USA* 91:4514–18

45. Zhu Z, Ma B, Homer RJ, Zheng T, Elias JA. 2001. Use of the tetracycline-controlled transcriptional silencer (tTS) to eliminate transgene leak in inducible overexpression transgenic mice. *J. Biol. Chem.* 276:25222–29

46. Clark JC, Tichelaar JW, Wert SE, Itoh N, Perl AK, et al. 2001. FGF-10 disrupts lung morphogenesis and causes pulmonary adenomas in vivo. *Am. J. Physiol. Lung Cell Mol. Physiol.* 280:L705–15

47. Whitsett JA, Clark JC, Picard L, Tichelaar JW, Wert SE, et al. 2002. Fibroblast growth factor 18 influences proximal programming during lung morphogenesis. *J. Biol. Chem.* 277:22743–49

48. Fisher GH, Wellen SL, Klimstra D, Lenczowski JM, Tichelaar JW, et al. 2001. Induction and apoptotic regression of lung adenocarcinomas by regulation of a K-ras transgene in the presence and absence of tumor suppressor genes. *Genes Dev.* 15:3249–62

49. Perl AK, Wert SE, Nagy A, Lobe CG, Whit-

sett JA. 2002. Early restriction of peripheral and proximal cell lineages during formation of the lung. *Proc. Natl. Acad. Sci. USA* 99:10482–87

50. Vegeto E, Allan GF, Schrader WT, Tsai MJ, McDonnell DP, O'Malley BW. 1992. The mechanism of RU486 antagonism is dependent on the conformation of the carboxy-terminal tail of the human progesterone receptor. *Cell* 69:703–13

51. Wang Y, Tsai SY, O'Malley BW. 2000. An antiprogestin regulable gene switch for induction of gene expression in vivo. *Adv. Pharmacol.* 47:343–55

52. Burcin MM, Schiedner G, Kochanek S, Tsai SY, O'Malley BW. 1999. Adenovirus-mediated regulable target gene expression in vivo. *Proc. Natl. Acad. Sci. USA* 96:355–60

53. Wang Y, DeMayo FJ, Tsai SY, O'Malley BW. 1997. Ligand-inducible and liver-specific target gene expression in transgenic mice. *Nat. Biotechnol.* 15:239–43

54. Zhao B, Chua SS, Burcin MM, Reynolds SD, Stripp BR, et al. 2001. Phenotypic consequences of lung-specific inducible expression of FGF-3. *Proc. Natl. Acad. Sci. USA* 98:5898–903

55. Wong EA, Capecchi MR. 1986. Analysis of homologous recombination in cultured mammalian cells in transient expression and stable transformation assays. *Somat. Cell Mol. Genet.* 12:63–72

56. Smithies O, Gregg RG, Boggs SS, Koralewski MA, Kucherlapati RS. 1985. Insertion of DNA sequences into the human chromosomal beta-globin locus by homologous recombination. *Nature* 317:230–34

57. Bradley A. 1987. Production and analysis of chimaeric mice. In *Teratocarcinomas and Embryonic Stem Cells: A Practical Approach*, ed. EJ Robertson, pp. 113–51. Oxford, UK: IRL

58. Smith AJ, Xian J, Richardson M, Johnstone KA, Rabbitts PH. 2002. Cre-loxP chromosome engineering of a targeted deletion in the mouse corresponding to the 3p21.3 region of homozygous loss

in human tumours. *Oncogene* 21:4521–29

59. Lerman MI, Minna JD. 2000. The 630-kb lung cancer homozygous deletion region on human chromosome 3p21.3: identification and evaluation of the resident candidate tumor suppressor genes. *Cancer Res.* 60:6116–33

60. Johnson L, Mercer K, Greenbaum D, Bronson RT, Crowley D, et al. 2001. Somatic activation of the K-*ras* oncogene causes early onset lung cancer in mice. *Nature* 410:1111–16

61. Ryding AD, Sharp MG, Mullins JJ. 2001. Conditional transgenic technologies. *J. Endocrinol.* 171:1–14

62. Soriano P. 1999. Generalized lacZ expression with the ROSA26 Cre reporter strain. *Nat. Genet.* 21:70–71

63. Sauer B, Henderson N. 1988. Site-specific DNA recombination in mammalian cells by the Cre recombinase of bacteriophage P1. *Proc. Natl. Acad. Sci. USA* 85:5166–70

64. Kilby NJ, Snaith MR, Murray JA. 1993. Site-specific recombinases: tools for genome engineering. *Trends Genet.* 9:413–21

65. Kwan KM. 2002. Conditional alleles in mice: practical considerations for tissue-specific knockouts. *Genesis* 32:49–62

66. Le Y, Sauer B. 2001. Conditional gene knockout using Cre recombinase. *Mol. Biotechnol.* 17:269–75

67. Kellendonk C, Tronche F, Monaghan AP, Angrand PO, Stewart F, Schutz G. 1996. Regulation of Cre recombinase activity by the synthetic steroid RU 486. *Nucleic Acids Res.* 24:1404–11

68. Feil R, Brocard J, Mascrez B, LeMeur M, Metzger D, Chambon P. 1996. Ligand-activated site-specific recombination in mice. *Proc. Natl. Acad. Sci. USA* 93:10887–90

69. Zhang Y, Riesterer C, Ayrall AM, Sablitzky F, Littlewood TD, Reth M. 1996. Inducible site-directed recombination in mouse embryonic stem cells. *Nucleic Acids Res.* 24:543–48

70. Brocard J, Feil R, Chambon P, Metzger D. 1998. A chimeric Cre recombinase inducible by synthetic, but not by natural ligands of the glucocorticoid receptor. *Nucleic Acids Res.* 26:4086–90

71. Danielian PS, White R, Hoare SA, Fawell SE, Parker MG. 1993. Identification of residues in the estrogen receptor that confer differential sensitivity to estrogen and hydroxytamoxifen. *Mol. Endocrinol.* 7:232–40

72. Minamino T, Gaussin V, DeMayo FJ, Schneider MD. 2001. Inducible gene targeting in postnatal myocardium by cardiac-specific expression of a hormone-activated Cre fusion protein. *Circ. Res.* 88:587–92

73. Kitayama K, Abe M, Kakizaki T, Honma D, Natsume R, et al. 2001. Purkinje cell-specific and inducible gene recombination system generated from C57BL/6 mouse ES cells. *Biochem. Biophys. Res. Commun.* 281:1134–40

74. Berton TR, Wang XJ, Zhou Z, Kellendonk C, Schutz G, et al. 2000. Characterization of an inducible, epidermal-specific knockout system: differential expression of lacZ in different Cre reporter mouse strains. *Genesis* 26:160–61

75. Arin MJ, Longley MA, Wang XJ, Roop DR. 2001. Focal activation of a mutant allele defines the role of stem cells in mosaic skin disorders. *J. Cell. Biol.* 152:645–49

76. Vasioukhin V, Degenstein L, Wise B, Fuchs E. 1999. The magical touch: genome targeting in epidermal stem cells induced by tamoxifen application to mouse skin. *Proc. Natl. Acad. Sci. USA* 96:8551–56

77. Brocard J, Warot X, Wendling O, Messaddeq N, Vonesch JL, et al. 1997. Spatiotemporally controlled site-specific somatic mutagenesis in the mouse. *Proc. Natl. Acad. Sci. USA* 94:14559–63

78. Kuhbandner S, Brummer S, Metzger D, Chambon P, Hofmann F, Feil R. 2000. Temporally controlled somatic mutagenesis in smooth muscle. *Genesis* 28:15–22

79. Schwenk F, Kuhn R, Angrand PO, Rajewsky K, Stewart AF. 1998. Temporally and

spatially regulated somatic mutagenesis in mice. *Nucleic Acids Res.* 26:1427–32

80. Imai T, Jiang M, Kastner P, Chambon P, Metzger D. 2001. Selective ablation of retinoid X receptor alpha in hepatocytes impairs their lifespan and regenerative capacity. *Proc. Natl. Acad. Sci. USA* 98:4581–86

81. Imai T, Jiang M, Chambon P, Metzger D. 2001. Impaired adipogenesis and lipolysis in the mouse upon selective ablation of the retinoid X receptor alpha mediated by a tamoxifen-inducible chimeric Cre recombinase (Cre-ERT2) in adipocytes. *Proc. Natl. Acad. Sci. USA* 98:224–28

82. Danielian PS, Muccino D, Rowitch DH, Michael SK, McMahon AP. 1998. Modification of gene activity in mouse embryos in utero by a tamoxifen-inducible form of Cre recombinase. *Curr. Biol.* 8:1323–26

83. Gautam A, Waldrep JC, Densmore CL. 2003. Aerosol gene therapy. *Mol. Biotechnol.* 23:51–60

84. Kanegae Y, Lee G, Sato Y, Tanaka M, Nakai M, et al. 1995. Efficient gene activation in mammalian cells by using recombinant adenovirus expressing site-specific Cre recombinase. *Nucleic Acids Res.* 23:3816–21

85. Wang Y, Krushel LA, Edelman GM. 1996. Targeted DNA recombination in vivo using an adenovirus carrying the cre recombinase gene. *Proc. Natl. Acad. Sci. USA* 93:3932–36

86. Stec DE, Davisson RL, Haskell RE, Davidson BL, Sigmund CD. 1999. Efficient liver-specific deletion of a floxed human angiotensinogen transgene by adenoviral delivery of Cre recombinase in vivo. *J. Biol. Chem.* 274:21285–90

87. Anton M, Graham FL. 1995. Site-specific recombination mediated by an adenovirus vector expressing the Cre recombinase protein: a molecular switch for control of gene expression. *J. Virol.* 69:4600–6

88. Aoki K, Barker C, Danthinne X, Imperiale MJ, Nabel GJ. 1999. Efficient generation of recombinant adenoviral vectors by Cre-lox recombination in vitro. *Mol. Med.* 5:224–31

89. Ueno T, Matsumura H, Tanaka K, Iwasaki T, Ueno M, et al. 2000. Site-specific integration of a transgene mediated by a hybrid adenovirus/adeno-associated virus vector using the Cre/loxP-expression-switching system. *Biochem. Biophys. Res. Commun.* 273:473–78

90. Ogasawara Y, Mizukami H, Urabe M, Kume A, Kanegae Y, et al. 1999. Highly regulated expression of adeno-associated virus large Rep proteins in stable 293 cell lines using the Cre/loxP switching system. *J. Gen. Virol.* 80 (Pt 9):2477–80

91. Kaspar BK, Vissel B, Bengoechea T, Crone S, Randolph-Moore L, et al. 2002. Adeno-associated virus effectively mediates conditional gene modification in the brain. *Proc. Natl. Acad. Sci. USA* 99:2320–25

92. Jackson EL, Willis N, Mercer K, Bronson RT, Crowley D, et al. 2001. Analysis of lung tumor initiation and progression using conditional expression of oncogenic K-*ras*. *Genes Dev.* 15:3243–48

93. Meuwissen R, Linn SC, van der Valk M, Mooi WJ, Berns A. 2001. Mouse model for lung tumorigenesis through Cre/lox controlled sporadic activation of the K-Ras oncogene. *Oncogene* 20:6551–58

Annu. Rev. Physiol. 2004. 66:665–88
doi: 10.1146/annurev.physiol.66.032102.150049
Copyright © 2004 by Annual Reviews. All rights reserved
First published online as a Review in Advance on September 15, 2003

BACTERIORHODOPSIN

Janos K. Lanyi

*Department of Physiology and Biophysics, University of California, Irvine,
California 92697; email: jlanyi@orion.oac.uci.edu*

Key Words   membrane proteins, protein structure, ion pump, retinal proteins

■ **Abstract**   Fourier transform infrared and Raman spectroscopy, solid-state NMR,
and X-ray crystallography have contributed detailed information about the structural
changes in the proton transport cycle of the light-driven pump, bacteriorhodopsin. The
results over the past few years add up to a step-by-step description of the configu-
rational changes of the photoisomerized retinal, how these changes result in internal
proton transfers and the release of a proton to the extracellular surface and uptake on
the other side, as well as the conservation and transformation of excess free energy
during the cycle.

INTRODUCTION

Since its discovery in the early seventies (1), the conceptual simplicity and experimental advantages of bacteriorhodopsin have made this light-driven proton pump
a testing ground for hypotheses of transport mechanisms and new experimental technologies. The existence of a pump for protons in the "purple membrane"
patches, not in direct contact with energy-transducing proteins in the other regions
of the membrane, generated much activity to explore the then newly proposed
chemiosmotic hypothesis of energy coupling. Detection of the intermediate states
of the reaction cycle (photocycle), after cryo-trapping or in real time by various
means, produced spectroscopic information with unprecedented signal/noise ratio. The kinetic schemes of great complexity that followed revealed the outlines of
the possible transport mechanism but also demonstrated the difficulties of understanding multistep reaction sequences. The naturally occurring two-dimensional
crystalline array of the purple membranes made the development of cryo-electron
microscopy possible. The primary structure, first from protein and then gene sequencing difference Fourier transform infrared (FTIR) spectra obtained by new
methods adapted to proteins, and the photocycle kinetics of a large number of
site-directed mutants, revealed that the protonated retinal Schiff base, and aspartate and glutamate residues, play the primary roles in the transport (as now seems
to be the case in other pumps) and identified the path of the transported proton. Assignment of the many vibrational bands of the protein with site-directed
mutations, and the FTIR and Raman bands of the retinal with isotope labeling, were

0066-4278/04/0315-0665$14.00

important advances that described the photocycle intermediates in specific molecular terms.

The wealth of spectroscopic data, together with the information from low-resolution electron diffraction maps, pointed to a well-defined transport mechanism. There was general agreement over the main steps (2–5). After photoisomerization of the retinal from all-*trans* to 13-*cis*,15-*anti*, the Schiff base proton is transferred to Asp-85 located on the extracellular side, and a proton is then released to the bulk from a site near the surface. Figure 1 (see color insert) shows the overall structure of the protein with the most important residues labeled. Reprotonation of the Schiff base is from Asp-96 located on the cytoplasmic side, aided by tilts of the cytoplasmic ends of helices F and G that were thought to result in increased hydration of this region. Reprotonation of the Schiff base through a proposed chain of water molecules is followed by reprotonation of Asp-96 from the cytoplasmic surface and reisomerization of the retinal to all-*trans*. Finally, transfer of a proton from Asp-85 to the vacant proton release site completes the cycle.

Although this was a more detailed model than available for other ionic pumps, it left many important questions unanswered. How is energy stored in the retinal and how is it transferred to the protein? What is the switch that gives the pump directionality? What is the cause of the changes in the pK_as of donor and acceptor groups? If a hydrogen-bonded chain of water develops to facilitate reprotonation of the Schiff base, where are these water molecules and how does the chain form? The answers are now at hand. FTIR spectra are providing more and more specific mechanistic insights (6–9). X-ray diffraction is yielding high-resolution maps for the protein and many of the intermediate states of the photocycle (10, 11). Solid-state NMR is producing interatomic distances and bond angles and their changes at functional regions (12, 13). Today the transport mechanism is understood at a level unimaginable only a few years ago.

As any fast-moving field, that of bacteriorhodopsin abounds in controversies, and not only hypotheses and interpretations but also data are intensely disputed. It is difficult to avoid taking sides in these debates. The reader is advised to take all review articles (including this one) with a grain of salt and to carefully read the original accounts of the research when confronted with differing points of view (11, 14, 15).

Bacteriorhodopsin Photocycle as a Reaction Sequence

It has been known from the earliest work that the photocycle contains the spectrally distinguishable quasi-stable states K, L, M, N, and O. Once K is produced from J, which arises from the excited state, the rest of the reaction cycle is a sequence of thermal reactions, as in other transport systems. Its kinetic description is therefore, of necessity, the point of departure for any proposed transport mechanism.

The intent of kinetic analyses has been to calculate the spectra of the intermediate states, define the reactions that connect them, and find their rate constants. The greatest problem is that the number of observable relaxation time-constants exceeds the number of identified spectral states (16). There is some agreement

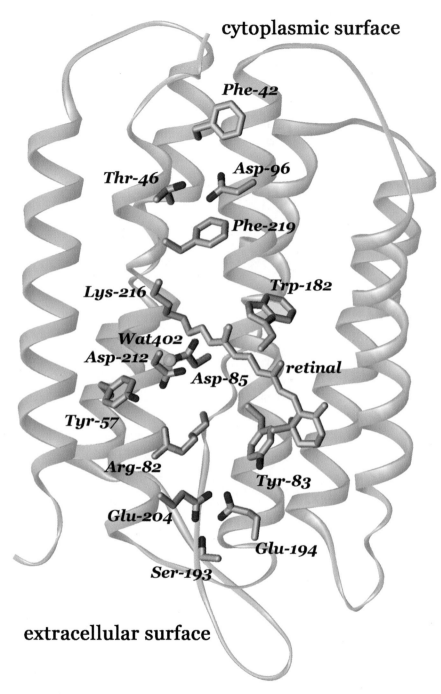

cytoplasmic surface

extracellular surface

Figure 1 Structure of bacteriorhodopsin, with functional side chains, the retinal, and Wat402 shown. Drawn from pdb coordinate file 1C3W.

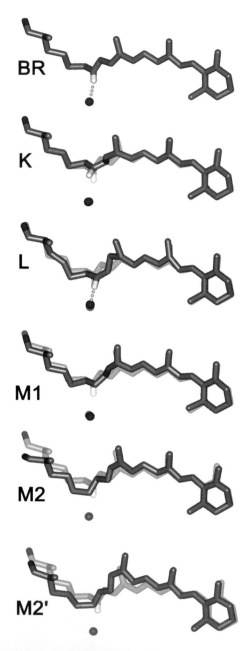

Figure 3 Structural changes of the retinal and the connected Lys-216 side chain in the first half of the photocycle. Wat402 is shown as hydrogen-bonded to the protonated Schiff base in BR and L only (see text). The images were constructed from the following coordinate sets: BR (1M0L); K (1M0K); L (1OOA); M_1 (1M0M); M_2 (1F4Z); M_2' (1C8S). Redrawn with permission from (47).

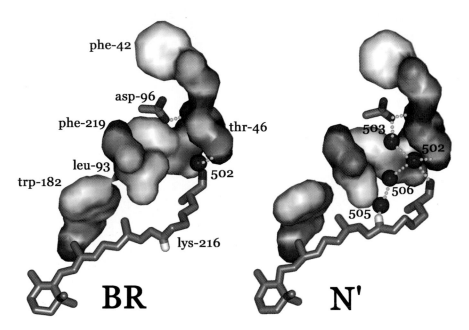

Figure 4 Structural changes in the cytoplasmic region of bacteriorhodopsin after reprotonation of the retinal Schiff base. The structures of the BR and N' states (see text) contain a hydrophobic barrier to proton exchange between Asp-96 and the retinal Schiff base in the former, and the formation of a cleft enclosing a chain of hydrogen-bonded water in the latter. Color coding for residues shown as surfaces is according to electrostatic potential.

that the discrepancy originates from substates with similar spectra and reversible reactions that lead to transient equilibrations of intermediate states (17–19). Others hold that bacteriorhodopsin is intrinsically heterogeneous and that the observed kinetics is from the summing of temporally overlapping photocycles, each from a different initial conformation (20, 21). Although these fundamentally differing models are mathematically equivalent, attempts have been made to decide the issue on other grounds. The best test for an equilibration reaction is the response of the reaction mixture to a perturbation. Such perturbation was applied as a heat pulse (22) or a second flash (23, 24), at suitable times after initiating the photocycle with a first light pulse. As expected from the equilibration model, depletion of the O or the M intermediate by the perturbation was followed by their partial recovery with the predicted kinetics. However, the controversy over single versus multiple photocycles, and unidirectional versus bi-directional photocycle reactions, continues (25, 26). An additional source of complexity is the dependence of the initial state of bacteriorodopsin on pH. For example, when the proton release site in the initial state is unprotonated the photocycle is altered (3, 27).

In any case, in spite of the conceptual simplicity of the photoisomerization of the retinal as the reaction that drives the transport, there is at present no kinetic scheme that fully accounts for the many and complex spectroscopic observations. Although the agreement between the predictions of the models of the best fit and the data is good, it is not perfect. As the quality of the kinetic traces improves, the small but real discrepancies become more evident. Perhaps it is naïve to expect that a system with as many degrees of freedom as a protein will exhibit kinetics as simple as reactions of small molecules. Indeed, with the reasonable assumption that the reaction barrier to each photocycle reaction consists of a range of activation energies within $\pm kT$, the fits improve greatly (23). There appears to be no way to test whether this is a valid, let alone unique, solution.

Today, most workers in the bacteriorhodopsin field have come to accept the lack of rigor in the kinetic fits, and regard the single photocycle scheme with reversible reactions as a workable approximation. Some degree of confidence in such a model has been gained from the fact that these kinds of models produce spectra of reasonable shape for all of the intermediates, and not only for bacteriorhodopsin, but also for the related halorhodopsin and proteorhodopsin.

The kinetic model that accounts for all (or most) states necessitated by mechanistic considerations, combines kinetic data from the wild-type and various mutant proteins. Figure 2 shows a single cycle, with K, L, M, N, and O in sequence. As much as possible to visualize, each reaction corresponds to a single defined physical event (discussed below). This necessitates substates for at least M and N, as shown. The three M substates, produced in successive equilibria, are suggested by the complex rise kinetics of M. The rationale is that M_1 arises from L by deprotonation of the retinal Schiff base, the M_1 to M_2 reaction disconnects the Schiff base from the extracellular side and connects it to the cytoplasmic side (the protonation switch), and the M_2 to M_2' reaction is linked to release of a proton to the extracellular surface (28, 29). The idea of two N substates is from the biphasic decay of N, observable under some conditions (30), where N arises from reprotonation of the

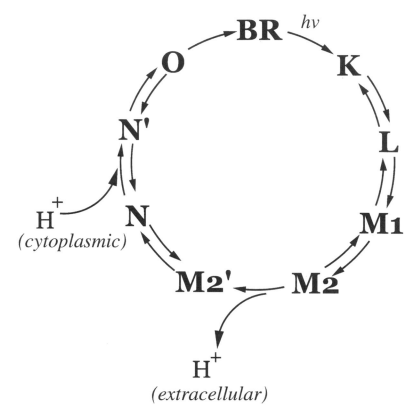

Figure 2 Photocycle scheme from kinetics in the wild-type and mutant bacteriorhodopsins. The K, L, M, N, and O states are distinguished by their spectral properties in the visible, infrared, Raman, and NMR, and by their crystal structures when available. The substates of M and N are identified from kinetic or spectral data or by their structural differences. Only the BR and O states contain all-*trans* retinal; in all other states the retinal is 13-*cis*,15-*anti*. Proton release to the extracellular surface and uptake from the bulk on the cytoplasmic side are shown as they occur under physiological conditions (pH > 7).

Schiff base by Asp-96, and the N to N' reaction corresponds to the reprotonation of Asp-96 from the cytoplasmic surface. There are likely substates also for O that arise after reisomerization of the retinal to all-*trans*, but they have not been widely studied.

Structure of the Protein

Bacteriorhodopsin is a small (24 kDa) integral membrane protein. Cryo-electron microscopy of two-dimensional crystals, at increasing resolutions (31, 32) to 3.0 Å, had described bacteriorhodopsin as the prototype of seven-transmembrane helical proteins. The retinal was seen to extend transversely into the interhelical space, and the positions of bulkier residues defined the orientation of most of the helices and therefore the pathway of the transported proton. The structure identified an interesting paradox: The region on the extracellular side of the retinal contains many polar residues known from mutagenesis to play essential roles in the deprotonation of the Schiff base and the release of a proton to the surface, but the region on the cytoplasmic side contains no obvious means to conduct a proton to reprotonate it. Transfer of a proton from Asp-96 to the Schiff base requires breaching a hydrophobic gap of 10–12 Å, and a chain of single-file hydrogen-bonded water molecules was postulated in this region. The electron diffraction at 3 Å (31) and a later X-ray structure at 2.95 Å resolution (33) did not allow its visualization, but several cavities in the cytoplasmic region that could contain a total of 7–8 water molecules (31) and later molecular dynamics calculations (34–36), suggested the existence of such a proton-wire.

The development of the cubic lipid phase crystallization method (37) made it possible to grow high-quality three-dimensional bacteriorhodopsin crystals, in the P6$_3$ space group for X-ray diffraction. Resolutions improved from 2.5 Å (38) to 2.30 Å (39), 1.90 Å (40), 1.55 Å (41), and then to 1.43 Å (42). The models described not only the protein monomers in atomic detail but also the lipids that form a single layer around them in the crystal in apparently the same way as in the two-dimensional P3 purple membrane lattice. It was a hindrance for the refinement that most of the crystals grown in this way turned out to be merohedral twins (39, 43). Although the twinning can be accounted for, either by detwinning the data before a conventional refinement (at low twin ratios) or by twinning the model in every calculation cycle (for twin ratios near 50:50), the effective resolutions are somewhat worse than for non-twinned crystals. The crystals in the cubic lipid phase are highly variable, and screening for a rare non-twinned crystal or for one with a low twin-ratio was a possible solution (44). However, crystals that diffract to better than about 2 Å are also rare, and there is only one report of a crystal with both low twin-ratio and exceptionally high resolution (41).

As anticipated (34, 36, 45), the high-resolution structures demonstrated that a water molecule, Wat402, separates the proton donor Schiff base and the proton acceptor, Asp-85. The extracellular region contains seven water molecules that, together with the polar sidechains of Asp-85, Asp-212, Thr-89, Trp-86, Tyr-83, Tyr-57, Tyr-185, Arg-82, Glu-194, Glu-204, and Ser-193, form a three-dimensional hydrogen-bonded network that reaches from the Schiff base to the surface. These are the residues that had been identified by mutagenesis to play roles in proton release to the surface. The structure gives a rationale for why Asp-85, and not the strongly hydrogen-bonded Asp-212, is the proton acceptor (41).

The structure also contains unexpected features. First, the cytoplasmic region does not have cavities or a chain of water molecules. If a proton is to be transferred through this region, an aqueous chain has to be created, transiently, during the photocycle. Second, four of the seven helices contain irregularities or kinks. In helices B, C, and F the discontinuity of the α-helix is from a proline residue, but in helix G, a new structural feature, a single turn of a π–helix termed a π–bulge, was identified at Lys-216 where the retinal is bound. It is associated with hydrogen bonds of the two main-chain C=O groups to water molecules. One of these, Wat501, bridges helices G and F by connecting O of Ala-215 to NE1 of Trp-182 near the retinal polyene chain. The other, Wat502, bridges helices G and B through a hydrogen-bond of O of Lys-216 with OG1 of Thr-46. Thr-46 is further connected to OD2 of Asp-96 on helix C, completing a chain of covalent and hydrogen bonds that links the region of the $C_{13}=C_{14}$ bond of the retinal, where the isomerization occurs, to the proton donor to the Schiff base, Asp-96.

The static structure of the protein thus contains clues to how events at the centrally located retinal might be linked to protonation changes at the two surfaces, but a real understanding of the mechanism of proton transport requires that structural information is also produced for the intermediate states. To date, all attempts to accomplish this have utilized photo-stationary states in the crystals at various temperatures and wavelengths of illumination, in the wild-type protein and mutants, followed by rapid cooling to 100 K for trapping the intermediates and the diffraction measurements. Although crystallography is usually regarded as the method of choice for structural studies, there are great difficulties in accurately describing the changes. At the $C_{13}=C_{14}-C_{15}=NZ$ segment of the retinal, where many of the changes of interest occur, the atomic displacements are small. In most cases, the photoconversion to the intermediate of interest is partial, and the refinement has to dissect the intermediate state from the overlapping unchanged state. The validity of such partial-occupancy refinement need be ascertained in each case. This was the reason for calculating omit maps, correlation coefficients of omit maps between data from illuminated and nonilluminated crystals, multiple independent diffraction measurements, and refinement of models from data of nonilluminated crystals with sham mixtures of conformations as controls (42, 46–48). In some studies (44, 49), the greater changes observed near the retinal than in more remote regions of the protein were cited as evidence that the changes are meaningful. Because the information content of diffraction data increases with resolution according to a cubic function, hard-won improvements in the quality of the crystals are of prime importance. In fact, in nearly all cases so far where the photoconversion is only partial, as for the K, L, and M in the wild-type protein, the structure calculated from data with resolutions 1.6 Å or better (42, 46–48) turned out to be quite different from earlier structures at 2.1–2.25 Å resolution (44, 49, 50).

Another difficulty is that the crystal lattice may resist large-scale conformational changes; either they will occur and disrupt the crystal, at least partly, (48, 51) or the intermediates in which they are produced will not be significantly populated (48). Less specific but more reliable information about such conformational shifts must be sought with probes. Spin-spin interactions at engineered residues revealed distance changes (52–54), and the hyperfine line-spacing of electron spectra of resonance (ESR) probes and the chemical reactivities of introduced cysteines provided information about changes in the exposure of groups at the surface (55–58).

Infrared spectroscopy provides crucial information about even subtle environmental changes at retinal and protein bonds. Assignment of the bands from such bonds with mutations and isotope labeling is generally not a problem (6), but assigning O-H bands to specific water molecules is a challenge (7). Resonance Raman spectra give very precise information about the geometry of the retinal (59). Because band shifts can be caused by very small structural changes, many features in the infrared and Raman do not have readily identified counterparts in the crystal structures. By comparing stationary spectra of cryo-trapped states and transient spectra at ambient temperature, these methods help to validate, and in some cases raise questions about, more direct structural information about the photocycle intermediates.

Solid-state NMR has been used to advantage to obtain atom-to-atom distances and angles in the nonilluminated states and in cryo-trapped intermediates (12, 13). With some notable exceptions, the numbers agree with those from the highest resolution X-ray structures. The disagreements are for some torsion angles, e.g., for the $C_{14}-C_{15}$ bond in the nonilluminated state and in two M states. The reason could be that in NMR the torsion angles refer to H$-$X$-$X$-$H, whereas in the X-ray structures they describe the geometry of C$-$X$-$X$-$C, and these do not need to be the same (42, 60). The difficulties of NMR are similar, and perhaps even more daunting, than for crystallography: how to find the conditions to trap the states of interest by illuminating an optically dense sample. The mass of an NMR sample is vastly greater than that of a single crystal, and the trapping conditions cannot be the same. The main advantage of NMR over crystallography is that the peaks from coexisting states do not usually overlap and thus partial occupancy is easier to handle.

Photoisomerization of the Retinal

The vibrational bands of the retinal have shown conclusively that in the K intermediate the retinal assumes a 13-*cis*,15-*anti* configuration twisted between C_{14} and the Schiff base NZ. The K state is produced either in a photostationary state at 100 K or transiently with a microsecond lifetime after a flash at ambient temperature. The hydrogen-out-of-plane (HOOP) bands that arise from displacement of the hydrogen atoms in $C_{14}-$H, $C_{15}-$H, and NZ$-$H out-of-the-plane of the retinal are observed in both (61–63), but the two forms of K are not quite equivalent. A strongly twisted retinal in K was predicted from molecular dynamics (64).

At 3.5 Å resolution and in projection, cryo-electron microscopy did not reveal any structural change in the K state (65), but the density map from X-ray diffraction of a three-dimensional crystal illuminated at 100 K contained numerous difference peaks (49). At 2.10 Å resolution, the retinal was modeled in a 13-*cis*,15-*anti* configuration relatively free of strain, with the NZ-H bond reoriented from the extracellular to the cytoplasmic side. Wat402 was absent, i.e., it either moved away from Asp-85 or became disordered, and Asp-85 moved closer to the Schiff base. The suggestion was made that these motions and those of Lys-216 and the peptide backbone of helix G, as well as nearby residues such as Tyr-57, initiate the subsequent photocycle reactions and the proton transfers that accompany them.

At 1.43 Å resolution, the same illumination regime yielded an entirely different structure for K (42). The changes were restricted to the C_{13} to NZ segment of the retinal, with no significant movements of protein atoms. The retinal is highly strained, apparently because a steric conflict with protein atoms prevents assumption of the normal sharp bend of the 13-*cis*,15-*anti* isomer. Increase of the bond angle at C_{13} from 112° to 145 ± 12° ($n = 4$) permits isomerization of the $C_{13}{=}C_{14}$ bond with virtually no change in the end-to-end distance of the polyene chain (Figure 3, see color insert). The rotation of the $C_{13}{=}C_{14}$ bond from 154° to –2 ± 39° is accompanied by counter-rotations of the $C_{14}{-}C_{15}$ and $C_{15}{=}NZ$ bonds with the result that the N–H bond does not reorient toward the cytoplasmic side. However, its direction is somewhat changed. Although Wat402 does not seem to have moved, its hydrogen-bond to NZ is weakened by decrease of the NZ–H–O angle from 147° to 116 ± 15°. A similarly twisted retinal in K (with Wat402 present) was refined independently from data for crystals grown by another method (66). The resolution was only 2.6 Å, but careful corrections were made for damage from exposure to the X-rays.

The strained retinal configuration in K has implications for the next photocycle steps because the excess free energy of K drives all subsequent photocycle reactions. If the isomerized retinal is to relax to an unhindered bent configuration, with NZ rotating toward the cytoplasmic side, it must become deprotonated and lose its electrostatic interaction with the neutralized Asp-85 (12). Although the retinal and its immediate surroundings are in a high-energy state in K, if proton transfer from the Schiff base to Asp-85 is to occur through Wat402 that connects them, it is not likely to happen directly. Formation of the M intermediate will require a reaction pathway that reorients the NZ–H bond toward Wat402. The sequence of retinal conformations shown in Figure 3, from the crystal structures of the BR, K, L, M_1, M_2, and M_2' states, describes the way this is achieved.

Deprotonation of the Retinal Schiff Base

Arguably, the first proton transfer in the cycle from the retinal Schiff base to Asp-85 is the most critical one. A large effort has been made to describe the L state by X-ray diffraction, FTIR, and NMR, with the expectation that it will reveal (*a*) why the protonation equilibrium between the Schiff base and Asp-85 shifts in the L to M reaction toward protonation of Asp-85 and (*b*) how the proton of the Schiff base is transferred to Asp-85. Is it over Wat402 or directly to Asp-85? Or does dissociation of Wat402 protonate Asp-85 and the OH⁻ moves to the cytoplasmic side where it receives the Schiff base proton, a mechanism that would make bacteriorhodopsin, in effect, a hydroxyl ion pump (67)? Alternatively, is the protonation of Asp-85 accompanied by transfer of a water molecule, perhaps Wat402, in the opposite cytoplasmic direction, in another version (13) of net transport of a hydroxyl ion?

In the first report of the crystal structure of L (44), to 2.10 Å resolution, there was little density change at the retinal, but the changes of the map in its vicinity were extensive. Wat402 was missing, and its dislocation was suggested to have triggered movements of the side chains of Ala-215 and Lys-216 on helix G, as well as Arg-82, Tyr-83, Asp-85, Trp-86, and Thr-89 on helix C. The breaking of the hydrogen bond between Asp-85 and Thr-89, was proposed to increase the pK_a of Asp-85 and make it the proton acceptor. A "bowing" of helix C toward helix G was suggested to decrease the distance between the Schiff base nitrogen and Asp-85, creating the conditions for proton transfer. [Inexplicably, however, in the deposited coordinate set (1E0P), the distance between NZ and Asp-85 increases rather than decreases.] There was controversy regarding whether the L state was actually produced with the illumination protocol, and if its yield, calculated from crystallographic data, could be as high as 70% (44, 68, 69). In a follow-up study (15), the absorption spectrum of the illuminated crystal corresponded more to what is expected for L, and the X-ray structure was claimed to confirm the first report.

Very few of the changes in these reports were observed in the structure of L at 1.62 Å resolution (47), however. Nearly all atomic displacements not near the retinal were found to be 0.2 Å or less, with the exception of a shift of the positions of the phenyl rings of Phe-219 and Phe-42 at Asp-96 that might account for the shift of the C=O stretch frequency of the aspartic acid in L (reviewed in 6). The Arg-82 side chain also showed larger than average displacement, but the distance between the Arg-82 CZ and the Asp-85 CG was unchanged. Wat402 was somewhat displaced but not absent. Significantly, the retinal was observed to undergo greater changes than in the K state (47). As expected, there are additional rotations in L around the $C_{13}=C_{14}$, $C_{14}-C_{15}$, and $C_{15}=NZ$ bonds that reorient the Schiff base N—H bond toward Wat402. The increased angle of the hydrogen-bond to 132 ± 9° (n = 6) provides a more effective pathway for transfer of the Schiff base proton to Asp-85 via Wat402 than exists in K. This could be the reason why L, and not K, leads to the further steps of the photocycle.

The changes at the retinal should be evaluated in light of the vibrational and NMR spectra of K and L. The coupling between the NZ-H in-plane bend and the $C_{15}=NZ$ stretch, that nearly completely disappears in K but reappears in L (62, 63), may be related to the breaking of the hydrogen-bond of NZ to Wat402 in K and its reestablishment in L. However, the vibrational coupling is stronger in L than in the nonilluminated state, and the downfield shift of the ^{15}N-labeled NZ indicates stronger interaction of the Schiff base with the counter-ion (70). Neither of these features of L would be predicted from the crystal structure. On the other hand, the strong HOOP band of C_{15}—H in K and its weaker amplitude in L (71) are consistent with the greater displacement of C_{15} from the plane of the retinal polyene chain in K than in L. From changes in the O—H stretch bands assigned to Wat401 and Wat402, it was argued that Asp-85 becomes the proton acceptor because in M the waters are more strongly coordinated by the carboxyl group of Asp-212 (72). The length changes of the relevant hydrogen bonds in M_1 are not seen in the X-ray structure (46) but might be too small to be detected with this method.

There is a more glaring discrepancy between FTIR and X-ray measurements over the presence of a new water molecule in L. Its location on the cytoplasmic side of the Schiff base near Leu-93 was inferred from the effects of various mutations on the frequency of an O—H stretch band assigned to water (73), but such a water molecule is not evident in the electron density maps.

Proton Release to the Extracellular Surface

During deprotonation of the Schiff base, i.e., during the rise of the M intermediate, a proton is released to the extracellular surface (74–76), but the released proton originates neither from the Schiff base nor Asp-85. Coupling between the protonation states of Asp-85 and the likely release site was demonstrated by the anomalous titration behavior of Asp-85 in the nonilluminated state (77). The biphasic titration suggested that at a physiological pH either Asp-85 or the site may be protonated, but not both. When applied to the photocycle, the model predicts well the observed pK_a of the release site, both before protonation of Asp-85 ($pK_a = 9$) and after ($pK_a = 5$–6) (28, 78). Furthermore, the coupling is abolished by the same mutations that block proton release during the rise of M (e.g., R82Q, E194Q, E204Q). Thus the direct cause of the proton release to the surface is identified as the protonation of Asp-85. The identity of the release site is not yet certain. Previously, it was suggested to be protonated Glu-204 (79), a complex that includes Glu-204 and Glu-194 (80) or a proton shared between Glu-194 and Glu-204 (33). From the depletion of a hydrogen-bonded continuum during the rise of M, however, it appears that the proton instead originates from the network of water molecules, in the vicinity of Arg-82, Glu-194, and Glu-204 (81). The site may be an $H_5O_2^+$ cluster (82).

According to the rules of the coupling (77, 83), the release of the proton at a pH higher than its pK_a will raise, in turn, the pK_a of Asp-85. This will shift the protonation equilibrium toward more complete deprotonation of the Schiff base, as indeed is found (28). Is the degree of coupling in the photocycle, where the retinal is 13-*cis*, the same as observed by titration in the dark, where it is all-*trans*? The pK_a of Asp-85 is expected to rise very strongly upon proton release. The resulting ΔpK_a between donor and acceptor can be decided from the ratio of L (protonated Schiff base) and M (deprotonated Schiff base) in the last of the quasi-equilibria during the rise of M. In the wild-type photocycle, this ratio is difficult to determine because the decay of the L state and the rise of the N state are temporally not well separated, and the spectra of L and N are similar. If N is not taken into account, it is easy to overestimate the residual L that coexists with M (84). A solution of this problem is to use the D96N mutant, where under most conditions the decay of M is so slow that N does not accumulate. In such systems, the kinetics suggested (28) two distinct steps in the deprotonation: (*a*) a pH-independent step labeled the M_1 to M_2 reaction, and (*b*) a pH-dependent step, with a pK_a of about 6 as predicted from the link between Schiff base deprotonation and proton release, labeled the M_2 to M_2' reaction. With increasing pH the L/M ratio was shifted by these two reactions to essentially zero (24, 28) in accordance with the parameters of the coupling in the dark.

Although proton release to the surface, followed with a pH indicator dye, is more or less coincident with the M_2 to M_2' phase of the formation of M (28), the kinetic coupling does not hold under all conditions (85). Furthermore, although the rise kinetics for M represents a relatively simple proton transfer, a very large and complex deuterium isotope effect suggests that it depends on other reactions (86). These observations imply that there is an intervening step between protonation of Asp-85 and proton release. The crystallographic structures of the M states produced in the wild-type (50), and the E204Q (29) and D96N (51) mutants indicated that this step could be the movement of the side chain of the positively charged Arg-82 away from Asp-85, approaching the Glu-194/Glu-204 pair. Thus shuttling of the positive charge of Arg-82 between the up and down positions appears to be the means of the coupling that drives proton release. The motion of Arg-82 occurs both when proton release is normal (in D96N) or blocked (in E204Q), arguing that its cause is the protonation of Asp-85 and not the proton release (29). The change of the hydrogen-bonding of the Arg-82 sidechain was detected by NMR, but it occurred in both early and late M (87). This would be consistent with the X-ray structures only if these were M_2 and M_2', a difficult question to answer because the M states for NMR were trapped at more extreme conditions (pH 10, with and without guanidinium hydrochloride).

Protonation Switch

The transport cycle of bacteriorhodopsin must contain a step in which the connectivity of the Schiff base changes from one side to the other, i.e., a switch. There may be more than one switch step in the cycle, but the minimal requirement for pumping is either that the Schiff base is first connected to Asp-85 and then disconnected from it or that the pK_a of Asp-85 is first lower and then becomes higher. The switch would thereby ensure that in the subsequent steps the Schiff base is reprotonated not by Asp-85 in a futile cycle but by the other donor, Asp-96.

The identity of the bacteriorhodopsin switch has been the subject of much experimental investigation and speculation. Over the past years the switch was proposed to reside in (*a*) a predicted change of the curvature of the retinal polyene chain upon deprotonation of the Schiff base that moves NZ away from Asp-85 (88); (*b*) reorientation of the Schiff base nitrogen from the extracellular to the cytoplasmic direction once its electrostatic interaction with Asp-85 is broken (12), as indeed was observed in the X-ray structures of M_1 and M_2 (46); (*c*) movement of the segment of helix C with Asp-85, first toward (in L) and then away (in M) from the Schiff base (44); (*d*) passage of OH$^-$ (from a proposed dissociation of Wat402 prior to deprotonation of the Schiff base) from the extracellular to the cytoplasmic side where it would function as proton acceptor (67); and (*e*) rise of the pK_a of Asp-85 upon proton release to the surface, which makes it less suited as proton donor than Asp-96 on the cytoplasmic side (77, 83).

Some of these proposals cannot be tested with currently available methods, and others seem to be contradicted by data. The most clear-cut experimental support is for a two-step switch that incorporates both possible ways to give direction to the transport: (*a*) reorientation of the Schiff base nitrogen from the extracellular to the cytoplasmic direction in the M_1 to M_2 reaction (Figure 3) and (*b*) rise of the pK_a of Asp-85 in the M_2 to M_2' reaction. The former establishes an alternating pathway for the proton transfer (24, 89), whereas the latter prevents reversal of the protonation of Asp-85.

Reprotonation of the Schiff Base

In the M_2' to N reaction that follows the switch steps, the protonated Asp-96 and the unprotonated Schiff base reach a new equilibrium in which both are partly protonated, and this equilibrium is later shifted to full protonation of the Schiff base. Transfer of proton from Asp-96 to the retinal Schiff base requires that the pK_a of Asp-96 be lowered from above 11 in the nonilluminated state (90) to not higher than the pK_a of the Schiff base (about 8) (91) at this state in the photocycle, and that a pathway be created for the proton in the rather hydrophobic cytoplasmic region.

It has been widely considered, on experimental (92–94) and theoretical (34–36) grounds, that these functions are accomplished by water in the cytoplasmic region. Indeed, there are no other obvious alternatives (95). Indirect but suggestive evidence in favor of the transient entry of water into the protein interior is that osmotic pressure (92) and hydrostatic pressure (94) have greater influence on the $M \leftrightarrow N$ equilibration than any other reaction in the photocycle. Conduction of a proton would be by a Grotthus mechanism that involves a sequence of proton transfers and rotations to reorient the hydrogen bonds along the chain (96).

In the M_2 state, produced by illumination of a crystal of the E204Q mutant at 1.9 Å resolution but nearly full occupancy, a cluster of hydrogen-bonded water appeared near Asp-96 (29). It contained Wat502, and two new water molecules connected to it. Of these, Wat504 was intercalated between Asp-96 and Thr-46. It seems reasonable that this and the generally more polar environment lowers the pK_a of Asp-96 as required, and the cluster itself is the beginning of the network that would extend, before the N intermediate is formed, to the retinal Schiff base. At this stage, however, a gap of 7.4 Å separates the nearest water in the cluster from the Schiff base.

The structure of an M state produced by illumination of a crystal of the wild-type protein was reported at 2.25 Å resolution and 35% occupancy (50). It contained a larger hydrogen-bonded network of water molecules in the cytoplasmic region: Wat501 and Wat502 (W721 and W720 in the nomenclature used) displaced from their positions in the nonilluminated structure, plus four new water molecules. As in the M of the E204Q mutant, one of these, W724, was intercalated between Asp-96 and Thr-46. The network extended from Asp-96 to within 5.8 Å of the Schiff base, and it was suggested that the remaining gap is breached by movements of a fluctuating water molecule. More recently, however, in M produced in the same way but at 1.52 Å resolution (and a similar occupancy), the four water molecules additional to Wat501 and Wat502 were not detected (48). Instead, a cluster near Asp-96 similar to that in E204Q mutant was seen. Asp-96 and Thr-46 were not bridged by a water molecule. The conformation of the retinal revealed this M to be an early rather than late M in the sequence. Thus there is currently no crystallographic information about the cytoplasmic region in M'_2 of the wild-type protein.

It is unlikely that the conditions for protonating the Schiff base by Asp-96 would exist before N is ready to form. The observation of large-scale conformational changes involving the cytoplasmic ends of helices F and G was a clue to what might be required. Most prominently, electron diffraction maps, in projection for M of the wild-type and the D96N mutant (97, 98) and in three dimensions for N of the F219L mutant (99), had shown that helix F tilts outward. The tilt was detected with neutron (100) and X-ray (101–105) diffraction, spin-spin broadening from pairs of spin-labels that revealed distance changes (52–54), and changes in chemical reactivity along the E-F interhelical loop in cysteine-scanning mutagenesis (58). The involvement of the E-F loop in the large-scale conformational transition from M to N is suggested additionally by the phenotypes of numerous single-site mutants obtained in random mutagenesis (106). Two reports also suggested a rotation of helix F, related to the unwinding of the kink in the E-F loop (53, 58). The appearance of only positive density in projection maps for M did not fully define the movement of helix G. From X-ray diffraction, the changed position of Hg label on a cysteine introduced at position 222 indicated that the cytoplasmic end of helix G tilts inward (107). Immobilization of helix F, but not the other helices, by cross-linking greatly slowed the reprotonation of the Schiff base (108).

The cause for the tilt of helix F away from helix G is to be found in the rearrangements of Lys-216 and the π–bulge of helix G, as well as side chains near the 13-methyl group of the retinal. As the 13-methyl group is gradually thrust upward in the K-L-M segment of the photocycle, it exerts force on the indole ring of Trp-182 and causes its move. Twists of the Lys-216 chain propagate to the π–bulge of helix G, and the peptide O of Ala-215 moves in the opposite direction. As a consequence, Wat501 changes its hydrogen bonding from Ala-215 to Thr-178 (29) and no longer bridges helices F and G. Movement of Ala-215 induces displacement of Leu-181 and Phe-219, i.e., repacking of residues between helices F and G. As a result Asp-96 and Thr-46 move apart to admit Wat504.

The D96G/F171C/F219L mutant contains three residue replacements that as single mutations are known to slow either the decay of M or the decay of N. Unexpectedly, the structure of this triple mutant, determined by cryo-electron microscopy first in projection (97) and then in three dimensions to 3.2 Å resolution (88), contains all the large-scale changes described for M (and N) but without illumination. Although there are no further structural changes in the M intermediate of this mutant, its photocycle and transport activity appear nearly normal (109). These observations raise the question of the relevance of the helical tilt in the transport mechanism. There are two separate issues: (*a*) the requirement for the large-scale structural change between the nonilluminated state and M, which is ruled out, and (*b*) the requirement for the helical tilts altogether.

The first N intermediate, after the protonation of the Schiff base but before reprotonation of Asp-96, is the only bacteriorhodopsin state with an anionic Asp-96. The cytoplasmic region must be greatly changed to accommodate this buried negative charge. If its structure could be determined at high resolution, it would very likely reveal the details of the full cytoplasmic proton conduction pathway. However, attempts to produce this N in crystals from the V49A mutant, where the lifetime of N is long (at pH > 7 in order to prevent protonation of Asp-96) (30), were not successful (48). On the other hand, the structure of N', produced in V49A at a pH low enough to ensure reprotonation of Asp-96, could be determined to 1.62 Å resolution and 37% occupancy (48).

As expected, in N' the retinal assumes a relaxed 13-*cis*,15-*anti* configuration, with the Schiff base NZ-H facing in the cytoplasmic direction as NZ in M_2'. Also as in M_2', there is no density where Wat402 would be. However, several new density peaks appear on the cytoplasmic side of the retinal, which could be modeled as water. They form a single-file hydrogen-bonded chain, Wat505–Wat506–Wat502–Wat503, which extends from the retinal Schiff base to Asp-96. This chain, shown in Figure 4 (see color insert), is either the pathway for proton transfer in the M to N reaction that persists until N' or a similar one. There is indication that the Schiff base is hydrogen-bonded after its deprotonation. The unexpected upfield position of its ^{15}N peak suggests that the deprotonated Schiff base is hydrogen-bonded in two M substates (110) and remains so after reprotonation in N (111), perhaps to a water molecule on the cytoplasmic side. The X-ray structures do not show evidence for such water at NZ after L until the N' state, however (48). There is no detectable tilt of any helices, including F and G, in the N' structure.

Intercalation of water into the tightly packed cytoplasmic region requires space, and the side chains of Leu-93, Thr-46, and Phe-219 are moved apart to form the required cavities. The displacement of Phe-219, in particular, already occurs in M_2 and M_2' (48). Did the water molecules enter this region in direct response to these movements? Cavities created deliberately by side chain replacement are not filled with water in every protein. However, when Phe-219 was replaced with Leu, the crystal structure of the F219L mutant contained two additional hydrogen-bonded water molecules where the Phe ring would have been (48). They were also hydrogen-bonded to Wat501, suggesting that the water molecules in the cluster were recruited by this existing buried water. One of the new water molecules is in the D85S/F219L mutant as well (112).

Proton Uptake at the Cytoplasmic Surface

The pK_a of the Schiff base is about 8 during its reprotonation in the M_2' to N reaction (at least in the D96N mutant) (91), and the pK_a of Asp-96 is about 7 during its reprotonation in the N to N' reaction (30, 113). Because these steps overlap in time and the Schiff base remains protonated during this time at pH well above 8, the protonation pathways between Asp-96 and the Schiff base and between the cytoplasmic surface and Asp-96 must be separate (27). Indeed, residue replacements near Asp-96, such as T46V (114), L100C, F171C, and L223C (108), have opposite effects on the two reactions: They accelerate the reprotonation of the Schiff base but slow the reprotonation of Asp-96. The pathway of the proton that reprotonates Asp-96 and the conformational changes that might facilitate the entry of water into the hydrophobic shield over Asp-96 are as yet uncertain (30), but in projection, the maps for the late M and the N state do show differences (104). The opening of a channel that might guide the proton from the cytoplasmic surface to Asp-96 was suggested from the structure of the triple mutant (88), but mutations that replaced bulky side chains with smaller ones at the opening of the putative pore (L100C, F171C, and L223C) block rather than accelerate the reprotonation of Asp-96 (30).

Deeper inside the protein, however, there appear to be few if any alternatives to the formation of an aqueous chain for conducting the proton from the surface to Asp-96. There is no such continuous chain in N' (48). Wat504, which was located between Asp-96 and Thr-46 in M_2 (29), has apparently moved to the cytoplasmic side of Asp-96. Reformation of its direct hydrogen bond with Thr-46 presumably raises the pK_a of Asp-96 at this time. A new water molecule in N', Wat520, is hydrogen bonded to the side chain of Asp-38 and the peptide O of Leu-99 at the surface. Wat504 and Wat520 might be the first and last members of a continuous chain between the surface and Asp-96, formed as a result of protein conformational fluctuation that existed with low occupancy during the N to N' reaction. If so, the conditions for creating this chain are distinct from the one between Asp-96 and the Schiff base that seems to persist for the duration of at least two photocycle steps. If the aqueous network in N' is the same as formed during the M_2' to N reaction and remains functional in N', it is not clear why the Schiff base proton does not equilibrate with the bulk.

There are six acidic residues at the cytoplasmic surface: Asp-36, Asp-38, Asp-102, Asp-104, Glu-161, and Asp-166. The responses of these groups to perturbation with a pulse of protons produced the hypothesis of a proton antenna on the surface (115). The interacting anionic groups would capture protons and funnel them into the protein. Asp-38, in particular, was proposed to be an essential residue (116) in conducting a proton because its replacement with Arg slowed down decay of M by about tenfold. Indeed, the appearance of Wat520 at Asp-38 in N' seems to implicate this residue in function, but replacing it, alone or in combination with other Asp residues with Asn, had only minor consequences (117). In the D38C mutant, proton uptake was delayed but only about twofold (118). Furthermore, of the dozen naturally occurring bacteriorhodopsins, five contain Arg at the position of Asp-38.

Reisomerization of the Retinal

Under most conditions, the reprotonation of Asp-96 and the thermal reisomerization of the retinal in the N $\rightarrow$ O transition occur at the same time in the photocycle, suggesting that these events are coupled to one another. Direct evidence for coupling is from the behavior of the D85N/F42C mutant (119). In this double mutant, residue 85 is neutral, and the pK_a of Asp-96 is lowered from >13 in the wild-type and 9.4 in D85N to 8.2. Raising the pH to 9 thus produces a charge state that resembles the N intermediate. Under these conditions, Raman spectra indicate that the retinal spontaneously isomerizes, without illumination, from a mixture of all-*trans* and 13-*cis*,15-*syn* (the isomeric composition in dark-adapted bacteriorhodopsin) to 13-*cis*,15-*anti*. The solid-state NMR spectrum of the retinal in D85N at high pH (120) confirmed this surprising finding. In pH jump experiments, the isomerization occurred on roughly the same time-scale as in the photocycle (119). Lowering the pH for the D85N/F42C mutant thus appears to be equivalent to the decay of N (with 13-*cis*,15-*anti* retinal and unprotonated Asp-96) that produces O (with all-*trans* retinal and protonated Asp-96).

Catalysis of the reisomerization of retinal by protonation of Asp-96 might be similar to the way the protonation of Asp-85 catalyzes reisomerization (27). Alternatively, protonation of Asp-96 might affect the isomeric state by breaking the hydrogen-bonded network of water molecules leading to the Schiff base, thereby destabilizing the cytoplasmic orientation of the N—H bond in the 13-*cis*,15-*anti* state in N (48).

On the other hand, in mutants with long-living N states, such as V49A or F171C, proton uptake (as followed with a pH indicator dye) and reprotonation of Asp-96 (as followed by the reappearance of its protonated C=O stretch) occur 10–100-times earlier then the reisomerization. Thus, at pH < 7 two N states follow one another: N with anionic Asp-96 and N' with protonated Asp-96 (30). Presumably, this is because the coupling between the Schiff base and the cytoplasmic surface, mediated perhaps by the aqueous network that links Asp-96 to the Schiff base in N, is perturbed.

Recovery of the Initial State

In the O state, prominent HOOP bands indicate that the newly formed all-*trans* retinal is twisted (121). It relaxes upon deprotonation of Asp-85, which restores the original charge state at the Schiff base. Near neutral pH, where proton release to the extracellular surface occurred during the rise of M, this proton is transferred to the vacant proton release site, but at pH < 6 the proton is released directly to the surface. The correlation of the decay time-constant of O in various mutants with the time-constant for deprotonation of Asp-85 in pH jump in the dark suggests that the rate-limiting step of this reaction is the loss of the Asp-85 proton (122). The latter is controlled by the protonation state of the proton release group, particularly at low pH (123).

The only clues to the structure of the O state are from X-ray diffraction of crystals of the D85S mutant (112) that contains all-*trans* retinal (in a mixture with 13-*cis*,15-*syn*) and a neutral residue 85 as O. In this mutant, the outward tilts of helices F and G at the cytoplasmic surface are replaced by tilts of helices A, B, D, and E on the extracellular side. How the more open conformation of the extracellular region might mediate the deprotonation of Asp-85 is not yet clear. The appearance of a new protonated C=O stretch band during the lifetime of the O state, which was assigned to Asp-212 (124, 125), suggests that the path of the proton is from Asp-85 to Asp-212 and then, perhaps over intervening water molecules, to the proton release group.

Energetics of the Pump

In nonilluminated bacteriorhodopsin, the retinal Schiff base titrates with a $pK_a >$ 13 (126), and Asp-85 with a pK_a of about 2.5 (127), but transfer of the Schiff base proton to Asp-85 during the photocycle does not need to overcome a barrier of this magnitude. The high pK_a of the Schiff base and the low pK_a of Asp-85 reflect the cost of an uncompensated negative charge in the first case, and a positive charge in the second, buried inside the protein, and these end points are not relevant to the internal proton transfer in the cycle. The free-energy gain that would match the pK_as of the donor and the acceptor was estimated to be equivalent to about 5 pH units (128). This represents 25–30 kJ/mol, or about half of the excess enthalpy of the K state (129). At this stage of the photocycle it must be stored in the retinal.

Ab initio Gaussian calculation of the energy level of all-*trans* and 13-*cis*,15-*anti* retinal as a function of the bond angle at C_{13} revealed that the 49 kJ/mol measured in K with calorimetry (129) predicts very nearly the same angle as found in the X-ray structure (42). Given the simplifying assumptions, the agreement means that at least a substantial part of the free-energy gain in K is in this bond distortion. The gradual decrease of the C_{13} bond angle in the K-L-M_1-M_2-M'_2 sequence to its initial value suggests (47) that the first half of the photocycle should be viewed as a relaxation of the strained retinal (Figure 3) and the protein matrix around it. If the displacements of Wat402, Asp-85, Arg-82, and other residues and water molecules in the extracellular region occur in a selective and timed way that can utilize the stored energy to drive proton transfers, this will transfer energy in a usable way from the retinal to the protein.

The spontaneous isomerization of the retinal to 13-*cis*,15-*anti* in the D85N/F42C mutant (119) suggests that when Asp-85 is protonated and Asp-96 is unprotonated, as in the N state, the retinal-binding site no longer easily accommodates all-*trans* retinal. In N, therefore, the 13-*cis*,15-*anti* retinal and the surrounding protein groups appear to be fully relaxed within the constraints of the binding site. The subsequent reactions to N', O, and back to the initial state are then driven by the recovery of the initial protein conformation including the Schiff base-Asp-85 ion pair.

The loss of the free energy occurs mainly in two of the photocycle steps. First, under physiological conditions proton release at the extracellular surface is at a pH well above the pK_a for proton release and dissipates about 10 kJ/mol (28). Second, when the low initial pK_a of Asp-85 (about 2.5) and the high initial pK_a of the extracellular proton release site (about 9) are regained in the last photocycle step, the proton transfer between them is strongly downhill. On the one hand, this fully recovers the initial state for a new round of transport, and on the other it provides thermodynamic force for uphill proton translocation across the membrane.

The *Annual Review of Physiology* is online at http://physiol.annualreviews.org

LITERATURE CITED

1. Oesterhelt D, Stoeckenius W. 1973. Functions of a new photoreceptor membrane. *Proc. Natl. Acad. Sci. USA* 70:2853–57

2. Mathies RA, Lin SW, Ames JB, Pollard WT. 1991. From femtoseconds to biology: mechanism of bacteriorhodopsin's light-driven proton pump. *Annu. Rev. Biophys. Biophys. Chem.* 20:491–518

3. Ebrey TG. 1993. Light energy transduction in bacteriorhodopsin. In *Thermodynamics of Membranes, Receptors and Channels*, ed. M Jackson, pp. 353–87. New York: CRC Press

4. Lanyi JK. 1993. Proton translocation mechanism and energetics in the light-driven pump bacteriorhodopsin. *Biochim. Biophys. Acta* 1183:241–61

5. Oesterhelt D. 1998. The structure and mechanism of the family of retinal proteins from halophilic archaea. *Curr. Opin. Struct. Biol.* 8:489–500

6. Maeda A, Kandori H, Yamazaki Y, Nishimura S, Hatanaka M, et al. 1997. Intramembrane signaling mediated by hydrogen-bonding of water and carboxyl groups in bacteriorhodopsin and rhodopsin. *J. Biochem.* 121:399–406

7. Kandori H. 2000. Role of internal water molecules in bacteriorhodopsin. *Biochim. Biophys. Acta* 1460:177–91

8. Dioumaev AK. 2001. Infrared methods for monitoring the protonation state of carboxylic amino acids in the photocycle of bacteriorhodopsin. *Biochemistry* 66:1269–76

9. Maeda A. 2001. Internal water molecules as mobile polar groups for light-induced proton translocation in bacteriorhodopsin and rhodopsin as studied by differ-

ence FTIR spectroscopy. *Biochemistry* 66:1256–68

10. Lanyi JK. 2003. Molecular mechanism of ion transport in bacteriorhodopsin: insights from crystallographic, spectroscopic, kinetic, and mutational studies. *J. Phys. Chem. B* 104:11441–48

11. Luecke H, Lanyi JK. 2003. Structural clues to the mechanism of ion pumping in bacteriorhodopsin. *Adv. Protein Chem.* 63:111–30

12. Herzfeld J, Tounge B. 2000. NMR probes of vectoriality in the proton-motive photocycle of bacteriorhodopsin: evidence for an 'electrostatic steering' mechanism. *Biochim. Biophys. Acta* 1460:95–105

13. Herzfeld J, Lansing JC. 2002. Magnetic resonance studies of the bacteriorhodopsin pump cycle. *Annu. Rev. Biophys. Biomol. Struct.* 31:73–95

14. Lanyi JK, Luecke H. 2001. Bacteriorhodopsin. *Curr. Opin. Struct. Biol.* 11:415–19

15. Neutze R, Pebay-Peyroula E, Edman K, Royant A, Navarro J, Landau EM. 2002. Bacteriorhodopsin: a high-resolution structural view of vectorial proton transport. *Biochim. Biophys. Acta* 1565:144–67

16. Lanyi JK, Váró G. 1995. The photocycles of bacteriorhodopsin. *Isr. J. Chem.* 35:365–86

17. Váró G, Lanyi JK. 1991. Thermodynamics and energy coupling in the bacteriorhodopsin photocycle. *Biochemistry* 30:5016–22

18. Zimanyi L, Lanyi JK. 1993. Deriving the intermediate spectra and photocycle kinetics from time-resolved difference spectra of bacteriorhodopsin. The simpler case

of the recombinant D96N protein. *Biophys. J.* 64:240–51

19. Chizhov I, Chernavskii DS, Engelhard M, Mueller KH, Zubov BV, Hess B. 1996. Spectrally silent transitions in the bacteriorhodopsin photocycle. *Biophys. J.* 71:2329–45

20. Eisfeld W, Althaus T, Stockburger M. 1995. Evidence for parallel photocycles and implications for the proton pump in bacteriorhodopsin. *Biophys. Chem.* 56: 105–12

21. Joshi MK, Bose S, Hendler RW. 1999. Regulation of the bacteriorhodopsin photocycle and proton pumping in whole cells of *Halobacterium salinarium*. *Biochemistry* 38:8786–93

22. Chernavskii DS, Chizhov IV, Lozier RH, Murina TM, Prokhorov AM, Zubov BV. 1989. Kinetic model of bacteriorhodopsin photocycle: pathway from M state to bR. *Photochem. Photobiol.* 49:649–53

23. Zimanyi L, Cao Y, Needleman R, Ottolenghi M, Lanyi JK. 1993. Pathway of proton uptake in the bacteriorhodopsin photocycle. *Biochemistry* 32:7669–78

24. Brown LS, Dioumaev AK, Needleman R, Lanyi JK. 1998. Connectivity of the retinal Schiff base to Asp85 and Asp96 during the bacteriorhodopsin photocycle: the local-access model. *Biophys. J.* 75:1455–65

25. Ludmann K, Gergely C, Váró G. 1998. Kinetic and thermodynamic study of the bacteriorhodopsin photocycle over a wide pH range. *Biophys. J.* 75:3110–19

26. Hendler RW, Bose S. 2003. Interconversions among four M-intermediates in the bacteriorhodopsin photocycle. *Eur. J. Biochem.* 270:3518–24

27. Balashov SP. 2000. Protonation reactions and their coupling in bacteriorhodopsin. *Biochim. Biophys. Acta* 1460: 75–94

28. Zimanyi L, Váró G, Chang M, Ni B, Needleman R, Lanyi JK. 1992. Pathways of proton release in the bacteriorhodopsin photocycle. *Biochemistry* 31: 8535–43

29. Luecke H, Schobert B, Richter HT, Cartailler J-P, Rosengarth A, Needleman R, Lanyi JK. 2000. Coupling photoisomerization of the retinal in bacteriorhodopsin to directional transport. *J. Mol. Biol.* 300: 1237–55

30. Dioumaev AK, Brown LS, Needleman R, Lanyi JK. 2001. Coupling of the reisomerization of the retinal, proton uptake, and reprotonation of Asp-96 in the N photointermediate of bacteriorhodopsin. *Biochemistry* 40:11308–17

31. Grigorieff N, Ceska TA, Downing KH, Baldwin JM, Henderson R. 1996. Electron-crystallographic refinement of the structure of bacteriorhodopsin. *J. Mol. Biol.* 259:393–421

32. Kimura Y, Vassylyev DG, Miyazawa A, Kidera A, Matsushima M, et al. 1997. Surface of bacteriorhodopsin revealed by high-resolution electron crystallography. *Nature* 389:206–11

33. Essen L, Siegert R, Lehmann WD, Oesterhelt D. 1998. Lipid patches in membrane protein oligomers: crystal structure of the bacteriorhodopsin-lipid complex. *Proc. Natl. Acad. Sci. USA* 95:11673–78

34. Zhou F, Windemuth A, Schulten K. 1993. Molecular dynamics study of the proton pump cycle of bacteriorhodopsin. *Biochemistry* 32:2291–306

35. Humphrey W, Logunov I, Schulten K, Sheves M. 1994. Molecular dynamics study of bacteriorhodopsin and artificial pigments. *Biochemistry* 33:3668–78

36. Roux B, Nina M, Pomes R, Smith JC. 1996. Thermodynamic stability of water molecules in the bacteriorhodopsin proton channel: a molecular dynamics free energy perturbation study. *Biophys. J.* 71:670–81

37. Landau EM, Rosenbusch JP. 1996. Lipidic cubic phases: a novel concept for the crystallization of membrane proteins. *Proc. Natl. Acad. Sci. USA* 93:14532–35

38. Pebay-Peyroula E, Rummel G, Rosenbusch JP, Landau EM. 1997. X-ray structure of bacteriorhodopsin at 2.5 angstroms

from microcrystals grown in lipidic cubic phases. *Science* 277:1676–81

39. Luecke H, Richter HT, Lanyi JK. 1998. Proton transfer pathways in bacteriorhodopsin at 2.3 angstrom resolution. *Science* 280:1934–37

40. Belrhali H, Nollert P, Royant A, Menzel C, Rosenbusch JP, Landau EM, Pebay-Peyroula E. 1999. Protein, lipid and water organization in bacteriorhodopsin crystals: a molecular view of the purple membrane at 1.9 Å resolution. *Struct. Fold. Des.* 7:909–17

41. Luecke H, Schobert B, Richter HT, Cartailler JP, Lanyi JK. 1999. Structure of bacteriorhodopsin at 1.55 Å resolution. *J. Mol. Biol.* 291:899–911

42. Schobert B, Cupp-Vickery J, Hornak V, Smith S, Lanyi J. 2002. Crystallographic structure of the K intermediate of bacteriorhodopsin: conservation of free energy after photoisomerization of the retinal. *J. Mol. Biol.* 321:715–26

43. Royant A, Grizot S, Kahn R, Belrhali H, Fieschi F, et al. 2002. Detection and characterization of merohedral twinning in two protein crystals: bacteriorhodopsin and p67phox. *Acta Crystallogr. D* 58:784–91

44. Royant A, Edman K, Ursby T, Pebay-Peyroula E, Landau EM, Neutze R. 2000. Helix deformation is coupled to vectorial proton transport in the photocycle of bacteriorhodopsin. *Nature* 406:645–48

45. Gat Y, Sheves M. 1993. A mechanism for controlling the pK_a of the retinal protonated Schiff base in retinal proteins. A study with model compounds. *J. Am. Chem. Soc.* 115:3772–73

46. Lanyi JK, Schobert B. 2002. Crystallographic structure of the retinal and the protein after deprotonation of the Schiff base: the switch in the bacteriorhodopsin photocycle. *J. Mol. Biol.* 321:727–37

47. Lanyi JK, Schobert B. 2003. Mechanism of proton transport in bacteriorhodopsin from crystallographic structures of the K, L, M1, M2, and M2′ intermediates of the photocycle. *J. Mol. Biol.* 328:439–50

48. Schobert B, Brown LS, Lanyi JK. 2003. Crystallographic structures of the M and N intermediates of bacteriorhodopsin: assembly of a hydrogen-bonded chain of water molecules between Asp96 and the retinal Schiff base. *J. Mol. Biol.* 330:553–70

49. Edman K, Nollert P, Royant A, Belrhali H, Pebay-Peyroula E, et al. 1999. High-resolution X-ray structure of an early intermediate in the bacteriorhodopsin photocycle. *Nature* 401:822–26

50. Sass HJ, Buldt G, Gessenich R, Hehn D, Neff D, et al. 2000. Structural alterations for proton translocation in the M state of wild-type bacteriorhodopsin. *Nature* 406:649–53

51. Luecke H, Schobert B, Richter HT, Cartailler JP, Lanyi JK. 1999. Structural changes in bacteriorhodopsin during ion transport at 2 angstrom resolution. *Science* 286:255–61

52. Thorgeirsson TE, Xiao W, Brown LS, Needleman R, Lanyi JK, Shin YK. 1997. Transient channel-opening in bacteriorhodopsin: an EPR study. *J. Mol. Biol.* 273:951–57

53. Xiao W, Brown LS, Needleman R, Lanyi JK, Shin YK. 2000. Light-induced rotation of a transmembrane alpha-helix in bacteriorhodopsin. *J. Mol. Biol.* 304:715–21

54. Radzwill N, Gerwert K, Steinhoff HJ. 2001. Time-resolved detection of transient movement of helices F and G in doubly spin-labeled bacteriorhodopsin. *Biophys. J.* 80:2856–66

55. Rink T, Riesle J, Oesterhelt D, Gerwert K, Steinhoff HJ. 1997. Spin-labeling studies of the conformational changes in the vicinity of D36, D38, T46, and E161 of bacteriorhodopsin during the photocycle. *Biophys. J.* 73:983–93

56. Pfeiffer M, Rink T, Gerwert K, Oesterhelt D, Steinhoff HJ. 1999. Site-directed spin-labeling reveals the orientation of the

amino acid side-chains in the E-F loop of bacteriorhodopsin. *J. Mol. Biol.* 287:163–71

57. Rink T, Pfeiffer M, Oesterhelt D, Gerwert K, Steinhoff HJ. 2000. Unraveling photoexcited conformational changes of bacteriorhodopsin by time resolved electron paramagnetic resonance spectroscopy. *Biophys. J.* 78:1519–30

58. Brown LS, Needleman R, Lanyi JK. 2002. Conformational change of the E-F interhelical loop in the M photointermediate of bacteriorhodopsin. *J. Mol. Biol.* 317:471–78

59. Althaus T, Eisfeld W, Lohrmann R, Stockburger M. 1995. Application of Raman spectroscopy to retinal proteins. *Isr. J. Chem.* 35:227–51

60. Lansing JC, Hohwy M, Jaroniec CP, Creemers AF, Lugtenburg J, et al. 2002. Chromophore distortions in the bacteriorhodopsin photocycle: evolution of the H-C$_{14}$-C$_{15}$-H dihedral angle measured by solid-state NMR. *Biochemistry* 41:431–38

61. Pande J, Callender RH, Ebrey TG. 1981. Resonance Raman study of the primary photochemistry of bacteriorhodopsin. *Proc. Natl. Acad. Sci. USA* 78:7379–82

62. Braiman M, Mathies R. 1982. Resonance Raman spectra of bacteriorhodopsin's primary photoproduct: evidence for a distorted 13-*cis* retinal chromophore. *Proc. Natl. Acad. Sci. USA* 79:403–7

63. Maeda A, Sasaki J, Pfefferle JM, Shichida Y, Yoshizawa T. 1991. Fourier transform infrared spectral studies on the Schiff base mode of all-*trans* bacteriorhodopsin and its photointermediates, K and L. *Photochem. Photobiol.* 54:911–21

64. Hayashi S, Tajkhorshid E, Schulten K. 2002. Structural changes during the formation of early intermediates in the bacteriorhodopsin photocycle. *Biophys. J.* 83:1281–97

65. Bullough PA, Henderson R. 1999. The projection structure of the low temperature K intermediate of the bacteriorhodopsin photocycle determined by electron diffraction. *J. Mol. Biol.* 286:1663–71

66. Matsui Y, Sakai K, Murakami M, Shiro Y, Adachi S, et al. 2002. Specific damage induced by X-ray radiation and structural changes in the primary photoreaction of bacteriorhodopsin. *J. Mol. Biol.* 324:469–81

67. Luecke H. 2000. Atomic resolution structures of bacteriorhodopsin photocycle intermediates: the role of discrete water molecules in the function of this light-driven ion pump. *Biochim. Biophys. Acta* 1460:133–56

68. Balashov SP, Ebrey TG. 2001. Trapping and spectroscopic identification of the photointermediates of bacteriorhodopsin at low temperatures. *Photochem. Photobiol.* 73:453–62

69. Royant A, Edman K, Ursby T, Pebay-Peyroula E, Landau EM, Neutze R. 2001. Spectroscopic characterization of bacteriorhodopsin's L-intermediate in 3D crystals cooled to 170 K. *Photochem. Photobiol.* 74:794–804

70. Hu JG, Sun BQ, Petkova AT, Griffin RG, Herzfeld J. 1997. The predischarge chromophore in bacteriorhodopsin: a ^{15}N solid-state NMR study of the L photointermediate. *Biochemistry* 36:9316–22

71. Fodor SPA, Pollard WT, Gebhard R, Van den Berg EMM, Lugtenburg J, Mathies RA. 1988. Bacteriorhodopsin's L550 intermediate contains a C14-C15 s-*trans* retinal chromophore. *Proc. Natl. Acad. Sci. USA* 85:2156–60

72. Tanimoto T, Furutani Y, Kandori H. 2003. Structural changes of water in the Schiff base region of bacteriorhodopsin: proposal of a hydration switch model. *Biochemistry* 42:2300–6

73. Maeda A, Tomson FL, Gennis RB, Balashov SP, Ebrey TG. 2003. Water molecule rearrangements around Leu93 and Trp182 in the formation of the L intermediate in bacteriorhodopsin's photocycle. *Biochemistry* 42:2535–41

74. Drachev LA, Kaulen AD, Skulachev VP. 1984. Correlation of photochemical cycle, proton release and uptake, and electric events in bacteriorhodopsin. *FEBS Lett.* 178:331–35

75. Grzesiek S, Dencher NA. 1986. Time-course and stoichiometry of light-induced proton release and uptake during the photocycle of bacteriorhodopsin. *FEBS Lett.* 208:337–42

76. Heberle J, Dencher NA. 1992. Surface-bound optical probes monitor proton translocation and surface potential changes during the bacteriorhodopsin photocycle. *Proc. Natl. Acad. Sci. USA* 89:5996–6000

77. Balashov SP, Imasheva ES, Govindjee R, Ebrey TG. 1996. Titration of aspartate-85 in bacteriorhodopsin: what it says about chromophore isomerization and proton release. *Biophys. J.* 70:473–81

78. Liu SY, Kono M, Ebrey TG. 1991. Effect of pH buffer molecules on the light-induced currents from oriented purple membrane. *Biophys. J.* 60:204–16

79. Brown LS, Sasaki J, Kandori H, Maeda A, Needleman R, Lanyi JK. 1995. Glutamic acid 204 is the terminal proton release group at the extracellular surface of bacteriorhodopsin. *J. Biol. Chem.* 270:27122–26

80. Balashov SP, Imasheva ES, Ebrey TG, Chen N, Menick DR, Crouch RK. 1997. Glutamate-194 to cysteine mutation inhibits fast light-induced proton release in bacteriorhodopsin. *Biochemistry* 36:8671–76

81. Rammelsberg R, Huhn G, Lubben M, Gerwert K. 1998. Bacteriorhodopsin's intramolecular proton-release pathway consists of a hydrogen-bonded network. *Biochemistry* 37:5001–9

82. Spassov VZ, Luecke H, Gerwert K, Bashford D. 2001. pK_a calculations suggest storage of an excess proton in a hydrogen-bonded water network in bacteriorhodopsin. *J. Mol. Biol.* 312:203–19

83. Richter HT, Brown LS, Needleman R, Lanyi JK. 1996. A linkage of the pK_as of Asp-85 and Glu-204 forms part of the reprotonation switch of bacteriorhodopsin. *Biochemistry* 35:4054–62

84. Althaus T, Stockburger M. 1998. Time and pH dependence of the L-to-M transition in the photocycle of bacteriorhodopsin and its correlation with proton release. *Biochemistry* 37:2807–17

85. Alexiev U, Marti T, Heyn MP, Khorana HG, Scherrer P. 1994. Covalently bound pH-indicator dyes at selected extracellular or cytoplasmic sites in bacteriorhodopsin. 2. Rotational orientation of helixes D and E and kinetic correlation between M formation and proton release in bacteriorhodopsin micelles. *Biochemistry* 33:13693–99

86. Brown LS, Needleman R, Lanyi JK. 2000. Origins of deuterium kinetic isotope effects on the proton transfers of the bacteriorhodopsin photocycle. *Biochemistry* 39:938–45

87. Petkova AT, Hu JG, Bizounok M, Simpson M, Griffin RG, Herzfeld J. 1999. Arginine activity in the proton-motive photocycle of bacteriorhodopsin: solid-state NMR studies of the wild-type and D85N proteins. *Biochemistry* 38:1562–72

88. Subramaniam S, Henderson R. 2000. Molecular mechanism of vectorial proton translocation by bacteriorhodopsin. *Nature* 406:653–57

89. Brown LS, Dioumaev AK, Needleman R, Lanyi JK. 1998. Local-access model for proton transfer in bacteriorhodopsin. *Biochemistry* 37:3982–93

90. Szaraz S, Oesterhelt D, Ormos P. 1994. pH-induced structural changes in bacteriorhodopsin studied by Fourier transform infrared spectroscopy. *Biophys. J.* 67:1706–12

91. Brown LS, Lanyi JK. 1996. Determination of the transiently lowered pK of the retinal Schiff base during the photocycle of bacteriorhodopsin. *Proc. Natl. Acad. Sci. USA* 93:1731–34

92. Cao Y, Váró G, Chang M, Ni B, Needleman R, Lanyi JK. 1991. Water is required for proton transfer from aspartate-96 to the bacteriorhodopsin Schiff base. *Biochemistry* 30:10972–79

93. Váró G, Lanyi JK. 1991. Distortions in the photocycle of bacteriorhodopsin at moderate dehydration. *Biophys. J.* 59:313–22

94. Varo G, Lanyi JK. 1995. Effects of hydrostatic pressure on the kinetics reveal a volume increase during the bacteriorhodopsin photocycle. *Biochemistry* 34:12161–69

95. Wikstrom M. 1998. Proton translocation by bacteriorhodopsin and heme-copper oxidases. *Curr. Opin. Struct. Biol.* 8:480–88

96. Nagel JF, Tristam-Nagle S. 1982. Hydrogen-bonded chain mechanisms for proton conduction and proton pumping. *J. Membr. Biol.* 74:1–14

97. Subramaniam S, Lindahl M, Bullough P, Faruqi AR, Tittor J, et al. 1999. Protein conformational changes in the bacteriorhodopsin photocycle. *J. Mol. Biol.* 287:145–61

98. Subramaniam S, Henderson R. 1999. Electron crystallography of bacteriorhodopsin with millisecond time resolution. *J. Struct. Biol.* 128:19–25

99. Vonck J. 2000. Structure of the bacteriorhodopsin mutant F219L N intermediate revealed by electron crystallography. *EMBO J.* 19:2152–60

100. Dencher NA, Dresselhaus D, Zaccai G, Bueldt G. 1989. Structural changes in bacteriorhodopsin during proton translocation revealed by neutron diffraction. *Proc. Natl. Acad. Sci. USA* 86:7876–79

101. Kamikubo H, Kataoka M, Váró G, Oka T, Tokunaga F, et al. 1996. Structure of the N intermediate of bacteriorhodopsin revealed by X-ray diffraction. *Proc. Natl. Acad. Sci. USA* 93:1386–90

102. Kamikubo H, Oka T, Imamoto Y, Tokunaga F, Lanyi JK, Kataoka M. 1997. The last phase of the reprotonation switch in bacteriorhodopsin: the transition between the M-type and the N-type protein conformation depends on hydration. *Biochemistry* 36:12282–87

103. Hendrickson FM, Burkard F, Glaeser RM. 1998. Structural characterization of the L-to-M transition of the bacteriorhodopsin photocycle. *Biophys. J.* 75:1446–54

104. Oka T, Yagi N, Fujisawa T, Kamikubo H, Tokunaga F, Kataoka M. 2000. Time-resolved X-ray diffraction reveals multiple conformations in the M-N transition of the bacteriorhodopsin photocycle. *Proc. Natl. Acad. Sci. USA* 97:14278–82

105. Oka T, Yagi N, Tokunaga F, Kataoka M. 2002. Time-resolved X-ray diffraction reveals movement of F helix of D96N bacteriorhodopsin during M-M_N transition at neutral pH. *Biophys. J.* 82:2610–16

106. Martinez LC, Turner GJ. 2002. High-throughput screening of bacteriorhodopsin mutants in whole cell pastes. *Biochim. Biophys. Acta* 1564:91–98

107. Oka T, Kamikubo H, Tokunaga F, Lanyi JK, Needleman R, Kataoka M. 1999. Conformational change of helix G in the bacteriorhodopsin photocycle: investigation with heavy atom labeling and X-ray diffraction. *Biophys. J.* 76:1018–23

108. Brown LS, Váró G, Needleman R, Lanyi JK. 1995. Functional significance of a protein conformation change at the cytoplasmic end of helix F during the bacteriorhodopsin photocycle. *Biophys. J.* 69:2103–11

109. Tittor J, Paula S, Subramaniam S, Heberle J, Henderson R, Oesterhelt D. 2002. Proton translocation by bacteriorhodopsin in the absence of substantial conformational changes. *J. Mol. Biol.* 319:555–65

110. Hu JG, Sun BQ, Bizounok M, Hatcher ME, Lansing JC, et al. 1998. Early and late M intermediates in the bacteriorhodopsin photocycle: a solid-state NMR study. *Biochemistry* 37:8088–96

111. Lakshmi KV, Farrar MR, Raap J, Lugtenburg J, Griffin RG, Herzfeld J. 1994. Solid state ^{13}C and ^{15}N NMR investigations of

the N intermediate of bacteriorhodopsin. *Biochemistry* 33:8853–57

112. Rouhani S, Cartailler JP, Facciotti MT, Walian P, Needleman R, et al. 2001. Crystal structure of the D85S mutant of bacteriorhodopsin: model of an O-like photocycle intermediate. *J. Mol. Biol.* 313:615–28

113. Zscherp C, Schlesinger R, Tittor J, Oesterhelt D, Heberle J. 1999. In situ determination of transient pK_a changes of internal amino acids of bacteriorhodopsin by using time-resolved attenuated total reflection Fourier-transform infrared spectroscopy. *Proc. Natl. Acad. Sci. USA* 96:5498–503

114. Brown LS, Yamazaki Y, Maeda A, Sun L, Needleman R, Lanyi JK. 1994. The proton transfers in the cytoplasmic domain of bacteriorhodopsin are facilitated by a cluster of interacting residues. *J. Mol. Biol.* 239:401–14

115. Checover S, Nachliel E, Dencher NA, Gutman M. 1997. Mechanism of proton entry into the cytoplasmic section of the proton-conducting channel of bacteriorhodopsin. *Biochemistry* 36:13919–28

116. Riesle J, Oesterhelt D, Dencher NA, Heberle J. 1996. D38 is an essential part of the proton translocation pathway in bacteriorhodopsin. *Biochemistry* 35:6635–43

117. Brown LS, Needleman R, Lanyi JK. 1999. Functional roles of aspartic acid residues at the cytoplasmic surface of bacteriorhodopsin. *Biochemistry* 38:6855–61

118. Schatzler B, Dencher NA, Tittor J, Oesterhelt D, Yaniv-Checover S, et al. 2003. Subsecond proton-hole propagation in bacteriorhodopsin. *Biophys. J.* 84:671–86

119. Dioumaev AK, Brown LS, Needleman R, Lanyi JK. 1998. Partitioning of free energy gain between the photoisomerized retinal and the protein in bacteriorhodopsin. *Biochemistry* 37:9889–93

120. Hatcher ME, Hu JG, Belenky M, Verdegem P, Lugtenburg J, et al. 2002. Control of the pump cycle in bacteriorhodopsin: mechanisms elucidated by solid-state NMR of the D85N mutant. *Biophys. J.* 82:1017–29

121. Smith SO, Pardoen JA, Mulder PPJ, Curry B, Lugtenburg J, Mathies R. 1983. Chromophore structure in bacteriorhodopsin's O640 photointermediate. *Biochemistry* 22:6141–48

122. Richter HT, Needleman R, Kandori H, Maeda A, Lanyi JK. 1996. Relationship of retinal configuration and internal proton transfer at the end of the bacteriorhodopsin photocycle. *Biochemistry* 35:15461–66

123. Balashov SP, Lu M, Imasheva ES, Govindjee R, Ebrey TG, et al. 1999. The proton release group of bacteriorhodopsin controls the rate of the final step of its photocycle at low pH. *Biochemistry* 38:2026–39

124. Dioumaev AK, Brown LS, Needleman R, Lanyi JK. 1999. Fourier transform infrared spectra of a late intermediate of the bacteriorhodopsin photocycle suggest transient protonation of Asp-212. *Biochemistry* 38:10070–78

125. Zscherp C, Schlesinger R, Heberle J. 2001. Time-resolved FT-IR spectroscopic investigation of the pH-dependent proton transfer reactions in the E194Q mutant of bacteriorhodopsin. *Biochem. Biophys. Res. Commun.* 283:57–63

126. Druckmann S, Ottolenghi M, Pande A, Pande J, Callender RH. 1982. Acid-base equilibrium of the Schiff base in bacteriorhodopsin. *Biochemistry* 21:4953–59

127. Fischer U, Oesterhelt D. 1979. Chromophore equilibria in bacteriorhodopsin. *Biophys. J.* 28:211–30

128. Brown LS, Bonet L, Needleman R, Lanyi JK. 1993. Estimated acid dissociation constants of the Schiff base, Asp-85, and Arg-82 during the bacteriorhodopsin photocycle. *Biophys. J.* 65:124–30

129. Birge RR, Cooper TM, Lawrence AF, Masthay MB, Zhang CF, Zidovetzki R. 1991. Revised assignment of energy storage in the primary photochemical event in bacteriorhodopsin. *J. Am. Chem. Soc.* 113:4327–28

Annu. Rev. Physiol. 2004. 66:689–733
doi: 10.1146/annurev.physiol.66.032102.150251
First published online as a Review in Advance on November 3, 2003

THE CYTOCHROME BC_1 COMPLEX: Function in the Context of Structure

Antony R. Crofts

Department of Biochemistry, and Center for Biophysics and Computational Biology, University of Illinois at Urbana-Champaign, Urbana, Illinois 61801; email: a-crofts@life.uiuc.edu

Key Words   antimycin, stigmatellin, myxothiazol, superoxide, mechanism

■ **Abstract**   The bc_1 complexes are intrinsic membrane proteins that catalyze the oxidation of ubihydroquinone and the reduction of cytochrome c in mitochondrial respiratory chains and bacterial photosynthetic and respiratory chains. The bc_1 complex operates through a Q-cycle mechanism that couples electron transfer to generation of the proton gradient that drives ATP synthesis.

Genetic defects leading to mutations in proteins of the respiratory chain, including the subunits of the bc_1 complex, result in mitochondrial myopathies, many of which are a direct result of dysfunction at catalytic sites. Some myopathies, especially those in the cytochrome b subunit, exacerbate free-radical damage by enhancing superoxide production at the ubihydroquinone oxidation site. This bypass reaction appears to be an unavoidable feature of the reaction mechanism. Cellular aging is largely attributable to damage to DNA and proteins from the reactive oxygen species arising from superoxide and is a major contributing factor in many diseases of old age. An understanding of the mechanism of the bc_1 complex is therefore central to our understanding of the aging process. In addition, a wide range of inhibitors that mimic the quinone substrates are finding important applications in clinical therapy and agronomy. Recent structural studies have shown how many of these inhibitors bind, and have provided important clues to the mechanism of action and the basis of resistance through mutation.

This paper reviews recent advances in our understanding of the mechanism of the bc_1 complex and their relation to these physiologically important issues in the context of the structural information available.

INTRODUCTION

Enzymes of the cytochrome (cyt) bc_1 complex family are central components of all the main energy transduction systems of the biosphere, accounting, through maintenance of the proton gradient, for $\sim$30% of energy transmission.[1] They all catalyze essentially the same reaction, the transfer of reducing equivalents from a quinol in the lipid phase (ubihydroquinone in the classical bc_1 complexes) to a higher-potential acceptor protein in the aqueous phase. This electron transfer is coupled to transfer of 1 H^+/e^- across the membrane. From an anthropocentric perspective, the mitochondrial complexes are of greatest interest, and the central role in the respiratory chain places them at the heart of cellular physiological function (1–4). The mitochondrial complexes arose from bacterial antecedents, and the catalytic core of three subunits [cyt c_1, cyt b and the Rieske iron sulfur protein (ISP)] is highly conserved (5–7). The latter two subunits span the archaeal/bacterial divide and are clearly more ancient than the cyt c_1 subunit, whose function is carried by a wide variety of different structures (3, 6). Because of the greater experimental flexibility, the bc_1 complexes from photosynthetic bacteria have proved useful models for the mitochondrial complexes, especially in kinetic studies. In these studies, the reactions of the complex in situ can be studied following photoactivation of the photosynthetic reaction center (RC), which generates both substrates in <1 ms, allowing better time resolution than is possible using rapid mixing.

Apart from the general philosophical interest, applications in the health and agronomical sciences arise from the central involvement of the complex in cellular energy metabolism and the many ways in which dysfunctions in this role impinge on physiological function.

A wide range of mitochondrial myopathies (using an extended definition that includes related disorders) have been located to mutations in genes encoding the proteins of the bc_1 complex (8–11). The general and specific symptoms of these myopathies provide abundant evidence of the central importance of the complex in energy metabolism. Many myopathies have been characterized at a molecular level, and many of the mutations have been reproduced in yeast or bacterial model systems for more detailed investigation (8, 9). In general, the mapping of lesions to the catalytic interfaces and demonstration of direct effects on catalytic function have provided a deeper understanding of the molecular basis of these important diseases.

---

[1]*Abbreviations*: bc_1 complex, ubiquinol:cytochrome c oxidoreductase (EC 1.10.2.2); b_L and b_H, low- and high-potential hemes of cytochrome b, respectively; cyt, cytochrome; $E_{m,(pH)}$, midpoint redox potential at pH indicated (pH 7 assumed if not indicated); $E_{h,(pH)}$, ambient redox potential at pH indicated; ESEEM, electron spin echo envelope modulation spectroscopy; HYSCORE, hyperfine sublevel correlation spectroscopy; HHDBT, 5-heptyl-6-hydroxy-4,7-dioxobenzothiazol; ISP, Rieske iron-sulfur protein; ISPH, reduced ISP; ISP_{ox}, oxidized, dissociated ISP; NQNO, 2-nonyl-4-hydroxyquinoline N-oxide; P^+/P, reaction center primary donor couple, $(BChl_2)^+/(BChl_2)$; P-phase, N-phase, aqueous phases in which the proton gradient is positive or negative, respectively; Q, oxidized form of quinone; QH_2, reduced form (hydroquinone, quinol) of quinone; $QH^\bullet$, $Q^{\bullet-}$, protonated and deprotonated forms of semiquinone; Qi-site (Q_o site), quinone reducing (quinol oxidizing) site of bc_1 complex; Rb., *Rhodobacter*; RC, photosynthetic reaction center; SQ, semiquinone (with protonation state unspecified); UHDBT, 5-undecyl-6-hydroxy-4,7-dioxobenzothiazol; UHNQ, 2-undecyl-3-hydroxy-1,4-naphthoquinone.

The Q_o-site of the complex is an important site for O_2 reduction to generate superoxide (12–18). The latter is a precursor of the reactive oxygen species (ROS) that cause damage to DNA and proteins, leading to cellular aging (19–24). The mitochondrial complexes and their ancestral bacterial counterparts share strong sequence and structural similarities at the level of the three catalytic subunits, and they operate through the same mechanism. This allows use of model systems from yeast and the photosynthetic bacteria to be used to explore at a molecular level how evolution has honed the mechanism so as to minimize this deleterious side reaction.

The two catalytic sites of the complex at which the redox reactions of quinones are processed are the sites of action of a wide range of inhibitors that mimic the quinones (25–31). Several of these are finding applications in medicine, based on differential species specificity, and in agronomy based on fungicidal properties, and an environmentally friendly nature (32–36). Elucidation of the mechanism of action of these inhibitors, and the availability of crystallographic structures showing their binding, promise to have wide applications in drug design and in understanding how resistant strains develop.

Over the past half dozen years or so, our understanding of the bc_1 complex has been greatly enhanced by the availability of crystallographic structures (37–48). These now include structures from seven species ranging across the animal, fungal, bacterial, and plant kingdoms, with occupancy of catalytic sites by a growing number of different inhibitors (at least 10 published at the present count), all interpreted against a background of a detailed mechanistic understanding, a sophisticated toolbox of biophysical and biochemical approaches, and the availability of molecular engineering protocols to allow detailed examination of the structure-function interface.

Structural aspects of the field have been extensively reviewed, and an additional essay would be superfluous if it were not for the rapid advancement of our understanding of the mechanistic interface, and the fact that some recent structures throw new light on these physiological aspects.

Overview of Mechanism

The bc_1 complex functions through a Q-cycle mechanism (49–54) (Figure 1, see color insert). The coupling to proton transfer depends on a bifurcated reaction at the Q_o-site of the complex, in which the two electrons from ubihydroquinone (quinol, QH_2) are passed to two different chains. The initial acceptor of the first electron from quinol is the [2Fe-2S] cluster of the Rieske iron sulfur protein (ISP), which feeds electrons via a bound c-type heme [cytochrome (cyt) c_1 in the bc_1 complex] to a mobile electron carrier protein (cyt c, or c_2 in bacteria) that then reduces a terminal acceptor (cytochrome oxidase in the mitochondrial chain, an oxidized photochemical reaction center in photosynthetic systems). In most mechanistic models, it is supposed from general principles of quinone chemistry (55) that a semiquinone intermediate is formed at the Q_o-site, although no such species sensitive to addition of Q_o-site inhibitors has been detected (52, 56). The second electron from quinol is passed to a lower-potential chain containing the low (cyt b_L) and a higher (cyt b_H) potential hemes of cyt b. This chain spans the membrane and delivers electrons to a second quinone processing site on the other side, the Q_i-site. At this site, quinone (Q) is reduced to quinol (QH_2) through a two-electron gate (57). Because this reaction requires two electrons, the Q_o-site has to turn over twice, oxidizing two QH_2, releasing four H^+, and delivering two electrons successively to each chain. If the complex starts with a quinone in the site, the first electron at the Q_i-site is stored on a bound semiquinone, which is reduced to QH_2 by the second electron. The two successive one-electron reactions lead to uptake of two H^+ on complete reduction of quinone. For the mitochondrial complex the overall reaction is shown in Equation 1; in α-proteobacteria, cyt c is replaced by cyt c_2.

$$QH_2 + 2\text{cyt}\,c^+ + 2H_N^+ \leftrightarrow Q + 2\text{cyt}\,c + 4H_P^+ \qquad\qquad 1.$$

Here subscripts N and P denote proton uptake from or release into the protochemically negative and positive aqueous phases, respectively.

The Q-cycle shown in Figure 1 is a generalized representation in the context of the structure of a variant of the Q-cycle that accounts for the ability of the complex to operate as an independent enzyme (58). It was proposed to explain apparent anomalies in the pre-steady-state kinetics of the bc_1 complex in chromatophores from the photosynthetic bacteria *Rhodobacter sphaeroides* and *Rb. capsulatus* (50–54), and of the isolated mitochondrial complexes (59–63). This modified version of Mitchell's original Q-cycle (64) was highly constrained by the set of reactions, by the stoichiometry of components, and by measured physicochemical parameters for rate and equilibrium constants, allowing discrimination from the many alternative models proposed at the time (53).

Although recent crystallographic structures have confirmed the main features expected from the Q-cycle, there were several unexpected findings. The most dramatic finding was the evidence for a large domain movement of the ISP to bring the tethered extrinsic domain containing the [2Fe-2S] cluster to two distinct reaction sites (39, 65–67). At the interface on cytochrome b near the Q_o-site (b-interface), the ISP oxidizes quinol and then moves through ~25 Å to deliver an electron to cytochrome c_1 (at the c-interface), as shown by the four different positions of the [2Fe-2S] cluster (a sample from the many different positions found in different crystals) in Figure 1. Mobility of the ISP extrinsic head has been the subject of much recent work, and the results have provided strong evidence that movement is required for catalysis (3, 68–72). The movement implies participation of five catalytic interfaces in turnover, the three sites (Q_o, Q_i, cyt c_1) expected from the modified Q-cycle model and two more for the catalytic interfaces at which ISP reacts. An important secondary conclusion is that the mobile extrinsic head acts as a second substrate at the Q_o-site (65–67, 73).

QH_2 Oxidation at the Q_o-Site

STRUCTURAL CONSIDERATIONS   None of the structures available to date shows any native occupant of the Q_o-site, so our discussions of mechanisms have been based on occupancy by inhibitors and extrapolation by inference to involvement of ligands that interact with them. The site for inhibitor binding is in the cyt b subunit, close to heme b_L. The site is spacious, reflecting the differential binding of inhibitors in two main domains (2, 3, 41, 67) (Figure 2, see color insert).

Inhibitors of class I, typified by the first example, 5-undecyl-6-hydroxy-4,7-dioxobenzothiazol (UHDBT) (25), bind in a domain distal from heme b_L (the distal domain). These are represented in structures by stigmatellin, HHDBT (the inhibitor has a heptyl instead of the undecyl tail of UHDBT), and 2-nonyl-4-hydroxyquinoline N-oxide (NQNO). They interact with the reduced ISP (ISPH), bound at the b-interface, through an H-bond to N_ε of the His-161 imidazole, which also binds one of the Fe-atoms of the cluster through N_δ. This interaction leads to an increase in E_m for ISP and to changes in the EPR spectrum in the g_x band. For stigmatellin, a ring $>C=O$ is the H-bond acceptor from the histidine $-NH$ of the ISPH. A second ligand, from the $-COO^-$ group of Glu-272, binds stigmatellin from cyt b, forming an H-bond to an $-OH$ group across the chromone ring from the carbonyl group (43, 67, 73) (see Figure 2). The inhibitor UHDBT was also reported to bind in this domain (41), but coordinate data for the structure are not available. However, a recent high-resolution structure by Palsdottir et al. (45) (PDB 1p84) of the yeast complex with HHDBT bound shows the configuration. Two interesting features of this new structure deserve attention. (*a*) Earlier experiments had suggested that the acid form of the inhibitor UHDBT was active because the titer increased as the pH was raised above the pK at 6.5 (74). However, unexpectedly, the structure shows that the group interacting with His-161 is the $-OH$ rather than one of the $>C=O$ groups, and spectrophotometric data suggest that this is in the dissociated form, which would be necessary for an H-bond acceptor function required for interaction with the reduced ISP, and would be in line with the fact that the pH for crystallization was well above the pK for this group measured at 6.1. (*b*) The second ligand, from Glu-272, is not involved in binding. Instead, the side chain points away from the inhibitor, in a configuration similar to that in native or in myxothiazol-containing structures (39, 67, 73). Other inhibitors showing similar binding behavior to that of UHDBT are 2-hydroxy-1–4-naphthoquinones like UHNQ (26, 75) and atovaquone (9, 33, 46, 76). The latter has been modeled as binding in the Q_o-site in a configuration similar to that for HHDBT (46).

Another recent structure of the bovine complex shows binding of NQNO at the Q_o-site (1nu1) (47). This inhibitor, and the n-heptyl analog HQNO, had generally been regarded as relatively specific reagents for the Q_i-site (27, 77), and the report by Gao et al. (47) studies mainly this interaction. However, the crystals also showed density at the Q_o-site, which was modeled as NQNO binding in a configuration similar to stigmatellin, with the $-OH$ group forming an H-bond with N_ε of His-161 of the ISP, and the —NO group forming a water-mediated H-bond with Glu-271 (equivalent to Glu-272 in chicken or yeast). Both these assignments are problematic since an —OH would not form an H-bond with the $-N_eH$ of the reduced ISP, and the $-N^{\delta+}O^{\delta-}$ group is generally considered to be highly polar in solution and would not be expected to form an H-bond with the carboxylate group. The authors speculate that the structure is tautomeric, with the H-atom shared between HO-(ring)-$N^{\delta+}O^{\delta-}$ and O=(ring)-N-OH forms. The latter form is found in the solid state and would provide more natural partners to the stigmatellin ligands.

The second class of inhibitors (class II), typified by myxothiazol, bind in the domain proximal to heme b_L (proximal domain) and do not interact with the ISP, which, in most structures, is found at the c-interface (41, 67). Other inhibitors in this class are MOA-type inhibitors such as MOA-stilbene, mucidin, and strobilurin (28), and also famoxadone, whose pharmacophore forms a similar H-bond (see below) (44). Anomalous features of the famoxadone-containing structure are discussed below. An important characteristic of all these structures was the finding that the Glu-272 side chain was rotated away from the liganding position in the stigmatellin-containing structures.

Both classes of inhibitor access the pocket from the lipid phase through the same relatively constricted tunnel, and, especially for those with hydrophobic tails, their binding volumes overlap substantially in the volume around the tunnel, accounting for the displacement of one class by another (67).

In general, the protein structure accommodates itself to the inhibitors and changes quite substantially, depending on occupancy. A significant factor in this accommodation is the configuration of Glu-272. In the stigmatellin-containing structures (43, 65–67, 73), the side chain occupies a major part of the volume in which the pharmacophore of myxothiazol (and other MOA-type inhibitors) would have to sit in that structure (Figure 2). To accommodate myxothiazol (or the other MOA-inhibitors), the side chain of Glu-272 rotates through at least 120° to a new configuration in which it contacts a water molecule at the end of a chain of H_2O leading to the P-phase water (67, 73), the proximal domain expands, and the distal domain closes up (67, 78). The rotation moves the Glu-272 side chain to allow the pharmacophoric group of each of the proximal domain inhibitors to form an H-bond to the backbone —NH (44, 67), a characteristic configuration for this class.

The structure containing the recently characterized inhibitor famoxadone (79) also shows occupancy of this proximal domain and rotational displacement of Glu-272 (44). As a novel feature, the hydrophobic tail of this inhibitor is on the opposite end of the structure compared with myxothiazol (the tail is attached to the pharmacophoric head), and accommodation of the extra volume distorts the protein further (compared with the distortion with myxothiazol). In a detailed analysis of the structure containing famoxadone, Gao et al. (44) suggested that, in addition to an H-bond to the backbone —NH of Glu-272 that famoxadone has in common with the MOA-stilbene and myxothiazol structures (67), aromatic interactions contribute strongly to the binding forces in this domain. Interestingly, the interaction of the rings of famoxadone with the aromatic side chains of the protein could be convincingly modeled by superposition of a crystal lattice of benzene (44). In contrast to the other class II inhibitors, the famoxadone-containing structure showed the ISP mobile domain at the b-interface. The distance from the His-161 of ISP to the inhibitor was $\sim$6.7 Å, too great to allow any specific liganding. Because occupancy of the b-interface could not be attributed to any direct interaction with famoxadone, the authors proposed that indirect forces arising from the structural accommodation were responsible, and they discussed some possible candidates.

MECHANISTIC CONSIDERATIONS   Critical features of the modified Q-cycle were the recognition that the complex operated as an independent enzyme (58) and that oxidation of two equivalents of QH_2 at the Q_o-site were necessary for complete turnover of the complex (50, 51). In the presence of antimycin as an inhibitor at the Q_i-site, turnover of the complex was limited because electrons could not leave the b-heme chain. Although four oxidizing equivalents, two in each of the chains (ISP$_{ox}$ and cyt c_1^+, hemes b_L and b_H), were available to accept electrons from QH_2, the two QH_2 oxidation reactions had different equilibrium constants. Consumption of the more favorable acceptors on oxidation of the first QH_2 constrained oxidation of the second QH_2 by the low value for the second of these ($K_{eq} \sim 1$) and effectively limited the antimycin-inhibited complex to oxidation of the first QH_2 (53, 54, 80). Furthermore, the first electron entering the oxidized high-potential chain was shared between available acceptors, predominantly ISP$_{ox}$ and ferricyt c_1, with a distribution at pH 7 that favored reduction of ISP$_{ox}$ as the higher-potential component. This fraction was therefore not seen in the kinetics measured at wavelengths appropriate for cyt c_1 and c_2 (cyt c_t). These features appeared to account for the kinetic mismatch between the rates of reduction of cyt c_t and b_H in the absence or presence of antimycin that had previously seemed contrary to the Q-cycle (reviewed in 53). Despite this simple explanation, several authors have recently discussed alternative scenarios in which rate limitation was suggested to occur at a step other than the first electron transfer reaction itself. In the simplest of these schemes, the kinetic disparity between rates of reduction of cyt c_1 and cyt b_H seen in mitochondrial complexes has been taken as evidence for a kinetic impediment at the level of ISP movement (81–83). Several groups have espoused more complex schemes involving conformational gating of electron transfer between the Q_o-site and cyt c_1 through a indirect coupling; the reduced ISP is held in position close to cyt b until released by conformational changes linked to electron transfer events elsewhere in the complex. These events have been variously suggested to be the electron transfer from a semiquinone-ISP$_{red}$ complex to heme b_L (84–86), electron transfer in the b-heme chain (83), rotation of Glu-272 associated with release of the second proton (87), and events at the Q_i-site (81, 86). Some support for gated mechanisms has come from structures. In several of these, marked asymmetries between the two halves of the dimeric complex have been observed (40, 88), and it has been suggested that these show short- or long-range conformational couplings that might be linked to dissociation of the ISP$_{red}$ (4, 86–90).

The structure from Gao et al. (44) with famoxadone in the Q_o-site showed the ISP locked at the b-interface, but without the liganding through His-161 assumed to provide the force for this locking with other inhibitors. Xiao et al. (91) investigated the kinetics of reduction of cyt c_1 by ISPH in the presence of famoxadone. They provided clear evidence for constraint on movement of the ISP mobile domain, based on a slowed electron donation to cyt c_1 in the presence of the inhibitor. In these experiments, the isolated complex from bovine mitochondria or *Rb. sphaeroides* was photoactivated through ionically bound ruthenium complexes. The slowed kinetics were taken as providing support for mechanisms in which the ISP movement was gated. From our own studies with famoxadone on the complex in situ in *Rb. sphaeroides*, this kinetic impediment was not apparent under conditions in which the quinone pool was oxidized or 30% reduced, when famoxadone behaved just like myxothiazol, so that ISPH reduced cyt c_1 more rapidly than in the absence of inhibitor. However, some inhibition was apparent when the pool was completely reduced (D.R.J. Kolling & A.R. Crofts, unpublished); possible explanations for this apparent discrepancy are currently being explored.

We have recently demonstrated that the assumptions used in derivation of the modified Q-cycle are sufficient to account for the data obtained in *Rb. sphaeroides* without the need to invoke alternative explanations based on allosteric control of the movement or dissociation of the mobile extrinsic ISP domain (54). We also showed that all reactions associated with association and dissociation of ISP at binding sites, and movement in either direction, occurred more than an order of magnitude faster ($\tau < 30~\mu$s) than the limiting reaction, and with low apparent activation barriers ($\sim$20 kJ mol^{-1}) (52, 54, 93, 94). The rate constants measured for the limiting reaction, the rate constant inferred for reduction of cyt c_1, and the activation energy for reduction of cyt c_1, have all been confirmed by direct measurement in the isolated *Rb. sphaeroides* complex (92). Because the simple explanation is sufficient in the case of the well-characterized reactions in *Rb. sphaeroides*, extension of these concepts to the mitochondrial complexes would likely be straightforward and should be explored before more complicated mechanisms are proposed. However, such an extension requires measurement of appropriate physicochemical parameters and the kinetics of all components. The extensive data from de Vries and colleagues (60–62) on the pre-steady-state kinetics of the bovine complex demonstrate many of the features expected from the simple hypothesis, but they were interpreted in the context of a double Q-cycle; more recent data using caged-UQH$_2$ (81) were interpreted in terms of an impediment to movement of ISP. The data set for the yeast complex are less complete and have also been differently interpreted (86, 87, 95).

FORMATION OF THE ES-COMPLEX AT THE Q_o-SITE   The overall reaction for oxidation of QH$_2$ at the Q_o-site of the oxidized bc_1 complex involves the [2Fe-2S] cluster of ISP$_{ox}$ and heme b_L of cyt b as the immediate acceptors.

$$QH_2 + ISP_{ox} + heme b_L \leftrightarrow Q + ISPH + heme b_L^- + H^+ \qquad 2.$$

The driving force for this reaction is calculated by summing the driving forces for the two partial electron transfer reactions, $\Delta G^{o'} = -F(E_{m, ISP} + E_{m, bL}) - 2E_{m, Q/QH2}) = -2.9\,kJ\,mol^{-1}$, giving a value of $K_{eq} = 3.2$ at pH 7.0, using $E_{m, ISP} = 310$ mV, $E_{m, bL} = -90$ mV, and $E_{m, Q/QH2} = 90$ mV. Under normal operation, the forward reaction is favored by removal of the electron from heme b_L by rapid transfer to b_H and the Q_i-site, with a more favorable equilibrium constant (50–54). Under physiological conditions, the equilibrium constants for the low potential chain are modulated by the proton gradient, so that under static head conditions, the complex comes into quasi-equilibrium (51).

A more complete description of the energy landscape requires partitioning of the driving force between a set of partial processes, including binding of substrates, activation barriers, electron and proton transfer reactions, and dissociation of products.

Two substrates contribute to formation of the ES-complex, QH_2 and ISP_{ox}, as shown in Scheme 1 (65–67, 73, 78, 96–99). Based on the structure of the stigmatellin-containing complex and following previous proposals for an involvement of ligands from ISP histidines (85, 100), we suggested that the ES-complex was formed with QH_2 in a position at the distal end similar to that found for stigmatellin (67, 73). A likely configuration would involve an H-bond between the ring $-OH$ of the quinol, and the N_c of ISP His-161, and between the other $-OH$ and the carboxylate of Glu-272. Because of the difference in pK_a values for QH_2 (pk > 11.5) and the ISP histidine (pK $\sim$ 7.6), the quinol $-OH$ is most likely the H-bond donor, requiring that the acceptor N_ε should be in the dissociated (imidazolate anion) form (95, 96). With this in mind, it is possible to account for the strong dependence on pH of the rate of electron transfer over the pH range below the pK_1 at 7.6 in terms of simple enzyme kinetics; the rate varies with [S] (the dissociated form) and approaches saturation (78, 93, 97, 99).

Scheme 1

$Eb_L.QH_2$

$QH_2 \nearrow \swarrow \quad \nwarrow \searrow ISP^{ox}$

$Eb_L \qquad Eb_L.QH_2.ISP^{ox} \leftrightarrows \{ES\}^{\#} \leftrightarrows Eb_L.QH^{\cdot} + HISP^{red} \leftrightarrows Eb_L^{-} + Q + HISP^{red} + H^{+}$

$ISP^{ox} \searrow \nwarrow \quad \swarrow \nearrow QH_2$

$Eb_L.ISP^{ox}$

THE BINDING CONSTANTS INVOLVED IN FORMATION OF THE ES-COMPLEX   A long history from several labs of work in photosynthetic bacteria has shown that QH_2 is preferentially bound compared with Q on oxidation at the Q_o-site, so that the dependence of rate on E_h is displaced from the E_m of the pool (at $\sim$90 mV) to an apparent $E_m \sim 130$–140 mV (reviewed in 57), but the molecular basis for this displacement of E_m was not understood. It had previously been observed that the steady-state rate of QH_2 oxidation in mitochondrial complexes showed a pH dependence over the range 5.5–9.5 (101, 102), recently attributed to three dissociable groups (103). The stimulation of electron transfer on raising the pH over the range 5.0–8 was attributed to the need for deprotonation of a group with a pK in the range 5.5–6.5. The rate of QH_2 oxidation during the first turnover of the site, seen in pre-steady-state kinetic measurements of the reactions in *Rb. sphaeroides* chromatophores, showed a similar stimulation, with an apparent pK of $\sim$6.3 (78, 93, 97, 99). A possible candidate for the dissociable group was His-161 of the ISP, but the pK value measured kinetically was displaced from the pK of 7.6 expected to determine the concentration of the dissociated ISP_{ox} form. A straightforward explanation for both these displacements is that they both reflect the same process, formation of the ES-complex, as shown in Scheme 1. Both equilibria are pulled over by the binding process through mass action; the binding of QH_2 will raise the apparent E_m for the oxidation reaction, and the binding of ISP_{ox} will pull the dissociated form of ISP_{ox} out of solution, giving an apparent shift in the pK (78). When the displacements were expressed as equilibrium constants for the binding processes, both sets of data showed similar values, with $K_{QH2} \approx K_{ISPox} \sim$ 17 ± 4 (78, 134). The similarity of these two values supports our suggestion that both reflect that same phenomenon: the liganding between QH_2 and ISP_{ox} involved in formation of the ES-complex. This depends on recognition that the complementary constants in the binding square (left-hand section of Scheme 1) are both small and of the same magnitude (66, 83). Although the thermodynamic equilibrium constants derived here are dimensionless, it seems appropriate to take the values as reflecting binding constants, following the rationale developed by Shinkarev et al. (118) for treatment of binding constants involving the ISP as a tethered substrate.

If this is the case, then the pH dependence over the range below pK_1 is accounted for by the properties of the ISP without invoking any controlling effect of other dissociable groups (see 103). (The slowing of the rate at pK above 8.0, likely determined by pK_2 of ISP_{ox}, is discussed more extensively below). The configuration of the ES-complex suggested requires specific properties of the histidine side chain involved; it has to be the group responsible for the pK_1 measured from redox titration as a function of pH. This assignment now seems well justified (104–107). The interpretation of a controlling role for this pK in determining the occupancy of the ES-complex is strongly supported by experiments with a mutant strain, Y156W, in which both the pK, and the whole curve for pH dependence, were shifted up by $\sim$1 pH unit (94).

ROLE OF Glu-272   The conformational change of the buried glutamate side chain (Glu-272) discussed above was revealed in structures from Berry's lab (39, 67, 73). In the presence of stigmatellin (2bcc), Glu-272 had rotated 120° away from a position seen in the native complex (1bcc), where it pointed toward heme b_L, to provide a second ligand to the inhibitor through an H-bond to an —OH group across the ring structure from the —C=O involved in interaction with the ISP. Molecular dynamics simulations (98) had predicted a relatively stable water chain leading from the aqueous phase on the cyt *c* side into the protein along the b_L heme edge to the Q_o-pocket. In the native structure, or that with myxothiazol bound (67), the

Glu-272 carboxylate contacted this water chain. We suggested that the two ligands that bind stigmatellin were also involved in formation of the ES-complex and that protonation of the Glu-272 carboxylate after formation of the SQ intermediate, followed by a rotation of the Glu-272 acidic side chain between the two positions, could provide a plausible pathway for transfer of a second proton from the site on oxidation of QH (67, 73). Consistent with this, mutant strains with the equivalent glutamate in *Rb. sphaeroides* (E295) modified to aspartate, glycine or glutamine, showed small increases (two- to threefold) in apparent K_m for QH_2, lowered rates of electron transfer, and resistance to stigmatellin (73). The two features required for this mechanism, orientation of Glu-272 seen in the stigmatellin structure (2bcc) and the water chain we predicted, have now been found in higher resolution structures. One of these is the yeast complex with stigmatellin from Hunte et al. (43, 108), from which these authors reached similar mechanistic conclusions. Both features are also well resolved in a recent 2.1 Å structure of the bovine complex (1pp9) (E.A. Berry, personal communication). The contribution of the H-bond from Glu-295 (Glu-272) to the binding of QH_2 is likely in the range ≤ 1 kJ mol^{-1}, as judged from the small increase in K_m for QH_2 in mutant strains (73). Palsdottir et al. (45) found the orientation of Glu-272 toward the water chain in their HHDBT structure, and Gao et al. (44) in their famoxadone structure. Palsdottir et al. (45) provide an interesting discussion of how secondary ligands might orient the QH_2 substrate in the site so as to provide a favorable configuration for binding to ISP_{ox} on formation of the ES-complex.

THE $g_x = 1.80$ COMPLEX   The $g_x = 1.80$ line in the CW X-band EPR spectrum of ISP is observed only when the quinone pool is oxidized and ISP reduced (109), and is attributed to interaction between Q and ISPH. As discussed in greater detail below, this likely involves an H-bond between the quinone $-C=O$ and the N_ε of His-161. Ding et al. (100, 110, 111) suggested a strong binding of a quinone by the ISPH to form the Q_{os} species of their double-occupancy model. However, a weakly bound complex seems to be required by the need for rapid dissociation of ISPH to allow participation in catalysis (65–67, 93). To explore the structure in greater detail, we used pulsed EPR to show that the g_x line position change involved an N-atom ligand to the [2Fe-2S] cluster (tentatively identified as N_δ of the histidine ring of His-161) with structural characteristics (as determined from the spin interaction) similar to those seen in the stigmatellin complex (112). In both the stigmatellin and quinone complexes, the involvement of N_ε in an H-bond with the occupant changed the spin interaction of the N_δ with the cluster in a similar fashion. The ESEEM and HYSCORE ^{14}N spectra of both these bound forms differed from that seen in the presence of myxothiazol, where the liganding histidines are likely exposed to the aqueous phase. This supported the view that a similar H-bond was formed [shown in the stigmatellin structure (39, 65) as between a ring $>C=O$ of the occupant and the $-N_\varepsilon H$ of His-161], and represented the first direct structural information about occupancy of the Q_o-site by a quinone species. The Q.ISPH complex is formally an EP-complex, and the strength of this bond is therefore a parameter of thermodynamic interest in defining the energy landscape.

THE STRENGTH OF THE BOND INVOLVED IN FORMATION OF THE $g_x = 1.80$ COMPLEX   Since the first demonstration in the case of UHDBT binding (25), a substantial literature on the change of E_m of the ISP in the presence of similar inhibitors has been interpreted in terms of a preferential binding of the reduced ISP by the inhibitor (31, 46, 75, 113). This has been challenged in a recent study (114), in which the kinetics of binding were measured by mixing dilute suspensions of inhibitor and isolated complex and then assaying the fraction in the active form by addition of excess substrate (DBH) and measurement of the steady-state turnover. Binding was shown to be second order [first order in (inhibitor)], with time constants in the 10 to 100 s range, depending on the concentrations used. The apparent binding rate constants for stigmatellin and myxothiazol were relatively independent of the redox state of the complex before mixing with inhibitor, and this was interpreted as showing that the binding of stigmatellin to oxidized and reduced ISP was at the same rate, i.e., no preferential binding to the reduced form. An alternative explanation is that the rates measured were limited by the probability of interaction of inhibitor and complex to form a bound micellar state, and that this (second-order) rate was independent of redox status. On reduction of the complex with DBH, the bound inhibitor would find the reduced ISP on the ms timescale, and this kinetic event would be undetected in the steady-state assay of turnover. This interpretation avoids the apparent conflict with the previous strong evidence for binding of stigmatellin or UHDBT to the reduced ISP.

In light of the change in E_m of ISP consequent on the binding of class I inhibitors, the interaction of ISPH with quinone might also be expected to induce an increase in E_m ISP, because the pulsed EPR data had shown that a similar bond is involved (112). Such a change in E_m has been measured directly by redox titration (115, 116) and indirectly by looking at the change in kinetics of cyt c on flash activation of chromatophores with and without addition of myxothiazol, over the E_h range around the E_m ISP (117, 118). Quantification of the changes showed that the E_m in the presence of myxothiazol was ~40 mV lower than that in the absence of inhibitor (118). Darrouzet et al. (116) investigated changes in E_m ISP in mutant strains with modified linker regions, and reported that myxothiazol induced a downward shift in the E_m ISP so that all strains showed a similar value. In wild type the shift was ~40 mV, comparable to the value found from our kinetic studies and to that with MOA-stilbene (115). From the structural data, no ligand is formed between myxothiazol and ISPH, rather the extrinsic domain is rotated away from its binding site to expose the histidine ligands to the aqueous phase (41, 67, 83). The E_m measured in the presence of myxothiazol therefore likely reflected the unliganded state, so that the change in E_m induced by addition of inhibitor would be due to loss of the bound state seen in the $g_x = 1.80$ complex. From the E_m change, a binding constant of $K_{ass} \sim 4$ could be calculated, showing that >80% of the ISP_{red} would be bound at $E_{h,7} \sim 200$ mV, where ISP is reduced and the quinone pool oxidized (118). Treatment of the thermodynamics of the binding reactions for the ISP required development of a new formalism appropriate to the peculiar status of the ISP as a tethered substrate (118).

The binding constant for formation of the $g_x = 1.80$ complex ($K_{ass} \sim 4$) is in the same range as that for the binding of QH_2 by Glu-272, as seen in the increased K_m in a mutant strain with glycine instead of glutamate [$K_m^G/K_m^E = 2.4$ (73)], and, following the arguments of Crofts et al. (66), explains why the apparent E_m for formation of the complex titrates with the same midpoint as the quinone pool (110).

IDENTIFICATION OF THE RATE-LIMITING STEP   From the above discussion, it seems likely that electron transfer from QH_2 to ISP_{ox} occurs through the H-bond by which the ES-complex is stabilized, and that it involves the dissociated ISP_{ox}. Because the pK on the reduced ISP is > 12 (107), electron transfer is coupled to H^+ transfer and the reaction is formally an H-transfer. Release of the proton occurs on oxidation of ISPH by cyt c_1 (above pK_1) or on rebinding of $ISP_{ox}H$ in the ES-complex (below pK_1).

Previously, Brandt & Okun (102), using a steady-state assay, confirmed the pH dependence for the steady-state rate of oxidation of ubiquinol by isolated bovine bc_1 complex observed by Link & von Jagow (101), and extended these studies to plastoquinol analogues and the yeast bc_1 complex. They also reported a novel dependence on pH of the activation energy for the steady-state turnover. From the pH dependence of the rate, they identified two controlling groups, one with a pK at 6.5 and a second at 9.5. The former group had to be in the dissociated state, and the latter protonated for rapid electron transfer, accounting for the bell-shaped pH dependence. Because of the pK values, they excluded any involvement of the ISP histidine group with its pK_1 at 7.6. On the other hand, the pH dependence of the activation energy was interpreted as showing that deprotonation of QH_2 was a necessary intermediate step preceding electron transfer. This followed earlier work by Rich & Bendall (172) on the non-enzymic oxidation of quinol by cyt c, which showed a strong acceleration of rate in the high pH range. This was interpreted in terms of a reaction mechanism dependent on QH^-, with the low probability of deprotonation providing the activation barrier. In contrast, the bc_1 complex catalyzed reaction showed a rate that decreased dramatically in this range. Because of this contradictory result, the mechanistic speculations had to be based on the pH dependence of the activation energy. Brandt & Okun (102) suggested that, following binding of the neutral quinol to the catalytic site, deprotonation to the quinol anion with direct release of the proton to the aqueous phase, with a pK in the range 14–16, was necessary before electron transfer could occur, and that this was responsible for most of the activation barrier. The deprotonation of QH_2 was incorporated into a detailed reaction scheme, the "proton-gated charge transfer mechanism," which included a double-occupancy model for electron transfer to heme b_L and control of the overall reaction by the redox state of heme b_L.

The earlier studies of Crofts & Wang (52), on the pre-steady-state kinetics of the complex in its native state in *Rb. sphaeroides*, and more recent extensions of this work by Hong et al. (93) have shown the following:

1. The reaction with the slowest rate under conditions of substrate saturation was the oxidation of QH_2 at the Q_o-site (52). This was also the reaction with the highest activation barrier (52, 93).

2. The rate of QH_2 oxidation showed the same bell-shaped pH dependence as seen in mitochondrial complexes, as discussed above (93, 94).

3. In contrast to the results of Brandt & Okun (102), the activation barrier was independent of pH (52, 93). Since varying pH below the pK_1 for ISP_{ox} varies the concentration of the dissociated ISP_{ox} species that is active in formation of the ES-complex, the activation barrier was independent of concentration for this substrate (93).

4. The activation barrier was also independent of the redox poise of the quinone pool (52, 93). Since reduction of the quinone pool increases the concentration of QH_2, the other substrate, these characteristics showed that the activation barrier was independent of substrate concentration, and therefore after formation of the ES-complex, as expected for an enzyme reaction.

5. The limiting reaction was transfer of the first electron from QH_2 to ISP_{ox} (93, 94).

These data were clearly at variance with those of Brandt & Okun (102) and precluded a mechanism of the sort they suggested. As shown by the pH dependence of the rate, and the acceleration of rate on reduction of the pool, the rate varied with concentration of either substrate, as expected from simple Michaelis-Menten considerations. The movement of the ISP could be assayed by measuring the lag times involved in reactions that incorporate it as a partial process. The time not accounted for by electron transfer events was always short (<30 μs), and the reactions had low activation barriers (54, 93). This was in line with weak association constants and the simple constrained diffusion mechanism suggested by the structures (65) and a steered MD simulation (98).

DEPENDENCE OF RATE ON DRIVING FORCE AND ON pH   Analysis of the kinetics in strains with mutations in ISP that lowered the E_m value had shown that the rate of QH_2 oxidation by the isolated complex depended on the driving force for the first electron transfer (119–121). We showed a similar dependence on driving force in pre-steady-state measurements assaying oxidation of the first QH_2 for the complex in situ (93, 94, 122). Similar mutations made in *Rb. sphaeroides* (94), *Paracoccus nitrificans* (121), and in yeast (120) all showed the same behavior. In contrast, the overall rate did not show changes consistent with a dependence on driving force on change in E_m of heme b_L (93). Mutant strains showed either a slowed electron transfer with increased driving force or no effect (123–125). The rate of reduction of heme b_L in the presence of antimycin after prior reduction of heme b_H showed the same rate constant and activation energy as the reduction of heme b_H when the latter was oxidized before flash activation (80, 93), although the redox potential of heme b_L is likely changed through coulombic interaction by ~-60 mV on reduction of heme b_H to give the $E_m \sim -90$ mV value measured through redox titration (123, 126). This strongly suggested that transfer of the first electron (from QH_2 to ISP_{ox}) was the rate-limiting step, and that of the second electron is intrinsically more rapid. The rate of the second electron transfer is gated [using the definition of Davidson (127); see Scheme 2] by the limiting rate of the first electron transfer. As a result, the change in rate expected on change of driving force is not seen experimentally.

Scheme 2

Definitions of *gated* and *coupled* electron transfer in terms of electron transfer reactions showing kinetic complexity. Adapted from Davidson (127)

Simple kinetic model

$$A_{ox} + B_{red} \overset{K_d}{\Longleftrightarrow} A_{ox} \cdots B_{red} \underset{-k_{lim}}{\overset{k_{lim}}{\Longleftrightarrow}} A_{red} \cdots B_{ox}$$

Kinetic complexity

$$A_{ox} + B_{red} \overset{K_d}{\Longleftrightarrow} A_{ox} \cdots B_{red} \underset{-k_x}{\overset{k_x}{\Longleftrightarrow}} [A_{ox} \cdots B_{red}]^{\#} \underset{-k_{ET}}{\overset{k_{ET}}{\Longleftrightarrow}} A_{red} \cdots B_{ox}$$

For true ET

$$k_{ET} \ll k_x \qquad K_x (k_x/k_{-x}) \gg 1 \qquad k_{lim} = k_{ET} \qquad \lambda_{obs} = \lambda_{et}$$

For *gated* ET

$$k_x \ll k_{ET} \qquad\qquad\qquad k_{lim} = k_x \qquad \lambda_{obs} = f(\Delta G^{\#}_x)$$

For *coupled* ET

$$k_{ET} \ll k_x \qquad K_x (k_x/k_{-x}) \ll 1 \qquad k_{lim} = K_x k_{ET} \qquad \lambda_{obs} = f(\lambda_{ET}, \lambda_x)$$

What controls the rate of the first electron transfer? The structures, and our speculation on the nature of the ES-complex, provided an opportunity to analyze the rate in the context of the distance for electron transfer (93). In the context of the proposed structure of the ES-complex, the rate of the first electron transfer was slower by about three orders of magnitude than that expected from the distance of ~7 Å suggested by our model (73, 93) and considerations by Moser et al. (128). The problem then was to account for this slow rate. The observed rate could be explained if a high value for reorganization energy ($\lambda \sim 2.0$ eV) was used, in line with the high activation barrier (93), but this value was much higher than that found in other electron transfer reactions occurring over similar distances (128), and no explanation was obvious from the structure.

PROTON-COUPLED ELECTRON TRANSFER AS A MODEL FOR QH_2 OXIDATION AT THE Q_o-SITE   Work on proton-coupled electron transfer in model compounds by

Nocera and colleagues (129) has emphasized the controlling effect of relative pK values on H$^+$ transfer through H-bonds. A detailed theoretical treatment by Cukier & Nocera (130) discussed the control of rate in terms of a Marcus treatment of proton-coupled electron transfer. Paddock and colleagues (131), in discussion of the proton-coupled electron transfer reaction at the Q_B-site in photochemical reaction centers, made simplifying assumptions that allowed separation of the role of the proton transfer from the electron transfer, by treating the former through a probability function. Combining these approaches led us to propose a treatment of the dependence of rate on driving force as applied to the Q_o-site reaction (99, 134), which avoids the difficulties arising from quantum mechanical considerations of the role of the proton (130).

(*a*) The electron transfer can occur only when the proton configuration is favorable. This requires that the proton be transferred through the H-bond before electron transfer can occur.

$$E \cdot b_L \cdot QH_2 \cdot ISP_{ox} \overset{\Delta G_{proton}}{\leftrightarrow} E \cdot b_L \cdot QH^- \cdot H^+ ISP_{ox}$$

$$\overset{\Delta G^{\#}_{electron}}{\leftrightarrow} ES^{\#} \rightarrow E \cdot b_L \cdot QH^\bullet \cdot ISPH. \qquad 3.$$

$$\Delta G_{proton} = -2.303\,RT\left(pK_{ISP_{ox}} - pK_{QH_2}\right). \qquad 4.$$

(*b*) The occupancy of the proton-transfer state needed for electron transfer is determined by the pK values of the H-bond donor and acceptor through a Brønsted term, ΔG_{proton}. Given the pK values for QH_2 (pK > 11.5) and ISP_{ox} (pK $\sim$ 7.6), the configuration is thermodynamically highly unfavorable, and the low probability of accessing the state represents a substantial part of the activation barrier. The step represented by $\{ES\}^{\#}$ in Scheme 1 is replaced by the two partial processes shown in Equation 3.

(*c*) Rates of H$^+$ transfer through H-bonds are inherently rapid ($\sim 2.10^{11}$ s^{-1}) (132, 133), $\sim$1000 faster than the maximal electron transfer rate at this distance. To a good approximation, the proton transfer contribution can therefore be treated as a separate probability function given by the Brønsted term. This treatment corresponds to the limit condition of a coupled reaction (127; see Scheme 2). In terms of reaction rate theory, this is equivalent to partitioning the activation barrier into the sum of terms, ΔG_{proton} and $\Delta G^{\#}_{electron}$ in this case, which can then be split out into separate terms in the Arrhenius equation and its derivatives. This allows for a great simplification in thermodynamic treatment. Using the pre-exponential terms suggested by Moser et al. (128), the following equation for the rate constant was proposed (99, 134):

$$\log_{10} k = 13 - \frac{\beta}{2.303}(R - 3.6) - \gamma \frac{\left(\Delta G^o_{ET} + \lambda_{ET}\right)^2}{\lambda_{ET}}$$

$$- \left(pK_{QH_2} - pK_{ISP_{ox}}\right). \qquad 5.$$

Here β is 1.4, the slope of the Moser-Dutton relationship between $\log_{10}k$ and distance, R is the distance in Å, ΔG^o_{ET} is the driving force for the electron transfer step, and λ_{ET} is the reorganization energy (both in V). The factor γ in the above equation has a value of $\sim$3.1 in the Moser-Dutton treatment, which differs from the value of 4.227 at 298 K ($\gamma = F/(4 \times 2.303\ RT)$) from classical Marcus treatment by incorporation of quantum mechanical factors contributing to a tunneling component (135–137).

The explanation involving the pK of QH_2 as a determinant in the activation barrier recalls the earlier suggestions that deprotonation of QH_2 might provide the activation barrier (102, 172). However, the reaction here is not a deprotonation, but formation of an improbable intermediate state. The electron transfer through the H-bond is coupled to proton transfer, and this provides a direct explanation for the pH independence of the activation barrier. The reaction mechanism proposed is well justified by theory and experiment (134). In contrast to the previous proposal (102), a substantial barrier to the electron transfer step is also necessary to explain the dependence on redox driving force demonstrated in mutant strains.

THE ODD BEHAVIOR OF STRAIN Y156W   Our own work on the dependence of reaction rate on driving force used mutant strains with modifications in ISP at Tyr-156 (Tyr-165 in bovine sequence), which forms an H-bond from the –OH to the $S\gamma$ of one of the cysteine ligands (Cys-139, bovine) (94). Measurement of the E_m and pK values of these strains showed decreases in E_m for all, but one strain (Y156W) also showed a substantial increase in pK (from 7.6 to 8.5). When $\log_{10}k$ for oxidation of QH_2 was plotted as a function of ΔE_m at pH 7.0, the points followed the dependence of rate on driving force expected from Marcus theory (135–137). However, a substantial part of the inhibition observed in strain Y156W could be attributed to the effect of the pK change on the concentration of the dissociated form as substrate; if this contribution was considered, the data point for this strain no longer fell on the same curve as those for other mutants. This anomaly might have called into question the validity of using the Marcus explanation for the inhibitory effect observed (94), but can now be explained by consideration of the role of pK_1 of ISP_{ox} in controlling several critical parameters:

1. In the formation of the ES-complex, the dissociated (imidazolate) form is the substrate (see above). The concentration of this form depends on pH and on pK, assumed to be pK_1 due to dissociation of His-161, leading to the dependence of rate on pH over the range 5.5–8 in wild type.

2. The pK_1 at 7.6 on the oxidized form also results in a dependence of $E_{m,\ ISP}$ on pH, the value decreases above pH 7.0, with a $\sim$59 mV/pH unit slope above the pK. Over this range (at pH > 8), the overall rate and the rate of the first electron transfer both decrease. The decrease in E_m of ISP above pK_1 might be expected to determine the overall driving force, but because the E_m of the Q/QH_2 couple also decreases by $\sim$59 mV/pH, the driving force is constant with pH over the range of pK_1. A second pK (pK_2) on the oxidized form at

~9.6 increases the slope of pH dependence at higher pH. As a consequence, pK_2 will be the critical determinant for the change in driving force; the fall-off in rate over the high pH range can be attributed to the loss of driving force as the effect of pK_2 comes into play.

3. The pK_1 also plays a critical role in determining the activation barrier through the Brønsted term discussed above. From Equation 5, an increase in pK_1 should lead to a lowering of the Brønsted barrier, and hence in increase in rate.

The curves showing the dependence of rate on driving force using mutant strains could be well fitted by using Equation 5 above (134). In particular, the anomalous behavior of strain Y156W with a shifted pK could be explained by recognizing that two effects of the pK_1 value offset each other. The lowered rate due to the substrate effect (depletion of the dissociated ISP_{ox} at pH $<$ pK_1, as in 1, above) was compensated by an increased rate constant associated with a lower activation barrier, due to the smaller contribution of the Brønsted term (as in 3). This simple explanation of the odd behavior provides strong support for the mechanism proposed.

The equilibrium constant between cyt c_1 and ISP is also determined by E_m ISP, because E_m cyt c_1 is independent of pH. This has important consequences in the kinetics of the high-potential chain measured in pre-steady-state experiments (92, 139) and can also be used to vary the driving force, so as to favor the forward electron transfer at higher pH. Engstrom et al. (92) used this effect, as well as mutant strains with modified $E_{m,ISP}$, to investigate the Marcus relationship for this reaction and found a rate independent of driving force. Since the observed rate was much slower than that calculated from the closest configuration suggested by the structures, they concluded that the reduction of cyt c_1 by ISPH was gated (127) by a conformational transition from a nonreactive state. From the high rate constant and the low activation barrier observed, this was a process different from the rate-limiting QH_2 oxidation reaction.

TRANSFER OF THE SECOND ELECTRON AT THE Q_o-SITE   It has long been recognized that the semiquinone intermediate formed after the first electron transfer has the potential for reduction of O_2, thus providing an explanation for the generation of superoxide observed at this site in mitochondria (12–15). This provides a background to discussion of the second electron transfer through which this potential hazard is minimized; the SQ is removed by rapid oxidation through the b_L heme. Evolution has honed the mechanism to limit the damage; the overall rate has to be optimized while keeping the species that reacts with O_2 at a minimal concentration and/or limiting access of the intermediate to O_2.

Hong et al. (93) discussed mechanisms for the second electron transfer, from SQ to heme b_L, that were compatible with the kinetic and thermodynamic properties, measured activation barriers, and structural information. They noted that the product of rate constant and fractional occupancy of the SQ intermediate state would be critical in determining the rate. The rate constant was determined by the distance to heme b_L and substitution of appropriate values into the equation of Moser et al. (128). Occupancy was determined in a complicated fashion by the parameters of the first electron transfer through the relation between E_m, driving force and semiquinone stability. Because the overall rate appeared to be determined by transfer of the first electron and was independent of the driving force for the second electron transfer, the gating of the latter by the former implied an intrinsic rate much faster than the rate-determining step. From these considerations, Hong et al. (93) ruled out single-occupancy mechanisms in which the occupancy of the intermediate state was very low and the intermediate was constrained to the distal domain. On the basis of these criteria, several sorts of hypothesis seem plausible:

1. Link (84) suggested that the product of the first electron transfer might be a complex between SQ and the reduced ISP in which a strong binding constant prevented oxidation of ISPH by raising the E_m value above that of the acceptor. Crofts and colleagues (67, 73, 93), with the benefit of the structural information by then available, have considered a modified version of this hypothesis, in which the strong binding constant prevented dissociation to products until after the second electron transfer from SQ to heme b_L. The intermediate complex would bind up the semiquinone, so that it was not available to O_2, and also bind up the ISPH, to prevent its movement, and oxidation down the high potential chain that would allow the ISP_{ox} to grab the second electron from the semiquinone and bypass the bifurcation. In this intermediate complex, the SQ could be relatively stable and therefore at high-enough occupancy to allow electron transfer at the observed rate from the distal domain in which the complex would have to be formed. Because neither SQ nor ISPH has been detected by EPR under conditions in which this complex might be favored (52, 56), it was necessary to introduce an additional ad hoc hypothesis, that both EPR signals were quenched through mutual magnetic interaction (56, 84). As Palsdottir et al. (45) noted, their recent structure of the complex containing HHDBT might be considered as mimicking an intermediate state in which the SQ is in the anionic form, complexed with the reduced ISP, after transfer out of second H^+ and before dissociation of the $SQ^- \cdot$ ISPH state. If this analogy is correct, their structure gives support for the Link (84) model. However, the stability of the inhibitor-ISPH complex might not necessarily reflect the stability of the $SQ^- \cdot$ ISPH complex. Since the side chain of Glu-272 is rotated away from liganding configuration in their structure, it might more conservatively be seen as providing support for the role of this residue in proton transfer.

2. An alternative mechanism was proposed by Crofts and colleagues (52, 67, 73, 93) in which the SQ was formed at very low concentration (thus minimizing reaction with O_2), owing to an unfavorable equilibrium constant. To overcome the kinetic barrier arising from the low occupancy suggested by the data, the semiquinone would have to move within the site closer to

heme b_L. From the distance obtained by modeling a quinone in the proximal domain to replace myxothiazol, the closer approach would provide a rate constant ~ 1000 times more rapid than from the distal domain. The movement proposed would require dissociation of the intermediate product complex before electron transfer to heme b_L could occur. On dissociation of the SQ.ISPH complex, the second H^+ would be transferred to the Glu-272 carboxylate, followed by a rotational displacement of the Glu-272 side chain to allow removal of the second H^+ from the site, and to open up the volume of the proximal domain for the $SQ \cdot {}^-$ (73). Earlier versions of this mechanism were based on the failure to detect either SQ or reduced ISP under conditions in which an intermediate complex would be maximal (52), so this experimental observation was naturally accounted for. The mechanism also accounted naturally for the large volume of the site, and explained the evidence from mutational studies showing that modifications that led to myxothiazol resistance also led to inhibition of electron transfer (67). If the resistance was interpreted in terms of an impediment to binding, then these mutations would also impede binding of SQ in this domain. The inhibition observed seemed to provide a clear indication that occupancy of the proximal domain during catalysis was important. Support for a dissociation of the SQ.ISPH intermediate has come from analysis of bypass reactions at the Q_o-site in the presence of myxothiazol (140, 141). To explain these, Muller et al. (141) interpreted their data as showing that dissociation of SQ.ISPH intermediate must occur, although under these circumstances, electron transfer to heme b_L was blocked (see more extensive discussion, below).

3. A third type of mechanism involves occupancy of the site by two quinones during catalysis, one tightly bound, the other readily dissociable as a substrate (85, 100, 110). The large volume of the site shown by the structures (39, 41, 67), the malleability of conformation suggested by differences in structures with different inhibitors in situ (39, 67, 78), the demonstration of a displacement of two quinones from the complex on binding either of the two classes of Q_o-site inhibitors (142), and different g_x line positions in the EPR spectrum of ISPH on partial extraction of ubiquinone [interpreted as reflecting two different quinone species (100, 110, 111, 115), both lost on addition of either class of inhibitors (100)], all favor double occupancy. Occupancy of the distal domain in the presence of proximal domain inhibitors (2, 111, 115) and the observation that some superoxide can be generated at the site in the presence of myxothiazol or other MOA-type inhibitors (100, 110, 143) demonstrate that occupancy of the two domains is not necessarily exclusive. However, there are major problems in understanding how such a double occupancy of quinone and proximal domain inhibitor might translate to a mechanism for normal turnover (2, 67, 144). Attempts to fit two quinones in existing structures lead to considerable distortion (142; A.R. Crofts, unpublished), and although a structural model has been presented (142), no evaluation of the energetic consequences was reported. The question of

energetic consequences is of importance, because, if this mechanism is to be accepted, double occupancy has to represent a favorable configuration compared to single occupancy. There remains the difficulty of explaining away the failure to find in any of the structures the tightly bound quinone expected (39, 41, 67). In the *Rb. sphaeroides* system, support for double occupancy was based on distinct EPR lines attributed to interaction of ISPH with weakly (Q_{ow}) and strongly (Q_{os}) binding species, or an empty site (100, 110), and the Q_{ow} species was predicted to have a binding constant substantially greater than those for the Q_B site of RC, or the Q_i-site, at both of which structurally well-defined quinones have been characterized. However, in the mitochondrial complexes where the structures show no occupancy, the changes in EPR spectra on extraction could be fit without invoking the additional spectral form associated with Q_{os} (P.R. Rich, personal communication); this could be interpreted as showing that there are substantial differences in binding between the bacterial and mitochondrial systems. Some features of earlier models need revision. The assignment of the quinone interacting with ISP as the tightly binding species (110, 115) must clearly be revised (67), because the ISP mobile domain has to be able to move. Double occupancy would provide an obvious explanation for the evidence indicating that occupancy of the proximal domain is important. However, this introduces a further problem. The differential effects of mutations on inhibitor binding were not reflected in differential effects on the EPR signals interpreted as showing two quinone species, although these would clearly have to be occupying the domains from which the inhibitors are selectively excluded (67, 144). These difficulties have so far precluded more general acceptance of this hypothesis.

4. As noted above, Crofts & Wang (52) considered mechanisms in which the SQ intermediate in the Q_o-site reaction was generated at low concentration, to give an up-hill reaction for the first electron transfer. Constraints on possible values for the equilibrium constant came from the detection limit of the SQ, and from the activation barrier. The former was based on unpublished experiments which showed that no myxothiazol-sensitive SQ could be detected under the "oxidant-induced reduction" conditions (145) expected to maximize [SQ] (K.A. Andrews, R.B. Gennis & A.R. Crofts, unpublished; see 146), leading to a maximal value for $K_{eq} \sim 10^{-4}$. Similar experiments in a mitochondrial complex yielded similar results, with a similar kinetic interpretation (56). The activation barrier represents the limiting value for instability of any intermediate state, giving a minimal value for K_{eq} determined by E^{act}. This lowest limit on K_{eq} represented a mechanism in which both electron transfers could occur simultaneously, with a common activation barrier and an activated state represented by the SQ intermediate; Crofts & Wang (52) discussed this as the most economical hypothesis. Two sets of data have since made this hypothesis untenable: (*a*) Hong et al. (93) and Guergova-Kuras et al. (94) pointed out that the slope of the curve of $\log_{10}k_{cat}$ v. driving force, obtained from experiments in which the driving force was

varied by using mutant strains in which the E_m of ISP was changed, was less steep ($\sim$0.009 mV^{-1}) than the value expected from this minimal hypothesis ($\sim$0.017 mV^{-1}), and therefore excluded such a mechanism; (*b*) Hong et al. (93) considered the constraints of occupancy and distance on rate for the single-occupancy case and concluded that if the rate constant was determined by the distance from a SQ occupant modeled in the distal pocket to heme b_L, the maximal rate possible, given the very low occupancy of the activated state from which the second electron would have to be transferred, would be too slow by several orders of magnitude to allow such a mechanism.

A concerted mechanism involving simultaneous transfer of two electrons to the two separate chains has recently been revisited by Kim et al. (41) and Trumpower and colleagues (3, 87), although they did not comment on the critical points discussed above. To overcome these arguments in the context of a single-occupancy model, it would be necessary to propose that the substrate QH$_2$ delivers both electrons simultaneously from a state bound in the distal domain with much higher occupancy than the activated state, through pathways that, although very different in terms of the distance involved, local structure, and reaction path, must function to give the same activation barrier and rate constant. A more detailed consideration must await justification of the physicochemical underpinnings of such a mechanism. Any alternative simultaneous model would require a faster rate constant than suggested by the distance. This could perhaps be achieved in a double-occupancy version of this type of mechanism, in which electron transfer was facilitated by a quinone species in the proximal domain. However, it would seem difficult in this context to account for the pattern of dependence on driving force discussed above, which had seemed to demonstrate control through the first electron transfer.

Some discrimination between single- and double-occupancy models might be made on the basis of the role of Glu-272 in liganding of QH$_2$ and in proton transfer out of the site. If Glu-272 is a ligand in the ES-complex, the volume occupied by the side chain and associated waters would largely fill the volume into which any second quinone would be likely to bind (Figure 2). Because the QH$_2$ in the distal domain has to be the substrate, there is no room for the strongly bound quinone in this proximal position without substantial distortion. As there is no evidence for any alternative binding domain close enough to heme b_L, this would argue against double-occupancy models being of mechanistic importance. If the glutamate does not serve this function, and a high degree of elasticity of structure without an energetic penalty is demonstrable, a wider set of possibilities remains open.

Discrimination between the first two models is more difficult. The question could clearly be resolved in favor of model 1 if the SQ.ISPH intermediate could be demonstrated by another spectroscopic approach than EPR. With a limiting first electron transfer and low occupancy of the SQ.ISPH state, it would be difficult to discriminate on the basis of the kinetics of reduction of heme b_H. However, the lag phase in the kinetics allows us to put limits on the population of intermediate states. In chromatophores poised with the quinone pool partly reduced, reduction of heme

b_H occurs after a lag of 120 μs. Of this, $\sim$100 μs is accounted for by identified electron transfer events (54). The unaccounted $\sim$20 μs would have to include the times for electron transfer through ISP to the Q_o-site (including movement), and the electron transfer through heme b_L. If the first electron transfer is rate determining, any accumulation of intermediate states (including centers in the SQ.ISPH state) would occur at the rate limit of $\sim$600 μs, at the expense of electrons delivered through heme b_L to b_H. Since no delay of this magnitude is seen, occupancy of any such state must be quite small. When heme b_H is reduced, the reduction of heme b_L follows similar kinetics (80, 93). Since the overall equilibrium constant under these latter conditions is close to 1, any significant stability of the intermediate state would represent an energy trough, but no kinetics that could be attributed to filling of such a trough are detected.

Hong et al. (93) discussed at length the interplay between SQ occupancy and rate for the second electron transfer. A value for $K_{eq} < 0.01$ for the first electron transfer, which seems a reasonable limit based on the lag, would be on the border-line of possible values compatible with electron transfer from the distal domain. A detailed kinetic model would allow exploration of the constraints on occupancy of intermediate states, but none is yet available. More detailed kinetic data, alternative spectroscopies for assaying ISPH [for example, by CD (96, 104)], SQ (by UV/VIS), and the Glu-272 changes (using FTIR), and better-defined models for cases 3 and 4, are needed before a definitive choice can be made between any of the above mechanisms. A computer model to allow exploration of the parameters determining Marcus theory constraints on the rates for the two electron transfers of the bifurcated reaction is available (http://www.life.uiuc.edu/crofts/Marcus_Bronsted/).

ELECTRON TRANSFER FROM CYT C_1 TO CYT C Berry's group had an early report of a bovine bc_1 complex dimer cocrystallized with cyt c, showing a weak occupancy of the latter, and an unresolved interaction domain (2; E.A. Berry, personal communication). Among recent structures, Lange et al. (88) have reported a complex between the yeast bc_1 complex and yeast cyt c at higher resolution that shows the interaction between cyt c_1 and its electron acceptor. Interestingly, only one of the two monomers had a cyt c attached, and asymmetry was also apparent elsewhere in the dimer, especially at the Q_i-site, where the occupancy by quinone was quite different in the two monomers. The authors speculated that these asymmetries were linked and might be of importance in allosteric control. From a mechanistic perspective, the structure shows the interface through which electron transfer between cyt c_1 and cyt c occurs, and Hunte et al. (147) presented a detailed analysis of possible electron transfer mechanisms in the light of the structure. The kinetics of electron transfer, the role of particular residues at the interaction interface, and the kinetic changes on mutation at the interface have been explored in detail using photoactivation of a yeast cytochrome c derivative labeled with ruthenium trisbipyridine at Lys-72 (148). The authors targeted acidic residues of the interface on cyt c_1 for mutation and concluded that acidic residues on opposite sides of the heme crevice of cyt c_1 are involved in binding with the positively charged cyt c, and that they direct the diffusion and binding of cyt c from the aqueous phase between the outer and inner mitochondrial membranes.

MECHANISTIC INFORMATION OF MEDICAL OR PHYSIOLOGICAL INTEREST   Detailed analysis of the structural basis of inhibitor function, and the mechanistic information available from mapping of mutations that affect binding to catalytic interfaces, are beyond the scope of this review, but have been discussed extensively elsewhere (1, 2, 8, 9, 41, 44–47, 65–67). Some more recent observations deserve further comment. New structures showing the binding configuration of HHDBT (45), NQNO (47), and famoxadone (44) have been mentioned above. The HHDBT structure has been used to model a likely configuration for the drug atovaquone (46), which is clinically important as an antimalarial agent, based on susceptibility to inhibition at the Q_o-site in *Plasmodium* species [also *Pneumocystis carinii* (pneumonia), *Toxoplasma gondii* (toxoplasmosis), and other fungal parasites] and resistance in vertebrates. The molecular basis for this susceptibility is revealed by mutant strains of *P. falciparum* and other *Plasmodium* species that develop resistance to atovaquone (33; reviewed in 9). Kessl et al. (46) modeled one of these mutations in yeast (L275F) and demonstrated substantial resistance, which was explained in terms of the increased volume of the Phe compared with Leu, which impinged on the binding volume. Figure 3 (see color insert) shows the position in the structure of residues at which resistant mutations are found in relation to the stigmatellin-binding volume.

BYPASS REACTIONS AT THE Q_o-SITE IN THE PRESENCE AND ABSENCE OF OXYGEN
As noted above, the well-studied reaction of the Q_o-site with O_2 to produce superoxide (SO) is generally attributed to one-electron reduction by the intermediate semiquinone species generated at the site when transfer of the second electron to the low potential chain is blocked (for recent reviews, see 14, 15). SO production occurs when electrons back up in the low-potential chain so that electron transfer from the intermediate SQ to heme b_L becomes limited. Under physiological conditions, this happens on development of a large back-pressure from Δp and in the lab when antimycin inhibits exit of electrons from the b-heme chain. The maximal rate of SO generation is $\sim 2\%$ of the normal turnover.

Several recent reports provide new insights as to the nature of the reaction. In mitochondrial complexes, myxothiazol inhibits the SO production in the presence of antimycin, but only by about 70%. In the absence of antimycin, production of SO is stimulated by addition of myxothiazol to give a similar rate ($\sim 0.6\%$ full turnover) (140, 141). Several groups have argued convincingly that the semiquinone at the Q_o-site is the reductant for O_2 (13–18). Because formation of semiquinone depends on oxidation of quinol, this could occur only if an ES-complex of some sort is formed in the presence of myxothiazol, with occupation of the distal domain of the site by ubiquinol and reaction with ISP_{ox} as the most likely configuration. Since most data seemed to indicate that myxothiazol eliminates all quinone species from the site (67, 100, 142), the possibility of such a double occupancy was surprising.

Muller et al. (141) found that the rate of the bypass reaction was not dependent on the presence of O_2. When the rate of quinol oxidation was assayed by reduction of cyt c, the bypass rates observed under anaerobic conditions in the presence of either antimycin or myxothiazol were the same as under aerobic conditions. This could be explained if generation of semiquinone under all conditions was rate-limiting, and independent of $[O_2]$, and if two pathways competed for the semiquinone, both with much higher rate constant, one to O_2 and the other indirectly to cyt c. Under aerobic conditions, O_2 out-competes the alternative pathway as acceptor for the second electron. A plausible pathway for the bypass under anaerobic conditions would be via the returning ISP_{ox}, which could oxidize the semiquinone after it had donated the first electron to cyt c via cyt c_1. Because the distance from the [2Fe-2S] cluster in the ISP_B position to cyt c_1 would preclude direct transfer from the complexed state (39), the reaction could occur only if the intermediate SQ.ISPH state dissociated to products. The bypass reaction therefore demonstrates that the SQ.ISPH state is not so tightly bonded as to preclude dissociation. A third observation of interest was that increasing the $[O_2]$ fivefold above ambient had no effect on the rate of SO generation in the presence of myxothiazol, but stimulated the rate in the presence of antimycin by $\sim 25\%$ (141). The saturation at ambient O_2 in the presence of myxothiazol indicates a high affinity compared with that in the absence of myxothiazol and shows that at least a fraction of the semiquinone was in a less reactive configuration when the Q_o-site was free of any proximal domain inhibitor (117).

The relation between occupancy of the distal domain and the size of the proximal domain inhibitor has been further explored by Muller et al. (143). They studied the K_m and V_{max} for ubiquinol substrates and showed that, whereas the V_{max} was independent of the nature of the inhibitor, the K_m was markedly dependent on the bulk of the inhibitor. The K_m varied from 36 μM for myxothiazol, the largest inhibitor, down to ~ 3 μM with mucidin, the smallest inhibitor. This latter value was the same as that observed in the absence of any Q_o-site inhibitor, either when antimycin alone was present to stimulate SO production or when turnover was measured in the absence of inhibitors, indicating that formation of the ES-complex was determined by the same kinetic parameters. This correlation between K_m and inhibitor size was extended by computational studies in which the volume overlap between the inhibitors and a UQ-1 test substrate was modeled using crystallographic structures. With a 3 Å adjustment of UQ-1 away from the position indicated by stigmatellin, which involved rotation of the Glu-272 ligand, it was possible to fit the inhibitor into the proximal domain with either no overlap (for mucidin) or partial overlap (increasingly for MOA-stilbene, famoxadone, and myxothiazol) with UQ-1, in line with the increasing K_m values. These structural considerations therefore support the hypothesis that the ES-complex can be formed in the presence of proximal domain inhibitors, since the overlaps in volume correlated well with the K_m values measured. However, the inhibitor volumes used were based on those in the structures containing inhibitors, which show substantial conformational changes compared with the native or stigmatellin-containing structures. Possible constraints from interaction with the protein were omitted from the modeling and were therefore not reflected in the overlap values.

The SO production that is insensitive to proximal domain inhibitors must indicate some double occupancy, but the question remains of how pertinent the data are to mechanism. It is by no means obvious that the occupancy of the ES-complex with the smaller inhibitors can be used to justify a double-occupancy mechanism under normal turnover. The 12-fold change in K_m for QH_2 with myxothiazol indicates a strong impediment to occupancy of the distal domain when the proximal domain is occupied by a species with a tail; this clear evidence for a constraint on double occupancy is more pertinent to occupancy by two quinone species.

These studies demonstrated that this physiologically important reaction is experimentally accessible and that our understanding of the mechanism is intimately linked to our understanding of the mechanism of QH_2 oxidation at the Q_o-site.

MITOCHONDRIAL MYOPATHIES MAPPING TO THE BC$_1$ COMPLEX   Mitochondrial myopathies have been extensively reviewed recently (8–11), and the relation between mitochondrial dysfunction and disease is well established. For those myopathies linked to mutation of the cyt b gene, many have been mapped through coding changes at the DNA level. For those leading to residue change rather than to gross truncation or deletion of sequence, the location in structural models provides a basis for detailed understanding of these diseases at the molecular level. (For a detailed discussion of the current field and of the interesting mechanistic and physiological insights provided by these mutations, see 8, 9).

Quinone Reduction at the Q_i-Site

STRUCTURAL CONSIDERATIONS   Several structures are available from different labs showing the native occupant, presumably ubiquinone (2, 39, 43, 47, 89, 149); others with the inhibitor antimycin as occupant (39, 47); and a recent structure shows NQNO at the site (47). This latter paper includes the fullest structural description of the Q_i-site yet available, either vacant (but with 4 water molecules forming an H-bonded network) (1ntz), or with quinone (1ntm), antimycin (1ntk), or NQNO (nu1) bound. Unfortunately, coordinates were not available at the time of writing. In general, the protein structure that Gao et al. (47) described was relatively indifferent to occupancy, with the vacant structure showing only 0.24 Å rmsd from the quinone-occupied structure. The inhibitor binding in all structures reported (2, 39, 47, 89, 149–151) showed volumes of occupancy that impinge on the surfaces at which mutational changes (152) lead to resistance.

All structures show that the quinone is liganded directly or indirectly through H-bonds involving three side chains—His-202, Asp-229, and/or Ser-206 (chicken or yeast numbering)—and/or by structurally defined waters (39, 43, 47, 89, 153, 154). In addition, some hydrophobic or aromatic interactions with side chains and the heme edge are recognized. Interestingly, four high-resolution structures show three different configurations of side chains involved in H-bonding of the

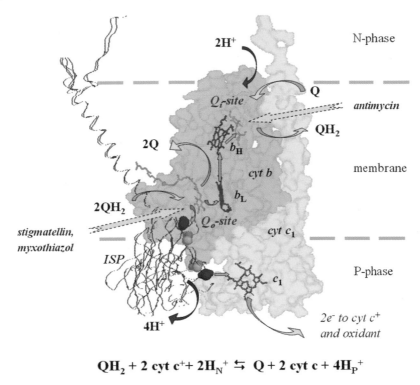

$$QH_2 + 2\ \text{cyt}\ c^+ + 2H_N^+ \leftrightarrows Q + 2\ \text{cyt}\ c + 4H_P^+$$

Figure 1   The Q-cycle mechanism shown in the context of a functional monomer. The three catalytic subunits of the bc_1 complex (PDB 2bcc) from chicken mitochondria are shown with a quinol modeled in the Q_o-site to replace stigmatellin, and quinone at the Q_i-site. The cyt b subunit is shown by its transparent cyan surface, and the cyt c_1 subunit by a transparent yellow surface. The Rieske iron sulfur protein is shown as a ribbon cartoon, docked at the b-interface (*blue*) or at the c-interface (*red*). The approximate position of the membrane is shown by the dashed lines. The b-hemes are shown in blue, heme c_1 in green. The [2Fe-2S] clusters are shown as space-filling models, colored as for the protein. Two additional positions for the cluster are shown (in *CPK colors*) to indicate the trajectory of movement between b- and c-interfaces. Binding and unbinding of quinone species, and docking of cyt c, are shown by broad curved arrows (*light-blue*). Electron transfers are shown by small green arrows, H^+ release and uptake are shown by curved blue arrows. The sites of action of inhibitors are shown by dotted light-blue arrows. For each turnover, two QH_2 molecules are oxidized to Q at the Q_o-site, and two successive electrons are passed down the pathways indicated by the green arrows. The two electrons going down the high-potential chain (ISP, cyt c_1, cyt c) are passed to a terminal oxidant. The two electrons going through the b-heme chain reduce Q to QH_2 at the Q_i-site through a two-electron gate, with storage of one electron as a SQ at the site.

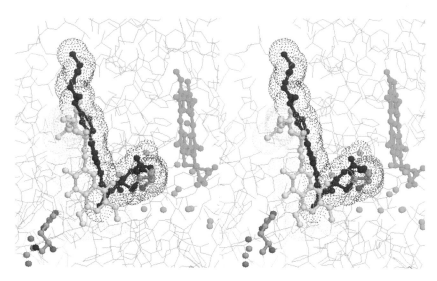

Figure 2 Stigmatellin and myxothiazol at the Q_o-site. The stigmatellin-containing structure of the bovine mitochondrial complex (PDB 1pp9) shows the positions of His-161 of ISP (*ball and stick model, bottom left*) and Glu-271 of cyt *b* (*center*, both in CPK colors) acting as ligands to the stigmatellin (yellow ball & stick model). The rest of the protein is represented by a wire-frame model, with cyt *b* in cyan and ISP in yellow. Two waters that H-bond to Glu-272 and waters contributing to the chain leading to the external P-phase are shown as small cyan spheres. Heme b_L is shown as a green ball and stick model. Superimposed on this structure is myxothiazol (*red ball and stick model*) to show the position relative to heme b_L in a separate structure. A dotted surface at the van de Waals radii is shown for stigmatellin (*yellow dots*), myxothiazol (*red dots*), and Glu-271 together with its two H-bonded waters (*blue dots*). The almost complete overlap of Glu-271 and its waters with the pharmacophore of myxothiazol (the horizontal arm of the "L" of myxothiazol) can best be appreciated by stereo viewing (stereo model is set up for crossed-eye viewing).

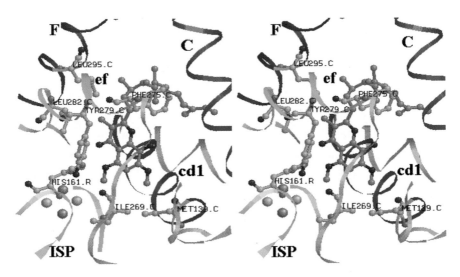

Figure 3 Residues modified in *P. falciparum* strains that are resistant to atovaquone. The residues are shown as ball and stick models in CPK coloring in the context of the stigmatellin structure, to show the binding of a similar occupant. The spans of cyt *b* contributing to the Q_o-site surface are shown as ribbons, with spans forming the Q_o-site binding domain identified as ribbon models of different colors. His-161 of ISP, the [2Fe-2S] cluster, and parts of the ISP protein (*green-blue ribbon, lower left*) are also shown. Stereo pair for crossed-eye viewing.

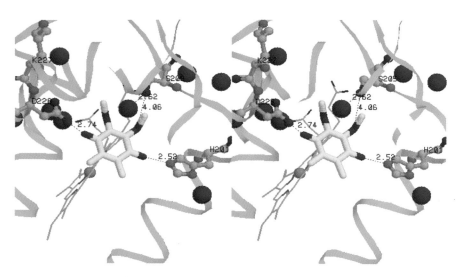

Figure 4   The liganding of quinone at the Q_i-site. The structure from Berry's bovine mitochondrial complex at 2.1 Å resolution (PDB 1pp9, 158) is shown, with potential H-bonding ligands to the quinone, and water molecules within 10.0 Å. The quinone is shown as a stick model, colored in CPK, but with C-atoms in yellow. Side chains involved as ligands are shown as ball and stick models in CPK colors. Distances shown are in Å. The rest of the protein close to the site is represented by a blue ribbon. Heme b_H is shown as a skinny stick model in CPK colors, with C-atoms in green. Water O-atoms are shown by larger red spheres. Stereo image for crossed-eyed viewing.

quinone carbonyls. Most notably, His-202 is directly H-bonded to the quinone in the bovine and chicken complexes from Berry (2, 39, 147, 154) (Figure 4, see color insert), but indirectly via bound water in the yeast structure described by Hunte et al. (43, 89) and the bovine complexes described by Gao et al. (47), whereas the H-bond to Asp-229 is direct in the Berry and Hunte structures, but via a water bridge in the Gao structure. An additional unpublished structure at 2.5 Å resolution from Iwata (153) is described as showing direct H-bonds with both the aspartate and histidine, as in the Berry structures. Differences are also apparent in the Ser-206 liganding, with the side chain within H-bonding distance in some structures, connected to either of the $-OCH_3$ O-atoms, and too far away in others. When considering these differences, note that the occupancy by quinone in all the structures is relatively low and that the B factors are relatively high, indicating a degree of uncertainty in assignment of the electron densities. The site is also quite spacious, allowing occupancy by a variety of inhibitors with overlapping volumes, and may allow alternative configurations for the quinone. The differences show an obvious structural plasticity, but beg the question of mechanistic relevance.

An alternative approach to mechanistically important structural features is through use of pulsed EPR or ENDOR to explore the structure around the semi-quinone bound at the Q_i-site through the magnetic interaction between neighboring nuclear magnets and the paramagnetic center (154–156). Exchangeable protons contributing to H-bonds with the SQ in the bovine mitochondrial complex were identified using ENDOR and H/D exchange (157). The spectra were interpreted as showing several H-bonds, but the spectral range limited the study to the proton region. Pulsed EPR has been used to study the bc_1 complex from *Rb. sphaeroides* (158), allowing exploration of neighboring ^{14}N nuclei. ESEEM and HYSCORE spectra provided an unambiguous demonstration of a N-ligand, interpreted as an H-bond to a histidine imidazole nitrogen, likely His-217, equivalent to His-202 of mitochondria complex. This supports the Berry structures as showing a functional configuration and opens the possibility of a dynamic role of the histidine. Spectra measured using ESEEM and HYSCORE, with H_2O or ^2H_2O as the medium solvent, resolved at least two (probably three) exchangeable protons affecting the SQ spin relaxation, likely contributing to H-bonds (D.R.J. Kolling, J.T. Holland, R.I. Samoilova, S.A. Dikanov, & A.R. Crofts, unpublished). The two strongest of these likely reflect the direct H-bonds to the ring carbonyl O-atoms observed in ^{14}N-ESEEM and the crystallographic structures, although firmer attribution to particular bonds must await further work using isotopic substitutions at other atoms.

OVERVIEW OF MECHANISM OF ELECTRON TRANSFER AT THE Q_i-SITE   Any mechanism must account for the fact that the Q_i-site operates as an interface between the one-electron chemistry of the b-heme chain, and the two-electron chemistry of the quinone pool. In addition, pathways for uptake of two protons and for exchange of substrate and product need to be identified. It is generally supposed that the site operates through a two-electron gate in which heme b_H reduces Q to SQ on one turnover of the Q_o-site, and SQ to QH_2 on a second turnover (50, 57). The structures provide important clues to these functions. In particular, they show the groups involved in binding, and accessibility to the site from aqueous and lipid phase.

THE ROLE OF THE SEMIQUINONE AT THE Q_i-SITE   Under normal turnover, the initial state of occupancy will depend on ambient redox potential and pH. A stable semiquinone is generated at the site on redox titration over the E_h range 30–60 mV higher than the E_m of the quinone pool ($E_{m, pool} \sim 90$ mV at pH 7), and is stabilized as the pH is raised over the range 7–9 (60, 154, 155). The SQ intermediate can also be generated by coulometric titration using a one-electron mediator (156), or by reversal of the second electron transfer (61, 145, 158). However, details of the equilibrium and rate constants, the interaction between semiquinone and heme b_H, and the specific role of ligands in this reverse reaction are unresolved (2, 125, 159–161). In each of these approaches, the yield of SQ, whether measured kinetically or thermodynamically, has always been substantially less (~ 0.5/heme b_H under conditions giving maximal yield) than the stoichiometry of bc_1 complex. Early work from Palmer's lab had suggested an explanation—this might reflect spin silencing from magnetic interaction with oxidized heme b_H (156). In this case, two populations of SQ might be expected, one with a fast, and the other with a slow relaxation. The thermodynamic and spectroscopic properties reported in the literature are those of a slowly relaxing species, easily saturated at temperatures <100 K.

The thermodynamic properties of the semiquinone have been explained in the context of the disproportionation reaction, and values for binding constants for the three species of the $Q/SQ/QH_2$ system, E_m values for the semiquinone couples for bound and unbound states, and pK values for the protolytic reactions have been proposed (60, 154, 155). In this discussion, it has been assumed that the disproportionation reactions of the bound and unbound couples reach equilibrium, but since structures became available, there has been little discussion of how this might occur at a site that appears too small to contain more than one species. The Q_i-site catalyzes a quinol-quinone transhydrogenase reaction that allows transfer from exogenous quinol to bound quinone or between exogenous quinones (162, 163). A plausible mechanism would be one in which two electrons from a donor Q_dH_2 are stored in the two b-hemes, allowing dissociation of the Q_d and its replacement by an acceptor Q_a, which would then leave the site as Q_aH_2 (162, 163). The rate of such a reaction would depend strongly on the probability of populating the $b_H^- b_L^-$ state. The unfavorable equilibrium constant for reduction of heme b_L through b_H (123, 126, 164) would account for the relatively slow rate observed, but this would likely be sufficient to allow equilibration of the Q and QH_2 components of the disproportionation reaction on the time scale of a redox titration. If equilibration through one-electron transfer processes occurs, it is presumably through direct or indirect interaction with mediators. The conditions used to assay semiquinone by EPR (high concentrations of redox mediators and a long equilibration time) are designed to encourage the equilibria of the disproportionation reaction, but they may not accurately reflect the kinetically important intermediate states. Furthermore, if the semiquinone signal at the Q_i-site is modulated by magnetic interaction with the spin of oxidized cytochrome b_H (156, 161), this might lead not only to loss of amplitude, but also to some displacement of the maximum of the bell-shaped titration curve. Since the latter data are important in calculation of thermodynamic parameters, a substantial miscounting of the SQ could lead to misleading values over portions of the pH/E_h titration surface.

In discussing the mechanistic role of the semiquinone, it is necessary to take account of the equilibration of the Q/QH$_2$ couple with heme b_H through one-electron transfer reactions. Reduction of heme b_H through the Q_i-site by exogenous quinones, with generation of a semiquinone, occurs on addition of quinol substrate to the oxidized complex in the presence of myxothiazol. In a definitive study by de Vries et al. (61), reduction by duroquinol was rapid, with the reaction $\sim$90% complete at the first point of measurement at 5 ms (using 300 μM QH$_2$). From this, a value of $\tau < 3$ ms seems appropriate. The measured rate was more rapid than the reaction through the Q_o-site monitored through the reduction of heme b_H in the presence of antimycin. The rate was also more rapid than that reported for the transhydrogenase reaction ($\tau \sim 170$ ms at 50 μM duroquinol, 163). From this, it seems likely that duroquinol, and presumably other quinols, could react directly at the Q_i-site by exchange with the initial occupant. In chromatophores, where the reactions using the native substrate can be followed *in situ*, the myxothiazol-insensitive reduction of heme b_H on generation of 1 QH$_2$ in the oxidized pool occurred with $\tau \sim 3$ ms (165, 166). This rate was similar to that for QH$_2$ oxidation through the Q_o-site under these conditions, suggesting a similar rate constant, but in both cases, the rate was limited by the delivery of QH$_2$ to the complex from the RC (50, 165).

For all these kinetically determined reactions, it is clear that the relevant equilibrium constants must include participation of heme b_H in the one-electron transfer reactions. One approach to determining these is through the phenomena associated with the cyt b_{150} component. This high-potential form of heme b_H has been recognized phenomenologically for over 30 years, having been observed in redox titrations of mitochondria and chromatophores in the early 1970s (164, 167). A hint as to the mechanism came from the observation that antimycin addition induced oxidation of this component (168) or eliminated it from redox titrations (125, 154, 155). In the presence of antimycin, the absorbance due to the cyt b_{150} form was lost and reappeared in the heme b_H titrating at $E_{m,7} \sim 50$ mV. The antimycin-induced oxidation of cyt b was explained through the following mechanism (168):

$$\text{cyt}b_H^- \cdot QH^{\bullet}(H^+) \overset{\overset{\displaystyle QH_2}{\displaystyle \curvearrowright}}{\longleftrightarrow} \text{cyt}b_H \cdot QH_2 \longleftrightarrow \text{cyt}b_H \cdot \text{vacant} \overset{\overset{\displaystyle AA}{\displaystyle \curvearrowright}}{\longleftrightarrow} \text{cyt}b_H \cdot AA. \qquad 6.$$

By analogy with the well-characterized inhibitor-induced electron transfers in the $Q_A Q_B$ sites of reaction centers and photosystem II (57), we suggested that the semiquinone at the Q_i-site accepts an electron from ferrocytochrome b_H and leaves as the quinol, so that the oxidized form of cyt b is stabilized by antimycin binding. Conversely, the cyt b_{150} form must arise from a reversal of the left-hand pair of reactions, as seen experimentally from the myxothiazol-insensitive reduction of heme b_H and generation of semiquinone on addition of QH_2 to the isolated complex (61, 145, 158), or the reduction of b_H on photoreduction of the pool in chromatophores (165, 166). For the reduction of heme b_H on flash activation in chromatophores, the observed equilibrium constant precluded mechanisms involving reduction of two heme b_H centers by the Q/QH_2 couple at the E_m values found by titration, and the alternative reaction, in which QH_2 reduced heme b_H with generation of SQ, was suggested (165).

The properties of cyt b_{150} in the mitochondrial complex have been explored in detail by Rich et al. (161), and discussed in terms of differential affects of occupancy by Q, SQ, or QH_2 on the E_m of heme b_H (see 160). A set of parameters that accounted successfully for the properties of both the SQ and cyt b_{150} was proposed; however, in deriving this set, Rich et al. (161) had to assume that SQ generated in the presence of oxidized heme b_H was EPR silent owing to magnetic quenching (see 156). An alternative approach to explaining the properties of cyt b_{150} has been through the equilibria associated with reversal of the forward reaction (165), as discussed below.

THE ROLE OF CYT B$_{150}$ IN THE MECHANISM OF THE Q$_i$-SITE   The properties of cyt b_{150} suggest that the Q$_i$-site catalyzes the following reaction, in which the cyt b_{150} is represented by the state cyt b_H^- QH$^•$ (125, 159, 165, 168). The reaction (Equation 7) can be considered as a reversal of the normal forward reaction by which ferroheme b_H reduces SQ to QH_2, and after release of the product, binds Q to re-establish the starting state for the Q$_i$-site.

$$QH_2(\text{pool}) + \text{cyt}b_HQ \leftrightarrow Q(\text{pool}) + \text{cyt}b_H^-QH^•(H^+), \qquad 7.$$

Partial reactions giving rise to the overall process of Equation 7 are

$$QH_2(\text{pool}) + \text{cyt}b_HQ \leftrightarrow Q(\text{pool}) + \text{cyt}b_HQH_2, \qquad 7a.$$

$$\text{cyt}b_HQH_2 \leftrightarrow \text{cyt}b_H^-QH^•(H^+), \qquad 7b.$$

At any defined pH, the equilibrium constant for formation of the cyt b_{150} form, obtained from the partial processes of Equations 7a and 7b, is given by

$$K_{eq} = K_{7a}K_{7b} = (K_{A(QH_2)}/K_{A(Q)})\exp\{(E_{m(\text{cyt}b_H)} - E_{m(QH^•/QH_2)}F/RT)\}, \qquad 8.$$

where K_A values are association (binding) constants for QH_2 and Q, and $E_{m(\text{cyt}bH)}$ and $E_{m(QH^•/QH_2)}$ refer to the mid-point potentials of cyt b_H, and of the SQ/quinol couple bound at the site, respectively.

Examination of Equation 8 shows that an increased value of K_{eq} could be expected if $E_{m(\text{cyt}bH)}$ is increased, if $E_{m(QH^•/QH_2)}$ is decreased, if the affinity of quinol is increased, or if the affinity of quinone is decreased. However, the affinities for Q and QH_2 have an importance that can be expressed in a different form. When expressed with respect to the potential of the quinone pool, the ratio of binding constants for Q and QH_2 contributes to the value for $E_{m(QH^{\bullet}/QH_2)}$ in such a way as to cancel the pre-exponential term. This gives

$$K_{eq} = \exp\{E_{m(cytb_H)} + E_{m(Q/QH^{\bullet})} - 2E_{m(Q_{pool})}\}F/RT, \qquad 9.$$

where $E_{m(Q/QH^{\bullet})}$ refers to the bound couple. This same expression can also be derived directly from Equation 6 using the conventional relationship between equilibrium constant and half-cell potentials. Incorporation of this equation in a computer model allows exploration of the variation of redox states of all components as a function of redox potential, pH, etc. (125). Interestingly, the properties of the cyt b_{150} phenomena are reproduced quite well using measured values for E_m and pK for heme b_H, reasonable values for binding constants (with a ratio of ~100 in favor of QH_2), and E_m values for the Q/SQ/QH_2 system derived from redox titrations and kinetic experiments (125). The properties of the SQ are also reproduced quite well, but are mechanistically linked to formation of SQ through reaction 6 above, so that the SQ observed is in the form cyt $b_H^- \cdot QH^{\bullet}$. An interesting feature of these simulations was that, because of the importance of E_m of heme b_H in Equation 9, the pH dependence of both b_{150} and the SQ were well explained, but are determined by the pK at ~7.8 on the heme b_H, without invoking a pK in the neutral range for the semiquinone (M. Guergova-Kuras & A.R. Crofts, unpublished).

Support for the above mechanisms also comes from analysis of strains with mutations in the Q_i-site. We have previously noted that in two mutants in which the E_m of cytochrome b_H was lowered, but by residue changes removed from the catalytic domain, the amount of cytochrome b_{150} was substantially decreased (124, 125, 169). This trend was apparent in mutants N223V, G220V, S218A, and N222A (all with lowered E_m for cyt b_H), which showed a reduced level of cyt$_{150}$ (169). Similarly, in T219A and in a double mutant K251D/D252K (which had higher E_m values) the level of cyt$_{150}$ was enhanced (125).

Although this simple model explains much of the data, in other mutants, including some showing higher levels of cytochrome b_{150} but a lower E_m for cyt b_H than wild type, the change in relative stability could not be attributed simply to changes in E_m of cytochrome b_H and would have to be explained in terms of changes in the other parameters of equation (125, 169). A lack of correlation between the amount of cytochrome b_{150} and the stability of the EPR-detectable semiquinone at the Q_i-site has also been noted in a series of mutants in the bc_1 complex of *Rb. capsulatus* (166). These departures from the simple model presented here are not unexpected, but suggest that the role of semiquinone stability needs to be explored more fully. Note that the role of protons in equilibrium with the partial reactions has not been explored. Changes in kinetics and in the equilibrium constants combined in K_{eq} in Equation (8) could reflect these protolytic equilibria, and the kinetics of the proton processing accompanying redox turnover. The relationship between cytochrome b_{150}, the semiquinone, and the protolytic equilibria clearly require further investigation.

THE FORWARD REACTION   As noted above, the operation of the two-electron gate at the Q_i-site is associated with a relatively stable SQ intermediate, and the mechanism is described by a set of equilibrium and rate constants defining the electron transfer reactions and the binding of Q and QH_2 from the pool. These also determine the properties of cyt b_{150}. In pre-steady-state kinetic experiments with the *Rb. sphaeroides* complex in situ, electron transfer rates in the forward direction through the *b*-heme chain to the Q_i-site are faster than the rate-limiting step. Since complete turnover delivers two electrons in succession to the Q_i-site (50, 54), this holds for electron transfer to both endogenous Q and SQ species. The fact that the observed rate is that of the limiting step at the Q_o-site restricts use of the kinetic data to estimation of upper limits for the forward rate constants. With the quinone pool initially oxidized (and the site presumably occupied by Q), cyt b_H remains partly reduced in the 100-ms range after generation of one QH_2 in the pool by flash excitation of the RC, suggesting an equilibrium constant for reduction of Q by heme b_H^- with $K_{eq} \sim 3$, in the range expected from the E_m of heme b_H, and the value for E_m Q/QH$^{\bullet}$ from simulation. This relatively stable level of heme b_H reduction is seen when the pool is initially oxidized, but titrates away as the pool starts to become reduced over the range below $E_{h,7}$ 150 mV, so that at 100 mV, reoxidation of heme b_H is complete. Over this same range, cyt b_{150} (likely the state cyt $b_H^- \cdot$ QH$^{\bullet}$, as discussed above) is formed in the pre-flash equilibrium mix, as determined by the equilibria shown in Equation 10. This brings up the question of the state of the Q_i-site under these conditions.

$$\text{cyt}b_L\text{cyt}b_H\text{Q} \overset{\text{Q to pool}}{\leftrightarrow} \text{cyt}b_L\text{cyt}b_H[\text{vac}] \underset{QH_2 \text{ from pool.}}{\leftrightarrow} \text{cyt}b_L\text{cyt}b_H\text{QH}_2 \leftrightarrow \text{cyt}b_L\text{cyt}b_H^-\text{QH}^{\bullet}. \quad 10.$$

At potentials around the midpoint of the Q-pool a substantial fraction of Q_i-sites (depending on pH) will be in the cyt b_{150} state, suggesting a forward reaction according to Equations 11 and 12:

$$\text{cyt}b_L\text{cyt}b_H^-\text{QH}^{\bullet} \overset{e^- \text{ from first } QH_2 \text{ at } Q_o\text{-site}}{\leftrightarrow} \text{cyt}b_L^-\text{cyt}b_H^-\text{QH}^{\bullet} \leftrightarrow \text{cyt}b_L\text{cyt}b_H^-\text{QH}_2 \underset{QH_2 \text{ to pool.}}{\leftrightarrow} \text{cyt}b_L\text{cyt}b_H^-[\text{vac}].$$

$$11.$$

$$\text{cyt}b_L\text{cyt}b_H^-[\text{vac}] \underset{Q \text{ from pool.}}{\leftrightarrow} \text{cyt}b_L\text{cyt}b_H^-\text{Q} \leftrightarrow \text{cyt}b_L\text{cyt}b_H\text{QH}^{\bullet} \overset{e^- \text{ from second } QH_2 \text{ at } Q_o\text{-site}}{\leftrightarrow} \text{cyt}b_L\text{cyt}b_H^-\text{QH}^{\bullet}. \quad 12.$$

If the rate constants for formation of the cyt b_{150} state (Equation 10) are rapid compared with the rate-limiting step, as seems quite possible from the data, then it is likely that most centers follow the pathway shown by Equations 11 and 12 under conditions in which QH_2 is initially available in the pool. In contrast to conventional schemes, the acceptor of the first electron at the Q_i-site would be the SQ, and the second electron would reduce Q to SQ to regenerate the starting state [see Slater (145) for an earlier discussion]. Since the bc_1 complex shows a maximal rate with the pool half reduced, and since under the coupled steady-state the pool is also partly reduced, this would represent the more physiologically important pathway for turnover.

The sequence of reactions 9 to 11, starting with the oxidized complex, also provides a basis for understanding the multiphasic reduction of heme b_H seen on addition of exogenous quinol reductants to the oxidized complex (61, 81, 95, 170). The initial rapid phase, which is antimycin sensitive and myxothiazol insensitive (61), would correspond to reaction 9. Then, from the perspective of heme b_H, the reaction of Equation 11 would represent a lag, and the reactions of Equation 11 would give an oxidative phase followed by a reductive phase with kinetics similar to the slower phase of reduction of cyt c_1, as observed (61).

LIGANDS OF MECHANISTIC IMPORTANCE   Consideration of two higher resolution crystallographic structures available at the time, and their EPR data, led Kolling et al. (158) to interpret the structural differences in the Q_i-site as reflecting changes in ligation for the different quinone species during the catalytic cycle, and a dynamic role for His-202 in transfer of protons and water into the site. This led to a scheme for the sequence of reactions involved in the forward electron transfer as shown in Figure 5. An alternative model has been suggested by Gao et al. (47), in which all bound quinone species are H-bonded through bridging waters (as shown by the quinone in their structure) during the catalytic cycle. This model is incompatible with the EPR data showing a direct H-bond between SQ and a histidine N. However, the structures proposed by Gao et al. (47) fit quite naturally into the scheme of Figure 5, providing a water-filled vacant site (F), and an intermediate between the vacant state (G) and quinone-bound states seen in the structures of Hunte (43, 89) (A) and Berry (2, 39, 147, 154) (B), prior to formation of the semiquinone in the environment seen by ESEEM (154) (C).

ACCESS TO THE QUINONE BINDING SITE OF SUBSTRATE, WATER, AND PROTONS   Structures for the vacant site show waters in the volume of the quinone, and all high-resolution structures have water in addition to the quinone within the site. For example, in addition to waters involved in H-bonds to the quinone, Ser-206 in the structures of both Hunte (43, 89) and Berry (154) forms an H-bond to a water that also forms an H-bond to a heme propionate, and both structures show waters that seem to be "caught in the act" of communication between the internal volume and the external phase through a small hole bordered by His-202.

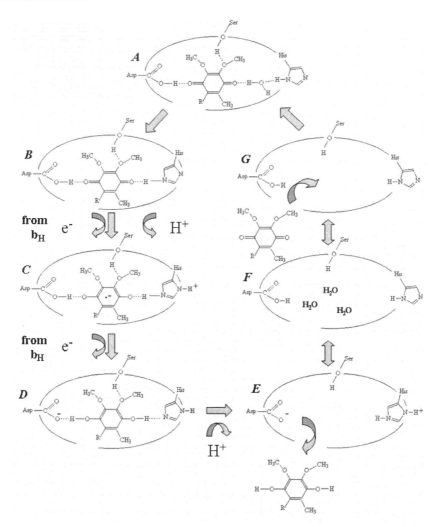

Figure 5   Scheme showing proposed involvement of liganding groups in the function of the Q_i-site (adapted from 158). The different occupants of the site are shown for different states of the cycle of reactions at the Q_i-site, shown in the direction of forward electron flow. State **A** might represent Hunte's structure (43), state **B** Berry's structure (see Figure 4), and state **C** the structure determined by Kolling et al. (158) using ESEEM. The vacant site with waters would correspond to the vacant site in Gao et al. (47), and their quinone-occupied site would be intermediate between **G** and **A**. The arrows show the direction of forward electron transfer. Reversal of the second electron transfer leading to formation of the semiquinone from QH_2 in addition to the oxidized complex would be represented by reversal of transitions leading to formation of state **A** from **C**. No attempt has been made to show the many different states of protonation of the residues involved, and the particular points of entry of H^+ in the scheme should therefore be regarded as flexible and would change with pH.

Lange et al. (89) discussed at length a possible role for waters around the site in transfer of H^+ into the site, and their relation to exterior waters and phospholipid. For the lipid phase, the quinone and inhibitor tails provide an Ariadne's thread, showing the path into the site from the lipid through a fairly wide access channel.

INTERACTIONS BETWEEN THE MONOMERS   A clear role for the dimeric structure lies in binding of the ISP N-terminal anchor so as to allow the cross-dimer interactions of the mobile head (65). The mechanistic discussions above have focused on models in which all reactions are within a monomer, but there have been many speculations about mechanisms involving dimeric Q-cycles and control through interactions between monomers in the dimeric complex (3, 38, 40, 60–62, 83, 86, 88, 145, 162, 171). Most of the tantalizing evidence for mechanistic interactions has come from work on mitochondrial complexes. Work from the author's lab in collaboration with Dr. Vlad Shinkarev to explore such interactions in the bacterial system has so far failed to provide any compelling evidence for electron transfer between monomers, or control by allosteric interactions across the dimer interface (unpublished). Possibly, the additional subunits of the mitochondrial complexes have evolved to provide a means for fine-tuning the mechanism through such interactions.

THE PHYSIOLOGICAL PERSPECTIVE   Although most attention has been paid to the Q_o-site, the Q_i-site has also been discussed as a possible site for SO generation, and a number of myopathies in cyt b have mapped to the site (8, 9). The well-established effect on SO production of blocking the Q_i-site, as in the antimycin stimulation, has led to some confusion, because antimycin is specific for the Q_i-site, and might have led to the conclusion that this site has an important role in SO production. More plausible is a mechanism based on a Q_o-site generation, stimulated by back-up in the b-heme chain. The relatively high redox potential of the semiquinone couples consequent on the stability of the Q_i-site SQ would make them unlikely donors in the production of superoxide.

Although the Q_o-site inhibitors have attracted most attention in clinical and agronomical applications, the species-selectivity of the Q_i-site is another attractive target that remains to be exploited.

ACKNOWLEDGMENTS

Support for the author's research from the following sources is gratefully acknowledged: NIH (GM 35438 to A.R.C., GM 62954 to Sergei Dikanov.), and a Fogarty grant (PHS 1 R03 TW01495, with Rimma Samoilova). The views presented are those of the author unless otherwise indicated. Helpful discussions with the following are gratefully acknowledged: Vlad Shinkarev, Ed Berry, Sergei Dikanov, Rimma Samoilova, Judy Hirst, Uli Brandt, Peter Rich, Hildur Palsdottir, Bernie Trumpower, Raúl Covián, Carola Hunte, David Kramer, and Florian Muller.

The *Annual Review of Physiology* is online at http://physiol.annualreviews.org

LITERATURE CITED

1. Crofts AR, Berry EA. 1998. Structure and function of the cytochrome bc_1 complex of mitochondria and photosynthetic bacteria. *Curr. Opin. Struct. Biol.* 8: 501–9

2. Berry E, Guergova-Kuras M, Huang L-S, Crofts AR. 2000. Structure and function of cytochrome bc complexes. *Annu. Rev. Biochem.* 69:1007–77

3. Hunte C, Palsdottir H, Trumpower BL. 2003. Protonmotive pathways and mechanisms in the cytochrome bc_1 complex. *FEBS Lett.* 545:39–46

4. Cramer WA, Soriano GM, Ponomarev M, Huang D, Zhang H, et al. 1996. Some new structural aspects and old controversies concerning the cytochrome b6f complex of oxygenic photosynthesis. *Annu. Rev. Plant Physiol. Plant Mol. Biol.* 47:477–508

5. Woese CR. 1987. Bacterial evolution. *Microbiol. Rev.* 51:221–71

6. Nitschke W, Muehlenhoff U, Liebl U. 1997. Evolution. In *Photosynthesis: A Comprehensive Treatise*, ed. A Raghavendra, pp. 285–304. Cambridge, UK: Cambridge Univ. Press

7. Degli Esposti M, de Vries S, Crimi M, Ghelli A, Patarnello T, Meyer A. 1993. Mitochondrial cytochrome b: evolution and structure of the protein. *Biochim. Biophys. Acta* 1143:243–71

8. Fisher N, Meunier B. 2001. Effects of mutations in mitochondrial cytochrome *b* in yeast and man—deficiency, compensation and disease. *Eur. J. Biochem.* 268:1155–62

9. Fisher N, Meunier B. 2002. The bc_1 complex: structure, function and dysfunction. In *Recent Research Developments in Human Mitochondrial Myopathies*, ed. J Garcia-Trejo, pp. 97–112. Kerala, India: Research Signpost

10. Rose MR. 1998. Mitochondrial myopathies. Genetic mechanisms. *Arch. Neurol.* 55:17–24

11. Vogel H. 2001. Mitochondrial myopathies and the role of the pathologist in the molecular era. *J. Neuropathol. Exp. Neurol.* 60:217–27

12. Boveris A. 1984. Determination of the production of superoxide radicals and hydrogen-peroxide in mitochondria. *Methods Enzymol.* 105:429–35

13. Turrens JF, Alexandre A, Lehninger AL. 1985. Ubisemiquinone is the electron donor for superoxide formation by complex III of heart mitochondria. *Arch. Biochem. Biophys.* 237:408–14

14. Skulachev VP. 1996. Role of uncoupled and non-coupled oxidations in maintenance of safely low levels of oxygen and its one-electron reductants. *Q. Rev. Biophys.* 29:169–202

15. Muller F. 2000. The nature and mechanism of superoxide production by the electron transport chain: its relevance to aging. *J. Am. Aging Assoc.* 23:227–53

16. Chen Q, Vazquez EJ, Moghaddas S, Hoppel CL, Lesnefsky EJ. 2003. Production of reactive oxygen species by mitochondria: central role of Complex III. *J. Biol. Chem.* 278: 36027–31

17. Kristal BS, Jackson CT, Chung HY, Matsuda M, Nguyen HD, Yu BP. 1997. Defects at center P underlie diabetes-associated mitochondrial dysfunction. *Free Radic. Biol. Med.* 22:823–33

18. Hoek TLV, Shao ZH, Li CQ, Schumacker PT, Becker LB. 1997. Mitochondrial electron transport can become a significant source of oxidative injury in cardiomyocytes. *J. Mol. Cell. Cardiol.* 29:2441–50

19. Harman D. 1992. Free radical theory of aging: history. *EXS* 62:1–10

20. Ames BN, Shigenaga MK, Hagen TM. 1993. Oxidants, antioxidants, and the

degenerative diseases of aging. *Proc. Natl. Acad. Sci. USA* 90:7915–21

21. Beckman KB, Ames BN. 1998. The free radical theory of aging matures. *Physiol. Rev.* 78:547–81

22. Raha S, Robinson BH. 2001. Mitochondria, oxygen free radicals, and apoptosis. *Am.J. Med. Genet.* 106:62–70

23. Orth M, Schapira AHV. 2001. Mitochondria and degenerative disorders. *Am. J. Med. Genet.* 106:27–36

24. Papa S, Skulachev VP. 1997. Reactive oxygen species, mitochondria, apoptosis and aging. *Mol. Cell. Biochem.* 174:305–19

25. Bowyer JR, Dutton PL, Prince RC, Crofts AR. 1980. The role of the Rieske iron-sulfur center as the electron donor to ferricytochrome c_2 in *Rhodopseudomonas sphaeroides. Biochim. Biophys. Acta* 592:445–60

26. Meinhardt SW, Crofts AR. 1982. The site and mechanism of action of myxothiazol as an inhibitor of electron transfer in *Rhodopseudomonas sphaeroides. FEBS Lett.* 149:217–22

27. von Jagow G, Link T. 1988. Use of specific inhibitors on the mitochondrial bc_1 complex. *Methods Enzymol.* 126:253–71

28. Brandt U, Schagger H, von Jagow G. 1988. Characterization of binding of the methoxyacrylate inhibitors to mitochondrial cytochrome c reductase. *Eur. J. Biochem.* 173:499–506

29. Tsai A-L, Kauten R, Palmer G. 1985. The interaction of yeast Complex III with some respiratory inhibitors. *Biochim. Biophys. Acta* 806:418–26

30. Link TA, Haase U, Brandt U, von Jagow G. 1993. What information do inhibitors provide about the structure of the hydroquinone oxidation site of ubiquinol:cytochrome c oxidoreductase. *J. Bioenerg. Biomembr.* 25:221–32

31. Slater EC. 1973. The mechanism of action of the respiratory inhibitor, antimycin. *Biochim. Biophys. Acta* 301:129–54

32. Srivastava IK, Morrisey JM, Darrouzet E,

Daldal F, Vaidya AB. 1999. Resistance mutations reveal the atovaquone-binding domain of cytochrome b in malaria parasites. *Mol. Microbiol.* 33:704–11

33. Beautement K, Clough JM, Defraine PJ, Godfrey CRA. 1991. Fungicidal Betamethoxyacrylates—from natural products to novel synthetic agricultural fungicides. *Pestic. Sci.* 31:499–519

34. Clough JM, Godfrey CRA. 1995. Growing hopes. *Chem. Br.* 31:466–69

35. Bartlett DW, Clough JM, Godwin JR, Hall AA, Hamer M, Parr DB. 2002. The strobilurin fungicides. *Pest-Manage. Sci.* 58:649–62

36. Sauter H, Ammermann E, Roehl F. 1996. Strobilurins—from natural products to a new class of fungicides. In *Crop Protection Agents from Nature*, ed. LG Copping, pp. 50–81. Cambridge, UK: Thomas Graham House, R. Soc. Chem.

37. Yu C-A, Xia J-Z, Kachurin AM, Yu L, Xia D, et al. 1996. Crystallization and preliminary structure of beef heart mitochondrial cytochrome-bc_1 complex. *Biochim. Biophys. Acta* 1275:47–53

38. Xia D, Yu C-A, Kim H, Xia J-Z, Kachurin AM, et al. 1997. Crystal structure of the cytochrome bc_1 complex from bovine heart mitochondria. *Science* 277:60–66

39. Zhang Z, Huang L-S, Shulmeister VM, Chi Y-I, Kim K-K, et al. 1998. Electron transfer by domain movement in cytochrome bc_1. *Nature* 392:677–84

40. Iwata S, Lee JW, Okada K, Lee JK, Iwata M, et al. 1998. Complete structure of the 11-subunit bovine mitochondrial cytochrome bc_1 complex. *Science* 281:64–71

41. Kim H, Xia D, Yu C-A, Xia JZ, Kachurin AM, et al. 1998. Inhibitor binding changes domain mobility in the iron-sulfur protein of the mitochondrial bc_1 complex from bovine heart. *Proc. Natl. Acad. Sci. USA* 95:8026–33

42. Yu C-A, Xia D, Kim H, Deisenhofer J, Zhang L, et al. 1998. Structural basis of functions of the mitochondrial

cytochrome bc_1 complex. *Biochim. Biophys. Acta* 1365:151–58

43. Hunte C, Koepke J, Lange C, Rossmanith T, Michel H. 2000. Structure at 2.3 Å resolution of the cytochrome bc_1 complex from the yeast *Saccharomyces cerevisiae* co-crystallized with an antibody Fv fragment. *Structure* 8:669–84

44. Gao X, Wen X, Yu C-A, Esser L, Tsao S, et al. 2002. The crystal structure of mitochondrial cytochrome bc_1 in complex with famoxadone: the role of aromatic-aromatic interaction in inhibition. *Biochemistry* 41:11692–702

45. Palsdottir H, Lojero CG, Trumpower BL, Hunte C. 2003. Structure of the yeast cytochrome bc_1 complex with a hydroxyquinone anion Q_o site inhibitor bound. *J. Biol. Chem.* 278: 31303–11

46. Kessl JJ, Lange BB, Merbitz-Zahradnik T, Zwicker K, Hill P, et al. 2003. Molecular basis for atovaquone binding to the cytochrome bc_1 complex. *J. Biol. Chem.* 278: 31312–18

47. Gao X, Wen X, Esser L, Quinn B, Yu L, Yu C-A, Xia D. 2003. Structural basis for the quinone reduction in the bc_1 complex: a comparative analysis of crystal structures of mitochondrial cytochrome bc_1 with bound substrate and inhibitors at the Q_i site. *Biochemistry.* 42:9067–80

48. Zhang HM, Kurisu G, Smith JL, Cramer WA. 2003. A defined protein-detergent-lipid complex for crystallization of integral membrane proteins: the cytochrome b_6f complex of oxygenic photosynthesis. *Proc. Natl. Acad. Sci. USA* 100:5160–63

49. Mitchell P. 1976. Possible molecular mechanisms of the protonmotive function of cytochrome systems. *J. Theor. Biol.* 62:327–67

50. Crofts AR, Meinhardt SW, Jones KR, Snozzi M. 1983. The role of the quinone pool in the cyclic electron-transfer chain of *Rhodopseudomonas sphaeroides*: a modified Q-cycle mechanism. *Biochim. Biophys. Acta* 723:202–18

51. Crofts AR. 1985. The mechanism of

ubiquinol: cytochrome c oxidoreductases of mitochondria and of *Rhodopseudomonas sphaeroides*. In *The Enzymes of Biological Membranes*, ed. AN Martonosi, 4:347–82. New York: Plenum

52. Crofts AR, Wang Z. 1989. How rapid are the internal reactions of the ubiquinol: cytochrome c_2 oxidoreductase? *Photosynth. Res.* 22:69–87

53. Crofts AR. 2003. The Q-cycle—a personal perspective. In The Millenium Issues on the History of Photosynthesis. *Photosynth. Res.* ed. Govindjee. In press

54. Crofts AR, Shinkarev VP, Kolling DRJ, Hong S. 2003. The modified Q-cycle explains the apparent mismatch between the kinetics of reduction of cytochromes c_1 and b_H in the bc_1 complex. *J. Biol. Chem.* 278:36191–201

55. Rich PR. 1981. Electron transfer reactions between quinols and quinones in aqueous and aprotic media. *Biochim. Biophys. Acta* 637:28–33

56. Junemann S, Heathcote P, Rich PR. 1998. On the mechanism of quinol oxidation in the bc_1 complex. *J. Biol. Chem.* 273:21603–7

57. Crofts AR, Wraight CA. 1983. The electrochemical domain of photosynthesis. *Biochim. Biophys. Acta* 726:149–86

58. Garland PB, Clegg RA, Boxer D, Downie JA, Haddock BA. 1975. Proton-translocating nitrate reductase of *Escherichia coli*. In *Electron Transfer Chains and Oxidative Phosphorylation*, ed. E Quagliariello, S Papa, F Palmieri, EC Slater, N Siliprandi, pp. 351–58. Amsterdam: North-Holland

59. King TE, Yu C-A, Yu L, Chiang Y. 1975. An examination of the components, sequence, mechanisms and their uncertainties involved in mitochondrial electron transport from succinate to cytochrome c. See Ref. 58, pp. 105–18

60. de Vries S, Berden JA, Slater EC. 1982. Oxidation-reduction properties of an antimycin-sensitive semiquinone anion bound to QH_2:cytochrome c

oxidoreductase. In *Function of Quinones in Energy Conserving Systems*, ed. BL Trumpower, pp. 235–46. New York: Academic

61. de Vries S, Albracht SPJ, Berden JA, Marres CAM, Slater EC. 1983. The effect of pH, ubiquinone depletion and myxothiazol on the reduction kinetics of the prosthetic groups of QH_2: cytochrome c oxidoreductase. *Biochim. Biophys. Acta* 723:91–103

62. de Vries S. 1983. *The pathway of electrons in QH_2 cytochrome c oxidoreductase*. PhD thesis. Amsterdam Univ., The Netherlands

63. Tsai A-L, Olson JS, Palmer G. 1983. The oxidation of yeast complex III. *J. Biol. Chem.* 258:2122–25

64. Mitchell P. 1975. Proton motive redox mechanism of the cytochrome b-c_1 complex in the respiratory chain: proton motive ubiquinone cycle. *FEBS Lett.* 56:1–6

65. Crofts AR, Guergova-Kuras M, Huang L-S, Kuras R, Zhang Z, Berry EA. 1999. The mechanism of ubiquinol oxidation by the bc_1 complex: the role of the iron sulfur protein, and its mobility. *Biochemistry* 38:15791–806

66. Crofts AR, Hong S, Zhang Z, Berry EA. 1999. Physicochemical aspects of the movement of the Rieske iron sulfur protein during quinol oxidation by the bc_1 complex. *Biochemistry* 38:15827–39

67. Crofts AR, Barquera B, Gennis RB, Kuras R, Guergova-Kuras M, Berry EA. 1999. The mechanism of ubiquinol oxidation by the bc_1 complex: the different domains of the quinol binding pocket, and their role in mechanism, and the binding of inhibitors. *Biochemistry* 38:15807–26

68. Tian H, White S, Yu L, Yu C-A. 1999. Evidence for the head domain movement of the Rieske iron-sulfur protein in electron transfer reaction of the cytochrome bc_1 complex. *J. Biol. Chem.* 274:7146–52

69. Tian H, Yu L, Mather MW, Yu C-A. 1998. Flexibility of the neck region of the Rieske iron-sulfur protein is functionally impor-

tant in the cytochrome bc_1 complex. *J. Biol. Chem.* 273:27953–59

70. Darrouzet E. Valkova-Valchanova M, Moser CC, Dutton PL, Daldal F. 2000. Uncovering the [2Fe2S] domain movement in cytochrome bc_1 and its implications for energy conversion *Proc. Natl. Acad. Sci. USA* 97:4567–72

71. Darrouzet E, Moser CC, Dutton PL, Daldal F. 2001. Large scale domain movement in cytochrome bc_1: a new device for electron transfer in proteins. *TIBS* 26:445–51

72. Darrouzet E, Daldal F. 2003. Protein-protein interactions between cytochrome b and the Fe-S protein subunits during QH_2 oxidation and large-scale domain movement in the bc_1 complex. *Biochemistry* 42:1499–507

73. Crofts AR, Hong SJ, Ugulava N, Barquera B, Gennis RB, et al. 1999. Pathways for proton release during ubihydroquinone oxidation by the bc_1 complex. *Proc. Natl. Acad. Sci. USA* 96:10021–26

74. Bowyer JR, Edwards CA, Ohnishi T, Trumpower BL. 1982. An analogue of ubiquinone which inhibits respiration by binding to the iron-sulfur protein of the cytochrome bc_1 segment of the mitochondrial respiratory chain. *J. Biol. Chem.* 257:8321–30

75. Matsuura K, Bowyer JR, Ohnishi T, Dutton PL. 1983. Inhibition of electron transfer by 3-alkyl-2-hydroxy-1,4-naphthoquinones in the ubiquinol-cytochrome c oxidoreductases of *Rhodopseudomonas sphaeroides* and mammalian mitochondria. Interaction with a ubiquinone-binding site and the Rieske iron-sulfur cluster. *J. Biol. Chem.* 258:1571–77

76. Fivelman QL, Butcher GA, Adagu IS, Warhurst DC, Pasvol G. 2002. Malarone treatment failure and in vitro confirmation of resistance of *Plasmodium falciparum* isolate from Lagos, Nigeria. *Malar. J.* 1:1–4

77. Van Ark G, Berden JA. 1977. Binding of HQNO to beef-heart sub-mitochondrial

particles. *Biochim. Biophys. Acta* 459: 119–27

78. Crofts AR, Guergova-Kuras M, Kuras R, Ugulava N, Li J, Hong S. 2000. Proton-coupled electron transfer at the Q_o-site: What type of mechanism can account for the high activation barrier? *Biochim. Biophys. Acta* 1459:456–66

79. Jordan DB, Livingston RS, Bisaha JJ, Duncan KE, Pember SO, et al. 1999. Mode of action of famoxadone. *Pestic. Sci.* 55:105–18

80. Meinhardt SW, Crofts AR. 1983. The role of cytochrome b_{566} in the electron transfer chain of *Rps. sphaeroides. Biochim. Biophys. Acta* 723:219–30

81. Hansen KC, Schultz BE, Wang G, Chan SI. 2000. Reaction of *Escherichia coli* cytochrome bo_3 and mitochondrial cytochrome bc_1 with a photoreleasable decylubiquinol. *Biochim. Biophys. Acta* 1456:121–37

82. Schultz BE, Chan SI. 2001. Structures and proton-pumping strategies of mitochondrial respiratory enzymes. *Annu. Rev. Biophys. Biomol. Struct.* 30:23–65

83. Yu C-A, Wen X, Xiao K, Xia D, Yu L. 2002. Inter- and intra-molecular electron transfer in the cytochrome bc_1 complex. *Biochim. Biophys. Acta* 1555:65–70

84. Link TA. 1997. The role of the "Rieske" iron sulfur protein in the hydroquinone oxidation (Q_p-) site of the cytochrome bc_1 complex: the "proton-gated affinity change" mechanism. *FEBS Lett.* 412: 257–64

85. Brandt U. 1998. The chemistry and mechanics of ubihydroquinone oxidation at center P (Q_o) of the cytochrome bc_1 complex. *Biochim. Biophys. Acta* 1365:261–68

86. Snyder CH, Gutierrez-Cirlos EB, Trumpower BL. 2000. Evidence for a concerted mechanism of ubiquinol oxidation by the cytochrome bc_1 complex. *J. Biol. Chem.* 275:13535–41

87. Trumpower BL. 2002. A concerted, alternating sites mechanism of ubiquinol oxidation by the dimeric cytochrome bc_1 complex. *Biochim. Biophys. Acta* 1555:166–73

88. Lange C, Hunte C. 2002. Crystal structure of the yeast cytochrome bc_1 complex with its bound substrate cytochrome *c. Proc. Natl. Acad. Sci. USA* 99:2800–5

89. Lange C, Nett JH, Trumpower BL, Hunte C. 2001. Specific roles of protein-phospholipid interactions in the yeast bc_1 complex structure. *EMBO J.* 23:6591–600

90. Nett JH, Hunte C, Trumpower BL. 2000. Changes to the length of the flexible linker region of the Rieske protein impair the interaction of ubiquinol with the cytochrome bc_1 complex. *Eur. J. Biochem.* 267:5777–82

91. Xiao K, Engstrom G, Rajagukguk S, Yu C-A, Yu L, et al. 2003. Effect of famoxadone on photoinduced electron transfer between the iron-sulfur center and cytochrome c_1 in the cytochrome bc_1 complex. *J. Biol. Chem.* 278:11419–26

92. Engstrom G, Xiao K, Yu C-A, Yu L, Durham B, Millett F. 2002. Photoinduced electron transfer between the Rieske iron-sulfur protein and cytochrome c_1 in the *Rhodobacter sphaeroides* cytochrome bc_1 complex: effects of pH, temperature, and driving force. *J. Biol. Chem.* 277:31072–78

93. Hong SJ, Ugulava N, Guergova-Kuras M, Crofts AR. 1999. The energy landscape for ubihydroquinone oxidation at the Q_o-site of the bc_1 complex in *Rhodobacter sphaeroides. J. Biol. Chem.* 274:33931–44

94. Guergova-Kuras M, Kuras R, Ugulava N, Hadad I, Crofts AR. 2000. Specific mutagenesis of the Rieske iron sulfur protein in *Rhodobacter sphaeroides* shows that both thermodynamic gradient and the pK of the oxidized form determine the rate of quinol oxidation by the bc_1 complex. *Biochemistry* 39:7436–44

95. Snyder C, Trumpower BL. 1998. Mechanism of ubiquinol oxidation by the cytochrome bc_1 complex—pre-steady-state

kinetics of cytochrome bc_1 complexes containing site-directed mutants of the Rieske iron-sulfur protein. *Biochim. Biophys. Acta* 1365:125–34

96. Ugulava NB, Crofts AR. 1998. CD-monitored redox titration of the Rieske Fe-S protein of *Rhodobacter sphaeroides*: pH dependence of the mid-point potential in isolated bc_1 complex and in membranes. *FEBS Lett.* 440:409–13

97. Crofts AR, Berry EA, Kuras R, Guergova-Kuras M, Hong S, Ugulava N. 1998. Structures of the bc_1 complex reveal dynamic aspects of mechanism. In *Photosynthesis: Mechanisms and Effects*, ed. G Garab, 3:1481–86. Dordrecht/Boston/London: Kluwer

98. Izrailev S, Crofts AR, Berry EA, Schulten K. 1999. Steered molecular dynamics simulation of the Rieske subunit motion in the cytochrome bc_1 complex. *Biophys. J.* 77:1753–68

99. Crofts AR, Guergova-Kuras M, Ugulava N, Kuras R, Hong S. 2002. Proton processing at the Q_o-site of the bc_1 complex of *Rhodobacter sphaeroides*. *Proc. Congr. Photosynth. Res., 12th, Brisbane, Australia*, S12-002. 6 pp.

100. Ding H, Moser CC, Robertson DE, Tokito MK, Daldal F, Dutton PL. 1995. Ubiquinone pair in the Q_o site central to the primary energy conversion reactions of cytochrome bc_1 complex. *Biochemistry* 34:15979–96

101. Link TA, von Jagow G. 1995. Zinc ions inhibit the Q_P center of bovine heart mitochondrial bc_1 complex by blocking a protonatable group. *J. Biol. Chem.* 270:25001–6

102. Brandt U, Okun JG. 1997. Role of deprotonation events in ubihydroquinone: cytochrome c oxidoreductase from bovine heart and yeast mitochondria. *Biochemistry* 36:11234–40

103. Covián R, Moreno-Sánchez R. 2001. Role of protonatable groups of bovine heart bc_1 complex in ubiquinol binding and oxidation. *Eur. J. Biochem.* 268:5783–90

104. Link TA. 1999. The structures of Rieske and Rieske-type proteins. *Adv. Inorg. Chem.* 47:83–157

105. Colbert C, Couture MM-J, Eltis LD, Bolin JT. 2000. A cluster exposed: structure of the Rieske ferredoxin from biphenyl dioxygenase and the redox properties of Rieske Fe-S proteins. *Structure* 8:1267–78

106. Ullmann GM, Noodleman L, Case DA. 2002. Density functional calculation of pK_a values and redox potentials in the bovine Rieske iron-sulfur protein. *J. Biol. Inorg. Chem.* 7:632–39

107. Zu Y, Manon M-J, Couture MM-J, Kolling DRJ, Crofts AR, et al. 2003. The reduction potentials of Rieske clusters: the importance of the coupling between oxidation state and histidine protonation state. *Biochemistry.* In press

108. Hunte C. 2001. Insights from the structure of the yeast cytochrome bc_1 complex: crystallization of membrane proteins with antibody fragments. *FEBS Lett.* 504:126–32

109. de Vries S, Albracht SPJ, Leeuwerik FJ. 1979. The multiplicity and stoichiometry of the prosthetic groups in QH_2: cytochrome c oxidoreductase as studied by EPR. *Biochim. Biophys. Acta* 546:316–33

110. Ding H, Robertson DE, Daldal F, Dutton PL. 1992. Cytochrome bc_1 complex [2Fe-2S] cluster and its interaction with ubiquinone and ubihydroquinone at the Q_o site: a double-occupancy Q_o site model. *Biochemistry* 31:3144–58

111. Sharp RE, Palmitessa A, Gibney BR, White JL, Moser CC, et al. 1999. Ubiquinone binding capacity of the *Rhodobacter capsulatus* cytochrome bc_1 complex: effect of diphenylamine, a weak binding Q_o site inhibitor. *Biochemistry* 38:3440–46

112. Samoilova RI, Kolling D, Uzawa T, Iwasaki T, Crofts AR, Dikanov SA. 2002. The interaction of the Rieske iron sulfur protein with occupants of the Q_o-site of the bc_1 complex, probed by 1D and 2D

electron spin echo envelope modulation. *J. Biol. Chem.* 277:4605–8

113. Brandt U, Haase U, Schägger H, von Jagow G. 1991. Significance of the "Rieske" iron-sulfur protein for formation and function of the ubiquinol-oxidation pocket of mitochondrial cytochrome c reductase (*bc*₁ complex) *J. Biol. Chem.* 266:19958–64

114. Covián R, Pardo JP, Moreno-Sánchez R. 2002. Tight binding of inhibitors to bovine *bc*1 complex is independent of the Rieske protein redox state—consequences for semiquinone stabilization in the quinol oxidation site. *J. Biol. Chem.* 277:48449–55

115. Sharp RE, Gibney BR, Palmitessa A, White J, Dixon JA, et al. 1999. Effect of inhibitors on the ubiquinone binding capacity of the primary energy conversion site in the *Rhodobacter capsulatus* cytochrome *bc*₁ complex. *Biochemistry* 38:14973–80

116. Darrouzet E, Valkova-Valchanova M, Daldal F. 2002. The [2Fe-2S] cluster Eₘ as an indicator of the iron-sulfur subunit position in the ubihydroquinone oxidation site of the cytochrome *bc*₁ complex. *J. Biol. Chem.* 277:2464–70

117. Crofts AR, Shinkarev VP, Dikanov SA, Samoilova RI, Kolling D. 2002. Interactions of quinone with the iron-sulfur protein of the *bc*₁ complex: Is the mechanism spring-loaded? *Biochim. Biophys. Acta* 1555:48–53

118. Shinkarev VP, Kolling DRJ, Miller TJ, Crofts AR. 2002. Modulation of the midpoint potential of the [2Fe-2S] Rieske iron sulfur center by Q₀ occupants in the *bc*₁ complex. *Biochemistry* 41:14372–82

119. Gatti DL, Meinhardt SW, Ohnishi T, Tzagoloff A. 1989. Structure and function of the mitochondrial *bc*₁ complex. A mutational analysis of the yeast Rieske iron-sulfur protein. *J. Mol. Biol.* 205:421–35

120. Denke E, Merbitzzahradnik T, Hatzfeld OM, Snyder CH, Link TA, Trumpower BL. 1998. Alteration of the midpoint potential of the Rieske iron-sulfur protein by changes of amino acids forming H-bonds to the iron-sulfur cluster. *J. Biol. Chem.* 273:9085–93

121. Schröter T, Hatzfeld OM, Gemeinhardt S, Korn M, Friedrich T, et al. 1998. Mutational analysis of residues forming hydrogen bonds in the Rieske [2Fe2S] cluster of the cytochrome *bc*₁ complex of *Paracoccus denitrificans*. *Eur. J. Biochem.* 255:100–6

122. Van Doren SR, Gennis RB, Barquera B, Crofts AR. 1993. Site-directed mutations of conserved residues of the Rieske iron-sulfur subunit of the cytochrome *bc*₁ complex of *Rhodobacter sphaeroides* blocking or impairing quinol oxidation. *Biochemistry* 32:8083–91

123. Yun C-H, Crofts AR, Gennis RB. 1991. Assignment of the histidine axial ligands to the cytochrome *b*_H and cytochrome *b*_L components of the *bc*₁ complex from *Rb. sphaeroides* by site-directed mutagenesis. *Biochemistry* 30:6747–54

124. Hacker B, Barquera B, Crofts AR, Gennis RB. 1993. Characterization of mutations in the cytochrome *b* subunit of the *bc*₁ complex of *Rhodobacter sphaeroides* that affect the quinone reductase site (Qc). *Biochemistry* 32:4403–10

125. Crofts AR, Barquera B, Bechmann G, Guergova M, Salcedo-Hernandez R, et al. 1995. Structure and function in the *bc*₁-complex of *Rb. sphaeroides*. In *Photosynthesis: From Light to Biosphere*, ed. P Mathis, 2:493–500. Dordrecht: Kluwer

126. Shinkarev VP, Crofts AR, Wraight CA. 2001. The electric field generated by photosynthetic reaction center induces rapid reversed electron transfer in the *bc*₁ complex. *Biochemistry* 40:12584–90

127. Davidson VL. 1996. Unraveling the kinetic complexity of inter-protein electron transfer reactions. *Biochemistry* 35:14036–39

128. Moser CC, Page CC, Farid R, Dutton PL. 1995. Biological electron transfer. *J. Bioenerg. Biomembr.* 27:263–74

129. Roberts JA, Kirby JP, Wall ST, Nocera

DG. 1997. Electron transfer within ruthenium(II) polypyridyl-(salt bridge)-dimethylaniline acceptor-donor complexes. *Inorg. Chim. Acta* 263:395–405

130. Cukier RI, Nocera DG. 1998. Proton-coupled electron transfer. *Annu. Rev. Phys. Chem.* 49:337–69

131. Graige MS, Paddock ML, Feher G, Okamura MY. 1999. Observation of the protonated SQ intermediate in isolated reaction centers from *Rhodobacter sphaeroides*: implications for the mechanism of electron and proton transfer in proteins. *Biochemistry* 38:11465–73

132. Kresge AJ, Silverman DN. 1999. Application of Marcus Rate Theory to proton transfer in enzyme catalyzed reactions. *Methods Enzymol.* 308:276–97

133. Pines E, Magnes B-Z, Lang MJ, Fleming GR. 1997. Direct measurement of intrinsic proton transfer rates in diffusion controlled reactions. *Chem. Phys. Lett.* 281:413–20

134. Crofts AR. 2003. Proton-coupled electron transfer at the Q_o-site of the bc_1 complex controls the rate of ubihydroquinone oxidation. *Biochim. Biophys. Acta.* In press

135. Marcus RA, Sutin N. 1985. Electron transfers in chemistry and biology. *Biochim. Biophys. Acta* 811:265–322

136. Page CC, Moser CC, Chen XX, Dutton PL. 1999. Natural engineering principles of electron tunnelling in biological oxidation-reduction. *Nature* 402:47–52

137. DeVault D. 1980. Quantum-mechanical tunnelling in biological systems. *Q. Rev. Biophys.* 13:387–564

138. Xiao K, Yu L, Yu C-A. 2000. Confirmation of the involvement of protein domain movement during the catalytic cycle of the cytochrome bc_1 complex by the formation of an inter-subunit disulfide bond between cytochrome b and the iron-sulfur protein. *J. Biol. Chem.* 275:38597–604

139. Zhang L, Tai CH, Yu L, Yu C-A. 2000. pH-induced intramolecular electron transfer between the iron-sulfur protein and cytochrome c_1 in bovine cytochrome bc_1 complex. *J. Biol. Chem.* 275:7656–61

140. Starkov AA, Fiskum G. 2001. Myxothiazol induces H_2O_2 production from mitochondrial respiratory chain. *Biochem. Biophys. Res. Commun.* 281:645–50

141. Muller F, Crofts AR, Kramer DM. 2002. Multiple Q-cycle bypass reactions at the Q_o-site of the cytochrome bc_1 complex. *Biochemistry* 41:7866–74

142. Bartoschek S, Johansson M, Geierstanger BH, Okun JG, Lancaster CRD, et al. 2001. Three molecules of ubiquinone bind specifically to mitochondrial cytochrome bc_1 complex. *J. Biol. Chem.* 276:35231–34

143. Muller FL, Roberts AG, Bowman MK, Kramer DM. 2003. Architecture of the Q_o site of the cytochrome bc_1 complex probed by superoxide production. *Biochemistry* 42:6493–99

144. Crofts AR, Barquera B, Gennis RB, Kuras R, Guergova-Kuras M, Berry EA. 1999. Mechanistic aspects of the Q_o-site of the bc_1 complex as revealed by mutagensis studies, and the crystallographic structure. In *The Phototrophic Prokaryotes*, ed. GA Peschek, W Loeffelhardt, G Schmetterer, pp. 229–39. New York: Plenum

145. Slater EC. 1981. The cytochrome b paradox, the BAL-labile factor and the Q-cycle. In *Chemiosmotic Proton Circuits in Biological Membranes*, ed. VP Skulachev, PV Hinkle, pp. 69–104. Reading, MA: Addison-Wesley

146. Andrews KM. 1984. *Purification and characterization of the cytochrome* bc_1 *complex from* Rhodobacter sphaeroides. PhD thesis, Univ. Illin. Urbana-Champaign

147. Hunte C, Solmaz S, Lange C. 2002. Electron transfer between yeast cytochrome bc_1 complex and cytochrome c: a structural analysis. *Biochim. Biophys. Acta* 1555:21–28

148. Tian H, Sadoski R, Zhang L, Yu C-A, Yu L, Durham B, Millett F. 2000. Definition of the interaction domain for cytochrome

c on the cytochrome bc_1 complex. Steady-state and rapid kinetic analysis of electron transfer between cytochrome c and *Rhodobacter sphaeroides* cytochrome bc_1 surface mutants. *J. Biol. Chem.* 275:9587–95

149. Berry EA, Zhang Z, Huang L-S, Kim SH. 1999. Structures of quinone-binding sites in bc complexes: functional implications. *Biochem. Soc. Trans.* 27:565–72

150. Berry EA, Huang L-S, Zhang Z, Kim SH. 1999. Structure of the avian mitochondrial cytochrome bc_1 complex. *J. Bioenerg. Biomembr.* 31:177–90

151. Zhang Z-L, Berry EA, Huang L-S, Kim S-H. 2000. Mitochondrial cytochrome bc_1 complex. *Subcell. Biochem.* 35:541–80

152. Brasseur G, Sami Saribas A, Daldal F. 1996. A compilation of mutations located in the cytochrome b subunit of the bacterial and mitochondrial bc_1 complex. *Biochim. Biophys. Acta* 1275:61–69

153. Iwata M. 2001. *Structural studies on cytochrome* bc$_1$ *complex from bovine heart mitochondria*. PhD diss. Uppsala Univ., Sweden

154. Ohnishi T, Trumpower BL. 1980. Differential effects of antimycin on ubisemiquinone bound in different environments in isolated succinate cytochrome c reductase complex. *J. Biol. Chem.* 255:3278–84

155. Robertson DE, Prince RC, Bowyer JR, Matsuura K, Dutton PL, Ohnishi T. 1984. Thermodynamic properties of the semiquinone and its binding site in the ubiquinol:cytochrome $c(c_2)$ oxidoreductase of respiratory and photosynthetic systems. *J. Biol. Chem.* 259:1758–63

156. de la Rosa FF, Palmer G. 1983. Reductive titration of CoQ-depleted Complex III from baker's yeast: evidence for an exchange-coupled complex between QH$^\bullet$ and low-spin ferricytochrome b. *FEBS Lett.* 163:140–43

157. Salerno JC, Osgood M, Liu Y, Taylor H, Scholes CP. 1990. Electron nuclear double resonance (ENDOR) of the Q-c.-ubiSQ radical in the mitochondrial electron transport chain. *Biochemistry* 29:6987–93

158. Kolling DRJ, Samoilova RI, Holland JT, Berry EA, Dikanov SA, Crofts AR. 2003. Exploration of ligands to the Q_i-site semiquinone in the bc_1 complex using high-resolution EPR. *J. Biol. Chem.* 278:39747–54

159. Hacker B, Barquera B, Gennis RB, Crofts AR. 1994. Site-directed mutagenesis of arginine-114 and tryptophan-129 in the cytochrome b subunit of the bc_1 complex of *Rhodobacter sphaeroides*: two highly conserved residues predicted to be near the cytoplasmic surface of putative transmembrane helices B and C. *Biochemistry* 33:13022–31

160. Salerno JC, Xu Y, Osgood P, Kim CH, King TE. 1989. Thermodynamic and spectroscopic characteristics of the cytochrome bc_1 complex. Role of quinone in the behavior of cytochrome b_{562}. *J. Biol. Chem.* 264:15398–403

161. Rich PR, Jeal AE, Madgwick SA, Moody AJ. 1990. Inhibitor effects on redox-linked protonations of the b hemes of the mitochondrial bc_1 complex. *Biochim. Biophys. Acta* 1018:29–40

162. Zweck A, Bechmann G, Weiss H. 1989. The pathway of the quinol/quinone transhydrogenation reaction in ubiquinol:cytochrome-c reductase of *Neurospora* mitochondria. *Eur. J. Biochem.* 183:199–203

163. Rich PR, Madgwick SA, Moss DA. 1991. The interactions of duroquinol, DBMIB and NQNO with the chloroplast cytochrome bf complex. *Biochim. Biophys. Acta* 1058:312–28

164. Dutton PL, Wilson DM. 1976. Redox potentiometry in biological systems. *Methods Enzymol.* 54:411–35

165. Glaser EG, Meinhardt SW, Crofts AR. 1984. Reduction of cytochrome b_{561} through the antimycin-sensitive site of the ubiquinol:cytochrome c_2 oxidoreductase

complex of *Rps. sphaeroides*. *FEBS Lett.* 178:336–42

166. Gray KA, Dutton PL, Daldal F. 1994. Requirement of histidine-217 for ubiquinone reductase activity (Q_i-site) in the cytochrome-bc_1 complex. *Biochemistry* 33:723–33

167. Dutton PL, Jackson JB. 1972. Thermodynamic and kinetics characterization of electron-transfer components in situ in *Rps. spheroides* and *Rhodospirillum rubrum*. *Eur. J. Biochem.* 30:495–510

168. Meinhardt SW, Crofts AR. 1984. A new effect of antimycin on the *b*-cytochromes of *Rps. sphaeroides*. In *Advances in Photosynthesis Research*, ed. C Sybesma, 1:649–52. The Hague: Nijhoff/Junk

169. Hacker B. 1994. *Mutational studies of the cytochrome bc_1 complex of* Rhodobacter sphaeroides. PhD thesis, Univ. Illinois, Urbana-Champaign. 94 pp.

170. Snyder CH, Trumpower BL. 1999. Ubiquinone at center n is responsible for triphasic reduction of cytochrome b in the cytochrome bc_1 complex. *J. Biol. Chem.* 274:31209–16

171. Bechmann G, Weiss H, Rich PR. 1992. Non-linear inhibition curves for tight-binding inhibitors of dimeric ubiquinol-cytochrome c oxidoreductases. Evidence for rapid inhibitor mobility. *Eur. J. Biochem.* 208(2):315–25

172. Rich PR, Bendall DS. 1980. The kinetics and thermodynamics of the reduction of cytochrome c by substituted p-benzoquinols in solution. *Biochim. Biophys. Acta* 592:506–18

Annu. Rev. Physiol. 2004. 66:735–69
doi: 10.1146/annurev.physiol.66.082602.092845
Copyright © 2004 by Annual Reviews. All rights reserved
First published online as a Review in Advance on October 20, 2003

INTERPRETING THE BOLD SIGNAL

Nikos K. Logothetis[1] and Brian A. Wandell[2]

[1]Max-Planck Institut für Biologische Kybernetik, Tübingen, Germany
[2]Department of Psychology, Stanford University, Stanford, California;
email: wandell@stanford.edu

Key Words visual cortex, blood-oxygen-level-dependent, fMRI, local field potential, multiunit activity

■ **Abstract** The development of functional magnetic resonance imaging (fMRI) has brought together a broad community of scientists interested in measuring the neural basis of the human mind. Because fMRI signals are an indirect measure of neural activity, interpreting these signals to make deductions about the nervous system requires some understanding of the signaling mechanisms. We describe our current understanding of the causal relationships between neural activity and the blood-oxygen-level-dependent (BOLD) signal, and we review how these analyses have challenged some basic assumptions that have guided neuroscience. We conclude with a discussion of how to use the BOLD signal to make inferences about the neural signal.

INTRODUCTION

Functional magnetic resonance imaging (fMRI) has greatly increased our ability to study localized brain activity in humans. Because humans are a cooperative and intelligent subject population, fMRI can be used to study problems that are very difficult to approach in animals. For example, what are the neural mechanisms that allow us to see objects, plan for the future, and recall ideas? How are feelings and thoughts represented in the brain?

The broad interest in these questions, coupled with the widespread availability of magnetic resonance (MR) scanners, brings together many investigators with differing perspectives on how to analyze brain computations. Also, each neuroscience tool, including the microscope, the microelectrode, and behavior, has limitations; each tool informs us about some, but not all, components of neural computation. Because of the diversity of investigators and the limited understanding of this new tool, fMRI data have been interpreted in a variety of ways.

As with other tools, fMRI will be integrated effectively into neuroscience when we can identify which aspect of the neural signals fMRI measures and how these measurements relate to those made with other instruments. Here, we describe our current understanding of the causal relationships between neural activity and the

0066-4278/04/0315-0735$14.00

blood-oxygen-level-dependent (BOLD) fMRI signals. We further analyze how to interpret the BOLD signal and indirect measure of brain activity in terms of the brain's synaptic activity and action potentials.

The process of understanding the relationship between the BOLD signal and neural events has challenged some basic assumptions that have guided neuroscience. For the past 40 years, measurements of cortical processing have been dominated by the analysis of action potentials. We now know that the BOLD signal does not correlate perfectly with action potentials, but rather measures a mix of continuous membrane potentials and action potentials. This complicates any comparison of it with single-unit physiology, but it also offers an opportunity to consider whether our understanding of neural information processing might extend beyond action potentials to include a range of signals that are an important part of neural computation.

With recent advances in understanding the physiological mechanisms that convert neural activity to BOLD signals, we have some answers about a question that is at the heart of the vast majority of experimental fMRI papers: What does the BOLD signal tell us about the neural signal? The reader should not be disappointed to learn that there are only partial answers as to how to design experiments that provide unambiguous interpretations from BOLD signals to neural signals. Methods of relating different neuroscience measurements have evolved over many decades in the areas of anatomy, microelectrodes, genomics, and behavior. Rather, we hope the reader will be encouraged to learn that good progress has been made and that it is possible to seriously address this important issue. The main portion of the review describes our view of the signaling pathway.

We conclude with a speculative discussion of what the BOLD signal can teach us about the neural signal. We note that some commonly used BOLD measures, for example the absolute size of the fMRI contrast signal, cannot be relied upon to measure the amplitude of the neural responses at two different cortical locations. On the other hand, measurements of stimulus selectivity at a single cortical location have a better logical foundation. Measurements of spatial maps are on a solid footing as well, although the spatial precision of map measurements may differ across cortex. We describe a preliminary mathematical framework that may be useful when extrapolating from BOLD to neural signals.

THE BOLD SIGNAL

The MRI Signal

Magnetic resonance measures how radio frequency electromagnetic waves act upon dipoles in a magnetic field. In the specific case of brain measurements, the MR signals arise mainly from the hydrogen nuclei in water, which are the only dipoles present in significant density to support measurements at high spatial resolution. The MR signal measures how these dipoles transition between different energy states. The now familiar MR images represent measurements of how these transitions differ depending upon the properties of the nearby tissue or the physiological state of the brain.

The hydrogen nuclei achieve a relatively organized, low-energy state when the subject enters the static (B0) field of the MR scanner. The strength of this magnetic field is specified in units of Tesla, and common field strengths range from 1.5 to 4.7 T. In the presence of a strong magnetic field, there is a tendency for the magnetic moment of these nuclei to align parallel or antiparallel to the main field—in much the same way as the dipole moment of a small bar magnet aligns itself with the local magnetic field.

The magnetic resonance measurement begins when the experimenter introduces a radio frequency (rf) pulse into the tissue. This pulse excites nuclei away from their resting state into a higher energy state. For any particular dipole, such as a hydrogen nucleus, this excitation is only effective at a resonance frequency known as the Larmor frequency. At this frequency, which is proportional to the magnetic field strength and the gyromagnetic constant of the hydrogen nucleus, the radio pulse contributes precisely the amount of energy needed to cause the nuclei to transition to a higher energy level and realign in the magnetic field.

The rf excitation pulses and the magnetic field gradients superimposed on the B0 field can be applied according to a variety of different timing and amplitude parameters. Adjustments to these parameters permit the investigator to obtain images that distinguish various properties of the brain, including structure (anatomical imaging), flow (perfusion imaging), or neural activity (functional imaging).

Information about the nearby tissue is derived from the rate at which the hydrogen nuclei return to the low-energy state following the excitation. Imagine the average initial orientation of a population of dipoles as a small vector pointed in three-dimensional space; this vector is aligned with the B0 field. The rf excitation pulse changes this population average. The relaxation back to the original state can be described as changes in two dimensions, longitudinal re-growth and transverse relaxation. Two exponential processes with time constants (T1 and T2) describe the relaxation back to the low-energy state. The T1 constant measures the relaxation in the direction of the B0 magnetic field (longitudinal re-growth). The T2 constant measures the transverse relaxation of the dipole in the x-y plane that is perpendicular to the B0 field. These changes in the local magnetic field are measured by special equipment (coils) that is placed within the scanner.

The transverse relaxation, also called spin dephasing or spin-spin interactions, is of special significance for fMRI. Any energy transition of a nucleus changes the local field at nearby nuclei; these exchanges are called spin-spin interactions. In an ideal homogeneous magnetic field, the transverse relaxation follows an exponential signal decay (free-induction decay, FID); the time constant of the decay is called T2. However, in physiological tissue the transverse relaxation is more rapid because of local field inhomogeneities including those caused by the tissue itself. When inhomogeneities are present, the decay constant is called T2*. Field variations randomly alter the frequency of the proton's precession, disturbing the phase coherence and speeding the transverse relaxation. In the brain, the size of these inhomogeneities depends upon the physiological state and in particular the composition of the local blood supply. This physiological state depends, in turn, on the neural activity. For this reason, measurement of the T2* parameter is an indirect measurement of neural activity.

MR signals are significant for neuroscience applications because it is possible to measure T2* at fairly high spatial resolution across the entire brain. These measurements are obtained by superimposing small gradients upon the main B0 magnetic field (for descriptions on image formation, see 1, 2). Much of the present review describes the chain connecting the neural activity and the resulting spatially varying field inhomogeneities that are measured by T2*.

The BOLD Contrast Mechanism

The mechanisms connecting neural activity to the measured T2* value are generally called the BOLD contrast mechanism; this measurement is currently the mainstay of brain fMRI studies.

As its name suggests, the BOLD contrast mechanism alters the T2* parameter mainly through neural activity–dependent changes in the relative concentration of oxygenated and deoxygenated blood. Deoxyhemoglobin (dHb) is paramagnetic (3) and influences the MR signal (4) unlike oxygenated Hb. In the presence of dHb, the T2 value decreases quadratically with field strength, as expected from the dynamic averaging owing to diffusion in the presence of field gradients (5, 6). The effects of dHb on T2* are even stronger, as first noticed by Ogawa et al. in their seminal studies on the rat brain in high fields (7–9). Specifically, Ogawa & Lee observed that blood vessel contrast varied with changes in blood oxygen demand or flow (9). They attributed the contrast increase to a magnetic susceptibility effect associated with the paramagnetic deoxyhemoglobin in red cells (7). They recognized the significance of their finding and concluded "BOLD contrast adds an additional feature to magnetic resonance imaging and complements other techniques that are attempting to provide positron emission tomography-like measurements related to regional neural activity" (8). Subsequently, the effect was demonstrated in the cat brain during the course of anoxia (10).

Given these observations, the reader may now be wondering about the origin of the signal increases typically reported in imaging experiments. Upon neural activation, any increases in dHb would be expected to enhance the field inhomogeneities and reduce rather than increase the BOLD contrast. The observed enhancement is due to an increase in cerebral blood flow (CBF) that overcompensates for the decrease in oxygen, delivering an oversupply of oxygenated blood (11, 12). At present, the reason for the mismatch between supply and consumption of blood oxygen is unclear. In a perfectly regulated system, the blood oxygen level would follow demand as neural activity waxed and waned. In fact, glucose supply does appear to match the consumption. Why should the glucose supply, but not the oxygen supply, match demand?

One possibility is that the vasculature delivers a fixed ratio of oxygen and glucose that is appropriate for an aerobic process. If both aerobic and anaerobic processes demand glucose, then the result would be an oxygen surplus (e.g., see 13). According to this hypothesis, the intensity of the BOLD signal depends on the relative proportion of local aerobic and anaerobic glucose metabolism, and understanding the distribution of anaerobic metabolism becomes an important consideration in interpreting the BOLD signal.

An alternative hypothesis is that oxygen extraction matches the metabolic needs and the excess oxygen present in the blood supply is due to an inefficient delivery process. Specifically, it has been proposed that this oversupply compensates for the inefficient, passive oxygen diffusion that occurs at high flow rates (14, 15). On this hypothesis, oxygenation supply is tightly coupled to neural activity, and the anaerobic processes in the astrocytes represent only a negligible amount of energy that is supplied by existing reserves (16, 17). There are some contradictory observations, however, so that the interpretation of these circulatory and metabolic changes remains uncertain (17, 18).

Physiological explanations of the BOLD signal can be tested by comparing theory with the amplitude and time course of the measured BOLD. The time course of the human BOLD response to a brief stimulus, the temporal impulse response function, is often called the hemodynamic response function (HRF). There is heterogeneity in the HRF across cortex of an individual observer (or animal), between observers, and possibly across different sensory, motor, and cognitive tasks (19–22). Figure 1a shows examples of the time course of the BOLD signals in visual and motor cortex after subjects viewed a 2-s stimulus or performed a motor act lasting 2 s (19, 23). The neural response to the sensory and motor acts ends fairly quickly, within a few hundred milliseconds past T = 0; the BOLD response begins roughly 2 s later, e.g., (24), rises to a plateau 6–9 s after stimulus onset, and then returns to baseline. In some instances the BOLD response has a post-stimulus undershoot, as can be seen in one of the four curves in Figure 1a (25–27).

There has been considerable focus on whether linear systems methods accurately model the BOLD signal. Specifically, a number of investigators have tried using the responses to brief stimuli (2 s) to predict the responses to visual or motor acts of longer duration (>8 s). Examples of measurements of responses to 8 s of visual stimulation and an 8-s motor act are shown in Figure 1b. The simple prediction of a linear time invariant (LTI) system, in which the response to the long-duration event is predicted from that to the short-duration event, fails significantly (19, 28, 29). It is possible, however, to make reasonably accurate predictions for responses to long-stimulus presentations on the basis of responses to stimuli of 6 s and greater (19, 28, 30) and to predict responses to brief stimuli from the responses to even briefer stimuli. Several authors have proposed modifications to the linear model to account for the inability to predict very long-duration responses from very short-duration responses (19, 28).

Despite the imprecision of the linear model predictions, a number of theorists have proposed formulae to describe the impulse response function. The formulae

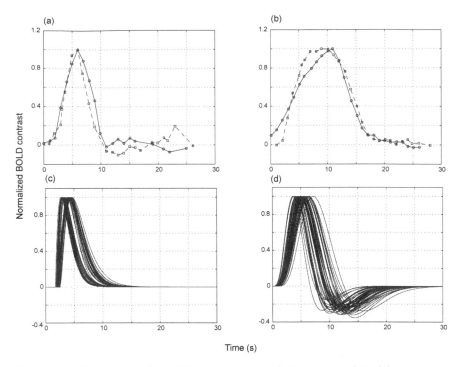

(a) (b) (c) (d)

Normalized BOLD contrast

Time (s)

Figure 1   Time course of the BOLD response. (*a,b*) Data are replotted from experiments in motor cortex (*open circles*) (19) and visual cortex (*open squares*) (161). The two panels show measurements in response to a visual stimulus or movement of 2 s (*a*) or 8 s duration (*b*). (*c,d*) Theoretical temporal impulse response functions (also called hemodynamic response functions) used in the statistical analyses of the BOLD response. These curves are intended to represent the BOLD response to brief stimuli of unit amplitude. The distributions of curves shown in these figures are derived from formulae in (162) (*c*) and (32) (*d*). The distributions of curves were created by randomly perturbing the parameters in the respective formulae by 10%.

for the HRF usually include parameters that permit the investigator to approximate a range of HRFs such as those in Figure 1*a,b*. Data analysis packages all include some assumption about the HRF, and two examples are shown in Figure 1*c,d*. The collection of curves in Figures 1*c,d* show the consequence of varying the parameters over a modest (10%) range (30–32).

The Significance of Human fMRI

Ogawa's groundbreaking work generated great interest in human BOLD fMRI. In 1992, three groups simultaneously and independently took the method a step further by measuring the intrinsic BOLD signal in humans (24, 33, 34). These reports initiated the flood of fMRI publications that have appeared in scientific journals ever since.

The success of fMRI stems partly from the availability of scanners, the noninvasive nature of the measurement, and its high spatiotemporal resolution compared with other methods available for human research. Prior to the BOLD fMRI, the only methods available to study the alert human brain were electro-encephalograms (EEG; scalp potentials), magneto-encephalograms (MEG), and positron emission tomography (PET). These methods suffer from either poor resolution or lack of spatial localization. The two-point spatial resolution of fMRI, on the other hand, at least in visual cortex, is on the order of two millimeters (35). Increased averaging can, in principle, yield comparable spatial resolution in PET, but this method requires the introduction of radioactive substances into the subject so that extensive averaging or even many repeated studies of a single brain are not practical. Using fMRI, one can study the same human brain for many hundreds of hours without significant health risk. This ability both improves the signal-to-noise ratio and permits the study of individuals whose brains are of particular interest, such as neurological case studies.

The Neural Signal

To form a more complete picture of the neural mechanisms underlying behavior, results from human fMRI experiments are often compared with those obtained in electrophysiology made in homologous regions of monkey cortex. The vast majority of such experiments in conscious animals report extracellular electrophysiological recordings. It is therefore worth considering the question: What do extracellular electrodes measure?

The Extracellular Field Potential

The extracellular medium surrounding neurons is a volume conductor with a resistivity (specific impedance) that ranges from 200–400 Ωcm depending on the neural site (36–39). This extracellular resistivity is higher than that of a saline bath, $\sim$65 Ωcm, 1 Hz to –10 kHz (N.K. Logothetis, unpublished measurements). The increased resistivity is due to limitations on the flow of ions in the extracellular medium compared with the saline. For the frequency range of electrophysiological signals (0 to about 2 kHz), inductive, magnetic, and propagative effects can be ignored, and the electric field can be considered homogeneous and purely resistive, obeying Ohm's law. In such a static description, the flow of positive ions (e.g., Na^+) into the active sites of a neuron appears as a current sink (inward current); the injected current flows down the core of the dendrites or axon. Because currents flow in closed loops, a current outflow from distant inactive membrane sites appears as a source (outward current) from which current flows back to the injection site through the extracellular medium. The finite resistance of the latter creates the extracellular field potentials (EFPs) that are measured by electrodes.

The weighted spatio-temporal sum of the sinks and sources from multiple cells is often called the mean extracellular field potential (mEFP; the black trace in

Figure 2*a*, see color insert). The weighting depends on several factors. Most importantly, the spatial alignment of elongated neural processes establishes an anisotropic conductivity across space and imparts a preferred directionality to the current flow. The directionality complicates the interpretation of the mEFP measurement: Cells whose current flow travels in opposing directions will cancel each other out. There are no practical methods for recovering information about the spatial conductance pattern because such information is rarely available, and there is no widely agreed upon model for interpreting the mEFP with respect to the factors that influence the cellular currents.

Despite its limitations the mEFP provides an important tool for system analysis. Depending on the recording site, on the choice of electrode size and type, on signal processing, and on the sophistication of mathematical analyses of its spatiotemporal distribution, the mEFP can be still used to explore the functional properties of neural tissues at a large range of spatial scales. It can be employed to study properties ranging from single neurons to neuronal ensembles on a scale of hundreds of microns. In neural structures with known anatomical organization the study of ensembles often provides insights in the cellular components involved in different computations, and the comparison of single- and multiple-unit activity is useful for defining functional modules. We suspect the same will prove true of the fMRI signal, especially as we learn more about how to constrain its interpretation by combining imaging with other invasive or non-invasive modalities.

Spiking Activity

If a microelectrode with a small tip is placed close to the soma or axon of a spiking neuron, spikes can be detected in the mEFP signal (Figure 2*b*). It is also possible to measure spikes simultaneously from a collection of cells in a small neighborhood using arrays of electrodes (40). Recent studies in rats, for instance, show that a group of four closely spaced electrodes (tetrodes) placed within 50–100 μm of pyramidal neurons in hippocampus provide accurate information on a number of spike parameters, including latency and amplitude; these spikes are measured simultaneously by intracellular recordings (41, 42).

Ever since the early development of microelectrodes, action potentials of well-isolated neurons have been used to understand the neural basis of behavior (43). The dominance of the single-unit action potential was strengthened by its suitability for studying the first steps of afferent information processing in sensory physiology (44–47).

Computation

There are compelling reasons for focusing on the role of the action potential; for example, the action potential is the only way neurons communicate over a distance. Hence, ganglion cell action potentials contain all of the information communicated from the retina to the rest of the nervous system. In general, information between major structures—such as the LGN to V1—is contained within the action potentials of the output neurons of these structures. Understanding this representation is essential to understanding neural computations.

Put this way, it is also easy to see the drawback of studying only action potentials: We learn little about the sub-threshold integrative processes that are essential for producing spikes within these structures. Within the retina, for example, ganglion cells are the only spiking neurons. If we restrict our study to ganglion cell spikes, we would miss the origins of trichromacy in the continuous photo-responses of the cone photoreceptors; we would miss how the rod and cone circuitry combine to represent signals spanning eight orders of magnitude; the formation of color opponency and the organization of the rod pathways would stay hidden from view. An analysis of the retina restricted to spikes is superficial, much like a view of the cortex that ends by declaring one region a face area and another a color area. Measuring only the spikes on large axons projecting between areas misses an essential question: What is the complete set of neural computations, including spikes and sub-threshold synaptic potentials, that create the output responses?

Sampling Bias

Single-unit recordings, in particular using large electrodes, do not sample cell sizes and types with equal probability (48, 49). Electrodes are more likely to sample large cells, and this bias, in turn, is partially responsible for the cell-type bias. The reason for the bias can be understood from basic signaling considerations. An action potential from a large neuron generates greater membrane current and a larger extracellular spike than an action potential from a small cell. Consequently, the extracellular field from a large cell remains above recording noise levels over a greater distance (50). For distances greater than about 140 μm, spikes become indistinguishable from background noise (42). Consequently, microelectrodes are likely to sample preferentially the somas and axons of large cells, a prediction supported by experimental work (49, 51). Further, in awake-behaving preparations, movement artifacts are likely to shift the microelectrode out of the range of a small cell's extracellular potential, making it difficult to hold the signals from such cells for the duration of an experiment. For all these reasons, we suggest that extracellular electrodes probably over-sample large cells (e.g., pyramidal cells in cerebral cortex and Purkinje neurons in cerebellar cortex).

An interesting study by Henze and colleagues provides more evidence of sampling bias. These investigators made simultaneous in vivo, intracellular, and extracellular measurements in the hippocampus of the anesthetized rat. They placed a tetrode array within the hippocampus and repeatedly advanced an intracellular electrode searching for a signal that could also be measured by the tetrode array. The 22 neurons that could be measured with the two types of electrodes were morphologically identified by a biocytin injection (42). Of these successful measurements, 21 were pyramidal cells and 1 was an inhibitory interneuron (42). This sampling distribution differs from any reasonable accounting of the distribution of cell types in hippocampus, which like the rest of cortex includes certain proportions of large- and medium-size pyramidal cells, different sizes of stellates, and many inhibitory neurons as small as a few microns. The measured sample from the tetrode array suggests a strong bias either in the intracellular sample or in the ability to measure across cell types with the extracellular tetrode array.

Separating Sub-Threshold Potentials and Spiking Activity

If a microelectrode with a relatively large tip (low impedance) is placed in the extracellular space, somewhat distant from large spiking neurons, the mEFP is dominated by the analog synaptic voltages (dendritic events) and summed action potentials from hundreds of nearby neurons (52, 53). This mEFP voltage can be subdivided into two parts that measure either the dendritic events or the summed spikes (for review see 54, 55). Specifically, applying a high-pass filter (cut-off 300–400 Hz) measures the multiple-unit spiking activity (MUA; Figure 2c, *upper panel*); applying a low-pass filter (cutoff < 200 Hz) produces a waveform that reflects the local synaptic voltages (LFP; Figure 2c, *lower panel*).

Multiunit Activity

Combined physiology-histology experiments demonstrate that the magnitude of MUA varies considerably across brain sites (e.g., neocortex versus hippocampus); within a region, however, the MUA magnitude is relatively constant. The local cell size is an important factor in determining the MUA magnitude (56, 57). Sites of large-amplitude and fast MUA signals generally have homogeneous populations of large cells (58). These experiments further show that amplitude of MUA signals covaries with the magnitude of extracellular spike potentials.

Local Field Potentials

The low-frequency range of the mEFP signal, the LFPs, represents slow electrical signals and sub-threshold activity. Until recently these signals were thought to represent synaptic events exclusively. Evidence for this came from combined electro-encephalographic (EEG) and intracortical recordings showing that the slow-wave activity in the EEG is largely independent of neuronal spiking (59–62). Unlike the MUA, the LFP magnitude is not correlated with cell size. Instead, the LFP amplitude reflects the extent and geometry of dendrites in each recording site. Cells in the so-called open field geometrical arrangement, in which dendrites face in one direction and somata in another, produce strong dendrite-to-soma dipoles when they are activated by synchronous synaptic input. Other cortical neurons are oriented horizontally and contribute less efficiently or not at all to the sum of potentials. The pyramidal cells with their apical dendrites running parallel to each other and perpendicular to the pial surface form an ideal open field arrangement and contribute maximally to both the macroscopically measured EEG and the local field potentials.

Thus LFPs reflect primarily a weighted average of synchronized dendro-somatic components of the synaptic signals of a neural population, most likely from within

0.5–3 mm of the electrode tip (63, 64). Yet, it has been further shown that LFPs may also measure other types of slow activity unrelated to synaptic events, including voltage-dependent membrane oscillations (e.g., 65) and spike afterpotentials. To be more specific, the soma-dendritic spikes in the neurons of the central nervous system are generally followed by afterpotentials, a brief delayed depolarization, the afterdepolarization, and a longer lasting afterhyperpolarization, which are thought to play an important role in the control of excitation-to-frequency transduction (e.g., 66–68). Afterpotentials, which are generated by calcium-activated potassium currents (e.g., 67, 69–72), have a duration on the order of 10 s of milliseconds and most likely contribute to the generation of the LFP signals, as first suggested by Buzsaki and his colleagues (73, 74).

In summary, MUA measures regional neuronal spiking, whereas the LFP measures slow waveforms, including synaptic potentials, afterpotentials of somato-dendritic spikes, and voltage-gated membrane oscillations. Hence, the LFPs are information-rich signals that may influence local neural excitations (75) and may reflect aspects of the input signal and the local intracortical processing mediated by the sub-threshold signals of interneurons. Unlike the MUA, the LFP does not reflect the action potentials carried by the principal (output) neurons.

BOLD IMAGES OF THE NEURAL SIGNAL

In the previous section, we reviewed a variety of ways to analyze the mEFP and described how each method could be an important measurement of the signals carried by the neuronal ensemble. In this section, we review the properties of the BOLD signal with the goal of finding its place within the array of measurements of the neural signal.

Sensory experiments frequently measure the relationship between stimulus energy and the BOLD response. In general, this relationship is nonlinear—the BOLD response increases according to a compressive, nonlinear, saturating function of stimulus energy (76). There are two stages of the path from stimulus energy to BOLD response that may introduce this nonlinearity: The neural signal may depend nonlinearly on the stimulus energy, and the BOLD response may depend nonlinearly on the neural signal. A number of studies using various techniques have presented data showing that in certain regimes there is a linear relationship between neural activity and the subsequent BOLD response (77–80). This hypothesis is important because if there is a linear relationship, there is a realistic possibility of inferring neural signals from the BOLD response.

Logothetis and colleagues recently examined the relationship between BOLD and neural activity by simultaneously acquiring electrophysiological and fMRI data from monkeys in a 4.7 T vertical scanner (27, 29, 81). The BOLD response reflected a local increase in neural activity as assessed by the mEFP signal. For the majority of recording sites, the amplitude of the BOLD response was a linearly increasing, but not a time invariant, function of LFPs, MUA, and the firing rate of small neural populations. A proportional increase in neural and hemodynamic signals was observed in experiments varying (within a limited range) the stimulus luminance contrast (81). This linear relationship over the measured range did not extend to zero, suggesting that the entire relationship is nonlinear. The nature of the nonlinearity in the low-contrast ranges, specifically, whether it is the tail-end or saturation point of a compressive function as indicated to us by A. Movshon (personal communication), remains to be investigated. The correlation between firing rate and BOLD was also confirmed with MRI/MRS studies in rats, whereby a linear relationship between the cerebral metabolic rate, glutamatergic neurotransmitter flux, and neural activity was demonstrated (82, 83).

The relative contribution of MUA and LFP signals to the BOLD response was examined by applying time-dependent frequency analysis to the raw data (81). In all these experiments, increases in the LFP range were both of greater spectral power and higher predictive reliability. After stimulus presentation, a transient increase in power was typically observed across all frequencies, followed by a lower power level that was maintained for the entire duration of stimulus presentation. A prominent characteristic in all of the spectrograms was a marked stimulus-induced increase in the magnitude of the LFP; this increase was always larger than that observed for MUA. A decrease in neural firing rates was also observed immediately after the termination of the stimulus.

Comparing LFPs and MUA as Predictors of the BOLD Response

Although the BOLD response is not linearly dependent on neural signals under all conditions, linear time-invariance (LTI) is a reasonable first approximation over restricted ranges (54, 55, 84, 85). In an LTI system, responses to arbitrary stimuli can be predicted from knowledge of the input signal and the system's impulse response function.

To evaluate which of the neural signals is the causal agent in creating the BOLD response, Logothetis et al. examined how well LTI methods could be used to predict BOLD fMRI signal from LFPs, MUAs, and spikes (81). The neural and BOLD measurements were obtained from a variety of cortical sites. Measurements included simultaneous recording of visually driven signals as well as spontaneous brain activity. The latter signals, when additionally prewhitened, offer an unbiased estimate of the hemodynamic impulse response. Impulse response functions (see Figure 2) were estimated from the responses to briefly presented stimuli, and the estimated impulse response function was convolved with several types of neural signals to evaluate how well each type of signal predicted the BOLD response.

Figure 3 (see color insert) shows the neural responses as well as the measured and estimated BOLD responses for four stimulus durations at one cortical site. Estimates of the time course of the neural response were relatively accurate over a range of stimulus durations (up to 12 s). As the stimulus duration increases, so do the errors. Deviations from linearity of this sort have also been described in the

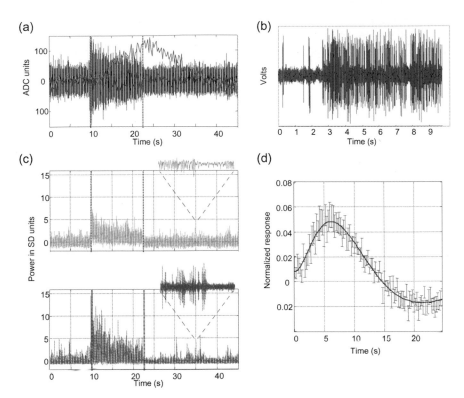

Figure 2 Simultaneous measurements of the neural and BOLD signals. (*a*) The black trace is a comprehensive (full-bandwidth; 0.050 Hz–22.3 kHz) mEFP signal. The red trace shows the average BOLD response measured by combining approximately 25 voxels (1 × 1 × 2 mm) near the electrode tip. The vertical red dotted lines delimit the stimulus presentation period. (*b*) Spike activity derived from the mEFP. (*c*) Frequency band separation of the mEFP. The green trace (*upper panel*) is the LFP. The signal is extracted by band pass filtering (10–150 Hz), rectification, and resampling (500 Hz). The blue trace (*lower panel*) is the multiunit activity (MUA). The signal is extracted from mEFP by band pass filtering (500–3000 Hz), rectification (absolute value), and resampling (1 Hz). The distinct characteristics of LFP and MUA signals before rectification are given in the expanded time scale of the insets, which show an interval of 1 s. (*d*) Estimated temporal impulse response function relating the neurophysiological and BOLD measurements in monkey (81). Very similar estimates are obtained using the LFP or MUA as the input neurophysiological signal.

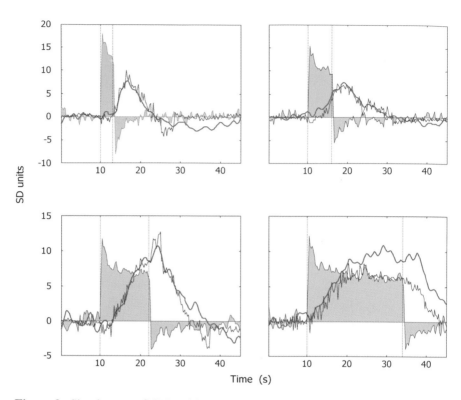

Figure 3   Simultaneous fMRI and intracortical recordings for visual stimuli of different duration. The four panels show responses to stimuli of 3, 6, 12, and 24 s. The blue-shaded trace is the LFP; the red trace is the BOLD response; the gray trace is the predicted BOLD response after convolution of the LFP with the temporal impulse response function. Linear time invariant (LTI) predictions are in good agreement over the duration range from 3 to 12 s, but the predictions of the latter portion of the 24-s stimulus are poor, suggesting the existence of nonlinearities not captured by the LTI analysis. The failure of accurate estimation of longer responses is in agreement with reports from human data, as described in the text.

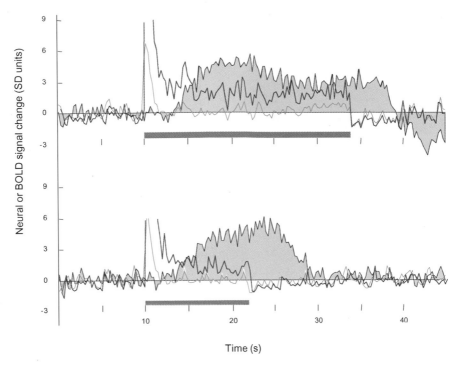

Figure 4 A cortical site in which the LFP and MUA predictions of the BOLD response differ significantly. Responses to a 24-s (*upper*) and 12-s (*lower*) stimulation are shown. The blue trace measures the LFP, the green trace measures the MUA, and the red-shaded trace measures the BOLD response. The MUA signal is brief and approximately the same in both experiments. The LFP and BOLD responses covary with the stimulus duration. At this cortical site and in 25% of the measurements, the LFP response matched the BOLD response but the MUA did not. There were no instances in which the MUA matched the BOLD response, but the LFP did not.

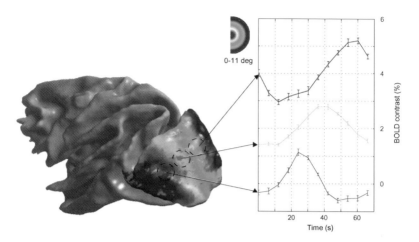

Figure 6 BOLD estimates of retinotopic organization in macaque V1 agree with single-unit estimates. The visual stimuli were a series of slowly expanding rings, each containing a collection of flickering squares (160, 163, 164). The stimulus begins as a small spot located at the center of the visual field; the spot becomes an expanding ring that grows to the edge of the stimulus display. As the ring disappears from view, a new spot, starting at the center replaces it. This stimulus causes a traveling wave of neural activity beginning in the foveal representation, several millimeters posterior to the lunate sulcus. Measuring the time course of the BOLD responses at a series of 3-mm radius regions-of-interest, we see that the BOLD response near the foveal representation is phase-advanced compared with the BOLD response measured in the more peripheral representations (*graph at right*). The phase of the BOLD response measures the visual field eccentricity that most effectively stimulates each cortical location. The most effective visual field eccentricity is indicated by pseudo-coloring the cortical surface according to the inset at the top (adapted from 142).

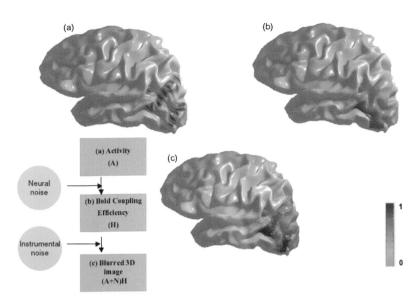

Figure 7 A computational model of the BOLD response. The inset in the lower left summarizes the sequence of steps used to create the three simulation images. Human cortex is segmented and approximated with a geometrically defined surface; the surface is presented as slightly inflated in order to expose the sulci. (*a*) Simulated neural activity, shown at its true spatial resolution. The neural activity varies at fine spatial resolution in the dorsal-ventral direction and at a coarser resolution in the posterior-anterior direction. (*b*) A representation of the coupling between the neural response and the BOLD response (HRE). In this simulation, the HRE is high on the lateral surface and low on the dorsal surface. (*c*) The product of the neural activity and HRE is blurred using a 5-mm (full-width half height) Gaussian kernel. The blurring is performed along the cortical surface (160). The pseudo-color overlay codes the relative size of the simulated measurements (arbitrary units; *key at right*).

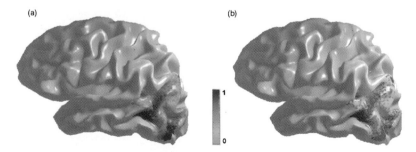

Figure 8 Simulations of the differential BOLD response. (*a*) The simulated differential BOLD response to the neural activity in Figure 6*a*. (*b*) The same BOLD response is shown after thresholding with a relatively high criterion. Other details as in Figure 7.

analysis of linear systems properties in the human literature (see above; e.g., 19, 30).

In general, LFPs and MUA vary in a similar manner. Hence, at those sites where the LFPs predicted the BOLD response, the MUA did too. Across cortical sites there was a tendency for the LFP-based estimate to perform slightly better than the MUA-based estimate: The LFP signal predicted 7.6% more of the variance than the MUA. The difference, although small, was statistically significant. The larger variability of MUA was mostly attributable to the stronger adaptation effects observed in this frequency range of the mEFP.

Spikes Can Be Dissociated from LFP Signals

There is a strong correlation between local field potentials and BOLD response. However, we still cannot be certain that the slightly greater predictive power of the LFP over the MUA implies that spikes are not a key variable. Because of the correlation between the LFP and MUA signals, either continuous potentials or spikes could be the key determinant of the BOLD response. Circumstances do exist, however, in which there is a dissociation between LFPs and spikes. Such a dissociation was observed in visually driven neuronal responses in striate cortex (81).

Figure 4 (see color insert) illustrates simultaneous measurements of LFP, MUA, and BOLD at a single cortical site in response to visual stimuli. The upper and lower panels show the responses to a 24- and 12-s stimuli, respectively. At this site, the MUA was of brief duration in response to both stimuli, returning to baseline approximately 2.5 s after stimulus onset. In contrast, the LFP duration was similar to that of the stimulus, remaining elevated for the entire stimulus duration. The BOLD response also varied with stimulus duration. Hence, at this site, where the MUA and LFP differ, only the LFP predicted the BOLD response. This dissociation could be observed in about 25% of the responses. There was no single observation period or recording site for which the opposite result was observed; that is, there was no measurement in which the MUA was correlated with the BOLD but the LFP was not.

An exquisite example of LFP-MUA dissociation was reported for cerebellar cortex by Lauritzen and collaborators (78, 86). In cerebellar cortex, climbing fiber stimulation causes monosynaptic excitation of the Purkinje cells, whereas simultaneous parallel fiber stimulation causes a disynaptic inhibition of the Purkinje cell spikes (via the basket cells). Lauritzen and colleagues measured LFPs, single-unit activity, and changes in cerebral flow by means of laser Doppler flowmetry (78). They demonstrated that despite the inhibition of Purkinje cell spikes, parallel fiber stimulation increases overall synaptic activity. Both LFPs and cerebral blood flow increased, following the synaptic activity, even though spiking activity ceased.

Finally, exploiting the differential effect of neuromodulators on different cellular sites may experimentally induce LFP-MUA dissociation. In preliminary results in macaques from the Logothetis laboratory, neuromodulators were injected during simultaneous electrophysiological and BOLD neuroimaging experiments. Using a triple pipette (electrode, saline, and a neuromodulator), 20 microliters of 0.01 mol 5HT (5-hydroxytryptoamine hydrochloride) were injected over a period of 10 min. Several minutes after the injection a profound suppression of the MUA was observed. The LFP signal showed a slight increase and returned to baseline within a few minutes. During this time, in which the MUA was silenced, no significant change was discernible in the BOLD response. Spectrograms obtained before and after the 5HT injection during visual stimulation confirmed that the stimulus-induced spikes were entirely eliminated, although LFP activity was moderately increased. These measurements show that it is possible to dissociate pharmacologically spiking activity and hemodynamic responses. On the basis of the several dissociations described in this section, we conclude that the LFP signal is the key variable for the BOLD response.

In summary, the BOLD response primarily reflects the input and local processing of neuronal information rather than the output signals, which are transmitted to other regions of the brain by the principal neurons. The long-range projection signals from these principal neurons are the measurements that are mainly accessible in single-cell recordings in the behaving animal.

NEURAL SIGNALS, METABOLISM, AND BLOOD FLOW

In the previous section, we reviewed the correlation between the neural signal and the BOLD response. Here we examine the cellular and molecular neurovascular coupling mechanisms that mediate this correlation. These mechanisms are the scaffolding for the neurometabolic coupling between energy demand and supply.

The brain, as does the heart, makes ongoing demands for energy. The mean level of energy required by the brain is very high, comparable to the energy required by the heart. On top of this mean level, the regional demand fluctuates over time and the delivery of oxygen and glucose modulates correspondingly. Experimental descriptions of the coupling between demand and supply were provided over a century ago in laboratory animals (87) and later verified with methods allowing local cerebral flow measurements. Although such methods had been used in conscious laboratory animals since the early 1960s (88), a precise quantitative assessment of the coupling between neural activity and regional blood flow was only possible after the introduction of the deoxyglucose autoradiographic technique that enabled spatially resolved measurements of glucose metabolism in laboratory animals (89). The results of a large number of experiments with the $[^{14}C]$deoxyglucose method have revealed a clear relationship between local cerebral activation and glucose consumption (90).

The first quantitative measurements of regional brain blood flow and oxygen consumption in humans were performed using the radiotracer techniques, which were followed by the introduction of PET (for a historical review see 91). With PET images, maps of activated brain regions were produced by detecting the indirect effects of neural activity on variables such as cerebral blood flow (92), cerebral blood volume (CBV) (11), and blood oxygenation (11, 12, 93). At the same time, optical imaging of intrinsic signals demonstrated the precision of neurovascular coupling by constructing detailed maps of cortical microarchitecture in both the anesthetized and the alert animal (94).

Although the existence of a regional coupling between neural activity, metabolism, and hemodynamic changes is now established, the nature of the link between these processes remains an area of study. If energy demand triggers CBF changes, which cellular processes and sites dominate the energy consumption? The structural and functional organization of the neuro-vascular system provides some insights to these questions.

Structural Neurovascular Coupling

One method for studying vascular structure is to infuse low viscosity resins into the vasculature and allow the resin to polymerize. Dissolving away the surrounding tissue with alkali leaves a cast of the three-dimensional distribution of vessels; the cast can then be sectioned and studied with scanning electron microscopy (SEM). Such corrosion cast studies reveal a vast vascular network. These casts also reveal several correlations between vascular density and neuronal structures.

Several investigators have shown that vascular density correlates with the number of synapses, rather than the number of neurons (95–97). For instance, Duvernoy and colleagues demonstrated that the human cortical vascular network could be subdivided based on density into four layers parallel to the surface. These layers systematically overlap with certain portions of the cytoarchitectonically defined Brodmann laminae. The first Duvernoy layer, consisting of vessels oriented approximately parallel to neural fibers, is entirely within the lower part of the molecular layer (Layer I). This layer had the lowest vascularization and IVc had the highest vascularization with a ratio of 3.3:1 (IVc:I, averaged across animals). In layers IVc and I in macaque striate cortex, the synaptic density ratio is 2.43:1, the astrocyte density is 1.2:1, and the neuronal density is 78.8:1. Hence, the vascular density parallels that of the perisynaptic elements rather than the neuronal somata.

A general principle proposed by a number of investigators is that primary sensory areas are characterized by a higher stimulus responsivity and capillary density than association areas and that the vascularization follows cortical plasticity (e.g., 95, 98–103). Figure 5b (see color insert) shows a clear example from the Duvernoy et al. study. In this section through human calcarine cortex, one can see a dense vascularization in the input layer (4C) that ends abruptly. The termination of the dense vascularization is in a position that would ordinarily fall near the human V1/V2 boundary.

Similar results were obtained in different cortical areas of rodents. For instance, endovascular casts revealed capillary densities resembling the whisker barrel pattern that characterizes the somatosensory cortex of rats (104) and the spatial patterns of stimulus-induced activation in the auditory cortex of chinchillas (98). It seems that an anatomical neurovascular association exists, and this relationship

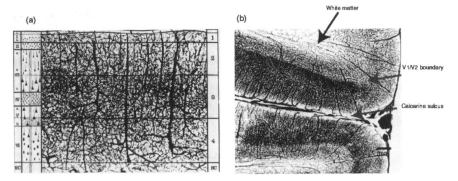

Figure 5   Cortical vascularization across layers and area boundaries. (*a*) Vascularization varies across the cortical layers, with the densest region falling near layers III and IV. Note that layer I is moderately well vascularized. The cell body density in layer I is two orders of magnitude lower than that of layer IVc. The synaptic density is about half that of IVc. Hence, the vascular density parallels synaptic density more closely than cell body density. (*b*) Vascular density varies across boundaries of functional areas. Dense vascularization is present in calcarine cortex, but the density ends abruptly at a location that is likely to be the V1/V2 boundary. These images were adapted from the work of Duvernoy and colleagues (95).

may be a source of the variations in regional coupling between neural activity and metabolism. The pattern of this anatomical neurovascular coupling suggests that blood supply is better correlated with the number of synapses rather than the number of neurons.

Functional Neurovascular Coupling

The cerebral metabolic rate (CMR) is commonly expressed in terms of oxygen consumption ($CMRO_2$). This is a convenient measure because about 90% of the glucose is aerobically metabolized so that the CMR parallels oxygen consumption (105, 106). The energy demands of different neural types probably reflect their electrical activity and physical shape, and size. Large projection neurons, which maintain energy-consuming processes over an extensive membrane surface, may have relatively large average energy requirements. Other cells, including glia and vascular endothelial cells, also play a role in metabolic function, and the specific mechanisms regulating the neurometabolic coupling constitute an area of active investigation. Several interesting cellular and molecular mechanisms and hypotheses have emerged concerning how the cerebral blood flow is coupled to energy consumption.

THE ROLE OF ASTROCYES IN COUPLING   There is a tightly regulated glucose metabolism in all brain cell types. An interesting case is the glial cell known as the astrocyte. The structural and functional characteristics of astrocytes make them ideal bridges between the neuropil and the intraparenchymal capillaries (for detailed references see 107). These specialized glia cells are massively connected with both neurons and the brain's vasculature.

It has been suggested that for each synaptically released glutamate molecule taken up with two to three Na^+ ions by an astrocyte, one glucose molecule enters the same astrocyte, two ATP molecules are produced through glycolysis, and two lactate molecules are released and consumed by neurons to yield 18 ATPs through oxidative phosphorylation. Recent studies established a quantitative relationship between imaging signals and the cycling of certain cerebral neurotransmitters (108–110) because synaptic activity is tightly coupled to glucose uptake (111, 112). Stoichiometric studies using NMR spectroscopy suggest that the utilization of glutamate (Glu), the dominant excitatory neurotransmitter of the brain (about 90% of the synapses in gray matter are excitatory) (113), is equal to the rate at which glutamate is converted to glutamine (Gln) (114). The Glu to Gln conversion occurs in the astrocytes, and the required energy is provided by glycolysis. Astrocytes are indeed enriched in glucose transporters. The transporters are driven by the electrochemical gradient of Na^+; for this reason there is a tight coupling between Glu and Na^+ uptake. Both Glu to Gln conversion and Na^+ restoration require ATP. Gln is subsequently released by astrocytes and taken up by the neuronal terminals to be reconverted to Glu (for review see 13). Calculations based on these findings suggest that the energy demands of glutamatergic neurons account for 80 to 90% of total cortical glucose usage in rats (114) and humans (115).

ENERGY CONSUMPTION   Which neural signaling events consume most of the energy? Two groups of investigators calculated an overall energy budget that estimates the energy requirements of neural signaling (16, 120). The budget is based on neural circuitry assumptions concerning the number of vesicles released per action potential, the number of post-synaptic receptors activated per vesicle released, the metabolic consequences of activating a single receptor and changing ion fluxes, and neurotransmitter recycling. The largest portion of energy consumption is attributed to the post-synaptic effects of glutamate (about 34% of the energy in rodents and 74% in humans).

An interesting possibility is that there is a dissociation between the mechanisms calling for the energy and those that use the energy; perhaps neurotransmitter-related signaling, rather than local energy use, drives the hemodynamic responses (121). Indeed, there is evidence that blood flow in a number of brain structures, including neocortex, cerebellum, and hippocampus, may be controlled directly by glutamate and GABA. In the cerebellar cortex, for example, the activation of parallel fibers releases glutamate and leads to the depolarization of Purkinje cells and interneurons. These cells, in turn, release GABA. Notably, the increased blood flow that typically follows the activation of parallel fibers is blocked by inhibitors of non-NMDA glutamate receptors, nitric oxide synthase, and adenosine receptors (122), whereas microinjections of glutamate have vascular effects similar to those observed during stimulation of the parallel fibers (123). In neocortex and hippocampus, microinjection of neurotransmitters dilates pial arterioles and/or precapillary microvessels, an effect attenuated by inhibitors of nitric oxide synthase (NOS) (124, 125).

These findings are in agreement with microstimulation experiments in which the increase in glucose utilization is assessed during orthodromic and antidromic stimulation. Orthodromic stimulation activates both pre- and post-synaptic terminals, whereas antidromic stimulation activates only post-synaptic terminals. Glucose utilization increased only during orthodromic stimulation (116–118; for review see 119).

Taken together, these results suggest that presynaptic activity (restoration of gradients) and neurotransmitter cycling are the main mechanisms that initiate brain energy production, e.g., via glial pathways. In this view, CBF is controlled by mechanisms that predict energy consumption, not by mechanisms that measure the consumption.

IMPLICATIONS FOR NEURAL CODING   While there are uncertainties concerning the mechanisms of CBF control, there is widespread agreement that cortical signaling consumes energy at high rates. This high level of energy consumption may have implications for neural coding strategies because, on the whole, it is better to conserve energy and reduce spike count. But before such a hypothesis can be evaluated, we must also have some measure of the benefits of these signals. The communicating brain, like the beating heart, is doing something worthwhile. A high cost (spike rate) may be worth the benefit for certain circuits and computations.

INTERPRETING THE BOLD SIGNAL

Many investigators use the BOLD response to estimate properties of neural signals. By describing the path from neural signals to the BOLD response, we hope to establish some principles for estimating neural signals from BOLD responses. In this section, we describe a framework to guide such inferences. First, we discuss two qualitative principles, and then we describe formulae and simulations specifically chosen to illustrate the limitations in using the BOLD response to estimate the amplitude and spatial distribution of neural signals.

Qualitative Principles

Given the complexity of the relationship between electrical activity and the BOLD signal, it will be some time before a complete model emerges. However, it is worthwhile to discuss some qualitative principles about the interpretation of the BOLD response. We describe two such principles concerning the interpretation of amplitudes and cortical maps.

COMPARISONS BETWEEN BOLD AND SINGLE-UNIT MEASUREMENTS: AMPLITUDES
A number of experiments demonstrate basic differences between the amplitude of
electrophysiological and BOLD responses. In one recent study Tolias and colleagues studied brain areas processing motion information (126) using an adaptation technique (127, 128). They repeatedly imaged a monkey's brain while the
animal viewed continuous motion in a single, unchanging direction. Under these
conditions, the BOLD response adapts. When the direction of motion abruptly
reverses, the measured activity immediately shows a partial recovery or rebound.
The extent of this rebound is an index of the average directional selectivity in the
activated area. The results confirmed previous electrophysiological studies revealing a distributed network of visual areas (V1, V2, V3, V5/MT) in the monkey that
process information about the direction of visual motion.

Surprisingly, strong BOLD activation was also observed in area V4; the neuroimaging rebound measurements and thus the inferred motion sensitivity of this
area is equal to that of area V5. Yet, single-unit recordings have repeatedly demonstrated very weak motion selectivity in area V4 (e.g., 129).

These results can be interpreted with respect to sub-threshold modulatory signals. Specifically, areas V4 and V5 are extensively interconnected (130–133).
Although the principal neurons of these areas may deliver long-range signals that
represent different stimulus properties, each area may influence the sensitivity of
the other by providing some kind of modulatory input. These short-range inputs
may be insufficient to drive the pyramidal cells recorded in a typical electrophysiology experiment. But the BOLD fMRI will measure this modulatory signal and
in this way provide measurements that do not match those of single-unit neurophysiology.

Similarly, measurements of visual attention, even measured using the same tasks
or stimulation conditions, can differ between fMRI and electrophysiology (134–
137; see review 138). A good example is the measurement of the effects of spatial
attention on neural activation. Attentional effects on the neurons of area V1 have
been very difficult to measure in monkey single-unit experiments (139, 140). Yet
using similar tasks strong attentional effects have been readily measurable with
fMRI in human V1 (135–137). In addition, attentional effects in area V4 were
found to be considerably larger in human fMRI than in monkey electrophysiology
(141).

As a general principle, we suggest that some of these differences may arise
because the BOLD response emphasizes different aspects of the neural signal:
Synaptic activity produced by short-range lateral or feedback input is often more
visible with imaging than with single-unit recordings.

COMPARISONS BETWEEN BOLD AND SINGLE-UNIT MEASUREMENTS: MAPS The identification of distinct visual areas, each containing a map-like representation of the
visual field, has been an important achievement in neuroscience. A number of
groups have developed methods for measuring visual field maps in the human

brain. We have been able to confirm that fMRI methods also measure accurate visual field maps in macaque brain (142). The basic method is to sweep a stimulus from fovea to periphery. As the stimulus travels eccentrically, a wave of activity passes across the visual field representation in cortex. The time at which the activity peaks at each cortical location is an indication of the most effective visual field eccentricity for stimulating that location.

Figure 6 (see color insert) shows an example of the foveal to peripheral representation that can be measured on the operculum (primary visual cortex in macaque). The colors indicate the visual field eccentricity that most strongly drives the fMRI signal at each cortical location.

The graphs in Figure 6 show the time course of the BOLD response in several cortical regions. These regions are each 3-mm radius disks drawn on the surface of the operculum. Although these regions are separated by only a few millimeters, the responses differ markedly, and the time of the peak BOLD response can be easily distinguished. The systematic shift of the time of peak activation reveals the well-known eccentricity map in primary visual cortex and confirms that in this region of the brain the BOLD response and the neural signals are colocalized.

There have been reports that colocalization is poor in somatosensory cortex. Measuring in macaque somatosensory cortex using a 1.5 T magnet and an echo-planar imaging (EPI) pulse sequence, Disbrow and colleagues reported somatosensory maps in areas 3a, 3b, 1, and 2 in separate fMRI and microelectrode recording sessions (143, 144). Although in half the cases the response fields overlapped, in the other half the fMRI signal was displaced from the electrical signal, sometimes by as much as a centimeter. Those measurements suggest that the ability to colocalize may turn out to depend on factors including the position within the cortical sheet, perhaps coupled to differences in the vascularization patterns, or to differences in experimental protocols (145).

Hemodynamic Response Efficiency

As we have reviewed here, the coupling between neural activity and the vascular response is significant in determining the amplitude and spatial resolution of the BOLD signal. We refer to the efficacy of this coupling as the hemodynamic response efficiency (HRE). Regions of sparse vascularization are likely to have low HRE and weak or absent BOLD response.

It is conceivable that below a certain vascularization density hemodynamic responses cannot be detected despite increases in neural activity and energy metabolism. White matter is a good example. The energy consumption in white matter is one fourth that of gray matter. This consumption has been measured in studies of glucose-metabolism and of blood flow changes using intracerebral O^{15}-H_2O microprobes (B. Weber, personal communication). White matter consumes energy for the restoration of ionic gradients perturbed by the spreading action potential at the Ranvier nodes where the Na^+/K^+-dependent ATPase is mainly responsible for the increased metabolic demands of brain tissue (146). In addition, neurotransmitter-mediated signaling may take place along fiber tracts lacking vesicular means of releasing neurotransmitters (for review see 147). In this receptor-mediated signaling mechanism, glutamate is released from axons via the reversal of a transporter to induce intracellular calcium spiking in glial cells by means of metabotropic glutamate receptors. The mechanism may serve the axon-glia interactions that are most likely involved in glutamate-induced glycolysis, such as that described for astrocytes in the vicinity of glutamatergic synapses (see for example 13). Yet, activation of the white matter has been rarely reported in the neuroimaging literature (148, 149), and a reasonable investigator may doubt the presence of a BOLD signal in white matter altogether.

An additional constraint for spatial resolution comes from the organization of the parenchymal blood supply. Pial arteries, often arising from arterial trunks, as well as capillary branching points very often have a slight rosary-like constriction at their point of origin (95, 98, 150, 151). These constrictions are widely believed to indicate the presence of muscular sphincters regulating the cortical arterial flow. Veins were never reported to show similar changes of diameter. We hypothesize that the density of such myogenic valves will partly determine the HRE and thus whether a cortical region exhibits powerful BOLD responses.

BOLD response measurements also may be influenced by precapillary shunts. These are anastomoses or arterio-venous shunts that serve as alternative channels and have been demonstrated in different tissues (152). When metabolic needs are low, the sphincters of the arterioles close and blood bypasses the capillary bed, flowing through the alternative channels directly to the venules. The pressure elevation that follows the increased metabolic demands opens the sphincters, and the blood flows again through the capillary bed. Precapillary arterio-venous shunts were rarely found in cortex (153) and were entirely absent in the corrosion cast studies (95). On the other hand, the vascular dilatation responsible for the increase in blood flow evoked by neural activity has been shown to propagate in a retrograde fashion to upstream arterioles located outside the activated area (154); a process quite possibly interfering with the spatial specificity of high-resolution imaging.

Finally, simultaneous recording of intrinsic signals and measurements of parenchymal flow with H_2 electrodes showed that activation increasing flow to a particular whisker barrel often leads to reduced flow to adjacent cortex (153). Such a vascular-contrast mechanism enhances local differences and may favorably influence the spatial resolution of a neuroimaging method. Caution is required here, too, because areas of activation may not always match the tissue region from which the signals are generated (98).

All in all, the anatomical substrate of the hemodynamic responses suggests a need for extreme caution when precise localization of function is required and quantitative comparisons must be made between signals coming from different areas. The HRE of each brain region may differ from other regions, even when the regions are nearby. It seems that with our current knowledge there is no secure way to determine a quantitative relationship between a hemodynamic response amplitude and its underlying neural activity in terms of either number of spikes per unit time per BOLD increase or amount of perisynaptic activity. This holds true even if BOLD is calibrated by taking into account the local perfusion of the tissue (156, 157).

The BOLD Signal Measures Circuit Properties

The data summarized above suggest that synaptic potentials are the strongest cause of the BOLD responses. An important consequence of this hypothesis is that the BOLD response will differ depending on the neural circuit properties within various regions of cortex.

Consider the implications in the case of two hypothetical neural circuits located in different regions of cortex. We suppose that in both regions a specific stimulus causes, on average, 10 action potentials above baseline. The two cortical regions differ, however, in that the first cortical region contains a circuit that measures the size of a purely excitatory input, whereas the circuitry in the second region measures the difference between opposing excitatory and inhibitory inputs. On the hypothesis that the BOLD response depends on synaptic potentials, the purely excitatory circuit usually will generate a smaller BOLD response than the circuit that compares opposing inputs. Hence, other factors being equal, the BOLD signal in the second region should exceed that in the first, despite the equal spiking activity.

In this example the amplitude of the BOLD response per action potential will differ across the two regions. Because we have assumed that both regions produce the same number of spikes, the region with the direct excitatory circuitry will have a smaller BOLD response/spike than the region with opposing inputs. Circuitry, then, is an additional reason for differential coupling between action potentials and BOLD responses.

A corollary of this reasoning is that the relationship between action potentials and BOLD responses contains useful information about local circuitry. For example, estimates of the number of spikes per 1% BOLD response in V1 and MT differ by an order of magnitude (80, 158; for a review see 159). If the incremental spikes in MT are produced by a direct excitatory connection, the activity associated with nine spikes might be enough to produce 1% BOLD response. If V1 signals are based on the comparisons of multiple signals, then a great deal of dendritic activity may be required to produce a small number of spikes. This hypothesis can be tested by an experimental analysis of the circuitry in these regions.

A Mathematical Framework

To increase the precision of the inferences we can make from BOLD responses to neural signals, it is useful to summarize the relationship in some simple equations and figures. In the following, we simulate the relationship between simple patterns of neural activity and the BOLD signal. We offer this simulation as a general guide to help reason about the relationship between the BOLD and neural signals.

BOLD Amplitude

We refer to the spatial pattern of stimulus-driven neural activity during an experiment as $A(x)$, and we describe the resulting spatial pattern of BOLD responses as $B(x)$. The spatial variable, x, refers to the position along the cortical surface. Our goal is to understand the relationship between these two quantities.

In addition to stimulus-driven activity, there will be uncontrolled neural responses that we summarize as a random variable, $\underset{\sim}{N}_N(x)$. These responses represent the ongoing neural noise. (We use the tilda to indicate that the term is a random variable). Hence, the total neural activity is $A(x) + \underset{\sim}{N}_N(x)$. An example of one pattern of neural activity and noise is shown in Figure 7a (see color insert).

At each location on the cortical surface, the neural activity is more or less efficiently coupled to the vascular response by the HRE. We express this efficiency as $H(x)$. The spatial distribution of the HRE is shown in Figure 7b. The vascular demand imposed at each point is $(A(x) + \underset{\sim}{N}_N(x))H(x)$.

The neural demand on the vasculature spreads across the cortical surface so that the BOLD signal is a blurred and noisy representation of this vascular demand. The degree of blurring, say on the order of 5 mm (full-width, half maximum) in primary visual cortex (160); of course, this quantity may not generalize across cortex. We use this number to illustrate the key issues for the moment. Mathematically, we can approximate the spatial spreading at x as the weighted sum of the demand in a nearby neighborhood, $n(x)$, on the cortical surface. The spread is measured by the pointspread function, $P(x)$.

$$\int_{n(x)} (A(u) + \underset{\sim}{N}_N(u))H(u)P(x - u)\, du.$$

Finally, as in all measurement systems, there is measurement noise, $\underset{\sim}{N}_M$. In this case, the noise can be attributed to factors such as the magnetic resonance instrument, brain pulsatility, and other uncontrollable experimental factors. We can express the relationship between the neural activity, $A(x)$, and the BOLD response, $B(x)$, by the equation

$$B(x) = \int_{n(x)} (A(u) + \underset{\sim}{N}_N(u))H(u)P(x - u)\, du + \underset{\sim}{N}_M(x).$$

The pseudo-color image in Figure 7c shows the measured BOLD response to the pattern of neural activity in Figure 7a.

The conventional subtraction methodology measures a BOLD response as the difference between an active stimulus condition and a control condition. Suppose that the stimulus-driven neural activity in the control condition is $\bar{A}$. The BOLD measurement described in most fMRI experiments is a spatial map of this difference.

$$\Delta B(x) = B(x) - \bar{B}(x)$$

$$= \int\limits_{n(x)} (A(u) + \underset{\sim}{N}_N(u))H(u)P(x-u)\,du + \underset{\sim}{N}_M(x)$$

$$- \int\limits_{n(x)} (\bar{A}(u) + \underset{\sim}{N}_N(u))H(u)P(x-u)\,du + \underset{\sim}{N}_M(x)$$

$$= \int\limits_{n(x)} (A(u) - \bar{A}(u) + 2\underset{\sim}{N}_N(u))H(u)P(x-u)\,du + 2\underset{\sim}{N}_M(x).$$

Figure 8a (see color insert) shows the differential BOLD response in a simulation that includes all types of noise. Figure 8b shows the locations where the differential BOLD response exceeds a relatively high threshold.

Consider a few basic features of the differential BOLD response formula. First, the expected values of the noise terms are zero (the factor of two arises because the two noise terms are subtracted). Second, the expected amplitude of the differential BOLD response is a spatially blurred version of the stimulus-driven differential activity multiplied by the HRE, $(A(x) - \bar{A}(x))H(x)$. It is only the stimulus-driven difference that is of experimental interest, but this difference cannot be directly estimated because the HRE is unknown. Consequently, a direct comparison of the amplitude of BOLD responses between well-separated cortical regions should only be made if one has some reason to expect that the HRE values in these two regions are equal, or at least known. It may be reasonable to compare the responses at nearby locations, assuming that the HRE varies smoothly across cortex or when we have other data about the HRE.

Even some simple conclusions are problematic when we consider this framework. For example, when the BOLD measurements at two locations are ordered $B(x) > B(y)$, without some information about the HRE, we cannot conclude securely that $A(x) > A(y)$.

BOLD Signal-to-Noise Ratio

Because of the heavy reliance on statistical reasoning in neuroimaging, many groups report a statistical map of the activity based on a variety of measures (z-score, p-value, etc.). In these maps, the statistical parameter represents the amplitude of the differential BOLD measurement relative to noise.

While there are many statistical models, and hence statistical parameter maps, all such models are based on some type of comparison of the size of the differential BOLD response to noise. The values in these parameter maps can be summarized by the signal-to-noise ratio, in which the signal term is in the numerator and the noise terms in the denominator,

$$SNR(x) \sim \frac{\displaystyle\int_{n(x)} (A(u) - \bar{A}(u))H(u)P(x - u)\, du}{2\underset{\sim}{N}_M + 2\displaystyle\int_{n(x)} H(u)\underset{\sim}{N}_N(u)P(x - u)\, du}.$$

In general, high signal-to-noise ratios are associated with a large z-score (smaller p-values). The precise relationship between these quantities depends upon the specific statistical model assumptions. Examining the SNR equation, we note a couple of simple properties.

First, suppose that (a) the measurement noise $\underset{\sim}{N}_M$ is small, and (b) the spatial spread, P(), is very localized. Then the signal-to-noise ratio simplifies to

$$SNR \sim \frac{(A - \bar{A})}{2\underset{\sim}{N}_N}.$$

This SNR model is implicit in many neuroimaging investigations; the hemo-dynamic response efficiency, H, disappears and the SNR is proportional to the signal difference divided by the noise.

Alternatively, consider the case in which (a) the instrumental noise dominates, and (b) the spatial spread is again modest. In that case the signal-to-noise ratio becomes

$$SNR \sim \frac{(A - \bar{A})H}{2\underset{\sim}{N}_M}.$$

In this case, the SNR measures the signal difference combined with the HRE.

In general, we do not know which noise dominates. In fact, the operating regime may depend on features of the instrument, pulse-sequence selection, voxel size, and so forth in complex ways.

The main purpose of working through these formulae is to alert us to some simple principles that should be kept in mind when inferring neural signals from BOLD responses. The simulation parameters were chosen to illustrate the effects of the HRE and spatial blurring of the differential BOLD response. In particular, the spatial periodicity of the neural activity in one direction (dorsal to ventral) was chosen to be too fine to be captured by the blurred BOLD response. The spatial variation in the anterior to posterior direction, however, is of sufficiently low spatial frequency that it passes through the blurring. Portions of this activity are unseen because the simulation uses a low HRE value on the dorsal surface. Hence, with these parameters the BOLD response simulation fails to mirror several aspects of the neural activity.

By choosing these parameters, which are possible but not necessarily true, we hope to encourage further research into measurements of the HRE and spatial blur-ring. Given the many successes of BOLD neuroimaging, the simulation parameters we have chosen may be too pessimistic. However, it would be far better to know the values of these parameters from direct measurements in specific regions of the brain. We hope, therefore, that this review and simulations will serve to motivate further investigation of these quantities.

CONCLUSIONS

In the short period of time since its introduction, fMRI has evolved to become the most important method for investigating human brain function. The BOLD response provides us unprecedented visibility of the neural activity in the human brain; a visibility that far exceeds previously available methods. Still, the visibility remains limited, and the data must be interpreted with these limitations in mind.

A number of important issues remain unresolved, and these issues require careful attention if we want to interpret neural mechanisms from this surrogate signal. The BOLD contrast mechanism reflects changes in cerebral blood volume, cerebral blood flow, and oxygen consumption. The interaction between neural activity and these variables involves a number of factors, including the cell types and circuitry driven during activation, and the processes that couple energy demand to its supply to the brain (HRE).

Research in a number of fields ranging from biochemistry, biophysics, and molecular biology to physics and engineering provides us daily with new information regarding cellular events that may be involved in the generation of responses or the technology that may be best applied to understand these aspects of the fMRI signals.

The first simultaneous fMRI and electrophysiological recordings clearly confirmed a longstanding assumption, i.e., the BOLD contrast mechanism reflects aspects of the neural responses elicited by a stimulus. The hemodynamic response primarily reflects the neuronal input to the relevant area of the brain and its processing there rather than the long-range signals transmitted by action potentials to other regions of the brain. It is reasonable to expect that output activity will usually correlate with neurotransmitter release and pre- and post-synaptic currents. But when input into a particular area plays a primarily modulatory role, fMRI experiments may measure activation that does not correlate well with single-unit measurements.

From examining the pathway between neural signals and BOLD response, we conclude that comparisons of the response amplitude at two locations may not reflect the neural signal amplitudes at these locations. Measuring (*a*) the relative tuning to various stimuli at a single cortical location and (*b*) spatial maps yields more secure conclusions. The relative tuning at a single location may be based on a single coupling, although even this is not certain because of laminar differences. And when one observes a map, it is likely to represent a neural map. Of course, the inability to measure a map using BOLD does not imply the absence of a neural map.

Finally, a great deal of research will be needed to elucidate the neurometabolic coupling and its subsequent hemodynamic response. Neuronal synaptic activity responses may be specific for stimuli and brain areas and are frequently nonlinearly related to the stimulus properties. Although glucose consumption is known to increase during the activity of inhibitory synaptic activity, the role of inhibition in the generation of BOLD remains elusive.

Despite these and other questions, fMRI has earned an indispensable position in neuroscience and, if combined with other invasive techniques, it promises a much better understanding of information processing in the central nervous system than any other technology alone.

ACKNOWLEDGMENTS

We thank Alyssa Brewer for reading and commenting on the manuscript and help with figure preparation. We thank Robert Dougherty, Edgar Galindo, Rainer Goebel, Peter Lennie, Marcus Raichle, David Ress, and Whitman Richards for their comments and suggestions, and Daniel Donoho for help in manuscript preparation. Supported by the Max-Planck Society and NEI RO1 EY03164.

The *Annual Review of Physiology* is online at http://physiol.annualreviews.org

LITERATURE CITED

1. Callaghan PT. 1991. *Principles of Nuclear Magnetic Resonance Microscopy.* New York: Oxford Univ. Press
2. Schmitt F, Stehling MK, Turner R. 1998. *Echo-Planar Imaging: Theory, Technique and Application.* Berlin: Springer
3. Pauling L, Coryell C. 1936. The magnetic properties and structure of hemoglobin. *Proc. Natl. Acad. Sci. USA* 22:210–16
4. Brooks RA, Battocletti JH, Sances A, Larson SJ, Bowman RL, Kudravcev V. 1975. Nuclear magnetic relaxation in blood. *IEEE Trans. Biomed. Eng.* 22:12–18
5. Thulborn KR, Waterton JC, Matthews PM, Radda GK. 1982. Oxygenation dependence of the transverse relaxation time of water protons in whole blood at high field. *Biochim. Biophys. Acta* 714:265–70
6. Thulborn KR, Waterton JC, Matthews PM. 1992. Dependence of the transverse relaxation time of water protons in whole blood at high field. *Biochem. Biophys. Acta* 714:265–72
7. Ogawa S, Lee TM, Nayak AS, Glynn P. 1990. Oxygenation-sensitive contrast in magnetic resonance image of rodent brain at high magnetic fields. *Magn. Reson. Med.* 14:68–78
8. Ogawa S, Lee TM, Kay AR, Tank DW. 1990. Brain magnetic resonance imaging with contrast dependent on blood oxygenation. *Proc. Natl. Acad. Sci. USA* 87:9868–72
9. Ogawa S, Lee TM. 1990. Magnetic resonance imaging of blood vessels at high fields: in vivo and in vitro measurements and image simulation. *Magn. Reson. Med.* 16:9–18
10. Turner R, Le Bihan D, Moonen CT, Despres D, Frank J. 1991. Echo-planar time course MRI of cat brain oxygenation changes. *Magn. Reson. Med.* 22:159–66
11. Fox PT, Raichle ME. 1986. Focal physiological uncoupling of cerebral blood flow and oxidative metabolism during somatosensory stimulation in human subjects. *Proc. Natl. Acad. Sci. USA* 83:1140–44
12. Fox PT, Raichle ME, Mintun MA, Dence C. 1988. Nonoxidative glucose consumption during focal physiologic neural activity. *Science* 241:462–64

13. Magistretti PJ, Pellerin L. 1999. Cellular mechanisms of brain energy metabolism and their relevance to functional brain imaging. *Philos. Trans. R. Soc. London Ser. B.* 354:1155–63

14. Hyder F, Shulman RG, Rothman DL. 1998. A model for the regulation of cerebral oxygen delivery. *J. Appl. Physiol.* 85:554–64

15. Buxton RB, Frank LR. 1997. A model for the coupling between cerebral blood flow and oxygen metabolism during neural stimulation. *J. Cereb. Blood Flow Metab.* 17:64–72

16. Attwell D, Laughlin SB. 2001. An energy budget for signaling in the grey matter of the brain. *J. Cereb. Blood Flow Metab.* 21:1133–45

17. Mintun MA, Lundstrom BN, Snyder AZ, Vlassenko AG, Shulman GL, Raichle ME. 2001. Blood flow and oxygen delivery to human brain during functional activity: theoretical modeling and experimental data. *Proc. Natl. Acad. Sci. USA* 98:6859–64

18. Shimojyo S, Scheinberg P, Kogure K, Reinmuth OM. 1968. The effect of graded hypoxia upon transient cerebral blood flow and oxygen consumption. *Neurology* 18:127–33

19. Glover GH. 1999. Deconvolution of impulse response in event-related BOLD fMRI. *Neuroimage* 9:416–29

20. Martindale J, Mayhew J, Berwick J, Jones M, Martin C, et al. 2003. The hemodynamic impulse response to a single neural event. *J. Cereb. Blood Flow Metab.* 23:546–55

21. Yang Y, Engelien W, Pan H, Xu S, Silbersweig DA, Stern E. 2000. A CBF-based event-related brain activation paradigm: characterization of impulse-response function and comparison to BOLD. *Neuroimage* 12:287–97

22. Pu Y, Liu HL, Spinks JA, Mahankali S, Xiong J, et al. 2001. Cerebral hemodynamic response in Chinese (first) and English (second) language processing revealed by event-related functional MRI. *Magn. Reson. Imaging* 19:643–47

23. Liu H, Gao JH. 2000. An investigation of the impulse functions for the nonlinear BOLD response in functional MRI. *Magn. Reson. Imaging* 18:931–38

24. Kwong KK, Belliveau JW, Chesler DA, Goldberg IE, Weisskoff RM, et al. 1992. Dynamic magnetic resonance imaging of human brain activity during primary sensory stimulation. *Proc. Natl. Acad. Sci. USA* 89:5675–79

25. Buxton RB, Wong EC, Frank LR. 1998. Dynamics of blood flow and oxygenation changes during brain activation: the balloon model. *Magn. Reson. Med.* 39:855–64

26. Frahm J, Kruger G, Merboldt KD, Kleinschmidt A. 1996. Dynamic uncoupling and recoupling of perfusion and oxidative metabolism during focal brain activation in man. *Magn. Reson. Med.* 35:143–48

27. Logothetis NK, Guggenberger H, Peled S, Pauls J. 1999. Functional imaging of the monkey brain. *Nat. Neurosci.* 2:555–62

28. Robson MD, Dorosz JL, Gore JC. 1998. Measurements of the temporal fMRI response of the human auditory cortex to trains of tones. *Neuroimage* 7:185–98. Erratum. *Neuroimage.* 1998. 8:228

29. Logothetis NK. 2002. The neural basis of the blood-oxygen-level-dependent functional magnetic resonance imaging signal. *Philos. Trans. R. Soc. London Ser. B* 357:1003–37

30. Boynton GM, Engel SA, Glover GH, Heeger DJ. 1996. Linear systems analysis of functional magnetic resonance imaging in human V1. *J. Neurosci.* 16:4207–21

31. Friston KJ, Fletcher P, Josephs O, Holmes A, Rugg MD, Turner R. 1998. Event-related fMRI: characterizing differential responses. *Neuroimage* 7:30–40

32. Worsley KJ. 2002. Statistical analysis of activation images. In *Functional Magnetic Resonance Imaging: An Introduction to Methods*, ed. P Jezzard, PM

Matthews, SM Smith, pp. 251–70. York: Oxford Univ. Press

33. Bandettini PA, Wong EC, Hinks RS, Tikofsky RS, Hyde JS. 1992. Time course EPI of human brain function during task activation. *Magn. Reson. Med.* 25:390–97

34. Ogawa S, Tank DW, Menon R, Ellermann JM, Kim SG, et al. 1992. Intrinsic signal changes accompanying sensory stimulation: functional brain mapping with magnetic resonance imaging. *Proc. Natl. Acad. Sci. USA* 89:5951–55

35. Engel SA, Glover GH, Wandell BA. 1997. Retinotopic organization in human visual cortex and the spatial precision of functional MRI. *Cereb. Cortex* 7:181–92

36. Ranck JBJ. 1963. Specific impedance of rabbit cerebral cortex. *Exp. Neurol.* 7:144–52

37. Ranck JBJ. 1966. Electrical impedance in the subicular area of rats during paradoxical sleep. *Exp. Neurol.* 16:416–37

38. Mitzdorf U. 1985. Current source-density method and application in cat cerebral cortex: investigation of evoked potentials and EEG phenomena. *Physiol. Rev.* 65:37–100

39. Nicholson C, Freeman JA. 1975. Theory of current source-density analysis and determination of conductivity tensor for anuran cerebellum. *J. Neurophysiol.* 38:356–68

40. Gray CM, Maldonado PE, Wilson M, McNaughton B. 1995. Tetrodes markedly improve the reliability and yield of multiple single-unit isolation from multi-unit recordings in cat striate cortex. *J. Neurosci. Methods* 63:43–54

41. Harris KD, Henze DA, Csicsvari J, Hirase H, Buzsaki G. 2000. Accuracy of tetrode spike separation as determined by simultaneous intracellular and extracellular measurements. *J. Neurophysiol.* 84:401–14

42. Henze DA, Borhegyi Z, Csicsvari J, Mamiya A, Harris KD, Buzsaki G. 2000. Intracellular features predicted by extracellular recordings in the hippocampus in vivo. *J. Neurophysiol.* 84:390–400

43. Adrian ED, Zotterman Y. 1926. The impulses produced by sensory nerve-endings, Part 2. The response of a single end-organ. *J. Physiol.* 61:151–71

44. Kuffler SW. 1953. Discharge patterns and functional organization of the mammalian retina. *J. Neurophysiol.* 16:37–68

45. Hubel DH, Wiesel TN. 1962. Receptive fields, binocular interaction and functional architecture in the cat's visual cortex. *J. Physiol.* 160:106–54

46. Mountcastle VB. 1957. Modality and topographic properties of single neurons of cat's somatic sensory cortex. *J. Neurophysiol.* 20:408–34

47. Lettvin JY, Maturana HR, McCulloch WS, Pitts WH. 1959. What the frog's eye tells the frog's brain. *Proc. Inst. Radio Engrs.* 47:1940–51

48. Stone J. 1973. Sampling properties of microelectrodes assessed in the cat's retina. *J. Neurophysiol.* 36:1071–79

49. Towe AL, Harding GW. 1970. Extracellular microelectrode sampling bias. *Exp. Neurol.* 29:366–81

50. Rall W. 1962. Electrophysiology of a dendritic neuron. *Biophys. J.* 2:145–67

51. Humphrey DR, Corrie WS. 1978. Properties of pyramidal tract neuron system within a functionally defined subregion of primate motor cortex. *J. Neurophysiol.* 41:216–43

52. Bishop GH, O'Leary JL. 1942. Factors determining the form of the potential record in the vicinity of the synapses of the dorsal nucleus of the lateral geniculate body. *J. Cell Comp. Physiol.* 19:315–31

53. Lorente de Nó R. 1947. Action potentials of the motorneurones of the hypoglossus nucleus. *J. Cell Comp. Physiol.* 29:207–88

54. Logothetis NK. 2003. The underpinnings of the BOLD functional magnetic resonance imaging signal. *J. Neurosci.* 23:3963–71

55. Logothetis NK. 2002. On the neural basis

of the BOLD fMRI signal. *Philos. Trans. R. Soc. London Ser. B.* 357:1003–37

56. Buchwald JS, Grover FS. 1970. Amplitudes of background fast activity characteristic of specific brain sites. *J. Neurophysiol.* 33:148–59

57. Nelson PG. 1966. Interaction between spinal motoneurons of the cat. *J. Neurophysiol.* 29:275–87

58. Grover FS, Buchwald JS. 1970. Correlation of cell size with amplitude of background fast activity in specific brain nuclei. *J. Neurophysiol.* 33:160–71

59. Fromm GH, Bond HW. 1964. Slow changes in the electrocorticogram and the activity of cortical neurons. *Electroencephalogr. Clin. Neurophysiol.* 17:520–23

60. Fromm GH, Bond HW. 1967. The relationship between neuron activity and cortical steady potentials. *Electroencephalogr. Clin. Neurophysiol.* 22:159–66

61. Ajmone-Marsan C. 1965. Electrical activity of the brain: slow waves and neuronal activity. *Isr. J. Med. Sci.* 1:104–17

62. Buchwald JS, Hala ES, Schramm S. 1965. A comparison of multi-unit activity EEG activity recorded from the same brain site in chronic cats during behavioral conditioning. *Nature* 205:1012–14

63. Mitzdorf U. 1987. Properties of the evoked potential generators: current source-density analysis of visually evoked potentials in the cat cortex. *Int. J. Neurosci.* 33:33–59

64. Juergens E, Guettler A, Eckhorn R. 1999. Visual stimulation elicits locked and induced gamma oscillations in monkey intracortical- and EEG-potentials, but not in human EEG. *Exp. Brain Res.* 129:247–59

65. Kamondi A, Acsady L, Wang XJ, Buzsaki G. 1998. Theta oscillations in somata and dendrites of hippocampal pyramidal cells in vivo: activity-dependent phase-precession of action potentials. *Hippocampus* 8:244–61

66. Granit R, Kernell D, Smith RS. 1963. Delayed depolarization and the repetitive response to intracellular stimulation of mammalian motoneurones. *J. Physiol.* 168:890–910

67. Harada Y, Takahashi T. 1983. The calcium component of the action potential in spinal motoneurones of the rat. *J. Physiol.* 335:89–100

68. Gustafsson B. 1984. Afterpotentials and transduction properties in different types of central neurones. *Arch. Ital. Biol.* 122:17–30

69. Chandler SH, Hsaio CF, Inoue T, Goldberg LJ. 1994. Electrophysiological properties of guinea pig trigeminal motoneurons recorded in vitro. *J. Neurophysiol.* 71:129–45

70. Walton K, Fulton BP. 1986. Ionic mechanisms underlying the firing properties of rat neonatal motoneurons studied in vitro. *Neuroscience* 19:669–83

71. Higashi H, Tanaka E, Inokuchi H, Nishi S. 1993. Ionic mechanisms underlying the depolarizing and hyperpolarizing afterpotentials of single spike in guinea-pig cingulate cortical neurons. *Neuroscience* 55:129–38

72. Kobayashi M, Inoue T, Matsuo R, Masuda Y, Hidaka O, et al. 1997. Role of calcium conductances on spike afterpotentials in rat trigeminal motoneurons. *J. Neurophysiol.* 77:3273–83

73. Buzsaki G. 1931. Theta oscillations in the hippocampus. *Neuron* 33:325–40

74. Buzsaki G, Bickford RG, Ponomareff G, Thal LJ, Mandel R, Gage FH. 1988. Nucleus basalis and thalamic control of neocortical activity in the freely moving rat. *J. Neurosci.* 8:4007–26

75. Bullock TH. 1997. Signals and signs in the nervous system: the dynamic anatomy of electrical activity is probably information-rich. *Proc. Natl. Acad. Sci. USA* 94:1–6

76. Wandell BA. 1999. Computational neuroimaging of human visual cortex. *Annu. Rev. Neurosci.* 22:145–73

77. Brinker G, Bock C, Busch E, Krep H, Hossmann KA, Hoehn-Berlage M. 1999.

Simultaneous recording of evoked potentials T2*-weighted MR images during somatosensory stimulation of rat. *Magn. Reson. Med.* 41:469–73

78. Mathiesen C, Caesar K, Akgoren N, Lauritzen M. 1998. Modification of activity-dependent increases of cerebral blood flow by excitatory synaptic activity and spikes in rat cerebellar cortex. *J. Physiol.* 512:555–66

79. Ogawa S, Lee TM, Stepnoski R, Chen W, Zhu XH, Ugurbil K. 2000. An approach to probe some neural systems interaction by functional MRI at neural time scale down to milliseconds. *Proc. Natl. Acad. Sci. USA* 97:11026–31

80. Rees G, Friston K, Koch C. 2000. A direct quantitative relationship between the functional properties of human and macaque V5. *Nat. Neurosci.* 3:716–23

81. Logothetis NK, Pauls J, Augath M, Trinath T, Oeltermann A. 2001. Neurophysiological investigation of the basis of the fMRI signal. *Nature* 412:150–57

82. Hyder F, Rothman DL, Shulman RG. 2002. Total neuroenergetics support localized brain activity: implications for the interpretation of fMRI. *Proc. Natl. Acad. Sci. USA* 99:10771–76

83. Smith AJ, Blumenfeld H, Behar KL, Rothman DL, Shulman RG, Hyder F. 2002. Cerebral energetics and spiking frequency: The neurophysiological basis of fMRI. *Proc. Natl. Acad. Sci. USA* 99:10765–70

84. Lauritzen M, Gold L. 2003. Brain function and neurophysiological correlates of signals used in functional neuroimaging. *J. Neurosci.* 23:3972–80

85. Lauritzen M. 2001. Relationship of spikes, synaptic activity, and local changes of cerebral blood flow. *J. Cereb. Blood Flow Metab.* 21:1367–83

86. Mathiesen C, Caesar K, Lauritzen M. 2000. Temporal coupling between neuronal activity and blood flow in rat cerebellar cortex as indicated by field potential analysis. *J. Physiol.* 523:235–46

87. Roy CS, Sherrington CS. 1890. On the regulation of the blood supply of the brain. *J. Physiol.* 11:85–108

88. Sokoloff L. 1981. Relationships among local functional activity, energy metabolism, and blood flow in the central nervous system. *Fed. Proc.* 40:2311–16

89. Sokoloff L, Reivich M, Kennedy C, DesRosiers MH, Patlak CS, et al. 1977. The C^{14}deoxyglucose method for the measurement of local cerebral glucose utilization: theory, procedure and normal values in the conscious and anesthetized albino rat. *J. Neurochem.* 28:897–916

90. Sokoloff L. 1977. Relation between physiological function and energy metabolism in the central nervous system. *J. Neurochem.* 29:13–26

91. Raichle ME. 2000. A brief history of human functional brain mapping. In *The Systems*, ed. AW Toga, JC Mazziotta, pp. 33–75. San Diego: Academic

92. Fox PT, Mintun MA, Raichle ME, Miezin FM, Allman JM, Van Essen DC. 1986. Mapping human visual cortex with positron emission tomography. *Nature* 323:806–9

93. Frostig RD, Lieke EE, Ts'o DY, Grinvald A. 1990. Cortical functional architecture and local coupling between neuronal activity and the microcirculation revealed by in vivo high-resolution optical imaging of intrinsic signals. *Proc. Natl. Acad. Sci. USA* 87:6082–86

94. Bonhoeffer T, Grinvald A. 1996. Optical imaging based on intrinsic signals. In *Brain Mapping, The Methods*, ed. AW Toga, JC Mazziotta, pp. 55–97. New York: Academic

95. Duvernoy HM, Delon S, Vannson JL. 1981. Cortical blood vessels of the human brain. *Brain Res. Bull.* 7:519–79

96. Schuz A, Palm G. 1989. Density of neurons and synapses in the cerebral cortex of the mouse. *J. Comp. Neurol.* 286:442–55

97. O'Kusky J, Colonnier M. 1982. A laminar analysis of the number of neurons,

glia, and synapses in the adult cortex (area 17) of adult macaque monkeys. *J. Comp. Neurol.* 210:278–90

98. Harrison RV, Harel N, Panesar J, Mount RJ. 2002. Blood capillary distribution correlates with hemodynamic-based functional imaging in cerebral cortex. *Cereb. Cortex* 12:225–33

99. Swain RA, Harris AB, Wiener EC, Dutka MV, Morris HD, et al. 2003. Prolonged exercise induces angiogenesis and increases cerebral blood volume in primary motor cortex of the rat. *Neuroscience* 117:1037–46

100. Argandona EG, Rossi ML, Lafuente JV. 2003. Visual deprivation effects on the s 100beta positive astrocytic population in the developing rat visual cortex: a quantitative study. *Brain Res. Dev. Brain Res.* 141:63–69

101. Argandona EG, Lafuente JV. 2000. Influence of visual experience deprivation on the postnatal development of the microvascular bed in layer IV of the rat visual cortex. *Brain Res.* 855:137–42

102. Black JE, Isaacs KR, Anderson BJ, Alcantara AA, Greenough WT. 1990. Learning causes synaptogenesis, whereas motor activity causes angiogenesis, in cerebellar cortex of adult rats. *Proc. Natl. Acad. Sci. USA* 87:5568–72

103. Toga AW. 1987. The metabolic consequence of visual deprivation in the rat. *Brain Res.* 465:209–17

104. Cox SB, Woolsey TA, Rovainen CM. 1993. Localized dynamic changes in cortical blood flow with whisker stimulation corresponds to matched vascular and neuronal architecture of rat barrels. *J. Cereb. Blood Flow Metab.* 13:899–913

105. Siesjo BoK. 1978. *Brain Energy Metabolism.* New York: Wiley & Sons

106. Ames A. 2000. CNS energy metabolism as related to function. *Brain Res. Brain Res. Rev.* 34:42–68

107. Araque A, Parpura V, Sanzgiri RP, Haydon PG. 1999. Tripartite synapses: glia, the unacknowledged partner. *Trends Neurosci.* 22:208–15

108. Magistretti PJ, Pellerin L, Rothman DL, Shulman RG. 1999. Neuroscience—energy on demand. *Science* 283:496–97

109. Rothman DL, Sibson NR, Hyder F, Shen J, Behar KL, Shulman RG. 1999. In vivo nuclear magnetic resonance spectroscopy studies of the relationship between the glutamate-glutamine neurotransmitter cycle and functional neuroenergetics. *Philos. Trans. R. Soc. London Ser. B* 354:1165–77

110. Shulman RG, Rothman DL. 1998. Interpreting functional imaging studies in terms of neurotransmitter cycling. *Proc. Natl. Acad. Sci. USA* 95:11993–98

111. Takahashi S, Driscoll BF, Law MJ, Sokoloff L. 1995. Role of sodium and potassium ions in regulation of glucose metabolism in cultured astroglia. *Proc. Natl. Acad. Sci. USA* 92:4616–20

112. Pellerin L, Magistretti PJ. 1994. Glutamate uptake into astrocytes stimulates aerobic glycolysis: a mechanism coupling neuronal activity to glucose utilization. *Proc. Natl. Acad. Sci. USA* 91:10625–29

113. Braitenberg V, Schuez A. 1998. *Cortex: Statistics and Geometry of Neuronal Connectivity.* Berlin: Springer

114. Sibson NR, Dhankhar A, Mason GF, Rothman DL, Behar KL, Shulman RG. 1998. Stoichiometric coupling of brain glucose metabolism and glutamatergic neuronal activity. *Proc. Natl. Acad. Sci. USA* 95:316–21

115. Pan JW, Stein DT, Telang F, Lee JH, Shen J, et al. 2000. Spectroscopic imaging of glutamate C4 turnover in human brain. *Magn. Reson. Med.* 44:673–79

116. Kadekaro M, Crane AM, Sokoloff L. 1985. Differential effects of electrical stimulation of sciatic nerve on metabolic activity in spinal cord and dorsal root ganglion in the rat. *Proc. Natl. Acad. Sci. USA* 82:6010–13

117. Kadekaro M, Vance WH, Terrell ML,

Gary HJ, Eisenberg HM, Sokoloff L. 1987. Effects of antidromic stimulation of the ventral root on glucose utilization in the ventral horn of the spinal cord in the rat. *Proc. Natl. Acad. Sci. USA* 84:5492–95

118. Nudo RJ, Masterton RB. 1986. Stimulation-induced [14]C2-deoxyglucose labeling of synaptic activity in the central auditory system. *J. Comp. Neurol.* 245:553–65

119. Jueptner M, Weiller C. 1995. Review—does measurement of regional cerebral blood flow reflect synaptic activity—implications for PET and fMRI. *Neuroimage* 2:148–56

120. Lennie P. 2003. The cost of cortical computation. *Curr. Biol.* 13:493–97

121. Attwell D, Iadecola C. 2002. The neural basis of functional brain imaging signals. *Trends Neurosci.* 25:621–25

122. Li J, Iadecola C. 1994. Nitric oxide and adenosine mediate vasodilation during functional activation in cerebellar cortex. *Neuropharmacology* 33:1453–61

123. Yang G, Iadecola C. 1996. Glutamate microinjections in cerebellar cortex reproduce cerebrovascular effects of parallel fiber stimulation. *Am. J. Physiol. Regul. Integr. Comp. Physiol.* 271:R1568–75

124. Faraci FM, Breese KR. 1993. Nitric oxide mediates vasodilatation in response to activation of N-methyl-D-aspartate receptors in brain. *Circ. Res.* 72:476–80

125. Fergus A, Lee KS. 1997. Regulation of cerebral microvessels by glutamatergic mechanisms. *Brain Res.* 754:35–45

126. Tolias AS, Smirnakis SM, Augath MA, Trinath T, Logothetis NK. 2001. Motion processing in the macaque: revisited with functional magnetic resonance imaging. *J. Neurosci.* 21:8594–601

127. Grill-Spector K, Kushnir T, Edelman S, Avidan G, Itzchak Y, Malach R. 1999. Differential processing of objects under various viewing conditions in the human lateral occipital complex. *Neuron* 24:187–203

128. Kourtzi Z, Kanwisher N. 2001. Representation of perceived object shape by the human lateral occipital complex. *Science* 293:1506–9

129. Desimone R, Schein SJ. 1987. Visual properties of neurons in area V4 of the macaque: sensitivity to stimulus form. *J. Neurophysiol.* 57:835–68

130. Felleman DJ, Van Essen DC. 1983. The connections of area V4 of macaque extrastriate cortex. *Soc. Neurosci. Abstr.* 9:153

131. Steele GE, Weller RE, Cusick CG. 1991. Cortical connections of the caudal subdivision of the dorsolateral area (V4) in monkeys. *J. Comp. Neurol.* 306:495–520

132. Maunsell JH, Van Essen DC. 1983. The connections of the middle temporal visual area (MT) and their relationship to a cortical hierarchy in the macaque monkey. *J. Neurosci.* 3:2563–86

133. Ungerleider LG, Desimone R. 1986. Cortical connections of visual area MT in the macaque. *J. Comp. Neurol.* 248:190–222

134. Polonsky A, Blake R, Braun J, Heeger DJ. 2000. Neuronal activity in human primary visual cortex correlates with perception during binocular rivalry. *Nat. Neurosci.* 3:1153–59

135. Gandhi SP, Heeger DJ, Boynton GM. 1999. Spatial attention affects brain activity in human primary visual cortex. *Proc. Natl. Acad. Sci. USA* 96:3314–19

136. Tong F, Engel SA. 1910. Interocular rivalry revealed in the human cortical blindspot representation. *Nature* 411:195–99

137. Kastner S, Ungerleider LG. 2000. Mechanisms of visual attention in the human cortex. *Annu. Rev. Neurosci.* 23:315–41

138. Blake R, Logothetis NK. 2002. Visual competition. *Nat. Rev. Neurosci.* 3:13–23

139. Luck SJ, Chelazzi L, Hillyard SA, Desimone R. 1997. Neural mechanisms of spatial selective attention in areas V1, V2, and V4 of macaque visual cortex. *J. Neurophysiol.* 77:24–42

140. McAdams CJ, Maunsell JH. 1999. Effects of attention on orientation-tuning

functions of single neurons in macaque cortical area V4. *J. Neurosci.* 19:431–41

141. Kastner S, De Weerd P, Desimone R, Ungerleider LG. 1998. Mechanisms of directed attention in the human extrastriate cortex as revealed by functional MRI. *Science* 282:108–11

142. Brewer AA, Press W, Logothetis NK, Wandell B. 2002. Visual areas in macaque cortex measured using functional MRI. *J. Neurosci.* 22:10416–26

143. Disbrow E, Roberts TP, Slutsky D, Krubitzer L. 1999. The use of fMRI for determining the topographic organization of cortical fields in human and nonhuman primates. *Brain Res.* 829:167–73

144. Disbrow EA, Slutsky DA, Roberts TP, Krubitzer LA. 2000. Functional MRI at 1.5 tesla: a comparison of the blood oxygenation level-dependent signal and electrophysiology. *Proc. Natl Acad. Sci. USA* 97:9718–23

145. Wandell B, Wade A. 2003. Functional neuroimaging of the visual pathway. *Neurol. Clin.* 21:417–43

146. Mata M, Fink DJ, Gainer H, Smith CB, Davidsen L, et al. 1980. Activity-dependent energy metabolism in rat posterior pituitary primarily reflects sodium pump activity. *J. Neurochem.* 34:213–15

147. Chiu SY, Kriegler S. 1994. Neurotransmitter-mediated signaling between axons and glial cells. *GLIA* 11:191–200

148. Mosier K, Bereznaya I. 2001. Parallel cortical networks for volitional control of swallowing in humans. *Exp. Brain Res.* 140:280–89

149. Tettamanti M, Paulesu E, Scifo P, Maravita A, Fazio F, et al. 2002. Interhemispheric transmission of visuomotor information in humans: fMRI evidence. *J. Neurophysiol.* 88:1051–58

150. Wolff HG. 1938. The cerebral blood vessels: anatomical principles. In *The Circulation of the Brain and Spinal Cord*, pp. 29–67. Baltimore: Williams & Wilkins

151. Reina-De L, Rodriguez-Baeza A, Sahuquillo-Barris J. 1998. Morphological characteristics and distribution pattern of the arterial vessels in human cerebral cortex: a scanning electron microscope study. *Anat. Rec.* 251:87–96

152. Chambers R, Zweifach BW. 1944. Topography and function of the mesenteric capillary circulation. *Am. J. Anat.* 75:173–205

153. Ravens JR. 1974. Anastomoses in the vascular bed of the human cerebrum. In *Pathology of Cerebral Microcirculation*, ed. J Cervos-Navarro, pp. 26–32. Berlin/New York: De Gruyter

154. Iadecola C, Yang G, Ebner TJ, Chen G. 1997. Local and propagated vascular responses evoked by focal synaptic activity in cerebellar cortex. *J. Neurophysiol.* 78:651–59

155. Woolsey TA, Rovainen CM, Cox SB, Henegar MH, Liang GE, et al. 1996. Neuronal units linked to microvascular modules in cerebral cortex: response elements for imaging the brain. *Cereb. Cortex* 6:647–60

156. Hyder F, Renken R, Kennan RP, Rothman DL. 2000. Quantitative multi-modal functional MRI with blood oxygenation level dependent exponential decays adjusted for flow attenuated inversion recovery (BOLDED AFFAIR). *Magn. Reson. Imaging* 18:227–35

157. Kida I, Kennan RP, Rothman DL, Behar KL, Hyder F. 2000. High-resolution CMR O_2 mapping in rat cortex: a multiparametric approach to calibration of BOLD image contrast at 7 Tesla. *J. Cereb. Blood Flow Metab.* 20:847–60

158. Heeger DJ, Huk AC, Geisler WS, Albrecht DG. 2000. Spikes versus BOLD: What does neuroimaging tell us about neuronal activity? *Nat. Neurosci.* 3:631–33

159. Heeger DJ, Ress D. 2002. What does fMRI tell us about neuronal activity? *Nat. Rev. Neurosci.* 3:142–51

160. Engel SA, Glover GH, Wandell BA. 1997. Retinotopic organization in human visual

cortex and the spatial precision of functional MRI. *Cereb. Cortex* 7:181–92

161. Liu HL, Pu Y, Nickerson LD, Liu Y, Fox PT, Gao JH. 2000. Comparison of the temporal response in perfusion BOLD-based event-related functional MRI. *Magn. Reson. Med.* 43:768–72

162. Boynton GM, Engel SA, Glover GH, Heeger DJ. 1996. Linear systems analysis

of functional magnetic resonance imaging in human V1. *J. Neurosci.* 16:4207–21

163. Engel SA, Rumelhart DE, Wandell BA, Lee AT, Glover GH, et al. 1994. fMRI of human visual cortex. *Nature* 369:525

164. Wandell BA, Press WA, Brewer A, Logothetis NK. 2000. *Soc. Neurosci. Abstr.* 26:821

Annu. Rev. Physiol. 2004. 66:771–98
doi: 10.1146/annurev.physiol.66.082602.095217
Copyright © 2004 by Annual Reviews. All rights reserved
First published online as a Review in Advance on November 3, 2003

LIVE OPTICAL IMAGING OF NERVOUS SYSTEM DEVELOPMENT

Cristopher M. Niell

*Neurosciences Program and Department of Molecular and Cellular Physiology,
Stanford University, Stanford, California 94305; email: cris@stanford.edu*

Stephen J Smith

*Department of Molecular and Cellular Physiology, Stanford University, Stanford,
California 94305; email: sjsmith@stanford.edu*

Key Words confocal, two-photon, migration, neurite growth, synaptogenesis

■ **Abstract** Although development of the nervous system is inherently a process
of dynamic change, until recently it has generally been investigated by inference from
static images. However, advances in live optical imaging are now allowing direct observation of growth, synapse formation, and even incipient function in the developing
nervous system, at length scales from molecules to cortical regions, and over timescales
from milliseconds to months. In this review, we provide technical background and
present examples of how these new methods, including confocal and two-photon microscopy, GFP-based markers, and functional indicators, are being applied to provide
fresh insight into long-standing questions of neural development.

INTRODUCTION

Much of modern neuroscience began with Ramon y Cajal's meticulous description of nervous system morphology (1). This work engendered the idea that by
understanding the structure of the nervous system, we can gain insight into its
function. The number of remarkably prescient hypotheses that Cajal was able to
make on the basis of light microscopy of fixed tissue, including the functional polarity of neurons, the nature of synaptic connections, and even the role of growth
cones in guidance of axon extension, provides support for such an approach to
understanding the brain.

Recent advances in live imaging have allowed us to go beyond Cajal's static
view to observe the dynamic structure of the nervous system. This vantage is invaluable in developmental neuroscience, where movement and change over time
are precisely what is interesting. In fact, just as understanding the cellular function
of the brain would be nearly impossible without structural description, to understand the development of the brain, one must know the dynamic steps by which it is

0066-4278/04/0315-0771$14.00

constructed. Live imaging has revealed new cellular structures and modes of growth and has generated hypotheses as to organizing principles of neural development. Furthermore, by leveraging optical imaging with modern molecular techniques, we have extraordinary power to understand these processes at the mechanistic level.

In this review, we focus on three broad approaches for applying live optical imaging to the development of the nervous system. In vitro study of cultured neurons provides a simplified system that is particularly amenable to understanding the cell biology of individual developmental processes. Imaging of neurons in vivo presents a more top-down approach by watching the process of development within its natural context. Finally, functional imaging has begun to fit these developmental processes into the final operation of the mature nervous system, as well as to reveal the role of neural activity in the process of development itself. Because we can not provide an account of every live imaging study, we have selected examples to illustrate how new techniques are being used to answer unresolved questions in neural development. Although we bring up some of the practical issues associated with various imaging methods, we refer the reader to several excellent texts for further details. (2–6)

THE NEED FOR LIVE IMAGING

Some developmental processes can be deduced from a series of static images of fixed tissue, taken from different individuals at various points in development. For instance, a stereotyped migratory pathway of a neuron could be pieced together from many images of cells at different points along the path. In these cases, live imaging simply streamlines the process, since imaging a single migrating neuron would directly delineate the pathway.

On the other hand, many processes cannot be understood from fixed specimens but must be observed directly by following a given neuron over time. In general, live imaging is necessary to determine dynamic parameters such as velocity of motion or stability of a structure, where measurements must be made on the same individual at multiple time points. For example, whereas Cajal observed the presence of filopodia on dendrites during development, it was not until recently that live imaging studies revealed that these filopodia are highly motile and have lifetimes of just minutes. Due to this rapid turnover, while fixed imaging may show ten filopodia on a branch at any given time, over the period of development hundreds of filopodia may appear.

Furthermore, live imaging is needed when properties of individual structures may be masked by pooling data from different static images. For example, averaging the number of synapses in many fixed specimens at multiple points in development can establish the number of total synapses formed per day, but determining the time course for a single synapse to form requires continual live imaging of an individual nascent synapse. Similarly, images of growing axonal arbors at different stages in development show a gradual increase in overall size. However, live imaging has revealed that this growth often occurs by a "two steps forward, one step backward" process, where many branches are actually retracted and the entire arbor undergoes continual remodeling. Fixed tissue shows the overall branch length increasing, but live imaging is necessary to realize that many individual branches are being eliminated.

IN VITRO IMAGING

There are many steps in the structural and functional development of a neuron: proliferation, fate specification, migration, projection and pathfinding of axons and dendrites to target regions, elaboration of axonal and dendritic arbors, and formation of a precise set of synaptic connections. Understanding the cell biological mechanisms underlying each of these steps provides a reductionist approach to the development of the nervous system.

In physics, one of the keys to such a bottom-up approach has been finding well-defined simplified systems, such as the abstraction of a basic pendulum or an object falling in a perfect vacuum. Although such a system lacks the complexity of real every-day objects, this simplicity makes it more tractable for study. In developmental neurobiology, cultured neurons can provide such a simple system (7). By isolating neurons from their natural surroundings, one removes much of the complexity and gains direct control over their environment, thereby allowing specific manipulations. Furthermore, there are now many reliable methods for transfecting DNA into cultured neurons (8), including calcium phosphate precipitation, lipofection, electroporation, and viral transfection, which allow manipulation at the genetic level.

For imaging as well, cultured neurons provide several advantages over intact specimens. First, they are easily accessible optically. Although many researchers use scanning confocal microscopy, it is possible to use a simple wide-field epifluorescence microscope for many experiments. Nomarski differential interference contrast (DIC) (3) provides a powerful method to visualize even transparent cell morphology in relation to fluorescently labeled structures (Figure 1D, see color insert). Also, the lower cell density of isolated or dispersed neuronal cultures allows easier interpretation of images and results by reducing overlap of neurons. Finally, it is easier to label neurons for imaging in vitro, particularly when indicators need to be directly loaded into or onto cells. Recently, the use of green fluorescent protein (GFP) (9) has revolutionized biological imaging by allowing cells and even specific molecules to be fluorescently tagged using genetic methods. In addition to enabling researchers to design specific markers and indicators with molecular biology, rather than synthetic chemistry, the use of GFP also reduces the problem of labeling cells to that of getting DNA into cells, for which there are several solutions (see above).

Cytoskeletal Mechanisms of Growth Cone Motility

To guide a growing axon, extracellular cues must be translated into modulation of growth cone motility. This motility is generated primarily by the cytoskeletal microtubule and actin filament systems. Beyond providing the structural underpinnings of the growth cone, the polymerization of these cytoskeletal structures, coupled with the action of molecular motors that bind to them, can generate forces that result in directed advance. Optical imaging provides a technique to "look under the hood" at the rearrangement and regulation of these cytoskeletal structures during growth.

In order to visualize cytoskeletal components, many studies use synthetic fluorescent molecules, conjugated either to cytoskeletal proteins themselves or to molecules that bind polymerized actin or tubulin, which must be loaded directly into cells (11). More recently, fluorescent protein fusions have been introduced as useful alternative methods for labeling actin and tubulin, as well as other molecules that interact with the cytoskeleton. Early fixed tissue studies revealed that the outer edge of a growth cone has a complex actin-based assembly, including a broad mesh in the lamellipodium and long thin filaments in filopodia behind which the microtubules provide structural support for the shaft of the axon. Presumably, rearrangements of these cytoskeletal structures results in advancement of the growth cone. However, although traditional fluorescence time-lapse imaging allows one to see accumulation of different structures, the combination of translocation by motors and polymerization makes it difficult to determine precisely how a structure moves. This requires some type of fiducial mark along the structure to track its motion.

There are two primary techniques for this. The first involves either activating or inactivating fluorescence at a particular point and tracking the motion of this point. Activation, or uncaging, takes advantage of fluorescent indicators that are bound to a molecule that quenches the fluorescence (12). Powerful UV illumination can sever this bond, resulting in activation of fluorescence in a spatially restricted area determined by the focus of the illumination. A similar effect can be obtained with a recently developed variant of GFP, which undergoes a conformational change upon UV illumination and results in activation of fluorescence (13). Conversely, fluorescence can be inactivated by photobleaching the label, leaving a dark region that can be tracked. Photobleaching is often simpler, as all fluorophores can be bleached (which is usually a drawback!), whereas uncaging requires specifically designed molecules. On the other hand, a localized bright signal on a dark background is generally easier to detect than the inverse. Both photobleaching and activation can be used in a wide variety of studies to determine the mobility of molecules.

The second technique, a relatively recent development, is speckle microscopy. This method uses low levels of indicator to produce irregular patterns of fluorescence along cytoskeletal structures (14). These speckles can be tracked to determine the movement of the cytoskeletal structures (Figure 1*E-H*). Because of the smaller number of fluorophores, more sensitive low-noise imaging methods are needed. Nevertheless, this may be simpler and less obtrusive than a photobleaching or photoactivation step. Furthermore, speckles are visible throughout a cell, rather than at a single location, allowing tracking of many points simultaneously. And speckle microscopy avoids the potential damage to the cell that can be incurred with photobleaching or activation. Computational tools are also being developed to easily extract kinetic parameters from speckle images (15).

Early live imaging studies, using wide-field DIC, were able to demonstrate the dynamics of actin transport in growth cones by ruffling of the overlying cell membrane. This revealed that even filaments that do not undergo net translocation are continually treadmilling away from the leading edge of the growth cone owing to the combined action of myosin motors and of polymerization on the ends (16). More recent studies have used photobleaching and speckle microscopy to elucidate the relative roles of translocation and polymerization of microtubules in extending the axon behind the growth cone *(17)*. Simultaneous imaging of actin and tubulin has shown that interaction between these two cytoskeletal components can lead to interstitial branching (18). A recent study (10), using speckle microscopy, demonstrated that the movement of microtubules in the leading edge of a growth cone is tightly coupled to actin-based arcs and filopodia (Figure 1).

Proliferation and Migration

Early in development, neurons are generated by repeated division of neural progenitors in a proliferative zone, and then migrate out to assume their final positions in the brain. Live imaging studies in cultured slices have given a glimpse of the cellular rearrangements involved in this process.

Because neurons in a vertical column of cortex share connections and many functional properties, it was initially proposed that progenitors give rise to clones, which migrate radially outward, preserving their initial mapping from the ventricular zone. O'Rourke et al. (19) tested this hypothesis by providing the first direct observation of migrating neurons in developing cortex. They labeled small groups of neurons with the fluorescent lipophilic membrane marker DiI in cultured brain slices from neonatal ferrets. These organotypic slice cultures can preserve structure well enough to allow formation of cortical layers and provide a three-dimensional substrate for migration. Although connectivity can change, and the extracellular environment is altered, slice cultures are a useful preparation when a neural structure is not optically accessible in vivo.

By tracking motile neurons for up to 45 h, O'Rourke et al. confirmed that most neurons migrate radially outward from the ventricular zone, as had been hypothesized. However, they also found a substantial fraction that migrated horizontally, thus disrupting a possible radial lineage. Furthermore, some neurons made turns or even reversed direction. Recent experiments using similar techniques have demonstrated multiple modes of radial movement (20), as well as a population of interneurons that actually returns to the ventricular zone before moving back out to their final location (21). These dynamics, and particularly the behavior and paths of individual neurons, could not be inferred from static images.

In order to determine how these migrating neurons arose, Chenn & McConnell (22) imaged progenitors while they were still proliferating in the ventricular zone. They found that when a progenitor divides, it can either create two more progenitors (symmetric division), or divide asymmetrically to produce a neuron and another progenitor that can continue dividing. Interestingly, they observed that symmetric division occurred when the progenitor cleaved vertically during mitosis, whereas in asymmetric divisions the neuron cleaved horizontally, with the upper daughter migrating out as a neuron and the lower daughter staying behind as a progenitor. These horizontal divisions also corresponded with disproportionate allocation of the protein Notch, which was localized to the apical half of the original progenitor and therefore would end up in the neuronal daughter. Thus their live imaging demonstrated that the functional asymmetry of division, resulting in two different cell fates, was accompanied by a physical asymmetry, both in the plane of division and the subcellular localization of Notch.

Synapse Assembly

Determining how synapse formation is initiated and how all of the molecular components are transported and assembled into a synapse is crucial for understanding both the development of neural connectivity and the function of the mature synapse. Although many molecules have been implicated as essential to synapse formation, it is difficult to understand their true functional role without knowing the overall sequence of events in synaptogenesis. In vitro imaging studies have proven extremely valuable in this regard by allowing researchers to track specific molecules as they arrive at the synapse.

Live imaging studies of synaptogenesis have taken advantage of GFP fusions to synaptic components to elucidate the order and time course of synaptic assembly. An advantage of the in vitro system is the ability to fix the preparation at the end of the time-lapse experiment and delve in with retrospective immunohistochemistry to gain a more complete inventory of the components present at the nascent synapse whose creation was just observed. One can also use retrospective electron microscopy to see the ultrastructure of the developing synapse. Another useful technique is the use of styryl dyes, such as FM-4–64, to observe vesicle turnover (23), one of the hallmarks of a functional synapse. These dyes are perfused onto the cells, during which time any synaptic vesicles that fuse with the membrane and are re-endocytosed in the presence of the dye will take it up. Once the dye is washed out of the medium, presynaptic boutons, which contain recycled vesicles, remain fluorescently labeled.

Using these techniques, Ahmari et al. (24) observed the dynamics of a GFP fusion to VAMP-2, a protein associated with presynaptic vesicles. Surprisingly, it was found that VAMP-GFP accumulated in moving puncta, which have become known as transport packets (Figure 2, see color insert). These packets travel along an axon until they arrive at a recent site of axo-dendritic contact, where they are stabilized. Retrospective immunohistochemistry revealed that the mobile transport packets actually contain many presynaptic components, suggesting that they may be complete or partial kits for building a synapse, just waiting for a site of cell-cell contact to initiate assembly.

Other studies also have explored the time course and mechanisms of synapse assembly. Ziv et al. (25) used the onset of FM-loading to mark the formation of new synapses. Once nascent synapses were time-stamped in this manner, they could wait a certain interval of time and use retrospective immunohistochemistry to determine how long it takes for other molecules to accumulate at the synapse. Studies of the postsynaptic active zone have shown that the scaffolding protein PSD-95 (SAP-90) is not transported in discrete packets but accumulates gradually at sites of contact (26), suggesting that presynaptic and postsynaptic active zone components may be transported by different mechanisms.

IN VIVO IMAGING

Although cultured neurons provide a powerful system for dissecting out cellular mechanisms underlying specific steps in development, they cannot be expected to provide definitive answers about how neurons differentiate and grow during brain development in vivo. Because the neurons have been removed from their natural surroundings and placed on the artificial substrate of a two-dimensional glass coverslip with an externally supplied medium, many processes that depend on normal interaction with neighboring cells through secreted molecules, intracellular signaling, cell-cell adhesion, patterns of synaptic activity, or simply the presence of a three-dimensional extracellular matrix may be perturbed in vitro. This makes some questions unanswerable—for example, one probably should not explore how Purkinje cells generate a planar dendritic arbor on a planar glass coverslip. Some processes can also occur very differently in vitro. For example, in dissociated cultures, contact events between dendrites and axons are rare, as most protrusions simply extend onto empty coverslip. In vivo, on the other hand, any dendritic protrusion within a dense neuropil will immediately contact something, which is quite likely to be an axon. Thus, in vivo, the question is not what happens when a contact occurs, because they are always in contact, but why a small fraction of contacts do lead to something.

Overall, although in vitro systems are invaluable for a reductionist approach, in vivo imaging is necessary to gain an overall picture of how a neuron actually develops in its native habitat (27), and how the individual steps fit together. Watching a process occur can often give insights available no other way. New labeling and imaging methods allow observations beyond simple cell morphology, including most of the molecular and functional markers of synaptic components and intracellular signaling used in in vitro studies sampled above.

Technical Advances Facilitating In Vivo Imaging

Of course, the reason that not all experiments are done in vivo is that it is often more challenging to label, image, and perturb a neuron that is buried in the brain of a living animal. Fortunately, great strides have been made recently in these areas. Early in vivo imaging studies required an invasive process to label cells with a vital dye, such as lipophilic carbocyanide dyes such as DiI or fluorescent molecules conjugated to dextrans, which could be transported throughout neuronal processes. Within the past decade, GFP-based markers have taken over, with the advantage of being able to use cell-type-specific promoters to restrict expression to defined populations. Although expressing a DNA reporter construct is also more difficult in a living animal, new methods are continuously being developed. The first GFP-based applications used viruses to infect cells (28). Recently, in vivo electroporation has provided a technique to insert DNA into localized populations or even single cells (29). Beyond these invasive techniques, methods to generate transgenic animals, either by transient mosaic expression (30) or creation of stable transgenic lines (31), allow cells to be labeled without disruption. An outstanding example of such technology is the numerous lines of transgenic mice generated by Feng et al. (32), which express GFP and spectral variants thereof in subsets of neurons throughout the nervous system.

A significant advance in in vivo imaging technology, beyond wide-field epifluorescence, was the development of the confocal microscope, which allowed three-dimensional imaging within specimens (5). A confocal microscope utilizes a point source to focus light down to a cone through the specimen. The emitted fluorescence light is then sent back through a pinhole, which corresponds to the focal point in the specimen, thus blocking out light from other focal planes. This creates a focal voxel within the specimen. By scanning the focal point within a plane, one can create an optical section, an image of a single plane within the tissue. Stepping the focus through the specimen builds up a three-dimensional image.

Although confocal microscopy has proven extremely useful for generating high-resolution, three-dimensional images, it has limitations for live imaging. In order to penetrate deeper into a turbid tissue, higher-intensity excitation is needed to allow excitation of the fluorophore and to generate enough photons that can escape the tissue and be detected. This results in damage both to the fluorophore, resulting in photobleaching, and to the tissue as a by-product of free radicals generated during excitation. The combined effects of photobleaching and photodamage can significantly limit the signal generated and length of time-lapse imaging that can be performed.

One way around these problems is the use of systems that increase the efficiency of detection, at some sacrifice of optical sectioning and resolution. Making a larger pinhole, either physically or effectively (by optics), allows more light to be detected (33). Multi-aperture confocals, such as the Nipkow disk or Yokogawa microlens array, use many pinholes simultaneously to collect images extremely rapidly (34). Whereas a standard confocal must scan a single point across the specimen, one rotation of the disk creates a complete optical section. Furthermore, light from many rotations can be summed to create an image over a longer period at lower intensity. In practice, these systems generate brighter images with lower rates of photodamage. Their improved sensitivity may be a consequence of relaxed optical sectioning (33), which allows collection of light from larger volumes, although it is also possible that some benefit comes from the lower peak intensity inducing fewer photodamaging reactions. Thus these systems can provide a powerful tool

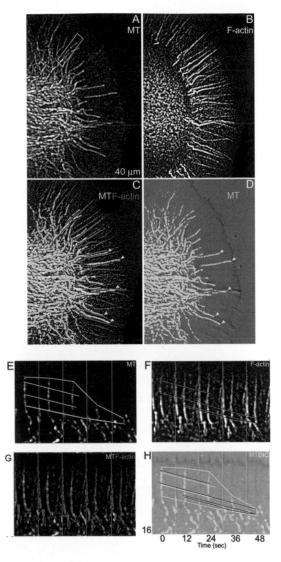

Figure 1 Fluorescence speckle microscopy reveals dynamics of cytoskeletal elements in the growth cone. MT, microtubules; DIC, Nomarski DIC transmitted light. (*A-D*) Organization of cytoskeletal elements in the growth cone. Many microtubules extend along actin filaments. (*E-H*) Consecutive speckle images reveal movement of microtubules is coupled to retrograde flow of actin filaments. Microtubules can also polymerize or depolymerize on either end. Reproduced by copyright permission of The Rockefeller University Press from Schaefer et al. (10).

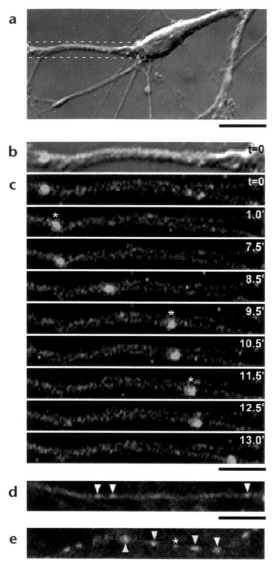

Figure 2 Dynamics of presynaptic transport packets. (*a*) A hippocampal pyramidal cell in culture, transfected with VAMP-GFP (*green*). Scale bar 10 μm. (*b*) Close-up of section of axon. (*c*) Consecutive images of a VAMP-GFP transport packet moving in a saltatory manner along an axon. Scale bar 5 μm. (*d,e*) Loading of a styryl dye, FM4-64 (*red*), indicates vesicle cycling. In immature neurons (*d*), moving VAMP-GFP puncta do not generally show recycling, indicating that they are not yet functional synapses, whereas in mature neurons (*e*) VAMP-GFP generally colocalizes with FM4-64 loading. Scale bar 5 μm. Reproduced by copyright permission of Nature Publishing Group from Ahmari et al. (24).

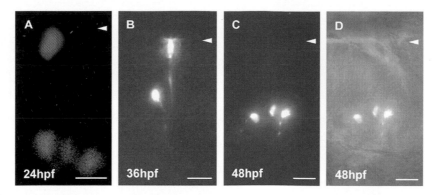

Figure 3   Neuronal proliferation in zebrafish hindbrain. (*A*) A progenitor cell (*red*) labeled with rhodamine dextran, and differentiated neurons (*green*) expressing GFP under the HuC neural promoter. Scale bar 10 μm. (*B,C*) Consecutive images of the progenitor show asymmetric division resulting in one neuron (*B*) and a symmetric division resulting in two neurons (*C*). (*D*) Image from (*C*) overlaid with brightfield to show location of neurons migrating away from ventricular zone. Scale bar 20 μm. Reproduced by copyright permission of The Company of Biologists, Ltd., from Lyons et al. (42).

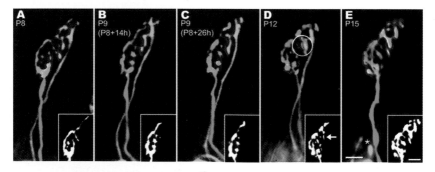

Figure 4   Consecutive images of competition between two motor neuron axons at a single neuromuscular junction. One neuron expresses YFP (*light green*) and the other expresses CFP (*blue*); the postsynaptic acetylcholine receptors are labeled by alpha-bungarotoxin (*red*). One axon slowly takes over all the territory, resulting in retraction of the other axon. However, no postsynaptic sites are eliminated or left vacated. Scale bar 10 μm. Reproduced from Walsh & Lichtman (50) with permission from Elsevier.

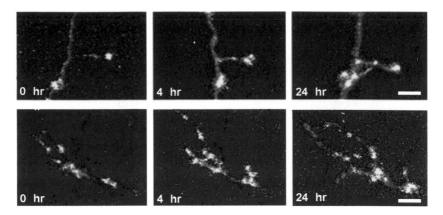

Figure 5 Time-lapse images of presynaptic assembly and axonal remodeling in *Xenopus*. Presynaptic sites are labeled by VAMP-GFP (*green*) and axons are labeled by DsRed (*top*) or DiI (*bottom*). Images after 4 or 24 h show extensive remodeling of both synapses and arbors. Scale bar 5 μm. Reproduced by copyright permission of Nature Publishing Group from Alsina et al. (52).

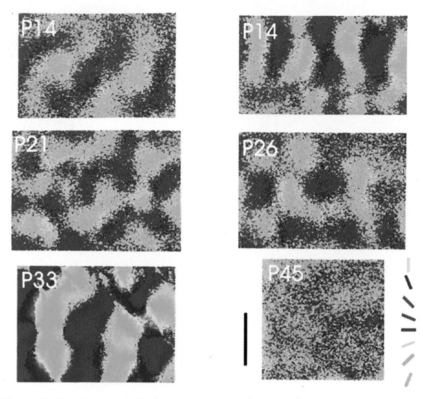

Figure 6   Development of orientation maps in ferret under normal and monocular deprivation revealed by intrinsic optical signal imaging. Direction of maximum orientation response is coded by color. In normal animals (*left*) an initially rough orientation map gradually becomes well defined. However, under monocular deprivation (*right*), the initial map deteriorates over time. Scale bar 1 mm. Reproduced by copyright permission of The Society for Neuroscience from Issa et al. (84).

when sensitivity, photobleaching, or acquisition speed are greater concerns than thin optical sections.

Possibly the most significant advance for in vivo imaging has been the development of two-photon microscopy (35). In a two-photon microscope, a fluorophore is excited by an extremely high-intensity laser at twice the usual excitation wavelength (i.e., half the energy per photon). The fluorophore effectively absorbs two photons simultaneously, each providing half the necessary excitation energy. Because the probability of such an event depends on the square of the intensity, excitation drops off rapidly away from the focal point of the laser, inherently isolating a voxel within the specimen. As the focus is scanned through the specimen, in a process similar to confocal scanning, a three-dimensional image can be constructed. This technique provides several advantages for in vivo imaging (36). First, the fluorescence excitation is at approximately twice the usual wavelength, in the near infrared instead of the visible. Longer-wavelength photons scatter more weakly and thus penetrate much deeper into tissue than visible light. Second, the optical sectioning in a two-photon microscope is provided by nonlinear excitation limited to the focus, rather than blocking out-of-focus photons as in a confocal. Therefore, all the photons that are emitted come from the focus, meaning that even if the light is scattered on the way out of the tissue, it can be collected to form an image. Finally, because fluorescence excitation is predominantly restricted to the focal point, the illumination does not cause photodamage at other parts of the tissue through which the laser passes. These three factors allow deeper penetration into tissue (up to 500 μm typically) with significantly reduced photodamage and photobleaching, which are crucial for extended live time-lapse imaging in vivo. As several examples below demonstrate, this has enabled remarkable feats of imaging in intact animals.

Furthermore, recent developments promise to extend the utility of two-photon microscopy (37). For example, by creating pulses of much higher peak intensity with a regenerative amplifier, Theer et al. (38) have imaged up to 1 mm deep in the cortex of an intact mouse. Helmchen et al. (39) have developed a fiber-optic-based head-mounted two-photon system that allows imaging to be performed even in an awake freely moving mouse.

Beyond imaging, perturbation of neurons or the environment in vivo can also be more challenging and the results more difficult to interpret. Many experiments rely on gross pharmacological manipulation, which can point to general effects, although lack of specificity can often be a problem. Recently, the techniques mentioned above for introducing DNA into individual cells, or generating knockout or knockin animals, are beginning to allow perturbation of gene function. One advantage of in vivo studies, though, is the ability to perform manipulations that are most relevant to the nervous system, such as altering sensory experience.

Proliferation and Migration in the Zebrafish

As discussed above, pioneering studies in cortical slices initially demonstrated mechanisms of asymmetric division and neuronal migration. Recently, these studies have been followed up by in vivo observations tracking new-born neurons for extended periods as they move through an intact nervous system.

Although in vivo imaging of migration in the cortex has not yet been performed, several studies have used the zebrafish to observe the birth and migration of neurons in other brain regions. (Alas, the zebrafish has no cortex.) The zebrafish is rapidly becoming a prime model organism for in vivo imaging studies of vertebrate development, because the larvae are nearly transparent; they develop extremely rapidly, three days from fertilization to free-swimming larvae, and they have been the focus of many mutant screens and other genetic analysis, which allows the role of candidate molecules to be tested in imaging experiments. Transient mosaic transgenics, which have small clusters or even individual neurons labeled, can be easily generated by injecting plasmid DNA into early embryos. Furthermore, stable transgenic lines that recapitulate the expression pattern of a specific gene can be made by raising a number of these fish to adulthood and screening for founders, which have incorporated the transgene into the germline (40). The examples in this section take advantage of specific markers, mutant strains, and transgenic lines to enable a more complete characterization of proliferation and migration in the zebrafish.

Das et al. (41) observed cell division in the early retina, using two-photon imaging and two transiently expressed GFP fusion proteins, one that localized to chromosomes, allowing dividing nuclei to be seen, and another localizing to the membrane to accentuate thin basal protrusions. Although they found very few of the horizontal plane divisions that Chenn & McConnell (22) observed, they did notice a shift from division along a central-peripheral plane to division along a circumferential plane. This shift correlated in time with the transition from proliferation and, presumably, symmetric divisions to neurogenesis. Furthermore, in mutants where neurogenesis was perturbed, this shift in division plane was similarly affected.

Lyons et al. (42) recently made a complete characterization of proliferation in the developing hindbrain of zebrafish. Their experiments were performed in a transgenic line expressing GFP in all differentiated neurons, which removed ambiguity about which progeny become neurons versus which remain progenitors. By injecting individual progenitor cells with different colored fluorescent marker, and imaging every eight hours (approximately the time for one round of division) through the entire course of proliferation, they were able to characterize the family tree of all eventual neurons. Surprisingly, they found that many divisions resulted in two neurons (Figure 3, see color insert), thus challenging the dogma that divisions result either in two progenitors or one progenitor and one neuron.

Finally, while imaging in culture is useful for looking at modes of movement, it is generally not possible to track a neuron all the way from its birthplace to its final location. However, because the zebrafish develops so quickly and the whole nervous system is visible, Koester & Fraser (43) have been able to follow neuronal precursors from the rhombic lip to the early cerebellum. Their observations resolved ambiguity about the route of migration, which had been proposed based on clonal analysis by directly watching individual cells traverse this route.

Axon Pathfinding

In vitro studies can provide much information about the response of growth cones to various extracellular cues or how cytoskeletal rearrangements result in motility of the growth cone. To understand how all this comes together to guide an axon to its actual target, however, one must observe the process in the normal complex environment with all the cues and substrates the growth cone naturally encounters. Live in vivo imaging has allowed researchers to follow axons throughout their course, revealing the dynamic changes associated with guidance. Furthermore, live imaging can reveal more complete explanations for the function of molecules implicated in guidance.

Early analysis in fixed tissue revealed that growth cones in different location have different morphologies, ranging from streamlined with few filopodia, to splayed-out with many extensive filopodia (44). Mason & Wang (45) performed time-lapse imaging demonstrating that these changes were associated with different guidance behaviors. First, growth cone advancement was shown to be discontinuous; axons would extend smoothly for short periods, pause, and begin advancing again. Pauses were particularly common at turning points, or at points where one set of extracellular guidance cues ended and another began. Growth cones that were rapidly moving along a straight path had a streamlined morphology, whereas growth cones that paused had more complex forms, suggesting that the highly motile filopodia are probing the environment to determine the future direction of growth.

Another interesting application of imaging to understanding the molecular basis of axon guidance is provided by Hutson & Chien's (46) analysis of pathfinding in the zebrafish optic chiasm. They studied the *astray* mutant, which had been found in a genetic screen as having misrouted retinal axons in the optic chiasm. The *astray* mutant is defective in the Robo2 receptor, which is believed to respond to repulsive ligands. To understand the mechanism of Robo at the optic chiasm, Hutson & Chien began by performing in vivo confocal time-lapse of wild-type axons crossing the midline. They found that even in the wild type, 17% of growth cones made errors in crossing the midline. However, these errant growth cones were retracted and continued upon the correct path. In *astray* mutants, 53% of axons left the normal pathway, suggesting that Robo is involved in keeping growth cones on the correct path. Furthermore, only 11% of these errant growth cones corrected their mistakes, indicating that Robo not only helps keep axons on the correct path but also permits growth cones to correct mistakes once they have left the proper path.

This experiment provides an excellent example of the power of live imaging in understanding molecular functions. Imaging the endpoint of the system simply revealed that all retinal axons project to the contralateral tectum, while mutant axons often end up in the wrong place. Live imaging showed that even wild-type axons make mistakes at the midline, and a significant contribution to the misrouting of *astray* axons is failure to correct these errors. Thus understanding the dynamic phenotype of the mutant, rather than simply the static endpoint phenotype, provides significant insight into the function of a gene.

Imaging Synaptogenesis In Vivo

In vitro studies of synaptogenesis are valuable for revealing the molecular steps in assembly once an axon and dendrite have come in contact. However, the artificial substrate of a culture dish, particularly in dissociated cultures, precludes normal patterns of interaction between axons and dendrites. Furthermore, issues of synaptic stability could be quite different in a culture dish where normal trophic factors and synaptic activity are perturbed. Live imaging is the obvious method to address such questions, as issues of stability can not be addressed from static images because the presence of a structure must be verified at multiple time points. Several lines of in vivo studies have begun to explore how synapse formation is initiated and subsequently stabilized under normal growth conditions.

The first live imaging studies of synapse formation in intact nervous system were the seminal experiments by Purves & Lichtman (47) on the mouse neuromuscular junction. Over the years, these studies have used a succession of labeling techniques, each advancing the state of the art at the time. Initial studies used vital dyes to label presynaptic terminals, and fluorescently tagged toxins, which bind postsynaptic acetylcholine receptors, to directly observe synapse formation and disassembly (48). More recent experiments have used GFP lines, which have a small subset of neurons labeled (49), to observe the competitive process whereby multiple axons innervating a muscle unit are eventually whittled down to a single afferent (Figure 4, see color insert). These studies have shown that in the peripheral nervous system, arborization and synapse stabilization are dominated by competition between axons and have demonstrated the importance of elimination processes in the specification of neural circuits.

An initial in vivo description of events leading to synaptogenic contact in the central nervous system was given by Jontes et al. (51). They imaged the interaction between the Mauthner axon and spinal cord motor neuron dendrites in the living zebrafish by labeling each member of the pair with lipophilic carbocyanide dyes and acquiring time-lapse images throughout the course of interaction. Imaging revealed several aspects of axonal extension associated with synapse formation. As the growth cone advanced, it paused briefly at regular intervals. Within an hour, axonal varicosities—presumably nascent synapses — appeared at the points where the growth cone had paused, suggesting that interaction with the postsynaptic partner slowed the growth cone. Furthermore, both the axonal growth cone and the dendrite had a large number of motile filopodia, each of which explored a volume sufficient to establish contact.

The study of Jontes et al. took advantage of the fact that the Mauthner axon forms synaptic contact with every motor neuron it passes, allowing identification of putative synapse locations. However, in general a marker is needed to indicate the presence of a nascent synapse. Alsina et al. (52) used VAMP-GFP, which had been used in vitro to demonstrate presynaptic transport packets, to look at the relation between axonal arborization and synapse formation in the *Xenopus* optic tectum. The projection of retinal ganglion cell axons to the optic tectum has a long history of live imaging, from O'Rourke & Fraser (53), who first observed the dynamic remodeling of axonal arbors during development. Alsina et al. showed that these dynamics are not limited to the arbor but involve formation and disassembly of synapses (Figure 5, see color insert). Furthermore, they were able to show that BDNF, a neurotrophin known to influence growth and receptive field formation, had a stabilizing effect on both the axonal arbor and the nascent synapses.

Effects of Synaptic Activity on Dendrite Growth

A crucial step in determining the connectivity of neural circuits is the elaboration of a dendritic arbor by a neuron. Live in vivo imaging has played a critical role in elucidating this process by revealing dynamic aspects of growth that cannot be deduced from static images and by demonstrating the role that the neural environment plays as a three-dimensional physical substrate and as a source of chemical cues and synaptic activity.

From static images of developing dendrites, one might assume that a dendritic arbor grows much like a tree, continuously elongating branches, which occasionally split off to form higher order branches. However, live imaging has shown that dendrites actually grow by a much more dynamic process, with many branches extending and quickly being retracted, in a process of remodeling that can continue over days (54). In addition, it has been shown that the lateral filopodia commonly observed on growing dendrites are highly motile and that most are transient (55).

Since these initial observations, in vivo experiments have continued to investigate how molecular mechanisms and synaptic activity affect this dynamic growth process to determine the final form of the dendritic arbor. Cline and colleagues have pursued this line of inquiry in the optic tectum of *Xenopus* tadpoles, using a combination of time-lapse imaging with both molecular and sensory perturbations. Initial experiments demonstrated high motility at early stages of development, which gradually tapers off as the dendrite grows and matures (56). To determine the molecular mechanism responsible for this maturation, they used a virus transfection method to test the role of CaMKII, which is known to be a signaling molecule downstream of synaptic activity, by overexpressing CaMKII along with its dominant-negative and constitutively active mutant forms. This revealed that CaMKII activity leads to increased growth early in development but stabilizes the growth later in development. Subsequent studies have tested the role of other activity-dependent molecules (57), including rho GTPase modulators of actin organization (58). Whereas studies of such molecules in fixed tissue can show the effect of the molecule on the final form of the dendrite, live imaging reveals how the molecule affects the process of growth. For example, many molecules affect the final size of the arbor, but some do this by altering the rate of formation of new branches, whereas others affect the stability of branches.

Although it is possible to show an effect of large perturbations, one wonders which effects are actually relevant to normal development. Because these experiments were performed in an intact living animal, the role of these activity-dependent mechanisms in development of the visual system can be tested directly by controlling visual input through the eye (59). Individual tectal cells of *Xenopus* tadpoles were labeled with GFP by single-cell electroporation. After imaging the structure of the dendritic arbor, the tadpoles were removed from the microscope and placed either in a visually deprived (dark) or visually enhanced (flashing LEDs) environment. After four hours in this environment, the same tectal cell was re-imaged. Remarkably, dendrites that received visual stimulus grew at nearly twice the rate during this interval. Furthermore, by overexpressing dominant-negative rho GTPases in the same cells that were imaged, it was shown that these molecules mediate the effect of light stimulation.

Long-Term Imaging of Synapse Stability

It has now been well established that during development, synaptic connections undergo drastic rearrangements accompanied by significant morphological rearrangements of axons and dendrites. Furthermore, recent two-photon studies in intact rat cortex suggest that sensory input can influence the rate of these rearrangements (60), particularly during the critical period when activity can affect the final receptive fields of neurons. It had long been assumed that this dynamic plasticity ends upon maturation of the neural circuits and that this stabilization could signal the closure of developmental critical periods. In fact, these short-term imaging studies have shown that morphological rearrangements are dramatically diminished after development. However, in order to test long-term stability, two separate groups have recently developed techniques to image neurons in living mice, not over hours, but over days or months (61, 62).

Grutzendler et al. (61) and Trachtenberg et al. (62) report long-term imaging in the cortex of mammals that required several technical advances. First, transgenic lines expressing GFP in a small subset of layer 5 pyramidal neurons were generated by Feng et al. (32), which provided both groups with animals with brightly labeled sets of pyramidal cell dendrites that could be observed at high resolution. Both groups used dendritic spines, upon which a large proportion of synapses are formed, as morphological proxies for synapses. Second, two-photon microscopy allowed imaging into the intact cortex, albeit limited to the more superficial layers. Finally, to repeatedly observe the same neurons, optical access to the brain lasting for weeks of repeated observation was necessary. Trachtenberg et al. (62) met this need by a craniotomy and cementing of a small coverslip window over the exposed brain. Grutzendler et al. (61) found that the skull did not need to be completely penetrated but could be thinned sufficiently to allow two-photon imaging without actually penetrating the cranium. Using these techniques, the investigators could place an anesthetized mouse under the microscope, use the pattern of blood vessels to return to the same neuron, and image fine structures as small as filopodia and spines, repeating this process on multiple days.

Both groups confirmed previous findings that in immature animals, dendritic protrusions were highly dynamic. Filopodia generally persisted for less than a day, and many spines turned over daily as well. However, in more mature animals, filopodia were nearly absent, and the overall pattern of dendritic branching was stabilized. One group (61) found that, in the visual cortex, spines became extremely stable as well, with 98% persisting over two weeks in the adult. However, the other group (62) found that spines in the whisker region of somatosensory cortex remained unstable, with up to 50% turning over in the course of a month. Moreover, this instability was increased when sensory activity was altered by trimming a subset of whiskers, accompanied by changes in the electrophysiological receptive field.

The differences between immature and mature animals suggest that stabilization of the dendritic arbor could explain the dramatic decrease in plasticity at the end of development, whereas rearrangement of synaptic connections can still allow for some degree of response to changes in sensory input. Although the source of discrepancy between these experiments, whether due to inherent differences between regions of cortex, different imaging methods, or different analysis, remains to be determined, both studies made important strides in our ability to observe structural changes in the intact mammalian cortex at the subcellular level.

FUNCTIONAL IMAGING

Watching the structural changes that result in the mature form of a neuron is a tremendous advance in understanding neural development. However, most neuroscientists would agree that what is fascinating about neurons is not just their structure but their ability to transmit and process signals that result in our sensory perception, thoughts, and behavior. Whereas systems neuroscientists hope to understand how a configuration of neurons processes information, developmental neuroscience attempts to explain how those neurons got to that configuration in the first place. Functional imaging of neural activity and signaling can bridge these two areas. Monitoring activity in developing networks, or in a network where developmental processes have been perturbed, provides an understanding of how the developmental program results in a properly functioning neural circuit. Furthermore, many studies have shown that incipient activity in immature networks can play a crucial role in development. Functional imaging allows us to look in on these signals that shape neural development.

Although electrophysiology is the most direct method of monitoring electrical signals within neurons, it is limited by the fact that a measurement is made at one point in space per electrode. On the other hand, optical imaging allows monitoring of signals throughout a specimen, while capturing morphology simultaneously. This is particularly useful in development because the patterns of activity and location of signals with respect to the changing structure of a neuron or network are important.

One of the greatest challenges in a functional imaging experiment is finding an indicator that gives a readout of activity. Because neurons transmit information over distance by action potentials, an indicator of membrane voltage seems like the obvious choice. Various voltage-sensitive indicators have been used to image action potentials from single dendrites to neuronal populations (63). However, voltage-sensitive dyes currently available have low fluorescence changes and are often difficult to load into tissues. Moreover, because action potentials are such brief events, extremely fast acquisition and high sensitivity are needed to pick up a signal at short sampling intervals.

Therefore, most functional imaging relies on indirect readouts of activity. The most common is fluorescent indicators of intracellular calcium concentration. Because neuronal membranes contain many voltage-gated calcium channels, action potential firing generally leads to increases in intracellular calcium. This increase can persist much longer than the action potential itself, decreasing the required scan rate and increasing the signal that can be detected. Also, calcium and other intracellular messengers are often more directly relevant to affecting cellular processes than the action potential itself, particularly in development. Furthermore, there are a number of fluorescent molecules, both synthetic and genetically encoded, that exhibit large changes in fluorescence upon binding calcium at physiological levels (64), and methods have been developed to load and image these indicators in both slices and intact living animals (65). In addition, there are many other indirect measures of activity that can be measured optically, such as changes in hemoglobin oxygenation and distribution (66); increases or translocation of intracellular signaling proteins, such as CaMKII (67, 68); or indicators of presynaptic vesicle recycling (23, 69).

Beyond choosing an appropriate indicator, the imaging system must be tailored to the specific type of activity to be studied. Because functional imaging implies a network of interacting neurons, most studies require at least an intact tissue such as a slice or explant, if not an in vivo preparation. The choice of imaging protocol then depends on the time course of the signals to be observed (milliseconds for action potentials, minutes for protein activation), as well as the desired spatial resolution (individual dendrites, cell bodies, or cortical columns). Imaging of slow signals, such as calcium transients at the low resolution of cell bodies, can be performed using a standard epifluorescence microscope and a standard charge-coupled device (CCD) camera. Photodiode arrays, as well as special high-speed CCD cameras, are available that allow extremely fast image acquisition in widefield modes. On the other hand, obtaining high resolution in thick tissue generally requires a confocal or two-photon system. Typical commercial scanning microscopes can collect image planes at rates only up to 10 Hz or so, but a few scanning microscopes have been designed for higher acquisition rates (70, 71), and even the standard scanners can collect line-scans at rates up to 1 KHz or thereabouts.

Calcium Transients and Branch Stabilization

Many studies have demonstrated that synaptic activity can play a role in the growth of dendritic arbors, as shown by the experiments of Cline discussed above. An elegant imaging experiment by Lohmann et al. (72) addressed the mechanism by which synaptic transmission can stabilize individual branches by looking at functional postsynaptic calcium responses to neurotransmitter release.

Lohmann et al. used an explant preparation from the chick retina that preserves the local neuronal circuitry, while allowing loading of indicators and bath application of pharmacological agents. To introduce calcium indicators into neurons, they used particle-mediated substance delivery, analogous to the "biolistic" or "gene gun" transfection method. In the original procedure, a pulse of high-pressure gas drives micron-sized gold beads, coated with DNA, into a tissue. A small number of the particles deliver their cargo into cell bodies, with minimal damage to the cells themselves, allowing expression of a transgene. Lohmann et al. adapted this method, coating beads with a calcium indicator (calcium-green dextran), instead of DNA, and calling the technique callistics (73). Incidentally, another variation of biolistics, known as DiOlistics, allows labeling of tissues with multiple colors of lipophilic dyes (74), which should prove valuable for studies of neuronal morphology.

By imaging dendrites of retinal ganglion cells loaded with the calcium indicator, they observed occasional brief localized elevations of intracellular calcium. Using bulk pharmacological manipulations, they determined that these signals originated with neurotransmission-dependent release of calcium from intracellular stores. Furthermore, blocking these calcium transients led to retraction of dendritic branches. To test the causal role of calcium signaling, they extended callistics again by adding a caged form of calcium to the beads. Like the caged indicators mentioned above, these molecules release calcium when a high-intensity UV beam breaks the bond between calcium-EGTA and the caging compound (75). In fact, they found that the localized release of calcium could stabilize a dendritic branch. Lohmann et al. hypothesized that these calcium signals were a result of transmission at sites of contact with synaptic partners. Although they were able to obtain only one direct example of contact between cells resulting in calcium influx, this provides an exciting lead in understanding the role of synaptic signaling in directing arbor formation.

Calcium Signaling in Differentiation and Growth Cone Guidance

Aside from being a useful readout of synaptic activity, calcium serves many roles as an intracellular signal. An ongoing series of experiments by Spitzer and colleagues has revealed dynamic patterns of calcium signals within neurons and linked these spikes and waves to aspects of neuronal differentiation and growth cone guidance.

By imaging cultured embryonic *Xenopus* neurons loaded with Fluo-3, Gu et al. (76) observed two types of calcium transients: spikes, which spread through the entire cell and occur at relatively low frequency (3 per h), and waves, which occur locally in growth cones at higher frequencies (8 per h). To investigate the possible role of these signals, they eliminated them by removing calcium from the medium. This resulted in increased axon growth and failure to differentiate fully, as measured by reduced potassium currents and decreased neurotransmitter levels. To test the role of spikes and waves separately, Gu & Spitzer (76) reintroduced calcium transients into calcium-deprived neurons by pulsing a high potassium/high calcium solution onto the neurons, which stimulated calcium influx with a time course and amplitude similar to spontaneous transients. Remarkably, stimulation at the spike frequency was sufficient to restore differentiation, and stimulation at the wave frequency restored axon extension to normal lengths.

A later study (77) in an exposed *Xenopus* spinal cord preparation confirmed that the growth cone waves in fact regulate axon pathfinding in vivo. Again, different observed calcium transient frequencies correlated with different growth cone behaviors, and by manipulating intracellular calcium Gomez & Spitzer were able to change the motion of growth cone correspondingly. Because rates of calcium transients were cell-type specific and dependent on growth cone location, these effects suggest that extracellular cues guide the growth cone through this frequency-dependent calcium signaling pathway. In fact, a recent study (78) has demonstrated that these effects extend down to individual filopodia, where extracellular cues can evoke turning by differential calcium signaling in filopodia on opposite sides of the growth cone.

Intrinsic Optical Signal Imaging of Cortical Map Development

A striking feature of the large-scale organization of the brain is the presence of topographic maps throughout the nervous system. Primary visual cortex contains several different forms of such spatial organization. Beyond the basic retinotopic map of the spatial visual field, V1 is further subdivided into regions receiving input from one eye versus the other (known as ocular dominance columns) and regions responding to input at a particular orientation (orientation maps). These maps are established during development, and the roles of genetic programming versus sensory activity have been actively studied. Methods for directly imaging these maps in developing animals, using intrinsic optical signals of neural response, are accelerating progress in this field.

Intrinsic optical signals can be defined as any measurable change in optical properties of a membrane, neuron, or tissue that reflects neural activity but does not depend on the presence of any dye or any exogenous gene, protein, or other substance. Many such signals have been identified in neurons, brain slices, and intact brains over the years (79), but most are extremely small and require heroic and inconvenient effort for their detection. One important exception to this rule of difficulty is a change in optical reflectance from cortical surfaces of the brain in vivo. Reflectance-change signals on the order of 0.1% or more can be visualized through open or windowed craniotomies or even through the skull. Technical requirements are not terribly demanding: Standard microscopic light sources and CCD cameras are quite useful. The largest components of such in vivo intrinsic signals result from changes in blood flow and hemoglobin oxygenation secondary to metabolic and signaling interaction between active neurons and the brain's rich vascular supply. Interestingly, the origins of this intrinsic optical signal appear to be similar to the signals that are measured by contemporary fMRI methods [see Logothetis & Wandell, this volume (80)].

Because intrinsic optical signals can be imaged in intact animals, with minimal surgery required, different sensory stimuli can be presented to the animal and the pattern of brain activation directly mapped onto the cortex (66). Resolution does not extend down to single cells, but permits studies of larger scale organization, such as visual retinotopy, ocular dominance columns, and orientation maps (Figure 6, see color insert), as well as chemotopic maps in the olfactory bulb (81) and somatotopic maps in cortex (82). Recent advances in methodology allow complete maps to be obtained in minutes, using Fourier analysis of the response to periodic stimuli in a continuous acquisition mode (83).

Intrinsic signal imaging has played a crucial role in experiments revealing the development of maps in visual cortex. Initial experiments described the gradual crystallization of maps, from a homogenous response field to sharp boundaries between ocular dominance columns and the corresponding emergence of a super-imposed orientation map (Figure 6) (85). Altering specific aspects of visual input, such as covering one eye (84) or creating a visual environment with stripes of just one orientation (86), results in perturbation of the corresponding map. These studies have provided sufficient detail to allow testing of theoretical models of input segregation (87). Experiments have also used these maps to guide electro-physiology, in order to trace the cellular changes underlying map formation. For example, Trachtenberg et al. (88) used intrinsic signal imaging to determine ocular dominance maps both before and after deprivation of vision in one eye for 24 h. These maps were used to guide single-cell recordings in regions that had switched dominance, revealing that the shift in mapping was due to changes in superficial layers of cortex, rather than deep layers.

Establishment of Local Circuit Connectivity

Underlying these topographic maps and all neural circuits are specific patterns of synaptic connectivity. Whereas imaging the structure of the neurons reveals the arborization field of possible synaptic contacts, eventually one wants to know the actual functional connections that form. Several functional imaging techniques have begun to elucidate the development of these circuits.

In order to trace the propagation of activity in the developing visual cortex, Nelson & Katz (89) imaged brain slices that had been loaded with a voltage-sensitive dye. By imaging with a high-speed photodiode array detector at a low-spatial resolution, thereby summing the activation of groups of neurons, they followed the depolarization that resulted from stimulation of afferent thalamic fibers. In the immature cortex, before normal visual activity begins, stimulation resulted in activation of neurons in a narrow radial column. However, in older animals, activity began to spread more laterally within layers of the cortex. They determined that this spread was due to horizontal connections within the more superficial layers. By correlating these time points with previous studies of map development, they concluded that these new intercolumn connections were forming during the period that orientation selectivity arises.

While sensory deprivation has long been known to result in gross disruption of development, it has yet to be determined how it plays its instructive role in directly shaping the normal development of a circuit. Most studies use large-scale changes in innervation area or lamination as their readout. However, if activity is presumed to be specifying the fine structure of connectivity, one should look for an effect on the synaptic inputs to individual neurons.

Callaway & Katz (90) pioneered a method that allows localization of functional inputs onto a single neuron by focally photostimulating uncaging of neurotransmitter. Rather than optically imaging the response to an electrical stimulus, this technique uses light to evoke the stimulus, with electrical recording as the readout. Caged neurotransmitters are applied to the bath of a brain slice, and a patch electrode is used to record from a single neuron. A focused laser beam can uncage the transmitter at specific points around the neuron, resulting in action potentials in neurons in the vicinity of the focus. If these neurons are synaptically connected to the patched neuron, a current will be recorded. Probing at thousands of sites maps out the location of neurons that are presynaptic to the given cell.

Shepherd et al. (91) used this technique to map the synaptic inputs onto layer 2/3 neurons in slices of barrel cortex from both normal animals and animals that had undergone sensory deprivation by trimming of whiskers. They found that deprived neurons above barrels had reduced synaptic input from layer 4 (which receives axons from thalamus) and increased local input from layer 2/3. They propose that this shift could account for previously demonstrated changes in receptive fields of these neurons. Thus beyond looking at the overall mapping on cortex, or the summed synaptic response of a neuron, they were able to map the effect of deprivation to a shift in the connectivity of the neural circuit.

Spontaneous Correlated Network Activity

Unlike computers, which are normally fully assembled before they are ever turned on, the nervous system is active even at early stages of development. In fact, it appears that such incipient activity may play a crucial role in shaping patterns of connectivity. In one particularly intriguing case, pioneering electrophysiological studies of Meister et al. (92) demonstrated the presence of spontaneous patterned activity in immature retina, occuring even before the development of functional photoreceptors. These activity patterns appear to play a substantial role in the early wiring of higher visual centers (93). Much of what we now know about the origins of these receptor-independent activity patterns has come from imaging studies.

In the first of these studies, Wong et al. (94) performed calcium imaging in explant preparations of immature retina. In order to load large populations of neurons, they utilized an AM-ester form of fura-2. This sidegroup makes the molecule hydrophobic, allowing it to cross the membrane and enter cells. However,

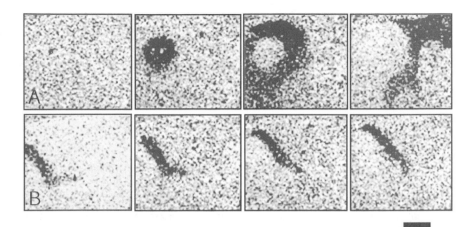

Figure 7 Waves of spontaneous activity spreading across a chick retina. Darker regions indicate an increase in activity, as measured by a fluorescent calcium indicator. Consecutive images are at 1.4-s intervals. Scale bar 1 mm. Reproduced by copyright permission of The Society for Neuroscience from Wong et al. (97).

once inside the cell, endogenous enzymes cleave the ester, trapping the indicator inside the cell. AM-esters provide a convenient way of loading large populations of neurons by bath application (95, 96). The low concentration in individual cells and the fact that nearly all neurons are loaded can make it difficult to discern fine structures such as dendrites due to lack of contrast. However, calcium signals in individual cell bodies or populations of neurons can be easily detected even with wide-field imaging. In immature retinal explants, Wong et al. (94) found large-scale waves of activation that moved across the retina (Figure 7). Furthermore, they were able to trace the source of these waves to a subcircuit in the retina, between amacrine cells and retinal ganglion cells, which led to the organized spontaneous activity.

Similar correlated activity has also been observed in both the spinal cord (98) and the hippocampus (99). To investigate factors that can influence the development of these network properties in the hippocampus, Aguado et al. (100) studied a line of transgenic mice that overexpressed BDNF, the same neurotrophin studied by Alsina et al. (52) (discussed above). When slices were made from BDNF over-expressing mice at embryonic day 18, a time when there is normally little correlated activity, they found increased overall spontaneous activity and, in particular, more neurons participating in coactivated ensembles. Elevated BDNF had increased the number of synapses in these immature slices and also affected the response to GABA transmission by elevating expression of the chloride transporter. Thus functional imaging revealed that the trophic effects of BDNF on synapse formation actually translate into shaping specific patterns of activity.

Although the exact role of this spontaneous correlated activity in directing development remains uncertain (101), imaging experiments can help develop perturbations to test their function (102). Furthermore, beyond their possible instructive role, patterns of spontaneous activity can give insight into the underlying circuit that is being developed. In the future, it will be interesting to understand how these immature spontaneous patterns solidify into mature network function.

FUTURE DIRECTIONS

While the rapid progress in live imaging has already provided new vistas for experiments, there are several areas where we believe further advances are likely to arise. First, live imaging can generate huge amounts of data, in the form of three-dimensional images at many time points. Extracting conclusions and, in particular, quantitative results from these pictures can be an overwhelming task. Most analysis consists of qualitative visual inspection of images or movies, followed by manual measurement of various parameters. In addition to the subjective aspect of this approach, such as defining structures or regions to be measured, the labor involved precludes all but the most straightforward measurements. Novel automated analysis techniques must be developed to enable us to extract all the useful information available from live imaging data, according to well-defined criteria. Two recent examples of such an approach are from Rodriguez et al. (103) and Koh et al. (104) who tackle the problems of analyzing dendritic arbor parameters and identifying dendritic spines, respectively.

Although amazing progress has been made in genetically encoded indicators in the decade since GFP was first expressed as a marker (105), we anticipate continuing advance in development of functional indicators. In particular, current genetically encoded calcium indicators are only beginning to be used in vertebrate neurons beyond the proof-of-principle stage. Likewise, the signal from GFP-based voltage sensors is not yet sufficient to take advantage of their potential for expression restricted to particular cell types. We also expect that sensors for other signaling pathways will be developed, such as cAMP indicators (106) and FRET-based readouts of GTPase activities (107), which have already been demonstrated.

New fluorophores may also be developed to take advantage of advances in imaging. In particular, because most current fluorescent molecules are not optimized for two-photon excitation, significant improvements in synthesized organic indicators are possible (108). Directed evolution techniques, which have been used to improve folding, stability, and spectral properties of fluorescent proteins (109, 110), could be used to increase their two-photon cross-sections as well. Furthermore, quantum dots, nanometer-sized semiconductor crystals that can be designed to have specific fluorescence properties, have two-photon absorption cross-sections up to several orders of magnitude greater than GFP or other current fluorophores (111) and have been shown to be applicable for in vivo imaging. If techniques can be developed to bind these to particular proteins, in a restricted pattern such as the way GFP is used, they would be a particularly useful tool.

Finally, techniques to interface optical imaging with electron microscopy studies of ultrastructure will prove extremely valuable. Optical imaging is generally limited by the wavelength of light to length scales of at least tenths of microns. However many subcellular structures, such as transport packets and synapses, are smaller than this. To verify the identity and makeup of such structures, techniques for retrospective electron microscopy must be expanded, particularly for use in vivo, where finding structures that have been observed at the light level and reconstructing them by serial section are almost prohibitively laborious. Techniques for identification, such as immunogold labeling and photoconversion, could be improved to facilitate such analysis. The development of the reAsH system, which uses a genetically encodable tag to specifically bind a photoconvertible fluorophore onto proteins of interest, provides an encouraging step in this direction (112).

CONCLUSION

While live imaging can be traced back to early cinematography studies at the beginning of the century, the field has accelerated rapidly in the past 20 years as a result of technological advances. New modes of microscopy permit us to peer deep into intact living nervous tissue, allowing long-standing questions in development to be addressed by directly watching the relevant processes. In addition to providing an initial description of the dynamic sequence of events in a developmental process, which often reveals unsuspected features such as transport packets and correlated spontaneous activity, live imaging is a vital tool in probing the mechanisms underlying these processes. Molecular techniques are facilitating these investigations by enabling advances in imaging such as GFP fusions and new functional indicators and by allowing perturbation of molecules in specific cells. These methods are generating profound insight into the development of the structures that Cajal observed over 100 years ago.

ACKNOWLEDGMENTS

We thank S. McConnell, Y. Hua, and J. Gilthorpe for comments on the manuscript, and members of the Smith lab for helpful discussions.

The *Annual Review of Physiology* is online at http://physiol.annualreviews.org

LITERATURE CITED

1. Cajal SRy. 1995. *Histology of the Nervous System of Man and Vertebrates*. New York: Oxford Univ. Press
2. Yuste R, Lanni F, Konnerth A, eds. 2000. *Imaging Neurons*. Cold Spring Harbor, NY: Cold Spring Harbor Lab. Press

3. Murphy DB. 2001. *Fundamentals of Light Microscopy and Electronic Imaging*. New York: Wiley-Liss
4. Diaspro A, ed. 2002. *Confocal and Two-Photon Microscopy*. New York: Wiley-Liss

5. Pawley JB. 1995. *Handbook of Biological Confocal Microscopy*. New York: Plenum

6. Marriott G, Parker I, eds. 2003. *Methods in Enzymology: Biophotonics*, Vols. 360, 361. San Diego: Academic

7. Banker G, Goslin K, eds. 1998. *Culturing Nerve Cells*. Cambridge, MA: MIT Press

8. Washbourne P, McAllister AK. 2002. Techniques for gene transfer into neurons. *Curr. Opin. Neurobiol.* 12:566–73

9. Tsien RY. 1998. The green fluorescent protein. *Annu. Rev. Biochem.* 67:509–44

10. Schaefer AW, Kabir N, Forscher P. 2002. Filopodia and actin arcs guide the assembly and transport of two populations of microtubules with unique dynamic parameters in neuronal growth cones. *J. Cell Biol.* 158:139–52

11. Dent EW, Kalil K. 2003. Dynamic imaging of neuronal cytoskeleton. *Methods Enzymol.* 361:390–407

12. Politz JC. 1999. Use of caged fluorochromes to track macromolecular movement in living cells. *Trends Cell Biol.* 9:284–87

13. Patterson GH, Lippincott-Schwartz J. 2002. A photoactivatable GFP for selective photolabeling of proteins and cells. *Science* 297:1873–77

14. Waterman-Storer CM, Danuser G. 2002. New directions for fluorescent speckle microscopy. *Curr. Biol.* 12:R633–40

15. Ponti A, Vallotton P, Salmon WC, Waterman-Storer CM, Danuser G. 2003. Computational analysis of f-actin turnover in cortical actin meshworks using fluorescent speckle microscopy. *Biophys. J.* 84:3336–52

16. Forscher P, Smith SJ. 1988. Actions of cytochalasins on the organization of actin filaments and microtubules in a neuronal growth cone. *J. Cell Biol.* 107:1505–16

17. Chang S, Svitkina TM, Borisy GG, Popov SV. 1999. Speckle microscopic evaluation of microtubule transport in growing nerve processes. *Nat. Cell Biol.* 1:399–403

18. Dent EW, Kalil K. 2001. Axon branching requires interactions between dynamic microtubules and actin filaments. *J. Neurosci.* 21:9757–69

19. O'Rourke NA, Dailey ME, Smith SJ, McConnell SK. 1992. Diverse migratory pathways in the developing cerebral cortex. *Science* 258:299–302

20. Nadarajah B, Brunstrom JE, Grutzendler J, Wong RO, Pearlman AL. 2001. Two modes of radial migration in early development of the cerebral cortex. *Nat. Neurosci.* 4:143–50

21. Nadarajah B, Alifragis P, Wong RO, Parnavelas JG. 2002. Ventricle-directed migration in the developing cerebral cortex. *Nat. Neurosci.* 5:218–24

22. Chenn A, McConnell SK. 1995. Cleavage orientation and the asymmetric inheritance of Notch1 immunoreactivity in mammalian neurogenesis. *Cell* 82:631–41

23. Cochilla AJ, Angleson JK, Betz WJ. 1999. Monitoring secretory membrane with FM1-43 fluorescence. *Annu. Rev. Neurosci.* 22:1–10

24. Ahmari SE, Buchanan J, Smith SJ. 2000. Assembly of presynaptic active zones from cytoplasmic transport packets. *Nat. Neurosci.* 3:445–51

25. Friedman HV, Bresler T, Garner CC, Ziv NE. 2000. Assembly of new individual excitatory synapses: time course and temporal order of synaptic molecule recruitment. *Neuron* 27:57–69

26. Bresler T, Ramati Y, Zamorano PL, Zhai R, Garner CC, Ziv NE. 2001. The dynamics of SAP90/PSD-95 recruitment to new synaptic junctions. *Mol. Cell Neurosci.* 18:149–67

27. Lichtman JW, Fraser SE. 2001. The neuronal naturalist: watching neurons in their native habitat. *Nat. Neurosci.* 4:1215–20 (Suppl.)

28. Wu GY, Zou DJ, Koothan T, Cline HT. 1995. Infection of frog neurons with vaccinia virus permits in vivo expression of foreign proteins. *Neuron* 14:681–84

29. Haas K, Jensen K, Sin WC, Foa L, Cline HT. 2002. Targeted electroporation in

Xenopus tadpoles in vivo—from single cells to the entire brain. *Differentiation* 70:148–54

30. Koster RW, Fraser SE. 2001. Tracing transgene expression in living zebrafish embryos. *Dev. Biol.* 233:329–46

31. Spergel DJ, Kruth U, Shimshek DR, Sprengel R, Seeburg PH. 2001. Using reporter genes to label selected neuronal populations in transgenic mice for gene promoter, anatomical, and physiological studies. *Prog. Neurobiol.* 63:673–86

32. Feng G, Mellor RH, Bernstein M, Keller-Peck C, Nguyen QT, et al. 2000. Imaging neuronal subsets in transgenic mice expressing multiple spectral variants of GFP. *Neuron* 28:41–51

33. Amos B, Reichelt S. 2001. SELS: a new method for laser scanning microscopy of living cells. *Microsc. Anal.* November:9–11

34. Maddox PS, Moree B, Canman JC, Salmon ED. 2003. Spinning disk confocal microscope system for rapid high-resolution, multimode, fluorescence speckle microscopy and green fluorescent protein imaging in living cells. *Methods Enzymol.* 360:597–617

35. Denk W, Strickler JH, Webb WW. 1990. Two-photon laser scanning fluorescence microscopy. *Science* 248:73–76

36. Denk W, Svoboda K. 1997. Photon upmanship: why multiphoton imaging is more than a gimmick. *Neuron* 18:351–57

37. Helmchen F, Denk W. 2002. New developments in multiphoton microscopy. *Curr. Opin. Neurobiol.* 12:593–601

38. Theer P, Hasan M, Denk W. 2003. Two-photon imaging to a depth of 1000 μm in living brains by use of a TI:Al2O3 regenerative amplifier. *Optics Lett.* 28:1002–4

39. Helmchen F, Fee MS, Tank DW, Denk W. 2001. A miniature head-mounted two-photon microscope. High-resolution brain imaging in freely moving animals. *Neuron* 31:903–12

40. Lin S. 2000. Transgenic zebrafish. *Methods Mol. Biol.* 136:375–83

41. Das T, Payer B, Cayouette M, Harris WA. 2003. In vivo time-lapse imaging of cell divisions during neurogenesis in the developing zebrafish retina. *Neuron* 37:597–609

42. Lyons DA, Guy AT, Clarke JD. 2003. Monitoring neural progenitor fate through multiple rounds of division in an intact vertebrate brain. *Development* 130:3427–36

43. Koster RW, Fraser SE. 2001. Direct imaging of in vivo neuronal migration in the developing cerebellum. *Curr. Biol.* 11:1858–63

44. Bovolenta P, Mason C. 1987. Growth cone morphology varies with position in the developing mouse visual pathway from retina to first targets. *J. Neurosci.* 7:1447–60

45. Mason CA, Wang LC. 1997. Growth cone form is behavior-specific and, consequently, position-specific along the retinal axon pathway. *J. Neurosci.* 17:1086–100

46. Hutson LD, Chien CB. 2002. Pathfinding and error correction by retinal axons: the role of astray/robo2. *Neuron* 33:205–17

47. Purves D, Lichtman JW. 1987. Synaptic sites on reinnervated nerve cells visualized at two different times in living mice. *J. Neurosci.* 7:1492–97

48. Balice-Gordon RJ, Lichtman JW. 1990. In vivo visualization of the growth of pre- and postsynaptic elements of neuromuscular junctions in the mouse. *J. Neurosci.* 10:894–908

49. Gan WB, Lichtman JW. 1998. Synaptic segregation at the developing neuromuscular junction. *Science* 282:1508–11

50. Walsh MK, Lichtman JW. 2003. In vivo time-lapse imaging of synaptic takeover associated with naturally occurring synapse elimination. *Neuron* 37:67–73

51. Jontes JD, Buchanan J, Smith SJ. 2000. Growth cone and dendrite dynamics in zebrafish embryos: early events in synaptogenesis imaged in vivo. *Nat. Neurosci.* 3:231–37

52. Alsina B, Vu T, Cohen-Cory S. 2001. Visualizing synapse formation in arborizing optic axons in vivo: dynamics and modulation by BDNF. *Nat. Neurosci.* 4:1093–101

53. O'Rourke NA, Fraser SE. 1986. Dynamic aspects of retinotectal map formation revealed by a vital-dye fiber-tracing technique. *Dev. Biol.* 114:265–76

54. Cline HT. 2001. Dendritic arbor development and synaptogenesis. *Curr. Opin. Neurobiol.* 11:118–26

55. Wong WT, Wong RO. 2000. Rapid dendritic movements during synapse formation and rearrangement. *Curr. Opin. Neurobiol.* 10:118–24

56. Wu GY, Cline HT. 1998. Stabilization of dendritic arbor structure in vivo by CaMKII. *Science* 279:222–26

57. Nedivi E, Wu GY, Cline HT. 1998. Promotion of dendritic growth by CPG15, an activity-induced signaling molecule. *Science* 281:1863–66

58. Li Z, Van Aelst L, Cline HT. 2000. Rho GTPases regulate distinct aspects of dendritic arbor growth in *Xenopus* central neurons in vivo. *Nat. Neurosci.* 3:217–25

59. Sin WC, Haas K, Ruthazer ES, Cline HT. 2002. Dendrite growth increased by visual activity requires NMDA receptor and Rho GTPases. *Nature* 419:475–80

60. Lendvai B, Stern EA, Chen B, Svoboda K. 2000. Experience-dependent plasticity of dendritic spines in the developing rat barrel cortex in vivo. *Nature* 404:876–81

61. Grutzendler J, Kasthuri N, Gan WB. 2002. Long-term dendritic spine stability in the adult cortex. *Nature* 420:812–16

62. Trachtenberg JT, Chen BE, Knott GW, Feng G, Sanes JR, et al. 2002. Long-term in vivo imaging of experience-dependent synaptic plasticity in adult cortex. *Nature* 420:788–94

63. Djurisic M, Zochowski M, Wachowiak M, Falk CX, Cohen LB, Zecevic D. 2003. Optical monitoring of neural activity using voltage-sensitive dyes. *Methods Enzymol.* 361:423–51

64. Rudolf R, Mongillo M, Rizzuto R, Pozzan T. 2003. Innovation: looking forward to seeing calcium. *Nat. Rev. Mol. Cell. Biol.* 4:579–86

65. Helmchen F, Waters J. 2002. Ca^{2+} imaging in the mammalian brain in vivo. *Eur. J. Pharmacol.* 447:119–29

66. Bonhoeffer T, Grinvald A. 1996. Optical imaging based on intrinsic signals—the methodology. In *Brain Mapping: The Methods*, ed. AW Toga, JC Mazziotta, pp. 550–97. New York: Academic

67. Shen K, Meyer T. 1999. Dynamic control of CaMKII translocation and localization in hippocampal neurons by NMDA receptor stimulation. *Science* 284:162–66

68. Gleason MR, Higashijima S, Dallman J, Liu K, Mandel G, Fetcho JR. 2003. Translocation of CaM kinase II to synaptic sites in vivo. *Nat. Neurosci.* 6:217–18

69. Yuste R, Miller RB, Holthoff K, Zhang S, Miesenbock G. 2000. Synapto-pHluorins: chimeras between pH-sensitive mutants of green fluorescent protein and synaptic vesicle membrane proteins as reporters of neurotransmitter release. *Methods Enzymol.* 327:522–46

70. Fan GY, Fujisaki H, Miyawaki A, Tsay RK, Tsien RY, Ellisman MH. 1999. Video-rate scanning two-photon excitation fluorescence microscopy and ratio imaging with cameleons. *Biophys. J.* 76:2412–20

71. Nguyen QT, Callamaras N, Hsieh C, Parker I. 2001. Construction of a two-photon microscope for video-rate Ca^{2+} imaging. *Cell Calcium* 30:383–93

72. Lohmann C, Myhr KL, Wong RO. 2002. Transmitter-evoked local calcium release stabilizes developing dendrites. *Nature* 418:177–81

73. Kettunen P, Demas J, Lohmann C, Kasthuri N, Gong Y, et al. 2002. Imaging calcium dynamics in the nervous system by means of ballistic delivery of indicators. *J. Neurosci. Methods* 119:37–43

74. Gan WB, Grutzendler J, Wong WT, Wong

RO, Lichtman JW. 2000. Multicolor "Di-Olistic" labeling of the nervous system using lipophilic dye combinations. *Neuron* 27:219–25

75. Ellis-Davies GC. 2003. Development and application of caged calcium. *Methods Enzymol.* 360:226–38

76. Gu X, Olson EC, Spitzer NC. 1994. Spontaneous neuronal calcium spikes and waves during early differentiation. *J. Neurosci.* 14:6325–35

77. Gomez TM, Spitzer NC. 1999. In vivo regulation of axon extension and pathfinding by growth-cone calcium transients. *Nature* 397:350–55

78. Gomez TM, Robles E, Poo M, Spitzer NC. 2001. Filopodial calcium transients promote substrate-dependent growth cone turning. *Science* 291:1983–87

79. Obrig H, Villringer A. 2003. Beyond the visible—imaging the human brain with light. *J. Cereb. Blood Flow Metab.* 23:1–18

80. Logothetis NK, Wandell BA. 2004. Interpreting the BOLD signal. *Annu. Rev. Physiol.* 66:735–69

81. Rubin BD, Katz LC. 1999. Optical imaging of odorant representations in the mammalian olfactory bulb. *Neuron* 23:499–511

82. Chen LM, Friedman RM, Ramsden BM, LaMotte RH, Roe AW. 2001. Fine-scale organization of SI (area 3b) in the squirrel monkey revealed with intrinsic optical imaging. *J. Neurophysiol.* 86:3011–29

83. Kalatsky VA, Stryker MP. 2003. New paradigm for optical imaging. Temporally encoded maps of intrinsic signal. *Neuron* 38:529–45

84. Issa NP, Trachtenberg JT, Chapman B, Zahs KR, Stryker MP. 1999. The critical period for ocular dominance plasticity in the Ferret's visual cortex. *J. Neurosci.* 19:6965–78

85. Crair MC, Gillespie DC, Stryker MP. 1998. The role of visual experience in the development of columns in cat visual cortex. *Science* 279:566–70

86. Sengpiel F, Stawinski P, Bonhoeffer T. 1999. Influence of experience on orientation maps in cat visual cortex. *Nat. Neurosci.* 2:727–32

87. Erwin E, Miller KD. 1998. Correlation-based development of ocularly matched orientation and ocular dominance maps: determination of required input activities. *J. Neurosci.* 18:9870–95

88. Trachtenberg JT, Trepel C, Stryker MP. 2000. Rapid extragranular plasticity in the absence of thalamocortical plasticity in the developing primary visual cortex. *Science* 287:2029–32

89. Nelson DA, Katz LC. 1995. Emergence of functional circuits in ferret visual cortex visualized by optical imaging. *Neuron* 15:23–34

90. Callaway EM, Katz LC. 1993. Photostimulation using caged glutamate reveals functional circuitry in living brain slices. *Proc. Natl. Acad. Sci. USA* 90:7661–65

91. Shepherd GM, Pologruto TA, Svoboda K. 2003. Circuit analysis of experience-dependent plasticity in the developing rat barrel cortex. *Neuron* 38:277–89

92. Meister M, Wong RO, Baylor DA, Shatz CJ. 1991. Synchronous bursts of action potentials in ganglion cells of the developing mammalian retina. *Science* 252:939–43

93. Katz LC, Shatz CJ. 1996. Synaptic activity and the construction of cortical circuits. *Science* 274:1133–38

94. Wong RO, Chernjavsky A, Smith SJ, Shatz CJ. 1995. Early functional neural networks in the developing retina. *Nature* 374:716–18

95. Yuste R. 2000. Loading brain slices with AM esters of calcium indicators. See Ref. 2, pp. 34.1–.9

96. Wong RO. 2000. Calcium imaging of retinal activity. See Ref. 2, pp. 41.1–.7

97. Wong WT, Sanes JR, Wong RO. 1998. Developmentally regulated spontaneous activity in the embryonic chick retina. *J. Neurosci.* 18:8839–52

98. O'Donovan M, Ho S, Yee W. 1994. Calcium imaging of rhythmic network activity in the developing spinal cord of the chick embryo. *J. Neurosci.* 14:6354–69

99. Garaschuk O, Hanse E, Konnerth A. 1998. Developmental profile and synaptic origin of early network oscillations in the CA1 region of rat neonatal hippocampus. *J. Physiol.* 507:219–36

100. Aguado F, Carmona MA, Pozas E, Aguilo A, Martinez-Guijarro FJ, et al. 2003. BDNF regulates spontaneous correlated activity at early developmental stages by increasing synaptogenesis and expression of the K^+/Cl^- co-transporter KCC2. *Development* 130:1267–80

101. Feller MB. 1999. Spontaneous correlated activity in developing neural circuits. *Neuron* 22:653–56

102. Huberman AD, Wang GY, Liets LC, Collins OA, Chapman B, Chalupa LM. 2003. Eye-specific retinogeniculate segregation independent of normal neuronal activity. *Science* 300:994–98

103. Rodriguez A, Ehlenberger D, Kelliher K, Einstein M, Henderson SC, et al. 2003. Automated reconstruction of three-dimensional neuronal morphology from laser scanning microscopy images. *Methods* 30:94–105

104. Koh IY, Lindquist WB, Zito K, Nimchinsky EA, Svoboda K. 2002. An image analysis algorithm for dendritic spines. *Neural Comput.* 14:1283–310

105. Zhang J, Campbell RE, Ting AY, Tsien RY. 2002. Creating new fluorescent probes for cell biology. *Nat. Rev. Mol. Cell. Biol.* 3:906–18

106. Zaccolo M, De Giorgi F, Cho CY, Feng L, Knapp T, et al. 2000. A genetically encoded, fluorescent indicator for cyclic AMP in living cells. *Nat. Cell Biol.* 2:25–29

107. Mochizuki N, Yamashita S, Kurokawa K, Ohba Y, Nagai T, et al. 2001. Spatio-temporal images of growth-factor-induced activation of Ras and Rap1. *Nature* 411:1065–68

108. Albota M, Beljonne D, Bredas JL, Ehrlich JE, Fu JY, et al. 1998. Design of organic molecules with large two-photon absorption cross sections. *Science* 281:1653–56

109. Campbell RE, Tour O, Palmer AE, Steinbach PA, Baird GS, et al. 2002. A monomeric red fluorescent protein. *Proc. Natl. Acad. Sci. USA* 99:7877–82

110. Crameri A, Whitehorn EA, Tate E, Stemmer WP. 1996. Improved green fluorescent protein by molecular evolution using DNA shuffling. *Nat. Biotechnol.* 14:315–19

111. Larson DR, Zipfel WR, Williams RM, Clark SW, Bruchez MP, et al. 2003. Water-soluble quantum dots for multiphoton fluorescence imaging in vivo. *Science* 300:1434–36

112. Gaietta G, Deerinck TJ, Adams SR, Bouwer J, Tour O, et al. 2002. Multicolor and electron microscopic imaging of connexin trafficking. *Science* 296:503–7

Annu. Rev. Physiol. 2004. 66:799–828
doi: 10.1146/annurev.physiol.66.052102.134444
First published online as a Review in Advance on September 22, 2003

CONTROL OF THE SIZE OF THE HUMAN MUSCLE MASS

Michael J. Rennie,[1,4] Henning Wackerhage,[1]
Espen E. Spangenburg,[3] and Frank W. Booth[2]

[1]Division of Molecular Physiology, School of Life Sciences, University of Dundee,
Dundee, DD1 4HN, Scotland, United Kingdom; [2]Department of Biomedical Sciences,
Medical Pharmacology and Physiology, and Dalton Cardiovascular Center, University
of Missouri-Columbia, Columbia, Missouri 65211; current addresses: [3]Exercise Biology
Program, University of California, Davis, California 95616; [4]University of Nottingham,
Graduate Entry Medical School, City Hospital, Derby, DE22 3NE, United Kingdom;
email: michael.rennie@nottingham.ac.uk; h.wackerhage@dundee.ac.uk;
spangenburge@missouri.edu (current: eespangenburg@ucdavis.edu),
boothf@missouri.edu

Key Words   adaptation, physical activity, nutrition, hypertrophy, protein turnover

■ **Abstract**   This review is divided into two parts, the first dealing with the cell and molecular biology of muscle in terms of growth and wasting and the second being an account of current knowledge of physiological mechanisms involved in the alteration of size of the human muscle mass. Wherever possible, attempts have been made to interrelate the information in each part and to provide the most likely explanation for phenomena that are currently only partially understood. The review should be of interest to cell and molecular biologists who know little of human muscle physiology and to physicians, physiotherapists, and kinesiologists who may be familiar with the gross behavior of human muscle but wish to understand more about the underlying mechanisms of change.

INTRODUCTION

Great strides have been made recently in understanding the regulation of the size of the muscle mass in humans. Dual X-ray absorptiometry (DEXA) (1, 2) and magnetic resonance imaging (MRI) (3–5), together with advances in immunohistochemical muscle fiber typing (6), have allowed the size of the human muscle mass and its components to be defined with previously unparalleled precision and sensitivity. We are now able to follow accurately small, relatively slow changes in muscle size during the extended timescales of sarcopenia (7, 8) and hypertrophy (9). The application of stable isotope tracer technology to the study of amino acids has markedly increased our knowledge of their transport and

0066-4278/04/0315-0799$14.00

intermediary metabolism and protein turnover in skeletal muscle (10–13), and positron emission spectroscopy (PET) promises to provide additional information (14). In parallel, ever more signal transduction pathways involved in the regulation of muscle growth are being elucidated, and powerful microarray methods enable identification of genes whose transcription is altered during muscle growth (15). The explosion of knowledge of the molecular cell biology of skeletal muscle (16, 17) has provided us with concepts, techniques, and reagents with which to probe the mechanisms underlying the observed changes in muscle mass in response to altered nutrition and physical activity.[1]

Hitherto, research on the human muscle mass and muscle growth signaling has usually been conducted by separate groups of researchers with limited mutual communication. The aim of this article is to review both areas and to show connections between them. We first discuss recent findings on muscle growth mechanisms; we then relate these findings to muscle growth in humans.

MUSCLE GROWTH MECHANISMS—SIGNAL TRANSDUCTION AND REGULATION OF PROTEIN TURNOVER

The specific aim of the first section is to review (*a*) the signal transduction pathways that sense the muscle's environment and respond to various factors within it, inputting upstream signals to the muscle growth regulation system; (*b*) the

---

[1] Abbreviations: ADP, adenosine diphosphate; AICAR, 5-aminoimidazole-4-carboxyamide ribonucleotide; AMP, adenosine monophosphate; AMPK, AMP-dependent protein kinase; ATP, adenosine triphosphate; 4E-BP1, initiation factor 4E-binding protein 1; CAIN, calcineurin inhibitor; cdk2, cyclin-dependent kinase 2; DEXA, dual X-ray absorptiometry; eIF2, eukaryotic translation initiation factor 4E; eIF4E, eukaryotic translation initiation factor 4E; ERK1/2, extracellular signal-regulated kinase 1/2; FAK, focal adhesion kinase; G_1, gap 1 phase of the cell cycle; Gasp-1, growth and differentiation factor-associated serum protein-1; GDF-8, growth differentiation factor 8/myostatin; GSK3β, glycogen synthase kinase 3β; IGF-1, insulin-like growth factor 1 (IGF-1Ea, MGF/IGF-1Eb, and IGF-1Ec are splice isoforms); JNK, c-Jun N-terminal protein kinase; LIF, leukemia-inhibitory factor; MEF2, myocyte enhancer factor 2; MGF, mechano-growth factor (synonymous with IGF-1 splice variant IGF-1Eb); MRI, magnetic resonance imaging; mTOR, mammalian target of rapamycin; MPS, muscle protein synthesis; MPB, muscle protein breakdown; MyoD, myoblast determination factor; NFAT, nuclear factor of activated T-cells; NF-κB, nuclear factor κB; p27^{Kip1}, p27 kinase inhibitor protein 1; p38, p38 stress-activated protein kinase; p70 S6k, p70 S6 kinase (S6 is a ribosomal protein); PI3K, phosphatidylinositol 3′-kinase; PET, positron emission spectroscopy; PKB/AKT, protein kinase B/AKT; PKC, protein kinase C; PPARγ, peroxisome proliferators-activated receptor γ; Raptor, mTOR-binding protein; rhGH, recombinant human growth factor; SHIP-2, SH2-containing inositol polyphosphate 5-phosphatase; SMAD3, human homologue from *Drosophila* Mad (mothers against decapentplegic) and *C. elegans* SMA gene; SRE1, serum response element 1; SRF, serum response factor; TGF-β, transforming growth factor-β; TNF-α, tumor necrosis factor-α.

transcriptional regulation of myostatin and IGF-1/MGF expression in response to growth-inducing stimuli; and (c) the specific regulation of muscle growth via regulation of mRNA translation and satellite cell proliferation in response to myostatin, IGF-1/MGF, and other factors. Most of the research has been carried out in nonhuman species, but the results are likely to be relevant to the observations made in human muscle.

Sensing of Growth Stimuli and Transcriptional Regulation

Strenuous, growth-inducing muscle activity is associated with changes in one or more of variables such as passive- and contraction-induced strain, sarcoplasmic calcium concentration, energy demand, intramuscular oxygen concentration, availability of hormones, growth factors and cytokines, temperature, and cellular damage (see Figure 1). A sufficient change in any of these variables will result in the altered activity of signal transduction pathways that regulate the transcription of genes differentially expressed during muscle growth. Signal transduction pathways shown to be activated in response to various forms of muscle contraction include those involving AMPK (18), calcineurin (19), ERK1/2 and p38 (20), JNK (20, 21) NF-κB (22), PI3K-PKB/AKT-mTOR (23), and PKC (24). In addition to these pathways, many of the transcription factors involved in myogenesis (25, 26) continue to be active in adaptive and regenerative processes in adult muscle. Thus, strenuous, growth-inducing muscle activity and other growth stimuli are likely to activate a signal transduction network rather than just one or two signal transduction pathways. Skeletal muscle hypertrophy signaling appears to mirror, to some extent, that observed during cardiac hypertrophy (27, 28). Here, we focus on the calcineurin and mechanical-chemical transduction pathways because these two signaling systems have been shown to be involved in muscle growth signaling.

CALCINEURIN-SIGNALING Ca^{2+} acts as a regulatory signal during skeletal muscle hypertrophy (29) and is of particular interest because during contraction there is a large transient change in cytosolic Ca^{2+}. Calcineurin is a Ca^{2+}-calmodulin-activated protein phosphatase that dephosphorylates the transcription factor NFAT, enabling its nuclear translocation and DNA binding. The calcineurin pathway has been linked not only to the regulation of skeletal muscle bulk growth but also to that of fast-to-slow phenotype conversion (30) and IGF-1 and Ca^{2+}-induced skeletal muscle hypertrophy, at least in cultured skeletal muscle (31).

However, regulation of nuclear Ca^{2+} concentrations may occur independently of transient changes in cytoplasmic Ca^{2+} calcium concentrations (32). In heart, Ca^{2+}-mediated cardiac muscle hypertrophy is induced partly through capacitative Ca^{2+} entry from the extracellular space by the way of transient receptor potential (Trp) proteins (33). It is not yet clear to what extent such a situation could occur in skeletal muscle, in which a much smaller proportion of Ca^{2+} flux arises via the sarcoplasmic reticulum. However, it seems likely that Ca^{2+} may be differentially routed toward nuclear Ca^{2+}-induced gene transcription, inducing activation of the Ca^{2+}-sensitive transcription factor (NFAT).

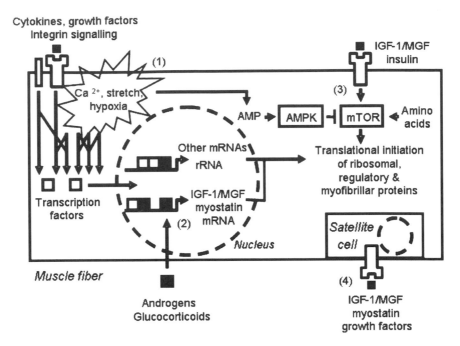

Figure 1 Overview of the main events during signal transduction and gene regulation leading to muscle hypertrophy. (1) Via receptor binding and cellular signals, cytokines and other growth factors are sensed and activate a network of signal transduction pathways that result (2) in the nuclear translocation or activation of transcription factors. Active, nuclear transcription factors (together with androgens and glucocorticoids via their soluble receptors) change the expression of the major muscle growth regulators IGF-1/MGF and myostatin or other muscle genes including ribosomal RNA (rRNA). Pathways that regulate translation or satellite cell function may also be activated by mechanisms other than IGF-1/MGF or myostatin (not shown). (3) IGF-1/MGF and insulin activate the PI3K-PKB/AKT-mTOR pathway, which enhances protein synthesis via increased translational initiation and the synthesis of ribosomal proteins for ribosome biogenesis. Availability of essential amino acids will activate mTOR signaling, whereas an increased energy demand sensed by AMPK will inhibit mTOR. (4) IGF-1/MGF, myostatin, and various other factors regulate an increased proliferation and differentiation of satellite cells.

Activation of NFAT transcriptional signaling is mediated by Ca^{2+}-induced increases in the phosphatase activity of calcineurin, which induces translocation of cytoplasmic NFAT to the nucleus (34). Overexpression of NFAT in transgenic mice results in cardiac hypertrophy and its knockout prevents it (35, 36). NFAT overexpression also results in inactivation of glycogen synthase kinase-3β, which mediates the nuclear location of NFAT (37) and possibly induces skeletal myotube hypertrophy (38).

With regard to skeletal muscle hypertrophy in animal models, the role of cal-cineurin remains controversial owing to the use of cyclosporin A, a nonspecific inhibitor of calcineurin. Inhibition of the calcineurin pathway in vivo with cy-closporin A prevents overload hypertrophy (39). Interestingly, overexpression of calcineurin in transgenic mice does not induce skeletal muscle hypertrophy (40, 41), which may indicate that skeletal muscle already contains sufficient calcineurin activity for muscle growth, and thus the addition of a constitutively active cal-cineurin is redundant (39). This contention is supported by the finding that cy-closporin blocked the growth of plantaris muscle after induced atrophy, which is noteworthy because the study clearly demonstrated that inhibition of calcineurin with cyclosporine is dependent upon the appropriate concentration of cyclosporine, the muscle, and selection of appropriate time points. Current thought suggests that the hyperactivation of calcineurin alone is not sufficient to induce skeletal mus-cle hypertrophy but that activation of various upstream or downstream regulators in conjunction with calcineurin activation may play a significant role in muscle hypertrophy (40).

MECHANICAL-CHEMICAL TRANSDUCTION   Nearly 30 years ago, Goldberg et al. summarized their work demonstrating that muscular activity appears to the fun-damental determinant of muscle mass (42). This concept has been extended and reinforced by more recent workers: Various sensors of mechanical strain seem to possess the ability to translate strain into chemical signals that induce the activation of the skeletal muscle α-actin promoter (43). The existence of a mechano-trans-duction mechanism in skeletal muscle is reinforced by the tensegrity hypothesis (44), which suggests that a protein framework within the cell maintains its overall cellular architecture; in response to mechanical forces, cell structural networks interact with gene and protein signaling networks to allow cytoskeletal proteins to reposition and renew themselves, permitting the cell to resist deformation from the applied forces. A possible candidate sensor of the increase in mechanical strain is focal adhesion kinase (FAK), a protein localized to the sacrolemma (45, 46); FAK autokinase activity increases during high mechanical loading. In addition, during mechanical loading of muscles, there are significant increases in both the total amount of FAK and its tyrosine phosphorylation status (47). The transcription factor, serum response factor (SRF), is a substrate of FAK, thereby providing a transcriptional link between membrane, the genome, and subsequent expression of muscle protein (43). Furthermore, binding of SRF to the serum response ele-ment (SRE1) within the chicken skeletal α-actin promoter is both necessary and sufficient for increased transcriptional activity of the skeletal α-actin promoter (46). The putative link between FAK and SRF has been strengthened by results indicating that SRF-mediated, skeletal α-actin promoter activity is dependent upon activation of β1D-integrin-RhoA signaling and that this activity can be completely abolished by cotransfection of a dominant-negative FAK, termed FRNK (48).

Different modes of exercise affect extracellular signal-regulated kinase (ERK1/2) and the 38-kDa stress-activated protein kinase (p38) in an almost universal way (20, 49) that seems to be intensity dependent. However, only those stimuli likely to result in hypertrophy, such as high-frequency electrical stimulation, increased p70S6 kinase and protein kinase B phosphorylation (49). In a study designed to untangle the effects of concentric and eccentric contractions on the phosphorylation of ERK1/2 and p38, Wretman and coworkers applied a panoply of stimuli to isolated rat extensor digitorum longus muscle in vitro: These included electrically stimulated concentric (shortening) or eccentric (lengthening) contractions or severe passive stretch, and application of antioxidants likely to counteract reactive oxygen species and induce intracellular acidosis (50). They concluded that mechanical activity, whether through contraction or stretch, increased the activity of both ERK1/2 and p38, whereas the ionic changes and increase in reactive oxygen species and acidosis exhibited after concentric contraction increased phosphorylation only of ERK1/2. This suggested that high mechanical stress was required for activation of p38. Because stretch per se does not appear to increase MPS in human muscle (see below), and because changes in ERK1/2 are common to types of exercise that differ markedly in their sequelae (hypertrophy or mitochondrial biogenesis), the likelihood of these signaling molecules being involved in hypertrophy is lessened.

TRANSCRIPTIONAL REGULATION   Most of the cellular signaling pathways discussed above control the location and activity of transcription factors that, in turn, regulate gene transcription. However, for many years the processes of protein turnover in skeletal muscle were thought to be modulated without requiring extensive gene expression, which, when it did occur, was considered a relatively sluggish process with a long latency, possibly up to days. These concepts are wrong: In addition to any translational regulation, metabolic alterations, such as an increase in the availability of insulin and glucose (51), or environmental stimuli such as exercise (52), can result in increases in gene transcription (not restricted to the so-called early response genes) within 1.5–3 h, even in adult human skeletal muscle. Thus the difference in the timescale between the adaptivity of transcription and the processes of protein turnover may be much less than previously thought. Skeletal muscle growth regulators such as IGF-1 and other regulatory factors, for instance cytokines, early genes, signal transduction proteins, and myogenic regulatory factors, are transcriptionally regulated in response to contractile overload induced by synergist ablation in rat. More microarray studies are needed to elucidate the behavior of genes during muscle growth in animals and humans. For more detailed reviews concerning the role of gene transcription in muscle hypertrophy, see Carson (53) and Baar et al. (54).

Regulation of the Expression of the Specific Muscle Growth Factors IGF-1/MGF and Myostatin

Changing the availability of the muscle growth factors IGF-1/MGF and myostatin appears to be a central regulatory process in adaptive muscle growth. Obviously,

IGF-1/MGF and myostatin are not directly regulated by stretch, overload, or muscle contraction but by the signal transduction pathways that sense these stimuli and consequently regulate the availability of these muscle growth factors for receptor binding. The protein availability depends on transcriptional regulation, translational regulation, splicing, localization, concentration of binding proteins, and proteolysis. The major regulatory step controlling the availability of IGF-1/MGF and myostatin in response growth-inducing stimuli appears to be transcriptional regulation. The transcriptional regulation of these factors involves some of the muscular signal transduction pathways mentioned above, as well as some developmental pathways and anabolic and catabolic steroid hormones.

REGULATION OF MYOSTATIN EXPRESSION Myostatin [growth/differentiation factor-8 (GDF-8)] is a member of the transforming growth factor-β (TGF-β) family (55). Depending upon the site of deletion in the myostatin gene, mice expressing the modified gene in muscle may exhibit either myofiber hyperplasia or hypertrophy (see 56 for references). Myostatin-null mice were, paradoxically, shown to be more susceptible than wild-type mice to hindlimb suspension-induced muscle atrophy (57). Myostatin expression is environmentally modifiable, i.e., it can be decreased during reloading but, oddly, it is unchanged during suspension-induced muscle atrophy hindlimb suspension (58). The regulation of myostatin expression appears to depend on major growth pathways. Binding sites for glucocorticoids, androgens, thyroid hormone receptors, myogenic differentiation factor 1, MEF2, PPARγ, and NF-κB, with appropriate positive and negative effects, have all been predicted for the myostatin promoter region and for glucocorticoids experimentally confirmed in human muscle (59, 60).

REGULATION OF IGF-1/MGF EXPRESSION In hypertrophying rodent muscle, IGF-1 mRNA rises nearly threefold within two days of functional overload and remains elevated thereafter (61), a phenomenon also observed in human skeletal muscle after a single resistance training bout (62). The increase in IGF-1 immunoreactivity was localized mostly within the fibers of rat anterior tibialis muscle four days after an eccentric-resistance training program that led eventually to hypertrophy (63), suggesting that pretranscriptional regulation is probably involved somehow in the exercise-induced increase. IGF-1 superfusion onto muscle in free-moving rats produces hypertrophy (64), and a similar maneuver rescues immobilized muscle from aging-associated sarcopenia (65), as does IGF-1 overexpression (66). The increase in IGF-1 and its splice variant MGF (IGF-1Eb) (see below) in skeletal muscle in response to mechanical loading may be regulated transcriptionally in rat muscle. However, the mechanisms that regulate MGF expression in response to an increase in muscle tension are currently unknown. Skeletal muscle IGF-1 expression increases in response to growth hormone and testosterone and decreases in response to glucocorticoid hormones, TNFα, and interleukin-1 (67).

Specific Muscle Growth Regulation

Muscle growth stimuli lead to the activation of a signal transduction network and to a changed availability of the major muscle growth factors IGF-1/MGF and myostatin. The activated signal transduction pathways and changed growth factor availability will then regulate the activity of "muscle growth executors," which are the translational or protein synthesis machinery and satellite cells.

TRANSLATIONAL REGULATION   The cellular and molecular mechanisms regulating the translation of mRNA in muscle have been elucidated to a much higher degree (68) than those regulating protein breakdown, partly because the protein/synthetic machinery forms a cohesive metabolic unit centered around the ribosome and the endoplasmic reticulum. There are a number of systems for achieving proteolysis, such as the ATP and ubiquitin-dependent proteasome (69), the lysosome, at least two cytoplasmic systems activated by various concentrations of Ca^{2+} (70), and even extracellular, lymphocyte-based systems, which act on muscle (71). There is good evidence that the myofibrillar apparatus is degraded by the proteasome; however, abundant amounts of proteasome mRNA or proteasome do not automatically produce an increase in proteolysis. For example, there are cases of paradoxical changes of proteasome activity in muscle in response to starvation and refeeding (72).

It is relatively easy to demonstrate regulatory changes in the machinery of protein synthesis, which are consonant with an increase in the synthesis of protein, e.g., ribosome, aggregation (73), whereas in the case of protein breakdown, changes in apparent activity of key components of the system may exist with no, or apparently opposite, changes in the extent of net protein balance. This has made it difficult to make much progress in understanding the physiological modulation of muscle protein breakdown, although some knowledge of the involvement of elements of signaling pathways also involved in regulation of protein breakdown is now being accumulated (70).

The mechanisms regulating translational regulation during muscle growth are becoming increasingly clear (see 74–76 for more details). IGF-1 is capable of promoting muscle growth by activating regulators of translational initiation or efficiency via the PI3K-PKB/AKT-mTOR pathway (76, 77). IGF-1 treatment leads to an increased phosphorylation of PKB/AKT, mTOR, GSK3β, and the translational regulators 4E-BP1 and p70S6k. When phosphorylated, 4E-BP1 detaches from eIF4E (78) (a translational initiation factor that mediates mRNA binding to the ribosome) and this initiates translation. Phosphorylated p70S6k promotes the increased translation of those mRNAs with a 5'-tract of pyrimidine (TOP), i.e., a series of cytosine or thymine repeats at the 5' gene terminus (79). All known ribosomal proteins have a TOP sequence, suggesting that mTOR regulates both ribosomal biogenesis and translation via p70S6k and 4E-BP1, respectively.

The main response of mTOR-dependent signaling occurs with a latency of only a few hours after growth-stimulating exercise. Hernandez et al. reported increases in PI3K and the translational regulator p70S6k occurring after 6 to 24 h and protein synthesis itself rising 12 to 24 h after resistance exercise in rat muscle (80). Similar results are available for human muscle (81). Rat muscles stimulated at high-frequency show p70S6k phosphorylation peaking at 3 to 6 h after stimulation, with some increased phosphorylation still apparent 36 h later in hypertrophying muscles (54). The delay in the activation of translational pathways and protein synthesis might be explained by the time necessary for strain-sensing and signaling, possibly via IGF-1 or MGF synthesis and secretion.

However, a recent paper shows that translational pathways and regulators are also transiently activated within 5–10 min after resistance exercise in rats (81a). This finding is interesting because a changed IGF-1 or MGF availability is unlikely to occur minutes after the stimulus and thus suggests other connections between the signal transduction pathways that sense resistance exercise signals and the translational regulators.

Essential amino acids stimulate protein synthesis via a nutrient-sensitive complex of two proteins, Raptor and mTOR, which, in humans, are expressed more in skeletal muscle than in other tissues (82, 83). It is likely that the Raptor-mTOR complex is destabilized and mTOR and the downstream translational regulators are activated when essential amino acid availability increases. An additional positive regulator, GβL, appears to be involved (83). The binding of GβL to mTOR strongly stimulates the kinase activity of mTOR toward S6K1 and 4E-BP1, an effect reversed by the stable interaction of Raptor with mTOR. The availability of essential amino acids sensed by this protein complex activates mTOR, as well as of the translational regulators eIF2, 4E-BP1, and p70S6k (84), which explains the observed stimulatory effect on protein synthesis.

In contrast, an increased energy demand (reflected by lowered ATP/ADP ratio, higher AMP, and lower creatine phosphate concentrations) leads to a depression of protein synthesis (85). A recently discovered interaction between AMPK and PKB-mTOR signaling in muscle has provided a possible mechanism for this effect: AMPK is activated by AMP and inhibited by ATP and creatine phosphate and is involved in the regulation of numerous cellular functions such as mitochondrial biogenesis and fuel metabolism (18, 86). Treatment of rats with the AMPK-activator, AICAR, resulted in a reduction in skeletal muscle protein synthesis. This was accompanied by a decreased activation of PKB-mTOR and its downstream targets p70S6k and 4E-BP1 (87).

Regulation of Satellite Cell Proliferation and Differentiation

Muscle fibers are permanently differentiated; therefore, they are incapable of mitotic activity to produce additional myonuclei in times of increased protein synthesis and muscle growth (88). Yet, myonuclear number increases during skeletal muscle hypertrophy, thereby maintaining the myonuclear domain (the amount of sarcoplasm managed by a single myonucleus) (89). The predominant source of the additional myonuclei is satellite cells that are localized in indentations in the sarcolemma beneath the basal lamina (90). Less commonly, satellite cells may possibly fuse with each other to form new fibers (hyperplasia) (90).

The requirement of satellite cell activation for muscle hypertrophy was first demonstrated by a "nontransgenic knockout" approach in which mild γ-irradiation (which damages DNA while leaving other cellular machinery intact) was employed to block satellite cell proliferation. In response to functional overload, myonuclear number or muscle size was not increased in irradiated rat muscles (91). Adams and coworkers found that most of the hypertrophy potentially achievable during mechanical overload was prevented by similar treatment for four months (92). Thus it is likely that neither endogenous mesenchymal stem cells nor extramuscular (e.g., bone marrow) stem cells contribute much to the stem cell population of overloaded muscles, and the proliferation and fusion of existing satellite cells are responsible for the full load-induced increases observed in the muscle mass.

The limited proliferative capacity and the decrease in satellite cell number during normal aging may be implicated in atrophy and poor regeneration in elderly subjects (93). The number of satellite cells is thought not to be limiting to hypertrophy in normal human skeletal muscle, even in the elderly, although it may be in chronic users of anabolic steroids (94). Numerous growth factors have been shown to increase satellite cell proliferation. Here we focus on the key muscle growth regulators IGF-1 and myostatin. The effects of IGF-1 on muscle growth are pleiotrophic, activating satellite cell proliferation by spurring progression through G_1 to S phase, increasing protein synthesis, decreasing protein degradation, and decreasing apoptosis. The mechanism by which IGF-I signals satellite cells to proliferate is by a decrease in $p27^{Kip1}$ protein concentrations via activation of the phosphatidylinositol $3'$-kinase (PI3K)/protein kinase B (PKB/Akt) signaling (95). As a result, increased $p27^{Kip1}$ inhibits cyclin-dependent kinase 2 (cdk2), producing a late G_1 arrest in the satellite cell cycle.

Myostatin inhibits both satellite cell proliferation and differentiation. Myostatin halts the satellite cell cycle by upregulating p21, which inactivates cyclin-dependent kinase activity so that retinoblastoma protein is particularly dephosphorylated (96). Myostatin also regulates satellite cell differentiation by inhibiting the expression of the myogenic growth factor, MyoD, via Smad 3 signaling (97). Reciprocally, MyoD upregulates myostatin to control myogenesis during the G_1 phase of the cell cycle (at least in C_2C_{12} myoblasts) (98).

SUMMARY OF REGULATORY MECHANISMS PRODUCING MUSCLE HYPERTROPHY   Resistance exercise and most other muscle growth factors lead to the activation of a signal transduction network that will regulate the expression of the muscle growth factors IGF-1, MGF, and myostatin. The activated signal transduction pathways and the changed receptor binding of IGF-1, MGF, and myostatin lead to the activation of translation or protein synthesis and satellite cell proliferation and differentation, resulting in muscle growth. Blocking any one of these signal transduction pathways or factors may limit hypertrophy. However, it is incorrect to conclude that the blocking of hypertrophy by a single pathway or factor supports the conclusion that only one mechanism accounts for all of the resistance exercise or strain-induced hypertrophy of skeletal muscle. Likewise, enhancement of resistance exercise or strain-induced hypertrophy by a single factor should not be interpreted to mean that this factor alone is the means whereby increased mechanical load signal are transmitted to muscle growth physiologically. Indeed, the normal physiological response to produce the full potential hypertrophy during a load-induced growth of skeletal muscle involves the orchestration of multiple, simultaneous, and temporarily related sequential signals.

ADAPTATION OF THE HUMAN MUSCLE MASS

The second part of this review is focused on observed changes in human MPS and breakdown and on translational and transcriptional control mechanisms, insofar as they act in adult human skeletal muscle, with particular reference to the effects of nutrition and physical activity.

EFFECTS OF NUTRITION ON SKELETAL MUSCLE PROTEIN MASS

The influences of food on protein metabolism are separable into two parts: those brought about through an increase in amino acid availability (e.g., the amino acid–activation of translational regulators via the Raptor-mTOR complex discussed above) and those resulting from increases in the concentration of hormones and growth factors (principally insulin and growth hormone/IGF-1) produced after stimulation by dietary secretogogues (e.g., glucose and amino acids).

Effects of Amino Acids

The discovery of the Raptor-mTOR complex and its likely function as an amino acid sensor (see above) has provided a likely explanation for the known stimulatory effect of amino acids on translation and protein synthesis. Here we review human studies in which the authors investigated this relationship in order to provide information that can be practically applied by those that wish to increase MPS in athletes, the elderly, or patients in whom muscle atrophy has occurred because of diminished protein synthesis. We aim to inform about effective amino acid concentrations, timings of ingestion, and the combination of amino acid feeding with resistance exercise.

MUSCLE PROTEIN SYNTHESIS   An increase in the supply of amino acids to skeletal muscle (in many in vitro and in vivo models) stimulates the incorporation of tracer amino acids into protein (74). This effect can be observed independently of any hormones, although insulin may enhance it (see below). In human subjects, intravenous infusion of mixed amino acids doubles the incorporation of stable-labeled tracer amino acid into anterior tibialis muscle without any increase in the availability of insulin (99). Modulation of MPS via availability of amino acids appears to show a sigmoidal relationship; rises and falls in amino acid availability cause rapid changes in MPS in the basal to postprandial range, with a shallower slope at the upper and lower concentration limits (100, 101). In humans, the upper limit of amino acid concentrations at which MPS appears to be saturated is about 50% greater than the blood amino acid concentrations normally achieved after a meal (101). These and other recent data (109–112) make a powerful point: The amount of amino acids necessary to stimulate MPS in the resting state and after exercise is in fact small (<10 g) compared with the accepted whole-body protein requirements (>70 g for most men).

In animal muscle, it is easily demonstrated that the branched chain amino acids (and leucine in particular) stimulate MPS in muscle cells in tissue culture, in perfused systems, and in intact mice and rats (102–106). A similar effect has been observed in whole human beings; administration of boluses of single essential amino acids (including threonine, valine, phenylalanine, and leucine) but not non-essential amino acids (such as proline, glycine, serine, and alanine) markedly stimulated the incorporation of tracer-labeled amino acids into muscle protein (107). Could such a stimulation could be sustained in vivo? When large amounts of leucine were infused in human subjects, the intramuscular concentrations of other amino acids fell (108), presumably owing to stimulation of MPS (together with the possible inhibition by leucine of MPB). However, synthesis of protein requires all 20 physiological amino acids, and those that have the highest concentration ratio between muscle protein and the free pool will have the largest fall under situations of net anabolism unless transport from the blood occurs sufficiently quickly. Unfortunately, the capacity for muscles to continue to produce protein when the supply of all 20 amino acids is limited is not known.

Studies of the latency and duration of the effect of amino acids on MPS suggest that it takes 30 min for a stimulatory effect to be detected; thereafter the rate of increase is rapid, and peak rates are obtained within 60 to 90 min (101). Then MPS falls back to basal levels despite the continued abundant availability of amino acids, suggesting that the system is full of protein and is no longer responsive to nutritional stimulation. The extent of the refractory period before restimulation can occur and the identity of the mechanisms involved in the switch-off are unknown. When the limb arterio-venous exchange methods are applied, after increases in blood amino acid concentration, the apparent increase in net amino acid balance (and in model-derived values for MPS) is greater and occurs more quickly than the increase of the rate of tracer incorporation into muscle protein (13, 109, 110). There may be two reasons for this: First, amino acids may in fact be inhibiting MPB more rapidly than simulating MPS (although this seems unlikely given the results of previous studies) (111); or second, the apparent increase in the net protein balance may be because the muscle amino acid pool is overfilled. This latter possibility is acknowledged in a recent paper from the Galveston group (112). In any case, attributing arterio-venous concentration difference × flow as signifying apparent increase in net protein balance should be regarded with caution under circumstances in which blood amino acid concentrations are changing; when arterial blood amino acid concentrations are rising (as after an oral dose of amino acids), there is a tendency to overestimate net muscle balance.

The cellular mechanisms involved are likely to be similar in human and animal muscle with stimulation of p70S6k and 4E-BP1 activation by amino acids (113).

MUSCLE PROTEIN BREAKDOWN   Although there is no doubt that increasing amino acid concentrations by intravenous infusion, meal feeding, or ingestion of free amino acids increases MPS, inhibitory effects of amino acids on MPB, which are relatively easy to detect in animal muscles (70, 114), are, in human, beings absent or at least much less evident than those on MPS (115, 116). Some of the protein anabolic effects of a protein meal in vivo may possibly be modulated through the stimulation of insulin, with consequent inhibitory effects on muscle protein breakdown, but in our experience they are slight. Certainly, in the postexercise period increased availability of amino acids enhances MPS without having an effect on protein breakdown (111).

Effects of Insulin

The effects of insulin on MPS in animals and humans may be different in terms of sensitivity and responsiveness. Much of the work demonstrating a marked stimulatory effect of insulin on MPS has been carried out in immature rodents or in only partially differentiated muscle cultured in vitro (117–119), and it has been much harder to obtain consistent results demonstrating a coherent pattern of responsiveness in adult (especially human) MPS to insulin. Two main areas of contention concern the question of the extent of the human MPS response to insulin and the dose response characteristics of the system.

MUSCLE PROTEIN SYNTHESIS   Barrett and coworkers (120, 121) first raised questions about the efficacy of insulin in human muscle when insulin was supplied to the forearm by close arterial infusion. No effects of insulin could be discerned on the disappearance of tracer into protein, i.e., protein synthesis, although there was a dose-dependent inhibition of protein breakdown. In this experimental model, protein metabolism in the forearm was assessed on the basis of arterio-venous balance of amino acids and the dilution across the arm of radio-labeled phenylalanine, an amino acid that is not subject to intermediary metabolism in muscle. However, the mathematical formula used by these workers produced results that may be underestimates of the rate of synthesis (see 122, 123 for discussion of this point). Also, Barrett and colleagues did not take muscle biopsies to check that the intramuscular concentration of amino acids was sufficient to sustain protein synthesis. Furthermore, as shown by Biolo and coworkers, when a different mathematical modeling approach was used for lysine and phenylalanine (a three-rather a than a two-pool model) (124), insulin could be shown to stimulate MPS. This conclusion was supported by independent data showing increased incorporation of stable tracer-labeled leucine into muscle protein sampled in the same period. Other workers have also demonstrated that insulin will stimulate MPS measured by incorporation or by limb arterio-venous difference exchange—but only when sufficient amounts of amino acids are present (116, 125–127).

Nevertheless, there are still no data that adequately describe the dose-response relationship between MPS, measured unequivocally by means of tracer incorporation into protein, and the availability of insulin in blood. There is a pressing need for construction of a dose-response curve (carried out using somatostatin to inhibit basal insulin and with insulin added back systematically) that will simultaneously measure amino acid balance across the limb and tracer incorporation into muscle protein.

MUSCLE PROTEIN BREAKDOWN   The effect of insulin on MPB has been well defined in terms of a decrease in the appearance of amino acids from preparations of muscle in tissue culture, in isolated whole muscles in perfused systems, and in measurement of arterio-venous tracer exchange in humans (116, 128). The major effect of insulin in inhibiting proteolysis appears to be modulated by effects on the proteasome, the ATP-ubiquitin-dependent proteolytic system that is responsible for myofibrillar protein breakdown in mammals (129). Despite a wealth of information describing alterations in mRNA and proteasome protein concentrations through manipulations of nutritional status, it has often been difficult to match up alterations in skeletal muscle balance or in measured protein breakdown with changes in the mRNA or proteasome content, e.g., in animals (72). In fact, no such parallel can be found from results of studies in human subjects as far as we are aware. In one well-designed study in which protein breakdown (measured as loss of amino acids from the limb) was elevated by three days of cortisol infusion, no changes occurred in the components of the proteasome pathway or their mRNA (130). A similar lack of correspondence between mRNA and protein for proteasome components and changes in net protein loss has been observed in muscle of lung cancer patients and in patients with acidosis due to renal disease (131, 132).

In short, after a meal and probably after exercise, the insulin-mediated decrease in MPB appears to be less important for the attainment of net anabolism than the stimulation of MPS.

Growth Hormone and Insulin-Like Growth Factor-1

Growth hormone has a number of metabolic actions on salt and water balance; fat metabolism; and, in growing animals and children, muscle and bone growth. Rennie recently reviewed the metabolic effects of growth hormone on human skeletal muscle and concluded that the balance of evidence suggests there are no major anabolic effects of exogenous rhGH in stimulating muscle protein accretion, muscle size, muscle strength, or muscle fiber characteristics in normal, healthy adult men or women, including the elderly (133). The published data, which have contributed to this conclusion, include information on incorporation of stable isotope-labeled amino acids into muscle and measurements of muscle mass and muscle fiber type using modern imaging and immunohistochemical methods.

There are, nevertheless, strong indications that IGF-1 involvement in metabolism may be locally important in skeletal muscle in humans and may modulate some of the effects of contractile activity in maintaining, or even increasing, muscle mass.

MGF is elevated in human muscle after exercise (134) but only in young (30-year-old) and not in old (75-year-old) subjects. One puzzling feature of this finding is that the MGF transcripts appear at concentrations that are very much lower than those of the IGF-1, so the effects of MGF must be because of different targeting or because the MGF is much more potent than IGF-1.

Although administration of IGF-1 seems to have acute stimulatory anabolic effects (135, 136) in muscle, long-term systemic administration of IGF-1 without its binding protein has no anabolic effect on lean body mass in elderly women (137), whereas a combination of IGF-1 with its binding protein 3 is markedly anabolic even in burn patients (138).

EFFECTS OF EXERCISE ON MUSCLE PROTEIN TURNOVER

It has long been known that regularly active muscle is able to maximize its mass despite a poor dietary protein intake; for example, hypertrophy of a muscle can occur after the ablation of a synergist muscle even in an undernourished animal (42). Any adaptive responses to increased physical activity that result in an increase in muscle mass or change in muscle composition must involve alterations of muscle protein turnover; thus muscle hypertrophy is always associated with increases in MPS, plus adaptive (likely remodeling related), increases in MPB (139, 140). Of course, during the upward swing of muscle mass, MPS has to exceed MPB, and this usually requires amino acids, either dietary or possibly those diverted from other body tissues.

There are no data available on the rate of MPS during (rather than after) resistance exercise in humans because of the poor time resolution of the methods available for its measurement. The likelihood that it is unchanged or depressed, but not raised, has been inferred from studies of walking and bicycling exercise (141–143). The probability of a depression of MPS during resistance exercise is reinforced by the observation of a fall in incorporation of tracer into protein during electrically stimulated maximal contractions in perfused rat hindlimb (144). The extent of the decrement was correlated with falls in creatine phosphate concentration and [ATP]/[ADP]. The associated rise in AMP is likely to activate AMPK, which in turn would inhibit mTOR-dependent translational regulators during acute exercise (see above). Bylund-Fellenius and colleagues (144) also reported a fall in MPB, indicated by a fall in 3-methylhistidine production during exercise, which is consonant with the fall in intramuscular 3-methylhistidine concentration observed during exercise in human subjects (142). This suggests that both areas of muscle protein turnover may be compromised during exercise by a fall in energy status.

Most studies on the responses to exercise of human muscle protein turnover have concentrated on postexercise changes after resistance exercise, likely to result in muscle hypertrophy. The first such study demonstrated that a single bout of high-intensity resistance exercise resulted in a doubling of the rate of biceps MPS within 4 h; an elevated but diminishing rate of MPS persisted for ~24 h (145). This time course was confirmed by later work on quadriceps (146), which also indicated a recovery to within 30% of basal rates by 48 h.

The extent (+80–100%) of the anabolic response of MPS to a single bout of exercise is surprising, because the rate of net accretion of muscle protein is very much slower than this: It may take 20 weeks of intense resistance exercise to increase muscle mass by 20%. The explanation is, of course, that MPB is also elevated after acute exercise (but probably not during it) and that, under circumstances in which exogenous amino acids are not provided, this usually exceeds MPS (146) so there is no net protein accretion. However, if food is eaten or mixed, free amino acids are ingested or infused after intense exercise, then MPS is stimulated beyond the rate achieved by exercise alone (111, 147–149). This effect is probably synergistic, rather than simply additive, which itself suggests that the effects of the contractile event(s) and of the amino acids act separately in activating pathways that eventually stimulate MPS.

Chronic Stretch and Immobilization

A major component of muscle activity in whole musculoskeletal systems is the stretching of muscles by their antagonists acting in the opposite sense across a joint; because this stretch itself has been implicated in activating mechanochemical transduction pathways, there has been a major interest in attempting to separate out the effects of passive and active tension development. Also, for centuries it has been recognized that any condition leading to inactivity in muscle is followed by muscle wasting—sarcopenia. The study of both condtions should provide information allowing us to understand the mechanisms between mechanical strain and muscle protein growth and wasting.

STRETCH   Passive stretch is well established to cause hypertrophy in animal muscle (150); possible mechanisms have been discussed above. Chronically stretched human multifidus muscle has a higher rate of protein synthesis than flaccid, effectively immobilized muscle (studied on either side of a scoliotic spinal column) (151). However, there is no evidence of an acute stimulatory effect on protein synthesis, measured as incorporation of ^{13}C leucine and 27 min of electromyographically silent, intermittent stretch in human muscle, despite isometric exercise generating the same force for the same period in the same subjects causing a marked stimulation of 50% (152).

IMMOBILIZATION   It has been known for ~15 years that immobilization of human muscle results in a fall in muscle mass caused by a substantial depression in MPS accompanied by a smaller fall in muscle protein breakdown so that muscle is in negative balance (153). This scenario has been confirmed by later studies of bed rest as a model for immobilization (154). The extent of the wasting is reduced by local electrical stimulation of muscle, probably through maintenance of translational efficiency that preserves muscle RNA (155). Stress probably acting via corticosteroids exacerbates muscle wasting in immobilized subjects, and anabolic steroids and resistance exercise lessen wasting (156).

Timing of Feeding in Relation to Exercise

The timing of the delivery of amino acids may be important in the extent of stimulation of MPS. Esmark and colleagues carried out a training study over 12 weeks with elderly men who were fed either immediately after exercise or 2 h later; the extent of hypertrophy was measured after 12 weeks (157). In the group of subjects who ate immediately after exercise, there was a bigger increase in skeletal muscle mass and fiber diameter. Flakoll and colleagues (158) used a similar maneuver but made measurements acutely after a single bout of intense bicycling rather than resistance exercise. The net balance of amino acids across the previously working legs was greater immediately after exercise than 3 h later. It has even been shown that amino acid feeding before resistance exercise increases the post-exercise response of MPS beyond that observed with identical feeding immediately after exercise (159), an effect the authors ascribe to the increased prior delivery of amino acids owing to exercise-induced increases in blood flow.

Nevertheless, some contradictory data exist. Rasmussen and colleagues found no difference in the net balance of amino acids across previously working legs when examined at 1 or 3 h post-exercise (160). Thus a possible "golden period" in which previous contractile activity predisposes muscle to accumulate amino acids as proteins remains a speculative, no matter how attractive, concept.

Effects of Resistance Training on the Response of Muscle Protein Turnover

Many of the muscle metabolic systems show adaptations with habitual physical activity. Whether habitual physical activity results in a chronically altered rate of muscle protein turnover is currently the subject of some interest. In diabetic rats trained to perform resistance exercise, Farrell and coworkers demonstrated a reduced response of MPS to exercise after training (161). However, obtaining a clear answer to this question for human muscle is difficult. First, the residual effects of a previous bout of exercise, which may last up to 72 h, depend on intensity. Second, there is the problem of the habitual dietary intake of athletes who are subjected to much marketing and coaching information suggesting that they need to eat large amounts of protein in order to maintain or build muscle mass; this is a problem because habitually high rates of dietary protein intake lead to the induction of amino acid catabolic enzymes (particularly of the branched chain and aromatic amino acids) that decrease the deposition of dietary protein (162, 163). Until this effect abates (after reducing protein intake), there will be a tendency to exhibit negative nitrogen balance, so studies should not be conducted with rapid variation in dietary protein contents.

There is, in fact, little data on the subject in respect to MPS or even muscle mass. Studies of military recruits undergoing intense physical training suggest that there is a loss of body protein over the first few days of training but that adaptation rapidly occurs and nitrogen balance is restored, all at the same rate of dietary protein intake (164). Butterfield & Calloway found that in young men undergoing physical training, exercise increased the efficiency of protein utilization (165), i.e., trained subjects would require less protein. Partial validation of this position was provided by the first of two studies by Phillips and colleagues (166, 167). When two groups of subjects, one strength-trained and the other sedentary, were compared, there were no differences in resting post-absorptive MPS or MPB; also when the post-exercise responses to a single bout of pleiometric exercise at 120% of each subjects concentric 1 RM were compared, the rise in MPS in the trained subjects was 50% less than in the sedentary group, and there was no rise in MPB, which increased by about 40% in the untrained group. Thus net muscle balance (MPS minus MPB) was improved to the same extent in each group. However, a different result was obtained in a second longitudinal study of the effects of 8 weeks of resistance training in young previously untrained men, studied in the fed state at rest and also after a bout of exercise at 80% of their pretraining 1 RM (166). These results suggested that there was no difference in the response of the subjects in the trained and untrained state to acute exercise; also, rather oddly, the trained subjects did now show a marked increase in MPB as a result of exercise. In addition, basal rates of MPS and MPB were in fact now higher in the trained state; one consequence of this was that the effect of training seemed to decrease the relative response to exercise, a result that was consonant with the earlier findings—but by a different mechanism! All in all, the data on net balance suggest that there was no effect of training tending to confirm the settled views of the present authors (143, 148) (although resisted by many athletes, their trainers, and, of course, sports nutrition companies) that habitual physical activity imposes no greater demands on protein requirements. As Phillips and coworkers (166) point out in their discussion, they did not test whether the same relative workload might affect protein turnover in trained and untrained subjects: It may be that if the above longitudinal studies had been conducted at the same relative intensity, a different result might have been obtained.

In the elderly, the rejuvenating effect of training may confound the issue. There is considerable controversy about whether aging is associated with a fall in muscle protein turnover [see (168) for review of this topic, which will not be dealt with further here]. However if it is true that the frail (as opposed to healthy) elderly show a fall in MPS, as seems likely, then exercise training may normalize it (169). The mechanism may be by decreasing the amount of TNF-α in muscle (170).

Effects of Creatine on Human Muscle Protein Turnover

Dietary supplements containing creatine have become popular with athletes and trainers hoping to promote greater increase in muscle mass and strength in resistance training programs (171–173). Measurements of myofibrillar protein synthesis (as incorporation of ^{13}C leucine) and forearm protein breakdown (as dilution of deuterated phenylalanine) were unable to discern any differences in subjects studied before and after creatine supplementation, either in the post-absorptive or the fed state, at rest, or immediately after acute exercise (174, 174a). These results appear to rule out any acute effect of creatine alone on translation of pre-existing mRNA or on MPB but do not invalidate the possibility of transcriptional changes or satellite cell activation stimulated by creatine and physical activity.

Effects of Intensity of Contraction and Metabolic Power Output on Muscle Protein Turnover

It seems clear that maneuvers resulting in a relatively rapid rise in muscle mass are all associated with substantial increases, albeit after a short latency, possibly of about one hour, in MPS as a result of translational stimulation produced by changes in 4E-BP1 and p70S6k phosphorylation (176). These changes are followed, probably shortly thereafter, by transcriptional changes associated with intense exercise. Thus questions of the extent and temporal pattern of disturbance need to be addressed.

In human muscle, our group (M.J. Rennie, D.J.R Cuthbertson, K. Esser & M. Fedele, unpublished work) consistently observe a long-lasting rise in p70S6k phosphorylation after acute, high-intensity exercise, with smaller transient rises in PKB (Akt) phosphorylation, which are associated with a consistent rise in incorporation of tracer-labeled amino acid into muscle protein, whether myofibrillar or sarcoplasmic. We find no difference in the extent of stimulation of p70S6k or MPS in different quadriceps in which the same amount of force is applied during stepping exercise (one leg up, one leg down, while carrying 20% of body weight) to exhaustion (81). Because concentric exercise is energetically much less efficient than eccentric exercise and normally requires a higher rate of ATP turnover, this suggests that the crucial factor in determining the extent of the rise of MPS is force or intensity rather than ATP turnover, unless there is some threshold effect beyond which the rise in MPS remains constant.

However, paradoxically, when ATP turnover and the extent of quadriceps motor unit recruitment is kept constant during exercise at 60, 75, and 90% of 1 RM for different numbers of repetitions, the stimulation of MPS is constant (175).

CONCLUSION

As we have seen, our current ability to describe the adaptive responses of skeletal muscle to a wide variety of circumstances with changes in mass, composition, and function is impressive. The time resolution of techniques for measuring changes in muscle mass and composition and rates of protein turnover have increased such that we can now make robust measurements of the time courses of, for example, the rate of myofibrillar protein synthesis, which was impossible 10 years ago. Much information about the interrelationships between signaling pathways, which are important for transcriptional and translational regulation, has been accrued, and we have a much better understanding of the importance of satellite cells for growth and regeneration of muscle. There are, however, a substantial number of gaps that need to be filled. We still have no clear idea of the temporal relationship between the components of amino acid sensing and signaling to the processes of protein synthesis and breakdown and how these are affected by individual amino acids, insulin, and IGF-1. The exact pathways by which anabolic and catabolic steroids affect gene transcription and translation of mRNA remain obscure in human muscle despite the existence of response elements predicted for the muscle genes; the commonality (if any) of the pathways between myofibers and satellite cells is not at all well understood. The nature of the dichotomy of the responses to short-term, high-intensity exercise leading to hypertrophy and long-term low-intensity exercise leading to mitochondriogenesis and fast-to-slow fiber type transition remains a mystery. We still require a good description of the dose-response relationship between exercise intensities and the observed changes in mass and protein composition, and until we have these, it will be difficult to sort out the relative contributions of signaling pathways, their commonality, additivity, or independence from each other in controlling the adaptive responses of muscle.

Nevertheless, the increasing power of post-genomic techniques, particularly the use of transcriptional profiling and subsequent bioinformatics, should enable us to identify previously unknown means of controlling transcriptional and translational events. Perhaps some time in the next 10 years, our view will suddenly snap into focus, and it will become obvious how, for example, changes in the concentrations of Ca^{2+} or AMP can modulate the size and shape of muscle.

ACKNOWLEDGMENTS

Supported by UK Medical Research Council, UK Biotechnology and Biological Sciences Research Council, The Wellcome Trust, World Anti-Doping Agency, Diabetes UK (all MJR), Royal Society (HW), and National Institutes of Health NIH AR19393 (FWB) and NIH AR48514 (EES).

The *Annual Review of Physiology* is online at http://physiol.annualreviews.org

LITERATURE CITED

1. Kohrt WM. 1998. Preliminary evidence that DEXA provides an accurate assessment of body composition. *J. Appl. Physiol.* 84:372–77

2. Wang W, Wang Z, Faith MS, Kotler D, Shih R, Heymsfield SB. 1999. Regional skeletal muscle measurement: evaluation of new dual-energy X-ray absorptiometry model. *J. Appl. Physiol.* 87:1163–71

3. Fuller NJ, Laskey MA, Elia M. 1992. Assessment of the composition of major body regions by dual-energy X-ray absorptiometry (DEXA), with special

reference to limb muscle mass. *Clin. Physiol.* 12:253–66

4. Lee RC, Wang Z, Heo M, Ross R, Janssen I, Heymsfield SB. 2000. Total-body skeletal muscle mass: development and cross-validation of anthropometric prediction models. *Am. J. Clin. Nutr.* 72:796–803

5. Scott SH, Engstrom CM, Loeb GE. 1993. Morphometry of human thigh muscles. Determination of fascicle architecture by magnetic resonance imaging. *J. Anat.* 182:249–57

6. Staron RS. 1997. Human skeletal muscle fiber types: delineation, development, and distribution. *Can. J. Appl. Physiol.* 22:307–27

7. Welle S. 2002. Cellular molecular basis of age-related sarcopenia. *Can. J. Appl. Physiol.* 27:19–41

8. Evans W. 1997. Functional metabolic consequences of sarcopenia. *J. Nutr.* 127:998S–1003

9. Bamman MM, Hill VJ, Adams GR, Haddad F, Wetzstein CJ, et al. 2003. Gender differences in resistance-training-induced myofiber hypertrophy among older adults. *J. Gerontol. A* 58:108–16

10. Hart DW, Wolf SE, Ramzy PI, Chinkes DL, Beauford RB, et al. 2001. Anabolic effects of oxandrolone after severe burn. *Ann. Surg.* 233:556–64

11. Biolo G, Iscra F, Bosutti A, Toigo G, Ciocchi B, et al. 2000. Growth hormone decreases muscle glutamine production and stimulates protein synthesis in hypercatabolic patients. *Am. J. Physiol. Endocrinol. Metab.* 279:E323–32

12. Biolo G, Gastaldelli A, Zhang XJ, Wolfe RR. 1994. Protein synthesis breakdown in skin and muscle: a leg model of amino acid kinetics. *Am. J. Physiol. Endocrinol. Metab.* 267:E467–74

13. Bohe J, Low JF, Wolfe RR, Rennie MJ. 2001. Latency duration of stimulation of human muscle protein synthesis during continuous infusion of amino acids. *J. Physiol.* 532:575–79

14. Fischman AJ, Yu YM, Livni E, Babich JW, Young VR, et al. 1998. Muscle protein synthesis by positron-emission tomography with L-[methyl-^{11}C]methionine in adult humans. *Proc. Natl. Acad. Sci. USA* 95:12793

15. Chen YW, Nader GA, Baar KR, Fedele MJ, Hoffman EP, Esser KA. 2002. Response of rat muscle to acute resistance exercise defined by transcriptional and translational profiling. *J. Physiol.* 545:27–41

16. Booth FW, Chakravarthy MV, Spangenburg EE. 2002. Exercise gene expression: physiological regulation of the human genome through physical activity. *J. Physiol.* 543:399–411

17. Baldwin KM, Haddad F. 2002. Skeletal muscle plasticity: cellular molecular responses to altered physical activity paradigms. *Am. J. Phys. Med. Rehabil.* 81:S40–51

18. Winder WW, Hardie DG. 1999. AMP-activated protein kinase, a metabolic master switch: possible roles in type 2 diabetes. *Am. J. Physiol. Endocrinol. Metab.* 277:E1–10

19. Meissner JD, Kubis HP, Scheibe RJ, Gros G. 2000. Reversible Ca^{2+}-induced fast-to-slow transition in primary skeletal muscle culture cells at the mRNA level. *J. Physiol.* 523:19–28

20. Widegren U, Wretman C, Lionikas A, Hedin G, Henriksson J. 2000. Influence of exercise intensity on ERK/MAP kinase signalling in human skeletal muscle. *Pflügers Arch.* 441:317–22

21. Aronson D, Boppart MD, Dufresne SD, Fielding RA, Goodyear LJ. 1998. Exercise stimulates c-Jun NH2 kinase activity and c-Jun transcriptional activity in human skeletal muscle. *Biochem. Biophys. Res. Commun.* 251:106–10

22. Hollander J, Fiebig R, Gore M, Ookawara T, Ohno H, Ji LL. 2001. Superoxide dismutase gene expression is activated by a single bout of exercise

in rat skeletal muscle. *Pflügers Arch.* 442:426–34

23. Turinsky J, Damrau-Abney A. 1999. Akt kinases 2-deoxyglucose uptake in rat skeletal muscles in vivo: study with insulin and exercise. *Am. J. Physiol. Regul. Integr. Comp. Physiol.* 276:R277–82

24. Richter EA, Derave W, Wojtaszewski JF. 2001. Glucose, exercise and insulin: emerging concepts. *J. Physiol.* 535:313–22

25. Arnold HH, Braun T. 2000. Genetics of muscle determination and development. *Curr. Top. Dev. Biol.* 48:129–64

26. Martin PT. 2003. Role of transcription factors in skeletal muscle and the potential for pharmacological manipulation. *Curr. Opin. Pharmacol.* 3:300–8

27. Molkentin JD, Olson EN. 1996. Defining the regulatory networks for muscle development. *Curr. Opin. Genet. Dev.* 6:445–53

28. Molkentin JD, Lu JR, Antos CL, Markham B, Richardson J, et al. 1998. A calcineurin-dependent transcriptional pathway for cardiac hypertrophy. *Cell* 93:215–28

29. Dunn SE, Burns JL, Michel RN. 1999. Calcineurin is required for skeletal muscle hypertrophy. *J. Biol. Chem.* 274:21908–12

30. Olson EN, Williams RS. 2000. Remodeling muscles with calcineurin. *BioEssays* 22:510–19

31. Semsarian C, Wu MJ, Ju YK, Marciniec T, Yeoh T, et al. 1999. Skeletal muscle hypertrophy is mediated by a Ca^{2+}-dependent calcineurin signalling pathway. *Nature* 400:576–81

32. Leite MF, Thrower EC, Echevarria W, Koulen P, Hirata K, et al. 2003. Nuclear cytosolic calcium are regulated independently. *Proc. Natl. Acad. Sci. USA* 100:2975–80

33. Hunton DL, Lucchesi PA, Pang Y, Cheng X, Dell'Italia LJ, Marchase RB. 2002. Capacitative calcium entry contributes to nuclear factor of activated T-cells nuclear translocation hypertrophy in cardiomyocytes. *J. Biol. Chem.* 277:14266–73

34. Crabtree GR, Olson EN. 2002. NFAT signaling: choreographing the social lives of cells. *Cell* 109(Suppl.):S67–79

35. Braz JC, Bueno OF, Liang Q, Wilkins BJ, Dai YS, et al. 2003. Targeted inhibition of p38 MAPK promotes hypertrophic cardiomyopathy through upregulation of calcineurin-NFAT signaling. *J. Clin. Invest.* 111:1475–86

36. Schubert W, Yang XY, Yang TT, Factor SM, Lisanti MP, et al. 2003. Requirement of transcription factor NFAT in developing atrial myocardium. *J. Cell Biol.* 161:861–74

37. Neilson J, Stankunas K, Crabtree GR. 2001. Monitoring the duration of antigen-receptor occupancy by calcineurin/glycogen-synthase-kinase-3 control of NF-AT nuclear shuttling. *Curr. Opin. Immunol.* 13:346–50

38. Vyas DR, Spangenburg EE, Abraha TW, Childs TE, Booth FW. 2002. GSK-3β negatively regulates skeletal myotube hypertrophy. *Am. J. Physiol. Cell Physiol.* 283:C545–51

39. Mitchell PO, Mills ST, Pavlath GK. 2002. Calcineurin differentially regulates maintenance and growth of phenotypically distinct muscles. *Am. J. Physiol. Cell Physiol.* 282:C984–92

40. Dunn SE, Chin ER, Michel RN. 2000. Matching of calcineurin activity to upstream effectors is critical for skeletal muscle fiber growth. *J. Cell Biol.* 151:663–72

41. Naya FJ, Mercer B, Shelton J, Richardson JA, Williams RS, Olson EN. 2000. Stimulation of slow skeletal muscle fiber gene expression by calcineurin in vivo. *J. Biol. Chem.* 275:4545–48

42. Goldberg AL, Etlinger JD, Goldspink DF, Jablecki C. 1975. Mechanism of work-induced hypertrophy of skeletal muscle. *Med. Sci. Sports Exerc.* 7:248–61

43. Carson JA, Wei L. 2000. Integrin signaling's potential for mediating gene expression in hypertrophying skeletal muscle. *J. Appl. Physiol.* 88:337–43

44. Ingber DE. 2003. Tensegrity II. How structural networks influence cellular information processing networks. *J. Cell Sci.* 116:1397–408

45. Fluck M, Carson JA, Schwartz RJ, Booth FW. 1999. SRF protein is upregulated during stretch-induced hypertrophy of rooster ALD muscle. *J. Appl. Physiol.* 86:1793–99

46. Carson JA, Schwartz RJ, Booth FW. 1996. SRF TEF-1 control of chicken skeletal α-actin gene during slow-muscle hypertrophy. *Am. J. Physiol. Cell Physiol.* 270:C1624–33

47. Gordon SE, Fluck M, Booth FW. 2001. Selected contribution: skeletal muscle focal adhesion kinase, paxillin, and serum response factor are loading dependent. *J. Appl. Physiol* 90:1174–83

48. Wei L, Wang L, Carson JA, Agan JE, Imanaka-Yoshida K, Schwartz RJ. 2001. β1 integrin and organized actin filaments facilitate cardiomyocyte-specific RhoA-dependent activation of the skeletal α-actin promoter. *FASEB J.* 15:785–96

49. Nader GA, Esser KA. 2001. Intracellular signaling specificity in skeletal muscle in response to different modes of exercise. *J. Appl. Physiol* 90:1936–42

50. Wretman C, Lionikas A, Widegren U, Lannergren J, Westerblad H, Henriksson J. 2001. Effects of concentric eccentric contractions on phosphorylation of MAPK(erk1/2) and MAPK(p38) in isolated rat skeletal muscle. *J. Physiol.* 535:155–64

51. Rome S, Clement K, Rabasa-Lhoret R, Loizon E, Poitou C, et al. 2003. Microarray profiling of human skeletal muscle reveals that insulin regulates approximately 800 genes during a hyperinsulinemic clamp. *J. Biol. Chem.* 278:18063–68

52. Pilegaard H, Ordway GA, Saltin B, Neufer PD. 2000. Transcriptional regulation of gene expression in human skeletal muscle during recovery from exercise. *Am. J. Physiol. Endocrinol. Metab.* 279:E806–14

53. Carson JA. 1997. The regulation of gene expression in hypertrophying skeletal muscle. *Exerc. Sport Sci. Rev.* 25:301–20

54. Baar K, Blough E, Dineen B, Esser K. 1999. Transcriptional regulation in response to exercise. *Exerc. Sport Sci. Rev.* 27:333–79

55. McPherron AC, Lawler AM, Lee SJ. 1997. Regulation of skeletal muscle mass in mice by a new TGF-β superfamily member. *Nature* 387:83–90

56. Nishi M, Yasue A, Nishimatu S, Nohno T, Yamaoka T, et al. 2002. A missense mutant myostatin causes hyperplasia without hypertrophy in the mouse muscle. *Biochem. Biophys. Res. Commun.* 293:247–51

57. McMahon CD, Popovic L, Oldham JM, Jeanplong F, Smith HK, et al. 2003. Myostatin-deficient (Mstn-/-) mice lose more skeletal muscle mass than wild-type controls during hindlimb suspension. *Am. J. Physiol. Endocrinol. Metab.* 285:E82–87

58. Kawada S, Tachi C, Ishii N. 2001. Content localization of myostatin in mouse skeletal muscles during aging, mechanical unloading and reloading. *J. Muscle Res. Cell Motil.* 22:627–33

59. Ma K, Mallidis C, Bhasin S, Mahabadi V, Artaza J, et al. 2003. Glucocorticoid-induced skeletal muscle atrophy is associated with upregulation of myostatin gene expression. *Am. J. Physiol. Endocrinol. Metab.* 285:E363–71

60. Ma K, Mallidis C, Artaza J, Taylor W, Gonzalez-Cadavid N, Bhasin S. 2001. Characterization of 5′-regulatory region of human myostatin gene: regulation by dexamethasone in vitro. *Am. J. Physiol. Endocrinol. Metab.* 281:E1128–36

61. DeVol DL, Rotwein P, Sadow JL, Novakofski J, Bechtel PJ. 1990. Activation

of insulin-like growth factor gene expression during work-induced skeletal muscle growth. *Am. J. Physiol. Endocrinol. Metab.* 259:E89–95

62. Bamman MM, Shipp JR, Jiang J, Gower BA, Hunter GR, et al. 2001. Mechanical load increases muscle IGF-I androgen receptor mRNA concentrations in humans. *Am. J. Physiol. Endocrinol. Metab.* 280:E383–90

63. Yan Z, Biggs RB, Booth FW. 1993. Insulin-like growth factor immunoreactivity increases in muscle after acute eccentric contractions. *J. Appl. Physiol* 74:410–14

64. Adams GR, McCue SA. 1998. Localized infusion of IGF-I results in skeletal muscle hypertrophy in rats. *J. Appl. Physiol.* 84:1716–22

65. Chakravarthy MV, Davis BS, Booth FW. 2000. IGF-I restores satellite cell proliferative potential in immobilized old skeletal muscle. *J. Appl. Physiol* 89:1365–79

66. Barton-Davis ER, Shoturma DI, Musaro A, Rosenthal N, Sweeney HL. 1998. Viral-mediated expression of insulin-like growth factor I blocks the aging-related loss of skeletal muscle function. *Proc. Natl. Acad. Sci. USA* 95:15603–7

67. Lang CH, Frost RA. 2002. Role of growth hormone, insulin-like growth factor-I, and insulin-like growth factor binding proteins in the catabolic response to injury and infection. *Curr. Opin. Clin. Nutr. Metab. Care* 5:271–79

68. Kimball SR, Farrell PA, Jefferson LS. 2002. Role of insulin in translational control of protein synthesis in skeletal muscle by amino acids or exercise. *J. Appl. Physiol.* 93:1168–80

69. Attaix D, Combaret L, Pouch MN, Taillandier D. 2001. Regulation of proteolysis. *Curr. Opin. Clin. Nutr. Metab. Care* 4:45–49

70. Kadowaki M, Kanazawa T. 2003. Amino acids as regulators of proteolysis. *J. Nutr.* 133:2052S–56

71. Watford M. 2003. Not all injury-induced muscle proteolysis is due to increased activity of the ubiquitin/proteasome system: evidence for up-regulation of macrophage-associated lysosomal proteolysis in a model of local trauma. *Nutr. Rev.* 61:34–38

72. Kee AJ, Combaret L, Tilignac T, Souweine B, Aurousseau E, et al. 2003. Ubiquitin-proteasome-dependent muscle proteolysis responds slowly to insulin release refeeding in starved rats. *J. Physiol.* 546:765–76

73. Morgan HE, Jefferson LS, Wolpert EB, Rannels DE. 1971. Regulation of protein synthesis in heart muscle. II. Effect of amino acid levels insulin on ribosomal aggregation. *J. Biol. Chem.* 246:2163–70

74. Kimball SR, Jefferson LS. 2002. Control of protein synthesis by amino acid availability. *Curr. Opin. Clin. Nutr. Metab. Care* 5:63–67

75. Rommel C, Bodine SC, Clarke BA, Rossman R, Nunez L, et al. 2001. Mediation of IGF-1-induced skeletal myotube hypertrophy by PI(3)/Akt/mTOR and PI(3)K/Akt/GSK3 pathways. *Nat. Cell Biol.* 3:1009–13

76. Bodine SC, Stitt TN, Gonzalez M, Kline WO, Stover GL, et al. 2001. Akt/mTOR pathway is a crucial regulator of skeletal muscle hypertrophy and can prevent muscle atrophy in vivo. *Nat. Cell Biol.* 3:1014–19

77. Pallafacchina G, Calabria E, Serrano AL, Kalhovde JM, Schiaffino S. 2002. A protein kinase B-dependent rapamycin-sensitive pathway controls skeletal muscle growth but not fiber type specification. *Proc. Natl. Acad. Sci. USA* 99:9213

78. Pause A, Belsham GJ, Gingras AC, Donze O, Lin TA, et al. 1994. Insulin-dependent stimulation of protein synthesis by phosphorylation of a regulator of 5′-cap function. *Nature* 371:762–67

79. Proud CG, Wang X, Patel JV, Campbell LE, Kleijn M, et al. 2001. Interplay between insulin nutrients in the

regulation of translation factors. *Biochem. Soc. Trans.* 29:541–47

80. Hernandez JM, Fedele MJ, Farrell PA. 2000. Time course evaluation of protein synthesis glucose uptake after acute resistance exercise in rats. *J. Appl. Physiol.* 88:1142–49

81. Cuthbertson DJR, Smith K, Babraj J, Waddell T, Watt PW, et al. 2002. Myofibrillar protein synthesis and the activity of p70S6 kinase in human quadriceps muscle after contractile activity with muscle shortening or stretching. *J. Physiol.* 539:P160 (Abstr.)

81a. Bolster DR, Kubica N, Crozier SJ, Williamson DL, Farrell PA, et al. 2003. Immediate response of mTOR-mediated signalling after acute exercise in rat skeletal muscle. *J. Physiol.* In press

82. Kim DH, Sarbassov DD, Ali SM, King JE, Latek RR, et al. 2002. mTOR interacts with raptor to form a nutrient-sensitive complex that signals to the cell growth machinery. *Cell* 110:163–75

83. Kim DH, Sarbassov dD, Ali SM, Latek RR, Guntur KV, et al. 2003. GβL, a positive regulator of the rapamycin-sensitive pathway required for the nutrient-sensitive interaction between raptor mTOR. *Mol. Cell* 11:895–904

84. Proud CG. 2002. Regulation of mammalian translation factors by nutrients. *Eur. J. Biochem.* 269:5338–49

85. Bylund-Fellenius AC, Ojamaa KM, Flaim KE, Li JB, Wassner SJ, Jefferson LS. 1984. Protein synthesis versus energy state in contracting muscles of perfused rat hindlimb. *Am. J. Physiol. Endocrinol. Metab.* 246:E297–305

86. Hardie DG, Hawley SA. 2001. AMP-activated protein kinase: the energy charge hypothesis revisited. *BioEssays* 23:1112–19

87. Bolster DR, Crozier SJ, Kimball SR, Jefferson LS. 2002. AMP-activated protein kinase suppresses protein synthesis in rat skeletal muscle through a down-regulated mammalian target of rapamycin (mTOR) signaling. *J. Biol. Chem.* 277:23977–80

88. Stockdale FE, Holtzer H. 1966. DNA synthesis myogenesis. *Exp. Cell Res.* 24:508–20

89. Allen DL, Roy RR, Edgerton VR. 1999. Myonuclear domains in muscle adaptation disease. *Muscle Nerve* 22:1350–60

90. Hawke TJ, Garry DJ. 2001. Myogenic satellite cells: physiology to molecular biology. *J. Appl. Physiol.* 91:534–51

91. Rosenblatt JD, Parry DJ. 1992. Gamma irradiation prevents compensatory hypertrophy of overloaded mouse extensor digitorum longus muscle. *J. Appl. Physiol.* 73:2538–43

92. Adams GR, Caiozzo VJ, Haddad F, Baldwin KM. 2002. Cellular molecular responses to increased skeletal muscle loading after irradiation. *Am. J. Physiol. Cell Physiol.* 283:C1182–95

93. Renault V, Rolland E, Thornell LE, Mouly V, Butler-Browne G. 2002. Distribution of satellite cells in the human vastus lateralis muscle during aging. *Exp. Gerontol.* 37:1513–14

94. Thornell LE, Lindstrom M, Renault V, Mouly V, Butler-Browne GS. 2003. Satellite cells and training in the elderly. *Scand. J. Med. Sci. Sports* 13:48–55

95. Chakravarthy MV, Abraha TW, Schwartz RJ, Fiorotto ML, Booth FW. 2000. Insulin-like growth factor-I extends in vitro replicative life span of skeletal muscle satellite cells by enhancing G1/S cell cycle progression via the activation of phosphatidylinositol 3′-kinase/Akt signaling pathway. *J. Biol. Chem.* 275:35942–52

96. Thomas M, Langley B, Berry C, Sharma M, Kirk S, et al. 2000. Myostatin, a negative regulator of muscle growth, functions by inhibiting myoblast proliferation. *J. Biol. Chem.* 275:40235–43

97. Langley B, Thomas M, Bishop A, Sharma M, Gilmour S, Kambadur R. 2002. Myostatin inhibits myoblast differentiation by down-regulating MyoD

expression. *J. Biol. Chem.* 277:49831–40

98. Spiller MP, Kambadur R, Jeanplong F, Thomas M, Martyn JK, et al. 2002. The myostatin gene is a downstream target gene of basic helix-loop-helix transcription factor MyoD. *Mol. Cell Biol.* 22:7066–82

99. Bennet WM, Connacher AA, Scrimgeour CM, Smith K, Rennie MJ. 1989. Increase in anterior tibialis muscle protein synthesis in healthy man during mixed amino acid infusion: studies of incorporation of [$^{1\text{-}13}$C]leucine. *Clin. Sci.* 76:447–54

100. Kobayashi H, Borsheim E, Anthony TG, Traber DL, Badalamenti J, et al. 2003. Reduced amino acid availability inhibits muscle protein synthesis and decreases activity of initiation factor eIF2B. *Am. J. Physiol. Endocrinol. Metab.* 284:E488–98

101. Rennie MJ, Bohe J, Wolfe RR. 2002. Latency, duration and dose response relationships of amino acid effects on human muscle protein synthesis. *J. Nutr.* 132:3225S–27

101a. Bohé J, Low A, Wolfe RR, Rennie MJ. 2003. Human muscle protein synthesis is modulated by extracellular not intramuscular amino acid availability: a dose response study. *J Physiol.* In press. DOI: 10.1113/jphysiol.2003.050674

102. Buse MG, Reid SS. 1975. Leucine: a possible regulator of protein turnover in muscle. *J. Clin. Invest.* 56:1250–61

103. Buse MG. 1981. In vivo effects of branched chain amino acids on muscle protein synthesis in fasted rats. *Horm. Metab Res.* 13:502–5

104. Buse MG, Weigand DA. 1977. Studies concerning the specificity of the effect of leucine on the turnover of proteins in muscles of control diabetic rats. *Biochim. Biophys. Acta* 475:81–89

105. Anthony JC, Yoshizawa F, Anthony TG, Vary TC, Jefferson LS, Kimball SR. 2000. Leucine stimulates translation initiation in skeletal muscle of postabsorptive rats via a rapamycin-sensitive pathway. *J. Nutr.* 130:2413–19

106. Dardevet D, Sornet C, Balage M, Grizard J. 2000. Stimulation of in vitro rat muscle protein synthesis by leucine decreases with age. *J. Nutr.* 130:2630–35

107. Smith K, Reynolds N, Downie S, Patel A, Rennie MJ. 1998. Effects of flooding amino acids on incorporation of labeled amino acids into human muscle protein. *Am. J. Physiol. Endocrinol. Metab.* 275:E73–78

108. Alvestrand A, Hagenfeldt L, Merli M, Oureshi A, Eriksson LS. 1990. Influence of leucine infusion on intracellular amino acids in humans. *Eur. J. Clin. Invest.* 20:293–98

109. Rasmussen BB, Tipton KD, Miller SL, Wolf SE, Wolfe RR. 2000. An oral essential amino acid-carbohydrate supplement enhances muscle protein anabolism after resistance exercise. *J. Appl. Physiol.* 88:386–92

110. Borsheim E, Tipton KD, Wolf SE, Wolfe RR. 2002. Essential amino acids muscle protein recovery from resistance exercise. *Am. J. Physiol. Endocrinol. Metab.* 283:E648–57

111. Biolo G, Tipton KD, Klein S, Wolfe RR. 1997. An abundant supply of amino acids enhances the metabolic effect of exercise on muscle protein. *Am. J. Physiol. Endocrinol. Metab.* 273:E122–29

112. Tipton KD, Borsheim E, Wolf SE, Sanford AP, Wolfe RR. 2003. Acute response of net muscle protein balance reflects 24-h balance after exercise and amino acid ingestion. *Am. J. Physiol. Endocrinol. Metab.* 284:E76–89

113. Rennie MJ, Hundal HS, Peyrollier K, Cuthbertson DJ, Smith K, et al. 2001. Myofibrillar protein synthesis (MPS) and the activity of P70^{s6} kinase in human skeletal muscle: the effects of contractile activity and essential amino acids (EAA). 2001. *J. Physiol.* 531P:39 (Abstr.)

114. Arnal M, Obled C, Attaix D, Patureau-Mirand P, Bonin D. 1987. Dietary control of protein turnover. *Diabetes Metab.* 13:630–42

115. Bennet WM, Connacher AA, Scrimgeour CM, Rennie MJ. 1990. The effect of amino acid infusion on leg protein turnover assessed by L-[^{15}N]phenylalanine and L-[$^{1-13}$C]leucine exchange *Eur. J. Clin. Invest.* 20:41–50. Erratum. 1990. *Eur. J. Clin. Invest.* 20:549

116. Nygren J, Nair KS. 2003. Differential regulation of protein dynamics in splanchnic skeletal muscle beds by insulin and amino acids in healthy human subjects. *Diabetes* 52:1377–85

117. Airhart J, Arnold JA, Stirewalt WS, Low RB. 1982. Insulin stimulation of protein synthesis in cultured skeletal cardiac muscle cells. *Am. J. Physiol. Cell Physiol.* 243:C81–86

118. Cameron CM, Kostyo JL, Adamafio NA, Brostedt P, Roos P, et al. 1988. The acute effects of growth hormone on amino acid transport and protein synthesis are due to its insulin-like action. *Endocrinology* 122:471–74

119. Dardevet D, Sornet C, Attaix D, Baracos VE, Grizard J. 1994. Insulin-like growth factor-1 insulin resistance in skeletal muscles of adult and old rats. *Endocrinology* 134:1475–84

120. Gelfand RA, Barrett EJ. 1986. Effect of physiological hyperinsulinemia on skeletal muscle protein synthesis and breakdown in man. *J. Clin. Invest.* 80:1–6

121. Louard RJ, Fryburg DA, Gelfand RA, Barrett EJ. 1992. Insulin sensitivity of protein glucose metabolism in human forearm skeletal muscle. *J. Clin. Invest* 90:2348–54

122. Biolo G, Declan Fleming RY, Wolfe RR. 1995. Physiologic hyperinsulinemia stimulates protein synthesis and enhances transport of selected amino acids in human skeletal muscle. *J. Clin. Invest.* 95:811–19

123. Wolfe RR, Volpi E. 2001. Insulin protein metabolism. In *Handbook of Physiology, Section 7. The Endocrine System*, ed. LS Jefferson, AD Cherrington, pp. 735–57. New York: Oxford Univ. Press

124. Wolfe RR. 1992. *Radioactive Stable Isotope Tracers in Biomedicine: Principles and Practice of Kinetic Analysis.* New York: Wiley-Liss

125. Bennet WM, Connacher AA, Scrimgeour CM, Jung RT, Rennie MJ. 1990. Euglycemic hyperinsulinemia augments amino acid uptake by human leg tissues during hyperaminoacidemia. *Am. J. Physiol. Endocrinol. Metab.* 259:E185–94

126. Bennet WM, Rennie MJ. 1991. Protein anabolic actions of insulin in the human body. *Diabetic Med.* 8:199–207

127. Bennet WM, Connacher AA, Jung RT, Stehle P, Rennie MJ. 1991. Effects of insulin amino acids on leg protein turnover in IDDM patients. *Diabetes* 40:499–508

128. Biolo G, Wolfe RR. 1993. Insulin action on protein metabolism. *Baillieres Clin. Endocrinol. Metab.* 7:989–1005

129. Jagoe RT, Lecker SH, Gomes M, Goldberg AL. 2002. Patterns of gene expression in atrophying skeletal muscles: response to food deprivation. *FASEB J.* 16:1697–712

130. Loftberg E, Gutierrez A, Wernerman J, Anderstam B, Mitch WE, et al. 2002. Effects of high doses of glucocorticoids on free amino acids, ribosomes and protein turnover in human muscle. *Eur. J. Clin. Invest.* 32:345–53

131. Jagoe RT, Redfern CP, Roberts RG, Gibson GJ, Goodship TH. 2002. Skeletal muscle mRNA levels for cathepsin B, but not components of the ubiquitin-proteasome pathway, are increased in patients with lung cancer referred for thoracotomy. *Clin. Sci.* 102:353–61

132. Roberts RG, Redfern CP, Graham KA, Bartlett K, Wilkinson R, Goodship TH. 2002. Sodium bicarbonate treatment and ubiquitin gene expression in acidotic

human subjects with chronic renal failure. *Eur. J. Clin. Invest* 32:488–92

133. Rennie MJ. 2003. Claims for the anabolic effects of growth hormone: a case of the emperor's new clothes? *Br. J. Sports Med.* 37:100–5

134. Hameed M, Orrell RW, Cobbold M, Goldspink G, Harridge SD. 2003. Expression of IGF-I splice variants in young and old human skeletal muscle after high resistance exercise. *J. Physiol* 547:247–54

135. Fryburg DA, Jahn LA, Hill SA, Oliveras DM, Barrett EJ. 1995. Insulin and insulin-like growth factor-I enhance human skeletal muscle protein anabolism during hyperaminoacidemia by different mechanisms. *J. Clin. Invest.* 96:1722–29

136. Butterfield GE, Thompson J, Rennie MJ, Marcus R, Hintz RL, Hoffman AR. 1997. Effect of rhGH and rhIGF-I treatment on protein utilization in elderly women. *Am. J. Physiol. Endocrinol. Metab.* 272:E94–99

137. Friedlander AL, Butterfield GE, Moynihan S, Grillo J, Pollack M, et al. 2001. One year of insulin-like growth factor I treatment does not affect bone density, body composition, or psychological measures in postmenopausal women. *J. Clin. Endocrinol. Metab.* 86: 1496

138. Herndon DN, Ramzy PI, DebRoy MA, Zheng M, Ferrando AA, et al. 1999. Muscle protein catabolism after severe burn: effects of age and IGF-1/IGFBP3 treatment. *Ann. Surg.* 229:713–20

139. Laurent GJ, Sparrow MP, Millward DJ. 1978. Changes in rates of protein synthesis breakdown during hypertrophy of the anterior and posterior latissimus dorsi muscles. *Biochem. J.* 176:407–17

140. Laurent GJ, Millward DJ. 1980. Protein turnover during skeletal muscle hypertrophy. *Fed. Proc.* 39:42–47

141. Carraro F, Stuart CA, Hartl WH, Rosenblatt J, Wolfe RR. 1990. Effect of exercise and recovery on muscle protein synthesis in human subjects. *Am. J. Physiol. Endocrinol. Metab.* 259:E470–76

142. Rennie MJ, Edwards RH, Krywawych S, Davies CT, Halliday D, et al. 1981. Effect of exercise on protein turnover in man. *Clin. Sci.* 61:627–39

143. Millward DJ, Bowtell JL, Pacy P, Rennie MJ. 1994. Physical activity, protein metabolism and protein requirements. *Proc. Nutr. Soc.* 53:223–40

144. Bylund-Fellenius A-C, Ojamaa KM, Flaim KE, Li JB, Wassner SJ, Jefferson LS. 1984. Protein synthesis versus energy state in contracting muscle of perfused rat hindlimb. *Am. J. Physiol. Endocrinol. Metab.* 246:E297–305

145. Chesley A, MacDougall JD, Tarnopolsky MA, Atkinson SA, Smith K. 1992. Changes in human muscle protein synthesis after resistance exercise. *J. Appl. Physiol.* 73:1383–88

146. Phillips SM, Tipton KD, Aarsland A, Wolf SE, Wolfe RR. 1997. Mixed muscle protein synthesis breakdown after resistance exercise in humans. *Am. J. Physiol. Endocrinol. Metab.* 273:E99–107

147. Tipton KD, Ferrando AA, Phillips SM, Doyle D, Jr., Wolfe RR. 1999. Postexercise net protein synthesis in human muscle from orally administered amino acids. *Am. J. Physiol. Endocrinol. Metab.* 276:E628–34

148. Rennie MJ, Tipton KD. 2000. Protein amino acid metabolism during and after exercise and the effects of nutrition. *Annu. Rev. Nutr.* 20:457–83

149. Levenhagen DK, Carr C, Carlson MG, Maron DJ, Borel MJ, Flakoll PJ. 2002. Postexercise protein intake enhances whole-body leg protein accretion in humans. *Med. Sci. Sports Exerc.* 34:828–37

150. Goldspink G. 1999. Changes in muscle mass phenotype and the expression of autocrine and systemic growth factors by muscle in response to stretch and overload. *J. Anat.* 194:323–34

151. Gibson JN, McMaster MJ, Scrimgeour CM, Stoward PJ, Rennie MJ. 1988. Rates

of muscle protein synthesis in paraspinal muscles: lateral disparity in children with idiopathic scoliosis. *Clin. Sci.* 75:79–83

152. Fowles JR, MacDougall JD, Tarnopolsky MA, Sale DG, Roy BD, Yarasheski KE. 2000. The effects of acute passive stretch on muscle protein synthesis in humans. *Can. J. Appl. Physiol.* 25:165–80

153. Gibson JNA, Halliday D, Watt PW, Stoward PJ, Morrison WL, Rennie MJ. 1987. Decrease in human quadriceps muscle protein turnover consequent upon leg immobilisation. *Clin. Sci.* 72:503–9

154. Ferrando AA, Lane HW, Stuart CA, Wolfe RR. 1996. Prolonged bed rest decreases skeletal muscle and whole-body protein synthesis. *Am. J. Physiol. Endocrinol. Metab.* 270:E627–33

155. Gibson JNA, Morrison WL, Scrimgeour CM, Smith KM, Stoward PJ, Rennie MJ. 1989. Effects of therapeutic percutaneous electrical stimulation of atrophic human quadriceps on muscle composition protein synthesis and contractile properties. *Eur. J. Clin. Invest.* 19:206–12

156. Ferrando AA, Wolfe RR, Tucker HN. 1997. Effects of bed rest with or without stress. In *Physiology, Stress, Malnutrition: Functional Correlates, Nutritional Interventions*, ed. JM Kinney, pp. 413–31. New York: Lippincott-Raven

157. Esmarck B, Andersen JL, Olsen S, Richter EA, Mizuno M, Kjaer M. 2001. Timing of postexercise protein intake is important for muscle hypertrophy with resistance training in elderly humans. *J. Physiol.* 535:301–11

158. Levenhagen DK, Gresham JD, Carlson MG, Maron DJ, Borel MJ, Flakoll PJ. 2001. Postexercise nutrient intake timing in humans is critical to recovery of leg glucose protein homeostasis. *Am. J. Physiol. Endocrinol. Metab.* 280:E982–93

159. Tipton KD, Rasmussen BB, Miller SL, Wolf SE, Owens-Stovall SK, et al. 2001.

Timing of amino acid-carbohydrate ingestion alters anabolic response of muscle to resistance exercise. *Am. J. Physiol. Endocrinol. Metab.* 281:E197–206

160. Rasmussen BB, Tipton KD, Miller SL, Wolf SE, Wolfe RR. 2000. An oral essential amino acid-carbohydrate supplement enhances muscle protein anabolism after resistance exercise. *J. Appl. Physiol.* 88:386–92

161. Farrell PA, Fedele MJ, Hernandez J, Fluckey JD, Miller JL, et al. 1999. Hypertrophy of skeletal muscle in diabetic rats in response to chronic resistance exercise. *J. Appl. Physiol.* 87:1075–82

162. Millward DJ, Price GM, Pacy PJH, Quevedo RM, Halliday D. 1991. The nutritional sensitivity of the diurnal cycling of body protein enables protein deposition to be measured in subjects at nitrogen equilibrium. *Clin. Nutr.* 10:239–44

163. Gibson NR, Fereday A, Cox M, Halliday D, Pacy PJ, Millward DJ. 1996. Influences of dietary energy protein on leucine kinetics during feeding in healthy adults. *Am. J. Physiol. Endocrinol. Metab.* 270:E282–91

164. Gontzea I, Sutzescu P, Dumitrache S. 1975. The influence of adaptation to physical effort on nitrogen balance in man. *Nutr. Reports Int.* 22:231–36

165. Butterfield GE, Calloway DH. 1984. Physical activity improves protein utilization in young men. *Br. J. Nutr.* 51:171–84

166. Phillips SM, Parise G, Roy BD, Tipton KD, Wolfe RR, Tamopolsky MA. 2002. Resistance-training-induced adaptations in skeletal muscle protein turnover in the fed state. *Can. J. Physiol. Pharmacol.* 80:1045–53

167. Phillips SM, Tipton KD, Ferrando AA, Wolfe RR. 1999. Resistance training reduces the acute exercise-induced increase in muscle protein turnover. *Am. J. Physiol. Endocrinol. Metab.* 276:E118–24

168. Dorrens J, Rennie MJ. 2003. Effects of ageing human whole body and muscle protein turnover. *Scand. J. Med. Sci. Sports* 13:26–33

169. Yarasheski KE, Pak-Loduca J, Hasten DL, Obert KA, Brown MB, Sinacore DR. 1999. Resistance exercise training increases mixed muscle protein synthesis rate in frail women and men ≥76 yr old. *Am. J. Physiol. Endocrinol. Metab.* 277:E118–25

170. Greiwe JS, Cheng B, Rubin DC, Yarasheski KE, Semenkovich CF. 2001. Resistance exercise decreases skeletal muscle tumor necrosis factor-α in frail elderly humans. *FASEB J.* 15:475–82

171. Earnest CP, Snell PG, Rodriguez R, Almada AL, Mitchell TL. 1995. The effect of creatine monohydrate ingestion on anaerobic power indices, muscular strength and body composition. *Acta. Physiol. Scand.* 153:207–9

172. Volek JS, Duncan ND, Mazzetti SA, Staron RS, Putukian M, et al. 1999. Performance muscle fiber adaptations to creatine supplementation and heavy resistance training. *Med. Sci. Sports Exerc.* 31:1147–56

173. Willoughby DS, Rosene J. 2001. Effects of oral creatine resistance training on myosin heavy chain expression. *Med. Sci. Sports Exerc.* 33:1674–81

174. Louis M, Poortmans JR, Francaux M, Hultman E, Berre J, et al. 2003. Creatine supplementation has no effect of creatine supplementation on human muscle protein turnover at rest in the postabsorptive or fed states. *Am. J. Physiol. Endocrinol. Metab.* 284:E764–70

174a. Louis M, Poortmans J, Francaux M, Berré J, Boisseau N, et al. 2003. No effect of creatine supplementation on human myofibrillar or sarcoplasmic protein synthesis after resistance exercise *Am. J. Physiol. Endocrinol. Metab.* DOI: 10.1152/ajpendo.00195.2003

175. Bowtell J, Park DM, Smith K, Cuthbertson DJR, Waddell T, Rennie MJ. 2003. Stimulation of human quadriceps protein synthesis after strenuous exercise: no effects of varying intensity between 60 and 90% of one repetition maximum (1RM). *J. Physiol.* 547.P (Abstr.)

176. Rennie MJ. 2001. Control of human muscle protein synthesis as a result of contractile activity and amino acid availability: implications for protein requirements. *Int. J. Sport Nutr. Exer. Metab.* 11:S168–74

Subject Index

and, 401–4

Synaptic physiology
  glutamate receptor ion
    channels and, 161–77
  learning mechanisms in
    addiction and, 447–67

Synaptic plasticity
  in ventral tegmental area
    due to drug abuse,
    447–67

Synaptogenesis
  live optical imaging of
    nervous system
    development, 771–83
  sleep and circadian
    rhythms in hibernation,
    278–79

Synchrotron radiation
  glutamate receptor ion
    channels and, 166

Syndrome X
  extracellular
    cAMP-adenosine
    pathway in renal
    physiology, 589–90

T

Tachyglossus aculeatus
  metabolic rate and body
    temperature during torpor,
    246, 257

Tadarida brasiliensis
  metabolic rate and body
    temperature during torpor,
    248

Tamarindus indica
  oral rehydration therapy
    and, 405

Tamias spp.
  metabolic rate and body
    temperature during torpor,
    247

Tamiasciurus hudsonicus
  field physiology and, 225

Tamoxifen
  corepressors in
    transcriptional regulation

by nuclear hormone
    receptors, 323, 334
  genetically engineered
    mouse models for lung
    cancer and, 657

Targeting
  corepressors in
    transcriptional regulation
    by nuclear hormone
    receptors, 315
  estrogens in the central
    nervous system and,
    297–98
  viral-based myocardial
    gene therapy in heart
    disease and, 52, 58–68
  voltage-gated ion channels
    in brain and, 499–506

Tarsipes rostratus
  metabolic rate and body
    temperature during torpor,
    245

TASK K$^+$ channels
  K$^+$ channels and metabolic
    regulation, 133

TATA-binding protein (TBP)
  corepressors in
    transcriptional regulation
    by nuclear hormone
    receptors, 316

Taxidea taxus
  metabolic rate and body
    temperature during torpor,
    242, 246, 256

TBL1-related protein 1
    (TBLR-1)
  corepressors in
    transcriptional regulation
    by nuclear hormone
    receptors, 316, 325

TBP-associated factor (TAF)
  corepressors in
    transcriptional regulation
    by nuclear hormone
    receptors, 316

Tbx4 gene
  epithelial-mesenchymal

interactions in developing
    lung and, 632

Tegula spp.
  biochemical indicators of
    stress and metabolism in
    marine ecological studies,
    198

Telomerase
  myocardial aging and
    senescence, 29, 37, 39

Temperature
  metabolic rate and body
    temperature during torpor,
    239–66

Tenrec ecaudatus
  metabolic rate and body
    temperature during torpor,
    248

Testosterone
  field physiology and, 228
  muscle mass size
    regulation and, 805

Tetanus toxin
  renal K transport regulation
    by dietary K intake, 557

tet genes
  genetically engineered
    mouse models for lung
    cancer and, 648, 652–53

Tetrabutylammonium
  inward rectifier
    K$^+$ channels and, 114

Tetracycline
  SP-B and SP-C expression
    in neonatal lung disease,
    607

Tetracycline-controlled
    silencer (rTS)
  genetically engineered
    mouse models for lung
    cancer and, 648

Tetracycline-transactivator
    (*tTA*-inducible system)
  genetically engineered
    mouse models for lung
    cancer and, 648, 652–53

Tetraethylammonium (TEA)

Cumulative Indexes

CONTRIBUTING AUTHORS, VOLUMES 62–66

CHAPTER TITLES, VOLUMES 62–66

Cardiovascular Physiology

Endocrinology

Gastrointestinal Physiology

Neurophysiology

Perspectives

Renal and Electrolyte Physiology

Respiratory Physiology

Special Chapter

Muscle Physiology

Special Topics

Circadian Rhythms

Proton and Electron Transporters

Transgenic Models

Transportopathies

Views and Overviews of the 20th Century